ORGANIC REACTION MECHANISMS · 1971

ORGANIC REACTION MECHANISMS · 1971

An annual survey covering the literature dated December 1970 *through November* 1971

Edited by

B. CAPON University of Glasgow
C. W. REES University of Liverpool

Interscience Publishers
A division of JOHN WILEY & SONS
London · New York · Sydney · Toronto

Copyright © 1972 by John Wiley & Sons Ltd.

All rights reserved. No part of this publication may be reproduced, stored in a retrieval system, or transmitted, in any form or by any means, electronic, mechanical photocopying, recording or otherwise, without the prior written permission of the Copyright owner.

Library of Congress Catalog Card Number 66-23143
ISBN 0 471 13472 4

Printed in Great Britain by
William Clowes & Sons Limited
London, Colchester and Beccles

Contributors

D. C. AYRES	Department of Chemistry, Westfield College, University of London
R. BAKER	Department of Chemistry, The University, Southampton
A. R. BUTLER	Department of Chemistry, St. Salvator's College, University of St. Andrews
B. CAPON	Department of Chemistry, The University, Glasgow
R. S. DAVIDSON	Department of Chemistry, The University, Leicester
T. L. GILCHRIST	Department of Chemistry, The University, Liverpool
M. J. P. HARGER	Department of Chemistry, The University, Leicester
A. C. KNIPE	Department of Chemistry, The New University, Ulster
A. LEDWITH	Donnan Laboratories, The University, Liverpool
P. J. RUSSELL	Donnan Laboratories, The University, Liverpool
I. D. R. STEVENS	Department of Chemistry, The University, Southampton
R. J. STOODLEY	Department of Organic Chemistry, The University, Newcastle-upon-Tyne
R. C. STORR	Department of Organic Chemistry, The University, Liverpool

Preface

This seventh volume of the series is a survey of the work on organic reaction mechanisms published in 1971. For convenience, the literature dated from December 1970 to November 1971, inclusive, was actually covered. The principal aim has again been to scan all the chemical literature and to summarize the progress of work on organic reaction mechanism generally and fairly uniformly, and not just on selected topics. Therefore, certain of the sections are somewhat fragmentary and all are concise. Nearly 5000 papers have been reported, and those which seemed at the time to be more significant are normally described and discussed, and the remainder are listed.

Our other major aim, second only to the comprehensive coverage, has been early publication since we felt that the immediate value of such a survey as this, that of current awareness, would diminish rapidly with time. In this we have been fortunate to have the expert cooperation of the English office of John Wiley and Sons.

July 1972

B.C.
C.W.R.

Contents

1. **Carbonium Ions** by R. Baker 1
   Bicyclic and Polycyclic Systems 1
   Participation by Aryl Groups 16
   Participation by Double and Triple Bonds 20
   Reactions of Small-ring Compounds 25
   Metallocenylmethyl Cations and Other Derivatives . . . 36
   Stable Carbonium Ions and their Reactions 38
   Other Reactions 43

2. **Nucleophilic Aliphatic Substitution** by I. D. R. Stevens . . 49
   Ion-pair Phenomena and Borderline Mechanisms . . . 49
   Solvent and Medium Effects 55
   Isotope Effects 58
   Neighbouring Group Participation 60
   Deamination and Related Reactions 79
   Reactions of Aliphatic Diazo-compounds 81
   Fragmentation Reactions 82
   Displacement Reactions at Elements Other than Carbon . 83
   Ambident Nucleophiles 95
   Substitution at Vinylic Carbon 97
   Reactions of α-Halogenocarbonyl Compounds 101
   S_N2 Processes and Other Reactions 103

3. **Carbanions and Electrophilic Aliphatic Substitution** by D. C. Ayres . . 111
   Carbanion Structure 111
   Reactions of Carbanions 113
   Proton Transfer, Hydrogen Isotope Exchange and Related Reactions . 119
   Electrophilic Reactions of Hydrocarbons 125
   Organometallics: Groups Ia, IIa, III 127
   Organometallics: Other Elements 130
   Miscellaneous Reactions 133

4. **Elimination Reactions** by A. C. Knipe 135
   Stereochemistry and Orientation in $E2$ Reactions . . . 135
   The $E1cB$ Mechanism 137
   The $E2C$ Mechanism 140
   Gas-phase Elimination Reactions 142
   Other Topics 145

5. **Addition Reactions** by R. C. Storr 151
   Electrophilic Additions 152
   Nucleophilic Additions 161
   Cycloadditions 164

6. **Nucleophilic Aromatic Substitution** by A. R. Butler 179
   The S_NAr Mechanism 179
   Heterocyclic Systems 181
   Meisenheimer and Related Complexes 184
   Substitution in Polyhaloaromatic Compounds 188
   Benzyne and Related Intermediates 188
   Other Reactions 192

7. **Electrophilic Aromatic Substitution** by A. R. Butler 195
   Sulphonation 197
   Nitration 198
   Nitrosation 200
   Azo Coupling 200
   Halogenation 201
   Metal Cleavage 204
   Decarboxylation 204
   Friedel–Crafts and Related Reactions 205
   Hydrogen Exchange 206
   Miscellaneous Reactions 208

8. **Molecular Rearrangements** by R. J. Stoodley 211
   Aromatic Rearrangements 212
   Sigmatropic Rearrangements 216
   Electrocyclic Reactions 228
   Rearrangements Involving Cycloreversions and Cycloadditions . . . 235
   Anionic Rearrangements 239
   Cationic Rearrangements 243
   Metal-catalysed Rearrangements 253
   Rearrangements Involving Electron-deficient Heteroatoms . . . 257
   Isomerizations 261
   Rearrangements Involving Ring Openings and Closures 264

9. **Radical Reactions** by A. Ledwith and P. J. Russell 275
   Introduction 275
   Structure and Stereochemistry 277
   Decomposition of Peroxides 281
   Decomposition of Azo-compounds 287
   Diradicals 290
   Atom-transfer Processes 296
   Additions 304
   Aromatic Substitution 313
   Rearrangements 320
   S_H2 Reactions 326
   Reactions Involving Oxidation or Reduction by Metal Salts . . . 331
   Radical Ions and Electron-transfer Processes 336
   Nitroxides 348

	Autoxidation	353
	Pyrolysis and Other Gas-phase Processes	355
	Radiolysis, ESR Spectroscopy and Miscellaneous	359

10. Carbenes and Nitrenes by T. L. GILCHRIST 367
Structure 367
Methods of Generation 368
Cycloadditions 373
Insertions and Abstractions 376
Rearrangements and Fragmentations 381
Reactions with Nucleophiles and Electrophiles 384
Carbenoids 387
Transition-metal Complexes 389

11. Reactions of Aldehydes and Ketones and their Derivatives
by B. CAPON 393
Formation and Reactions of Acetals and Ketals 393
Hydrolysis and Formation of Glycosides 397
Hydration of Aldehydes and Ketones and Related Reactions . . 402
Reactions with Nitrogen Bases 404
Hydrolysis of Enol Ethers and Esters 408
Enolization and Related Reactions 409
Aldol and Related Reactions 413
Other Reactions 414

12. Reactions of Acids and their Derivatives by B. CAPON . . . 419
Carboxylic Acids 419
Non-carboxylic Acids 457

13. Photochemistry by R. S. DAVIDSON 467
Physical Aspects 469
Carbonyl Compounds 475
Carboxylic Acids and Related Compounds 490
Olefins 493
Aromatic Hydrocarbons 501
Heterocyclic Compounds 504
Nitrogen-containing Compounds 508
Halogen-containing Compounds 516
Carbonium Ions and Carbanions 517
Miscellaneous Compounds 518
Other Photoreactions 520

14. Oxidation and Reduction by M. J. P. HARGER 527
Ozonation and Ozonolysis 527
Oxidation by Metallic Ions 529

Oxidation by Molecular Oxygen 537
Other Oxidations 541
Reductions 546
Hydrogenation and Hydrogenolysis 555

Author Index, 1971 561

Subject Index, Cumulative, 1970–1971 611

CHAPTER 1

Carbonium Ions[1]

R. BAKER

Chemistry Department, The University, Southampton

Bicyclic and Polycyclic Systems	1
Derivatives of Norbornane and Related Compounds	1
Other Bicyclic Systems	5
Polycyclic Systems	13
Participation by Aryl Groups	16
Arylalkyl Compounds	16
Benzonorbornene Derivatives	18
Other Reactions involving Phenyl Participation	19
Participation by Double and Triple Bonds	20
Double-bond Participation	20
Triple-bond Participation	25
Reactions of Small-ring Compounds	25
Cyclopropylmethyl Derivatives	25
Participation by More Remote Cyclopropyl Rings	29
Reactions of Cyclopropyl Derivatives	32
Reactions of Cyclobutyl Derivatives	34
Protonated Cyclopropane Intermediates	34
Metallocenylmethyl Cations and Other Derivatives	36
Stable Carbonium Ions and their Reactions	38
Other Reactions	43

Bicyclic and Polycyclic Systems

Derivatives of Norbornane and Related Compounds

^{13}C and ^{1}H NMR and Raman spectroscopic investigations of the 1,2-dimethylnorbornyl cation have been reported.[2] In SbF_5–SO_2 and FSO_3H–SbF_5–SO_2, the amount of 6,2 σ delocalization is nearly identical with that in the 2-methylnorbornyl cation, but in the former the barrier to C_1,C_2 Wagner–Meerwein shift is lowered so far that it cannot be frozen out on the NMR time scale even at –140°C. This low barrier results from a degenerate tertiary–tertiary rearrangement. The C—$H_{6\,exo}$ bond is involved to a major extent in the delocalization.

---

[1] S. Winstein and M. Sakai, "Non-Classical Ions and Homoaromaticity", *Kagaku No. Ryoiki*, **25**, 127 (1971); R. E. Leone and P. von R. Schleyer, "Degenerate Carbonium Ions", *Angew. Chem. Int. Ed.*, **9**, 860 (1970); W. R. Dolbier, "Mechanisms of Solvolytic Spirane Rearrangements", *Mech. Mol. Migr.* **3**, 1 (1971); R. C. Bingham and P. von R. Schleyer, *Fortschr. Chem. Forsch.*, **18**, 1 (1971); M. J. Goldstein and R. Hoffman, "Symmetry, Topology and Aromaticity", *J. Am. Chem. Soc.*, **93**, 6193 (1971).

[2] G. A. Olah, J. R. DeMember, C. Y. Lui, and R. D. Porter, *J. Am. Chem. Soc.*, **93**, 1442 (1971); see *Org. Reaction Mech.*, **1970**, 2.

The relative rates of 6,2- and 3,2-hydride shifts ($k_6:k_3$) are <100 and <200 in the acetolysis of Δ^3-[1-^{14}C]cyclopentenylethyl toluene-p-sulphonate and 2-exo-[4-^{14}C]-norbornyl toluene-p-sulphonate, respectively.[3] ^{14}C studies have also shown that in the solvolysis of exo-norbornyl p-bromobenzenesulphonate, isotopic scrambling occurs in the starting ester owing to internal return and also in the initial product as well as in the reaction intermediates.[4]

Deuterium tracer studies have added further evidence that classical ions generated in deaminations can survive several Wagner–Meerwein rearrangements or hydride shifts.[5] An $endo$–$endo$ hydride shift to a secondary carbonium ion has been observed in the deamination of 3-exo-aminobornane-2-exo-ol,[6] and a 3,2-$endo$,$endo$-methyl migration in the decomposition of a triazoline.[7] The results were interpreted in terms of open classical intermediates. β-Deuterium isotope effects have been reported to be inconsistent with anchimeric assistance in the solvolysis of exo-norbornyl derivatives. Differences in the force-constant change associated with the hydrogens at C-2, C-3 and C-6 in formation of the transition states for the exo- and $endo$-derivatives are suggested as the origin of the observed isotope effects.[8] The observation that β-deuterium isotope effects in the solvolysis of 1,2-dimethyl-exo-2-norbornyl p-nitrobenzoate in solvents varying from 50 to 70 vol. % ethanol are directly related to the amount of elimination product is suggested as being due to rate-determining elimination in competition with substitution and internal return.[9]

Acetolysis of the epimeric exo-6-methoxycarbonyl-2-norbornyl p-bromobenzene-sulphonates proceeds through classical carbonium ions with $k_{exo}/k_{endo} = 4.4$, although some Wagner–Meerwein rearrangement occurs.[10] Little positive charge was believed to reside on the migrating carbon atom in the transition state for migration. $endo$-6-Methoxycarbonyl-exo-2-norbornyl p-bromobenzenesulphonate solvolyses with anchimeric assistance and lactone formation.

In the solvolysis of α-fenchyl toluene-p-sulphonate (**1**) in acetic acid, a mixture of products is obtained, but the main component is 4-methylsantenyl acetate (**2**) resulting from methyl migration, Wagner–Meerwein rearrangement and hydride shift.[11] (**3**) is also produced by a process involving solvent assistance and hydride shift.

Rates of displacement reactions on (**4**)–(**7**) have been measured, the relative rates being found to be very similar to the reactivities observed for solvolyses. Steric interactions between the leaving group and other groups are again the determining factor.[12] Rearrangements of a series of dihydrodicyclopentadiene derivatives in orthophosphoric acid have also been studied.[13]

Low $exo:endo$ rate ratios have been observed in the solvolysis of a number of exo-2,3-o-arylene-5-norbornyl toluene-p-sulphonates (**8**) and (**9**).[14] A small increase in

---

[3] C. J. Collins and C. E. Harding, *Ann. Chem.*, **745**, 124 (1971).
[4] C. C. Lee, B.-S. Hahn, L. K. M. Lam, and D. J. Woodcock, *Can. J. Chem.*, **48**, 3831 (1970).
[5] V. F. Raaen, B. M. Benjamin, and C. J. Collins, *Tetrahedron Letters*, **1971**, 2613; see *Org. Reaction Mech.*, **1970**, 6.
[6] P. Wilder, Jr., and W. C. Hsieh, *J. Org. Chem.*, **36**, 2552 (1971).
[7] S. Rengaraju and K. D. Berlin, *Tetrahedron*, **27**, 2399 (1971).
[8] S. E. Scheppele, *Chem. Comm.*, **1971**, 592.
[9] K. Humski, *Croat. Chem. Acta*, **42**, 501 (1971).
[10] G. W. Oxer and D. Wege, *Tetrahedron Letters*, **1971**, 457.
[11] A. Coulombeau, C. Coulombeau, and A. Rassat, *Bull. Soc. Chim. France*, **1970**, 4389.
[12] I. Rothberg and R. V. Russo, *J. Chem. Soc.* (B), **1971**, 1214.
[13] P. Wilder, D. J. Cash, R. C. Wheland, and G. W. Wright, *J. Am. Chem. Soc.*, **93**, 791 (1971).
[14] R. Baker and T. J. Mason, *J. Chem. Soc.* (B), **1971**, 1144.

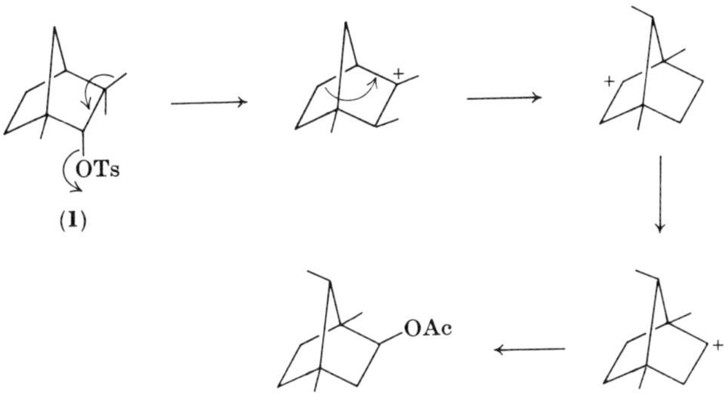

exo/endo rate ratio with change in solvent from acetic acid to formic acid was interpreted as resulting from an increasing carbon–carbon bond participation. The reactions were discussed in terms of k_Δ and k_s pathways but it was suggested that, in these systems, some "leakage" could occur.

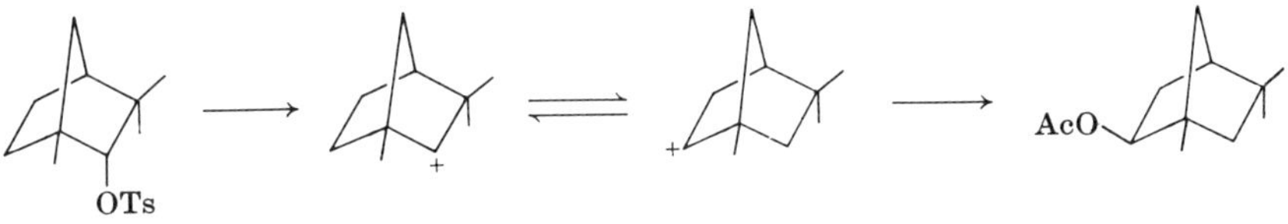

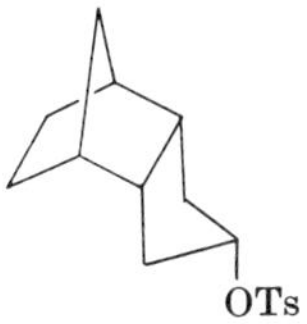

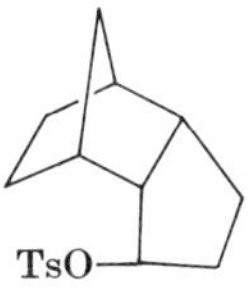

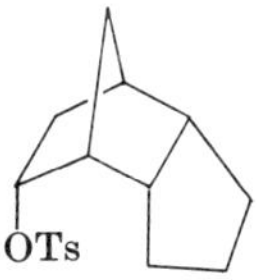

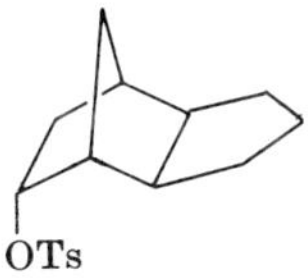

In the addition of bromine and chlorine to *exo,exo-* and *endo,endo-*5,6-dideuterionorbornene, proton loss to form a tricyclic product occurs from C-6 with *exo* and *endo* stereoselectivity, respectively.[15] It was suggested that elimination could occur from (10) in bromination and from (11) in chlorination. Elimination from unsymmetrical bridged ions was also considered. Greater than 50% of *exo-cis* addition was found for hydrogen chloride and bromide to 2,3-dideuterionorbornene, and a classical norbornyl carbonium ion was suggested.[16] In formic acid, methanol and hydrogen fluoride, Wagner–Meerwein rearrangement became equally important.

A detailed re-investigation has been made of the bromination of norbornene and the products have been found to consist of a mixture of five dibromides, bromonortricylene and 2-*exo*-bromonorbornane.[17] From ^{14}C studies, 2-*exo*,3-*endo*-dibromonorbornane was

[15] N. H. Werstiuk and I. Vancas, *Can. J. Chem.*, **48**, 3963 (1970).
[16] J. K. Stille and R. D. Hughes, *J. Org. Chem.*, **36**, 340 (1971).
[17] D. R. Marshall, P. Reynolds-Warnhoff, E. W. Warnhoff, and J. R. Robinson, *Can. J. Chem.*, **49**, 885 (1971).

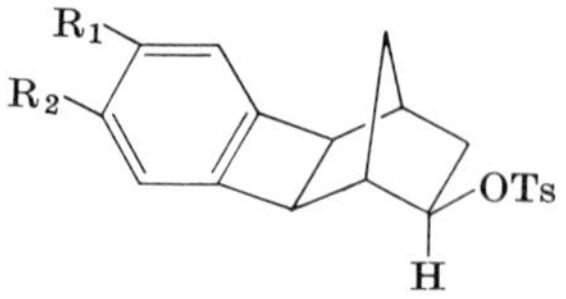

(8)

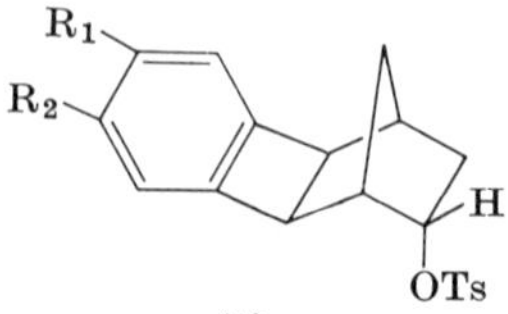

(9)

$R_1 = R_2 = H$
$R_1 = R_2 = OMe$
R_1 or $R_2 = NO_2$, R_1 or $R_2 = H$

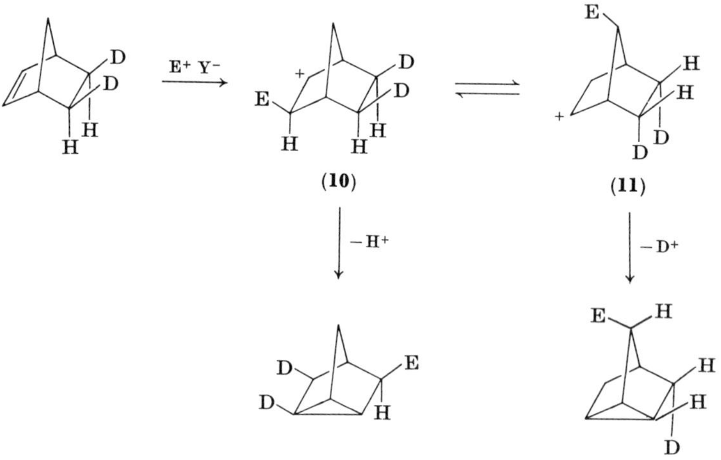

(10) (11)

shown to arise from bromonium ion (12) and also carbonium ion (13). *endo*-Attack has been observed in the reaction of bromine with *anti*-7-bromo-5-phenylbenzonorbornadiene and *syn*-7-bromo-2-phenylnorbornene.[18] It was suggested that the reversal from normally observed *exo* attack is due to the reduction in the σ delocalization in the transition state for this pathway.

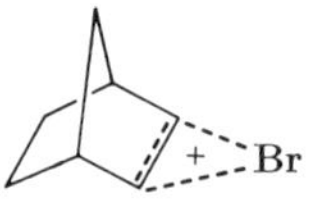

(12) 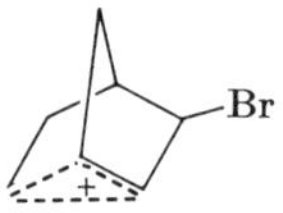 (13)

Palladium chloride–copper chloride catalysed addition reactions to tricyclo[4.2.1.0^{2,5}] derivatives have been studied.[19] Products from two competitive reactions involving a copper(II) complex and palladium(II) complex were obtained, but neither process appeared to proceed through free carbonium ions. Reaction of camphene, norbornene, benzonorbornene and dibenzobarrelene with Pb(OAc)$_{4-n}$(N$_3$)$_n$ appears to proceed through carbonium ion intermediates.[20]

[18] R. Caple, G. M. S. Chen, and J. D. Nelson, *J. Org. Chem.*, **36**, 2870 (1971).
[19] C. J. R. Adderley, J. W. Nebzydoski, M. A. Battiste, R. Baker, and D. E. Halliday, *Tetrahedron Letters*, **1971**, 3545.
[20] E. Zbiral and A. Stutz, *Tetrahedron*, **27**, 4953 (1971).

Carbonium Ions

The 7-norbornyl cation has been generated from α-deuterio-2-bicyclo[3.2.0]heptyl *p*-bromobenzenesulphonate in acetic acid and products have been studied.[21] Observed ^{13}C chemical shifts in norbornyl derivatives have been interpreted in terms of bond lengths and inductive and steric effects.[22] Special kinetic salt effects have been measured in the solvolysis of t-butyl bromide, *exo*-2-chloro-1- and -2-methylnorbornane, isobornyl chloride and camphene hydrochloride.[23]

An intermediate diazotic acid (14) is formed in the reaction of 1,1-disubstituted hydrazines with nitrous acid, which decomposes by two alternative routes.[24] One involves formation of a nitrenium ion (15) which undergoes rearrangement and reaction with solvent and, in the other, nitrous oxide is lost with hydrogen transfer to the nitrogen.

Other studies include assessments of torsional effects in the [2.2.1] system;[25] equilibration reactions of norborneols and 1-methylnorborneols;[26] dependence of rates on sodium hydroxide concentration in the hydrolysis of optically active *exo*- and *endo*-norbornyl toluene-*p*-sulphonates;[27] sulphur–oxygen bond cleavage of secondary and tertiary toluene-*p*-sulphonates under nucleophilic solvolytic conditions;[28] properties of 2-arylnorbornene oxides and the dimer formed by dehydration of the 2-*p*-anisyl derivative;[29] reactions of *endo*-tricyclo[3.2.1.0^{2,4}]oct-6-ene *exo*-oxide;[30] formation of norbornylene-mercurinium ions;[31] sulphuric acid and formic acid catalysed hydration of *endo*- and *exo*-norbornene-5-carboxylic acids;[32] lactone formation from 2-*endo*-cyano-, 2-*endo*-cyanomethyl- and 2-*endo*-carboxymethyl-5-norbornenes;[33] and deamination of *endo*- and *exo*-bornylamines.[34]

Other Bicyclic Systems

Further studies on "memory effects" have appeared. Magnification of a memory effect by substitution of an alkyl group for hydrogen of the non-migrating group R_3 in (16) γ to the original cationic charge, providing a more stable doubly rearranged ion, is reported.[35] In deamination reactions of (17) and (18) the effect of the γ-methyl substitution on the multiplicative memory effect was a factor of 6 for the C-1 ("near") versus C-4 ("far") ring expansion comparison and about 13 in the comparison with the parent system. However, limits ranging from 1450 to 64,000 for the multiplicative memory effects, indicating very large enhancements of selectivity by methyl substitution, were

---

[21] B. Funke and S. Winstein, *Tetrahedron Letters*, **1971**, 1477.
[22] J. B. Grutzner, M. Jautelat, J. B. Dence, R. A. Smith, and J. D. Roberts, *J. Am. Chem. Soc.*, **92**, 7107 (1970).
[23] C. A. Bunton, T. W. Del Pesco, A. M. Dunlop, and K. U. Yang, *J. Org. Chem.*, **36**, 887 (1971).
[24] P. G. Gassman and K. Shudo, *J. Am. Chem. Soc.*, **93**, 5899 (1971).
[25] S. P. Jindall, S. S. Sohoni, and T. T. Tidwell, *Tetrahedron Letters*, **1971**, 779; S. P. Jindall and T. T. Tidwell, *Tetrahedron Letters*, **1971**, 787; J. M. Mellor and C. F. Webb, *Tetrahedron Letters*, **1971**, 4025.
[26] A. Coulombeau and A. Rassat, *Bull. Soc. Chim. France*, **1970**, 4393.
[27] P. Hirsjarvi, T. Kiutamo, J. Korvenranta, E. Tenhunen, and M. Vilen, *Suom. Kemistilehti*, **43**, 519 (1970).
[28] P. G. Gassman, J. M. Hornback, and J. M. Piescone, *Tetrahedron Letters*, **1971**, 1425.
[29] T. J. Gerteisen, D. C. Kleinfelter, G. C. Brophy, and S. Sternhell, *Tetrahedron*, **27**, 3013 (1971).
[30] B. C. Henshaw, D. W. Rome, and B. L. Johnson, *Tetrahedron*, **27**, 2255 (1971).
[31] G. A. Olah and P. R. Clifford, *J. Am. Chem. Soc.*, **93**, 1261 (1971).
[32] H. Geiger, *Tetrahedron*, **27**, 165 (1971).
[33] T. Sasaki, S. Eguchi, and M. Sugimoto, *Bull. Chem. Soc. Japan*, **44**, 1382 (1971).
[34] D. V. Banthorpe, D. G. Morris, and C. A. Bunton, *J. Chem. Soc.* (B), **1971**, 687.
[35] J. A. Berson, J. M. McKenna, and H. Junge, *J. Am. Chem. Soc.*, **93**, 1296 (1971).

R = Me, H
X = NH$_2$

Carbonium Ions

found in deamination and solvolysis of the 1-methyl-2-norbornylcarbinyl systems (**19** and **20**).

Evidence has been obtained that the free-energy benefit ($\Delta\Delta F^{+}$) of replacing hydrogen by methyl monotonically approaches zero at zero overall ΔF^{+}.[36] Thus, in reactions of 1-methyl-7-norbornylcarbinyl derivatives (**21**), despite the extra stability of tertiary cation (**22**) over secondary cation (**23**), the ratio of products from "far" and "near" branches is close to unity. The data suggest that $\Delta\Delta F^{+}$ of methyl for hydrogen replacement at the site of a developing carbonium ion (C-β) in these rearrangements (**24**) is about 0.4–1.1 kcal mole^{-1} compared to 6 kcal mole^{-1} for overall ΔF^{+}.

$$\begin{array}{c} R_1 \\ R_2-C-C^+ \\ R_3 \quad H \end{array} \xrightarrow{} \begin{array}{c} R_1 \\ R_2-C^+-C-R_3 \\ R_3 \quad H \end{array}$$

(**24**)

Whilst the relatively unselective nature of deamination reactions compared with solvolytic processes has been accepted for some time, in some deaminations the reverse has been demonstrated.[37] Competing paths used to examine selectivity are migration of a β-substituent (R_1) and migration of a β ring member (G, J) in carbonyl derivatives of general formula (**25**); the series of structures (**26**)–(**31**) was examined. Deamination invariably produces an increase in preference for ring expansion over the solvolytic reactions; in some cases a selectivity ratio as large as 100 is found. The results were rationalized in terms of ground-state conformations, which in the bicyclic systems places the leaving group as far as possible from the bulky bicyclic system. In this situation migration is not favoured in the solvolytic processes, so that the migration in deamination has enhanced relative importance. Normally, in systems such as 3-phenyl-2-butyl the energies of transition states for deamination do not lie far above the barriers for internal rotation and the choice of phenyl or methyl migration depends on the distribution in the diazonium ion ground state.

"Memory effects" in formation of (**33**) and (**34**) resulting from hydride shift, and rearrangement in carbonium ion reactions of (**32**), have been shown not to result from ion-pairing effects.[38] The hydride shift process results in ca. 20% preservation of optical purity in p-bromobenzenesulphonate hydrolysis, but in deamination the hydroxyl group in the ring-expanded product replaces the migrating carbon with inversion of configuration. The results are consistent with either a slow interconversion of conformational isomers or non-classical bonding.

A number of studies on trans-fused cyclopropanes have been published. In the solvolysis of the trans-bicyclo[6.1.0]non-2-yl derivatives (**35** and **36**), the trans,trans isomer was found to be greater than 10^4 times as reactive as the trans,cis isomer.[39] This difference arises because in (**36**) the leaving group lies over the cyclopropane ring so that only a

---

[36] J. A. Berson and J. W. Foley, J. Am. Chem. Soc., **93**, 1297 (1971).
[37] J. A. Berson, J. W. Foley, J. M. McKenna, H. Junge, D. S. Donald, R. T. Luibrand, N. G. Kundu, W. J. Libbey, M. S. Poonian, J. J. Gajewski, and J. B. E. Allen, J. Am. Chem. Soc., **93**, 1299 (1971).
[38] J. A. Berson, R. T. Luibrand, N. G. Kundu, and D. G. Morris, J. Am. Chem. Soc., **93**, 3075 (1971).
[39] K. B. Wiberg and T. Nakahira, J. Am. Chem. Soc., **93**, 5193 (1971); P. G. Gassman, E. A. Williams, and F. J. Williams, J. Am. Chem. Soc., **93**, 5199 (1971).

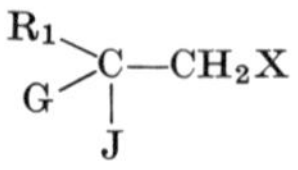

poor interaction exists between the developing p-orbital and the cyclopropane ring. In reaction of (35) the developing p-orbital has its lobe directly over the bridging C—C bond. No "cross-over" in products was found for the epimeric derivatives, and the stereochemistry was the same for the two reactions. Only the *trans,trans* epimer showed scrambling in the solvolysis when the material was labelled α-D (37).

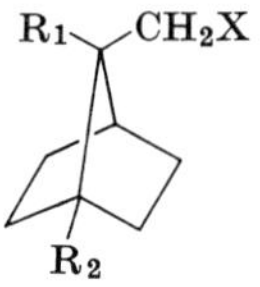

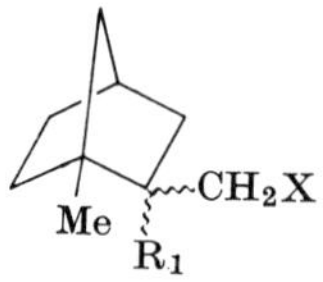

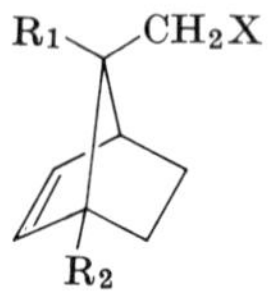

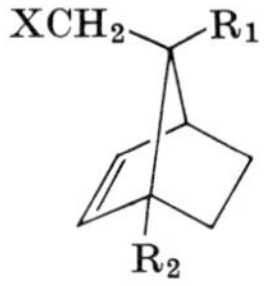

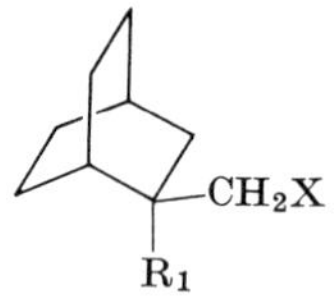

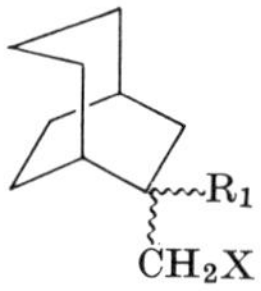

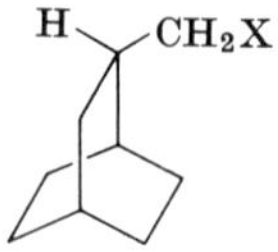

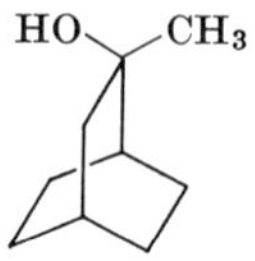

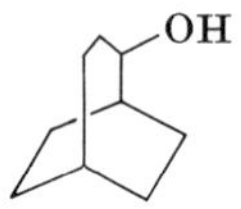

No evidence for cyclopropane participation was found in the solvolysis of 4-*trans*-bicyclo[5.1.0]octane p-bromobenzenesulphonate.[40] It was concluded that requirements for participation, which involve either the interaction of a single bent cyclopropyl bond with an orthogonal p-orbital, or the interaction of the orbitals of two different cyclopropyl carbon–carbon single bonds with two ends of the p-orbital, are not satisfied with this system. Acid-catalysed addition to *trans*-bicyclo[5.1.0]oct-3-ene occurs with product-determining protonation at the bridgehead.[41] Products indicate that protonation of the cyclopropane ring is preferred over that of the double bond, which is consistent with the reactive nature of the former, as shown in solvolytic studies of analogous systems.

A bisected bishomoallylic intermediate has been suggested on the basis of similar rates and products found in the solvolysis of *endo*- and *exo*-2-bicyclo[3.1.0]hexyl 3,5-dinitrobenzoates; substitution of a 5-methyl group had a similar effect on the solvolysis of the two epimers.[42]

[40] P. G. Gassman, J. Seter, and F. J. Williams, *J. Am. Chem. Soc.*, **93**, 1673 (1971).
[41] P. G. Gassman and F. J. Williams, *J. Am. Chem. Soc.*, **93**, 2704 (1971).
[42] E. C. Friedrich and M. A. Saleh, *Tetrahedron Letters*, **1971**, 1373.

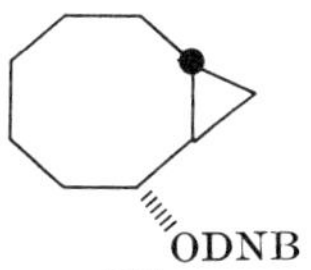

Studies of the bicyclo[3.1.0]hex-3-en-1-yl cation are reported. On the basis of the solvolysis of 2-*exo*-chloro-*exo*-bicyclo[3.1.0]hex-3-ene (**38**) which gives predominantly 2-*exo* derivatives, and photolysis of benzene in deuteriophosphoric acid, benzovalene (**39**) is suggested as the key intermediate in the photolytic hydration of benzene.[43] All the deuterium is incorporated into the 6-*endo* position (see **40**) in the photolysis, which excludes the benzenonium ion (**41**) as a possible intermediate.

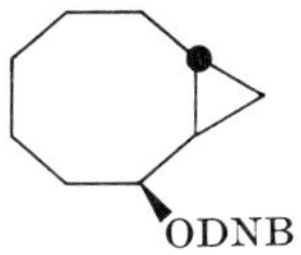

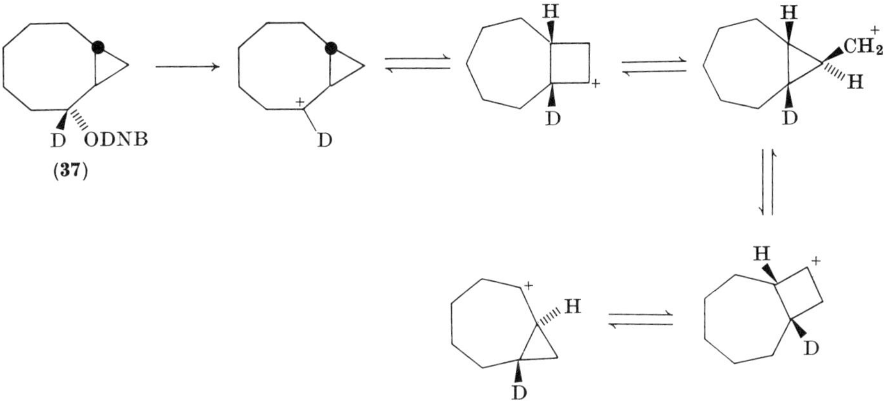

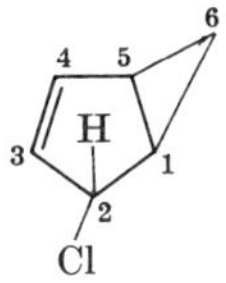

The bicyclo[3.1.0]hex-3-en-2-yl cation has also been generated in SbF_5–SO_2ClF at −100° and in FSO_3H–SO_2ClF at −120°.[44] NMR results indicate that the carbonium ion has a "closed" cyclopropane ring with conjugation from the C-1—C-6 and C-5—C-6 bonds to the allyl system. A slow sigmatropic rearrangement in the ion was observed

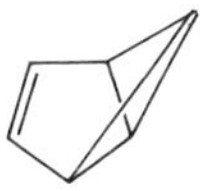

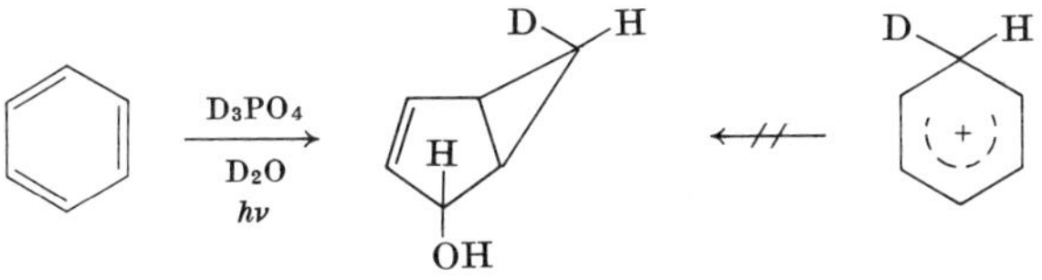

[43] J. A. Berson and N. M. Hasty, Jr., *J. Am. Chem. Soc.*, **93**, 1549 (1971).
[44] P. Vogel, M. Saunders, N. M. Hasty, Jr., and J. A. Berson, *J. Am. Chem. Soc.*, **93**, 1551 (1971).

with $\Delta F^* = 15 \pm 1$ kcal mole^{-1} at $-90°$: this compares with ring opening to benzenonium ion with $\Delta F^* = 19.8$ kcal mole^{-1}. The activation for the sigmatropic rearrangement is substantially higher than that found for the analogous heptamethyl ion ($\Delta F^* = 9$ kcal mole^{-1} at $-89°$).

In line with the probable antihomoaromatic character of the four-electron cationic intermediate, solvolysis of the bicyclo[3.2.1]octa-2,6-dienyl p-nitrobenzoates (**42**) is 235 times slower in aqueous acetone at 100° than that of the monoene analogue.[45] No evidence for the intermediacy of a 1,4-bishomotropylium ion was found in the solvolysis of (**43**) and (**44**), and both epimers reacted more slowly than cyclopent-2-enyl p-nitrobenzoate, an allylic model compound.[46] The results are consistent with intervention of the allylic intermediate (**45**). Solvolysis of bicyclo[4.3.1]deca-2,4,8-trien-*exo*-7-yl p-nitrobenzoate (**46**) in aqueous acetone is, however, 10³ faster than that of a comparable allylic system, this being consistent with the intervention of the bishomotropylium ion (**47**).[47]

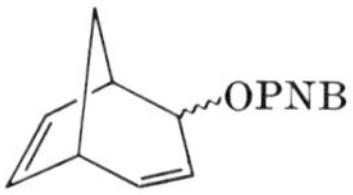

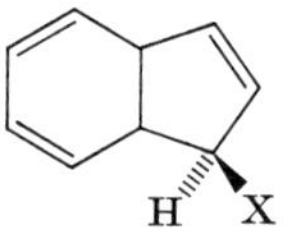

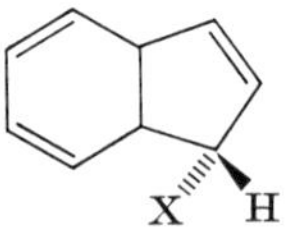

4-Acetoxy- and 4-hydroxy-1-methylbicyclo[2.2.2]oct-2-yl toluene-p-sulphonates rearrange on reaction with methyl-lithium or methylmagnesium iodide in ether, to yield bicyclo[3.2.1]octene derivatives.[48] It has been suggested that *cis*- and *trans*-2-pinanyl p-nitrobenzoates (**48** and **49**) solvolyse through delocalized ions that can be interconverted.[49]

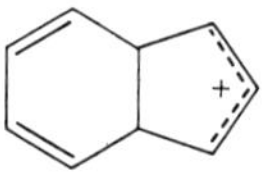

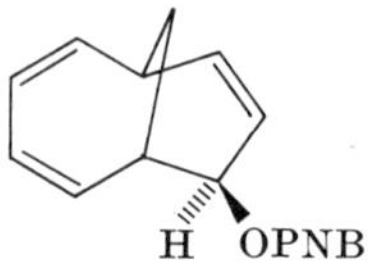

Common intermediates involving equilibrating ions are suggested for acid-catalysed hydration of α- and β-pinene.[50] With sulphuric acid in anhydrous acetic acid, α- and β-pinene appear to give an intimate ion pair that is stabilized by the counter-ion against attack by external nucleophiles;[51] olefin is formed by proton loss to the counter-ion. Support for the hypothesis was found by use of 10% aqueous acetic acid, in which the differences in product formation from the two olefins are reduced owing to greater dissociation of ion pairs, and also in the easier attack by external nucleophiles.

[45] A. F. Diaz, M. Sakai, and S. Winstein, *J. Am. Chem. Soc.*, **92**, 7477 (1970).
[46] D. Cook, A. Diaz, J. P. Dirlam, D. L. Harris, M. Sakai, J. C. Barborak, P. von R. Schleyer, and S. Winstein, *Tetrahedron Letters*, **1971**, 1405.
[47] P. Seidl, M. Roberts, and S. Winstein, *J. Am. Chem. Soc.*, **93**, 4089 (1971).
[48] W. Kraus and C. Chassin, *Ann. Chem.*, **747**, 98 (1971).
[49] J. R. Salmon and D. Whittaker, *J. Chem. Soc.* (B), **1971**, 1249.
[50] C. M. Williams and D. Whittaker, *J. Chem. Soc.* (B), **1971**, 668.
[51] C. M. Williams and D. Whittaker, *J. Chem. Soc.* (B), **1971**, 672.

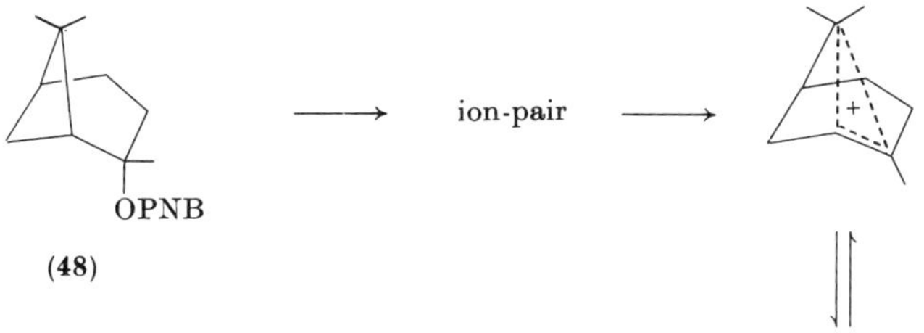

(48)

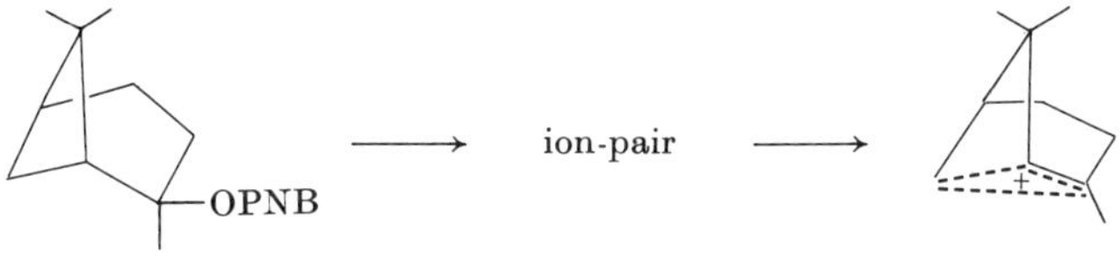

(49)

The stereochemistry of the addition of acid to bicyclo[1.1.0]butane has been examined.[52] In addition of chlorosulphonyl isocyanate to a number of bicyclo[1.1.0]-butanes and bicyclo[2.1.0]pentanes, the mechanism appears to involve initial S_E2-like attack at the least hindered bridgehead carbon atom followed by either ring inversion or rearrangement of a cyclobutyl to a cyclopropylcarbinyl carbonium ion.[53] Choice of the latter pathway is determined by the extent of substitution.

A number of reports have appeared of Ag- and other metal-catalysed rearrangement of bicyclobutanes although the mechanism still appears unclear. Thus, with tricyclo-[4.1.0.0^{2,7}]heptane (50) and 1-methyltricyclo[4.1.0.0^{2,7}]heptane stepwise bond cleavage is suggested, with the various metal catalysts acting as specific Lewis acids.[54] In this case the mechanism is considered to involve hybrid (51) of a metal-complexed carbene and a metal-bonded carbonium ion, and the products obtained from reaction of (50) with a number of metals can be envisaged as arising through (52a) and (52b). Although the way in which the different catalysts control the nature of the products remains to be elucidated, species such as (52) were trapped with formation of methyl ethers when the reaction was performed in the presence of the dicarbonylchlororhodium dimer in methanol. In similar studies it has been suggested that isomerizations of bicyclo[1.1.0]-butanes[55] and tricyclo[4.1.0.0^{2,7}]heptanes[56] in the presence of silver catalysts involve rupture of two strained bonds to produce a cationic species with Ag$^+$ bonded to an sp^2-hybridized electron pair (argentocarbonium ion). An experiment with tricyclo[4.1.0.0^{2,7}]-heptane has, however, indicated that initial attack of Ag(I) on bicyclobutane involves a one-bond rupture since again methyl ethers were obtained that were envisaged as arising from an intermediate similar to (52).[57] The intermediacy of carbenoid-Ag(I) or argento-carbonium ions in all Ag(I)-catalysed reactions has, however, been questioned. In comparison of Pd(II)- and Ag(I)-catalysed reactions of bicyclobutanes somewhat different

---

[52] G. Szeimies and A. Schlober, *Tetrahedron Letters*, **1971**, 3631.
[53] L. A. Paquette, G. R. Allen, Jr., and M. J. Broadhurst, *J. Am. Chem. Soc.*, **93**, 4503 (1971).
[54] P. G. Gassman and T. J. Atkins, *J. Am. Chem. Soc.*, **93**, 4597 (1971); see also P. G. Gassman and T. Nakai, *J. Am. Chem. Soc.*, **93**, 5897 (1971).
[55] L. A. Paquette, R. P. Henzel, and S. E. Wilson, *J. Am. Chem. Soc.*, **93**, 2335 (1971).
[56] L. A. Paquette and S. E. Wilson, *J. Am. Chem. Soc.*, **93**, 5934 (1971).
[57] M. Sakai, H. H. Westberg, H. Yamaguchi, and S. Masamune, *J. Am. Chem. Soc.*, **93**, 4611 (1971).

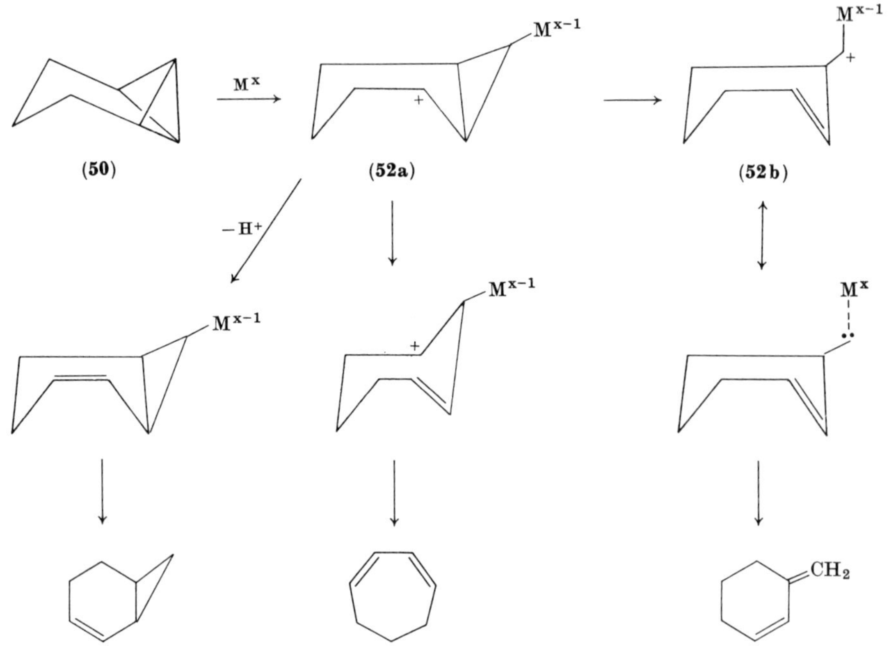

products were obtained.[58] In the Pd(II) reactions, products resembled those obtained from the corresponding diazo compounds, but products from the Ag(I) reactions resembled those of carbonium ion rearrangements. A suggested hypothesis for these processes was heterolytic cleavage of the C-1—C-2 bond followed by a rearrangement of cyclopropylcarbinyl–allylcarbinyl type.

Conformationally isomeric carbonium ions are formed in solvolysis of cis- and trans-bridgehead p-nitrobenzoates of the bicyclo[4.4.0]decane, bicyclo[4.3.0]nonane and bicyclo[3.3.0]octane series.[59] Rates of reaction of the epimeric pairs differed substantially and this, together with product variations, requires different intermediates. Ion pairs do not appear to play an important role but the carbonium ions are suggested as retaining some of the ring geometry of their precursors.

6-Oxabicyclo[3.2.1]oct-1-ylmethyl p-bromobenzenesulphonate has been synthesized and allowed to react in acetic acid; the large amount of unrearranged product was suggested as due to dipole-induced resistance to C—C bond migration.[60] The reactivities of 9-substituted 9-chlorobicyclo[3.3.1]nonanes have been examined.[61] Other studies

[58] M. Sakai and S. Masamune, J. Am. Chem. Soc., **93**, 4610 (1971).
[59] R. C. Fort, Jr., R. E. Hornish, and G. A. Liang, J. Am. Chem. Soc., **92**, 7558 (1970).
[60] E. J. Grubbs, R. A. Froehlich, and H. Lathrop, J. Org. Chem., **36**, 504 (1971).
[61] L. Baiocchi, M. Giannangeli, and G. Palazzo, Tetrahedron Letters, **1970**, 5025; **1971**, 2077.

include acid-catalysed rearrangements of tetrafluoro-1-methoxy-3,5-dimethylbenzobarrelene;[62] 4-hydroxycyclohexa-2,5-dienones[63] and exo- and endo-1,2,4,4-tetramethylbicyclo[3.2.0]hept-6-en-2-ols;[64] thermal rearrangement of bicyclo[6.1.0]nonatrienyl chloride;[65] and solvolysis of 7-methylenebicyclo[3.2.1]oct-1-yl toluene-p-sulphonate.[66]

Polycyclic Systems

Calculations of bridgehead reactivities have been extended and improved.[67] Slight modifications have been made to the computer program of Westheimer's classical treatment. New "ideal" bond angles (112.4° for CCH_2C, 111.3° for CCHC, 106.1° for HCH, 110.7° for HCHC and 107.8° for HCC) have been used to conform with those of simple alkanes and "harder" non-bonded potentials have been adopted. Results obtained were less satisfactory for halides than for sulphonate esters, possibly owing to the greater conformational flexibility of the former. Some of the variations in k_{OTs}/k_{Br} in tertiary systems are, it is suggested, the result of ground-state interactions in the esters which would be unimportant with halides; this might also be a factor with secondary systems. The extremely low reactivity of tricyclo[5.2.1.0^{4,10}]dec-10-yl toluene-p-sulphonate was interpreted as resulting from the absence of *anti*-periplanar C—C or C—H bonds to afford hyperconjugative stabilization of the transition state.

Solvolyses of 2-adamantyl toluene-p-sulphonate and 1-adamantyl bromide in 80% ethanol and 75% dioxan show only minor rate enhancement on addition of sodium azide. No correlation is found between relative rate and product data, indicating the limiting character of the reactions.[68] The Sneen, Carter and Kay linear free-energy relationship between carbonium ion stability and azide incorporation was extended to include a number of less reactive substrates.

The non-variance of the α-D effects in solvolysis of 2-adamantyl 2,2,2-trifluoroethanesulphonate in a number of solvents indicates the lack of solvent participation in the rate-determining step.[69] Solvolysis rates of 1-adamantyl toluene-p-sulphonate have been correlated against Y values,[70] and silver-ion assisted reactions of 1-adamantyl halides in ethanol have been studied.[71]

Rates of solvolysis of (53)–(55) have been compared.[72] (54) and (55) react ca. 10^3 times slower than (53) as a result of non-bonded interactions: no rearrangement was observed.

The [3,5,7-$^2H_1$]-1-adamantyl carbonium has been generated in SbF_5–SO_2ClF.[73] No proton scrambling was seen after heating at 105° for 90 min, setting a lower limit for the

---

[62] H. Heaney and S. V. Ley, *Chem. Comm.*, **1971**, 1342.
[63] G. F. Burkinshaw, B. R. Davis, E. G. Hutchinson, P. D. Woodgate, and R. Hodges, *J. Chem. Soc.* (C), **1971**, 3002.
[64] L. B. Jones and V. K. Jones, *J. Org. Chem.*, **36**, 1017 (1971).
[65] J. C. Barborak, T. M. Su, P. von R. Schleyer, G. Boche, and G. Schneider, *J. Am. Chem. Soc.*, **93**, 279 (1971).
[66] F. E. Ziegler and J. A. Kloek, *Tetrahedron Letters*, **1971**, 2201.
[67] R. C. Bingham and P. von R. Schleyer, *J. Am. Chem. Soc.*, **93**, 3189 (1971); *Tetrahedron Letters*, **1971**, 23.
[68] D. J. Raber, J. M. Harris, R. E. Hall, and P. von R. Schleyer, *J. Am. Chem. Soc.*, **93**, 4821 (1971).
[69] J. M. Harris, R. E. Hall, and P. von R. Schleyer, *J. Am. Chem. Soc.*, **93**, 2551 (1971); V. J. Shiner, Jr., and R. D. Fischer, *J. Am. Chem. Soc.*, **93**, 2553 (1971).
[70] D. N. Kevill, K. C. Kolwyck, and F. L. Weitl, *J. Am. Chem. Soc.*, **92**, 7300 (1970).
[71] D. N. Kevill and V. M. Horvath, *Tetrahedron Letters*, **1971**, 711.
[72] R. C. Bingham, P. von R. Schleyer, Y. Lambert, and P. Deslongchamps, *Can. J. Chem.*, **48**, 3739 (1970).
[73] P. Vogel, M. Saunders, W. Thielecke, and P. von R. Schleyer, *Tetrahedron Letters*, **1971**, 1429.

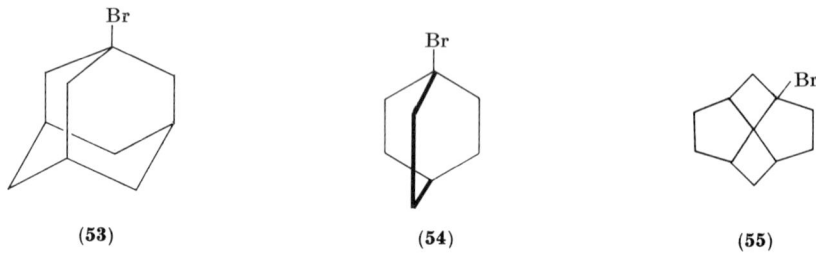

(53)   (54)   (55)

rate of 2×10^{-5} sec^{-1} and leading to the minimum energy barrier of $E_A > 29$ or 30.5 kcal mole^{-1}. This compares with $E_A = 15$ kcal mole^{-1} found for 1,2 hydride shifts in acyclic ions.

Reaction of bromine with protoadamantene has led to the formation of 2,4-diaxial-dibromoadamantane,[74] and 1,2- and 2,4-disubstituted adamantanes have been prepared by a reaction sequence beginning with 4-protoadamantanone.[75] Protoadamantane has been shown to undergo ionic bromination in the 6-position as predicted by conformational-analysis calculations;[76] the solvolysis rate of 6-bromoprotoadamantane in 80% ethanol was also shown to agree to within a factor of three with the value predicted on the basis of strain calculations.

Under Koch–Haaf carboxylation conditions 2-(1-adamantyl)propan-2-ol yields 3-isopropyladamantane-1-carboxylic acid as a consequence of rearrangement by inter-molecular hydride shifts.[77] Use of dilute conditions in the Koch–Haaf reaction has been demonstrated to provide a synthetic route to 2-(1-adamantyl)-2-methylpropionic acid. Rearrangements also take place in the bromination of 2-methyladamantane; 4-eq- and 4-ax-bromo-2-(dibromomethylene)adamantane are formed together with 1-bromo-adamantane and cis- and trans-1-bromo-4-methyladamantane.[78] Homoadamantyl methyl ketone is a rearrangement product from addition of 1-adamantyl cation to prop-2-ynyl alcohol.[79]

78% of the total scrambling possible on a statistical basis is found after heating [2-^{14}C]adamantane (56) with aluminium bromide in CS$_2$ at 110° for 8 hr; suggested rearrangements are indicated in the attached formulae.[80] Rearrangement of 2-methyl-adamantane to 1-methyladamantane proceeds by a similar skeletal rearrangement but is facilitated by the methyl group. Oligomerization of 2,4-dehydroadamantane by aluminium chloride in carbon disulphide appears to proceed by initial hydride abstraction from the tertiary position.[81] Evidence has been presented for formation of a small amount of 1,2'-biadamantane in this reaction.[82]

Reaction of spiro[adamantane-2,4'-homoadamantan-5'-ol] in concentrated H$_2$SO$_4$ gave adamantylideneadamantane and 2,2'-biadamantane which arise from the same cation by proton extrusion and intermolecular hydride abstraction from the tertiary

---

[74] D. Lenoir, P. von R. Schleyer, C. A. Cupas, and W. E. Heyd, *Chem. Comm.*, **1971**, 26.
[75] B. D. Cuddy, D. Grant, and M. A. McKervey, *Chem. Comm.*, **1971**, 27.
[76] A. Karim, M. A. McKervey, E. M. Engler, and P. von R. Schleyer, *Tetrahedron Letters*, **1971**, 3987.
[77] D. J. Raber, R. C. Fort, E. Wiskott, C. W. Woodworth, P. von R. Schleyer, J. Weber, and H. Stetter, *Tetrahedron*, **27**, 3 (1971).
[78] J. R. Alford, D. Grant, and M. A. McKervey, *J. Chem. Soc.* (C), **1971**, 880.
[79] J. K. Chakrabarti and A. Todd, *Chem. Comm.*, **1971**, 556.
[80] Z. Majerski, S. H. Liggero, P. von R. Schleyer, and A. P. Wolf, *Chem. Comm.*, **1971**, 1596.
[81] H. J. Storesund, *Tetrahedron Letters*, **1971**, 3911.
[82] H. J. Storesund, *Tetrahedron Letters*, **1971**, 4353.

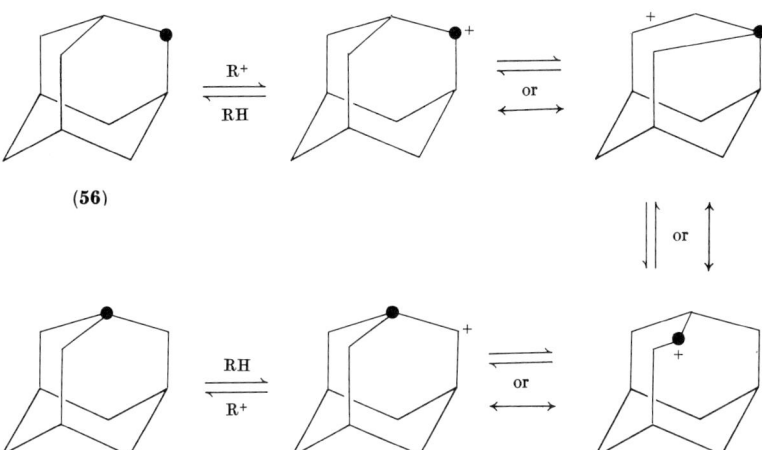

(56)

position of another molecule.[83] Oxidation of adamantane in fuming sulphuric acid is considered to involve hydride transfer from the adamantane nucleus to the acid.[84] A pinacol–pinacolone rearrangement and other reactions of highly hindered spiro-adamantane[85] and pyrolysis of 2-adamantyl methanesulphonate have been reported.[86]

In the acetolysis of exo-tetracyclo[3.3.0.0.^{2,8}.0^{4,6}]oct-3-yl p-nitrobenzoate (57), deuterium labelling implicates stepwise, stereospecific cleavage of first one, followed by the second cyclopropane to yield ultimately the allylic cation (58); degeneracy arising from equilibration of the C-3, C-2 and C-1 bridges was excluded.[87] Products arising from the three distinct carbonium ions were identified and an approximate energy diagram for the reaction was constructed.

Product studies on the acetolysis of (59) and (60) have indicated that steric or strain effects could not have been responsible for the high degree (>95%) of stereochemical

[83] E. Boelema, H. Wynberg, and J. Strating, *Tetrahedron Letters*, **1971**, 4029.
[84] H. W. Geluk and J. L. M. A. Schlatmann, *Rec. Trav. Chim.*, **90**, 516 (1971).
[85] G. B. Gill and D. Hands, *Tetrahedron Letters*, **1971**, 181.
[86] J. Boyd and K. H. Overton, *Chem. Comm.*, **1971**, 211.
[87] W. Lotsch and A. S. Kende, *Angew. Chem. Int. Ed.*, **10**, 559 (1971).

retention observed in the solvolysis of the secondary toluene-p-sulphonates. Nearly equal acetate distribution was found in the reactions of (59) and (60).[88]

Treatment of 1-(N,N-dichloroamino)adamantane (61) with $AlCl_3$–CH_2Cl_2 gave a rearrangement product that was converted into endo-7-(aminomethyl)bicyclo[3.3.1]-nonan-3-one by aqueous acid. The mechanism is considered to involve electron-deficient nitrogen (62) rearrangement and collapse of the carbonium ion with nucleophiles.[89]

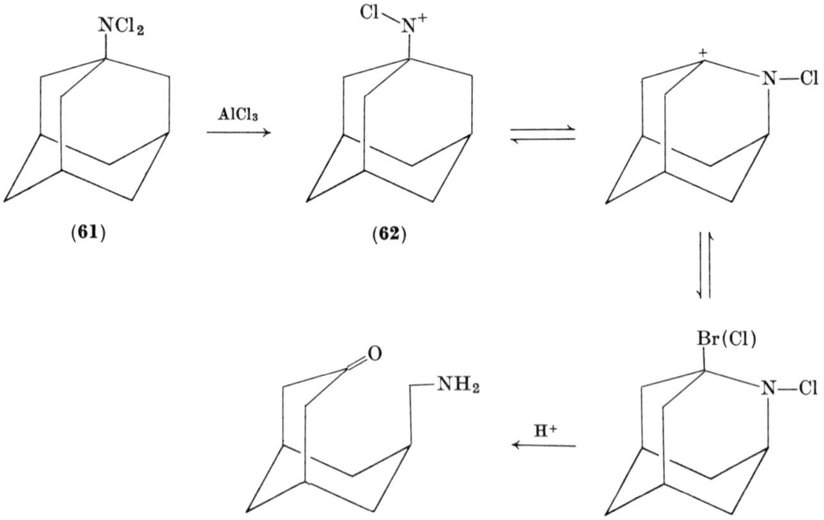

Sativene has been isomerized by treatment with $Cu(OAc)_2$ in refluxing acetic acid to a mixture of isomeric sesquiterpenes consisting of recovered sativene (7%), cyclosativene (32%) and isosativene (61%).[90] Neoclovene has been prepared.[91]

Participation by Aryl Groups

Arylalkyl Compounds

The effects of sodium azide on the solvolysis of a series of 1-aryl-2-propyl and 1-aryl-1-propyl toluene-p-sulphonates has been investigated.[92] Reactions involving higher amounts of the k_s component show greater rate enhancement due to added azide and higher amounts of azide product. Solvolysis of the β-aryl derivatives were dissected into their component k_s and k_Δ pathways. A plot of log k_s against log (azide ion-water competition ratios) shows a marked deviation and no correlation is found against the carbonium ion–selectivity relationship.[93] Although this has been previously suggested as

---

[88] W. L. Dilling and J. A. Alford, *Tetrahedron Letters*, **1971**, 761; see *Org. Reaction Mech.*, **1969**, 18.
[89] P. Kovacic, J. H. Liu, E. M. Levi, and P. D. Roskos, *J. Am. Chem. Soc.*, **93**, 5801 (1971).
[90] J. E. McMurry, *J. Org. Chem.*, **36**, 2826 (1971).
[91] T. F. W. McKillop, J. Martin, W. Parker, J. S. Roberts, and J. R. Stevenson, *J. Chem. Soc. (C)*, **1971**, 3375.
[92] D. J. Raber, J. M. Harris, and P. von R. Schleyer, *J. Am. Chem. Soc.*, **93**, 4829 (1971); see *Org. Reaction Mech.*, **1969**, 21.
[93] R. A. Sneen, J. V. Carter, and P. S. Kay, *J. Am. Chem. Soc.*, **88**, 2594 (1966); see also ref. 68.

evidence for the intermediacy of ion pairs, direct displacement on the neutral substrate was favoured in the present case.

Results from acetolysis of *threo*-3-aryl-2-butyl *p*-bromobenzenesulphonates have been discussed in terms of k_Δ and k_S processes. A suggestion was made that the k_S pathway involves initial ionization to an ion-pair intermediate.[94]

The ethylenephenonium ion and ethylene-*p*-toluenium ions have been prepared in SbF_5–SO_2ClF at $-78°$: styryl carbonium ions are also formed.[95]

^{14}C isotope effects for the solvolysis of phenethyl and *p*-methoxyphenethyl *p*-nitrobenzenesulphonate in formic and trifluoroacetic acid have provided further evidence for aryl participation.[96] Deuterium isotope effects have been measured in the formolysis of *threo*-1-methyl-2-*p*-tolylpropyl toluene-*p*-sulphonate and the results are considered to be consistent with an unsymmetrical transition state.[97]

Neighbouring aryl group participation has been observed in the deamination of 3-arylalanine ethyl esters in trifluoroacetic acid and is more important when a *p*-methoxy rather than a nitro group is a substitutent in the aromatic ring.[98] In the solvolysis of 2,2,2-triphenylethyl toluene-*p*-sulphonate in alcohol–dioxan solution, substitution is more favourable than elimination under high pressure, owing to the decrease in volume of the system.[99]

2-(1-Azulenyl)ethyl toluene-*p*-sulphonates (**63**) undergo acetolysis 68,000 times faster than phenethyl toluene-*p*-sulphonate. A series of compounds substituted in the

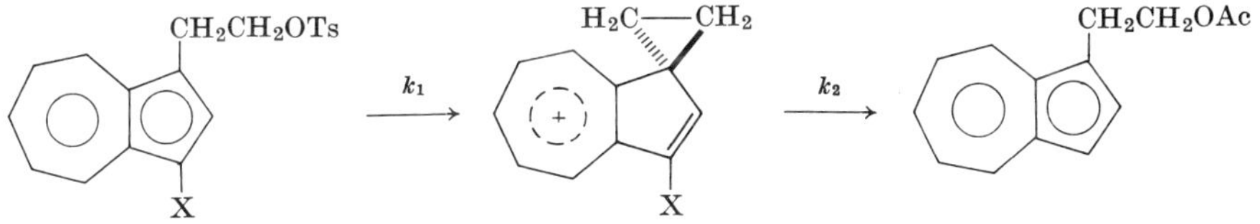

(**63**)

X = H, NO_2, $COCH_3$, Br, SCH_3

3-position was used and, from the $\rho = 3.745$ and the absence of substantial deuterium scrambling (0 when X = H and 12% when X = NO_2), ion-pair return from the intermediate was discounted. The processes, it was concluded, proceed totally by the k_Δ pathway.[100] Neighbouring-group participation has also been observed in mass-spectral fragmentation of some azulenes.[101]

4-(*p*-Methoxyphenyl)[2,2-$^2H_1$]butyl *p*-bromobenzenesulphonate, in formic acid, yielded dideuterated 6-methoxytetralin (36%) and 4-(*p*-methoxyphenyl)butyl alcohol after LiAlH$_4$ reduction; 74% of the tetralin was shown to arise by an Ar_1-5 pathway. The relative rate of Ar_1-5 to Ar_1-6 pathway was calculated as 5.6 and the selectivity was noted as low for an electrophilic aromatic substitution by a poor nucleophile.[102]

[94] H. C. Brown and C. J. Kim, *J. Am. Chem. Soc.*, **93**, 5765 (1971).
[95] G. A. Olah and R. D. Porter, *J. Am. Chem. Soc.*, **92**, 7627 (1970).
[96] Y. Yukawa, T. Ando, M. Kawada, K. Token, and S. G. Kim, *Tetrahedron Letters*, **1971**, 847.
[97] S. L. Loukas, F. S. Varveri, M. R. Velkou, and G. A. Gregoriou, *Tetrahedron Letters*, **1971**, 1803.
[98] K. Koga, C. C. Wu, and S. Yamada, *Tetrahedron Letters*, **1971**, 2283.
[99] Y. Okamoto and T. Yano, *Tetrahedron Letters*, **1971**, 919.
[100] R. N. McDonald and J. R. Curtis, *J. Am. Chem. Soc.*, **93**, 2530 (1971).
[101] R. G. Cooks, N. L. Wolfe, J. R. Curtis, H. E. Petty, and R. N. McDonald, *J. Org. Chem.*, **35**, 4048 (1970).
[102] V. R. Haddon and L. M. Jackman, *J. Am. Chem. Soc.*, **93**, 3832 (1971).

It is suggested that cyclized products from acid-catalysed dehydration of 3-methyl-3-(p-tolyl)butan-1-ol arise through Ar_1-4 and Ar_1-5 participation.[103]

No participation was observed in the base-promoted reaction of 3-(p-hydroxyphenyl)-propyl p-bromobenzenesulphonate.[104] MO calculations have been made on the phenethyl carbonium ion to phenonium ion transformation.[105]

Benzonorbornene Derivatives

β-Deuterium isotope effects have been measured for solvolysis of *exo*- and *endo*-2-benzonorbornenyl p-bromobenzenesulphonates.[106] Small isotope effects were found for for *exo*-derivatives in acetic and formic acid but the value obtained after substitution of two nitro groups into the 6- and the 7-position of the benzene ring was slightly greater than for the parent compound. The small deuterium isotope effects were ascribed to aryl participation but, as this decreased with the dinitro derivative, hyperconjugation with neighbouring C—H bonds could increase with a corresponding increase in isotope effect. The larger isotope effect with the dinitro derivative was also discussed in terms of greater release of steric compression in the transition state than in the parent compound. Larger isotope effects were found in solvolysis of the *endo*-derivatives, involving no aryl participation. A larger effect of the 3-*exo*-proton in solvolysis of *exo*-derivatives was considered to result from the favourable alignment with the vacant p orbital on C-2, compared to the 60° dihedral angle between the 3-*endo*-proton and the p orbital. No geometric dependence of 3-*exo*- or 3-*endo*-protons was found in solvolysis of the *endo*-isomers, owing to an identical alignment of both protons with the vacant p orbital of the classical ion.

Effects of substituents on solvolysis of benzobicyclo[2.2.2]octen-2(*exo* and *endo*)-yl toluenesulphonates have also been studied, and many of the features of the analogous [2.2.1]system were exhibited.[107]

Solvolysis of a series of *endo*- and *exo*-chromium tricarbonyl complexes of benzonorbornenyl methanesulphonates (**64–66**) have been examined.[108] Acetolysis of (**66**)

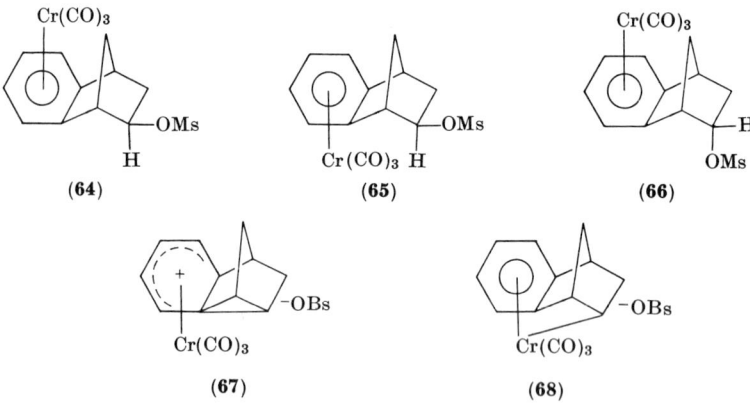

[103] A. A. Khalaf and R. M. Roberts, *J. Org. Chem.*, **36**, 1040 (1971).
[104] W. le Noble and B. Gabrielsen, *Tetrahedron Letters*, **1971**, 3417.
[105] E. I. Snyder, *J. Am. Chem. Soc.*, **92**, 7529 (1970).
[106] H. Tanida and T. Tsushima, *J. Am. Chem. Soc.*, **93**, 3011 (1971).
[107] H. Tanida and S. Miyazaki, *J. Org. Chem.*, **36**, 425 (1971).
[108] R. S. Bly and R. C. Strickland, *J. Am. Chem. Soc.*, **92**, 7459 (1970); D. K. Wells and W. S. Trahanovsky, *J. Am. Chem. Soc.*, **92**, 7461 (1970).

is 14 times slower than that of the uncomplexed analogue, reflecting the electron-withdrawing inductive effect of the tricarbonylchromium. Complex (**65**) undergoes reaction 266 times faster than (**64**), indicating the additional driving force of the *endo*-tricarbonylchromium over that due to the benzene ring itself; a part of this difference might be explained in terms of steric effects. It was concluded that the additional stabilization in the transition state for solvolysis of (**65**) could be the result of either homoconjugation (**67**) or direct interaction (**68**); the latter of these was favoured because of the slower rate of acetolysis of (**64**).

Aryl participation has been demonstrated to be less important in the solvolysis of benzo[7,8]bicyclo[4.2.1]nonen-9(*exo*)-yl *p*-bromobenzenesulphonate (**69**) than in that of the analogous [2.1.1], [2.2.1] and [3.2.1] systems.[109] Molecular models indicate that

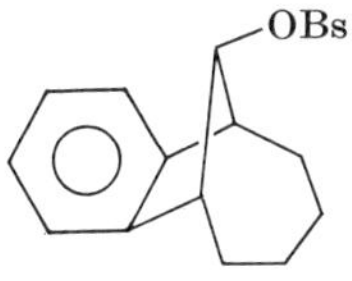

(**69**)

the distance between the reacting carbon and the junctive carbons of the aromatic and aliphatic part increase in these bicyclo[*n*.2.1] systems as *n* increases. Thus in the solvolysis of (**69**) solvent competes with aryl assistance.

Addition of hydrogen chloride in acetic acid to benzonorbornadiene involves two competing mechanisms, one involving a non-classical carbonium ion and the other a *syn*-concerted molecular addition.[110]

Other Reactions involving Phenyl Participation

Complete retention of configuation was found in acetolysis, methanolysis and trifluoroacetolysis of an optically active 1-tosyloxy[2.2]paracyclophane (**70**).[111] Acetolysis rates

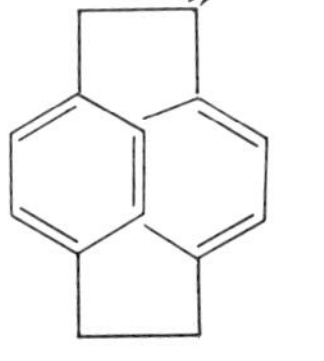

(**70**)

were about 10^2 times faster than for isopropyl toluene-*p*-sulphonate. This, together with the almost total absence of olefin products, points to a predominant k_Δ process, and a highly strained bridged ion is suggested as an intermediate with positive charge distributed over both benzene rings. In one ring, charge can be delocalized by direct benzene-to-α-carbon bonding and in the other by relayed conjugation between the benzene rings.

[109] H. Tanida and T. Irie, *J. Org. Chem.*, **36**, 2777 (1971).
[110] S. J. Cristol and J. M. Sullivan, *J. Am. Chem. Soc.*, **93**, 1967 (1971).
[111] R. E. Singler and D. J. Cram, *J. Am. Chem. Soc.*, **93**, 4443 (1971); see also R. E. Singler, R. C. Helgeson, and D. J. Cram, *J. Am. Chem. Soc.*, **92**, 7625 (1970).

Although this would involve molecular distortion some decreases in $\pi-\pi$ repulsion of the benzene rings strain was expected.

Some aryl participation has been observed in displacement reactions of substituted phenylcyclohexyl toluene-p-sulphonates,[112] and in the cyclization of diazo-ketones.[113] No phenyl migration was observed in acetolysis or trifluoroacetolysis of 5,5-diphenylcyclononyl toluene-p-sulphonate.[114] Migration of the ethyl group from C-9 to C-10 in 9,9,10-triethylphenanthrenium ion catalysed by HCl–AlCl$_3$ in CH$_2$Cl$_2$ has been studied.[115]

Participation by Double and Triple Bonds

Double-bond Participation

Evidence has been presented for a symmetrically bridged structure of the 7-methoxy-norbornadienyl carbonium ion (**71**).[116] When the solution was warmed a degenerate *syn–anti* isomerization was observed for the bridgehead protons [(**72**) to (**73**)] (these are

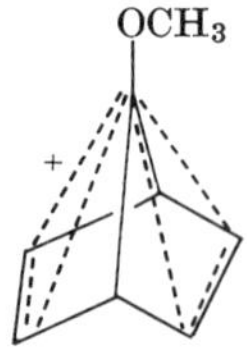

(**71**)

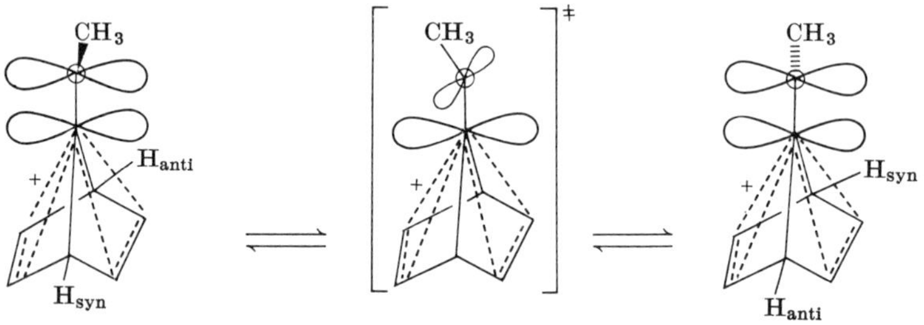

(**72**) ⇌ ⇌ (**73**)

unsymmetrical since the methyl function must lie in the C-1–C-7–C-4 plane to enable $p-p$ π overlap to occur between C-7 and oxygen. The activation energy for this isomerization is much lower than for the 7-methoxynorbornyl or norbornenyl carbonium ions and was interpreted as due to greater delocalization of charge into the double bonds, thus decreasing the requirement for $p-p$ π overlap between C-7 and the oxygen atom. Observed chemical shifts were best explained in terms of a symmetrically bridged transition state.

[112] S. K. Core and F. J. Lotspeich, *J. Org. Chem.*, **36**, 499 (1971).
[113] D. J. Beames, T. R. Klose, and L. N. Mander, *Chem. Comm.*, **1971**, 773.
[114] J. M. McIntosh, M. H. Miskow, and G. Wood, *Can. J. Chem.*, **49**, 3243 (1971).
[115] D. V. Korchagina, B. G. Derendyaev, and V. G. Shubin, *Izv. Akad. Nauk SSSR, Ser. Khim*, **1971**, 441; see also V. G. Shubin, D. V. Korchagina, B. G. Derendyaev, V. I. Mamatyuk, and V. A. Koptyug, *J. Org. Chem. USSR*, **6**, 2074 (1970).
[116] R. K. Lustgarten, M. Brookhart, and S. Winstein, *Tetrahedron Letters*, **1971**, 141.

The direct observation of the 7-hydroxy- and 7-methoxy-quadricyclyl carbonium ions in FSO_3H-SO_2, together with their rates and products of rearrangement, have been reported.[117]

In the thermal isomerizations of *anti*-7-norbornenyl and 7-norbornadienyl thiocyanates in polar aprotic solvents, isothiocyanates of retained structure but no tricyclic products were formed; this was attributed to the greater stability of the bicyclic skeleton than of the tricyclic structure.[118] The nature of the carbonium ions produced from *exo*-5-norbornenyl and nortricyclyl thiocyanates has been discussed in terms of ion pairs and in relation to the position of the anion relative to the carbonium ion.[119]

Substantial participation was found in the solvolysis of *endo*-bicyclo[3.2.1]oct-6-en-8-yl toluene-*p*-sulphonate in acetic acid and ethanol–water mixtures. Product of retained configuration was obtained and the rate of solvolysis was ca. 10^5 times faster than for the saturated analogue. The effectiveness of double-bond participation in this system is less than in the analogous [2.2.1] and [2.1.1] structures owing to the greater distance between the carbon–carbon double bond and the developing carbonium ion.[120]

Enhanced double-bond participation was found in the acetolysis of (**74**) compared with that of the parent norbornyl system.[121] The rate of solvolysis was 3×10^4 times faster than for the saturated analogue, and the product acetate had completely retained configuration. An explanation was advanced that in the solvolysis the benzene ring forces the C—O bond ionization to follow a more vertical reaction coordinate (**75**) than

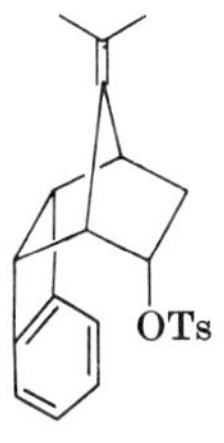

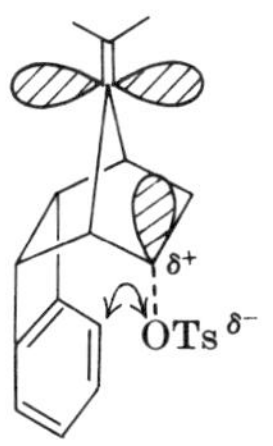

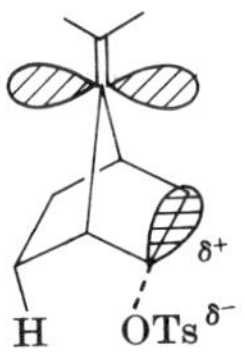

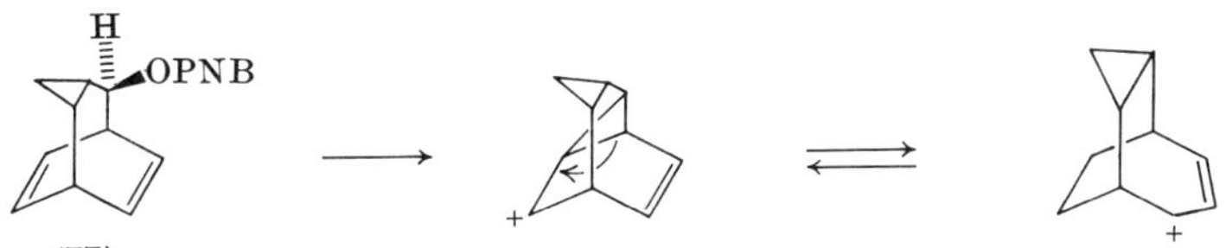

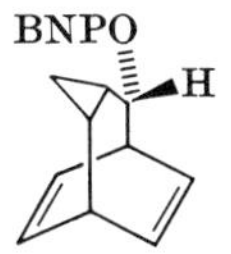

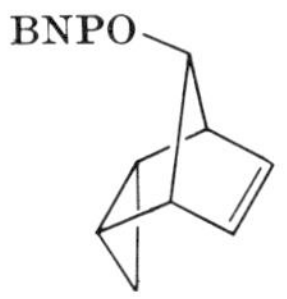

[117] M. Brookhart, R. K. Lustgarten, D. L. Harris, and S. Winstein, *Tetrahedron Letters*, **1971**, 943.
[118] L. A. Spurlock and Y. Mikuriya, *J. Org. Chem.*, **1971**, 1549.
[119] L. A. Spurlock and W. G. Cox, *J. Org. Chem.*, **93**, 146 (1971).
[120] B. A. Hess, Jr., *J. Am. Chem. Soc.*, **93**, 1000 (1971).
[121] R. Baker and J. C. Salter, *J. Chem. Soc.* (B), **1971**, 757.

for the norbornyl system (76). Better overlap of the double bond with the developing p orbital would then be found in (75).

It has been suggested that (77) solvolyses in acetic acid with homoallylic π-participation followed by rearrangement. Greater than 95% of the acetate with retained configuration was found in the acetolysis of (78), and cyclopropane participation was postulated since in this case a good alignment can be found for overlap involving the cyclopropane ring and the developing orbital. (78) was more reactive than (77).[122]

Solvolysis of syn-endo-tricyclo[3.2.1.0^{2,4}]oct-6-en-8-yl p-nitrobenzoate (79) in dioxan–water is 10 times faster than that of anti-norbornen-7-yl p-nitrobenzoate. Formation of mainly product with retained configuration indicates the absence of bridge flipping in the reaction.[123]

Cyclization to a [2.2.1] system occurs in the acetolysis of (3-methoxy-4-methylcyclohex-3-enyl)methyl p-bromobenzenesulphonate (80) owing to the additional nucleophilicity

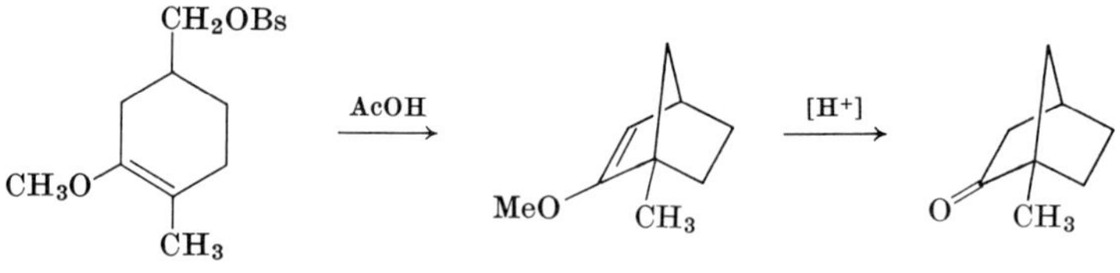

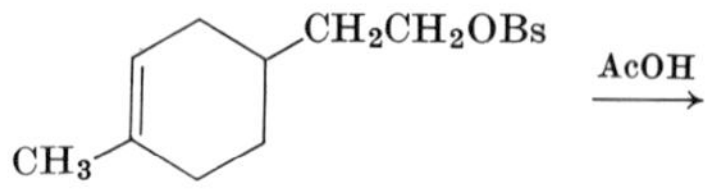

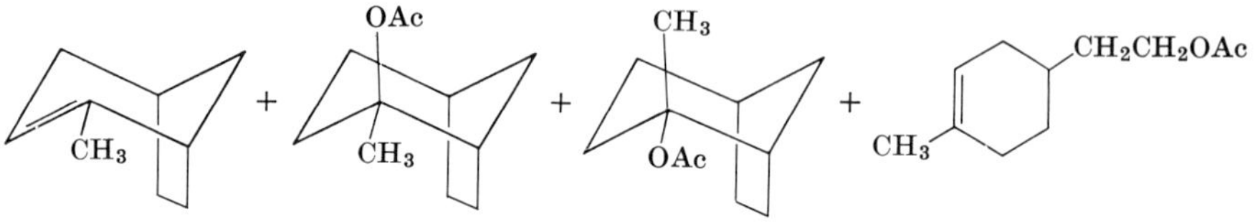

rendered to the double bond by the substituents.[124] 2-(4-Methylcyclohex-3-enyl)ethyl p-bromobenzenesulphonate (81) undergoes acetolysis at 80° 6.8 times faster than 2-(cyclohex-3-enyl)ethyl p-bromobenzenesulphonate; a number of cyclized products are obtained from (81).[125] A good yield of bicyclo[3.2.1]octan-2-one can be obtained after

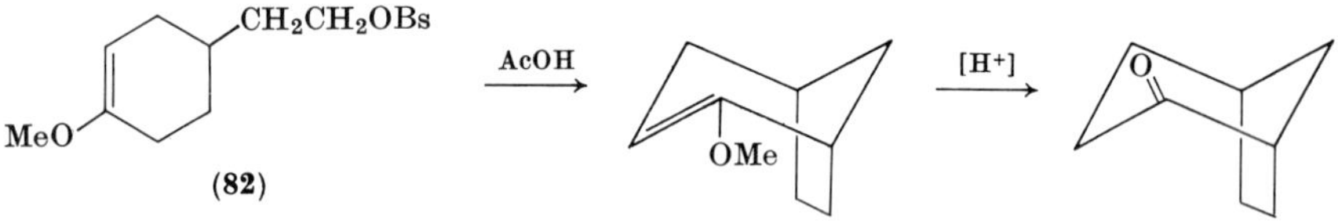

[122] A. S. Kende, J. K. Jenkins, and L. E. Friedrich, *Chem. Comm.*, **1971**, 1215.
[123] A. P. Jovanovich and J. B. Lambert, *J. Chem. Soc.* (B), **1971**, 1129.
[124] H. Felkin and C. Lion, *Tetrahedron*, **27**, 1403 (1971).
[125] H. Felkin and C. Lion, *Tetrahedron*, **27**, 1375 (1971).

acid hydrolysis of the products of acetolysis of 2-(4-methoxycyclohex-3-enyl)ethyl p-bromobenzenesulphonate (**82**); this undergoes reaction at 80° 97 times faster than 2-(cyclohex-3-enyl)ethyl p-bromobenzenesulphonate.[126]

The acetolysis of 2-(cyclobut-2-enyl)ethyl toluene-p-sulphonate has been shown to involve initial opening to hexa-trans-3,5-dienyl toluene-p-sulphonate.[127]

Norborn-5-ene-cis-2,exo-3-diyl bis(toluene-p-sulphonate) reacts in acetic acid 500 times more rapidly than the related saturated bis(toluene-p-sulphonate). It was suggested that homoallylic participation becomes more important in charge delocalization when an electron-withdrawing group is adjacent to the leaving group.[128] A similar situation was found in the solvolysis of bicyclo[2.2.2]oct-5-ene-cis-2,exo-3-diyl bis-(toluene-p-sulphonate) and no evidence was found for dicarbonium formation.[129]

A number of allenic compounds have been studied. 2-Methylpenta-3,4-dienyl toluene-p-sulphonate (**83**) and 1-methylpenta-3,4-dienyl toluene-p-sulphonate (**84**) yield similar product mixtures on acetolysis; a common bisected cyclopropyl-

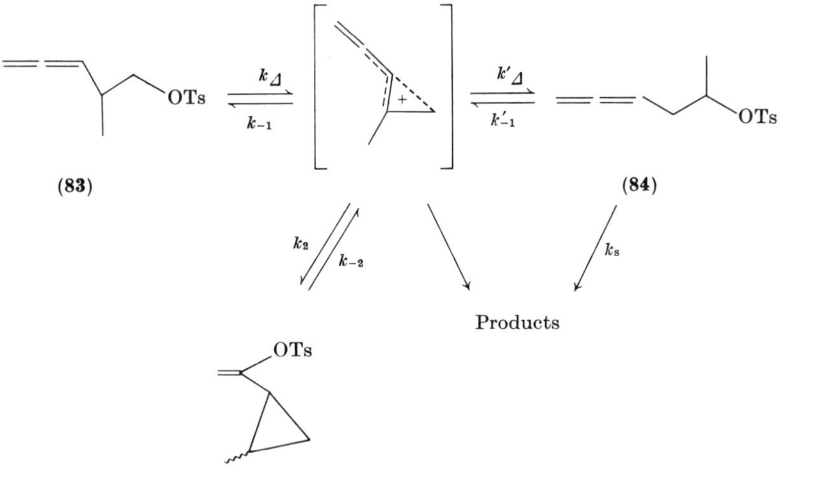

carbinyl carbonium ion intermediate was suggested. Double-bond participation is indicated since (**83**) and (**84**) react 40 and 160 times faster than 2-methylpentyl toluene-p-sulphonate.[130] Similar studies are reported for 1,6-dimethylhepta-4,5-dienyl toluene-p-sulphonate[131] and a number of other methyl-substituted allenic derivatives and cyclopropylvinyl chloride.[132]

Solvolysis of 3-(2-tosyloxyethyl)cyclopentanone (**85**) yields only 3% of norcamphor (**86**) in acetic acid but in trifluoroacetic acid only norcamphor was observed; participation of the enol is required.[133] Interaction of a double bond with a protonated ketone group in (**87**) is reported.[134]

[126] H. Felkin and C. Lion, *Tetrahedron*, **27**, 1387 (1971).
[127] R. N. McDonald and E. P. Lyznicki, Jr., *J. Am. Chem. Soc.*, **93**, 5920 (1971).
[128] J. B. Lambert and A. G. Holcomb, *J. Am. Chem. Soc.*, **93**, 2994 (1971).
[129] J. B. Lambert and A. G. Holcomb, *J. Am. Chem. Soc.*, **93**, 3952 (1971).
[130] R. S. Macomber, *J. Am. Chem. Soc.*, **92**, 7101 (1970).
[131] B. Ragonnet, M. Santelli, and M. Bertrand, *Tetrahedron Letters*, **1971**, 955.
[132] M. Santelli and M. Bertrand, *Tetrahedron Letters*, **1971**, 3767.
[133] J. L. Marshall, *Tetrahedron Letters*, **1971**, 753.
[134] H. Hart and G. M. Love, *Tetrahedron Letters*, **1971**, 2267.

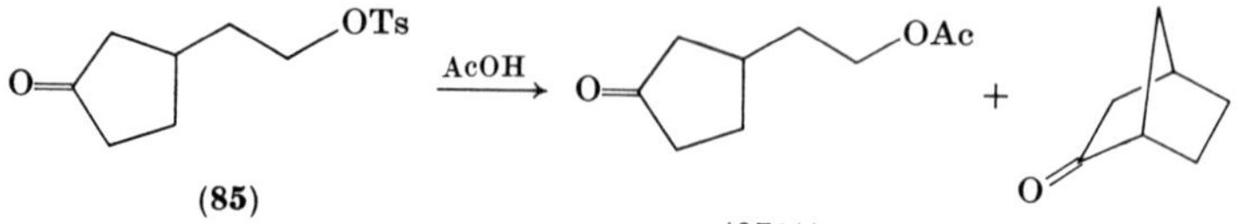

(85) —AcOH→ (97%) + (86) 3%

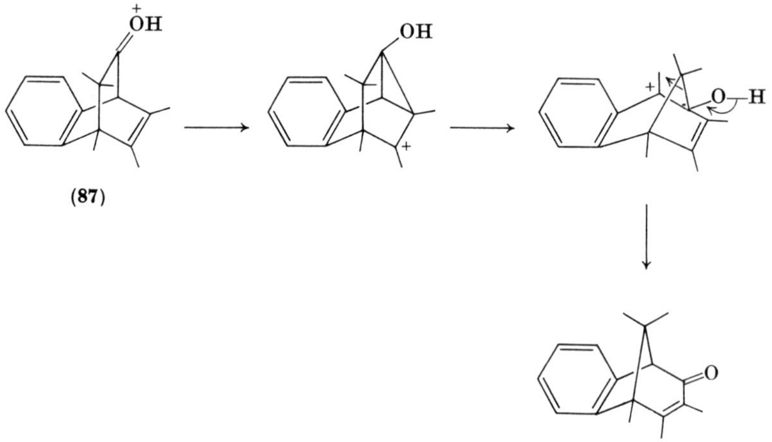

(87)

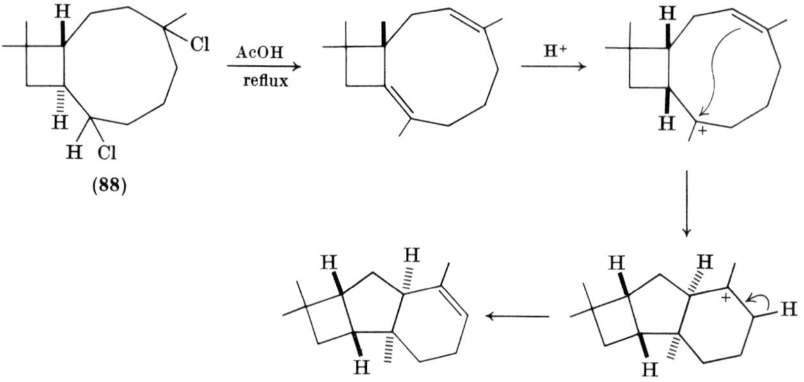

(88) —AcOH reflux→ —H⁺→

(89)

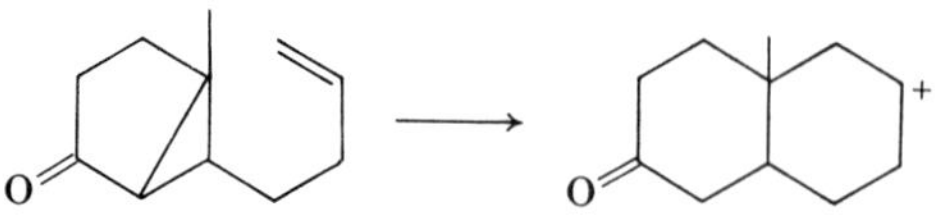

(90) → (91)

2,6,10,10-Tetramethyltricyclo[7.2.0.0^{2,7}]undec-5-ene (**89**) is formed by cyclization after elimination of HCl from caryophyllene dihydrochloride (**88**).[135] Participation of the double bond in the reaction of 6-(but-3-enyl)-5-methyl-*endo*-bicyclo[3.1.0]hexan-2-one (**90**) in presence of stannic chloride leads to efficient generation of a cyclohexyl carbonium ion (**91**).[136]

Formation of a lanosterol system through a non-enzymic cyclization[137] and a discussion of the enzymic cyclization process of partially cyclized squalene 2,3-oxide to the lanosterol system have been reported.[138]

MINDO/2 calculations have been made on the 7-norbornyl, 7-norbornenyl and 7-norbornadienyl ions and parent hydrocarbons. A barrier of 26 kcal mole^{-1} was calculated for the "flipping" of 7-norbornadienyl carbonium ion, which is consistent with experimental evidence.[139] Interaction of orbitals in 7-norbornenyl carbonium ion has been discussed.[140]

An investigation has been made of the π route to 2,4-disubstituted adamantanes,[141] and the addition of hydrogen chloride and deuterium chloride to bicyclo[2.2.1]heptadiene and quadricyclene.[142] Cyclization of 4-(2,6,6-trimethylcyclohexenyl)-2-methylbutanal gives 1,2,3,4,4a,7,8a-octahydro-2α,4$\alpha\beta$,5,8$\alpha\beta$-tetramethylnaphthalene-1β-ol, involving a Wagner–Meerwein shift.[143]

Triple-bond Participation

Cyclodecyn-6-one, cyclodecyn-6-ol and 6-methylcyclodecyn-6-ol have been shown to cyclize transannularly to yield bicyclo[4.4.0]decanones, highly selectively.[144]

Acetylenic bond participation has been demonstrated in the conversion of (**92**) into (**93**),[145] and this procedure has formed part (**94** to **95**) of a synthesis of ($\pm$)-progesterone.[146]

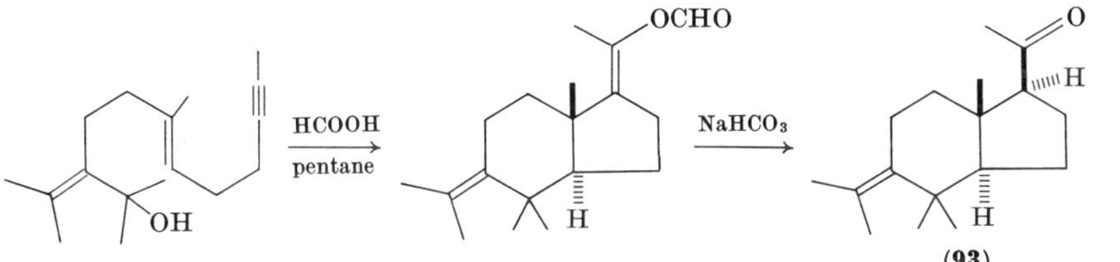

A method of generating the D-ring of 20-keto steroids involving cyclization at a triple bond, as in (**96**) to (**97**), has been described.[147]

---

[135] K. Gollnick, G. Schade, A. F. Cameron, C. Hannaway, and J. M. Robertson, *Chem. Comm.*, **1971**, 46.
[136] G. Stork and P. A. Grieco, *Tetrahedron Letters*, **1971**, 1807.
[137] E. E. van Tamelen and J. W. Murphy, *J. Am. Chem. Soc.*, 92, 7204 (1970).
[138] E. E. van Tamelen and J. H. Freed, *J. Am. Chem. Soc.*, 92, 7207 (1970).
[139] M. J. S. Dewar and W. W. Schoeller, *Tetrahedron*, 27, 4401 (1971).
[140] R. Hoffman, *Accounts Chem. Res.*, 4, 1 (1971).
[141] T. Sasaki, S. Eguchi, and T. Toru, *Tetrahedron Letters*, **1971**, 1109.
[142] T. C. Morrill and B. E. Greenwald, *J. Org. Chem.*, 36, 2769 (1971).
[143] G. Saucy, R. E. Ireland, J. Bordner, and R. E. Dickerson, *J. Org. Chem.*, 36, 1195 (1971).
[144] R. J. Balf, B. Rao, and L. Weiler, *Can. J. Chem.*, 49, 3135 (1971).
[145] W. S. Johnson, M. B. Gravestock, R. J. Parry, R. F. Myers, T. A. Bryson, and D. H. Miles, *J. Am. Chem. Soc.*, 93, 4330 (1971).
[146] W. S. Johnson, M. B. Gravestock, and B. E. McGarry, *J. Am. Chem. Soc.*, 93, 4332 (1971).
[147] P. T. Lansbury and G. E. Dubois, *Chem. Comm.*, **1971**, 1107.

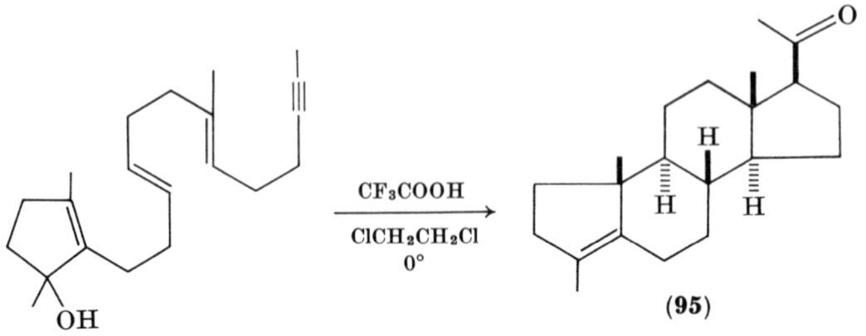

Reactions of Small-ring Compounds

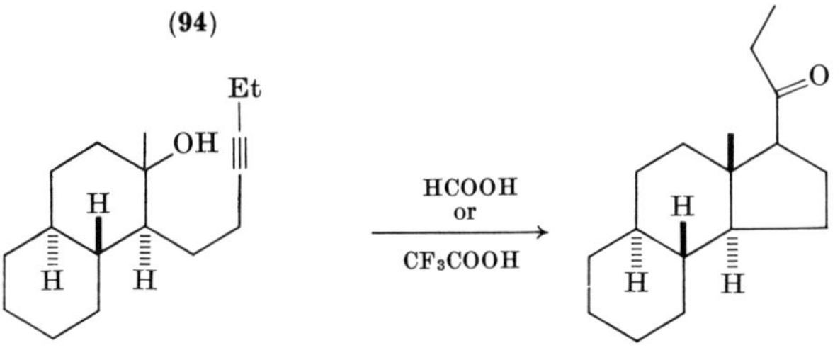

Cyclopropylmethyl Derivatives

In studies on the three different rearrangement processes involving the cyclopropyl cation, the cyclopropylcarbinyl–cyclopropylcarbinyl, the cyclopropylcarbinyl–cyclobutyl, and the cyclopropylcarbinyl-allylcarbinyl have all been confirmed as completely stereospecific. In these experiments, the substrate was labelled with six deuterium atoms, leaving only a single hydrogen atom which facilitated NMR interpretations.[148]

No deuterium scrambling was found in solvolysis and borohydride products of 1-[^{2}H$_1$]cyclobutyl methanesulphonate; all the deuterium was found at the methine positions in cyclopropylmethanol, cyclobutanol, methylcyclopropane and but-1-ene, indicating absence of hydrogen shift between the methine and methylene positions. Products from [1,1-^{2}H$_2$]cyclopropylmethyl and [2,2,4,4-^{2}H$_4$]cyclobutyl methanesulphonates indicated absence of hydrogen shift among the three methylene positions.[149] These results were explained in terms of rate-determining formation of two intimate ion pairs formed from the cyclopropylmethyl derivative and one intimate ion pair from cyclobutyl methanesulphonate. Reaction with a nucleophile, isomerization to another ion pair or further dissociation to a solvent-separated ion pair could then take place. Borohydride, a strong nucleophile, would attack at the intimate ion-pair stage, allowing less scrambled methylene positions than in attack by water.

Further details on the solvolysis of 2-methylene-1-adamantyl and spiro[cyclopropane-1,2'-(1'-adamantyl)] derivatives have been published.[150]

Essentially no deuterium scrambling was observed in solvolysis of cis- or trans-bicyclo[4.1.0]hept-2-yl p-nitrobenzoate, confirming the earlier conclusion of Goering

[148] Z. Majerski and P. von R. Schleyer, J. Am. Chem. Soc., **93**, 665 (1971); see Org. Reaction Mech., **1970**, 13.
[149] Z. Majerski, S. Borcic, and D. E. Sunko, Chem. Comm., **1970**, 1636.
[150] V. Buss, R. Gleiter, and P. von R. Schleyer, J. Am. Chem. Soc., **93**, 3927 (1971); see Org. Reaction Mech., **1970**, 33; **1969**, 49.

and Rubenstein that each isomer gave a common cyclopropylcarbinyl carbonium ion intermediate.[151]

Hydrolysis rates and products for *exo-* and *endo-*tricyclo[3.2.0.0^{2,7}]hept-3-yl *p*-nitrobenzoates (**98** and **99**) are almost identical, indicating equally efficient overlap of the cyclopropane ring with the developing *p* orbital from either conformation. Absence of cyclopropylcarbinyl–cyclopropylcarbinyl isomerization was indicated by the absence of protium incorporation at the α-position in the solvolysis of α-[^{2}H$_1$] derivatives.[152]

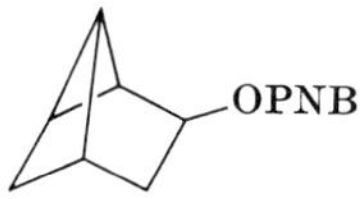

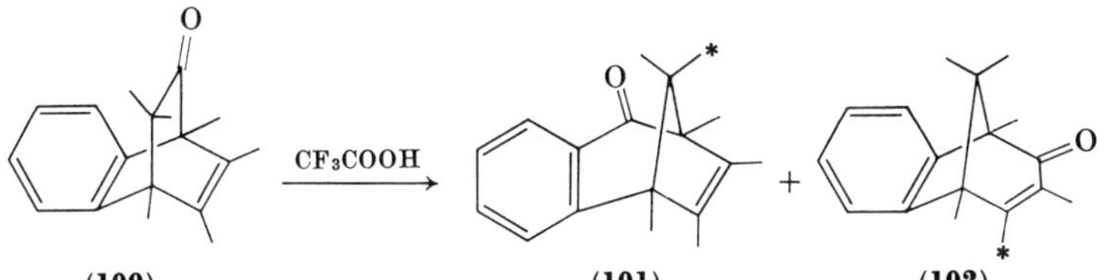

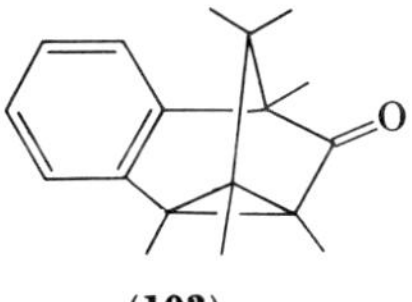

---

[151] L. E. Friedrich and G. B. Schuster, *Tetrahedron Letters*, **1971**, 3171; H. L. Goering and K. E. Rubenstein, Abstr. Papers, 151st Amer. Chem. Soc. Meeting, March 28–31, 1966, p. K011; see also ref. 42.

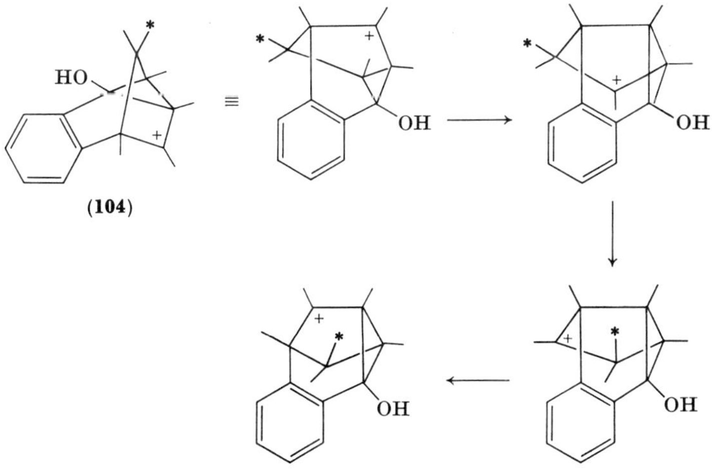

[152] R. K. Lustgarten, *J. Am. Chem. Soc.*, **93**, 1275 (1971).

A new degenerate cyclopropylcarbinyl carbonium ion rearrangement has been suggested as occurring in the addition of trifluoroacetic acid to (**100**) which led to the formation of (**101**)–(**103**). On treatment of (**101**) with CF_3CO_2D at 60°, five of the six methyl groups underwent deuterium exchange and only the 8-*anti*-methyl group (asterisk) was unexchanged. Protonation of (**101**) to give (**104**) was suggested, followed by a series of rearrangements in which the carbon atoms constituting the three-carbon bridge in (**101**) "revolve" with respect to the five-membered ring. Stereospecific methyl migration is required in this mechanism and all the intermediates are cyclopropylcarbinyl carbonium ions except the immediate product of methyl migration.[153]

The hydroazulenyl acetate (**106**), an intermediate for a projected guaiol synthesis, has been prepared by solvolysis of the hydroindanyl methanesulphonate (**105**), the mechanism involves a cyclopropylcarbinyl rearrangement.[154]

Rates and products of solvolysis of a series of cycloalkylcarbinyl toluene-*p*-sulphonates of ring size five to twelve have been determined.[155] The major product, except for the cyclopentyl and cyclohexyl derivatives, was the 1-methylcycloalkene arising by a 1,2-hydride shift. Hydrogen participation (k_Δ) was proposed as being directly related to the relief of strain on formation of the sp^2 centre in the ring and also partially to the inductive effect of the ring. k_s, the solvent assisted pathway, was competitive with the k_Δ process. Ring expansion also occurred on solvolysis of the cyclopentyl derivative, and carbon–carbon bond participation was suggested.

Activation parameters for solvolysis of cyclopropylcarbinyl and cyclobutyl 2-naphthalenesulphonates have been determined in acetic acid and ethanol. ΔS^* for acetolysis of the cyclopropylcarbinyl derivative is much more negative than that for related substrates. A good correlation of rates of solvolysis for crotyl against cholesteryl derivatives was found in a series of solvents of varying ionizing strength.[156]

Hydration of (+)-sabinene to terpinen-4-ol proceeds with a high degree of stereospecificity (>80%). Hydration of (−)-α-thujene is slower but yields the same products in fairly similar proportions; products are derived from cyclopropylcarbinyl intermediates.[157] (−)-Chrysanthemone and (+)-2,4,4-trimethylbicyclo[3.1.1]hept-2-en-6-one undergo BF_3-catalysed rearrangement; a mechanism of product formation is suggested, involving sequential 1,2-alkyl shifts and intermediate cyclopropylcarbinyl carbonium ions.[158]

Formulation of squalene from presqualene pyrophosphate appears to involve conversion of a cyclopropylcarbinyl pyrophosphate to a cyclobutyl carbonium ion followed by

---

[153] H. Hart and G. M. Love, *J. Am. Chem. Soc.*, **93**, 6264 (1971); see also ref. 134.
[154] J. A. Marshall and A. E. Greene, *Tetrahedron Letters*, **1971**, 859.
[155] A. P. Krapcho and R. G. Johanson, *J. Org. Chem.*, **36**, 146 (1971).
[156] D. D. Roberts, *J. Org. Chem.*, **35**, 4059 (1970).
[157] T. Norin and L. A. Smedman, *Acta Chem. Scand.*, **25**, 2010 (1971).
[158] W. F. Erman, R. S. Treptow, P. Bakuzis, and E. Wenkert, *J. Am. Chem. Soc.*, **93**, 657 (1971).

conversion into an allyl carbonium ion.[159] Conversion of presqualene alcohol into squalene is suggested as involving equilibration of bicyclobutonium ions.[160]

Rearrangement of the 1-cyclopropylvinyl carbonium ion, obtained by uncatalysed and silver-catalysed solvolysis of 1-cyclopropyl-1-iodoethylene, appears to proceed through an unsymmetrical 2-methylenecyclobutyl carbonium ion.[161]

Decomposition of N-nitroso-N-(exo-bicyclo[3.1.0]hex-2-en-6-yl)urea which leads to formation of bicyclo[2.1.1]hexane and tricyclo[3.2.0.0^{2,6}]hexane derivatives,[162] and methanolysis of 1,1-bis(diazomethyl)cyclopropane[163] have been studied. Products from deoxygenation of cyclopropylcarbinol with KOH–CHBr$_3$ were found to contain cyclopropylcarbinyl, cyclobutyl and allylcarbinyl alcohols, bromides and ethers. Deoxygenation of cyclopropyl[^{14}C]carbinol gave isotopically scrambled cyclopropylcarbinol containing 22% of the ^{14}C rearranged from α-C to the cyclopropyl group. On the basis of the ^{14}C results it was concluded that both deoxidation and deamination involve similar rearrangement processes.[164]

Participation by More Remote Cyclopropyl Rings

Homocyclopropylcarbinyl rearrangement appears to occur in the solvolysis of pentacyclo[4.4.0.0^{2,4}0^{3,8}0^{5,7}]dec-9-yl toluene-p-sulphonate (**107**) in acetic acid.[165] Whereas the infrared spectrum of the corresponding ketone would predict a rate 0.2 times that of cyclohexyl toluene-p-sulphonate, the observed rate is 10^3 times faster. This enhancement is, however, considerably less than that found for systems that can yield degenerate carbonium ions. 97% of 2-acetoxytetracyclo[5.3.0.0^{3,5}0^{4,8}]dec-9-ene derived from (**108**) was formed in the solvolysis.

Evidence from solvolysis studies on the exo-syn-tricyclo[3.2.1.0^{2,4}]oct-6-en-8-yl derivatives (**109**) and (**110**) in aqueous acetone and acetic acid, respectively, suggests

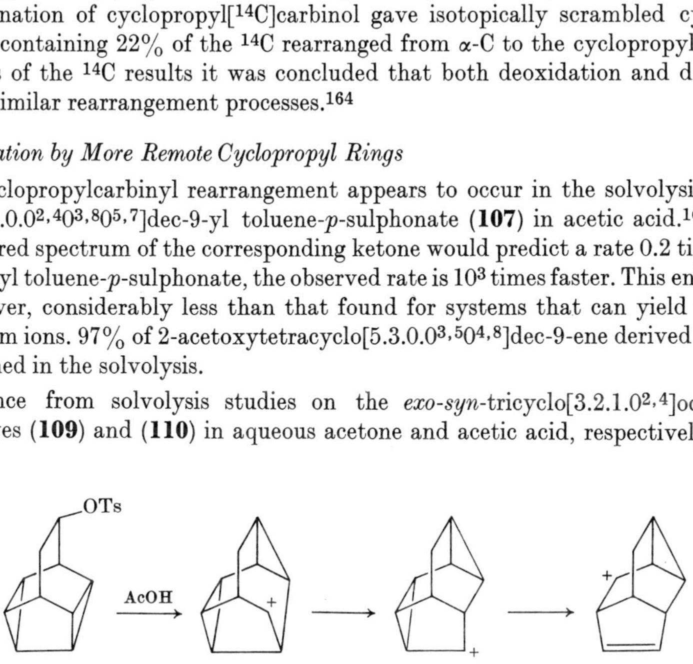

(**107**) $\xrightarrow{\text{AcOH}}$ → → (**108**)

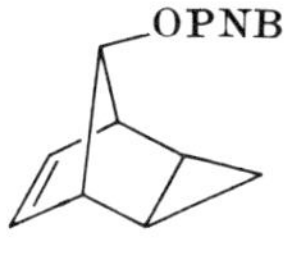

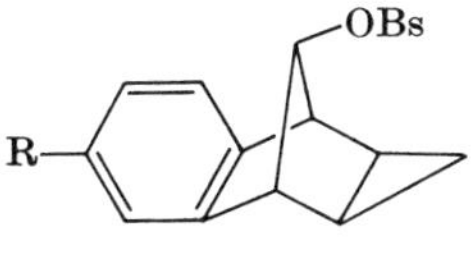

(**109**)   (**110**)
R = H, Br or CH$_3$

---

[159] E. E. van Tamelen and M. A. Schwartz, *J. Am. Chem. Soc.*, **93**, 1780 (1971).
[160] L. J. Altman, R. C. Kowerski, and H. C. Rillings, *J. Am. Chem. Soc.*, **93**, 1782 (1971).
[161] S. A. Sherrod and R. G. Bergman, *J. Am. Chem. Soc.*, **93**, 1925 (1971).
[162] W. Kirmse and F. Scheidt, *Angew. Chem. Int. Ed.*, **10**, 263 (1971).
[163] W. Kirmse and B. Brinkman, *Chem. Comm.*, **1971**, 259.
[164] A. J. Cessna and C. C. Lee, *Can. J. Chem.*, **48**, 3953 (1970).
[165] W. G. Dauben and C. H. Schallhorn, *J. Am. Chem. Soc.*, **93**, 2254 (1971), see *Org. Reaction Mech.*, **1970**, 40.

that rate enhancement is best attributed to relief of non-bonded interactions in the transition state rather than to homoconjugative electron release by cyclopropane. Retained alcohol and acetate were the sole products in nearly quantitative yield on solvolysis of (109) and (110).[166]

Products (112)–(114) were obtained from 3'-diazospiro[cyclopropane-1,7'-norbornan-3'-one] (111) on decomposition in aqueous tetrahydrofuran. Formation of the bridged ion (116) from participation of the *syn*-C-7'-cyclopropyl bond with loss of nitrogen from (115) resulting from *exo*-protonation of (111) was suggested. The occurrence of this participation was considered to be favoured in this case by relief of an unfavourable electrostatic interaction between a positive charge at C-3' and the adjacent carbonyl group. Formation of (118) on decomposition of 3'-diazospiro[cyclopropane-1,5'-norbornan-2'-one] (117) was explained by participation of the C-4'—C-7' bond.[167]

Solvolysis of (119) was found to be 15,929 times faster than that of (121) in aqueous ethanol at 70°, whilst the rate of solvolysis of (120) was only 10.3 times faster. This result shows the relative importance of "vertical" stabilization of a carbonium ion by an adjacent cyclopropyl group compared to a double bond. (122) and (123) were obtained on solvolysis of (119) in 50% aqueous acetone containing calcium carbonate, and this product mixture could be converted into (122) and the ketone (124) under unbuffered solvolytic conditions.[168]

Substitution of a cyclopropyl group at C-7 of benzo-2-norbornyl toluene-*p*-sulphonates (125) had little effect on the solvolysis reactions but substitution at C-3 (126) produced a marked decrease in the *exo/endo* rate ratio. These results are consistent with the non-classical nature of the benzo-2-norbornenyl carbonium ion.[169]

No significant interaction occurs between the β-cyclopropyl group and the allylic cation to form the four π-electron antiaromatic trishomocyclopentadienyl carbonium ion in the solvolysis of (127). In 80% aqueous acetone at 100°, (127) solvolyses 3.1 times faster than (128), reflecting the rate-retarding effect of the β-cyclopropyl group. (127) yields product with predominant (96%) retention of configuration.[170]

Participation by the carbon–carbon bond of the oxiran ring in the solvolysis of (128) has been suggested to account for the greater reactivity (259) of (128) compared to (129) (1).[171] The dialdehyde (130) was produced in 52% yield on reaction of (128) in aqueous acetone, together with a number of other products. Any carbon–carbon bond participation in the solvolysis of (129) must have occurred after the ionization step, and the formation of 17% of cyclohept-1-enecarboxaldehyde (131) was evidence for this.

Cyclopropane participation has been suggested as the major reason for the high reactivity of *endo*-tricyclo[3.2.1.0^{2,4}]oct-6-en-8-one in thermal decarbonylation.[172] On the basis of NMR and UV spectra, cyclopropyl conjugation was suggested in unprotonated benzo[*a*]spiro[2,5]octa-1,4-diene-3-one.[173] Electrostatic interaction of small rings with charge has been used to explain the reactivity of a number of cyclopropyl derivatives in solvolysis reactions.[174]

---

[166] M. A. Battiste, P. F. Ranken, and R. Edelman, *J. Am. Chem. Soc.*, **93**, 6276 (1971).
[167] P. Yates and J. D. Fenwick, *J. Am. Chem. Soc.*, **93**, 4618 (1971).
[168] B. A. Howell and J. G. Jewett, *J. Am. Chem. Soc.*, **93**, 798 (1971).
[169] D. Lenoir, P. von R. Schleyer, and J. Ipaktschi, *Ann. Chem.*, **750**, 28 (1971).
[170] A. F. Diaz, D. L. Harris, M. Sakai, and S. Winstein, *Tetrahedron Letters*, **1971**, 303.
[171] D. L. Whalen, *J. Am. Chem. Soc.*, **92**, 7619 (1970).
[172] S. C. Clarke and B. L. Johnson, *Tetrahedron*, **27**, 3555 (1971).
[173] P. Rys, P. Skrabal, and H. Zollinger, *Tetrahedron Letters*, **1971**, 1797.
[174] I. J. Miller, *Austral. J. Chem.*, **24**, 2013 (1971).

Carbonium Ions

Reactions of Cyclopropyl Derivatives

Heating the dichlorocyclopropane derivative (**132**) at 160° for 4 hr gave a mixture of (**133**) and (**134**) in the ratio 91:9; similarly, (**135**) gave the same allyl chlorides in the ratio 14:86. The rearrangement was shown to be largely stereospecific since, when heated with sodium methoxide in methanol, both (**132**) and (**133**) gave the methyl ethers (**136**) and (**137**) in the same ratio 87:13. Similarly, (**135**) and (**134**) gave the same ethers in the ratio 15:85. It was concluded that the external nucleophile did not participate in the ring-opening step and the rearrangement of cyclopropyl chlorides to allyl chlorides was at least 95% stereospecific.[175]

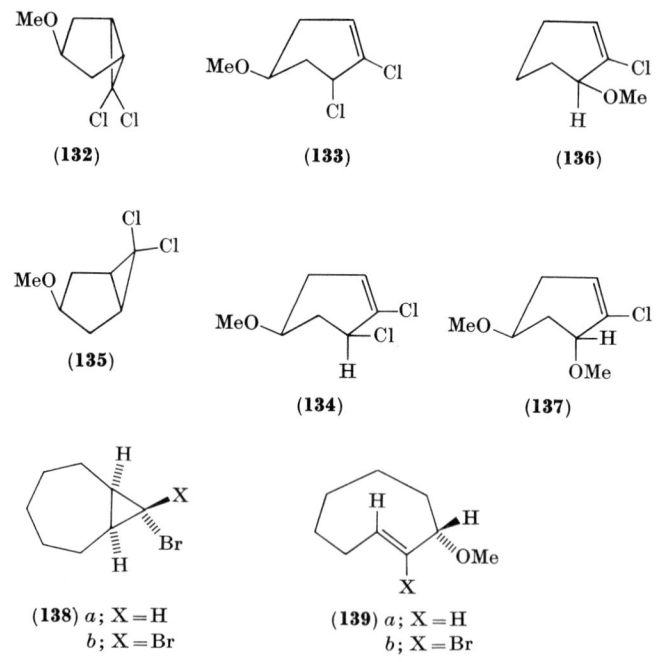

[175] I. Fleming and E. J. Thomas, *Tetrahedron Letters*, **1971**, 2485.

(140), a; X=H
    b; X=Br

(141), a; X=H
    b; X=Br

(142), a; X=H
    b; X=Br

High yields of 3-methoxy- and 2-bromo-3-methoxy-*trans*-cyclo-octene (**139a** and **139b**) were obtained when *exo*-8-bromo- and 8,8-dibromo-bicyclo[5.1.0]octane (**138a** and **138b**) were treated with silver perchlorate in methanol solution at 20°.[176] Silver-catalysed methanolysis of *exo*-8-bromo[5.1.0]oct-2-ene (**140a**) has been shown to yield approximately a 1:1 mixture of 3-methoxycyclo-octa-*cis*-1,*trans*-4-diene (**141a**) and 5-methoxycyclo-octa-*cis*-1,*trans*-3-diene (**142a**). Similarly, a 1:1 mixture of 4-bromo-3-methoxycyclo-octa-*cis*-1,*trans*-4-diene (**141b**) and 4-bromo-5-methoxycyclo-octa-*cis*-1,*trans*-3-diene (**142b**) was obtained from 8,8-dibromobicyclo[5.1.0]oct-2-ene (**140b**).[177] The results are consistent with methanol attack at both ends of a free *trans,trans*-allylic carbonium ion.

Methanolysis of 1-chloro-1-(phenylthio)cyclopropane has been studied.[178] Anodic oxidation of methyl-substituted cyclopropanecarboxylic acids has been discussed in terms of cationic intermediates.[179]

Addition of hydrogen chloride to tetracyclo[3.2.0.0^{2,7}.0^{4,6}]heptane-1,5-dicarboxylic acid results in formation of two products, *exo*-5-chlorotricyclo[2.2.1.0^{2,6}]heptane-2,*endo*-3-dicarboxylic acid and its 2-*exo* epimer. Cyclopropane cleavage, in this case, appears to proceed by inversion with the nucleophile and addition of the electrophile by a combination of retention and inversion. The mechanism involves protonation to give a delocalized carbonium ion followed by attack by nucleophile.[180] Cyclopropyl ring opening has been observed in addition of dichloro- and dibromo-carbene to bicyclo[4.1.1]oct-2-ene.[181]

A number of MINDO/2 calculations have been made involving the cyclopropyl carbonium ion. The ring opening is predicted to be disrotatory but also to require activation; differences from observed solvolysis results and other calculations were assumed to arise from geometrical differences.[182] Antiaromatic pathways for ring opening of the cyclopropyl carbonium ion have been calculated and compared with aromatic pathways. The differences in activation energy between the two pathways were 30.6, 35.0 and 27.6 kcal mole^{-1} for the cyclopropyl carbonium ion, carbanion and radical, respectively. Calculations of the reverse reactions by the "forbidden" antiaromatic pathways uncovered problems. The allyl carbonium ion failed to cyclize to the cyclopropyl carbonium ion and it was concluded that no reliance should be placed on calculations based on the simple transition-state picture unless it has been established that the forward and the backward reaction follow the same reaction path.[183] In a further investigation into potential surfaces it was concluded that pericyclic reactions can be divided into two classes, those involving aromatic transition states that have unique geometries of

---

[176] M. S. Baird and C. B. Reese, *Chem. Comm.*, **1970**, 1644; see *Org. Reaction Mech.*, **1970**, 42.
[177] M. S. Baird and C. B. Reese, *Tetrahedron Letters*, **1971**, 4637.
[178] U. Schollkopf, E. Ruban, P. Tonne, and K. Riedel, *Tetrahedron Letters*, **1970**, 5077.
[179] L. B. Rodewald and M. C. Lewis, *Tetrahedron*, **27**, 5273 (1971).
[180] S. J. Cristol, J. K. Harrington, T. C. Morrill, and B. E. Greenwald, *J. Org. Chem.*, **36**, 2773 (1971).
[181] J. Hatem and B. Waegell, *Tetrahedron Letters*, **1971**, 2069.
[182] M. J. S. Dewar and S. Kirschner, *J. Am. Chem. Soc.*, **93**, 4290 (1971).
[183] M. J. S. Dewar and S. Kirschner, *J. Am. Chem. Soc.*, **93**, 4291 (1971).

minimum energy and those with antiaromatic transition states that have two distinct geometries that minimize the total energy. The former reactions may be slower than their antiaromatic counterparts if steric effects or other considerations are important.[184]

Symmetry rules, derived in a different manner to those of Woodward–Hoffmann, have been presented for ring opening of cyclopropyl carbonium ion and other reactions.[185] CNDO/2 calculations for the thermal opening of the aziridinyl cation to a 2-azaallyl cation are in accord with solvolysis results for N-chloroaziridines: the linear 2-azaallyl cation was predicted to be the most stable.[186]

Reactions of Cyclobutyl Derivatives

Cyclobutyl 2-naphthalenesulphonate has been shown to undergo solvolysis in a range of solvents with little nucleophilic assistance by solvent but considerable anchimeric assistance. Correlation of rates of solvolysis against those of neophyl, 2-adamantyl and pinacolyl toluene-p-sulphonates confirm the dominance of a $k_\varDelta$ route and the unimportance of a k_s process in solvolysis of the cyclobutyl derivative.[187]

Steric factors in carbonium ion rearrangements of cyclobutyl derivatives have been investigated. In solvolysis and deamination of *cis*- and *trans*-isomers of 2-methyl- and 3-isopropyl-cyclobutyl derivatives the mechanism involves rate-determining stereospecific rearrangement of the cyclobutyl derivative to an intermediate 2-alkylcyclopropylcarbinyl ion pair; steric effects have been demonstrated to be extremely important in both product and rate determination.[188]

For solvolyses of spiro[2.3]hexane derivatives it has been found[189] that generation of a nitrenium ion adjacent to a cyclobutane ring results in expansion to a five-membered ring containing nitrogen and that products are formed on generation of a nitrenium ion from an azetidine precursor consistent with ring contraction to an aziridine.[190]

Protonated Cyclopropane Intermediates

Ab initio molecular-orbital calculations with complete geometry optimization have been made for the 2-propyl, methyl-eclipsed 1-propyl and methyl-staggered 1-propyl carbonium ions and corner-, edge- and face-protonated cyclopropanes. The 2-propyl carbonium ion was predicted to be the most stable of the series of $C_3H_7^+$ cations; the methyl-eclipsed 1-propyl carbonium was the next stable and appeared to be the only other potential minimum on the $C_3H_7^+$ surface. Neither the edge- nor corner-protonated cyclopropane coincided with an energy minimum and the latter of these was only slightly less stable than the methyl-eclipsed 1-propyl carbonium ion. It was suggested that this ion could be important as a reaction intermediate.[191]

An accurate NMR method has been reported for the measurement of equilibrium deuterium isotope effects in carbonium ions undergoing rapid degenerate rearrangements **(143a)** $\rightleftharpoons$ **(144a)**.[192] In SbF_5–SO_2ClF, a single doublet at τ 7.07 ppm is attributed to the six protons of the methyl groups attached to C^+ averaged with the six protons of the isopropyl group. For **(143b)** $\rightleftharpoons$ **(144b)**, however, two doublets at ca. τ 7 appeared; the

[184] M. J. S. Dewar and S. Kirschner, *J. Am. Chem. Soc.*, **93**, 4292 (1971).
[185] R. G. Pearson, *Accounts Chem. Res.*, **4**, 152 (1971).
[186] R. G. Weiss, *Tetrahedron*, **27**, 271 (1971).
[187] D. D. Roberts, *J. Org. Chem.*, **36**, 1913 (1971).
[188] I. Lillien and L. Handloser, *J. Am. Chem. Soc.*, **93**, 1682 (1971).
[189] R. Maurin and M. Bertrand, *Tetrahedron Letters*, **1970**, 5065.
[190] P. G. Gassman and A. Carrasquillo, *Tetrahedron Letters*, **1971**, 109.
[191] L. Radom, J. A. Pople, V. Buss, and P. von R. Schleyer, *J. Am. Chem. Soc.*, **93**, 1813 (1971).
[192] M. Saunders, M. H. Jaffe, and P. Vogel, *J. Am. Chem. Soc.*, **93**, 2558 (1971).

lower-field doublet was attributed to the averaging of six methyl protons attached to C⁺ of (**143b**) with the six protons of the methyls attached to the methylene group of (**144b**). The higher-field doublet was due to the averaging of the five protons of the methyls attached to CH of (**143b**) with the five protons of the methyls on C⁺ of (**144b**). The lower-field six-proton doublet indicates that the carbonium ion prefers to be substituted by unlabelled methyl groups and that β-deuterium destabilizes the carbonium centre more than γ-deuterium. K was calculated as 1.132 ± 0.0007 and the enthalpy difference 54 ± 3 cal mole^{-1}, which is very close to the $\Delta\Delta F^{\ne} = 55$–$59$ cal mole^{-1} obtained for the β-deuterium isotope effect on the reaction rate in solvolysis of t-butyl chloride. The energy surface involving the three "t-hexyl" ions (**145**), (**146**) and (**147**) was calculated and it was postulated that the reaction between these passes through corner- or edge-protonated cyclopropane intermediates in which protons migrate from corner to corner yielding transition states or intermediates (**148**), (**149**) and (**150**).[192]

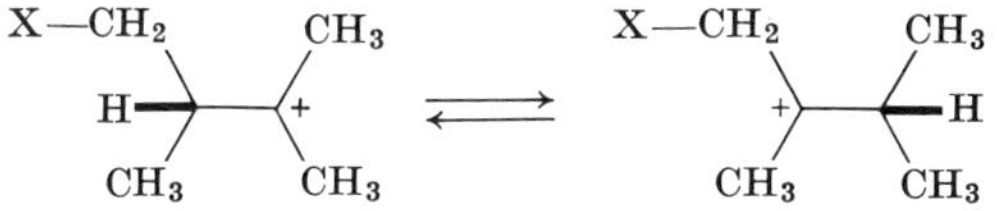

(**143**) a; X=H
   b; X=D

(**144**) a; X=H
   b; X=D

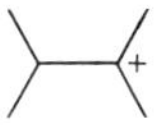

(**145**)    (**146**) 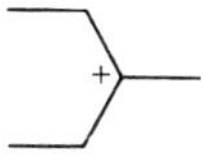   (**147**)

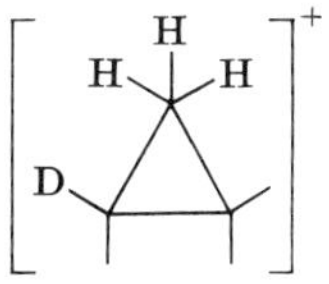

(**148**) 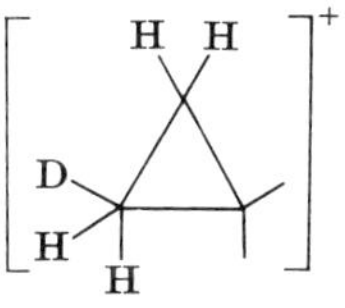   (**149**) 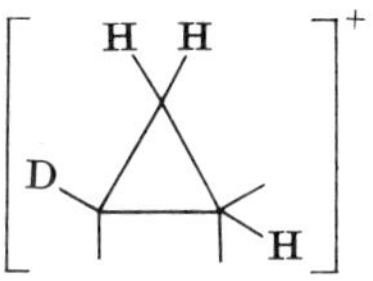   (**150**)

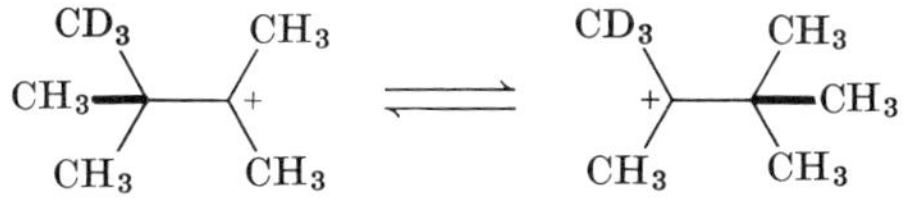

(**151**)   (**152**)

A (β–γ) differential isotope effect, such that (**151**) is more stable than (**152**), has been observed when a 1:3 mixture of 2,2,3-trimethylbut-3-ylium ion and deuteriated 1,1,1-trideuterio-2,2,3-trimethylbut-3-ylium ion (**151**)⇌(**152**) was prepared from the corresponding chlorides in SbF₅–SO₂ClF at $-100°$. A value of 50–60 cal mole^{-1} was obtained for the differential isotope effect for interchange of H and D between methyls with C-H's β and γ to tertiary carbonium ions; this value is comparable to that obtained in solvolysis of alkyl systems. When the mixture (**151**)⇌(**152**) is warmed six other isomeric tertiary ions are formed by a scrambling process amongst the protons, and the suggested mechanism for rearrangement involved corner-protonated cyclopropane intermediates.[193]

[193] M. Saunders and P. Vogel, *J. Am. Chem. Soc.*, **93**, 2559 (1971).

Rapid interchange of methyl groups has been observed in the dimethylisopropyl carbonium ion by an NMR technique using multi-labelled substrate mixtures. Protonated cyclopropanes were again implicated, but edge-protonated appeared to be less stable than corner-protonated cyclopropanes.[194]

The stereochemistry of the formation of 1,2-dimethylcyclopropane in the deamination of optically active 3-methylbut-2-ylamine has been investigated; the intervention of the 3-methylbut-2-yl carbonium ion and corner-protonated cyclopropanes was discussed.[195]

Norcarane has been obtained by pyrolytic decomposition of cyclohexylmethyl and cycloheptyl chloroformates and by deamination of the corresponding amines. The wide variety of products was considered to be consistent with intervention of protonated cyclopropane intermediates.[196]

In the solvolysis of [1-^{14}C]propylmercuric perchlorate in aqueous dioxan, acetic acid or formic acid, 0.6, 2.5 or 3.5% isotopic scrambling was observed; the ^{14}C was distributed equally over C-2 and C-3. Since the leaving group is neutral mercury, a high-energy intermediate is produced in these reactions. The small amount of scrambling in aqueous dioxan is consistent with the importance of a direct displacement process in this solvent. The intervention of protonated cyclopropanes was discussed but differentiation into corner- or edge-protonated species was not possible.[197]

The mechanism of protolytic cleavage of arylcyclopropanes is consistent with rate-determining protonation; methyl substituent effects are consistent with corner-protonated intermediates. A smaller amount of positive charge is generated at the benzylic carbon in these reactions than in analogous styrene hydration reactions, as shown from ρ values.[198]

Protonated cyclopropanes have been suggested for involvement in Koch–Haaf carboxylations of t-alkyl-substituted cycloalkanols.[199] Exclusive retention of the nucleophile has been demonstrated in the cleavage process of a cyclopropane derivative with hydrogen.[200]

Metallocenylmethyl Cations* and Other Derivatives

The *trans*-tricarbonyl(π-pentadienyl)iron cation (**153**) has been generated in FSO_3H at $-78°$ and the *trans*-carbonium ion observed in equilibrium with the corresponding *cis*-form (**154**).[201]

σ-2-Acetoxyethyl- and σ-2-acetoxypropyl-bis(biacetyl dioximato)pyridinecobalt (**155**) have been shown to be very reactive in solvolytic reactions; the mechanism suggested involved the intermediates (**156**).[202]

---

* See page 60 for discussion of "vertical" stabilization.

[194] M. Saunders and P. Vogel, *J. Am. Chem. Soc.*, **93**, 2561 (1971).
[195] M. S. Silver and A. G. Meek, *Tetrahedron Letters*, **1971**, 3579.
[196] W. E. Dupuy, H. R. Hudson, and P. A. Karam, *Tetrahedron Letters*, **1971**, 3193.
[197] C. C. Lee and J. Law, *Can. J. Chem.*, **49**, 2746 (1971).
[198] M. A. McKinney, S. H. Smith, S. Hempelman, M. M. Gearen, B. U. M. Pearson, and L. Pearson, *Tetrahedron Letters*, **1971**, 3657.
[199] J. A. Peters and H. van Bekkum, *Rec. Trav. Chim.*, **90**, 65 (1971).
[200] J. B. Hendrickson and R. K. Boeckman, Jr., *J. Am. Chem. Soc.*, **93**, 4491 (1971).
[201] T. S. Sorensen and C. R. Jablonski, *J. Organomet. Chem.*, **25**, C62 (1970); C. P. Lillya and R. A. Sahatjian, *J. Organomet. Chem.*, **25**, C67 (1970).
[202] B. T. Golding, H. L. Holland, U. Horn, and S. Sakrikar, *Angew. Chem. Int. Ed.*, **9**, 959 (1970).

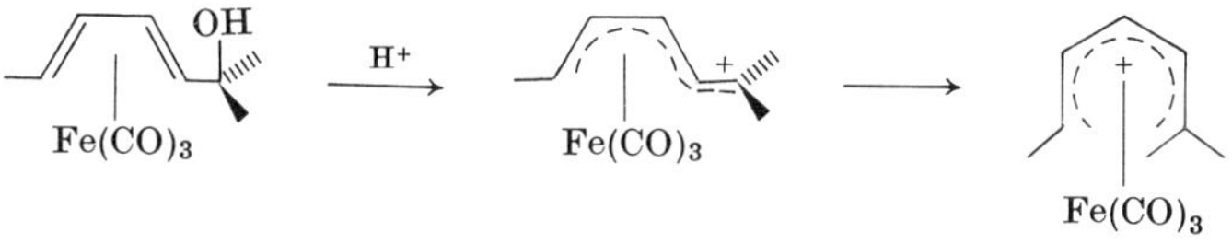

(153) (154)

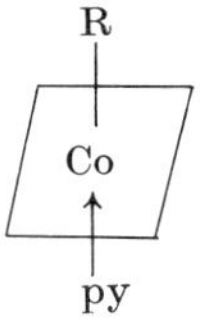

(155) (156)

R = CH₂CH₂OCOCH₃
R = CH₂CH(OCOCH₃)CH₃

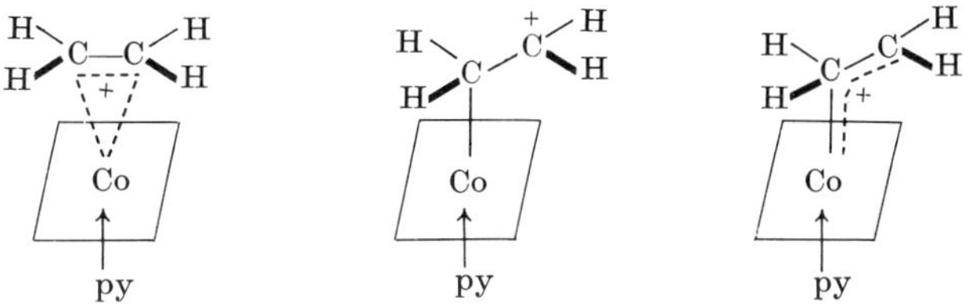

Strongly basic amines react with tricarbonyl(*syn,syn*-1,5-dimethyl pentadienyl)iron carbonium ion to give the tricarbonyl(*cis,trans*-dienylamine)iron complexes by *exo*-attack of nucleophile on the coordinated pentadienyl group. Weakly basic amines react with inversion to give tricarbonyl(*trans,trans*-dienylamine)iron complexes.[203] Allyltetracarbonyliron carbonium ions have been prepared by electrophilic addition to tetracarbonyl(tetramethyl)alleneiron.[204]

Substitution of a dimethylamino group for an acetate has been shown to proceed with retention in reactions of 1-ferrocenylethyl acetate.[205] Retention of configuration has also been found in the reaction of (157) with dimethylamine in the presence of aluminium chloride.[206]

Neighbouring-group participation and ring expansion have been observed on formation of the tricarbonyl-(1-methylcyclohexadienyl)iron carbonium ion (159) on heterolysis of tricarbonyl-[5-*endo*-methyl-5-*exo*-(tosyloxymethyl)cyclopentadiene]iron (158) in (CH₃CO)₂O/aq. HBF₄. The corresponding 5-*endo*-ester is slowly decomposed under the same conditions.[207] Formation of dicyclopentadienylcobalt carbonium ions on heterolysis of 2-oxoalkyl derivatives has been studied.[208]

α-Ferrocenyl carbonium ions, generated from ferrocenylethylenes dissolved in 90% formic acid, dimerize and then cyclize by homo- or hetero-annular pathways.[209] In solvolytic reactions of ferrocenyldihalogenocyclopropanes, products are formed from allyl carbonium ions obtained by opening of the cyclopropane ring.[210]

The ferrocenophane 6- and 7-carbonium ions have been prepared from 6- and 7-hydroxyferrocenophanes in trifluoroacetic acid. Unusual behaviour in NMR spectra

---

[203] G. Maglio, A. Museo, and R. Palumbo, *J. Organomet. Chem.*, **32**, 127 (1971).
[204] D. H. Gibson, R. L. Vonnahme, and J. E. McKiernan, *Chem. Comm.*, **1971**, 720.
[205] G. W. Gokel and I. K. Ugi, *Angew. Chem. Int. Ed.*, **10**, 191 (1971).
[206] P. Dixneuf, *Tetrahedron Letters*, **1971**, 1561.
[207] G. E. Herberich and H. Müller, *Chem. Ber.*, **104**, 2781 (1971); see *Org. Reaction Mech.*, **1970**, 50.
[208] G. E. Herberich and G. Greiss, *J. Organomet. Chem.*, **27**, 113 (1971).
[209] W. M. Horspool, R. G. Sutherland, and J. R. Sutton, *Can. J. Chem.*, **48**, 3542 (1970).
[210] W. M. Horspool, R. G. Sutherland, and B. J. Thompson, *J. Chem. Soc.* (C), **1971**, 1563.

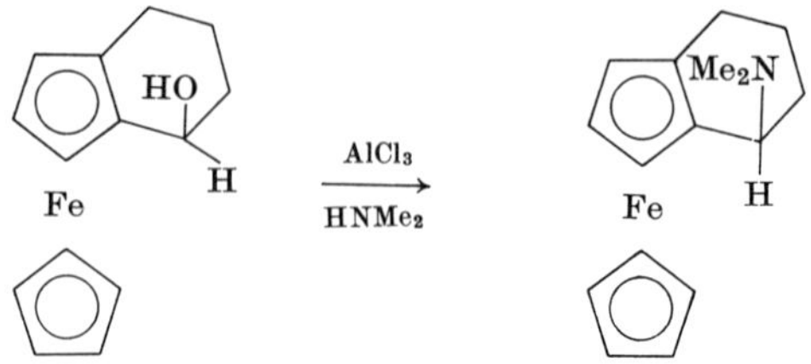

(157)

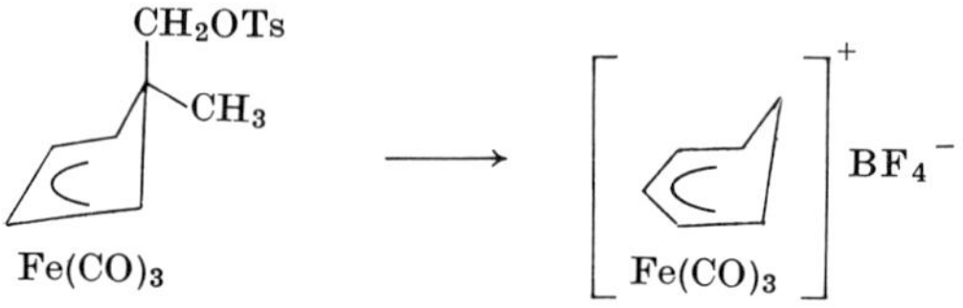

(158) (159)

reflects molecular deformations that result from interaction of the positive charge of the carbonionic centre and the π-electrons of the cyclopentadiene ring and of the charge and the Fe atom.[211]

A symmetry-controlled process has been postulated as responsible for formation of a single product in the addition of cyclopentadiene to ferrocenyl carbonium ion. Either a [6 + 4] cycloaddition followed by rapid proton elimination or a [4 + 2] cycloaddition of the diene to the exocyclic double bond followed by 1,5-sigmatropic rearrangement takes place.[212]

Substituent effects in the ESR spectra have been studied,[213] and the NMR spectra[214] and stability of α-ferrocenyl carbonium ions[215] have been investigated.

Modified extended Hückel (S.C.C.) calculations have been made on the ferrocenylmethyl carbonium ion.[216]

Stable Carbonium Ions and their Reactions

A monocyclic 1,3-bishomotropylium ion (161) has been prepared from cis-bicyclo-[6.1.0]nona-2,4,6-triene (160) in FSO_3H–SO_2ClF at ca. −125°. Deuteration of (160) was shown to occur fairly stereoselectively (70%) on the exo-side, to yield deuterium mainly on the outside of (161).[217]

The 1,6-methano-bridged bishomotropylium ion (163) has been prepared from bicyclo[4.2.2]decatetraene (162), and (163) was suggested as the intermediate (164) involved in electrophilic addition to (162).[218]

[211] M. Hisatome and K. Yawakawa, *Tetrahedron Letters*, **1971**, 3533.
[212] T. D. Turbitt and W. E. Watts, *Chem. Comm.*, **1971**, 631.
[213] R. Prins and A. R. Korswagen, *J. Organomet. Chem.*, **25**, C74 (1971).
[214] M. Hisatome and K. Yawakawa, *Tetrahedron*, **1971**, 2101.
[215] A. N. Nesmeyanov, L. I. Kazakova, M. D. Reshelova, L. A. Kazitsina, and E. G. Perevalova, *Izv. Akad. Nauk SSSR, Ser. Khim.*, **1970**, 2804.
[216] R. Gleiter and R. Seeger, *Helv. Chim. Acta*, **54**, 1217 (1971).
[217] P. Warner and S. Winstein, *J. Am. Chem. Soc.*, **93**, 1284 (1971).
[218] G. Schröder, U. Prange, B. Putze, J. Thio, and J. F. M. Oth, *Chem. Ber.*, **104**, 3406 (1971).

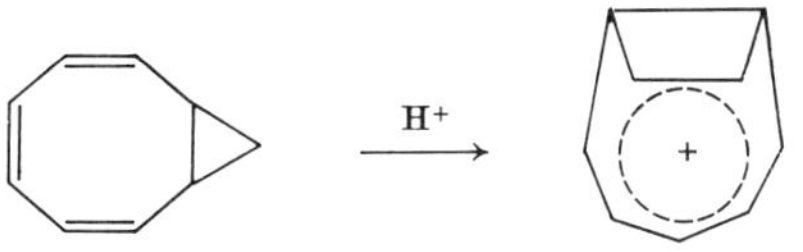

Carbonium Ions

(160) → **(161)** (H⁺)

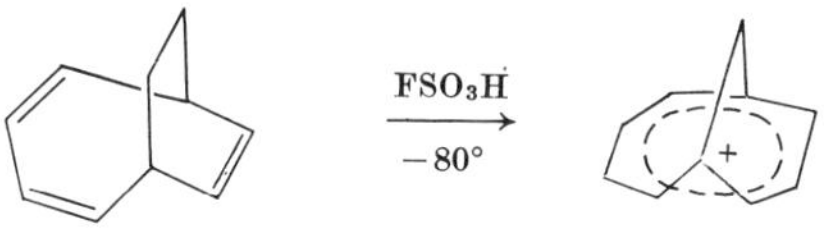

(162) → **(163)** (FSO₃H, −80°)

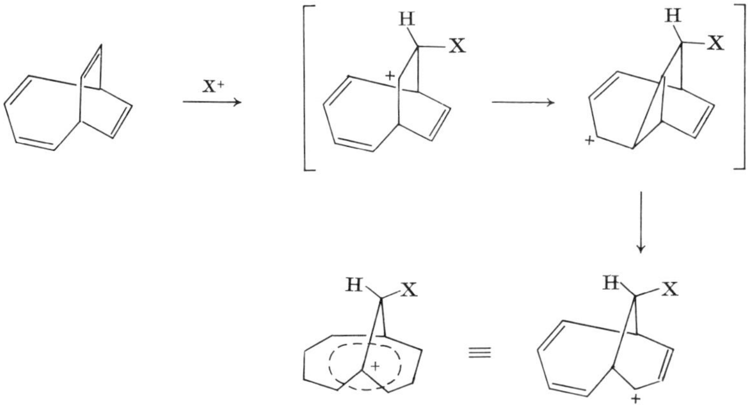

(164)

4,5-Benzohomotropone has been prepared and in D_2SO_4 appears to exist as a hydroxyhomotropylium ion.[219]

At −120° in FSO_3H–SO_2F_2 the tricarbonylcyclooctatetraeneiron complex **(165)** yields the tricarbonylcyclooctatrienyliron complex **(166)**, which undergoes electrocyclic ring closure at −60° to the bicyclo[5.1.0]octadienyltricarbonyliron complex **(167)**.[220] It has also been shown that protonation of tricarbonylmethylcyclooctatetreneiron **(168)** occurs at C-6 and C-7, the internal positions of the free diene portion, to form **(169)** and **(170)** in about a 2:1 ratio. At −60° both complexes undergo ring closure to **(171)** and **(172)** at rates of 3.5×10^{-4} sec^{-1} and 7.0×10^{-4} sec^{-1}, corresponding to $\Delta F^{\ne} = 15.7$ and 15.3 kcal mole^{-1}, respectively. It was suggested that protonation of **(165)** would occur in similar fashion.[221]

Carbon-protonation **(174)** of tricarbonyltroponeiron **(173)** in CH_2Cl_2 with concentrated H_2SO_4 has been observed; quenching in methanol containing sodium carbonate

---

[219] Y. Sugimura, N. Soma, and Y. Kishida, *Tetrahedron Letters*, **1971**, 91.
[220] M. Brookhart and E. R. Davis, *J. Am. Chem. Soc.*, **92**, 7622 (1970).
[221] M. Brookhart and E. R. Davis, *Tetrahedron Letters*, **1971**, 4349.

(165) **(166)** **(167)**

(168) **(169)** **(170)** **(171)** **(172)**

(173) **(174)** **(175)**

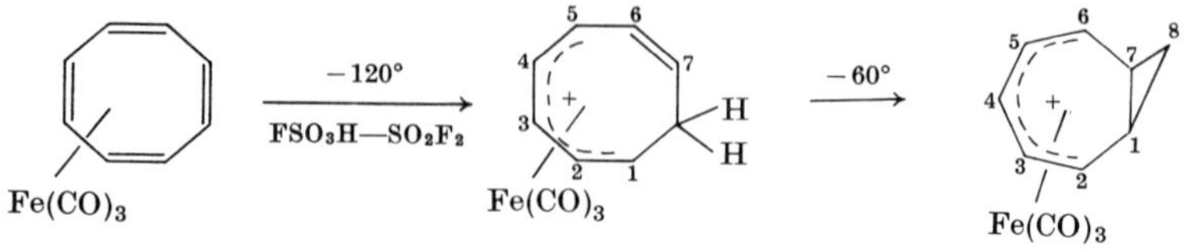

gave (**175**) in greater than 90% yield. An estimate of the rate of ring fluxional isomerism of the $Fe(CO)_3$ component was obtained by deuteriation experiments.[222]

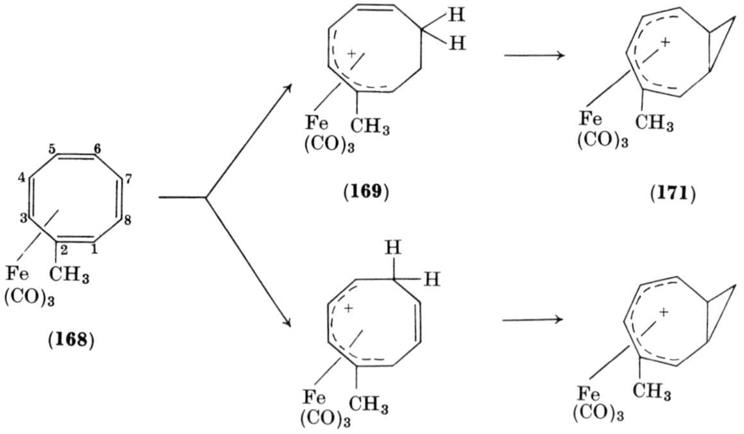

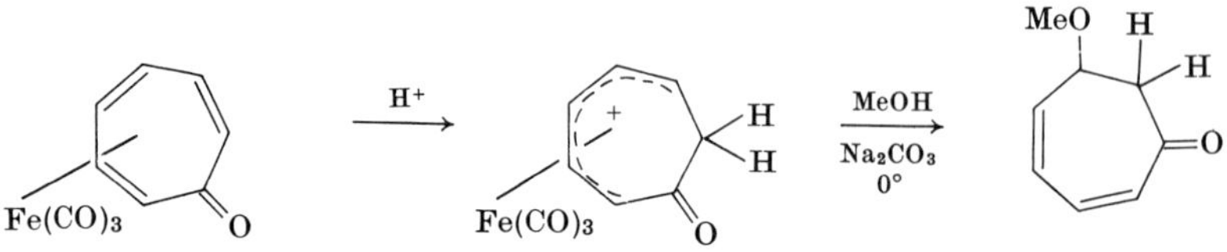

Reactions of cyclopropylallyl carbonium ions have been studied in H_2SO_4, FSO_3H and $FSO_3H–SbF_5$. In $FSO_3H–SbF_5$, acid-catalysed ring opening leads almost entirely to cyclopentenyl carbonium ions derived from dienylic carbonium ions. In the other two solvents internal structural rearrangements predominate. The mechanism has been shown to involve rearrangements leading to more stable carbonium ions beginning from 30 kcal mole^{-1} of strain energy contained in the cyclopropane ring.[223] Rearrangement of *cis*-1,2,3,4,5-pentamethylcyclopentenyl carbonium ion to the *trans*-isomer follows H_0 with unit slope. Intermediacy of 1,2,3,4,5-pentamethylcyclopentadiene, followed by reprotonation to give the *trans*-ion product, was suggested.[224]

[222] A. Eisenstadt and S. Winstein, *Tetrahedron Letters*, **1971**, 613.
[223] T. S. Sorensen and K. Rajeswari, *J. Am. Chem. Soc.*, **93**, 4222 (1971).
[224] T. S. Sorensen, I. J. Miller, and C. M. Urness, *Can. J. Chem.*, **48**, 3374 (1971).

Kinetics of rearrangement of protonated aldehydes and ketones have been measured and the relative rates found to be in the ratios $10^7:10^2:10^2:1$ for α-dibranched aldehydes, α-monobranched aldehydes, α-dibranched ketones and α-monobranched ketones, respectively. The first three groups rearrange by a two-step process involving 1,2-alkyl and/or hydride shifts, whereas the fourth group rearranges by way of tertiary α-hydroxyalkyl carbonium ions and protonated α-dibranched aldehydes. Protonated aldehydes were found to be about 9 kcal mole^{-1} less stable than isomeric protonated ketones.[225]

Conditions have been found for protonation of olefins in superacids to form stable carbonium ions without polymerization. Specific conditions vary with the substrates.[226] Protonation of 3-, 4-, 5- and 6-membered protonated cyclic ethers and sulphides[227] and the formation of α-keto carbonium ions have been studied.[228]

NMR spectra for four halogenoacetones in SO_2–SbF_5–FSO_3H show long-range coupling and intramolecular hydrogen bonding directly related to the electronegativity of the heteroatom. The high coupling was held to be due to the high contribution of the hydroxycarbonium ion mesomeric with the oxonium ion.[229]

^{13}C-NMR spectroscopy indicates little 1,3-interaction in most allyl carbonium ions. When steric constraints make overlap of the p orbitals on C-2 and C-3 more favourable, as in the cyclobutenyl ion produced by reaction of 4-chloro-1,2,3,4-tetramethylcyclobutene in FSO_3H–SO_2, significant 1,3-interaction is apparent. Generation of allyl cation by hydride abstraction from an olefin was also reported.[230]

Evidence for the planarity of the t-butyl carbonium ion has been obtained by Raman and infrared spectroscopy; close similarities were found for the structures of the t-pentyl dimethylisopropyl, and pentamethylethyl carbonium ions.[231] A linear relationship of the ^{13}C NMR chemical shift against σ^+ has been found for a series of *para*-substituted styryl carbonium ions in SbF_5–SO_2ClF solution.[232] A series of donor–acceptor complexes has been prepared from methylfluorosilanes in SbF–SO_3ClF. In SbF_5–HF–SO_2ClF, protolytic cleavage of the Si—C single bond took place, to give methane and the higher homologous methylfluorosilanes.[233]

General electrophilic reactivity of covalent C—C and C—H single bonds of alkanes has been discussed; it involves the σ-donor ability of shared electron pairs via two-electron, three-centre bond formation. Protolytic attack is suggested to take place on the C—H or C—C bonds where the major part of the electron density resides; the transition states are of three-centre pentacoordinated carbonium ion nature (**176**). Strong evidence for

$$\left[\begin{array}{c} R \\ | \\ C \\ R\ \ R \end{array} \diagup \begin{array}{c} H \\ H \end{array} \right]^+ \equiv \left[R_3C \cdots \diagup \begin{array}{c} H \\ H \end{array} \right]^+$$

(**176**)

[225] D. M. Brouwer and J. A. van Doorn, *Rec. Trav. Chim.*, **90**, 1010 (1971).
[226] G. A. Olah and Y. Halpern, *J. Org. Chem.*, **36**, 2354 (1971).
[227] G. A. Olah and P. J. Szilagyi, *J. Org. Chem.*, **36**, 1121 (1971).
[228] J. P. Bégué, M. Charpentier-Morize, and C. Pardo, *Tetrahedron Letters*, **1971**, 4736; see also J. P. Bégué and M. Charpentier-Morize, *Angew. Chem. Int. Ed.*, **10**, 327 (1971).
[229] R. Jost, P. Rimmelin, and J. M. Sommer, *Chem. Comm.*, **1971**, 1243.
[230] G. A. Olah, P. R. Clifford, Y. Halpern, and R. G. Johanson, *J. Am. Chem. Soc.*, **93**, 4219 (1971).
[231] G. A. Olah, J. R. DeMember, A. Commeyras, and J. L. Bribes, *J. Am. Chem. Soc.*, **93**, 459 (1971).
[232] G. A. Olah, R. D. Porter, and D. P. Kelly, *J. Am. Chem. Soc.*, **93**, 464 (1971).
[233] G. A. Olah and Y. K. Mo, *J. Am. Chem. Soc.*, **93**, 4942 (1971).

the three-centre-bond transition state was found in the formation of the t-butyl carbonium ion, rather than isobutene, on treatment of isobutane with DSO_3F–SbF_5 or DF–SbF_5 at $-78°$ and also by hydrogen–deuterium exchange of adamantane in DF–SbF_5. Isobutane yielded methane and isopropyl cation and C—C bond protolysis in n-alkanes was shown to be a general reaction. An empirical reactivity order in alkanes of tertiary CH > C—C > secondary CH ≫ primary CH was found.[234]

Alkylation of alkanes by alkyl carbonium ion salts in SO_2ClF has been reported. Intermolecular hydride transfer between tertiary and secondary carbonium ions was shown to be faster than alkylation, so that the alkylation products are also those derived from the new alkanes and carbonium ions formed in the hydride transfer reactions. Products are determined by the relative rates of hydrogen transfer and alkylation, and some intramolecular rearrangements also take place. Propylation of propane by isopropyl fluoroantimonate gave 2,3-dimethybutane (26%), 2-methylpentane (28%), 3-methylpentane (14%) and n-hexane (32%). Reaction of other alkanes such as isobutane and n-butane with t-butyl and sec-butyl salts were studied and the reaction was shown to be general and can proceed without involvement of olefins.[235]

Electrophilic nitration and nitrolytic cleavage of alkanes and cycloalkanes have been demonstrated with stable nitronium salts, such as $NO_2^+PF_6^-$, $NO_2^+SbF_6^-$ or $NO_2^+BF_4^-$ in methylene chloride–tetramethylene sulphone solution. It was suggested that the reactions occur by three-centre-bond transition states, and the nitronium ion was considered bent (sp^2) with an empty p orbital on nitrogen. Tertiary C—H bonds were shown to be most reactive and C—C bonds were generally more reactive than secondary or primary C—H bonds. Examples investigated were nitration of methane, cyclohexane and adamantane to yield nitromethane, nitrocyclohexane and nitroadamantane, respectively. In some cases the nitrations are accompanied by protolytic denitrations and *tert*-butyl carbonium ion can be obtained from 2-nitro-2-methylpropane. More conventional nitrating mixtures (HNO_3–H_2SO_4) were also successful; nitration of isobutane at $50°$ gave CH_3NO_2 (28%), $CH_3CH_2NO_2$ (4.9%), $(CH_3)_2CHNO_2$ (0.5%) and $(CH_3)_3CNO_2$ (66.6%).[236]

Thermal ring expansion of octachlorobicyclo[3.2.0]hepta-2,6-diene in presence of two moles of anhydrous aluminium chloride gave heptachlorotropylium heptachlorodialuminate, which yielded octachlorocycloheptatriene on quenching with water.[237] Benzo[3,4]cyclobuta[1,2-*a*]benzo[*d*]tropylium cation has been prepared and converted into the methyl ether by quenching with methanol. pK measurements indicate that fusion of the benzocyclobutadiene group has a destabilizing effect on the tropylium cation.[238] Ditropylium bis(hexachloroantimonate) has been prepared in 70% yield by reaction of $SbCl_5$ with heptafulvalene,[239] and formation of a stable cyclohepta-2,4,6-trienylmethyl cation has been described.[240] Metal-catalysed reactions of the homotropylium ion have been discussed.[241]

Tetra-arylcyclobutenyl carbonium ions have been formed in quantitative yield by

---

[234] G. A. Olah, Y. Halpern, J. Shen, and Y. K. Mo, *J. Am. Chem. Soc.*, **93**, 1251 (1971).
[235] G. A. Olah and J. A. Olah, *J. Am. Chem. Soc.*, **93**, 1256 (1971).
[236] G. A. Olah and H. C. Lin, *J. Am. Chem. Soc.*, **93**, 1259 (1971).
[237] K. Kusuda, R. West, and V. N. M. Rao, *J. Am. Chem. Soc.*, **93**, 3627 (1971).
[238] P. J. Garrett and K. P. C. Vollhardt, *Chem. Comm.*, **1971**, 1143.
[239] H. Volz and M. Volz de Lecca, *Ann. Chem.*, **750**, 136 (1971).
[240] W. Betz and J. Daub, *Angew. Chem. Int. Ed.*, **10**, 269 (1971).
[241] P. Warner, *Tetrahedron Letters*, **1971**, 723.

reaction of diarylacetylenes with strong proton acids.[242] The triaminocyclopropenyl carbonium has been reported.[243]

Stable arylalkoxy carbonium ions have been directly observed on treatment of an aromatic acetal with BF_3; the carbonium ion showed a high rotational barrier about the C_{aryl}—$^+CHOCH_3$ bond with a coalescence temperature at $+20°$.[244] Reduction potentials of the tropylium, dibenzotropylium and homotropylium ions in sulphuric acid have been measured.[245] Isotopic exchange in arenonium ions is reported.[246]

Other studies include reactions of phthalides and phthaleins;[247] kinetics of hydrolysis of 2-t-butyloxaziridine in strong acid;[248] and generation of carbonium ions from substituted π-allylpalladium chloride complexes in strong acids.[249]

Other Reactions

Correlations between S_N1 reactivity and carbonium ion stability has been shown to be poor if structures vary widely.[250] Relative carbonium ion stabilities have been determined by NMR methods involving direct observation of the equilibrium between two carbonium ions and their covalent precursors.[251] Relative heats of formation (ΔH_R^+) of acylium ions in SbF_5–FSO_3H have been compared with gas-phase values. The normal inductive order observed in the gas phase becomes a Baker–Nathan order in solution.[252] Relative heats of formation of a series of 2-substituted 4,4-dimethyl-1,3-dioxolium ions have been measured calorimetrically.[253]

An extensive study has been made of solvolysis reactions of alkyl picrates. A smaller amount of positive charge appears to be developed in acetolysis of picrates than in equivalent reactions of arenesulphonates; the consequence is a smaller amount of hydride shift. A slightly higher positive ΔS^* for picrate acetolysis was suggested as due to the delocalized nature of the picrate anion. The variable toluene-p-sulphonate–picrate rate ratio was attributed to the smaller charge development in acetolysis of the picrates.[254]

A series of 2,2,2-trifluoroethanesulphonates has been prepared and a linear free-energy relationship found for a series of X—SO_3R derivatives and σ_n with ρ (= +10.3). The trifluoroethanesulphonate–methanesulphonate rate ratio was suggested as a measure of nucleophilic character of transition states.[255] The nucleophilicity of 2,2,2-trifluoroethanol is equivalent to that of formic acid and this substance has, therefore, application as a "limiting solvent".[256]

---

[242] A. E. Lodder, H. M. Buck, and L. J. Oosterhoff, *Rec. Trav. Chim.*, **89**, 1229 (1970).
[243] Z. Yoshida and Y. Tawara, *J. Am. Chem. Soc.*, **93**, 2573 (1971).
[244] M. Rabinovitz and D. Bruck, *Tetrahedron Letters*, **1971**, 245.
[245] M. Feldman and W. C. Flythe, *J. Am. Chem. Soc.*, **93**, 1547 (1971).
[246] V. G. Shubin, A. A. Tabatskaya, B. G. Derendyaev, D. V. Korgehagina, and V. A. Koptyug, *Zhue. Org. Khim.*, **6**, 2072 (1970); *J. Org. Chem. USSR*, **6**, 2079 (1970).
[247] A. C. Hopkinson, *J. Chem. Soc. (B)*, **1971**, 1752.
[248] A. R. Butler and B. C. Challis, *J. Chem. Soc. (B)*, **1971**, 778.
[249] J. Lukas and P. A. Kramer, *J. Organomet. Chem.*, **31**, 111 (1971).
[250] J. Hine, *J. Am. Chem. Soc.*, **93**, 3701 (1971).
[251] S. V. McKinley, J. W. Rakshys, Jr., A. E. Young, and H. H. Freedman, *J. Am. Chem. Soc.*, **93**, 4715 (1971).
[252] J. W. Larsen, P. A. Bouis, M. W. Grant, and C. A. Lane, *J. Am. Chem. Soc.*, **93**, 2067 (1971).
[253] J. W. Larsen and S. Ewing, *J. Am. Chem. Soc.*, **93**, 5107 (1971).
[254] M. L. Sinnott and M. C. Whiting, *J. Chem. Soc. (B)*, **1971**, 965.
[255] R. K. Crossland, W. E. Wells, and V. J. Shiner, Jr., *J. Am. Chem. Soc.*, **93**, 4217 (1971).
[256] M. D. Bentley and J. A. Lacadie, *Tetrahedron Letters*, **1971**, 741.

The extent of fragmentation in the reaction of the series of tertiary alcohols (**177**) increases with the size of R (Table 1). It was concluded that the absence of a strong base for proton abstraction was important, together with stability of fragmentation product.[257]

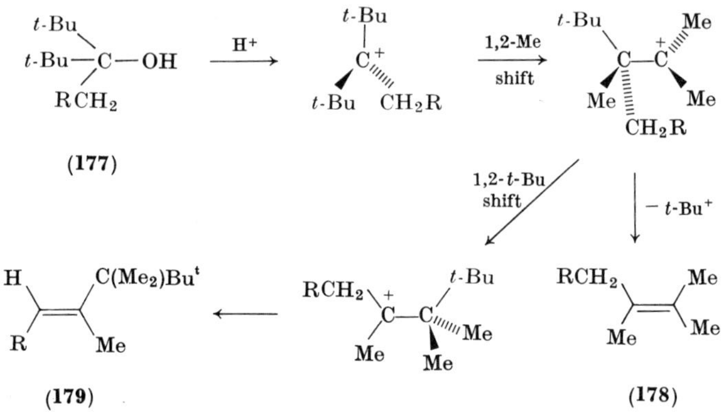

Table 1. Products from the reaction $RCH_2(t\text{-}Bu)_2COH$ with concentrated H_2SO_4 at 25°

R	(**178**) (%)	(**179**) (%)	Other fragmentation products (%)
H	22	66	2
Me	31	52	14
Et	40	39	15
i-Pr	51	25	16
t-Bu	79	—	17

Acid-catalysed rearrangements in 1-allyl-2,4-di-t-butyl-3,6-dimethylbenzene and similar crowded systems involve protonation, migration of a methyl group and finally loss of a t-butyl group.[258]

Optically active 2-phenylbutane has been formed by capture of 2-phenyl-2-butyl carbonium ion by a chiral organosilicon hydride.[259]

Optimum conditions have been found for [4 + 3] cycloadditions of simple allyl carbonium ions and conjugated dienes, they involve stirring the diene with the allyl iodide and silver trifluoroacetate in isopentane at −78°. Butadiene and 2-methylallyl carbonium ion give 1-methylcyclohepta-1,4-diene in small yield, and reactions with isoprene, 2,3-dimethylbutadiene and cyclopentadiene are also described. Results were parallel to those of Diels–Alder additions of benzyne at ca. 30°. The utility of silver trifluoroacetate–isopentane as an electrophilic system for generation of carbonium ions was emphasized.[260]

---

[257] J. E. Dubois, J. S. Lomas, and D. S. Sagatys, *Tetrahedron Letters*, **1971**, 1349.
[258] K. H. Lai and B. Miller, *Tetrahedron Letters*, **1971**, 3575.
[259] J. L. Fry, *J. Am. Chem. Soc.*, **93**, 3558 (1971).
[260] H. M. R. Hoffman, G. F. P. Kernaghan, and G. Greenwood, *J. Chem. Soc.* (B), **1971**, 2257.

A complex mixture (**180**) of alcohols was obtained from the reaction of tropylium perchlorate and cyclopentadiene in 66% aqueous dioxan.[261]

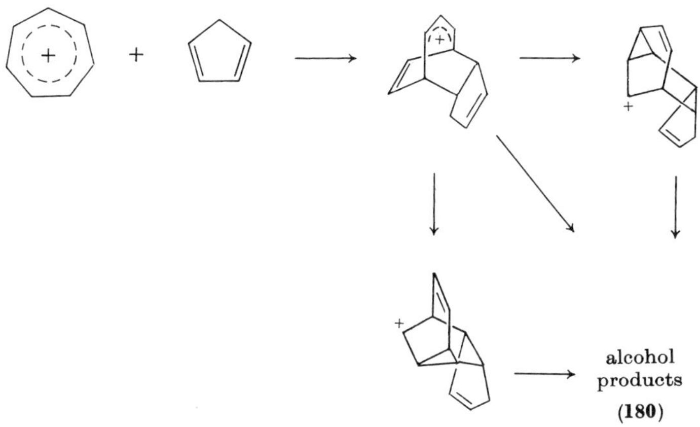

The stereochemistry of product formation from 4-t-butylcyclohexyl carbonium ions, generated by decarboxylation, anodic oxidation and lead tetra-acetate oxidative decarboxylation, has been investigated; axial bond formation is favoured when the stereochemistry of the leaving group is not influential and in reactions in which there is substantial bond-making in the transition state.[262]

It has been proposed that hydride shift from hydrocarbon to aryl carbonium ions is a two-step process; an electron-transfer process followed by hydrogen-atom transfer.[263] 1,3 Hydride shifts have been observed in reactions of phenol and 5,5-dimethyl-N-nitrosooxazolidone.[264] Rates of solvolysis of N-(p-nitrophenyl)-o- and -p-nitrobenzhydrazonyl bromides in 70% dioxan have been studied as a function of pH; the rate-determining step is 1,3 dipolar ion formation.[265]

Stability of polyenyl carbonium ions increases with conjugation, but the effect is smaller than that associated with methyl hyperconjugation.[266]

A kinetic study of the β-fission of dialkyhydroxy carbonium ions has shown the reaction rate to depend on the solvent, largely owing to differences in interaction of the hydroxyl group with solvent in the initial and the transition state. No fission was found for 3,3-dimethylpentanone and hexanone owing to unfavourable orbital orientation.[267] Secondary and tertiary oxocarbonium ions, $C_5H_{11}CO^+$, interconvert readily at room temperature in FSO_3H-SbF_5 by recarbonylation, rearrangement and recarbonylation.[268] Carbonylation of ethylene in FSO_3H and FSO_3H-SbF_5 has been studied.[269]

[261] S. Ito and I. Itah, *Tetrahedron Letters*, **1971**, 2969.
[262] S. D. Elakovich and J. G. Traynham, *Tetrahedron Letters*, **1971**, 1435.
[263] M. Ballester, J. Riera-Figueras, J. Castaner, and A. Rodriguez-Siurana, *Tetrahedron Letters*, **1971**, 2079.
[264] M. S. Newman and C. D. Beard, *J. Am. Chem. Soc.*, **92**, 7564 (1970).
[265] A. F. Hegarty, M. Cashman, J. B. Aylward, and F. L. Scott, *J. Chem. Soc.* (B), **1971**, 1879.
[266] N. C. Deno and P. C. Scholl, *J. Am. Chem. Soc.*, **93**, 2702 (1971).
[267] D. M. Brouwer and J. A. van Doorn, *Rec. Trav. Chim.*, **90**, 535 (1971).
[268] H. Hogeveen and C. F. Roobeck, *Rec. Trav. Chim.*, **89**, 1121 (1970).
[269] J. Lukas and H. Hogeveen, *Chem. Ber.*, **104**, 2964 (1971).

In contrast to solvolysis of the 20β-epimer, formolysis of 3β-acetoxy-5α-pregnan-20α-yl toluene-p-sulphonate has been shown to proceed without ring enlargement of the D ring but forms predominantly 17β-methyl-18-nor-5α,17α-pregn-13-en-3β-yl acetate; a favourable conformation is found between the 17α-hydrogen atom and the 20α-group.[270]

Benzenonium ions have been formed by hydride abstraction from a series of methyl-cyclohexa-1,4-dienes arising from alkylation of polycyclic hydrocarbons.[271]

Rate enhancements in a series of pinacolyl-type toluene-p-sulphonates are, it is suggested, steric in origin.[272]

^{13}C NMR spectra of para-substituted triphenylmethanols have been interpreted in terms of charge densities; results are in accord with simple resonance picture with relative large positive charges at positions α and meta to the substituent. The amount of positive charge delocalized from the central carbon is proportional to σ^+. Charge densities correlate well with those calculated by the CNDO method.[273] Stabilization of the 1,3 orbital overlap in 1,3-diphenylcyclobutenyl carbonium ions has been studied by comparison of the NMR spectra of cycloalkenyl carbonium ions.[274] Substantial secondary β-deuterium isotope effects were found on F NMR chemical shifts of p-fluorophenyl-methyl carbonium ions.[275] Other studies include proton resonance spectra of steroids in sulphuric acid;[276] identification of $C_3H_6^+$ radical cation by ion cyclotron resonance spectroscopy;[277] and the X-ray structure determination of a dicarbonium ion containing two triphenylmethylium groups.[278]

A study has been made of t-butyl halide solvolysis under action of metal ions.[279] Exchange between t-butyl carbonium ion and isobutane in SO_2 solution is of the first order in both species, $E_A = 3.6$ kcal mole^{-1}, $\Delta S^* = -27$ e.u. The isopropyl chloride–arsenic pentachloride complex reacts rapidly with n-butane to yield propane and t-butyl carbonium ions but does not react with neopentane or cyclopropene.[280]

Hydride transfer to the trityl carbonium ion has been demonstrated to be an extremely mild and neutral procedure for the deprotection of benzyl ethers and benzyloxycarbonyl esters and for oxidation of benzyl ethers to aldehydes.[281]

In acid-catalysed rearrangements of androst-5-ene and D-homoandrost-5-ene, mixtures of 8(9)-enes are obtained; equilibration of configurations at C-5, C-10, C-13 and C-14 in the latter compound leads to complete loss of optical activity.[282] Acid-catalysed conversion of germacrone-type sesquiterpenes into cadinene-type compounds by 80% aqueous AcOH or $AlCl_3$ in dry ether,[283] and carbonium ion rearrangements in janusene, hemiisojanusene and isojanusene,[284] have been studied.

[270] F. B. Hirshmann, D. M. Kantz, S. S. Deshmane, and H. Hirshmann, Tetrahedron, 27, 2041 (1971).
[271] D. F. Lindow and R. G. Harvey, J. Am. Chem. Soc., 93, 3786 (1971).
[272] A. P. Krapcho, B. S. Bak, R. G. Johanson, and N. Rabjohn, J. Org. Chem., 35, 3722 (1970).
[273] G. J. Ray, R. J. Kurland, and A. K. Colter, Tetrahedron, 27, 735 (1971).
[274] D. G. Farnum, A. Mostashari, and A. A. Hagedorn, J. Org. Chem., 36, 698 (1971).
[275] J. W. Timberlake, J. A. Thompson, and R. W. Taft, J. Am. Chem. Soc., 93, 274 (1971).
[276] H. A. Jones, J. Chem. Soc. (B), 1971, 99.
[277] M. L. Gross and F. W. McLafferty, J. Am. Chem. Soc., 93, 1267 (1971).
[278] J. S. McKechnie and I. C. Paul, J. Chem. Soc. (B), 1971, 918.
[279] E. S. Rudakov and I. V. Kozhevnikov, Tetrahedron Letters, 1971, 1333.
[280] S. Brownstein and J. Bornais, Can. J. Chem., 49, 7 (1971).
[281] D. H. R. Barton, P. D. Magnus, G. Streckert, and D. Zurr, Chem. Comm., 1971, 1109.
[282] D. N. Kirk and P. M. Shaw, Chem. Comm., 1971, 948.
[283] M. Iguchi, M. Niwa, and S. Yamahura, Chem. Comm., 1971, 974.
[284] S. J. Cristol and M. A. Imhoff, J. Org. Chem., 36, 1854 (1971).

Investigations have been made into solvolytic dimerization of 3,3,3-trifluoropropene,[285] acid-catalysed isomerization of *cis*-1-phenylbuta-1,3-diene and *cis*-1-methyl-3-phenylallyl alcohol[286] and the mechanism of addition of HF to cholest-14-en-7-ols.[287]

Four distinct types of reaction have been observed in reactions of silylcarbonium ions (Si—C+): (*a*) 1,2-migration of aryl groups from Si to C, yielding fluorosilanes; (*b*) Friedel–Craft substitution of carbonium ions on aromatic substrate; (*c*) isomerization to β-silyl carbonium ions by hydrogen migration; and (*d*) elimination of water to give vinylsilane.[288] A number of t-alkyl carbonium ions have been generated from alcohols or alkanes, and their reactivities towards hydride-transfer reactions with organosilicon hydride have been studied.[289] On solvolysis of 2-bromo-2-(trimethylsilyl)propane, only isopropenyltrimethylsilane was obtained.[290]

Formation of (**182**) from (**181**) in 95% yield in acetic acid has been interpreted as involving migration of a diphenylphosphinyl group.[291]

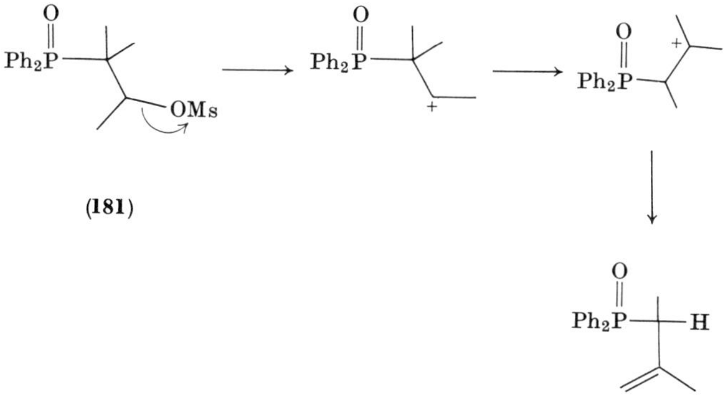

It has been suggested that the mechanism of copper-catalysed decomposition of bis-5-(cyclohex-1-enyl)pentanoyl peroxide in ionizing media involves organocopper intermediates and carbonium ions.[292] Rearrangements in cobaltic fluoride fluorinations of bridged-ring hydrocarbons have been investigated.[293]

Carbonium ions have been implicated in the formation of acetates on oxidation of propylmercury chloride with lead tetra-acetate.[294] The mechanism of electrochemical oxidation of phenylethylenes has been discussed in terms of cation radicals.[295] Cathodic reduction of the tropylium cation,[296] polarographic reduction of organic cations in

---

[285] P. C. Myhre and G. D. Andrews, *J. Am. Chem. Soc.*, **92**, 7595 (1970).
[286] Y. Pocker and M. J. Hill, *J. Am. Chem. Soc.*, **93**, 690 (1971).
[287] J. C. Jacquesy, R. Jacquesy, and S. Moreau, *Bull. Soc. Chim. France*, **1971**, 3609.
[288] A. G. Brook and K. H. Pannell, *Can. J. Chem.* **48**, 3679 (1970).
[289] F. A. Carey and H. S. Tremper, *J. Org. Chem.*, **36**, 758 (1971).
[290] F. K. Cartledge and J. P. Jones, *Tetrahedron Letters*, **1971**, 2193.
[291] P. F. Cann, D. Howells, and S. Warren, *Chem. Comm.*, **1971**, 1149.
[292] D. L. Struble, A. L. J. Beckwith, and G. E. Gream, *Tetrahedron Letters*, **1970**, 4794.
[293] J. Battersby, R. Stephens, and J. C. Tatlow, *Tetrahedron Letters*, **1970**, 5041.
[294] V. T. Kampel, K. A. Bilevich, and O. Y. Okhlobystin, *Dokl. Akad. Nauk SSSR*, **197**, 95 (1971).
[295] J. D. Stuart and W. E. Ohnesorge, *J. Am. Chem. Soc.*, **93**, 4531 (1971).
[296] P. H. Plesch and A. Stasko, *J. Chem. Soc.* (B), **1971**, 2052.

aqueous sulphuric acid,[297] anodic oxidation of arylcyclopropanes,[298] and anodic cyclization of olefinic alcohols to cyclic ethers,[299] have been studied.

1,2,3-Tri-t-butylcyclopropene was found to be oxidized by m-chloroperoxybenzoic acid through the tri-t-butylcyclopropenyl carbonium ion.[300] 5-t-Butyl-2,3-dihydrofuran derivatives have been prepared by reaction of perchloric acid with α,β-unsaturated ketones.[301] Reactions of lithium tetra-alkylborates with alkylating reagents depend on the carbonium ion activity of the alkylating reagent.[302] Mixtures of hydrocarbons are obtained by acid-catalysed reaction of o-, m- and p-cymene with isobutene, diisobutene and triisobutene.[303]

Other studies reported include the photochemistry of protonated eucarvone;[304] reaction of the chlorotropylium cation with nucleophiles;[305] addition of carbonium ions to trichloroethylene;[306] acid-catalysed decomposition of aliphatic hydroperoxide in the presence of alcohols;[307] reaction of Malachite Green with primary amines, methoxylamine and hydrazines;[308] generation of the cyclopropylidenecarbinyl carbonium ion;[309] photolysis of triphenylmethyl carbonium ion in 99% sulphuric acid, yielding 9-phenylfluoren-9-yl and 9-hydroxyfluoren-9-yl carbonium ions;[310] photooxygenation of triphenyl-pyrylium and -thiopyrylium cations;[311] preparation of 8-cyano-8-(2,3-diphenylcyclopropenyl)heptafulvenylium fluoroborate;[312] reaction of chloroplatinic acid with 1,1-dimethylcyclopropane;[313] reaction of 2-methylthio-1,3-dithiolanylium ion with N,N-dimethyldithiocarbamate;[314] and syntheses of the dithiopyrylium dicarbonium ion[315] and 3,4-diaminotriafulvenes.[316]

Basicity constants of some ethers[317] and ion-pair dissociation constants for a number of hexachloroantimonate salts[318] have been measured.

Ab initio MO calculations for C_1 and C_2 hydrocarbons,[319] CNDO/2 evaluations of substituent effects on charge distribution of conjugated ions[320] and calculations on ^{13}C coupling constants in a model carbonium ion[321] have been made.

[297] P. H. Plesch and I. Sestakova, *J. Chem. Soc.* (B) **1971**, 1337.
[298] T. Shono and Y. Matsumura, *J. Org. Chem.*, **35**, 4157 (1970).
[299] T. Shono, A. Ikeda, and Y. Kimura, *Tetrahedron Letters*, **1971**, 3599.
[300] J. Ciabattoni and J. P. Kocienski, *J. Am. Chem. Soc.*, **93**, 4902 (1971).
[301] W. Randel and K. Besserer, *Ann. Chem.*, **751**, 168 (1971).
[302] V. A. Hogg and G. Hesse, *Ann. Chem.*, **751**, 95 (1971).
[303] D. E. Boone, E. J. Eisenbraun, P. W. Flanagan, and R. D. Grigsby, *J. Org. Chem.*, **36**, 2042 (1971).
[304] K. E. Hine and R. F. Childs, *J. Am. Chem. Soc.*, **93**, 2323 (1971).
[305] B. Fohlisch and E. Hang, *Chem. Ber.*, **104**, 2324 (1971).
[306] K. Bott, *Chem. Ber.*, **103**, 3850 (1970).
[307] J. O. Turner, *Tetrahedron Letters*, **1971**, 887.
[308] J. E. Dixon and T. C. Bruice, *J. Am. Chem. Soc.*, **93**, 3248 (1971).
[309] M. L. Poutsma and P. A. Ibarbia, *J. Am. Chem. Soc.*, **93**, 441 (1971).
[310] D. M. Allen and E. D. Owen, *Chem. Comm.*, **1971**, 848.
[311] Z. Yoshida, T. Sugimoto, and S. Yoneda, *Tetrahedron Letters*, **1971**, 4259.
[312] M. Oda, K. Tamate, and Y. Kitahara, *Chem. Comm.*, **1971**, 347.
[313] D. B. Brown and M. J. Strauss, *Chem. Comm.*, **1971**, 128.
[314] T. Nakai and M. Okawara, *Bull. Chem. Soc. Japan*, **43**, 3882 (1970).
[315] Z. Yoshida, S. Yoneda, T. Sugimoto, and O. Kikukawa, *Tetrahedron Letters*, **1971**, 3999.
[316] Z. Yoshida and T. Tawara, *Tetrahedron Letters*, **1971**, 3603.
[317] A. M. Avedikian and A. Kergomard, *Bull. Soc. Chim. France*, **1971**, 87.
[318] P. M. Bowyer, A. Ledwith, and D. C. Sherrington, *J. Chem. Soc.* (B), **1971**, 1511.
[319] W. A. Lathan, W. J. Hehre, and J. A. Pople, *J. Am. Chem. Soc.*, **93**, 808 (1971).
[320] J. R. Grunwell and J. F. Sebastian, *Tetrahedron*, **27**, 4387 (1971).
[321] G. A. Maciel, *J. Am. Chem. Soc.*, **93**, 4375 (1971).

CHAPTER 2

Nucleophilic Aliphatic Substitution

I. D. R. STEVENS

Chemistry Department, University of Southampton

Ion-pair Phenomena and Borderline Mechanisms	49
Solvent and Medium Effects	55
Isotope Effects	58
Neighbouring Group Participation	60
Participation by Ether and Hydroxyl Groups	61
Participation by Thioether and Thiol Groups	62
Participation by Halogens	64
Participation by Carbonyl Groups	65
Participation by Ester, Carboxyl and Amide Groups	67
Participation by Amino, Silyl and Phosphinyl Groups	70
Participation by Neighbouring Carbanion and Carbon—Metal Bonds	72
Neighbouring Carbon and Hydrogen	74
Deamination and Related Reactions	79
Reactions of Aliphatic Diazo-compounds	81
Fragmentation Reactions	82
Displacement Reactions at Elements Other than Carbon	83
Silicon, Germanium, Tin and Lead	83
Phosphorus, Arsenic and Antimony	85
Sulphur	89
Other Elements	94
Ambident Nucleophiles	95
Substitution at Vinylic Carbon	97
Reactions of α-Halogenocarbonyl Compounds	101
S_N2 Processes and Other Reactions	103

Ion-pair Phenomena and Borderline Mechanisms

A full kinetic analysis of the ion-pair return situation has been carried out by Macomber.[1] He has shown that the situation of reactions (1) is kinetically the same as the triangular scheme of reactions (2). His results show that: (a) if $k_1 = k_3$, then, even if $k_2 \gg k_1$, a normal first-order kinetic plot is obtained and that the rearrangement is not detectable from the kinetics; (b) plots of $\log [(\text{HOTs})_\infty] - \log [(\text{HOTs})_\infty - (\text{HOTs})_t]$ against time are curved upwards if $k_3 > k_1$ and $k_1 \gg k_2$, and are curved downwards if $k_1 > k_3$ and $k_1 \ll k_2$. The important point is that, for deviation to be detectable, k_1 and k_2 must be of the same order of magnitude. He has a computer program available for loan, which solves for all the rate constants.

The Sneen postulate,[2] reactions (3), has received further attention. McLennan[3] has analysed the data for $E2$ reactions and has decided that the evidence is only consistent with a classical $E2$ reaction and that this is probably also true for S_N2.[3] Kohnstam

---

[1] R. S. Macomber, *J. Org. Chem.*, **36**, 2182 (1971).
[2] See *Org. Reaction Mech.*, **1970**, 63; **1969**, 72; **1966**, 44.
[3] D. J. McLennan, *Tetrahedron Letters*, **1971**, 2317.

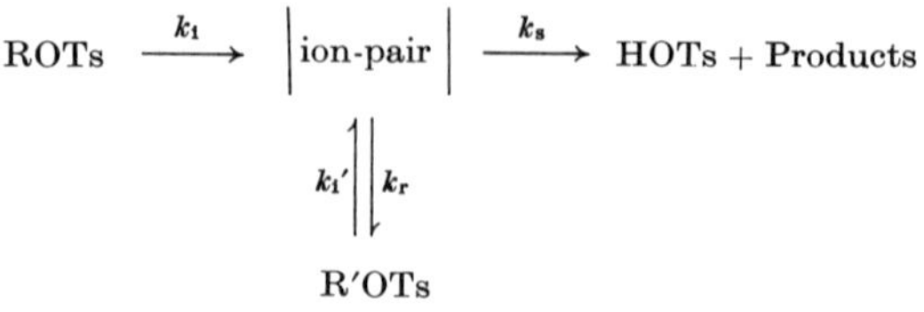

$$\text{ROTs} \xrightarrow{k_1} |\text{ion-pair}| \xrightarrow{k_s} \text{HOTs} + \text{Products} \qquad \ldots (1)$$

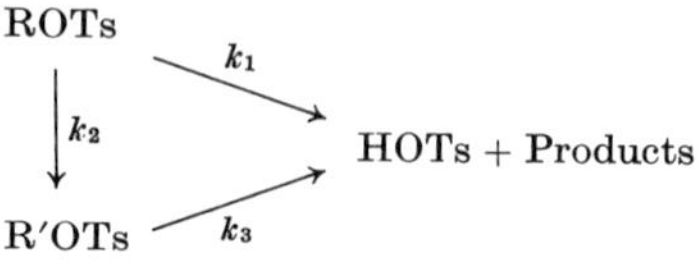

$$k_1' \updownarrow k_r$$

$$\text{R'OTs}$$

$$\begin{array}{c} \text{ROTs} \xrightarrow{k_1} \\ \downarrow k_2 \qquad \searrow \\ \text{R'OTs} \xrightarrow{k_3} \end{array} \text{HOTs} + \text{Products} \qquad \ldots (2)$$

$$\text{RX} \underset{k_{-1}}{\overset{k_1}{\rightleftharpoons}} \text{R}^+ \ \text{X}^- \begin{array}{c} \xrightarrow{k_2} S_N1, E1 \\ \xrightarrow{k_{2n}} S_N2, E2 \end{array} \qquad \ldots (3)$$

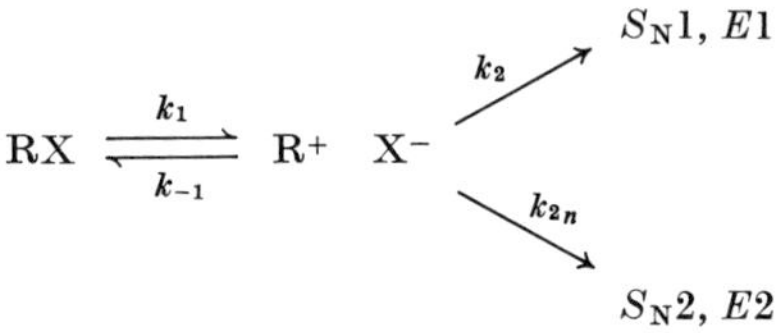

and his co-workers[4] have returned to the attack with an analysis of the data on solvolysis of arylmethyl chlorides. They feel that their analysis of the substituent effects on the rate compared to the same substituents on the rate of solvolysis of p-$NO_2C_6H_4CHClAr$ and of the salt effects on the two reactions rules out the Sneen mechanism and that "any general discussion of the mechanism of nucleophilic substitution *must* therefore include S_N1 and S_N2 processes". Schleyer and his colleagues[5] have compared the effect of added azide on the reactions of isopropyl, 2-adamantyl and 1-adamantyl toluene-p-sulphonates in 80% ethanol and in 75% dioxan. The two adamantyl compounds show little rate enhancement with added azide and react by an LIM mechanism. Furthermore, the amount of azide product is much less than one would expect if the rate enhancement were due to nucleophilic attack by azide on the ion-pair. On the other hand, the isopropyl (and 2-octyl) compounds give exactly the percentage of alkyl azide that would be expected from the large rate enhancement, viz: $(\%RN_3/\%ROS) + 1 = k(\text{observed})/k(\text{solvolysis})$. They also show that Sneen's results in 25% and 30% dioxan are fitted quite accurately by an S_N2 mechanism if one assumes a small negative salt effect ($b = -1.0$) instead of the small positive salt effect used by Sneen ($b = +1.0$). They therefore conclude that the reactions proceed by concurrent S_N1 and S_N2 reactions of the classical type. On the other hand, Scott,[6] in a long paper, has analysed the available data on the reactions of the four methyl halides (as a limiting case) with nucleophiles in water. By using a standard linear free-energy approach and making the assumption that for attack at a carbonium ion the ratio of the ρ values for one nucleophile against a standard nucleophile (ρ_n/ρ_{ns}) should be constant, he evaluates the two constants for equation (4) from the reactions of trityl carbonium ions with water and lyate ion. This

$$\log k^{\text{obs}} - \log K_e = (\rho_n/\rho_{ns})(\log k_s^{\text{obs}} - \log K_e) + C \qquad \ldots (4)$$

enabled him to evaluate $K_e (=k_1/k_{-1})$ for each methyl halide, and hence, using the relation $k^{\text{obs}} = K_e k_{2n}$, which is true when $k_{2n} \gg K_e$ (the situation to be expected when return is

---

[4] B. J. Gregory, G. Kohnstam, A. Queen, and D. J. Reid, *Chem. Comm.*, **1971**, 797.
[5] D. J. Raber, J. M. Harris, R. E. Hall, and P. von R. Schleyer, *J. Am. Chem. Soc.*, **93**, 4821 (1971).
[6] J. M. W. Scott, *Can. J. Chem.*, **48**, 3807 (1970).

much faster than attack), he evaluates k_{2n} for each halide/nucleophile pair. The log k_{2n} values he arrives at (Table 1) give a linear correlation against the force constants of the methyl halides, the gas-phase heats of ionization and the Swain–Scott correlation (ρ_r/ρ_{rs}, the ratio of the ρ values of one methyl halide with different nucleophiles against the ρ value for methyl bromide with the same nucleophiles). On the basis of these linear

Table 1. Equilibrium constants and nucleophilicity rates for methyl halides[6]

Halogen	F	Cl	Br	I
log K_e	−13.90	−8.81	−7.19	−6.68
log $k_{2n}(H_2O)$	3.06	−0.55	−0.94	−2.19

correlations, he feels that Sneen is right for the limiting case of the methyl halides and hence also in general. One particular aspect of his results is that they reproduce the observed order of reactivity of the methyl halides with nucleophiles, MeF < MeCl < MeBr > MeI, which has not previously been satisfactorily accounted for.[6] Scott's arguments seem convincing and Schleyer's use of a negative salt effect to "straighten out" Sneen's curve requires more justification. Clearly the field needs further work and one hopes that protagonists will enter the fray on both sides and not only to preserve the *status quo*. (See also ref. 68.)

Goering and his colleagues[7,8] have reported further work on the stereochemistry of ion-pair return for phenethyl, α-*p*-methoxyphenethyl and substituted benzhydryl *p*-nitrobenzoates in 70% and 90% aqueous acetone. Titrimetric (k_t), racemization (k_{rac}) and ^{18}O equilibration (k_{eq}) rates have been determined. In the phenethyl system, $k_{eq} = 0.76 k_t$ and $k_{rac} = 0.03 k_{eq}$, showing that return occurs with very little racemization, whereas for the *p*-methoxy compound, which reacts 3×10^4 faster, $k_{eq} = 0.57 k_t$ and $k_{rac} = 0.71 k_{eq}$; i.e. return occurs with much racemization, suggesting that the stereochemistry of return depends on the lifetime of the intermediate ion pair. The ratio k_{rac}/k_{eq} for α-*p*-methoxyphenethyl is not affected by added azide, which does, however, accelerate k_t with a special salt effect. These workers use the standard Winstein scheme to explain their results with (III) intercepted by azide and (II) not. The azide experiments show that (III) accounts for 46% and (II) for 54% of return. Strangely, both give

$$RX \rightleftharpoons R^+ X^- \rightleftharpoons R^+ \| X^- \rightarrow \text{Products}$$

(I)   (II)   (III)

the same (k_{rac}/k_{eq}) ratio.[7] Similar results were found in the benzhydryl series, with $k_{rac}/k_{eq} < 1$, and racemization increases with increasing reactivity (carbonium ion stability). As in the *p*-methoxyphenethyl case, azide ion gives rise to a special salt effect on the solvolysis of the *p*-tolyl compound but does not alter the racemization-to-equilibration ratio. Capture of the solvent-separated ion pair by azide gives 4-methylbenzhydryl azide of inverted configuration, while the configuration of the alcohol formed in the

[7] H. L. Goering, R. G. Briody, and G. Sandrock, *J. Am. Chem. Soc.*, **92**, 7401 (1970).
[8] H. L. Goering and H. Hopf, *J. Am. Chem. Soc.*, **93**, 1224 (1971).

presence of added 2,6-lutidine is 4% retained. They suggest that this occurs by internal water capture in (**III**), solvation to anion making it somewhat more nucleophilic than

$$HO^{\delta-}\cdots H\cdots^-OpNB$$

those outside. 1-p-Methoxyphenethyl p-nitrobenzoate gives 8% retention and p-chlorobenzhydryl p-nitrobenzoate 10% under similar conditions.[8] Similar work on the mercuric chloride-catalysed ionization of p-chlorobenzhydryl chloride suggests that, in acetone, ionization proceeds only as far as the intimate ion pair (**II**) (X = HgCl$_3^-$) and that racemization and chlorine exchange occur at this stage, but only within the counter-ion, whereas, in acetonitrile, solvent-separated ion pairs (**III**) are formed with scrambling of labelled chloride into the entire chloride ion pool. By using the dissociation constant of trityl mercuritrichloride as a model and setting k_{-2} equal to the diffusion rate, they have evaluated k_2 and hence k_{-1} for the two solvents acetone and acetonitrile. The former differ by more than 10^3 and the latter by less than ten.[9]

Further work on the *trans*-α,γ-methylphenylallyl p-nitrobenzoates in aqueous acetone has shown that less return occurs as the solvent becomes more aqueous, and also that return occurs with less racemization. This is consistent with a shorter ion-pair lifetime. The ratio of racemization to equilibration in the rearrangement of the α-phenyl-γ-methyl to the γ-phenyl-α-methyl isomer is 0.67, showing that return occurs with inversion about one-third of the time, although no *cis* ⇌ *trans* leakage occurs. This should be contrasted with the 5-methylcyclohex-2-enyl p-nitrobenzoate case where the stereospecificity is now ascribed to special conformational factors.[10]

The solvolysis of the α,α- and γ,γ-dimethylallyl chlorides in the presence of nucleophiles (NaOEt, NaBH$_4$) in water, ethanol and aqueous ethanol has been studied. The primary/tertiary product ratio rises with increase in nucleophilicity of the reagent, but it rises faster for the γ,γ- than for the α,α-dimethylallyl compound. The authors suggest, therefore, that the γ,γ-compound suffers S_N2 attack, and, from the difference in the ratio of the products derived from each isomer, that water and ethanol react only at the solvent-separated stage (**III**), while borohydride attacks the intimate ion-pairs (**II**).[11]

In the solvolysis of the epimeric p-nitrobenzoates (**1**) in 70% acetone, rearrangement in the recovered p-nitrobenzoates is very different from that in the product alcohols and hence at least two types of ion pair must be formed[12] (see also Chapter 1, p. 27).

Crampton and Grunwald[13,14] have published further details of their work on ion-pair exchange in acetic acid. They find that p-toluidinium toluene-p-sulphonate undergoes proton exchange *via* prior exchange with p-toluidinium acetate, which then exchanges with acetic acid. The former exchange, they suggest, occurs through the ion quartet (**2**), and they ascribe the increased rate for exchange over that for tertiary anilines as being due to the hydrogen bonding indicated in (**2**).[13] The exchange of the anilinium salt with metal acetate [reaction (5) (B = p-toluidine)] shows a variation for k_{+M} with

---

[9] A. Diaz, I. L. Reich, and S. Winstein, *J. Am. Chem. Soc.*, **92**, 7598 (1970).
[10] H. L. Goering, G. S. Koermer, and E. C. Linsay, *J. Am. Chem. Soc.*, **93**, 1230 (1971); cf. *Org. Reaction Mech.*, **1970**, 62.
[11] M. Eckert-Maksić, Z. Majerski, S. Borčić, and D. E. Sunko, *Tetrahedron*, **27**, 2119 (1971).
[12] R. K. Lustgarten, *J. Am. Chem. Soc.*, **93**, 1275 (1971).
[13] M. R. Crampton and E. Grunwald, *J. Am. Chem. Soc.*, **93**, 2987 (1971).
[14] M. R. Crampton and E. Grunwald, *J. Am. Chem. Soc.*, **93**, 2990 (1971).

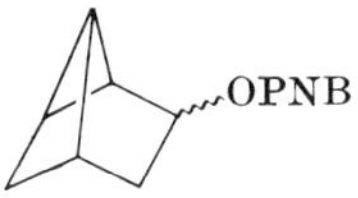

(1) [norbornyl-OPNB structure]

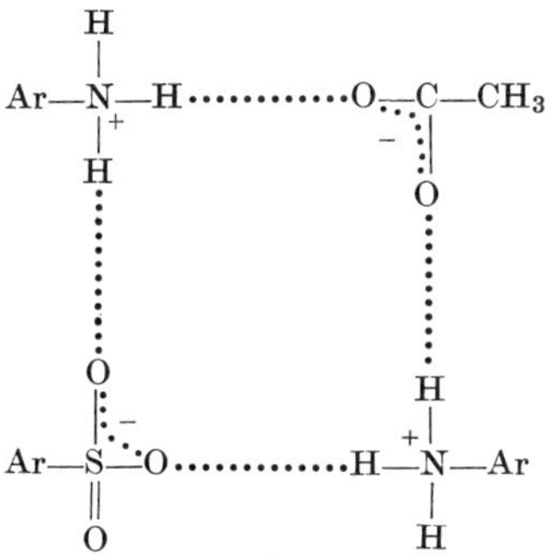

(2) [hydrogen-bonded ion pair structure with ArNH₃⁺···OAc⁻ and ArSO₃⁻···H₃N⁺Ar]

$$BH^{\oplus}OTs^- + M^+OAc^- \underset{k_{-M}}{\overset{k_{+M}}{\rightleftharpoons}} M^+OTs^- + BH^+OAc^{\ominus} \quad \ldots(5)$$

metal of a factor of seven, while k_{-M} is sensibly constant. They suggest that this is due, in particular, to a strong (but variable) interaction between metal cations of Lewis acid character and acetate anions, which must be weakened in the exchange process.[14]

In further work on the solvolytic isomerization of *cis*-4-t-butylcyclohexyl toluene-*p*-sulphonate, Okamoto and his co-workers[15] have found that recovered toluene-*p*-sulphonate shows isomerization to the *trans*-4-t-butyl epimer, not only in phenol–benzene as previously reported, but also in acetic acid and 60% aqueous acetone (Table 2). From the fact that a plot of percent *trans*-4-toluene-*p*-sulphonate (*trans*-4-ROTs) versus toluene-*p*-sulphonate anion concentration shows an intercept when extrapolated to zero anion concentration, they argue that epimerization is occurring in an ion pair, and that retained product arises as a result of solvolysis of *trans*-4-ROTs. However,

Table 2. Percent *trans*-4-ROTs and *cis*-4-ROS in the solvolysis of *cis*-4-t-butylcyclohexyl toluene-*p*-sulphonate[15]

Solvent	Phenol	Acetic acid	60% Acetone
% *trans*-4-ROTs	1.1	2.2	9.4
% *cis*-4-ROS	20.4	5.8	2.7

the percentage of *trans*-4-ROTs increases with increasing solvent ionizing power, but the amount of *cis*-4-ROS (acetate, alcohol or ether) decreases with increasing solvent nucleophilicity, which does not seem to accord with the authors' explanation.

The reaction of cyclo-octyl and cyclodecyl arylsulphenates with chlorine in acetic acid gives rise to intimate and solvent-separated chlorosulphenate ion pairs which decompose to cycloalkyl ion pairs [reactions (6)]. Added lithium perchlorate changes the ratio of halide to acetate products but does not affect the amount of deuterium scrambling, which is the same in both the chloride and the acetate products in each case. Scrambling experiments were done on the [1-²H₁]- and the [1,2,2,n,n-²H₅]-compounds in each case and showed that mainly 1,5-shifts occurred.[16]

---

[15] K. Okamoto, S. Saitô, and H. Shingu, *Bull. Chem. Soc. Japan*, **43**, 3008 (1970); see *Org. Reaction Mech.*, **1970**, 87; ref. 177.

[16] J. G. Traynham and A. W. Foster, *J. Am. Chem. Soc.*, **93**, 6216 (1971).

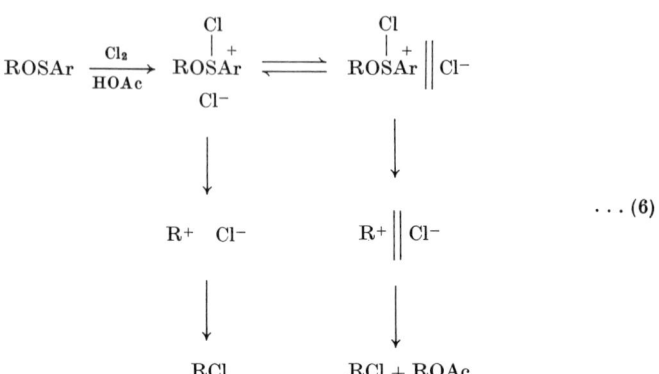

... (6)

The S_Ni reaction of 1-adamantyl chloroformate shows an m value of between 0.38 and 0.61 (depending on solvent).[17] This may be compared with the value of 1.20 which Schleyer reported for 1-adamantyl bromide[18] and suggests that the S_Ni is not truly LIM. Other work on alkyl chloroformate decomposition has also been interpreted as involving ion pairs, due to extensive rearrangement found in the products. This rearrangement is almost completely suppressed by the addition of ca. 10% of pyridine hydrochloride, even neopentyl chloroformate giving 42% of neopentyl chloride.[19]

The alcoholysis of benzhydryl chloride with six normal alcohols and three iso-alcohols has been studied,[20] and the methanolysis of the o-, m-, and p-phenylbenzhydryl chlorides has also been examined.[21] The kinetics of the hydrolysis of p-alkylbenzyl chlorides in aqueous acetone have been measured, and the effect of catalysis by mercuric chloride has been examined.[22] The hydrolysis of para-substituted α,α-dichloro- and α,α,α-trichloro-toluene has also received attention.[23]

Details of Fleury's work on 2-azabicyclo[2.2.1]hept-2-yl toluene-p-sulphonates have appeared.[24]

The effect of ion-pair structure on electron and proton-transfer,[25] and of pressure on the tight ion-pair ⇌ loose ion-pair equilibrium of sodium and lithium fluorenide ion-pairs[26] have been reported.

Further spectroscopic studies on metal–carbanion ion pairs have been carried out.[27a] Other studies on ion pairs have also been reported.[27b]

[17] D. N. Kevill and F. L. Weitl, *Tetrahedron Letters*, **1971**, 707.
[18] See *Org. Reaction Mech.*, **1970**, 67.
[19] P. W. Clinch and H. R. Hudson, *J. Chem. Soc.* (B), **1971**, 747.
[20] I. G. Murgulescu and I. Demetrescu, *Rev. Roum. Chim.*, **16**, 781 (1971); I. Demetrescu, *Rev. Roum. Chim.*, **15**, 1721 (1970).
[21] G. W. Gribble and M. S. Smith, *J. Org. Chem.*, **36**, 2724 (1971).
[22] R. Anantaraman, A. Balasubramanian, and K. Saramma, *Indian J. Chem.*, **9**, 605 (1971); *Chem. Abs.*, **75**, 75661 (1971).
[23] F. Quemeneur, B. Bariou, and M. Kerfanto, *Compt. Rend.* (C), **272**, 497 (1971).
[24] J.-M. Biehler and J.-P. Fleury, *Tetrahedron*, **26**, 3171 (1971).
[25] Y. Karasawa, G. Levin, and M. Szwarc, *J. Am. Chem. Soc.*, **93**, 4614 (1971).
[26] S. Claesson, B. Lundgren, and M. Szwarc, *Trans. Faraday Soc.*, **66**, 3053 (1971).
[27a] L. Lee, R. Adams, J. Jagur-Grodzinski, and M. Szwarc, *J. Am. Chem. Soc.*, **93**, 4149 (1971); J. W. Burley and R. N. Young, *Chem. Comm.*, **1971**, 1649; *J. Chem. Soc.* (B), **1971**, 1018; R. H. Cox, H. W. Terry, and L. W. Harrison, *J. Am. Chem. Soc.*, **93**, 3297 (1971).
[27b] M. Szwarc, *Suomen Kem.* (A), **43**, 173 (1971); U. Miotti and M. Padovan, *Boll. Sci. Fac. Chim. Ind. Bologna*, **27**, 269 (1970).

Solvent and Medium Effects

The use of a simple dipole–dipole model for statistical-mechanical calculations of normal salt effects has been reported.[28] The parameters necessary are the dipole moments and molecular radii of the species involved, and ways of estimating these are discussed. Perrin and Pressing's semi-empirical calculations[28] give b-values (in acetic acid and acetone) in quite good agreement with experimental results. They suggest that normal salt effects are due to salt-assisted ionization of the partly polar transition state and that the observed linear dependence on salt concentration, instead of the quadratic form expected from such a process, is due to the fact that the first salt molecule is much more effective than the others because of dipole–dipole repulsion between it and the second and other salt ion-pairs. They agree with Winstein's interpretation of the special salt effect and estimate lifetimes for ion pairs in the range 10^{-5} to 10^{-7} sec. Bunton[29] and his co-workers have also discussed normal salt effects in solvolysis of t-butyl bromide, 1- and 2-methyl-*exo*-2-chloro-norbornane, isobornyl chloride and camphene hydrochloride in a number of solvents. They find the anion order $ClO_4^- > OTs^-$, NO_3^-, $Br^- > Cl^- \approx$ No salt $> F^- > OH^-$, and use Perrin and Pressing's explanation, suggesting that the effects are due to changes in solvent structure interactions induced between the transition state for ionization and the salt anion (particularly when this is large).

The use of quaternary ammonium or phosphonium salts to overcome the problem of reaction heterogeneity has been beautifully exemplified by Starks.[30] He calls the process phase-transfer-catalysis and it is shown for the reaction of alkyl bromide with aqueous sodium cyanide in reactions (7). The quaternary salt acts to transport cyanide

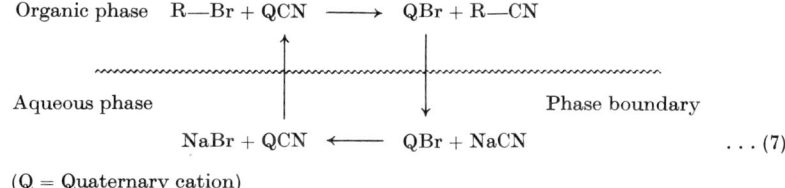

(Q = Quaternary cation)

ion into the organic phase and bromide ion into the aqueous phase. Suitable salts are tris(decyl)methylammonium chloride and tributylhexadecylphosphonium bromide which are used at concentrations of 1–10% of the organic phase, with a three-fold excess of the nucleophile salt in the (usually saturated) aqueous phase. A wide variety of anions can be used.

A wide-ranging survey of solvent polarity indices from various sources, e.g. E-values, Z-values, A-values, etc. have been compared and cross-correlated and correlated with other data by using multiparameter expressions incorporating them. In general, these expressions are little better than use of Dimroth's E-values alone.[31] The influence of solvent on the kinetics of the diphenyldiazomethane–benzoic acid reaction has been studied. The rate constants are fitted by equation (8), with a correlation coefficient of 0.983, for reaction in 20 alcohol solvents (K is the Kirkwood dielectric function and the

$$\log k = -1.72 + 4.5K + 2.8\sigma^* \qquad \ldots(8)$$

[28] C. L. Perrin and J. Pressing, *J. Am. Chem. Soc.*, **93**, 5705 (1971).
[29] C. A. Bunton, T. W. Del Pesco, A. M. Dunlop, and K.-U. Yang, *J. Org. Chem.* **36**, 887 (1971).
[30] C. M. Starks, *J. Am Chem. Soc.*, **93**, 195 (1971).
[31] F. W. Fowler, A. R. Katritzky, and R. J. D. Rutherford, *J. Chem. Soc.* (B), **1971**, 460.

σ^* are those of Taft). The introduction of a third parameter (the number of γ-hydrogen atoms in the alcohol) gives a marginal improvement $(r = 0.991)$. A similar correlation for 22 aprotic solvents is fitted by equation (9), where Qm is the heat of mixing with

$$\log k_0 = -0.47 + 3.6K - 0.0034Qm \qquad \ldots(9)$$

chloroform $(r = 0.92)$. As can be seen from the magnitude of its constant, this parameter (they also tried Inoue's $\Delta\nu_D$ function) makes only a marginal difference.[32]

The hypsochromic shift in the ultraviolet spectrum of tetrazolium and thiadiazolium salts has been shown to correlate with Y and Z values.[33] However, the azo $\rightleftharpoons$ hydrazone tautomerism in the azo-dyes (3) shows no correlation with dielectric constant or Z value.

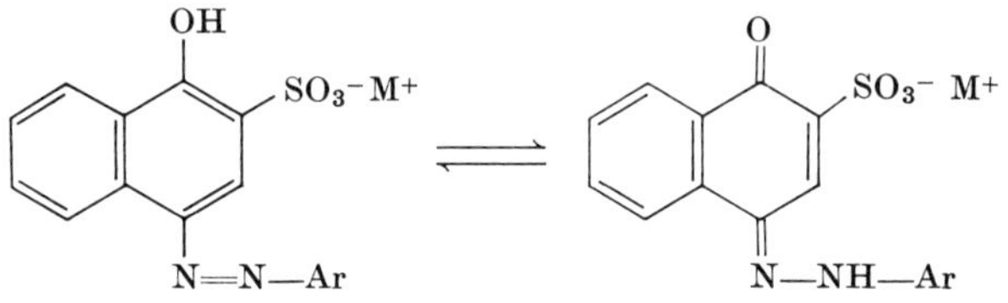

The authors suggest that, in aqueous solvents, changes in the equilibrium constant are due to breakdown of the three-dimensional water structure and specific solvation of the dye molecule.[34]

1-Adamantyl toluene-p-sulphonate solvolysis in water–ethanol, –methanol, –acetone and –dioxan mixtures and in isopropanol, t-butanol and acetic acid has been shown to correlate against Y values, with an m of 1.098 ± 0.057 (cf. ref. 18). The authors[35] conclude that a scale of Y values based on their results shows no improvement on Winstein and Fainberg's original or Schleyer's[18] (see also Chapter 1, p. 13).

The rates of solvolysis of arylmethyl chlorides in trifluoroethanol correlate against those in formic acid with a slope of 1.03, confirming other results that trifluoroethanol is a very non-nucleophilic solvent.[36]

Solvent effects on Hammett ρ-values have been examined. An attempt to correlate them with a solvent polarity index ξ_0 (a function of the activity coefficients of reagents and transition state) gives a fair amount of scatter.[37]

Abraham[38] has reported further work on the free energy of transfer from ethyl benzoate to 17 other solvents and from methanol to 11 other solvents for the reactions of Et_3N with p-$NO_2C_6H_4CH_2Cl$, of Me_3N with RCl, for solvolysis of t-butyl chloride and for the ion pairs $Me_4N^+Cl^-$ and $Et_4N^+I^-$. As might be expected transition states are stabilized by polar and destabilized by non-polar solvents. Dipolar aprotic solvents are more effective at this than alcohols. The results also show that the transition state for t-butyl chloride solvolysis strongly resembles the quaternary ions with a charge of ca. 0.8 unit, whereas that for the Menschutkin reaction is much less polar.[38] Similarly,

[32] N. B. Chapman, M. R. J. Dack, and J. Shorter, *J. Chem. Soc.* (B), **1971**, 834.
[33] A. M. Kiwan and H. M. N. H. Irving, *J. Chem. Soc.* (B), **1971**, 898.
[34] R. L. Reeves and R. S. Kaiser, *J. Org. Chem.*, **35**, 3670 (1970).
[35] D. N. Kevill, K. C. Kolwyck, and F. L. Weitl, *J. Am. Chem. Soc.*, **92**, 7300 (1970).
[36] M. D. Bentley and J. A. Lacadie, *Tetrahedron Letters*, **1971**, 741.
[37] T. Matsui and N. Tokura, *Bull. Chem. Soc. Japan*, **44**, 756 (1971).
[38] M. H. Abraham, *J. Chem. Soc.* (B), **1971**, 299; *Tetrahedron Letters*, **1970**, 5233; see *Org. Reaction Mech.*, **1970**, 68; **1969**, 77.

Haberfield and his colleagues[39] have shown that the enthalpy change in the transfer of pyridine quaternization from methanol to DMF is due solely to greater solvation of the transition state in the latter solvent and not to ground-state effects.

Cation and cavity selectivity of "cryptates" (**4**) ($m = 0$, 1 or 2, $n = 1$, 0 or 2; $m = n = 1$ or 2) towards the alkali and alkaline-earth metal cations shows that cation selectivity

$$\begin{array}{c}
\text{CH}_2\text{CH}_2\text{-O-(CH}_2\text{CH}_2\text{-O-)}_m\text{CH}_2\text{CH}_2 \\
\text{N-CH}_2\text{CH}_2\text{-O-(CH}_2\text{CH}_2\text{-O-)}_m\text{CH}_2\text{CH}_2\text{-N} \\
\text{CH}_2\text{CH}_2\text{-O-(CH}_2\text{CH}_2\text{-O-)}_n\text{CH}_2\text{CH}_2
\end{array}$$

(**4**)

is very high within a column of the Periodic Table.[40] The solvation of sodium tetraphenylborate by macrocyclic polyethers has also been studied.[41]

Dickson and Hyne have reported[42] further studies on the effect of pressure on the solvolysis of benzyl chloride in t-butyl alcohol–water mixtures and interpret their results as due to changes in solvent structure. The effect of pressure on the hydrolysis of benzyl chloride in pure water at temperatures from 0° to 15 °C has been reported. $\Delta V^{\ddagger}$ changes from -4 ml. to $+4$ ml. over the temperature range $+6°$ to $0°$C, indicating a large change in entropy in this region. It is suggested that this is due to the vastly increased structure of water, with a clathrated transition state, leading to rapid proton-transfer by a "tunnelling" process.[43] The effect of pressure and solvent on the solvolysis of 2,2,2-triphenylethyl toluene-p-sulphonate has been examined. In dioxan, 8 M in alcohol, with one equivalent of alkoxide, the ether/alkene ratio increases with increasing pressure.[44]

The activity coefficients of transfer of cations and anions from water to DMSO have been determined.[45] The effect of added water on ion-pair relaxation processes in DMSO have been studied by ultrasonic absorption.[46] The viscosities of salts in DMSO have been measured; sodium ions are solvated to the same size as butyltri(isopentyl)ammonium ions, and perchlorate appears to be a solvent structure-breaker.[47] The solvation of $AlCl_3$ by DMSO has been studied,[48] and the thermal pressure and energy–volume coefficients of DMSO–water mixtures have been measured.[49]

Thermal pressure and energy–volume coefficients have also been reported for methanol–water and t-butyl alcohol–water mixtures.[50]

[39] P. Haberfield, A. Nudelman, A. Bloom, R. Romm, and H. Ginsberg, *J. Org. Chem.*, **36**, 1792 (1971).
[40] J. M. Lehn and J. P. Sauvage, *Chem. Comm.*, **1971**, 440.
[41] A. M. Grotens, J. Smid, and E. de Boer, *Chem. Comm.*, **1971**, 759.
[42] S. J. Dickson and J. B. Hyne, *Can. J. Chem.*, **49**, 2394 (1971).
[43] G. J. Hills and C. A. N. Viana, *Nature*, **229**, 194 (1971).
[44] Y. Okamoto and T. Yano, *Tetrahedron Letters*, **1971**, 919.
[45] M. Le Demezet, *Bull. Soc. Chim. France*, **1970**, 4550.
[46] D. R. Dickson and P. Kruus, *Can. J. Chem.*, **49**, 3107 (1971).
[47] N.-P. Yao and D. N. Bennion, *J. Phys. Chem.*, **75**, 1727 (1971).
[48] J. C. Boubel, J. J. Delpuech, M. R. Khaddar, and A. Peguy, *Chem. Comm.*, **1971**, 1265.
[49] D. D. Macdonald and J. B. Hyne, *Can. J. Chem.*, **49**, 611 (1971).
[50] D. M. Macdonald and J. B. Hyne, *Can. J. Chem.*, **49**, 2636 (1971).

Solvation studies on alkali-metal salts in tetramethylurea have been carried out. As with DMSO, cations are extensively and anions very little solvated.[51] Other solvent studies are reported in references 52–56.

Isotope Effects

Model calculations on the relative magnitudes of deuterium and tritium isotope effects have been reported, with the assumption that there are no tunnelling effects. These show that, provided the individual effects are normal and reasonably large and exhibit "regular" temperature-dependence, the ratio r of deuterium to tritium effects lies within the values 1.33–1.58 for the temperature range 20–1000°K. This compares well with the Swain–Schaad value of 1.44. However they show that deviations are not rare, but that r can be used to calculate k_H/k_D from k_H/k_T for primary isotope effects with reliability if care is taken. For secondary isotope effects, such as are found in, e.g., S_N1, S_N2, $E2$ reactions, the median value of r is ca. 1.44 at zero temperatures and ca. 1.8 at infinite temperature; but the limits of variation are from −2.92 to +3.29 at zero temperature and 1.45 to 1.98 at infinite temperature, the differences depending on the type of potential barrier associated with the transition state.[57]

Shiner and Dowd[58] have further studied the effect of changing the leaving group on the α-deuterium isotope effect in limiting solvolysis. Using the system, $CH_3-C{\equiv}C-CHXCH_3$ (X = OTs, Br or I) they measured the effects of α-, β- and γ-deuterium substitution. They conclude that their results show that the reactions in 70% trifluoroethanol are LIM and that the maximum values for $k_H/k_{\alpha D}$ are OTs 1.226, Br 1.123 and I 1.089 and they compare these to their earlier theoretical values.[59] A product study showed that the leaving group has an effect (ROTs gives 20.6% of ether while RBr gives 25.4%) even in LIM solvolysis and they therefore suggest that nucleophilic attack on the second of two ion-pairs is the product-determining step.[58] Values for the α-k_H/k_D ratio in the solvolysis of 2-adamantyl trifluoroethanesulphonate in 70% and 97% trifluoroethanol at 25° (1.225 and 1.228) are also in agreement with occurrence of this reaction by a LIM mechanism. In other words, there is no nucleophilic solvent

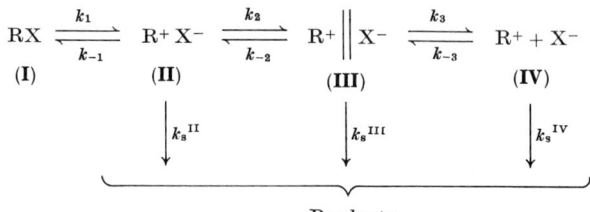

participation in the rate-determining step, which may refer to k_1, k_2 or k_3 on the Winstein scheme (annexed). In these cases where the maximum α-isotope effect is observed, k_2 is rate-determining and $k_{-1} > k_2$, i.e. one is not measuring the true ionization rate for 2-adamantyl trifluoroethanesulphonate. From their work on pinacolyl p-bromobenzene-

---

[51] B. J. Barker and J. A. Caruso, *J. Am. Chem. Soc.*, **93**, 1341 (1971).
[52] R. D. Guthrie, *J. Am. Chem. Soc.*, **92**, 7219 (1970).
[53] J. Ducom, *Bull. Soc. Chim. France*, **1971**, 3523, 3529.
[54] C. V. Krishnan and H. L. Friedman, *J. Phys. Chem.*, **74**, 3900 (1970).
[55] J. Macau, L. Lamberts, and P. Huyskens, *Bull. Soc. Chim. France*, **1971**, 2387.
[56] P. Murlo, A. Kivinen, and L. Strandman, *Suomen Kem.* (B), **44**, 308 (1971).
[57] M. J. Stern and P. C. Vogel, *J. Am. Chem. Soc.*, **93**, 4664 (1971).
[58] V. J. Shiner and W. Dowd, *J. Am. Chem. Soc.*, **93**, 1029 (1971).
[59] See *Org. Reaction Mech.*, **1969**, 79.

sulphonate, reported last year,[60] they estimate $k_{-1}/k_2 = 50$ for this case. They conclude that the α-deuterium isotope effect is the best probe for solvent participation in the rate-determining step (minimum substrate perturbation) but that, since mechanisms are complex, attempts to understand them must use all available techniques because of their synergic effects.[61] The last point is borne out by the results of Harris, Hall and Schleyer[62] who have also measured the α-deuterium isotope effect in the 2-adamantyl system (toluene p-sulphonate at 75.1°) and find a value of 1.23 (corrected to 25°); they contrast this with the value of 1.22 for cyclohexyl toluene-p-sulphonate in acetic acid and point to the fact that it shows an m value of 0.44 as compared with the 2-adamantyl compound where it is 0.91, indicating that nucleophilic solvent participation has altered drastically.

The magnitude of the α-deuterium isotope effect observed in the reverse Menschutkin reaction of N-benzyl-N,N-dimethylanilinium bromide (deuteriated at the benzylic position) in chloroform has been used to show that the reaction is of the S_N1 type.[63]

Equilibrium isotope effects in the 2,3-dimethyl-2-butyl (**5**, **6**) and 2,3,3-trimethyl-2-butyl (**7**) cations have been measured in SbF_5/SO_2ClF. The equilibrium constant for

$$DH_2C\diagdown\underset{H_3C\diagup}{\overset{+}{C}}-\underset{CH_3}{\overset{CH_3}{\diagup}}CH \quad \rightleftharpoons \quad DH_2C\diagdown\underset{H_3C\diagup}{HC}-\underset{CH_3}{\overset{+}{\overset{CH_3}{\diagup}}}C$$

(**5a**)   (**5b**)

$$D_3C\diagdown\underset{D_3C\diagup}{\overset{+}{C}}-\underset{CH_3}{\overset{CH_3}{\diagup}}CH \quad \rightleftharpoons \quad D_3C\diagdown\underset{D_3C\diagup}{HC}-\underset{CH_3}{\overset{+}{\overset{CH_3}{\diagup}}}C$$

(**6a**)   (**6b**)

$$D_3C\diagdown\underset{H_3C\diagup}{\overset{+}{C}}-\underset{CH_3}{\overset{CH_3}{\diagup}}C-CH_3 \quad \rightleftharpoons \quad D_3C\diagdown\underset{H_3C\diagup}{H_3C-C}-\underset{CH_3}{\overset{+}{\overset{CH_3}{\diagup}}}C$$

(**7a**)   (**7b**)

the reaction (**5a**) ⇌ (**5b**) is 1.132 (at −56°)[64] and that for (**6a**) ⇌ (**6b**) is 2.584 (at −79°).[65] These may be compared to the values for [$^2H_9$]-t-butyl chloride solvolysis of 2.30–2.46, equivalent to 1.097–1.105 per deuterium atom,[66] and clearly represent limiting values for the β-deuterium isotope effect. The reaction (**7a**) ⇌ (**7b**) has an equilibrium constant of 1.40 (at −53°), representing a value of 1.12 per deuterium atom.[67] Further discussion of these reactions is in Chapter 1, p. 34.

Schubert and Henson[68] have re-examined the isotope effects in the neopentyl system in trifluoroacetic acid, for all the possible γ-deuterated compounds:

$$CD_3C(CH_3)_2-CH_2OSO_2Ar \;(\mathbf{8}), \quad (CD_3)_2C(CH_3)-CH_2OSO_2Ar \;(\mathbf{9})$$

[60] See *Org. Reaction Mech.*, **1970**, 59.
[61] V. J. Shiner and R. D. Fisher, *J. Am. Chem. Soc.*, **93**, 2553 (1971).
[62] J. M. Harris, R. E. Hall, and P. von R. Schleyer, *J. Am. Chem. Soc.*, **93**, 2551 (1971).
[63] E. C. F. Ko and K. T. Leffek, *Can. J. Chem.*, **49**, 129 (1971).
[64] M. Saunders, M. H. Jaffe, and P. Vogel, *J. Am. Chem. Soc.*, **93**, 2558 (1971).
[65] M. Saunders and P. Vogel, *J. Am. Chem. Soc.*, **93**, 2561 (1971).
[66] See *Org. Reaction Mech.*, **1968**, 70.
[67] M. Saunders and P. Vogel, *J. Am. Chem. Soc.*, **93**, 2559 (1971).
[68] W. M. Schubert and W. L. Henson, *J. Am. Chem. Soc.*, **93**, 6299 (1971).

and $(CD_3)_3C-CH_2OSO_2Ar$ (10), where $Ar = 2,4-(NO_2)_2C_6H_3$. No kinetic isotope effect is observed $(k_H/k_D = 1.07)$;[69] but a substantial one on methyl migration is found, namely $k_{CH_3}/k_{CD_3} = 1.31$ (when the non-migrating groups are CH_3, CD_3 and CH_3, CH_3, respectively) and 1.22 (when they are CD_3, CD_3 and CH_3, CD_3). Since there is no kinetic effect, but nevertheless a distinct preference for CH_3 to migrate faster than CD_3, it is highly probable that the migration step occurs after the rate-determining one. They use the Sneen scheme to explain this and the fact that the ratio of the rates of solvolysis of neopentyl to ethyl arenesulphonates changes from 480:1 in trifluoroacetic acid to 1:1750 in ethanol. Thus for the ethyl compounds the value of k_{2n} changes enor-

$$(CH_3)_3CCH_2OSO_2Ar \underset{k_{-1}}{\overset{k_1}{\rightleftharpoons}} (CH_3)_3C\overset{+}{C}H_2 \quad \overset{-}{O}SO_2Ar \xrightarrow{k_2} (CH_3)_2\overset{+}{C}CH_2CH_3$$
$$\overset{-}{O}SO_2Ar$$

$\downarrow k_{2n}$

Nucleophilic attack and/or dissociation

$\downarrow$

Products

mously, whereas for neopentyl k_{2n} is always (negligibly) small and rearrangement to tertiary cation provides a constant escape hatch $(k_2 > k_1$ and $k_2 > k_{2n})$ to products.[68]

β-Deuterium isotope effects for acetolysis and formolysis have been measured for 5,6-benzo-2-norbornyl p-bromobenzenesulphonates (see Chapter 1, p. 18).[70]

The deuterium isotope effect in the attacking group in the quaternization of N,N-dimethylaniline and the corresponding [$^2H_6$]-compound with methyl toluene-p-sulphonate in nitrobenzene is (k_H/k_D) 0.86–0.92. The major effect is on $\Delta H^{\ddagger}$, which has been interpreted as showing that non-bonded interactions are the important factor.[71]

The solvent isotope effect on the mercury(II)-catalysed aquation of octahedral cobalt(III) complexes has been studied.[72]

Neighbouring Group Participation

Traylor and his colleagues[73] have further elaborated their σ–π conjugation theory of neighbouring group effects in a wide-ranging review covering participation by carbon σ-bonds, carbon–metal bonds and ferrocenyl systems. The important point is that such stabilization is "vertical" and occurs without change of bond length and bond angle, in contrast to the ND_I or bridged-ion theory where such changes are important. He argues that for a molecule $R_3ABCR'_2-X$ the stereochemistry of σ–π conjugation will be such that an antiperiplanar arrangement of the A—B and C—X bonds will be more favourable than a synperiplanar one, and this in turn will be much more favourable than any other orientation. The exo stereoselectivity of the 2-norbornyl cation is thus accounted for by microscopic reversibility: σ–π conjugation from C-1 to C-6 favours C^2–X bond heterolysis and by microscopic reversibility exo attack, without the need for either bridging or the 'windscreen wiper effect' of Brown. Other results that follow are

[69] Cf. M. J. Blandamer and R. E. Robertson, *Can. J. Chem.*, **42**, 2137 (1964).
[70] H. Tanida and T. Tsushima, *J. Am. Chem. Soc.*, **93**, 3011 (1971).
[71] K. T. Leffek and A. F. Matheson, *Can. J. Chem.*, **49**, 439 (1971).
[72] S. C. Chan and S. F. Chan, *Austral. J. Chem.*, **24**, 895 (1971).
[73] T. G. Traylor, W. Hanstein, H. J. Berwin, N. A. Clinton, and R. S. Brown, *J. Am. Chem. Soc.*, **93**, 5715 (1971); see also *Org. Reaction Mech.*, **1970**, 70.

that rate-acceleration by ND_I will lead to a change of $n\varDelta$ for n participating groups [e.g. $(ArCH_2)_2C(CH_3)X$ ionization], whereas $\sigma-\pi$ conjugation leads to an acceleration of $(\varDelta)^n$ for n groups [e.g. $(cyclopropyl)_2C(CH_3)X$ ionization], thus accounting for the observed differences between the two systems. It also follows that hyperconjugation should have no effect on ground-state properties. In conclusion, Traylor et al. do not deny that cationic solvolysis intermediates are bridged, but state that, because all the stabilization of cations is also seen in vertical processes, such bridging contributes very little to stabilization of the transition state.

Participation by Ether and Hydroxyl Groups

Dehydration of the cyclohexa-2,5-diene-1,4-diols (**11**) under acid conditions results either in migration of R^1 to give phenol or in ring-expansion to oxepin (**12**), by O–3 participation as shown. The product ratio is controlled by the nature of the groups

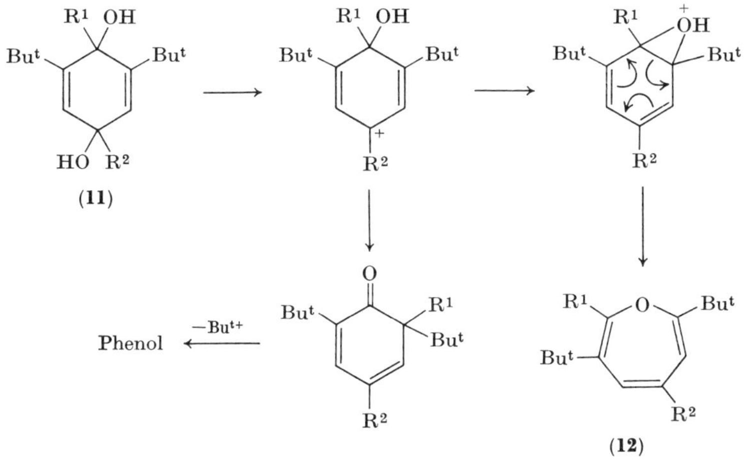

R^1 and R^2. With R^2 = t-butyl and R^1 either methyl or $p\text{-}Me_2NC_6H_4$, only phenol is formed, and with R^1 = p-tolyl and R^2 = phenyl, mesityl or p-methoxyphenyl, oxepin is formed to the extent of 17, 17, and 70%, respectively.[74]

Reaction of the sodium salt of 2-methylcyclohexane-1,3-dione with 1,4-dichlorobutan-2-one in ether–dimethoxyethane with one equivalent of ethoxide gives the decalindione epoxide (**13**).[75] Treatment of monotoluene-p-sulphonate (**14**) with t-butoxide–t-butyl alcohol affords the *exo*-epoxide (**15**), presumably by capture of the intermediate ion pair. The corresponding *endo*-,*endo*-isomer reacts similarly to afford the *endo*-epoxide.[76]

Solvolysis of the glucosides (**16**; Z = N_3 or OAc) gives the bicyclic acetals (**17**; Z = N_3 or OAc) by O–3 participation of the ring-oxygen, followed by closure to the alkyl-oxygen of the acetate group.[77] The silver perchlorate-catalysed ethanolysis of 2-chloro[1-$^{14}C^1$]-cyclohexanone gives 2-ethoxycyclohexanone with 20–25% of the label at C-2. This has been interpreted as involving addition of alcohol to the carbonyl group and rearrangement *via* an epoxide.[78] Electrochemical reduction of $TsO(CH_2)_nOTs$ gives good

---
[74] S. Berger, G. Henes, and R. Rieker, *Tetrahedron Letters*, **1971**, 1257.
[75] S. Danishefsky and G. A. Koppel, *Chem. Comm.*, **1971**, 367.
[76] J. M. Coxon, M. P. Hartshorn, and A. J. Lewis, *Chem. Comm.*, **1970**, 1607.
[77] C. Bullock, L. Hough, and A. C. Richardson, *Chem. Comm.*, **1971**, 1276.
[78] T. Masuike, N. Furukawa, and S. Oae, *Bull. Chem. Soc. Japan*, **44**, 448 (1971).

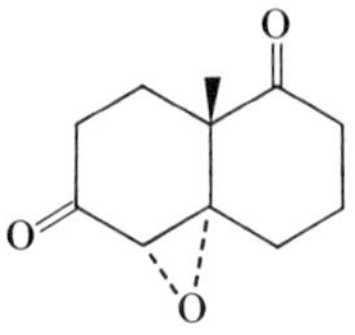

(13) 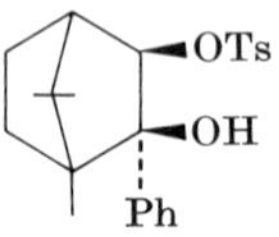   (14) 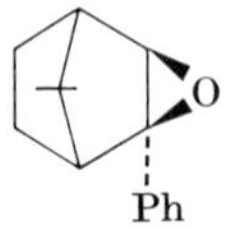   (15)

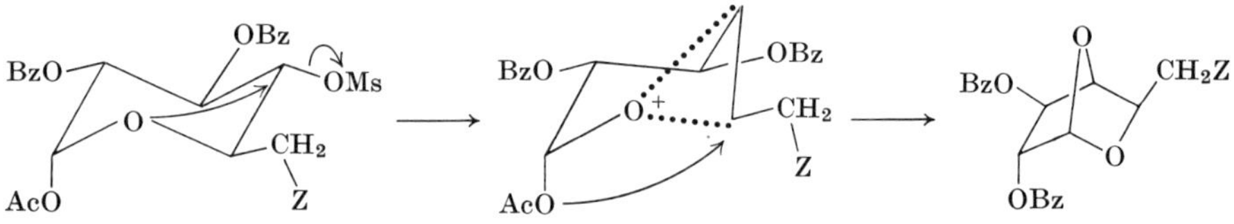

(16) → (17)

yields of epoxide for $n = 2$, 46% of THF for $n = 4$ and 14% of tetrahydropyran for $n = 5$. No oxetan was found in the reaction of the compound with $n = 3$.[79] A kinetic and product study of the alkaline hydrolysis of $ClCH_2C(CH_3)_2OOH$ showed that it reacted about 10^4 times faster than neopentyl chloride and 37 times faster than $ClCH_2C(CH_3)_2CH_2OH$ (18). The acceleration compared to neopentyl has been ascribed to O–4 participation and that compared to (18) to operation of the α-effect. Product studies showed that the four-membered cyclic peroxide could be isolated.[80] Richardson et al. have also reported the kinetics of oxetan formation for (18), 3-chloropropanol and 3-chloro-2,2-dimethylpropanol,[81] and yet another example of O–4 participation has been reported.[82]

Neighbouring methoxy participation in the *trans*-2-arylcyclohexyl toluene-*p*-sulphonate series has been studied[83] as a function of the number of methoxyl groups on the aromatic ring and of nucleophile concentration. With the dipotassium salt of thioglycollic acid in methanol, the *o*-methoxyphenyl compound shows 82% retention of stereochemistry at low nucleophile/toluene-*p*-sulphonate ratios (2:1) and 85% inversion at 50:1 ratio. The 2′,3′,4′-trimethoxyphenyl compound gives only product of retained stereochemistry.

Other examples of O–5 and O–6 participation have also been reported.[84–88]

Participation by Thioether and Thiol Groups

The rates of hydrolysis of the 2α,5-epithio-5α-cholestan-3β-yl bromide and the 3α-yl methanesulphonate, of the 3β-bromide of the corresponding sulphoxide, and of the 2α,5-epoxy-5α-cholestan-3α- and -3β-yl methanesulphonates and the 3α-yl bromide in 70, 80 and 90% aqueous dioxan have been measured. The partial structures are shown as (19). The 3β-compounds have the leaving group *endo* on the bicyclo[2.2.1] ring, and the

[79] R. Gerdil, *Helv. Chim. Acta*, **53**, 2097 (1970).
[80] W. H. Richardson and V. F. Hodge, *J. Am. Chem. Soc.*, **93**, 3996 (1971).
[81] W. H. Richardson, C. M. Golino, R. H. Wachs, and M. B. Yelvington, *J. Org. Chem.*, **36**, 943 (1971).
[82] P. F. Hudrlik and A. M. Hudrlik, *Tetrahedron Letters*, **1971**, 1361.
[83] S. K. Core and F. J. Lotspeich, *J. Org. Chem.*, **36**, 399 (1971).
[84] T. Sasaki, S. Eguchi, and T. Kiriyama, *Tetrahedron Letters*, **1971**, 2651.
[85] J. A. Marshall and A. E. Greene, *Tetrahedron Letters*, **1971**, 859.
[86] R. D. H. Murray, M. Sutcliffe, and P. H. McCabe, *Tetrahedron*, **27**, 4901 (1971).
[87] W. H. Richardson and V. F. Hodge, *Tetrahedron Letters*, **1971**, 749.
[88] O. A. Ching Puente, *Bol. Soc. Quím. Peru*, **36**, 13, 45, 60 (1970); *Chem. Abs.*, **73**, 110017 (1970); **74**, 64368, 64372 (1971); I. Ondvus, *Chem. Prumysl*, **21**, 168 (1971).

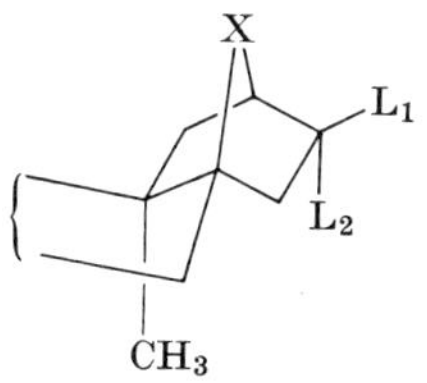

(19) $\begin{pmatrix} \text{a } X = O, L_1 = H, L_2 = OMs \\ \text{b } X = S, L_1 = H, L_2 = Br \end{pmatrix}$

3α-compounds have the *exo* orientation. The relative rates are given in Table 3. Using the rate of the 3β-oxygen compound (**19a**) as a model for the inductive retardation,

Table 3. Relative rates of solvolysis of (**19**) in 70% dioxan at 25°

X =	S	SO	O	Norbornyl
exo	1.0	0.79	9.0	1.1×10^4
endo	1.2×10^8	—	1.1×10^{-2}	13

Tanida and his colleagues[89] suggest that the acceleration is due to S–3 participation and amounts to 1.1×10^{10}. However, both (**19a**) and (**19b**) give products with retained stereochemistry, 100% for the sulphur compound and 95% for the oxygen. The latter figure suggests that even in (**19a**) some participation occurs and that the figure for acceleration by sulphur should be somewhat larger than 10^{10}.

S–3 participation has been invoked[90] to account for the equilibration of (**20**) and (**21**) and for the reaction of either with terminal alkenes to give the *p*-chlorobenzenesulphenyl chloride adduct of the alkene and *trans*-propenylbenzene. Since equilibration of (**20**) and (**21**) is much faster than formation of propenylbenzene, attack by chloride on the intermediate thiironium ion is much faster at carbon than at sulphur. The alkaline

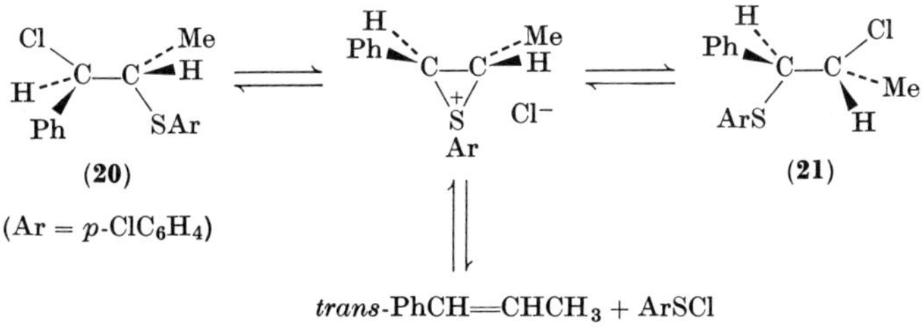

(**20**) (**21**)

(Ar = *p*-ClC$_6$H$_4$)

trans-PhCH=CHCH$_3$ + ArSCl

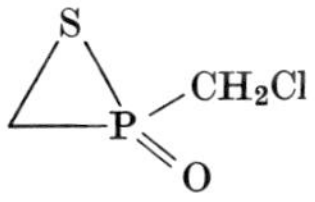

(**22**) 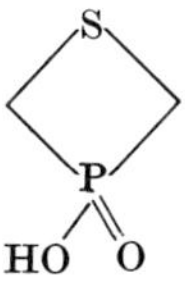 (**23**)

[89] T. Tsuji, T. Komeno, H. Itani, and H. Tanida, *J. Org. Chem.*, **36**, 1648 (1971).
[90] G. H. Schmid and P. H. Fitzgerald, *J. Am. Chem. Soc.*, **93**, 2547 (1971).

hydrolysis of $(ClCH_2)_2P(:O)SH$ goes with S–3 participation to give (23) through (22).[91]

S–6 participation occurs in solvolysis of the thioacetal of 5-tosyl-L-arabinose to give a perhydrothiophen derivative.[92]

Modena, Tonellato and their colleagues have reported further on sulphur participation in vinyl cations and this is discussed below (refs. 347–349).

Participation by Halogens

Trifluoroacetolysis of the trifluoromethanesulphonates (24) leads to products of halogen migration (25), amounting to 5%, 55% and 58% of the product for X = Cl, Br and I, respectively. In the case of (24c), 6% of 1-(iodomethyl)propyl trifluoroacetate is formed, and it is suggested that the intermediate four-membered iodonium ion undergoes the ring contraction shown.[93] Br–4 participation has also been observed in the rapid con-

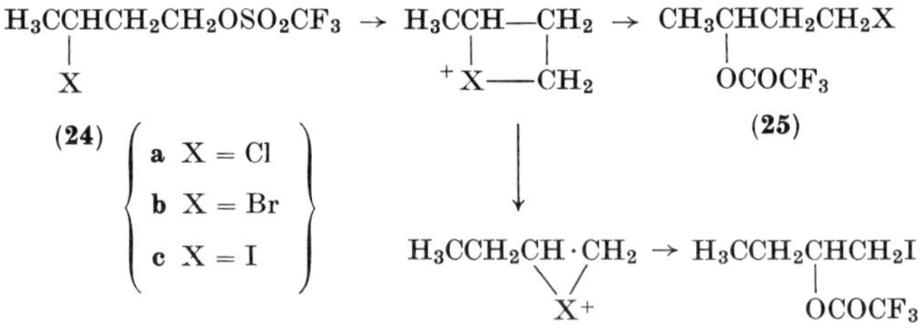

$$H_3CCHCH_2CH_2OSO_2CF_3 \rightarrow H_3CCH—CH_2 \rightarrow CH_3CHCH_2CH_2X$$
$$\quad\quad\quad | \quad\quad\quad\quad\quad\quad\quad | \quad\quad | \quad\quad\quad\quad\quad\quad\quad |$$
$$\quad\quad\quad X \quad\quad\quad\quad\quad\quad\quad\quad ^+X——CH_2 \quad\quad\quad\quad\quad OCOCF_3$$

(24) a X = Cl
      b X = Br
      c X = I

$$H_3CCH_2CH \cdot CH_2 \rightarrow H_3CCH_2CHCH_2I$$
$$\quad\quad\quad \backslash \ / \quad\quad\quad\quad\quad\quad\quad |$$
$$\quad\quad\quad\quad X^+ \quad\quad\quad\quad\quad\quad\quad OCOCF_3$$

$$CH_2—CHCH_2CH_2CHR$$
$$\ \backslash\ /\ \quad\quad\quad\quad\quad |$$
$$\quad O \quad\quad\quad\quad\quad\quad Cl$$

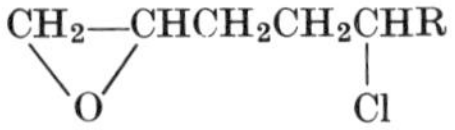

(26)
(R = H or CH_3)

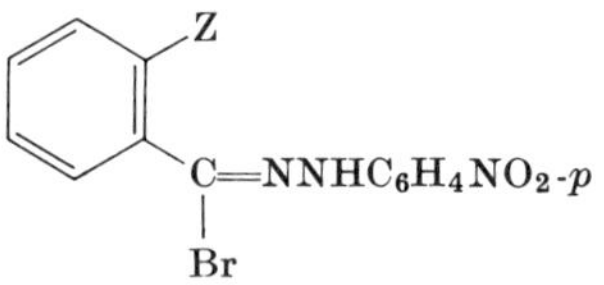

(27) (with Z on aromatic ring, C=NNHC_6H_4NO_2-*p*, Br substituent)

(Z = F, Cl, Br or NO_2)

version of $(BrCH_2)_3CCH_2OH$ into $(BrCH_2)_3CCH_2OSO_3F$ and $(BrCH_2)_2C(CH_2OSO_3F)_2$ in sulphur dioxide solutions of SbF_5–HSO_3F.[94]

The chloro-epoxides (26) have been shown to undergo ring-opening in trifluoroacetic acid with Cl–5 participation.[95]

Scott and his co-workers have ascribed the acceleration by all *ortho*-groups in the solvolyses of (27) to a field (dipole repulsion) effect.[96]

Deamination of 2-chloro[1-$^{14}C_1$]ethylamine leads to 2-chloroethanol with 27% of the label at the chlorine-bearing carbon, showing that Cl–3 participation occurs.[97]

[91] V. E. Bel'skii, N. V. Isvasyuk, S. V. Povarenkina, and I. M. Shermergorn, *Izv. Akad. Nauk SSSR, Ser. Khim.*, **1970**, 1407.
[92] J. Harness and N. A. Hughes, *Chem. Comm.*, **1971**, 811.
[93] P. E. Peterson and W. F. Boron, *J. Am. Chem. Soc.*, **93**, 4076 (1971).
[94] J. H. Exner, L. D. Kershner, and E. R. Larsen, *Chem. Comm.*, **1971**, 1174; cf. *Org. Reaction Mech.*, **1970**, 77.
[95] P. E. Peterson, J. M. Indelicato, and B. R. Bonazza, *Tetrahedron Letters*, **1971**, 13.
[96] A. F. Hegarty, M. Cashman, J. B. Aylward, and F. L. Scott, *J. Chem. Soc.* (B), **1971**, 1879.
[97] O. A. Reutov, T. A. Smolina, and O. Y. Polevaya, *Doklady Akad. Nauk SSSR*, **191**, 366 (1970); *Proc. Acad. Sci. USSR*, **191**, 205 (1970).

Participation by Carbonyl Groups

The hydrolysis of (**28**) in aqueous base proceeds as shown,[98] while reaction of the triterpenoid ketones (**29**) and (**30**) (partial structures) in t-BuOH–t-BuOK gives the cyclobutanone derivatives shown.[99] The same groups of workers re-examined the reaction of (**31**) and suggest similar mechanisms for the formation of (**32**) and (**33**). The American workers have treated the methyl enol ether of (**31**) which also gives the same products and have suggested that ions (**34**) and (**35**) are formed by independent

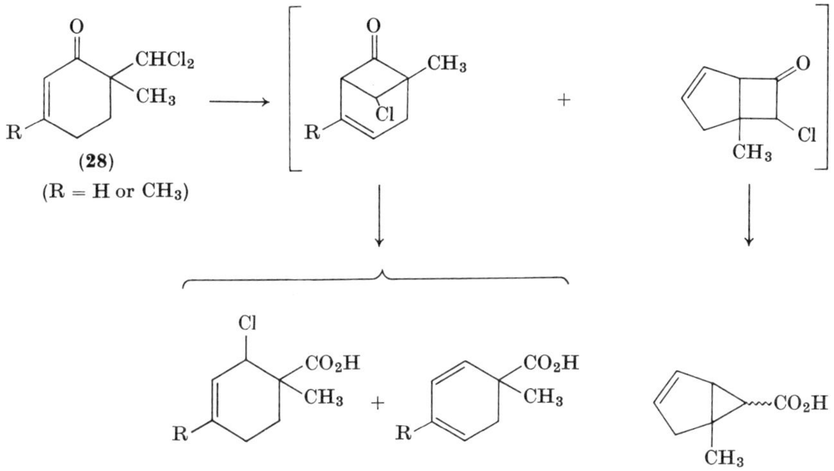

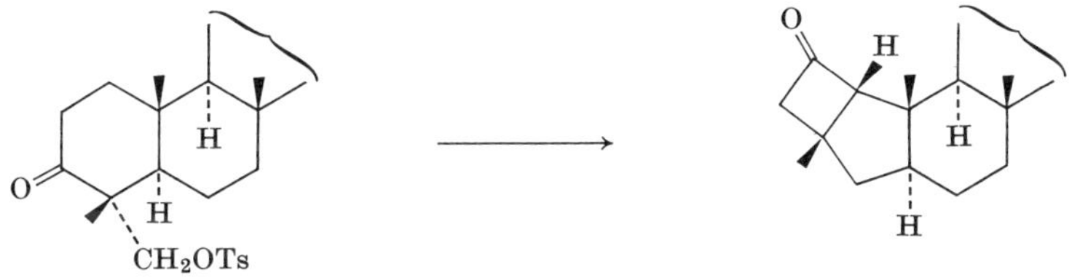

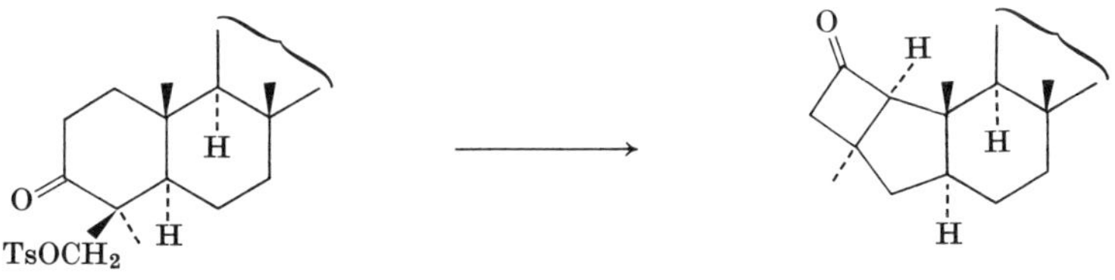

[98] E. Wenkert, P. Bakuzis, R. J. Baumgarten, C. L. Leicht, and H. P. Schenk, *J. Am. Chem. Soc.*, **93**, 3208 (1971).
[99] Y. Tsuda, T. Tanno, A. Ukai, and K. Isobe, *Tetrahedron Letters*, **1971**, 2009.

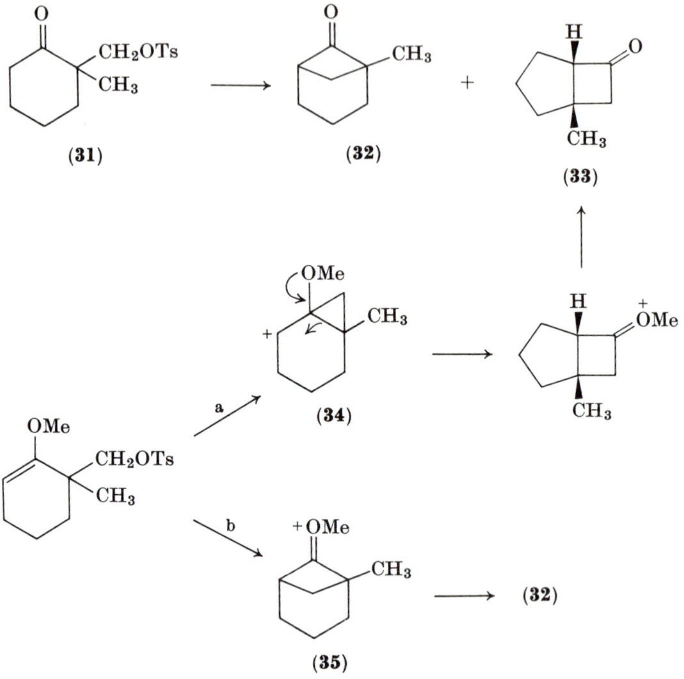

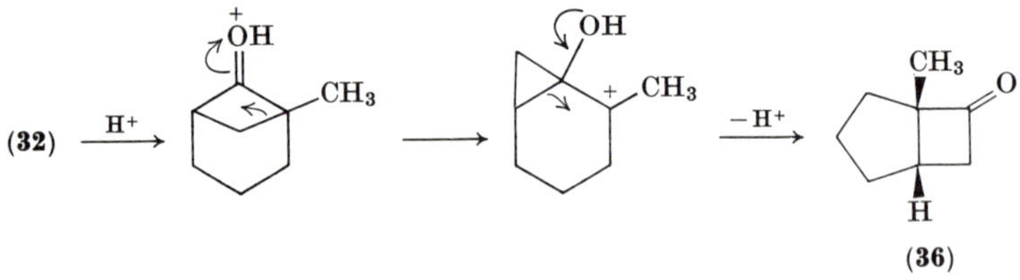

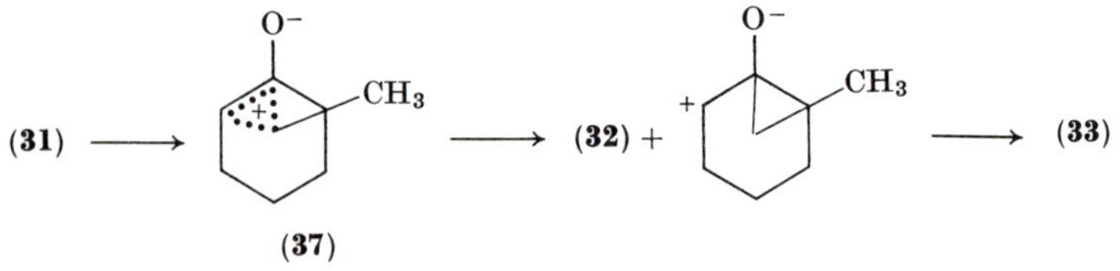

routes. The Japanese workers[99] have examined the acid-catalysed rearrangement of (32), which gives (36) presumably by the route shown, and they therefore suggest that the enolate from (31) gives the bridged ion (37), which then partitions to (32) or the zwitterion shown.

Acid chlorides of the type RCH$_2$COCl react with N-benzylideneaniline in carbon tetrachloride to give chloro-amides of the type RCH$_2$CONPhCHClPh, which are converted by triethylamine via the enols into azetidinones (38).[100] Enol participation in the solvolysis of (39) gives only norbornanone and never bicyclo[3.2.0]heptan-2-one.[101]

[100] A. K. Bose, G. Spiegelman, and M. S. Manhas, *Tetrahedron Letters*, **1971**, 3167.
[101] J. L. Marshall, *Tetrahedron Letters*, **1971**, 753.

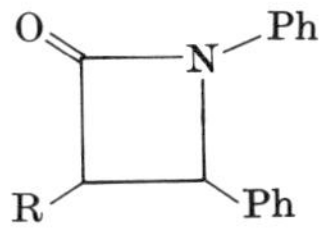

[Structures (38), (39), (40) → (41) shown]

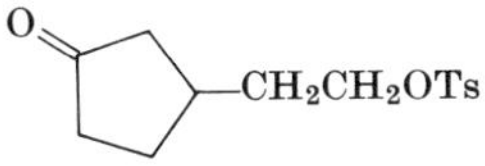

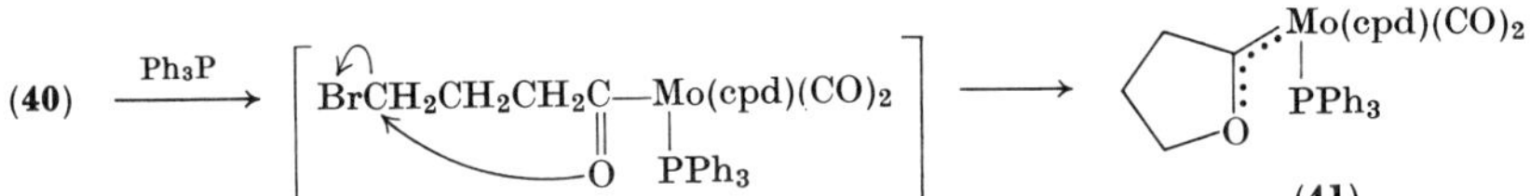

Allenyldimethylsulfonium salts react with ketones to give furans by ring-closure through carbonyl oxygen with both S_N2 and S_N2' displacement in the intermediate zwitterion.[102]

The reaction of (3-bromopropyl)tricarbonyl(*pentahapto*-cyclopentadienyl)molybdenum (40) with triphenylphosphine affords the intermediate complex (41) by the route shown.[103]

Participation by Ester, Carboxyl and Amide Groups

Migrations of alkoxycarbonyl groups have been reviewed.[104] Also, Paulsen has reviewed his work on acetoxonium ion rearrangement[105] and has reported studies of these in the amino-sugar series,[106-109] where he finds that the nitrogen-containing ion is much more stable than the oxygen ones and hence disturbs the equilibrium. Thus 3-amino-3-deoxy-D-glucose penta-acetate gives the 2,3-oxazoline-derivative of D-*manno*-configuration, and the corresponding 6-amino-6-deoxy-D-glucose compound is converted into the 5,6-dihydro-oxazine derivative of D-*ido*-configuration.[106] Treatment of 1,2,4/3,5-penta-acetoxycyclopentane with anhydrous HF affords the acetoxonium ion (42) which undergoes a 10-step migration around the ring equivalent to pseudorotation with a free energy of activation of 18.0–18.3 kcal mol^{-1}, depending on solvent.[107] A similar equilibrium is set up when the triesters of glycerol, 1,3/2-cyclopentanetriol and 1,3/2-

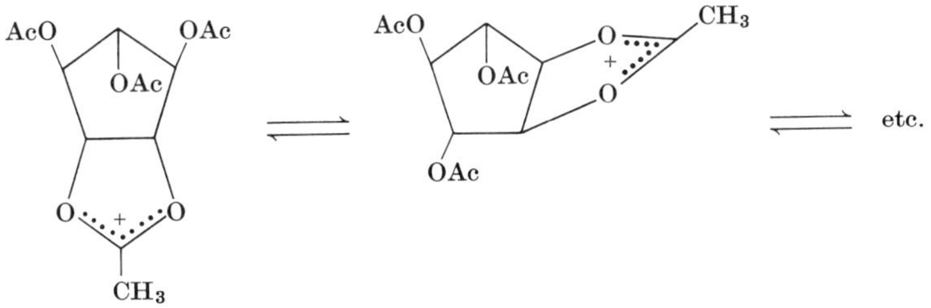

[Structure (42) shown in equilibrium]

[102] J. W. Batty, P. D. Howes, and C. J. M. Stirling, *Chem. Comm.*, **1971**, 534.
[103] F. A. Cotton and C. M. Lukehart, *J. Am. Chem. Soc.*, **93**, 2672 (1971).
[104] R. M. Acheson, *Accounts Chem. Res.*, **4**, 177 (1971).
[105] H. Paulsen, *Chimia (Aarau)*, **24**, 290 (1970); H. Paulsen, H. Behre, and C.-P. Herold, *Fortschr. Chem. Forsch.* **14**, 473 (1970).
[106] H. Paulsen and C.-P. Herold, *Chem. Ber.*, **104**, 1311 (1971); cf. *Org. Reaction Mech.*, **1970**, 79.
[107] H. Paulsen and H. Behre, *Chem. Ber.*, **104**, 1299 (1971).
[108] H. Paulsen and H. Behre, *Chem. Ber.*, **104**, 1281 (1971).
[109] H. Paulsen and H. Behre, *Chem. Ber.*, **104**, 1264 (1971).

cyclohexanetriol are treated with antimony pentachloride in nitromethane. The free energies of activation are all close to 18 kcal mol^{-1} with entropies of activation of ca. 3 eu.[108] Other work on diesters has shown that only *trans*-1,2- or *trans*-1,3-diesters react with $SbCl_5$ to give cyclic salts, but that only *cis*-diesters give similar salts in liquid HF.[109]

The anhydro-sugar derivative (**43**) reacts with chlorosulphonyl isocyanate with S_N2' participation by acetoxyl as shown in the reaction sequence (10).[110]

(**43**)

(R¹ = CO₂Et, R² = SO₂Cl) . . . (10)

(**44**)

(**45**)

(**46**)

(**47**)

. . . (11)

(R = H, CH_3; R′ = H, CH_3; R″ = H, CH_3, CO_2Et)

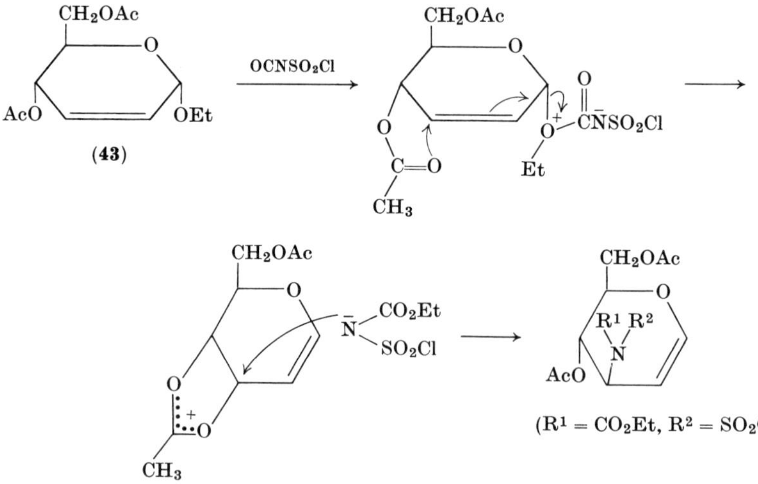

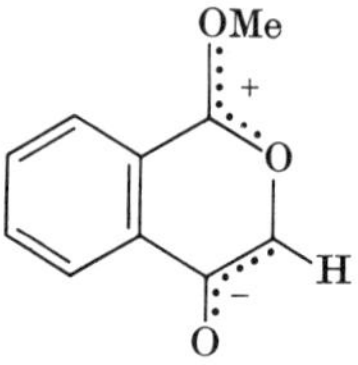

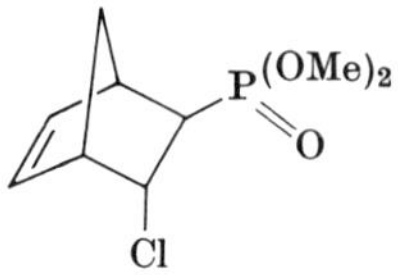

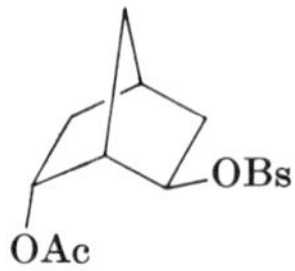

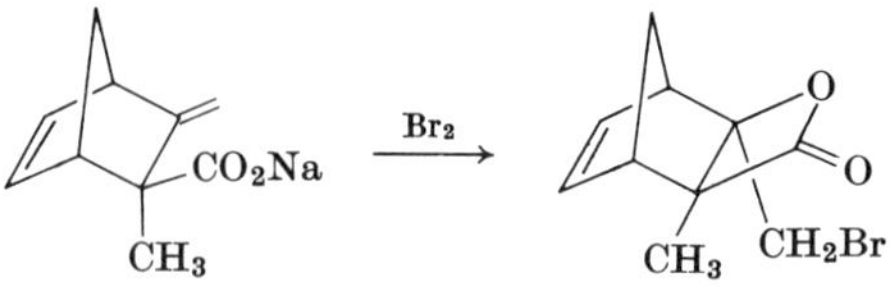

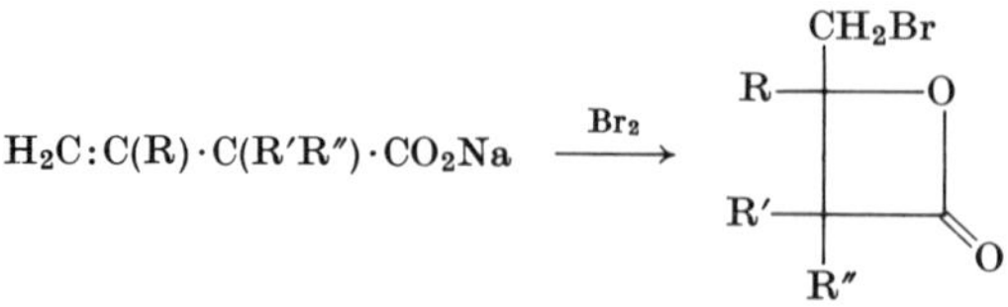

[110] A. Jordaan and G. J. Lourens, *Chem. Comm.*, **1971**, 581.

o-Methoxycarbonyldiazoacetophenone reacts with copper acetylacetonate with neighbouring carboxyl participation to afford the zwitterion (**44**), which may be trapped by dienes.[111] Participation by the phosphorus ester in (**45**) gives a rate acceleration of 4.6×10^4 over the corresponding *exo*-chloride, which is itself anchimerically accelerated.[112] (**46**) also shows significant acceleration by acetoxyl participation, reacting about 600 times more rapidly than the *exo*-6-acetoxy isomer.[113]

Other reports of neighbouring ester participation in the sugar field have also appeared.[114–116]

The formation of β-lactones in the bromination of β,γ-unsaturated acids has been shown to be a general reaction [see (11)] and not confined to the bicyclic compound (**47**). However, iodo-lactonization proceeds to give the γ-lactone.[117] Deamination of phenylalanine and *p*-methoxyphenylalanine occurs with either aryl or carboxyl participation; the ratio of one to the other depends on the aryl group and the solvent, but with *p*-nitrophenylalanine only carboxyl participation is observed.[118] The disproportionation of *o*-carboxyphenyl disulphides involves anchimeric acceleration by the *o*-carboxylate anion.[119] The intramolecular catalysis of phosphate triester hydrolysis by neighbouring carboxyl has been studied.[120]

The stereochemistry of ring-closure and ring-opening of α-lactams has been studied by Sarel and his co-workers.[121] Using the optically pure bisnorchlolanyl compounds (**48**; side chain only shown) they find that, as expected, ring closure occurs with inversion at C-23 and that ring-opening by attack at the carbonyl group occurs with retention. However, ring-opening with attack at C-23 also occurs with retention of configuration, which they suggest occurs via the four-centre planar transition complex (**49**). An α-lactam has also been implicated in the reaction of $Ph_2C(Cl)CONH_2$ with cyanamide.[122]

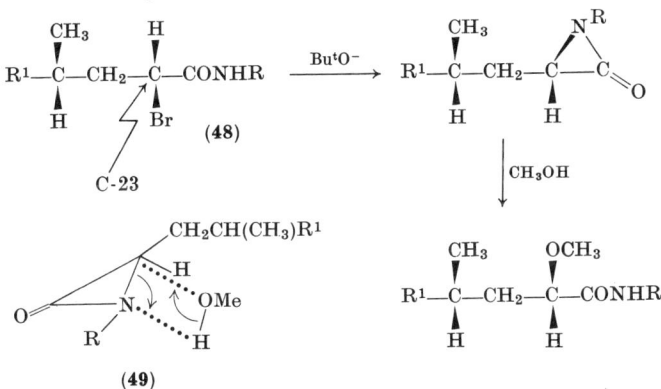

(R^1 = Steroid, R = Bu^t or 1-Adamantyl)

[111] M. Takebayashi, T. Ibata, K. Ueda, and T. Ohashi, *Bull. Chem. Soc., Japan*, **43**, 3964 (1971).
[112] C. E. Griffin, *Colloq. Int. Cent. Nat. Rech. Sci.*, **1970**, No. 185, 95; *Chem. Abs.*, **74**, 52679 (1971).
[113] G. W. Oxer and D. Wege, *Tetrahedron Letters*, **1971**, 457.
[114] J. S. Brimacombe and A. M. Mofti, *Chem. Comm.*, **1971**, 241.
[115] J. G. Buchanan, J. Conn, A. R. Edgar, and R. Fletcher, *J. Chem. Soc.* (C), **1971**, 1515.
[116] I. Ziderman, *Carbohydrate Res.*, **17**, 220 (1971).
[117] W. E. Barnett and J. C. McKenna, *Tetrahedron Letters*, **1971**, 2595; *Chem. Comm.*, **1971**, 551.
[118] K. Koga, C. C. Wu, and S. Yamada, *Tetrahedron Letters*, **1971**, 2287.
[119] L. Field, P. M. Giles, and D. L. Tuleen, *J. Org. Chem.*, **36**, 623 (1971).
[120] R. H. Bromilow, S. A. Khan, and A. J. Kirby, *J. Chem. Soc.* (B), **1971**, 1091.
[121] S. Sarel, B. A. Weissman, and Y. Stein, *Tetrahedron Letters*, **1971**, 373.
[122] K. Lempert, J. Puskas, and S. Vezer, *Acta Chim. Acad. Sci. Hung.*, **67**, 369 (1971).

The effect of solvent on O-5 ring closures of $BzNHCH_2CH_2Br$ and $BzOCH_2CH_2Br$ in 30–100% aqueous ethanol has been studied. A linear correlation against Y with an m of 0.13 and 0.18, respectively, suggests an internal S_N2 mechanism.[123] Full details have appeared on the dehydration of o-$RNHCOC_6H_4CH_2CR^1R^2OH$ in sulphuric acid. The change in participation of O versus N with R^1 and R^2 has been interpreted as showing that N participation only occurs if the amide group does not suffer protonation before water is lost from the hydroxyl group.[124] There has also been another study of this system.[125] For the analogous closure of o-$RNHCOC_6H_4CR^1R^2OH$ it has been suggested that participation by oxygen results in reversible formation of the imino-phthalide, whereas closure on nitrogen is irreversible.[126]

The diazonium ion (**50**), formed by protonation of the corresponding diazo-compound, decomposes by the competitive participation of each of the three centres indicated by arrows.[127] The nine-membered chloro-lactam (**51**) cyclizes more rapidly than the

(**50**)   (**51**)   (**52**)

eight-membered one (**52**); this has been ascribed to conformational effects.[128] Participation through oxygen and sulphur occurs in the formation of oxazolinium and thiazolinium ions by the protonation of N-allyl(and substituted allyl)-amides, -urethanes, -ureas and -thioureas.[129] Other examples of participation involving amide groups are reported in references 130–135.

Participation by Amino, Silyl and Phosphinyl Groups

The acetolysis of $ArN(CH_2CD_2Cl)_2$ occurs with complete scrambling of deuterium; but in the reaction with a toluene-p-thiolate anion, no scrambling occurs when Ar is phenyl but occurs when Ar is p-methoxyphenyl. This suggests that an aziridinium ion is formed by a process competitive with S_N2 attack.[136] The kinetics of aziridine ring-closure has been studied.[137] N-t-Butyl-2-(chloromethyl)aziridine (**53**) and N-t-butyl-3-

---

[123] F. L. Scott, E. J. Flynn, and D. F. Fenton, *J. Chem. Soc.* (B), **1971**, 277; cf. *Org. Reaction Mech.*, **1970**, 79.
[124] D. M. Bailey and C. G. De Grazia, *J. Org. Chem.*, **35**, 4088 (1970); cf. *Org. Reaction Mech.*, **1970**, 80.
[125] C.-L. Mao and C. R. Hauser, *J. Org. Chem.*, **35**, 3704 (1970).
[126] D. M. Bailey and C. G. De Grazia, *J. Org. Chem.*, **35**, 4093 (1970).
[127] R. Scarpati and M. L. Graziano, *Tetrahedron Letters*, **1971**, 2085.
[128] O. E. Edwards, J. M. Paton, M. H. Benn, R. E. Mitchell, C. Watanatada, and K. N. Vohra, *Can. J. Chem.*, **49**, 1648 (1971).
[129] S. P. McManus, J. T. Carroll, and C. U. Pittman, *J. Org. Chem.*, **35**, 3768 (1970).
[130] T. A. Foglia, L. M. Gregory, and G. Maerker, *J. Org. Chem.*, **35**, 3779 (1970).
[131] E. E. Smissman, R. A. Robinson, J. B. Carr, and A. J. B. Matuszak, *J. Org. Chem.*, **35**, 3821 (1970).
[132] Y. S. Rao and R. Filler, *Chem. Comm.*, **1970**, 1622.
[133] L. A. Paquette and J. F. Kelly, *J. Org. Chem.*, **36**, 442 (1971).
[134] S. K. Talapatra, B. C. Maiti, and B. Talapatra, *Tetrahedron Letters*, **1971**, 2683.
[135] P. Shanmugam and P. Lakshminarayana, *Tetrahedron Letters*, **1971**, 2323.
[136] M. H. Benn, P. Kazmaier, and C. Watanatada, *Chem. Comm.*, **1970**, 1685.
[137] V. N. Bochenkov and V. F. Borodkin, *Tr. Ivanov. Khim.-Teknol. Inst.*, **1969**, No. 11, 113; *Chem. Abs.*, **74**, 140428 (1971).

Nucleophilic Aliphatic Substitution

chloroazetidine (54) are equilibrated in the presence of HCl (with concurrent ring-opening). Reaction of either with alkanethiolate, alkoxide or amine gives product of retained structure, suggesting an S_N2 process; but on hydrolysis or reaction with cyanide ion both (53) and (54) give the substituted azetidines. In acetolysis, (53) gives t-BuNHCH(CH$_2$OAc)$_2$, while (54) gives both that and t-BuNHCH$_2$CH(OAc)CH$_2$OAc. The results are interpreted in terms of the unsymmetrical ions (55) and (56).[138]

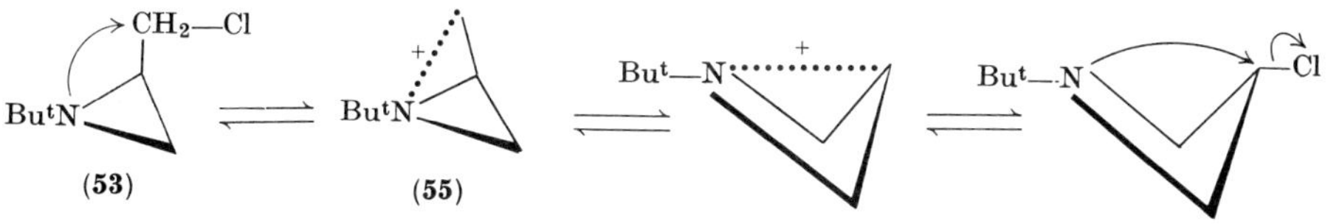

A common aziridinium ion (57; R^1 = OMe, R^2 = Me) has been implicated in the solvolysis of halogeno-codeinone and -neopinone acetals and has been isolated by treatment of the 8β-iodoneopinone derivative with AgClO$_4$ in benzene at room temperature.[139] However, in the presence of a free 6α-hydroxyl group in the codeine series, O–5 participation precedes N–3 participation, leading to product of retained configuration at C-14, as shown in sequence (12).[140]

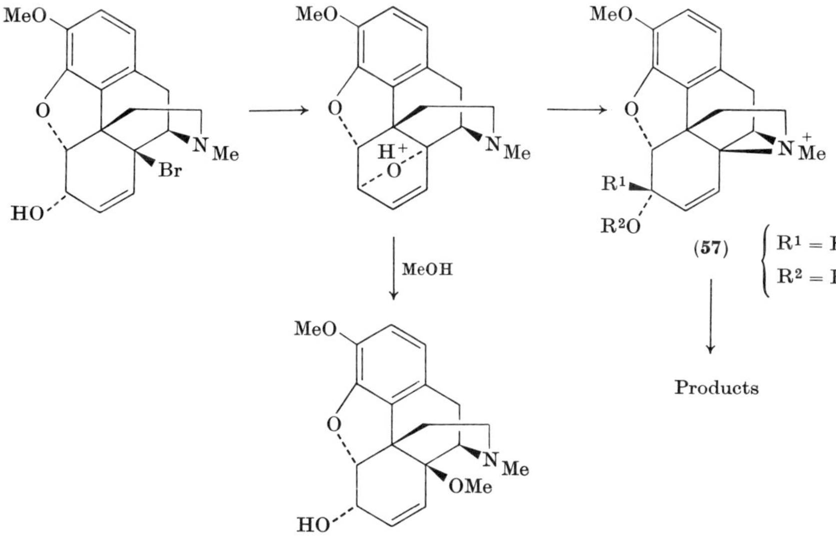

... (12)

An mY treatment of the rates of solvolysis of tropan-3α-yl methanesulphonates gives an m of 0.33 in aqueous ethanol, indicating significant nucleophilic participation. However, reaction with phenoxide and thiophenoxide is some 900 times faster than with benzylamine, and the product is of 3β-configuration, showing that, with active nucleophiles, a S_N2 mechanism competes successfully with N–4 participation.[141]

[138] V. R. Gaertner, *J. Org. Chem.*, **35**, 3952 (1970).
[139] R. M. Allen and G. W. Kirby, *Chem. Comm.*, **1971**, 1121.
[140] K. Abe, Y. Nakamura, M. Onda, and S. Okuda, *Tetrahedron*, **27**, 4495 (1971).
[141] P. Scheiber, G. Kraiss, and K. Nádor, *J. Chem. Soc.* (B), **1971**, 2154.

α,β-Epoxyketone hydrazones afford pyrazolinols if the β-carbon is benzylic and the carbonyl and epoxide are cisoid.[142] Another example of N–5 participation is reported in ref. 143. 1-Thio-3,4,6-tri-O-acetyl-S-(o-aminophenyl)-2-O-methanesulphonyl-thio-β-D-glucopyranoside undergoes ring-closure to the dihydro-D-*manno*-tetrahydroxypyranobenzothiazine triacetate.[144]

In the reactions of four chloromethyldisilanes with aluminium chloride, silicon migrates to carbon after the ionization step.[145]

Acetolysis of 1,2-dimethyl-2-(diphenylphosphinyl)propyl methanesulphonate affords 3-(diphenylphosphinyl)-2-methylbut-1-ene by migration of the diphenylphosphinyl group, and 2-(diphenylphosphinyl)-2-methylpropyl toluene-*p*-sulphonate similarly gives 1-(diphenylphosphinyl)-2-methylpropene. Another example is shown in reaction (13).[146]

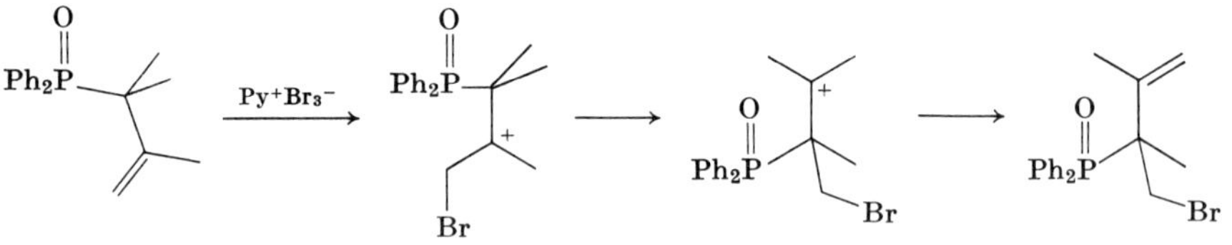

... (13)

Participation by Neighbouring Carbanion and Carbon—Metal Bonds

Wolff–Kishner reduction of 2,3-dihydro-5,6-benzopyran gives 40% of 2-cyclopropylphenol by C⁻–3 participation of the intermediate anion.[147] Crombie and his co-workers[148] have found that, in the synthesis of presqualane alcohol, the enolate (**58**), formed by Michael addition at the indicated bond, closes to form a cyclopropane derivative in high yield.[148] Arylidenemalononitriles react with the cyclic enol phosphate (**59**) in a similar fashion to give good yields of cyclopropanes.[149] *meso*-1,2-Dithian-3,6-dicarboxylic acid (**60**) in 0.1N-sodium hydroxide undergoes the ring-contraction shown about 100 times faster than the (*R,S*)-isomer. The reaction appears to be general for dithianes with enolizable groups at C-3.[150] The ring-closure–elimination of the dibromo-diesters, $EtO_2CCHBr(CH_2)_nCHBrCO_2Et$, to give diethylcycloalkene-1,2-dicarboxylates has been studied as a function of chain length in dimethylformamide; good yields are obtained, with sodium hydride as base, for formation of cyclo-butene, -pentene and -hexene, and 21% of cycloheptene-1,2-dicarboxylic ester is obtained but none of the cyclic C_8 compound; when $n = 1$ and t-butoxide is used as base, (**61**) is formed, and one equivalent of base leads to (**62**) in 77% yield.[151]

[142] D. L. Coffen and D. G. Korzan, *J. Org. Chem.*, **36**, 390 (1971).
[143] J. J. Tufariello and J. P. Tette, *Chem. Comm.*, **1971**, 469.
[144] M. Sekiya and S. Ishiguro, *Tetrahedron Letters*, **1971**, 431.
[145] K. Tamao and M. Kumada, *J. Organomet. Chem.*, **30**, 339 (1971).
[146] P. F. Cann, D. Howells, and S. Warren, *Chem. Comm.*, **1971**, 1148.
[147] N. Bellinger, D. Cagniant, and P. Cagniant, *Tetrahedron Letters*, **1971**, 49.
[148] R. V. M. Campbell, L. Crombie, and G. Pattenden, *Chem. Comm.*, **1971**, 218.
[149] E. Corre and A. Foucaud, *Chem. Comm.*, **1971**, 570.
[150] J. P. Danehy and V. J. Elia, *J. Org. Chem.*, **36**, 1394 (1971).
[151] R. N. McDonald and R. R. Reitz, *Chem. Comm.*, **1971**, 90.

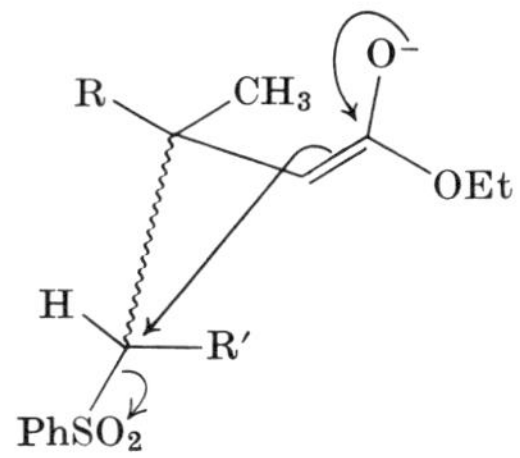

(58)
(R = $C_{11}H_{19}$, R' = $C_{14}H_{23}$)

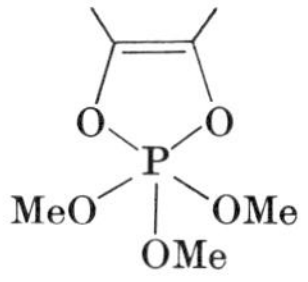

(59)

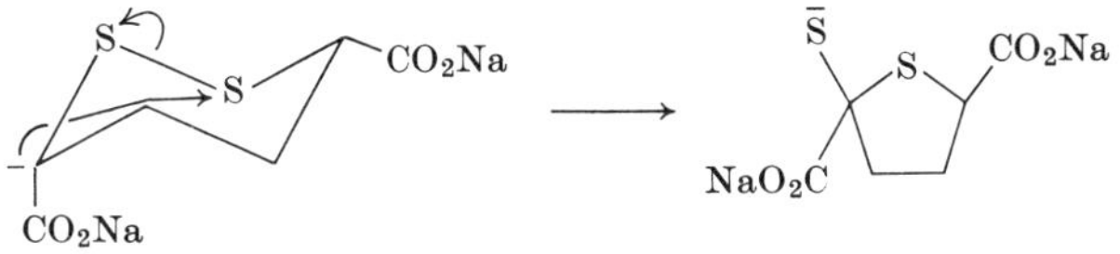

(60)

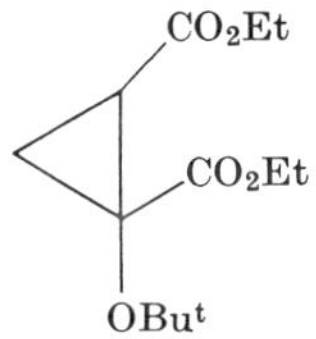

(61)

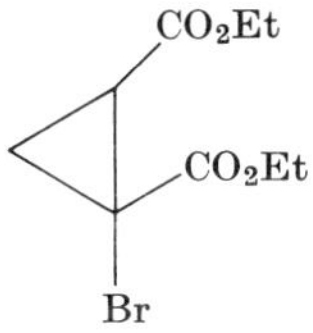

(62)

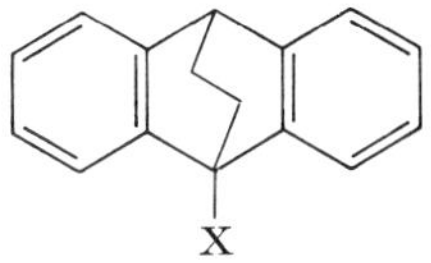

(63)

Hanstein, Berwin and Traylor[152] have reported data relevant to the σ–π conjugation of carbon—metal bonds. By comparing the effect of metal-containing groups on the charge-transfer (CT) frequency of compounds of the type $PhCH_2X$ (X = $SnMe_3$, HgCl, $SiMe_3$, etc.) with tetracyanoethylene against that for toluene they evaluated a set of σ^+ values (see *Org. Reaction Mech.*, **1970**, 71). A similar comparison using (**63**) shows that in this orientation all groups give the same CT frequency, i.e. that rotation into the nodal plane of the π-system removes the effect, and also that any inductive or direct field effect is negligibly small. But where correct orientation is possible, as in the benzyl compounds, a large effect is found due to the ability of the carbon—metal bond to σ–π conjugate; e.g. for the series $PhCH_2X$, the CT frequency is given by $\nu = 4300\sigma + 24,250$, whereas for the corresponding compounds $PhCH_2HgX$ it is given by $\nu = 11,200\sigma + 15,700$.

Olah and Clifford[153] have examined the ^{1}H-NMR spectra of $CH_3OCH_2CH_2HgCl$ and (**64**) in FSO_3H–SbF_5–SO_2. The former gives a spectrum consistent with a bridged mercurium ion and so does the latter at $-70°$. However, the spectrum of (**64**) collapses to a single line at $-30°$, indicating more extensive equilibration processes at this temperature.

Further examples of the ring closure of stannanes, $(CH_3)_3SnCH_2CH_2CR_2OH$, to cyclopropanes on treatment with thionyl chloride have appeared. The norbornyl

[152] W. Hanstein, H. J. Berwin, and T. G. Traylor, *J. Am. Chem. Soc.*, **92**, 7476 (1970).
[153] G. A. Olah and P. R. Clifford, *J. Am. Chem. Soc.*, **93**, 1261 (1971).

compound (**65**), however, gives only norbornene and no tricyclic hydrocarbon.[154] Both *cis*- and *trans*-1,4-dibromocyclo-octane on treatment with magnesium afford *cis*-cyclo-octene; participation by the intermediate Grignard compound as in (**66**) has been suggested as a route. *cis*-1,5-Dibromocyclononane similarly gives *cis*-cyclononene as well as some of the more expected *cis*-bicyclo[4.3.0]nonane.[155]

Neighbouring Carbon and Hydrogen

Pople and his colleagues[156] have reported the results of their *ab initio* calculations, with full geometrical optimization, for the $C_3H_7^+$ cation. They find that the isopropyl cation is more stable than any other arrangement by some 17 kcal mol^{-1}, and that the most stable geometry for the n-propyl cation is as shown in (**67**), although the other structures (**68**), (**69**), (**70**) are only ca. 0.5 kcal mol^{-1} of higher energy. This shows that methyl shift requires very little energy, and that methyl rotation in the corner-protonated cyclopropane requires none [(**68**) ⇌ (**69**)]. Edge-protonated cyclopropane is some

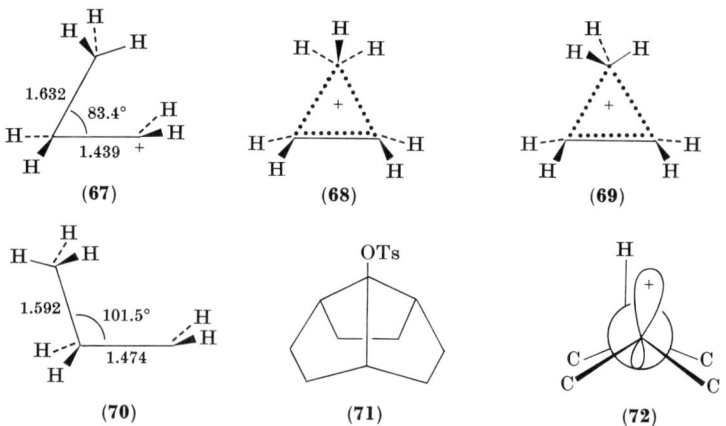

10 kcal mol^{-1} higher in energy than (**68**); and the face-protonated structure is some 120 kcal mol^{-1} higher. From the difference between edge- and corner-protonated cyclopropanes they conclude that the 1,2-methyl shift is much easier than a 1,3-hydrogen shift. It should be noted that neither (**67**) nor (**70**) has a geometry like the "classical"

---

[154] D. D. Davis, R. L. Chambers, and H. T. Johnson, *J. Organomet. Chem.*, **25**, C13 (1970); cf. *Org. Reaction Mech.*, **1970**, 81.
[155] M. S. Baird, C. B. Reese, and M. R. D. Stebles, *Chem. Comm.*, **1971**, 1340.
[156] L. Radom, J. A. Pople, V. Buss, and P. von R. Schleyer, *J. Am. Chem. Soc.*, **93**, 1813 (1971).

one for $C_2H_5CH_2^+$ and in particular that the "sp^2" atom is bent with the hydrogen atoms moved towards the methyl group.

Bingham and Schleyer[157] have reported that the tosylate (71) solvolyses 10^9 times more slowly than one would predict (see Chapter 1, p. 13) and have ascribed this to lack of hyperconjugation in the intermediate carbonium ion, which must have the eclipsed geometry shown in the Newman projection (72). This accords well with the calculations of Pople, Schleyer and their co-workers reported last year.[158]

The rates of solvolysis of $Me_3SiC(CH_3)_2Br$ (73) and $Me_3CC(CH_3)_2Br$ (74) in aqueous ethanol have been compared. The former gives an mY correlation with an m of 0.70. The only product is olefin, which may be due to an $E2$ process; however, if we assume that the solvolysis rate is due to an ionization process, the minimum rate ratio (73):(74) is 1:38,000. The authors suggest that the low rate of ionization of (73) is due to lack of C—C hyperconjugation because of the longer C—Si bond.[159] The solvolysis of a number of other α-(trimethylsilyl)-substituted halides has also been found to be very slow, showing that this is a general phenomenon.[160]

The solvolysis of eleven t-butyl- and methyl-substituted cyclohexane chlorohydrins has been studied, as regards both kinetics and products.[161] Isotope effects have been measured for (75) and (76) and are (k_H/k_D) 2.8 and 2.9, respectively. While the latter

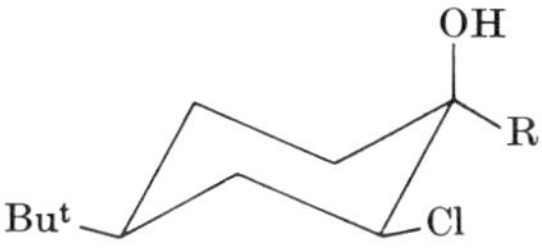

(75)   (76)   (77)

(R = H or Me)

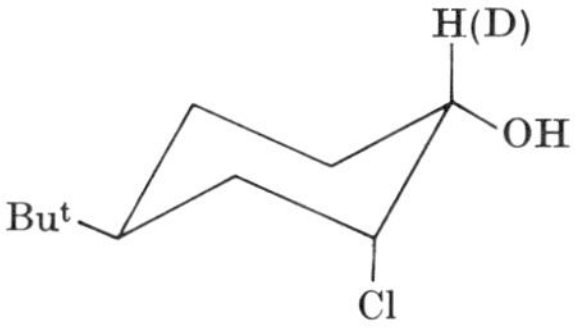

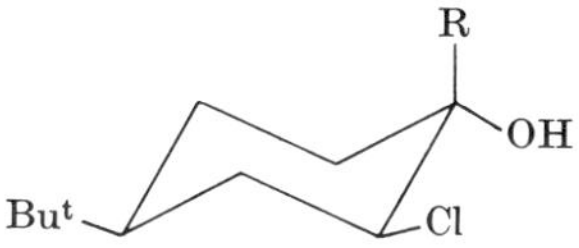

value is what might be expected, the former strongly suggests H-participation in a twist-boat transition state, as does the product, which is entirely 4-t-butylcyclohexanone from each of (75) and (76). The diequatorial chlorohydrin (77; R = H) also reacts as a twist-boat to give diol of retained stereochemistry, while the corresponding 1-methyl compound (77; R = Me) gives 4-t-butyl-2-methylcyclohexanone and 3-t-butylcyclopentyl methyl ketone by methyl participation and ring-contraction, respectively. In contrast to the hydrogen case, the compound (75; R = Me) reacts exclusively by ring-contraction to give the cyclopentyl methyl ketone. Deamination of the amino alcohol corresponding to (75; R = Me) also goes exclusively by ring-contraction to give only cis-3-t-butylcyclopentyl methyl ketone.[162]

Treatment of the acids (78) and (79) with BF_3 in acetic acid affords the same (1:1) mixture of products shown. Refluxing in formic acid gives additionally (80), and use of deuterium-labelled (79) affords labelled products indicating that the reaction proceeds by ring-contraction and expansion rather than by methyl migration and epimerization.[163a]

---

[157] R. C. Bingham and P. von R. Schleyer, *Tetrahedron Letters*, **1971**, 23; *J. Am. Chem. Soc.*, **93**, 3189 (1971).
[158] See *Org. Reaction Mech.*, **1970**, 83.
[159] F. K. Cartledge and J. P. Jones, *Tetrahedron Letters*, **1971**, 2193.
[160] M. A. Cook, C. Eaborn, and D. R. M. Walton, *J. Organomet. Chem.*, **29**, 389 (1971).
[161] D. Dicko and H. Bodot, *Tetrahedron*, **27**, 1761 (1971).
[162] P. H. Lewis, S. Middleton, and M. J. Rosser, *Austral. J. Chem.*, **24**, 865 (1971).
[163a] D. J. Dunham and R. G. Lawton, *J. Am. Chem. Soc.*, **93**, 2075 (1971).

. . . (14)

A similar preference for ring-contraction is exhibited in the *trans-syn-cis*-steroid series, reaction (14), where reaction occurs 150 times more rapidly than for the *trans-anti-trans*-isomer. Ring-contraction occurs with retention of configuration at C-5.[163b] On the other hand, dehydration with thionyl chloride in pyridine results in partial backbone rearrangement as well as ring-contraction.[164] Other studies on backbone rearrangements have also been reported.[165,166] Racemization during the backbone rearrangement of D-homoandrost-5-ene has been shown to occur by rearrangement to the 8(14)-unsaturated compound followed by protonation from both the α- and the β-face.[167] Other rearrangements in the steroid series have been reported.[168]

The zinc–acetic acid reduction of tabersonine gives vincamsonine,[169] and similarly akuammicine gives decarbonylgeissoschizine in 66% yield.[170,171]

In contrast to the 20β-epimer, 3β-acetoxy-5α-pregnan-20α-yl toluene-*p*-sulphonate formolyses almost without ring-enlargement, presumably owing to conformational effects on the side-chain orientation [see reaction (15)].[172]

^{13}C-NMR has been used to examine the scrambling in cyclopentyl[1-^{13}C]methanol and the corresponding chloride in zinc chloride–HCl. It is suggested[173] that the primary cation is formed and gives the primary product, as well as rearranging to a cyclohexyl carbonium ion in an irreversible step; complete scrambling of the label then occurs before capture by chloride ion. The irreversibility of the cyclopentylcarbinyl to cyclohexyl cation step is called seriously into question by the results on the electrochemical oxidation of cyclohexane in fluorosulphonic–acetic acid mixture, which gives (**81**) presumably by the route shown; under the same conditions cyclohexene also gives (**81**).[174]

The rate of acetolysis of the compounds (**82**) is nearly identical for Z = H and Z = OMe, and the products arise by preferential migration of the *anti*-group. Similar effects are reported for the solvolyses of (**83**) and the results have been interpreted in terms of

---

[163b] J. M. Coxon, M. P. Hartshorn, and C. N. Muir, *Chem. Comm.*, **1970**, 1591.
[164] J. M. Coxon, M. P. Hartshorn, and C. N. Muir, *Chem. Comm.*, **1971**, 659.
[165] F. Frappier, J. Thierry, and F.-X. Jarreau, *Tetrahedron Letters*, **1971**, 1887.
[166] C. Berrier, J. C. Jacquesy, and R. Jacquesy, *Tetrahedron Letters*, **1971**, 4567.
[167] D. N. Kirk and P. M. Shaw, *Chem. Comm.*, **1971**, 948.
[168] C. Monneret, P. Choay, Q. Khoung-Huu, and R. Goutarel, *Tetrahedron Letters*, **1971**, 3223; J. Libman and Y. Mazur, *Chem. Comm.*, **1971**, 729, 730.
[169] P. Maupérin, J. Lévy, and J. Le Men, *Tetrahedron Letters*, **1971**, 999.
[170] W. B. Hinshaw, J. Lévy, and J. Le Men, *Tetrahedron Letters*, **1971**, 995.
[171] J. Lévy, P. Maupérin, M. D. de Maindreville, and J. Le Men, *Tetrahedron Letters*, **1971**, 1003.
[172] F. B. Hirschmann, D. M. Kautz, S. S. Deshmane, and H. Hirschmann, *Tetrahedron*, **27**, 2041 (1971).
[173] A. N. Lovtsova, T. N. Shatkina, O. A. Reutov, E. T. Lipmaa, and T. I. Pehk, *Doklady Akad. Nauk SSSR*, **192**, 346 (1970); *Proc. Acad. Sci. USSR*, **192**, 354 (1970).
[174] J. Bertram, M. Fleischmann, and D. Pletcher, *Tetrahedron Letters*, **1971**, 349.

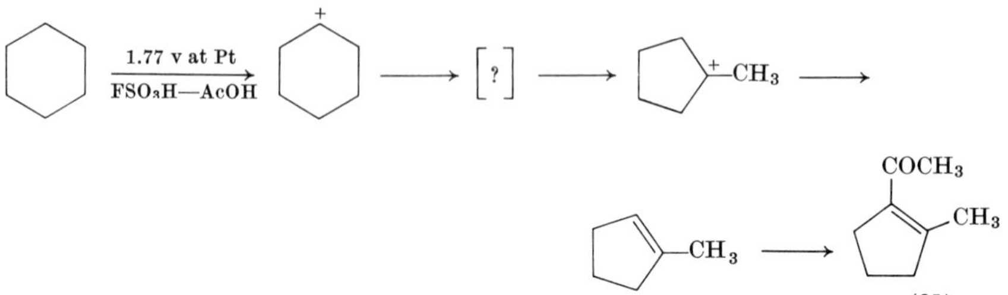

(81)

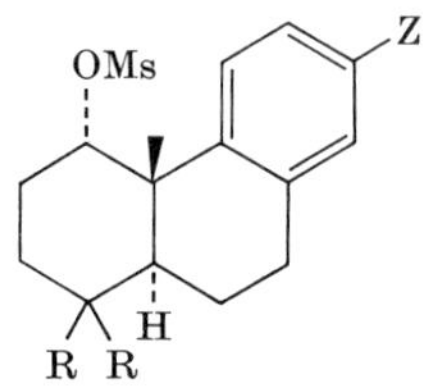

(82)
(R = H or Me; Z = H or OMe)

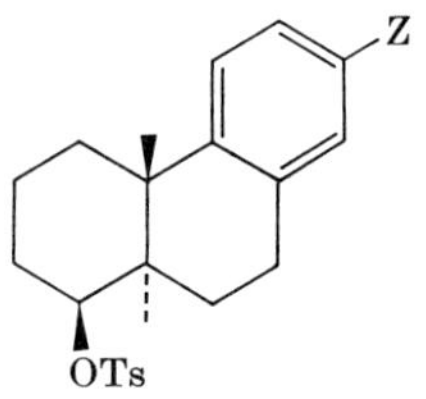

(83)
(Z = H or OMe)

equilibrating classical ions. No kinetic effect of methyl participation is observed.[175,176] A study of the equilibrating cyclopropylcarbinyl cations produced by protonation of thujopsene has been carried out.[177] The dehydration of 2-methyl-2-substituted-propane-1,3-diols in sulphuric acid shows that methyl migrates preferentially over other alkyl groups and that the rate is affected by steric acceleration by the non-migrating group.[178]

The reaction of norbornanone with diazomethane affords only ring-expanded ketones and no epoxide products; the relative reactivity of the ketonic products towards diazomethane has been examined.[179] The change in the ratio of ring-expansion to epoxide formation in 4-substituted cyclohexanones and N-substituted 4-piperidones has been attributed to a field effect of the substituent.[180,181] Similar effects are observed in the deamination of the corresponding 1-hydroxy-4-substituted-cyclohexylmethylamines and 4-hydroxy-N-substituted-4-piperidylmethylamines.[181,182]

Other studies involving neighbouring carbon have also been reported.[183–185]

The rates of acetolysis and formolysis of the cycloalkylcarbinyl toluene-p-sulphonates of ring size 5–12 have been measured. A linear plot of the rates of formolysis versus the ring-strain energy (from heats of combustion) is obtained; but it must be noted that the maximum rate ratio observed was only 10 in formolysis and 5.5 in acetolysis. That

[175] H. W. Whitlock, P. B. Reichardt, and F. M. Silver, *J. Am. Chem. Soc.*, **93**, 485 (1971).
[176] H. W. Whitlock and L. E. Overman, *J. Am. Chem. Soc.*, **93**, 2247 (1971).
[177] S. Ito, M. Yatagai, K. Endo, and M. Kodama, *Tetrahedron Letters*, **1971**, 1153.
[178] T. Yvernault and M. Mazet, *Bull. Soc. Chim. France*, **1971**, 2652; M. Mazet and M. Desmaison-Brut, *Bull. Soc. Chim France*, **1971**, 2656.
[179] G. Fachinetti, F. Pietra and A. Marsili, *Tetrahedron Letters*, **1971**, 393.
[180] H. Favre, D. Gravel, Z. Hamlet, M. Ménard, and J. Temler, *Can. J. Chem.*, **49**, 3097 (1971).
[181] H. Favre, Z. Hamlet, R. Lanthier, and M. Ménard, *Can. J. Chem.*, **49**, 3075 (1971).
[182] H. Favre, Z. Hamlet, M. Ménard, G. Roblot, and J. Temler, *Can. J. Chem.*, **49**, 3086 (1971).
[183] E. V. Brandt, D. Ferreira, and D. G. Roux, *Chem. Comm.*, **1971**, 116; N. M. K. Ng Ying Kin and J. M. Williams, *Chem. Comm.*, **1971**, 1123.
[184] Y. G. Bundel, I. Y. Levina, N. Y. Smordina, and O. A. Reutov, *Zhur. Org. Khim.*, **6**, 409 (1970); *J. Org. Chem. USSR*, **6**, 407 (1970); S. Smolinski, B. Dybek, and A. Mroczek, *Rocz. Chem.*, **44**, 527 (1970).
[185] N. Bosworth and P. D. Magnus, *Chem. Comm.*, **1971**, 618.

some H-participation occurs was shown for the cyclo-octyl case by deuterium labelling.[186] Formolysis and acetolysis have also been studied for α-deuteriated and α-tritiated cycloheptyl, 4-methylcycloheptyl and 3-methylcyclopentyl toluene-p-sulphonates. Less than 12% of isomerization was observed, and greater than half of this was due to 1,2-shifts.[187]

1,5-Hydrogen shifts have been studied for the 80% ethanolysis of eight substituted longifolene derivatives.[188]

Oestrone undergoes a retro-dienone–phenol isomerization on treatment with hydrogen fluoride–antimony pentafluoride.[189] Rearrangement, with halide exchange, is observed when n-alkyl chlorides and bromides are passed through a $SnCl_2$–KCl melt.[190]

Deamination and Related Reactions

Further work on the importance of micelles in deamination processes has been reported by Moss and Talkowski.[191] Using trimethyl-(1-methylheptyl)ammonium ions to favour micelle formation, they have shown that the rates of deamination are 6–16 times faster within the micelle than in solution. This has been ascribed to effects on the dissociation constants of nitrous acid and RNH_3^+ and to the higher local concentration of nitrite anions near the positively charged micelle.

The reaction of primary amines with N_2O_4 in THF affords high yields of the corresponding nitrates. 1-Phenylethylamine reacts with 60–70% retention of configuration. The Moss scheme (16) reported last year is used to explain the results. The high substitution/elimination ratio observed is due, it is suggested, to the low basicity of the nitrate anion.[192]

$$RNH_2 + N_2O_4 \longrightarrow R-N=N-OH \cdots H \atop O_2NO \longrightarrow R^+ + N_2 + OH \atop | \atop H \atop ONO_2^- \longrightarrow RONO_2 \quad \ldots (16)$$

Decomposition of the nitrosourea (84) gives the products shown, by protonation of the intermediate diazotate and reaction of methoxide within the ion-triplet formed as a result of its decomposition (as in the Moss scheme). The same products are formed in the same ratios when the intermediate is formed by protonation of the preformed diazo-compound. However, when the nitrosourea (85) is the substrate, the products (shown) are formed in different ratios than those from protonation of (86). It is suggested that the diazo-compound protonates at both possible sites, as shown.[193] Kirmse and his colleagues[194–196] have published work on the deamination of (1R,2S)-2-cyclohexyl-1-methylpropylamine, (S)-2-cyclohexyl-1-butylamine, (1R,2R)-2-phenyl-1-methylpropyl-

---
[186] A. P. Krapcho and R. G. Johanson, *J. Org. Chem.*, **36**, 146 (1971).
[187] Y. G. Bundel, K. G. Pankratova, M. Reali, N. M. Przheval'skii, and O. A. Reutov, *Zhur. Org. Khim.*, **7**, 425 (1971); *Chem. Abs.*, **75**, 4944 (1971).
[188] L. Stéhelin, J. Lhomme, and G. Ourisson, *J. Am. Chem. Soc.*, **93**, 1650 (1971).
[189] J. P. Gesson, J. C. Jacquesy, and R. Jacquesy, *Tetrahedron Letters*, **1971**, 4733.
[190] R. A. Bailey and S. F. Prest, *Can. J. Chem.*, **49**, 1 (1971).
[191] R. A. Moss and C. J. Talkowski, *Tetrahedron Letters*, **1971**, 703.
[192] F. Wudl and T. B. K. Lee, *J. Am. Chem. Soc.*, **93**, 271 (1971); cf. *Org. Reaction Mech.*, **1970**, 88.
[193] W. Kirmse and J. Heese, *Chem. Comm.*, **1971**, 258.
[194] W. Kirmse and W. Gruber, *Chem. Ber.*, **104**, 1789, 1795 (1971).
[195] W. Kirmse and H. Arold, *Chem. Ber.*, **104**, 1800 (1971).
[196] W. Kirmse, H. Arold, and B. Kornrumpf, *Chem. Ber.*, **104**, 1783 (1971).

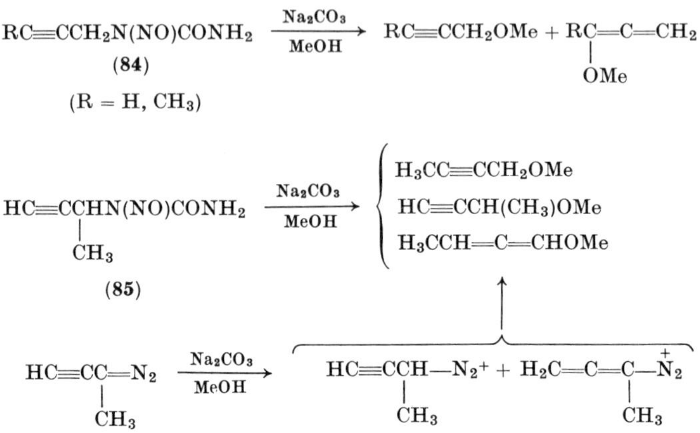

$$RC{\equiv}CCH_2N(NO)CONH_2 \xrightarrow[\text{MeOH}]{Na_2CO_3} RC{\equiv}CCH_2OMe + \underset{\underset{OMe}{|}}{RC}{=}C{=}CH_2$$

(84)

(R = H, CH$_3$)

$$\underset{\underset{CH_3}{|}}{HC{\equiv}CCHN(NO)CONH_2} \xrightarrow[\text{MeOH}]{Na_2CO_3} \begin{cases} H_3CC{\equiv}CCH_2OMe \\ HC{\equiv}CCH(CH_3)OMe \\ H_3CCH{=}C{=}CHOMe \end{cases}$$

(85)

$$\underset{\underset{CH_3}{|}}{HC{\equiv}CC{=}N_2} \xrightarrow[\text{MeOH}]{Na_2CO_3} \overbrace{\underset{\underset{CH_3}{|}}{HC{\equiv}CCH{-}N_2^+} + \underset{\underset{CH_3}{|}}{H_2C{=}C{=}C{-}\overset{+}{N_2}}}$$

(86)

amine, and (S)-2-phenyl-1-butylamine and on (1'–S)-4-methyl-1-(1'-methylpropyl)-pentylamine and (S)-2-ethyl-6-methylheptylamine. Although the major product from the 2-phenyl-1-methylpropylamine is alcohol of unchanged configuration, some 40% of hydrogen migration occurs to give α-ethyl-α-methylbenzyl alcohol with 6% inversion of configuration, and a nearly identical stereochemical result is observed in the 2-phenyl-1-butylamine reaction. On the other hand, the product of hydrogen migration in the wholly alkyl systems is completely racemic for the cyclohexylbutylamine and there is 41–60% retention of configuration in the open-chain compounds. Essentially identical results to those from deamination were obtained by solvolyses of the corresponding toluene-p-sulphonates in each case, Kirmse has argued that, if hydrogen migration takes place with inversion at the leaving group centre [as has been shown; sequence (17)],

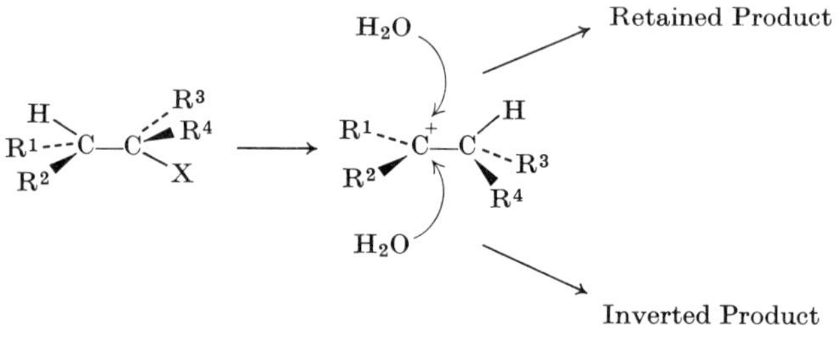

... (17)

the intermediate carbonium ion must rotate to become achiral and steric approach control will normally ensure retention of configuration where R^3 and/or R^4 are alkyl. However, this leaves unanswered the problem for $R^3 = R^4 = H$ and also the difference shown between the cyclohexyl and the phenyl cases. He therefore suggests that micellar effects may be the most significant factor, as these will vary with the structure of the starting material but not with its mode of formation.[194,195] Deamination of (R)-2-amino-2-methylbutan-1-ol gives 2-methylbutanal with 30% inversion of configuration, while dehydration of the corresponding diol gives completely racemic aldehyde. Ring-opening of (R)-2-ethyl-2-methyloxiran gives aldehyde with 6% inversion. Noting that least racemization occurs with the best leaving group, the authors[196] suggest that migration of hydrogen is not synchronous with ionization and that unsymmetrically solvated carbonium ions account for the results.

Deamination of 3-aminopropanol with sodium nitrite–sulphuric acid in the presence of sodium chloride has been studied at 0°, 5° and 50°. The ratio of 3-chloropropan-1-ol to propane-1,3-diol (the major products) goes from 0.45 to 42 over this temperature range, and this has been interpreted as representing a change of diazotizing agent from nitrous acid to nitrosyl chloride.[197]

Other work on deamination has been reported.[198,199]

Reactions of Aliphatic Diazo-compounds

Protonation of the bisdiazo-compound (**87**) forms the monodiazonium ion which decomposes by competitive reaction with solvent, ring expansion and ring-closure as shown.

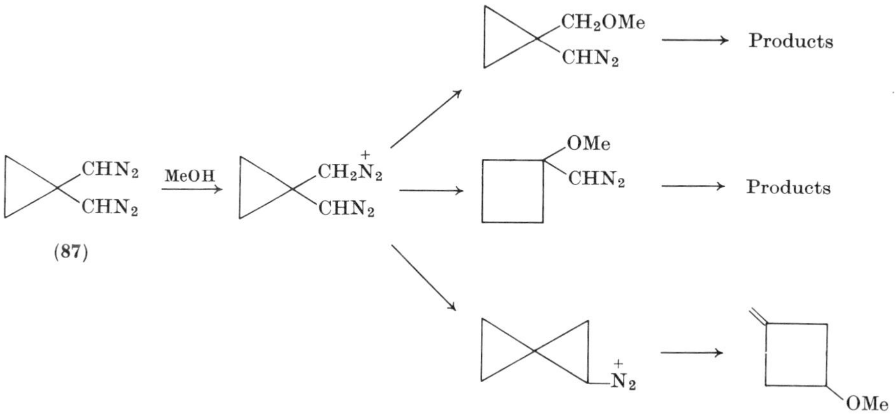

Most of the products arise by the first two routes, but up to 15% goes by way of the spiropentanediazonium ion.[200]

The acid-catalysed hydrolysis of aryldiazomethanes in 60% dioxan–water shows general acid-catalysis obeys the Brønsted relation with a slope of 0.69 and has a solvent isotope effect k_{H_2O}/k_{D_2O} of 2.6. No deuterium exchange occurs into the starting materials and an A-S_E2 mechanism is therefore proposed.[201]

Further work on hydrolysis of aryl diazoketones (**88**), (**89**) and (**90**) has been carried out. Both the series (**88**) and (**89**) give correlations against σ^+ with ρ values of -0.68 and -2.05, respectively. General acid-catalysis is observed and the solvent isotope effect shows a correlation against σ^+ with ρ values of -0.15 for (**88**) and $+0.31$ for (**89**). A mixed $A1/A$-S_E2 mechanism has therefore been proposed with a smooth change-over

$$\text{ArCO} \cdot \text{CN}_2 \cdot \text{Ph} \qquad \text{PhCO} \cdot \text{CN}_2 \cdot \text{Ar} \qquad \text{H}_3\text{CCO} \cdot \text{CN}_2 \cdot \text{Ar}$$
$$(\mathbf{88}) \qquad\qquad (\mathbf{89}) \qquad\qquad (\mathbf{90})$$

[197] O. A. Reutov, T. A. Smolina, and O. Y. Polevaya, *Izv. Akad. Nauk SSSR, Ser. Khim.*, **1970**, 1912.
[198] D. Fărcasiu and L. H. Schwartz, *Compt. Rend.* (C), **273**, 168 (1971).
[199] See ref. 37 of Chapter 1.
[200] W. Kirmse and B. Brinkman, *Chem. Comm.*, **1971**, 259.
[201] H. Dahn and G. Diderich, *Helv. Chim. Acta*, **54**, 1950 (1971).

as the substituent on the benzene ring varies from p-MeO to p-nitro.²⁰² Similar studies in the series p-NO₂C₆H₄C(N₂)R, where R = H, CH₃, Ph, CO₂Et, COPh or COCH₃) show a correlation of rate with Taft's σ^* ($\rho = -1.87$); an A-S_E2 mechanism is proposed.²⁰³

The hydrolysis of potassium diazoacetate has been studied²⁰⁴ as a function of pH. The reaction goes by an A-1 route in strongly basic solution and changes over to an A-S_E2 mechanism in less basic or acid solution. The rate of loss of nitrogen from ⁺N₂CH₂CO₂⁻ has been estimated to have an upper limit of 10^7 sec⁻¹.

Fragmentation Reactions

1,3-Dithianyl-substituted compounds of the type (**91**) (prepared from the α,β-unsaturated ketone) undergo the fragmentation shown when treated with phenyl-lithium.²⁰⁵ A silacyclohexane (**92**) fragments with proton transfer as the major course of solvolysis [reaction (18)].²⁰⁶

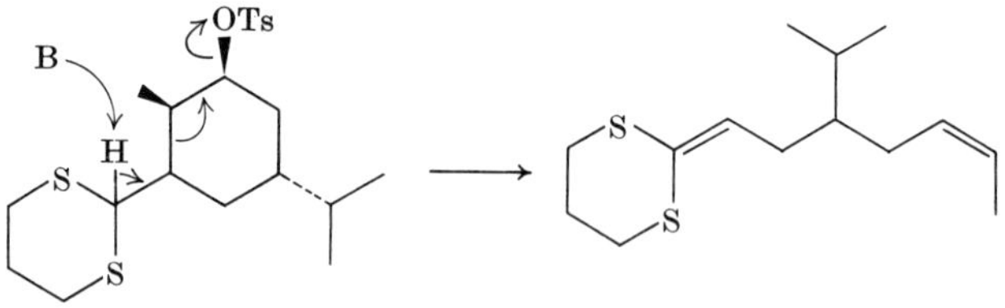

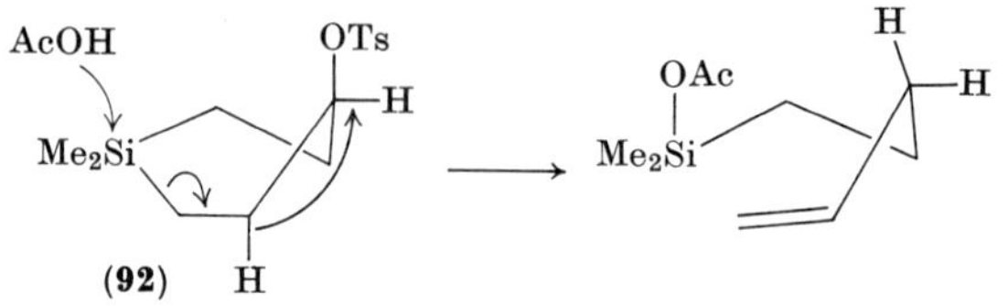

$$(Bu^t)_2C(OH)CH_2R \rightarrow Bu^t_2\overset{+}{C}-CH_2R \rightarrow (CH_3)_2\overset{+}{C}-\underset{Bu^t}{\overset{CH_3}{\underset{|}{\overset{|}{C}}}}-CH_2R$$

(**93**) (**94**)

(R = H, Me, Et, i-Pr or Buᵗ)

Dehydration of the alcohol (**93**) proceeds mainly through cation (**94**), which either loses a proton or fragments with loss of a t-butyl cation. Dehydration with SOCl₂–pyridine goes mainly by the former route with little fragmentation, while in concentrated sulphuric acid the amount of fragmentation rises with the size of R, varying from 22% for R = Me to 79% for R = t-Bu.²⁰⁷

---

²⁰² W. Jugelt and L. Berseck, *Tetrahedron*, **26**, 5557 (1970).
²⁰³ W. Jugelt and L. Berseck, *Tetrahedron*, **26**, 5581 (1970).
²⁰⁴ M. M. Kreevoy and D. E. Konasewich, *J. Phys. Chem.*, **74**, 4464 (1970).
²⁰⁵ J. A. Marshall and J. L. Belletire, *Tetrahedron Letters*, **1971**, 871.
²⁰⁶ S. S. Washburne and R. R. Chawla, *J. Organomet. Chem.*, **31**, C20 (1971).
²⁰⁷ J. S. Lomas, D. S. Sagatys, and J. E. Dubois, *Tetrahedron Letters*, **1971**, 599, 1349.

4,6-O-Benzylidene-2-O-toluene-p-sulphonyl-α-D-ribo-hexopyranosid-3-ulose reacts with Et_3N in methanol both to give an α-methoxyoxiran and by fragmentation through the enolate anion (**95**).[208]

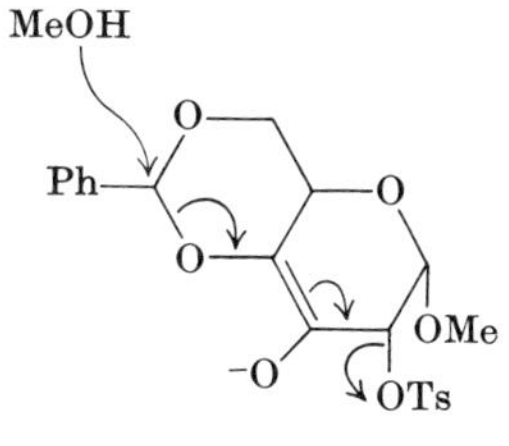

(**95**) 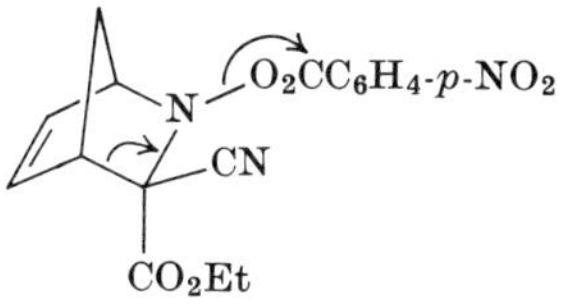 (**96**)

The epoxide of α-pinene undergoes a two-step fragmentation when treated with HBr in ether.[209] The azabicycloheptene (**96**) fragments when treated with trifluoroacetic acid, although the corresponding N-toluene-p-sulphonate does not do so in aqueous dioxan.[24]

Other reactions involving fragmentation are the solvolysis of arylsulphonylmethyl nitrates,[210] and the electro-oxidation of benzylic ketones.[211]

Displacement Reactions at Elements Other than Carbon

Silicon, Germanium, Tin and Lead

The racemization of chlorosilanes has received further attention. The kinetics of the racemization of menthoxy-(1-naphthyl)phenylchlorosilane with hexamethylphosphoric triamide (HMPTA) in carbon tetrachloride is of the first order in silane and second order with respect to HMPTA. DMF, DMSO and 1-methylpyrrolidone also give a second-order dependence. Corriu and Leard therefore suggest a six-coordinate silicon species as responsible for the racemization, k_2 of reaction (19) being rate-determining.[212] Sommer

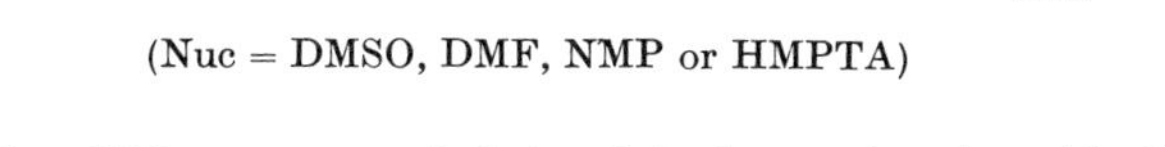

$$Nuc + R_3SiCl \underset{}{\overset{k_1}{\rightleftharpoons}} Nuc\text{—}SiR_3Cl \overset{k_2}{\rightleftharpoons} \begin{array}{c} Nuc \\ | \\ R'\text{—}Si\text{—}R'' \\ | \\ R \quad Cl \\ | \\ Nuc \end{array} \quad \ldots (19)$$

(Nuc = DMSO, DMF, NMP or HMPTA)

and his co-workers[213] have retracted their original assertion that chloride exchange and racemization of aryl(methyl)-1-naphthylchlorosilanes occurs *via* siliconium ions. The rate of exchange is now reported to show a linear correlation with σ ($\rho = +0.4$), and the ratio of racemization to exchange rates varies with the substituents ($k_{rac}/k_{exch} = 1.3$,

---

[208] A. Dmytraczenko, W. A. Szarek, and J. K. N. Jones, *Chem. Comm.*, **1971**, 1220.
[209] P. H. Boyle, W. Cocker, D. H. Grayson, and P. V. R. Shannon, *Chem. Comm.*, **1971**, 395.
[210] A. Bruggink, B. Zwanenburg, and J. B. F. N. Engberts, *Tetrahedron*, **27**, 4571 (1971).
[211] L. L. Miller, V. R. Koch, M. E. Larscheid, and J. F. Wolf, *Tetrahedron Letters*, **1971**, 1389.
[212] R. J. P. Corriu and M. Leard, *Chem. Comm.*, **1971**, 1087.
[213] L. H. Sommer, G. D. Homer, A. W. Messing, J. L. Kutschinski, F. O. Stark, and K. W. Michael, *J. Am. Chem. Soc.*, **93**, 2093 (1971).

1.8 and 2.0 for Ar = Ph, Ph$_3$Si and C$_6$F$_5$, respectively). The perfluorophenyl group increases the exchange rate (over the phenyl compound) by 1.1 but the racemization rate by a factor of 1.7. Racemization in the presence of perchlorate salts is 10^3 times *slower* than with the corresponding chlorides and they therefore accept the suggestions made by Grant and Prince[214] that exchange occurs by competitive inversion and retention mechanisms.

Corriu and his colleagues have reported the synthesis of asymmetric (optically active) silanes.[215] The reaction of racemic (1-naphthyl)phenylchlorosilane with menthol in the presence of HCl occurs with asymmetric induction to give a 70% yield of one diastereoisomer. The processes of (20), with $k_R \neq k_S$ would account for this, together with the fact

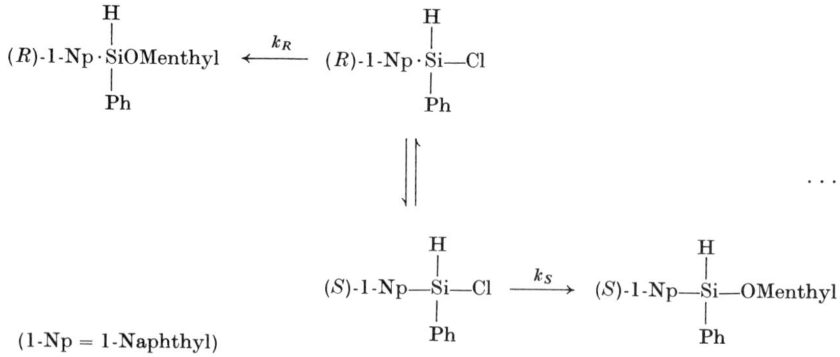

$$(R)\text{-1-Np}\cdot\underset{\underset{Ph}{|}}{\overset{\overset{H}{|}}{Si}}\text{OMenthyl} \xleftarrow{k_R} (R)\text{-1-Np}\cdot\underset{\underset{Ph}{|}}{\overset{\overset{H}{|}}{Si}}-\text{Cl}$$

$$\Updownarrow \qquad\qquad\qquad \dots(20)$$

$$(S)\text{-1-Np}-\underset{\underset{Ph}{|}}{\overset{\overset{H}{|}}{Si}}-\text{Cl} \xrightarrow{k_S} (S)\text{-1-Np}-\underset{\underset{Ph}{|}}{\overset{\overset{H}{|}}{Si}}-\text{OMenthyl}$$

(1-Np = 1-Naphthyl)

that the racemization at silicon is more rapid in the starting chlorosilane than in the product menthoxysilane.[216]

The reaction of (**97**) with Grignard reagents occurs with inversion of configuration in diethyl ether and with retention of configuration in dimethoxyethane, except for allylmagnesium bromide, which gives the inverted product. The reactions of (**98**) with Grignard reagents show similar behaviour without the exception of allyl compounds; reaction of (**98**) with dimethylmagnesium in ether occurs with retention of configuration

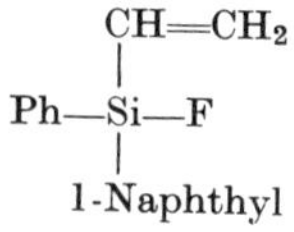

$$\text{Ph}-\underset{\underset{\text{1-Naphthyl}}{|}}{\overset{\overset{\text{CH}=\text{CH}_2}{|}}{Si}}-\text{F}$$

(**97**)

[structure of (**98**): tetrahydronaphthalene ring fused with Si bearing 1-Naphthyl and X]

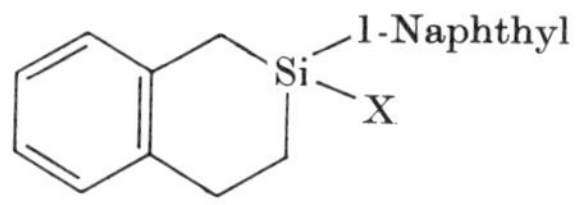

(**98**)

(X = F, OMe)

and therefore the change in stereospecificity is ascribed to electrophilic catalysis by magnesium bromide. In agreement with this conclusion, Corriu *et al.*[217] find that reaction of (**98**; X = OMe) with crotylmagnesium bromide in ether occurs with retention of configuration (92%), but that in the presence of 1 equivalent of magnesium bromide the product is formed with 57% inversion of configuration. They therefore conclude,

---

[214] See *Org. Reaction Mech.*, **1969**, 104.
[215] R. J. P. Corriu, G. F. Lanneau, and M. Leard, *Chem. Comm.*, **1971**, 1365.
[216] R. J. P. Corriu and G. F. Lanneau, *Tetrahedron Letters*, **1971**, 2771.
[217] R. J. P. Corriu, J. P. Masse, and G. Royo, *Chem. Comm.*, **1971**, 252; cf. *Org. Reaction Mech.*, **1967**, 88.

contrary to earlier reports, that electrophilic catalysis is not an essential part of the S_Ni-Si mechanism.[217] The rearrangement of (methyl-1-naphthylphenylsilyl)acetic acid to (methyl-1-naphthylphenylsilyl) acetate, catalysed by triethylamine, occurs with retention of configuration. An intramolecular 4-membered cyclic transition state has been proposed.[218]

A further report on the reaction of optically active silanes R_3SiH with R'_3SnOR'' confirms that retention of stereochemistry at silicon is observed.[219] A Hammett treatment for the reactions of $ArSiHMe_2$ with $n\text{-}Bu_3SnOMe$ and of $PhSiHMe_2$ with $n\text{-}Bu_3SnOAr$ gives linear correlations for each with ρ values of $+0.90$ and -1.32, respectively. On the basis of this and the stereochemical work, a four-centre transition state with charge separation is proposed for this S_Ni-Si reaction.[220]

The acid-catalysed hydrolysis of the 5-co-ordinate nitrilotriphenoxysilanes occurs by protonation on oxygen and rate-determining cleavage of the Si—O bond, except when the silicon also bears a chlorine; for the latter case it is suggested that rate-determining ionization of the Si—Cl bond occurs without prior protonation. The corresponding nitrilotriphenoxychlorogermane does not react in the same way, and this is used as evidence for $p\pi$–$d\pi$ bonding from chlorine to germanium.[221]

A further report[222] on the hydrolysis of aryltrimethyl-silanes and -stannanes by alkali in aqueous DMSO, in which eight new aryl compounds have been studied, shows that the p-methoxyphenyl and p-N,N-dimethylanilino-compounds fall well off the line of a ρ–σ plot. The authors therefore suggest that a concerted electrophilic–nucleophilic attack occurs.

A correction[223] has appeared in connection with last year's report on the hydrolysis of $PhNHSiMe_3$, and contains a masterly use of tautology to de-emphasize an error. It is now found that reactions are inhibited at low methoxide concentration but are very fast at high base concentration.[223]

The kinetics of the hydrolysis of 12 trisubstituted chlorosilanes have been measured at $-45°$ in acetone containing 1.6–9.6 moles of water per mole of silane.[224]

Phosphorus, Arsenic and Antimony

Ugi, Ramirez and their co-workers[225–227] have reviewed the mechanisms of permutational isomerization of 5-co-ordinate phosphorus compounds. They discuss the accepted mechanism, proposed by Berry, of pseudorotation (BPR) and suggest a valid alternative which they call "turnstile rotation" (TR). The molecular motions involved in BPR and TR are summarized in (**99**) for the former and (**100**) for the latter. A Newman diagram

---

[218] A. G. Brook, J. M. Duff, and D. G. Anderson, *J. Am. Chem. Soc.*, **92**, 7567 (1970).
[219] M. Pereyre and J. Pijselman, *J. Organomet. Chem.*, **25**, C27 (1970); cf. *Org. Reaction Mech.*, **1970**, 93.
[220] J. Pijselman and M. Pereyre, *J. Orgnomet. Chem.*, **32**, C72 (1971).
[221] R. E. Timms, *J. Chem. Soc.* (A), **1971**, 1969.
[222] A. R. Bassindale, C. Eaborn, R. Taylor, A. R. Thompson, D. R. M. Walton, J. Cretney, and G. J. Wright, *J. Chem. Soc.* (B), **1971**, 1155.
[223] A. R. Bassindale, C. Eaborn, and D. R. M. Walton, *J. Organomet. Chem.*, **27**, C24 (1971); cf. *Org. Reaction Mech.*, **1970**, 94, ref. 237.
[224] V. P. Mileshkevich, G. A. Nikolaev, V. F. Evdokimov, and A. V. Karlin, *Zhur. Obsch. Khim.*, **41**, 634 (1971).
[225] F. Ramirez and I. Ugi, *Adv. Phys. Org. Chem.*, **9**, 26 (1971).
[226] P. Gillespie, P. Hoffman, H. Klusacek, D. Marquarding, S. Pfohl, F. Ramirez, E. A. Tsolis, and I. Ugi, *Angew. Chem. Int. Ed.*, **10**, 687 (1971).
[227] I. Ugi, D. Marquarding, H. Klusacek, P. Gillespie, and F. Ramirez, *Accounts Chem. Res.*, **4**, 288 (1971).

(99)

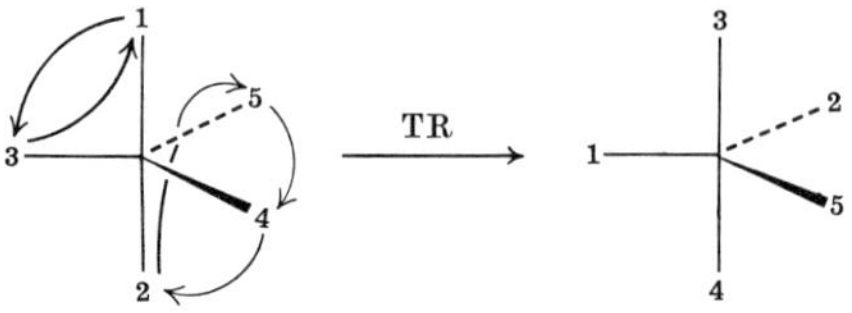

(100)

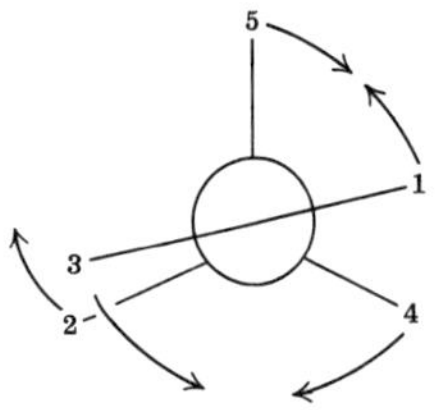

(101)

of TR is shown (101) looking along the three-fold axis of groups 2, 4, 5. While the overall result of the two processes is the same, it should be noted that in TR all the ligands participate in the process, while in BPR one of them remains fixed. Because in each process exchange occurs between two pairs of apical and equatorial groups, the graphs that connect the various isomers are identical. CNDO/2 calculations of the energy barriers are not very different, e.g. for PF_5, values of 3.5 kcal mol^{-1} and 11.1 kcal mol^{-1} are predicted for BPR and TR, respectively. However, a number of important differences should be mentioned: (i) for each BPR, there are four different TR processes that produce the same result; (ii) TR is essentially an internal rotation while this motion is entirely absent in BPR; (iii) the motions leading to the transition state are different involving C_{4v} symmetry in BPR and C_{2v} plus C_{3v} local symmetries in TR; (iv) TR has a higher probability than BPR as a result of (i); (v) the ligand movement and the barriers to such movement vary in a different way as a function of the ligand set and its distribution (apical and equatorial) between the two processes. Ugi and his colleagues suggest that trigonal-bipyramidal phosphorus compounds with apical-equatorial rings *must* rearrange by a TR process, for example (102) for which BPR is not possible. For acyclic compounds exchange may, however, be easier by BPR or occur by both processes.

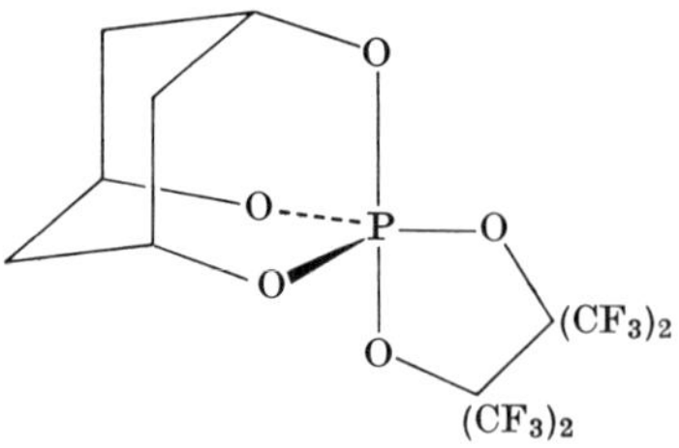

(102) 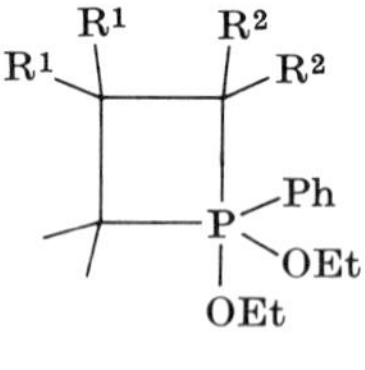 (103) 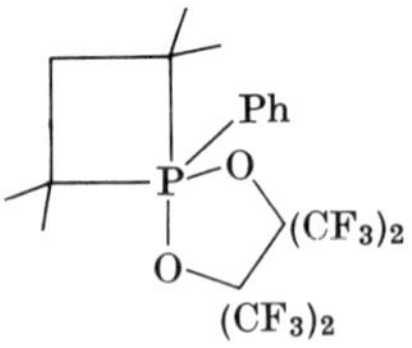 (104)

$$\begin{Bmatrix} R^1 = H \text{ or } Me \\ R^2 = Me \text{ or } H \end{Bmatrix}$$

The rates of pseudorotation of PF_5 and AsF_5 have been measured by infrared and NMR methods.[228] Pseudorotation in cyclic phosphoranes (**103**) and (**104**) with four-membered rings has been examined by NMR methods:[229] (**103**) shows non-equivalence of the methylene-protons of the ethoxy groups at low temperatures, which disappears on raising the temperature, and a ΔG of ca. 15 kcal mol^{-1} is estimated. (**104**) shows non-equivalence of the CF_3 groups and ΔG for the equilibration process is 20 kcal mol^{-1}. In each case BPR to place the four-membered ring diequatorial is invoked to account for the equilibration; but in view of the foregoing discussion a TR process must be seriously considered as this avoids the diequatorial four-membered ring. Other work on positional isomerization in cyclic phsophorus(v) compounds has also been discussed only in terms of BPR.[230,231]

The hydrolysis of some di-(t-butyl)phosphonium salts has been found[232] to be extremely slow, between 10^3 and 10^4 times slower than for the corresponding methyl compounds (Reviewer's estimate from product data in the paper). The product is that of Hofmann elimination, viz. isobutene. Mono-t-butylphosphonium salts react normally, to give phosphine oxide, and only 50 times slower than the methyl analogues. This effect has been attributed to steric crowding. Other crowded phosphonium salts also show abnormal behaviour in hydrolysis.

An *an initio* treatment of the rotation barrier in methylphosphine shows that the eclipsed conformer is the most stable.[233]

The alkaline hydrolysis of phosphonium salts (**105**) and (**106**) results in loss of phenyl from (**105**) and ring-opening for (**106**), as expected from loss of the most electronegative group. Change of base from NaOH in DMSO to aqueous sodium carbonate causes a change in products such that (**105**) reacts mainly by ring-cleavage to give the same products as (**106**). It is suggested that this is due to a decrease in nucleophilicity of hydroxide-ion and that, as this decreases, equilibration of (**105**) and (**106**) takes place, via the common ylid, in competition with cleavage.[234]

(**105**)   (**106**)   (**107**)

The stereochemistry of the oxidation of (**107**) to the corresponding phosphonate by *m*-chloroperbenzoic acid is affected by hydrogen ion concentration but not by solvent or temperature. Addition of trifluoroacetic acid causes a change from 97% retention of configuration to 89% inversion at 0.07M. It is suggested that in the absence of strong acid oxidation occurs at sulphur with retention via a three-membered ring displacement, whereas at high acid concentration sulphur is preferentially protonated and attack by

---

[228] J. Brickmann, *Ber. Bunsenges. Phys. Chem.*, **75**, 747 (1971).
[229] D. Z. Denney, D. W. White, and D. B. Denney, *J. Am. Chem. Soc.*, **93**, 2066 (1971).
[230] R. E. Duff, R. K. Oram, and S. Trippett, *Chem. Comm.*, **1971**, 1011; M. Eisenhut and R. Schmutzler, *Chem. Comm.*, **1971**, 1452.
[231] B. C. Chang, W. E. Conrad, D. B. Denney, D. Z. Denney, R. Edelman, R. L. Powell, and D. W. White, *J. Am. Chem. Soc.*, **93**, 4004 (1971).
[232] J. R. Corfield, N. J. De'ath, and S. Trippett, *J. Chem. Soc.* (C), **1971**, 1930.
[233] I. Absar and J. R. Van Wazer, *Chem. Comm.*, **1971**, 611.
[234] H. M. Priestley and J. P. Snyder, *Tetrahedron Letters*, **1971**, 2433.

peracid occurs at phosphorus such that the SH group is placed in an apical position. Fragmentation to sulphur, phosphinate and carboxylic acid then results in inversion of configuration at phosphorus.[235] The reverse reaction, conversion of a phosphonothioate into a thiophosphonothioate by P_4S_{10} is reported to occur with 80–100% retention of configuration. However, hydrolysis with potassium hydroxide to regenerate the initial phosphonothioate gives completely racemic material, even though the potassium salt of the phosphonothioic acid is optically stable.[236] In contrast to the reduction of phosphine oxides by $LiAlH_4$, which affords racemization of both starting material and product, the $LiAlH_4$ reduction of phosphine thio-oxides occurs with complete retention of configuration. As with reduction by hexachlorodisilane, it is suggested that this is due to attack at the soft electrophile sulphur.[237]

The absolute stereochemistry of O-menthyl S-methyl phenylphosphonothioate (108) has been determined by X-ray crystallography. As a result the stereochemistry of its reactions with methyl- and phenyl-magnesium bromide has been reassessed. It is now concluded that they proceed with retention of configuration at phosphorus, and also that the reaction of menthyl phenylphosphinate with sulphur and a secondary amine and then methyl iodide, which forms (108), also proceeds with retention.[238] Similar conclusions have also been arrived at as a result of chemical correlations.[239,240] The reactions of ethyl phenylphosphonate to give O-trimethylsilyl derivatives and the ethyl alkyl phenylphosphonates all proceed with retention of stereochemistry.[240] The hydrolysis of (109) occurs with complete lack of stereospecificity to give a 3:1 ratio of ethyl to methyl ester. This result casts serious doubt on last year's report on the reaction of the methyl ester with CD_3O^-.[241] The ester (110) hydrolyses in aqueous

(109)   (110)   (R = Me, Menthyl)

alkaline dioxan as shown, with complete retention of stereochemistry at phosphorus,[242] but phosphinate esters react with lithium anilide with inversion of configuration.[243]

The rates of alkaline hydrolysis of 3,4-disubstituted 1-alkoxyphosphole 1-oxides have been compared with those of the corresponding phosphol-3-ene 1-oxides and the phosphol-2-ene 1-oxides.[244] No satisfactory explanation for the differences was found.

An unusual, base-induced rearrangement from phosphorus to carbon of a phospholanium salt has been described.[245]

[235] A. W. Herriott, *J. Am. Chem. Soc.*, **93**, 3304 (1971).
[236] L. P. Rieff, L. J. Szafraniec, and H. S. Aaron, *Chem. Comm.*, **1971**, 366.
[237] R. Luckenbach, *Tetrahedron Letters*, **1971**, 2177; cf. *Org. Reaction Mech.*, **1970**, 99.
[238] J. Donohue, N. Mandel, W. B. Farnham, R. K. Murray, K. Mislow, and H. P. Benschop, *J. Am. Chem. Soc.*, **93**, 3792 (1971).
[239] W. B. Farnham, R. K. Murray, and K. Mislow, *Chem. Comm.*, **1971**, 605.
[240] G. R. Van den Berg, D. H. J. M. Platenburg, and H. P. Benschop, *Chem. Comm.*, **1971**, 606.
[241] K. E. De Bruin and M. J. Jacobs, *Chem. Comm.*, **1971**, 59; *Org. Reaction Mech.*, **1970**, 98.
[242] N. J. De'ath, K. Ellis, D. J. H. Smith, and S. Trippett, *Chem. Comm.*, **1971**, 714.
[243] A. Nudelman and D. J. Cram, *J. Org. Chem.*, **36**, 335 (1971).
[244] F. B. Clarke and F. H. Westheimer, *J. Am. Chem. Soc.*, **93**, 4541 (1971).
[245] A. N. Hughes and C. Srivanavit, *Can. J. Chem.*, **49**, 879 (1971).

The photolysis of 1-azidophosphetane 1-oxides has been reported.[246]

Chlorodiphenylphosphine forms salts reversibly with trimethylphosphine and dimethylphenylphosphine; methyldiphenylphosphine fails to react.[247]

A structure–reactivity correlation for the hydrolysis of phosphoramidate monoanions suggests that at the transition state the bond to the leaving group is broken to a greater extent than that to the nucleophile is formed, and that no five-covalent intermediate is formed in the hydrolysis of such mono- and di-esters.[248]

Dichlorobis(triphenylphosphine)nickel reacts with methylmagnesium bromide by attack at phosphorus with the formation of the corresponding (methyldiphenylphosphine)nickel complex and phenylmagnesium bromide.[249]

The intermediate in the Michaelis–Arbusov reaction of $MeC(CH_2O)_3P$ with bromine, i.e. $MeC(CH_2O)_3PBr^+Br^-$, has been isolated.[250] Benzoyl chloride reacts with $(MeO)_2PNEt_2$ in the normal Arbusov manner; however, reaction with the corresponding cyclic phosphorus amide, $(CH_2O)_2PNEt_2$ occurs by attack at nitrogen to give $(CH_2O)_2PCl$ and N,N-diethylbenzamide.[251]

The use of shift reagents allows assignment of configuration to 2,2,3,4,4-pentamethylphosphetanium salts by ^{1}H-NMR spectroscopy.[252] ^{13}C–^{31}P coupling constants have been reported for phosphetanium salts.[253]

Sulphur

"A Commentary on the Optical Activity of Sulphur"[254] and a review[255] on the stereochemistry of organic sulphur compounds have appeared.

The preparation of stable tetraco-ordinate sulphur compounds has been described.[256]

Full details on the proposed S_N1 reaction at sulphur, reported last year, have now been published.[257]

Ciuffarin and his group[258,259] have published further work on the reaction of Ph_3CSX with nucleophiles. For X = Cl, Br, I or SCN, they argue that the Brønsted coefficients for the leaving group suggest that bond-breaking at the transition state is far advanced, but that the leaving-group order suggests the reverse. They therefore feel that, even in aqueous dioxan, the reversible formation of a 3-co-ordinate intermediate followed by rate-determining loss of X^- is the best mechanism.[258] For reactions where X = p-$NO_2C_6H_4O$, nitrogen nucleophiles show larger Brønsted coefficients than do oxygen nucleophiles (0.75–1.5 versus 0.24). Anilines show the largest values (1.5) and hence have the highest bond orders to sulphur at the transition state. It is suggested that this is due to the softness of anilines attacking the soft electrophile sulphur.[259]

---

[246] M. J. P. Harger, *Chem. Comm.*, **1971**, 442.
[247] F. Ramirez and E. A. Tsolis, *J. Am. Chem. Soc.*, **92**, 7553 (1970).
[248] S. J. Benkovic and E. J. Sampson, *J. Am. Chem. Soc.*, **93**, 4009 (1971).
[249] M. L. H. Green, M. J. Smith, H. Felkin, and G. Swierczewski, *Chem. Comm.*, **1971**, 158.
[250] G. K. McEwen and J. G. Verkade, *Chem. Comm.*, **1971**, 668.
[251] C. Brown and R. F. Hudson, *Tetrahedron Letters*, **1971**, 3191.
[252] J. R. Corfield and S. Trippett, *Chem. Comm.*, **1971**, 721.
[253] G. A. Gray and S. E. Cremer, *Tetrahedron Letters*, **1971**, 3061.
[254] C. K. Ingold, *Int. J. Sulfur Chem.*, **6**(B), 65 (1971).
[255] K. K. Andersen, *Int. J. Sulfur Chem.*, **6**(B), 69 (1971).
[256] J. C. Martin and R. J. Arhart, *J. Am. Chem. Soc.*, **93**, 2339, 2341 (1971).
[257] A. M. Kiwan and H. M. N. H. Irving, *J. Chem. Soc.* (B), **1971**, 901; cf. *Org. Reaction Mech.*, **1970**, 101.
[258] L. Senatore, E. Ciuffarin, and L. Sagramora, *J. Chem. Soc.* (B), **1971**, 2191; cf. *Org. Reaction Mech.*, **1970**, 101.
[259] E. Ciuffarin, L. Senatore, and M. Isola, *J. Chem. Soc.* (B), **1971**, 2187.

The reaction of arene(1)sulphenyl chlorides with substituted benzhydryl aryl(2) sulphides to give diaryl disulphides and benzhydryl chlorides has been studied.[260] Linear correlations against σ are observed for arene(1) and for aryl(2) [ρ(1) is very small; ρ(2) = -1.29], and against σ^+ for the benzhydryl group ($\rho = -2.98$). The reaction is thought to proceed by way of a sulphonium salt (probably formed in a pre-equilibrium) followed by rate-determining ionization of the benzhydryl component.[260]

The rates of sulphur—sulphur bond cleavage have been measured by variable-temperature ^{1}H-NMR spectroscopy. For the reactions (21) and (22) the observed rates are

$$(CH_3)_2\overset{+}{S}SCH_3 + S(CH_3)_2 \underset{}{\overset{k_1}{\rightleftharpoons}} (CH_3)_2S + CH_3S\overset{+}{S}(CH_3)_2 \quad \ldots (21)$$
$$(\mathbf{111})$$

$$(\mathbf{111}) + CH_3SSCH_3 \underset{k_3}{\overset{k_2}{\rightleftharpoons}} (CH_3)_2S + CH_3S\cdot\underset{\underset{SCH_3}{|}}{\overset{+}{S}CH_3} \quad \ldots (22)$$

$k_1 = 1.5 \times 10^5$, $k_2 = 1.06 \times 10^4$ and $k_3 = 1.4 \times 10^7$ mol^{-1} sec^{-1}. (**111**) catalyses the exchange of S-alkyl groups between dimethyl disulphide and other dialkyl and diaryl disulphides. On the other hand, (**111**) does not give an observable reaction with dimethyl ether on the NMR time scale.[261] It is suggested that this is due to the hardness of oxygen relative to sulphur. It should be noted that methylation or S-methylation on sulphur increases the rate of attack at sulphur by greater than 10^6.

The equilibrium reaction of ArSNHPh with 2,4-dichloroaniline has been studied. A linear correlation with σ^- was found ($\rho = -1.71$) and suggests electron-donation from nitrogen to sulphur.[262]

o-t-Butylbenzenethiolate reacts with ethyl 2,4-dinitrophenyl disulphide faster than the *para*-isomer owing to ground-state factors.[263a]

Dialkyl disulphides react with (Et$_2$N)$_3$P to give dialkyl sulphides with inversion of stereochemistry at one alkyl group.[263b] Attack at sulphur, with concomitant ring-cleavage, has been reported for 3-chloro-4,5-benzothiazole.[264]

The reactions of sulphoxides with N,N'-bis(toluene-p-sulphonyl)sulphur diimide to give sulphilimines proceeds with 94% inversion of stereochemistry in pyridine,[265,266] and with 95% retention of stereochemistry in benzene.[266] The suggested mechanisms are outlined in reactions (23) and (24). The alkaline hydrolysis of the sulphilimines [reaction (25)] gives the sulphoxides with inversion of configuration.[265,267] Cram's group also report that the electrophilic substitution at the sulphilimine-sulphur by peracid involves retention of configuration.[265]

The stereochemical cycle $(+)$-(S)-(**113**) $\to$ (**114**) $\to$ (**116**) $\to$ (**117**) $\to$ (**113**) plus the reaction (**114**) $\to$ (**115**) $\to$ $(-)$-(R)-(**113**) has been used to show that displacement by nucleophiles at 4-co-ordinate hexavalent sulphur proceeds with inversion of stereo-

[260] U. Quintily and G. Scorrano, *Chim. Ind.* (*Milan*), **53**, 36 (1971).
[261] S. H. Smallcombe and M. C. Caserio, *J. Am. Chem. Soc.*, **93**, 5826 (1971).
[262] F. A. Davis, S. Divald, and A. H. Confer, *Chem. Comm.*, **1971**, 294.
[263a] D. S. Garwood and D. C. Garwood, *Tetrahedron Letters*, **1970**, 4959.
[263b] D. N. Harpp and J. G. Gleason, *J. Am. Chem. Soc.*, **92**, 2437 (1971).
[264] D. E. L. Carrington, K. Clarke, and R. M. Scrowston, *Tetrahedron Letters*, **1971**, 1075.
[265] D. J. Cram, J. Day, D. R. Rayner, D. M. von Schriltz, D. J. Duchamp, and D. C. Garwood, *J. Am. Chem. Soc.*, **92**, 7369 (1970).
[266] B. W. Christensen, *Chem. Comm.*, **1971**, 597.
[267] R. E. Cook, M. D. Glick, J. J. Rigau, and C. R. Johnson, *J. Am. Chem. Soc.*, **93**, 924 (1971).

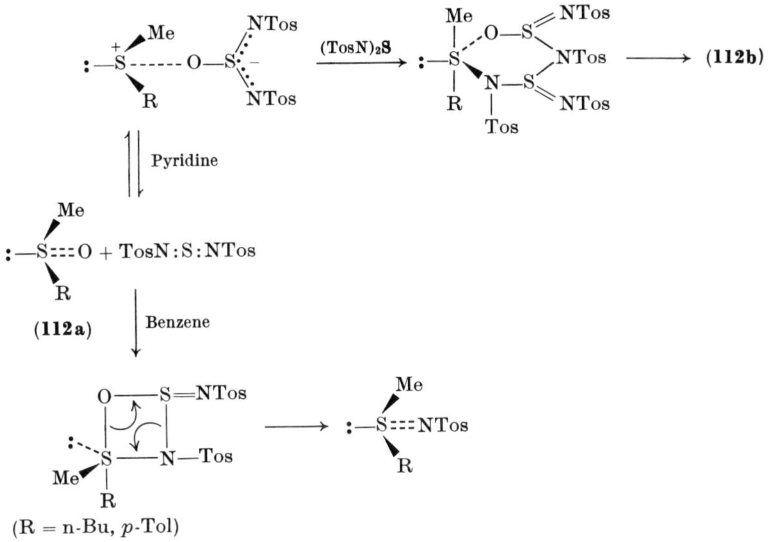

chemistry. Since the reduction step (**117**) → (**113**) is known to proceed with retention, and the electrophilic substitution also goes with retention (see above), the stereochemistry of the nucleophilic displacement steps follows.[268] This confirms the result that displacement in sulphonate esters proceeds with inversion.[269]

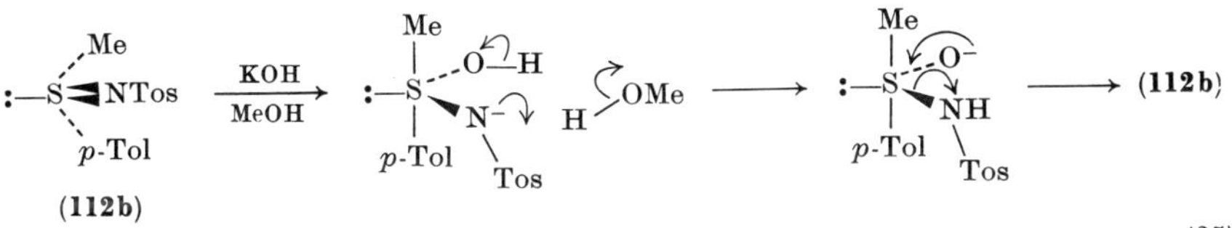

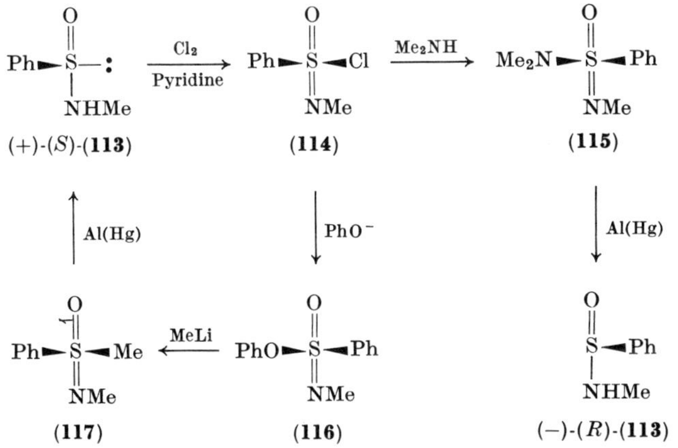

[268] E. U. Jonsson and C. R. Johnson, *J. Am. Chem. Soc.*, **93**, 5308 (1971).
[269] *Org. Reaction Mech.*, **1969**, 110.

Reduction of (+)-2-methyl-2,3-dihydro-4,5-[2-^{2}H$_1$]benzothiophen 1,1-dioxide with LiAlH$_4$ to the dihydrothiophen and re-oxidation to the dioxide proceeds without loss of optical activity or loss of deuterium. Oxidation of the dihydrothiophen to the sulphoxide gives both diastereomers.[270]

In contrast to the work of Cram, Oae and his group[271-273] have reported that the hydrolysis of the cyclic sulphilimines (118) gives the 2-alkoxy cyclic sulphides, by elimination in the intermediate ylid [reaction (26)]. The mechanism was based on the

$$\begin{array}{c}\text{(CH}_2)_n\\ \diagup\hspace{1cm}\diagdown\\ \text{S—CH}_2\\ \|\\ \text{NTos}\end{array} \xrightleftharpoons[]{\text{KOH—ROH}} \begin{array}{c}\text{(CH}_2)_n\\ \diagup\hspace{1cm}\diagdown\\ \overset{+}{\text{S}}\text{—}\underline{\text{CH}}\\ |\\ {}^-\text{NTos}\end{array} \longrightarrow \begin{array}{c}\text{(CH}_2)_n\\ \diagup\hspace{1cm}\diagdown\\ \overset{+}{\text{S}}\text{=CH}\end{array} \longrightarrow \begin{array}{c}\text{(CH}_2)_n\\ \diagup\hspace{1cm}\diagdown\\ \text{S—CH}\\ |\\ \text{OR}\end{array}$$

(118)

(n = 3, 4 or 5; Tos = toluene-p-sulphonyl)   ... (26)

low deuterium isotope effect in the α,α'-[^{2}H$_4$] compound and ρ value of +2.0 for the arenesulphonyl group. The sulphilimine from 7-thiabicyclo[2.2.1]heptane did not react.[271] In DMSO and in DMF, sulphilimines react with cyanide ion to give the sulphide and N-(toluene-p-sulphonyl)urea,[272] and with arenethiolates by attack at carbon with displacement of (RSNTos)$^-$.[273]

1- and 7-Bicyclo[2.2.1]heptyl toluene-p-sulphonates are cleaved to the corresponding alcohols by sodium methoxide or cyanide in methanol. For the 7-compound, the reaction has been shown to occur with >92% retention of configuration.[274]

The rates of hydrolysis of N-mesitylarenesulphinamides correlate with σ (ρ = +1.3). A Taft σ^0 treatment shows that the p-nitro-group has no resonance contribution and the p-methoxy-group a very small one. A direct displacement without the intervention of an intermediate 4-co-ordinate species is therefore suggested.[275]

The hydrolysis of p-methoxyphenyl p-methoxybenzenesulphinyl sulphone in aqueous dioxan or dimethoxyethane shows general base catalysis by N,N'-diethylbenzylamine, but pyridine and N-benzylpyrrolidine are nucleophilic catalysts, showing that even very slight changes in steric hindrance can cause changes from general to specific catalysis.[276] The reaction of p-tolyl toluene-p-sulphinyl sulphone with butane-1-thiol does not require acid-catalysis, in contrast to the corresponding reaction with di-n-butyl sulphide.[277]

The reduction and racemization of sulphoxides by halide ions has received wide attention.[278-281] Similar mechanisms have been proposed by all the groups concerned.

[270] T. A. Whitney and D. J. Cram, *J. Org. Chem.*, **35**, 3964 (1970).
[271] H. Kobayashi, N. Furukawa, T. Aida, K. Tsujihara, and S. Oae, *Tetrahedron Letters*, **1971**, 3109.
[272] S. Oae, T. Aida, K. Tsujihara, and N. Furukawa, *Tetrahedron Letters*, **1971**, 1145.
[273] T. Aida, N. Furukawa, and S. Oae, *Tetrahedron Letters*, **1971**, 4255.
[274] P. G. Gassman, J. M. Hornback, and J. M. Pascone, *Tetrahedron Letters*, **1971**, 1425.
[275] J. B. Biasotti and K. K. Andersen, *J. Am. Chem. Soc.*, **93**, 1178 (1971).
[276] J. L. Kice and J. D. Campbell, *J. Org. Chem.*, **36**, 2291 (1971).
[277] J. L. Kice and J. D. Campbell, *J. Org. Chem.*, **36**, 2288 (1971).
[278] D. Landini, G. Modena, F. Montanari, and G. Scorrano, *J. Am. Chem. Soc.*, **92**, 7168 (1970).
[279] D. Landini, G. Modena, U. Quintily, and G. Scorrano, *J. Chem. Soc.* (B), **1971**, 2041.
[280] H. Yoshida, T. Numata, and S. Oae, *Bull. Chem. Soc. Japan*, **44**, 2875 (1971).
[281] R. H. Rynbrandt, *Tetrahedron Letters*, **1971**, 3553.

That shown in reactions (27)–(30) is basically that of Modena and his colleagues. It was proposed on the observations that the rates of the two processes are proportional to

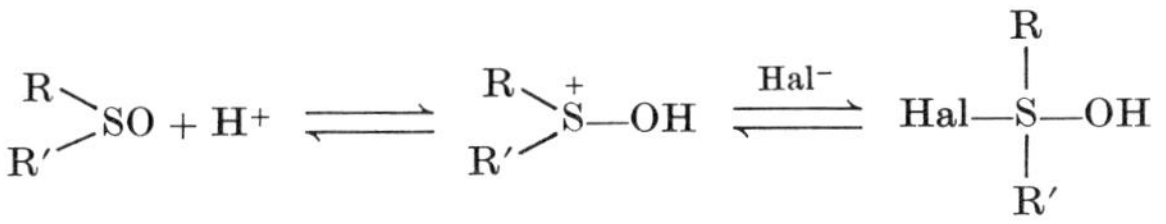

$$\underset{R'}{\overset{R}{>}}SO + H^+ \rightleftharpoons \underset{R'}{\overset{R}{>}}\overset{+}{S}-OH \underset{}{\overset{Hal^-}{\rightleftharpoons}} Hal-\underset{R'}{\overset{R}{\underset{|}{S}}}-OH \qquad \ldots (27)$$

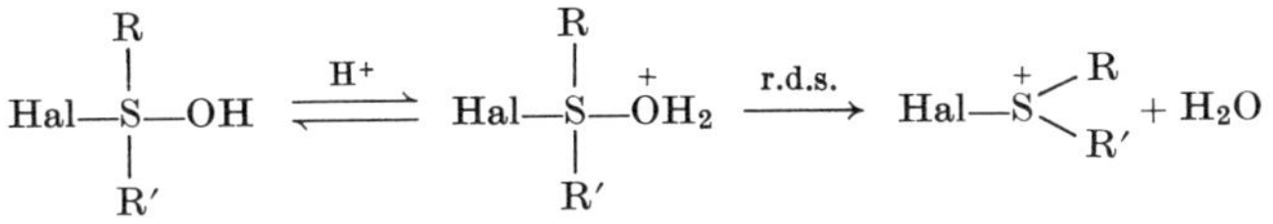

$$Hal-\underset{R'}{\overset{R}{\underset{|}{S}}}-OH \overset{H^+}{\rightleftharpoons} Hal-\underset{R'}{\overset{R}{\underset{|}{\overset{+}{S}}}}-OH_2 \overset{r.d.s.}{\longrightarrow} Hal-\overset{+}{S}\underset{R'}{\overset{R}{<}} + H_2O \qquad \ldots (28)$$

(Hal = Cl, Br or I)

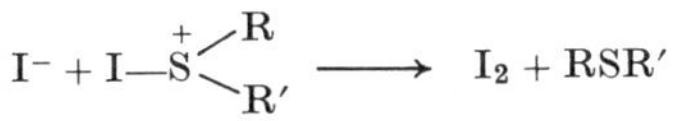

$$I^- + I-\overset{+}{S}\underset{R'}{\overset{R}{<}} \longrightarrow I_2 + RSR' \qquad \ldots (29)$$

$$Hal-\overset{+}{S}\underset{R'}{\overset{R}{<}} + Hal^- \rightleftharpoons \underset{R'}{\overset{R}{>}}S(Hal)_2 \overset{H_2O}{\longrightarrow} \underset{R'}{\overset{R}{>}}SO \qquad \ldots (30)$$

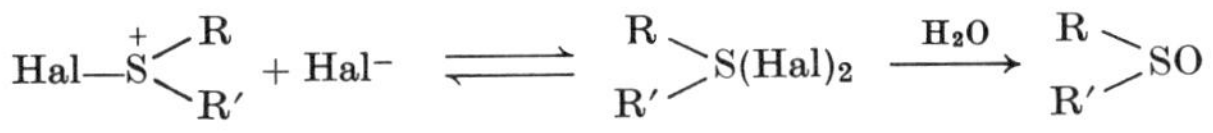

(Hal = Cl or Br) Racemic

sulphoxide and halide concentration, that the acid-dependence was such that both a pre-equilibrium protonation and another proton were involved. The tendency for racemization of isopropyl phenyl sulphoxide to show deviations towards general acid-catalysis in sulphuric acid led to the above choice, as this mechanism would involve general acid-catalysis if the two steps of (28) were to collapse into one.[278,279] Reduction is observed only with iodide ion. Rynbrandt[281] has suggested that the racemization step is really that of reaction (31), because he finds that, in the presence of molecular sieve,

$$RR'SCl_2 \rightleftharpoons RR'S + Cl_2 \qquad \ldots(31)$$

sulphoxides of the type $RS(O)CH_2R'$ are converted into the α-chloro-sulphides, $RSCHClR'$, by HCl. This mechanism was also considered by Modena and amounts to allowing that the reaction of equation (29) is reversible. Reaction (31) has been studied by Wilson and Chang,[282] using ^{19}F-NMR where R and R' are aryl, and their evidence shows it to be a direct equilibrium and not one involving an intermediate diarylsulphonium chloride. Oae's group have also studied the effects of ring size on the rates of reduction of cyclic sulphoxides.[283] The acid-catalysed racemization and disproportionation of t-butyl phenyl sulphoxide and α-methylbenzyl phenyl sulphoxide have been studied.[284]

Triphenylsulphonium tetrafluoroborate reacts with allyl-, phenyl- and vinyl-lithium to give allylbenzene, biphenyl and styrene, respectively. The other product is diphenyl sulphide. A mechanism involving a sulphurane has been proposed. This collapses by ligand coupling, which is only possible when the coupling ligands have π-orbitals. Thus the reaction is a general one for aryl and vinyl compounds, but reaction with t-butyl-lithium affords biphenyl and t-butyl phenyl sulphide.[285]

[282] G. E. Wilson and M. M. Y. Chang, *Tetrahedron Letters*, **1971**, 875.
[283] S. Tamagaki, M. Mizuno, H. Yoshida, H. Hirota, and S. Oae, *Bull. Chem. Soc. Japan*, **44**, 2456 (1971).
[284] G. Scorrano, U. Quintily, and G. Modena, *Chim. Ind. (Milan)*, **53**, 155 (1971).
[285] R. W. La Rochelle and B. M. Trost, *J. Am. Chem. Soc.*, **93**, 6077 (1971).

The reaction of toluene-p-sulphonyl chloride and cyanide with cyanide ion has been studied.[286]

Alkane- and arene-sulphenyl chlorides, -sulphinyl chlorides and -sulphonyl chlorides react with tri-n-propylamine and trichlorosilane to give the corresponding dialkyl and diaryl disulphides in good yield.[287]

Other Elements

Details of extensive work on the reaction of alkanethiolates with 1-aryl-2-haloacetylenes have been published.[288] Reaction occurs with nucleophilic displacement at halogen, exclusively when it is bromine and iodine, but in competition with attack at carbon of chloroacetylenes. The rates of attack are correlated by a Hammett plot with ρ values of 1.15–1.28 depending on the alkyl group. Using sec-BuSC≡CX (X = Cl, Br or I), which reacts much faster than the arylacetylenes, and SO_3^{2-} as standard nucleophile, Swain–Scott nucleophilicities have been determined for a large number of nucleophiles. These vary from 1.82, for RS^-, to –4.9 for $S_2O_3^{2-}$ (for attack at bromine). Values for attack at iodine parallel those at bromine and both differ greatly from those for attack at other elements, e.g. carbon, platinum or sulphur. A good correlation of these nucleophilicities is found when the Edwards oxibase scale is used. Table 4 gives a summary of the findings

Table 4. Attack at halogen (X) versus attack at carbon (C) in displacements in ArC≡CX[288]

Halogen	S^{2-}	EtS^-	SO_3^{2-}	HS^-	CN^-	$S_2O_3^{2-}$	OH^-	I^-
Cl	C + X	C + X	C	C			C	
Br	X	X	X	X	X	C + X	C + X	(No reaction)
I	←—————————————— Attack at X only ——————————————→							

for attack at carbon versus attack at halogen for a selection of nucleophiles. The HSAB principle explains the pattern quite well.[288]

The kinetics of the displacement at halogen have been measured for the reaction of potassium iodide with $(NO_2)_2CFX$ (X = Cl, Br or I) in 70% ethanol. Displacement at chlorine is about 10^8 slower than with chlorotrinitromethane.[289] Further reports on the reaction of acetylenic halides with trialkyl phosphites have also appeared.[290,291]

Preparative use of the displacement at halogen has been made in the reduction of twelve α-chloro- and α-bromo-ketones with lithium iodide and boron trifluoride etherate. Good yields (95–100%) of the reduced ketones are formed. α-Bromocamphor gives camphor in 85% yield, but α-chloronorbornanone fails to react.[292] Tris(dimethylamino)-

---

[286] F. P. Corson and R. G. Pews, *J. Org. Chem.*, **36**, 1654 (1971).
[287] T. H. Chan, J. P. Montillier, W. F. Van Horn, and D. N. Harpp, *J. Am. Chem. Soc.*, **92**, 7224 (1970).
[288] M. C. Verploegh, L. Donk, H. J. T. Bos, and W. Drenth, *Rec. Trav. Chim.*, **90**, 765 (1971).
[289] M. S. L'vova, V. I. Slovetskii, L. V. Okhlobystina, V. M. Khutoretskii, and A. A. Fainzil'berg, *Izv. Akad. Nauk SSSR, Ser. Khim.*, **1971**, 95.
[290] D. W. Burt and P. Simpson, *J. Chem. Soc.* (C), **1971**, 2872.
[291] A. Fujii and S. I. Miller, *J. Am. Chem. Soc.*, **93**, 3694 (1971).
[292] J. M. Townsend and T. A. Spencer, *Tetrahedron Letters*, **1971**, 137.

phosphine reacts with CCl_4 and aldehydes to give 1,1-dichloroalk-1-enes; the mechanism involves displacement at chlorine.[293]

The reaction of fluorotrinitromethane with iodide ion involves displacement at nitrogen,[294] and this is also involved in the reaction of aryl methyl sulphides with chloramine-T.[295]

In a series of papers[296-298] on displacement at boron, Hawthorne and his co-workers find that compounds of the type $Ar_2BH \cdot NH_2Me$ react with 2,2-diphenylethylamine by prior ionization (S_N1-B),[296] that compounds of the type $RBH_2 \cdot NMe_3$ and $ArBH_2 \cdot NMe_3$ react with tri-n-butylphosphine by an S_N2-B route, except for the cases where R = t-Bu (S_N1-B) and Ar = p-methoxyphenyl and mesityl, which show small S_N1-B components.[297] $Me_3N \cdot BH_3$ and $Et_3N \cdot BH_3$ also react by the S_N2-B route.[298]

The kinetics of the displacement of carbon monoxide from allyldicarbonylnitrosyliron by trialkylphosphines suggest an association–elimination mechanism.[299] 4,4'-Disubstituted diphenyl sulphides react with *trans*-dichlorodipyridineplatinum(II) to displace chloride; the logarithms of the rates of reaction give a linear plot against the sum of the σ_p constants, and it is suggested that bond-making to platinum is the driving force.[300] An association–elimination route has also been proposed for the reaction of phosphines with tricarbonylcyclo-octatetraenylruthenium.[301]

Triphenylphosphine reacts with peroxycarbonates by nucleophilic attack at peroxy-oxygen.[302]

Ambident Nucleophiles[303]

Reutov and his group[304-307] have reviewed much of their work on the alkylation of acetoacetic ester and of acetylacetone in hexamethylphosphorotriamide (HMPT). They find that for substituted acetoacetates, $RCOCH_2CO_2Et$ (R = Me, CF_3, i-Pr or t-Bu), reacting with ethyl toluene-p-sulphonate, the ratio of oxygen to carbon-alkylation is independent of counter-ion (Li, Na, K or Cs). The variation of rates of reaction with concentration is fitted by the Acree equation; only free-ions are reactive. The rates of both C-alkylation (k_C) and O-alkylation (k_O) increase with inductive electron-release by the R group; but, with the exception of CF_3, where $k_C = 0$, the ratio k_C/k_O is almost unaffected.[304,305] The variation in the ratio k_C/k_O with leaving group decreases in the order I > Br > Cl > OTs with all alkyl groups. The amount of variation in k_C/k_O is

---

[293] G. Lavielle, J.-C. Combret, and J. Villieras, *Bull. Soc. Chim. France*, **1971**, 2047.
[294] V. I. Slovetskii, M. S. L'vova, A. A. Fainzil'berg, and T. I. Chaeva, *Izv. Akad. Nauk SSSR, Ser. Khim.*, **1970**, 2690.
[295] F. Ruff and A. Kucsman, *Acta Chim. (Budapest)*, **65**, 107 (1970).
[296] F. J. Lalor, T. Paxson, and M. F. Hawthorne, *J. Am. Chem. Soc.*, **93**, 3156 (1971).
[297] D. E. Walmsley, W. L. Budde, and M. F. Hawthorne, *J. Am. Chem. Soc.*, **93**, 3150 (1971).
[298] W. L. Budde and M. F. Hawthorne, *J. Am. Chem. Soc.*, **93**, 3147 (1971).
[299] G. Cardaci and S. M. Murgia, *J. Organomet. Chem.*, **25**, 483 (1970).
[300] J. R. Gaylor and C. V. Senoff, *Can. J. Chem.*, **49**, 2390 (1971).
[301] F. Faraone, F. Cusmano, and R. Pietropaolo, *J. Organomet. Chem.*, **26**, 147 (1971).
[302] W. Adam and A. Rios, *J. Org. Chem.*, **36**, 407 (1971).
[303] S. A. Shevelev, *Uspekhi Khim.*, **39**, 1773 (1970); *Russ. Chem. Rev.*, **39**, 844 (1970).
[304] A. L. Kurts, A. Macias, I. P. Beletskaya, and O. A. Reutov, *Tetrahedron*, **27**, 4759 (1971).
[305] A. L. Kurts, P. I. Dem'yanov, A. Macias, I. P. Beletskaya, and O. A. Reutov, *Tetrahedron*, **27**, 4769 (1971); *Doklady Akad. Nauk USSR*, **195**, 1117 (1970); *Proc. Acad. Sci. USSR*, **195**, 920 (1970).
[306] A. L. Kurts, N. K. Genkina, A. Macias, I. P. Beletskaya, and O. A. Reutov, *Tetrahedron*, **27**, 4777 (1971).
[307] A. L. Kurts, A. Macias, I. P. Beletskaya, and O. A. Reutov, *Tetrahedron Letters*, **1971**, 3037.

large with MeX (10^3 for acetylacetone and 160 for acetoacetate) and almost negligible for sec-BuX (3.6 and 5, respectively).[306] Rates of reaction for the free ions have been measured for acetoacetic ester with EtX (X = OTs, Br or I) in DMF (Na⁺ salt). The rate of C-alkylation increases by nearly 10^3 as X goes from OTs to I whereas k_O only changes by a factor of 10. These authors use the HSAB principle to rationalize the changes.[304,306] The stereochemistry of the O-alkyl product has also been examined in both HMPT and DMF. The results show that for the sodium, potassium and caesium enolates of acetoacetic ester, a W conformation is favoured over U or S ones, but that the lithium enolate exists in a ca. 4:3 ratio of W to U + S forms.[307]

The kinetics of the reaction of the sodium salt of 9-fluorenone oxime with methyl iodide and toluene-p-sulphonate have been measured.[308] By complexing the sodium ions with dibenzo-18-crown-6-polyether, the rate for the free ion was measured, and, by using sodium tetraphenyl borate as a common-ion source, the rate for the ion pair was measured for each of the two reactive sites. The effect of leaving group, OTs/I, was then available for each site for both free ions and ion pairs. For the free ions the rate ratio $k(\text{MeOTs})/k(\text{MeI})$ was 1.0 for O-alkylation and 0.067 for N-alkylation. For the ion pairs, the ratio for O-alkylation was also 1.0, but that for N-alkylation was 1.2. This ratio is often used as the basis for a hard–soft classification, but it appears that this classification depends on the state of ion aggregation.

The ratio of O- to C-acetylation of benzoylacetone and its sodium salt varies with solvent, from 0.1 in benzene to ∞ in diglyme and DMF; in the presence of pyridine the two O-acetates are readily equilibrated and are converted slowly into the C-acetyl compound.[309]

The alkylation of the sodium enolates of acetylacetone and ethyl acetoacetate in DMSO with $(-)$-(R)-sec-butyl bromide has been studied.[310] O-Alkylation proceeds mainly with inversion (ca. 35% racemization), but C-alkylation goes with 10–20% retention of configuration (80–90% racemization). The authors suggest that this may arise by front-side attack, but their main conclusion is that the O- and C-alkylation processes should not be regarded as simple competitive ones with analogous transition states.

The rates and products of alkylation of the enolates of a number of alkylphenones have been studied.[311] The ratio of C- to O-acylation of the magnesium enolates of 4,4-diphenylbutan-2-one has been shown to vary with the acylating agent and the solvent.[312] Moreover, another study of this type has been reported.[313]

The formation and alkylation (almost exclusively on carbon) of specific isomers of lithium enolates has been studied by two groups.[314,315]

The alkylation of the sodium salts of imines (to give enamines or C-alkylimines) in diglyme, "Diethylcarbitol," xylene and HMPT has been examined. The ratio of N- to C-alkylation increases with the hardness of the leaving group.[316]

---

[308] S. G. Smith and M. P. Hanson, *J. Org. Chem.*, **36**, 1931 (1971).
[309] M. Suama, Y. Nakao, and K. Ichikawa, *Bull. Chem. Soc. Japan*, **44**, 2811 (1971).
[310] M. Suama, T. Sugita, and K. Ichikawa, *Bull. Chem. Soc. Japan*, **44**, 1999 (1971).
[311] H. D. Zook and J. A. Miller, *J. Org. Chem.*, **36**, 1112 (1971).
[312] P. Anguibeaud and M.-J. Lagrange, *Compt. Rend.* (C), **272**, 1506 (1971).
[313] B. Cardillo, G. Casnati, V. Malatesta, and A. Pochini, *Rend. Inst. Lomb. Sci. Lett.* A, **104**, 404 (1970); *Chem. Abs.*, **75**, 19451 (1971).
[314] H. O. House, M. Gall, and H. D. Olmstead, *J. Org. Chem.*, **36**, 2361 (1971).
[315] I. J. Borowitz, E. W. R. Casper, and R. K. Crouch, *Tetrahedron Letters*, **1971**, 105.
[316] G. J. Heiszwolf and H. Kloosterziel, *Rec. Trav. Chim.*, **89**, 1217 (1970).

Lithium N-isopropylcyclohexylamide in THF at $-78°$ gives the enolate anions from esters. These are alkylated in 50–95% yield exclusively at carbon, the reaction being carried out by adding the enolate to the alkylating agent in DMSO at room temperature; quenching of the initial enolate with heavy water results in the incorporation of only 50–75% of one atom of deuterium![317]

The methylation of $PhCH\!\cdots\!CH\!\cdots\!CH_2{}^- Na^+$ and of $Ph_2C\!\cdots\!CH\!\cdots\!CH_2{}^- Na^+$ in liquid ammonia has been studied. The ratio of alkylation at C-1 to that at C-3 varies with leaving group, the former being the harder centre.[318]

The variation in the percentage of nitrile to isocyanide with temperature has been examined for the reaction of trityl chloride and perchlorate with $Ph_4As^+CN^-$ in acetonitrile.[319]

It has been reported[320] that the sodium enolate of the tetrahydropyranyl ether of $PhCH(OH)CO_2Me$ in DMSO reacts with alkyl halides to give C-alkylation at 25° and O-alkylation at 80°. Details of the reasons are awaited with interest.

A number of studies of the alkylation and hydroxyalkylation of phenoxide anions have been reported.[321–324]

The methylation of thallous salts of 2-hydroxythiophens under heterogeneous conditions has been examined,[325] and the alkylation of the benzothiazole-2-thiolate anion has been considered in terms of the oxibase scale.[326]

A study of the isomerization of benzyl and p-bromobenzyl thiocyanates to the isothiocyanates in fused quaternary ammonium salts has been carried out.[327]

Substitution at Vinylic Carbon[328–330]

Extended Hückel calculation[331] on the 1-cyclopropylvinyl cation show that the most stable conformation is (**119**). This is the same as that reported last year[158] for the cyclopropylmethyl cation, but the barrier to rotation is much less than in the cyclopropylmethyl case, being only 4.5–6.5 kcal mol^{-1} as against 17.5. The energetics of the S_N2 displacements of H$^-$ by H$^-$ at sp^3 and sp^2 centres has also been calculated. The transition-state energies differ by only 14.5 kcal mol^{-1} ($sp^2 > sp^3$); but the free energy of activation for addition of hydride to ethylene was calculated to be some 60 kcal mol^{-1} less than that for the S_N2 process. Thus an addition–elimination mechanism of substitution is predicted to be markedly favoured over direct substitution even though the latter is not much worse than S_N2 at saturated carbon.[331] Non-empirical LCAO–MO–SCF calculations on the protonation of acetylene show that the bridged ion (**120**) is less stable

---

[317] M. W. Rathke and A. Libert, *J. Am. Chem. Soc.*, **93**, 2318 (1971).
[318] W. S. Murphy, R. Boyce, and E. A. O'Riordan, *Tetrahedron Letters*, **1971**, 4157.
[319] T. Austad, J. Songstad, and L. J. Stangeland, *Acta Chem. Scand.*, **25**, 2327 (1971).
[320] K. L. Shepard and J. I. Stevens, *Chem. Comm.*, **1971**, 951.
[321] B. Miller and K.-H. Lai, *Chem. Comm.*, **1971**, 334.
[322] W. J. le Noble, T. Hayakawa, A. K. Sen, and Y. Tatsukami, *J. Org. Chem.*, **36**, 193 (1971).
[323] G. Casiraghi, G. Casnati, and G. Sartori, *Tetrahedron Letters*, **1971**, 3969.
[324] K. D. Sears and R. J. Engen, *Chem. Comm.*, **1971**, 612.
[325] E. B. Pedersen and S.-O. Lawesson, *Tetrahedron*, **27**, 3861 (1971).
[326] A. F. Halasa and G. E. P. Smith, *J. Org. Chem.*, **36**, 636 (1971).
[327] D. Mravec, J. Kalamár, and J. Hrivňák, *Coll. Czech. Chem. Comm.*, **35**, 3274 (1970).
[328] G. Modena, *Accounts Chem. Res.*, **4**, 73 (1971).
[329] C. A. Grob, *Chimia*, **25**, 87 (1971).
[330] G. Modena and U. Tonellato, *Adv. Phys. Org. Chem.*, **9**, 185 (1971).
[331] D. R. Kelsey and R. G. Bergman, *J. Am. Chem. Soc.*, **93**, 1953 (1971).

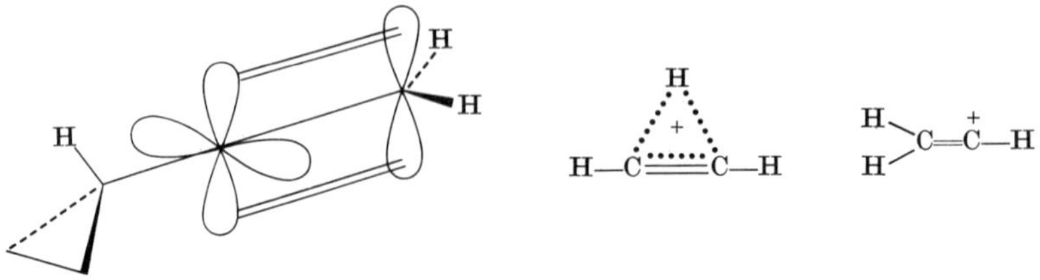

than the vinyl ion (**121**) and that no barrier exists to the conversion of (**120**) into (**121**).[332] The reaction of but-2-yne with carbon monoxide in FSO_3H–SbF_5 affords the acylium ion (**123**) when the ratio of alkyne to CO is 1:25 and the ratio of FSO_3H to SbF_5 is 1:10. Only the E-isomer (**123**) is isolated and no evidence for $trans \rightleftharpoons cis$ equilibration was found. As a result the authors suggest that the intermediate vinyl cation (**122**) may be non-linear. At 1:1 ratios of acid to SbF_5, the reaction affords (**124**), and it is suggested that this is formed by the rearrangement shown.[333] The addition of the 1-adamantyl cation to propargyl alcohol affords 2-acetylhomoadamantane.[334] The acid-catalysed addition of water, methanol, and acetic and trifluoroacetic acid to (**125**) goes through the cyclopropylidenyl cation; the major products are the acetylenes formed by ring opening.[335]

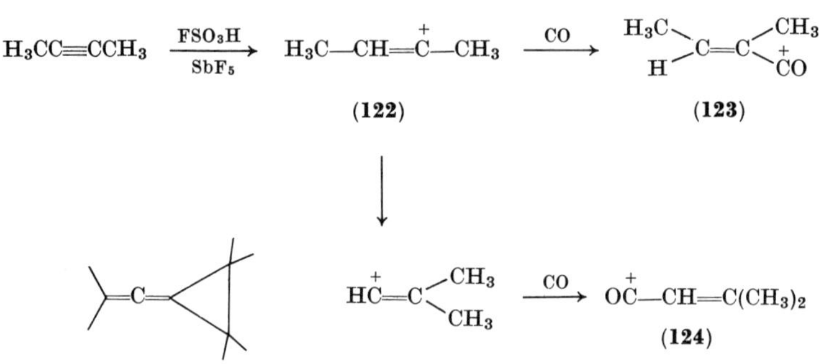

Schleyer, Hanack and their groups[336] have reported the rates of solvolysis of a number of cyclic vinyl trifluoromethanesulphonates. The cyclopentenyl and cyclohexenyl compounds react at 1.1×10^{-5} and 3×10^{-4} times the rate of the 1-methylpropenyl compound, and this is attributed to the destabilization of the vinyl cation by the enforced bending, the result being in agreement with their *ab initio* calculations that the linear geometry is the most favourable one. Rate and product studies on some substituted cyclohexenyl compounds show quite large rate variations and that ring-contraction and/or methyl-migration are quite favourable; the rate variations are not explained but are not due to anchimeric acceleration.[336]

[332] A. C. Hopkinson and I. G. Csizmadia, *Chem. Comm.*, **1971**, 1291.
[333] H. Hogeveen and C. F. Roobeek, *Tetrahedron Letters*, **1971**, 3343.
[334] J. K. Chakrabarti and A. Todd, *Chem. Comm.*, **1971**, 556.
[335] M. L. Poutsma and P. A. Ibarbia, *J. Am. Chem. Soc.*, **93**, 440 (1971).
[336] W. D. Pfeifer, C. A. Bahn, P. von R. Schleyer, S. Bocher, C. E. Harding, K. Hummel, M. Hanack, and P. J. Stang, *J. Am. Chem. Soc.*, **93**, 1513 (1971).

Grob and Spaar have published the details of their work on the 2-bromobuta-1,3-diene system reported in 1969.[337] Bending of the diene into a ring results in complete loss of reactivity; thus 2-bromocyclohexa-1,3-diene and the 4,6-dimethyl analogue are stable up to 180° in 80% ethanol and to silver–ion at 70°.[338]

Full details of Kelsey and Bergman's work on *cis*- and *trans*-1-cyclopropyl-1-iodopropene have been published. 6-Iodohexa-2,3-diene reacts with silver acetate to give the same ions. Internal return is observed for the cyclopropanes and ring-expansion to 2-ethylidenecyclobutyl acetates also occurs.[339] The similar work on 1-cyclopropyl-1-iodoethylene has also been published in full.[340]

Bässler and Hanack[341] have attempted to form a "primary" vinyl cation by solvolysis of bromomethylenecyclopropane in aqueous ethanol. An m value of 0.52, plus the isolation of cyclobutanone as the only product, is adduced as evidence of their success.[341] The solvolysis of bromoallene has been studied in a similar attempt and showed an m value of 0.44 (cf. allyl bromide, $m = 0.455$). The allene solvolysed 4×10^3 times more slowly than propargyl bromide, and as both should give the same ion this was ascribed to ground-state differences.[342]

Styryl trifluoromethanesulphonate and the [β,β-^{2}H$_2$] isomer have been solvolysed in 80% ethanol. Other evidence points to the intervention of a vinyl cation, and so the isotope effect (k_H/k_D) of 1.42 is of interest. This is appreciably larger than that observed in the solvolysis of [2.2.2-^{2}H$_3$]-1-phenylethyl p-bromobenzenesulphonate (1.220) and would indicate a much greater involvement of the β-hydrogen atoms at the transition state.[343]

Rappoport has reported on the trifluoroacetolysis of the 1-(p-methoxyphenyl)-2-methylprop-1-enyl p-bromobenzene- and toluene-p-sulphonates.[344]

The preparation of 2-chloroethylenediazonium hexachloroantimonate has been reported,[345] and the intermediacy of an ethylenediazonium ion has been inferred for the reaction of (**126**) with base. The major product is the enol ether of isobutyraldehyde, with some methallyl ether. Deuterium-labelling studies suggest that this ether is formed by isomerization of the ethylenediazonium ion to the methallyldiazonium ion.[346]

Modena and Tonellato[347] have now published full details of their work on the system (**127**). The rates show a dependence on solvent polarity and not on solvent nucleophilicity and, in acetone, common-ion depression was observed. In acetic acid, (**127**) (Ar^1 = Ar^2 = p-tolyl, Ar^3 = Ph) shows a special salt effect; in other solvents, normal salt effects were observed. The rates are correlated by σ for Ar^1 and Ar^3 and by σ^+ for Ar^2 with an overall correlation by $\log k = \log k_0 - 2.85(\sigma_2^+ + 0.44\sigma_1 + 0.52\sigma_3)$. All the effects are consistent with rate-determining ionization to give a relatively long-lived ion pair. The rates are

---

[337] C. A. Grob and R. Spaar, *Helv. Chim. Acta*, **53**, 2119 (1971); cf. *Org. Reaction Mech.*, **1969**, 115.
[338] C. A. Grob and H. R. Pfaendler, *Helv. Chim. Acta*, **53**, 2130 (1970).
[339] D. R. Kelsey and R. G. Bergman, *J. Am. Chem. Soc.*, **93**, 1941 (1971); cf. *Org. Reaction Mech.*, **1970**, 108.
[340] S. A. Sherrod and R. G. Bergman, *J. Am. Chem. Soc.*, **93**, 1925 (1971); cf. *Org. Reaction Mech.*, **1969**, 117.
[341] T. Bässler and M. Hanack, *Tetrahedron Letters*, **1971**, 2171.
[342] C. V. Lee, R. J. Hargrove, T. E. Dueber, and P. J. Stang, *Tetrahedron Letters*, **1971**, 2519.
[343] R. J. Hargrove, T. E. Dueber, and P. J. Stang, *Chem. Comm.*, **1970**, 1614.
[344] Z. Rappoport and J. Kaspi, *Tetrahedron Letters*, **1971**, 4039.
[345] K. Bott, *Tetrahedron Letters*, **1971**, 2227.
[346] M. S. Newman and C. D. Beard, *J. Am. Chem. Soc.*, **92**, 7564 (1970).
[347] G. Modena and U. Tonellato, *J. Chem. Soc. (B)*, **1971**, 374, 381, 1569; cf. *Org. Reaction Mech.*, **1970**, 75, 107.

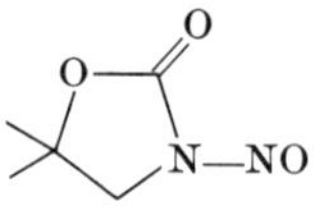

(126) [structure: oxazolidinone with N—NO]

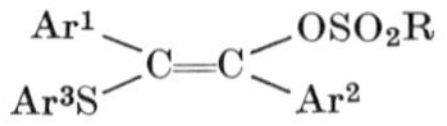

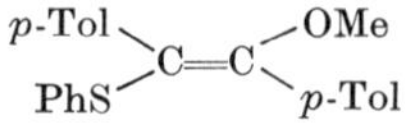

(127) $Ar^1\!\!\!\diagdown\!\!\!C\!\!=\!\!C\!\!\diagup\!\!\!OSO_2R$ / $Ar^3S\diagup \quad \diagdown Ar^2$

(128) $p\text{-Tol}\diagdown C\!\!=\!\!C\diagup OMe$ / $PhS\diagup \quad \diagdown p\text{-Tol}$

(R = $C_6H_2(NO_2)_3$, C_6H_4Br or $C_6H_4CH_3$)

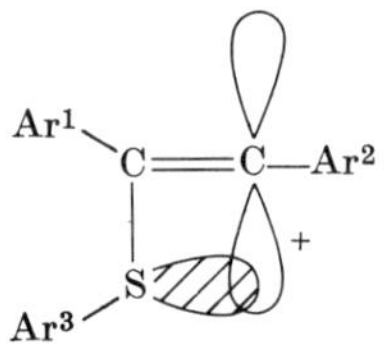

(129) [bridged sulfonium ion structure]

faster than those of the corresponding triarylvinyl arenesulphonates by factors of 15.6–33 depending on solvent, and this, together with the fact that (127; $Ar^1 = Ar^2 = p$-tolyl, Ar^3 = Ph), in acetone/methanol, gives the methyl enol ether (128) with $\geqslant 97\%$ stereospecificity, leads them to conclude that reaction occurs by way of the unsymmetrically bridged ion (129). Their work on (127; $Ar^1 = p$-tolyl, Ar^2 = Ph and Ar^1 = Ph, Ar^2 = p-tolyl) reported last year then leads them to suggest that (129) is in rapid equilibrium with the isomer with Ar^1 and Ar^2 reversed, rather than that a symmetrically bridged ion is formed.[347] Further work on the alkyl-substituted system (130) has shown that the rate accelerations due to sulphur participation range from 3.3×10^2 (for R^1 = Ph, R^2 = Me, R^3 = Ph), through 3.6×10^4 (for $R^1 = R^2 = R^3$ = Me) to ca. 7×10^5 (for $R^1 = R^2$ = n-Pr, R^3 = Me).[348,349] While these are undoubtedly large accelerations, their claim, that S-3 participation gives bigger effects in vinyl than in saturated systems, seems slightly exaggerated in view of the example in ref. 89.

Vinyl cations have also been implicated in the electro-oxidation of α-morpholinostilbene.[350]

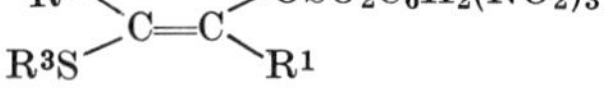

(130) $R^2\diagdown C\!\!=\!\!C\diagup OSO_2C_6H_2(NO_2)_3$ / $R^3S\diagup \quad \diagdown R^1$

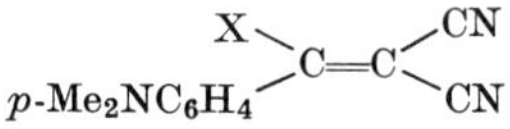

(131) $X\diagdown C\!\!=\!\!C\diagup CN$ / $p\text{-Me}_2NC_6H_4\diagup \quad \diagdown CN$

(132) $Ar\diagdown C\!\!=\!\!C\diagup F$ / $R_f\diagup \quad \diagdown Cl$

(X = Cl, F)

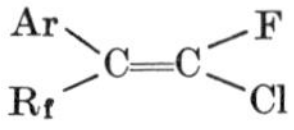

The displacement of chloride and fluoride by aromatic amines from (131) goes by an addition–elimination route. The reactions have large negative entropies of activation and are of the second-order in amine, always for X = F and generally for X = Cl. The mechanism suggested is pre-equilibrium amine addition, followed by either halide elimination with solvent assistance and/or amine-mediated proton transfer and halide loss.[351]

Other addition–elimination reactions are the displacement of chloride from (132; R_F = perfluoroalkyl) by methoxide ion, which takes place with retention of configura-

[348] G. Capozzi, G. Modena, and U. Tonellato, *J. Chem. Soc.* (B), **1971**, 1700.
[349] A. Burighel, G. Modena, and U. Tonellato, *Chem. Comm.*, **1971**, 1325.
[350] S. J. Huang and E. T. Hsu, *Tetrahedron Letters*, **1971**, 1385.
[351] Z. Rappoport and R. Ta-Shma, *J. Chem. Soc.* (B), **1971**, 871, 1461.

tion,[352] and the displacement of fluoride from cis- and trans-1-fluoro-2-(phenylsulphonyl)-ethylene by methoxide and benzenethiolate, which is non-stereospecific.[353] The reaction of cyclopentadienylidenetriphenylphosphorane with tetracyanoethylene also has an addition–elimination mechanism.[354]

Details of the proof of the cyclohexyne mechanism of replacement of chlorine in 1-chlorocyclohexenes have now been published.[355]

A general discussion of nucleophilic displacement of halogen at ethynyl carbon may be found in reference 288, and other work on this subject has also been reported.[356,357]

Reactions of α-Halogenocarbonyl Compounds

The adduct (**133**; R = H) of dichloroketene with cyclopentadiene has attracted attention, for its products of solvolysis depend on the reaction conditions. Acetolysis and hydrolysis lead to tropolone,[358] and alkaline methanolysis to methyl benzoate. The latter reaction probably goes by the route shown, for specifically labelled benzoate is formed in [O-^{2}H]-methanol.[359] The mechanism of the acetolysis has been clarified by Bartlett and Ando[360]

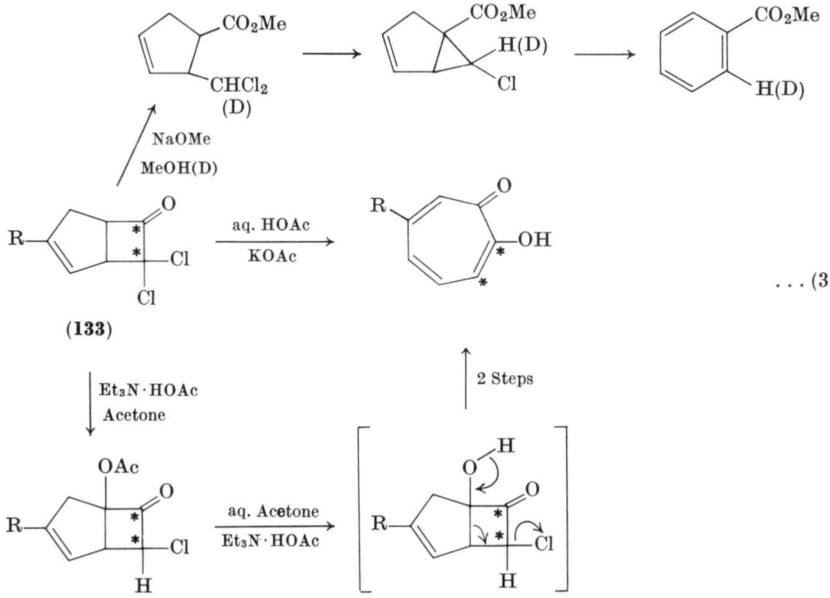

... (32)

[352] D. J. Burton and H. C. Krutsch, *J. Org. Chem.*, **36**, 2351 (1971).
[353] G. Marchese and F. Naso, *Chim. Ind. (Milan)*, **53**, 760 (1971).
[354] C. W. Rigby, E. Lord, M. P. Naan, and C. D. Hall, *J. Chem. Soc.* (B), **1971**, 1192.
[355] P. Caubère and J.-J. Brunet, *Tetrahedron*, **27**, 3515 (1971); cf. *Org. Reaction Mech.*, **1969**, 119.
[356] R. Tanaka, M. Rodgers, R. Simonaitis, and S. I. Miller, *Tetrahedron*, **27**, 2651 (1971).
[357] R. Tanaka and S. I. Miller, *Tetrahedron Letters*, **1971**, 1753.
[358] H. C. Stevens, J. K. Rinehart, J. M. Lavanish, and G. M. Trenta, *J. Org. Chem.*, **36**, 2780 (1971).
[359] P. R. Brook and A. J. Duke, *J. Chem. Soc.* (C), **1971**, 1764.
[360] P. D. Bartlett and T. Ando, *J. Am. Chem. Soc.*, **92**, 7518 (1970).

by the steps shown in the sequences (32), the isolation of (**134**; R = H) being most important, after labelling (R = t-Bu, and $^{14}C_2$ from labelled dichloroketene) had shown that rearrangement had occurred. This mechanism also accounts for the observations of Asao et al., who had postulated a 1-chloronorcaradienone as intermediate.[360a] The same rearrangement has also been reported for (**134**; R = i-Pr), which afforded β-thujaplicin and for the 2-isopropyl compound (**134**; R = H) which gives γ-thujaplicin.[361] The step (**133**) → (**134**) has previously been reported in the bicyclo[4.2.0]octane series.[362]

Further work on the solvolysis of the *trans*-2-haloacetyl-3-phenyl-1,4-benzodioxans has shown that they too react by enolization followed by ionization. The products are entirely those of attack at the tertiary centre.[363]

1-Halo-1,1-diphenylacetones and their 3-halo-isomers react with an excess of base to give, cleanly, the Favorskii products; but at low (10^{-5}M) base concentration 1-phenylindan-2-one is formed by the mechanism reported last year for acetolysis.[364]

4β-Bromo-5β-androstan-3-one acetolyses to the 2β-acetoxy-3-ketone via the Δ^2-enol.[365] This appears to be general for *cis*-AB-steroids, for 4β-bromo-5β-cholestan-3-one also gives the 2β-acetate; this has been shown to arise by epimerization of the initial 2α-acetate, and it was suggested that it was itself formed by a *trans*-S_N2' attack on the enol.[366]

The silver ion-catalysed solvolysis of α-bromoisobutyrophenone has been studied. Three concurrent processes are postulated to account for the dependence on silver ion and hydrogen ion, one first-order and one second-order in silver ion, and the third first-order in both. Reversible addition of solvent to the carbonyl group and reversible complexing of this with silver ion was followed by rate-determining ionization, which may be either uncatalysed, silver ion- or hydrogen ion-catalysed.[367]

Silver hexafluoroantimonate-catalysed reaction of the 1-acyl-1-bromocyclohexanes (**135**) leads to the bridged ion (**136**). It is suggested that this arises by formation of the α-keto carbonium ion, followed by 1,4-hydrogen shift and participation by the carbonyl group. The hydrogen shift probably occurs as a series of 1,2-steps, because 1-benzoyl-1-bromocyclohexane gives the 6-oxa-7-phenylbicyclo[3.2.1]octenium ion.[368,369]

(**135**)   (**136**)

(R = Ph, *p*-MeOC$_6$H$_4$, OMe or NMe$_2$)

[360a] T. Asao, T. Machiguchi, and Y. Kitahara, *Bull. Chem. Soc. Japan*, **43**, 2662 (1970); cf. *Org. Reaction Mech.*, **1970**, 287, ref. 285.
[361] K. Tanaka and A. Yoshikoshi, *Tetrahedron*, **27**, 4889 (1971).
[362] V. R. Fletcher and A. Hassner, *Tetrahedron Letters*, **1970**, 1071.
[363] V. Rosnati, F. Sannicolo, and G. Zecchi, *Gazzetta*, **101**, 344 (1971); cf. *Org. Reaction Mech.*, **1969**, 120.
[364] F. G. Bordwell and R. G. Scamehorn, *J. Am. Chem. Soc.*, **93**, 3410 (1971); cf. *Org. Reaction Mech.*, **1970**, 110.
[365] R. B. Warnebolt and L. Weiler, *Tetrahedron Letters*, **1971**, 3413.
[366] J. Y. Satoh and T. T. Takahashi, *Chem. Comm.*, **1970**, 1714.
[367] D. J. Pasto and J. P. Sevenair, *J. Am. Chem. Soc.*, **93**, 711 (1971).
[368] J. P. Bégué and M. Charpentier-Morize, *Angew. Chem. Int. Ed.*, **10**, 327 (1971).
[369] J. P. Bégué, M. Charpentier-Morize, and C. Pardo, *Tetrahedron Letters*, **1971**, 4737.

Alkyl lithium cuprates react with α-bromo-ketones to give good yields of displacement products if the bromine is tertiary.[370]

The reactions of $Ph_2CBr-COAr$ and $PhArCBr-COPh$ have been studied.[371,372]

The kinetics of the Perkow reaction of phenacyl chlorides with triethyl phosphite have been measured,[373] as have those of the reaction of phenacyl bromide with secondary amines.[374]

The reaction of erythro-α,β-dibromobutyrophenone with azide ion in DMF gives α-azidocrotonophenone. Other acyl dibromides react similarly.[375]

Displacement of halide in α-chloro-sulphones has been examined.[376]

S_N2 Processes and Other Reactions

Reviews have appeared on nucleophilic substitution by elimination–addition,[377] and on the hard–soft acid–base principle.[378]

A number of theoretical treatments of the S_N2 process have been carried out. Lowe[379] has pointed out that, for the reaction of fluoride ion on methyl fluoride, eleven molecular orbitals are involved, and of these the two most important are the highest unoccupied and the eighth. Hence calculations that neglect high-level empty MO's are likely to be seriously in error. His treatment shows that approach of a charged base causes charge polarization in the substrate that is bond-weakening (to the leaving group) only if attack is along the C_3 axis. Neutral bases must approach much more closely to carbon to cause equivalent bond-weakening.[379] Fukui and his co-workers have reported that an *ab initio* treatment of the chloride–methyl chloride reaction shows that rear-side approach is more favourable than front-side.[380] Extended Hückel calculations on the hydride-ion–ethane reaction suggest a positive charge on carbon of 0.35 unit, and a 0.7 unit charge on each hydrogen at the transition state. The geometry is such that the $C-C-H^{\delta-}$ angle is 96°.[331] A CNDO/2 treatment for the anions $CH_3F_2^-$, CH_3FOH^- and CH_3FCN^- shows that the trigonal-bipyramidal structure is a local minimum and therefore should be regarded as an intermediate, and that for such a structure apical entry and apical loss are favoured (cf. displacements at phosphorus). Gillespie and Ugi[381] also considered the possibility of attack at a face of the tetrahedral carbon that was not opposite to the leaving group [reaction (33)]. Such an intermediate (**137**) can only collapse by apical loss and hence returns to starting material unless pseudorotation (or turnstile rotation)[225] converts it into (**138**) which then collapses to product with *retention* of configuration. They suggest that this might occur in S_N2 reactions at cyclobutyl-carbon where rear-side

---

[370] J. E. Dubois, C. Lion, and C. Moulineau, *Tetrahedron Letters*, **1971**, 177.
[371] V. S. Karavan, L. A. Timofeeva, and T. I. Temnikova, *Zhur. Org. Khim.*, **6**, 958 (1970); *J. Org. Chem. USSR*, **6**, 962 (1970).
[372] V. S. Karavan, L. A. Timofeeva, Y. A. Patrushkin, and T. I. Temnikova, *Zhur. Org. Khim.*, **6**, 742 (1970); *J. Org. Chem. USSR*, **6**, 746 (1970).
[373] A. Arcoria and S. Fisichella, *Tetrahedron Letters*, **1971**, 3347.
[374] L. M. Litvinenko, L. A. Perel'man, A. F. Popov, and L. I. Voroshilova, *Zhur. Org. Khim.*, **6**, 2090 (1970); *J. Org. Chem. USSR*, **6**, 2096 (1970).
[375] A. Hassner, G. L'abbé, and M. J. Miller, *J. Am. Chem. Soc.*, **93**, 981 (1971).
[376] K. Ogura and G. Tsuchihashi, *Chem. Comm.*, **1970**, 1689.
[377] W. Pritzkow, *Wiss. Z. Tech. Hochsch. Chem. "Carl Schorlemmer" Leuna-Merseburg*, **12**, 213 (1970); *Chem. Abs.*, **74**, 111124 (1971).
[378] P. J. Majewski, *Wiad. Chem.*, **24**, 841 (1970); *Chem. Abs.*, **74**, 87001 (1971).
[379] J. P. Lowe, *J. Am. Chem. Soc.*, **93**, 301 (1971).
[380] H. Fujimoto, S. Yamabe, and K. Fukui, *Tetrahedron Letters*, **1971**, 439, 443.
[381] P. D. Gillespie and I. Ugi, *Angew. Chem. Int. Ed.*, **10**, 503 (1971).

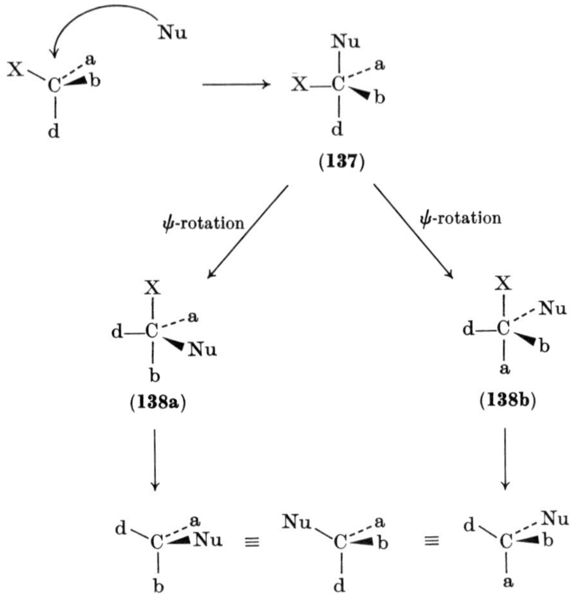

$$\ldots (33)$$

$$(R^1 = R^2 = H, R^3, R^4 = -(CH_2)_4-; R^1, R^2 = -(CH_2)_4-, R^3 = R^4 = H)$$

$$\ldots (34)$$

attack is hindered. Their calculations account for the acceleration of S_N2 processes by carbonyl (and similar) groups, as these would stabilize the trigonal-bipyramidal intermediate through occupying an equatorial position.[381]

The reduction of (**139**) by lithium aluminium hydride [reaction (34)], may be an example of the process suggested by Gillespie and Ugi. The Japanese authors, however, propose a cyclic four-centre mechanism.[382]

Firestone[383] has continued his propaganda on behalf of the Linnett, double-quartet theory and has applied it to the S_N2 transition state. He has quantitized his L-strain concept to allow a correct prediction of the order of alkyl group reactivity, the effect of α-unsaturation and the α-heteroatom effect. He has also discussed its application to the problem of kinetic and thermodynamic acidity and to ions such as HF_2^- and trihalide anion, which he describes as "frozen" S_N2 transition states.

Eschenmoser and his co-workers[384] have described intramolecular S_N2 reactions as exocyclic when they have the geometry of (**140**) and endocyclic when like (**141**). Careful crossing experiments have shown that the transformation (**142**) → (**143**) is inter-

[382] H. Yamanaka, T. Yagi, K. Teramura, and T. Ando, *Chem. Comm.*, **1971**, 380.
[383] R. A. Firestone, *J. Org. Chem.*, **36**, 702 (1971); cf. *Org. Reaction Mech.*, **1969**, 352, 372.
[384] L. Tenud, S. Farooq, J. Seibl, and A. Eschenmoser, *Helv. Chim. Acta*, **53**, 2059 (1970).

molecular. Another example, with methyl-transfer to sulphur, together with the above one, leads them to conclude that endocyclic S_N2 is unfavourable, and hence that the nucleophile must approach from the direction opposite to the leaving group, i.e. it must place both entering and leaving groups apical in the trigonal bipyramid. This conclusion is the same as that for nucleophilic displacement at sulphur reported last year.[385]

The change in the enthalpy of solution, between methanol and DMF, has been used as a probe for transition-state differences. For S_N2 reactions, this indicates that increasing basicity in the nucleophile moves the transition state closer to reactants. The effect of electron-withdrawing groups on the central carbon fits well with the theoretical treatment of Harris and Kurz.[386]

Ion-cyclotron resonance spectroscopy has allowed a study of gas-phase ion-molecule reactions. From a study of molecule–cation reactions (e.g. $HCl + MeFH^+$, $H_2O + EtClH^+$), methyl cation affinities have been evaluated for a number of nucleophiles, and this, together with the proton affinities, has enabled the relative displacing abilities to be predicted for the general reaction. This order is similar to the normal nucleophilicity order but does contain some surprises. For example, CO is more nucleophilic than HI, water falls between this and HBr, and molecular nitrogen will displace HF.[387] Anion–molecule reactions have received theoretical as well as experimental attention. The ratio of experimental to theoretical rates has been used as a measure of reaction probability. These probabilities fall in the range 0.2–0.32 for charge-localized anions but are less than 0.03 for charge-delocalized anions (e.g. $PhCH_2^-$). Solvated alkoxides, $RO^- \cdot HOR$, react about 3 times less rapidly than unsolvated ones.[388]

The nucleophilicities of ten anions towards n-propyl toluene-p-sulphonate have been determined in DMSO. The halogens fall in the opposite order to that in water, fluoride being 10^3 times more reactive than iodide.[389] The same halide order, Cl, Br, I, is found from the pyrolysis of molten tetra-n-pentylammonium halides.[390] The reactivity of imide and sulphonamide anions towards methyl iodide correlates with basicity, with the exception of benzenesulphonamide.[391] The nucleophilicity of N-halogenosulphonamide anions is comparable to that of azide.[392] Relative nucleophilicities have been determined for eight α-substituted phenylacetonitrile enolate anions in liquid ammonia.[393]

---

[385] cf. *Org. Reaction Mech.*, **1970**, 102.
[386] P. Haberfield, *J. Am. Chem. Soc.*, **93**, 2091 (1971); cf. *Org. Reaction Mech.*, **1970**, 112, ref. 376.
[387] D. Holtz, J. L. Beauchamp, and S. D. Woodgate, *J. Am. Chem. Soc.*, **92**, 7484 (1970).
[388] D. K. Bohme and L. B. Young, *J. Am. Chem. Soc.*, **92**, 7354 (1970).
[389] R. Fuchs and K. Mahendran, *J. Org. Chem.*, **36**, 730 (1971).
[390] J. E. Gordon and P. Varughese, *Chem. Comm.*, **1971**, 1160.
[391] J. F. Bunnett and J. H. Beale, *J. Org. Chem.*, **36**, 1659 (1971).
[392] F. E. Hardy, *J. Chem. Soc.* (B), **1971**, 1899.
[393] H. A. Smith, R. L. Bissell, W. G. Kenyon, J. W. MacClarence, and C. R. Hauser, *J. Org. Chem.*, **36**, 2132 (1971).

Shiner and his colleagues[394] have described the preparation and reactivity of 2,2,2-trifluoroethanesulphonates. These compounds are intermediate in reactivity between toluene-*p*- and trifluoromethane-sulphonates. They have shown that, for LIM reactions, a linear correlation of rate against σ_m is obtained for the substituent on sulphur in the sulphonate. Cobalt(I) acts as leaving group in the reaction of cobalt alkyls with rhodium.[395] The reaction of *threo*-t-BuCHDCHDFe(CO)$_2$cyclopentadiene with bromine gives *erythro*-t-BuCHDCHDBr. Reaction with mercuric chloride gives the alkyl-mercurichloride with retention.[396] The kinetics of methyl-transfer from aryldimethylsulphonium salts to a number of nucleophiles in water and acetonitrile has been studied.[397]

Neopentyl toluene-*p*-sulphonate undergoes direct displacement with ethoxide and hydroxide.[398] The rates of displacement, by iodide ion in acetone, have been measured for hindered toluene-*p*-sulphonates in the tricyclo[5.2.1.0^{2,6}]decane system.[399] The kinetics of the reaction of sodium *p*-nitrophenoxide with ten aryl-substituted α-bromoacetanilides have been studied.[400] The reactions of 2,2-dinitroalkyl toluene-*p*-sulphonates with nucleophiles have been examined.[401] The thermodynamic parameters for the reaction of chloromethylated polystyrene with 2-aminobutanol are nearly identical to those for benzyl chloride.[402] The kinetics of the reactions of benzyl iodide with sodium alkoxides have been measured.[403] Thermodynamic parameters have been determined for bromide exchange in α-D-glycosyl bromides.[404] Halogen exchange (Cl, Br or I) has been studied for cinnamyl chloride and bromide.[405] The solvolysis of sulphonium ylids involves protonation followed by S_N2 reaction.[406] The mechanism of the isomerization of dimethyl sulphite to methyl methanesulphonate has been reinvestigated.[407]

The ratio of $E2$ to S_N2 in the reaction of sodium and potassium cyanide with α-ethylbenzyl toluene-*p*-sulphonate in DMF, DMSO, 1-methylpyrrolidone and hexamethylphosphorotriamide has been investigated. It is observed that the substitution to elimination ratio increases at low salt concentration. This has been interpreted as the free anions giving more S_N2 reaction than do the ion-pairs.[408]

The paper by de la Mare and Vernon[409] refuting Bordwell's attack on the S_N2' process is convincing, which Bordwell's attack never was. An investigation of the chloride exchange and rearrangement of crotyl and α-methylallyl chlorides in acetonitrile has shown that the S_N2/S_N2' ratio is 6×10^4 and 2×10^2, respectively, for the primary and secondary positions.[410] The lithium aluminium hydride reduction of β-phenylallylic alcohols takes place by an S_N2' process.[411] Similar reduction of allylic halides in the

---

[394] R. K. Crossland, W. E. Wells, and V. J. Shiner, *J. Am. Chem. Soc.*, **93**, 4217 (1971).
[395] D. Dodd and M. D. Johnson, *Chem. Comm.*, **1971**, 1371.
[396] G. M. Whitesides and D. J. Boschetto, *J. Am. Chem. Soc.*, **93**, 1529 (1971).
[397] J. K. Coward and W. D. Sweet, *J. Org. Chem.*, **36**, 2337 (1971).
[398] Y. Okamoto and T. Yano, *Tetrahedron Letters*, **1971**, 4285.
[399] I. Rothberg and R. V. Russo, *J. Chem. Soc.* (B), **1971**, 1214.
[400] H. W. Johnson and Y. Iwata, *J. Org. Chem.*, **36**, 1921 (1971).
[401] H. G. Adolph, *J. Org. Chem.*, **36**, 806 (1971).
[402] H. Kawabe and M. Yanagita, *Bull. Chem. Soc. Japan*, **44**, 896 (1971).
[403] G. Murgulescu and D. Oancea, *Rev. Roum. Chim.*, **15**, 1635 (1970).
[404] M. J. Duffy, G. Pass, and G. O. Phillips, *J. Chem. Soc.* (B), **1971**, 785.
[405] B.-S. Lee and I. Lee, *Wonjąryok Hakhoeji*, **1**, 87 (1969); *Chem. Abs.*, **74**, 87112 (1971).
[406] W. Ando, T. Toyama, and T. Migita, *Chem. Comm.*, **1971**, 756.
[407] A. J. W. Brook and R. K. Robertson, *J. Chem. Soc.* (B), **1971**, 1161.
[408] A. Loupy and J. Seyden-Penne, *Bull. Soc. Chim. France*, **1971**, 2306.
[409] P. B. D. de la Mare and C. A. Vernon, *J. Chem. Soc.* (B), **1971**, 1699.
[410] J. A. Hemmington and B. D. England, *J. Chem. Soc.* (B), **1971**, 1347.
[411] W. T. Borden and M. Scott, *Chem. Comm.*, **1971**, 387.

bicyclo[3.2.1]oct-2-enyl system has been shown to occur with synfacial stereochemistry.[412] 2-[(α-Substituted amino)benzyl]acrylophenones react with morpholine and t-butylamine by an S_N2' process.[413]

The exchange of 1,1,2,3,3-pentachloropropane with aluminium radiotrichloride takes place exclusively at the primary positions.[414]

Rate differences in the ethanolyses of *cis*- and *trans*-5-substituted cyclo-octyl toluene-*p*-sulphonates are due to steric effects and not to participation.[415]

The reactions of 4-t-butyl-1-methylcyclohexanols with alkyl and aryl nitriles (Ritter reaction), cyanide and chloride ions, carbon monoxide and aromatic ethers (catalysed by sulphuric acid) have been shown to be S_N1 in nature.[416]

Tertiary amines are cleaved by refluxing acetic anhydride to give secondary amides and acetates if one of the groups gives a sufficiently stable cation.[417] A correction has been made to the rates of ether cleavage by acetyl perchlorate previously reported.[418] The alcoholysis of methallyl and isobutyl arenesulphonates has been studied in six alcohols.[419] Rudakov has reported extensively on the work of his group on the metal ion-catalysed hydrolysis of alkyl halides.[420] The kinetics of the reaction of $CH_3N(CH_2CH_2Cl)_2 \cdot HCl$ with thiols have been measured.[421]

The hydrolysis of chloroacetic acid and its sodium salt have received much attention. Salt effects have been studied in aqueous solution,[422] kinetics measured in methanol–water mixtures,[423] and rates of halide displacement on its esters examined.[424]

The effect of solvent composition on the H^--dependence of the rate of alkaline hydrolysis of chloroform in ethanol–water has been studied.[425]

Quaternization continues to receive a great deal of attention. The thermodynamic parameters for methylation of the *cis*- and *trans*-4-t-butyl-*N*,*N*-dimethylcyclohexylamines are such as to suggest that both react in conventional chair forms and not through a twist-boat.[426]

Further work on the stereochemistry of tropane quaternization has been reported.[427] The rates of reaction of *N*,*N*-dimethylanilines with methyl iodide have been measured.[428]

---

[412] C. W. Jefford, A. Sweeney, D. T. Hill, and F. Delay, *Helv. Chim. Acta*, **54**, 1691 (1971).
[413] N. H. Cromwell, K. Matsumoto, and A. D. George, *J. Org. Chem.*, **36**, 272 (1971).
[414] H. Khalaf, *Tetrahedron Letters*, **1971**, 4239.
[415] N. L. Allinger, C. L. Neumann, and H. Sugiyama, *J. Org. Chem.*, **36**, 1360 (1971).
[416] P. J. Beeby and S. Sternhell, *Austral. J. Chem.*, **24**, 809 (1971).
[417] P. Mariella and K. H. Brown, *Can. J. Chem.*, **49**, 3349 (1971).
[418] A.-M. Avedikian, S. Coffi Nketsia, and A. Kergomard, *Bull. Soc. Chim. France*, **1971**, 435.
[419] R. V. Sendega, M. K. Mikhalevech, and R. V. Vizgert, *Organic Reactivity (Tartu)*, **7**, 512, 635 (1970).
[420] I. V. Kozhevnikov and E. S. Rudakov, *Organic Reactivity (Tartu)*, **7**, 761 (1970); E. S. Rudakov and I. V. Kozhevnikov, *Organic Reactivity (Tartu)*, **7**, 770, 771 (1970); E. S. Rudakov, *Organic Reactivity (Tartu)*, **7**, 779 (1970); E. S. Rudakov, V. V. Zamashchikov, E. G. Gushchina, V. P. Tret'yakov, and V. D. Belayev, *Organic Reactivity (Tartu)*, **7**, 788 (1970); E. S. Rudakov, V. V. Zamashchikov, and R. I. Rudakova, *Organic Reactivity (Tartu)*, **7**, 804 (1970).
[421] H. M. Rauen and H. Schriewer, *Arzneim.-Forsch.*, **21**, 693 (1971); *Chem. Abs.*, **85**, 62783 (1971).
[422] E. V. Sergeev and S. I. Rad'ko, *Kinet. Katal.*, **12**, 555, 877 (1971).
[423] W. Köhler and L. Neuheiser, *Z. Phys. Chem. (Leipzig)*, **245**, 272 (1971).
[424] J. Ashworth and B. A. W. Coller, *Trans. Faraday Soc.*, **67**, 1077 (1971).
[425] J. Konecny, *Z. Phys. Chem. (Frankfurt)*, **73**, 157 (1970).
[426] N. L. Allinger and J. C. Graham, *J. Org. Chem.*, **36**, 1688 (1971).
[427] G. Fodor, R. V. Chastain, D. Frehel, M. J. Cooper, N. Mandava, and E. L. Gooden, *J. Am. Chem. Soc.*, **93**, 403 (1971).
[428] V. Baliah and V. M. Kanagasabapathy, *Indian J. Chem.*, **9**, 182 (1971); *Chem. Abs.*, **74**, 140690 (1971).

Quaternization studies on pyridine derivatives include measurement of activation volume for alkylation,[429] activation energies for reaction with α-chloro-ethers,[430] reaction with α,ω-dibromides,[431] and reaction of pyridine with n-dodecyl bromide.[432] Pyridine, quinoline and isoquinoline have been methylated in benzene and chloroform and the rates measured.[433] Methylation (and demethylation of the products) has been studied for substituted tetrahydroisoquinolines.[434] Activation parameters have been measured for phenanthridine methylation in nitrobenzene,[435] and N-substituted α- and β-perhydroacridines have been alkylated in acetone.[436] Methylation of cinnolines has been studied.[437]

Rates of quaternization have been compared with pK_a values for 3- and 6-substituted N-methyl-cis-hexahydrocarbazoles.[438] Ethylation of 5,6-disubstituted 1-ethylbenzimidazole has been examined in acetone solution.[439]

The reaction of diethylamine with ω-bromoalkyl ethers of 1,4-benzodioxans and phenols has been studied,[440] and its reaction with alkyl halides in ether, and that of ethanol with sec-butyl bromide, have been examined.[441]

The opening of epoxides continues to be a subject for extensive study. It has been suggested that the rearrangement of bicyclo[2.2.1]heptadiene monoepoxide is a sigmatropic one.[442] Perhydrolysis of epoxides affords vic-hydroperoxy-alcohols.[443] Other epoxide reactions have been with organometallic reagents[444] and with nucleophiles;[445] acid-catalysed opening has received the most attention.[446]

---

[429] H.-D. Brauer and H. Kelm, *Z. Phys. Chem. (Frankfurt)*, **76**, 98 (1971).

[430] J. Jonas, M. Kratochvíl, J. Mikula, and J. Pichler, *Coll. Czech. Chem. Comm.*, **36**, 202 (1971).

[431] J. G. Leipoldt, L. D. C. Bok, S. S. Basson, and G. F. S. Wessels, *Z. Phys. Chem. (Frankfurt)*, **72**, 116 (1971).

[432] K. Murai and C. Kimura, *Asahi Garasu Kogyo Gijutsu Shoreikai Kenkyu Hokoku*, **17**, 117 (1970); *Chem. Abs.*, **75**, 87799 (1971).

[433] J. M. Cachaza and M. A. Herráez, *Ann. Quím.*, **67**, 11 (1971); *Chem. Abs.*, **74**, 140470 (1971).

[434] J. Volford, G. Toth, G. Bernáth, and J. Kóbor, *Tetrahedron Letters*, **1971**, 4019.

[435] B. R. T. Keene and G. L. Turner, *Tetrahedron*, **27**, 3405 (1971).

[436] N. Bărbulescu and F. Potmischil, *Annalen*, **752**, 22 (1971).

[437] M. H. Palmer and P. S. McIntyre, *Tetrahedron*, **27**, 2913 (1971).

[438] A. Smith and J. H. P. Utley, *J. Chem. Soc.* (B), **1971**, 1201.

[439] T. N. Chegolya and A. F. Pozharskii, *Sb. Nauch. Tr. Vladimir. Politekh. Inst.*, **1969**, No. 7, 138; *Chem. Abs.*, **74**, 22366 (1971).

[440] V. Dauksas, I. Dembinskiene, and J. Smitaile, *Let. Tsk. Ankst. Mokyklu Mokslo Darb., Chem. Chem. Technol.*, **11**, 205 (1970); *Chem. Abs.*, **75**, 87943 (1971).

[441] G. Geiseler and H. Tilgner, *Wiss. Z. Tech. Hochsch. Chem. "Carl Schorlemmer" Leuna-Merseburg*, **12**, 236 (1970); G. Geiseler, J. Hoffmann, and H. Teske, *Wiss. Z. Tech. Hochsch. Chem. "Carl Schorlemmer" Leuna-Merseburg*, **12**, 238 (1970); *Chem. Abs.*, **74**, 124468, 124469 (1971).

[442] R. Grigg and G. Shelton, *Chem. Comm.*, **1971**, 1247.

[443] W. Adam and A. Rios, *Chem. Comm.*, **1971**, 822.

[444] R. Grigg, R. Hayes, and A. Sweeney, *Chem. Comm.*, **1971**, 1248; J. Staroscik and B. Rickborn, *J. Am. Chem. Soc.*, **93**, 3046 (1971); D. M. Wieland and C. R. Johnson, *J. Am. Chem. Soc.*, **93**, 3047 (1971); J. L. Namy, D. Abenhaim, and G. Boireau, *Bull. Soc. Chim. France*, **1971**, 2943.

[445] W. N. Marmer and D. Swern, *J. Am. Chem. Soc.*, **93**, 2719 (1971); B. Rickborn and R. M. Gerkin, *J. Am. Chem. Soc.*, **93**, 1693 (1971); M. A. Khuddus and D. Swern, *Tetrahedron Letters*, **1971**, 411; Y. V. Lykov and V. F. Shvets, *Tr. Mosk. Khim.-Tekhnol. Inst.*, **1970**, No. 66, 33; *Chem. Abs.*, **75**, 87776 (1971); Y. Aleksandrov, B. I. Taranin, and V. A. Shushunov, *Kinet. Katal.*, **12**, 883 (1971).

[446] J. Biggs, N. B. Chapman, A. F. Finch, and V. Wray, *J. Chem. Soc.* (B), **1971**, 55; J. Biggs, N. B. Chapman, and V. Wray, *J. Chem. Soc.* (B), **1971**, 63, 66, 71; S. W. Pelletier and D. L. Herald, *Chem. Comm.*, **1971**, 10; J. Grimaldi and M. Bertrand, *Bull. Soc. Chim. France*, **1971**, 973; S. A. Reines, J. R. Griffith, and J. G. O'Rear, *J. Org. Chem.*, **36**, 1209 (1971); V. F. Shvets and Y. V. Lykov, *Kinet. Katal.*, **12**, 347 (1971); N. Ikekawa, M. Morisaki, H. Ohtaka, and Y. Chiyoda, *Chem.*

The ring-opening of aziridines has also been studied.[447]

An α-effect is observed in the reaction of the Malachite Green cation with hydrazines and hydroxylamines.[448]

The rearrangement of α-bromoboranes has been reported. The half-life for rearrangement depends on the nucleophilicity of the reagent attacking boron.[449]

β-Alkylamino-sulphone hydrochlorides and methiodides react with thiophenols by an elimination–addition mechanism.[450]

Further work on the mechanism of the reaction of alcohols with triphenylphosphine

$$ROH + Ph_3P + CCl_4 \rightarrow RCl + Ph_3PO \qquad \ldots(35)$$

and carbon tetrahalides [reaction (35)] has been carried out. Applied to active [1-$^2H_1$]-neopentyl alcohol it gives [1-$^2H_1$]neopentyl chloride of $[\alpha]_D = +0.13°$, whereas lithium chloride in DMF reacts with neopentyl toluene-p-sulphonate to give chloride of $[\alpha]_D = +0.085°$. Normal S_N2 reaction of the neopentyl toluene-p-sulphonate is observed with sodium acetate, affording neopentyl acetate, which on hydrolysis gives neopentanol of $85 \pm 17\%$ inverted configuration. The reaction (35) has therefore some S_N2 characteristics, but external nucleophile does not compete for any intermediate phosphorane, and the rate for pentan-1-ol is only 14 times greater than for neopentanol. 2-exo-Bicyclo-[3.2.0]heptanol gives a mixture of inverted chloride and 7-chloronorbornane, but this last is the sole product formed (with inversion) from 7-norbornanol. The postulated mechanism involves the formation of Ph$_3$P(Cl)OR with subsequent four-centre rearrangement.[451] Further work on this is awaited with interest.

Comm., **1971**, 1498; J. R. Hanson, *Chem. Comm.*, **1971**, 1343; I. G. Guest, J. G. Ll. Jones, and B. A. Marples, *Tetrahedron Letters*, **1971**, 1979; R. D. H. Murray, R. W. Mills, and J. M. Young, *Tetrahedron Letters*, **1971**, 2393; E. Staude and U. Staude, *Z. Phys. Chem. (Frankfurt)*, **74**, 1 (1971); P. L. Barili, G. Bellucci, B. Macchia, F. Macchia, and G. Parmigiani, *Gazzetta*, **101**, 300 (1971); T. Trinka and M. Černý, *Coll. Czech. Chem. Comm.*, **36**, 2216 (1971); M. Prystaš, H. Gustafsson and F. Šorm, *Coll. Czech. Chem. Comm.*, **36**, 1487 (1971).

[447] V. R. Gaertner, *J. Heterocyclic Chem.*, **8**, 177 (1971); S. Fujita, T. Hiyama, and H. Nozaki, *Bull. Chem. Soc. Japan*, **43**, 3239 (1970).
[448] J. E. Dixon and T. C. Bruice, *J. Am. Chem. Soc.*, **93**, 3248 (1971).
[449] H. C. Brown and Y. Yamamoto, *J. Am. Chem. Soc.*, **93**, 2796 (1971).
[450] A. S. Angeloni, P. De Maria, A. Fini, and G. Salvadori, *Tetrahedron*, **26**, 5601 (1970).
[451] R. G. Weiss and E. I. Snyder, *J. Org. Chem.*, **35**, 1627 (1970); **36**, 403 (1971).

CHAPTER 3

Carbanions and Electrophilic Aliphatic Substitution

D. C. AYRES

Department of Chemistry, Westfield College, University of London

Carbanion Structure	111
Reactions of Carbanions	113
Proton Transfer, Hydrogen Isotope Exchange and Related Reactions	119
Electrophilic Reactions of Hydrocarbons	125
Organometallics: Group Ia, IIa, III	127
Organometallics: Other Elements	130
Miscellaneous Reactions	133

Carbanion Structure

Transannular interaction involving the allylic anion was evident[1] in the enhancement of base-catalysed exchange in bicyclo[3.2.1]octa-2,6-diene but the cyclopropane system of the analogue (**1**) was ineffective; nor was any appreciable effect detected at an adjacent centre in a study[2] of relative carbon acidity by deuteration in various media.

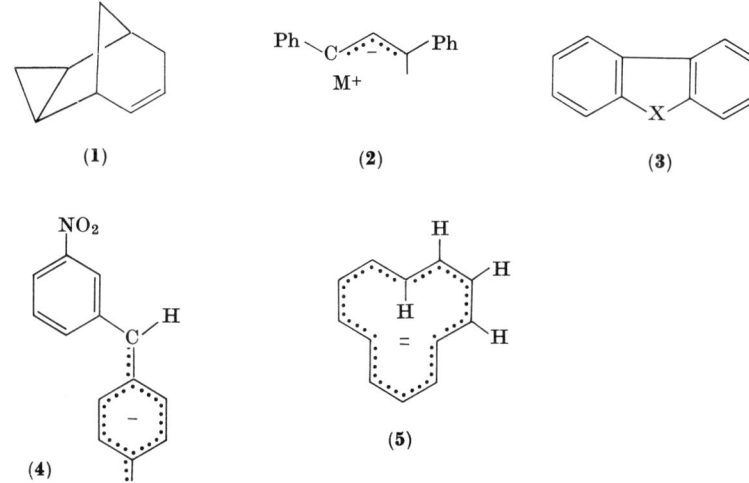

An investigation[3] of the electronic spectra of the salts of 1,3-diphenylbut-1-ene (**2**) evealed maxima due to two kinds of solvated ion-pairs in addition to that arising from he solvent-separated pair. Spectral changes due to solvation processes were independent

---

[1] P. K. Freeman and T. A. Hardy, *Tetrahedron Letters*, **1971**, 3939.
[2] M. J. Perkins and P. Ward, *Chem. Comm.*, **1971**, 1134.
[3] J. W. Burley and R. N. Young, *J. Chem. Soc.* (B), **1971**, 1018.

of concentration and the power of 15 solvents was determined from the appearance, or non-appearance, of the solvent-separated absorption and from the shifts induced by the counter-ions (M = Na, K or Cs). The relation between the chemical shifts of the carbazole nitranion [(3), X = N$^-$] and the carbanion from 4,5-methylenephenanthrene showed that the nitrogen atom of the former was strongly associated with the counter-ion, whilst the carbanionic charge of the latter was extensively delocalized;[4] the separation of ion-pairs was detected by NMR spectroscopy, which also showed[5] control of delocalization in nitrophenylmethide ions (4) by nitro-group conjugation. In the latter, line-broadening in less basic media indicated radical-ion formation and this was confirmed in the visible range. Pulse radiolysis work[6] established that reduction of tetranitromethane by electron capture led to initial production of the nitroform anion $^-$C(NO$_2$)$_3$ in solvents of lower dielectric constant than water where nitroform dissociation is limited.

The two-stage addition of electrons during reduction of [12]-21-annulene by lithium–THF was demonstrated by polarography and ESR spectra. The ring current and charge in the dianion (5) shifted the 3 inner PMR signals to high field (τ 14.6) and the peripheral protons, now non-equivalent, were detectably deshielded.[7] Full details on the formation of dianions from azocines have appeared;[8] PMR evidence of aromatic character was presented and protonation without isomerization was demonstrated. The inductive and steric effects of ring substituents pointed to C-4 as the centre for alkylation and the first protonation step. The major alkylation product was 2,3-pyridocyclobutene, and the following mechanism was suggested for its formation from benzophenone.

The PMR spectra of α-picolyl-lithium[9] and of 1,1,4,4-tetraphenylbutadiene dianion[10] in THF solution have been recorded. Spectroscopic measurements of the tetraphenylethylene dianion (6) have also been made[11,12] but they do not lead to a distinction between twisted and planar structures; other work [13,13a] on the PMR spectra of carbanions assigns a twist angle of 30° to (6) on the assumption that it is sp^2 hybridized.

The postulation of a favoured equatorial lobe in a carbanion adjacent to sulphur has been illustrated,[14] since *cis*-dimethyldithians (7) on lithiation exchange *stereospecifically* at H$_e$. Equatorial alkylation was observed and the conformational preference for

---

[4] R. H. Cox, *Can. J. Chem.*, **49**, 1377 (1971).
[5] C. A. Fyfe, A. Albagli, and R. Stewart, *Can. J. Chem.*, **48**, 3721 (1970).
[6] K.-D. Asmus, S. A. Chaudhri, N. B. Nazhat, and W. F. Schmidt, *Trans. Faraday Soc.*, **67**, 2607 (1971).
[7] J. F. M. Oth and G. Schröder, *J. Chem. Soc.* (B), **1971**, 904.
[8] L. A. Paquette, J. F. Hansen, and T. Kakihana, *J. Am. Chem. Soc.* **93**, 168, 174 (1971).
[9] K. Konishi, K. Takahashi, and R. Asami, *Bull. Chem. Soc. Japan*, **44**, 2281 (1971).
[10] M. Ushio, M. Takaki, K. Takahashi, and R. Asami, *Bull. Chem. Soc. Japan*, **44**, 2559 (1971).
[11] D. H. Eargle, *J. Am. Chem. Soc.*, **93**, 3859 (1971).
[12] J. F. Garst, *J. Am. Chem. Soc.*, **93**, 6312 (1971).
[13] K. Takahashi, M. Takaki, and R. Asami, *J. Phys. Chem.*, **75**, 1062 (1971).
[13a] K. Takahashi, Y. Inoue, and R. Asami, *Org. Mag. Res.*, **3**, 349 (1971).
[14] A. A. Harlmann and E. L. Eliel, *J. Am. Chem. Soc.*, **93**, 2572 (1971).

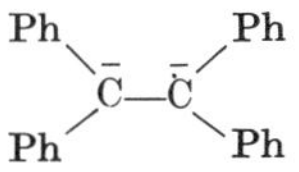

exchange was also found in epimeric 2-methyldithian. Despite the failure to detect ring currents in related compounds[15] the greatly enhanced acidity of cyclic thiopyran S,S-dioxides compared to that of their open-chain allylic analogues was ascribed[16] to aromatic character in anions of type (8) and the necessary trigonal configuration at α-carbon was justified.

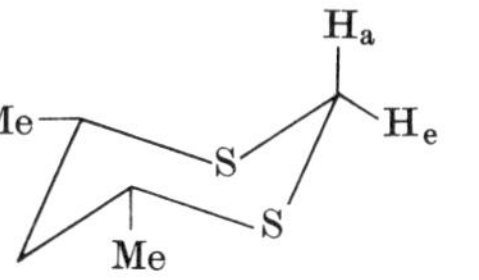

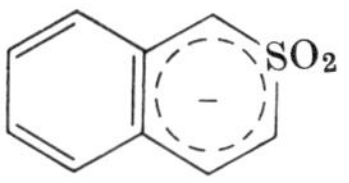

The ionization of alkylmagnesiumbromides has been studied[17] for hexamethylphosphorotriamide (HMPT) solutions; their behaviour contrasts with that of the chlorides which are largely undissociated. Calculated[18] dipole moments in alkyltin compounds $R_n SnCl_{4-n}$ were consistent with the relative electronegativities of the alkyl groups. There has been an *ab initio* calculation[19] of the electronic structure of cyclopropane, and it has been suggested[20] that anions are best evaluated comparatively with assumption of their molecular geometry.

Reactions of Carbanions

A comment on 1,2 shifts in carbanions has been published,[21] and electrocyclic reactions of cyclopropyl ions and cyclobutane have been studied[22] by the MINDO/2 method; calculations were also made[23] for 7-norbornyl ions and radicals.

Base-catalysed addition of dichloromethyl thioethers to olefins[24] and reactions of the methylsulphinyl anion with stilbenes[25] and unsaturated carbonyl compounds[26] have been investigated. In the general base-catalysed fission of arylsulphonylmethyl nitrates a concerted mechanism was preferred[27] to one involving a sulphonyl carbanion. Two-fold

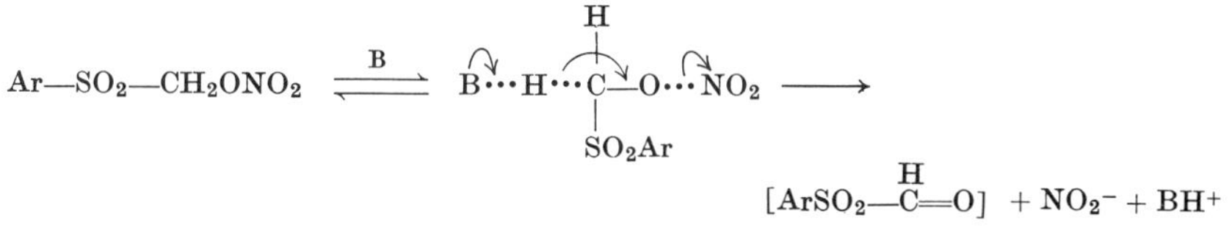

---

[15] *Org. Reaction Mech.*, **1970**, 121.
[16] S. Bradamante, S. Maiorana, A. Mangia, and G. Pagani, *J. Chem. Soc.* (B), **1971**, 74.
[17] J. Ducom and B. Denise, *J. Organomet. Chem.*, **26**, 305 (1971).
[18] R. Gupta and B. Majee, *J. Organomet. Chem.*, **33**, 169 (1971).
[19] H. Marsmann, J.-B. Robert, and J. R. van Wazer, *Tetrahedron*, **27**, 4377 (1971).
[20] P. H. Owens and A. Streitwieser, *Tetrahedron*, **27**, 4471 (1971).
[21] R. W. Alder, *Tetrahedron Letters*, **1971**, 193 (1971).
[22] M. J. S. Dewar and S. Kirschner, *J. Am. Chem. Soc.*, **93**, 4290, 4291, 4292 (1971).
[23] M. J. S. Dewar and W. W. Schoeller, *Tetrahedron*, **27**, 4401 (1971).
[24] M. Makosza and E. Bialecka, *Tetrahedron Letters*, **1971**, 4517.
[25] B. G. James and G. Pattenden, *Chem. Comm.*, **1971**, 1015.
[26] G. A. Russell and R. L. Blankespoor, *Tetrahedron Letters*, **1971**, 4573.
[27] A. Bruggink, B. Zwanenburg, and J. B. F. N. Engberts, *Tetrahedron*, **27**, 4571 (1971).

alkylation of *vic*-dibromoalkanes is not observed on treatment with methyl lithio- or sodio-isobutyrate: the product from (±)- and *meso*-3,4-dibromohexane afforded only *trans*-hex-3-ene. This was explained[28] in terms of bromine abstraction and electron transfer through a bridged radical (9) which equilibrates to the markedly more stable (10) when addition of the second electron is relatively slow.

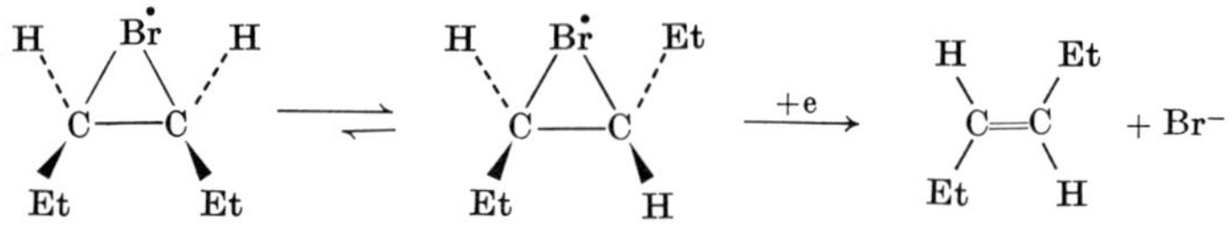

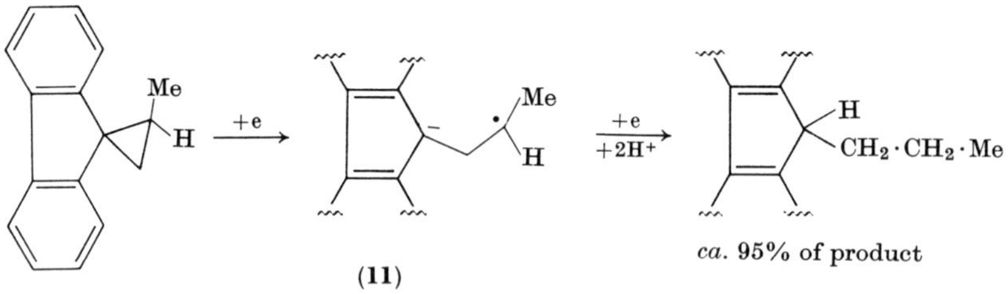

There remains uncertainty about the steps leading to the reduction of cyclopropanes by dissolving metals, but it has been shown[29] that when steric factors are minimal a mesomeric anion (11) is a preferred intermediate. The rate of reaction of methyl phenylacetylene, following electron transfer from sodium biphenylide, increased on addition of another carbon acid (RH) in a manner that was only consistent with the participation of a dianion in the rate-determining step:[30] in general, a low concentration of the very strong A^{2-} base formed by the exchange $2A^{\cdot -} \rightleftharpoons A^{2-} + A$ provides an easier route than direct protonation of $A^{\cdot -}$. The reaction of the two-electron addition product of *cis*-bicyclo[6.1.0]nona-2,4,6-triene in carbon tetrachloride leads to the addition of a dichloromethylene group;[31] in more polar solvents C-1–C-8 cleavage to (12) occurs and protonation afforded *all-cis*-cyclonona-1,3,6-triene, whilst 5-substituted derivatives were obtained on methylation. A radical-coupling mechanism is preferred[32] for the latter reaction and for the dimerization induced by an excess of oxygen. The synthesis of alkali-metal salts of another 10π-electron anion (13) from 2,3-dimethylpyrrole has been described.[33] Suprafacial [1,4] shifts of hydrogen and carbon did not occur in monoanions derived from but-1-enes [(14), R = benzyl or cyclopropyl], but double-bond participation led to a dianion (15) in which these shifts took place: the results are compatible[34] with an orthogonal model of the allylic systems which on rotation gave a planar 8π-electron dianion (16). The sequential metallation of 1,4-enynes by n-butyl-

---

[28] D. G. Korzan, F. Chen, and C. Ainsworth, *Chem. Comm.*, **1971**, 1053.
[29] H. M. Walborsky, M. S. Aronoff, and M. F. Schulman, *J. Org. Chem.*, **36**, 1037 (1971).
[30] G. Levin and M. Szwarc, *Chem. Comm.*, **1971**, 1029.
[31] M. Ogliaruso, *J. Am. Chem. Soc.*, **92**, 7490 (1970).
[32] T. I. Ito, F. C. Baldwin, and W. H. Okamura, *Chem. Comm.*, **1971**, 1317, 1440.
[33] H. Volz and R. Draese, *Tetrahedron Letters*, **1970**, 4917.
[34] R. M. Magid and S. E. Wilson, *Tetrahedron Letters*, **1971**, 19.

$$Ph \cdot C \equiv C \cdot Me^{2-}, 2Na^+ + Ph \cdot C \equiv C \cdot Me$$
$$\downarrow$$
$$Ph \cdot \underline{C} = C = CH_2, Na^+ + Ph\underline{C} = CH - CH_3, Na^+$$

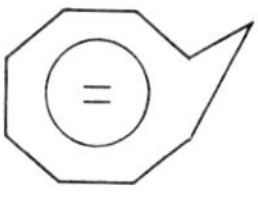

(12) 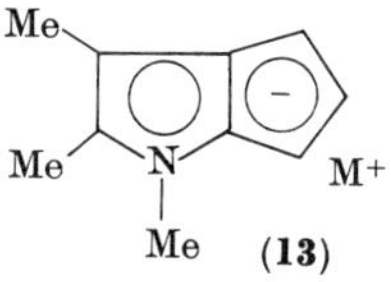   (13)

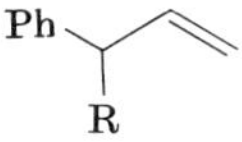

(14)

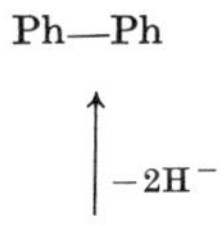

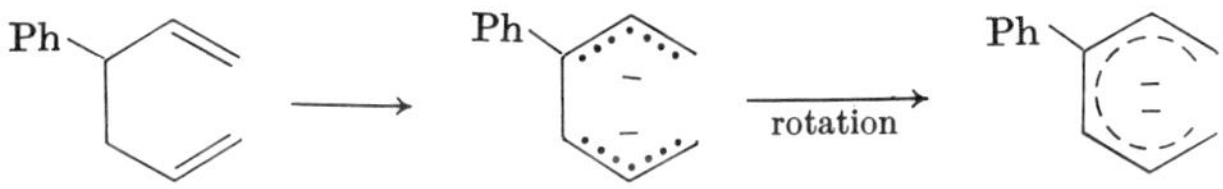

(15) → (rotation) → (16) Ph—Ph $-2H^-$

lithium–ether was followed by PMR and UV spectra; the products of protonation of the dianion (**17**) were determined[35] quantitatively and the rapid isomerization catalysed by methylsulphinyl carbanion afforded an allene (**18**) as a precursor to the vinylacetylene end product. Linolenyl ethers on metallation gave monoanions (**19**) which undergo[36] 1,8 hydrogen shifts and cyclize to a mixture of isomers that correspond to the rearrangement products of (**20**).

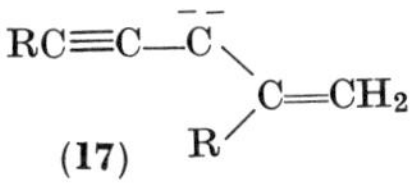

$$RC \equiv C - \overset{--}{C} \diagdown_{C=CH_2} \diagup R$$
(17)

$$R-CH=C=CH-CR=CH_2$$
(18)

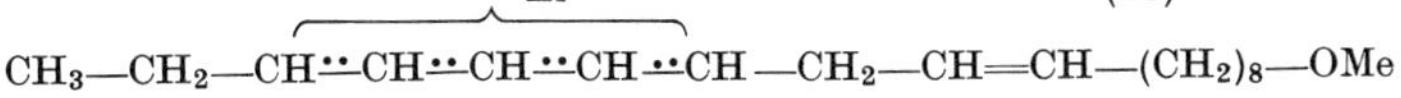

$$CH_3-CH_2-CH \cdots CH \cdots CH \cdots CH \cdots CH - CH_2 - CH = CH - (CH_2)_8 - OMe \quad -Li^+$$
(19)

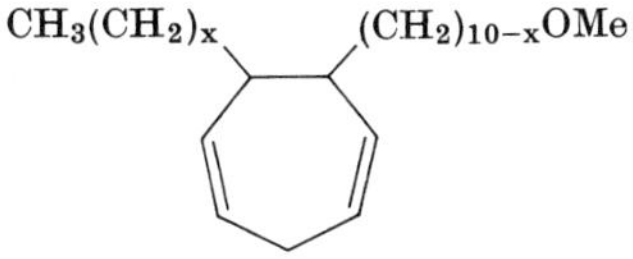

$$CH_3(CH_2)_x \qquad (CH_2)_{10-x}OMe$$
(20)

[35] J. Klein and S. Brenner, *Tetrahedron*, **26**, 5807 (1970).
[36] J. Klein and S. Glily, *Tetrahedron*, **27**, 3477 (1971).

Trapping techniques showed that, in the presence of sodium naphthalene, dibenzyl is formed from benzyl halides by a carbanion route,[37] and the nature of the carbanionoid rearrangement products of alkyl sulphides was established[38] by alkylation. The balance of C- and O-alkylation of esters in DMSO is temperature-dependent.[39] Solvent effects on the O-alkylation of alkali-metal acetoacetates have been investigated;[40] O-alkylation was dominant in an aprotic solvent (HMPT) but C-alkylation was more competitive in DMF.[41] The rates of alkylation of enolates of acetoacetates decrease in the order of their basicity;[42] $Me_3C \cdot CO \cdot CH_2 \cdot CO_2Et > Me_2CH \cdot CO \cdot CH_2 \cdot CO_2Et > CH_3 \cdot CO \cdot CH_2 \cdot CO_2Et$. CNDO/2 calculations[43] indicate that a methyl group may stabilize ions, relative to hydrogen, by electron withdrawal; in consequence it would follow that the most substituted enolate anion of an unsymmetrical ketone is the *most* stable. The calculations could not take account of the effects of the solvent and counter-ion. It has been observed[44] that the mixture of isomers produced on alkylation of allylic carbanions is controlled by the hardness of the leaving group.

$$\begin{array}{c} Ph \\ \\ Ph \end{array} C \cdots \overset{H}{C} \cdots CH_2 \quad \xrightarrow[\text{(b) MeOTs}]{\text{(a) MeI}} \quad \begin{array}{c} Ph \\ \\ Ph \end{array} C-CH=CH_2 \quad \text{yields (a) 64\%} \\ \text{(b) 96\%}$$

Carbon is a leaving group in the base-induced cleavage of 3,7-dimethyltricyclo-[3.3.0.0^{3,7}]octan-1-ol.[45] The relation between the nucleophilicity and basicity of 1,1-dinitrocarbanions in the Michael addition has been studied[46] and the trapping of carbanions by electron transfer to nitro compounds has been used[47] to measure the primary isotope effect for abstraction of protons from solvent molecules. There has been an investigation[48] of the effect of solvent composition on carboxylate-catalysed halogenation of nitroalkanes and it has been reported[49] that the usual second-order kinetics do not apply to the reaction between hypochlorite and nitroethane. In the Darzens reaction with chloroacetonitrile the slow step is carbanion formation followed by non-selective aldolization.[50] Synthetic and mechanistic features of radical-anion intermediates have been reviewed.[51]

An allylic carbanion rearrangement (21) is proposed for the aconitate-isomerase-catalysed interconversion of *cis*- and *trans*-aconitate; (*pro-S*)-hydrogen is activated at methylene-carbon atoms and intramolecular transfer was demonstrated.[52] It was

---

[37] J. W. Rakshys, *Tetrahedron Letters*, **1971**, 4745.
[38] J. F. Biellmann and J. B. Ducep, *Tetrahedron Letters*, **1971**, 33; cf. ref. 44.
[39] K. L. Shepard and J. I. Stevens, *Chem. Comm.*, **1971**, 951.
[40] A. L. Kurts, A. Macias, I. P. Beletskaya, and O. A. Reutov, *Dokl. Akad. Nauk SSSR*, **197**, 1088 (1971); *Chem. Abs.*, **75**, 5124 (1971).
[41] A. L. Kurts, A. Macias, I. P. Beletskaya, and O. A. Reutov, *Tetrahedron*, **27**, 4759 (1971).
[42] A. L. Kurts, P. I. Dem'yanov, A. Macias, I. P. Beletskaya and O. A. Reutov, *Tetrahedron*, **27**, 4769 (1971).
[43] J. R. Grunwell and J. F. Sebastian, *Tetrahedron*, **27**, 4387 (1971).
[44] W. S. Murphy, R. Boyce, and E. A. O'Riordan, *Tetrahedron Letters*, **1971**, 4157.
[45] W. T. Borden, V. Varma, M. Cabell, and T. Ravindranathan, *J. Am. Chem. Soc.*, **93**, 3800 (1971).
[46] V. K. Krylov, I. V. Tselmskii, L. I. Bagal, and M. F. Kozlova, *Organic Reactivity (Tartu)*, **7**, 612, 616 (1971); *Chem. Abs.*, **75**, 5057 (1971).
[47] R. O. Guthrie, A. T. Young, and G. W. Pendygraft, *J. Am. Chem. Soc.*, **93**, 4947 (1971).
[48] T. Tenno and A. Talvik, *Organic Reactivity (Tartu)*, **7**, 1194, 1206 (1970); *Chem. Abs.*, **75**, 19569, 19581 (1971).
[49] R. R. Lii and S. I. Miller, *J. Chem. Soc. (B)*, **1971**, 2271.
[50] B. Deschamp and J. Seyden-Penne, *Tetrahedron*, **27**, 3959 (1971).
[51] N. L. Holy and J. D. Marcum, *Angew. Chem. Int. Ed.*, **10**, 115 (1971).
[52] J. P. Klinman and I. A. Rose, *Biochemistry*, **10**, 2259 (1971).

(21)

(22) $(3S, 4R)$-[4,4-^{2}H, ^{3}H] citrate

$(2S)$-[2,2-^{2}H, ^{3}H] acetyl CoA

$(4R)$-*cis*-[4-^{3}H] aconitate

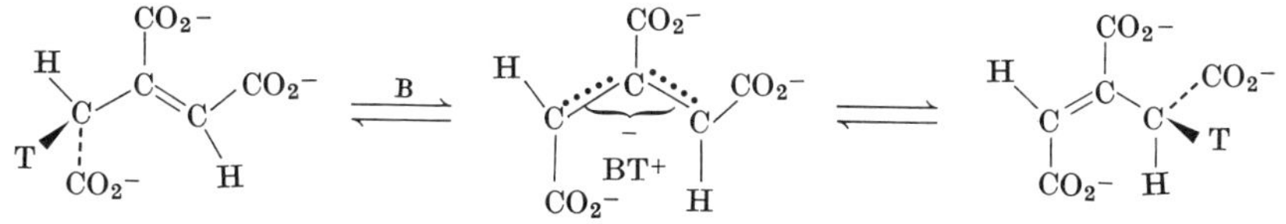

further shown[53a] that aldol and retroaldol reactions (**22**) catalysed by citrate enzymes proceeded with inversion. Substantial amounts of the 6β-isomer (**23**) are retained in aldimino derivatives of methoxymethyl penicillanate on base-catalysed equilibration of the 6α-compounds.[53b]

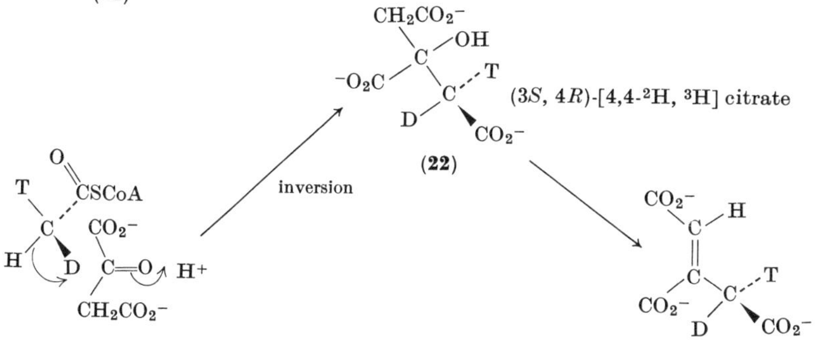

A route to 2-pyrones employing sulphonium ylids has been devised[54] and speculative mechanisms have been suggested; ylids have also been used[55] in the synthesis of vinyl sulphides. The disulphoxide (**24**) underwent a Pummerer rearrangement in acetic anhydride in which the 5-acetoxy-2-acetylthieno[3,2-*b*]thiophen [(**25**), X = Ac] was obtained,[56] rather than the expected product (**25**, X = H): direct acetylation of [(**25**), X = H] did not occur under the conditions used and this group was therefore inserted by trapping of the acetyl cation at C-1 or C-6 by an ylid intermediate. The transition energies

(23)

(24)

(25)

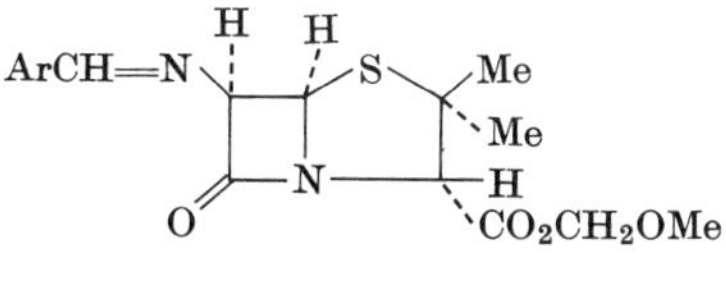

calculated[57] from the electronic spectra show that the ylid structure describes the ground state of sulphonium and phosphonium cyclopentadienylids. As predicted, electrophiles

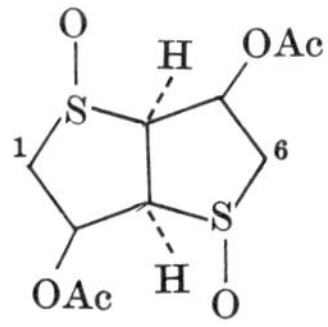

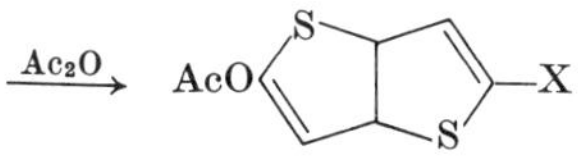

[53a] J. P. Klinman and I. A. Rose, *Biochemistry*, **10**, 2267 (1971).
[53b] J. R. Jackson and R. J. Stoodley, *Chem. Comm.*, **1971**, 647.
[54] Y. Hayasi and H. Nozaki, *Tetrahedron*, **27**, 3085 (1971).
[55] T. Mukaiyama, T. Kumamoto, S. Kukuyama, and T. Taguchi, *Bull. Chem. Soc. Japan*, **43**, 2870 (1969).
[56] J. Kuszmann, P. Sohar, and G. Horvath, *Tetrahedron*, **27**, 5055 (1971).
[57] Z. Yoshida, K. Iwata, and S. Yoneda, *Tetrahedron Letters*, **1971**, 1519.

are substituted in the dienide ring[58] of the phosphonium ylids, which show no tendency to undergo Michael addition since strong dienophiles also react with net substitution at C-2 (see Chapter 7). In the reaction[59] of arsenic and phosphorus ylids ($Ph_3Z=CH \cdot CO \cdot Ar$) with aromatic aldehydes, betaine formation is rate-determining: the arsenic compounds are more nucleophilic and are suitable for Wittig syntheses.

Zero-order bromination of the disulphone $RCH(SO_2Et)_2$ was observed[60] when re-protonation of the derived carbanion was slow, and ionization rates were determined in the presence of a number of bases. Dissociation constants were evaluated for this carbon acid (R = Me) and a number of related compounds; they showed the expected dependence on electron-withdrawing and -donating substituents. In contrast to the acidity the velocity constants for the reactions of the carbanions with [Br_2] and [Br_3^-] were little affected by substituents, although a reduction in rate for R = Ph indicated delocalization of charge. Factors that govern the chlorination of thiapropellanes have been studied,[61] and conditions have been given[62] for the conversion of the derived sulphones into cyclobutenes by the Ramberg–Bäcklund reaction. The bromosulphone [(**26**), R = Me] undergoes this rearrangement with rapid capture of the carbanion relative to proton exchange at the methinyl position: however, a preference for exchange was found in [(**26**), R = H] where the intermolecular process is not hindered; this suggests[63] that 1,3 intramolecular proton capture competes in some instances with ion-pair return. It has also been proposed[63a] that the greater rate of elimination relative to the *observed* rate of exchange in tosylates (**28**) may be the result of a carbanion mechanism in which expulsion of tosylate [cf. Br^- in (**27**)] competes more favourably with internal return than does the exchange reaction ($k_3 \gg k_2$).

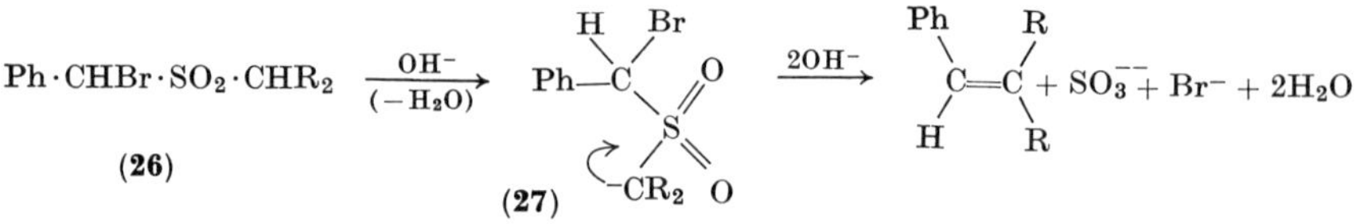

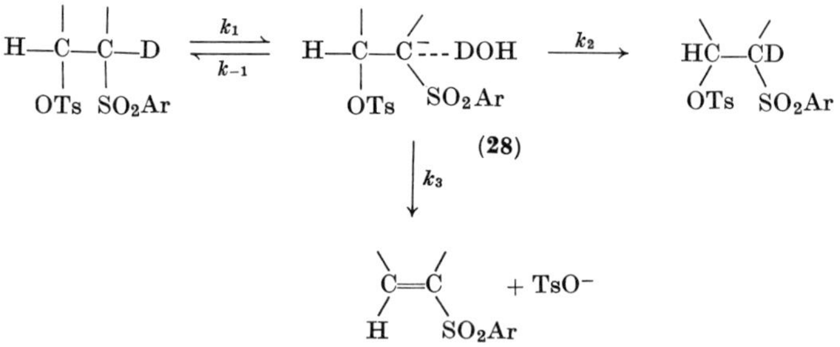

---

[58] Z. Yoshida, S. Yoneda, Y. Murata, and H. Hashimoto, *Tetrahedron Letters*, **1971**, 1523, 1527.
[59] N. A. Nesmeyanov, E. V. Binshtok, and O. A. Reutov, *Dokl. Akad. Nauk SSSR*, **1971**, 198, 1102; *Chem. Abs.*, **75**, 62756 (1971).
[60] R. P. Bell and B. G. Cox, *J. Chem. Soc.* (B), **1971**, 652.
[61] L. A. Paquette, R. E. Wingard, J. C. Phillips, G. L. Thompson, L. K. Read, and J. Clardy, *J. Am. Chem. Soc.*, **93**, 4508 (1971).
[62] L. A. Paquette, J. C. Phillips, and R. E. Wingard, *J. Am. Chem. Soc.*, **93**, 4516 (1971).
[63] F. G. Bordwell and M. D. Wolfinger, *J. Am. Chem. Soc.*, **93**, 6303 (1971).
[63a] F. G. Bordwell, J. Weinstock, and T. F. Sullivan, *J. Am. Chem. Soc.*, **93**, 4728 (1971).

Proton Transfer, Hydrogen Isotope Exchange and Related Reactions

The geometry of sulphinyl and sulphonyl carbanions remains an open question,[64] but conflicting accounts[65,66] of the stereochemistry of alkylation and exchange in α-lithio sulphoxides have been resolved,[67] and it is most probable that the lithio salt (**29**) retains its configuration on deuteration but that inversion occurs on methylation. The reaction with acetone occurred with retention and the products [as (**30**)] were employed[68] in the synthesis of epoxides of known absolute configuration. The preferred configuration

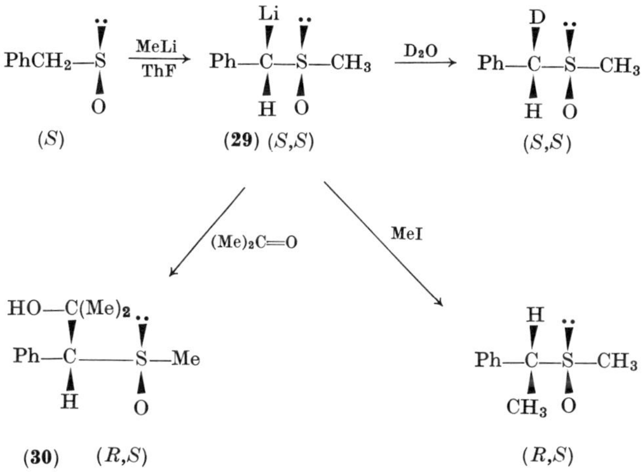

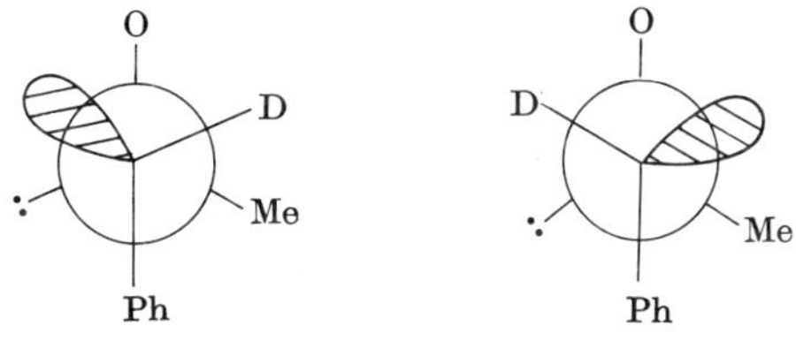

(**31**) of the anion in the above reactions is not firm and the alternative (**32**) leading to deuteration with inversion is possible. A calculation has been made[69] of the preferred conformation in 5-membered-ring sulphoxides, but in studies[70] of exchange rates in sulphones of the type (**33**) by a PMR technique there was only a small preference for

---

[64] *Org. Reaction Mech.*, **1970**, 135.
[65] K. Nishikata and K. Nishio, *Chem. Comm.*, **1971**, 958.
[66] T. Durst, R. Viau, and M. R. McClory, *J. Am. Chem. Soc.*, **93**, 3077 (1971).
[67] T. Durst and M. Nishio, personal communications.
[68] T. Durst, R. Viau, R. Van Den Elzen, and C. H. Nguyen, *Chem. Comm.*, **1971**, 1334.
[69] R. Lett, S. Bory, B. Moreau, and A. Marquet, *Tetrahedron Letters*, **1971**, 3255.
[70] M. D. Brown, M. J. Cook, B. J. Hutchinson, and A. R. Katritzky, *Tetrahedron*, **27**, 593 (1971).

equatorial exchange and trigonal intermediates could not be excluded; this symmetrical model is adopted[71] in the constrained sulphonyl carbanion (**34**) where, in the absence of tight hydrogen bonding $k_e = k_\alpha$. Partial racemization of sulphones [e.g. (**35**) in [O-^{2}H$_1$]-methanol with potassium methoxide, re-resolution and deuterium analysis[72] gave rate constants for net inversion with exchange, inversion without exchange (isoinversion) and racemization with exchange. The acyclic sulphone (**35**) largely retains its configuration, whilst in (**36**) (cf. ref. 71) inversion occurs most frequently and isoinversion infrequently. These findings are accounted for in terms of an energetically preferred carbanion (**37**) [cf. (**31**), in which there are similar gauche interactions of C$^-$], and stabilizing solvent molecules which exchange hydrogen bonds between three adjacent sites; retention follows collapse by capture of a proton or deuteron. In (**36**) the geometry constrains the carbanion and two conformers (A and B) are readily interconverted, leading to the observed preference for inversion and exchange by a cyclic movement of solvent molecules: the small isotope effect is consistent with a process of this kind. The behaviour of (**36**) was independent of the counter-ion and was little affected by the addition of crown ether. In contrast, retention is normal in exchange with (**38**) and depends on

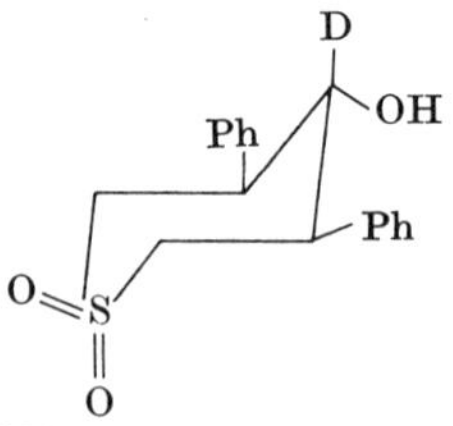

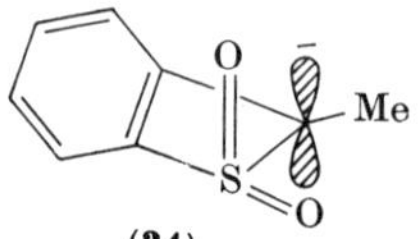

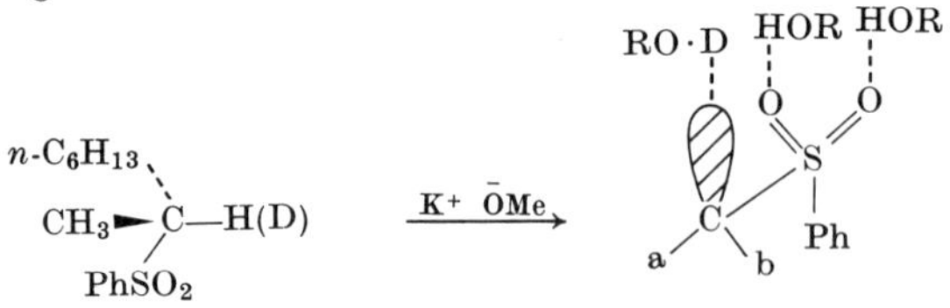

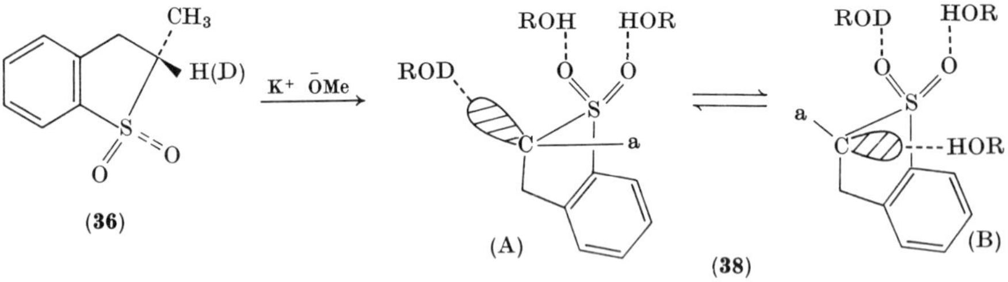

assembly of hydrogen donor solvents and carbanion on one face of a contact ion pair: capture of potassium ion by crown ether gave the characteristics of an ordinary solvent-separated ion pair, increasing both rates but relatively enhancing that of racemization.

[71] L. A. Paquette, J. P. Freeman, and M. J. Wyvratt, *J. Am. Chem. Soc.*, **93**, 3216 (1971).
[72] J. N. Roitman and D. J. Cram, *J. Am. Chem. Soc.*, **93**, 2225, 2231 (1971).

A review of the acidity of carbon acids has appeared[73] and the acidity of fluorinated compounds R_FH has been determined[74] from the polarographic parameters of their mercury derivatives $(R_F)_2Hg$; the results for sp^3 hybridized compounds were not accountable on the basis of the inductive effect of fluorine. The rates of exchange in p-trifluoromethyltoluene are in keeping[75] with a π-inductive effect rather than a fluorine no-bond resonance mechanism. The relative rates of exchange at the bridgehead in triptycene and structures modified by fusion of another benzene ring and by simplification to 9,10-dihydro-9,10-ethenoanthracene, did correlate with changes in the inductive effect. This was quantified[76] as $\Sigma 1/r$ where r is the distance between the acidic centre and each sp^2 hybridized carbon in the molecule. Steric inhibition of exchange and the destabilizing effect of alkyl groups on pyramidal carbanions have been observed[77] for α-alkyltoluenes. The overall rate of base-catalysed exchange of 6,7-benzobicyclo-

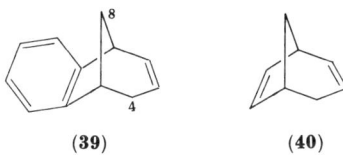

(39)   (40)

[3.2.1]octa-2,6-diene (39) is slower by an order of magnitude than that of its alkene analogue (40); the relative rate at the 4-*exo*-position was greater in both models. There was no evidence here[78] of the marked interaction between the alkene and the 8-position characteristic of carbonium ion reactions.

Acidity functions have been given[79] for DMSO–water mixtures that correspond well with the scale based on the ionization of aniline indicators, save for crossing in the more aqueous region. The activity coefficients of nitrogen acids (N–H dissociates) probably vary relative to those of carbon acids owing to the H-bonding interactions. The equilibrium acidities of phenyl and t-butylacetylene towards lithium cyclohexylamide have been determined.[80] Great differences were found in the acidity relative to an extensively conjugated carbon acid under conditions affecting ion-pair stabilization; fair correlation of equilibrium constants with carbanion basicity has been reported[81] for competitive metallation of arylalkanes, whose acidity has also been compared[82] with that of phosphine oxides. The kinetic acidity of arylalkanes can be correlated[83] with the intensities of the

---

[73] J. R. Jones, *Quart. Rev. (London)*, **3**, 365 (1971).
[74] K. P. Butin, A. P. Kashin, I. P. Beletskaya, L. S. German, and V. R. Polishchulk, *J. Organomet. Chem.*, **25**, 11 (1970).
[75] A. Streitwieser and D. Holtz, *J. Org. Chem.*, **35**, 4288 (1970).
[76] A. Streitwieser, M. J. Maskornick, and G. R. Ziegler, *Tetrahedron Letters*, **1971**, 3927.
[77] A. Streitwieser, P. C. Mowery, and W. R. Young, *Tetrahedron Letters*, **1971**, 3931; cf. ref. 43.
[78] J. M. Brown, E. N. Cain, and M. C. McIvor, *J. Chem. Soc.* (B), **1971**, 730.
[79] A. F. Cockerill and J. E. Lamper, *J. Chem. Soc.* (B), **1971**, 503.
[80] A. Streitwieser and D. M. E. Reuben, *J. Am. Chem. Soc.*, **93**, 1794 (1971).
[81] E. S. Petrov, M. I. Terekhova, and A. I. Shatenshtein, *Organic Reactivity (Tartu)*, **7**, 1234 (1970); *Chem. Abs.*, **75**, 26089 (1971).
[82] E. S. Petrov, E. N. Tsvetkov, M. I. Kabachnik, and A. I. Shatenshtein, *Zhur. Obsch. Khim.*, **41**, 1172 (1971); *Chem. Abs.*, **75**, 80951 (1971).
[83] N. N. Zatsepina, A. V. Kirova, N. S. Kolodina, and I. F. Tupitsyn, *Organic Reactivity (Tartu)*, **7**, 667 (1970); *Chem. Abs.*, **75**, 34960 (1971).

C—H vibrations in their IR spectra. The kinetic isotope effect for exchange in diphenylmethane varies with binary solvent composition,[84] and the maximum isotope effect for the ionization rate (k_H/k_D) of a given carbon acid (nitroethane) and base (OH^-) occurs[85] near the equivalence point in the two pK values of the acid–base system ($\Delta pK = 0$); difficulties in correlating Brønsted β values and kinetic isotope effects with the transition state for proton transfer have been discussed.[86] The kinetics of D-exchange at aromatic ring positions and at substituent methyl groups have been studied[87] with potassium–DMSO, and the structure of the intermediate carbanion has been related to acidity; solvent and isotope effects on the metallation rates of carbon acids by the sodium and potassium adducts of naphthalene and biphenyl have also been investigated.[88] The rate of proton exchange in trimethylsulphoxonium nitrate is 10^9 times that found in trimethylsulphonium iodide, illustrating the difference in the stability of the sulphur ylids.[89]

The relative basicities of a group of polynitroalkanes were correlated[90] with the ^1H chemical shift of dissolved dinitrofluoromethane; and intramolecular proton transfer in dinitrocarbanions has been investigated.[91] The rate of exchange between aliphatic amines and 4-nitrophenylnitromethane was insensitive to steric factors and there was no definite evidence of proton tunnelling. In acetonitrile solution the C—H(D) bonds were largely broken in the transition state.[92] The formation of nitronate was limited in nitrosugars by steric factors, particularly the interaction with an adjacent equatorial substituent in a rigid ring [e.g. (41A)]. In these circumstances[93] retroaldolization was relatively fast and epimerization occurred at C-2 on recyclization; it is worth noting that compression of the anion (42) is not a factor once the ring (B) is opened. The solvent isotope

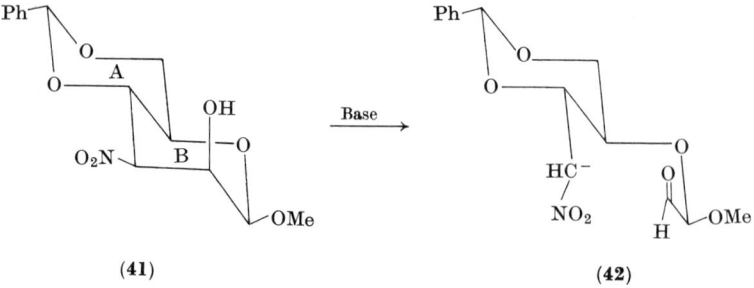

[84] F. S. Yakushin, Y. I. Ranneva, V. A. Marchenko, I. A. Rowanskii, and A. I. Shatenshtein, *Kinet. Katal.*, **12**, 591 (1971); *Chem. Abs.*, **75**, 87791 (1971).
[85] R. P. Bell and B. G. Cox, *J. Chem. Soc.* (B), **1971**, 783.
[86] F. G. Bordwell and W. J. Boyle, *J. Am. Chem. Soc.*, **93**, 512 (1971).
[87] A. I. Shatenshtein, I. A. Romanskii, and I. O. Shapiro, *Organic Reactivity* (*Tartu*), **7**, 337 (1970); *Chem. Abs.*, **73**, 130522 (1970).
[88] E. S. Petrov, M. I. Terekhova, and A. I. Shatenshtein, *Dokl. Akad. Nauk SSSR*, **192**, 1053 (1970); *Chem. Abs.*, **73**, 119845 (1970).
[89] A. F. Relkasheva, L. A. Kiprianova, and I. P. Samchenko, *Dokl. Akad. Nauk SSSR*, **197**, 396 (1971); *Chem. Abs.*, **75**, 48182 (1971).
[90] V. M. Khuloretskii, L. V. Okhlobystina, A. D. Naumov, A. V. Kessenikh, and A. D. Fainzil'berg, *Izv. Akad. Nauk SSSR, Ser. Khim.*, **1970**, 2374; *Chem. Abs.*, **75**, 48239 (1971).
[91] V. N. Dronov and I. V. Tselinsky, *Organic Reactivity* (*Tartu*), **7**, 594 (1970); *Chem. Abs.*, **74**, 124558 (1970).
[92] E. F. Caldin, A. Jarczewski, and K. T. Leffek, *Trans. Faraday Soc.*, **67**, 110 (1971).
[93] H. H. Baer and W. Rank, *Can. J. Chem.*, **49**, 3197 (1971); H. H. Baer and J. Kouar, *Can. J. Chem.* **49**, 1940 (1971).

effect on proton transfer from 2-nitropropane and phenyl 3-methyl-1-phenylbutyl ketone to methoxide ion has been explained[94] in terms of its multiple solvation.

The rate of bromination and detritiation of malononitrile is determined by a slow, general base-catalysed transfer from the carbon acid; the small isotope effect (k_H/k_D = 1.48) and a Brønsted coefficient near unity show[95] that the carbanion is highly developed in the transition state. The conjugate bases of the nitriles resemble oxyacids in that their protonation is almost diffusion-controlled. The same conclusion was reached[96] concerning the transition state for hydrogen transfer in 2-methyl-3-phenylpropionitrile, and similar properties were forecast for other carbanions that lack resonance stabilization; a detailed study[97] of factors affecting proton transfer in H_2O–D_2O mixtures has been made for the more acidic nitriles. The re-resolution method[72] has also been used to investigate isoinversion in (−)-2-methyl-3,3,3-triphenylpropionitrile, a proton tour being favoured[98] for catalysis by organic bases. pK values which correlated with σ-values have been obtained for phenylacetic acid derivatives.[99]

The mechanism of the base-catalysed methylene–azomethine rearrangement, a prototype of *in vivo* transamination, has been examined.[100] Stereospecificity in the reaction was shown for (**43**), where the bulky phenyl and t-butyl groups made for conformational purity in an anion, (**44**), whose structure was retained by contact

$$\text{Me}-\underset{\text{Ph}}{\overset{\text{H}}{\text{C}}}-\text{N}=\text{C}\underset{\text{D}}{\overset{\text{Bu}^t}{}} \rightleftharpoons \underset{\text{Ph}}{\overset{\text{Me}}{\text{C}}}\cdots\underset{\text{N}^-}{}\cdots\underset{\text{Bu}^t}{\overset{\text{D}}{\text{C}}} \rightleftharpoons \underset{\text{Ph}}{\overset{\text{Me}}{\text{C}}}=\text{N}-\underset{\text{D}}{\overset{\text{H}}{\text{C}}}-\text{Bu}^t$$

(**43**)   (**44**)   (**45**)

ion-pairing with K^+ and by solvation on the side on which the proton was suprafacially transferred. Addition of crown ether also showed that retention during the faster exchange reaction was the result of contact ion-pairing. Isotope effects on the collapse of (**44**) to (**43**) and (**45**) were determined, and stereospecific transamination, without enzymic mediation, was demonstrated[101] for the related ketimine [(**43**), Me for D]. Isoracemization of peptides has been reported,[102] and the more acidic α-positions of azlactones are racemized more rapidly than in the open-chain benzylidene derivatives, $\text{ArCH}=\text{N}-\text{CHR}-\text{CO}_2\text{R}'$, which are further stabilized by bulky ester groups (R').[103] Acid catalysis leads to proton exchange of the methyl group attached to the pyrimidine ring of thiamine and some of its derivatives.[104] The rate of exchange at C-5 of uridine was proportional to glutathione (GSH) concentration and reached maximum near pD 9; Michael addition of GS^- ions was suggested[105] as a mechanism and an analogy with

---

[94] V. Gold and S. Grist, *J. Chem. Soc.* (B), **1971**, 2282.
[95] F. Hibbert, F. A. Long, and E. A. Walters, *J. Amer. Chem. Soc.*, **93**, 2829 (1971).
[96] L. Melander and N.-A. Bergman, *Acta Chem. Scand.*, **25**, 2264 (1971).
[97] F. Hibbert and F. A. Long, *J. Am. Chem. Soc.*, **93**, 2836 (1971).
[98] S. M. Wong, H. P. Fischer, and D. J. Cram, *J. Am. Chem. Soc.*, **93**, 2235 (1971).
[99] K. Zahorieva, J. N. Stefanovsky, and I. G. Pojorlieff, *Tetrahedron Letters*, **1971**, 1663.
[100] R. D. Guthrie, D. A. Jaeger, W. Meister, and D. J. Cram, *J. Am. Chem. Soc.*, **93**, 5137 (1971).
[101] D. A. Jaeger and D. J. Cram, *J. Am. Chem. Soc.*, **93**, 5153 (1971).
[102] J. Kovacs, H. Cortegiano, R. E. Cover, and G. L. Mayers, *J. Am. Chem. Soc.*, **93**, 1541 (1971).
[103] I. Z. Siemon and L. Wilschowitz, *Z. Naturforsch.*, **26b**, 762 (1971).
[104] D. W. Hutchinson, *Biochemistry*, **10**, 542 (1971).
[105] T. I. Kalman, *Biochemistry*, **10**, 2567 (1971).

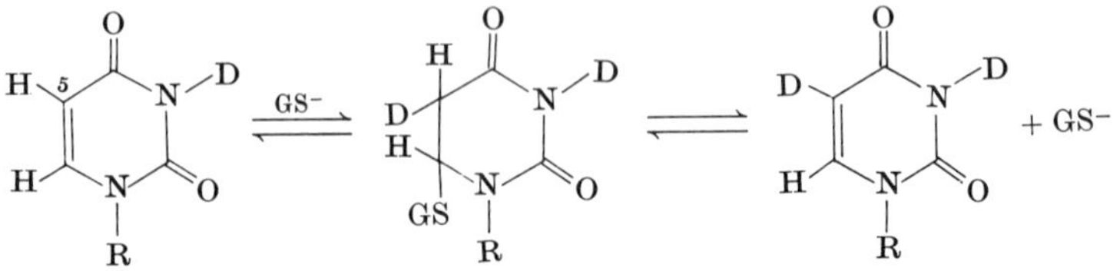

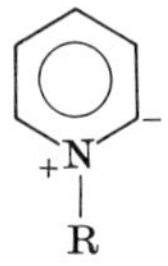

(46) 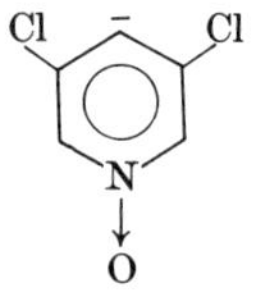   (47)

C-5 alkylations was indicated. Deuteroxide ion catalysed exchange[106] in N-substituted pyridinium ions via the ylid (46), whose localized charge led to competing internal proton return; varying the R group had a predominantly inductive effect on the rate, and this could be used to control ylid reactivity in synthetic applications. General base catalysis was ineffective during exchange involving similar localized anions [as (47)]; an internal return mechanism was preferred[107] in which reprotonation was faster than substitution:

$$\text{CH} + \text{B}^- \rightleftharpoons \text{C}^-\text{---HB} \xrightarrow[\text{DB}]{} \text{C}^-\text{---BD} \rightarrow \text{CD} + \text{B}^-$$

The rate of exchange at C-8 in purines is 2000 times that at C-2 and is little affected by substituents in the pyrimidine ring;[108] in benzimidazole, with no electronegative atoms in the fused ring, the rate of exchange at C-2 *increases* significantly.

The principle of microscopic reversibility has been examined[109] on the basis of three-dimensional potential-energy surfaces that allow the possibility of exchange between symmetrical systems by unsymmetrical pathways.

A preliminary account has appeared[110,111] of the σ-basicity of C—C and C—H bonds of alkanes and the transition states for H-exchange, alkylation and nitration[112] that involve triangular three-centre bonds. One example which substantiates this mode (48)

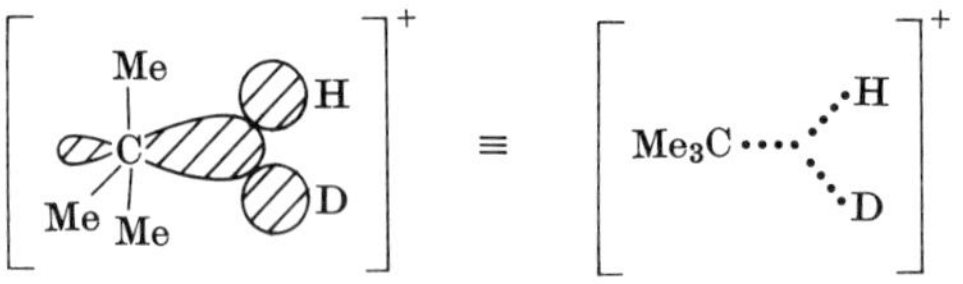

(48)

[106] J. A. Zoltewicz and L. S. Helmick, *J. Am. Chem. Soc.*, **92**, 7547 (1970).
[107] J. A. Zoltewicz, C. L. Smith, and G. M. Kauffman, *J. Heterocyclic Chem.*, **8**, 337 (1971).
[108] J. A. Elvidge, J. R. Jones, C. O'Brien, and E. A. Evans, *Chem. Comm.*, **1971**, 394; M. Maeda, M. Saneyoshi, and Y. Kawazoe, *Chem. Pharm. Bull.*, **19**, 1641 (1971).
[109] M. H. Abraham, D. Dodd, M. D. Johnson, E. S. Lewis, and R. A. More O'Ferrall, *J. Chem. Soc.* (B), **1971**, 762.
[110] G. A. Olah, Y. Halpern, J. Shen, and Y. K. Mo, *J. Am. Chem. Soc.*, **93**, 1251 (1971).
[111] G. A. Olah and J. A. Olah, *J. Am. Chem. Soc.*, **93**, 1256 (1971).
[112] G. A. Olah and H. C. Lin, *J. Am. Chem. Soc.*, **93**, 1259 (1971).

is the significant D-exchange at the methine proton of isobutane in DF–SbF$_5$, under conditions that excluded the alternative addition to isobutene and quenching of H–D by the t-butyl cation. Direct alkylation in the absence of alkenes was demonstrated by using RHal–SbF$_5$ in SO$_2$ClF solution. Nitration (**49a**) and nitrolysis (**49b**) occurred with nitronium salts in the absence of free radicals. The solvent and nitrated products

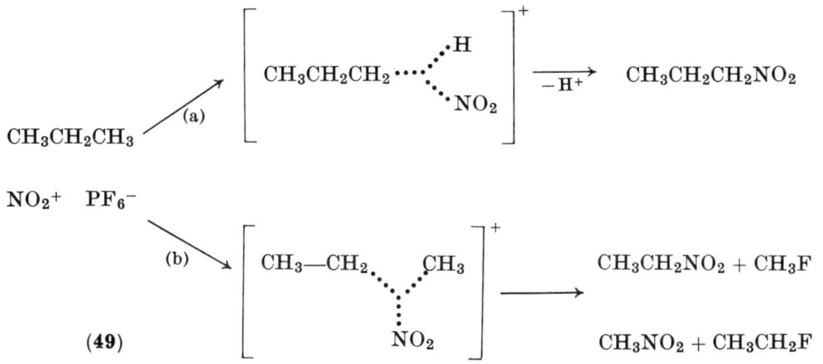

$$CH_3CH_2CH_3 \xrightarrow{(a)} [CH_3CH_2CH_2 \cdots \overset{H}{\underset{NO_2}{\cdots}}]^+ \xrightarrow{-H^+} CH_3CH_2CH_2NO_2$$

$$NO_2^+ \ PF_6^-$$

$$\xrightarrow{(b)} [CH_3-CH_2 \underset{NO_2}{\overset{CH_3}{\cdots}}]^+ \longrightarrow \begin{array}{l} CH_3CH_2NO_2 + CH_3F \\ CH_3NO_2 + CH_3CH_2F \end{array}$$

(**49**)

limit the forward reaction, being stronger bases than the alkanes themselves. Electrophilic α-chlorination and solvolysis follow the formation of σ-adducts from alkylbenzenes.[113]

Electrophilic Reactions of Hydrocarbons

The uniparticulate electrophile chlorosulphonyl isocyanate has been used to show[114] that in bicyclobutanes the more strained bond breaks and the carbonium ion formed may isomerize rapidly [(**50**) to (**51**)] before it is trapped by intramolecular reaction; alternatively, when rearrangement is less favoured electronically, the initially formed ion will be trapped by the competing ring inversion.

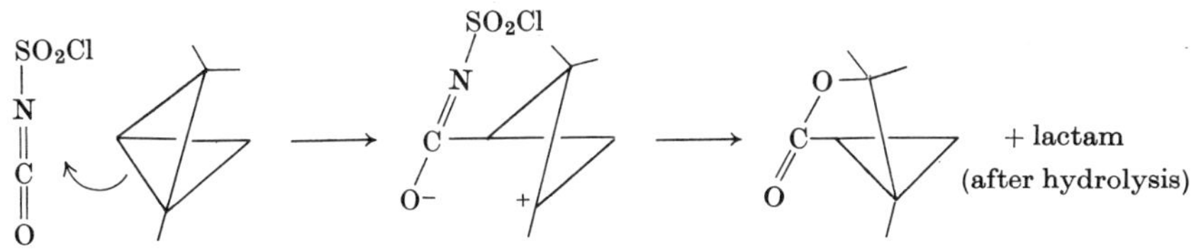

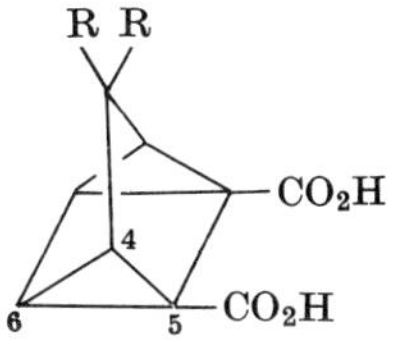

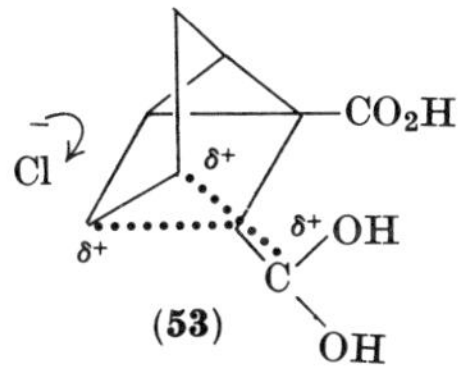

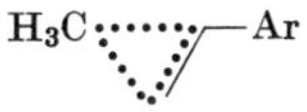

[113] G. Illuminati, L. Mandolini, E. M. Arnett, and R. Smoyer, *J. Chem. Soc.* (B), **1971**, 2206.
[114] L. A. Paquette, G. R. Allen, and M. J. Broadhurst, *J. Am. Chem. Soc.*, **93**, 4503 (1971).

The products of cleavage of *trans*-bicyclo[5.1.0]oct-3-ene by acetic acid were determined by protonation of the cyclopropane ring, whose high reactivity was attributed[115] to "twist-bending". Fission of quadricyclenedicarboxylic acid [(**52**), R = H] gave normal nucleophilic coordination with inversion at the most strained 6-position. The isolation of products generated by both retention and inversion at C-5 was accounted for[116] by homoconjugate addition, final protonation occurring from either side of the enol tautomer (**53**); a related ester [(**52**), RR = isopropylidene] underwent fission of the strained ring system with a similar steric result.[117] Quadricyclene undergoes ionic cleavage only by inversion.[118]

In the Pt(II) insertion reaction of cyclopropanes[119] the donor capacity of the centre

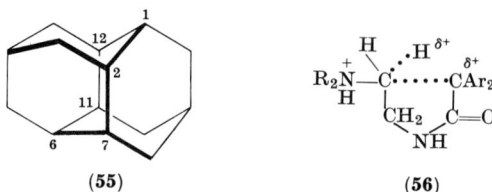

combining with platinum is critical and the reactivity order for various R groups is alkyl > benzyl > phenyl. There has been an *ab initio* calculation of the electronic structure of cyclopropane.[120] Internal and external cleavage of cyclopropane rings by hydrogen chloride has been studied[121] and anti-Markownikoff fission by diborane was shown[122] to be regioselective but not sterically analogous to its reactions with alkenes. Markownikoff addition of hydrogen bromide occurs[123] in the absence of electrostatic interactions with neighbouring carbonyl groups, and nucleophilic retention was observed.

The acid-catalysed cleavage of caranones[124] and solvolysis of bicyclobutanones[125] was reported, and a corner-protonated structure [(**54**); cf. (**48**)] was proposed[126] for the protolytic cleavage of arylcyclopropanes. The selective halogenation of adamantanone and 1-methyladamantanone has been accomplished;[127] the belt positions (1, 2, 6, 7, 11, 12) of diamantane (**55**) are brominated in preference to apical positions in the absence of Lewis acids.[128]

(**55**)   (**56**)

---

[115] P. G. Gassman and F. J. Williams, *J. Am. Chem. Soc.*, **93**, 2704 (1971).
[116] S. J. Cristol, J. K. Harrington, J. G. Morrill, and B. E. Greenwald, *J. Org. Chem.*, **36**, 2773 (1971).
[117] S. F. Nelsen, J. B. Gillespie, and P. J. Hintz, *Tetrahedron Letters*, **1971**, 2361.
[118] T. C. Morrill and B. E. Greenwald, *J. Org. Chem.*, **36**, 2769 (1971).
[119] K. G. Powell and F. J. McQuillin, *Tetrahedron Letters*, **1971**, 3313.
[120] H. Marsmann, J.-B. Robert, and J. R. van Wazer, *Tetrahedron*, **27**, 4377 (1971).
[121] R. S. Boikess and M. Mackay, *J. Org. Chem.*, **36**, 901 (1971).
[122] B. Rickborn and S. E. Wood, *J. Am. Chem. Soc.*, **93**, 3940 (1971).
[123] J. B. Hendrickson and R. K. Boeckman, *J. Am. Chem. Soc.*, **93**, 4491 (1971).
[124] F. Fringuelli and A. Taticchi, *J. Chem. Soc.* (C), **1971**, 297.
[125] G. Szeimies and A. Schlosser, *Tetrahedron Letters*, **1971**, 3631.
[126] M. A. McKinney, S. H. Smith, S. Hempelman, M. M. Gearen, and L. Pearson, *Tetrahedron Letters*, **1971**, 3657.
[127] H. Hamill, A. Karim, and M. A. McKervey, *Tetrahedron*, **27**, 4317 (1971).
[128] T. M. Gund, P. von R. Schleyer, and C. Hoogzand, *Tetrahedron Letters*, **1971**, 1583.

A base (H_2O or HSO_4^-) participates in the nitration of the strong carbon acid dinitroacetonitrile in nitrosulphuric acid.[129] A transition state (**56**) was proposed[130] for the intramolecular cyclization of *N,N*-disubstituted amides of diarylglycollic acids; the reaction rate was reduced by electron-donating aryl substituents and by bulky *N*-substituents.

Organometallics: Groups Ia, IIa, III

Aminomethyl lithium compounds useful in enamine synthesis have been prepared[131] from the favoured exchange with butyl-lithium;

$$R_2NCH_2SnBu_3 + BuLi \rightarrow R_2NCH_2Li + SnBu_4$$

Rates of exchange between enamines and deuterioacetone have been studied.[132] The reaction[133] of lithium aryls and ^{14}C-labelled bromobenzene in ether is inhibited by lithium salts[134] and accelerated by ammonium salts, behaviour which was ascribed[135] to different structures of the respective complexes for exchange. The rate of reaction of *n*-propylzinc with benzaldehyde in ether was also increased by the addition of tetrabutylammonium salts, and the proportion of reduced product fell in consequence.[136] Metal halides were also effective[137] in alkyl exchanges of triethylaluminium, which reacted[138] with oct-1-ene in hydrocarbons as the monomer. Inter- and intra-molecular induction of asymmetrical lithiation of ferrocene has been demonstrated,[139] a cyclic intermediate being invoked[140] to account for the initial preference for 2-lithiation of dimethylaminoethylferrocene. There has been a kinetic study[141] of the fission of tetrahydrofuran by *n*-butyl-lithium, but the reviewer feels that the reaction order of 2.5 in THF was not assigned with sufficient accuracy to require a revision of suggested mechanisms. 7-Alkylnorbornen-7-ols can be prepared[142] by using lithium alkyls which give both *syn*- and *anti*-alkylation with little reduction, whereas the Grignard reaction proceeds largely by *syn*-attack: control by steric and polar interactions is discussed.

The Schlenk equilibrium was studied[143] for solutions of methyl- and *tert*-butylmagnesium halides in ether and THF by 100 MHz NMR spectroscopy to a lower limiting temperature of ca. −100°. Chemical-shift changes indicated that polymeric Me_2Mg

---

[129] E. S. Mints, E. L. Golod, and L. I. Bagol, *Zhur. Org. Khim.*, **6**, 1137 (1970); *J. Org. Chem. USSR*, **6**, 1142 (1970).
[130] V. S. Shklyaev and A. Z. Koblowa, *Zhur. Org. Khim.*, **6**, 2545 (1970); *J. Org. Chem. USSR*, **6**, 2558 (1970).
[131] D. J. Peterson, *J. Am. Chem. Soc.*, **93**, 4027 (1971).
[132] W. H. Daly, J. G. Underwood, and S. C. Kuo, *Tetrahedron Letters*, **1971**, 4375.
[133] A. P. Batalov, A. A. Kvasov, G. A. Rostokin, and I. A. Korshunov, *Tr. Khim. Khim. Tekhnol.*, **1970**, 40; *Chem. Abs.*, **75**, 75898 (1971).
[134] A. P. Batalov and M. A. Skvortsova, *Kinet. Katal.*, **1971**, 326; *Chem. Abs.*, **75**, 80761 (1971).
[135] A. P. Batalov and G. A. Rostokin, *Dokl. Akad. Nauk SSSR*, **198**, 1334 (1971); *Chem. Abs.*, **75**, 87837 (1971).
[136] R. Chastrette and R. Amouroux, *Tetrahedron Letters*, **1970**, 5165.
[137] A. P. Batalov, *Zhur. Obsch. Khim.*, **41**, 1143 (1971); *Chem. Abs.*, **75**, 87920 (1971).
[138] P. E. M. Allen and A. E. Byers, *Trans. Faraday Soc.*, **67**, 1718 (1971).
[139] T. Aratani, T. Gonda, and H. Nozaki, *Tetrahedron*, **26**, 5453 (1970).
[140] D. W. Slocum, C. A. Jennings, T. R. Engelmann, B. W. Rockett, and C. R. Hauser, *J. Org. Chem.*, **36**, 377 (1971).
[141] S. C. Honeycutt, *J. Organomet. Chem.*, **29**, 1 (1971).
[142] F. R. S. Clark and J. Warkentin, *Can. J. Chem.*, **49**, 2223 (1971).
[143] G. E. Parris and E. C. Ashby, *J. Am. Chem. Soc.*, **93**, 1206 (1971).

was precipitated from ether at low temperatures and that MeMgBr predominates at 30°. In THF it was difficult to assign shifts to MeMgHal but a statistical distribution is probable in this solvent. t-Butylmagnesium halides also predominate at ambient temperature, the concentration of di-t-butylmagnesium increased as the temperature fell; exchange was slower for the bulky group but was accelerated by alkoxides. The mixed bridge structure (**57**) is important[143] in exchange reactions and it has been noted[144] that transfer of a

$$\begin{array}{c} \text{Solv} \diagdown \quad \diagup \text{R} \diagdown \quad \diagup \text{R} \\ \text{Mg} \quad \text{Mg} \\ \text{Hal} \diagup \quad \diagdown \text{Y} \diagup \quad \diagdown \text{Solv} \end{array}$$

Y = R, Hal, or $^-$OR

(**57**)

bridging alkyl group to another aggregate leads to inversion in primary Grignard compounds. Multiplicity in the spectra of *gem*-dimethylcycloalkylmagnesium halides is attributed[145] to a relatively slow carbanion inversion, but the rate of α-inversion of cyclic secondary reagents is fast enough for detection[146] by PMR and is both solvent- and concentration-dependent, whilst cyclohex-3-enylmagnesium bromide rearranges slowly to form[147] the primary cyclopent-2-enylmethylmagnesium bromide.

Pinacol formation and 1,2-, 1,4- and 1,6-additions have been detected[148] in the reactions of benzophenones with Grignard reagents. A Hammett correlation was satisfactory with t-butylmagnesium chloride, despite varying proportions of these four pathways; this is explained by a common rate-limiting step:

$$\text{Ar}_2\text{C}=\text{O} + \text{Me}_3\text{CMgCl} \rightarrow \text{Ar}_2\overset{-}{\text{C}}\text{—}\overset{+}{\text{O}}\text{MgCl} + \text{Me}_3\text{C}\cdot$$

This electron-transfer mechanism is indicated when the rate of reaction of a given acceptor with an α-branched Grignard is greater than that of an unbranched reagent. For primary Grignard reagents that react through the dialkylmagnesium the rate was determined by the transfer:

$$\begin{array}{c} \text{R} \diagdown \quad \quad \quad \quad \quad \quad \quad \quad \diagup \text{R} \\ \text{Mg} + \text{Ph}_2\text{C}=\text{O} \rightarrow \text{Ph}_2\text{—}\overset{\cdot}{\text{C}}\text{—O—Mg} \\ \text{R} \diagup \quad \quad \quad \quad \quad \quad \quad \quad \cdot \text{R} \end{array}$$

(**58**)

A subsequent rapid exchange of alkyl groups between Grignard and the biradical (**58**) followed by decomposition to *H*- and *D*-benzhydrol, accounts[149] for observed isotope effects. The rate of reaction of methylmagnesium bromide with benzophenone was followed[150] in the UV from the rate of disappearance of its complex with the ketone; it was of the first order when the concentration of the more reactive dimethylmagnesium

---

[144] G. Fraenkel, C. E. Cottrell, and D. T. Dix, *J. Am. Chem. Soc.*, **93**, 1704 (1971).
[145] E. Pechhold, D. G. Adams, and G. Fraenkel, *J. Org. Chem.*, **36**, 1368 (1971).
[146] A. Maercker and R. Geuss, *Angew. Chem. Int. Ed.*, **10**, 270 (1971).
[147] A. Maercker and R. Geuss, *Angew. Chem. Int. Ed.*, **9**, 909 (1970).
[148] T. Holm and I. Crossland, *Acta Chem. Scand.*, **25**, 59 (1971).
[149] T. Holm, *J. Organomet. Chem.*, **29**, C45 (1971).
[150] E. C. Ashby, J. Laemmle, and H. M. Neumann, *J. Am. Chem. Soc.*, **93**, 4601 (1971).

was reduced to a low level. The rate of reaction of dipropylmagnesium with pinacolone in *n*-heptane was dependent on the polarity and steric constants of added donor solvents,[151] and related work on propylmagnesium bromide has also been published.[152]

The similarity in the rates of reaction of allylic Grignard reagents with unhindered epoxides and ketones has been attributed[153] to an S_E2' mechanism:

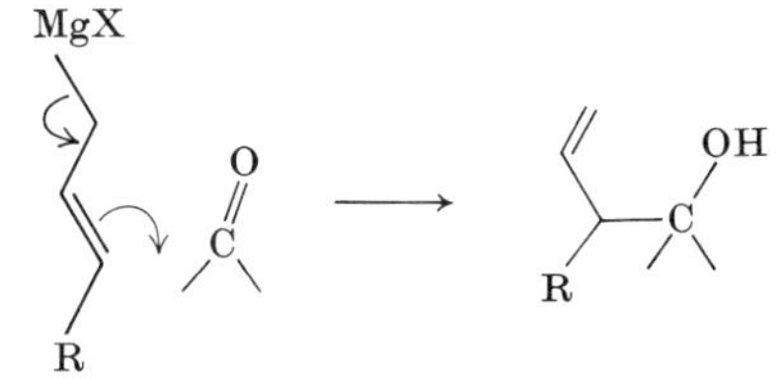

BrMgCH$_2$—CH=CHBu^t $\rightleftharpoons$ CH$_2$=CH—CHBu^t
(59) |
                                  MgBr

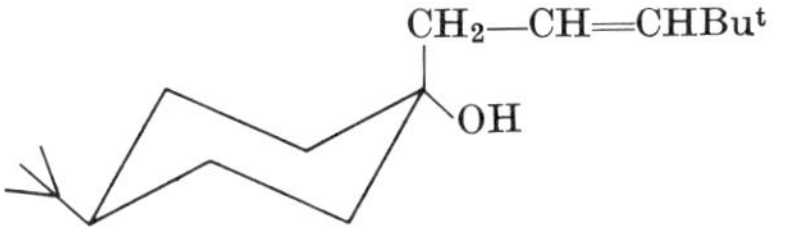

(60) 41%
(61) minor product 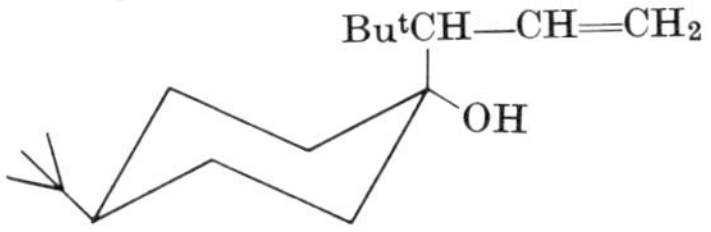
(62) 33% 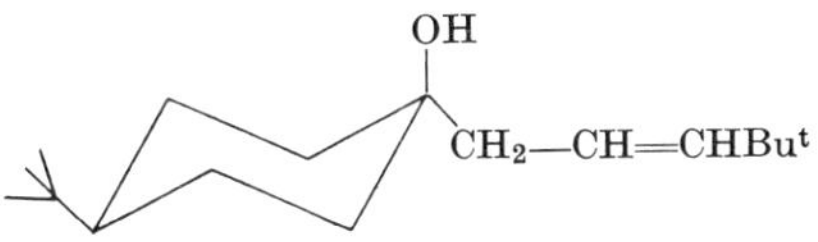
(63) 26% 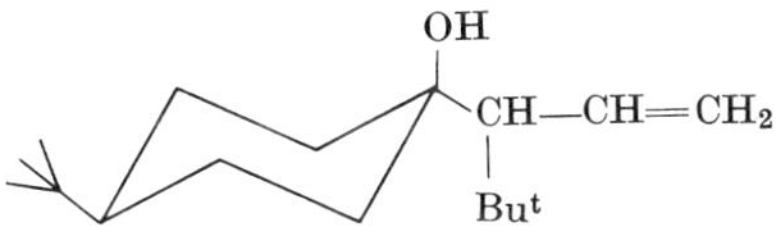

With ambident t-butylallylmagnesium bromide (**59**) this type of reaction gives a measure[154] of steric hindrance at the carbonyl centre. Of the two isomeric equatorial alcohols (**60**) and (**61**) formed from t-butylcyclohexanone, the former predominates as a result of steric strain in a reactant-like transition state; the comparable yields of axial alcohols (**62**) and (**63**) show that torsional strain is effective in the transition state leading to their formation.[155] It has been argued[156] that polar factors must be considered when the method is used to evaluate the steric effects of α-deuteration. It should be stressed that when these reactions are slowed by steric and/or polar factors a later transition state is to be expected; this point has been further discussed[157] and adamantanone has been used[158] as a model to test the nature of transition states. Dipropylcadmium and dipropylzinc in reaction[159] with t-butylcyclohexanone gave more equatorial attack than did

---

[151] J. Koppel, S. Vaiga, and A. Tuulmets, *Organic Reactivity (Tartu)*, **7**, 898 (1970); *Chem. Abs.*, **74**, 140569 (1970); J. Koppel and A. Tuulmets, *Organic Reactivity (Tartu)*, **7**, 911 (1970); *Chem. Abs.*, **74**, 140568 (1970).
[152] J. Koppel and A. Tuulmets, *Organic Reactivity (Tartu)*, **7**, 1178, 1187 (1970); *Chem. Abs.*, **74**, 19401, 19403 (1970).
[153] M. Felkin, C. Frajerman, and G. Roussi, *Ann. Chim. (Paris)*, **6**, 17 (1971).
[154] M. Chérest, H. Felkin, and C. Frajerman, *Tetrahedron Letters*, **1971**, 379.
[155] M. Chérest and H. Felkin, *Tetrahedron Letters*, **1971**, 383.
[156] A. J. Kresge and V. Nowlan, *Tetrahedron Letters*, **1971**, 4297.
[157] P. Geneste, G. Lamaty, and J. Roque, *Tetrahedron Letters*, **1970**, 5007.
[158] P. Geneste, G. Lamaty, C. Moreau, and J. P. Roque, *Tetrahedron Letters*, **1970**, 5011.
[159] P. R. Jones, W. J. Kauffman, and E. J. Goller, *J. Org. Chem.*, **36**, 186 (1971).

their methyl analogues; the competing reduction afforded axial alcohol with dipropylzinc, and the marked effect of added iodide on the balance of products revealed a complex transition state. The formation of axial alcohol on addition of trimethylaluminium rises with its concentration relative to the ketone in benzene solution.[160] Chelation in the transition state for the reaction of β-hydroxy ketones with Grignard reagents leads to stereoselective synthesis of cis-diols;[161] meso-diols are formed[162] by the selective addition of benzylmagnesium chloride to the α-face of the complex (64) and a number of isomeric methyl-1-phenylcyclohexanols has been prepared.[163] The reaction between Grignard reagents and aroyl peroxides has been studied.[164]

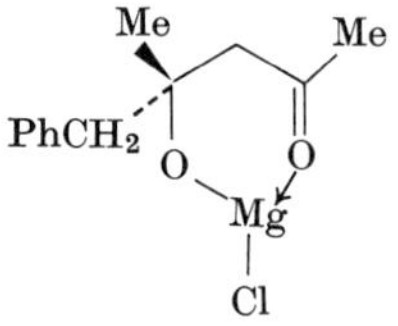

(64)

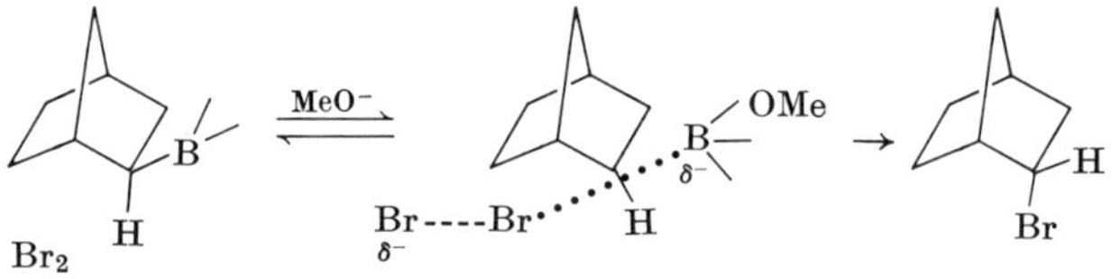

(65)

The formation of an "ate" complex (65) in the methoxide-catalysed bromination of tri-exo-norbornylborane accounts for the high yield of endo-bromo compound, rather than the usual reaction, with retention, through a four-centre transition state.[165]

Organometallics: Other Elements

Organosilicon compounds (X = H, F, or OMe) of known absolute configuration were

$$(+)\text{-}\alpha\text{Np}\cdot\text{Ph}\cdot\underset{\underset{\text{Me}}{|}}{\text{Si}}\text{—X} + \text{EtLi} \rightarrow \alpha\text{-Np}\cdot\text{Ph}\cdot\underset{\underset{\text{Me}}{|}}{\text{Si}}\text{—Et-(+)}$$

$$(+)\text{-}\alpha\text{Np}\cdot\text{Ph}\cdot\underset{\underset{\text{Et}}{|}}{\text{Si}}\text{—X} + \text{MeLi} \rightarrow \alpha\text{-Np}\cdot\text{Ph}\cdot\underset{\underset{\text{Et}}{|}}{\text{Si}}\text{—Me-(−)}$$

shown[166] to react with Grignard reagents and lithium alkyls with retention of configuration: on the other hand, chlorosilanes reacted with inversion. α-Chloroalkylsilanes react with sodium in the gaseous phase faster than the sterically analogous neopentyl chloride. The difference did not correspond to the C—Cl bond energies and was ascribed[167] to trimethylsilylmethide (66) participation in the transition state. Considerable interaction is possible between carbon–metal bonds and neighbouring π-electron systems in tin and mercury alkyls by σ–π conjugation;[168] quantitative evidence of π-bond participation

---

[160] E. C. Ashby and S. Yu, *Chem. Comm.*, **1971**, 351.
[161] E. Ghera and S. Shoua, *Chem. Comm.*, **1971**, 398.
[162] A. Barabas and A. T. Balaban, *Tetrahedron*, **27**, 5495 (1971).
[163] J. R. Luderer, J. E. Woodall, and J. L. Pyle, *J. Org. Chem.*, **36**, 2909 (1971).
[164] M. Okubo, K. Maruyama, and J. Osugi, *Bull. Chem. Soc. Japan*, **44**, 125 (1971).
[165] H. C. Brown and C. F. Lane, *Chem. Comm.*, **1971**, 521.
[166] R. Corriu and G. Royo, *Tetrahedron*, **27**, 4289 (1971).
[167] D. S. Boak and B. G. Gowenlock, *J. Organomet. Chem.*, **29**, 385 (1971).
[168] W. Hanstein, H. J. Berwin, and T. G. Traylor, *J. Am. Chem. Soc.*, **93**, 7476 (1970).

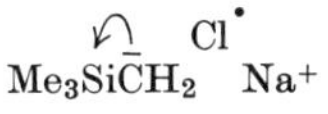

$$\text{Me}_3\text{SiCH}_2 \overset{\curvearrowleft}{} \overset{\cdot}{\text{Cl}} \quad \text{Na}^+ \qquad \left[R \cdots \underset{\text{Br}}{\overset{\overset{\text{Br}}{|}}{\text{Hg}}} \cdots \text{Br} \right] \qquad \left[\overset{\delta-}{\text{Br}} \cdots \text{Br} \cdots R \cdots \overset{\delta+}{\text{MX}} \right]$$

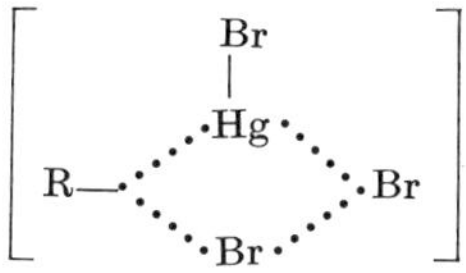

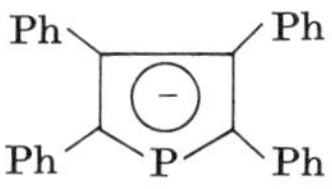

(66)   (67)   (68)

(69) norbornene···Hg^{++}

(70) 1,2,3,4-tetraphenyl phosphole anion (Ph, Ph, Ph, Ph on cyclopentadienyl ring with P, ⊖)

in the iododestannylation and mercuridestannylation of tin norbornenyls has been obtained.[169] The charge-transfer maxima of tetracyanoethylene complexes of trimethylsilylbenzenes show[170] that donor substituents transfer charge to the silicon atom; charge transfer from silicon is limited for dimethylphenyl-, methyldiphenyl-, and triphenyl-silyllithium, for their absorption spectra and exchange constants with fluorene are similar.[171]

Rates of base-induced cleavage of compounds of the type $\text{XC}_6\text{H}_4-(\text{C}\equiv\text{C})_n-\text{MEt}_3$ (M = Si or Ge) increased with electron withdrawal and for greater values of n, being most marked for silicon. Plots of log k_{rel} to σ values indicated[172] that separation of inductive and mesomeric substituent effects was not justified; calculations of σ_{ij} for a substituent atom i and reaction at a remote side j were made but proved inferior to simple σ values. Optically active derivatives of Group IV elements have been prepared:[173]

$$\text{Ph}_3\text{M}^- + (S)\text{-}(+)\text{-s-BuBr} \rightarrow (R)\text{-}(-)\text{-s-BuMPh}_3$$

Here inversion of configuration was confirmed by the observation[174] of inversion in the cleavage of tin tetra-alkyls with bromine, which involves an open transition state (68), rather than retention through one (67) closed by attachment to a metal (e.g. mercury) with vacant orbitals. The mechanism of protodemercuration of mercury dialkyls has been correlated[175] with the strength of the corresponding carbon acids (RH); the rate of iododemercuration in alkylmercury halides increased with the stability of the substrate carbanion[176] but did not correspond with the nucleophilic character of the organic group in mercury dialkyls.[177] The extent of nuclear deuteration of benzylchloromercury

---

[169] R. M. G. Roberts, *J. Organomet. Chem.*, **32**, 323 (1971).
[170] H. Alt and H. Bock, *Tetrahedron*, **27**, 4965 (1971).
[171] A. G. Evans, M. A. Hamid, and N. H. Rees, *J. Chem. Soc.* (B), **1971**, 1110.
[172] C. Eaborn, R. Eastwood, and D. R. M. Walton, *J. Chem. Soc.* (B), **1971**, 127.
[173] F. R. Jensen and D. D. Davis, *J. Am. Chem. Soc.*, **93**, 4047 (1971).
[174] F. R. Jensen and D. D. Davis, *J. Am. Chem. Soc.*, **93**, 4048 (1971).
[175] I. P. Beletskaya, L. V. Savinykh, and O. A. Reutov, *Dokl. Akad. Nauk SSSR*, **197**, 1325 (1971); *Chem. Abs.*, **75**, 36284 (1971).
[176] I. P. Beletskaya, L. V. Savinykh, and O. A. Reutov, *Izv. Akad. Nauk SSSR, Ser. Khim.*, **1971**, 1585; *Chem. Abs.*, **75**, 87749 (1971).
[177] I. P. Beletskaya, L. V. Savinykh, and O. A. Reutov, *J. Organomet. Chem.*, **26**, 13 (1971).

in the reaction with deuterium chloride was controlled by the specificity of the solvent–metal association.[178] A number of bis(perfluoroalkyl) mercurials have been synthesized[179] from mercury(II) halides and perfluoroalkyl carbanions; mercurinium ions showing metal–proton coupling have been characterized[180] by PMR spectroscopy; the ion (**69**) differs from the hydroxynorbornyl cation generated under the same conditions.

Alkali heterocyclopentadienides are formed[181] by removal of a 1-phenyl group by the metal in ethereal or hydrocarbon solvents; the heteroatom is strongly nucleophilic in phosphadienides (**70**), which react with alkyl halides, including methylene chloride. Depending on the conditions the arsenic analogues react similarly or by extrusion of the heteroatom and formation of 1,2,3-triphenylnaphthalene.

The vitamin B_{12}-catalysed methylmalonyl-isomerase reaction has been simulated[182] *in vitro*, and models for enzyme-bound corrinoids have been prepared[183] by electrochemical reduction of organometallic Co(III) chelates; the ions formed strongly nucleophilic organic ligands:

$$[PhCo^{I}\{(do)(doh)pn\}]^{-} \xrightarrow{H^{+}/DMF} [Co^{III}\{(do)(doh)pn\}(dmf)_2]^{+} + benzene$$

Alkylcobalt(III) species exchange the alkyl group with cobalt(I) anions and equilibria are also established[184] by transfer between different metal ions. Electrophilic reactions of σ-bonded organo-transition-metal ions have been reviewed,[185] and an extensive study[186] has been made of dialkylmanganese compounds formed in THF by metathesis of Grignard reagents. They react by β-elimination to form alkenes:

$$(CH_3CH_2CH_2)_2Mn \rightarrow CH_3CH=CH_2 + HMnCH_2CH_2CH_3$$
$$(\mathbf{71})$$

These alkenes are hydrogenated in a second step; the necessary hydrogen is derived indirectly and in part from the hydrido species (**71**), whilst the remainder is obtained by dehydrogenation of the solvent. Catalysis of the alkylmanganese decomposition, and reactions with alkyl halides, were also described. Mercury(II) and thallium(III) species displace[187] the dicarbonyl-π-cyclopentadienyliron cation from (**72**) where the rate coefficients for an S_E2 mechanism varied over a wide range. The substitution of tetraethyltin by mercury(II) halides is similar in alcoholic solution[188] and in acetonitrile;[189] steric effects in a group of tetra-alkyltins on the reaction rate were consistent with an S_E2 mechanism in these solvents.[190] The most probable mechanism for pyridiniomethylthallium(III)[191] and pyridiniomethylchromium(III)[192] species with chloride ion is an S_N2 reaction (**73**) at the methylene group.

---

[178] Y. G. Bundel, V. I. Rosenberg, I. N. Krokhina, and O. A. Reutov, *Zhur. Org. Khim.*, **6**, 1531 (1970); *J. Org. Chem. USSR*, **6**, 1531 (1970).
[179] B. L. Dyatkin, S. R. Sterlin, B. I. Martynov, E. I. Mysov, and I. L. Knunyants, *Tetrahedron*, **27**, 2843 (1971).
[180] G. A. Olah and P. R. Clifford, *J. Am. Chem. Soc.*, **93**, 1261 (1971).
[181] E. H. Braye, I. Caplier, and R. Saussez, *Tetrahedron*, **27**, 5523 (1971).
[182] J. N. Lowe and L. L. Ingraham, *J. Am. Chem. Soc.*, **93**, 3801 (1971).
[183] G. Costa, A. Puxeddu, and E. Reisenhofer, *Chem. Comm.*, **1971**, 993.
[184] M. D. Johnson, *Record Chem. Progr.*, **31**, 143 (1971).
[185] D. Dodd and M. D. Johnson, *Chem. Comm.*, **1971**, 1371.
[186] M. Tamura and J. Kochi, *J. Organomet. Chem.*, **29**, 111 (1971).
[187] D. Dodd and M. D. Johnson, *J. Chem. Soc.* (B), **1971**, 662.
[188] M. H. Abraham and F. Behbahany, *J. Chem. Soc.* (A), **1971**, 1469.
[189] M. H. Abraham and M. J. Hogarth, *J. Chem. Soc.* (A), **1971**, 1474.
[190] M. M. Abraham, F. Behbahany, and M. J. Hogarth, *J. Chem. Soc.* (A), **1971**, 2566.
[191] M. D. Johnson and D. Vamplew, *J. Chem. Soc.* (B), **1971**, 507.
[192] D. Dodd, M. D. Johnson, and D. Vamplew, *J. Chem. Soc.* (B), **1971**, 1841.

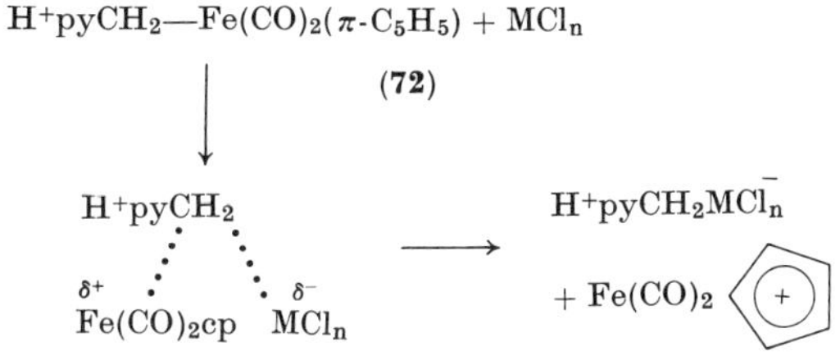

H^+pyCH_2—$Fe(CO)_2(\pi$-$C_5H_5) + MCl_n$

(72)

H^+pyCH_2
$\overset{\delta^+}{Fe(CO)_2cp} \cdots \overset{\delta^-}{MCl_n}$ $\longrightarrow$ $H^+pyCH_2MCl_n^- + Fe(CO)_2 \langle + \rangle$

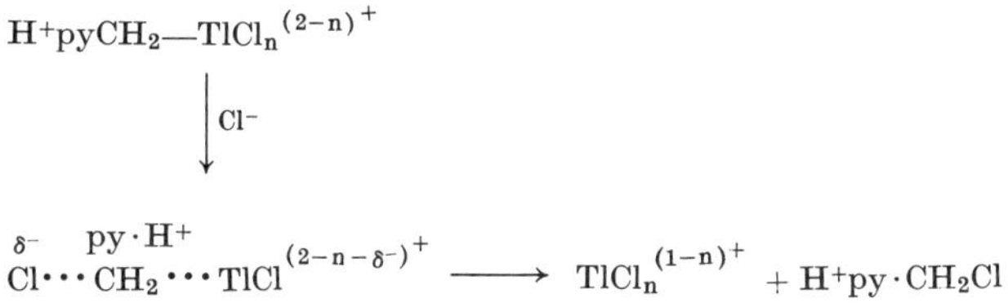

H^+pyCH_2—$TlCl_n^{(2-n)+}$

$\downarrow Cl^-$

$\overset{\delta^-}{Cl} \cdots \overset{py \cdot H^+}{CH_2} \cdots TlCl^{(2-n-\delta^-)+}$ $\longrightarrow$ $TlCl_n^{(1-n)+} + H^+py \cdot CH_2Cl$

(73)

There has been further work on the regioselectivity and mechanism of the reactions of lithium dialkylcoppers with dienones[193] and epoxides.[194,195] Vinylpalladium intermediates have been obtained[196] by exchange between vinylsilanes and palladium salts.

Miscellaneous Reactions

Mercury alkyls reduce trityl bromide.[197] Further examples of cyclopropane synthesis from α-halocarbonyl compounds and sulphonium ylids have been given;[198] there has been a comparative review[199] of ylid chemistry.

Cyclohexynes are intermediates in the synthesis[200] of thioenol ethers and enamines by the reaction of bases and chlorocyclohexenes; a chelating ring effect[201] in thiothiazolinyllithium derivatives (74) leads to improvements in alkylations.

The rate of the zero-order bromination of N,N-disubstituted hydrazones is determined[202] by isomerization to the more compressed *anti*-isomer (75) in which the methine-hydrogen atom is more accessible. Benzaldoximes react in solution with chlorine to give

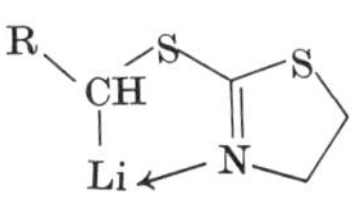

(74)

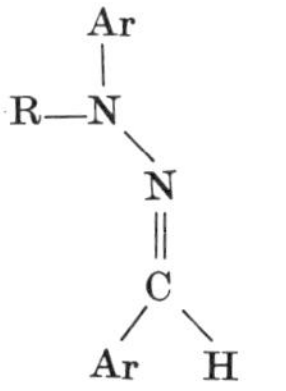

(75)

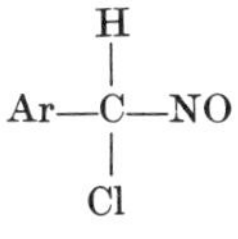

(76)

[193] J. A. Marshall, R. A. Ruden, L. K. Hirsh, and M. Phillipe, *Tetrahedron Letters*, **1971**, 3795.
[194] J. Staroscik and B. Rickborn, *J. Am. Chem. Soc.*, **93**, 3046 (1971).
[195] D. M. Wieland and C. R. Johnson, *J. Am. Chem. Soc.*, **93**, 3047 (1971).
[196] W. P. Weber, R. A. Feux, A. K. Willard, and K. E. Koenig, *Tetrahedron Letters*, **1971**, 4701.
[197] O. A. Reutov, E. V. Uglova, V. D. Makhaev, and V. S. Petrosyan, *J. Org. Chem. USSR*, **6**, 2164 (1970); *Chem. Abs.*, **74**, 53959 (1970).
[198] P. Bravo, G. Fronza, G. Gaudiano, C. Ticozzi, and M. G. Zubiani, *Tetrahedron*, **27**, 3563 (1971).
[199] R. F. Hudson, *Chem. in Britain*, **7**, 287 (1971).
[200] P. Caubere and J. J. Brunet, *Tetrahedron*, **27**, 3515 (1971).
[201] K. Hirai, H. Matsuda, and Y. Kishida, *Tetrahedron Letters*, **1971**, 4359.
[202] F. L. Scott, F. A. Groeger, and A. F. Hegarty, *J. Chem. Soc. (B)*, **1971**, 1411.

benzylidene chlorides and/or hydroxamoyl chlorides: the intermediate C-nitroso compound (**76**) is thought[203] to decompose to the hydroxamoyl chloride by proton transfer catalysed by alcohol or an amine. The presence of nitro substituents encourages benzylidene chloride formation through the conjugate base of (**76**).

The exchange between dimethylarsenic deuteride and diethylarsine has been studied[204] by PMR spectroscopy and the chiroptical properties of $\beta\gamma$-unsaturated sulphoxides have been investigated.[205]

A nitroxide free radical was coordinated to cobinamide co-enzymes as a probe to demonstrate[206] that homolysis of the Co—C bond occurs during catalysis.

[203] Y. H. Chiang, *J. Org. Chem.*, **36**, 2146 (1971).
[204] W. R. Cullen and W. R. Leeder, *Can. J. Chem.*, **48**, 3757 (1970).
[205] D. N. Jones, E. E. Helmy, R. J. K. Taylor, and A. C. F. Edmonds, *Chem. Comm.*, **1971**, 1401.
[206] P. Y. Law, D. G. Brown, E. L. Lien, B. M. Babior, and J. M. Wood, *Biochemistry*, **10**, 3428 (1971).

CHAPTER 4

Elimination Reactions

A. C. KNIPE

Chemistry Department, The New University of Ulster

Stereochemistry and Orientation in *E*2 Reactions	135
The *E1cB* Mechanism	137
The *E2C* Mechanism	140
Gas-phase Elimination Reactions	142
Other Topics	145

Stereochemistry and Orientation in *E*2 Reactions

The steric course of *E*2 eliminations of bicyclo[2.2.2]octyl 'onium bases and tosylates has been studied,[1] by deuterium labelling techniques, for a range of base–solvent systems including Bu^tO^-–(DMSO or C_6H_6) and RO^-–ROH. Only with Bu^tO^- in aprotic solvents, and on pyrolysis of the quaternary hydroxide, can substantial yields of olefin be obtained from the quaternary bases and in each case the mechanism is almost exclusively *syn*-*E*2. By investigation of *trans*-3-methylbicyclo[2.2.2]oct-2-yltrimethylammonium base and open-chain 'onium compounds an alternative *syn*-mechanism through cyclic intramolecular fragmentation of the corresponding ylid has been shown to compete only if there is enforced *syn*-planarity of the bonds to be broken, the approach of an external base to the β-H is made difficult, the β-H closely approaches the 'onium group and an aprotic solvent is employed. As with bicycloheptyl analogues[2] the contribution of *syn*-*E*2 elimination for the *exo*-bicyclo-octyl tosylates is smaller than for the 'onium bases, yet favoured relative to *anti*-elimination in non-polar solvents. This solvent-dependence has previously been attributed to a "pseudocyclic" mechanism facilitated by ion pairing of the attacking base.[3]

It has generally been found that the tendency for *syn*-elimination in acyclic systems increases with vicinal steric interaction, particularly if a bulky base is employed.[4] *syn*-Elimination has now been reported[5] for reaction of the sterically unencumbered systems *threo*- or *erythro*-PhCHDCHDX with Bu^tOK in benzene and to a much smaller extent in Bu^tOH and DMSO. This is also in accord with the suggestion[3,6] that the effective reagent in *syn*-elimination is not the alcoholate anion as in *anti*-elimination but the RO^-K^+ ion pair. For the tosylates, deuterium kinetic isotope effects (k_H/k_D) for *syn*- and *anti*-eliminations in benzene are ca. 2.2 and 4.0, respectively, and indicate that the reactions are mechanistically distinct. In benzene *syn*-elimination accounts for up to 20% of the total elimination pathway in this system where steric interactions

---

[1] J. Sicher, M. Pánková, J. Závada, L. Kniezo, and A. Orahovats, *Coll. Czech. Chem. Comm.*, **36**, 3128 (1971).
[2] Cf. *Org. Reaction Mech.*, **1970**, 150.
[3] Cf. *Org. Reaction Mech.*, **1968**, 146.
[4] See *Org. Reaction Mech.*, **1970**, 157.
[5] W. F. Bayne and E. I. Snyder, *Tetrahedron Letters*, **1971**, 571.
[6] Cf. J. Závada, M. Pánková, and J. Sicher, *Chem. Comm.*, **1968**, 1145.

are minimal; thus it is clear that a rationale for *syn*-elimination cannot be based on repulsive steric interactions alone.

Investigation of stereospecifically β-deuteriated and $\beta,\beta,\beta',\beta'$-tetradeuteriated 5-decyl and 5-nonyl trimethylammonium bases and tosylates has revealed that 'onium bases with Bu^{t}OK in protic or aprotic solvents (Bu^{t}OH, DMSO or C$_6$H$_6$), and also on pyrolysis, give the *trans*- and *cis*-olefin (dec-5-ene or non-4-ene) mainly by *syn*- and *anti*-elimination, respectively, and that *syn*-elimination predominates.[7] Only in MeO$^-$–MeOH, or for reactions of the tosylates under any of the conditions studied, is *trans*-olefin formed mainly by *anti*-elimination. Thus earlier findings[8] that $E2$ eliminations of C$_8$–C$_{18}$-cycloalkyl 'onium bases give *cis*- and *trans*-cycloalkenes by *anti*- and *syn*-eliminations, respectively, whereas for tosylates and bromides[3] *trans*-cycloalkene formation by the *syn*-mechanism is less general, are also features of reaction of open-chain compounds. The contribution of *syn*-elimination is, however, less pronounced and in MeO$^-$–MeOH *anti*-elimination predominates. The contribution of *syn*-elimination to Bu^{t}O$^-$-induced formation of *trans*-dec-5-ene from the corresponding tosylate decreases from 27 to 14.5 to 3% as the solvent is changed from C$_6$H$_6$ to Bu^{t}OH and DMF. This marked solvent-dependence has also been reported for bicyclo-octyl tosylates[1] and no doubt, in part, accounts for the almost exclusive *anti*-$E2$ eliminations reported for 2-butyl tosylates in EtO$^-$–EtOH and Bu^{t}O$^-$–DMSO. The significance of *trans/cis*-olefin ratios has been discussed.

Also consistent with these trends are observations[9] that both *cis*- and *trans*-but-2-ene are formed by exclusive *anti*-elimination from ($\pm$)-*erythro*- and ($\pm$)-*threo*-2-bromo-3-deuteriobutane [(**1**) and (**2**), respectively] with EtOK–EtOH, Bu^{s}OK–Bu^{s}OH, Bu^{t}OK–Bu^{t}OH, Bu^{t}OK–DMSO, Bu^{t}OK–THF and Bu$^n_4$NF–DMF. The primary deuterium

isotope effects are equal for formation of *trans*-but-2-ene from (**1**) and *cis*-but-2-ene from (**2**) in a given medium. Since it is unlikely that the stereochemistry of reaction of 2-hexyl halides with Bu^{t}OK–DMSO differs markedly from that of 2-butyl bromide, these results suggest that *syn*-elimination is not responsible for the high *trans/cis*-hex-2-ene ratios previously observed. Such high ratios have now been found to be characteristic of 2-alkyl halide eliminations promoted by a wide variety of bases in dipolar aprotic solvents.[10]

For reactions of 2-butyl halides with several bases (including carboxylate, cyanide and halide ions as well as alkoxide and phenoxide) an inverse dependence of the *trans/cis*-but-2-ene ratio and the percentage of but-1-ene has been found[10] in alcoholic solvents, whereas in dipolar aprotic solvents (DMF and DMSO) high *trans/cis* ratios of 3.0–4.0 are independent of the percentage of but-1-ene which varies from 2 to 40%. It has been concluded, by virtue of the high *trans/cis* ratio, that the eliminations in dipolar aprotic solvents have a high degree of double-bond character in the transition state, although

---

[7] J. Sicher, J. Závada, and M. Pánková, *Coll. Czech. Chem. Comm.*, **36**, 3140 (1971).
[8] Cf. *Org. Reaction Mech.*, **1969**, 155.
[9] R. A. Bartsch, *J. Am. Chem. Soc.*, **93**, 3683 (1971).
[10] R. A. Bartsch, C. F. Kelly, and G. M. Pruss, *J. Org. Chem.*, **36**, 662 (1971).

the reason for this divergence of orientation from that observed in alcoholic solvents is not clear (see also discussion in ref. 1). Only small changes in butene proportions are effected upon variation of alcoholic solvent for a common base, and the much larger effects of variation of alkoxides in the respective alcohols have therefore been attributed primarily to differing base strengths and not to solvent effects. Plots of log (% cis-but-2-ene/% but-1-ene) against log (% trans-but-2-ene/% but-1-ene) are rectilinear for eliminations in dipolar aprotic solvents and in alcohols, respectively, and attest to regularities in the effects of base, solvent and leaving group for the 55 combinations represented. The linear free-energy relationships suggest that the stereochemistry of elimination is invariant for each plot and is most probably exclusively *anti* by comparison with the known stereochemistry of eliminations of (**1**) and (**2**). There is no discontinuity to suggest unusual steric interaction with bulky bases,[11] such as Bu^{t}OK and Et$_3$COK. Orientation in eliminations from 2-butyl toluene-p-sulphonate induced by 13 base–solvent combinations[12] did not correlate with these results.

Products of elimination have been determined for reaction of a series of 2-pentyl *para*-substituted phenyl sulphones with sodium ethylene glycollate in refluxing ethylene glycol and with Bu^{t}OK in refluxing pyridine.[13] In either medium the percentage of pent-1-ene (ca. 77 and 96%, respectively) was independent of the phenyl substituent and indicated a more Hoffmann-like pattern for the latter case. The results are consistent with an $E1cB$-like transition state. The ratio of *trans*- to *cis*-pent-2-ene from the mesityl sulphone is anomalous in both media, although the proportion of pent-1-ene is as for other aryl sulphones. This is clear indication that *trans/cis* ratios can be influenced by the steric requirements of the leaving group, although the stereochemistry of these eliminations has not yet been determined. A five-centre cyclic transition state has been proposed to account for the ready elimination reaction of the p-dimethylamino sulphone in the absence of base.

The generation of *trans*-cinnamate from phenylalanine by the action of L-phenylalanine ammonia-lyase has been found[14] to involve antiperiplanar elimination of ammonia and the *pro-S* proton from C-3. The kinetics and distribution of products from reactions of 2-halogeno-2,3,3-trimethylbutane with NaSEt or NaOMe in methanol containing NaClO$_4$ could not be satisfactorily explained.[15] Base-induced dehydrohalogenation[16] of R$_F$CH$_2$CHICH$_2$CO$_2$H (R$_F$ = perfluoroalkyl) gives R$_F$'CF=CHCH=CHCO$_2$H through R$_F$CH$_2$CH=CHCO$_2$H and not R$_F$CH=CHCH$_2$CO$_2$H. High yields (>99%) of unrearranged alk-1-enes have been obtained[17] upon reaction of n-alkyl bromides with hexamethylphosphorotriamide at 180–210°.

The $E1cB$ Mechanism

It has been suggested that an $E1cB$ mechanism generally applies for reaction of β-oxy ketones possessing leaving groups in the pK_a range 4–16 and that for good leaving groups k_1 is rate-determining; for poor leaving groups k_2 is rate-determining and for leaving

---

[11] Cf. *Org. Reaction Mech.*, **1970**, 156–157.
[12] D. H. Froemsdorf and M. D. Robbins, *J. Am. Chem. Soc.*, **89**, 1737 (1967).
[13] A. K. Colter and R. E. Miller, *J. Org. Chem.*, **36**, 1898 (1971).
[14] K. R. Hanson, R. H. Wightman, J. Staunton, and A. R. Battersby, *Chem. Comm.*, **1971**, 185.
[15] J. F. Bunnett and D. L. Eck, *J. Org. Chem.*, **36**, 897 (1971).
[16] N. O. Brace, *J. Org. Chem.*, **36**, 1904 (1971).
[17] R. S. Monson, *Chem. Comm.*, **1971**, 113.

groups of intermediate pK_a partitioning of the enolate anion is important.[18] Investigation of the formation of but-3-en-2-ones by elimination of *para*-substituted phenols from 4-(*p*-substituted phenoxy)butan-2-ones was undertaken in order to test the feasibility of $E1cB$ mechanisms that have previously been proposed to account for reactions of butan-2-ones bearing 4-methoxy or 4-*p*-substituted benzoyloxy leaving groups:

$$\text{SH} \underset{k_{-1}[\text{BH}^+]}{\overset{k_1[\text{B}]}{\rightleftarrows}} \text{S}^- \overset{k_2}{\longrightarrow} \text{Products}$$

Use of tertiary amine buffers made possible kinetic detection of S^-, since the presence of a significant amount of sufficiently strong general acid favoured collapse to SH, and choice of poor leaving groups (pK_a ca. 9–10) kept k_2 not much greater than k_{-1}. General base catalysis approaching saturation kinetics at high amine buffer concentrations was observed. The rate constant k_2 is markedly affected by the nature of the leaving group and relative to $\rho' = 1$ for the ionization of *para*-substituted phenols, $\rho' = 0.066 \pm 0.002$ for formation of the enolate ions and $\rho' = 0.67 \pm 0.08$ for their decomposition to products in 2-dimethylaminoethanol buffers. It has therefore been estimated that in the transition state ca. 67% of the charge resides on the phenoxy portion of S^-. (These results closely resemble those[19] for aryl 2-aryloxyethyl sulphones, sulphoxides and sulphonium salts in EtONa–EtOH.) Of particular interest is that k_2 has been found to depend on the pK_a of the general acid in such a way that S^- is stabilized towards elimination in low relative to high pK_a buffers. This has been attributed to ion-pair association with BH^+ and finds analogy elsewhere.[20,21]

Further support for $E1cB$ reactions[19] of XCH_2CH_2OPh has been gained by comparison[22] of their relative rates of base-catalysed elimination of PhO^- with those for nucleophilic vinylic substitution on $XCH=CHR$ and with ionization rate and equilibrium constants (k_{ion} and K_{eq}) for the carbon acids MeX.

Small leaving group effects for both *syn* and *anti* base-initiated eliminations from cyclohexanes (**3**) and (**4**), wherein the β-proton is activated by an $ArSO_2$ group have been interpreted in terms of a common $E1cB$ mechanism.[23] Retardation of proton

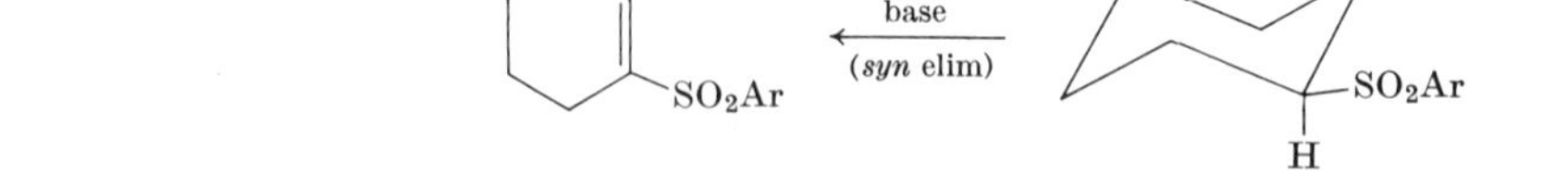

[18] L. R. Fedor and W. R. Glave, *J: Am. Chem. Soc.*, **93**, 985 (1971).
[19] See *Org. Reaction Mech.*, **1970**, 158.
[20] T. I. Crowell, R. T. Kemp, R. E. Lutz, and A. A. Wall, *J. Am. Chem. Soc.*, **90**, 4638 (1968).
[21] Z. Rappoport and E. Shohamy, *J. Chem. Soc.* (B), **1971**, 2060.
[22] Z. Rappoport, *J. Chem. Soc.* (B), **1971**, 171.
[23] F. G. Bordwell, J. Weinstock, and T. F. Sullivan, *J. Am. Chem. Soc.*, **93**, 4728 (1971).

abstraction due to chair deformation in cyclohexane systems bearing 1,2-diequatorial substituents has been invoked to explain high *anti/syn* rate ratios which may therefore be attributed to retardation of *syn*-elimination. For reactions of (**3**) and (**4**) with OH⁻ and with Me₃N in 50% aqueous dioxan, general base catalysis has been observed and Hammett ρ values are small (0.33–0.59) for both *syn*-elimination of (**4**) and *anti*-elimination of (**3**). The close correspondence of ρ values (ca. 1.5–1.9) for *syn*- and *anti*-elimination is also a feature of reactions of analogous cyclohexyl and cyclopentyl chlorides with base in 80% EtOH. By comparison of rate data, activation parameters and ρ values it has been concluded that base-promoted eliminations activated by a sulphone group in acyclic, C₅-cyclic and C₆-cyclic systems all proceed by the same mechanism. Even though elimination is generally faster than the expected rate of deuterium exchange and reactions of *threo*- and *erythro*-ArSO₂CHMeCHMeOBs are highly stereoselective, it has been argued that the common mechanism is carbanionic. Thus it has been suggested that exchange with deuterated solvent may be less able to compete with internal return of a sulphonyl carbanion than is loss of OTs⁻ from a comparable β-tosyl sulphonyl carbanion. The stereoselective preference for *anti*-elimination reactions of the acyclic *p*-bromobenzenesulphonates with tertiary amines (reagents known to promote internal return) can also be rationalized if it is assumed that elimination competes effectively with internal return only when the carbanion is generated from a conformation where *anti*-elimination can occur. As the sulphonyl group is one of the weaker electron-withdrawing groups the conclusions are consistent with a postulate that most 1,2-elimination reactions proceed through carbanion or carbonium ion intermediates and that truly concerted eliminations are relatively rare.[24]

The competitive elimination and H–D exchange reactions[25] of 1-methoxyacenaphthenes (**6a–d**) (p. 140) change from exclusively *cis*- to preferential *trans*-stereoselectivity as the cation in Bu^{t}O⁻–M⁺–Bu^{t}OH is changed through Li⁺, K⁺, Cs⁺ to (CH₃)₄N⁺. The kinetic isotope effects (k_H/k_D) for elimination and exchange reactions of (**6c**) and (**6d**) fall in the range 1.6–1.8 for cations Cs⁺, K⁺ and K⁺ crown ether and suggest that both reactions proceed through a carbanion intermediate. The strong preference for *syn*-reactions of Bu^{t}O⁻Li⁺ has been interpreted in terms of a strong coordination between Li⁺ and the ether oxygen linkage, i.e. formation of (**9**) is greatly preferred over that of (**8**). The coordinating ability is progressively reduced for K⁺ and Cs⁺. Intermediates (**8**) and (**9**) become almost degenerate in the absence of coordinating cations and this has been dramatically demonstrated by the large decrease in *syn*-elimination and exchange upon addition of dicyclohexyl 18-crown ether to Bu^{t}OK or use of the tetramethylammonium cation. In contrast, crown ether has no effect on reactions in MeOK–MeOH in which the base is already fully dissociated.

It has been shown that amine-catalysed elimination of HCN from 2,6-dimethyl-4-($\alpha,\alpha,\beta,\beta$-tetracyanoethyl)aniline (HAX) in chloroform goes to completion even if the amine is in deficit.[21] This special case arises since HAX is more acidic than HX and X⁻ is therefore neutralized by BH⁺ with regeneration of the catalyst B:

$$\text{ArC(CN)}_2\text{CH(CN)}_2 + \text{R}_3\text{N} \underset{k_{-1}}{\overset{k_1}{\rightleftharpoons}} \text{R}_3\text{NH}^+ + \text{ArC(CN)}_2\overset{-}{\text{C}}\text{(CN)}_2 \longrightarrow \text{ArC(CN)}{=}\text{C(CN)}_2 + \text{CN}^-$$

(HAX)　　(B)　　　　　(BH⁺)　　　　(AX⁻)　　　　　　　(A)

Where R = Et or n-Bu and [B] < [HAX] the reaction is of zero order in both B and HAX within a run and of first order in B for different runs. This behaviour is typical of "*E1cB*

[24] F. G. Bordwell, *Accounts Chem. Res.*, **3**, 281 (1970).
[25] D. H. Hunter and D. J. Shearing, *J. Am. Chem. Soc.*, **93**, 2348 (1971).

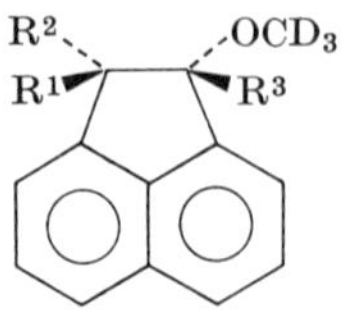

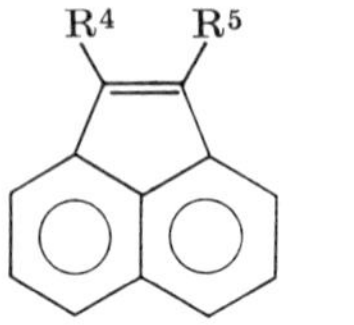

(6a) $R^2 = D; R^1 = R^3 = H$
(6b) $R^1 = D; R^2 = R^3 = H$
(6c) $R^1 = R^2 = R^3 = D$
(6d) $R^1 = R^2 = R^3 = H$

(7a) $R^4 = R^5 = H$
(7b) $R^4 = D; R^5 = H$
(7c) $R^4 = R^5 = D$

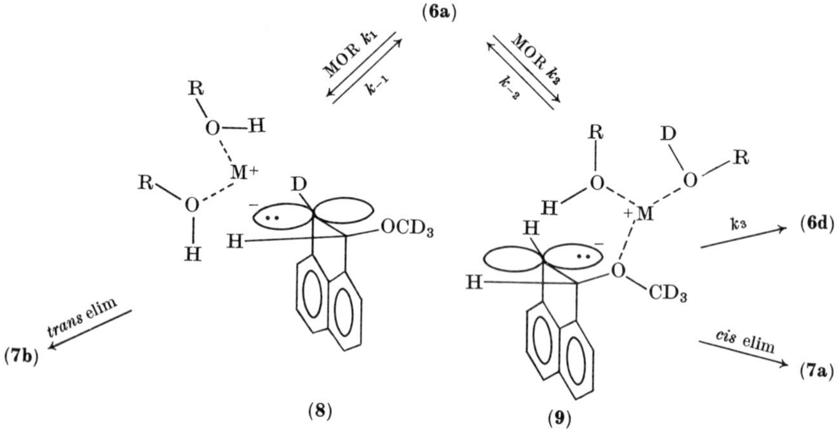

reaction of the second type", wherein neutralization of HAX by B is almost complete and k_2 is rate-determining. For amines of lower pK_b the reaction is of second order, there being less complete neutralization of HAX and onset of the "pre-equilibrium $E1cB$ mechanism", as indicated by deuterium exchange and the absence of a deuterium isotope effect ($k_H/k_D \simeq 1.0$ for all amines studied). The inverse rate dependence on [BH$^+$] expected of the rate expression $k_{obs} = k_1 k_2/k_{-1}$[BH$^+$] was not found and this has been attributed to CN$^-$ expulsion electrophilically assisted by BH$^+$, most probably within an ion pair (cf. ref. 18 and 20). The $(E1cB)_{ip}$ mechanism has also been proposed for R = Et or n-Bu and by other workers[26] to account for kinetics of the analogous reaction of triphenyl[2-(1-phenyl-1,2,2-tricyanoethyl)cyclopentadienylidene]phosphorane with tertiary amines in C_6H_6 and C_6H_6–CH_3CN. Base-catalysed reactions of methyl 6β-phthalimidopenicillanate have been studied.[27]

The *E2C* Mechanism

The generation of olefins by elimination promoted by weak bases has been reviewed.[28] Further definition of the spectrum of transition states ranging from *E2H*-like to *E2C*-like has been realized by detailed analysis of effects of alkyl, aryl, benzyl, bromine and

[26] E. Lord, M. P. Naan, and C. D. Hall, *J. Chem. Soc.* (B), **1971**, 220.
[27] B. G. Ramsay and R. J. Stoodley, *Chem. Comm.*, **1971**, 450.
[28] A. J. Parker, *Chem. Technol.*, **1971**, 297.

```
   B----H                  H--B                    H  B
   R^β   H                R^β  H   R^α            R^β  H  H  R^α
     \C==C/R^α              \C==C/                   \C==C/
   R^β/    \R^α           R^β/    \R^α            R^β/    \R^α
       X                      X                       X
      E2H                                            E2C
```

methoxycarbonyl substituents ($R^α$ and $R^β$) on rates of bimolecular $β$-elimination promoted by a wide range of base–solvent systems.[29] It is clear that $E2C$-like reactions give high yields of the most stable olefin provided that the requirement of *anti*-geometry of $β$-hydrogen and leaving group is not violated. Although the $E2C$ transition state is believed to be very product-like, substituent effects on rate suggest that $α$- and $β$-aryl groups are not well conjugated with the incipient double bond and this has not yet been adequately explained. Product distribution can be predicted from the generalization that $E2C$-like transition states are very product-like whereas $E2H$-like are carbanion-like. Also of synthetic application is the general tendency for halide ions in acetone to induce much more rapid elimination than do alkoxides in alcohol with very weakly acidic substrates, whereas for more acidic substrates the converse is true. Tetrabutylammonium acetate in dipolar aprotic solvents is a particularly effective base for promoting $E2$ reactions.

The effect of bulky $α$-substituents is to accelerate $E2C$ yet retard S_N2 reactions, and this has recently been claimed as evidence against any association between base and $C_α$ in the $E2C$-like transition state.[30] The arguments in favour of this association have therefore been further outlined and it has been suggested that, with respect to changes in hybridization at $C_α$, $E2C$-like transition states are S_N1-like, a difference being that there is negligible positive charge at $C_α$. The effects of *para*-substituents on the rates of S_N2- and $E2C$-like reactions of 1-aryl-1-bromopropanes with NBu_4Cl in DMF have been measured and compared with their effect on the corresponding solvolysis in 90% acetone in order to test this hypothesis.[31] The S_N2 reactions show much the same features as S_N2 reactions of benzyl bromide, and replacement of the aryl group by hydrogen has little effect. The rates of $E2C$-like reaction are also extremely insensitive to aryl substituents; however, there is a substantial rate decrease upon removal of the aryl group. There is no correlation between substituent effects for the S_N1 and $E2C$ reactions, for which $ρ$ values -5.30 and <-0.6, respectively, have been determined. It has been concluded that there is negligible charge development at $C_α$ in these $E2C$-like reactions and that the concept of a spectrum of $E2$ transition states bounded at one extreme by a paenecarbonium ion for elimination by weak bases can be discarded.

The differences in Hoffmann–Saytzeff and *trans/cis*-olefin proportions from $E2C$- and $E2H$-like reactions have been applied to synthetic advantage.[32-34] Thus reaction of (**10**) with strong hydrogen but weak carbon bases[33] (e.g. Bu^tOK–Bu^tOH) gave (**11a**) (1.5%) and (**11b**) (96%) by elimination of the most acidic hydrogen H_a. Conversely, under $E2C$-like conditions (e.g. Bu_4NBr in acetone) elimination of H_b gave (**11c**) (90%). Both $E2C$- and $E2H$-like reactions gave cleaner products in higher yield than $E1$ or

---

[29] G. Biale, D. Cook, D. J. Lloyd, A. J. Parker, J. D. R. Stevens, J. Takahashi, and S. Winstein, *J. Am. Chem. Soc.*, **93**, 4735 (1971).
[30] Cf. *Org. Reaction Mech.*, **1970**, 162.
[31] D. J. Lloyd and A. J. Parker, *Tetrahedron Letters*, **1970**, 5029.
[32] D. J. Lloyd and A. J. Parker, *Tetrahedron Letters*, **1971**, 637.
[33] D. J. Lloyd, D. M. Muir, and A. J. Parker, *Tetrahedron Letters*, **1971**, 3015.
[34] J. Avraamides and A. J. Parker, *Tetrahedron Letters*, **1971**, 4043.

$$\underset{\substack{| \quad | \\ H_a \quad OTs \\ (\mathbf{10})}}{\overset{\substack{H_a \quad H_b \\ | \quad |}}{Ph-C-CH-C}}\begin{smallmatrix}CH_3 \\ \diagdown \\ CH_3\end{smallmatrix} \xrightarrow{\text{Elimination}} \begin{array}{ll} cis & PhCH{=}CH{-}CH(CH_3)_2 \quad (\mathbf{11a}) \\ trans & PhCH{=}CH{-}CH(CH_3)_2 \quad (\mathbf{11b}) \end{array} \Bigg\} \text{Hofmann}$$

$$PhCH_2CH{=}C(CH_3)_2 \quad (\mathbf{11c}) \quad \text{Saytzeff}$$

$$\left.\begin{array}{l} PhCH_2CH_2{-}C{\diagup}\begin{smallmatrix}CH_2 \\ \diagdown \\ CH_3\end{smallmatrix} \quad (\mathbf{11d}) \\ PhC(CH_3){=}C(CH_3)_2 \quad (\mathbf{11e}) \end{array}\right\} \text{Rearranged}$$

equilibration reactions and with less than 1% of rearrangement to (**11d**) or (**11e**). It has been demonstrated, for reactions of *erythro*-1,2-dibromo-1-(*p*-nitrophenyl)-2-phenylethane with NBu₄CN, that an *E2C*-like dehydrobromination is slowed less than a competing *E2Br*-debromination upon transfer from dipolar aprotic to protic solvent since the loose *E2C*-like transition state is a better H-bond acceptor than its tighter *E2Br*-like counterpart.[34] In DMF *anti*-debromination gave 4-nitro-*trans*-stilbene (>99%) and the proportion of α-bromo-4-nitro-*cis*-stilbene, the product of *anti*-dehydrobromination, increased to 90% as the solvent composition approached pure EtOH.

Elimination reactions of phenethyl bromide in CH₃CN are promoted by tetraethylammonium fluoride, which acts as a strong base, whereas chloride and bromide ion effect only substitution under the same conditions.[35] The kinetic effect of ring substituents has now been studied[36] for reactions of ArCH₂CH₂X with F⁻ at 5°. Hammett ρ values (2.033 and 1.879) and 2,2-²H₂ isotope effects (3.99 and 5.03) have been determined for X = Cl and Br, respectively. The ρ values are considered too small and the isotope effects too large for *E1cB*-reaction[37] but are in the range expected for the *E2* reaction. The isotope effect is smaller than for reaction of phenethyl bromide with a conventional base, yet the small k_{OTs}/k_{Br} ratio (0.027) suggests that bond breaking is not well advanced. It has been concluded that the reaction proceeds through a tight reactant-like transition state where the β-proton is less than half transferred to base and that the more *E2C*-like reactions of other halide ions are consistent with the HSAB principle. Fluoride ion promises to be a useful reagent for dehydrohalogenation in dipolar aprotic solvents, and reaction features including the orientation rule for olefin formation from secondary alkyl derivatives have been discussed in detail[38] after a preliminary report that was reviewed last year.[39] The influence of dipolar aprotic solvents on reactions of NaCN and KCN with 2-phenylpropyl toluene-*p*-sulphonate has been interpreted in terms of ion-pair association which retards substitution more than elimination.[40]

Gas-phase Elimination Reactions

Structure–reactivity relationships in homogeneous gas-phase reactions, including elimination, have been reviewed,[41] and heterogeneous catalytic β-eliminations have been compared with their heterolytic liquid-phase analogues.[42]

[35] J. Hayami, N. Ono, and A. Kaji, *Tetrahedron Letters*, **1968**, 1385; *Nippon Kagaku Zasshi*, **92**, 87 (1971).
[36] J. Hayami, N. Ono, and A. Kaji, *Bull. Chem. Soc. Japan*, **44**, 1628 (1971).
[37] See D. J. McLennan, *Quart. Rev. (London)*, **21**, 490 (1967); Z. Rappoport, *Tetrahedron Letters*, **1968**, 3601.
[38] N. Ono, *Bull. Chem. Soc. Japan*, **44**, 1369 (1971).
[39] See *Org. Reaction Mech.*, **1970**, 163; J. Hayami, N. Ono, and A. Kaji, *Tetrahedron Letters*, **1970**, 2727.
[40] A. Loupy and J. Seyden-Penne, *Bull. Soc. Chim. France*, **1971**, 2306.
[41] G. G. Smith and F. W. Kelly, *Progr. Phys. Org. Chem.*, **8**, 75 (1971).
[42] H. Noller, P. Andréu, and M. Hunger, *Angew. Chem. Int. Ed.*, **10**, 172 (1971).

Techniques that have successfully identified ionic intermediates in solution have been applied to gas-phase pyrolyses.[43-46] No such species could be detected during formation of butenes by unimolecular gas-phase pyrolysis of (−)-s-butyl thiocyanate.[43] No racemization occurred and s-butyl isothiocyanate (<3%) was formed only through back-addition to butenes and not by "internal return" of the ambident leaving group. The product composition differed markedly from that of the solvolysis reaction. A six-centre transition state has been proposed that is consistent with these findings and with the activation parameters. Pyrolyses of ethyl, [1,1-^{2}H$_2$]ethyl and [^{2}H$_5$]ethyl thiocyanate have also been studied.[45] The molecular isotope effects, formation of $CH_2=CD_2$ (95%) from the 1,1-^{2}H$_2$ isomer and absence of radical intermediates likewise favour a six-centre over an alternative four-centre transition state which might have been expected to yield detectable thiocyanic acid (rather than isothiocyanic acid, the only isomer found so far). In sharp contrast a four-centre transition state is favoured for s-butyl isothiocyanate pyrolyses.[44] Rate effects of α-alkyl substitution suggest that isothiocyanates eliminate through a transition state of comparable polarity to that for carboxylic esters, although product distributions suggest that there is more incipient double-bond character in the former case. Only minor scrambling of the oxygen of unpyrolysed ^{18}O-labelled ethyl acetate could be detected at pyrolysis temperatures.[46] This result supports a cyclic elimination mechanism and places the ion-pair mechanism[47] in question since equilibration between the ester and such an intermediate might be expected.

Rates of pyrolyses (330–410°) of 1-arylethyl acetates have given a measure of the electronic effects of *ortho*-substituents on this reaction which proceeds by partial formation of a carbonium ion at the α-sidechain carbon.[48] For *m*-CF$_3$ the value $\sigma^+ = 0.563$ has been evaluated and found to correlate existing data on electrophilic aromatic substitutions better than the value 0.52 usually employed. That this group is no more deactivating in the *ortho*- than in the *meta*-position has been attributed to stabilization of the electron-deficient reaction centre, through space, by the fluorine lone pair. The effects of *o*-nitro and *o*-halogeno substituents fit a linear correlation of kinetic data for 1-arylethyl acetates and the corresponding methyl carbonates. The use of such correlations to derive o-σ^+ values is, however, invalid since for reactions at the α-position through electron-deficient transition states, where the solvent is absent or of low polarity, these substituents deactivate more than from the *meta*-positions, probably through a direct field effect. Effects of halogeno substituents are paralleled in the abnormally low activation by *o*- relative to *p*-O- and *p*-S-containing substituents for similar reactions, and the high activation by the *o*-benzyl group has been attributed to interaction between the substituent π-electrons and the incipient cation.[49] It is therefore clear that, although pyrolysis of 1-arylethyl acetates is a useful model reaction for predicting the effects of *meta*- and *para*-substituents in electrophilic aromatic substitutions,[50] the model cannot be simply extended to *ortho*-substituents. Relative rates of pyrolysis of 1-phenyl- and 1-pentafluorophenyl-ethyl acetates have revealed exalted reactivity of the perfluoro group.[51]

[43] N. Barroeta, A. Maccoll, M. Cavazza, L. Congiu, and A. Fava, *J. Chem. Soc.* (B), **1971**, 1264.
[44] N. Barroeta, A. Maccoll, M. Cavazza, L. Congiu, and A. Fava, *J. Chem. Soc.* (B), **1971**, 1267.
[45] N. Barroeta and A. Maccoll, *J. Am. Chem. Soc.*, **93**, 5787 (1971).
[46] G. G. Smith, K. J. Voorhees, and F. M. Kelly, *Chem. Comm.*, **1971**, 789.
[47] Cf. A. Maccoll and P. J. Thomas, *Progr. Reaction Kinetics*, **4**, 119 (1967).
[48] R. Taylor, *J. Chem. Soc.* (B), **1971**, 622.
[49] R. Taylor, *J. Chem. Soc.* (B), **1971**, 1450.
[50] R. Taylor *et al.*, *Tetrahedron*, **19**, 937 (1963); *J. Am. Chem. Soc.*, **84**, 4817 (1962).
[51] R. Taylor, *J. Chem. Soc.* (B), **1971**, 255.

The method of O'Neal and Benson[52] has been used effectively[53] to estimate Arrhenius A-factors and permit evaluation of activation energies (E_a) from experimental rate constants for the six-centre decompositions of pentan-2-one (59), methoxyacetone (58), pentane-2,4-dione (51), methyl butyrate (≫70), hept-1-ene (54) and 4-methylhex-1-ene (54–55). Experimental results for ethyl formate were in accord with the predictions.

There has been much theoretical debate over the interpretation of k_H/k_D ratios and controversy has been heightened by the recent publication of Bordwell and Boyle.[54] In order to obtain further experimental data with which to resolve conflicting interpretations, the isotope effects k_H/k_D for six-centre thermolyses of pent-4-en-2-ol, but-3-enoic acid and but-3-yn-1-ol have been determined as a function of temperature.[55] The substrates can be broadly classified as β-hydroxy olefins although there is significant variation in the nature of the hydroxyl bond and the hybridization of both C-1 and C-4 (terminal carbon to which H is transferred). Consistently with the proposed concerted sigmatropic mechanism, all three reactions occur at almost the same rate despite the range of structural variation, occur with equal facility in gas or liquid phase, and are insensitive to medium dielectric effects. Over a wide temperature range (100°) k_H/k_D in each case corresponded to that estimated by presuming zero-point energy difference (O–H versus O–D) alone to determine the rate ratio. Therefore, it has been claimed that the validity of the kinetic deuterium isotope criterion is not affected by such changes in hybridization and valence geometry and that the ionic character of the O—H bond is of minor importance in determining the magnitude of k_H/k_D for a concerted H-transfer. It is, to the Reviewer, surprising that the three transition states are by implication each fully symmetrical. Clearly some demonstration of the sensitivity of k_H/k_D to the symmetry of the transition state for sigmatropic hydrogen migration is required to enhance these results. Also consistent with a concerted six-centre elimination with little charge development at the 3-position are kinetics of formation of formaldehyde and the corresponding α-methylstyrene upon pyrolysis of seven 3-arylbut-3-en-1-ols, for which the ρ value -0.59 has been estimated.[56]

A four-centre decomposition of 2,2-difluoroethyltrimethoxysilane has been described,[57] and diarylketenes have been prepared by pyrolysis of their alkyl trialkylsilyl acetals.[58] Pyrolysis of (12) gave exclusively (14); this is consistent with a four-centre transition state (13) which is probably accommodated by the ability of silicon to increase its

$$Ph_2C=C\begin{matrix}OMe\\O^{18}-SiMe_3\end{matrix} \xrightarrow{\Delta} \left[Ph_2C\cdots\cdots C\begin{matrix}OMe\\\vdots\\\vdots\\O^{18}\end{matrix}SiMe_3\right] \longrightarrow Ph_2C=C=O^{18}$$

(12) (13) (14)

$+ \text{MeOSiMe}_3$

$$\begin{matrix}CH_3HC—CHEt\\H\quad\quad O\\Ph_2C=C\\\quad\quad OSiEt_3\end{matrix} \xrightarrow{\Delta} Ph_2CHCO_2SiEt_3 + \begin{matrix}Me\\H\end{matrix}C=C\begin{matrix}H\\Et\end{matrix}$$

(15)

[52] H. E. O'Neal and S. W. Benson, *J. Phys. Chem.*, **71**, 2903 (1967).
[53] A. T. Blades and H. S. Sandhu, *Int. J. Chem. Kinetics*, **3**, 187 (1971).
[54] F. G. Bordwell and W. J. Boyle, Jr., *J. Am. Chem. Soc.*, **93**, 512 (1971).
[55] H. Kwart and M. C. Latimore, *J. Am. Chem. Soc.*, **93**, 3770 (1971).
[56] K. J. Voorhees and G. G. Smith, *J. Org. Chem.*, **36**, 1755 (1971).
[57] D. Graham, R. N. Haszeldine, and P. J. Robinson, *J. Chem. Soc.* (B), **1971**, 611.
[58] Y.-N. Kuo, F. Chen, and C. Ainsworth, *J. Am. Chem. Soc.*, **93**, 4604 (1971).

coordination number through 3d-orbital participation. A variety of analogous ketene acetals have been studied, including (15) which took a different course, probably by the mechanism indicated.

Homogeneous pyrolyses of 1,1-dichlorobutane and 1,1-dichloropropane have been found[59] to give 1-chloroalkenes in the ratios $cis/trans = 1.00$ and 1.2 (cf. equilibrium values 1.6 and 1.9). This slight preference for *trans*-product is expected for a planar four-centred activated complex. On comparison with data for EtCl, n-PrCl, n-BuCl and 1,1-dichloroethane, rate acceleration by α-chlorination and β-alkylation is apparent. There was no evidence for a radical mechanism for the primary pyrolysis of 1,2-dichloropropane for which rates and product distributions have also been interpreted.[60] In contrast, radical chain mechanisms have been suggested for dehydrochlorination of 1,2-dichloroethane[61] and dehydrobromination of 1,1- and 1,2-dibromoethane[62] in the absence of inhibitors. In the presence of inhibitors unimolecular decomposition of the dibromoethanes occurs and rate acceleration and retardation by α- and β-halogen, respectively, again qualitatively reflect substituent effects on solvolyses of the alkyl halides. Rate differences between ethyl fluoride and 1,2-difluoroethane have also been explained in terms of the polar transition state,[63] and results for HF elimination from 1,1,2,2-tetrafluoroethane in a single-pulse shock tube further demonstrate, within the context of previous work, the increase in activation energy with increasing fluorination.[64] Kinetics of unimolecular decomposition of cycloheptyl and cyclo-octyl chloride have been found[65] to fit the elimination rate sequence cyclo-hexyl < -pentyl < -heptyl < -octyl, in accord with that for the corresponding bromides[66] and for S_N1 solvolyses of the toluene-p-sulphonates. A relationship between activation energy and the alkene–alkane strain energy differences has been derived, in support of the analogy between solvolyses and gas-phase elimination reactions.

Hydrogen bromide catalysed thermal decomposition of cyclohexanecarboxylic acid to CO, H_2O and cyclohexene;[67] and the kinetics of formation of ketene and acetic acid from acetic anhydride[68] have been investigated. The concept[69] of a tight polar four-centre transition state for thermolysis of triisobutylaluminium is further substantiated by results for unimolecular elimination of but-1-ene from an equilibrating mixture of methyldi-n-butylaluminium and n-butyldimethylaluminium.[70]

Other Topics

Quantitative treatment of the ortho effect with respect to elimination reactions has been reviewed.[71]

[59] K. A. Holbrook and K. A. W. Parry, *J. Chem. Soc.* (B), **1971**, 1762.
[60] K. A. Holbrook and J. S. Palmer, *Trans. Faraday Soc.*, **67**, 80 (1971).
[61] K. A. Holbrook, R. W. Walker, and W. R. Watson, *J. Chem. Soc.* (B), **1971**, 577.
[62] P. T. Good and A. Maccoll, *J. Chem. Soc.* (B), **1971**, 268.
[63] J. A. Kerr and D. M. Timlin, *Trans. Faraday Soc.*, **67**, 1376 (1971).
[64] G. E. Millward, R. Hartig, and E. Tschuikow-Roux, *Chem. Comm.*, **1971**, 465.
[65] M. Dakubu and J. L. Holmes, *J. Chem. Soc.* (B), **1971**, 1040.
[66] W. C. Herndon and J. M. Sullivan, *J. Phys. Chem.*, **74**, 995 (1970).
[67] S. I. Ahonkhai and E. U. Emovon, *J. Chem. Soc.* (B), **1971**, 2031.
[68] P. G. Blake and A. Speis, *J. Chem. Soc.* (B), **1971**, 1877.
[69] K. W. Egger, *Int. J. Chem. Kinet.*, **1**, 459 (1969).
[70] K. W. Egger and A. T. Cocks, *Trans. Faraday Soc.*, **67**, 2629 (1971).
[71] M. Charton, *Progr. Phys. Org. Chem.*, **8**, 235 (1971).

In an attempt to decide whether electron withdrawal in the leaving group influences mainly C_β—H or C_α—X bond-breaking, the rates of elimination from a series of sulphonates ZC_6H_4—CH_2CH_2—$OSO_2C_6H_4Y$ with Bu^tOK–Bu^tOH have been correlated by the Hammett equation.[72] For the most electron-withdrawing substituents, ρ_Z decreases as σ_Y increases and a shift to a transition state with less carbanion character is implied. As σ_Z increases, ρ_Y becomes less positive (1.24–0.94), suggesting that increased C_β—H bond-breaking is accompanied by decreased C_α—X bond-breaking and a shift to a transition state of greater carbanion character. These results and those of earlier studies of secondary deuterium isotope effects have been made the basis for a new interpretation of the reactivity ratio[73] k_{OTs}/k_{Br}. For elimination from Z-phenethyl derivatives in Bu^tOK–Bu^tOH at 30°, k_{OTs}/k_{Br} increases from 0.2 to 1.6 as Z is changed from p-OMe through H, p-Cl, m-Br to p-NO_2; thus it has previously been suggested that increased C_β—H bond-breaking induces greater C_α—X bond-breaking. This conclusion, which makes the $E1cB$ mechanism paradoxical, is the exact opposite of that derived[72] from ρ_Y values. It has therefore been suggested[73] that, for cases when no C—X bond-breaking occurs in the transition state, k_{OTs} should be slightly greater than k_{Br} (since the reactivity is dependent on the electrophilic nature of the α-carbon) and as the extent of bond-breaking increases there should be a decrease in the ratio (until it is dominated by polarizability considerations) followed by a further increase (as the delocalization factor gains importance). Thus trends in the values of the ratio that are small or less than unity can be interpreted in terms of either increasing or decreasing C—X bond-breaking.

It has been shown that for reactions of $ZC_6H_4CHBrCH_3$ with Bu^tOK–Bu^tOH, Bu^tOK–Bu^tOH–DMSO and EtONa–EtOH (in which there are kinetic complications), ρ values successively decrease but are in all cases positive (although r values were low) and the β-deuterium isotope effects are 5.0, 3.5 and 3.4, respectively.[74] Only in EtONa–EtOH is a β-Me group rate-enhancing (1.7-fold). The $E2$ transition-state structure is therefore much more sensitive to the nature of solvent and base for 1- than for 2-arylethyl bromides and is believed to vary from nearly carbanionic in Bu^tOK–Bu^tOH–DMSO to nearly $E1$ in EtONa–EtOH. Controversy surrounding the 1-arylethyl system has been heightened by a re-examination of reaction of phenethyl bromide with EtONa–EtOH which has suggested that explanation in terms of the competitive S_N2–$E2$ scheme remains a viable alternative to the "unified" S_N2–$E2$ mechanism.[75]

Reaction scheme (2) is favoured[76] over scheme (1) for conversion of vicinal dihalides into olefins[77] by stabilized carbanions: however, (**17a**) and (**17b**) with diphenylacetonitrile (or methyl isobutyrate) in KNH_2–liquid NH_3 gave only (**18b**) rather than the thermodynamic mixture of (**18b**) (60%) and (**18a**) (40%). This stereoselectivity has been attributed to transfer of a second electron to (**22**), the more stable of the equilibrated bridged radicals (**21**) and (**22**). With potassium in liquid NH_3 the thermodynamic mixture of olefins is obtained, presumably through fast second-electron transfer before equilibration of the radicals. Studies[77] of debromination and dehydrobromination of *meso*- and *dl*-stilbene dibromides in DMF have been extended[78] in an attempt to con-

---

[72] J. Banger, A. F. Cockerill, and G. L. O. Davies, *J. Chem. Soc.* (B), **1971**, 498.
[73] Cf. *Org. Reaction Mech.*, **1967**, 121.
[74] T. Yoshida, Y. Yano, and S. Oae, *Tetrahedron*, **27**, 5343 (1971).
[75] D. J. McLennan, *Tetrahedron Letters*, **1971**, 2317.
[76] D. G. Korzan, F. Chen, and C. Ainsworth, *Chem. Comm.*, **1971**, 1053.
[77] Cf. *Org. Reaction Mech.*, **1970**, 151–152.
[78] W. K. Kwok and S. I. Miller, *J. Org. Chem.*, **35**, 4034 (1970).

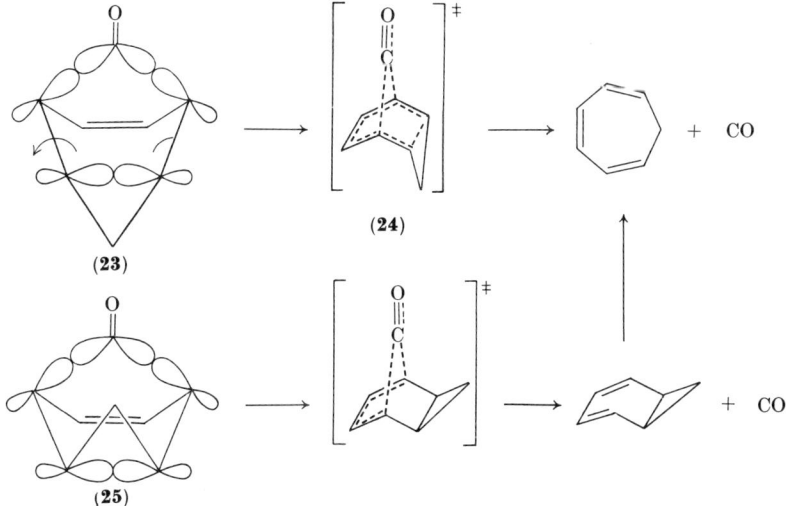

struct a scale of nucleophilicity towards halogen. In DMF nucleophilicities are much greater than in protic solvents and fall in the order $F^- > Cl^- > DMF$ and $I^- > Br^- > Cl^- > Sn^{2+} > DMF$ towards hydrogen and bromine, respectively. The magnitude and direction of substituent effects on debromination of cinnamic acid dibromides with I^- in 80% aqueous acetone are similar to those for decarboxylative debromination of the acid anions and suggest that attack of I^- is on the α- rather than the benzylic bromine.[79]

The critical effect of the configuration of component cyclopropane rings on relative rates of thermal decarbonylation of carbonyl bridge compounds has been demonstrated.[80] Differences in ground-state energies may contribute to the fact that (**23**) is 10^7 times more

[79] K. M. Ibne-Rasa, A. Ahmad, and Amir-Ud-Din, *J. Sci. Res. (Lahore)*, **3**, 1 (1968); *Chem. Abs.*, **74**, 31474 (1971); K. M. Ibne-Rasa and A. Ahmad, *J. Sci. Res.*, **3**, 1 (1968); *Chem. Abs.*, **73**, 119813 (1970).

[80] S. C. Clarke and B. L. Johnson, *Tetrahedron*, **27**, 3555 (1971).

reactive than (**25**) or norborn-2-en-7-one; however, it is felt that the major factor is the favourable interaction of the *endo*-cyclopropane ring during cleavage of the bridgehead-to-carbonyl bond;[81] thus the transition state (**24**) resembles a monohomobenzene with loosely bound carbon monoxide.

The product ratios for BuLi-induced elimination and fragmentation of (**26a–c**) have been found to vary markedly with steric hindrance to approach of the nucleophile at sulphur;[82] thus fragmentation increases in the order (**26c**) < (**26a**) < (**26b**). An $E2$

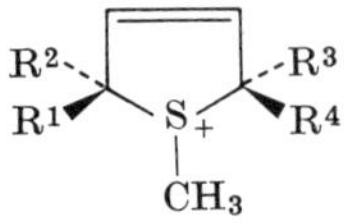

(**26a**) $R^1 = R^3 = CH_3$; $R^2 = R^4 = H$
(**26b**) $R^1 = R^4 = CH_3$; $R^2 = R^3 = H$
(**26c**) $R^1 = R^2 = R^3 = R^4 = CH_3$

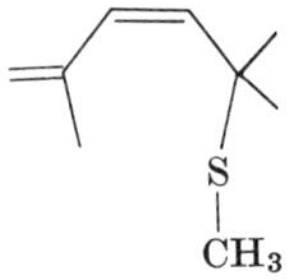

(**27**)

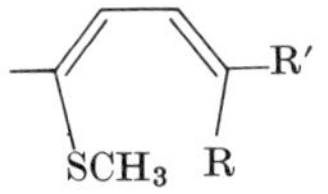

(**28**) $R = H$, $R' = Me$
(**29**) $R = Me$, $R' = H$

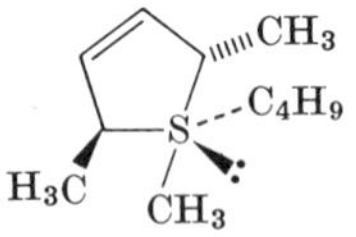

(**30**)

elimination probably accounts for formation of (**27**) from (**26c**) whereas the 4:1 mixture of (**28**) and (**29**) obtained from both (**26a**) and (**26b**) may be produced by a zwitterion mechanism. Stereospecific formation of hexa-*cis*-2, *trans*-4-diene from (**26a**) and of only *trans,trans* and *cis,cis* isomers from (**26b**) has been attributed to concerted fragmentation of intermediate sulphuranes (**30**) rather than to stepwise fragmentation of intermediate zwitterions.

Although dehydration of (**31a**) in thionyl chloride–pyridine has been reported[83] to give (**32a**) and little rearranged product (**33a**), it has been found that the major product

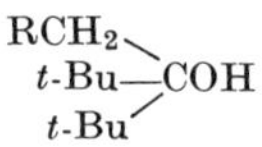

(**31a**) $R = H$
(**31b**) $R \neq H$

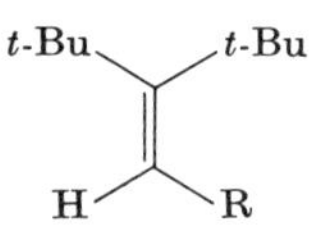

(**32a**) $R = H$
(**32b**) $R \neq H$

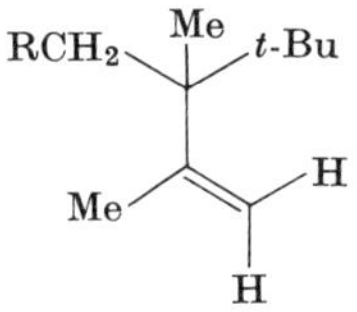

(**33a**) $R = H$
(**33b**) $R \neq H$

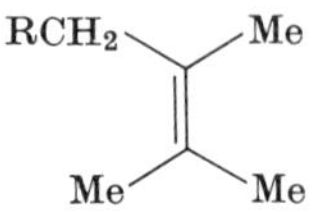

(**34a**) $R = H$
(**34b**) $R \neq H$

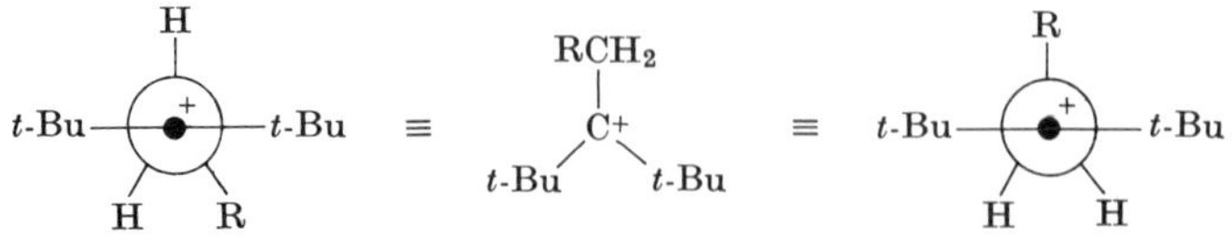

(**35A**)

(**35B**)

[81] Cf. *Org. Reaction Mech.*, **1967**, 39.
[82] B. M. Trost and S. D. Ziman, *J. Am. Chem. Soc.*, **93**, 3825 (1971).
[83] M. S. Newman, A. Arkell, and T. Fukunaga, *J. Am. Chem. Soc.*, **82**, 2498 (1960).

from (**31b**) is always the rearranged olefin (**33b**) along with (**32b**) and (**34b**) in that order of importance.[84] These findings confirm the carbonium character of the intermediate formed upon ionization of the chlorosulphite. When R = H, conformer (**35A**) favours elimination of H[+] whereas when R = H the preferred conformer (**35B**) precludes coplanar alignment of the unoccupied p-orbital with H[+], and 1,2 methyl shift occurs. The distributions of butenes from dehydration of the four isomeric butanols by strong acid at 160° have been reported,[85] and the catalytic activities of anhydrous metal bromides for dehydration of Bu[t]OH and *tert*-pentyl alcohol have been related to the standard potential and ionic radius of the metal.[86] Hexamethylphosphorotriamide (HMPT) has been used as an effective solvent catalyst for dehydrohalogenation of primary alkyl halides[17] and for dehydration of primary or secondary alcohols[87] at 220–240°. Normal alcohols give modest yields of unrearranged alk-1-enes and 1-dimethylaminoalkanes, whereas secondary alcohols give only unrearranged olefins. Reaction of *trans*-4-t-butylcyclohexanol is equally rapid under these conditions; thus a *trans*-diaxial orientation of OH and β-H is unnecessary. The Taft equation has been applied to dehydration and dehydrogenation reactions of alcohols over alkaline-earth silicates in order to clarify the cause of selectivity of such acid–base bifunctional catalysts.[88]

A cyclic transition state has been proposed for formation of ethylene and tri-n-butyltin acetate upon decomposition of (2-acetoxyethyl)tri-n-butyltin.[89]

Rates of fragmentation of sulphonates (**36**)→(**37**), etc., in 10% dioxan are interpreted in terms of a more or less concerted decarboxylative dearylsulphonation with substantial positive charge development at C_β in the transition state.[90] The magnitudes of ρ values for variation of X and Y ($\rho^+ = -3.1$) and Ar ($\rho = +1.17$) were considered insufficient for either pre-equilibrium or irreversible vinylcarbonium ion formation although such intermediates have now been identified for a variety of solvolyses.[91] The m value (0.1)

is also very low; however, in the opinion of the Reviewer this might well be a feature of heterolysis at a site neighbouring a charged group.[92] Competing $E1cB$, $E1cB$–HBA and $E2$ reactions are believed to account for the kinetics of formation of arylacetylenes by alkoxide-induced elimination reactions of *cis*- and *trans*-styryl chlorides,[93] whereas with organolithium compounds *cis*- and *trans*-styryl fluorides yield phenylacetylene by

---

[84] J. S. Lomas, D. S. Sagatys, and J. E. Dubois, *Tetrahedron Letters*, **1971**, 599.
[85] J. Warkentin and K. E. Hine, *Can. J. Chem.*, **48**, 3545 (1970).
[86] E. Tsukurimichi, *Kogyo Kagaku Zasshi*, **74**, 130 (1971); *Chem. Abs.*, **74**, 111210 (1971).
[87] R. S. Monson, *Tetrahedron Letters*, **1971**, 567.
[88] H. Niiyama and E. Echigoya, *Bull. Chem. Soc. Japan*, **44**, 1739 (1971).
[89] J. Tsurugi, M. Iida, R. Nakao, T. Fukumoto, and N. Murata, *Bull. Chem. Soc. Japan*, **44**, 777 (1971).
[90] I. Fleming and C. R. Owen, *J. Chem. Soc.* (B), **1971**, 1293.
[91] Cf. *Org. Reaction Mech.*, **1969**, 115.
[92] Cf. F. G. Bordwell and A. C. Knipe, *J. Org. Chem.*, **35**, 2956, 2959 (1970).
[93] M. Schlosser and V. Ladenberger, *Chem. Ber.*, **104**, 2873 (1971).

the $E2cB$ pathway and also *cis*- and *trans*-alkylphenylethylenes, respectively, by a stereoretentive addition–elimination sequence.[94] DMSO has been found[95] to cause dramatic increase in the rate of formation of phenylacetylene from *cis*-β-chloro- and β-bromo-styrenes and their 4-NO_2 analogues in MeONa–MeOH. The values k_H/k_D were in the range 1.6–3.7, no exchange of the α-H could be detected (in contrast with results in ref. 93) and a concerted paenecarbanion elimination is favoured. The effect of [DMSO] on the extent of C—X and C—H bond-breaking in the transition state has been discussed.

The epoxides (**40**) and (**41**) have been obtained by a novel *syn*-elimination ring closure

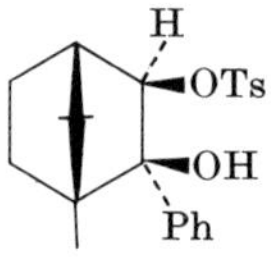

(**38**) 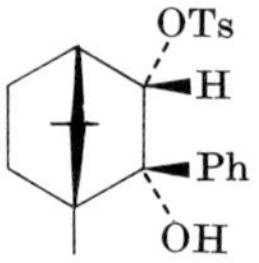   (**39**) 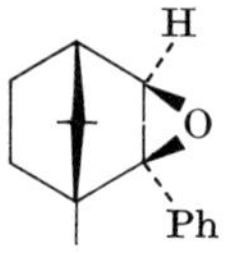   (**40**) 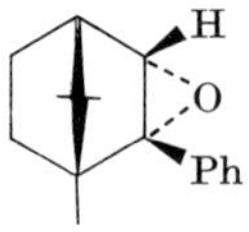   (**41**)

between Bu^tOK–Bu^tOH and (**38**) or (**39**), respectively.[96] Nematic solvents enhance the yield of olefin obtained upon pyrolysis of xanthates in a closed system[97] and may have synthetic application. Primary salt effects of poly(diallyldimethylammonium chloride) have been interpreted for reaction between OH^- and the like charged chloromaleate or chlorofumarate ions,[98] and kinetics of reaction between 4,5-dibromo-2-cyclohexyl-2-phenylvaleronitrile and Et_2NH have been interpreted.[99] Avoidance of ferrocenyl–ferrocenyl non-bonded interaction during formation of 1,2-diferrocenyl-1,2-diphenyl-ethane probably accounts for exclusive formation of the *trans*-isomer by Clemmensen reduction of benzoylferrrocene.[100] Products of reaction of *n*-propyl and *n*-dodecyl phosphates with Bu^tOK–DMSO are in accord with the following mechanism:[101]

$$\underset{\underset{OR}{|}}{\overset{\overset{O}{\|}}{PhOPOPh}} + Bu^tO^- \longrightarrow PhO^- + \underset{\underset{OR}{|}}{\overset{\overset{O}{\|}}{PhOPOBu^t}} \xrightarrow{Bu^tO^-} \underset{\underset{OR}{|}}{\overset{\overset{O}{\|}}{PhOPO^-}} + (CH_3)_2C\!=\!CH_2$$

Model calculations have been used to estimate relative tritium–deuterium kinetic isotope effects and their temperature-dependences for five basic reaction types including $E2$ elimination from Bu^sX with amide ion.[102]

[94] M. Schlosser and M. Zimmerman, *Chem. Ber.*, **104**, 2885 (1971).
[95] G. Marchese, F. Naso, and V. Sgherza, *Gazzetta*, **101**, 251 (1971); *Chem. Abs.*, **75**, 87913 (1971).
[96] J. M. Coxon, M. P. Hartshorn, and A. J. Lewis, *Chem. Comm.*, **1970**, 1607.
[97] W. E. Barnett and W. H. Sohn, *Chem. Comm.*, **1971**, 1002.
[98] T. Ueda, S. Harada, and N. Ise, *Chem. Comm.*, **1971**, 99.
[99] E. de Hoffmann, J. P. Schmit, and J. J. Charette, *J. Org. Chem.*, **35**, 4016 (1970).
[100] S. I. Goldberg, W. D. Bailey, and M. L. McGregor, *J. Org. Chem.*, **36**, 761 (1971).
[101] R. A. Bartsch and D. G. Wallin, *J. Org. Chem.*, **36**, 1013 (1971).
[102] M. J. Stern and D. C. Vogel, *J. Am. Chem. Soc.*, **93**, 4664 (1971).

CHAPTER 5

Addition Reactions

R. C. STORR

Department of Organic Chemistry, University of Liverpool

Electrophilic Additions	152
Halogen and Related Additions	152
Addition of Hydrogen Halides	155
Hydration and Related Reactions	156
Miscellaneous Electrophilic Additions	159
Nucleophilic Additions	161
Cycloadditions	164
2 + 4-Cycloadditions	164
2 + 3-Cycloadditions	168
2 + 2-Cycloadditions	172
Other Cycloadditions and Cyclizations	174
Metal-catalysed Cycloadditions	177

There is little that is fundamentally new in addition reactions this year.

A review of the application of Perturbation MO theory to the development of symmetry rules for (mainly acyclic) chemical reactions has appeared.[1] Dewar has presented an excellent review of the application of the concept of aromaticity to transition states for all types of pericyclic reactions.[2]

Fukui has reviewed his Frontier MO approach to organic reactions.[3] Application of the orbital phase continuity principle and the use of GI orbitals leads to selection rules for concerted reactions without recourse to symmetry considerations. The selection rules agree with those of Woodward and Hoffmann.[4]

Hoffmann and his co-workers have turned their attention to gaining a better understanding of non-concerted disallowed reactions. Investigation of the potential surface for the "tetramethylene diradical"—the species commonly considered in classical terms as the diradical intermediate in cyclobutane cleavage or 2 + 2-cycloaddition— shows no real potential minimum between cyclobutane and two ethylene molecules. There is, however, a large region of coordinate space where the energy of the tetramethylene species varies little with conformation. Such a flat region may well be indistinguishable from a true minimum, since the species will, on average, spend a relatively long time exploring the surface, so allowing the possibilities for interception or diversion that are normally associated with a classical diradical. The term "twixtyl" is proposed for such a species. If these results are confirmed by more sophisticated calculations, our view of non-concerted reactions and diradical intermediates will have to be re-examined.[5a]

[1] R. G. Pearson, Accounts Chem. Res., **4**, 152 (1971).
[2] M. J. S. Dewar, Angew. Chem. Int. Ed., **10**, 761 (1971).
[3] K. Fukui, Fortschr. Chem. Forsch., **15**, 3 (1970).
[4] W. A. Goddard, J. Am. Chem. Soc., **92**, 7520 (1970).
[5a] R. Hoffmann, S. Swaminathan, B. G. Odell, and R. Gleiter, J. Am. Chem. Soc., **92**, 7091 (1970).

Stereochemical observations in the thermal fragmentation of [4.4.2]propella-2,4-dienes are considered to be more consistent with a "conventional" semi-stabilized 1,4-diradical than with a twixtylic intermediate.[5b]

Electrophilic Additions

Reviews of electrophilic additions, involving neighbouring-group participation,[6] and of the additions of halogen azides,[7] have appeared.

Halogen and Related Additions

Dubois and Huynh[8] have presented evidence that there are two mechanisms involved for the "tribromide ion" term in the rate expressions for addition of bromine to olefins. These are considered to be electrophilic attack by Br_3^- on the substrate (this predominates for highly reactive olefins) and rate-determining attack by Br^- on the charge-transfer complex between Br_2 and olefin. The exceptionally high value for the solvent isotope effect ($k_{MeOH}/k_{MeOD} = 1.40$) for the addition of Br_2 to pent-1-ene indicates that H-bonding solvation of the developing Br^- is important in the rate-determining collapse of the intermediate charge-transfer complex to a bromonium ion–bromide ion pair; solvent effects show that the transition state for this collapse closely resembles the bromonium ion.[9] A comparison of heats of combustion and activation parameters for Br_2 addition to pairs of cis- and trans-olefins indicates that ground-state steric effects are retained in the transition state, which therefore resembles a bridged bromonium ion rather than an open ion where steric interactions would be relieved.[10]

The driving force for exo attack of Br_2, which overrides the unfavourable steric effect of an anti-7-substituent, in the addition to benzonorbornadienes does not apply when a 2-phenyl group is introduced. It therefore seems likely that σ-delocalization rather than steric or torsional effects is responsible.[11]

Five dibromides in addition to exo-2-bromonorbornene and bromonortricyclene have been identified in the reaction of Br_2 with norbornene. The origin of these dibromides is consistent with exo addition of electrophilic bromine followed by 6,1- or 6,2-hydride shifts in the resulting non-classical ion or Wagner–Meerwein related classical cations (Scheme 1). Labelling confirms that (**1**) is formed by the two routes shown.[12] A study of the mechanism of formation of nortricyclyl bromide and chloride in the addition of Br_2 and Cl_2 to norbornene indicates that proton elimination occurs mainly from the rearranged ion with Br_2 and the unrearranged ion with Cl_2 or from the unsymmetrical norbornonium ions (**2**) and (**3**).[13]

Detailed studies of the products, kinetics and effect of added salts on the addition of Br_2 to acetylenes in acetic acid suggest that an open vinyl cation–bromide ion pair which collapses to cis- and trans-dibromide or reacts with solvent to give Markovnikov

---

[5b] L. A. Paquette and G. L. Thompson, *J. Am. Chem. Soc.*, **93**, 4920 (1971).
[6] V. I. Staninets and E. A. Shilov, *Uspekhi Khim.*, **1971**, 272; *Russ. Chem. Rev.*, **1971**, 272.
[7] A. Hassner, *Accounts Chem. Res.*, **4**, 9 (1971).
[8] J.-E. Dubois and X. Q. Huynh, *Tetrahedron Letters*, **1971**, 3369.
[9] F. Garnier, R. H. Donnay, and J.-E. Dubois, *Chem. Comm.*, **1971**, 829.
[10] K. Yates and R. S. McDonald, *J. Am. Chem. Soc.*, **93**, 6297 (1971).
[11] R. Caple, G. M.-S. Chen, and J. D. Nelson, *J. Org. Chem.*, **36**, 2870 (1971).
[12] D. R. Marshall, P. Reynolds-Warnhoff, E. W. Warnhoff, and J. R. Robinson, *Can. J. Chem.*, **49**, 885 (1971).
[13] N. H. Werstiuk and I. Vancas, *Can. J. Chem.*, **48**, 3963 (1970).

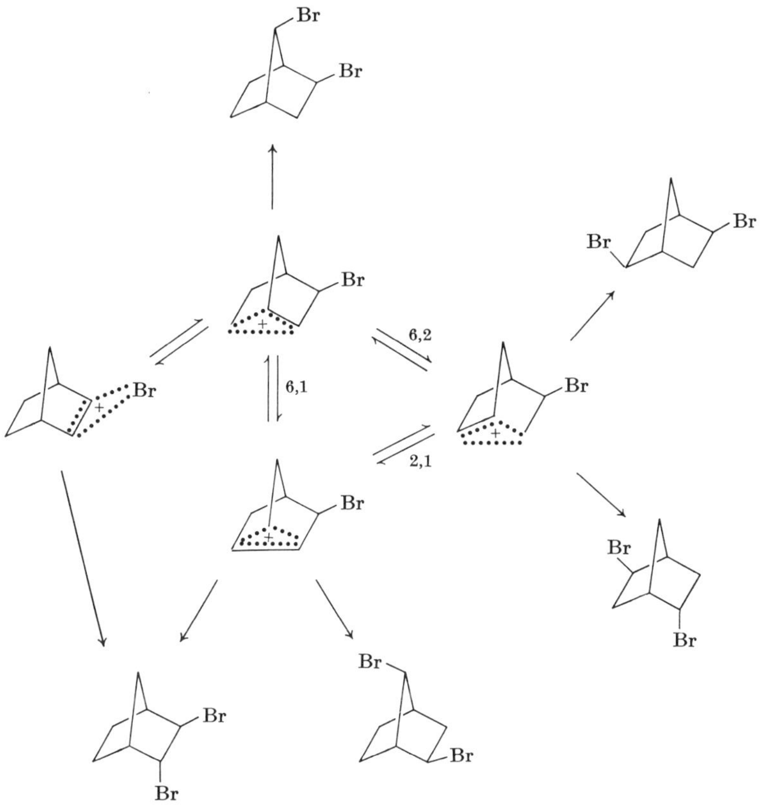

Scheme 1

bromoacetate is involved for phenylacetylenes. For alkylacetylenes a cyclic bromonium ion intermediate leads to *trans* addition.[14]

Identical stereospecificity is observed for addition of Br_2 or BrCl to optically active penta-2,3-diene in methanol. For iodination, the specificity (ICl > IBr > I_2) depends on the nucleophilicity of the counter-ion which can cause racemization of the initial adducts by competitive $E2$ and S_N2 reactions.[15] The regiospecificity and *trans*-stereospecificity of bromohydrin formation from olefins and *N*-bromosuccinimide (NBS) in moist DMSO are in accord with a bromonium ion intermediate; the precise nature

[14] J. A. Pincock and K. Yates, *Can. J. Chem.*, **48**, 3332 (1970).
[15] M. C. Findlay, W. L. Waters, and M. C. Caserio, *J. Org. Chem.*, **36**, 275 (1971).

of the brominating agent is not yet clear but NBS itself seems most likely.[16] Phenyl migration occurs in ionic but not free-radical addition of BrN_3 to 3,3,3-triphenylpropene; competing radical and ionic pathways account for the products from Br_2 addition to this olefin. Phenyl migration also occurs in BrN_3 addition to 3,3-diphenylpropene but it is less evident in addition of IN_3.[17]

Through conjugation of enol-oxygen with a carbonyl group as in β-keto-enol ethers does not cause any great diminution in reactivity of the enol function towards addition of Br_2.[18]

The first cases of β-lactone formation in the addition of Br_2 to β,γ-unsaturated acids have been reported.[19] Participation by allylic and homoallylic double bonds in the reactions of NBS with norbornenyl enol acetates[20] and by chloromethyl-chlorine in the reaction of NBS with exo-2-cyano-endo-3-(chloromethyl)bicyclo[2.2.1]hept-5-ene[21] has also been reported. Transannular addition products are observed in the addition of Br_2 to cyclonona-1,2-diene[22] but not to cycloocta- or cyclodeca-1,2-diene.[23] Use of partially resolved cyclonona-1,2-diene shows that intermediate dissymmetric species (most probably bridged bromonium ions) are involved for both 1,2- and transannular adducts.[24]

The stereochemistry of Br_2 addition of cyclohex-4-ene-1,2-dicarboxyic acids[25] and additions of dipyridine–bromine perchlorate (a mild bromide-free source of electrophilic bromine)[26] have been investigated. Asymmetric induction has been observed in the addition of Br_2 to an optically active vinyl sulphoxide.[27] Stereospecific *trans* addition of N-bromo- and N-chloro-bis(trifluoromethyl)amines to butenes at $-78°$ in the dark suggests that these reagents add by an ionic rather than by a 4-centre mechanism under such conditions.[28] Several kinetic studies of Br_2 additions have been reported.[29] Formation of Au^{III}–carbon σ-bonds occurs in the addition of Br_2 to linear Au^I complexes of o-styryl- and o-allyl-phenyldiphenylphosphine.[30]

The isomerization and chlorination of decachlorobi(cyclopenta-2,4-dienyl),[31] the kinetics and mechanism of the reaction of chlorine with 2-methylallyl halides (addition

---

[16] D. R. Dalton and V. P. Dutta, *J. Chem. Soc.* (B), **1971**, 85.
[17] A. Hassner and J. S. Teeter, *J. Org. Chem.*, **36**, 2176 (1971).
[18] D. R. Marshall and T. R. Roberts, *J. Chem. Soc.* (B), **1971**, 797.
[19] W. E. Barnett and J. C. McKenna, *Chem. Comm.*, **1971**, 551; *Tetrahedron Letters*, **1971**, 2595.
[20] J. J. Riehl and F. Jung, *Tetrahedron Letters*, **1971**, 325.
[21] H. Christol, J. Coste, and F. Plénat, *Bull. Soc. Chim. France*, **1971**, 3001.
[22] M. S. Baird, C. B. Reese, and A. Shaw, *Tetrahedron*, **27**, 231 (1971).
[23] C. B. Reese and A. Shaw, *Tetrahedron Letters*, **1971**, 4641.
[24] L. R. Byrd and M. C. Caserio, *J. Am. Chem. Soc.*, **93**, 5758 (1971).
[25] J. Klein and E. Dunkelblum, *J. Org. Chem.*, **36**, 142 (1971).
[26] C. F. Hammer and C. E. Costello, *Tetrahedron Letters*, **1971**, 1743.
[27] D. J. Abbott and C. J. M. Stirling, *Chem. Comm.*, **1971**, 472.
[28] M. G. Barlow, G. L. Fleming, R. N. Haszeldine, and A. E. Tipping, *J. Chem. Soc.* (C), **1971**, 2744.
[29] E. A. Shilov, Y. A. Serguchev, G. B. Sergeev, V. V. Smirnov, and V. A. Kofman, *Dokl. Akad. Nauk SSSR*, **197**, 1096 (1971); *Chem. Abs.*, **75**, 4930 (1971); G. Heublein and D. Stodermann, *J. Prakt. Chem.*, **312**, 1121 (1971); R. Gelin and D. Pigasse, *Bull. Soc. Chim. France*, **1971**, 2186; G. D. Mel'nikov and Y. I. Porfir'eva, *Zhur. Org. Khim.*, **6**, 1953 (1970); *J. Org. Chem. USSR*, **6**, 1967 (1970); L. P. Zalukaev, G. B. Sergeev, V. V. Smirnov, and Z. V. Shmyreva, *Kinet. Katal.*, **12**, 754 (1971).
[30] M. A. Bennett, K. Hoskins, W. R. Kneen, R. S. Nyholm, P. B. Hitchcock, R. Mason, G. B. Robertson, and A. D. C. Towl, *J. Am. Chem. Soc.*, **93**, 4591 (1971).
[31] V. Mark and E. D. Weil, *J. Org. Chem.*, **36**, 676 (1971).

and substitution)[32] and the reaction of but-2-yne with chlorine and benzonitrile[33] have been studied. cis-1,2-Dichlorides are formed predominantly when olefins are treated with $SbCl_5$ in CCl_4.[34]

Three concurrent pathways have been detected for the addition of iodine to sodium phenylpropiolate. A termolecular Ad_E3 mechanism predominates at high I^- concentration. A bimolecular reaction, possibly involving rate-determining attack of free I_2 to form a cation, occurs at lower I^- concentration, together with a third reaction which probably corresponds to iodination by hydrated iodine cation (H_2OI^+).[35]

The stereo- and regio-specificity of iodo- and bromo-acetate formation observed in the addition of I_2 and Br_2 in acetic acid to the 22(23)-ethylenic bond of the ergosterol side chain is surprising considering that this double bond is flanked only by chiral centres that have no obvious directing functions.[36] Norborn-5-ene-*exo*-2-carboxylic acids with I_2 in weakly alkaline solution give di-iodo compounds by *cis* addition and iodohydroxy compounds by a process involving intramolecular rearrangement.[37]

The characteristics of the addition of iodonium nitrate to alkenes and dienes accord with a bridged iodonium ion.[38] Acetyl hypoiodite seems to be the active reagent in the reaction of I_2 and peracetic acid with diphenylacetylene in acetic acid; *trans*-α-iodo-α-acetoxystilbene is the primary product which is further oxidized to benzil.[39] The formation of *vic*-chloroiodoalkanes from alkenes with $CuCl_2$ and I_2 or an iodine donor has been discussed.[40]

Addition of Hydrogen Halides

The kinetics and activation parameters for the ionic or possibly polar four-centre addition of HF to vinyl fluorides have been studied.[41]

The addition of HCl to 1,2-dimethylcyclohexene in acetic acid is essentially similar to that reported previously for cyclohexene in that it can be rationalized by competing Ad_E2 and Ad_E3 mechanisms. The former involves a carbonium ion–chloride ion pair which collapses mainly to *cis*-HCl adduct. The latter is termolecular and involves *trans* addition of HCl or HOAc and is generally more important in polar solvents.[42]

Greater capture by nucleophile at C_2 than at C_1 occurs in the addition of HCl and HBr but not of HF, HCO_2H or MeOH to 2,3-dideuterionorbornene. This implies that a classical norbornyl cation can be intercepted by reactive nucleophiles before irreversible leakage to a non-classical ion occurs or complete equilibrium with the rearranged

---

[32] I. V. Bodrikov, S. V. Spiridonova, Z. S. Smolyen, and A. I. Subbotin, *Zhur. Org. Chim.*, **6**, 684 (1970); *J. Org. Chem. USSR*, **6**, 686 (1970).
[33] K.-W. Thiem, *Chem. Ber.*, **103**, 3842 (1970).
[34] S. Uemura, O. Sasaki, and M. Okano, *Chem. Comm.*, **1971**, 1064.
[35] M. H. Wilson and E. Berliner, *J. Am. Chem. Soc.*, **93**, 208 (1971).
[36] D. H. R. Barton, J. P. Poyser, P. G. Sammes, M. B. Hursthouse, and S. Neidle, *Chem. Comm.*, **1971**, 715.
[37] H. Geiger, *Tetrahedron*, **27**, 157 (1971).
[38] U. E. Diner and J. W. Lown, *Can. J. Chem.*, **49**, 403 (1971); U. E. Diner, M. Worsley, and J. W. Lown, *J. Chem. Soc.* (C), **1971**, 3131.
[39] Y. Ogota and I. Urasaki, *J. Org. Chem.*, **36**, 2164 (1971).
[40] W. C. Baird, J. H. Surridge, and M. Buza, *J. Org. Chem.*, **36**, 2088 (1971).
[41] L. O. Moore, *Can. J. Chem.*, **49**, 2471 (1971).
[42] R. C. Fahey and C. A. McPherson, *J. Am. Chem. Soc.*, **93**, 2445 (1971).

classical ion is established.[43] The addition of HCl and DCl to norbornadiene and quadricyclene cannot involve a symmetrically bridged cation as the sole precursor to *exo*-dehydronorbornyl chloride.[44]

Addition of HCl to the allylidenecyclopropane (**4**) is mainly 1,4 and does not involve cleavage of the cyclopropane ring.[45] The observed *trans* addition of HCl to 1-cyanocyclohexene in anhydrous alcohol, which proceeds largely with concomitant hydrolysis

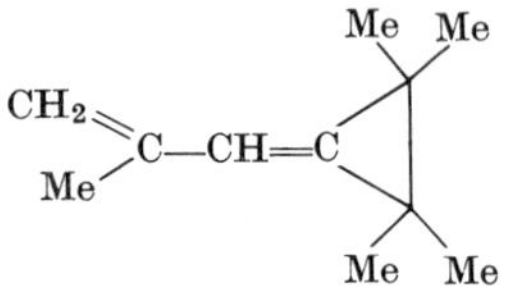

and esterification of the CN group, is explained in terms of protonation from the least hindered side of the chloro-methylidene imine intermediate (**5**) through the transition state (**6**) which minimizes chlorine lone-pair and C≡N π-orbital interactions.[46]

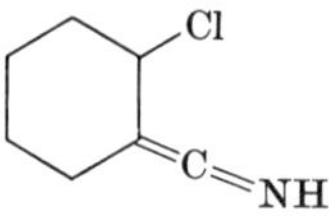

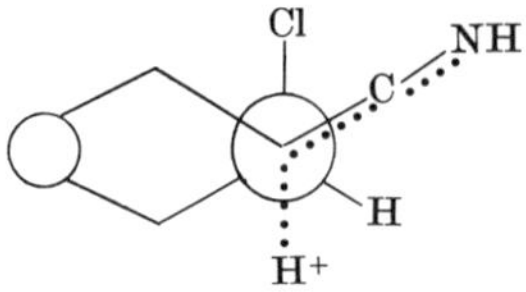

Kinetic studies of the Hg^{II}-catalysed additions of HCl to acetylenes have been reported.[47]

Hydration and Related Reactions

General acid catalysis has now been demonstrated in the acid-catalysed hydration of simple olefins such as *trans*-cyclo-octene and 2,3-dimethylbut-2-ene, so establishing that rate-determining protonation to give a carbonium ion, rather than pre-equilibrium π-complex formation followed by rate-determining collapse to a carbonium ion, is involved.[48] The reasons why general catalysis was not observed in other studies[49] have been discussed.

Further evidence that the counter-ion can influence the fate of a carbonium ion comes from a study of the acid-catalysed hydration of α- and β-pinene in acetic acid.[50] The acid-catalysed hydration of *endo*- and *exo*-norborn-5-ene-2-carboxylic acid has been studied and the accompanying rearrangements have been discussed.[51]

Studies of the mechanism of acid-catalysed hydration and isomerization of *tert*- and iso-hexene,[52] the high-pressure acid-catalysed liquid-phase hydration of crotonaldehyde[53]

---

[43] J. K. Stille and R. D. Hughes, *J. Org. Chem.*, **36**, 340 (1971).
[44] T. C. Morrill and B. E. Greenwald, *J. Org. Chem.*, **36**, 2769 (1971); see also S. J. Cristol, J. K. Harrington, T. C. Morrill, and B. E. Greenwald, *J. Org. Chem.*, **36**, 2773 (1971).
[45] M. L. Poutsma and P. A. Ibarbia, *Tetrahedron Letters*, **1970**, 4967.
[46] S. N. Balasubrahmanyam and M. Balasubramanian, *J. Chem. Soc.* (C), **1971**, 827.
[47] R. Gelin and D. Pigasse, *Bull. Soc. Chim. France*, **1971**, 1840; O. N. Temkin, N. F. Alekseeva, R. M. Flid, O. L. Kaliya, G. K. Shestakov, and T. I. Dolgina, *Dokl. Akad. Nauk SSSR*, **196**, 836 (1971); *Chem. Abs* , **74**, 124448 (1971).
[48] A. J. Kresge, Y. Chiang, P. H. Fitzgerald, R. S. McDonald, and G. H. Schmid, *J. Am. Chem. Soc.*, **93**, 4907 (1971).
[49] See, for example, J. L. Jensen, *Tetrahedron Letters*, **1971**, 7.
[50] C. M. Williams and D. Whittaker, *J. Chem. Soc.* (B), **1971**, 668, 672.
[51] H. Geiger, *Tetrahedron*, **27**, 165 (1971).
[52] Y. G. Osokin and S. I. Kryukov, *Nauch. Korf. Yaroslav. Tekhnol. Inst. 21st*, **1969**, 160; *Chem. Abs.*, **74**, 111229 (1971).
[53] S. K. Bhattacharyya and G. B. Purohit, *Proc. Indian Nat. Sci. Acad.* (*A*), **36**, 154 (1970); *Chem. Abs.*, **74**, 22297 (1971).

and the effect of pressure on the acid-catalysed hydration of propene[54] have been reported.

Addition of DCl to benzonorbornadiene leads to complete Wagner–Meerwein scrambling, whereas addition of DOAc leads to an excess of unrearranged adduct. Since it was shown that Cl⁻ is not a poorer nucleophile than acetate, it seems that capture of a non-classical ion or a rapidly equilibrating pair of classical ions is in competition with concerted 4- or 6-centre addition of HOAc.[55] *syn*-Addition, possibly concerted, of acetic acid to (1*RS*,2*RS*)-*trans*-cyclo-oct-2-enyl acetate has been observed.[56]

Very little selectivity between cyclohexene and oct-1-ene is found for the acid-catalysed addition of phenols and acetic acid and this may be compared with the higher selectivity normally found for halogen addition or epoxidation. This is reasonable, however, since initial attack by H⁺ will give an open secondary carbonium ion for both olefins, whereas stability of the bridged ions involved in halogen addition or epoxidation will be greater for cyclohexene.[57]

Perdeuteration of norbornene occurs at 250° with 10% CD_3CO_2D/D_2O.[58] The acid-catalysed addition of alcohols to α-methylstyrene[59] has been studied, and further mechanistic studies have been reported for the Prins reaction of styrene derivatives.[60]

The small and apparently random variations, observed with the introduction of bridgehead methyl groups, in *exo* selectivity for hydroboration of norbornenes and base-catalysed H-exchange in camphor offer no confirmatory evidence for the importance of torsional effects in determining *exo* selectivity.[61] The reaction pathway and structures of hydroboration products from buta-1,3-diene and borane in THF have been clarified.[62] Hydroboration of protoadamantane (obtained together with 2,4-didehydroadamantane by pyrolysis of 2-adamantyl methanesulphonate) gives all four possible adamantanols.[63] Directive effects in the monohydroboration of alkynes and enynes[64] and the use of 1,3,2-benzodioxaborole as a convenient monofunctional hydroborating agent[65] for converting olefins into alkylboronic acid derivatives have been discussed.

Detailed studies of the addition of triphenylaluminium to *para*-substituted diphenylacetylenes indicate that rate-determining electrophilic attack by monomeric AlPh₃ through a four-centre transition state (7) is involved.[66] Mixed alkenyl–hydride-bridged alkylaluminium dimers have been observed, as well as the already established alkenyl-bridged dimers in the monohydroalumination of alkynes.[67]

---

[54] H. Takaya, N. Todo, T. Hosoya, and T. Minegishi, *Bull. Chem. Soc. Japan*, **44**, 1175, 1179 (1971).
[55] S. J. Cristol and J. M. Sullivan, *J. Am. Chem. Soc.*, **93**, 1967 (1971).
[56] G. H. Whitham and M. Wright, *J. Chem. Soc.* (C), **1971**, 891.
[57] D. T. Dalgleish, D. C. Nonhebel, and P. L. Pauson, *J. Chem. Soc.* (C), **1971**, 1174.
[58] N. H. Werstiuk and T. Kadai, *Chem. Comm.*, **1971**, 1349.
[59] S. Sekiguchi, K. Abe, J. Takenaka, and K. Matsui, *Kogyo Kogaku Zasshi*, **73**, 2325 (1970); *Chem. Abs.*, **74**, 124439 (1971).
[60] C. Bocard, M. Hellin, M. Davidson, and F. Coussemant, *Bull. Soc. Chim. France*, **1971**, 490, 877; M. Karpaty, M. Hellin, M. Davidson, and F. Coussemant, *Bull. Soc. Chim. France*, **1971**, 1731, 1736; J.-P. Durand, M. Davidson, M. Hellin, and F. Coussemant, *Bull. Soc. Chim. France*, **1970**, 4355.
[61] S. P. Jindal and T. T. Tidwell, *Tetrahedron Letters*, **1971**, 783.
[62] H. C. Brown, E. Negishi, and P. L. Burke, *J. Am. Chem. Soc.*, **93**, 3400 (1971).
[63] J. Boyd and K. H. Overton, *Chem. Comm.*, **1971**, 211.
[64] G. Zweifel, G. M. Clark, and N. L. Polston, *J. Am. Chem. Soc.*, **93**, 3395 (1971).
[65] H. C. Brown and S. K. Gupta, *J. Am. Chem. Soc.*, **93**, 1816 (1971).
[66] J. J. Eisch and C. K. Hordis, *J. Am. Chem. Soc.*, **93**, 2974, 4496 (1971).
[67] G. M. Clark and G. Zweifel, *J. Am. Chem. Soc.*, **93**, 527 (1971).

$$\begin{array}{c} Ph\diagdown\diagup Ph \\ C\!=\!\overset{+}{C} \\ \vdots\vdots \\ \underset{Ph}{\overset{Ph}{|}}\!-\!Al\!-\! \end{array}$$

(7)

Direct NMR spectral evidence has been obtained for mercurinium ions in the mercuration of olefins.[68]

The close parallel in the stereochemistries of oxymercuration of, and addition of Br_2 and BrX but not HBr to, substituted cyclohexenes is put forward as evidence that cyclic mercurinium ions are involved in oxymercuration.[69] This contrasts with Brown and Liu's conclusions drawn from studies of the stereochemical course of the oxymercuration of norbornene and 7,7-dimethylnorbornene.[70]

The mechanism of *trans*-oxymercuration, including both olefin and alkoxy exchange reactions, has been further examined. A reaction pathway involving deoxymercuration, induced by attack of olefin on the oxymercurial, is considered important for olefin exchange.[71]

Solvent attack on an unsymmetrical mercurinium ion now seems more acceptable than a four-centre mechanism for *cis*-alkoxymercuration of such olefins as *trans*-cyclooctene and norbornene with mercuric acetate. The situation is less clear for the acetoxymercuration that accompanies oxymercuration.[72]

Whereas mercuric chloride or bromide give mainly N-mercuration with enamines, mainly C-mercuration occurs with the more electrophilic mercuric acetate; this forms the basis of a general conversion of enamines into tertiary amines.[73] Intramolecular participation by amino groups in mercuration,[74] and asymmetric induction in olefin hydration via oxymercuration-demercuration by chiral Hg^{II} carboxylates[75] have been observed. The oxymercuration of phenyl cinnamate[76] and the effect of solvent on the stereospecificity of oxymercuration of $(-)$-*trans*-cyclo-octene[77] have been studied.

Predominant attack of peracid *cis* to the secondary methyl group in 1,6-dimethylcyclohexene is attributed to torsional effects.[78] The slower epoxidation, in benzene, of 3-alkylcyclohexenes than cyclohexene indicates that the effect of these substituents is predominantly steric, not inductive.[79] The effect of substituents on the position of epoxidation of vinylallenes has been studied. α-Allenic epoxides and cyclopentenones result from epoxidation of the vinyl and allene groups, respectively.[80] Epoxidation of

[68] G. A. Olah and P. R. Clifford, *J. Am. Chem. Soc.*, **93**, 2320 (1971); see also G. A. Olah and P. R. Clifford, *J. Am. Chem. Soc.*, **93**, 1261 (1971).
[69] D. J. Pasto and J. A. Gontarz, *J. Am. Chem. Soc.*, **92**, 7480 (1970).
[70] H. C. Brown and K. T. Liu, *J. Am. Chem. Soc.*, **92**, 3502 (1970).
[71] R. D. Bach, R. N. Brummel, and R. F. Richter, *Tetrahedron Letters*, **1971**, 2879.
[72] R. D. Bach and R. F. Richter, *Tetrahedron Letters*, **1971**, 3915.
[73] R. D. Bach and D. K. Mitra, *Chem. Comm.*, **1971**, 1433.
[74] J. Perie, J. P. Laval, J. Roussel, and A. Lattes, *Tetrahedron Letters*, **1971**, 4399.
[75] R. M. Carlson and A. H. Funk, *Tetrahedron Letters*, **1971**, 3661.
[76] P. L. Nayak and M. K. Rout, *J. Inst. Chem. Calcutta*, **43**, 32 (1971); *Chem. Abs.*, **75**, 48082 (1971).
[77] V. I. Sokolov, L. L. Troitskaya, and O. A. Reutov, *Izv. Akad. Nauk SSSR, Ser. Khim.*, **1970**, 2648; *Chem. Abs.*, **74**, 142010 (1971).
[78] P. M. McCurry, *Tetrahedron Letters*, **1971**, 1841.
[79] D. B. Inglis, *Chem. Ind. (London)*, **1971**, 1268.
[80] J. Grimaldi and M. Bertrand, *Bull. Soc. Chim. France*, **1971**, 957.

cyclonona-1,2-diene with peracetic acid in CH_2Cl_2 gives cyclononane-1,2-dione and cyclo-octene oxide, but mechanistic details are not yet available.[81] Oxabicyclobutanone is not involved in the peracid oxidation of cyclopropenones.[82] The products from 1-hydroperoxy-1-isopropylnaphthalen-2(1H)-one under basic conditions are those arising from competing inter- and intra-molecular epoxidations with subsequent rearrangements.[83] Epoxidation and *trans*-diol formation by means of *o*-sulphoperbenzoic acid,[84] studies of the stereochemistry of epoxidation of *cis*-*p*-menthene,[85] the kinetics of epoxidation of *trans*-fused bicyclic olefins with *p*-nitroperbenzoic acid,[86] and the epoxidation of halo-olefins with permaleic and monochloroperacetic acid[87] have been reported.

Kinetic studies of the epoxidations of styrenes in dilute solution in benzene with t-butyl hydroperoxide in the presence of molybdenum naphthenate are consistent with a scheme involving reversible association between peroxide and catalyst, followed by rate-determining electrophilic attack of this complex on the olefin (Scheme 2).

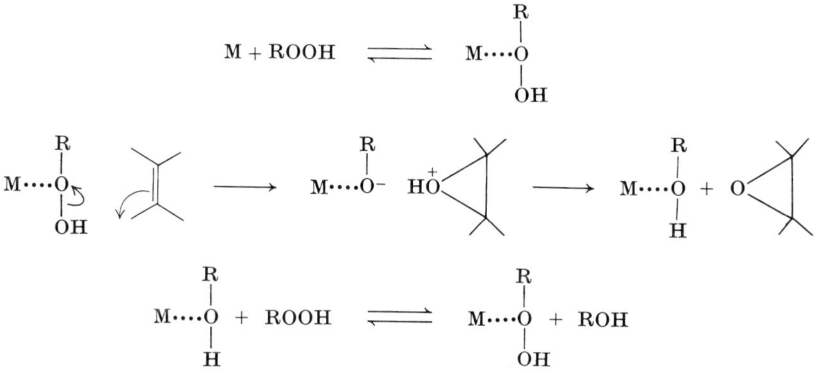

Scheme 2

This is basically the same as the mechanism for vanadium acetylacetonate and $Mo(CO)_6$-catalysed epoxidations, differences between V and Mo being accommodated by a smaller equilibrium constant for complex formation for Mo.[88]

Miscellaneous Electrophilic Additions

The steric and electronic effect of substituents on the rate of addition of 2,4-dinitrobenzenesulphenyl chloride to aliphatic alkenes in AcOH at 25° have been determined; it is concluded that bridging by S in these additions is more important than that by halogen in halogen additions and that this makes sulphenyl halide addition less selective and more dependent than the latter on steric factors.[89] The rates of exchange of *p*-chlorobenzenesulphenyl chloride from its adducts with various cyclic olefins to oct-1-ene

---

[1] W. P. Reeves and G. S. Stroebel, *Tetrahedron Letters*, **1971**, 2945.
[2] J. K. Crandall and W. W. Conover, *Tetrahedron Letters*, **1971**, 583.
[3] J. Carnduff and D. G. Leppard, *Chem. Comm.*, **1971**, 975.
[4] J. M. Bachhawat and N. K. Mathur, *Tetrahedron Letters*, **1971**, 691.
[5] J. Kuduk-Jaworska and Z. Chabudzinski, *Roczn. Chem.*, **44**, 1987 (1971); *Chem. Abs.*, **74**, 112225 (1971).
[6] J. Itier, M. Tournaire, and A. Casadevall, *Compt. Rend.*, **271**, 878 (1970).
[7] V. G. Dryuk, M. S. Malinovskii, and A. F. Kurochkin, *Zhur. Org. Khim.*, **6**, 2361 (1970); *J. Org. Chem. USSR*, **6**, 2373 (1970).
[8] G. R. Howe and R. R. Hiatt, *J. Org. Chem.*, **36**, 2493 (1971).
[9] G. M. Beverly and D. R. Hogg, *J. Chem. Soc.* (B), **1971**, 175.

allow a discussion of the factors that govern C and S attack by nucleophiles in episulphonium ions.[90] The stereochemistries of addition of 2,4-dinitrobenzenesulphenyl chloride to, and oxymercuration of, dimethyl cyclohex-4-ene-*trans*-1,2-dicarboxylate have been studied.[91]

Electrophile-assisted ionization to give $F_2C\cdots\overset{+}{C}H\cdots CH_2$ rather than conventional addition to give a carbonium ion seems to be a key step in the reactions of $CH_2=CH-CF_3$ with electrophiles. This is illustrated for dimerization[92] in Scheme 3.

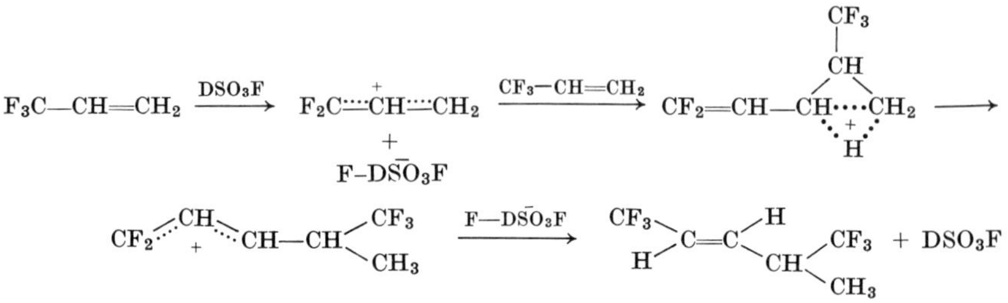

$$F_3C-CH=CH_2 \xrightarrow{DSO_3F} F_2C\cdots\overset{+}{C}H\cdots CH_2 \xrightarrow{CF_3-CH=CH_2} CF_2=CH-\underset{H}{\overset{CF_3}{\underset{|}{CH}}}\cdots\overset{+}{CH_2} \longrightarrow$$

$$CF_2\cdots\overset{CH}{\underset{+}{\cdots}}\cdots CH-CH\overset{CF_3}{\underset{CH_3}{}} \xrightarrow{F-DSO_3F} \underset{H}{\overset{CF_3}{}}C=C\underset{CH}{\overset{H}{\underset{CH_3}{\overset{CF_3}{}}}} + DSO_3F$$

Scheme 3

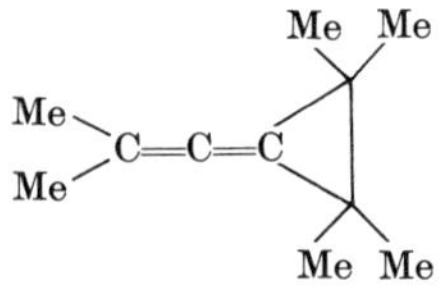

(8) 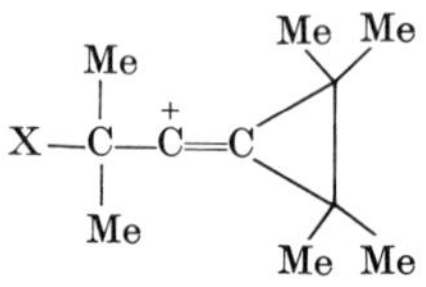   (9)

Electrophilic additions to the allene (8) give products resulting from terminal attack of the electrophile because of the special stabilization of the intermediate vinyl cation (9) by the cyclopropane ring. However, when bridging in the intermediate is sufficiently important to change the product-determining step to nucleophilic opening of a bridged ion, reversal of this orientation is observed.[93]

The directional orientation for a variety of additions to 1,2,3,4,7,7-hexachloronorborna-2,5-diene cannot be accommodated solely by the simple steric model proposed by Brown for 7,7-dimethylnorbornene.[94]

Several electrophilic additions and 1,3-dipolar cycloadditions to bicyclo[2.2.0]hexa-2,5-diene (Dewar benzene) have been accomplished without skeletal rearrangement.[95] The first simple electrophilic additions of strongly polarized cationoid complexes of type X^+Y^- ($X^+ = RCO^+$, R^+, NO_2^+, RS^+, $Y^- = BF_4^-$) to olefins and acetylenes to give monomeric products, rather than polymerization, have been achieved.[96] Bridged

---

[90] G. H. Schmid and P. H. Fitzgerald, *J. Am. Chem. Soc.*, **93**, 2547 (1971).
[91] N. S. Zefirov, V. N. Chekulaeva, N. D. Antonova, and L. G. Gurvich, *Dokl. Akad. Nauk SSSR*, **192**, 567 (1970).
[92] P. C. Myhre and G. D. Andrews, *J. Am. Chem. Soc.*, **92**, 7595, 7596 (1970).
[93] M. L. Poutsma and P. A. Ibarbia, *J. Am. Chem. Soc.*, **93**, 440 (1971).
[94] D. I. Davies, P. Mason, and M. J. Parrott, *J. Chem. Soc.* (C), **1971**, 3428.
[95] E. E. van Tamelen and D. Carty, *J. Am. Chem. Soc.*, **93**, 6102 (1971).
[96] W. A. Smit, A. V. Semenovsky, V. F. Kucherov, T. N. Chernova, M. Z. Krimer, and O. V. Lubinskaya, *Tetrahedron Letters*, **1971**, 3101.

olefins of the norbornene type give products with $Pb(OAc)_{4-n}(N_3)_n$ that apparently result from an ionic reaction initiated by electrophilic azide.[97] Transannular cyclizations of 6-substituted cyclodecynes have been studied. These are generally highly selective and lead to products with the bicyclo[4.4.0]decane skeleton.[98, 99] Similarly, only the six-membered ring product is observed in the acid-catalysed intramolecular cyclization of oct-6-yn-2-one.[99]

Intramolecular addition of a carbonium ion to an acetylene forms the basis of a new stereoselective approach to *trans*-fused steroid D rings.[100] Transannular π-bond participation is observed in the electrophilic additions of cyclo-octa-*cis*-1-*cis*-5-diene.[101] Intramolecular lactone formation occurs when norborn-5-enes with *endo*-2-CN, -CH_2CN and -CO_2Me substituents are treated with polyphosphoric acid.[102]

The effect of substituents on silver–olefin complex formation has been studied and correlated by a modified Kirkwood–Westheimer cavity model.[103] Free carbonium ions are not involved in $PdCl_2$- and $CuCl_2$-catalysed additions to bicyclic olefins.[104] Other aspects of electrophilic additions that have received attention are various additions to dehydrojanusene,[105] 2-phenylnorbornene,[106] 5-ethylidene- and 5-vinyl-bicyclo[2.2.1]-hept-2-ene,[107] cyano ynamines (giving cyclobutene derivatives),[108] the addition of the 1-adamantyl cation to prop-2-ynyl alcohol,[109] a comparison of the reactivity of allenes and acetylenes,[110] investigations of the modes of reaction of the diphenyl-diazenium cation with olefins,[111] and the effect of $AlCl_3$ on the addition of the prenyl cation to prenyl thiolacetate.[112]

The kinetics and mechanism of the gas-phase triethylaluminium-catalysed dimerization of ethylene to but-1-ene[113] have been studied, and activation parameters for the addition of S and other Group VIA atoms in their ground triplet state[114] have been determined.

Nucleophilic Additions

The mechanism of addition of $R\bar{C}(NO_2)_2$ ions to methyl acrylate varies with the nature of R. When R is capable of resonance-stabilizing negative charge, acid-catalysis is

[97] E. Zbiral and A. Stütz, *Tetrahedron*, **27**, 4953 (1971).
[98] B. Rao and L. Weiler, *Tetrahedron Letters*, **1971**, 927; J. Balf, B. Rao, and L. Weiler, *Can. J. Chem.*, **49**, 3135 (1971).
[99] C. E. Harding and M. Hanach, *Tetrahedron Letters*, **1971**, 1253.
[100] P. T. Lansbury and G. E. Dubois, *Chem. Comm.*, **1971**, 1107.
[101] H. J. Franz, W. Höbold, R. Höhn, G. Müller-Hagen, R. Müller, W. Pritzkow, and H. Schmidt, *J. Prakt. Chem.*, **312**, 622 (1970).
[102] T. Sasaki, S. Eguchi, and M. Sugimoto, *Bull. Chem. Soc. Japan*, **44**, 1382 (1971).
[103] C. F. Wilcox and W. Gaal, *J. Am. Chem. Soc.*, **93**, 2453 (1971).
[104] C. J. R. Adderley, J. W. Nebzydoski, M. A. Battiste, R. Baker, and D. E. Halliday, *Tetrahedron Letters*, **1971**, 3545.
[105] S. J. Cristol and M. A. Imhoff, *J. Org. Chem.*, **36**, 1849 (1971).
[106] J. M. Coxon, M. P. Hartshorn, and A. J. Lewis, *Austral. J. Chem.*, **24**, 1017 (1971).
[107] T. C. Shields, *Can. J. Chem.*, **49**, 1142 (1971).
[108] T. Sasaki and A. Kojima, *J. Chem. Soc.* (C), **1971**, 3056.
[109] J. K. Chakrabarti and A. Todd, *Chem. Comm.*, **1971**, 556.
[110] T. Okamoto, K. Takagi, Y. Sakakibara, and S. Kunichika, *Bull. Inst. Chem. Res. Kyoto Univ.*, **48**, 96 (1970).
[111] G. Cauquis and M. Genies, *Tetrahedron Letters*, **1971**, 3959.
[112] K. Takabe, T. Katagiri, and J. Tanaka, *Tetrahedron Letters*, **1971**, 1503.
[113] K. W. Egger, *Trans. Faraday Soc.*, **67**, 2638 (1971).
[114] O. P. Strausz, W. B. O'Callaghan, E. M. Lown, and H. E. Gunning, *J. Am. Chem. Soc.*, **93**, 559 (1971); see also J. Connor, A. Van Roodselaar, R. W. Fair, and O. P. Strausz, *J. Am. Chem. Soc.*, **93**, 560 (1971).

observed and protonation of the intermediate carbanion is rate-determining; in other cases acid-catalysis is not observed. From a detailed kinetic study in 50% dioxan and in water this change in mechanism is attributed mainly to a lowering of transition-state energy for dissociation of the initial adduct anion by such R groups, although in borderline cases the rate of protonation of this anion may also be a critical factor. The effect of R on the specific rate of addition of $\bar{RC}(NO_2)_2$ is also discussed.[115]

Rate-determining polarization of a π-complex between ylid and olefin by acid is proposed for the acid-catalysed nucleophilic addition of cyclopentadienyltriphenylphosphorane to a series of benzylidenemalononitriles in benzene[116] (Scheme 4).

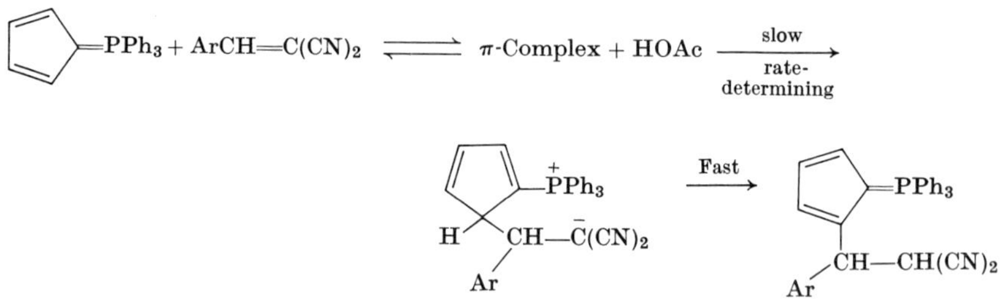

$$\text{Cp=PPh}_3 + \text{ArCH=C(CN)}_2 \rightleftharpoons \pi\text{-Complex} + \text{HOAc} \xrightarrow[\text{rate-determining}]{\text{slow}}$$

Scheme 4

Steric acceleration observed for the *o*-t-butylbenzenethiolate anion in its addition to *N*-ethylmaleimide (and S_N2 displacement of 2,4-dinitrobenzenethiolate anion from 2,4-dinitrophenyl disulphide) is attributed to steric inhibition of solvation of the thiolate anion, so making it more reactive than the *para*-isomer.[117]

The predominantly (possibly exclusively) *trans*-addition of Grignard reagents to alkynols suggests that this involves concerted addition of external and co-ordinated molecules (10) and not concerted addition of R and Mg from the same co-ordinated molecule (11).[118] Co-ordination via a primary or tertiary amine substituent promotes the addition of Grignard reagents to alkenes or alkynes.[119]

Nucleophiles attack the methylene side chain of 4-(aminomethylene)oxazole-5(4*H*)-ones (12) rather than the ring carbonyl. The analogous penicillenic acid (13) behaves similarly when intramolecular addition of the thiol group is prevented; this has implications for the mechanism of penicillin carcinogenicity, since it shows that, when the side chain sulphur is oxidized, binding to other thiol sites can be by alkylation rather than acylation.[120]

Studies of the kinetics and effects of different dipolar aprotic solvents on the homogeneous base-catalysed addition of alkyl-aromatic compounds to conjugated unsaturated hydrocarbons have been reported and discussed.[121] As expected, 3-alkylpyridines react less readily than their 2- and 4-isomers in such additions. The intermediate carbanion

---

[115] L. A. Kaplan and H. B. Pickard, *J. Am. Chem. Soc.*, **93**, 3447 (1971).
[116] E. Lord, M. P. Naan, and C. D. Hall, *J. Chem. Soc.* (B), **1971**, 213.
[117] D. S. Garwood and D. C. Garwood, *Tetrahedron Letters*, **1970**, 4959.
[118] F. W. Von Rein and H. G. Richey, *Tetrahedron Letters*, **1971**, 3777, 3781; H. G. Richey and S. S. Szacs, *Tetrahedron Letters*, **1971**, 3785.
[119] H. G. Richey, W. F. Erickson, and A. S. Heyn, *Tetrahedron Letters*, **1971**, 2183.
[120] J. L. Longridge and D. Timms, *J. Chem. Soc.* (B), **1971**, 848, 852.
[121] H. Pines, W. M. Stalick, T. G. Holford, J. Golab, H. Lazar, and J. Simonik, *J. Org. Chem.*, **36**, 2299 (1971).

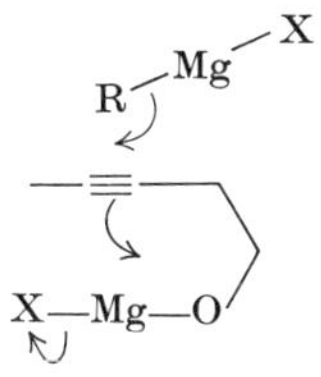

(10) 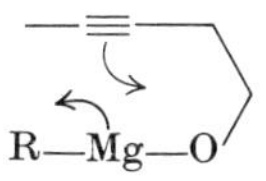 (11) 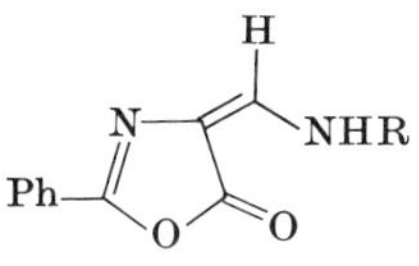 (12) 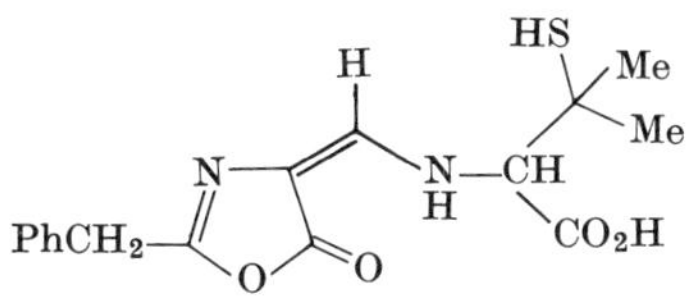 (13)

formed from 3-ethyl- or 3-sec-butylpyridine and ethylene, but not conjugated olefins, undergoes a novel cyclization to the pyridine 2-position.[122] The dipolar aprotic solvents N-methylpyrrolidin-2-one and N-methylpiperid-2-one are not inert and themselves undergo base-catalysed additions to conjugated diolefins (the 3-position is the point of proton removal).[123]

The ambident anion from benzothiazole-2-thiol undergoes Michael additions via N; this is rationalized by application of the oxibase scale.[124] Base-catalysed addition of methanol to hexafluorobut-2-yne and (trifluoromethyl)acetylene is predominantly *trans*, as expected.[125] The base-catalysed addition of thiophenol to *cis*- and *trans*-2-fluorovinyl phenyl sulphones is stereospecifically *cis*.[126]

Asymmetric induction has been observed in the addition of piperidine to an optically active vinyl sulphone.[127] The position of nucleophilic addition of lithium dimethylcuprate to a variety of cyclohexadienones is overwhelmingly determined by the relative steric accessibility of the 4- and the 6-position.[128] That phosphonium cyclopentadienylide forms Michael rather than Diels–Alder adducts with a variety of dienophiles demonstrates its ylid rather than ylene character.[129] Bisulphite-catalysed incorporation of deuterium at the 5-position of cytidine 5-phosphate without deamination has been reported.[130] Nona-2,7-diynedioic acid derivatives undergo a novel addition–cyclization with methylamine to give isoquinoline-1,3-dione derivatives.[131]

The kinetics of the additions of thiocyanic acid and hydriodic acid to propiolic ester in DMF,[132] of methanol to α,β-unsaturated ketones[133] and of imidazole to acrylonitrile[134]

---

[122] S. V. Kannan and H. Pines, *J. Org. Chem.*, **36**, 2304 (1971); H. Pines, S. V. Kannan, and W. M. Stalick, *J. Org. Chem.*, **36**, 2308 (1971).
[123] H. Pines, S. V. Kannan, and J. Simonik, *J. Org. Chem.*, **36**, 2311 (1971).
[124] A. F. Halasa and G. E. P. Smith, *J. Org. Chem.*, **36**, 636 (1971).
[125] E. K. Raunio and T. G. Frey, *J. Org. Chem.*, **36**, 345 (1971).
[126] G. Marchese, F. Naso, L. Schenetti, and V. Sciacovelli, *Chim. Ind. (Milan)*, **53**, 843 (1971).
[127] D. J. Abbott, S. Colonna, and C. J. M. Stirling, *Chem. Comm.*, **1971**, 471.
[128] J. A. Marshall, R. A. Ruden, L. K. Hirsch, and M. Phillippe, *Tetrahedron Letters*, **1971**, 3795.
[129] Z. Yoshida, S. Yoneda, H. Hashimoto, and Y. Murata, *Tetrahedron Letters*, **1971**, 1527.
[130] K. Kai, Y. Wataya, and H. Hayatsu, *J. Am. Chem. Soc.*, **93**, 2089 (1971).
[131] N. J. McCorkindale, D. S. Magrill, R. A. Raphael, and J. L. C. Wright, *J. Chem. Soc. (C)*, **1971**, 3620.
[132] G. V. Dvorko and T. F. Karpenko, *Ukr. Khim. Zhur.*, **37**, 41 (1971); *Chem. Abs.*, **74**, 140445 (1971).
[133] R. Luft, S. Delattre, and T. F. Arnaudo, *Bull. Chim. Soc. France*, **1971**, 1317.
[134] F. Yamada and Y. Fujimoto, *Bull. Chem. Soc. Japan*, **44**, 557 (1971).

have been reported. Full details of pyrazoline formation from Mannich bases and hydrazines have now appeared.[135] Addition of enolate anions to allenic sulphonium salts provides a simple new synthesis of furans.[136]

Cycloadditions

2 + 4-Cycloadditions

Simple models which explain qualitatively the effect of substituents on the rate of Diels–Alder reactions in terms of their effects on the energies of HO and LVMO have appeared,[137, 138] Interaction between reactant orbitals of the same symmetry leads to 'transition state orbitals' whose stabilization is greatest when the energy gap between the interacting orbitals is smallest. Electron-withdrawing substituents lower the energy of orbitals, but electron-releasing ones have the opposite effect. The orbital diagram for a conventional Diels–Alder reaction between electron-rich diene and electron-deficient dienophile is as in Figure 1. The dominant interaction is between π_A and ϕ_2, and this interaction increases as the electron-withdrawing effect of dienophile and the electron-releasing effect of diene substituents is increased. The situation is changed for the

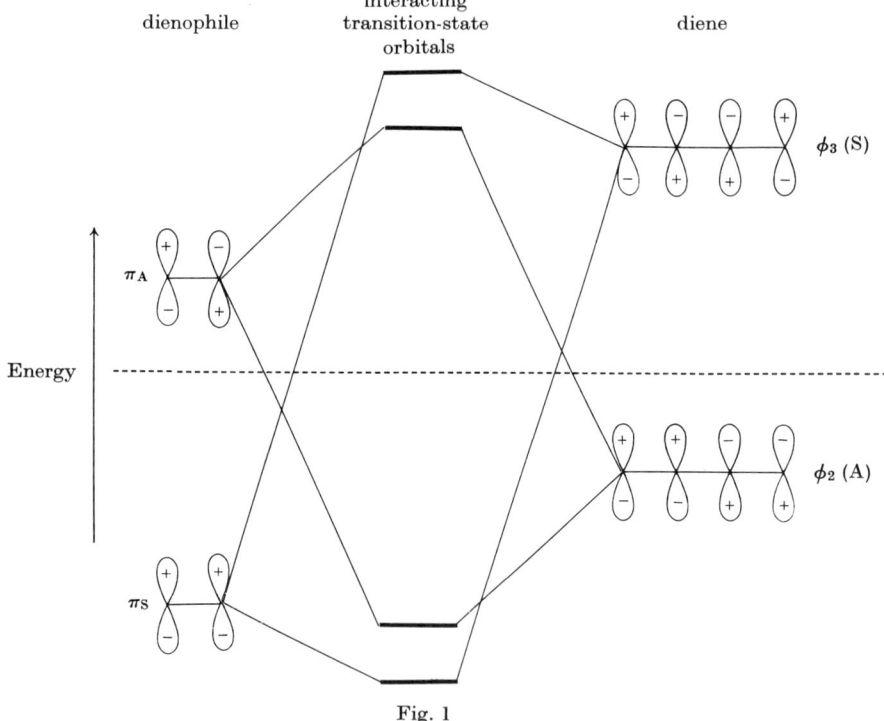

Fig. 1

[135] F. L. Scott, S. A. Houlihan, and D. F. Fenton, *J. Chem. Soc.* (C), **1971**, 80; *Org. Reaction Mech.*, **1970**, 189.
[136] J. W. Batty, P. D. Howes, and C. J. M. Stirling, *Chem. Comm.*, **1971**, 534.
[137] O. Eisenstein and N. T. Anh, *Tetrahedron Letters*, **1971**, 1191.
[138] R. Sustmann, *Tetrahedron Letters*, **1971**, 2721.

Diels–Alder reaction with inverse electron demand (Figure 2), where the dominant interaction is now between π_S and ϕ_3; in this case withdrawing groups in diene and the releasing groups in the dienophile increase the interaction.[138]

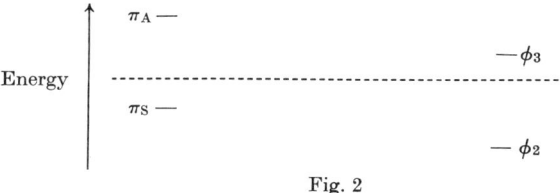

Fig. 2

Explanation of regiospecificity remains one of the outstanding problems associated with the Diels–Alder reaction. Application of the concept of hard and soft centres leads to simple, apparently successful predictions; it is assumed that bonds are formed preferentially between the softest centres—that is those centres that have the greatest coefficients in the HO and LVMO.[139]

The suggestion that attractive van der Waals interactions between methyl substituents in acrylic esters and sp^2 carbons in cyclopentadienes are responsible for the low proportion of adducts with an *endo*-ester group[140] has been refuted since the same pattern is not observed for the addition of acrylic esters to 2,5-dimethyl-3,4-diphenylcyclopentadienone. The results with cyclopentadiene are reinterpreted in terms of steric effects due to the methylene-CH_2 interfering with the normal secondary interactions that would otherwise lead to *endo* orientation of the ester group.[141]

anti-Orientation of methylene-bridge chlorine in Diels–Alder adducts of pentachlorocyclopentadiene is unexpected on steric grounds and is rationalized by attractive interactions due to dipole–dipole, dipole–induced-dipole and London dispersion forces between the polarizable chlorine and the dienophile. The amount of *anti* orientation increases with the polarity of the dienophile and with added Lewis acid catalysts which increase the polarity of the dienophile by coordination.[142]

In order to account for anomalies in the reactivity order of perhalocyclopropenes in Diels–Alder additions to cyclopentadiene and furan,[143] a mechanism involving prior ionization of the perhalocyclopropene followed by addition of a cyclopropenium ion to the diene has been considered. However, a careful search for evidence to substantiate this failed.[144]

Normally, *trans*-dienophiles are more reactive than *cis*, probably because steric interactions prevent *cis*-substituents from exerting full conjugative activation and because steric compression becomes more severe in the transition state. Maleonitrile is, however, more reactive than fumaronitrile; it is suggested that when the above steric effects are minimized, as with small $C{\equiv}N$ groups, favourable secondary interactions, which are greater for two *cis*-groups capable of π-overlap, take over.[145]

[139] O. Eisenstein, J.-M. Lefour, and N. T. Anh, *Chem. Comm.*, **1971**, 969.
[140] Y. Kobuke, T. Fueno, and J. Furukawa, *J. Am. Chem. Soc.*, **92**, 6548 (1970).
[141] K. N. Houk and L. J. Luskus, *J. Am. Chem. Soc.*, **93**, 4606 (1971).
[142] K. L. Williamson and Y.-F. Li Hsu, *J. Am. Chem. Soc.*, **92**, 7385 (1970).
[143] D. C. F. Low and S. W. Tobey, *J. Am. Chem. Soc.*, **90**, 2376 (1968); *Org. Reaction Mech.*, **1968**, 175.
[144] R. M. Magid and S. E. Wilson, *J. Org. Chem.*, **36**, 1775 (1971).
[145] T. Miki and T. Matsuo, *Chem. Pharm. Bull.*, **19**, 858 (1971).

In the additions of *trans*-β-nitrostyrenes to *trans,trans*-1,4-diphenylbutadiene the aryl group shows a markedly increased tendency relative to the nitro group to be *endo* as electron-withdrawing groups are introduced into its *para*- but not *meta*-position.[146] Further details of the activation-volume data from high-pressure kinetic studies of Diels–Alder reactions have appeared. These fully support a concerted mechanism and the fact that, for the reactions of maleic anhydride, the transition state is smaller than the ultimate adduct is attributed to secondary interactions.[147]

Detailed studies of Diels–Alder reactions of nitrosobenzenes with dienes indicate a concerted reaction through a polar transition state.[148] Several studies of the effect of substituents on the regiospecificity of Diels–Alder reactions have appeared.[149] Kinetic investigations have been reported for the Diels–Alder reactions of unsymmetrically 1,4-disubstituted dienes with 2,6-dimethyl-*p*-benzoquinone[150] (concerted with an asymmetric transition state), of hepta-3,5-dien-2-one with maleic anhydride[151] and of 1-alkoxydienes with glyoxylic esters.[152] Hammett correlations have been observed in the Diels–Alder reactions of 6-substituted fulvenes with maleic anhydride.[153] The influence of chlorine substituents on the dienophilic activity of cyclic carbonates and anhydrides has been discussed,[154] and the enthalpy of addition of isoprene to tetracyanoethylene (TCNE) has been determined calorimetrically.[155]

Semiempirical SCF MO CNDO/2 theory has been applied to the Diels–Alder reaction of ethylene and *cis*-butadiene,[156] and HMO calculations have been reported for the Diels–Alder reactions of fulvenes.[157]

The major proportion of 4 + 2 adducts from itaconic acid and methyl 5-(*p*-methoxyphenyl)penta-*trans*-2,*trans*-4-dienoate (160°, 8 hr) involve loss of configuration of the diene component but this is possibly the result of isomerization to relieve steric strain in the initial adducts.[158]

An example of the rarely observed Diels–Alder addition of a homodiene has been reported in the addition of bicyclopentene to TCNE.[159]

In line with the picture built up by Bartlett,[160] the rates and product distribution for the dimerization of chloroprene and related dienes is consistent with a competing concerted Diels–Alder reaction requiring a *cisoid*-diene, and a diradical process leading mainly to 2 + 2 adducts. Normally the first process is favoured except where substituents

---

[146] P. C. Jain, Y. N. Mukerjee, and N. Anand, *Chem. Comm.*, **1971**, 303.
[147] R. A. Grieger and C. A. Eckert, *J. Am. Chem. Soc.*, **92**, 7149 (1970).
[148] G. Kresze and W. Kosbahn, *Tetrahedron*, **27**, 1931 (1971); G. Kresze, H. Saitner, J. Firl, and W. Kosbahn, *Tetrahedron*, **27**, 1941 (1971).
[149] C. Schmidt, S. D. Sabnis, E. Schmidt, and D. K. Taylor, *Can. J. Chem.*, **49**, 371 (1971); M. F. Ansell and A. H. Clements, *J. Chem. Soc.* (C), **1971**, 269, 275; T. Inukai and T. Kojima, *J. Org. Chem.*, **36**, 924 (1971).
[150] M. T. H. Liu and C. Schmidt, *Tetrahedron*, **27**, 5289 (1971).
[151] V. S. Markevich, V. A. Terent'ev, N. K. Shtivel, and A. P. Gryaznov, *Zhur. Org. Khim.*, **6**, 2011 (1970); *J. Org. Chem. USSR*, **6**, 2020 (1970).
[152] S. D. Yablonovskaya, S. V. Bogathov, and S. M. Makin, *Zhur. Org. Khim.*, **6**, 2005 (1970); *J. Org. Chem. USSR*, **6**, 2015 (1970).
[153] N. Takeno and M. Morita, *Nippon Kagaku Zasshi*, **91**, 1182 (1970); *Chem. Abs.*, **75**, 19515 (1971).
[154] H.-D. Scharf, W. Küsters, and J. Fleischhauer, *Chem. Ber.*, **104**, 3030 (1971).
[155] F. E. Rogers, *J. Phys. Chem.*, **75**, 1734 (1971).
[156] O. Kikuchi, *Tetrahedron*, **27**, 2791 (1971).
[157] N. Takeno, K. Suzuki, and M. Morita, *Nippon Kagaku Zasshi*, **91**, 877 (1970).
[158] A. J. Baker and A. J. Goudie, *Chem. Comm.*, **1971**, 180.
[159] J. E. Baldwin and R. K. Pinschmidt, *Tetrahedron Letters*, **1971**, 935.
[160] P. D. Bartlett, *Quart. Rev. (London)*, **24**, 473 (1970).

specially stabilize a diradical intermediate.[161] A common diradical intermediate has been suggested to account for the simultaneous formation of 2 + 2 and 2 + 4 cyclodimers in the borohydride reduction of 1-methyl-4-cyanopyridinium iodide.[162] The tropylium cation functions as a diene in its highly stereospecific endo-4 + 2-addition to cyclopentadiene.[163] 5-Alkenylcyclohexa-1,3-dienes (**14**; $n = 0-3$) give intramolecular Diels–Alder adducts with the stereochemistry shown.[164]

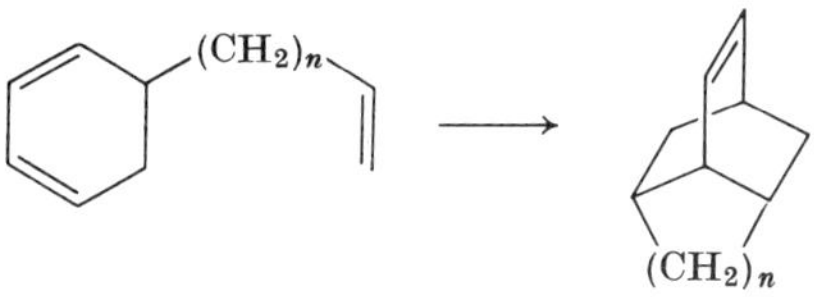

(**14**) 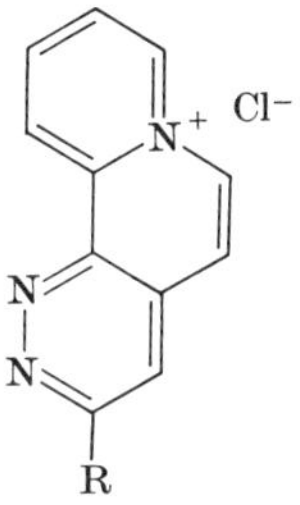   (**15**)

R = 2-pyridyl

Deamination has been observed in the Diels–Alder adducts of isoindole derivatives with benzyne and dimethyl acetylenedicarboxylate, but the fate of the R—N̈: (nitrene) fragment has not been established.[165]

Vinylallenes undergo exclusive (concerted) Diels–Alder reactions with α,β-unsaturated ketones, no 2 + 2 adducts of the allene function being observed.[166]

Initial Diels–Alder reactions involving an aryl bond in the diene component are involved in the cycloadditions of diarylmethylenecyclopropanes with TCNE and in the thermal dimerization of phenylallene[167] (see also other cycloadditions). Attempted Diels–Alder reaction of cis-3,4-dichlorocyclobutene with 3,6-di-(2-pyridyl)-s-tetrazine gives (**15**).[168]

Full details of the Diels–Alder reactions of hexafluorobicyclo[2.2.0]hexa-2,5-diene[169] and a preliminary report of the dienophilic reactivity of polyfluorobicyclo[2.2.0]hex-2-enes[170] have appeared.

The novel 2,5-diphenyl-3,4-diazacyclopentadienone (diene with inverse electron-demand)[171] has been reported, and the highly reactive diene isobenzofuran has been isolated.[172] A new cis-azodienophile is obtained by oxidation of 5-aryl-s-triazolin-3-ones.[173]

---

[161] C. A. Stewart, J. Am. Chem. Soc., **93**, 4815 (1971).
[162] F. Liberatore, A. Casini, and V. Carelli, Tetrahedron Letters, **1971**, 3829.
[163] S. Itô and I. Itoh, Tetrahedron Letters, **1971**, 2969.
[164] A. Krantz and C. Y. Lin, Chem. Comm., **1971**, 1287.
[165] L. J. Kricka and J. M. Vernon, Chem. Comm., **1971**, 942.
[166] M. Bertrand, J. Grimaldi, and M. Waegell, Bull. Soc. Chim. France, **1971**, 962.
[167] J. E. Baldwin and R. E. Peavy, J. Org. Chem., **36**, 1441 (1971); J. E. Baldwin and L. E. Walker, J. Org. Chem., **36**, 1440 (1971).
[168] J. A. Elix, M. Sterns, W. S. Wilson, and R. N. Warrener, Chem. Comm., **1971**, 426.
[169] M. G. Barlow, R. N. Haszeldine, and R. Hubbard, J. Chem. Soc. (C), **1971**, 90.
[170] W. J. Feast, W. K. R. Musgrave, and R. G. Weston, Chem. Comm., **1971**, 709.
[171] P. J. Fagan and M. J. Nye, Chem. Comm., **1971**, 537.
[172] R. N. Warrener, J. Am. Chem. Soc., **93**, 2346 (1971); D. Wege, Tetrahedron Letters, **1971**, 2337.
[173] B. T. Gillis and J. G. Dain, J. Org. Chem., **36**, 518 (1971).

2 + 4-Cycloaddition of chlorosulphonyl isocyanate to vinyldihydronaphthaline involves addition to C=O under kinetic control but to C=N under thermodynamic control.[174] Cycloaddition of the 1,4-dipole resulting from interaction of 4-phenyltriazolin-3,5-dione and vinyl ethers to the C=O of alkyl ketones has been reported.[175] 1,4-Dipolar cycloadditions of the extremely stable 1,4-dipoles, e.g. (**17**), from the electron-rich olefin (**16**) and sulphonyl isocyanates and isothiocyanates have been investigated.[176] See also p. 173 for further examples of this type of reaction. Substituent effects give further support for the stepwise nature of the 4 + 2-addition of acridizinium ions to olefins.[177]

$$CH_2=C\begin{smallmatrix}NMe_2\\SMe\end{smallmatrix} \quad \xrightarrow{PhSO_2N=C=X} \quad \overset{+}{C}H_2-C\begin{smallmatrix}SMe\\NMe_2\end{smallmatrix}$$
$$\underset{X\cdots NSO_2Ph}{\overset{C}{|}}$$

X = O, S

(**16**)   (**17**)

Arrhenius parameters for the gas-phase retro-Diels–Alder reaction of bicyclo[2.2.2]oct-2-ene suggest a concerted rather than a diradical mechanism.[178] Full details of the retro-Diels–Alder reactions of tricyclo[6.2.1.0^{2,7}]undeca-3,5,9-trienes and tricyclo-[6.2.1.0^{2,7}]undeca-3,5,7,9-tetraenes[179] have appeared.

2 + 3-Cycloadditions

A review, in Polish, of 1,3-dipolar cycloadditions has appeared.[180] The simple picture explaining the effect of substituents in cycloadditions in terms of their influence on the relative energies of HO and LV orbitals, as described for the Diels–Alder reaction (p. 164) has been applied to 1,3-dipolar cycloadditions.[181]

Additions of 2-azaallyl anions to dienes[182] (only 2 + 3 adducts are found) and to C=N and N=N systems[183] have been reported. The stereospecificity observed supports the probable concerted nature of these allowed reactions for which the term 1,3-anionic cycloaddition is proposed. The additions of 2-azaallyl anions involve favourable redistribution of charge from C to N as the reaction proceeds. The situation should be somewhat less favourable for a 1,2-diazaallyl anion, since charge is already extensively localized on N. Such additions have been observed, but only to strained π-bonds.[184]

New studies on the mechanism of ozonolysis contradict the generally accepted view that ozone addition to an olefinic bond is of the 1,3-dipolar type (see Chapter 14).[185]

Although it is entirely reasonable that a zwitterionic intermediate (**18**) should be involved in the 3 + 2-cycloaddition of (methyleneamino)phosphanes $X_2P-N=CPh_2$

---

[174] R. J. P. Barends, W. N. Speckamp, and H. O. Huismann, *Tetrahedron Letters*, **1970**, 5301.
[175] S. R. Turner, L. J. Guilbault, and G. B. Butler, *J. Org. Chem.*, **36**, 2838 (1971).
[176] R. Gompper and B. Wetzel, *Tetrahedron Letters*, **1971**, 529.
[177] I. J. Westerman and C. K. Bradsher, *J. Org. Chem.*, **36**, 969 (1971).
[178] A. T. Cocks and H. M. Frey, *J. Chem. Soc.* (A), **1971**, 1661.
[179] W. P. Lay, K. Mackenzie, and J. R. Telford, *J. Chem. Soc.* (C), **1971**, 3199.
[180] W. Sliva, *Wiad. Chem.*, **25**, 1 (1971); *Chem. Abs.*, **74**, 99087 (1971).
[181] R. Sustmann, *Tetrahedron Letters*, **1971**, 2717.
[182] T. Kauffmann and R. Eidenschink, *Angew. Chem. Int. Ed.*, **10**, 739 (1971).
[183] T. Kauffmann, M. Berg, E. Ludorff, and A. Woltermann, *Angew. Chem. Int. Ed.*, **9**, 960 (1970).
[184] T. Kauffmann, D. Berger, B. Scheerer, and A. Woltermann, *Angew. Chem. Int. Ed.*, **9**, 961 (1970).
[185] P. R. Story, J. A. Alford, J. R. Burgess, and W. C. Ray, *J. Am. Chem. Soc.*, **93**, 3042, 3044 (1971).

$$X_2\overset{+}{P}-N=CPh_2$$
$$\underset{\underset{CHR}{\searrow}}{|}$$
$$CH_2$$

(18)

to electrophilic olefins, it is possible that this system may be a 1,3-dipole by virtue of $d_\pi-p_\pi$ overlap.[186]

Kinetic investigations of the reaction of nitrile oxides with arylacetylenes have been carried out.[187,188] These establish that isoxazole and oxime are formed concurrently. The absence of an isotope effect in oxime formation implies that this is formed through an intermediate and, since activation energies for both oxime and isoxazole formation are similar[187] and there is little variation in product distribution with temperature and solvent, it is suggested that both products are derived from a common intermediate, possibly in a ratio that is determined by the conformational distribution of this species.[188] The evidence against stepwise oxime and concerted isoxazole formation is, however, not yet compelling.

Bent 1,3-dipoles, such as N-methyl-C-phenylnitrone, show anomalously low $k_{\text{norbornene}}/k_{\text{cyclohexene}}$ rate ratios when compared with linear 1,3-dipoles because of steric inhibition in the transition state. For less hindered nitrones the ratio increases. The isomeric adducts from N-methyl-C-phenylnitrone and norbornene result from addition of cis- and trans-forms of the nitrone, which are interconverted under the reaction conditions. Steric interactions lead to the major adduct's being formed from the less stable cis-nitrone.[189]

Substituent and solvent effects and activation parameters for the cycloadditions of benzonitrile oxides to thiobenzophenones to give 1,4,2-oxathiazoles,[190] of azides to α-ketophosphorus ylids to give triazoles and triphenylphosphine oxide,[191] and of ethyl diazoacetate to isobutylidene- and 3-methyl-2-(ethoxycarbonyl)butylidene-Meldrum acids to give thermally labile 1-pyrazolines,[192] are consistent with concerted processes with polar transition states. Characteristics for the cycloadditions of munchnone imines (20), derived from the hydrofluoroborate salts of Reissert compounds (19), to aryl propiolates similarly suggest a concerted process. These additions are therefore like those of munchnones, sydnones and sydnone imines, except that in this case the initial adduct (21) can be isolated and loses HNCO only slowly.[193]

The electronic structure of the open forms of three-membered rings have been discussed.[194] Huisgen has continued his studies of azomethine ylids derived from aziridines. These substantiate the established scheme of conrotatory ring opening of the aziridine

---

[186] A. Schmidpeter and W. Zeiss, *Angew. Chem. Int. Ed.*, **10**, 396 (1971).
[187] A. Battaglia, A. Dondoni, and A. Mangini, *J. Chem. Soc.* (B), **1971**, 554.
[188] P. Beltrame, P. Sartirana, and C. Vintani, *J. Chem. Soc.* (B), **1971**, 814.
[189] L. W. Boyle, M. J. Peagram, and G. H. Whitham, *J. Chem. Soc.* (B), **1971**, 1728.
[190] A. Battaglia, A. Dondoni, G. Maccagnani, and G. Mazzanti, *J. Chem. Soc.* (B), **1971**, 2096.
[191] P. Ykman, G. L'abbe, and G. Smets, *Tetrahedron*, **27**, 845 (1971).
[192] F. Nierlich, P. Schuster, and O. E. Polansky, *Monatsh.*, **102**, 438 (1971).
[193] W. E. McEwen, K. B. Kanitkar, and W. M. Hung, *J. Am. Chem. Soc.*, **93**, 4484 (1971); W. E. McEwen, I. C. Mineo, and Y. H. Shen, *J. Am. Chem. Soc.*, **93**, 4479 (1971).
[194] E. F. Hayes and A. K. Q. Siu, *J. Am. Chem. Soc.*, **93**, 2090 (1971).

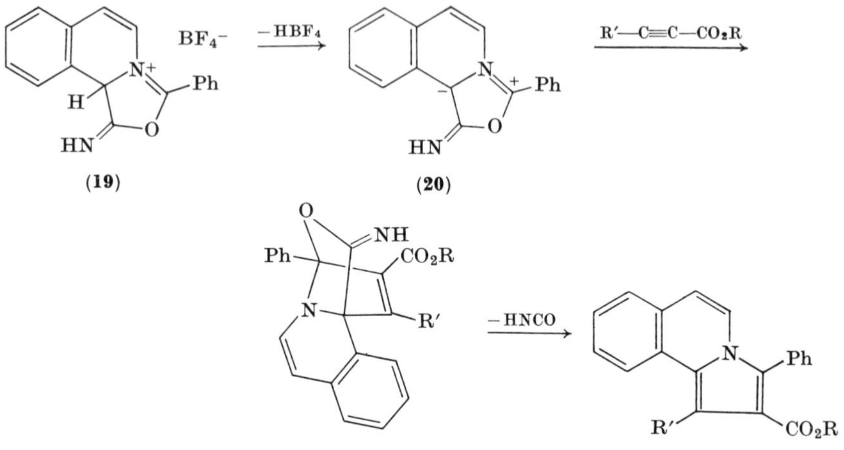

and isomerization of the open dipoles in competition with trapping by 1,3-dipolarophiles.[195] The scope of such additions to aromatic compounds,[196] to C-heteromultiple bonds[197] and to azo compounds[198] has been demonstrated.

Cyanostilbene oxides behave similarly to aziridines, isomerizing and adding via carbonyl ylids formed by conrotatory ring-opening.[199] Formation of dioxolans from gem-dicyanostilbene oxides and aldehydes or ketones involves concerted addition of the carbonyl ylid through a polar transition state.[200]

4,5-Dihydro-1,3,5-oxazaphosph(v)oles (22) react with alkynes and alkenes with (presumably prior) elimination of phosphoric acid esters to give the 3 + 2 adducts expected from nitrile ylids.[201] Azomethine oxides give 3-pyrrolidinones rather than the expected adducts with allenes.[202] N-Methoxycarbonyl-2,3-homopyrrole (methyl 2-azabicyclo[3.1.0]hex-3-ene-2-carboxylate) (23) undergoes cycloadditions that can be explained by thermally allowed disrotatory opening to give the 1,3-dipole (24).[203] The heterocyclic ylid (25) can be trapped in situ by 1,3-dipolarophiles, otherwise it gives formal dimers of the carbene (26).[204]

The steric course and regioselectivity of the addition of diazoacetic ester to cis- and trans-cinnamic esters and of aliphatic diazo compounds to α,β-unsaturated carboxylic esters has been discussed in terms of the steric and electronic effects of substituents on the transition state for concerted addition. Normally, large substituents are preferentially

---

[195] R. Huisgen and H. Mäder, J. Am. Chem. Soc., **93**, 1777 (1971); H. Hermann, R. Huisgen, and H. Mäder, J. Am. Chem. Soc., **93**, 1779 (1971); J. H. Hall and R. Huisgen, Chem. Comm., **1971**, 1187; J. H. Hall, R. Huisgen, C. H. Ross, and W. Scheer, Chem. Comm., **1971**, 1188.
[196] R. Huisgen and W. Scheer, Tetrahedron Letters, **1971**, 481.
[197] R. Huisgen, V. Martin-Ramos, and W. Scheer, Tetrahedron Letters, **1971**, 477.
[198] E. Brunn and R. Huisgen, Tetrahedron Letters, **1971**, 473 (1971).
[199] H. Hamberger and R. Huisgen, Chem. Comm., **1971**, 1190; A. Dahman, H. Hamberger, R. Huisgen, and V. Markowski, Chem. Comm., **1971**, 1192.
[200] A. Robert, J.-J. Pommeret, and A. Foucaud, Tetrahedron Letters, **1971**, 231.
[201] K. Burger and J. Fehn, Angew. Chem. Int. Ed., **10**, 728, 729 (1971).
[202] M. C. Aversa, G. Cum, and N. Uccella, Chem. Comm., **1971**, 156.
[203] F. W. Fowler, Angew. Chem. Int. Ed., **10**, 135 (1971).
[204] M. Caprosu, M. Petrovanu, I. Druță, and I. Zugrăvescu, Bull. Soc. Chim. France, **1971**, 1834.

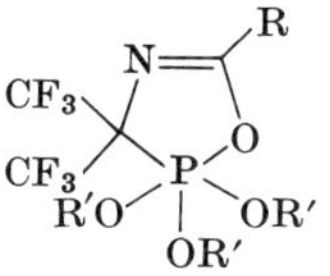

oriented *anti* in the transition state, but in some cases π-overlap between substituents, e.g. CO_2Me and Ph, can lead to a favoured *syn*-orientation.[205]

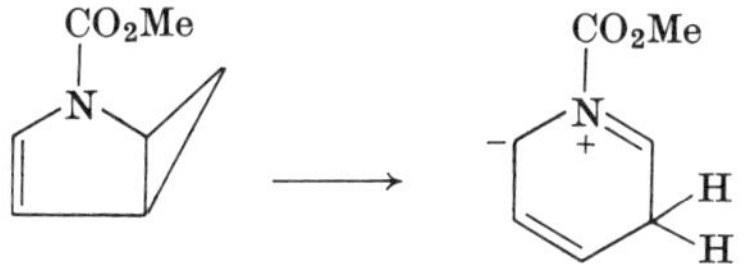

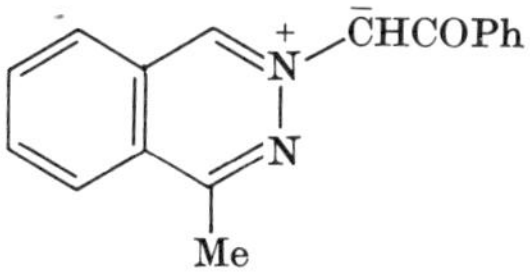

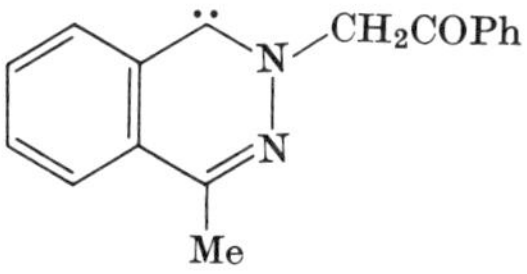

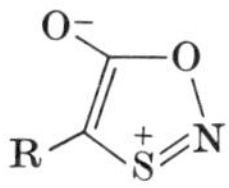

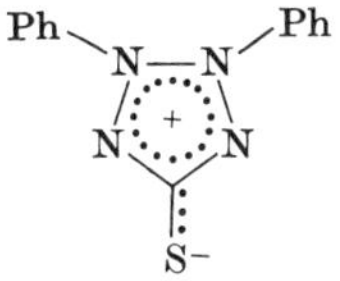

Other aspects of 1,3-dipolar cycloadditions that have received attention are: the additions of mesoionic 1,3,2-oxathiazolium 5-oxides (27),[206] the dehydrodithizone system (28),[207] pyridazinium and pyridinium methylides,[208] 3,4-diazacyclopentadienone N-oxides,[209] of glycosyl azides to electron-rich olefins,[210] and nitronate esters to activated olefins;[211] the orientation of addition of diazo compounds to acetylenes[212] and of vinyl azides to phenylacetylene,[213] and the intramolecular cycloadditions of several methyl substituted N-methyl-C-(hept-6-enyl)nitrones.[214]

Pyridinium N- and C-ylids with diphenylcyclopropenone give 1,3-oxazines and 2-pyrones, respectively, rather than the expected cycloadducts.[215]

Photochemical cycloadditions of N,C-diphenylsydnone and 2,5-diphenyltetrazole to dimethyl acetylenedicarboxylate apparently involve initial photo-elimination of CO_2 and N_2 from the heterocycle and rearrangement of the potential 1,3-dipolar fragment to N,C-diphenylnitrile imine, which is trapped.[216]

---

[205] P. Eberhard and R. Huisgen, *Tetrahedron Letters*, **1971**, 4337, 4343.
[206] H. Gotthardt. *Tetrahedron Letters*, **1971**, 1281.
[207] P. Rajagopalan and P. Penev, *Chem. Comm.*, **1971**, 490.
[208] T. Sasaki, K. Kanematsu, Y. Yukimoto, and S. Ochiai, *J. Org. Chem.*, **36**, 813 (1971).
[209] J. P. Freeman and M. J. Hoare, *J. Org. Chem.*, **36**, 19 (1971).
[210] R. E. Harmon, R. A. Earl, and S. K. Gupta, *Chem. Comm.*, **1971**, 296.
[211] R. Grée and R. Carrié, *Tetrahedron Letters*, **1971**, 4117.
[212] G. Manecke and H.-U. Schenck, *Chem. Ber.*, **104**, 3395 (1971).
[213] G. L'abbe and A. Hassner, *Bull. Soc. Chim. Belges*, **80**, 209 (1971).
[214] N. A. LeBel and E. G. Banucci, *J. Org. Chem.*, **36**, 2440 (1971).
[215] T. Sasaki, K. Kanematsu, and A. Kakehi, *J. Org. Chem.*, **36**, 2451 (1971).
[216] C. S. Angadiyavar and M. V. George, *J. Org. Chem.*, **36**, 1589 (1971); H. Gotthardt and F. Reiter, *Tetrahedron Letters*, **1971**, 2749.

Benzonitrile oxide undergoes $4\pi + 2\pi$-addition to 6,6-dimethyl- and 6,6-diphenyl-fulvene but gives the first reported $4\pi + 6\pi$ adduct for a nitrile oxide with 6-(dimethylamino)fulvene.[217] The first clear example of a thermally allowed $1 + 3$ cheletropic addition to a 1,3-dipole is provided by the reaction of an isocyanide with an azomethine imine.[218]

2 + 2-Cycloadditions

Extended Huckel and MINDO calculations for the addition of vinyl cations to olefins, acetylenes and allenes suggest that the stepwise mode (reaction 1) is more reasonable than concerted $_\pi 2_s + _\pi 2_a$ addition.[219]

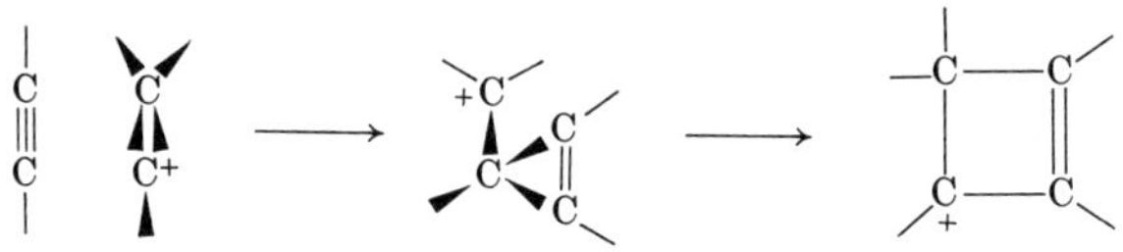

$$\ldots (1)$$

Failure to observe a secondary deuterium isotope effect in the addition of diphenylketene to α-methylstyrene ($k_{\text{Ph(CH}_3)\text{C=CH}_2} = k_{\text{Ph(CD}_3)\text{C=CH}_2}$) supports a concerted mechanism since formation of a zwitterion by attack at the β-carbon atom of the styrene should be affected by D-substitution in the methyl group.[220] A concerted $_\pi 2_s + _\pi 2_a$ mechanism is thought most likely for addition of dichloroketene to a variety of olefins. Norbornene is surprisingly inert to this highly electrophilic ketene because of steric hindrance by the bridge-methylene group in the orthogonal $_\pi 2_s + _\pi 2_a$ approach.[221] The predominant regioisomers formed in the addition of dichloroketene to cholest-2-ene and to 4-t-butylcyclohexene are also rationalized in terms of concerted $_\pi 2_s + _\pi 2_a$ addition.[222] The steric effects that cause predominant *endo* orientation of the larger of two ketene substituents in their adducts with cyclopentadiene have been observed to extend to other olefins.[223] An intramolecular ketene cycloaddition is involved in the photolysis of compounds of type (**29**).[224] (see also Chapter 13, under Carbonyl Compounds). There is some slight evidence that the additions of aldóketenes to chloral are concerted.[225] The stereochemistry of addition of chloroketene to imines has been investigated.[226]

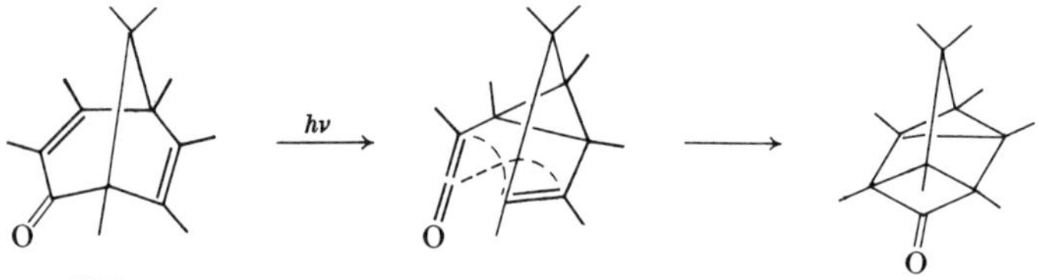

(**29**)

[217] P. Caramella, P. Frattini, and P. Grünanger, *Tetrahedron Letters*, **1971**, 3817.
[218] J. A. Deyrup, *Tetrahedron Letters*, **1971**, 2191.
[219] H.-U. Wagner and R. Gompper, *Tetrahedron Letters*, **1971**, 4061, 4065.
[220] E. I. Snyder, *J. Org. Chem.*, **35**, 4287 (1970).
[221] L. Ghosez, R. Montaigne, A. Roussel, H. Vanlierde, and P. Mollet, *Tetrahedron*, **27**, 615 (1971).
[222] A. Hassner, V. R. Fletcher, and D. P. G. Hamon, *J. Am. Chem. Soc.*, **93**, 264 (1971).
[223] W. T. Brady and R. Roe, *J. Am. Chem. Soc.*, **91**, 1662 (1971); W. T. Brady, F. H. Parry, and J. D. Stockton, *J. Org. Chem.*, **36**, 1486 (1971).
[224] H. Hart and G. M. Love, *J. Am. Chem. Soc.*, **93**, 6266 (1971).
[225] W. T. Brady and L. Smith, *J. Org. Chem.*, **36**, 1637 (1971).
[226] D. A. Nelson, *Tetrahedron Letters*, **1971**, 2543.

Concerted $_\pi 2_s + {}_\pi 2_a$ addition through a partially charged transition state is suggested to account for the exclusive 2 + 2, Markovnikov addition of chlorosulphonyl isocyanate to dienes under conditions of kinetic control.[227] On the other hand, the reversible but highly stereoselective addition of chlorosulphonyl isocyanate to *cis*- and *trans*-stilbene has been interpreted in terms of a dipolar intermediate.[228] Stereochemistry, activation parameters and the effect of solvent polarity on the rate indicate that the 2 + 2-addition of toluene-*p*-sulphonyl isocyanate to enol ethers is concerted through a polar transition state.[229] The stereochemistry of the low-temperature, reversible 2 + 2-addition of *N*-sulphinyltosylamide to vinyl ethers has been studied.[230] A transition-state model based on concerted $_\pi 2_s + {}_\pi 2_a$ addition accounts for the asymmetric induction observed in the 2 + 2-additions of sulphenes to optically active enamines.[231]

2:1(2 + 2 + 2) Adducts are formed together with 1:1(2 + 2) adducts via a zwitterionic intermediate in the reactions of ketene imines with ynamines.[232]

A stepwise reaction via an allyl-stabilized diradical has been suggested for the 2 + 2-addition of polyfluoro-olefins and chlorofluoro-olefins to allenes. This is based mainly on the regiospecificity of addition for chlorotrifluoroethylene and the relatively small variation with temperature in the distribution of adducts formed by addition to the two different double bonds in 3-methylbuta-1,2-diene. A larger variation was expected for two competing concerted $_\pi 2_s + {}_\pi 2_a$ additions with relatively different transition states.[233] A two-step mechanism has been suggested for the dimerization of polyhalogenoallenes.[234]

Although the $_\pi 2_s + {}_\pi 2_a$ dimer is the major isomer formed in the dimerization of cyclo-octa-*cis*-1-*trans*-3-diene, *trans,trans*- and *cis,cis*-fused products are also formed. A mechanism involving a diradical intermediate is favoured to account for all products since the ratio of all dimers is consistent over a range of temperatures and in the presence of dienes and dienophiles which divert a major portion of the reaction.[235]

2 + 2-Cycloadditions of tetramethoxyethylene to a variety of electrophilic double bonds[236] and of phenyl vinyl ethers to TCNE have been reported.[237] Addition of phenols to the charge-transfer complexes involved in the latter reaction has been observed.[238]

In contrast to other di- and tri-cyclopropylethylenes which undergo ionic 2 + 2-additions with TCNE, dicyclopropyldiphenylethylene and tetracyclopropylethylene give cyclopropane-cleaved trisubstituted vinylcyclopentane derivatives, probably through a radical-cation intermediate.[239]

With enamines, *N*-substituted oxindolideneacetophenones (**30**) give 2 + 2 adducts when the substituent is electron-releasing (alkyl) but 2 + 4 adducts when it is withdrawing

---

[227] E. J. Moriconi and W. C. Meyer, *J. Org. Chem.*, **36**, 2841 (1971).
[228] H. J. Friedrich, *Tetrahedron Letters*, **1971**, 2981.
[229] F. Effenberger, P. Fischer, G. Prossel, and G. Kiefer, *Chem. Ber.*, **104**, 1987 (1971); F. Effenberger, G. Prossel, and P. Fischer, *Chem. Ber.*, **104**, 2002 (1971).
[230] W. Wucherpfennig, *Tetrahedron Letters*, **1971**, 1891.
[231] L. A. Paquette, J. P. Freeman, and S. Maiorana, *Tetrahedron*, **27**, 2599 (1971).
[232] L. Ghosez and C. de Perez, *Angew. Chem. Int. Ed.*, **10**, 184 (1971).
[233] D. R. Taylor, M. R. Warburton, and D. B. Wright, *J. Chem. Soc.* (C), **1971**, 385; D. R. Taylor and D. B. Wright, *J. Chem. Soc.*, (C), **1971**, 391.
[234] N. Detzer and A. Roedig, *Tetrahedron*, **27**, 5697 (1971).
[235] A. Padwa, W. Koehn, J. Masaracchia, C. L. Osborn, and D. J. Trecker, *J. Am. Chem. Soc.*, **93**, 3633 (1971).
[236] R. W. Hoffmann, U. Bressel, J. Gehlhaus, and H. Hauser, *Chem. Ber.*, **104**, 873 (1971).
[237] J. R. Dombroski, M. L. Hallensleben, and W. Regel, *Tetrahedron Letters*, **1971**, 3881.
[238] M. L. Hallensleben, *Tetrahedron Letters*, **1971**, 3883.
[239] S. Nishida, I. Moritani, and T. Teraji, *Chem. Comm.*, **1971**, 36.

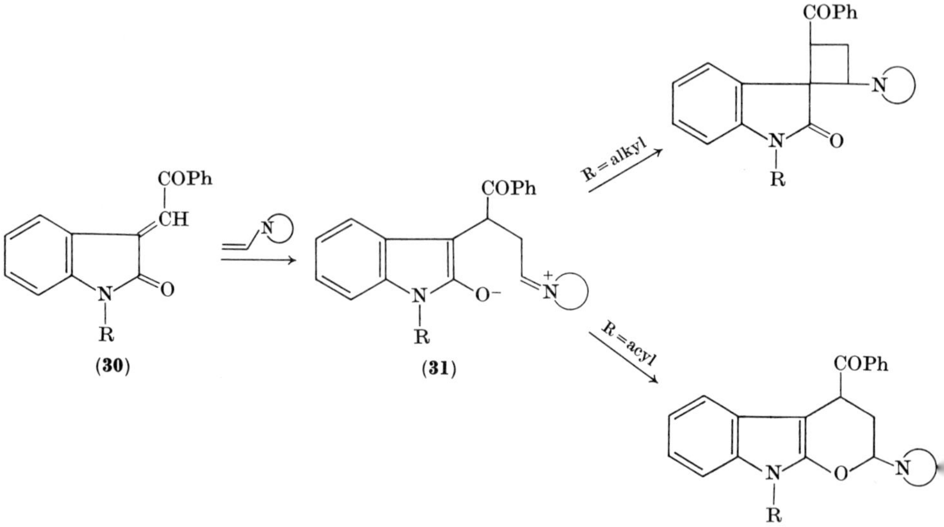

(acyl). This is rationalized in terms of the zwitterionic intermediate (**31**) in which releasing groups reinforce the nucleophilicity of the oxindole 3-position.[240]

The first tetraquinocyclobutane has been synthesized by thermal dimerization of a diquinoethylene.[241]

Other Cycloadditions and Cyclizations

A review of thermal additions of C—C multiple bonds to strained σ-bonds in carbocycles has appeared.[242]

In the reaction of chlorosulphonyl isocyanate with bicyclobutanes and bicyclopentanes a mechanism involving initial S_E2-like attack at the less hindered bridgehead carbon atoms, followed by cyclobutyl-to-cyclopropyl carbonium ion rearrangement or conformational ring inversion and collapse of the zwitterion, is proposed to account for the stereochemistry and effect of methyl substituents.[243]

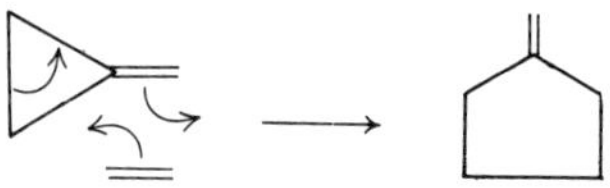

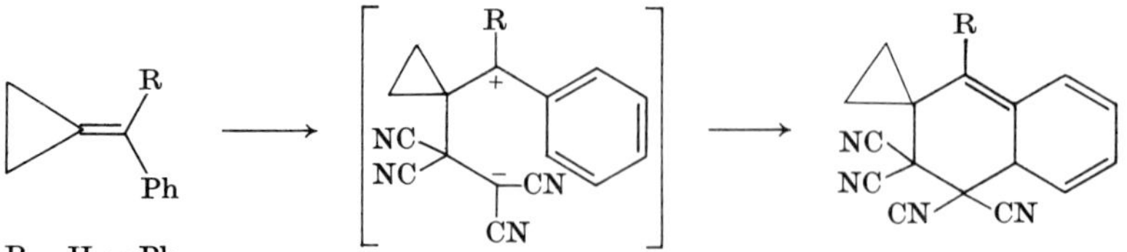

. . . (2)

[240] G. Tacconi, A. Gamba, F. Marinone, and G. Desimoni, *Tetrahedron*, **27**, 561 (1971).
[241] S. Koster and R. West, *Chem. Comm.*, **1971**, 1380.
[242] P. G. Gassman, *Accounts Chem. Res.*, **4**, 128 (1971).
[243] L. A. Paquette, G. R. Allen, and M. J. Broadhurst, *J. Am. Chem. Soc.*, **93**, 4503 (1971).

Optimum conditions for the generation and trapping of allyl cations with dienes, including the first tentative example of trapping with an acyclic diene, have been reported.[244] Examples of, probably concerted, $\sigma^2 + \pi^2 + \pi^2$ cycloadditions (**32**) of methylenecyclopropanes[245] and cyclopropylallenes[246] to dienophiles have appeared. In the case of phenyl-substituted methylenecyclopropanes with TCNE the reaction takes a different course [reaction (2)] to give (**33**) apparently through a zwitterion.[245] The adduct (**33**; R = Ph) was suggested as an intermediate, but was not isolated, in the same reaction carried out at higher temperatures by other workers.[247]

An allowed 10π retro-$_\pi 2_a + _\pi 2_a + _\pi 2_s + _\pi 2_s + _\pi 2_s$ cycloaddition in which Cu^I does not change the occupancy of orbitals but merely lowers the energy of activation is suggested for the conversion of the Cu^I complex of azo compound (**34**) into cyclooctatetraene.[248]

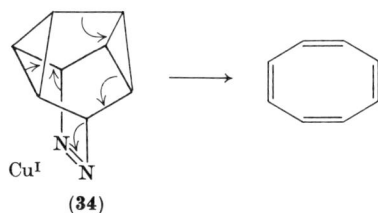

(**34**)

Clear examples of allowed 8 + 2-cycloaddition have been observed between 1,6-dimethylenecycloheptadiene and TCNE or azocarboxylic ester.[249] A new synthesis of 4,5-disubstituted azulenes from 6-[2-(dimethylamino)vinyl]fulvene and dimethyl acetylenedicarboxylate possibly also involves an initial $_\pi 8_s + _\pi 2_s$ cycloaddition.[250] Intramolecular 6 + 2-cycloaddition of 6-vinylfulvene to give dihydropentalene has been reported.[251] Further details of the reactions of *cis*-bicyclo[6.1.0]nonatriene, which normally reacts as its valence isomer tricyclo[4.3.0.0^{7,9}]nona-2,4-diene, with dienophiles, but which gives formal 2 + 2(homo-8 + 2) cycloadducts with TCNE and benzyne,[252] and of the 4 + 4-dimerization of diphenylinden-2-one[253] have appeared. The structure of the double 6 + 4 adduct of tropone and dimethylfulvene has been reported.[254] Attempted Diels–Alder addition of maleic anhydride to nitrosocyclohexene is claimed to give the novel adduct (**35**).[255] The dimer from 8-cyanoheptafulvene has been rationalized in terms of 1,5-hydrogen shifts in an initially formed 8 + 8 cycloadduct.[256]

[244] H. M. R. Hoffmann, G. F. P. Kernaghan, and G. Greenwood, *J. Chem. Soc.*, (B), **1971**, 2257.
[245] R. Noyori, N. Hayashi, and M. Katô, *J. Am. Chem. Soc.*, **93**, 4948 (1971); see also F. W. Fowler, *Angew. Chem. Int. Ed.*, **10**, 135 (1971); J. E. Baldwin and R. K. Pinschmidt, *Tetrahedron Letters*, **1971**, 935.
[246] D. J. Pasto and A. Chen, *J. Am. Chem. Soc.*, **93**, 2562 (1971).
[247] J. E. Baldwin and R. E. Peavy, *J. Org. Chem.*, **36**, 1441 (1971).
[248] R. M. Moriarty, C.-L. Yeh, and N. Ishibi, *J. Am. Chem. Soc.*, **93**, 3085 (1971).
[249] G. C. Farrant and R. Feldmann, *Tetrahedron Letters*, **1970**, 4979.
[250] R. W. Alder and G. Whittaker, *Chem. Comm.*, **1971**, 776.
[251] J. J. Gajewski and C. J. Cavender, *Tetrahedron Letters*, **1971**, 1057.
[252] C. S. Baxter and P. J. Garratt, *Tetrahedron*, **27**, 3284 (1971); J. E. Baldwin and R. K. Pinschmidt, *Chem. Comm.*, **1971**, 820.
[253] J. M. Holland and D. W. Jones, *J. Chem. Soc.* (C), **1971**, 608.
[254] N. S. Bhacca, L. J. Luskus, and K. N. Houk, *Chem. Comm.*, **1971**, 109.
[255] G. Just and W. Zehetner, *Chem. Comm.*, **1971**, 81.
[256] M. Oda, Y. Kayama, and Y. Kitahara, *Chem. Comm.*, **1971**, 505.

The 1,4-addition of dicyanocarbene to cyclo-octatraene involves the triplet carbene and is not an example of $_\pi 4_s + {}_\omega 2_s$-cycloaddition.[257] Phosphinothioylidene, $R—\ddot{P}{=}S$, from the dechlorination of phenylphosphonothioic dichloride, gives the product of such an addition with benzil but acts as a dienophile (P=S) with 2,3-dimethylbutadiene.[258] The cycloaddition of ethoxyacetylene to diphenylketene to give 1-ethoxy-3,3a-dihydro-3-oxo-3a-phenylazulene can be rationalized as an allowed $_\pi 2_s + {}_\pi 2_a + {}_\pi 2_a$ addition followed by ring-opening of the norcaradiene intermediate (**36**).[259]

The addition of Schiff bases to maleimides, giving 1:2 spiro adducts (**37**),[260] the reaction of ketene–SO$_2$ adducts with ketene imines (to give a different type of adduct

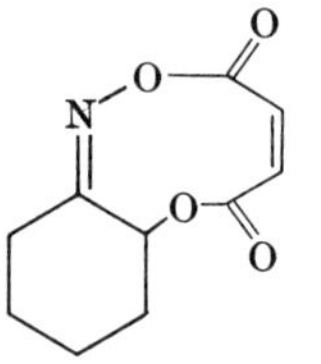

(**35**)

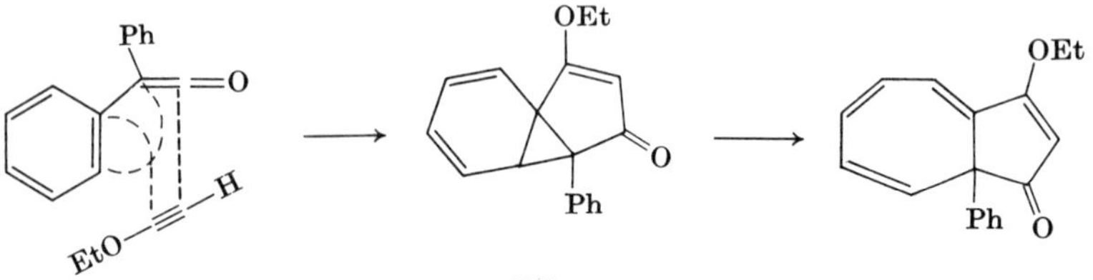

(**36**)

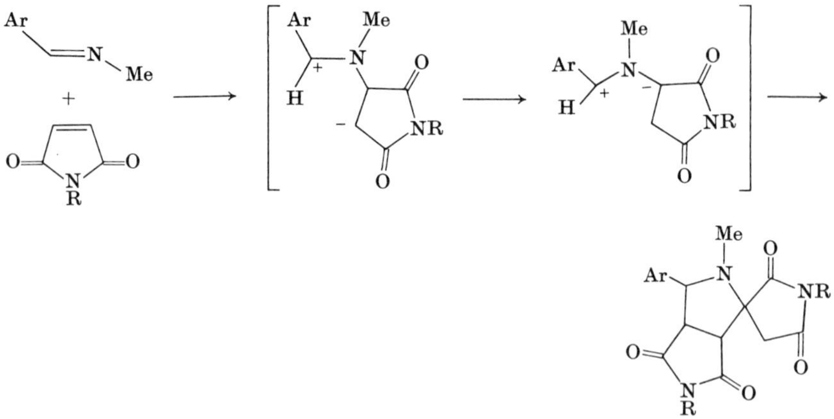

(**37**)

[257] A. G. Anastassiou, R. P. Cellura, and E. Ciganek, *Tetrahedron Letters*, **1970**, 5267.
[258] S. Nakayama, M. Yoshifuji, R. Okazaki, and N. Inamoto, *Chem. Comm.*, **1971**, 1186.
[259] H. Teufel and E. F. Jenny, *Tetrahedron Letters*, **1971**, 1769.
[260] L. Zirngibl, T. Wagner-Jauregg, E. Pretsch, D. J. Stage, N. J. Hales, and C. W. Paris, *Tetrahedron*, **27**, 2203 (1971).

than with imines)[261] and various aspects of the addition reactions of heterocyclic compounds with acetylenedicarboxylic ester[262] have been reported.

Primary and secondary deuterium isotope effects indicate that the "ene" reaction between allene and perfluorocyclobutanone is concerted.[263] On the other hand, a common diradical intermediate is suggested for the formation of both "ene" and 2 + 2 adducts of tetramethylallene and electron-deficient acetylenes.[264] Of relevance to reactions such as the "ene" reaction is the proposal of a general isotopic method for differentiating $S2$, $S'2$ and $S1$ (SR) substitution reactions and tautomerism in systems of the type A=B—A—X; this is based on the fact that each process has a particular and predictable pattern of reactant and product isotope distribution if A is labelled. Application to the additive substitution reactions of azodicarboxylic ester with 1,3-diphenylpropene and N-benzylidenebenzylamine indicates that both reactions proceed by at least two competing mechanisms.[265]

Metal-catalysed Cycloadditions

In a reply to van der Lugt's criticism,[266] Mango has reaffirmed that interaction with metal orbitals can change a disallowed into an allowed process.[267]

The mechanism of this "forbidden-to-allowed" catalytic process in 2 + 2-cycloadditions has been discussed in some detail.[268,269] The spatial redistribution of metal valence electrons which occurs in the process is affected by the ligand fields due to other non-reacting ligands, since such fields may remove the degeneracy of the d orbitals to which electron density is transferred. This can restrict certain modes of transformation; for example, the co-ordinated cyclobutane (**38**) in which the shaded d_{zx} orbital is preferentially occupied can only cleave without restriction in the direction (a). The suitability of metal complexes for catalytic processes was discussed.[268]

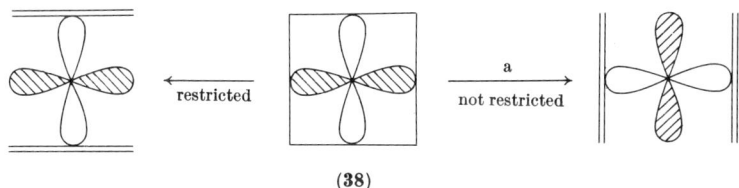

(**38**)

Catalysts for the valence isomerization of prismane are all moderate to weak, easily reduced Lewis acids. Neither of the current mechanistic explanations, "forbidden-to-allowed" catalysis or stepwise reaction through a σ-bonded intermediate, is considered adequate and it is suggested that the process involves formation of a very short-lived

---

[261] J. M. Bohen and M. M. Joullié, *Tetrahedron Letters*, **1971**, 1815.
[262] R. M. Acheson and J. N. Bridon, *Chem. Comm.*, **1971**, 1225; F. Fried, J. B. Taylor, and R. Westwood, *Chem. Comm.*, **1971**, 1226; R. M. Acheson and N. D. Wright, *Chem. Comm.*, **1971**, 1421; R. M. Acheson and J. K. Stubbs, *J. Chem. Soc.* (C), **1971**, 3285; R. M. Acheson and D. F. Nisbet, *J. Chem. Soc.* (C), **1971**, 3291; R. M. Acheson and J. McK. Woolard, *J. Chem. Soc.* (C), **1971**, 3296.
[263] W. R. Dolbier and S.-H. Dai, *Chem. Comm.*, **1971**, 166.
[264] H. A. Chia, B. E. Kirk, and D. R. Taylor, *Chem. Comm.*, **1971**, 1144.
[265] M. M. Shemyakin, L. A. Neiman, S. V. Zhukova, Y. S. Nekrasov, T. T. Pehk, and E. T. Lippmaa, *Tetrahedron*, **27**, 2811 (1971).
[266] W. T. A. M. van der Lugt, *Tetrahedron Letters*, **1970**, 2281; *Org. Reaction Mech.*, **1970**, 208.
[267] F. D. Mango, *Tetrahedron Letters*, **1971**, 505.
[268] F. D. Mango and J. H. Schachtschneider, *J. Am. Chem. Soc.*, **93**, 1123 (1971).
[269] G. L. Caldow and R. A. MacGregor, *J. Chem. Soc.* (A), **1971**, 1654.

charge-transfer complex between catalyst and prismane which gives a prismane radical cation (**39**) that rearranges and collapses as shown.[270]

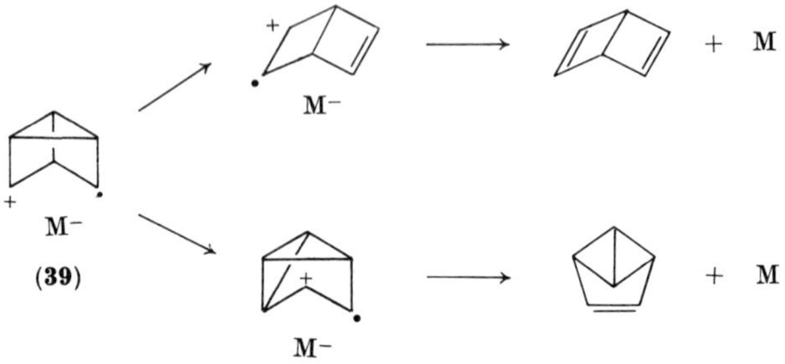

A general scheme has been advanced to account for the Ag^+-catalysed additions of benzyne to cyclo-octatetraene and benzene. The key step in the catalytic sequence is formally the reverse of that in the oxidative addition of Ag^+ to strained σ-bonds.[271] The suggestion[272] that charge-transfer complexes involving Ag^{2+} are involved in the case of cyclooctatetraene was rejected.[271] Further studies of the stepwise Ni-catalysed dimerization of butadiene have been reported.[273] Ni-catalysed addition of bicyclobutanes[274] and bicyclopentane[275] to olefins possibly involves oxidative addition as key step. Cycloadditions of cyclopentadieneiron–allyl and –cyclopropylmethyl complexes to TCNE, for example reaction (3), are best interpreted in terms of dipolar intermediates.[276] Cycloaddition of cyclopentadiene to ferrocenylcarbonium ions has been reported.[277]

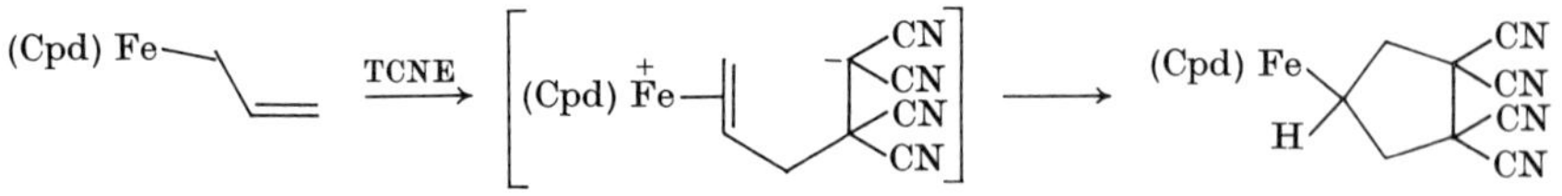

... (3)

[270] K. L. Kaiser, R. F. Childs, and P. M. Maitlis, *J. Am. Chem. Soc.*, **93**, 1270 (1971).
[271] L. A. Paquette, *Chem. Comm.*, **1971**, 1076.
[272] P. Warner, *Tetrahedron Letters*, **1971**, 723.
[273] J. M. Brown, B. T. Golding, and M. J. Smith, *Chem. Comm.*, **1971**, 1240; P. W. Jolly, I. Tkatchenko, and G. Wilke, *Angew. Chem. Int. Ed.*, **10**, 328, 329 (1971).
[274] R. Noyori, T. Suzuki, Y. Kumagai, and H. Takaya, *J. Am. Chem. Soc.*, **93**, 5894 (1971).
[275] R. Noyori, T. Suzuki, and H. Takaya, *J. Am. Chem. Soc.*, **93**, 5896 (1971).
[276] W. P. Giering and M. Rosenblum, *J. Am. Chem. Soc.*, **93**, 5299 (1971).
[277] T. D. Turbitt and W. E. Watts, *Chem. Comm.*, **1971**, 631.

CHAPTER 6

Nucleophilic Aromatic Substitution

A. R. BUTLER

Department of Chemistry, St. Salvator's College, University of St. Andrews

The S_NAr Mechanism	179
Heterocyclic Systems	181
Meisenheimer and Related Complexes	184
Substitution in Polyhaloaromatic Compounds	188
Benzyne and Related Intermediates	188
Other Reactions	192

ortho-Effects in nucleophilic aromatic substitution have been reviewed[1] and Miller's semiempirical method of calculating rates of nucleophilic substitution has been successfully applied to halogen-exchange rates for chloro-, bromo-, and iodo-2,4-dinitrobenzene in an aprotic solvent (N,N-dimethylformamide).[2]

The S_NAr Mechanism

The rates of alkaline hydrolysis of a number of substituted nitrobenzenes and 2,4-dinitrobenzenes in aqueous DMSO correlate well with the J_- function for that medium. The rate-determining step depends upon the proportion of DMSO in the solvent.[3] In cases where addition of the hydroxide ion is slow, the ratio $k(H_2O)/k(D_2O)$ is ca. unity for those reactions with a transition state resembling the reactants but smaller for more advanced transition states.[4] The Swain–Scott measure of nucleophilicity has been discussed with reference to the reaction of amines with 1-chloro-2,4-dinitrobenzene and picryl chloride.[5] Aniline does not catalyse the reaction between aniline and 2,4,6-trinitroanisole[6] but does catalyse the reaction with various 2,4-disubstituted halobenzenes.[7] Steric factors do not account for the low reactivity of diethylamine with fluoronitrobenzenes in benzene.[8]

Solvent effects have been measured for a number of nucleophilic substitution reactions[9,10] but no clear pattern emerges, as the effects appear to be rather specific. For

[1] M. Charton, *Progr. Phys. Org. Chem.*, **8**, 235 (1971).
[2] F. H. Kendall, J. Miller, and R. Wong, *J. Chem. Soc.* (B), **1971**, 1521.
[3] K. Bowden and R. S. Cook, *J. Chem. Soc.* (B), **1971**, 1765, 1771.
[4] K. Bowden, R. S. Cook, and M. J. Price, *J. Chem. Soc.* (B), **1971**, 1778.
[5] R. Minetti and A. Bruylants, *Bull. Classe Sci., Acad. Roy. Belg.*, **56**, 904 (1971); *Chem. Abs.*, **75**, 36962 (1971).
[6] J. Lelieuve, R. Gaboriaud, and R. Schaal, *Compt. Rend* (C), **272**, 1780 (1971).
[7] S. M. Shein and L. A. Suchkova, *Organic Reactivity (Tartu)*, **7**, 732 (1970).
[8] N. S. Nudelman and J. A. Brieux, *An. Asoc. Quím. Argent.*, **58**, 217 (1970); *Chem. Abs.*, **74**, 75910 (1971).
[9] S. M. Shein, N. K. Danilova, and N. I. Kuznetsova, *Organic Reactivity (Tartu)*, **7**, 458 (1970); N. K. Danilova and S. M. Shein, *Organic Reactivity (Tartu)*, **7**, 476 (1970); N. K. Danilova and S. M. Shein, *Organic Reactivity (Tartu)*, **7**, 718 (1970); P. S. Radhakrishnamurti and J. Sahu, *Proc. Indian Acad. Sci.* (A), **73**, 192 (1971); *Chem. Abs.*, **75**, 87927 (1971); J. Kaválek, J. Socha, J. Urbánek, and M. Večeřa, *Coll. Czech. Chem. Comm.*, **36**, 209 (1971); K. K. Satpathy and P. L. Nayak, *J. Indian Chem. Soc.*, **48**, 531 (1971).
[10] S. M. Shein and L. Suchkova, *Organic Reactivity (Tartu)*, **7**, 748 (1971).

example, addition of methanol lowers the rate of reaction between thiocyanate ion and 1-iodo-2,4-dinitrobenzene owing to strong solvation of the thiocyanate ion by methanol.[11] Charge-transfer complex formation has been thought to affect the rate of reaction between aniline and 1-chloro-2,4-dinitrobenzene but the effect is better explained by solvation of the aniline.[10]

Imidazole reacts readily with picryl chloride to give 2,4,6-trinitrophenylimidazole,[12] and generally the carbanions derived from phenylalkylacetonitriles displace chloride ion from a number of chloronitrobenzenes.[13] In the reaction of various nitrophenyl trifluoromethyl sulphones with sodium methoxide the nitro group may be replaced.[14] There has been a kinetic study of the reactions of esters of some nitrobenzenesulphonic acids with n-butylamine. In the reaction with 2,4-dinitrobenzenesulphonyl chloride SO_2 is eliminated.[15] The reactions of various diamines with 1-chloro-2,4-dinitrobenzene in ethanol have been examined.[16]

The reaction of 4-chloronitrobenzene with sodium thiophenoxide in methanol is 2–3 times faster than that of 4-chlorophenyl trifluoromethyl sulphone, although the σ-constant for CF_3SO_2 is greater than that for NO_2.[17] Substituents at the 4-position affect the rate of the latter reaction in the order $SO_2CF_3 > NO_2 > F > Cl$.[18] The transmission of polar effects between two rings has been examined for the reaction of 3'- and 4'-substituted 4-bromo-3-nitrobiphenyls with piperidine in methanol. The Hammett ρ-constant is +1.00, but strongly electron-withdrawing groups do not lie on this plot.[19]

Fluorine at the *para*-position in nitrobenzene is replaced by methoxide ion in methanol faster than at the *ortho*-position. However, with 2,4-difluoronitrobenzene it is fluorine at the 2-position that is replaced first. The analysis of this situation by Bamkole and Hirst[20] shows that it is not really anomalous. The reactivity of each fluorine in 2,4-difluoronitrobenzene is influenced by the second fluorine at the *meta*-position but in one case it is a *para*-activated system and, in the other, *ortho*-activated and the response is different. Linked with this is the observation that the effect of disubstitution at the *meta*-positions in fluorobenzene on the free energies of activation for the reaction with methoxide ion is not additive.[21] An attempt has been made to detect an α-effect in the reaction of 1-chloro-2,4-dinitrobenzene with various nucleophiles. Four primary amines give a linear Brønsted plot, with a high β value, but the exalted positions of hydrazine and methoxylamine are no greater than those of other nucleophiles that have no lone pair α to the position of nucleophilic attack (e.g. aniline and morpholine).[22] The relative reactivities of hydroxide and alkoxide ions towards 1-fluoro-2,6-dinitrobenzene have

---

[11] Y. Kondo, K. Uosaki, and N. Tokura, *Bull. Chem. Soc. Japan*, **44**, 2548 (1971).

[12] R. Minetti and A. Bruylants, *Bull. Classe Sci., Acad. Roy. Belg.*, **56**, 1047 (1971); *Chem. Abs.*, **75** 87917 (1971).

[13] M. Makoszu, J. M. Jagusztyn-Grochowska, and M. Jaurdosiuk, *Rocz. Chem.*, **45**, 858 (1971).

[14] V. N. Boiko and L. M. Yagupol'skii, *Zhur. Org. Khim.* **6**, 1874 (1970); *J. Org. Chem. USSR*, **6**, 1885 (1970).

[15] R. V. Vizgert, E. N. Ozdrovsky, I. V. Kozak, and I. M. Ozdrovskaya, *Organic Reactivity (Tartu)*, **7**, 1093 (1970).

[16] T. I. Petrenko and I. V. Smimov-Zamkov, *Organic Reactivity (Tartu)*, **8**, 107 (1971).

[17] K. V. Solodova and S. M. Shein, *Zhur. Org. Khim.*, **6**, 1461 (1970); *J. Org. Chem. USSR*, **6**, 1475 (1970).

[18] S. M. Shein and K. V. Solodova, *Zhur. Org. Chim.*, **6**, 1465 (1970); *J. Org. Chem. USSR*, **6**, 1479 (1970).

[19] C. Dell'Erba, G. Guanti, and G. Garbarino, *Tetrahedron*, **27**, 1807 (1971).

[20] T. L. Bamkole and J. Hirst, *Chem. Comm.*, **1971**, 69.

[21] J. Hirst and S. J. Una, *J. Chem. Soc.* (B), **1971**, 2221.

[22] G. Biggi and F. Pietra, *J. Chem. Soc.* (B), **1971**, 44.

been evaluated and compared with the reactions of 1-fluoro-2,4-dinitrobenzene.[23] The reaction of 2-halotropones results in replacement of halogen but with 2-fluorotropone the reaction is faster than expected. On the other hand, quinuclidine does not displace fluorine at all, although there is reaction with the other halotropones. Removal of fluoride ion is assisted by protonation, which is possible with piperidine but not with quinuclidine.[24]

Two explanations have been offered for the very fast acetylation of benzamidine by p-nitrophenyl acetate in chlorobenzene: the bifunctional reactivity of benzamidine[25] or extensive delocalization of the positive charge of the transition state.[26] If the latter is the correct explanation, it should apply equally to the reaction of benzamidine and 1-chloro-2,4-dinitrobenzene. In fact, this reaction is much slower than that with p-nitrophenyl acetate and this lends more support to the former explanation.[27]

The reactivities of hydrogen peroxide and the hydroperoxide ion towards 3'-substituted 3,4-benzotropolone-1',2'-quinone (1) and the dianion of purpurogalloquinone indicate that charge repulsion decides which species is the more reactive.[28] Chlorine may be

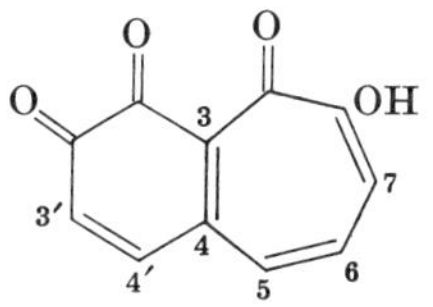

(1)

substituted for various other groups on polynitrobenzenes by reaction with $POCl_3$ and DMF. The reactive species appears to be $Me_2\overset{+}{N}=CHCl$.[29]

The reaction of 2-halo-3,5-dinitrobenzoic acids with piperidine in water or benzene containing cuprous chloride involves formation of a complex cuprous salt.[30] In the methylation of halotoluenes and haloanisoles by $LiCuMe_2$ the order of reactivities is $m > o > p$. For reaction with phenoxide ion the order is changed to $m > p > o$ owing to steric factors.[31]

Heterocyclic Systems

Nucleophilic substitution on nitrogen heterocycles has been reviewed (in Rumanian),[32] and several workers have compared the activating effects of nitro and aza groups. The reactions of hydroxide ion with various alkoxynitropyridines indicate the absence of

---

[23] J. Murto, I. Wavtiovaara, and J. Janhoven, *Suomen Kem.* (B), **44**, 312 (1971).
[24] F. Pietra and F. Del Cima, *J. Chem. Soc.* (B), **1971**, 2224.
[25] F. M. Menger, *J. Am. Chem. Soc.*, **88**, 3081 (1966).
[26] H. Anderson, C. Su, and J. W. Watson, *J. Am. Chem. Soc.*, **91**, 482 (1969).
[27] G. Biggi, F. Del Cima, and F. Pietra, *Tetrahedron Letters*, **1971**, 2811.
[28] P. D. Collier, *J. Chem. Soc.* (B), **1971**, 637.
[29] V. L. Zbarskii, G. M. Shutov, V. F. Zhilin, R. G. Chirkova, and E. Y. Orlova, *Zhur. Org. Khim*, **7**, 310 (1971); *Chem. Abs.*, **74**, 111267 (1971).
[30] V. N. Lisitsyn and A. D. Tishchenko, *Izv. Vyssh. Uchebn. Zaved. Khim. Khim. Tekhnol.*, **13**, 992 (1970); *Chem. Abs.*, **74**, 63690 (1971).
[31] V. N. Drozd and O. I. Tribonova, *Zhur. Org. Khim.*, **6**, 2493 (1970); *Chem. Abs.*, **74**, 64001 (1971).
[32] K. Bacaloglu, *Stud. Cerc. Chim.*, **19**, 819 (1971).

steric factors with the aza group.[33] Aza activation is much smaller in five-membered rings than in six, and in thiazoles all positions (2-, 4-, and 5-) are equally activated, contrary to previously held views. There is no addition–elimination mechanism in the reactions of these compounds with methoxide ion.[34] In the reactions of 2-chlorothiazole with benzenethiolate ion, the 4- and 5-positions act as *meta*- and *para*-positions in benzene with respect to a Hammett plot, and the high ρ value (+5.3) indicates how very sensitive thiazoles are to substituent effects.[35] The reaction of halopyridines, quinolines and quinoxalines with piperidine is very similar to the reaction with halonitrobenzenes but the element effect is smaller.[36]

The triazolium salt (**2**) reacts readily with methoxide ion with replacement of bromine[37]

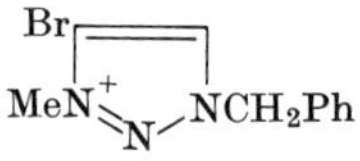

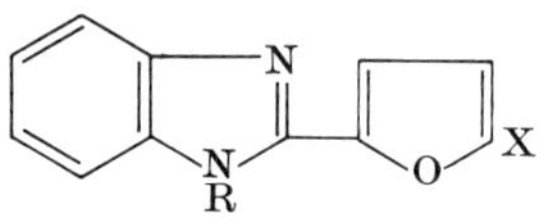

and the kinetics of halogen (X) replacement by piperidine and dimethylamine on (**3**) and (**4**) have been reported.[38] The enhanced reactivity of 6-chlorophenanthridine (**5**), substituted at the 1- and 10-positions by methyl groups, to nucleophiles has been

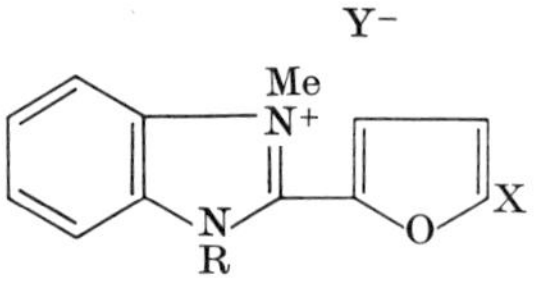

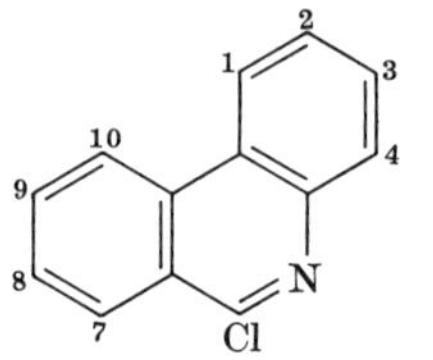

explained in terms of the relief of steric strain.[39] Replacement of the trimethylammonium group on purines (**6**) and pyrimidines by hydroxide ion has been reported.[40] A kinetic study of the reactions of substituted 2- and 4-chloropyrimidines with piperidine in isooctane shows that substituents at the 6-position have a stronger effect on the 4- than the 2-position.[41] Changes of solvent have a bigger effect on the reactions of 4-chloropyrimidines.[42a] 5-Chloro-4-nitro- and 4-chloro-5-nitro-benzofurazan (**7**) show an unexpectedly

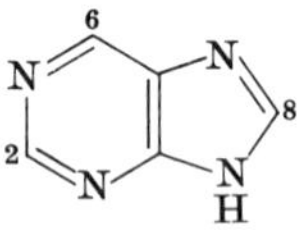

---

[33] J. Murto, L. Nummela, M.-L. Hyvönen, and I. Wartiovaara, *Suomen Kem.* (B), **43**, 517 (1970).
[34] M. Bosco, L. Forlani, P. E. Todesco, and L. Troisi, *Chem. Comm.*, **1971**, 1093.
[35] M. Bosco, L. Forlani, V. Liturri, P. Riccio, and P. E. Todesco, *J. Chem. Soc.* (B), **1971**, 1373.
[36] G. B. Bressan, I. Giardi, G. Illuminati, P. Linda, and G. Sleiter, *J. Chem. Soc.* (B), **1971**, 225.
[37] M. Begtrup, *Acta Chem. Scand.*, **25**, 795, 803 (1971).
[38] V. N. Novikov and V. T. Bukhaeva, F. T. Pozharskii, and A. M. Simonov, *Khim. Geterotsikl Soedin*, **7**, 252 (1971); *Chem. Abs.*, **75**, 48073 (1971).
[39] B. R. T. Keene and G. L. Turner, *Tetrahedron*, **27**, 3405 (1971).
[40] G. B. Barlin and A. C. Young, *J. Chem. Soc.* (B), **1971**, 821, 1675.
[41] V. P. Mamayev, O. A. Zagulayeva, and V. P. Krivopalov, *Dokl. Akad. Nauk SSSR*, **193**, 600 (1970); *Chem. Abs.*, **74**, 3250 (1971).
[42a] V. A. Zagulayeva, S. M. Shein, A. I. Shvets, V. P. Mamayev, and V. P. Krivopalov, *Organic Reactivity (Tartu)*, **7**, 1133 (1970).

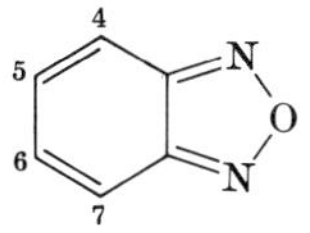

high reactivity towards methoxide ion, and fluorine is replaced more readily than chlorine or bromine.[42b] With 4-methoxy-5-nitrobenzofurazan an orange addition compound is formed. The sulphur analogue of benzofurazan is even more reactive.[43] Reaction between thiophenol and 2-chloro-1,3-dimethylbenzimidazolium ion (8) results in replacement of chlorine by way of an addition intermediate. Both breaking of the S—H bond and formation of a C—S bond are involved in the rate-determining step and so the reaction rate depends on both the acidity and the nucleophilicity of thiophenol.[44]

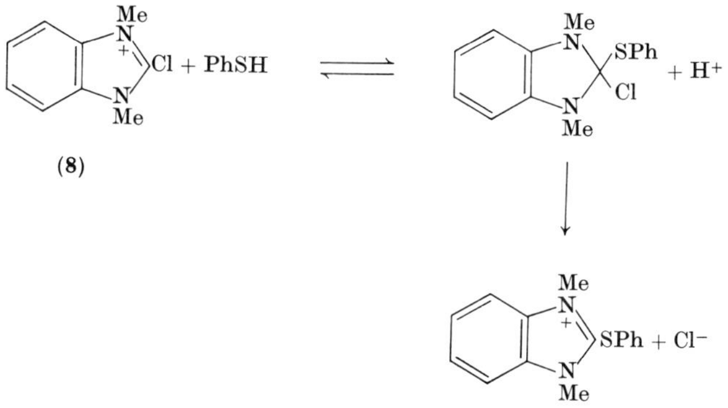

The replacement of one chlorine in 2,4-dichloro-6-(phenylamino)-s-triazine by water, methanol or phenol in the presence of a tertiary amine involves formation of (9) as the slow step.[45] In the alkaline hydrolysis of ammelide the rate-determining step is elimination of amide ion from the dianion (10); the slow step in acid hydrolysis is also loss of an amide ion.[46] There has been a very full kinetic study of the alkaline hydrolysis of chloro-s-triazines; the initial reaction is attack by hydroxide ion and protonation to give the intermediate (11), formed in a non-steady equilibrium state, which undergoes deprotonation and release of chloride ion.[47]

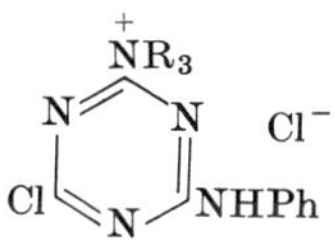

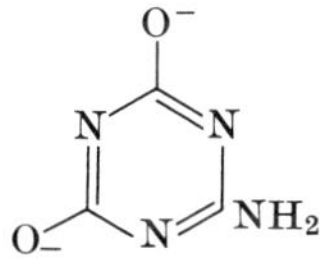

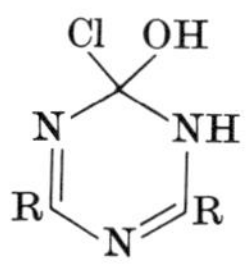

[42b] D. Dal Monte, E. Sandri, L. Di Nunno, S. Florio, and P. E. Todesco, *J. Chem. Soc.* (B), **1971**, 2209.
[43] J. J. K. Boulton and P. Kirby, *Chem. Comm.*, **1970**, 1618.
[44] P. Dembech, A. Ricci, G. Seconi, and P. Vivarelli, *J. Chem. Soc.* (B), **1971**, 557.
[45] G. Ostrogovich, E. Fliegl, and R. Bacaloglu, *Tetrahedron*, **27**, 2885 (1971).
[46] G. Ostrogovich, E. Fliegl, and R. Bacaloglu, *Tetrahedron*, **27**, 3869 (1971).
[47] P. Rys, A. Schmitz, and H. Zollinger, *Helv. Chim. Acta*, **35**, 163 (1971).

Replacement of the fluorines of phosphonitrilic fluorides by butyl-lithium is geminal for the first two, but the third and fourth enter antipodally. This is attributed to the π-inductive effect of the substituent on a delocalized homomorphic π system.[48]

Meisenheimer and Related Complexes

Application of techniques for studying rapid reactions, and NMR studies, have contributed significantly to an understanding of Meisenheimer complex formation. The complex (12) results from reaction of hydroxide ion and 1,3,5,8-tetranitronaphthalene, and the equilibrium constant for complex formation is reduced by a factor of 0.3 with deuteroxide in D_2O.[49] Reaction of methoxide ion and 2,4,6-tricyanoanisole gives the 1,1-complex (13) but the equilibrium constant is much smaller than that for formation of

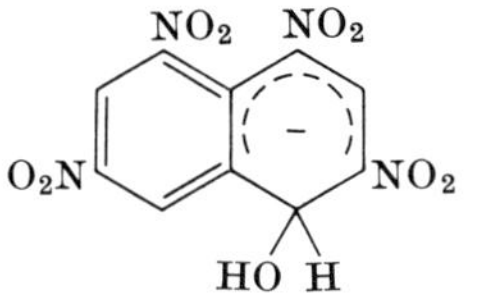

(12) 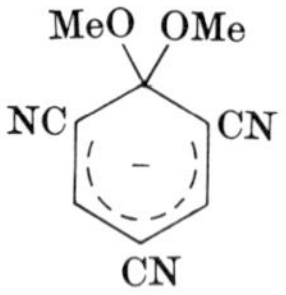   (13)

the equivalent complex from trinitroanisole. A complex could not be observed in the reaction of 1-bromo-2,4,6-tricyanobenzene and hydroxide ion, although bromine is displaced, and this means that nucleophilic attack must be the slow step.[50] The rates of formation of 1,3-complexes, and their conversion into the more stable 1,1-complexes, have been measured in the reaction of methoxide ion with various nitro- and cyano-anisoles. Addition of DMSO has a profound effect on all the rate constants involved.[51] The same workers[53] reported a comprehensive study of the kinetics of Meisenheimer complexes formed from nitronaphthalenes[52] in DMSO solution and also the formation of (14) from 3,5-dinitrobenzonitrile. Interaction of methoxide ion and N-*tert*-butyl-2,4,6-trinitrobenzamide gives the complex (15) and this is followed by loss of nitrite ion.[54]

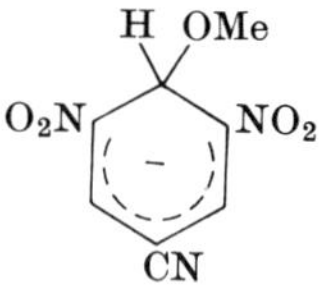

(14)

---

[48] N. L. Paddock, T. N. Ranganathan, and S. M. Todd, *Can. J. Chem.*, **49**, 164 (1971).
[49] J. H. Fendler, E. J. Fendler, and L. M. Casilio, *J. Org. Chem.*, **36**, 1749 (1971).
[50] E. J. Fendler, W. Ernsberger, and J. H. Fendler, *J. Org. Chem.*, **36**, 2333 (1971).
[51] C. Dearing, F. Terrier, and R. Schaal, *Compt. Rend.* (C), **271**, 349 (1970); F. Terrier, J.-C. Hallé, M.-P. Simonnin, and M.-J. Lecourt, *Organic Mag. Resonance*, **3**, 361 (1971); F. Terrier and M.-P. Simonnin, *Bull. Soc. Chim. France*, **1971**, 677.
[52] F. Millot and F. Terrier, *Bull. Soc. Chim. France*, **1971**, 3897.
[53] F. Terrier, F. Millot, and M.-P. Simonnin, *Tetrahedron Letters*, **1971**, 2933.
[54] E. J. Fendler, D. M. Camioni, and J. H. Fendler, *J. Org. Chem.*, **36**, 1544 (1971).

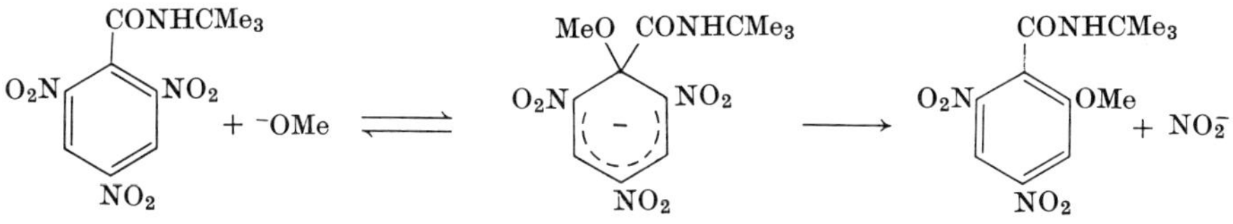

(15)

The initial principal reaction between trinitrotoluene and methoxide ion is removal of a proton from the methyl group to give an anion, and this process has been studied by stopped-flow and temperature-jump techniques. Subsequently there is reaction between the anion and trinitrotoluene to form a complex, probably with structure (16).[55] Anion

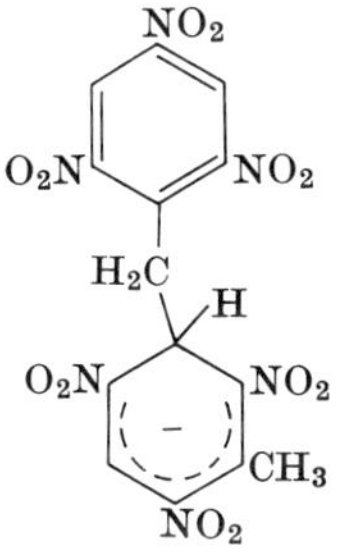

(16) 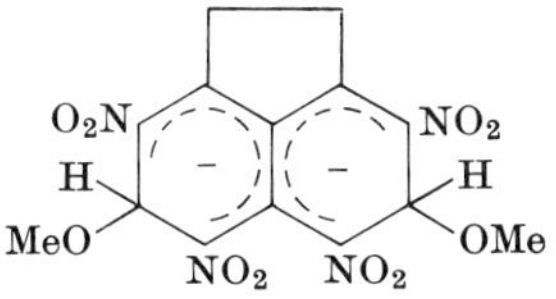   (17)

formation does not occur with 3,5,6,8-tetranitroacenaphthene, and the sodium salt of the complex (17) may be isolated.[56]

The kinetics of the formation of the complex (18) from 2,4,6-dinitroanilino-N-methylpropionamide and methoxide ion have studied. Acidification of an alcoholic solution of this complex gives the conjugate acid.[57] There is hydrogen bonding between

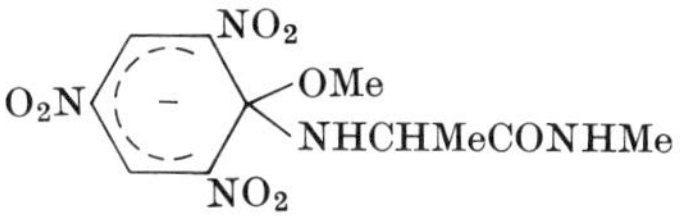

(18)

the amido group and the o-NO_2, which is maximized in the conformation adopted by the Meisenheimer complex, but if the amido group is alkylated, and hydrogen bonding is not possible, the anilino group ionizes and no complex is formed. If the anilino group is alkylated, no 1,1-complex can be formed for steric reasons, and instead a 1,3-complex results.[58] A previous report[59] of a stable intermediate resulting from the reaction of diethyl malonate and 1-fluoro-2,4-dinitrobenzene in the presence of Et_3N was not confirmed.[60] However, a compound with an active methylene group (e.g. sodiomalonic

[55] C. F. Bernasconi, *J. Org. Chem.*, **36**, 1671 (1971).
[56] C. H. J. Wells and J. A. Wilson, *Tetrahedron Letters*, **1971**, 4521.
[57] J. J. K. Boulton and N. R. McFarlane, *J. Chem. Soc.* (B), **1971**, 925.
[58] J. J. K. Boulton, P. J. Jewess, and N. R. McFarlane, *J. Chem. Soc.* (B), **1971**, 928.
[59] P. Baudet, *Helv. Chim. Acta*, **49**, 545 (1966).
[60] K. T. Leffek and P. H. Tremaine, *Can. J. Chem.*, **49**, 1979 (1971).

ester) does form a coloured adduct (19) with 9-nitroanthracene, which on acidification gives the conjugate acid (20).[61]

Full details have been given of the formation of the first Meisenheimer complex (21) of thiophen, resulting from the interaction of 2-methoxy-3,5-dinitrothiophen and

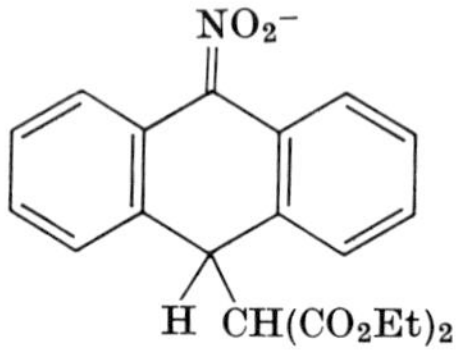

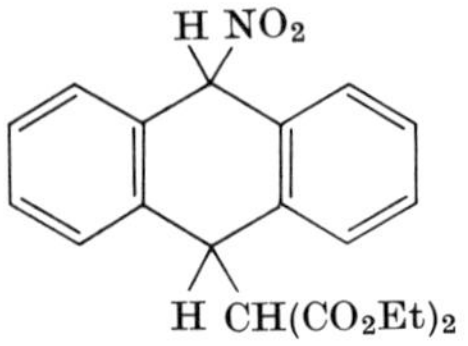

methoxide ion. This complex is formed more readily than that for trinitroanisole, and related intermediates are formed during nucleophilic attack on thiophen compounds, but reactions other than displacement may occur (e.g. ring opening).[62] Complexes of the type (22), formed from thienylcopper and trinitrobenzene, may play a part in copper-promoted reactions leading to biaryls; there is probably initial co-ordination between copper and the NO_2 group.[63] Phenylethynylcopper reacts with trinitrobenzene in pyridine to give a similar complex (23).[64]

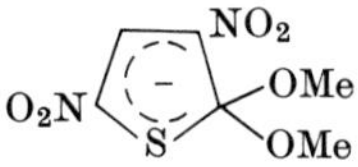

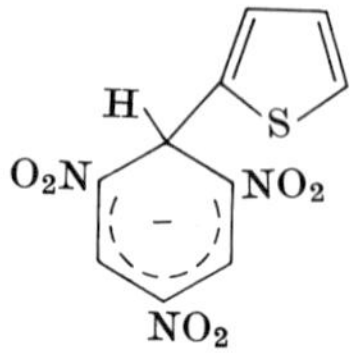

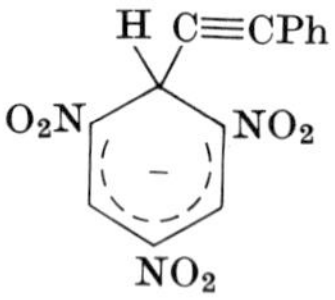

Conversion of (24) into (25) involves ionization at the C* atom and cyclization, as indicated by a kinetic study.[65] Reaction of octahydrotriborate and 1-chloro-2,4,6-

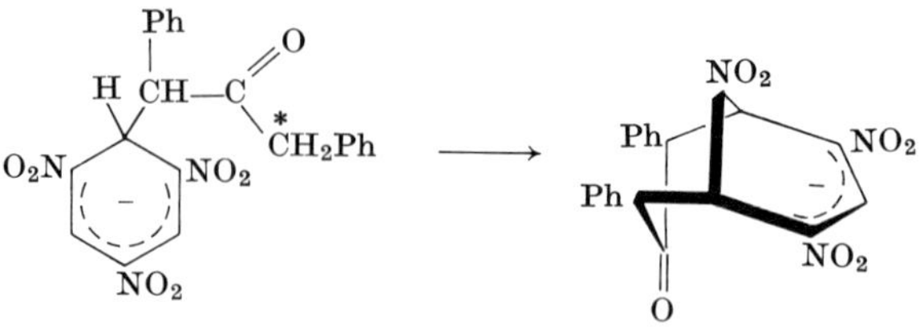

[61] R. H. Williams and H. R. Snyder, *J. Org. Chem.*, **36**, 2327 (1971).
[62] G. Doddi, G. Illuminati, and F. Stegel, *J. Org. Chem.*, **36**, 1918 (1971).
[63] M. Nilsson, C. Ullenius, and O. Wennerström, *Tetrahedron Letters*, **1971**, 2713.
[64] O. Wennerström, *Acta Chem. Scand.*, **26**, 789 (1971).
[65] M. J. Strauss and H. Schran, *Tetrahedron Letters*, **1971**, 2349.

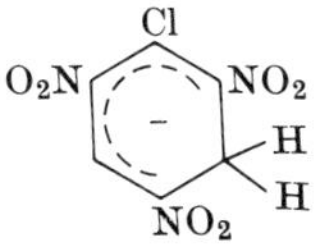

(26)

trinitrobenzene results in transfer of a hydride ion to give the complex (26). Hydride ion will also displace chlorine from this compound, but no evidence could be obtained for formation of the 1,1-complex.[66] The conjugate acid of a Meisenheimer complex (27)

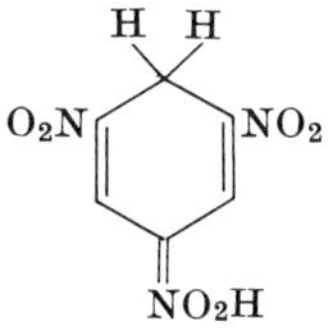

(27) 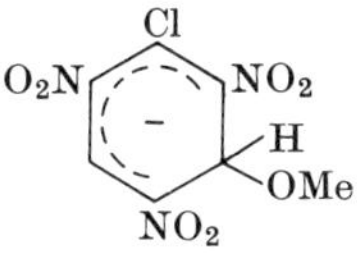   (28) 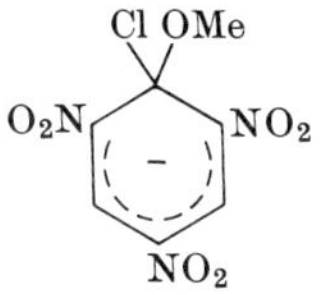   (29)

decomposes in acid solution to a variety of products.[67] The initial reaction between 4-substituted 1-chloro-2,6-dinitrobenzenes and methoxide ion is addition at the 3-position to give (28), and not at the 1-position, which would give (29), although eventually displacement of chlorine occurs.[68]

Nucleophilic aromatic substitution occurs only with great difficulty on rings not activated by electron-withdrawing groups. However, ethoxide ion will displace chlorine from 5-chloroacenaphthylene, and it has been suggested that the addition intermediate formed during this reaction is stabilized by formation of a cyclopentadienide ion (30).[69]

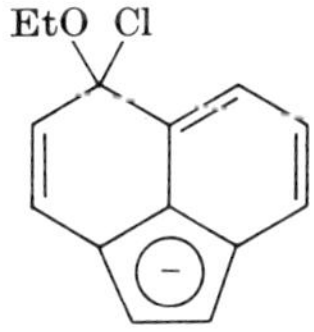

(30)

A kinetic study of complex formation between trinitrobenzene and the lyate anion in aqueous ethanol and aqueous methanol has shown that a change from pure alcohol to aqueous alcohol has very little effect on leaving-group departure.[70] Solvation effects in complex formation have been discussed;[71,72] the reaction between methoxide ion and trinitroanisole is more exothermic in DMSO than in methanol.[73] Isotopic substitution

---

[66] L. A. Kaplan and A. R. Siedle, *J. Org. Chem.*, **36**, 937 (1971).
[67] C. Mobergans and O. Wennerström, *Acta Chem. Scand.*, **25**, 2355 (1971).
[68] M. R. Crampton, M. El Ghariani, and H. A. Khan, *Chem. Comm.*, **1971**, 834.
[69] M. J. Perkins, *Chem. Comm.*, **1971**, 231.
[70] C. F. Bernasconi and R. G. Bergstrom, *J. Org. Chem.*, **36**, 1325 (1971).
[71] G. Illuminati, G. Sleiter, and M. Speranza, *J. Org. Chem.*, **36**, 1723 (1971).
[72] L. G. Gan and A. R. Norris, *Can. J. Chem.*, **49**, 2490 (1971).
[73] J. W. Larsen, K. Amin, and J. H. Fendler, *J. Am. Chem. Soc.*, **93**, 2910 (1971).

of ring hydrogen atoms does not affect complex formation between trinitrobenzene and ethoxide ion but the rate of complex formation is higher in EtOD.[72] Thiolate ions and trinitrobenzene give both 1:1 (**31**) and 2:1 (**32**) adducts; the equilibrium constant for formation of the 2:1 adduct is much bigger in water than alcohol owing, in part, to improved solvation of the doubly charged species.[74]

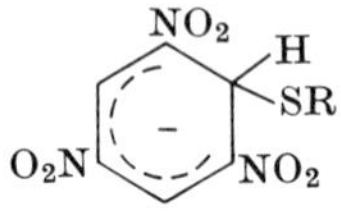

(**31**) 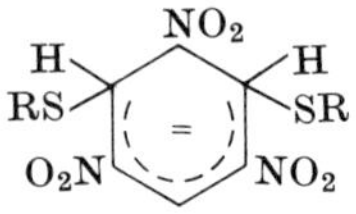   (**32**)

Further work has shown that the effect of surfactants and electrolytes on the rate of formation and decomposition of Meisenheimer complexes follows a complicated pattern and cannot be explained by simple electrolyte theory. Some micelles have a very large effect on spiro complexes.[75]

Substitution in Polyhaloaromatic Compounds

Piperidine will displace fluorine from hexafluorobenzene. The reaction is catalysed by piperidine but the extent depends upon the solvent. Fluorine atoms exert no *ortho*-effect.[76] A variety of products results from nucleophilic attack of polyfluoroalkyl anions on pentafluoropyridine and tetrafluoropyridazine. With $CF_3CF_2^-$ there is kinetic control of the products but, with increasing bulk, there is a gradual change to thermodynamic control and this is complete with $(CF_3)_3C^-$.[77]

Benzyne and Related Intermediates

Ab initio SCF and configuration interaction calculations on *o*-, *m*-, and *p*-benzyne indicate that, in the ground state, *o*-benzyne is singlet but for the others the singlet–triplet energy separations are too small to fix the ground state.[78] The addition of benzyne to cycloheptatriene is said to yield (**33**) and (**34**), and formation of (**33**) by [2 + 6] addition

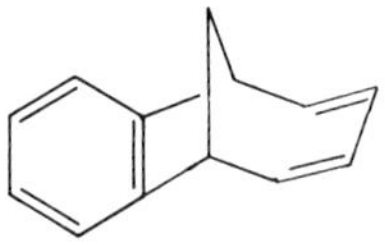

(**33**) 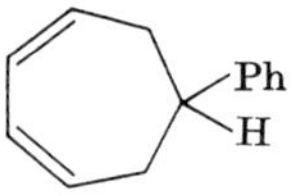   (**34**)

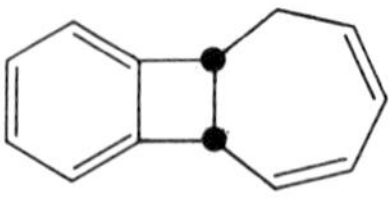

(**35**)

---

[74] M. R. Crampton and M. El Ghariani, *J. Chem. Soc.* (B), **1971**, 1043.
[75] J. H. Fendler, E. J. Fendler, and M. V. Merritt, *J. Org. Chem.*, **36**, 2172 (1971); L. M. Casilio, E. J. Fendler, and J. H. Fendler, *J. Chem. Soc.* (B), **1971**, 1377.
[76] S. M. Shein and P. P. Rodionov, *Organic Reactivity (Tartu)*, **7**, 1150, 1168 (1970); *Kinet. Katal.*, **11**, 1378 (1971); *Chem. Abs.*, **74**, 52724 (1971).
[77] R. O. Chambers, R. P. Corbally, M. Y. Gribble, and W. K. R. Musgrave, *Chem. Comm.*, **1971**, 1345.
[78] D. L. Wilhite and J. L. Whitten, *J. Am. Chem. Soc.*, **93**, 2858 (1971).

requires, for the conservation of orbital symmetry, that benzyne must be present in the antisymmetric form.[79] However, the structure of (**33**) has been questioned and the alternative suggestion is (**35**).[80] Several possible structures for *m*-benzyne have been suggested and "resonance energy per electron" calculations indicate that (**36**)

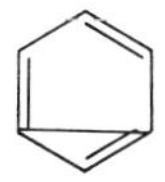

(**36**)

should have considerable stability.[81] There has been a comprehensive review of highly halogenated arynes.[82]

Benzyne may be generated photochemically by irradiation of phthaloyl peroxide and undergoes stereospecific [2 + 4] and non-stereospecific [2 + 2] cycloaddition reactions, showing that it is identical (i.e. a singlet) with thermally generated benzyne. The photochemical generation has advantages for reaction of benzyne with low-boiling compounds.[83] The relative mobilities of halogens in *o*-, *m*-, and *p*-dihalobenzenes in benzyne formation have been examined. The pattern is fairly complex as either loss of a proton or expulsion of halide ion may be the rate-determining step but, in general, the heavier halide is lost preferentially. For example, *m*-chloroiodobenzene gives chlorobenzyne.[84] This paper indicates that Professor Bunnett has ability as a poet as well as a chemist. Thermal decomposition of the diazocyclopentadiene-2-carboxylate anion (**37**) gives the dehydrocyclopentadienyl anion (**38**), which is more reactive than benzyne as it

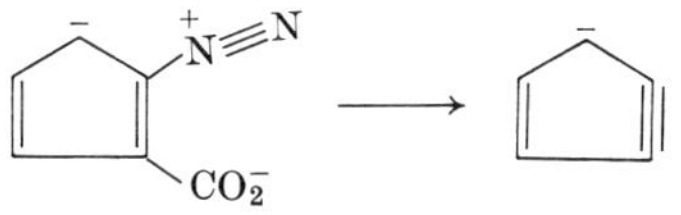

(**37**) → (**38**)

is more strained.[85] Formation of butyl ethers from potassium *tert*-butoxide in DMSO and halonaphthalenes proceeds through an aryne intermediate, except for the fluoronaphthalenes, which undergo direct replacement.[86] However, *p*-fluorotoluene and lithium diethylamide react to give a benzyne and it is the iodo-compound that undergoes direct replacement of the halogen.[87]

An aryne mechanism may occur concurrently with a free-radical pathway. The formation of 1,3-benzyne is only the minor route in the decomposition of 2-carboxy-4-nitrobenzenediazonium chloride, suspended in an inert solvent in an effort to repress

[79] I. Tabushi, H. Yamada, Z. Yoshida, and H. Kuroda, *Tetrahedron Letters*, **1971**, 1093.
[80] L. Lombardo and D. Wege, *Tetrahedron Letters*, **1971**, 3981.
[81] B. A. Hess and L. J. Schood, *Tetrahedron Letters*, **1971**, 17.
[82] H. Heaney, *Fortschr. Chem. Forsch.* **16**, 35 (1971).
[83] M. Jones and M. R. DeCamp, *J. Org. Chem.*, **36**, 1536 (1971).
[84] J. F. Bunnett and F. J. Kearley, *J. Org. Chem.*, **36**, 184 (1971).
[85] J. C. Martin and D. R. Block, *J. Am. Chem. Soc.*, **93**, 451 (1971).
[86] R. H. Hales, J. S. Bradshaw, and D. R. Pratt, *J. Org. Chem.*, **36**, 314, 318 (1971).
[87] G. Wittig, C. N. Rentzea, and M. Rentzea, *Ann. Chem.*, **744**, 8 (1971).

the explosive nature of the reaction.[88] Formation of the same aryne is the only pathway in the reaction of both 5- and 6-chloro- and -bromo-pseudocumene [(**39**) and (**40**)] with

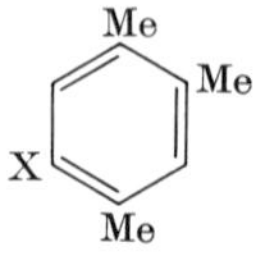

(**39**) 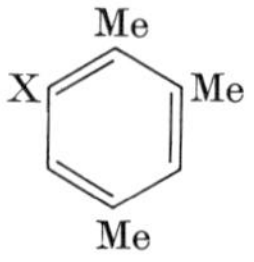 (**40**)

potassium amide, but the iodo-compounds react partially by a radical mechanism involving reaction of an electron donor with iodopseudocumene to give a radical anion. Addition of an improved electron-donor (potassium metal) completely removes the aryne pathway.[89] A full account of aryne participation in the decomposition of substituted N-nitrosoacetanilides has appeared; the results indicate that addition of an arynophile diverts the mechanism from free radical to aryne, and the aryne is formed by removal of a proton *ortho* to the diazonium function to give (**41**) and subsequent loss of

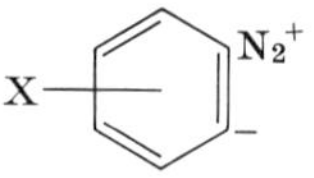

(**41**)

nitrogen. The main puzzle with this reaction is that addition of furan gives 2-phenylfuran rather than the aryne adduct, but this is explained by the fact that furan also reacts so readily with free radicals that the percursory benzenediazonium ion is diverted along that pathway rather than undergoing conversion into an aryne.[90] Although diazotization of aniline in the presence of acetate may yield N-nitrosoacetanilide, presence of water prevents subsequent conversion into an aryne and the reaction of 2,5-di-*tert*-butylaniline, which does give an aryne, appears to be a special case. However, addition of acetic anhydride removes water as it is formed, and this leads to production of arynes under simple, mild conditions.[91] Production of an arynyl free radical (**42**) by the decomposition of 4-nitrophthalic anhydride (**43**) at 650° has been described;[92] photolysis of 1,2,3,4-tetrafluoro-5,6-di-iodobenzene gives tetrafluorobenzyne as well as 2,3,4,5-tetrafluoro-6-iodophenyl radicals.[93]

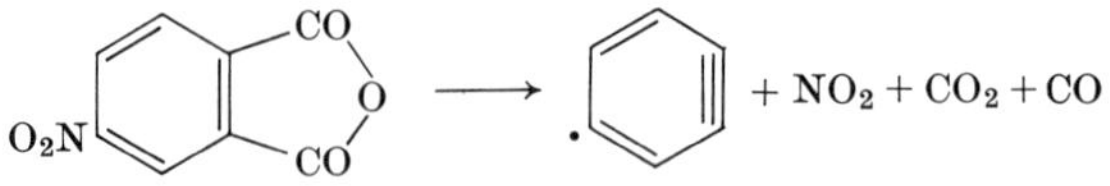

(**42**) (**43**) $+ NO_2 + CO_2 + CO$

[88] R. A. Rossi, R. Hoyos de Rossi, and H. E. Bertorello, *J. Org. Chem.*, **36**, 2905 (1971).
[89] J. K. Kim and J. F. Bunnett, *J. Am. Chem. Soc.*, **92**, 7463, 7465 (1970).
[90] D. L. Brydon, J. I. G. Cadogan, J. Cook, M. J. P. Harger, and J. T. Sharp, *J. Chem. Soc.* (B), **1971**, 1996.
[91] J. I. G. Cadogan, J. R. Mitchell, and J. T. Sharp, *Chem. Comm.*, **1971**, 1.
[92] E. K. Fields and S. Meyerson, *Tetrahedron Letters*, **1971**, 719.
[93] J. P. N. Brewer, I. F. Eckhard, H. Heaney, M. G. Johnson, B. A. Marples, and T. J. Ward, *J. Chem. Soc.* (C), **1970**, 2569.

Reactions involving benzynes may be used to prepare N-methyl-N-(phenethyl)-aniline,[94] m-substituted phenylacetonitriles,[95] and arylcyclohexanones.[96] The formation of 4H-3,1-benzoxazolin-4-ones from a diazonium ion does not involve a benzyne.[97] The reaction of benzyne with various substituted tropones yields a number of 1,4- and 1,6-cycloaddition products.[98] Different mixtures of isomers are obtained from the addition of methoxide ion and methanol to 4-chlorobenzyne, but the products ratio is the same in MeOH as in MeOD, so that breaking of the O—H or O—D bond cannot occur in the rate-determining step.[99]

Tetrachlorobenzyne readily undergoes addition reactions with furan and 1,3-diphenylisobenzofuran.[100] Tetrafluorobenzyne adds normally to furan but with thiophen gives 1,2,3,4-tetrafluoronaphthalene and not the expected episulphide.[101]

The reactions of benzyne towards transition-metal complexes parallels the behaviour of alkenes and alkynes,[102] but attempts to prepare a complex of benzyne and platinum, analogous to the complex of acetylene and platinum, were unsuccessful although the fate of the benzyne was drastically affected by the presence of the metal.[103] The presence of silver ions also affects the product distribution of cycloaddition reactions of benzyne but Warner's explanation,[104] in terms of charge-transfer complex formation with Ag^{++}, has been rejected for one involving orbital symmetry considerations.[105]

The production of hetarynes is a topic of considerable current interest and has been excellently reviewed.[106] The phenothiazyne (**44**) has been suggested as an intermediate

(**44**)   (**45**)   (**46**)   (**47**)

in the conversion of 1- or 2-chloro-10-methylphenothiazines into 2-amino-10-methylphenothiazine,[107] and (**45**) reacts with hydroxide ion to give (**47**) by way of a pyridazyne intermediate (**46**).[108] Reaction of 3-halopyridines with aqueous KOH at 300° to give 3-hydroxypyridine and 4-pyridone occurs by both direct substitution and formation of a

---

[94] A. R. Lepley, R. H. Becker, and A. G. Giumanini, *J. Org. Chem.*, **36**, 1222 (1971).
[95] T. Kametani, K. Kigasawa, M. Hiiragi, T. Aoyama, and O. Kusama, *J. Org. Chem.*, **36**, 327 (1971).
[96] T. Kametani, S. Noguchi, I. Agata, T. Aono, K. Kigasawa, M. Hiiragi, T. Hayasaka, and O. Kusama, *J. Chem. Soc.* (C), **1971**, 1047; T. Kametani, K. Kigasawa, M. Hiiragi, T. Hayasaha, and O. Kusama, *J. Chem. Soc.* (C), **1971**, 1051.
[97] R. R. Schmidt and W. Schneider, *Tetrahedron Letters*, **1970**, 5095.
[98] M. Kato, Y. Okamoto, and T. Miwa, *Tetrahedron*, **27**, 4013 (1971).
[99] J. F. Bunnett, C. Pyun, and J. K. Kim, *Amer. Chem. Soc., Div. Petrol Chem. Reprints*, **14**, C76 (1969); *Chem. Abs.*, **74**, 3178 (1971).
[100] G. A. Moser, F. E. Tibbetts, and M. D. Rausch, *Organomet. Chem. Syn*, **1**, 99 (1971); *Chem. Abs.*, **75**, 35546 (1971).
[101] S. Hayashi and N. Ishikawa, *Nippon Kagaku Zasshi*, **91**, 1000 (1970); *Chem. Abs.*, **74**, 75787 (1971).
[102] R. G. Miller and D. P. Kuhlman, *J. Organomet. Chem.*, **26**, 401 (1971).
[103] T. L. Gilchrist, F. J. Graveling, and C. W. Rees, *J. Chem. Soc.* (C), **1971**, 977.
[104] P. Warner, *Tetrahedron Letters*, **1971**, 723.
[105] L. A. Paquette, *Chem. Comm.*, **1971**, 1076.
[106] T. Kauffmann and R. Wirthwein, *Angew. Chem. Int. Ed.*, **10**, 20 (1971).
[107] D. H. Jones, *J. Chem. Soc.* (C), **1971**, 132.
[108] Y. Maki, G. P. Beardsley, and M. Takaya, *Tetrahedron Letters*, **1971**, 1507.

pyridyne.[109] In the gas-phase, pyridines, unlike benzynes, do not readily dimerize.[110] No hetaryne could be detected in the reaction of 5-halo-1-methylimidazole and lithium piperidide[111] or the conversion of 4-bromo-6-phenylpyrimidine into the 4-amino-compound.[112]

Other Reactions

A previous study[113] suggesting that iodine exchange between KI and 1-iodo-2,4-dinitrobenzene is of zero order in iodide has been shown to be incorrect. The mechanism is concurrent S_N2 and a heterogeneous reaction.[114] Base-catalysed hydrogen exchange with the 1-methylthiolanium ion (48) causes rapid replacement of the three methyl and

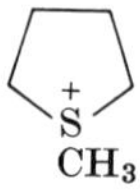

(48)

two of the four α-methylene hydrogen atoms.[115] Reaction of 1,5-naphthyridine 1,6-dioxide and $POCl_3$ involves nucleophilic substitution by chloride ion and loss of oxygen,[116] and reductive dehalogenation may occur alongside nucleophilic substitution in the reaction of bulky amines with iodo-nitro-compounds.[117] The mechanism of the Ullmann reaction has been discussed.[118]

An addition reaction has been suggested as a step in the reaction of the quinoline (49) with phenyl-lithium[119] and of chlorophosphazenes with amines.[120] In the former case

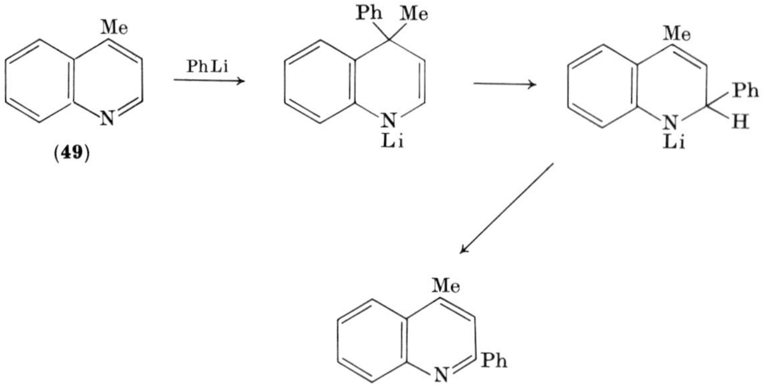

(49)

[109] J. Zoltewicz and A. A. Sale, *J. Org. Chem.*, **36**, 1455 (1971).
[110] J. M. Kramer and R. S. Berry, *J. Am. Chem. Soc.*, **93**, 1303 (1971).
[111] D. A. de Bie, H. C. van der Plas, and G. Geurtsen, *Rec. Trav. Chim.*, **90**, 594 (1971).
[112] J. de Valk and H. C. van der Plas, *Rec. Trav. Chim.*, **90**, 1239 (1971).
[113] C. A. Marcopoulos, *Z. phys. Chem. (Leipzig)*, **236**, 64 (1967).
[114] C. H. Bovington, D. F. Maundrell, and B. Dacre, *J. Chem. Soc.* (B), **1971**, 767.
[115] G. Barbarella, A. Garbesi, and A. Fava, *Helv. Chim. Acta*, **35**, 341 (1971).
[116] E. V. Brown and A. C. Plasz, *J. Org. Chem.*, **36**, 331 (1971).
[117] F. Pietra, M. Bartolozzi, and F. Del Cima, *Chem. Comm.*, **1971**, 1232.
[118] T. O. Tuong and M. Hida, *Bull. Chem. Soc. Japan*, **44**, 765 (1971).
[119] Y. Otsuji, K. Yutani, and E. Imoto, *Bull. Chem. Soc. Japan*, **44**, 520 (1971).
[120] J. M. E. Goldschmidt and E. Licht, *J. Chem. Soc.* (A), **1971**, 2429.

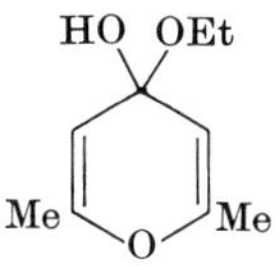

(50)

addition is followed by rearrangement. The addition complex (50) is an intermediate in one pathway for the hydrolysis of 4-ethoxy-2,6-dimethylpyrylium salts,[121] but hydrolysis of the 2-methyl-4,6-diphenylpyrylium ion results in ring opening.[122] Rate-determining attack by hydroxide ion at the 2-position occurs in the hydrolysis of 2-alkoxytropones and the reaction rates are correlated well by the Yakawa–Tsuno equation with $r = 0.56$.[123] Hydrolyses of 1-tosylimidazole and 1-tosyl-3-methylimidazolium chloride[124] and pyrimidine sulphonates have been studied.[125] The reaction of 4-halo-3-nitrobenzenediazonium ions with water in the presence of a base yields 4-diazo-2-nitrophenol; the first step is nucleophilic attack by the base to give an intermediate which reacts, in a second step, with water.[126]

Nucleophilic attack has been proposed in the cyclization of 2-acyl-2-(2-hydroxyethyl)-5-nitropyrrole to 5-acyl-2,3-dihydropyrrolo[2,1-b]oxazole (51),[127] and of *trans*-penta-

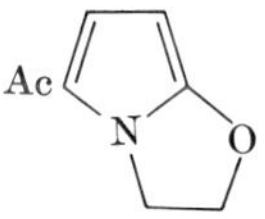

(51) 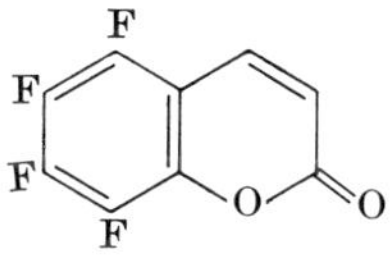   (52)

fluorocinnamic acid to 5,6,7,8-tetrafluorocoumarin (52).[128] The latter reaction can be brought about thermally or photochemically. Reaction of diazomethane with trinitrobenzene at low temperatures results in formation of addition products.[129]

---

[121] G. Salvadori and A. Williams, *J. Am. Chem. Soc.*, **93**, 2727 (1971).
[122] A. Williams, *J. Am. Chem. Soc.*, **93**, 2733 (1971).
[123] K. Bowden and M. J. Price, *J. Chem. Soc.* (B), **1971**, 1784.
[124] P. Voisin, P. Monjoint, M. Laloi-Diard, and M. Vilkas, *Compt. Rend.* (C), **272**, 332 (1971).
[125] D. J. Brown and J. A. Hoskins, *J. Chem. Soc.* (B), **1971**, 2214.
[126] A. Mačháčková and V. Štěrba, *Coll. Czech. Chem. Comm.*, **36**, 3197 (1971).
[127] V. Vechietti, E. Dradi, and F. Lauria, *J. Chem. Soc.* (C), **1971**, 2554.
[128] H. Heaney and A. P. Price, *Chem. Comm.*, **1971**, 894.
[129] J. C. van Velzen, C. Kruk, and T. J. de Boer, *Rec. Trav. Chim.*, **90**, 842 (1971).

CHAPTER 7

Electrophilic Aromatic Substitution

A. R. BUTLER

Department of Chemistry, St. Salvator's College, University of St. Andrews

Sulphonation	197
Nitration	198
Nitrosation	200
Azo Coupling	200
Halogenation	201
Metal Cleavage	204
Decarboxylation	204
Friedel–Crafts and Related Reactions	205
Hydrogen Exchange	206
Miscellaneous Reactions	208

In a general review of electrophilic aromatic substitution Olah[1] discusses the nature of the transition state and defends his views on the role of π-complexes. After a general historical introduction to the mechanism of aromatic nitration, Ridd[2] argues for the importance of diffusion-controlled reactions in explaining some of Olah's results. There has been another general review of electrophilic substitution (in Rumanian),[3] and Cram[4] has given more details of the reactions of cyclophanes. An intermediate similar to the σ-complex in electrophilic aromatic substitution, but without solvation, has been detected in the gas phase by ion cyclotron resonance.[5]

It has been shown that CNDO/2 calculations are far better than either HMO or PPP calculations in predicting qualitatively the activation energies for the reactions of substituted benzenes,[6] but simple HMO calculations for protonated pyridine do provide a guide to the reactions of the methylpyridinium cation.[7]

The Mills–Nixon effect can be rationalized by simple perturbation of Hückel theory,[8] and the results are consistent with Streitwieser's model.[9] In biphenylene the 1-position is activated towards hydrogen exchange but deactivated towards protodesilylation, indicating different resonance demands of the two reactions in the transition state.[10]

---

[1] G. A. Olah, *Accounts Chem. Res.*, **4**, 240 (1971).
[2] J. H. Ridd, *Accounts Chem. Res.*, **4**, 248 (1971).
[3] I. Demetrescu, *Stud. Cerc. Chim.*, **19**, 947 (1971).
[4] D. J. Cram and J. M. Cram, *Accounts Chem. Res.*, **4**, 204 (1971); D. T. Hefelfinger and D. J. Cram, *J. Am. Chem. Soc.*, **93**, 4754 (1971).
[5] S. A. Benezra, M. K. Hoffman, and M. M. Bursey, *J. Am. Chem. Soc.*, **92**, 7501 (1970).
[6] G. R. Howe, *J. Chem. Soc.* (B), **1971**, 981, 984.
[7] J. Kuthan, N. V. Koshmina, and M. Ferles, *Coll. Czech. Chem. Comm.*, **35**, 3825 (1970).
[8] R. D. Rieke, *J. Org. Chem.*, **36**, 227 (1971).
[9] A. Streitwieser, G. R. Ziegler, P. C. Mowery, A. Lewis, and R. G. Lawler, *J. Am. Chem. Soc.*, **90**, 1357 (1968).
[10] R. Taylor, *J. Chem. Soc.* (B), **1971**, 536.

Three new semiempirical relationships, based largely on the Dewar–Grisdale treatment,[11] have emerged from a general analysis of linear free-energy relationship.[12] Taylor has used the pyrolysis of 1-arylethyl acetates to great effect in the evaluation of σ^+ constants. An improved value of +0.565 for the m-CF_3 group has been suggested[13] and the exalted reactivity of pentafluorobenzene, observed in demetallation reactions, has been confirmed and appears to be electronic in origin.[14] The pyrolysis reaction cannot be used to measure σ_0^+ constants.[15] Studies of protodesilylation and hydrogen exchange of some aromatic compounds with oxygen- and sulphur-containing substituents have adduced evidence in favour of hyperconjugation.[16]

In an interesting study[17] an attempt has been made to assess the leaving ability of some electrophiles by a study of model compounds. For example, reaction of (**1**) with

(**1**)

HCl results in migration and loss of NO_2^+, rather than Cl^+. However, Br^+ is lost from the bromo analogue. As well as being an intrinsic quality, leaving ability depends on whether the electrophile is lost in a unimolecular process or removed by a nucleophile. In the first case the order is $NO_2^+ < $ i-$Pr^+ \sim SO_3 < $ t-$Bu^+ \sim ArN_2^+ < ArCHOH^+ < NO^+ < CO_2 < B(OH)_3$ and in the latter case $CH_3^+ < Cl^+ < Br^+ < D^+ < RCO^+ < H^+ < I^+ < Hg^{++} < Me_3Si^+$. Nitration of p-haloanisoles may result in displacement of the halogen and this reaction has been used to measure *ipso* partial rate factors, i.e. the activating or deactivating effect of a substituent on attack at the carbon atom bearing the substituent. For H, I, Br and Cl the values are 1, 0.119, 0.077 and 0.069, respectively, values which show surprisingly little variation.[18]

The most reactive sites in 2,3-dihydrofluoroanthene (**2**) are the 1- and 9-positions.[19]

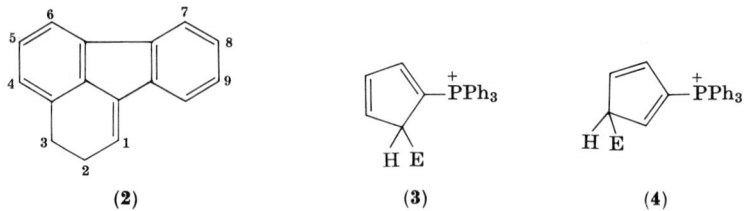

(**2**)   (**3**)   (**4**)

[11] M. J. S. Dewar and P. Grisdale, *J. Am. Chem. Soc.*, **84**, 3548 (1962).
[12] K. C. C. Bancroft and G. R. Howe, *J. Chem. Soc.* (B), **1971**, 1221.
[13] R. Taylor, *J. Chem. Soc.* (B), **1971**, 622.
[14] R. Taylor, *J. Chem. Soc.* (B), **1971**, 255.
[15] R. Taylor, *J. Chem. Soc.* (B), **1971**, 1450.
[16] F. P. Bailey and R. Taylor, *J. Chem. Soc.* (B), **1971**, 1446.
[17] C. L. Perrin, *J. Org. Chem.*, **36**, 420 (1971).
[18] C. L. Perrin and G. A. Skinner, *J. Am. Chem. Soc.*, **93**, 3389 (1971).
[19] C. Finger, *Chem. Ber.*, **103**, 2567 (1971).

Electrophilic attack occurs at the 2-position of phosphonium cyclopentadienylide to give the intermediate (3), whereas attack at the 3-position would give an intermediate with cross-conjugation (4).[20] Tellurophen is more reactive towards electrophiles than is thiophen or selenophen, but less than furan,[21] while thiophen, because of its different geometry, displays smaller steric effects in acylation than benzene.[22] A study of electrophilic substitution on benzothiophen and benzofuran indicates that annelation deactivates both rings and does not affect orientation.[23] The ρ value for ionization potentials of substituted furans, thiophens, selenophens and pyrroles is more negative than for electrophilic substitution.[24] Orientation effects with other heterocycles,[25] tricarbonyl-(cyclo-octatetraene)iron[26] and the cyclopentadienide ion[27] have been considered. Formation of substituted benzenes from phenylmercury salts and Pd(II) in the presence of an oxidizing agent and a nucleophile follows the same pattern as electrophilic substitution.[28] Most electrophilic reagents react with 2H-heptafluoronaphthalene with replacement of the single hydrogen, but with nitric acid the addition product (5) is formed.[29]

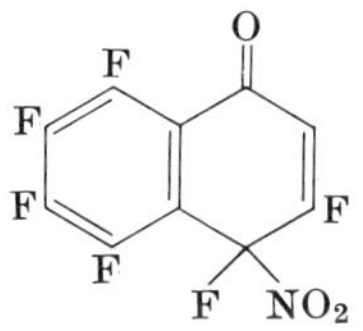

(5)

Sulphonation

The various sulphonating agents present in oleum have been discussed,[30] and kinetic studies of the sulphonation of naphthalene, thionaphthalene and 2-methylnaphthalene,[31] 5,6-dihydro-1,2-dimethyl-4H-pyrrolo[3,2,1-ij]quinoline[32] and polymethylbenzenes[33]

---

[20] D. Lloyd and M. I. C. Singer, Chem. Ind. (London), 1971, 786; Z. Yoshida, S. Yoneda, Y. Murata, and H. Hashimoto, Tetrahedron Letters, 1971, 1523.
[21] F. Friryuelli, G. Marino, G. Sovelli, and A. Taticchi, Chem. Comm., 1971, 1441.
[22] S. Clementi, P. Linda, and M. Vergoni, Tetrahedron, 27, 4667 (1971).
[23] S. Clementi, P. Linda, and G. Marino, J. Chem. Soc. (B). 1971, 79.
[24] P. Linda, G. Marino, and S. Pignataro, J. Chem. Soc. (B), 1971, 1585.
[25] Z. I. Aksel'rod and V. M. Berezovskii, Uspek. Khim., 39, 1337 (1970); Russ. Chem. Rev., 39, 627 (1970); J. P. Kutney, H. W. Hanssen, and G. V. Naiv, Tetrahedron, 27, 3323 (1971).
[26] B. F. G. Johnson, J. Lewis, and G. L. P. Randall, J. Chem. Soc. (A), 1971, 422.
[27] M. I. Rybinskaya and L. M. Korneva, Uspek. Khim., 1971, 444; Russ. Chem. Rev., 1971, 247.
[28] P. M. Henry, J. Org. Chem., 36, 1886 (1971).
[29] V. D. Shteingarts, O. I. Osina, N. G. Kostina, and G. G. Yakobson, Zhur. Org. Khim., 6, 833 (1970); J. Org. Chem. USSR, 6, 835 (1970).
[30] A. Koeberg-Telder and H. Cerfontain, Rec. Trav. Chim., 90, 193 (1971).
[31] M. K. Paktev and Z. Y. Gorokhova, Zhur. Prikl. Khim (Leningrad), 44, 589 (1971); Chem. Abs., 74, 140497 (1971).
[32] M. I. Vinnik, L. D. Abramovich, L. G. Yudin, and V. A. Budylin, Zhur. Org. Khim., 6, 1061 (1970); J. Org. Chem. USSR, 6, 1064 (1970).
[33] A. A. Spryskov and Z. A. Yakovleva, Izv. Vyssh. Uchebn. Zaved., Khim. Khim. Tekhnol, 13, 1139 (1970); Chem. Abs., 74, 3187 (1971).

have been reported. The reaction of sulphuric acid with p-hydroxylbenzenesulphonic acid[34] and chlorobenzenedisulphonic acids[35] results in hydrolysis.

The curvature in a plot of log $a_{H_2S_2O_7}$ against log k for the sulphonation of 4-hydroxyazobenzene and 4'-hydroxyazobenzene-4-sulphonic acid is explained by a change in rate-determining step from σ-complex formation to proton abstraction.[36] Sulphonation of p-phenylenediamine can be effected by anodic oxidation in an aqueous solution containing sodium sulphite; the mechanism is probably concerted charge transfer and sulphonation.[37]

Nitration

Olah's review[1] has already been referred to and there is also a Rumanian review of aromatic nitration.[38] A book on aromatic nitration by well-known workers in this field[39a] and a second edition of Ingold's classic treatise, with a large section on the mechanism of aromatic nitration,[39b] have also appeared.

Measurement of the kinetics of nitration in HNO_3–Ac_2O gives a somewhat different value for the relative reactivity of benzene and toluene from that obtained by the competitive method and both values are very different from that obtained with nitrating agents known to contain NO_2^+, indicating a unique mechanism for nitration by HNO_3–Ac_2O.[40] It is also reported that with HNO_3–H_2SO_4 the relative reactivities of some aromatic compounds decrease with increasing H_2SO_4 concentration.[41] The importance of the direct field effect in the deactivating properties of positive poles has been shown by comparing the rates of nitration of (6) and (7); the electronic effects are the same but (7) nitrates 200 slower than (6) owing to the different positions of the positive poles;[42] if the positive pole is attached directly to the ring then the inductive effect is relatively more important.[43] There is evidence of $\pi(d-p)$ interaction between the aromatic ring

(6)   (7)

---

[34] A. Spryskov and Z. A. Yakovleva, *Izv. Vyssh. Uchebn. Zaved., Khim. Khim. Tekhnol.*, **13**, 1625 (1971); *Chem. Abs.*, **74**, 140430 (1971).

[35] V. V. Kharitonov, A. A. Spryskov, and V. P. Leshchev, *Izv. Vyssh. Uchebn. Zaved., Khim. Khim. Tekhnol.*, **13**, 1151 (1970); *Chem. Abs.*, **74**, 3171 (1971).

[36] F. Buncel, W. M. J. Strachan, and H. Cerfontain, *Can. J. Chem.*, **49**, 152 (1971).

[37] K. Sasaki, H. Imai, Y. Tanimizu, and H. Shiba, *Nippon Kagaku Zasshi*, **91**, 1030 (1970); *Chem. Abs.*, **74**, 63707 (1971).

[38] I. Demetrescu, *Stud. Cerc. Chim.*, **19**, 377 (1971).

[39a] J. Hoggett, R. B. Moodie, J. R. Penton, and K. Schofield, "Nitration and Aromatic Reactivity", Cambridge University Press, London, 1971.

[39b] C. K. Ingold, "Structure and Mechanism in Organic Chemistry", G. Bell and Sons, Ltd., London, 1969.

[40] S. R. Hartshorn, R. B. Moodie, and K. Schofield, *J. Chem. Soc.* (B), **1971**, 1256.

[41] A. Vyhidy, M. Magyar, and R. Berbes, *Acta Chim. Acad. Sci. Hung.*, **69**, 107 (1971); A. Vyhidy, O. Répásy, and J. Tahnács, *Acta Chim. Acad. Sci. Hung.*, **69**, 349 (1971).

[42] G. Mossa, A. Ricci, and J. H. Ridd, *Chem. Comm.*, **1971**, 332.

[43] F. De Sarlo, G. Grynkiewicz, A. Ricci, and J. H. Ridd., *J. Chem. Soc.* (B), **1971**, 714.

and positive poles containing —SMe$_2^+$ and —SeMe$_2^+$, as there is with positive poles of P, As and Sb, but not $\pi(p-p)$ overlap.[44] Nitration of N-benzylaniline in a non-acidic medium (HNO$_3$–Ac$_2$O) results in attack of the aniline component, but in an acid medium (HNO$_3$–H$_2$SO$_4$) protonation of the nitrogen diverts attack to the benzyl ring; with acetic acid as solvent the main product is benzyl-N,p-dinitroaniline.[45]

Steric hindrance prevents nitration at the 2'-position in 2,4,5-trinitrobiphenyl but there is a fair amount of attack at the 3'-position, so previous classification of the picryl group as o/p-directing is incorrect.[46] On the other hand, o-nitration in a series of primary alkylbenzenes is not much affected by the size of the alkyl group; with secondary alkyl groups steric hindrance is reflected in the enthalpy of activation.[47] From a study of the nitration of benzotrichloride and (2,2,2-trichloroethyl)benzene there is no support for a previous suggestion that there is a hyperconjugative component in the substituent effect of CCl$_3$, but there may be one with CF$_3$.[48]

Products of nitration of 4-phenylpyridine and 4-benzylpyridine,[49] various pyrimidones,[50] quinoline,[51a] diphenyl sulphoxide[51b] and 4-hydroxycinnoline[52] have been reported. Nitration of 4-(2,4-dinitrophenyl)-1-methylpyrazole (**8**) occurs at the 3- and 5-positions but the 5-nitro compound slowly decomposes, so the ratio of the two products

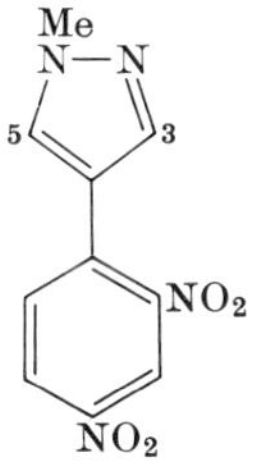

(**8**)

changes with time.[53] The explosive nature of the nitration of thiophen is due to the intrusion of nitrosation, an autocatalytic reaction; if traces of nitrous acid are removed by the addition of urea and the concentration of thiophen is kept low, the reaction proceeds normally and a kinetic study is reported.[54]

Nitration (HNO$_3$–Ac$_2$O) of 5-bromohemimellitene (**9**) results in a 60% yield of 3,4,5-trimethyl-2-nitrophenol owing to decomposition of the dienone (**10**) during work-up of the products.[55] A very comprehensive investigation using, among other methods, isotopic tracer studies, has shown that nitration of 1,3,5-tri-*tert*-butyl-2-nitrobenzene

---

[44] H. M. Gilow, M. De Shazo, and W. C. Van Cleve, *J. Org. Chem.*, **36**, 1745 (1971).
[45] T. A. Modro, *Roczn. Chem.*, **45**, 825 (1971).
[46] E. V. Condon and J. P. Trivedi, *J. Org. Chem.*, **36**, 1926 (1971).
[47] J. M. A. Bass and B. M. Wepster, *Rec. Trav. Chim.*, **90**, 1081, 1089 (1971).
[48] G. Grynkiewicz and J. H. Ridd., *J. Chem. Soc.* (B), **1971**, 716.
[49] F. De Sarlo and J. H. Ridd., *J. Chem. Soc.* (B), **1971**, 712.
[50] C. D. Johnson, A. R. Katritzky, M. Kingsland, and E. F. V. Swiven, *J. Chem. Soc.* (B), **1971**, 1.
[51a] D. H. G. Crout, J. R. Penton, and K. Schofield, *J. Chem. Soc.* (B), **1971**, 1254.
[51b] N. C. Marziano, E. Maccarone, and R. C. Passerini, *J. Chem. Soc.* (B), **1971**, 745.
[52] R. B. Moodie, J. R. Penton, and K. Schofield, *J. Chem. Soc.* (B), **1971**, 1493.
[53] M. D. Coburn, *J. Heterocyclic Chem.*, **8**, 293 (1971).
[54] A. R. Butler and J. B. Hendry, *J. Chem. Soc.* (B), **1971**, 102.
[55] D. J. Blackstock, M. B. Hartshorn, A. J. Lewis, K. E. Richards, and J. Vaughan, *J. Chem. Soc.* (B), **1971**, 1212.

yields certain amounts of 3,5-di-*tert*-butyl-2,4- and -2,6-dinitrotoluene; there is intramolecular rearrangement of the intermediate cyclohexadienyl carbonium ion (**11**) and alkyl fragmentation of the 4-*tert*-butyl group.[56]

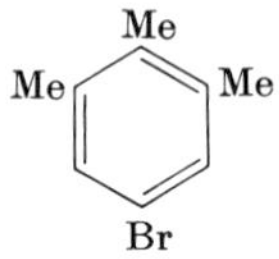

(9) 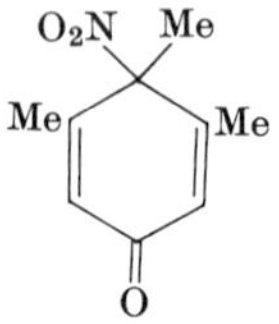 (10) 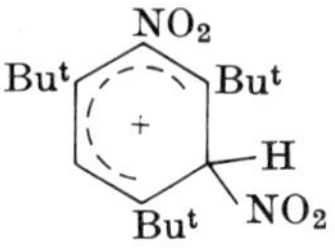 (11)

Nitrosation

Full details have been given of a study of the nitrosation of phenol and anisole in perchloric acid. For phenol the rate is constant below 1 M acid and then increases to reach a maximum in 7.5 M acid. For anisole there is no acid-independent region but with both loss of H^+ is a slow step. Phenol reacts by way of a dienone intermediate, not available to anisole and, although demethylation occurs readily with the latter, this process follows nitrosation. Identification of the nitrosating species is difficult but, above 5 M acid, it is probably NO^+.[57]

Azo Coupling

The reactions of substituted benzenediazonium ions with aniline,[58] acetone,[59] acetylacetone,[60] acetoacetanilide and naphtholate anions[61] all accord with the Hammett relationship. There is some deviation with highly activated ions and this may be due to non-additivity of substituent effects and the incursion of diffusion-controlled reaction rates. Studies of the reaction of benzenediazonium ions with amines have been reported.[62]

Unlike nitration by nitronium tetrafluoroborate, diazo coupling does not proceed faster in tetramethylene sulphone.[63] The rate of reaction between bisdiazotized benzidine and 2-naphthol-3,6-disulphonic acid is lower than expected from a knowledge of the electrophilicity of the monomeric diazonium ion owing to the formation of dimeric and oligomeric aggregates, which may be broken by addition of urea.[64]

A detailed kinetic analysis has shown the presence of specific acid catalysis in the diazotization of aniline in MeOH–CCl$_4$, that had not been detected previously.[65]

[56] P. C. Myhre, M. Beug, K. S. Brown, and B. Östman, *J. Am. Chem. Soc.*, **93**, 3452 (1971).
[57] B. C. Challis, and A. J. Lawson, *J. Chem. Soc.* (B), **1971**, 770.
[58] V. Beránek and M. Večeřa, *Coll. Czech. Chem. Comm.*, **35**, 3402 (1970).
[59] V. Macháček, O. Macháčková, and V. Štěrba, *Coll. Czech. Chem. Comm.*, **35**, 2945 (1970); *Coll. Czech. Chem. Comm.*, **36**, 3187 (1971).
[60] V. Macháček, J. Panchartek, and V. Štěrba, *Coll. Czech. Chem. Comm.*, **35**, 3410 (1971).
[61] H. Kropáčová, J. Panchartek, V. Štěrba, and K. Valter, *Coll. Czech. Chem. Comm.*, **35**, 3287 (1970); J. Kaválek, J. Panchartek, and V. Štěrba, *Coll. Czech. Chem. Comm.*, **35**, 3470 (1971).
[62] L. M. Rozhdestvenskaya, I. L. Bagal, and B. A. Porai-Koshits, *Organic Reactivity* (*Tartu*), **8**, 11 (1971); B. V. Passet, T. M. Timofeyeva, and V. A. Sidorov, *Organic Reactivity* (*Tartu*), **8**, 297 (1971).
[63] J. R. Penton and H. Zollinger, *Helv. Chim. Acta* **54**, 573 (1971).
[64] B. Gloor and H. Zollinger, *Helv. Chim. Acta*, **54**, 553, 563 (1971).
[65] Z. A. Schelly, *J. Phys. Chem.*, **64**, 4062 (1970).

Halogenation

Azobenzene forms a complex with Pd(II) and this reacts with halogen to give exclusively *o*-halogenation.[66] Under acid conditions bromination and chlorination of quinolin-8-ol (**12**) and its copper chelate occurs at the 5- and iodination at the 7-position but under basic conditions these are reversed.[66a]

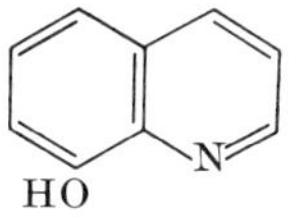

(**12**)

Ring fluorination occurs in preference to side-chain attack in the reaction of toluene with xenon fluorides, suggesting an ionic mechanism, but the relative reactivity with respect to benzene is more consistent with a radical mechanism.[67]
Chlorination by HOCl in acidified aqueous dioxan probably involves formation of $ClOH_2^+$ and addition of $AgClO_4$ removes any traces of free chlorine and also catalyses the reaction, possibly by formation of $AgCl_2^+$.[68] Full details have been given of the chlorination of anisole by HOCl in the presence of α-cyclodextrin.[69] Chlorination of α-methyl-α-isopropenylbenzyl alcohol (**13**) by *tert*-butyl hypochlorite results in a rearranged product (**14**), and the bridged carbonium ion (**15**) is a probable intermediate.[70]

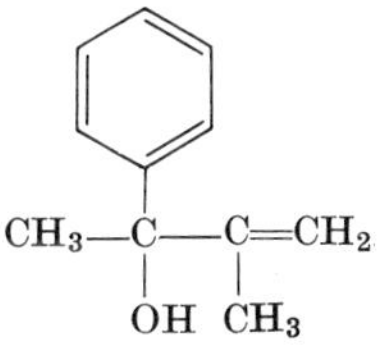

(**13**) 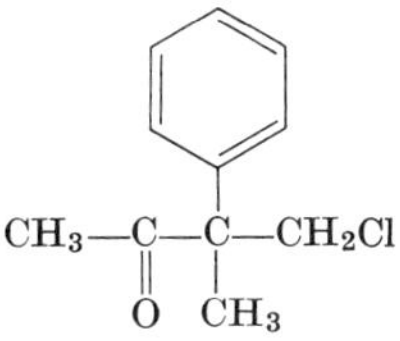   (**14**) 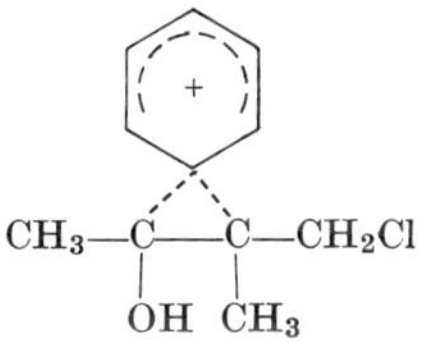   (**15**)

A certain amount of chlorination occurs during the bromination of alkylbenzenes in the presence of LiCl and this was thought to be due to formation of BrCl but it now appears more likely to be due to the equilibrium:[71]

$$2Cl^- + Br_2 \rightleftharpoons 2Br^- + Cl_2$$

although iodine is thought to catalyse aromatic bromination by the formation of IBr.[72] Chlorination of amines involves initial *N*-chlorination and cleavage of the N—Cl bond to give arylnitrenium and chloride ions.[73]

---

[36] D. R. Fakey, *J. Organomet. Chem.*, **27**, 283 (1971).
[36a] H. Gershon, M. W. McNeil, and S. G. Schulman, *J. Org. Chem.*, **36**, 1616 (1971).
[37] T. C. Shieh, E. D. Feit, C. L. Chernik, and N. C. Yang, *J. Org. Chem.*, **35**, 4020 (1970).
[38] P. B. D. de la Mare and L. Main, *J. Chem. Soc.* (B), **1971**, 90.
[39] R. Breslow and P. Campbell, *Bioorganic Chemistry*, **1**, 140 (1971).
[70] V. R. Kartashov, I. V. Bodrikov, and V. P. Pushkarev, *Kinet. Katal.*, **11**, 1334 (1970).
[71] E. Baciocchi and C. Mandolini, *Tetrahedron Letters*, **1971**, 1659.
[72] G. Kothandaraman, J. Rajaram, and J. C. Kuriacose, *Proc. Indian Acad. Sci., Sect. A*, **72**, 144 (1970); *Chem. Abs.*, **74**, 78106 (1971).
[73] P. G. Gassman and G. A. Campbell, *J. Am. Chem. Soc.*, **93**, 2567 (1971).

Studies of the chlorination of anisole,[74] p-fluoroanisole,[75] phenyl ethers,[76] hexaethylbenzene, dodecahydrotriphenylene and trindane[77] have been reported.

Aromatic bromination may not always be irreversible, for heating HBr and o-bromo-N,N-dimethylaniline causes partial rearrangement to the p-bromo compound.[78] There has been a further discussion of relative rates of bromination of toluene and tert-butylbenzene in solvents rich in trifluoroacetic acid. It is argued that formation of significant amounts of the meta-isomer with t-butylbenzene indicates that the relative reactivity is governed by ground-state solvation effects.[79] The directive effect of non-aromatic annelated rings on the bromination of phenol depends on the size of the ring. Generally attack is at the β-position (16) but for a six-membered ring the preferred position is α.[80]

(16)

Bromination at the ortho-position occurs only once in each ring with 4,4',4"-trisubstituted triphenylamines, but hydrogen exchange occurs readily at both ortho-positions; the bulky bromine atom prevents rotation of the ring to a position correct for attack at the other ortho-position. However, for the sterically less demanding hydrogen exchange this restriction does not apply.[81] Pyrazole, 1-methylpyrazole, and 3,5-dimethylpyrazole are all brominated at the 4-position and there is a secondary isotope effect of 1.39.[82] Bromination of the 2,3-dihydro-1,4-diazepinium ion occurs exclusively at the 6-position and if this position is already substituted there is an addition reaction rather than substitution.[83]

The kinetics of aromatic bromination have been analysed in detail.[84] Studies of the bromination of alkyl phenyl ethers and sulphides,[85] anisole,[86] 1-methylnaphthalene,[87] p-toluidine,[88] N,N-dimethylaniline[89] and p-nitrophenol[90] have been reported. In the

[74] K. Seguchi, T. Asano, A. Sera, and R. Gotô, *Bull. Chem. Soc. Japan*, **43**, 3368 (1970).
[75] R. Ganesan and S. Katachandra, *Z. Phys. Chem. (Frankfurt)*, **75**, 212 (1971).
[76] R. Ganesan and S. Katachandra, *Z. Phys. Chem. (Frankfurt)*, **72**, 269 (1970).
[77] G. Illuminati, C. Mandolini, E. M. Arnett, and R. Smoyer, *J. Chem. Soc. (B)*, **1971**, 2206.
[78] F. Effenberger and P. Menzel, *Angew. Chem. Int. Ed.*, **10**, 493 (1971).
[79] L. M. Stock and M. R. Wasielewski, *J. Org. Chem.*, **36**, 1002 (1971).
[80] J. L. G. Nilsson, H. Selander, H. Sievertsson, I. Skånberg, and G.-G. Svensson, *Acta Chem. Scand.*, **25**, 94 (1971).
[81] W. S. Kelley, L. Monack, P. T. Rogge, R. N. Schwartz, S. P. Varimbi, and R. I. Walter, *Ann. Chem.*, **744**, 129 (1971).
[82] B. E. Boulton and B. A. W. Coller, *Austral. J. Chem.*, **24**, 1413 (1971).
[83] C. Barnett, D. R. Marshall, L. A. Mulligan, and D. Lloyd, *J. Chem. Soc. (B)*, **1971**, 1529.
[84] P. Alcais, R. Uzan, J. J. Aaron, F. Rothenberg, and J.-E. Dubois, *Bull. Soc. Chim. France*, **1971**, 612.
[85] S. Ahmed and J. L. Wardell, *Tetrahedron Letters*, **1971**, 3089.
[86] D. C. Rao, J. Rajaram, and J. Kuriacose, *Chem. Comm.*, **1971**, 754; N. S. Granapragsam and N. Joseph, *Curr. Sci. (India)*, **40**, 83 (1971); *Chem. Abs.*, **74**, 99210 (1971); J. J. Aaron and J.-E. Dubois, *Bull. Soc. Chim. France*, **1971**, 603.
[87] J.-M. Bonnier and J. Rinaudo, *Bull. Soc. Chim. France*, **1971**, 2092.
[88] V. P. Kudesia, *Bull. Soc. Chim. Belges*, **80**, 469 (1971).
[89] R. Uzan and J.-E. Dubois, *Bull. Soc. Chim. France*, **1971**, 598.
[90] N. Sridhar, J. Rajaram, and J. Kuriacose, *Curr. Sci. (India)*, **40**, 10 (1971).

presence of AlCl$_3$, bromination of 2-formylfuran gives 4-bromo-2-formylfuran and 3-bromo-2-chloro-5-formylfuran.[91]

Thallic bromide will effect aromatic bromination and thallium salts may also be used in catalytic amounts.[92] Sulphur dioxide is a good solvent for bromination and probably assists polarization of the molecular bromine to give an electrophilic species.[93]

A full account has been given of the use of thallium trifluoroacetate and KI in effecting aromatic iodination. The first step is metallation to give (**17**) and this reacts with

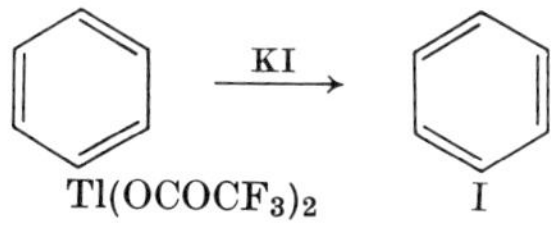

$$\underset{(17)}{\underset{Tl(OCOCF_3)_2}{\bigcirc}} \xrightarrow{KI} \underset{I}{\bigcirc}$$

KI, iodine entering at the same position as the thallium. The method is simple, yields are good, and it has wide applicability, extending to heterocycles. With o/p-directing substituents attack is normally at the *para*-position, but if the reaction mixture is heated before reaction with KI the *m*-iodo compound results owing to a change from kinetic to thermodynamic control.[94] If there is a basic site in the substituent the thallium salt complexes with this initially and thallation occurs at the *ortho*-position.[95]

The iodinating agent present in iodine dissolved in HNO$_3$–H$_2$SO$_4$ is thought to be I$^+$, formed from NO$_2{}^+$ and I$_2$,[96] but in the absence of H$_2$SO$_4$ the situation is more complicated. Iodination of polymethylbenzenes by iodine and HNO$_3$ in acetic acid as solvent is catalysed by N$_2$O$_4$ and hydrogen ions. A kinetic study has indicated that the iodinating agent is protonated NO$_2$I which reacts in a slow step with the aromatic compound.[97]

Several unusual pathways involving dienones occur in the chlorination of 3,4-dimethylphenol and related compounds.[98] The intermediacy of such compounds has been shown by a study of the acid-catalysed rearrangement of (**18**) to (**19**). The rate of reaction is directly proportional to H_0 and substitution by deuterium at the 4-position depresses the rate by a factor of four.[99]

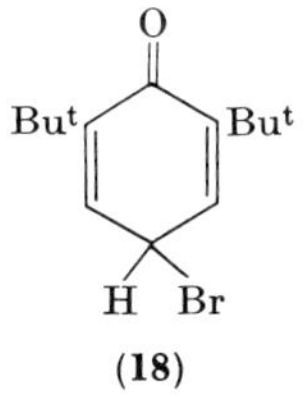

(**18**) 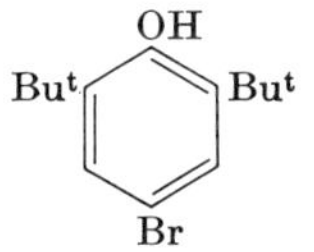 (**19**)

[91] B. Roques, M.-C. Zaluski, and M. Dutheil, *Bull. Soc. Chim. France*, **1971**, 238; B. Rocques, M.-C. Zaluski, M. Bonhomme, and M. Robba, *Bull. Soc. Chim. France*, **1971**, 242.
[92] S. Uemura, K. Sohma, M. Okano, and K. Ichikawa, *Bull. Chem. Soc. Japan*, **44**, 2490 (1971).
[93] J. P. Canselier, *Bull. Soc. Chim. France*, **1971**, 1785.
[94] A. McKillop, J. D. Hunt, M. J. Zelesko, J. S. Fowler, E. C. Taylor, G. McGillivray, and F. Kienzle, *J. Am. Chem. Soc.*, **93**, 4841 (1971).
[95] E. C. Taylor, F. Kienzle, R. L. Robey, A. McKillop, and J. D. Hunt, *J. Am. Chem. Soc.*, **93**, 4845 (1971).
[96] A. M. Sedov and A. N. Novikov, *Zhur. Org. Khim.*, **7**, 517 (1971); *Chem. Abs.*, **75**, 5601 (1971).
[97] A. R. Butler and A. P. Sanderson, *J. Chem. Soc.* (B), **1971**, 2265.
[98] P. B. D. de la Mare and B. N. B. Hannan, *Chem. Comm.*, **1971**, 1324.
[99] P. B. D. de la Mare, A. Singh, J. G. Tillett, and M. Zeltner, *J. Chem. Soc.* (B), **1971**, 1122.

Metal Cleavage

This subject is discussed in a book on organometallic chemistry.[100]

Cleavage of the C—M bond in X—C_6H_4—MMe_3 (M = Si or Sn) is a concerted reaction involving electrophilic attack at the carbon atom and breaking of the C—M bond. This explains the anomalous effect of some substituents (X).[101] A study of the effect of water on the cleavage of substituted phenyltrimethylstannanes in ethanolic $HClO_4$ has been reported.[102] Sulphur dioxide undergoes an insertion reaction with these compounds to give (**20a**). The mechanism is thought to be electrophilic sulphidestannylation.[103]

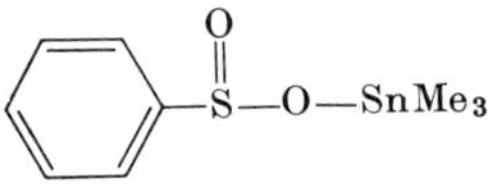

(**20a**) 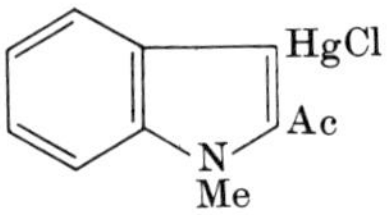   (**20b**)

Protolysis of 2-acetyl-3-chloromercuri-1-methylindole (**20b**) leads to demercuration,[104] and a full kinetic study of substituent effects on the rate of iododemercuration of several organomercury compounds has been reported.[105] Mercury exchange with pentafluorophenylmercury bromide requires the presence of a nucleophile: for this reason the reaction proceeds in DMSO but not in benzene.[106] Iododeboronation of naphthaleneboronic acids has been studied.[107]

Decarboxylation

The anion of 5-amino-1-β-D-ribofuranosylimidazole-4-carboxylic acid 5′-phosphate is stable but both the cation and the zwitterion are decarboxylated fairly readily. At low pH there is acid catalysis of cation decarboxylation.[108] With increasing acid concentration the rate-determining step in the decarboxylation of pyrrole-2-carboxylic acid changes from protonation of the anion to give (**21**) to decarboxylation of (**21**).[109]

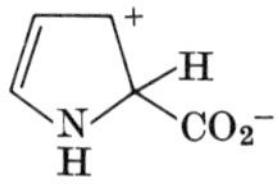

(**21**)

---

[100] C. Eaborn and R. W. Bott, in "Organometallic Compounds of the Group IV Elements", A. G. MacDiarmid, ed., Marcel Dekker, New York, 1968, Vol. 1, Part 1, p. 407.
[101] A. R. Bassindale, C. Eaborn, R. Taylor, A. R. Thompson, D. R. M. Walton, J. Cretney, and G. J. Wright, *J. Chem. Soc.* (B), **1971**, 1155.
[102] C. Eaborn, A. R. Thompson, and D. R. M. Walton, *J. Organomet. Chem.*, **29**, 257 (1971).
[103] C. W. Fong and W. Kitching, *J. Am. Chem. Soc.*, **93**, 3791 (1971).
[104] L. V. Pepekina, L. G. Yudin, A. I. Pavlyuchenko, M. I. Vinnik, and A. N. Kost, *Zhur. Org. Khim.*, **7**, 839 (1971); *Chem. Abs.*, **75**, 48197 (1971).
[105] I. P. Beletskaya, L. V. Savinykh, and O. A. Reutov, *J. Organomet. Chem.*, **26**, 13 (1971); I. P. Beletskaya, L. V. Savinykh, V. N. Gulyachkino, and O. A. Reutov, *J. Organomet. Chem.*, **26**, 23 (1971).
[106] I. P. Beletskaya, I. I. Zakharcheva, and O. A. Reutov, *Dokl. Akad. Nauk SSSR*, **195**, 837 (1971); *Chem. Abs.*, **75**, 48089 (1971).
[107] R. L. Bruce and A. A. Humffray, *Austral. J. Chem.*, **24**, 1085 (1971).
[108] G. J. Litchfield and G. Shaw, *J. Chem. Soc.* (B), **1971**, 1474.
[109] G. E. Dunn and G. K. J. Lee, *Can. J. Chem.*, **49**, 1032 (1971).

Friedel–Crafts and Related Reactions

Acetylation of 2,3,4,5-tetrahydroacenaphthene (**22**) yields the 8-, 6- and 7-isomers and the results may be rationalized in terms of strain in the molecule.[110] Acetylation of

(**22**)   (**23**)

1-methylnaphthalene gives almost exclusively the 4-isomer, but 2-methylnaphthalene is less selective and more sensitive to solvent changes.[111] Localization-energy calculations predict that the most reactive position in 2-methyl-2H-cyclopenta[d]pyridazine (**23**) is 7 and this has been confirmed by a study of the product of trifluoroacylation.[112] Reaction of pyrazine with alkyl-lithium causes alkylation at the 2- and 3-positions.[113] In common with other five-membered heterocycles, the β-position of furan is very unreactive towards acetylation,[114] and furan, thiophen and pyrrole show considerable differences in their sensitivity to substituent effects in trifluoroacetylation. Although furan is intermediate in reactivity, ρ is more negative than for thiophen.[115]

Acylation of anisole by acid chlorides is acid-catalysed,[116] as is the reaction between cymene and isobutene, di-isobutene, and tri-isobutene;[117] a complex mixture of hydrocarbons results from the cymene reactions. Other Friedel–Crafts reactions involve cyclohexyl chloride, bromide, and iodide,[118] benzoylation of phenanthrene[119] and toluene,[120] alkylation by propene,[121] and reactions with 3-chloropropionyl chloride.[122] The chloromethylation of mesitylene, catalysed by $ZnCl_2$,[123] is acid-catalysed and alkylation is faster than sulphonation in a solution of H_2SO_4 and di-isobutene in octane.[124] Studies of the cycloalkylation reactions of 3-alkylpyridines and ω-(3-pyridyl)alk-1-enes have been reported.[125]

---

[110] F. Perin, P. Jacquignon, and N. P. Buu-Hoï, *Bull. Soc. Chim. France*, **1971**, 1371.
[111] J.-M. Bonnier and J. Rinaudo, *Bull. Soc. Chim France*, **1971**, 2094.
[112] H. L. Ammon, P. H. Watts, A. G. Anderson, D. M. Forkey, L. D. Grina, and Q. Johnson, *Tetrahedron*, **26**, 5707 (1970).
[113] W. Schwaiger and J. P. Ward, *Rec. Trav. Chim.*, **90**, 513 (1971).
[114] G. Ciranni and S. Clementi, *Tetrahedron Letters*, **1971**, 3833.
[115] S. Clementi and G. Marino, *Chem. Comm.*, **1971**, 1642.
[116] R. Corriu and G. Dabosi, *Bull. Soc. Chim France*, **1971**, 1666, 1670.
[117] D. E. Boone, E. J. Eisenbraun, P. W. Flanagan, and R. D. Grigsby, *J. Org. Chem.*, **36**, 2042 (1971).
[118] S. D. Lyushin, P.-W. Wu, N. P. Kurnosova, and M. S. Shakhgel'diev, *Dokl. Akad. Nauk Azerb. SSSR*, **26**, 36 (1970); *Chem. Abs.*, **74**, 52725 (1971).
[119] P. H. Gore, C. K. Thadani, S. Thorburn, and M. Yusuf, *J. Chem. Soc.* (C), **1971**, 2329.
[120] L. R. Pettiford, *J. Chem. Soc.* (B), **1971**, 1089.
[121] N. I. Plotkina, N. V. Gein, and V. S. Plyusnin, *Zhur. Fiz. Khim.*, **44**, 1943 (1970); *Chem. Abs.*, **73**, 119869 (1970).
[122] P. O. I. Virtanen, H. Malo, and H. Ruotsalainen, *Suomen Kem.*, (B), **43**, 512 (1971).
[123] M. M. Lyushin, S. D. Mekhtiev, and S. N. Guseinova, *Zhur. Org. Chim.*, **6**, 1432 (1970); *J. Org. Chem. USSR*, **6**, 1445 (1970).
[124] M. K. Paktev and Z. Y. Gorokhova, *Zhur. Prikl. Khim.* (*Leningrad*), **44**, 1116 (1971); *Chem. Abs.*, **75**, 62871 (1971).
[125] S. V. Kannan and H. Pines, *J. Org. Chem.*, **36**, 2301 (1971); H. Pines, S. V. Kannan, and W. M. Stalick, *J. Org. Chem.*, **36**, 2308 (1971).

An NMR study has indicated formation of intermediates of the type [Me$_2$NCHCl]$^+$-OSOCl$^-$ during Vilsmeier formylation in DMF.[126]

Gas-phase alkylation of benzene and toluene by γ-radiolysis is an electrophilic reaction but, to explain the statistical distribution of isomers at low pressures, low substrate selectivity, and the small kinetic isotope effect, initial formation of a π-complex has been proposed,[127] as is found in the isopropylation of benzene and toluene.[128] In the presence of AlBr$_3$ lactones will alkylate benzene. The stereochemistry of the reaction is consistent with complex formation between the lactone and AlBr$_3$ and subsequent reaction of the complex with Al$_2$Br$_6$ to give an ion pair.[129] Oxonium complexes have been detected by IR spectroscopy in aromatic benzoylation catalysed by AlCl$_3$.[130]

Friedel–Crafts reactions are catalysed by the half-sandwich compounds ArMo(CO)$_3$ and proceed by carbonium ion generation.[131] There is a redox reaction between an aryl iodide and tri-iodoborane to give iodine and aryldi-iodoborane.[132]

Hydrogen Exchange

Kresge and his collaborators[133] have reported a very full study correlating equilibrium protonation and hydrogen exchange for benzene-1,3,5-triol and its ethyl and methyl ethers. In fairly concentrated acid σ-complexes are formed and even anisole is protonated at the *para*-position in HClO$_4$ more concentrated than 70%.[133] The extent of protonation has been studied as a function of acid concentration and the results correlate well with the H_c acidity function.[134] Measurement of substituent effects, assuming they are additive, predict a pK_a value for the conjugate acid of benzene of -23.[135] The rates of hydrogen exchange for these compounds correlate equally well with H_c and H_0 but the slopes of plots vary from compound to compound. Clearly the acidity-dependence of the rate of exchange and equilibrium protonation are not the same, and individual variation of the exchange reaction probably indicates differing extent of proton transfer in the transition state. This study casts doubt on the value of kinetic acidity-dependence as a criterion for reaction mechanism in view of the variation found for reactions with the same mechanism.[136] Two sets of molecular-orbital calculations for the benzenonium ion, the intermediate in hydrogen exchange, have been reported[137] and protonation of some substituted hemimellitenes in fluorosulphuric acid occurs significantly at a ring carbon atom bearing a methyl group (**24**).[138] A new acidity scale has been proposed, based on the protonation of azulenes.[139]

---

[126] G. J. Martin, S. Poignant, M. L. Filleux, and M. T. Quemeneur, *Tetrahdron Letters*, **1970**, 5061.
[127] S. Takamuku, K. Iseda, and H. Sakurai, *J. Am. Chem. Soc.*, **93**, 2420.
[128] R. Nakane, O. Kurihara, and A. Takematsu, *J. Org. Chem.*, **36**, 2753 (1971).
[129] D. A. Waples and J. I. Brauman, *Chem. Comm.*, **1971**, 1075.
[130] R. Corriu, M. Dore, and R. Thomassin, *Tetrahedron*, **27**, 5601 (1971).
[131] M. F. Farona and J. F. White, *J. Am. Chem. Soc.*, **93**, 2826 (1971).
[132] W. Siebert, F. R. Rittig, and M. Schmidt, *J. Organomet. Chem.*, **25**, 305 (1970); W. Siebert, K.-J. Schaper, and M. Schmidt, *J. Organometal. Chem.*, **25**, 315 (1970).
[133] A. J. Kresge, Y. Chiang, and L. E. Hakka, *J. Am. Chem. Soc.*, **93**, 6167 (1971).
[134] M. T. Reagan, *J. Am. Chem. Soc.*, **91**, 5506 (1969).
[135] A. J. Kresge, H. J. Chen, L. E. Hakka, and J. E. Kouba, *J. Am. Chem. Soc.*, **93**, 6174 (1971).
[136] A. J. Kresge, S. G. Mylonakis, Y. Sato, and V. P. Vitullo, *J. Am. Chem. Soc.*, **93**, 6181 (1971).
[137] W. Jakubetz and P. Schuster, *Angew. Chem. Int. Ed.*, **10**, 497 (1971), N. S. Isaacs and D. Cvitas, *Tetrahedron*, **47**, 4139 (1971).
[138] M. P. Hartshorn, K. E. Richards, J. Vaughan, and G. J. Wright, *J. Chem. Soc.* (B), **1971**, 1629.
[139] T. K. Rodima and V. Haldna, *Organic Reactivity* (*Tartu*), **7**, 391 (1971).

Because of the ease with which it may be followed and the absence of steric effects, hydrogen exchange continues to be used to compare the reactivities of different positions in aromatic compounds. Such reactions with heteroaromatics have been reviewed.[140] Benzothiophen and thienothiophens react more readily than thiophen,[141] and five-membered heterocycles show the expected relative reactivities (thiophen < furan < pyrrole).[142] The effect of substituents on hydrogen exchange at the 2-, 3- and 8-positions of fluoroanthenes (**25**) has been analysed in terms of atom–atom polarizabilities, σ-inductive effects, and direct field effects.[143] Only the 6-position undergoes detectable

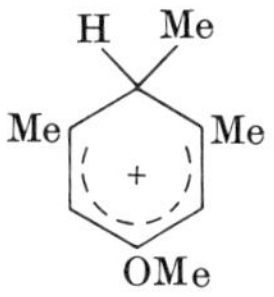

(24) 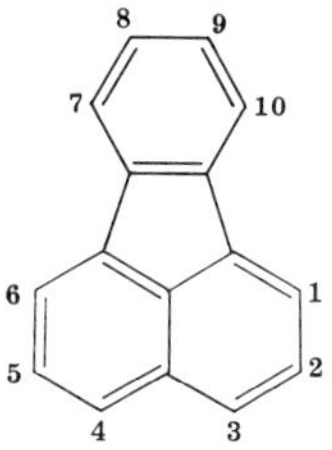 (25) 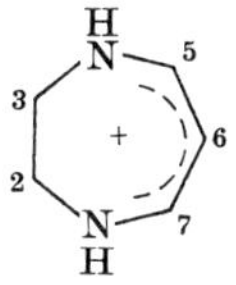 (26)

hydrogen exchange on the 2,3-dihydro-1,4-diazepinium ion (**26**) and this is due to the location of the positive charge in the intermediate on the two N atoms.[144] The relative reactivities of the 2- and 3-positions of pyrrole invert as the acidity is changed and it is suggested that this is due to an earlier transition state in a more acid medium.[145] Localization energies predict the relative reactivities of the non-equivalent positions of

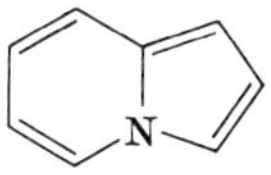

(27)

indolizine (**27**) incorrectly, but better correlation is obtained with the π-electron densities.[146] Studies of hydrogen exchange with quinoline, 2-naphthol, and some corresponding aza, thia and oxo derivatives,[147] substituted phenanthrenes,[148] 1,8-dimethylnaphthalene, acenaphthene and perinaphthane,[149] various organometallic carbonyls[150] and ferrocene[151] have been reported.

[40] Y. Kawazoe and M. Maeda, *Kagaku No Ryoiki, Zokan*, **92**, 39 (1970); *Chem. Abs.*, **73**, 108897 (1971); Z. I. Aksel'rod and V. M. Berezovskii, *Uspek. Khim.*, **39**, 1337 (1970).
[41] T. A. Yakushina, E. N. Zvyagintseva, V. P. Litinov, S. Ozolins, Y. L. Gol'dfarb, and A. I. Shatenshstein, *Zhur. Obshch. Khim.*, **40**, 1622 (1970); *Chem. Abs.*, **74**, 63692 (1971).
[42] K. Schwetlick and K. Unverferth, *Wiss. Z. Tech. Hochsch. Chem. "Carl Schorlemmer" Leuna-Merseburg*, **12**, 230 (1970); *Chem. Abs.*, **74**, 124421 (1971).
[43] K. C. C. Bancroft and G. R. Howe, *J. Chem. Soc.* (B), **1971**, 400.
[44] A. R. Butler, D. Lloyd, and D. R. Marshall, *J. Chem. Soc.* (B), **1971**, 795.
[45] G. P. Bean, *Chem. Comm.*, **1971**, 421.
[46] W. Engewald, M. Mühlstädt, and C. Weiss, *Tetrahedron*, **27**, 851, 4171 (1971).
[47] U. Bressel, A. R. Katritzky, and J. R. Lea, *J. Chem. Soc.* (B), **1971**, 4, 11.
[48] C. Eaborn, A. Fischer, and D. R. Killpack, *J. Chem. Soc.* (B), **1971**, 2142 (1971).
[49] M. C. A. Opie, G. J. Wright, and J. Vaughan, *Austral. J. Chem.*, **24**, 1205 (1971).
[50] M. N. Nefedova, V. N. Setkina, A. I. Khatami, A. G. Ginzburg, and D. N. Kursanov, *Dokl. Akad. Nauk SSSR*, **198**, 889 (1971); *Chem. Abs.*, **74**, 75711 (1971).
[51] D. N. Kursanov, V. N. Setkina, E. I. Fedin, M. N. Nefedova, and A. I. Khatami, *Dokl. Akad. Nauk SSSR*, **192**, 347 (1970).

A previous report[152] that hydrogen exchange in porphyrins is accelerated by the presence of a metal has now been denied.[153] Studies of the deuteriation of benzyldicarbonyl-π-cyclopentadienyliron indicate that the $-CH_2Fe(CO)_2-\pi-C_5H_5$ group activates the benzene ring more than methoxy does. The transition state is stabilized either by neighbouring-group participation via a π-complex (28), or by vertical stabilization brought about by overlap of an appropriately filled orbital of the iron with the outer p-orbital of the partially quinonoid organic group (29).[154]

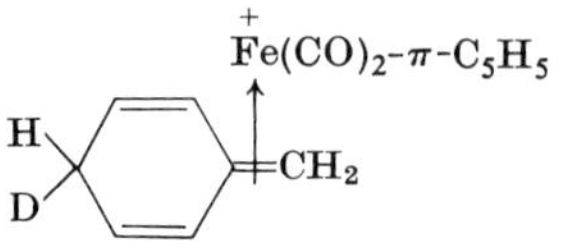

(28) 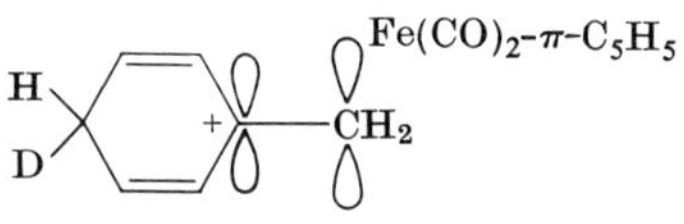   (29)

Acid-catalysed hydrogen exchange of resorcinol involves attack of the hydrogen ion on the neutral molecule. In alkaline solution the situation is much more complex and five substrate–catalyst pairs may be involved.[155] Exchange on the methyl group of 2-methylbenzo[b]thiophen (30) is thought to occur via the intermediates (31) and (32).[156]

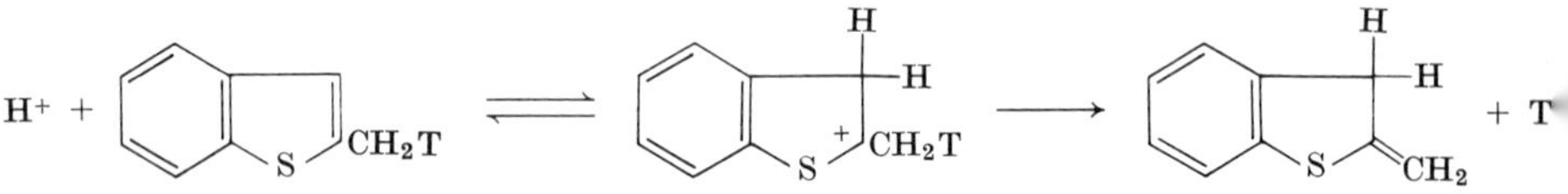

(30)   (31)   (32)

Absorption or exchange may occur in the reaction of deuterium and an anthracene dianion:alkali-metal complex.[157]

The helium tritiide ion (HeT+) is a powerful electrophile and will react in the gas phase with halobenzenes with hydrogen exchange. The reaction is analogous to electrophilic substitution in solution and appears to be a powerful method of studying this reaction without the complication of solvation effects. The reagent has low selectivity. With anisole an excited oxonium ion may be formed.[158]

Miscellaneous Reactions

Electrophilic cyclization is thought to occur in the reaction of substituted benzoic acids with formaldehyde and SO_3 to give phthalide products,[159] the conversion of 2-halo-1,1,4,4-tetraphenylbuta-2,3-dien-1-ol (33) to give a fulvene (34) or a furan (35) in the presence of $HgBr_2$,[160] and in the formation of (36) from 2-anilino-1,4-naphthoquinone.[161]

[152] R. Grigg, A. Sweeney, and A. W. Johnson, *Chem. Comm.*, **1970**, 1237.
[153] J. B. Paine and D. Dolphin, *J. Am. Chem. Soc.*, **93**, 4080 (1971).
[154] S. N. Anderson, D. H. Ballard, and M. D. Johnson, *Chem. Comm.*, **1971**, 779.
[155] V. Gold, J. R. Lee, and A. Gitter, *J. Chem. Soc.* (B), **1971**, 32.
[156] C. Eaborn and G. J. Wright, *J. Chem. Soc.* (B), **1971**, 2262.
[157] M. Ichikawa and K. Tamaru, *J. Am. Chem. Soc.*, **93**, 2079 (1971).
[158] F. Cacace and G. Perez, *J. Chem. Soc.* (B), **1971**, 2086; F. Cacace, R. Cipollini, and G. Ciranni, *J. Chem. Soc.* (B), **1971**, 2089.
[159] L. S. Forney and A. T. Jurewicz, *J. Org. Chem.*, **36**, 689 (1971).
[160] F. Toda, N. Ooi, and K. Akagi, *Bull. Chem. Soc. Japan.* **144**, 1050 (1971).
[161] G. Schroeder and W. Lüttke, *Chem. Ber.*, **104**, 2908 (1971).

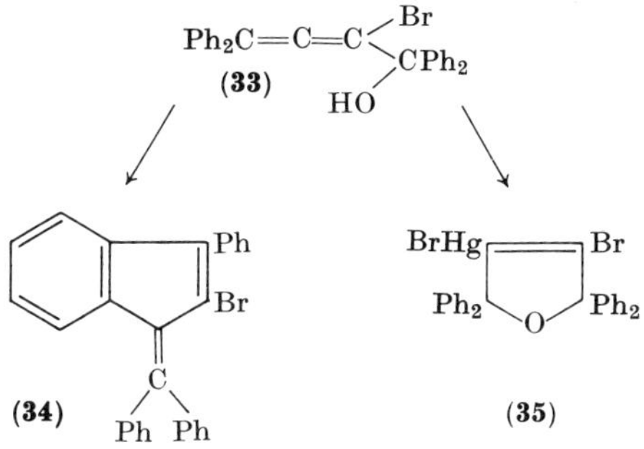

2-(Chloromercury)-1,3-dimethylbenzimidazolium ion (**37**) undergoes the unusual reaction of unimolecular acidolysis catalysed by the chloride ion; substituents in the benzene ring have only a small effect on the rate.[162] The formation of dicyclopentadienyl-cobalt cations from the 5-*exo*-(3-oxoalkyl) derivatives of cyclopentadienyl(cyclopentadiene)cobalt,[163] acid-catalysed racemization of optically active acylferrocenes,[164] and the automerization of 9-ethyl-9,10-dimethylphenanthrenonium ion[165] have been described. Reaction of pyrrylmagnesium bromide and methyl pyrrole-1-carboxylate gives 1,2′-dipyrrolyl ketone (**38**) and this has been shown not to occur by initial attack at the 1-position to give (**39**) and subsequent rearrangement.[166]

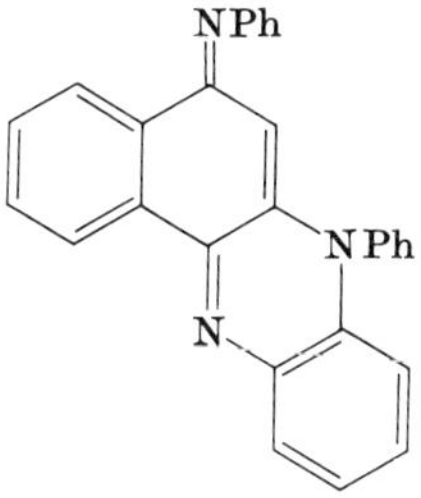

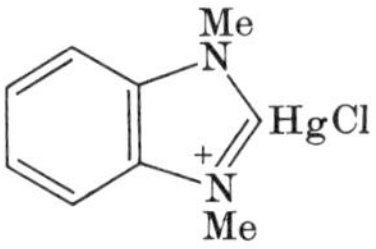

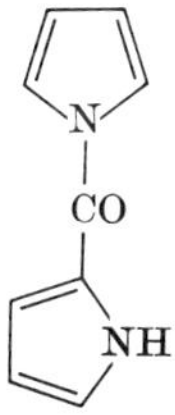

Polyfluorinated arenonium ions are formed when fluorinated halocyclohexadienes are dissolved in SbF_5.[167]

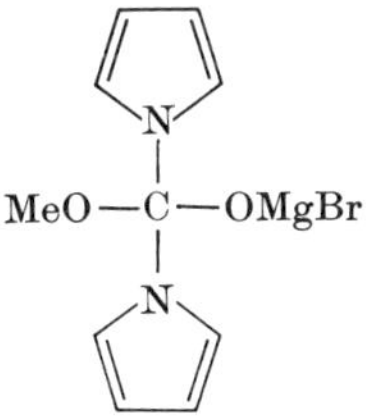

---

[162] C. J. Cooksey, D. Dodd, and M. D. Johnson, *J. Chem. Soc.* (B), **1971**, 1380.
[163] G. E. Herberich and G. Greiss, *J. Organomet. Chem.*, **27**, 113 (1971).
[164] H. Falk, H. Lehner, J. Paul, and U. Wagner, *J. Organomet. Chem.*, **28**, 115 (1971).
[165] V. G. Shubin, D. V. Korchagina, B. G. Derendyaev, V. I. Mamatyuk, and V. A. Koptjug, *Zhur. Org. Chim.*, **6**, 2072 (1970); *J. Org. Chem. USSR*, **6**, 2074 (1970).
[166] C. E. Loader and H. J. Anderson, *Can. J. Chem.*, **49**, 1064 (1971).
[167] V. D. Steingarts and Y. V. Pozdnyakovich, *Zhur. Org. Khim.*, **7**, 734 (1971); *Chem. Abs.*, **75**, 48006 (1971).

Palladium acetate effects nuclear acetoxylation of alkylbenzenes but the normal orientation rules are not obeyed as the reaction involves oxypalladation to give (**40**)

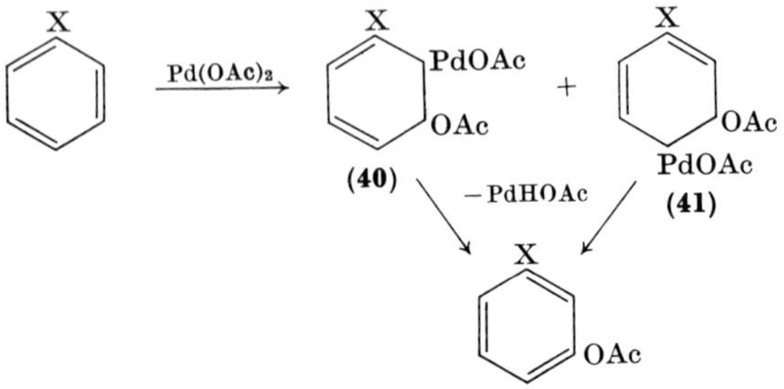

$$\text{C}_6\text{H}_5\text{X} \xrightarrow{\text{Pd(OAc)}_2} \underset{(\mathbf{40})}{\text{[X-C}_6\text{H}_5\text{(PdOAc)(OAc)]}} + \underset{(\mathbf{41})}{\text{[X-C}_6\text{H}_5\text{(OAc)(PdOAc)]}} \xrightarrow{-\text{PdHOAc}} \text{X-C}_6\text{H}_4\text{-OAc}$$

and (**41**) and then elimination of PdHOAc.[168] The reaction of benzene and an ethylene in the presence of Pd(II) involves formation of a σ-complex[169] and there is no hydride shift.[170]

[168] L. Eberson and L. Gomez-Gonzales, *Chem. Comm.*, **1971**, 263.
[169] S. Danno, I. Moritani, Y. Fujiwara, and S. Teranishi, *Bull. Chem. Soc. Japan*, **47**, 3966 (1970); I. Moritani, S. Danno, Y. Fujiwara, and S. Teranishi, *Bull. Chem. Soc. Japan.*, **44**, 578 (1971); I. Moritani, Y. Fujiwara, and S. Danno, *J. Organomet. Chem.*, **27**, 279 (1971).
[170] S. Danno, I. Moritani, Y. Fujiwara, and S. Teranishi, *J. Chem. Soc. (B)*, **1971**, 196.

CHAPTER 8

Molecular Rearrangements

R. J. STOODLEY

Department of Organic Chemistry, The University, Newcastle-upon-Tyne

Aromatic Rearrangements	212
Benzene Derivatives	212
Cyclohexadiene Derivatives	215
Sigmatropic Rearrangements	216
[3,3]-Migrations	216
Claisen and Related Rearrangements	216
Cope Rearrangements	219
[2,3]-Migrations	222
[1,3]-Migrations	224
[1,5]-Migrations	226
Electrocyclic Rearrangements	228
Aromatic Processes	229
Antiaromatic Processes	233
Rearrangements Involving Cycloreversions and Cycloadditions	235
Anionic Rearrangements	239
[1,2]-Migrations	239
[1,5]-Migrations	241
Enolate-induced Reorganizations	241
Cationic Rearrangements	243
[1,2]-Migrations Involving Departures of Leaving Groups	243
Shifts to Primary Centres	243
Shifts to Secondary Centres	244
Shifts to Tertiary Centres	246
[1,2]-Migrations to Unsaturated Centres	248
[1,3]-Migrations	252
Other Migrations	253
Metal-catalysed Rearrangements	253
Rearrangements Involving Electron-deficient Heteroatoms	257
[1,2]-Migrations	257
[1,3]-Migrations	260
Isomerizations	261
Interconversions of Diastereoisomers	261
Prototropic Shifts	262
Alkenes	262
Alkynes	264
Tautomerizations	264
Rearrangements Involving Ring Openings and Closures	264
Three-membered Rings	264
Four-membered Rings	266
Five-membered Rings	267
Six-membered Rings	270
Seven-membered and Larger Rings	273

Aromatic Rearrangements

Benzene Derivatives

The arylnitramine rearrangement has been reviewed.[1]

The rate-determining step in the HCl-catalysed rearrangement of N-substituted N-nitrosoanilines to p-nitrosoanilines is considered to involve the formation of NOCl and the secondary aromatic amine.[2] However, in the case of N-methyl-N-nitrosoaniline, the observed isotope effects coupled with the influence of acid concentration and of added N-methylaniline on the reaction rate are inconsistent with such a pathway; the evidence suggests that the rearrangement is intramolecular and that the denitrosation is a separate reversible reaction.[3]

There have been further studies of the benzidine rearrangement[4] and, in the case of 2,2′-dimethoxyhydrazobenzene,[5] the reaction is markedly accelerated by an increase in pressure. Related reorganizations are probably involved in the HCl-catalysed conversion of 2-amino-1-anilino-4-phenylimidazole into 2-amino-4-(p-aminophenyl)-5-phenylimidazole[6] and in the KOBu^t-induced reaction of (**1**) and 1-fluoro-2-nitrobenzene to give (**2**).[7]

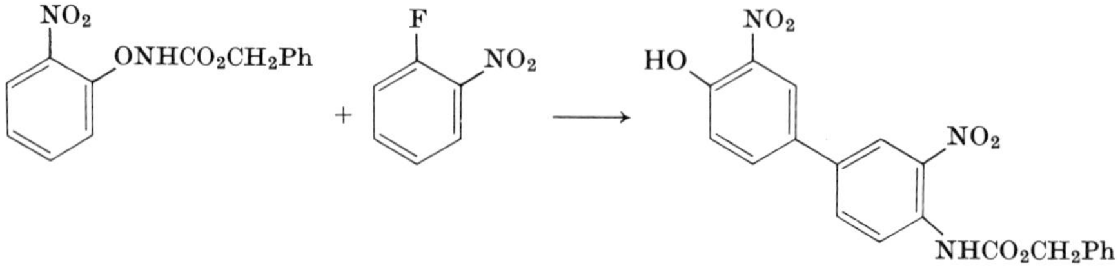

NO$_2$ ONHCO$_2$CH$_2$Ph   +   F   NO$_2$ ⟶   HO—(NO$_2$)—(NO$_2$)—NHCO$_2$CH$_2$Ph

(**1**)   (**2**)

Normally, the acid-catalysed Wallach rearrangement of azoxybenzenes leads to p-hydroxyazobenzenes; however, preferential migration of the hydroxyl group to the *ortho*-position of the naphthyl ring occurs with α- and β-phenylazonaphthalenes.[8] Arenesulphonyl chlorides also react with azoxybenzenes to give *o*- and *p*-(arenesulphonyloxy)azobenzenes.[9] The product ratio is sensitive to the nature of the sulphonyl chloride: electron-withdrawing substituents in the aryl ring promote the *ortho*-rearrangement and electron-releasing groups favour the *para*-rearrangement. ^{18}O-Labelling experiments suggest that the *ortho*-migration is intramolecular, occurring through an O-bridged ion-pair, while an intermolecular process, possibly involving a solvent-separated ion-pair, is implicated in the *para*-rearrangement. When 1,2-dihydroxyindole is treated with toluene-p-sulphonyl chloride at $-78°$ the unstable tosylate (**3**) is formed; however, in the presence of water this derivative affords a mixture of (**5**: R^1 = OH, R^2 = H) and

---

[1] W. N. White, *Mech. Mol. Migr.*, **3**, 109 (1971).
[2] *Org. Reaction Mech.*, **1970**, 243.
[3] T. D. B. Morgan and D. L. H. Williams, *Chem. Comm.*, **1970**, 1671.
[4] D. V. Banthorpe, A. Cooper, and M. O'Sullivan, *J. Chem. Soc.* (B), **1971**, 2054; Z. J. Allan, *Tetrahedron Letters*, **1971**, 4225; Z. J. Allan and J. Podstata, *Coll. Czech. Chem. Comm.*, **36**, 3053 (1971).
[5] J. Osugi, M. Sasaki, and I. Onishi, *Nippon Kagaku Zasshi*, **91**, 714 (1970); *Chem. Abs.*, **73**, 119909 (1970); *Rev. Phys. Chem. Japan*, **40**, 39 (1970); *Chem. Abs.*, **74**, 63758 (1971).
[6] H. Honeck and H. Beyer, *Chem. Ber.*, **104**, 407 (1971).
[7] T. Sheradsky and G. Salemnick, *Tetrahedron Letters*, **1971**, 645.
[8] E. Buncel and A. Dolenko, *Tetrahedron Letters*, **1971**, 113.
[9] S. Oae, T. Maeda, S. Kozuka, and M. Nakai, *Bull. Chem. Soc. Japan*, **44**, 2495 (1971).

(**5**: R¹ = H, R² = OH), in accord with the intervention of (**4**).[10] The detection of a CIDNP effect in the acetic anhydride-induced rearrangement of 4-methylpyridine 1-oxide to 4-(acetoxymethyl)pyridine and 3-acetoxy-4-methylpyridine suggests that a radical-pair mechanism is involved.[11] However, in cases where cleavage of the N—O bonds occurs during the slow step, the rearrangement rate is influenced by the polarity of the solvent.[12]

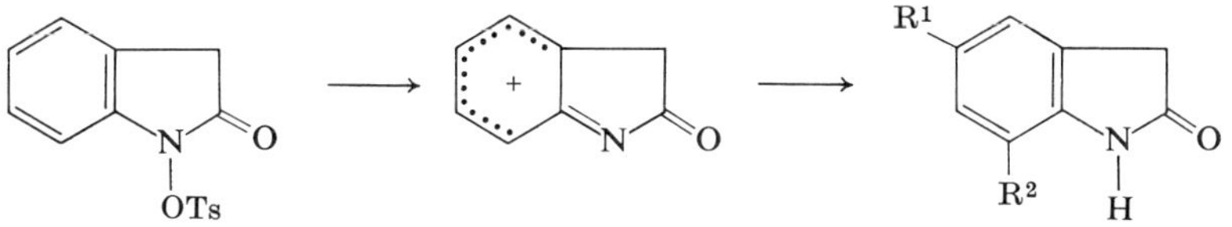

The acid-promoted rearrangement of (**6**) to (**9**), in which the cyclohexadienyl cations (**7**) and (**8**) probably intervene, is stereospecific and the S-configuration is retained.[13]

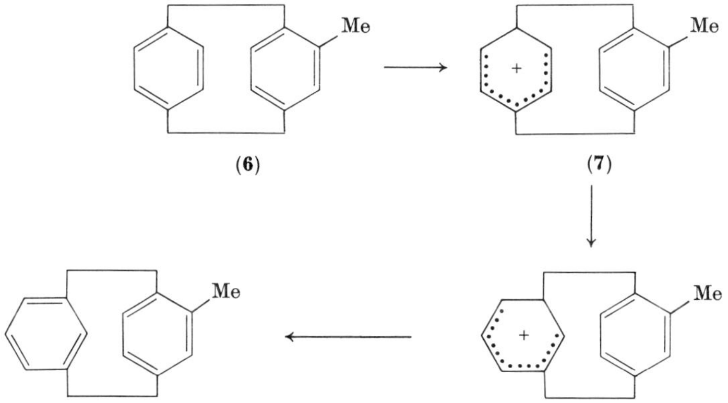

The TiCl₄-promoted Fries rearrangement of t-butylphenyl propionates has been examined.[14]

Several studies have been concerned with the intramolecular nucleophilic displacement of substituents attached to an aromatic ring. Thus, the Smiles rearrangement has been surveyed[15] and further examples of it have been reported.[16] ¹⁸O-Labelling experiments reveal that such a process is involved in the LiBH₄ reduction of ethyl 2-methyl-2-(p-nitrophenoxy)propionate to 2-hydroxy-1-(p-nitrophenoxy)propane.[17] Aryl thionobenzoates (**10**) rearrange to aryl thiolbenzoates (**12**) when heated; the first-order reaction is facilitated by the presence of electron-withdrawing substituents, R, in the migrating aromatic ring (ρ + 2.11) and possesses a large negative $\Delta S^{\ddagger}$, concordant with a transition

[10] P. G. Gassman and G. A. Campbell, *Chem. Comm.*, **1971**, 1437.
[11] H. Iwamura, M. Iwamura, T. Nishida, and S. Sato, *J. Am. Chem. Soc.*, **92**, 7474 (1970).
[12] K. Ogino and S. Oae, *Kogyo Kagaku Zasshi*, **73**, 2300 (1970); *Chem. Abs.*, **74**, 124566 (1971).
[13] M. H. Delton, R. E. Gilman, and D. J. Cram, *J. Am. Chem. Soc.*, **93**, 2329 (1971).
[14] R. Martin and J. M. Betoux, *Bull. Soc. Chim. France*, **1971**, 3648.
[15] W. E. Truce, E. M. Kreider, and W. W. Brand, *Org. Reactions*, **18**, 99 (1970).
[16] H.-H. Otto, *Tetrahedron Letters*, **1970**, 5189.
[17] M. Harfenist and E. Thom, *J. Org. Chem.*, **36**, 1171 (1971).

state leading to the formation of (11).[18] A similar mechanism is involved in the corresponding thermal reorganization of aryl thionocarboxylates.[19] Aryl hydrazonates (13) afford aryl hydrazides (14) when heated; since the reaction rate is accelerated by electron-withdrawing substituents, R, in the migrating aromatic ring, the reaction is probably analogous to the Chapman rearrangement.[20] Above pH 7 the s-triazinyl group of (15)

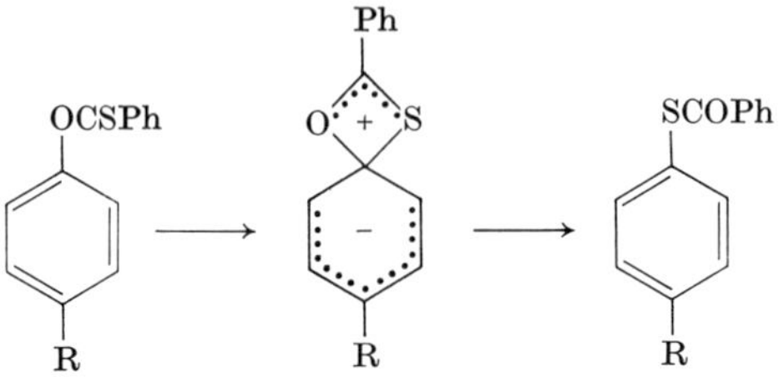

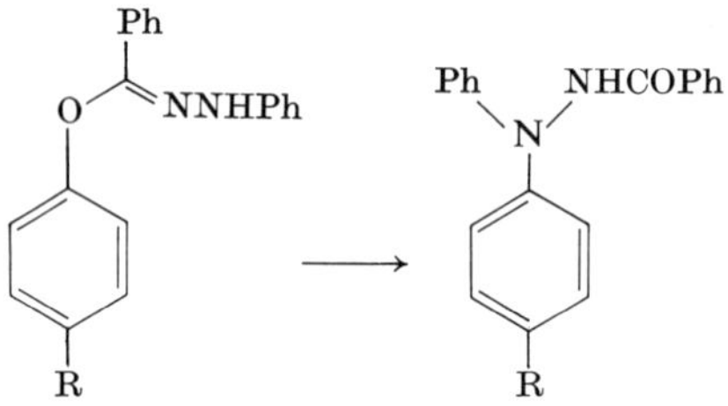

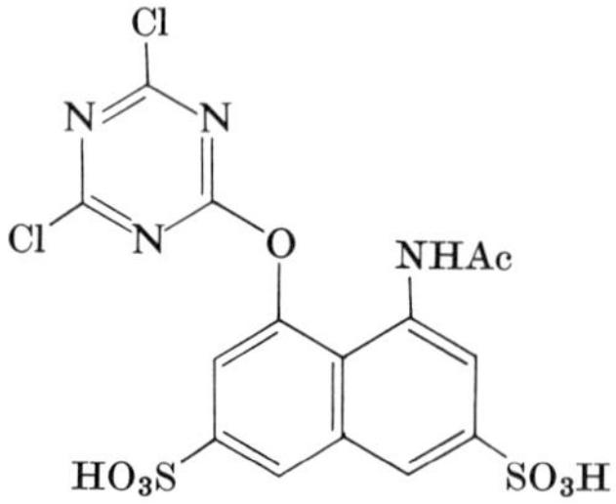

migrates to the amide-N atom which then undergoes deacetylation.[21] Certain N-aryloxypyridinium salts undergo interesting base-catalysed reorganizations;[22] thus (16) affords (18) in the presence of triethylamine, possibly via (17).

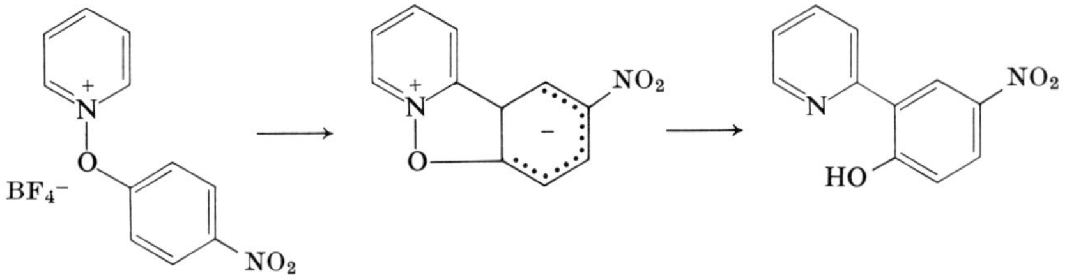

[18] Y. Araki and A. Kaji, Bull. Chem. Soc. Japan, **43**, 3214 (1970).
[19] A. Kaji, Y. Araki, and K. Miyazaki, Bull. Chem. Soc. Japan, **44**, 1393 (1971).
[20] A. F. Hegarty, J. A. Kearney, M. P. Cashman, and F. L. Scott, Chem. Comm., **1971**, 689.
[21] R. Budziarek, J. Chem. Soc. (C), **1971**, 74.
[22] R. A. Abramovitch, S. Kato, and G. M. Singer, J. Am. Chem. Soc., **93**, 3074 (1971).

The rearrangements that occur when nitrobenzenesulphenanilides are heated,[23] N-acetyl-2-nitro-2'-(methylthio)diphenylamines are reduced with P(OEt)$_3$,[24] and aryl 2-hydroxyaminoaryl sulphones are treated with NaOH[25] have been examined.

Cyclohexadiene Derivatives

The first-order rate-coefficient for the HClO$_4$-catalysed isomerization of 4-bromo-2,6-di-t-butylcyclohexa-2,5-dienone to 4-bromo-2,6-di-t-butylphenol correlates with H_0 and the reaction exhibits a relatively large primary isotope effect ($k_H/k_D = 4.2$ in acetic acid). These results indicate that the slow step of the reaction involves removal of the 4-proton from the protonated substrate by a neutral base.[26] A similar rate-determining step is probably involved in the acid-catalysed rearrangement of 1-diphenylmethylene-4-tritylcyclohexa-2,5-diene to 1-diphenylmethyl-4-tritylbenzene.[27] Neutral alumina at 70° promotes the isomerization of quinone methides to alkenylphenols; for instance, 2,6-di-t-butyl-4-ethylidenecyclohexa-2,5-dienone is converted into 2,6-di-t-butyl-4-vinylphenol.[28]

The behaviour of alkylcyclohexa-2,5-dienones in acid has been examined;[29] for example, protonated 3-ethyl-4,4-dimethylcyclohexa-2,5-dienone rearranges to 3-ethyl-4,5-dimethylphenol ca. twice as rapidly as protonated 3,3,4-trimethylcyclohexa-2,5-dienone affords 3,4,5-trimethylphenol. For the first time a spiran intermediate has been isolated from the rearrangement of a bicyclic dienone; for instance, (**20**) is obtained when (**19**) is treated with BF$_3$.[30] Such species are implicated in the aromatizations of a number of steroidal substrates.[31] In certain cases, however, appropriate substitution can lead to a change in the aromatization mechanism; thus, in the presence of HBr (**21**) affords (**23**), probably via the spiran intermediate (**22**), while the 6-oxo derivative of (**21**) yields (**24**).[32] Evidently, the carbonyl group inhibits carbonium ion formation at position 5 and aromatization proceeds by methyl migration. A number of other aromatizations are accompanied by methyl migrations.[33] The reaction of (**25**) with HF–SbF$_5$ to give (**26**) provides an interesting example of a steroidal dearomatization.[34]

Isomerization of 3-allyl-1,2,3,4,5-pentamethyl-6-methylenecyclohexa-1,4-diene to 4-pentamethylphenylbut-1-ene occurs at room temperature; a homolytic pathway is implicated on the basis of D-labelling experiments and the observation that the reaction is slowed down by radical inhibitors.[35] 1-Allyl-2,4-di-t-butyl-3,6-dimethylbenzene is the initial product when 4-allyl-2,6-di-t-butyl-4,6-dimethylcyclohexa-2,5-dien-1-ol and

---

[23] F. A. Davis, R. B. Wetzel, T. J. Devon, and T. F. Stackhouse, *J. Org. Chem.*, **36**, 799 (1971).
[24] Y. Maki, T. Hosokami, and H. Suzuki, *Tetrahedron Letters*, **1971**, 3509.
[25] M. F. Grundon, D. J. Maitland, and W. L. Matier, *J. Chem. Soc.* (C), **1971**, 654.
[26] P. B. D. de la Mare, A. Singh, J. G. Tillett, and M. Zeltner, *J. Chem. Soc.* (B), **1971**, 1122.
[27] H. Takeuchi, T. Nagai, and N. Tokura, *Bull. Chem. Soc. Japan*, **44**, 753 (1971).
[28] D. Braun and B. Meier, *Angew. Chem. Int. Ed.*, **10**, 566 (1971).
[29] K. L. Cook and A. J. Waring, *Tetrahedron Letters*, **1971**, 1675, 3359.
[30] G. F. Burkinshaw, B. R. Davis, E. G. Hutchinson, P. D. Woodgate, and R. Hodges, *J. Chem. Soc.* (C), **1971**, 3002.
[31] T. Wolff and H. Dannenberg, *Tetrahedron*, **27**, 3417 (1971); J. Libman and Y. Mazur, *Chem. Comm.*, **1971**, 730, 1146; J. R. Hanson, *Chem. Comm.*, **1971**, 1343; J. R. Hanson and T. D. Organ, *J. Chem. Soc.* (C), **1971**, 1313.
[32] J. R. Hanson, *Chem. Comm.*, **1971**, 1119.
[33] D. F. Lindow and R. G. Harvey, *J. Am. Chem. Soc.*, **93**, 3786 (1971); J. Meney, Y.-H. Kim, R. Stevenson, and T. N. Margulis, *Chem. Comm.*, **1970**, 1706; W. Brown and A. B. Turner, *J. Chem. Soc.* (C), **1971**, 2566.
[34] J. P. Gesson, J. C. Jacquesy, and R. Jacquesy, *Tetrahedron Letters*, **1971**, 4733.
[35] B. Miller and K.-H. Lai, *Tetrahedron Letters*, **1971**, 1617.

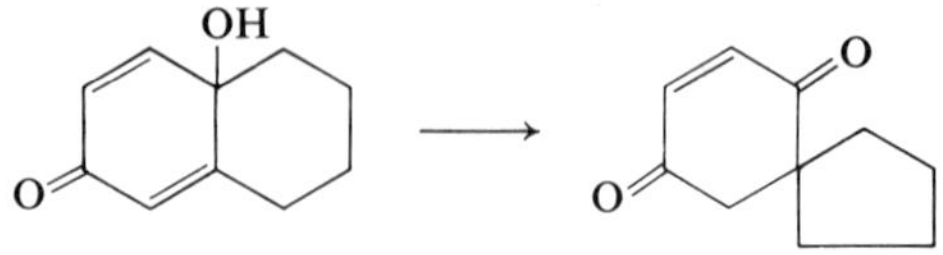

(19) (20)

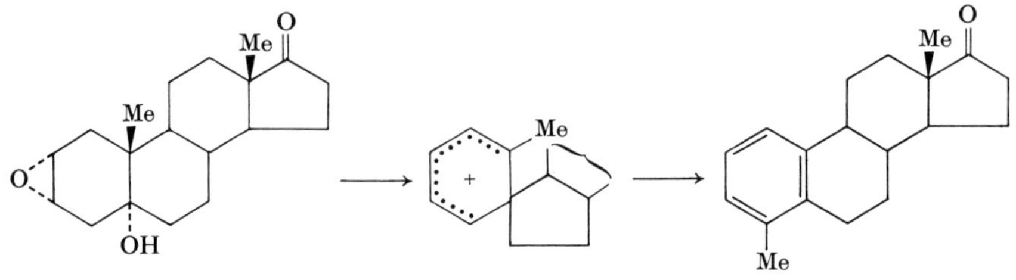

(21) (22) (23)

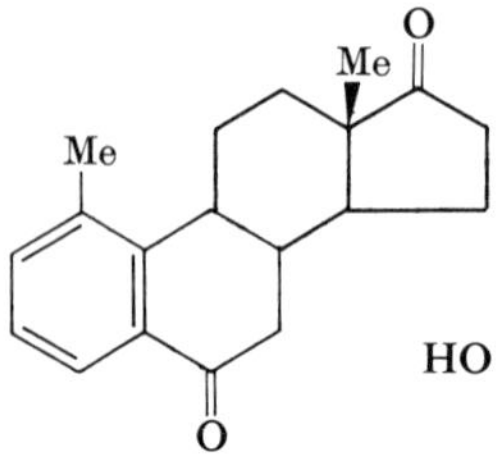

(24) 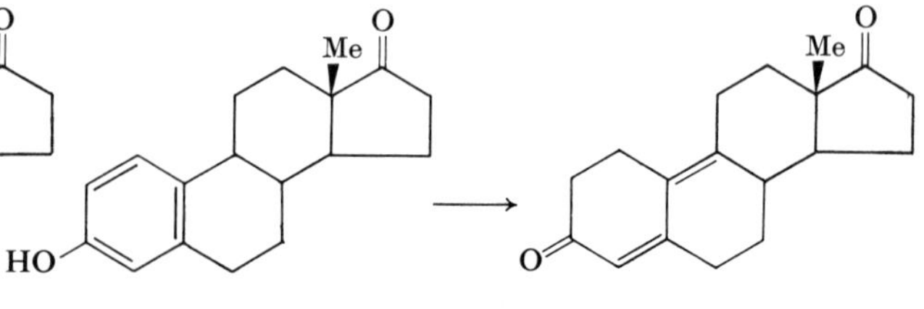 (25) (26)

3-allyl-1,5-di-t-butyl-3-methyl-6-methylenecyclohexa-1,4-diene are treated with acid; under more forcing conditions, when the product loses a t-butyl group, methyl migration surprisingly represents the major reaction pathway.[36]

Sigmatropic Rearrangements

Several reviews dealing with pericyclic processes, which present an alternative theoretical approach to that of Woodward and Hoffmann, have been published.[37] Narcissistic reactions have also been discussed.[38]

[3,3]-*Migrations*

Claisen and Related Rearrangements. The thermal conversion of tetrahydro-2-methyl-5-methylene-2-vinylfuran into 4-methylcyclohept-4-enone is a first-order reaction; the

---

[36] K.-H. Lai and B. Miller, *Tetrahedron Letters*, **1971**, 3575.
[37] K. Fukui, *Fortschr. Chem. Forsch.*, **15**, 3 (1970); *Accounts Current Chem. Res.*, **4**, 57 (1971); R. G. Pearson, *Accounts Current Chem. Res.*, **4**, 152 (1971); H. E. Zimmerman, *Accounts Current Chem. Res.*, **4**, 272 (1971); M. J. S. Dewar, *Angew. Chem. Int. Ed.*, **10**, 761 (1971); C. W. Jefford and V. Burger, *Chimia*, **25**, 297 (1971).
[38] L. Salem, *Accounts Current Chem. Res.*, **4**, 322 (1971).

activation parameters indicate that a concerted [3,3]-sigmatropic shift is involved.[39] When heated, (**27**) was believed to afford the bicyclo[2.1.0]pentene (**28**); however, the latter formulation is shown to be incorrect and (**29**) is the product.[40] It therefore appears that the Claisen equilibrium between (**28**) and (**29**) lies entirely on the side of the enol ether. The isomerization of 7,8-diethyl-1,3-dimethyl-1-oxabicyclo[4.2.0]octa-3,7-dien-5-one to 2-acetyl-3,4-diethyl-5-methylphenol probably involves the intermediacy of 3-acetyl-4,5-diethyl-6-methylbicyclo[2.2.0]hex-5-en-2-one, formed by a Claisen rearrangement.[41]

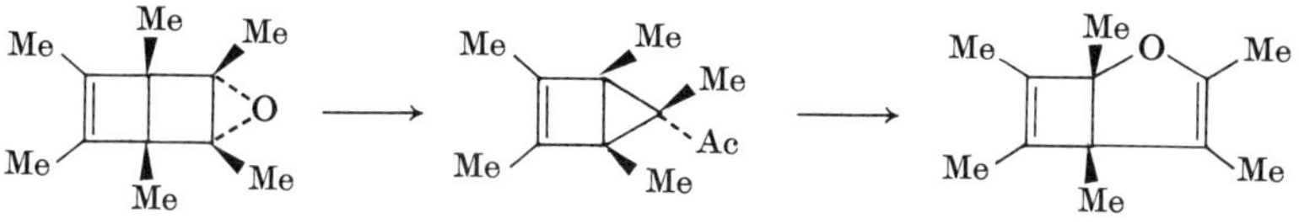

(**27**) (**28**) (**29**)

6-Allylcyclohexa-2,4-dienones are well-established intermediates in the thermal reorganization of O-allylphenols. In certain cases it is possible to trap such species by intramolecular [4 + 2]-cycloadditions since (**31**) is obtained when (**30**) is heated in decalin.[42] Tropolone ethers provide related examples; thus, (**34**) is produced when (**32**) is heated, presumably by the intervention of (**33**).[43] When allyl pentafluorophenyl ether is pyrolysed it is possible to isolate 4-allyl-2,3,4,5,6-pentafluorocyclohexa-2,5-dienone.[44] 4,6-Disubstituted 6-allylcyclohexa-2,4-dienones undergo reverse Claisen rearrangements when heated;[45] for example, (**35**) affords (**36**).

A number of reactions that probably involve [3,3]-sigmatropic rearrangements of allyl vinyl ethers generated *in situ* have been investigated. For example, N,N-dimethylpent-4-enamides are formed from allylic alcohols and either cyano-N,N-dimethylaminoacetylene[46] or 1-methoxy-1-(N,N-dimethylamino)prop-1-ene.[47] The intermediates, in which the keten O,N-acetal double bonds possess the Z-configuration, probably rearrange via chair-like transition states[47,48] since N,N-dimethyl-*erythro*-2,3-dimethylpent-4-enamide is obtained from *trans*-but-2-enol and 1-methoxy-1-(N,N-dimethylamino)prop-1-ene. Sigmatropic shifts are also implicated in the reactions of 3-hydroxyhex-1-en-4-ynes with vinyl ethers[49] and ethyl Z-β-(benzylamino)crotonate with crotonyl chloride.[50]

---

[39] S. J. Rhoads and C. F. Brandenburg, *J. Am. Chem. Soc.*, **93**, 5805 (1971); S. J. Rhoads and J. M. Watson, *J. Am. Chem. Soc.*, **93**, 5813 (1971).
[40] B. C. Baumann, M. Rey, J. Markert, H. Prinzbach, and A. S. Dreiding, *Helv. Chim. Acta*, **54**, 1589 (1971).
[41] J. N. Hanifin and E. Cohen, *J. Org. Chem.*, **36**, 910 (1971).
[42] A. J. Quillinan and F. Scheinmann, *Chem. Comm.*, **1971**, 966.
[43] R. M. Harrison, J. D. Hobson, and M. M. Al Holly, *J. Chem. Soc.* (C), **1971**, 3084.
[44] G. M. Brooke, *Tetrahedron Letters*, **1971**, 2377.
[45] B. Miller, *J. Org. Chem.*, **35**, 4262 (1970); M. F. Ansell and V. J. Leslie, *J. Chem. Soc.* (C), **1971**, 1423.
[46] T. Sasaki, A. Kojima, and M. Ohta, *J. Chem. Soc.* (C), **1971**, 196.
[47] N. Sucrow and W. Richter, *Chem. Ber.*, **104**, 3679 (1971).
[48] W. Sucrow, B. Schubert, W. Richter, and M. Slopianka, *Chem. Ber.*, **104**, 3689 (1971).
[49] P. Cresson, *Compt. Rend.* (C), **273**, 1382 (1971).
[50] P. W. Hickmott and G. Sheppard, *J. Chem. Soc.* (C), **1971**, 2112.

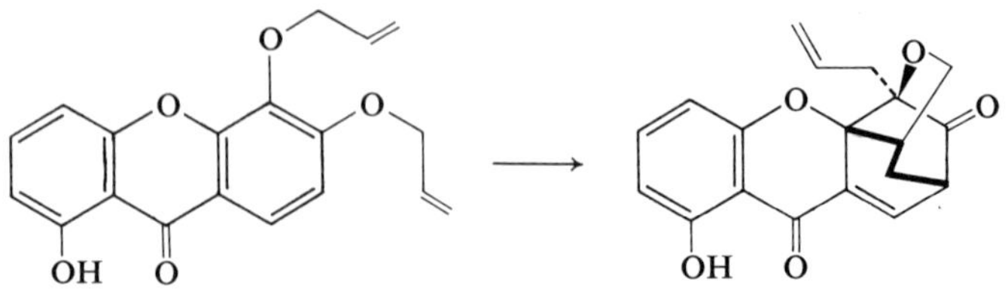

(30) (31)

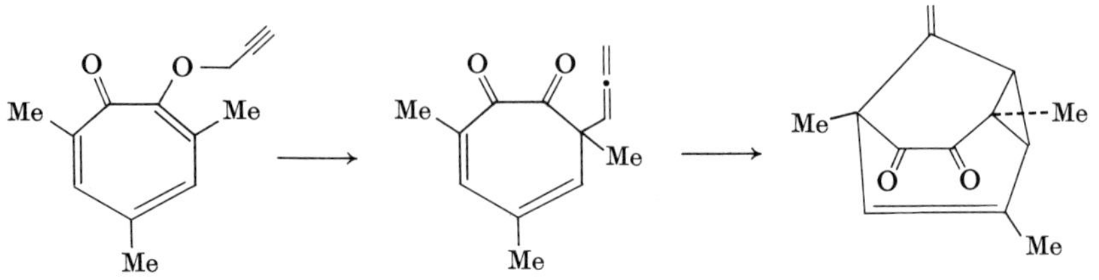

(32) (33) (34)

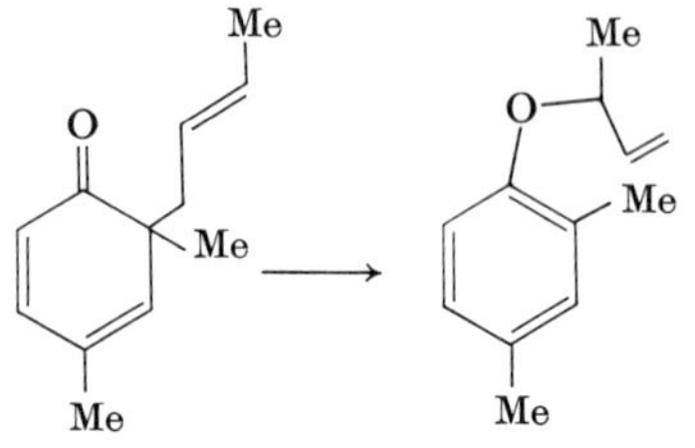

(35) (36)

The thermal behaviour of allyloxy derivatives of indoles,[51] coumarins,[52] uracils,[53] quinolones,[54] cyclohexenones,[55] and propenylbenzene[56] has been examined and the rates of Claisen rearrangement of *para*-substituted allyl phenyl ethers have been determined in nematic and isotropic solvents.[57]

Thio- and amino-Claisen rearrangements have been reviewed.[58] Thermolyses of allyl thienyl sulphides[59] and allyl vinyl sulphones[60] illustrate the former process while the formation of quinolones by heating the adducts of N-allylanilines and dimethyl acetylenedicarboxylate[61] is an example of the latter rearrangement. An intermolecular mechanism is implicated in the $ZnCl_2$-catalysed reorganization of N-(3,3-dimethylallyl)anilines.[62]

Related [3,3]-sigmatropic shifts in which two heteroatoms participate include the

[51] H. Plieninger, H. Sirowej, and D. Rau, *Chem. Ber.*, **104**, 1863 (1971).
[52] M. M. Ballantyne, P. H. McCabe, and R. D. H. Murray, *Tetrahedron*, **27**, 871 (1971); R. D. H. Murray, M. M. Ballantyne, and K. P. Mathai, *Tetrahedron*, **27**, 1247 (1971).
[53] B. A. Otter, A. Taube, and J. L. Fox, *J. Org. Chem.*, **36**, 1251 (1971).
[54] T. R. Chamberlain and M. F. Grundon, *J. Chem. Soc.* (C), **1971**, 910.
[55] Y. Tamura, Y. Kita, M. Shimagaki, and M. Terashima, *Chem. Pharm. Bull. (Japan)*, **19**, 571 (1971).
[56] H. M. Okely and M. F. Grundon, *Chem. Comm.*, **1971**, 1157.
[57] W. E. Bacon and G. H. Brown, *Mol. Cryst. Liq. Cryst.*, **12**, 229 (1971); *Chem. Abs.*, **75**, 48167 (1971).
[58] E. Winterfeldt, *Fortschr. Chem. Forsch.*, **16**, 75 (1971).
[59] J. Z. Mortensen, B. Hedegaard, and S.-O. Lawesson, *Tetrahedron*, **27**, 3831 (1971).
[60] J. F. King and D. R. K. Harding, *Chem. Comm.*, **1971**, 959.
[61] G. Schmidt and E. Winterfeldt, *Chem. Ber.*, **104**, 2483 (1971).
[62] N. Takamatsu, S. Inoue, and Y. Kishi, *Tetrahedron Letters*, **1971**, 4661.

isomerizations of allylic thiocyanates to allylic isocyanates[63] and of 6-acyloxycyclohexa-2,4-dienones to 4-acyloxycyclohexa-2,5-dienes,[64] and the conversion of O-(2-pyridyl)-oximes into 3-(2-oxoalkyl)-2-pyridones.[65] Three heteroatoms are involved in the isomerization of allylic azides[63] and the rearrangement of the adducts derived from amide oximes and dimethyl acetylenedicarboxylate or methyl propiolate;[66] for example, (38) is formed when (37) is heated in diphenyl ether.

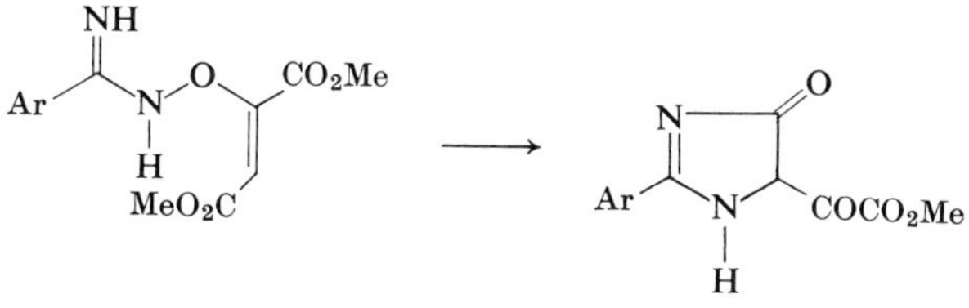

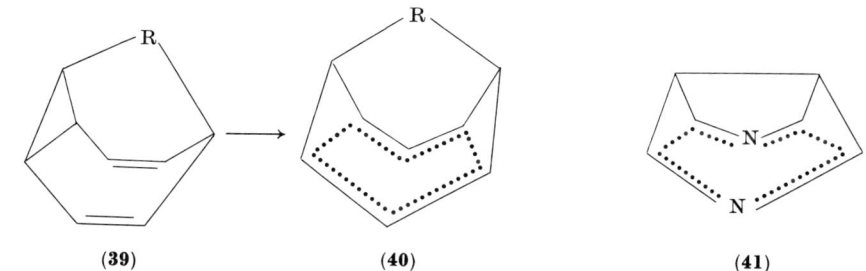

(37)　　　　　　　　　　　　(38)

Cope Rearrangements. There have been further calculations of the course of the Cope rearrangement. The MINDO/2 method accounts for the large rate increase in passing from hexa-1,5-diene to rigid derivatives and predicts that the activation energy for the conversion of (39) into (40) will rapidly diminish as the size of R decreases.[67] Indeed, it has been suggested that in certain semibullvalenes the bishomoaromatic structure will possess a lower free energy than the classical structure; thus, 3,7-diazasemibullvalene is predicted to exist as the bishomopyrazine (41).[68] Calculations using the CNDO/2 method[69] indicate that the six-centre boat-like transition state is favoured in the degenerate Cope rearrangement of cyclopentadiene dimer.

(39)　　　　　　(40)　　　　　　(41)

Whereas (±)-3,4-diphenylhexa-1,5-diene rearranges to *trans,trans*-1,6-diphenylhexa-1,5-diene on heating, the *meso*-isomer affords a mixture of the *cis,trans*- (63%) and the *trans,trans*-diene (37%).[70] Evidently in the latter case the six-centre transition state (in which both phenyl groups occupy boat-equatorial positions) is only slightly higher in energy than the four-centre chair-like counterpart (in which one phenyl substituent

---

[63] R. J. Ferrier and N. Vethaviyaser, *J. Chem. Soc.* (C), **1971**, 1907.
[64] D. H. R. Barton, P. D. Magnus, and M. J. Pearson, *J. Chem. Soc.* (C), **1971**, 2231.
[65] T. Sheradsky and G. Salemnick, *J. Org. Chem.*, **36**, 1061 (1971).
[66] N. D. Heindel and M. C. Chun, *Chem. Comm.*, **1971**, 664; *Tetrahedron Letters*, **1971**, 1439.
[67] M. J. S. Dewar and W. W. Schoeller, *J. Am. Chem. Soc.*, **93**, 1481 (1971).
[68] M. J. S. Dewar, Z. Náhlovská, and B. D. Náhlovský, *Chem. Comm.*, **1971**, 1377.
[69] P. Beltrame, A. Gamba, and M. Simonetta, *Chem. Comm.*, **1970**, 1660.
[70] R. P. Lutz, S. Bernal, R. J. Boggio, R. O. Harris, and M. W. McNicholas, *J. Am. Chem. Soc.*, **93**, 3985 (1971).

is axially oriented). The activation parameters ($\Delta H^{\neq}$ 35.5 kcal mole^{-1}, $\Delta S^{\neq}$ −13.1 cal deg^{-1} mole^{-1}) for the rearrangement of 1,1-dideuteriohexa-1,5-diene are consilient with a concerted reorganization.[71] A [3,3]-sigmatropic shift to give 3,4-diethylhexa-1,2,4,5-tetraene probably represents the initial step in the thermal isomerization of deca-3,7-diene to 1,2-diethyl-3,4-bis(methylene)cyclobut-1-ene.[72] D-Labelling experiments reveal that hexa-1,2-dien-5-yne undergoes a degenerate Cope rearrangement before conversion into 3-methylenehex-1-en-5-yne.[73]

Further examples of molecules that exhibit fluxional behaviour due to rapid Cope rearrangements have been studied.[74] D-Labelling experiments establish that (**42**) is undergoing a degenerate reorganization at room temperature, although at 80° it is converted into (**43**), probably by a diradical pathway.[75] The equilibration of (**44**) and (**45**) reveals that the former derivative is preferred by at least 3.7 kcal mole^{-1} at 140°,[76] suggesting that the system is stabilized by the presence of the π-electron acceptor adjacent to the cyclopropane ring.

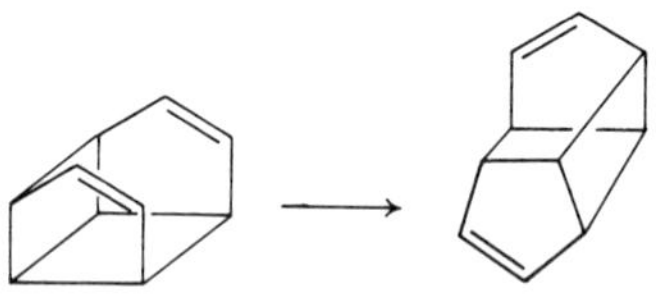

(**42**)   (**43**)

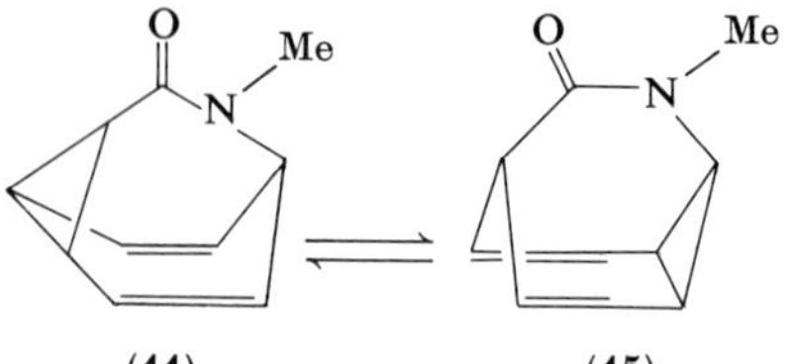

(**44**)   (**45**)

The acid-initiated rearrangement of 4-allyl-4-methylcyclohexa-2,5-dienone tosylhydrazone involves a [3,3]-sigmatropic shift of the allyl group followed by aromatization.[77] A retro-Diels–Alder reaction to give (**46**) is considered to be involved in the conversion of basketene into (**47**);[78] an independent synthesis of (**46**) confirms that it undergoes the Cope rearrangement to (**47**) at 70°.[79] The first step in the thermal transformation of (**48**) into (**50**) was postulated as involving opening of the cyclobutene ring;[80] however, such a route is now eliminated and an initial Cope rearrangement of (**48**) to (**49**) is suggested.[81] A similar mechanism accounts for the conversion of (**51**) into (**52**).[82]

[71] W. von E. Doering, V. G. Toscano, and G. H. Beasley, *Tetrahedron*, **27**, 5299 (1971).
[72] H. A. Brune, H. P. Wolff, and H. Hüther, *Tetrahedron*, **27**, 3949 (1971).
[73] H. Hopf, *Chem. Ber.*, **104**, 1499 (1971).
[74] J. F. M. Oth, E. Machens, H. Röttele, and G. Schröder, *Ann. Chem.*, **745**, 112 (1971).
[75] J. S. McKennis, L. Brener, J. S. Ward, and R. Pettit, *J. Am. Chem. Soc.*, **93**, 4957 (1971).
[76] G. R. Krow and K. C. Ramey, *Tetrahedron Letters*, **1971**, 3141.
[77] M. Schmid, H.-J. Hansen, and H. Schmid, *Helv. Chim. Acta*, **54**, 937 (1971).
[78] *Org. Reaction Mech.*, **1970**, 275.
[79] E. Vedejs, *Chem. Comm.*, **1971**, 536.
[80] *Org. Reaction Mech.*, **1970**, 269.
[81] E. Vedejs, *Tetrahedron Letters*, **1970**, 4963.
[82] L. A. Paquette and J. C. Stowell, *J. Am. Chem. Soc.*, **93**, 5735 (1971).

Bicyclo[3.2.2]nona-2,6,8-triene is equilibrated at 160° with 7-vinylcycloheptatriene, which rapidly isomerizes to the 3-vinyl derivative by a [1,5]-H shift: *cis*-7-vinylnorcaradiene (**54**), formed by Cope rearrangement of the benzo-derivative (**53**) at 270°, affords a dihydrobenzindene by a vinylcyclopropane–cyclopentene reorganization.[83] Similar processes are deemed likely in the thermolysis of 1-methoxy-6,7-bis(methoxycarbonyl)bicyclo[3.2.0]nona-3,6,8-trien-2-one to yield 2-methoxy-1,6-bis(methoxycarbonyl)bicyclo[4.3.0]nona-2,4,7-trien-9-one.[84] Vapour-phase pyrolysis of the *endo*-derivative (**55**) produces a mixture of the *exo*-isomer and (**56**); kinetic and D-labelling studies indicate that the 3,4-bond moves to position 8,[85] possibly representing a [3,3]-sigmatropic shift in which the cyclobutane ring acts in place of a double bond.

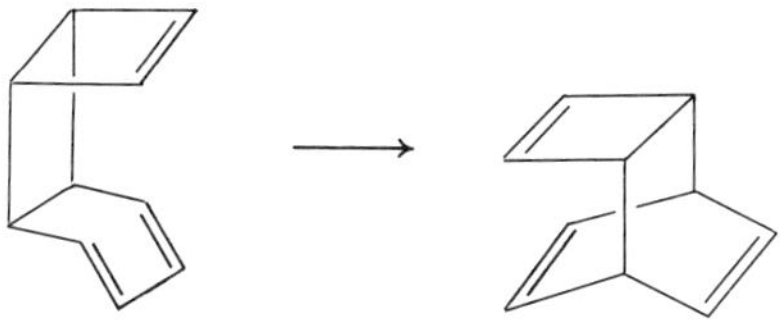

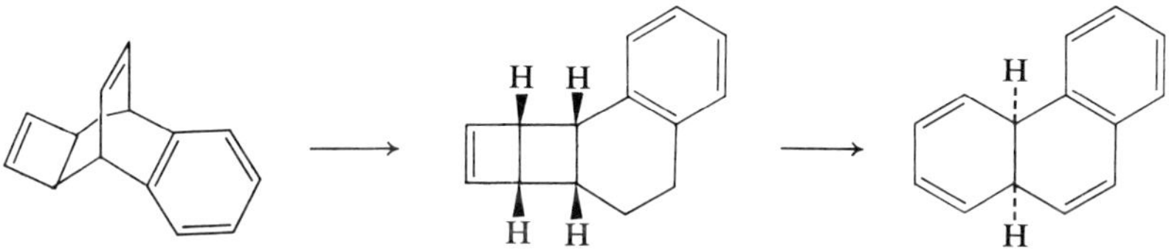

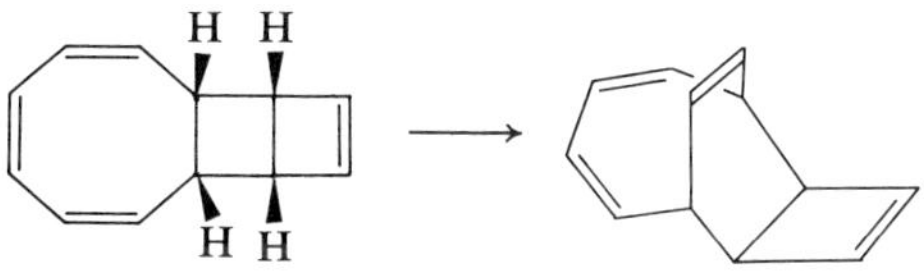

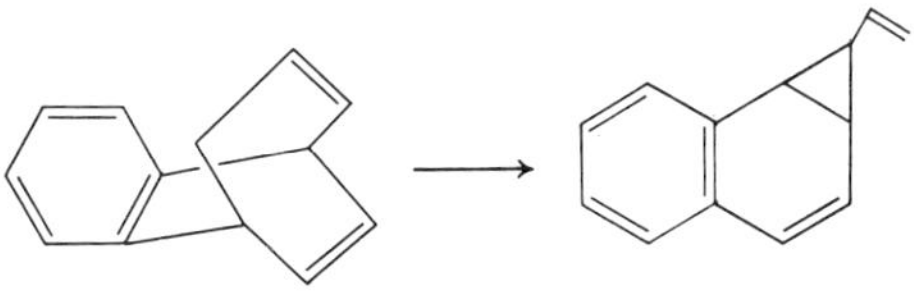

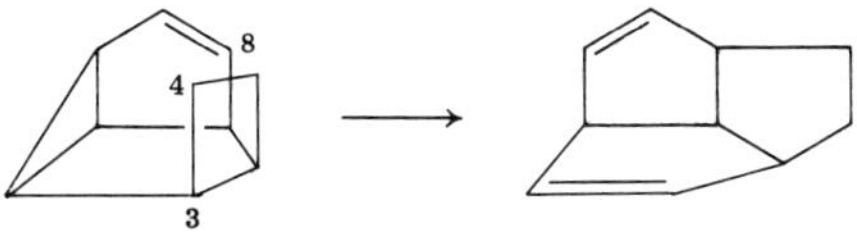

[83] T. Tsuji, H. Ishitobi, and H. Tanida, *Bull. Chem. Soc. Japan*, **44**, 2447 (1971).
[84] T. Uyehara, M. Funamizu, and Y. Kitahara, *Chem. Ind. (London)*, **1971**, 486.
[85] R. Srinivasan, *J. Am. Chem. Soc.*, **92**, 7542 (1970).

A Cope rearrangement of intermediate divinyl episulphones is probably involved in the formation of dihydrothiepin 1,1-dioxides from vinyldiazomethanes and SO_2.[86] The Cope rearrangements of *trans*-divinylcyclopropanes,[87] and of *trans,trans*-[88] and *cis,trans*-cyclodeca-1,5-dienes[89] have been examined.

There have been further studies of the oxy-Cope reaction. For example, the rate constants for the thermolyses of the 3-hydroxy derivatives of hexa-1,5-diene, hex-1-en-5-yne and hexa-1,5-diyne have been measured.[90] The reaction has been utilized to enlarge cyclic derivatives by two or four C atoms[91] and, in certain cases, improved yields result if the hydroxyl group is converted into the trimethylsilyl ether.[92]

[2,3]-*Migrations*

The transition-state geometry for the [2,3]-sigmatropic rearrangement of allylic ammonium ylids has been investigated; thus, (**57**) is rapidly converted into (**58**) at $-50°$, and (**59**) affords (**60**) at $120°$, but (**61**) fails to rearrange.[93] In the last case the participating orbitals of the carbanion and the N atom are orthogonal and, consequently, severe twisting of the allyl group would be necessary for migration to occur. When treated with NaH, the N-allyl-N,N-dimethylammonium iodide of methyl 6β-aminopenicillanate rearranges stereospecifically to methyl 6α-allyl-6β-(N,N-dimethylamino)-penicillanate.[94] A mixture of (**63**) and (**64**) is formed when the quaternary ammonium salt (**62**) is treated with a base, indicating that the ylid undergoes a [2,3]-sigmatropic shift and a cyclization.[95] On oxidation allylic hydrazines yield rearranged allylic azo compounds, probably by way of allylic diazenes; for example, (**66**) is produced when (**65**) is oxidized with $Pb(OAc)_4$.[96]

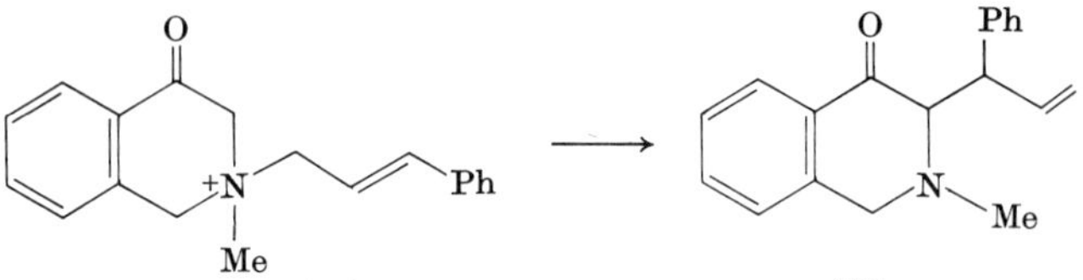

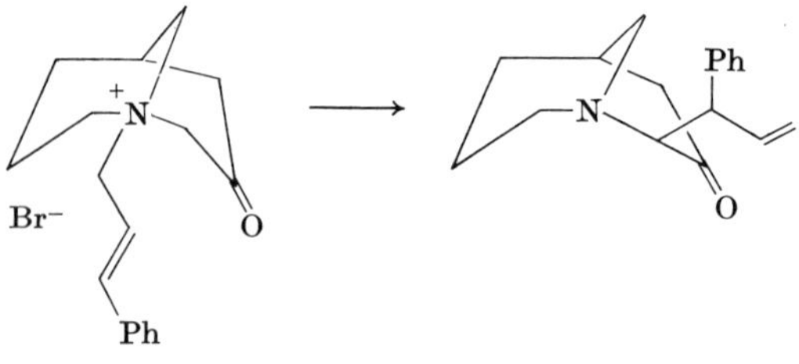

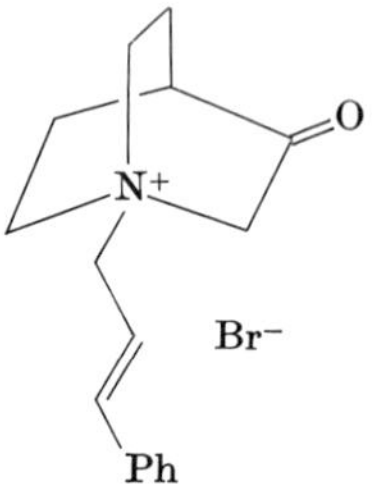

[86] L. A. Paquette and S. Maiorana, *Chem. Comm.*, **1971**, 313.
[87] J. A. Pettus, Jr., and R. E. Moore, *J. Am. Chem. Soc.*, **93**, 3087 (1971).
[88] K. Takeda, K. Tori, I. Horibe, M. Ohtsuru, and H. Minato, *J. Chem. Soc.* (C), **1970**, 2697.
[89] K. Takeda, I. Horibe, and H. Minato, *J. Chem. Soc.* (C), **1970**, 2704; *Chem. Comm.*, **1971**, 88.
[90] A. Viola, J. H. MacMillan, R. J. Proverb, and B. L. Yates, *Chem. Comm.*, **1971**, 936.
[91] R. C. Cookson and P. Singh, *J. Chem. Soc.* (C), **1971**, 1477.
[92] R. W. Thies, *Chem. Comm.*, **1971**, 237.
[93] S. Mageswaran, W. D. Ollis, I. O. Sutherland, and Y. Thebtaranonth, *Chem. Comm.*, **1971**, 1494.
[94] G. V. Kaiser, C. W. Ashbrook, and J. E. Baldwin, *J. Am. Chem. Soc.*, **93**, 2342 (1971).
[95] S. Mageswaran, W. D. Ollis, and I. O. Sutherland, *Chem. Comm.*, **1971**, 1493.
[96] J. E. Baldwin, J. E. Brown, and G. Höfle, *J. Am. Chem. Soc.*, **93**, 788 (1971).

(62) **(63)** **(64)**

(65) **(66)**

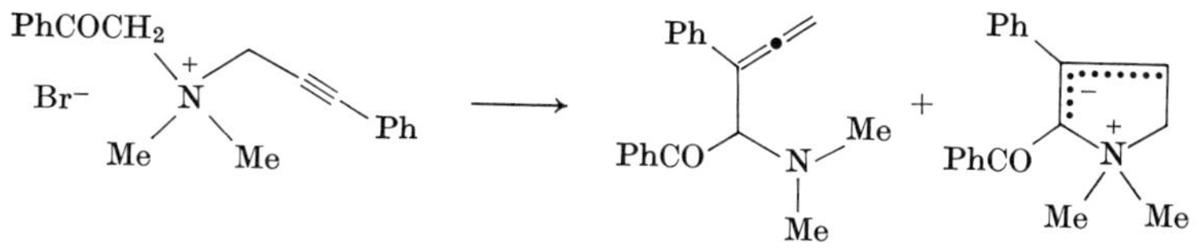

Almost complete stereospecificity is observed in the Bu^{n}Li-initiated conversion of (**67**) into (**68**) and the *cis*-isomer of (**68**), demonstrating that the [2,3]-shift involves the doubly suprafacial transition state.[97]

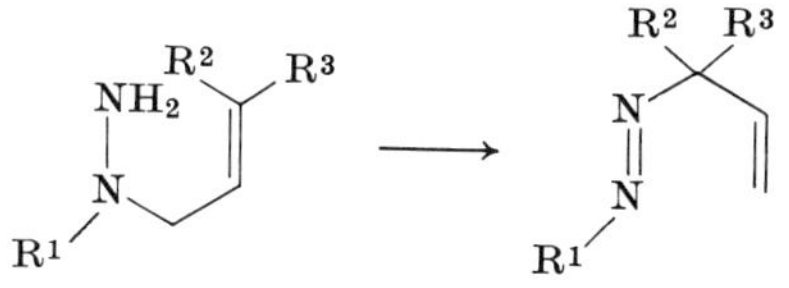

The base-induced reactions of diallyl and allyl benzyl sulphides have been investigated;[98] for example, 2,2-dimethyl-1-methylthio-1-phenylbut-3-ene is produced when benzyl 3,3-dimethylallyl sulphide is treated with Bu^{n}Li and methyl iodide.

Attempts to isolate the stabilized ylid derived from (**69**) have been unrewarding; in the presence of Bu^{n}Li, (**69**) rearranges to 2,6-di-t-butyl-4-(1,1-dimethylallyl)-4-(methylthio)cyclohexa-2,5-dienone even at $-40°$.[99] Moreover, the product of the reaction of

(67) **(68)**

(69)

(70) **(71)** **(72)**

(73)

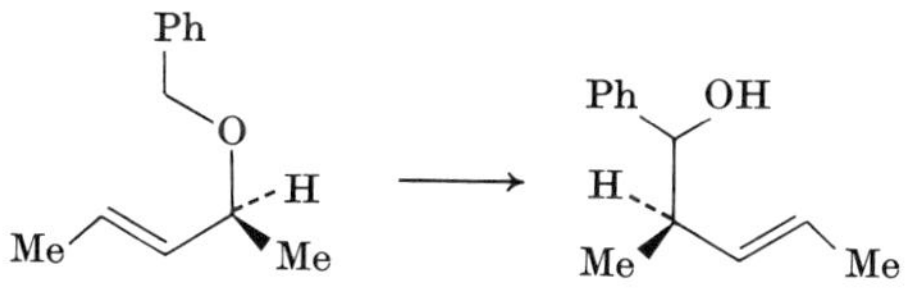

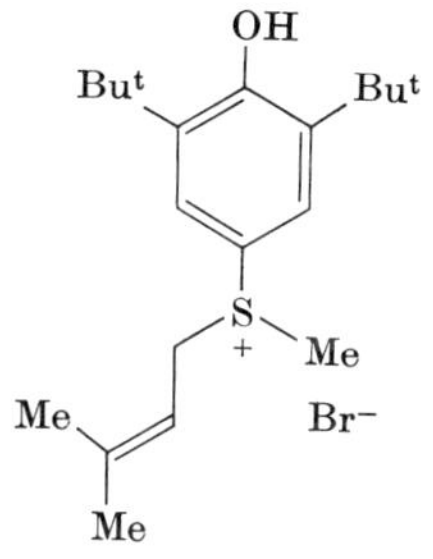

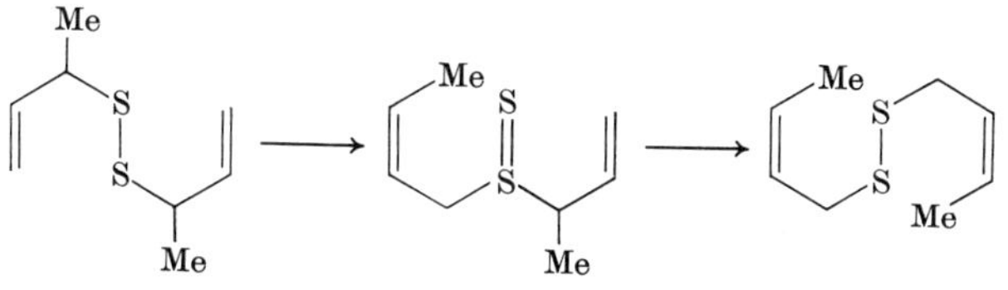

97 J. E. Baldwin and J. E. Patrick, *J. Am. Chem. Soc.*, **93**, 3556 (1971).
98 J. F. Biellmann and J. B. Ducep, *Tetrahedron Letters*, **1971**, 33; V. Rautenstrauch, *Helv. Chim. Acta*, **54**, 739 (1971).
99 J. E. Baldwin and W. F. Erickson, *Chem. Comm.*, **1971**, 359.

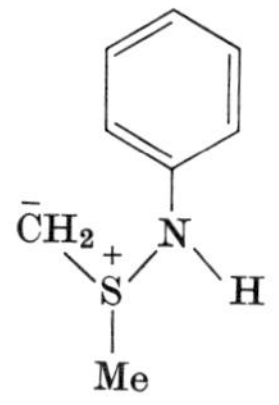

[2,3-bis(methoxycarbonyl)prop-2-enyl]dimethylsulphonium bromide with mild base is not the stabilized ylid but is tetramethyl 1-(methylthio)hexa-1,5-diene-2,3,4,5-tetracarboxylate.[100] Reduction of the S—O bonds of allylic sulphenates with P(OMe)$_3$ provides a useful procedure for the conversion of allylic sulphoxides, which are in thermal equilibrium with allylic sulphenates, into rearranged allylic alcohols.[101] While alkyl allyl disulphides are thermally stable, α-substituted allylic disulphides rearrange quantitatively at room temperature with double allylic inversion. The activation parameters ($\Delta H^{\ne}$ 19.8 kcal mole^{-1}, $\Delta S^{\ne}$ −8.9 cal deg^{-1} mole^{-1}) for the conversion of (**70**) into (**72**) are consilient with the intermediacy of (**71**).[102]

In the presence of triethylamine N-aryl-S,S-dimethylsulphimides isomerize to o-(methylthiomethyl)arylamines, probably by a [2,3]-sigmatropic shift of the ylid (**73**).[103]

[1,3]-*Migrations*

According to orbital-symmetry rules, thermal [1,3]-C shifts should occur with inversion of configuration. However, in the thermal reorganization of 3,3-dicyano-2-methyl-4-phenylpent-1-ene to the 1,1-dicyano-isomer the 1-phenylethyl group migrates with at least 95% retention of configuration.[104] The strongly polarized nature of the allyl group may account for this anomalous behaviour.

Two consecutive suprafacial [1,3]-shifts have been postulated to account for the thermal behaviour of quinuclidine derivatives;[105] for example, (**74**) is transformed into (**75**) by heat. The gas-phase pyrolysis of bicyclo[3.2.0]hept-2-ene has been scrutinized: one mode of reaction involves its isomerization to bicyclo[2.2.1]hept-2-ene by a [1,3]-C shift although it is not certain if the process is concerted.[106]

A number of [1,3]-shifts involving norbornadiene derivatives have been reported. For example, vapour-phase thermolysis of benzonorbornadiene yields 1,2-benzotropilidene; D-labelling studies are concordant with the intervention of benzonorcaradiene.[107] A similar intermediate is probably involved in the rearrangement of (**76**) to (**77**).[108] 7-Acetoxy- and 7-hydroxy-norbornadiene are converted into 7-methoxycycloheptatriene in the presence of methanolic NaOMe; it has not been established when ether formation occurs but a [1,3]-sigmatropic C shift is implicated.[109] A study of the thermolysis of the 9-*exo*-cyano-9-*endo*-methyl- and 9-*endo*-cyano-9-*exo*-methyl derivatives of *cis*-bicyclo-[6.1.0]nona-2,4,6-triene suggests that a homolytic pathway is involved in the [1,3]-C shift.[110]

There have been further investigations of the vinylcyclopropane–cyclopentene reorganization. A concerted mechanism is not operating in the thermolysis of the isomeric

---

[100] J. E. Baldwin, J. A. Walker, A. J. H. Labuschagne, and D. F. Schneider, *Chem. Comm.*, **1971**, 1382.
[101] D. A. Evans, G. C. Andrews, and C. L. Sims, *J. Am. Chem. Soc.*, **93**, 4956 (1971).
[102] G. Höfle and J. E. Baldwin, *J. Am. Chem. Soc.*, **93**, 6307 (1971).
[103] P. Claus, W. Vycudilik, and W. Rieder, *Monatsh.*, **102**, 1571 (1971).
[104] R. C. Cookson and J. E. Kemp, *Chem. Comm.*, **1971**, 385.
[105] J. Dolby, R. Dahlbom, K.-H. Hasselgren, and J. L. G. Nilsson, *Acta Chem. Scand.*, **25**, 735 (1971); K.-H. Hasselgren, J. Dolby, and J. L. G. Nilsson, *Tetrahedron Letters*, **1971**, 2917; J. Dolby, K.-H. Hasselgren, and J. L. G. Nilsson, *J. Heterocyclic Chem.*, **8**, 663 (1971).
[106] A. T. Cocks and H. M. Frey, *J. Chem. Soc.* (A), **1971**, 2564.
[107] M. Pomerantz, T. H. Witherup, and W. C. Schumann, *J. Org. Chem.*, **36**, 2080 (1971).
[108] R. Criegee and B. Bastani, *Chem. Ber.*, **103**, 3942 (1970); J. Ipaktschi, *Chem. Ber.*, **103**, 3944 (1970)
[109] B. Franzus, W. C. Baird, Jr., R. E. Felty, J. C. Smith, and M. L. Scheinbaum, *Tetrahedron Letters* **1971**, 295.
[110] F.-G. Klärner, *Tetrahedron Letters*, **1971**, 3611.

ethyl 2-methyl-3-(*trans*-propenyl)cyclopropanecarboxylates; the results favour the diradical route.[111] 1-Isopropyl-4-methylbicyclo[3.1.0]hex-3-ene undergoes a degenerate rearrangement when heated; D-labelling experiments suggest that there are two competitive processes leading to homolysis of the cyclopropane ring, one involving retention and the other inversion of the original conformation.[112] A common diradical intermediate is implicated in the conversion of ethyl bicyclo[3.1.0]hex-2-ene-6-*endo*-carboxylate into an equilibrium mixture of the 6-*exo*-isomer and ethyl bicyclo[3.1.0]hex-2-ene-4-carboxylate.[113] However, the *exo*- and *endo*-isomers of 1,6-diphenylbicyclo[3.1.0]hex-4-ene display contrasting behaviour: thus, (**79**) equilibrates with (**80**) while (**81**) affords a mixture of (**83**), (**78**), and (**80**); since the first product (**83**) is only obtained in 7% yield, the radical (**82**), formed from (**81**), shows a strong tendency to flip to (**79**) before bond reformation.[114]

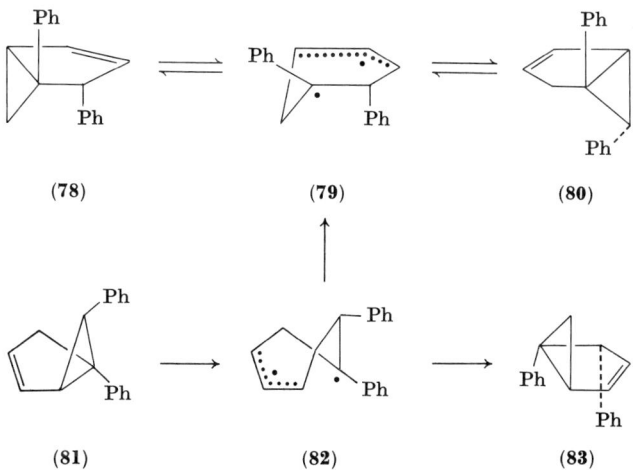

[111] P. H. Mazzocchi and H. J. Tamburin, *J. Am. Chem. Soc.*, **92**, 7220 (1970).
[112] W. von E. Doering and E. K. G. Schmidt, *Tetrahedron*, **27**, 2005 (1971).
[113] R. A. Clark, *Tetrahedron Letters*, **1971**, 2279.
[114] J. S. Swenton and A. Wexler, *J. Am. Chem. Soc.*, **93**, 3066 (1971).

Interest in methylenecyclopropane rearrangements continues and the experimental evidence[115] points to the intervention of non-planar diradicals. Calculations reinforce this picture since the ground state of singlet trimethylene is predicted as possessing an open-shell structure with one methylene group orthogonal.[116] Thermolysis of allylidenecyclopropanes yields methylenecyclopent-2-enes probably by way of non-planar diradicals; there are some differences in the products obtained when (84) and (86) are heated at 120° although in each case the major derivative is (85).[117] Diradical intermediates are also implicated in the isomerization of 2-phenylalkenylidenecyclopropanes to 3-phenylbis(methylene)cyclopropanes.[118] However, concerted processes have been postulated for the rearrangements of (dimethylvinylidene)benzobicyclo[n.1.0]alkenes; thus, the conversion of (87) into (88) by low-pressure vapour-phase pyrolysis may take place by a thermally allowed ($\pi^2a + \sigma^2a + \sigma^2s$) process involving disrotation, double-bond formation, and H-transfer.[119]

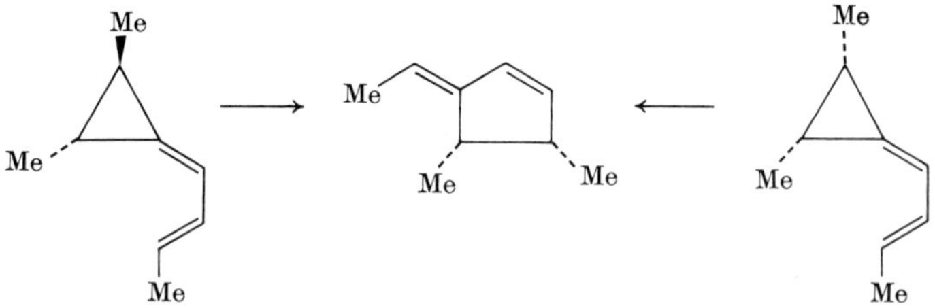

The degenerate rearrangements of methylene- and 1,2-dimethylene-cyclobutanes have been reviewed,[120] and the kinetics of the gas-phase equilibration of ethylidenecyclobutane and 2-methylmethylenecyclobutane have been studied.[121]

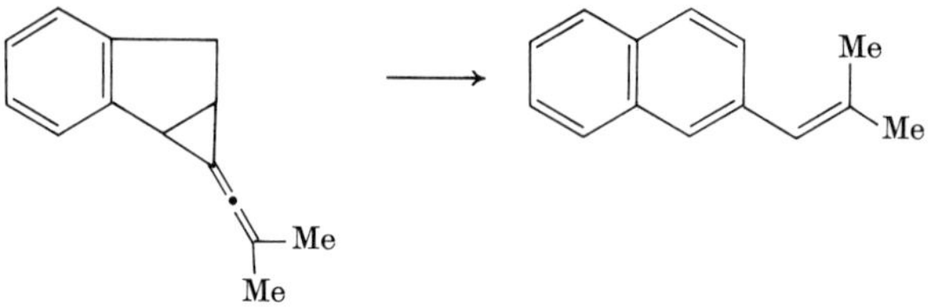

[1,5]-*Migrations*

Kinetic studies indicate that the vapour-phase rearrangement of *cis*-2,3-dimethylpenta-1,3-diene to 3,4-dimethylpenta-1,3-diene involves a [1,5]-H shift.[122] Similar migrations

[115] *Org. Reaction Mech.*, **1970**, 258; J. J. Gajewski, *J. Am. Chem. Soc.*, **93**, 4450 (1971); W. R. Dolbier, Jr., K. Akiba, J. M. Riemann, C. A. Harmon, M. Bertrand, A. Bezaguet, and M. Santelli, *J. Am. Chem. Soc.*, **93**, 3933 (1971).
[116] M. J. S. Dewar and J. S. Wasson, *J. Am. Chem. Soc.*, **93**, 3081 (1971).
[117] W. R. Roth and T. Schmidt, *Tetrahedron Letters*, **1971**, 3639.
[118] T. B. Patrick, E. C. Haynie, and W. J. Probst, *Tetrahedron Letters*, **1971**, 423.
[119] I. H. Sadler and J. A. G. Stewart, *Chem. Comm.*, **1970**, 1588.
[120] J. E. Baldwin and R. H. Fleming, *Fortschr. Chem. Forsch.*, **15**, 281 (1970).
[121] M. C. Flowers and A. R. Gibbons, *J. Chem. Soc.* (B), **1971**, 362.
[122] H. M. Frey, A. M. Lamont, and R. Walsh, *J. Chem. Soc.* (A), **1971**, 2642.

are probably involved in the isomerizations of other diene derivatives.[123] The vacuum-pyrolysis of (89) to give (91) is postulated as proceeding by cyclization of (90); the keten may arise from (89) by a [1,5]-H shift.[124] Details of related rearrangements of penicillin sulphoxides,[125] in which the sulphoxide function acts as the diene component, have been published. The activation energy of such rearrangements is dramatically lowered in the case of (92), which affords (93) at 35°; the latter derivative rapidly equilibrates with the thiosulphoxylate (94) by a [2,3]-sigmatropic shift.[126] Aziridine N-oxides undergo similar [1,5]-H shifts under even milder conditions; thus, 1-t-butyl-2-methylaziridine 1-oxide yields N-allyl-N-t-butylhydroxylamine at −30°.[127] Evidently relief of ring strain provides a substantial driving force to such reactions.

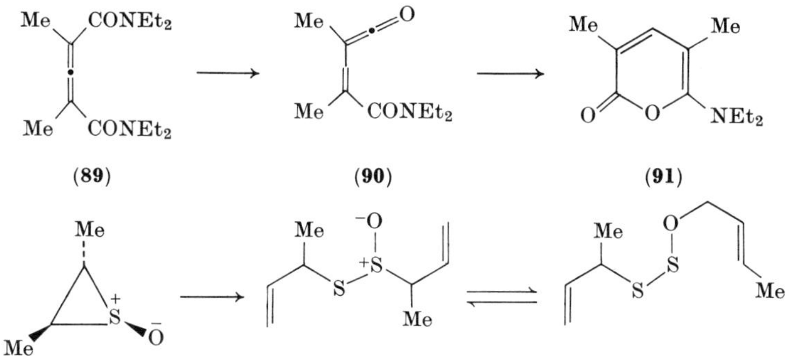

The abnormal Claisen rearrangement, which involves homodienyl [1,5]-H shifts, has been reviewed[128] and further examples of the reaction have been studied.[54-56] Homodienyl [1,5]-H shifts are also implicated in the thermal transformations of chrysanthemumdicarboxylic acid[129] and cis- and trans-13-p-nitrobenzoyl-13-azabicyclo[10.1.0]-tridecanes.[130]

There have been further studies of [1,5]-C shifts. The activation parameters for the pyrolytic rearrangement of 5,5-dimethylcyclopentadiene have been measured and the results indicate that a concerted [1,5]-methyl shift is involved.[131] Similar alkyl shifts are entailed in the thermal reorganizations of alkylated palladium and nickel corroles.[132] A sigmatropic shift of an ester group is postulated to occur in the conversions of dimethyl 4-acetoxybicyclo[3.1.0]hex-3-ene-6,6-dicarboxylate[133] and dimethyl 3-acetoxycyclohexene-4,4-dicarboxylate[134] into dimethyl isophthalate. [1,5]-Aryl shifts have been

---

[123] T. Sasaki, S. Eguchi, and H. Yamada, *Tetrahedron Letters*, **1971**, 99; E. J. Corey and D. K. Herron, *Tetrahedron Letters*, **1971**, 1641; K. W. Egger and T. Ll. James, *J. Organomet. Chem.*, **26**, 335 (1971).
[124] J. Ficini, J. Pouliquen, and J.-P. Paulme, *Tetrahedron Letters*, **1971**, 2483.
[125] *Org. Reaction Mech.*, **1970**, 264; D. H. R. Barton, F. Comer, D. G. T. Greig, P. G. Sammes, C. M. Cooper, G. Hewitt, and W. G. E. Underwood, *J. Chem. Soc.* (C), **1971**, 3540.
[126] J. E. Baldwin, G. Höfle, and S. C. Choi, *J. Am. Chem. Soc.*, **93**, 2810 (1971).
[127] J. E. Baldwin, A. K. Bhatmagar, S. C. Choi, and T. J. Shortridge, *J. Am. Chem. Soc.*, **93**, 4082 (1971).
[128] H. J. Hansen, *Mech. Mol. Migr.*, **3**, 177 (1971).
[129] L. Crombie, C. F. Doherty, G. Pattenden, and D. K. Woods, *J. Chem. Soc.* (C), **1971**, 2739.
[130] I. J. Burnstein, P. E. Fanta, and B. S. Green, *J. Org. Chem.*, **35**, 4084 (1970).
[131] S. McLean and D. M. Findlay, *Can. J. Chem.*, **48**, 3107 (1970).
[132] R. Grigg, A. W. Johnson, and G. Shelton, *J. Chem. Soc.* (C), **1971**, 2287; *Ann. Chem.*, **746**, 32 (1971).
[133] J. A. Berson and R. G. Salomon, *J. Am. Chem. Soc.*, **93**, 4620 (1971).
[134] R. A. Baylouny, *J. Am. Chem. Soc.*, **93**, 4621 (1971).

noted in the reduction and thermolysis of substituted indenes[135] and in the thermal rearrangement of the piperidine enamine of Pummerer's ketone.[136]

The Me_3Si group undergoes [1,5]-migrations with extreme ease and, on the basis of NMR spectroscopy, 5-trimethylsilylcyclopentadiene is a fluxional moleule.[137] However, this behaviour is not displayed in DMSO solution, since the trimethylsilyl derivative decomposes to the ylid (**95**).[138] According to NMR spectroscopy, which indicates that the methoxy and methoxycarbonyl groups are equivalent, (**96**) is undergoing rapid [1,5]-shifts of the Me_3Si group even at $-60°$.[139] A similar situation prevails with 4-(trimethylsilyloxy)pent-3-en-2-one[140] while, in the case of the 7-triphenyltin derivative of cycloheptatriene, rapid [1,5]-shifts of the Ph_3Sn group are implicated.[141]

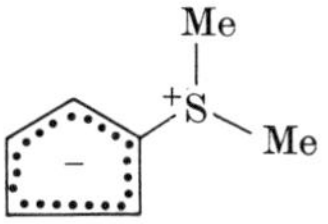

The acid-catalysed rearrangement of 1,1,1-trichloro-2-hydroxyalkan-4-ones to 1,1,5-trichloroalk-1-en-4-ones is postulated as involving the intermediacy of (**97**), which undergoes a sigmatropic [1,5]-Cl shift.[142]

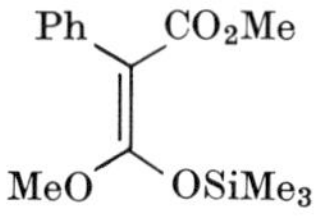

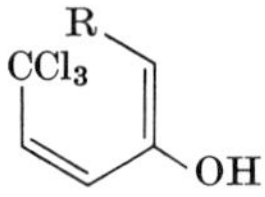

Electrocyclic Rearrangements

Valence isomerizations of heterocyclic systems[143] and electrocyclic reactions of 5-membered heterocycles[144] have been reviewed.

Calculations using the MINDO/2 procedure indicate that the cyclopropyl cation opens in a disrotatory manner and the cyclopropyl anion in a conrotatory fashion,[145] in agreement with the EH method. However, in contrast to the latter method, the SCF calculations predict that the cyclopropyl radical opens in a disrotatory manner.[145,146] The activation-energy differences between the aromatic and the antiaromatic ring-openings of the cyclopropyl cation, anion and radical are estimated to be 30.6, 35.0, and 27.6 kcal mole^{-1}, respectively. In retracing the ring-closures by the forbidden route, cyclization of the allyl anion and radical occurred by the reverse pathway but the allyl cation failed to cyclize. Consequently, the potential surface of a reaction that involves an antiaromatic transition state need not possess a continuous valley between the reactant and product.[147] Indeed, two distinct minimum-energy geometries, that are separated by a maximum, are available for the transition state of such a reaction;

---

[135] L. L. Miller and R. F. Boyer, *J. Am. Chem. Soc.*, **93**, 646, 650 (1971).
[136] D. Beck and K. Schenker, *Helv. Chim. Acta*, **54**, 734 (1971).
[137] *Org. Reaction Mech.*, **1970**, 265.
[138] S. McLean and G. W. B. Reed, *Can. J. Chem.*, **48**, 3110 (1970).
[139] Y.-N. Kuo, F. Chen, and C. Ainsworth, *Chem. Comm.*, **1971**, 137.
[140] H. Shanan-Atidi and Y. Shvo, *Tetrahedron Letters*, **1971**, 603.
[141] R. B. Larrabee, *J. Am. Chem. Soc.*, **94**, 1510 (1971).
[142] E. Kiehlmann, P.-W. Loo, B. C. Menon, and N. McGillivray, *Can. J. Chem.*, **49**, 2964 (1971).
[143] L. A. Paquette, *Angew. Chem. Int. Ed.*, **10**, 11 (1971).
[144] J. Elguero, *Bull. Soc. Chim. France*, **1971**, 1925.
[145] M. J. S. Dewar and S. Kirschner, *J. Am. Chem. Soc.*, **93**, 4290 (1971).
[146] E. Haselbach, *Helv. Chim. Acta*, **54**, 2257 (1971).
[147] M. J. S. Dewar and S. Kirschner, *J. Am. Chem. Soc.*, **93**, 4291 (1971).

the valley from a reactant will lead to one of these forms while the depression from a product to the other.[148] No reliance should therefore be placed upon calculations involving antiaromatic processes unless the same path is followed by the forward and the reverse reactions. Calculations on the electrocyclic transformation of cyclobutene to *cis*-butadiene suggest that a stepwise conrotation, in which the methylene groups rotate after the ring-cleavage, is much preferred to a mechanism whereby the two processes occur simultaneously.[149]

A linear correlation between the Arrhenius activation parameters for the rearrangement of strained cyclic olefins and $\bar{\nu}$ of the $\pi \rightarrow \pi^*$ electronic transition has been noted.[150]

Dielectrocyclic reactions, in which two π-bonds are transformed into two σ-bonds, have been recognized and the advantages of this description over the cycloaddition classification have been discussed.[151]

Aromatic Processes

Optically active *trans*-2,3-di-t-butylcyclopropanone racemizes at 80° by a first-order process; the reaction, which is relatively insensitive to the polarity of the solvent and occurs in Bu^{t}OD without isotope incorporation, probably involves disrotatory openings to give a dipolar intermediate.[152] There have been several further studies of the disrotatory ring-opening of cyclopropyl halides;[153] (**98**, R = OMe) is formed in addition to the allylic ether in the methanolysis of (**98**; R = Cl), attesting to the intervention of the cyclopropyl cation.[154] A similar species is possibly involved in the acid-catalysed methanolysis of 4-isopropylidene-2,2,5,5-tetramethyl-1-oxaspiro[2.2]pentane to give 2,5-dimethyl-3-(1-methoxy-1-methylethyl)hexa-3,4-dien-2-ol as the initial product.[155] An interesting difference in the thermal behaviour of (**99**) and its *trans*-isomer is noted; the former derivative affords 1,1-dicyano-3-ethylidene-2-methylcyclobutane, possibly by disrotation to (**100**) since the reaction rate is at least 10 times faster in acetonitrile than in benzene; in contrast, the *trans*-isomer of (**99**) yields dicyanomethylene-*trans*-2,3-dimethylcyclobutane, probably by a diradical intermediate.[156]

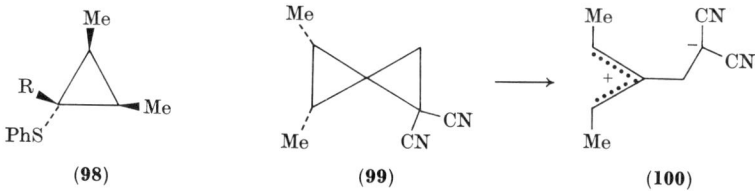

(98)   (99)   (100)

[148] M. J. S. Dewar and S. Kirschner, *J. Am. Chem. Soc.*, **93**, 4292 (1971).
[149] K. Hsu, R. J. Buenker, and S. D. Peyerimhoff, *J. Am. Chem. Soc.*, **93**, 2117 (1971); W. J. Engelbrecht and M. J. DeVries, *J. S. Afr. Chem. Inst.*, **1971**, 24, 38; *Chem. Abs.*, **74**, 124555, 124556.
[150] J. E. Baldwin and A. H. Andrist, *J. Am. Chem. Soc.*, **93**, 3289 (1971).
[151] E. C. W. Scheuneman and W. G. Laidlaw, *J. Am. Chem. Soc.*, **93**, 5731 (1971).
[152] D. B. Sclove, J. F. Pazos, R. L. Camp, and F. D. Greene, *J. Am. Chem. Soc.*, **92**, 7488 (1970).
[153] I. Fleming and E. J. Thomas, *Tetrahedron Letters*, **1971**, 2485; M. S. Baird and C. B. Reese, *Tetrahedron Letters*, **1971**, 4637; W. M. Horspool, R. G. Sutherland, and B. J. Thompson, *J. Chem. Soc.* (C), **1971**, 1558; E. V. Dehmlow and J. Schönefeld, *Ann. Chem.*, **744**, 42 (1971); F. Nerdel, P. Hentschel, W. Brodowski, and J. Buddrus, *Ann. Chem.*, **746**, 6 (1971); W. Tochtermann, D. Schäfer, and C. Rohr, *Chem. Ber.*, **104**, 2923 (1971); T. Asahara, K. Ono, and K. Tanaka, *Bull. Chem. Soc. Japan*, **44**, 1130 (1971).
[154] U. Schöllkopf, E. Ruban, P. Tonne, and K. Riedel, *Tetrahedron Letters*, **1970**, 5077.
[155] J. K. Randall and D. R. Paulson, *J. Org. Chem.*, **36**, 1184 (1971).
[156] J. J. Gajewski and L. T. Burka, *J. Am. Chem. Soc.*, **93**, 4592 (1971).

The equilibria between cycloheptatrienes and norcaradienes have been accounted for theoretically by using the Walsh cyclopropane model,[157] and further examples have been studied experimentally.[158] For example, on the basis of ^{13}C- and ^{1}H-NMR spectroscopy, (**101**; R = H) is in rapid equilibrium with the cycloheptatriene isomer whereas (**101**; R = Cl) exists solely as the norcaradiene derivative.[159] In hot ethanol 1-chlorocycloheptatriene rearranges to a mixture of benzyl chloride and benzyl ethyl ether; the first step of the reaction probably involves valence tautomerization to 1-chlorobicyclo[4.1.0]hepta-2,4-diene, which then undergoes electrocyclic opening of the cyclopropane ring.[160] A similar sequence is implicated in the thermal conversion of octachlorocycloheptatriene into octachlorotoluene.[161] The adduct (**102**), derived from 2,5-dimethyl-3,4-diphenylcyclopentadienone and 1,2,3-triphenylcyclopropene, rearranges to 1,5-dimethyl-2,3,4,6,7-pentaphenylcycloheptatriene when heated in toluene, probably through (**103**).[162] X-ray crystallography suggests that the thermolysis product of 1,2,6,7-tetraphenyl-3,4-diazabicyclo[4.1.0]hepta-2,4-diene is 3,4,5,6,7-pentaphenyl-4H-1,2-diazepine;[163] therefore, [1,5]-H shifts accompany the ring-opening reaction. The sulphoxide (**104**) gives benzocinnoline when heated, probably by SO extrusion from the valence tautomer (**105**).[164]

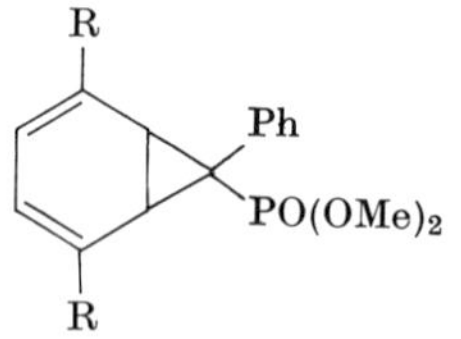

(**101**) 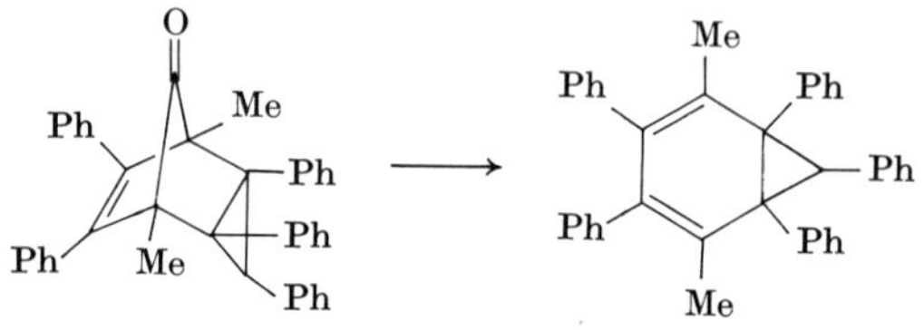   (**102**)   (**103**)

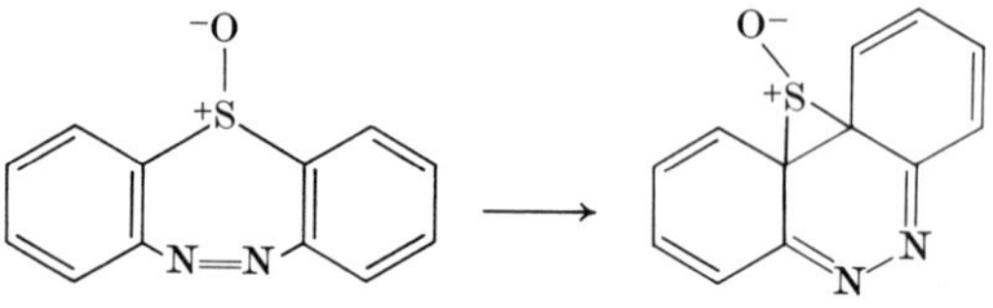

(**104**)   (**105**)

The kinetics of the Diels–Alder addition of tetracyanoethylene, maleic anhydride and dicyanomaleimide to cyclo-octatetraene (which reacts in its valence-tautomeric form) have been studied.[165] In the case of 1-fluorocyclo-octatetraene, the course of the cyclo-

---

[157] *Org. Reaction Mech.*, **1970**, 272; H. Gunther, *Tetrahedron Letters*, **1970**, 5173.
[158] H. Dürr and H. Kober, *Angew. Chem. Int. Ed.*, **10**, 342 (1971); G. E. Hall and J. D. Roberts, *J. Am. Chem. Soc.*, **93**, 2203 (1971); E. Ciganek, *J. Am. Chem. Soc.*, **93**, 2207 (1971).
[159] H. Gunther, B. D. Tunggal, M. Regitz, H. Scherer, and T. Keller, *Angew. Chem. Int. Ed.*, **10**, 563 (1971).
[160] B. Föhlisch and W. Vodrazka, *Tetrahedron Letters*, **1971**, 1207.
[161] K. Kusuda, R. West, and V. N. M. Rao, *J. Am. Chem. Soc.*, **93**, 3627 (1971).
[162] D. J. Anderson and A. Hassner, *J. Am. Chem. Soc.*, **93**, 4339 (1971).
[163] J. N. Brown, R. L. Towns, and L. M. Trefonas, *J. Am. Chem. Soc.*, **92**, 7436 (1970).
[164] Y. L. Chow, J. N. S. Tam, J. E. Blier, and H. H. Szmant, *Chem. Comm.*, **1971**, 1604.
[165] R. Huisgen, *Mod. Sviluppi Sin. Org. Corso. Estivo Chem.* 10*th*, **1967**, 281; *Chem. Abs.*, **74**, 111370 (1971).

addition depends on the dienophile; thus, tetracyanoethylene reacts with 7-fluoro-bicyclo[4.2.0]octa-2,4,6-triene while 4-methyl-1,2,4-triazoline-3,5-dione affords the adduct of 1-fluorobicyclo[4.2.0]octa-2,4,6-triene.[166] The equilibrium between (106) and (107) is very much in favour of the former derivative; however, (107) can be trapped with N-phenylmaleimide and it probably intervenes in the KOBu^t-induced conversion of (106) into benzonitrile.[167] In contrast, 7-azabicyclo[5.2.0]octa-2,5-dien-8-one predominates in the equilibrium with 1-azacyclo-octa-2,4,6-trien-8-one.[168] The dianion (108) reacts with benzophenone to give (111; R = Ph$_2$CO$^-$), in accord with a disrotatory electrocyclization of (109) to (110); support for this pathway comes from the observation that (111; R = H) is produced when 2-methoxy-3,8-dimethyl-1-azacyclo-octa-1,5,7-triene is treated with KOBu^t.[169] A disrotatory ring-opening is involved in the thermolysis of (112) to 3,4,5,6-tetrachlorobicyclo[6.1.0]nona-2,4,6-triene, although the *endo*-isomer of (112) is stable to heat.[170]

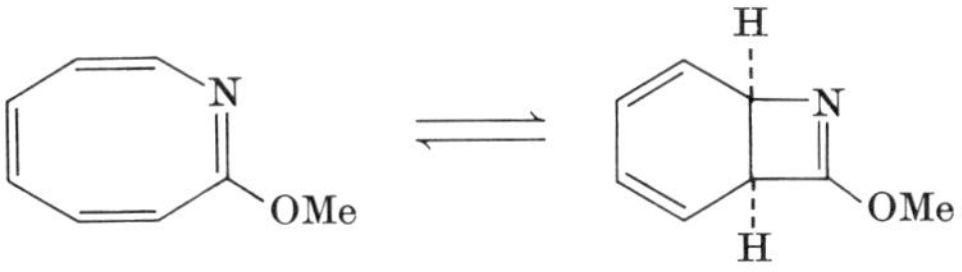

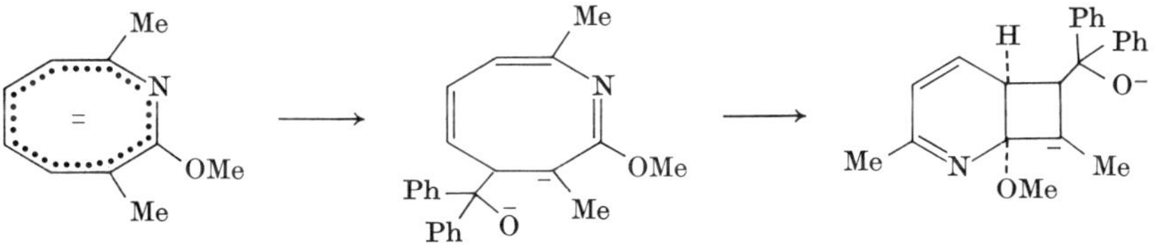

The syntheses of *cis*- and *trans*-9,10-dihydronaphthalenes have been accomplished but there is no evidence to suggest that they interconvert with the appropriate cyclodecapentaenes.[171] The thiaannulene (113) undergoes disrotatory electrocyclization at

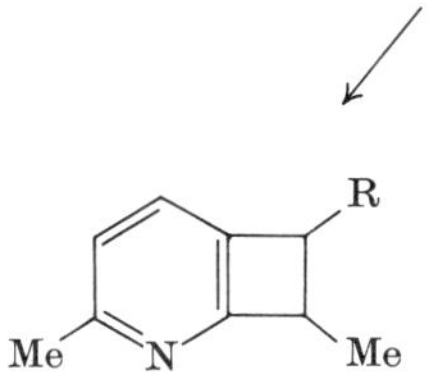

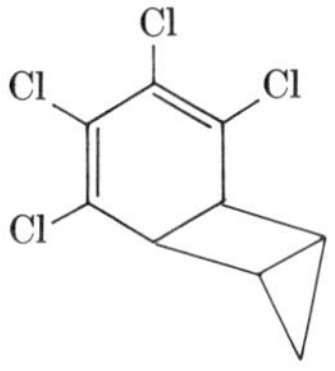

[166] G. Schröder, G. Kirsch, J. F. M. Oth, R. Huisgen, W. E. Konz, and U. Schnegg, *Chem. Ber.*, **104**, 2405 (1971).
[167] L. A. Paquette, T. Kakihana, J. F. Hansen, and J. C. Philips, *J. Am. Chem. Soc.*, **93**, 152 (1971).
[168] L. A. Paquette, T. Kakihana, and J. F. Kelly, *J. Org. Chem.*, **36**, 435 (1971).
[169] L. A. Paquette and T. Kakihana, *J. Am. Chem. Soc.*, **93**, 174 (1971).
[170] W. P. Lay, K. MacKenzie, and J. R. Telford, *J. Chem. Soc.* (C), **1971**, 3199.
[171] E. E. van Tamelen and B. C. T. Pappas, *J. Am. Chem. Soc.*, **93**, 6111 (1971); E. E. van Tamelen, T. L. Burkoth, and R. H. Greenley, *J. Am. Chem. Soc.*, **93**, 6120 (1971).

room temperature to (114).[172] Related electrocyclic processes probably occur in the ring-closures of aldazines,[173] azomethine ylids,[174] and 2-azidopyridines,[175] and the ring-opening of 2,3,4,5,6-pentaphenyl-2H-pyran.[176] The rate-determining step in the thermal rearrangement of (115; $R^1 = R^2 = H$) to 2-methyl-2H-chromene is postulated to involve homoelectrocyclic ring-opening to (116), which, on the basis of D-labelling studies, affords the product by sequential [1,5]- and [1,7]-H shifts and electrocyclization. Moreover, since (115; $R^1 = H$, $R^2 = Me$) opens much faster than (115; $R^1 = Me$, $R^2 = H$), only one of the two possible disrotatory processes is probably involved.[177]

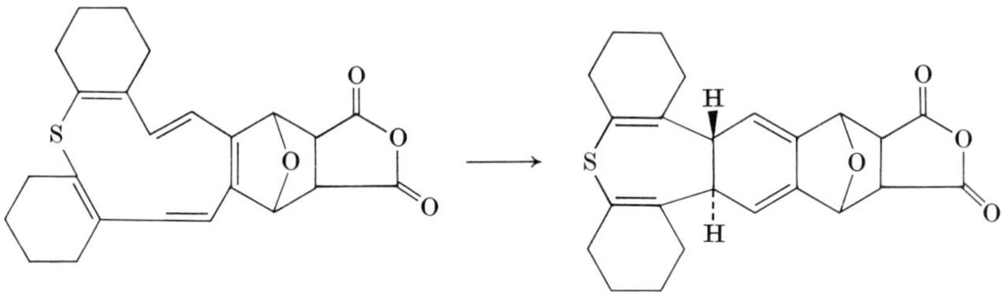

(113)   (114)

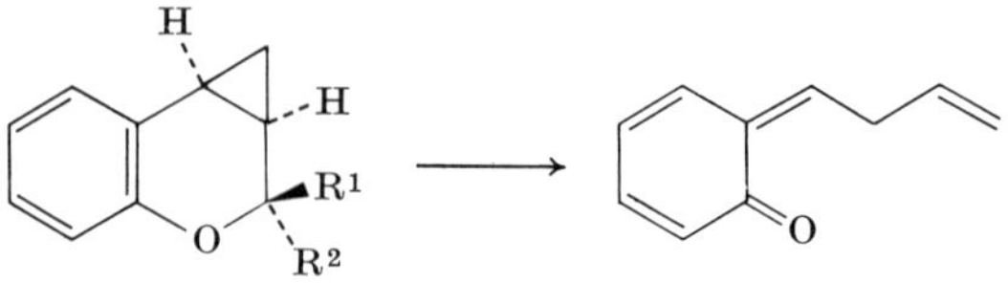

(115)   (116)

In spite of numerous studies, the thermal rearrangement of cis-bicyclo[6.1.0]nona-2,4,6-triene (117; $R^1 = R^2 = H$) to cis-8,9-dihydroindene (120; $R^1 = R^2 = R^3 = H$) is not fully understood. The product almost certainly arises by electrocyclization of (119); however, the reactant is expected to afford cis,cis,trans,cis-cyclononatetraene, either by a direct electrocyclic ring-opening or by a sigmatropic [3,3]-shift to (118) which then undergoes conrotation. Trapping experiments suggest that the latter tetraene and (121; $R^1 = R^2 = H$), the product of disrotatory closure of the starting material, are formed although neither of these species is a likely intermediate in the rearrangement.[178] 1,2,7,8,9,9 - Hexadeuterio - cis - bicyclo[6.1.0]nona - 2,4,6-triene yields 5,6,7,7,8,9 - hexadeuterio-cis-bicyclo[4.3.0]nona-2,4,8-triene; this result, while not excluding nonconcerted processes, is concordant with the intervention of cis-bicyclo[5.2.0]nona-1,4,7-triene which reorganizes to (119) by a concerted ($\sigma^2 s + \sigma^2 a$) pathway.[179] The activation

[172] A. B. Holmes and F. Sondheimer, Chem. Comm., **1971**, 1434.
[173] J. Elguero, R. Jacquier, and C. Marzin, Bull. Soc. Chim. France, **1970**, 4119.
[174] Y. Tamura, N. Tsujimoto, and M. Ikeda, Chem. Comm., **1971**, 310.
[175] T. Sasaki, K. Kanematsu, and M. Murata, Tetrahedron, **27**, 5121 (1971).
[176] J.-P. LeRoux, G. Letertre, P.-L. Desbere, and J.-J. Basselier, Bull. Soc. Chim. France, **1971**, 4059.
[177] R. Hug, G. Fráter, H.-J. Hansen, and H. Schmid, Helv. Chim. Acta, **54**, 306 (1971).
[178] A. G. Anastassiou and R. C. Griffith, J. Am. Chem. Soc., **93**, 3083 (1971); C. S. Baxter and P. J. Garratt, Tetrahedron, **27**, 3285 (1971).
[179] J. E. Baldwin and A. H. Andrist, J. Am. Chem. Soc., **93**, 4005 (1971).

parameters for the rearrangement of (**117**; $R^1 = H$, $R^2 = Me$ and $R^1 = R^2 = H$) are similar, while those of the *syn*-isomer (**117**; $R^1 = Me$, $R^2 = H$), which reacts ca. 100 times more slowly, parallel those of (**117**; $R^1 = R^2 = Me$). These results imply that the reactants require to adopt tub conformations in order to undergo the initial Cope rearrangement.[180] On being heated (**117**; $R^1 = H$, $R^2 = Cl$) and its *syn*-isomer afford (**120**; $R^1 = R^3 = H$, $R^2 = Cl$) at similar rates and the reactions are insensitive to the polarity of the solvent. However, D-labelling experiments dramatically reveal that the reactions are mechanistically different since (**117**; $R^1 = D$, $R^2 = Cl$) affords (**120**; $R^1 = D$, $R^2 = Cl$, $R^3 = H$) while (**120**; $R^1 = H$, $R^2 = Cl$, $R^3 = D$) is formed from (**117**; $R^1 = Cl$, $R^2 = D$). The rearrangement of *anti*-9-chlorobicyclo[6.1.0]nona-2,4,6-triene therefore parallels that of (**117**; $R^1 = R^2 = H$), while that of the *syn*-isomer is postulated as involving a rate-determining disrotatory closure to (**121**; $R^1 = Cl$, $R^2 = H$) which then undergoes disrotatory opening of the cyclopropane ring.[181] 2,7-Diphenylbicyclo-[6.1.0]nona-2,4,6-triene, which is readily derived from 2,6-diphenylbicyclo[5.2.0]nona-2,5,8-triene at $-25°$ by a Cope rearrangement, rearranges at $120°$ to 2,7-diphenylbicyclo-[4.3.0]nona-2,4,7-triene of undefined stereochemistry.[182]

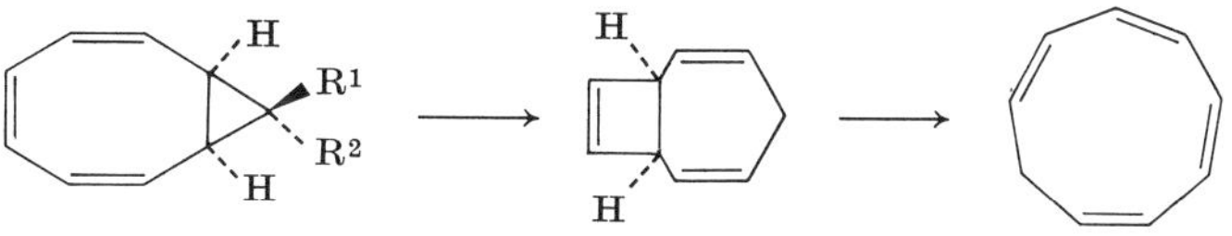

(**117**)   (**118**)   (**119**)

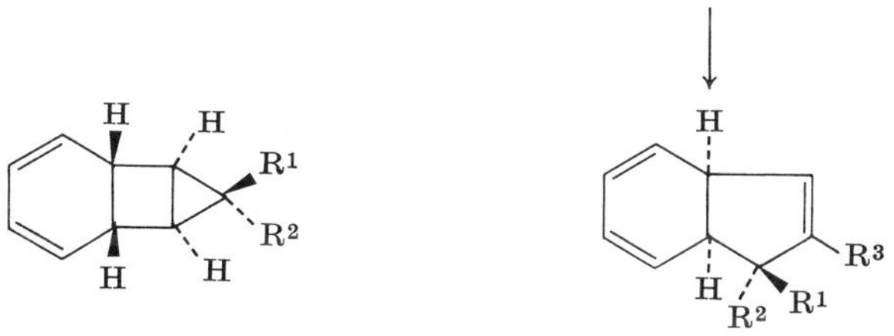

(**121**)   (**120**)

Tetracarbonyl-μ-dichlorodirhodium effectively catalyses the rearrangement of (**117**; $R^1 = R^2 = H$) to (**120**; $R^1 = R^2 = R^3 = H$) at $35°$, in contrast to the thermal reorganization which requires heating at $90°$. Moreover, whereas *trans*-7,7-dimethylbicyclo[4.3.0]-nona-2,4,8-triene is formed by heating (**117**; $R^1 = R^2 = Me$), the *cis*-derivative is the predominant product in the Rh(I)-catalysed reaction.[183]

Antiaromatic Processes

There have been further studies of the electrocyclic ring-openings of aziridines to azomethine ylids,[184] and there is evidence to suggest that oxirans can interconvert with

[180] A. G. Anastassiou and R. C. Griffith, *Chem. Comm.*, **1971**, 1301.
[181] J. C. Barborak, T.-M. Su, P. von R. Schleyer, G. Boch, and G. Schneider, *J. Am. Chem. Soc.*, **93**, 279 (1971).
[182] L. A. Paquette and M. J. Epstein, *J. Am. Chem. Soc.*, **93**, 5936 (1971).
[183] R. Grigg, R. Hayes, and A. Sweeney, *Chem. Comm.*, **1971**, 1248.
[184] J. A. Deyrup and S. C. Clough, *Chem. Comm.*, **1970**, 1620; J. H. Hall and R. Huisgen, *Chem. Comm.*, **1971**, 1187; J. H. Hall, R. Huisgen, C. H. Ross, and W. Scheer, *Chem.. Comm.*, **1971**, 1188; R. Huisgen and H. Mäder, *J. Am. Chem. Soc.*, **93**, 1777 (1971); J. W. Lown and K. Matsumoto, *J. Org. Chem.*, **36**, 1405 (1971).

carbonyl ylids.[185] Furthermore, the stereochemistry of the cycloadducts obtained from *cis*- and *trans*-cyanostilbene oxides and suitable dipolarophiles indicates that the thermal ring-openings occur in a conrotatory manner.[186] In the case of vinyl epoxides it is possible to trap the carbonyl ylids by intramolecular cycloaddition; thus, gas-phase pyrolysis of (**122**) produces a mixture of the *cis*-epoxide and (**125**). The results imply that (**123**), which may be formed by conrotatory ring-opening of (**122**), interconverts with (**124**) which then either undergoes conrotatory closure to the *cis*-epoxide or disrotation to the *cis*-dihydrofuran.[187] Carbonyl ylids have also been postulated as intermediates in the decomposition of phenyl(bromodichloromethyl)mercury in the presence of benzaldehyde.[188] For the first time the thermal conversion of a nitrone into an oxaziridine has been realized.[189]

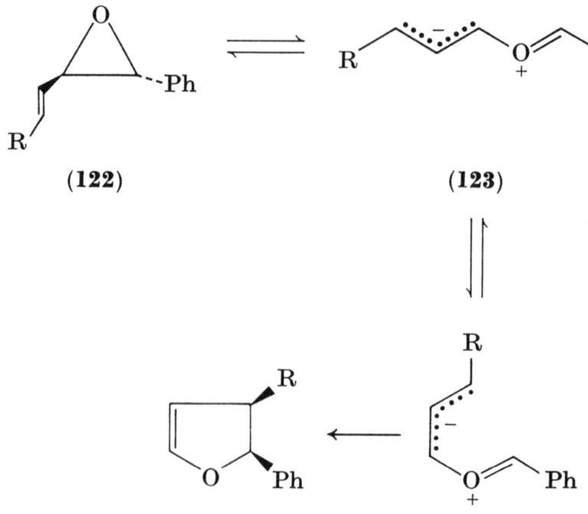

Thermolysis of (**126**) furnishes a mixture of (**128**) and (**130**) in the ratio of 47.7 : 52.3; evidently, the conrotatory cleavage involving (**127**) is preferred to that involving (**129**), a result in keeping with a steric D-isotope effect.[190] The failure of 1,5-dimethylbicyclo-[3.3.0]octa-2,6-diene and 3,7-dideuteriobicyclo[3.3.0]octa-2,6-diene to undergo an antarafacial–antarafacial Cope rearrangement[191] casts doubt upon the proposal that such a mechanism operates in the reorganization of 1-methoxybicyclo[3.2.0]hepta-3,6-dien-2-one.[192] Moreover, since 1,6-dideuteriobicyclo[4.2.0]octa,2,7-diene equilibrates with the 3,8-dideuterio-isomer on being heated,[191] neither the oxo nor the methoxy groups are obligatory. The results can be explained by postulating the intervention of *cis,trans,cis*-cyclic trienes.

Electrocyclic ring-openings are probably involved in the thermal isomerizations of

---

[185] H. Hamberger and R. Huisgen, *Chem. Comm.*, **1971**, 1190; J. W. Lown and K. Matsumoto, *Can. J. Chem.*, **49**, 3443 (1971).
[186] A. Dahmen, H. Hamberger, R. Huisgen, and V. Markowski, *Chem. Comm.*, **1971**, 1192.
[187] J. C. Paladini and J. Chuche, *Tetrahedron Letters*, **1971**, 4383.
[188] C. W. Martin, J. A. Landgrebe, and E. Rapp, *Chem. Comm.*, **1971**, 1438.
[189] H. O. Larson, K. Y. W. Ing, and D. L. Adams, *J. Heterocyclic Chem.*, **7**, 1227 (1970).
[190] R. E. K. Winter and M. L. Honig, *J. Am. Chem. Soc.*, **93**, 4616 (1971).
[191] J. E. Baldwin and M. S. Kaplan, *J. Am. Chem. Soc.*, **93**, 3969 (1971).
[192] T. Miyashi, M. Nitta, and T. Mukai, *J. Am. Chem. Soc.*, **93**, 3441 (1971).

1-chloro-,[193] 1-ethyl-2-vinyl-,[194] cis-3,4-diphenyl-[195] and benzo-cyclobutenes,[196] 2,3,4,4-tetramethyloxetene,[197] and thiete 1,1-dioxide.[198] A stereochemical study of the thermolyses of the 2,3-dimethylbicyclo[2.1.0]pentanes to hepta-2,5-dienes indicates the preference for conrotation over disrotation to be at least 10^5 times less than with the corresponding cyclobutenes.[199]

trans,cis,cis,trans-Cyclodeca-1,3,5,7-tetraene and *trans*-bicyclo[6.2.0]deca-2,4,6-triene rapidly interconvert at 48° and there is a slight preference for the latter derivative ($K = 1.48$); in contrast the 9,9-dichloro-10,10-difluoro derivative exists exclusively in the bicyclic form.[200] *cis,cis,cis,trans*-Cyclodeca-1,3,5,7-tetraene is probably an intermediate in the thermal conversion of *cis*-bicyclo[6.2.0]deca-2,4,6-triene into *trans*-bicyclo[4.4.0]deca-2,4,7-triene.[201] An 8π-electron electrocyclization followed by [1,5]-H shifts is implicated in the thermal conversions of 6-vinylfulvenes into dihydropentalenes,[202] while a conrotatory closure of a 16π-electron cation is postulated in the oxidative cyclization of 1,19-dialkyl-1,19-dideoxybiladiene-*ac* dibromides in the presence of nickel salts.[203]

Rearrangements Involving Cycloreversions and Cycloadditions

Pyrolysis of methyl 2,2,4,4-tetramethylbicyclo[1.1.0]butane-1-carboxylate gives methyl *trans*-3-isopropenyl-2,2-dimethylcyclopropane-1-carboxylate, in accord with a ($\sigma^2 s + \sigma^2 a$) process; the concomitant formation of methyl 2-isopropenyl-4-methylpent-3-enoate

[193] D. Dickens, H. M. Frey, and J. Metcalfe, *Trans. Faraday Soc.*, **67**, 2328 (1971).
[194] H. M. Frey, J. Metcalfe, and B. M. Pope, *Trans. Faraday Soc.*, **67**, 750 (1971).
[195] J. I. Brauman and W. C. Archie, Jr., *Tetrahedron*, **27**, 1275 (1971).
[196] W. Oppolzer, *J. Am. Chem. Soc.*, **93**, 3824, 2833 (1971).
[197] L. E. Friedrich and G. B. Schuster, *J. Am. Chem. Soc.*, **93**, 4602 (1971).
[198] J. F. King, P. De Mayo, C. L. McIntosh, K. Piers, and D. J. H. Smith, *Can. J. Chem.*, **48**, 3704 (1970).
[199] J. A. Berson, W. Bauer, and M. M. Campbell, *J. Am. Chem. Soc.*, **92**, 7515 (1970).
[200] S. W. Staley and T. J. Henry, *J. Am. Chem. Soc.*, **92**, 7612 (1970).
[201] S. W. Staley and T. J. Henry, *J. Am. Chem. Soc.*, **93**, 1292 (1971).
[202] J. J. Gajewski and C. J. Cavender, *Tetrahedron Letters*, **1971**, 1057.
[203] R. Grigg, A. P. Johnson, A. W. Johnson, and M. J. Smith, *J. Chem. Soc.* (C), **1971**, 2457.

and methyl 2-isopropylidene-4-methylpent-3-enoate is the result of an acid-catalysed rearrangement.[204] 1,3-Dimethylcyclobutadiene is intercepted as its maleic anhydride adduct in the thermolysis of (131).[205]

The rates of cycloreversion of 4-substituted 1-chlorobicyclo[2.2.0]hexanes to 4-substituted 2-chlorohexa-1,5-dienes is sensitive to the nature of the 4-substituent but is not influenced by the solvent polarity, in accord with the intervention of diradical intermediates.[206] 1,2,5,6-Tetramethyl-3,4,7,8-tetramethylenecyclo-octa-1,5-diene (133), formed by thermolysis of (132), undergoes an internal cycloaddition to afford (134) on being heated.[207]

There have been a number of investigations of the thermal behaviour of 6,6-dimethylbicyclo[3.1.1]heptane derivatives. The normal cleavage mode involves homolysis of the 1–6 bond;[208] for example, (136) is the initial product of the gas-phase pyrolysis of (135; R^1 = Me, R^2 = OH). However, rupture of other bonds of the cyclobutane ring may also occur since nopinol (135; R^1 = OH, R^2 = H) yields (137).[209] That even more substantial reorganization can take place is illustrated by the formation of 3,6-dimethyl-2-methylenehept-5-enal during the thermolysis of cis-pinocarveol (138); D-labelling experiments suggest that a dissociation–recombination pathway is involved.[210] The thermal cyclization of trichloroacrylic anhydride to (140) is probably triggered by an intramolecular [2 + 2]-cycloaddition to give (139).[211]

Tricycloheptenes undergo an interesting series of thermal transformations. For example, (141; R = Ph) yields (143) at 100°, probably by internal cycloaddition of the diradical (142; R = Ph); at 175° the quadricyclane isomerizes to the norbornadiene (144). In contrast, the diradical (142; R = Me) derived from (141; R = Me) suffers a different fate and rearranges to 1,2,2,3-tetramethylbicyclo[3.2.0]hepta-3,6-diene and 1,6,7,7-tetramethylcycloheptatriene.[212]

Retro-Diels–Alder reactions are implicated in the thermal racemization of bicyclo-[2.2.1]hept-5-ene-trans-2,3-dicarboxylic acid[213] and the thermolysis of basketene[214] and diazabasketene (145).[215] Thus, flash-vacuum-pyrolysis of the latter derivative gives HCN and 1-azacyclo-octatetraene, probably through (147) which may be formed from the cycloreversion product (146) by a Cope rearrangement. An analogous sequence of reactions is likely to be involved in the BF_3-induced reorganization of diazabasketene N-oxide to benzaldehyde oxime.[216] The reduction of (148) affords dimethyl cyclo-octa-1,4,6-triene-1,2-dicarboxylate, implying that the intermediate (149) readily undergoes cycloreversion.[217] A similar reorganization occurs when (150; R = CH_2, O, or NTs) is

---

[204] D. P. G. Hamon and C. F. Lill, Austral. J. Chem., 24, 1667 (1971).
[205] H. Ona, H. Yamaguchi, and S. Masamune, J. Am. Chem. Soc., 92, 7495 (1970).
[206] E. N. Cain, Tetrahedron Letters, 1971, 1865.
[207] W. T. Borden and A. Gold, J. Am. Chem. Soc., 93, 3830 (1971).
[208] K. H. Schulte-Elte, M. Gadola, and G. Ohloff, Helv. Chim. Acta, 54, 1813 (1971); J. Tanaka, T. Katagiri, and K. Ozawa, Bull. Chem. Soc. Japan, 44, 130 (1971); J. M. Coxon, R. P. Garland, and M. P. Hartshorn, Austral. J. Chem., 23, 2531 (1971); Chem. Comm., 1971, 1131.
[209] J. M. Coxon, R. P. Garland, and M. P. Hartshorn, Chem. Comm., 1970, 1709.
[210] J. M. Coxon, R. P. Garland, and M. P. Hartshorn, Austral. J. Chem., 24, 1481 (1971).
[211] A. Roedig, H. H. Bauer, B. Heinrick, and D. Kubin, Chem. Ber., 104, 3525 (1971).
[212] L. A. Paquette and L. M. Leichter, J. Am. Chem. Soc., 93, 5128 (1971).
[213] R. E. Pincock, M.-M. Tong, and K. R. Wilson, J. Am. Chem. Soc., 93, 1669 (1971).
[214] Org. Reaction Mech., 1970, 275; K.-W. Shen, Chem. Comm., 1971, 391.
[215] D. W. McNeil, M. E. Kent, E. Hedaya, P. F. D'Angelo, and P. O. Schissel, J. Am. Chem. Soc., 93, 3817 (1971).
[216] J. P. Snyder, L. Lee, and D. G. Farnum, J. Am. Chem. Soc., 93, 3816 (1971).
[217] H. W. Whitlock, Jr., and P. F. Schatz, J. Am. Chem. Soc., 93, 3837 (1971).

(131) (132) (133) (134) (135) (136) (137) (138) (139) (140)

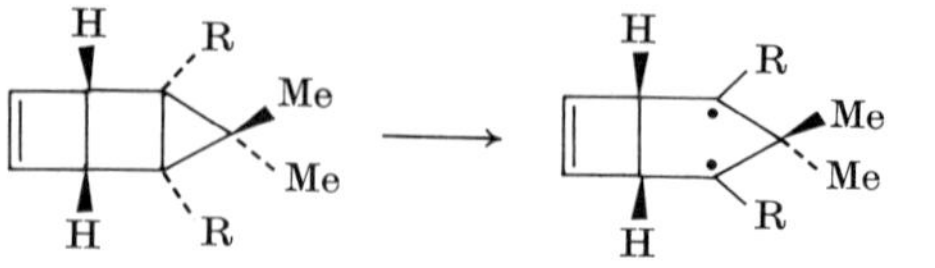

(141) (142)

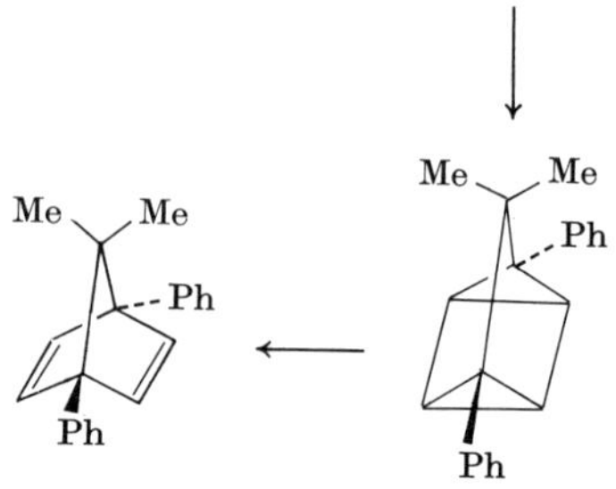

(144) (143)

heated.[218] The conversion of bicyclo[4.2.2]deca-2,4,7,10-tetraene into *cis*-9,10-dihydronaphthalene is postulated as involving (151);[78] the intermediate can be trapped as its iron carbonyl complex (152), which, on the basis of NMR spectroscopy, undergoes rapid valence tautomerism at 45°.[219]

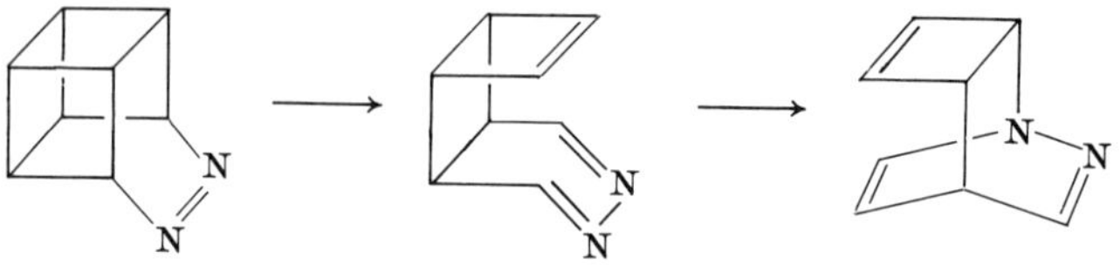

(145) (146) (147)

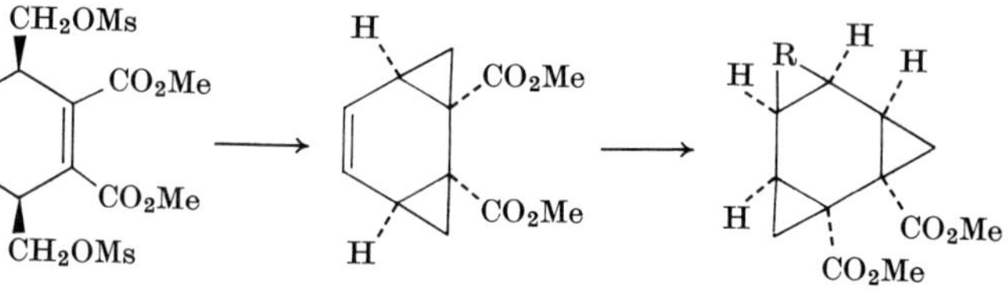

(148) (149) (150)

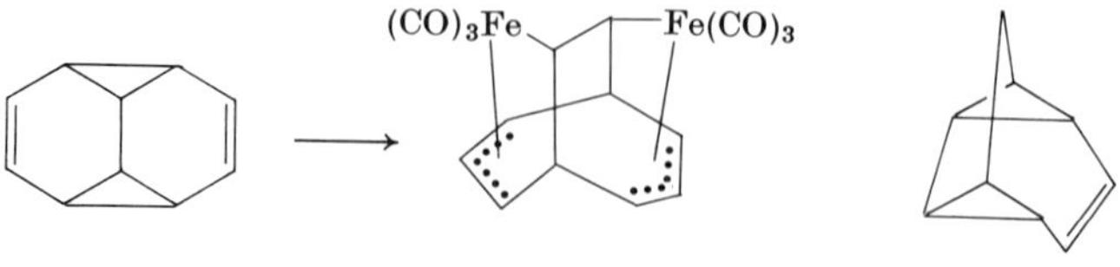

(151) (152) (153)

[218] H. Prinzbach and D. Stusche, *Helv. Chim. Acta*, **54**, 755 (1971).
[219] R. Aumann, *Angew. Chem. Int. Ed.*, **10**, 189 (1971).

Rearrangements, resulting from intramolecular [4 + 2]-cycloadditions, are observed when 2-(but-3-enyl)cycloheptadienones,[220] 5-(pent-4-enyl)-[221] and 5-vinyl-cyclohexa-1,3-dienes[222] are heated. The rearrangement of bicyclo[4.2.1]nona-2,4,7-triene to cis-bicyclo[4.3.0]nona-2,4,7-triene at 290° is postulated as involving an intramolecular Diels–Alder reaction in the first step, to give (**153**) which, by two successive homodienyl [1,5]-H shifts, affords the product; D-labelling experiments are in accord with this pathway.[223]

Intramolecular ene reactions are implicated in the thermal cyclizations of 2-methyl-3-(pent-4-enyl)cyclohex-2-enone.[224]

Anionic Rearrangements

Reactions involving the migrations of alkoxycarbonyl groups have been reviewed.[225]

[1,2]-Migrations

Reviews discussing the Stevens, Sommelet–Hauser, Meisenheimer, and Wittig rearrangements have been published.[226]

The sulphonium ylid (**154**), generated from dibenzyl sulphide and benzyne, undergoes a Stevens rearrangement with migration of the benzyl group; the reaction is accompanied by CIDNP signals in accord with a radical-pair intermediate.[227] Stevens-like reorganizations ensue when 2,4-diphenylthietan 1-oxides and 1,1-dioxides are treated with strong bases,[228] and when aminimides are heated.[229] The base-promoted reactions of quaternary benzylammonium salts have been investigated[230] and, in the case of N,N,N-trimethyl-α-phenylneopentylammonium halides,[231] Stevens and ortho- and para-Sommelet–Hauser rearrangements occur. The former process is favoured in non-planar solvents, while an increase in base concentration promotes the ortho-rearrangement. Optically active α-ethyl-α-methylbenzyl isocyanide isomerizes when heated to the almost racemic cyanide, in accordance with a radical-pair mechanism.[232] The detection of CIDNP signals during the Martynoff rearrangement of nitrones to oxime ethers suggests that the reaction proceeds by way of a caged radical pair.[233] In the presence of two molar

---

[20] C. A. Cupas, W. E. Heyd, and M.-S. Kong, *J. Am. Chem. Soc.*, **93**, 4623 (1971).
[21] A. Krantz and C. Y. Lin, *Chem. Comm.*, **1971**, 1287.
[22] P. Heinbach, K.-J. Ploner, and F. Thömel, *Angew. Chem. Int. Ed.*, **10**, 276 (1971).
[23] J. A. Berson, R. R. Boettcher, and J. J. Vollmer, *J. Am. Chem. Soc.*, **93**, 1540 (1971).
[24] R. Ramage and A. Sattar, *Tetrahedron Letters*, **1971**, 649.
[25] R. M. Acheson, *Accounts Chem. Res.*, **4**, 177 (1971).
[26] A. R. Lepley and A. G. Giumanini, *Mech. Mol. Migr.*, **3**, 297 (1971); S. H. Pine, *Org. Reactions*, **18**, 403 (1970); *J. Chem. Educ.*, **48**, 99 (1971); G. Wittig, *Bull. Soc. Chim. France*, **1971**, 1921.
[27] H. Iwamura, M. Iwamura, T. Nishida, M. Yoshida, and J. Nakayama, *Tetrahedron Letters*, **1971**, 63.
[28] R. M. Dodson, P. D. Hammen, and J. Y. Fan, *J. Org. Chem.*, **36**, 2703 (1971); R. M. Dodson, P. D. Hammen, and R. A. Davis, *J. Org. Chem.* **36**, 2693 (1971); R. M. Dodson, P. D. Hammen, E. H. Jancis, and G. Klose, *J. Org. Chem.*, **36**, 2698 (1971).
[29] H. P. Benecke and J. H. Wikel, *Tetrahedron Letters*, **1971**, 3479.
[30] P. E. Iverson, *Tetrahedron Letters*, **1971**, 55; E. A. Sedor, *Tetrahedron Letters*, **1971**, 323; A. R. Lepley and A. G. Giumanini, *J. Org. Chem.*, **36**, 1217 (1971).
[31] S. H. Pine, E. M. Munemo, T. R. Philips, G. Bartolini, W. D. Cotton, and G. C. Andrews, *J. Org. Chem.*, **36**, 984 (1971).
[32] S.-I. Yamada, M. Shibasaki, and S. Terashima, *Chem. Comm.*, **1971**, 1008.
[33] D. G. Morris, *Chem. Comm.*, **1971**, 221.

equivalents of Bu^nLi allyl phenyl ether is converted into propiophenone via the Wittig-rearrangement product, lithium 1-phenylallyl oxide.[234] Contrary to previous claims, the silylcarbinol to silyl ether rearrangement proceeds with inversion of configuration at C;[235] thus, in the presence of triethylamine (R)-phenyl(triphenylsilyl)methan-[^{2}H$_1$]ol yields [α-^{2}H$_1$]benzyl triphenylsilyl ether, which is reduced to (S)-[α-^{2}H$_1$]benzyl alcohol. Reversal of the above reorganization has been achieved for the first time: benzyl triethylsilyl ether gives α-(triethylsilyl)benzyl alcohol in the presence of Bu^{t}Li.[236]

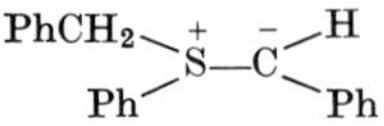

(154)  (155) (156)

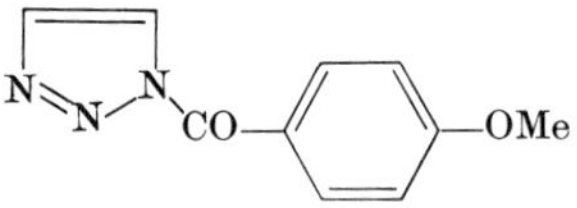

(157) 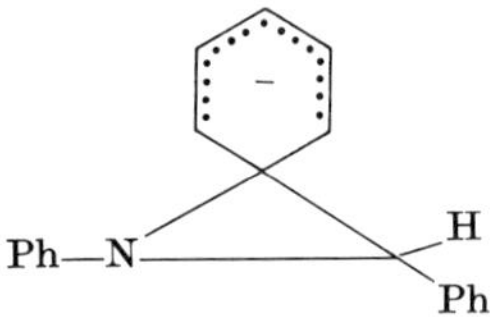 (158) 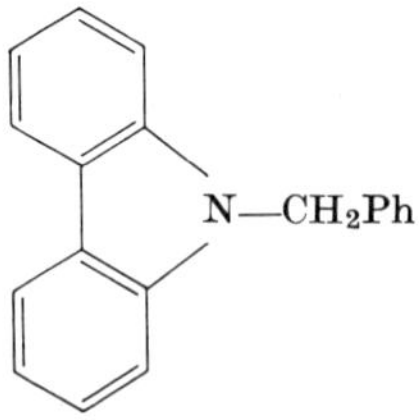 (159)

Reduction of benzophenone anil with Na in THF affords the dianion (155), which when treated at −60° with ethyl chloroformate and then water furnishes ethyl N,2,2-triphenylglycinate; if the addition of water is delayed until the solution warms to room temperature N-diphenylmethyl-N-phenylcarbamate is the product. Trapping experiments with methyl iodide indicate that acylation initially occurs at the more nucleophilic carbanionic centre to give (156), which undergoes a [1,2]-shift of the ethoxycarbonyl group on warming.[237] A similar migration of the acyl group of (157) occurs in the presence of triethylamine, probably by an intermolecular route.[238] N,N-Diphenylbenzylamine affords N-phenylbenzhydrylamine when treated with Bu^nLi, probably via the cyclic intermediate (158); such a species is precluded with (159) which fails to rearrange under corresponding conditions.[239] Migration of the pyridyl ring occurs when 2-(2-pyridyl)ethyl chloride reacts with Li in THF although a phenyl shift takes place with 2-(4-pyridyl)ethyl chloride; radical intermediates are implicated by ESR measurements.[240]

The KOBu^t-induced ring expansions of (bromomethylene)cyclobutanes to 1-bromocyclopentenes is triggered by the formation of the vinyl anions;[241] in the case of 1-bromomethylene-2,2-dimethylcyclobutane the [1,2]-shift of the primary C atom is preferred to that of the tertiary one although the product ratio is influenced by the stereochemistry of the double bond.[242]

[234] D. R. Dimmel and S. B. Gharpure, J. Am. Chem. Soc., 93, 3991 (1971).
[235] M. S. Biernbaum and H. S. Mosher, J. Am. Chem. Soc., 93, 6221 (1971); A. G. Brook and J. D. Pascoe, J. Am. Chem. Soc., 93, 6224 (1971).
[236] R. West, R. Lowe, H. F. Stewart, and A. Wright, J. Am. Chem. Soc., 93, 282 (1971).
[237] J. G. Smith and G. E. F. Simpson, Tetrahedron Letters, 1971, 3295.
[238] P. Ykman, G. L'abbé, and G. Smets, Tetrahedron Letters, 1971, 5225.
[239] J. J. Eisch and C. A. Kovacs, J. Organomet. Chem., 30, C97 (1971).
[240] J. J. Eisch and C. A. Kovacs, J. Organomet. Chem., 25, C33 (1970).
[241] K. L. Erickson, J. Markstein, and K. Kim, J. Org. Chem., 36, 1024 (1971).
[242] K. L. Erickson, J. Org. Chem., 36, 1031 (1971).

The Pummerer rearrangement of thioanhydrohexitol sulphoxides has been studied.[243]

The KNHPh-induced isomerization of 1,2,4-tribromobenzene to the 1,3,5-derivative formally involves a [1,2]-shift of the bromine atom; kinetic and radiolabelling studies indicate that the intermolecular reaction is initiated by the formation of the aryl anion (**160**).[244] An intermolecular pathway is also implicated in the KF-induced isomerization of (**161**; $R^1 = F$, $R^2 = C_3F_7$) to (**161**; $R^1 = C_3F_7$, $R^2 = F$).[245]

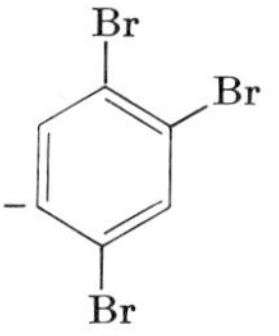

(160) 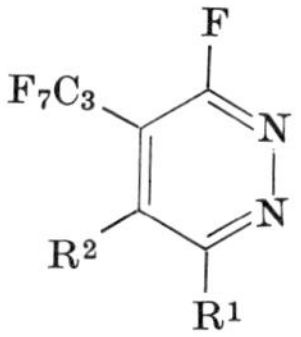   (161)

[1,5]-*Migrations*

NMR spectroscopy indicates that (**162**) rapidly interconverts with (**163**) above 170° by a [1,5]-acyl transfer ($\Delta G^{\ddagger}$ 22.3 kcal mole^{-1}).[246] When methyl 2-mercaptobenzoate is heated in benzylamine, 2-(methylthio)benzanilide is produced;[247] the claim that an intramolecular $S_N i$ process is involved would seem to be unlikely in view of Eschenmoser's studies.[248] Intermolecular [1,5]-shifts of a methyl group occur when (**164**) is thermolysed.[249]

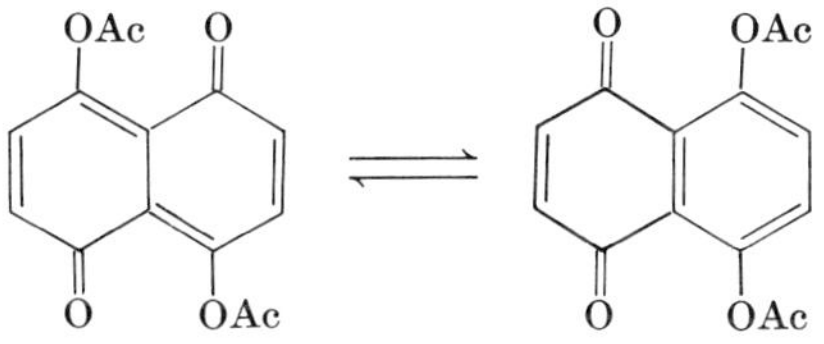

(162)   (163) 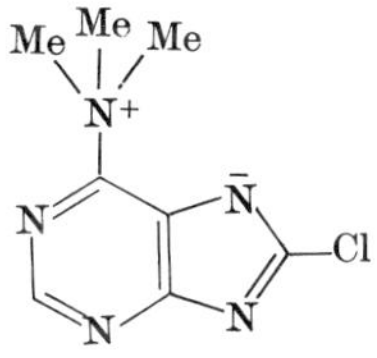   (164)

Enolate-induced Reorganizations

The Favorskii rearrangement has been reviewed and studied further.[250] 1-Halo-1,1-diphenylpropanone and 1-halo-3,3-diphenylpropanone react with methanolic NaOMe

---

[243] J. Kuszmann, P. Sohár, and G. Horvath, *Tetrahedron*, **27**, 5055 (1971).
[244] J. F. Bunnett and C. E. Mayer, Jr., *J. Am. Chem. Soc.*, **93**, 1183 (1971); J. F. Bunnett and G. Scorrano, *J. Am. Chem. Soc.*, **93**, 1190 (1971); D. J. McLennan and J. F. Bunnett, *J. Am. Chem. Soc.*, **93**, 1198 (1971); J. F. Bunnett and I. N. Feit, *J. Am. Chem. Soc.*, **93**, 1201 (1971).
[245] R. D. Chambers, Yu. A. Cheburkov, J. A. H. MacBridge, and W. K. R. Musgrave, *Chem. Comm.*, **1970**, 1647; *J. Chem. Soc.* (C), **1971**, 532.
[246] I. C. Calder, D. W. Cameron, and M. D. Sidell, *Chem. Comm.*, **1971**, 360.
[247] J. S. Grivas and K. C. Navada *J. Org. Chem.*, **36**, 1520 (1971).
[248] L. Tenud, S. Farooq, J. Seibl, and A. Eschenmoser, *Helv. Chim. Acta*, **53**, 2059 (1970).
[249] J. Kiburis and J. H. Lister, *J. Chem. Soc.* (C), **1971**, 1587.
[250] A. A. Akhrem, T. K. Ustynyuk, and Yu. A. Titov, *Russian Chem. Rev.*, **1970**, 732; A. V. Shchelkunov, Z. M. Muldakhmetov, N. A. Rakhimzhanova, and T. A. Favorskaya, *Zhur. Org. Khim.*, **6**, 936 (1970); *J. Org. Chem. USSR*, **6**, 1530 (1970).

to give a similar mixture of methyl 3,3-diphenylpropionate and 1-phenylindan-2-one; the former product probably originates from 2,2-diphenylcyclopropanone and the latter from its dipolar tautomer. In the corresponding reactions of 1-halo-1,1,3-triphenyl- and 1-halo-1,3,3-triphenyl-propanones, 1,3-diphenylindan-2-one is the exclusive product.[251] A Favorskii-like rearrangement occurs with α-bromoimines; thus, (166) is formed in the reaction of (165) with KOBu^t.[252]

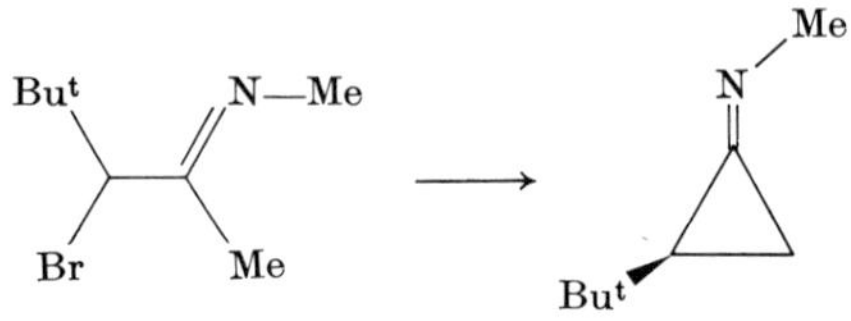

The homo-Favorskii rearrangement of 6-dichloromethylcyclohex-2-enones has been examined; for instance, NaOH converts (167) into a mixture of (168), (169) and (170), probably by the route shown in Scheme 1.[250] An unusual enolate rearrangement is implicated in the reaction of 2-methyl-2-(toluene-p-sulphonyloxymethyl)cyclohexane with base since, in addition to the expected 1-methylbicyclo[3.1.0]heptan-6-one, 1-methylbicyclo[3.2.0]heptan-6-one is produced.[253,254]

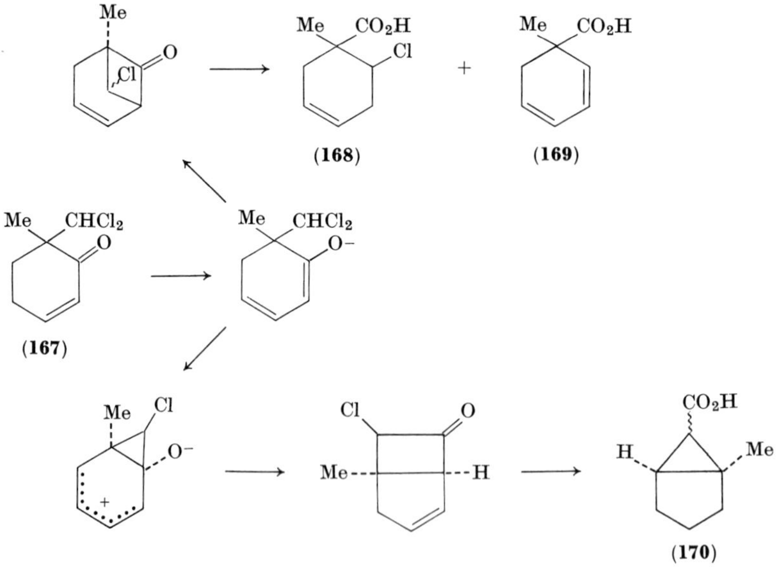

Scheme 1

[251] F. G. Bordwell and R. G. Scamehorn, *J. Am. Chem. Soc.*, **93**, 3410 (1971).
[252] H. Quast, E. Schmitt, and R. Frank, *Angew. Chem. Int. Ed.*, **10**, 651 (1971).
[253] E. Wenkert, P. Bakuzis, R. J. Baumgarten, C. L. Leicht, and H. P. Schenk, *J. Am. Chem. Soc.*, **93**, 3208 (1971).
[254] Y. Tsuda, T. Tanno, A. Ukai, and K. Isobe, *Tetrahedron Letters*, **1971**, 2009.

The Ramberg–Bäcklund rearrangement is applicable to the synthesis of highly strained unsaturated propellanes;[255] thus, (**172**) is formed, albeit in low yield, when (**171**) is treated with Bu[n]Li. However, in the case of (**173**) the major product is (**175**), which suggests that the carbanion undergoes cyclization to (**174**).[256] In the triethylamine-induced conversion of αα-dichlorodibenzyl sulphone into diphenylacetylene, it is possible to isolate the intermediate 2,3-diphenylthiiren 1,1-dioxide in over 90% yield.[257]

Enol(ate) intermediates are implicated in the equilibrations of mandelaldehyde to 2-hydroxyacetophenone,[258] menthone to isomenthone,[259] cis- to trans-4-t-butyl-2-halocyclohexanones,[260] and the α-ketols of bicyclo[2.2.1]heptanes,[261] and in the isomerization of 2-hydroxy-3,4-bis(diphenylmethylene)cyclobutanone to 3,4-bis(diphenylmethyl)-cyclobutene-1,2-dione.[262]

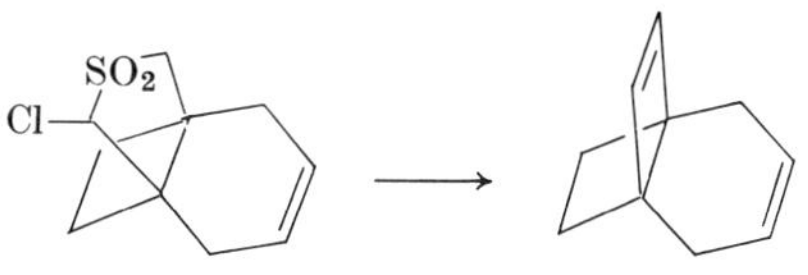

(**171**)   (**172**)

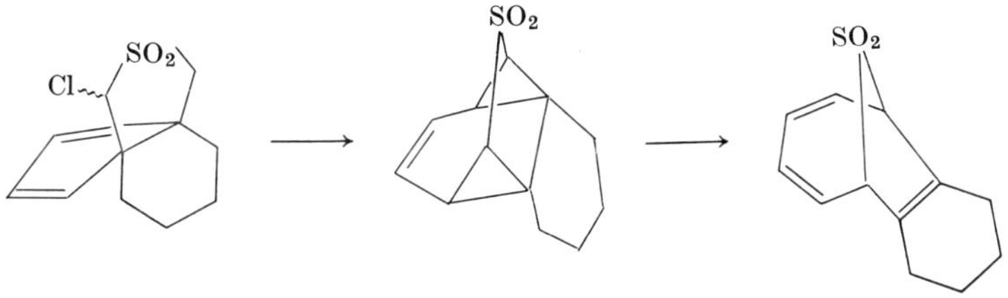

(**173**)   (**174**)   (**175**)

Cationic Rearrangements

[1,2]-*Migrations Involving Departures of Leaving Groups*

Rearrangements involving α-halo-epoxides[263] and amino-ketones[264] have been reviewed.

Shifts to Primary Centres. 2,5,5-Trimethylhepta-1,3,6-triene is produced when (**176**) is pyrolysed; the [1,2]-shift of the electron-pair is probably initiated by heterolysis of the ester linkages.[265]

The reductive ring-expansion of cyclopropenecarboxylic esters to cyclobutenes, which occurs in the presence of LiAlH$_4$–AlCl$_3$, is a synthetically useful reaction; thus,

---

[255] L. A. Paquette and R. W. Houser, *J. Am. Chem. Soc.*, **93**, 4522 (1971); L. A. Paquette, J. C. Philips, and R. E. Wingard, Jr., *J. Am. Chem. Soc.*, **93**, 4516 (1971).
[256] L. A. Paquette, R. E. Wingard, Jr., and R. H. Meisinger, *J. Am. Chem. Soc.*, **93**, 1047 (1971).
[257] J. C. Philips, J. V. Swisher, D. Haidukewych, and O. Morales, *Chem. Comm.*, **1971**, 22.
[258] D. W. Griffiths and C. D. Gutsche, *J. Am. Chem. Soc.*, **93**, 4788 (1971).
[259] Y. Pocker and R. F. Buchholz, *J. Am. Chem. Soc.*, **93**, 2905 (1971).
[260] P. Moreau, A. Casadevall, and E. Casadevall, *Bull. Soc. Chim. France*, **1971**, 3973.
[261] C. Coulombeau and A. Rassat, *Bull. Soc. Chim. France*, **1971**, 505.
[262] F. Toda, N. Ooi, Y. Takehira, and K. Akagi, *Bull. Chem. Soc. Japan*, **44**, 1998 (1971).
[263] R. N. McDonald, *Mech. Mol. Migr.*, **3**, 67 (1971).
[264] C. L. Stevens, P. M. Pillai, and M. E. Munk, *Mech. Mol. Migr.*, **3**, 237 (1971).
[265] T. Sasaki, S. Eguchi, M. Ohmo, and T. Umemura, *J. Org. Chem.*, **36**, 1968 (1971).

1,2-dipropylcyclobutene is obtained from ethyl 2,3-dipropylcycloprop-2-enecarboxylate.[266] Phenyl migration accompanies the acid-catalysed dehydration of 4-hydroxymethyl-4-phenylpiperidines[267] and, surprisingly, the methyl group preferentially migrates in the corresponding reaction of 2-n-butyl-, 2-isobutyl-, and 2-t-butyl-2-methylpropane-1,3-diols.[268] In the presence of acetic anhydride and HBF_4 the *exo*-derivative (**177**) undergoes ring-expansion.[269] The deamination of 7β-aminomethyl-6,7,8,14-tetrahydro-7α-methyl-6,14-*endo*-ethenothebaine also involves an initial ring-enlargement.[270]

A [1,2]-hydride shift is implicated in the acetic anhydride-induced conversion of 1,2-(*N*-aminoepimino)-3,4,5,6-diethylidene-1,2-dideoxy-L-iditol into 1-deoxy-3,5:4,6-diethylidene-L-*xylo*-hex-2-ulose *N*-acetylhydrazone.[271]

Examples of [1,2]-heteroatom shifts include the room-temperature ring-expansion of 1-t-butyl-2-(methylsulphonyloxymethyl)azetidine,[272] the formic acid-catalysed conversion of (**178**) into (**179**),[273] and the formation of 1-arylthio-2,3-bis(methylsulphonyloxy)propanes by mesylation of 2-arylthiopropane-1,3-diols.[274] *N*-Chlorosuccinimide converts 2-benzamido-2-(benzylthio)propanoic acid into 2-benzamido-3-(benzylthio)propenoic acid;[275] the reaction, which formally involves an oxidative [1,2]-shift, presumably involves the intermediacy of *N*-benzoyl-*S*-benzylcysteine formed by an elimination–addition sequence.

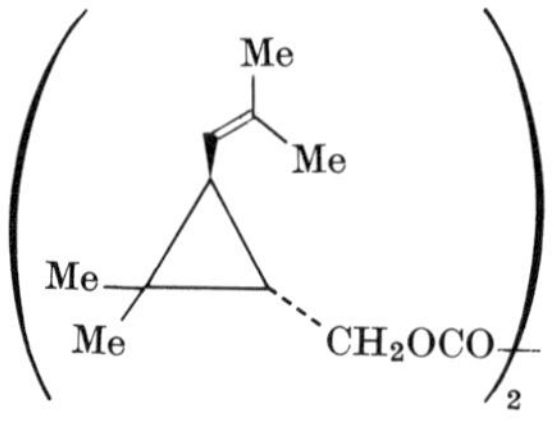

(**176**) 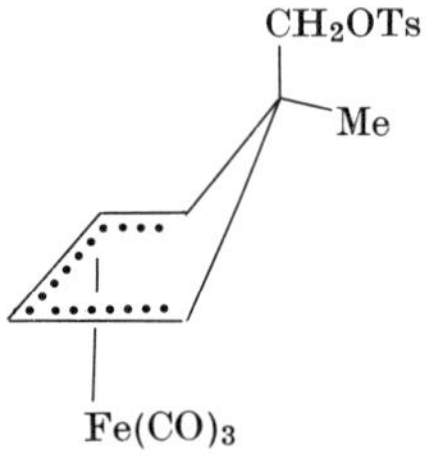   (**177**)

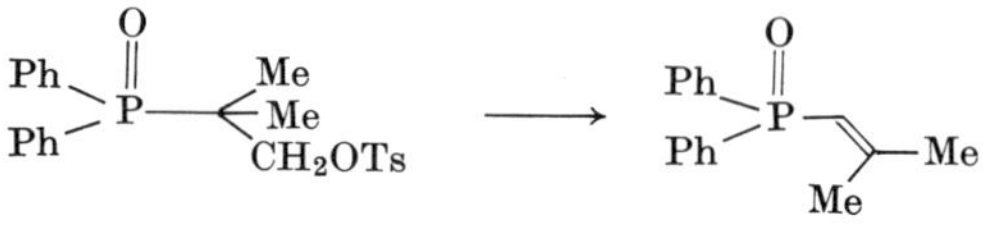

(**178**)   (**179**)

Shifts to Secondary Centres. There have been further studies of the Th(III)-induced rearrangements of olefins.[276] Cyclohexene is converted into cyclopentanecarbaldehyde

---

[266] W. J. Gensler, J. J. Langone, and M. B. Floyd, *J. Am. Chem. Soc.*, **93**, 3829 (1971).
[267] M. A. Iorio and M. Miraglia, *Tetrahedron*, **27**, 4983 (1971).
[268] T. Yvernault and M. Mazet, *Bull. Soc. Chim. France*, **1971**, 2652.
[269] G. E. Herberich and H. Müller, *Chem. Ber.*, **104**, 2781 (1971).
[270] J. W. Lewis and M. J. Readhead, *J. Chem. Soc.* (C), **1971**, 2298.
[271] H. Paulsen and M. Budzis, *Chem. Ber.*, **103**, 3794 (1970).
[272] T. Masuda, A. Chinone, and M. Ohta, *Bull. Chem. Soc. Japan*, **43**, 3287 (1970).
[273] P. F. Cann, D. Howells, and S. Warren, *Chem. Comm.*, **1971**, 1148.
[274] M. S. Khan and L. N. Owen, *J. Chem. Soc.* (C), **1971**, 1442.
[275] P. M. Pojer and I. D. Rae, *Tetrahedron Letters*, **1971**, 3081.
[276] A. McKillop, B. P. Swann, and E. C. Taylor, *Tetrahedron Letters*, **1971**, 5281.

in 85% yield by methanolic Th(NO$_3$)$_3$;[277] the stereoselectivity of the ring-contraction is illustrated by the formation of (**181**) from (**180**).[278] [1,2]-Alkyl shifts occur during the reactions of 20β-acetoxy-5α-pregnanes with BF$_3$,[279] the solvolyses of sulphonates from fenchane,[280] pinane,[281] tricycloundecane,[282] cholestane[283] and octahydrophenanthrene,[284] the ring-openings of epoxides of diterpene,[285] tricyclo-octene,[286] yomogi alcohol,[287] protoadamantene[288] and tropylidene,[289] and the AlHCl$_2$ reduction of the methanesulphonate of hydroxy-12β-conanine.[290] The initial product from the reaction of LiI and the *exo*-oxide of bicyclo[3.2.0]hept-6-ene is bicyclo[2.1.0]hexane-6-*endo*-carbaldehyde.[291] Hexamethylphosphoramide and LiBr in benzene is also an effective catalyst in promoting the rearrangements of epoxides to aldehydes and/or ketones.[292] The product of the thermal rearrangement of (**182**) and its *anti*-isomer is (**183**), indicating that complete inversion of configuration occurs at the migration terminus; the reaction is therefore non-concerted.[293] Pyrolysis of 2-adamantyl methanesulphonate at 520° affords a mixture of protoadamantene and 2,4-dehydroadamantane.[294] ^{14}C-Labelling experiments have established that adamantane undergoes a degenerate isomerization when heated with AlBr$_3$[295] in a manner analogous to that of 2-methyladamantane.[296] There have been further studies of the pinacol rearrangement,[297] and the effect of pressure on such processes has been examined.[298] A [1,2]-phenyl shift is implicated in the reaction of 2-bromo-1,3-diphenylpropanedione with *N*-methylhydrazine to give (**184**).[299]

A Wagner–Meerwein rearrangement occurs when the 2α,5-epithio-5α-cholestane derivative (**185**) is solvolysed; in contrast, the *endo*-isomer of (**185**), which reacts ca. 2.3 × 10^7 times more rapidly, affords a product with retained configuration.[300] S-Participation is also implicated in the mesylation of 1-arylthiopropan-2-ols[301] and the BF$_3$-promoted cyclization of arylthiovinyl sulphonates to benzo[*b*]thiophens.[302] The carbonyl-oxygen atom can also participate in the displacement of a leaving group on an

---

[277] A. McKillop, J. D. Hunt, E. C. Taylor, and F. Kienzle, *Tetrahedron Letters*, **1970**, 5275.
[278] E. J. Corey and T. Ravindranathan, *Tetrahedron Letters*, **1971**, 4753.
[279] S. Aoyama, K. Kamata, and T. Komeno, *Chem. Pharm. Bull. (Japan)*, **19**, 1329 (1971).
[280] A. Coulombeau, C. Coulombeau, and A. Rassat, *Bull. Soc. Chim. France*, **1970**, 4389.
[281] T. Hirata and T. Suga, *J. Org. Chem.*, **36**, 412 (1971).
[282] D. F. MacSweeney and R. Ramage, *Tetrahedron*, **27**, 1481 (1971).
[283] A. K. Bose and N. G. Steinberg, *J. Org. Chem.*, **36**, 2400 (1971).
[284] H. W. Whitlock, Jr., and L. E. Overman, *J. Am. Chem. Soc.* **93**, 2247 (1971); H. W. Whitlock, Jr., P. B. Reichardt, and F. M. Silver, *J. Am. Chem. Soc.*, **93**, 485 (1971).
[285] R. D. H. Murray, R. W. Mills, and J. M. Young, *Tetrahedron Letters*, **1971**, 2393.
[286] B. C. Henshaw, D. W. Rome, and B. L. Johnson, *Tetrahedron*, **27**, 2255 (1971).
[287] A. F. Thomas and W. Pawlak, *Helv. Chim. Acta*, **54**, 1822 (1971).
[288] D. Lenoir, P. von R. Schleyer, C. A. Cupas, and W. E. Heyd, *Chem. Comm.*, **1971**, 26.
[289] P. W. Schiess and M. Wisson, *Tetrahedron Letters*, **1971**, 2389.
[290] G. Lukacs, P. Longevialle, and X. Lusinchi, *Tetrahedron*, **27**, 1891 (1971).
[291] D. L. Garin, *J. Org. Chem.*, **36**, 1697 (1971).
[292] B. Rickborn and R. M. Gerkin, *J. Am. Chem. Soc.*, **93**, 1693 (1971).
[293] J. Salaun and J. M. Conia, *Tetrahedron Letters*, **1971**, 4023.
[294] J. Boyd and K. H. Overton, *Chem. Comm.*, **1971**, 211.
[295] Z. Majerski, S. H. Liggero, P. von R. Schleyer, and A. P. Wolf, *Chem. Comm.*, **1970**, 1596.
[296] *Org. Reaction Mech.*, **1970**, 287.
[297] G. Dana and J. Wiemann, *Bull. Soc. Chim. France*, **1970**, 3894.
[298] T. Moriyoshi, and K. Tamura, *Rev. Phys. Chem. Japan*, **40**, 48 (1970).
[299] M. J. Nye and W. P. Tang, *Chem. Comm.*, **1971**, 1395.
[300] T. Tsuji, T. Komeno, H. Itani, and H. Tanida, *J. Org. Chem.*, **36**, 1648 (1971).
[301] M. S. Khan and L. N. Owen, *J. Chem. Soc.* (C), **1971**, 1448.
[302] G. Capozzi, G. Melloni, and G. Modena, *J. Chem. Soc.* (C), **1970**, 2621; **1971**, 3018.

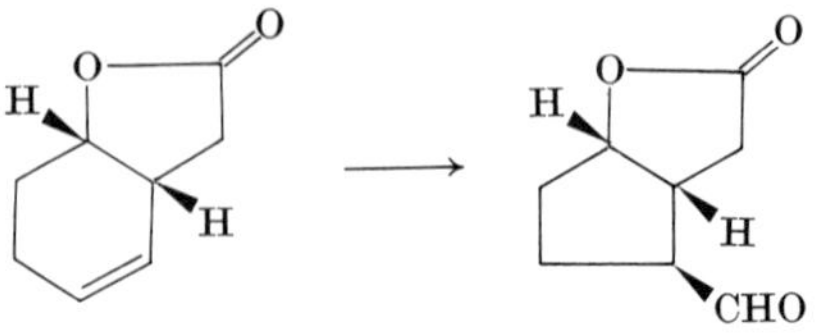

(180) (181)

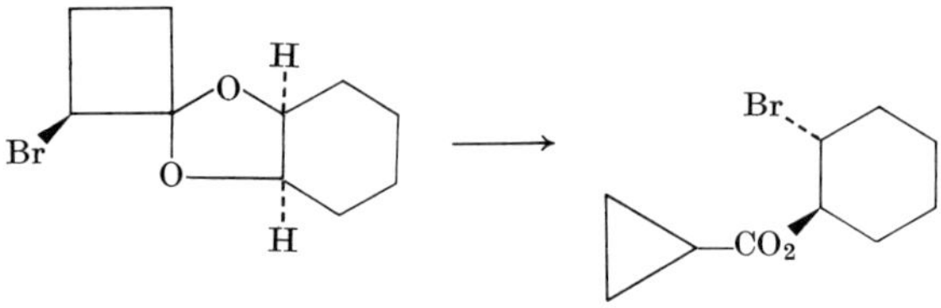

(182) (183) 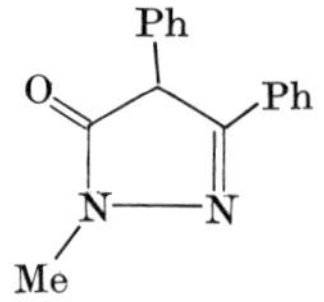 (184)

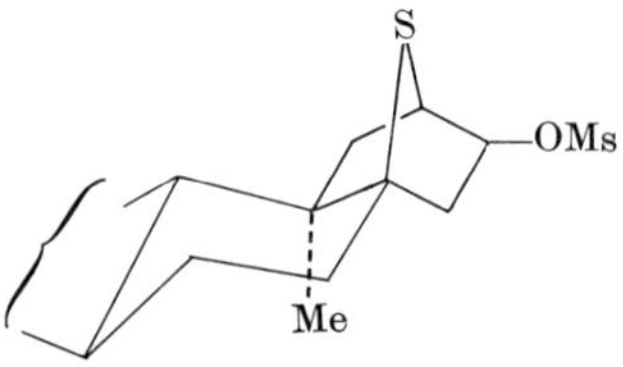

(185)  (186)

adjacent C atom;[303] thus, ^{14}C-labelling studies suggest that (186) intervenes in the AgClO$_4$-promoted ethanolysis of 2-chlorocyclohexanone. A [1,2]-shift of the ring-oxygen atom occurs in the reaction of α-D-glucopyranose 4-methanesulphonates with NaN$_3$.[304]

Shifts to Tertiary Centres. The formation of 2-acetyl-4-t-butyl-1-methylcyclohexyl chloride in the reaction of pinane (187) with acetyl chloride and AlCl$_3$ probably takes place by way of (188), formed by hydride abstraction and cleavage of the cyclobutyl ring.[305]

Details of the protoadamantane route to disubstituted adamantanes have been published.[306] In addition to undergoing a pinacol rearrangement, biadamantane-2,2'-

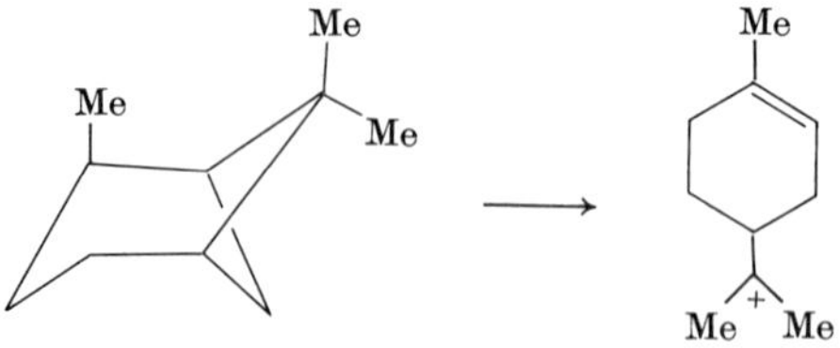

(187) (188)

[303] T. Masuike, N. Furukawa, and S. Oae, *Bull. Chem. Soc. Japan*, **44**, 448 (1971); A. Dmytraczenko, W. A. Szarek, and J. K. N. Jones, *Chem. Comm.*, **1970**, 1220.
[304] C. Bullock, L. Hough, and A. C. Richardson, *Chem. Comm.*, **1971**, 1276.
[305] R. F. Tavares, J. Dorsky, and W. M. Easter, *J. Org. Chem.*, **36**, 2434 (1971).
[306] D. Lenoir, R. Glaser, P. Mison, and P. von R. Schleyer, *J. Org. Chem.*, **36**, 1821 (1971).

diol affords the epoxide in the presence of acid.³⁰⁷ With $ZnCl_2$ spiro[adamantane-2,4′-homoadamantan-5′-ol] yields adamantylideneadamantane and 3-bishomoadamantene, although spiro[adamantane-2,5′-homoadamant-2′-ene] and spiro[adamantane-2,5′-(2′,4′-dehydrohomoadamantane)] are the major products in the presence of $SOCl_2$.³⁰⁸ Alkyl shifts take place during the acid-catalysed dehydrations of steroidal tertiary alcohols³⁰⁹ and the ring-openings of steroidal,³¹⁰ terpene,³¹¹ cyclohexane,³¹² and bicyclodecane³¹³ epoxides. The acid-catalysed dehydration of monoterpene alcohols often leads to complex mixtures of products; in certain cases these can be avoided by the use of SiO_2 and, for example, 2-*endo*-phenylisoborneol is specifically rearranged to 1-phenylcamphene.³¹⁴ Examples of Al_2O_3-induced oxiran rearrangements have been investigated.³¹⁵ The epoxide (**189**) undergoes an acid-induced hydride shift to give 2-*endo*-phenylbornan-3-one, whereas under corresponding conditions the *exo*-isomer of (**189**) gives predominantly 2,2,3-trimethylcyclohex-3-enecarbaldehyde.³¹⁶

Rearrangement of (**190**) to (**191**) occurs with BF_3-acetic anhydride,³¹⁷ while, in the presence of $LiAlH_4$–$AlCl_3$, (**192**) undergoes a stereospecific ring-contraction to (**193**).³¹⁸

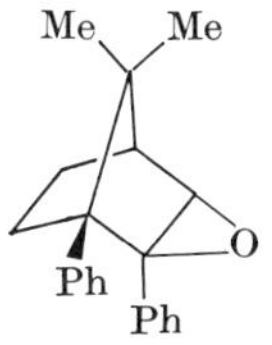

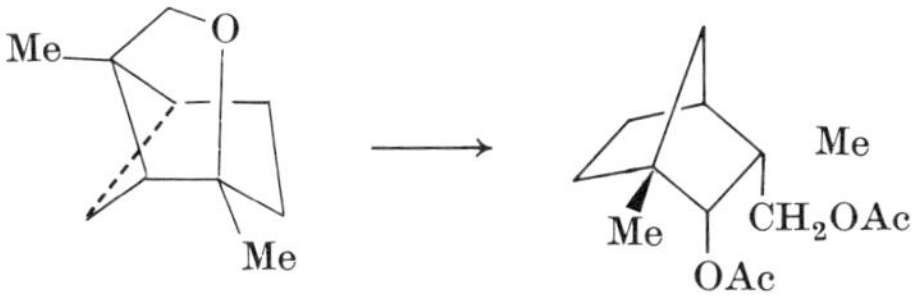

³⁰⁷ G. B. Gill and D. Hands, *Tetrahedron Letters*, **1971**, 181.
³⁰⁸ E. Boelema, H. Wynberg, and J. Strating, *Tetrahedron Letters*, **1971**, 4029.
³⁰⁹ J. M. Coxon, M. P. Hartshorn, and C. N. Muir, *Chem. Comm.*, **1970**, 1591; **1971**, 659; J. G. Ll. Jones and B. A. Marples, *J. Chem. Soc.* (C), **1971**, 572; F. Kohen, R. A. Mallory, and I. Scheer, *J. Org. Chem.*, **36**, 716 (1971); C. Monneret and Q. Khuong-Huu, *Bull. Soc. Chim. France*, **1971**, 623.
³¹⁰ D. Lavie and E. C. Levy, *Tetrahedron*, **27**, 3941 (1971); I. G. Guest, J. G. Ll. Jones, and B. A. Marples, *Tetrahedron Letters*, **1971**, 1979; I. G. Guest and B. A. Marples, *J. Chem. Soc.* (C), **1971**, 576; V. Tortorella, L. Toscano, C. Vetushi, and A. Romeo, *J. Chem. Soc.* (C), **1971**, 2422.
³¹¹ L. J. Ames, A. F. H. Baines, J. M. Coxon, and M. P. Hartshorn, *Austral. J. Chem.*, **24**, 1899 (1971); G. Berti, F. Bottari, A. Marsili, I. Morelli, and A. Mandelbaum, *Tetrahedron*, **27**, 2143 (1971); G. Berti, A. Marsili, I. Morelli, and A. Mandelbaum, *Tetrahedron*, **27**, 2217 (1971); L. K. Lala, *J. Org. Chem.*, **36**, 2560 (1971); P. A. Gunn, R. McCrindle, and R. G. Roy, *J. Chem. Soc.* (C), **1971**, 1018; H. Watanabe, J. Katsuhara, and N. Yamamoto, *Bull. Chem. Soc. Japan*, **44**, 1328 (1971).
³¹² G. Berti, B. Macchia, F. Macchia, and L. Monti, *J. Chem. Soc.* (C), **1971**, 3371.
³¹³ R. A. Kretchmer and W. J. Frazee, *J. Org. Chem.*, **36**, 2855 (1971).
³¹⁴ C. R. Hughes, D. F. MacSweeney, and R. Ramage, *Tetrahedron*, **27**, 2247 (1971).
³¹⁵ V. S. Joshi, N. P. Damodaran, and S. Dev, *Tetrahedron*, **27**, 459 (1971).
³¹⁶ J. M. Coxon, M. P. Hartshorn, and A. J. Lewis, *Austral. J. Chem.*, **24**, 1009 (1971).
³¹⁷ N. Bosworth and P. D. Magnus, *Chem. Comm.*, **1971**, 618.
³¹⁸ E. Demole, P. Enggist and M. C. Borer, *Helv. Chim. Acta*, **54**, 1845 (1971).

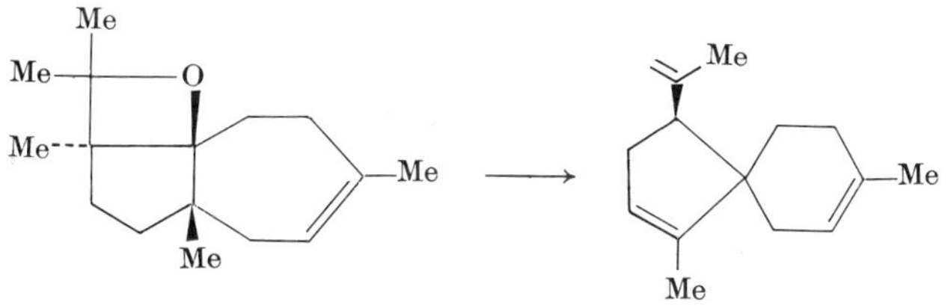

A ring-contraction also occurs when 1-bromo-7,7-dimethylnorbornanone is treated with KOH under forcing conditions, probably by a semi-benzylic mechanism.[319] Similar pathways are implicated in the reactions of other α-halo-ketones.[320] [1,2]-Phenyl shifts ensue when janusene dibromide is treated with silver acetate.[321]

[1,2]-*Migrations to Unsaturated Centres*

A common carbonium-ion intermediate is implicated in the acid-catalysed hydration of α- and β-pinene: migration of the methylene bridge of this species leads to fenchane derivatives, while migration of the isopropylidene bridge affords bornane derivatives.[322] The acid-induced interconversions of sativene, cyclosativene and isosativene[323] and the SiO_2–P_2O_5-catalysed isomerization of camphene[324] have been studied. A variety of rearrangements involving [1,2]-alkyl shifts, which are initiated by the protonation of steroidal olefins[325] and bicyclodecenes,[326] have been reported. An equilibrium mixture of (**195**) and (**196**) is obtained when (**194**) is treated with formic acid; D-labelling experiments, which establish that epimerization of the carboxyl group is not involved, are congruent with the route shown in Scheme 2.[327] X-ray crystallographic analysis demonstrates that (**197**) is the product of H_3PO_4-catalysed cyclization of 4-(2,6,6-trimethylcyclohexenyl)-2-methylbutanal.[328] The acid-promoted reorganizations of 1-methoxybenzobarrelenes have been examined;[329] thus, (**200**; R = H) is the major and (**201**) the minor product when (**198**; R^1 = Me, R^2 = R^3 = H) is treated with H_2SO_4. The rearrangement probably involves the intermediary of (**199**), formed by a [1,2]-phenyl shift from position 1 to position 6. In contrast, (**200**; R = Me) is obtained from (**198**; R^1 = H, R^2 = R^3 = Me); D-labelling studies implicate the intervention of (**202**), derived by a [1,2]-aryl migration from position 4 to position 5.

The addition of bromine to alkenes can induce skeletal rearrangements; for example, a 2:3 mixture of 2*ax*,4*eq*- and 2*eq*,4*eq*-dibromoadamantane is obtained from protoadamantene.[330] In the reactions of dehydrojanusene[331] and 2,6,6-trimethyl-7-oxabicyclo[3.2.1]oct-2-ene[332] with bromine, respectively [1,2]-phenyl and [1,2]-oxygen migrations occur. [1,2]-Shifts to vinyl cations are probably involved in the formation of homoadamantanone methyl ketone from prop-2-ynyl alcohol and the 1-adamantyl

---

[319] W. C. Fong, R. Thomas, and K. V. Scherer, Jr., *Tetrahedron Letters*, **1971**, 3789.
[320] D. Bawdry, J. P. Bégué, and M. Charpentier-Morize, *Bull. Soc. Chim. France*, **1971**, 1416.
[321] S. J. Cristol and M. A. Imhoff, *J. Org. Chem.*, **36**, 1854 (1971).
[322] C. M. Williams and D. Whittaker, *J. Chem. Soc.* (B), **1971**, 668, 672.
[323] J. E. McMurry, *J. Org. Chem.*, **36**, 2826 (1971).
[324] F. Petit, M. Evrard, and M. Blanchard, *Bull. Soc. Chim. France*, **1971**, 4176.
[325] F. Frappier, J. Thierry, and F.-X. Jarreau, *Tetrahedron Letters*, **1971**, 1887; C. Monneret, P. Choay, Q. Khuong-Huu, and R. Goutarel, *Tetrahedron Letters*, **1971**, 3223; H. Irie, S. Uyeo, and K. Kuriyama, *Tetrahedron Letters*, **1971**, 3467; R. J. Chambers and B. A. Marples, *Tetrahedron Letters*, **1971**, 3747; C. Berrier, J.-C. Jacquesy, and R. Jacquesy, *Tetrahedron Letters*, **1971**, 4567; P. Sengupta, J. Mukherjee, and M. Sen, *Tetrahedron*, **27**, 2473 (1971); D. N. Kirk and P. M. Shaw, *Chem. Comm.*, **1971**, 948; P. Bourguignon, J.-C. Jacquesy, R. Jacquesy, J. Levisalles, and J. Wagnon, *Bull. Soc. Chim. France*, **1971**, 269; J.-P. Berthelot, and J. Levisalles, *Bull. Soc. Chim. France*, **1971**, 1888.
[326] J.-P. Berthelot, J.-C. Jacquesy, and J. Levisalles, *Bull. Soc. Chim. France*, **1971**, 1896.
[327] D. J. Dunham and R. G. Lawton, *J. Am. Chem. Soc.*, **93**, 2075 (1971).
[328] G. Saucy, R. E. Ireland, J. Bordner, and R. E. Dickerson, *J. Org. Chem.*, **36**, 1195 (1971).
[329] H. Heaney and S. M. Ley, *Chem. Comm.*, **1971**, 224, 1342.
[330] B. D. Cuddy, D. Grant, and M. A. McKervey, *J. Chem. Soc.* (C), **1971**, 3173.
[331] S. J. Cristol and M. A. Imhoff, *J. Org. Chem.*, **36**, 1849 (1971).
[332] J. Wolinsky, R. O. Hutchins, and J. H. Thorstenson, *Tetrahedron*, **27**, 753 (1971).

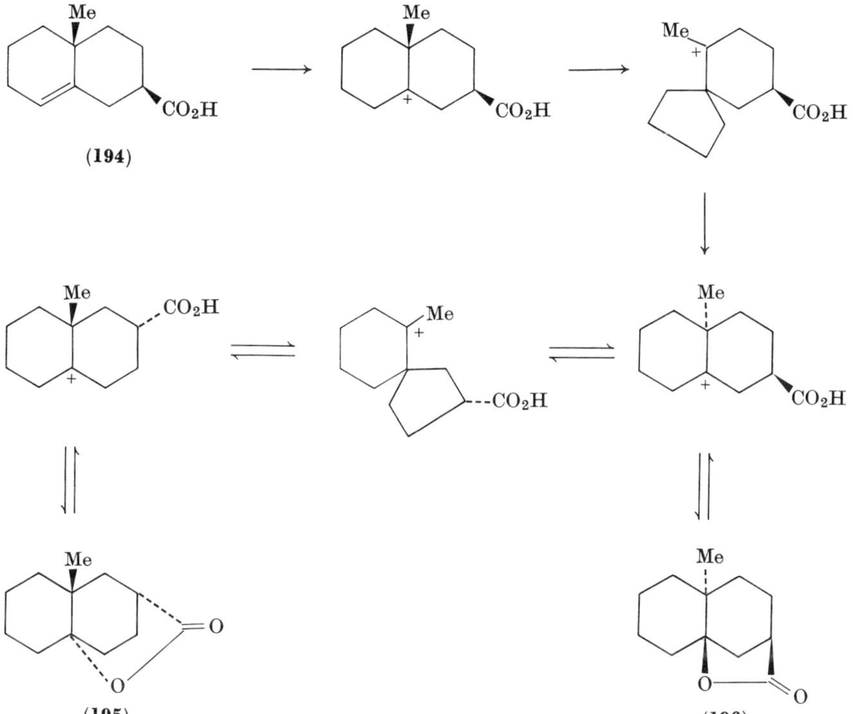

Molecular Rearrangements

249

Scheme 2

(194), (195), (196)

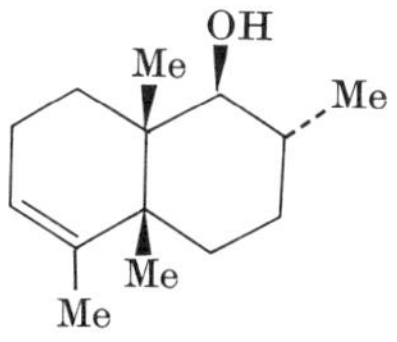

(197), 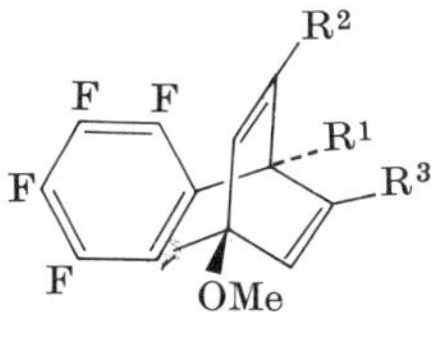 (198), 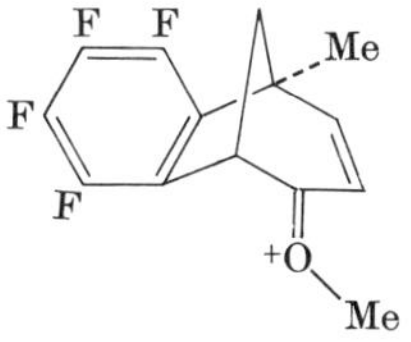 (199)

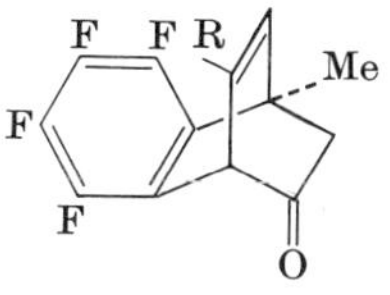

(200), 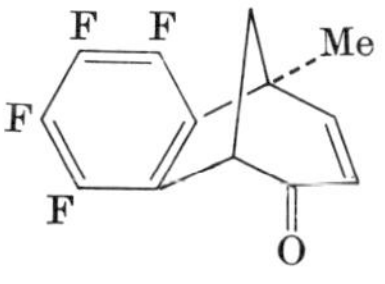 (201), 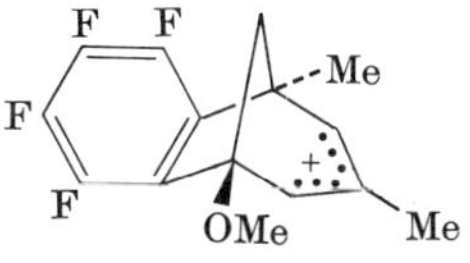 (202)

cation in strong H_2SO_4,[333] and the vapour-phase isomerization of 2,2-dimethylbut-3-yne to 2,3-dimethylbutadiene in the presence of SiO_2–Al_2O_3.[334] 1-Aryl-2-(diphenylphosphinyl)-2-phenylethanes are formed from triphenylphosphine, an arylacetylene and water; the [1,2]-phenyl shift occurs more readily when electron-withdrawing groups are present in the phenyl ring of the arylacetylene.[335] The NaCN-induced rearrangement of 7β-iodoneopinone (**203**) to (**205**) involves the intermediacy of (**204**), which can be isolated from the reaction as its perchlorate salt.[336]

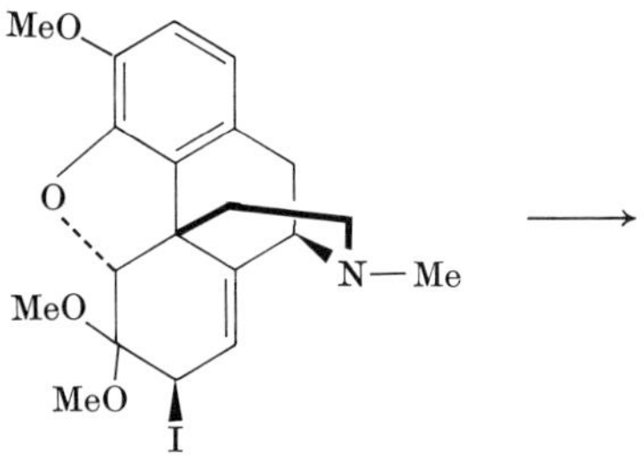

Surprisingly, T-labelling studies indicate that a [1,2]-hydride shift represents the major route for the alkali-induced rearrangement of D-ribose.[337] The requirement of a small amount of water for the toluene-*p*-sulphonic acid-promoted conversion of α-hydroxy-αα-diphenylacetaldehyde into benzoin may indicate that (**206**) is involved.[338]

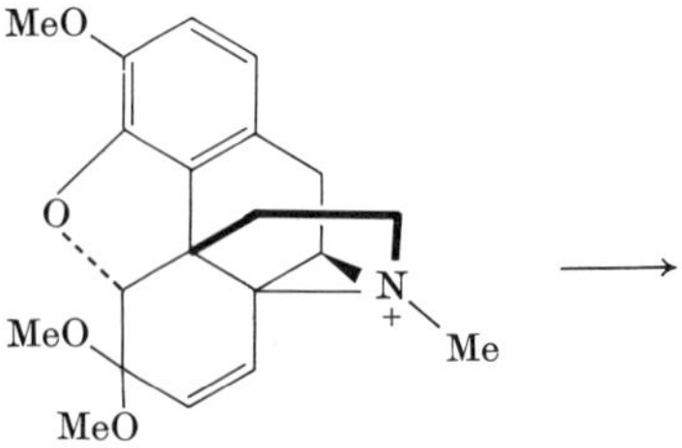

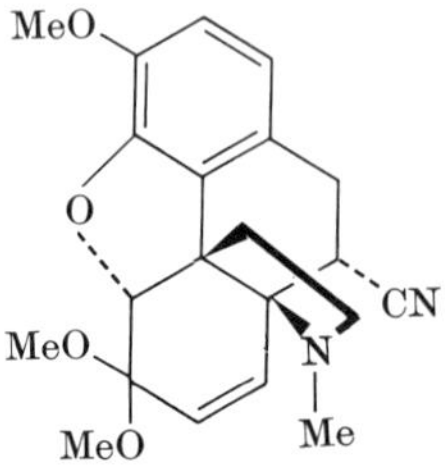

[1,2]-Shifts to a ketonic carbonyl group occur in the reactions of terpenes with BF_3,[339] bicyclo[2.2.2]octadienones with acid,[340] and protoadamantanone with PCl_5.[306,330,341] There have been further studies of the α-ketol rearrangement.[342] For example, the alkali-induced equilibration of 1-hydroxy-3,3-dimethylnorbornan-2-one and its 7,7-dimethyl isomer shows that the latter derivative is slightly favoured; moreover,

---

[333] J. K. Chakrabarti and A. Todd, *Chem. Comm.*, **1971**, 556.
[334] A. Mortreux and M. Blanchard, *Bull. Chem. Soc. France*, **1970**, 4035.
[335] E. M. Richards and J. C. Tebby, *J. Chem. Soc.* (C), **1971**, 1059.
[336] R. M. Allen and G. W. Kirby, *Chem. Comm.*, **1971**, 1121.
[337] W. B. Gleason and R. Barker, *Can. J. Chem.*, **49**, 1433 (1971).
[338] T. D. Inch, P. Watts, and N. Williams, *Chem. Comm.*, **1971**, 174.
[339] W. F. Erman, R. S. Treptow, P. Bakuzis, and E. Wenkert, *J. Am. Chem. Soc.*, **93**, 657 (1971); J. Lhomme and G. Ourisson, *Bull. Soc. Chim. France*, **1970**, 3935.
[340] H. Hart and G. M. Love, *Tetrahedron Letters*, **1971**, 2267.
[341] B. D. Cuddy, D. Grant, and M. A. McKervey, *Chem. Comm.*, **1971**, 27.
[342] R. W. Alder, *Tetrahedron Letters*, **1971**, 193; M. J. Nye and W. P. Tang, *Chem. Comm.*, **1971**, 1394.

NMR spectroscopy reveals that the isomers are in rapid equilibrium at 180°.[343] 1-Hydroperoxy-1-isopropylnaphthalen-2(1*H*)-one rearranges to a mixture of (**207**) and (**208**); the former product probably arises by an intramolecular epoxidation followed by an α-ketol rearrangement and the latter by an intermolecular epoxidation.[344] Water induces the ring-contraction of cyclobutane-1,2-dione to 1-hydroxycyclopropanecarboxylic acid.[345] The corresponding Ba(OH)$_2$-promoted ring-contractions of 5α-oestrane-3,4- and 2,3-diones have been studied; ^{14}C-labelling experiments reveal that there is a 3:1 preference for migration of the 2—3 bond over the 4—5 bond in the former case, and an equal probability for shifts of the 1—2 and 3—4 bonds in the latter.[346] The mechanism of the thenilic acid rearrangement has been investigated kinetically.[347] Kinetic measurements of the AlCl$_3$-promoted isomerizations of 2-cyclohexyl-2-phenylpentan-3-one, cyclohexyl α-ethyl-α-methylbenzyl ketone, and 2-cyclohexyl-2-methylbutyrophenone have also been carried out.[348] [1,2]-Aryl shifts are implicated in the reactions of 5,5-diaryldithiohydantoins with AlCl$_3$,[349] and in the KSCN-promoted transformation of α-benzoyl-α-chlorobenzyl benzyl sulphide into (**209**).[350] A double

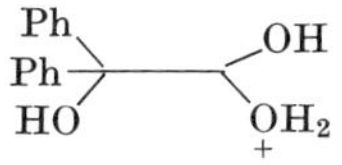

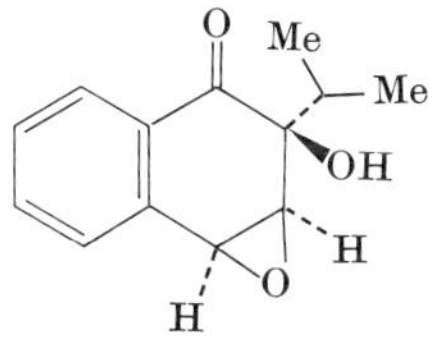

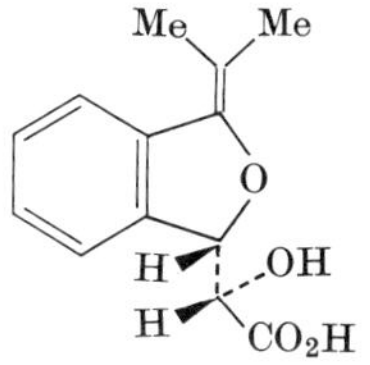

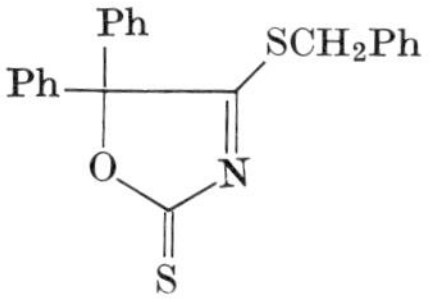

ketol rearrangement involving the [1,2]-shift of a heteroatom is exemplified in the NaOMe-induced conversion of (**210**) into (**211**); the structure of the product has been secured by X-ray analysis.[351]

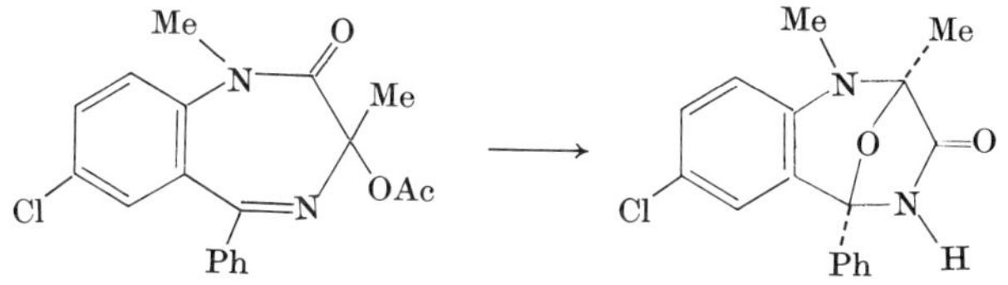

A number of reactions have been reported which involve [1,2]-shifts to the imino group. Alkaloids provide examples in which bond fragmentations result;[352] thus, tabersonine

[343] J. V. Paukstelis and D. N. Stevens, *Tetrahedron Letters*, **1971**, 3549; A. Nickon, T. Nishida, J. Frank, and R. Muneyuki, *J. Org. Chem.*, **36**, 1075 (1971).
[344] J. Carnduff and D. G. Leppard, *Chem. Comm.*, **1971**, 975.
[345] J. M. Conia and J. M. Denis, *Tetrahedron Letters*, **1971**, 2845; H. G. Heine, *Chem. Ber.*, **104**, 2869 (1971).
[346] J. Alais and J. Levisalles, *Bull. Soc. Chim. France*, **1971**, 3731, 3737.
[347] G. P. Nilles and R. D. Schuetz, *J. Org. Chem.*, **36**, 2486, 2489 (1971).
[348] B. Calas and L. Giral, *Bull. Soc. Chim. France*, **1971**, 2629.
[349] E. Koltai, J. Nyitrai, K. Lempert, and L. Bursics, *Chem. Ber.*, **104**, 290 (1971).
[350] D. N. Harpp and P. Mathiaparam, *J. Org. Chem.*, **36**, 2886 (1971).
[351] A. Walser, G. Silverman, J. Blount, R. I. Fryer, and L. H. Sternbach, *J. Org. Chem.*, **36**, 1465 (1971).
[352] C. Pierron, J. Garnier, J. Lévy, and J. Le Men, *Tetrahedron Letters*, **1971**, 1007; M.-J. Hoizey, L. Olivier, J. Lévy, and J. Le Men, *Tetrahedron Letters*, **1971**, 1071; R. T. Brown, J. S. Hill, G. F. Smith, and K. S. J. Stapleford, *Tetrahedron*, **27**, 5217 (1971).

(212) is reduced by formic acid to (213). In contrast, Zn–acetic acid reduction of (212) affords (214) a [1,2]-alkyl shift taking place;[353] under corresponding conditions related alkaloids undergo similar reorganizations.[354] Migration of the phenyl or methyl group ensues when (215) is heated; the reaction is undoubtedly intramolecular since full retention of activity in the ketonic products is observed with an optically active reactant.[355] A [1,2]-methyl shift is implicated in the thermally induced conversion of 2,3,3,4,5-pentamethyl-3H-pyrrolenine into 2,3,4,5,5-pentamethyl-5H-pyrrolenine since the reaction is acid-catalysed and is speeded up with an increase in solvent polarity.[356]

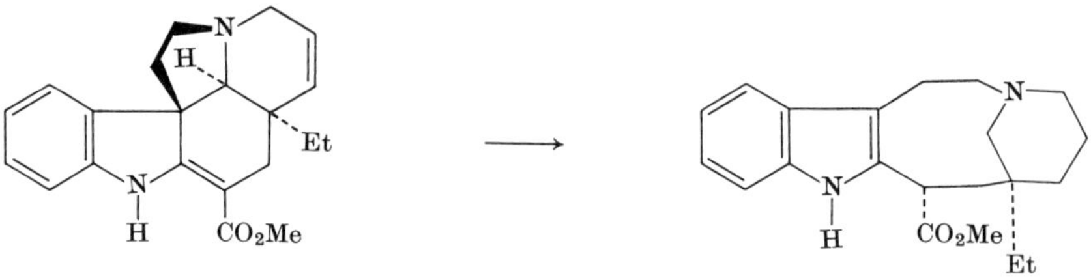

[1,3]-*Migrations*

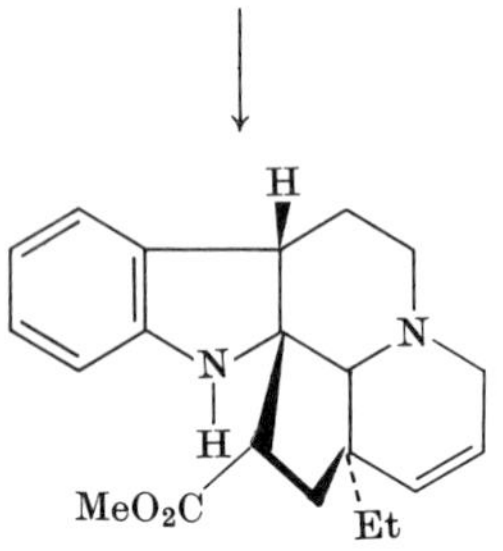

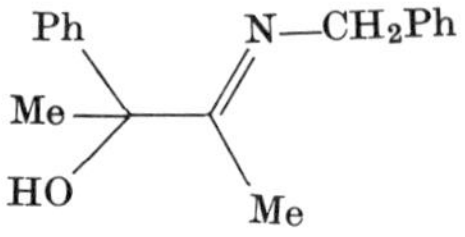

The Rupe and Meyer–Schuster rearrangements have been reviewed.[357]

Kinetic studies suggest that the thermally induced isomerization of N-arylaziridine-1-carboximidoyl chlorides to N-aryl-N-(2-chloroalkyl)carbodiimides involves heterolysis of the C—Cl bond in the slow step.[358] The rate of thermal equilibration of 3-chloro-2-ethyl-3-methyl- and 3-chloro-3-ethyl-2-methyl-3H-azirines is sensitive to solvent polarity, suggesting that the azacyclopropenyl ion–chloride ion pair intervenes.[359] ^{18}O-Labelling experiments reveal that the Pd(II)-catalysed allylic rearrangement of but-2-enyl propionate involves a cyclic six-membered transition state.[360] Allylic rearrangements take place with hex-2-enopyranosylpurine nucleosides[361] and in the

[353] P. Maupérin, J. Lévy, and J. Le Men, *Tetrahedron Letters*, **1971**, 999.
[354] W. B. Hinshaw, Jr., J. Lévy, and J. Le Men, *Tetrahedron Letters*, **1971**, 995; J. Lévy, P. Maupérin, M. D. de Maindreville, and J. Le Men, *Tetrahedron Letters*, **1971**, 1003.
[355] H. Mizuno, S. Terashima, and S. Yamada, *Chem. Pharm. Bull. (Japan)*, **19**, 227 (1971).
[356] J. L. Wong, M. H. Ritchie, and C. M. Gladstone, *Chem. Comm.*, **1971**, 1093.
[357] S. Swaminathan and K. V. Narayanan, *Chem. Rev.*, **71**, 429 (1971).
[358] D. A. Tomalia, T. J. Giacobbe, and W. A. Sprenger, *J. Org. Chem.*, **36**, 2142 (1971).
[359] J. Ciabattoni and M. Cabell, Jr., *J. Am. Chem. Soc.*, **93**, 1482 (1971).
[360] P. M. Henry, *Chem. Comm.*, **1971**, 328.
[361] R. J. Ferrier and M. M. Ponpipom, *J. Chem. Soc. (C)*, **1971**, 553.

oxidation of 1-arylprop-2-enes with 2,3-dichloro-5,6-dicyanobenzoquinone.[362] The LiAlH$_4$ reduction of (**216**) affords (**217**), congruent with a synfacial displacement.[363] On reaction with NaOMe 8-chloro-8-methylbicyclo[4.2.0]octan-7-one affords 1-

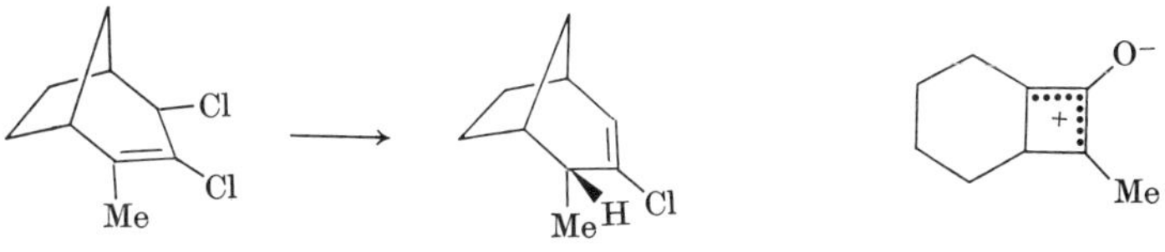

(**216**)   (**217**)   (**218**)

methoxy-8-methylbicyclo[4.2.0]octan-7-one, probably through (**218**).[364] The first step in the potassium acetate-induced conversion of 6,6-dichlorobicyclo[3.2.0]hept-2-en-8-one into tropolone involves a similar rearrangement and the intermediate 1-acetoxy-6-chlorobicyclo[3.2.0]hept-2-en-8-ones can be isolated.[365] Related reorganizations occur in the reaction of 4β-acetoxy-1,2-dihydrosantonine with acetic anhydride[366] and the sodium acetate-promoted conversion of 4-bromo- into 2-acetoxy-3-keto-5β-steroids.[367]

Other Migrations

(2-Benzyloxyphenyl)diphenylmethyl chloride rearranges to (α-chlorobenzyloxyphenyl)-diphenylmethane when heated in benzene; the reaction probably involves an intramolecular [1,5]-hydride shift.[368] When 2-(1-adamantyl)propan-2-ol is treated with H$_2$SO$_4$–formic acid 3-isopropyladamantane-1-carboxylic acid is produced, suggesting that the initially formed carbonium ion rearranges by an intermolecular hydride shift.[369]

Metal-catalysed Rearrangements

There have been extensive studies of the Ag(I) ion-catalysed rearrangements of molecules containing strained σ-bonds and the subject has been reviewed.[370]

The isomerization of cubane to cuneane (**222**) is considered to involve the oxidative addition of Ag(I) ion to a σ-bond to give (**219**), which rearranges to the product via (**220**) and (**221**).[371] The rate constants of such isomerizations correlate with the Taft σ* constants of the substituents, implying that the metal–σ-bond interactions are similar to those in Ag(I) ion–olefin π-complexes; thus, electron-withdrawing substituents weaken the σ-complex and slow down the rearrangement rate.[372] Related reorganizations occur

---

[362] F. E. Lutz and E. F. Kiefer, *Chem. Comm.*, **1970**, 1722.
[363] C. W. Jefford, A. Sweeney, D. T. Hill, and F. Delay, *Helv. Chim. Acta*, **54**, 1691 (1971).
[364] W. T. Brady and J. P. Hieble, *J. Org. Chem.*, **36**, 2033 (1971).
[365] P. D. Bartlett and T. Ando, *J. Am. Chem. Soc.*, **92**, 7518 (1970).
[366] P. S. Aumeer and T. B. H. McMurry, *Chem. Comm.*, **1971**, 641.
[367] R. B. Warneboldt and L. Weiler, *Tetrahedron Letters*, **1971**, 3413.
[368] W. H. Starness, Jr., *J. Org. Chem.*, **36**, 2508 (1971).
[369] D. J. Raber, R. C. Fort, E. Wiskott, C. W. Woodworth, P. von R. Schleyer, J. Weber, and H. Stetter, *Tetrahedron*, **27**, 3 (1971).
[370] L. A. Paquette, *Accounts Current Chem. Res.*, **4**, 280 (1971).
[371] J. E. Byrd, L. Cassar, P. E. Eaton, and J. Halpern, *Chem. Comm.*, **1971**, 40.
[372] G. F. Koser, *Chem. Comm.*, **1971**, 388.

with homocubane (and can also be induced by copper bronze)[373] and 1,1-dihomocubane derivatives.[374] The stereochemical integrity of substituents is maintained in the corresponding rearrangements of *seco*-cubane derivatives;[374,375] for example, (**224**) is obtained from (**223**). Moreover, the rates of these reactions are only slightly affected by changing the 2- and 5-substituents, which suggests that an Ag(I) ion-induced cleavage of a C—C bond on the cubic surfaces is involved in the rate-determining step.[375] The major product of the AgBF$_4$-induced reaction of (**225**) is (**226**) although the *anti*-isomer of (**225**) is unreactive.[376] Rapid isomerization of the prismane (**227**) ensues in the presence of AgNO$_3$ to give a mixture of 1,2,5-tri-t-butylbicyclo[2.2.0]hexa-2,5-diene and (**231**); the intervention of (**228**), which gives the Dewar benzene via (**229**) and the benzvalene via (**230**), is postulated. Both products rearrange to 1,2,4-tri-t-butylbenzene in slower reactions.[377]

The rearrangements of bicyclobutanes to butadienes have been scrutinized, it being shown that the course of such reactions is sensitive to the bicyclobutane substituents and to the metal-ion catalyst. For example, cleavage of the 1—2 and 3—4 bonds results when (**232**; $R^1 = R^3 =$ Me, $R^2 = R^4 =$ H) and (**232**; $R^1 = R^4 =$ Me, $R^2 = R^3 =$ H) are treated with AgClO$_4$.[378] In contrast, rupture of the 1—2 and 1—3 bonds is a competing reaction when bis(benzonitrile)dichloropalladium(II) is employed as the catalyst.[379] The reactions of 1,2,2-trimethylbicyclobutane with Ag(I) ion,[380] pentafluorophenylcopper tetramer,[381] and di-μ-chlorobis[π-allylpalladium(II)][382] involve 1—2 and 1—3 bond-breakages, whereas cleavage of the 1—3 and 2—3 bonds is observed with bis(benzonitrile)dichloropalladium(II);[382] these results demonstrate that the reaction outcome depends not only on the metal ion but also on the ligand. However, 2,5-dimethylhexa-2,4-diene is the common product of the reaction of (**232**; $R^1 = R^2 = R^3 = R^4 =$ Me) with both AgBF$_4$[380] and tetracarbonyl-μ-dichlorodirhodium.[383] The latter catalyst is also effective in cleaving the 2—3 and 1—4 bonds of 1-methyl-2,2-diphenylbicyclobutane; the formation of 3,4-dihydro-2-methyl-1-phenylnaphthalene and 3,4-dihydro-2-methyl-1-phenylazulene is concordant with the intermediacy of (**233**).[384]

Dramatic differences are also observed in the opening of (**234**; R = H):[385] thus, Ag(I) ion affords *cis,cis*-cyclohepta-1,3-diene,[386] tetracarbonyl-μ-dichlorodirhodium gives 3-methylenecyclohexene,[383] bis(benzonitrile)dichloropalladium(II) furnishes a mixture of *cis,cis*-cyclohepta-1,3-diene and 3-methylenecyclohexene,[387] and *trans*-chlorocar-

---

[373] N. B. Chapman, J. M. Key, and K. J. Toyne, *Tetrahedron Letters*, **1970**, 5211.
[374] W. G. Dauben, C. H. Schallhorn, and D. L. Whalen, *J. Am. Chem. Soc.*, **93**, 1446 (1971); L. A. Paquette and J. C. Stowell, *J. Am. Chem. Soc.*, **93**, 2459 (1971).
[375] L. A. Paquette, R. S. Beckley, and T. McCreadie, *Tetrahedron Letters*, **1971**, 775; H. H. Westberg and H. Ona, *Chem. Comm.*, **1971**, 248.
[376] J. Wristers, L. Brener, and R. Pettit, *J. Am. Chem. Soc.*, **92**, 7499 (1970).
[377] K. L. Kaiser, R. F. Childs, and M. M. Maitlis, *J. Am. Chem. Soc.*, **93**, 1270 (1971).
[378] M. Sakai, H. Yamaguchi, H. H. Westberg, and S. Masamune, *J. Am. Chem. Soc.*, **93**, 1043 (1971); L. A. Paquette, S. E. Wilson, and R. B. Henzel, *J. Am. Chem. Soc.*, **93**, 1288 (1971).
[379] M. Sakai, H. Yamaguchi, and S. Masamune, *Chem. Comm.*, **1971**, 486.
[380] L. A. Paquette, R. P. Henzel, and S. E. Wilson, *J. Am. Chem. Soc.*, **93**, 2335 (1971).
[381] P. G. Gassman and F. J. Williams, *J. Am. Chem. Soc.*, **92**, 7631 (1970); *Tetrahedron Letters*, **1971**, 1409.
[382] P. G. Gassman, G. R. Meyer, and F. J. Williams, *Chem. Comm.*, **1971**, 842.
[383] P. G. Gassman, T. J. Atkins, and F. J. Williams, *J. Am. Chem. Soc.*, **93**, 1812 (1971).
[384] P. G. Gassman and T. Nakai, *J. Am. Chem. Soc.*, **93**, 5897 (1971).
[385] P. G. Gassman and T. J. Atkins, *J. Am. Chem. Soc.*, **93**, 4597 (1971).
[386] *Org. Reaction Mech.*, **1970**, 276.
[387] M. Sakai and S. Masamune, *J. Am. Chem. Soc.*, **93**, 4610 (1971).

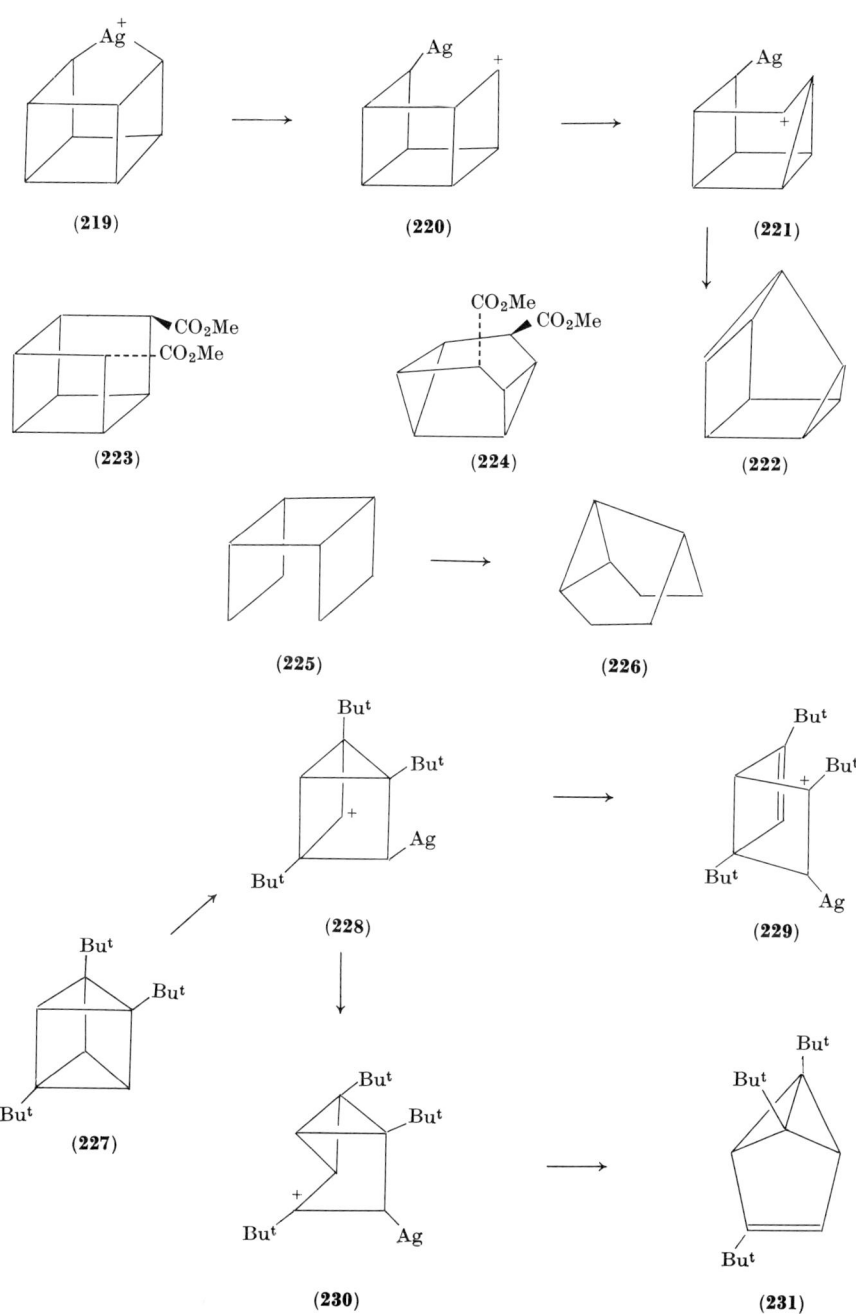

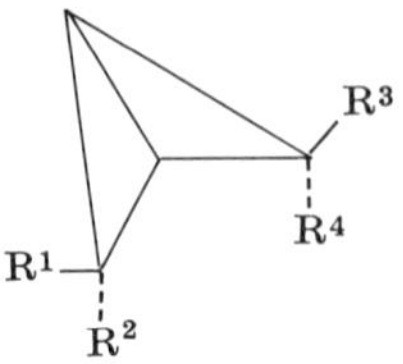

(232) 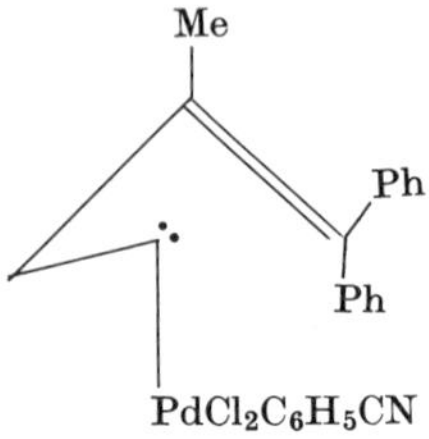 (233)

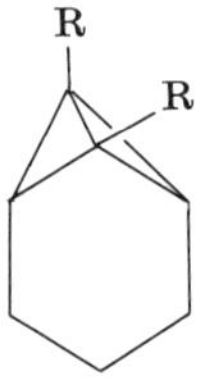

(234) 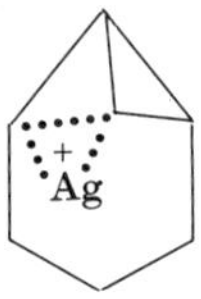 (235) 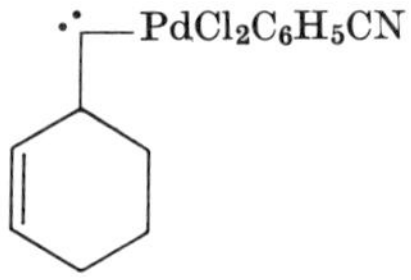 (236)

bonylbis(triphenylphosphine)rhodium yields a mixture of 3-methylenecyclohexene and bicyclo[4.1.0]hept-2-ene.[388] The Ag(I) ion-catalysed rearrangement of (**234**; R = D) proceeds more rapidly than that of (**234**; R = H); the inverse D-isotope effect (k_H/k_D = 0.85) is consistent with the intervention of the co-ordinated cyclopropyl cation (**235**) which rearranges to the product by electrocyclic ring-opening and expulsion of the Ag(I) ion.[389] Moreover, when the reaction of (**234**; R = H) with AgClO$_4$ is performed in methanol the 2-methoxybicyclo[4.1.0]heptanes are produced, providing further evidence for the intermediacy of (**235**).[390] Since bis(benzonitrile)dichloropalladium reacts with (cyclohex-3-enyl)diazomethane to give a mixture of 3-methylenecyclohexene and cis,cis-cyclohepta-1,3-diene, which is similar to that obtained in the corresponding reactions of (**234**; R = H), the common metal-complexed carbene intermediate (**236**) is implicated.[387] The Hg(OAc)$_2$-induced oxidative cleavage of 1,3,3-trimethylcyclopropane to 1,1-diacetoxy-2,3-dimethylbut-2-ene is also postulated as involving a metal–carbene complex.[391]

When heated at 300° bicyclo[2.1.0]pentane is isomerized to cyclopentene; in the presence of tetracarbonyl-μ-dichlorodirhodium the reaction occurs at room temperature.[392] This catalyst also cleaves the propellane (**237**) to give a mixture of 4- and 5-methylenecyclopentenes.[393] A low yield of bicyclo[3.2.1]octa-2,6-diene (**239**; R = H) is formed when (**238**; R = H) is heated with chlorotris(triphenylphosphine)rhodium; isotopic labelling studies reveal that (**238**; R = D) affords (**239**; R = D) and therefore a H atom is transferred stereospecially across the face of the molecule.[394]

Nickel derivatives that are particularly effective in catalysing the isomerization of cis-hexa-1,4-diene to trans-2-methylpenta-1,3-diene have been prepared.[395] Isotopic

[388] P. G. Gassman and T. J. Atkins, *J. Am. Chem. Soc.*, **93**, 1042 (1971).
[389] L. A. Paquette and S. E. Wilson, *J. Am. Chem. Soc.*, **93**, 5934 (1971).
[390] M. Sakai, H. H. Westberg, H. Yamaguchi, and S. Masamune, *J. Am. Chem. Soc.*, **93**, 4611 (1971).
[391] T. Shirafugi, Y. Yamamoto, and H. Nozaki, *Tetrahedron Letters*, **1971**, 4713.
[392] P. G. Gassman, T. J. Atkins, and T. J. Lumb, *Tetrahedron Letters*, **1971**, 1643.
[393] P. G. Gassman and E. A. Armour, *Tetrahedron Letters*, **1971**, 1431.
[394] T. J. Katz and S. A. Cerefice, *J. Am. Chem. Soc.*, **93**, 1049 (1971).
[395] L. W. Gosser and G. W. Parshall, *Tetrahedron Letters*, **1971**, 2555; R. G. Miller, P. A. Pinke, R. D. Stauffer, and H. J. Golden, *J. Organomet. Chem.*, **29**, C42 (1971).

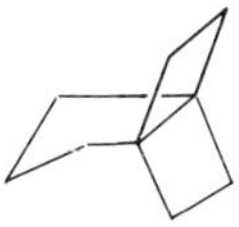

(237) 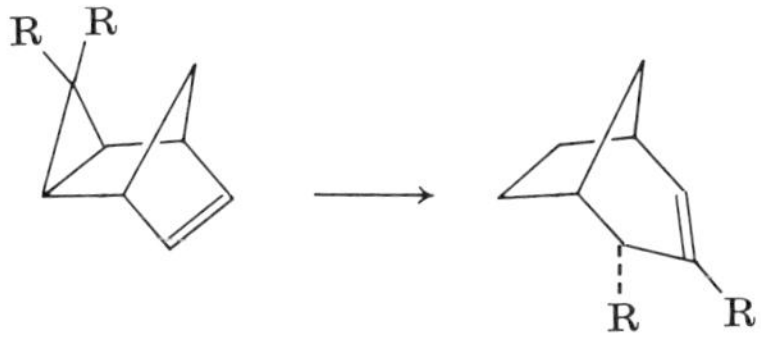　　　(238)　　　(239)

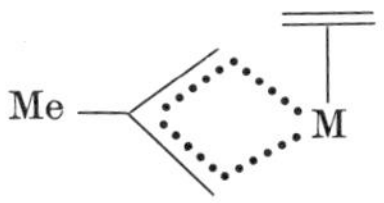

(240)

labelling studies reveal that 2-methyl[1,1-²H₂]penta-1,4-diene is converted into 2-methyl[3,3-²H₂]penta-1,4-diene in the presence of such catalysts; the intermediacy of (240) is postulated.[396]

The role of the metal in catalysing symmetry-forbidden valence isomerizations, such as the disproportionation of olefins, has been discussed.[397] 1,2,3,4,5,6-Hexamethylbicyclo[2.2.0]hexa-2,5-diene undergoes metal-catalysed ring-opening to hexamethylbenzene,[398] whereas the monoepoxide of the former derivative affords predominantly methyl [5-(1,2,3,4,5-pentamethylcyclopentadienyl)] ketone in the presence of tetracarbonyl-μ-dichlororhodirhodium.[399] Dewar benzene derivatives are formed when 1,1'-bicyclopropenes are heated with AgClO₄.[400]

Rearrangements Involving Electron-deficient Heteroatoms

[1,2]-*Migrations*

A review of the Beckmann rearrangement has been published.[401] The rearrangements of the oximes of homoadamantan-4-one,[402] cyclopentyl phenyl ketone,[403] 5-methylhex-3-en-2-one,[404] cyclic αβ-unsaturated ketones[405] and 3-o-carboranes[406] have been examined. 4-Methoxy-1-azabicyclo[3.2.0]hept-3-ene-2,7-dione is formed from (241) at 60°, presumably via 4-methoxyazacyclohepta-3,5-diene-2,7-dione.[407] Complexes of SO₃ and Lewis bases have been examined as reagents for promoting the Beckmann

---

[396] R. G. Miller, H. J. Golden, D. J. Baker, and R. D. Stauffer, *J. Am. Chem. Soc.*, **93**, 6308 (1971).
[397] F. D. Mango, *Tetrahedron Letters*, **1971**, 505; G. S. Lewandos and R. Pettit, *Tetrahedron Letters*, **1971**, 397.
[398] C. J. Attridge and S. J. Maddock, *J. Organomet. Chem.*, **26**, C25 (1971).
[399] R. Grigg and G. Shelton, *Chem. Comm.*, **1971**, 1247.
[400] R. Weiss and C. Schlierf, *Angew. Chem. Int. Ed.*, **10**, 811 (1971).
[401] H. Mukamal, *Nuova Chem.*, **47**, 79 (1971); *Chem. Abs.*, **75**, 75559 (1971).
[402] T. Sasaki, S. Eguchi, and T. Toru, *J. Org. Chem.*, **36**, 2454 (1971).
[403] C. Beaute, N. Thoai, and J. Wiemann, *Bull. Soc. Chim. France*, **1971**, 3327.
[404] N. Thoai, N. Chieu, and J. Wiemann, *Ann. Chim. (France)*, **6**, 235 (1971).
[405] T. Sato, H. Wakatsuka, and K. Amano, *Tetrahedron*, **27**, 5381 (1971); Y. Tamura, Y. Kita, and M. Terashima, *Chem. Pharm. Bull. (Japan)*, **19**, 529 (1971); A. P. Schroff and C. J. Shaw, *Anal. Chem.*, **43**, 454 (1971).
[406] L. I. Zakharkin, V. N. Kalinin, and V. V. Gedymin, *Tetrahedron*, **27**, 1317 (1971).
[407] N. Hatanaka, H. Ohta, O. Simamura, and M. Yoshida, *Chem. Comm.*, **1971**, 1364.

rearrangement.[408] Alkyl nitrones, e.g. (**242**), readily rearrange to N-alkylamides, e.g. (**243**), on treatment with toluene-p-sulphonyl chloride in pyridine; the stereochemistry of the nitrone does not influence the course of reaction which provides a useful alternative to the Beckmann rearrangement.[409] Acyclic amidoximes, e.g. (**244**), are converted into

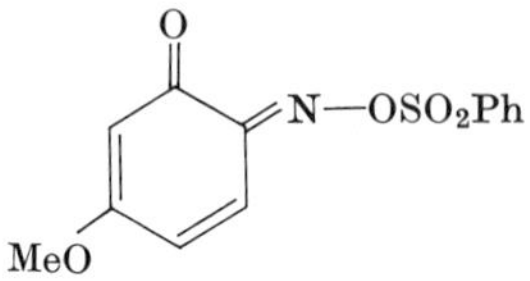

carbodiimides by P_2O_5; migration of the aryl group is facilitated when an electron-releasing *para*-substituent is present.[410] Cyclic amidoxime toluene-p-sulphonates, e.g. (**245**), undergo NaOH-initiated ring-expansions to the cyclic ureas.[411] An interesting reorganization of 3,4-diphenylbut-3-en-2-one oxime benzoate to 1-phenyl-3-methylisoquinoline is reported to occur on heating; an initial Beckman rearrangement is postulated.[412]

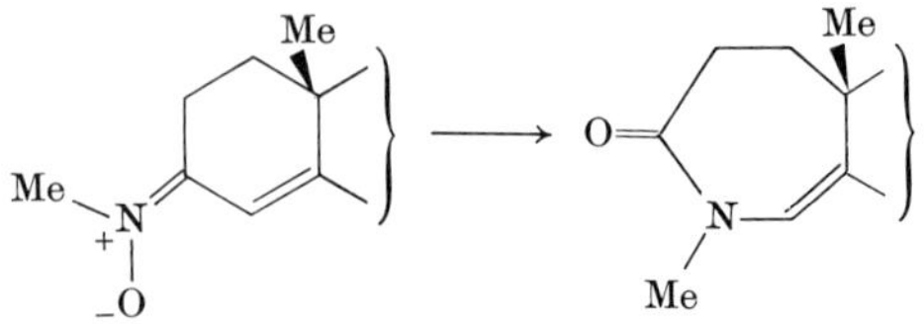

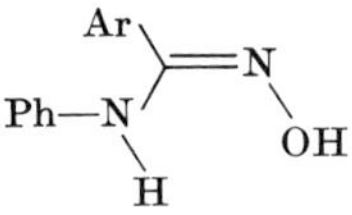

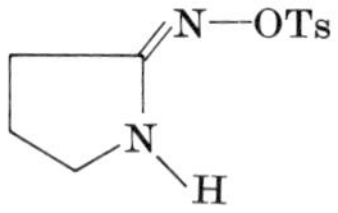

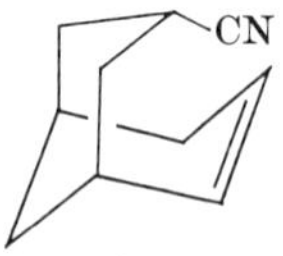

There have been further studies of the Schmidt rearrangement[413] and hexan-2-one and acetophenone have been found to react 3–6 times faster in D_2O–D_2SO_4 than in the corresponding undeuteriated medium.[414] Adamantan-2-one affords 4eq-(methanesulphonyloxy)adamantan-2-one as the major product in the presence of HN_3–methanesulphonic acid, implying that fragmentation to (**246**) is the preferred reaction pathway.[415] The Schmidt rearrangements of 3-chlorocyclohex-2-enones,[416] isoflavanones,[417] and 3-o-carborane derivatives[406] have been examined.

The kinetics of the Hofmann rearrangement of substituted N-bromo- and N-chlorobenzamides have been studied; the shift of the aryl group is considered to be concerted with the loss of the halogen.[418] The Hofmann rearrangement of 3-o-carborane derivatives

---

[408] K. K. Kelly and J. S. Matthews, *J. Org. Chem.*, **36**, 2159 (1971).
[409] D. H. R. Barton, M. J. Day, R. H. Hesse, and M. M. Pechet, *Chem. Comm.*, **1971**, 945.
[410] J. Garapon, B. Sillion, and J. M. Bonnier, *Tetrahedron Letters*, **1970**, 4095.
[411] A. LeBerre, C. Renault, and P. Giraudeau, *Bull. Soc. Chim. France*. **1971**, 3245.
[412] S. Goszczyński and E. Salwińska, *Tetrahedron Letters*, **1971**, 3027.
[413] G. F. Tereshchenko, G. I. Koldobskii, A. S. Yenin, and L. I. Bagal, *Organic Reactivity (Tartu)*, **7**, 1116 (1970).
[414] G. F. Tereshchenko, G. I. Koldobskii, and L. I. Bagal, *Zhur. Org. Khim.*, **6**, 2633 (1970); *Chem. Abs.*, **74**, 63781 (1971).
[415] T. Sasaki, S. Eguchi, and T. Toru, *Tetrahedron Letters*, **1971**, 1109; *J. Org. Chem.*, **35**, 4109 (1970).
[416] V. Tamura and Y. Kita, *Chem. Pharm. Bull. (Japan)*, **19**, 1735 (1971).
[417] D. Misiti and V. Rimatori, *Gazzetta*, **101**, 167 (1971).
[418] T. Imamoto, Y. Tsuno, and Y. Kukawa, *Bull. Chem. Soc., Japan*, **44**, 1632, 1639, 1644 (1971).

has also been investigated.[406] A homolytic cleavage of the N—Br bond is considered to trigger the thermal rearrangement of N-bromo-β-lactams to 2-bromoalkyl isocyanates.[419] The generation of a nitrenium ion adjacent to a small ring results in ring-expansion; for example, 1-methyl-2-phenylpyrrolidine is obtained when N-chloro-N-methyl-(1-phenylcyclobutyl)amine is treated with silver trifluoroacetate followed by NaBH$_4$. Formation of a nitrenium ion within a four-membered ring leads to degradation, probably by an initial ring-contraction; thus, 1-chloro-2-phenylazetidine affords ethanolamine and benzaldehyde.[420] 2-Azabicyclo[2.2.1]heptanes with a leaving group attached to the N atom also rearrange via nitrenium ion intermediates.[421] When (**247**; R = H or Me) is treated with POCl$_3$ followed by a reducing agent, (**248**; R = H or Me) is formed; however, no ring-contraction is observed in the corresponding reaction of (**247**; R = Ac).[422] A [1,2]-phenyl shift results when (**249**) is treated with acetic acid,[423] but the acyl group of (**250**) migrates in the presence of acetic anhydride.[424]

In acyclic ketones there is an equal preference for the migration of the cyclobutyl and isopropyl groups during Baeyer–Villiger oxidation; by contrast in this oxidation (**251**) affords >98% of the lactone derived by shift of the cyclobutane ring.[425] ^{18}O-Labelling experiments are concordant with the intervention of (**253**) in the rearrangement of (**252**)

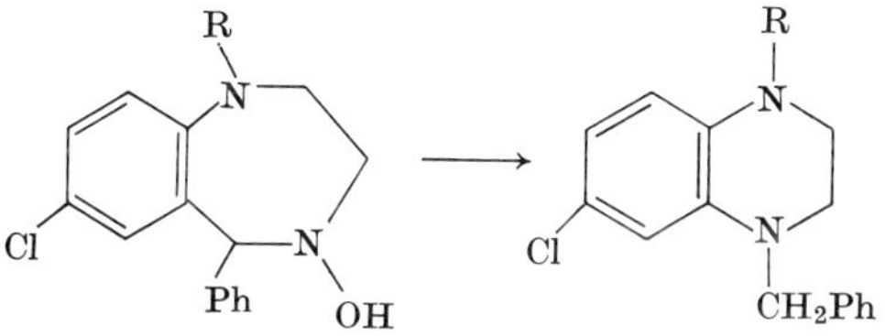

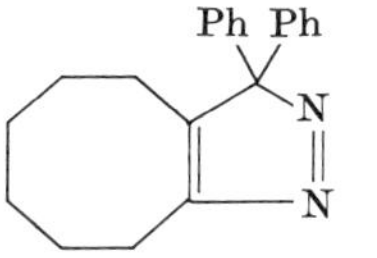

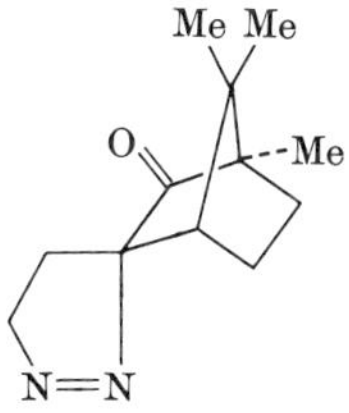

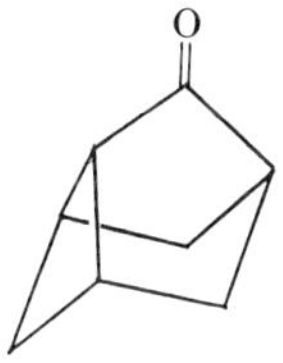

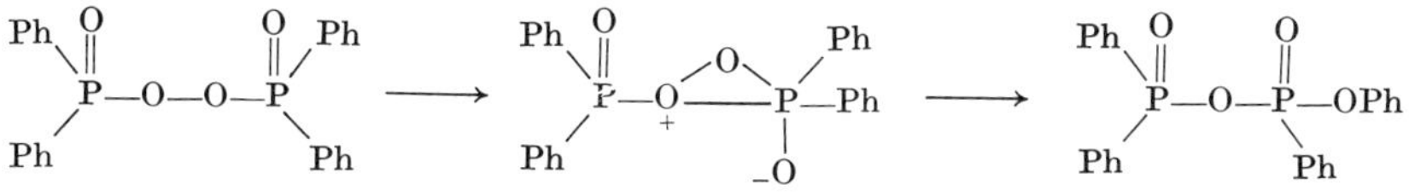

[419] K.-D. Kampe, *Ann. Chem.*, **752**, 142 (1971).
[420] P. G. Gassman and A. Carrasquillo, *Tetrahedron Letters*, **1971**, 109.
[421] P. G. Gassman and K. Shudo, *J. Am. Chem. Soc.*, **93**, 5899 (1971).
[422] A. Waber, G. Silverman, R. I. Fryer, L. H. Sternbach, and J. Hellerbach, *J. Org. Chem.*, **36**, 1248 (1971).
[423] G. Wittig and J. J. Hutchinson, *Ann. Chem.*, **741**, 89 (1970).
[424] G. Snatzke, H. Langen, and J. Himmelreich, *Ann. Chem.*, **744**, 142 (1971).
[425] S. A. Monti and C. K. Ward, *Tetrahedron Letters*, **1971**, 697.

to (**254**).[426] The rearrangements of tertiary alicyclic peroxides[427] and di-isopropylbenzene hydroperoxides[428] have been studied.

[1,3]-*Migrations*

The rates of thermal rearrangement of acetophenone oxime thionocarbamates to the thiolcarbamates are not appreciably influenced either by solvent polarity or by *para*-substitution in the benzene ring; moreover, free-radical species may be detected by ESR measurements in accordance with a homolytic dissociation–recombination mechanism.[429] Radical-pairs are also implicated in the fast intramolecular migration of the arylazo group of (**255**) since CIDNP signals may be detected.[430] However, kinetic studies of substituent effects in the aromatic ring on the rate of thermal rearrangement of (**256**)

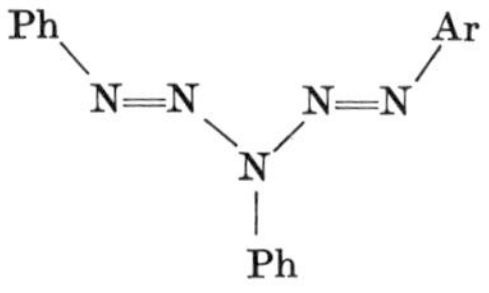

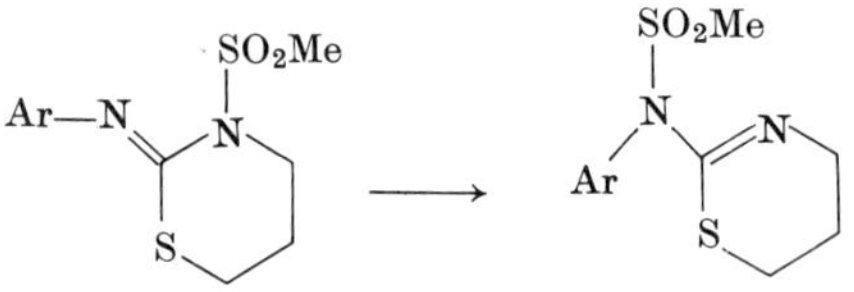

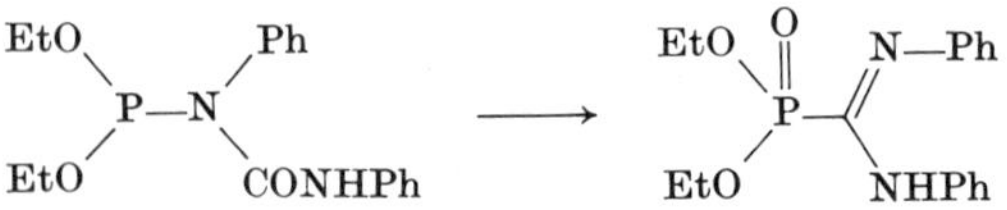

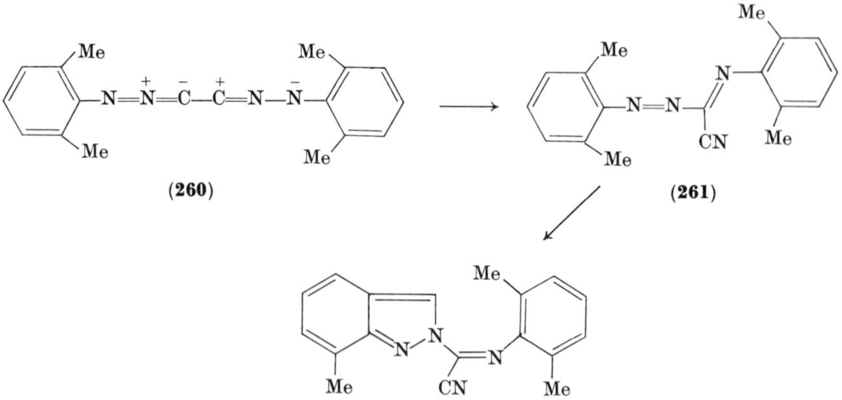

[426] R. L. Dannley, R. L. Waller, R. V. Hoffmann, and R. F. Hudson, *Chem. Comm.*, **1971**, 1362.
[427] R. D. Bushick, *Tetrahedron Letters*, **1971**, 579.
[428] V. A. Yablokov, S. Petrova, and V. A. Shushunov, *Kinet. Katal.*, **11**, 1054 (1970); *Chem. Abs.*, **73**, 119919 (1970).
[429] R. F. Hudson, A. J. Lawson, and E. A. C. Lucken, *Chem. Comm.*, **1971**, 807; B. Cross, R. J. G. Searle, and R. E. Woodall, *J. Chem. Soc.* (C), **1971**, 1833.
[430] J. Hollaender and W. P. Neumann, *Angew. Chem. Int. Ed.*, **9**, 804 (1970).

to (**257**) suggest that the methylsulphonyl group migrates as a positively charged ion.[431] The primary step in the thermally induced conversion of (**258**) into (**259**) probably involves a [1,3]-shift of the P to the carbonyl-oxygen atom; the derived species is postulated to yield the product in a bimolecular process.[432] An interesting oxidative intramolecular rearrangement of (**260**) to (**262**) has been reported; the initial step possibly involves a [1,3]-shift in which (**261**) is formed.[433] The Et_3OBF_4-induced conversion of thionesters into thiolesters proceeds by an intermolecular mechanism.[434]

Isomerizations

Interconverions of Diasteroisomers

The nomenclature for intramolecular exchange processes has been discussed and it is suggested that the process which leads to the exchange of position of identical ligands should be known as topomerization.[435] The energy barriers for the topomerization of simple imines by the inversion pathway, calculated by the EH method, are in reasonable agreement with the experimentally determined values.[436] The rates of *syn–anti* topomerization of substituted *N*-aryliminocarbonates and *N*-aryliminodithiocarbonates show linear Hammett correlations ($\rho = +1.44$ and $+1.30$, respectively) in accord with planar inversion of N in the rate-determining step.[437] It has been estimated that (**263**) undergoes less than one rotation during 10^8 inversions,[438] although the transition state for *syn–anti*-isomerization of *N*-(1,2,3-trimethylbut-2-enylidene)benzenesulphonamide is postulated to possess some torsional character.[439] The rates of brominations of *N*,*N*-disubstituted *N'*-(arylmethylene)hydrazines,[440] 5-(arylmethylenehydrazino)tetrazoles[441] and arylidenesemicarbazones[442] are independent of the bromine concentration, in accord with slow *syn–anti* isomerizations; Hammett correlations favour the rotational mechanism. Similar pathways are implicated in the thermal isomerization of *para*-donor- and *para*-acceptor-substituted azobenzenes[443] and in the topomerization of *para*-substituted hexafluoroacetone *N*-phenylimines (when the *para* group = H, Cl, F, OMe, or Me);[444] however, in the case of hexafluoroacetone *N*-(*p*-nitrophenyl)imine the inversion route is implicated.[444] Thio-oxime ethers are estimated to undergo *syn–anti* isomerization ca. 10^{11} times faster than the corresponding oxime ethers, although the mechanism of this process has not been established.[445]

---

[431] G. Toth and L. Toldy, *Tetrahedron*, **27**, 5025 (1971).
[432] R. F. Hudson and A. Mancuso, *Tetrahedron Letters*, **1971**, 3821.
[433] C. Grundmann, S. K. Datta, and R. F. Sprecher, *Ann. Chem.*, **744**, 88 (1971).
[434] T. Oishi, M. Mori, and Y. Ban, *Tetrahedron Letters*, **1971**, 1777.
[435] G. Binsch, E. L. Eliel, and H. Kessler, *Angew. Chem. Int. Ed.*, **10**, 570 (1971).
[436] F. Kerek, G. Ostrogovich, and Z. Simon, *J. Chem. Soc.* (B), **1971**, 541; C. H. Warren, G. Wettermark, and K. Weiss, *J. Am. Chem. Soc.*, **93** 4658 (1971).
[437] H. Kessler, P. F. Bley, and D. Leibfritz, *Tetrahderon*, **27**, 1687 (1971); A. Liden and J. Sandström, *Tetrahedron*, **27**, 2893 (1971).
[438] H. Kessler and D. Leibfritz, *Chem. Ber.*, **104**, 2143 (1971).
[439] M. Raban and E. Carlson, *J. Am. Chem. Soc.*, **93**, 685 (1971).
[440] F. L. Scott, F. A. Groeger, and A. F. Hegarty, *J. Chem Soc* (B), **1971**, 1411.
[441] J. C. Tobin, A. F. Hegarty, and F. L. Scott, *J. Chem.Soc.* (B), **1971**, 2198.
[442] F. L. Scott, T. M. Lambe, and R. N. Butler, *Tetrahedron Letters*, **1971**, 2909.
[443] P. D. Wildes, J. G. Pacifici, G. Irick, Jr., and D. G. Whitten, *J. Am. Chem. Soc.*, **93**, 2004 (1971).
[444] G. E. Hall, W. J. Middleton, and J. D. Roberts, *J. Am. Chem. Soc.*, **93**, 4778 (1971).
[445] C. Brown, B. T. Grayson, and R. F. Hudson, *Tetrahedron Letters*, **1970**, 4925.

The interconversions of the conformers of N-(benzyloxycarbonyl)proline[446] and N-methyl(thioformamide)[447] have been studied by NMR spectroscopy; in the latter case the isomers can be separated by thin-layer chromatography. Rotational isomerism has also been observed with aromatic guanidinium salts[448] and 5-oxidovinylphenanthridinium salts.[449] The rate-determining step in the conversion of S-benzoyl-1,1,3-trimethylisothiouronium bromide into 1-benzyl-1,3,3-trimethylthiourea involves a *syn–anti* isomerization.[450]

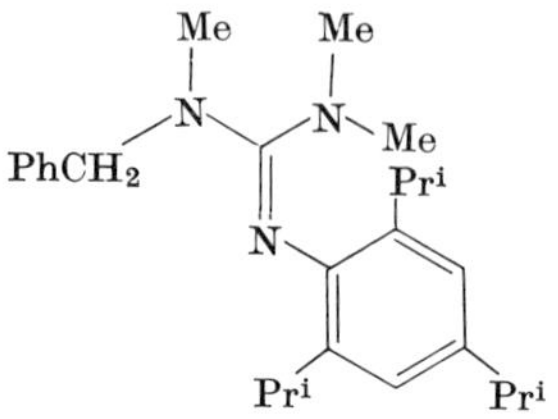

(263)

(264)

The H_2SO_4-catalysed *cis–trans* isomerization of 1-phenylbuta-1,3-diene is slowed down by 2.24–3.15 times in D_2SO_4 and D is incorporated only at the terminal C atom, suggesting that (264) is formed in the slow step of the reaction; loss of water occurs in the rate-determining step of the corresponding *cis–trans* isomerization of 1-methyl-3-phenylallyl alcohol.[451] The *cis–trans* isomerizations of enol propionates and vinyl halides catalysed by Pd(II),[452] crotyl acetates in the presence of acid,[453] 4-oxopent-2-enoic acid in the presence of KSCN,[454] vinyl halides,[455] 1,2,3,3-tetrachloroprop-1-ene,[456] 2-(N-methylanilino)but-2-ene,[457] 1-ethyl-4-(4-hydroxystyryl)quinolinium chloride[458] and diethyl p-bromoanilinomaleate[459] have been examined.

The thermal conversion of the *endo*-derivative of N-toluene-p-sulphonyl-2-thia-5-azabicyclo[2.2.1]heptane 2-oxide into the thermodynamically favoured *exo*-isomer[460] and the equilibration of the diastereoisomers of 3-acetoxy-*trans*-cyclo-octene[461] have also been studied.

Prototropic Shifts

Alkenes. The prototropic isomerization of olefins has been reviewed.[462]

4-Vinylcyclohexene undergoes a $RuCl_3$-induced isomerization to 4-ethylidenecyclohexene, which then disproportionates to a mixture of ethylbenzene and 1- and 3-ethyl-

[446] H. L. Maia, K. G. Orrell, and H. N. Rydon, *Chem. Comm.*, **1971**, 1209.
[447] W. Walter and E. Schaumann, *Chem. Ber.*, **104**, 3361 (1971).
[448] H. Kessler and D. Leibfritz, *Chem. Ber.*, **104**, 2158 (1971).
[449] R. M. Acheson and I. A. Selby, *J. Chem. Soc.* (C), **1971**, 691.
[450] T. C. Bruice and R. F. Pratt, *Chem. Comm.*, **1971**, 1259.
[451] Y. Pocker and M. J. Hill, *J. Am. Chem. Soc.*, **93**, 691 (1971).
[452] P. M. Henry, *J. Am. Chem. Soc.*, **93**, 3547 (1971).
[453] W. G. Young, H. E. Green, and A. F. Diaz, *J. Am. Chem. Soc.*, **93**, 4782 (1971).
[454] C. Santiago and S. Seltzer, *J. Am. Chem. Soc.*, **93**, 4546 (1971).
[455] G. J. Martin and N. Naulet, *Bull. Soc. Chim. France*, **1970**, 4001.
[456] H. Khalaf and K. Kirchhoff, *Tetrahedron Letters*, **1971**, 3861.
[457] M. Riviere and A. Lattèrs, *Tetrahedron Letters*, **1971**, 4563.
[458] D. Schulte-Frohlinde and H. Güsten, *Ann. Chem.*, **749**, 49 (1971).
[459] S. Toppet, E. Van Loock, G. L'abbe, and G. Smets, *Chem. Ind. (London)*, **1971**, 703.
[460] P. S. Portoghese and V. G. Telang, *Tetrahedron*, **27**, 1823 (1971).
[461] G. H. Whitham and M. Wright, *J. Chem. Soc.* (C), **1971**, 891.
[462] L. A. Yanovskaya and Kh. Shakhidayatov, *Russ. Chem. Rev.*, **1970**, 859.

cyclohexene.[463] Hydroperoxides greatly accelerate the rates of dichlorotris(triphenylphosphine)rhodium-catalysed rearrangements of olefins,[464] and such catalysts selectively convert vinylcycloalkenes and vinylcycloalkanes into the exocyclic-double-bond isomers.[465] Evacuated CaO is an effective catalyst for the isomerization of but-1-ene into cis- and trans-but-2-ene at 30°,[466] while Cu_2O induces the rearrangement of allyl isocyanide to propenyl isocyanide at room temperature.[467] 1,2,3,4,5-Pentamethylcyclopentadiene is implicated in the conversion of cis-1,2,3,4,5-pentamethylcyclopentyl cation into the trans-isomer in aqueous H_2SO_4.[468] The isomerizations of 1-methyl-4-(ethoxycarbonylmethylene)phosphorinane,[469] nitro-olefins,[470] enamines,[471] steroidal olefins,[472] decalin enol acetates,[473] methylcyclohexadienes,[474] hepta-1,3,5-trienes[475] and octa-1,3,6-trienes[476] have been examined.

1,2-Diphenylbut-3-ene affords 1,2-diphenylbut-2-ene when treated with two molar equivalents of Bu^{n}Li followed by water; when the reaction is quenched with D_2O two D atoms are incorporated, one at position 1 and the other at position 4, in accord with the intervention of (265). Under similar conditions 3-phenylhexa-1,5-diene affords a mixture of 3-phenylhexa-2,4- and -2,5-diene; quenching with D_2O implicates the intermediacy of (266). Furthermore, heating the dianion at 60° yields biphenyl in 77% yield, suggesting that the planar 8π-electron dianion (267) cyclizes to (268) which then eliminates LiH.[477] Metallation of linolenyl alcohol (269) with Bu^{n}Li for 3–4 hours and subsequent carboxylation with CO_2 indicates that protons are abstracted from the two activated methylene

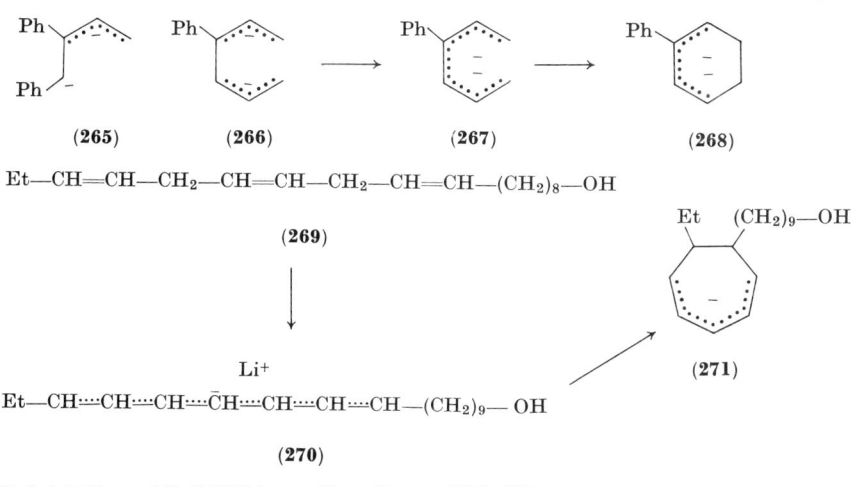

463 C. J. Attridge and P. J. Wilkinson, Chem. Comm., 1971, 620.
464 J. E. Lyons, Chem. Comm., 1971, 562.
465 J. E. Lyons, J. Org. Chem., 36, 2497 (1971).
466 K. Tanabe, N. Yoshii, and H. Hattori, Chem. Comm., 1971, 464.
467 T. Saegusa, I. Murase, and Y. Ito, Tetrahedron, 27, 3795 (1971).
468 T. S. Sorensen, I. J. Miller, and C. M. Urness, Can. J. Chem., 48, 3374 (1970).
469 L. D. Quin, J. W. Russell, Jr., R. D. Prince, and H. E. Shook, Jr., J. Org. Chem., 36, 1495 (1971).
470 L. Lešeticky and M. Procházka, Coll. Czech. Chem. Comm., 36, 307 (1971).
471 H. Mazarguil and A. Latters, Tetrahedron Letters, 1971, 975.
472 N. Ikekawa, Y. Honma, N. Morisaki, and K. Sakai, J. Org. Chem., 35, 4145 (1970).
473 H. Favre, F. Huet, and L. Varfalvy, Can. J. Chem., 49, 1776 (1971).
474 C. W. Spangler and R. P. Hennis, J. Org. Chem., 36, 917 (1971).
475 K. W. Egger and T. Ll. James, J. Chem. Soc. (B), 1971, 348.
476 D. McHale, Tetrahedron, 27, 4843 (1971).
477 R. M. Magid and S. E. Wilson, Tetrahedron Letters, 1971, 19.

groups at a similar rate; however, if the metallation is prolonged for >20 hr the cyclic products that are formed suggest that (**270**) isomerizes to (**271**) which then undergoes [1,6]-H shifts;[478] according to orbital symmetry predictions the latter process should involve an antarafacial H transfer although this pathway is precluded in the above example.

The interconversions of tricarbonyliron complexes of polyenes,[479] tetracyclodecadiene[219,480] and tricyclodecatriene[481] have been investigated.

Alkynes. The base-catalysed isomerization of acetylenes has been reviewed.[482] Zinc oxide is considered to act as a base in catalysing the methylacetylene–allene isomerization.[483] The base-induced isomerizations of pent-4-en-1-ynes[484] and hexa-1,5-diynes,[485] and the kinetics of the gas-phase iodine-catalysed isomerization of methylacetylene to allene[486] and of the interconversions of 3-(N,N-dimethylamino)prop-1-yne, 1-(N,N-dimethylamino)allene and (dimethylamino)methylacetylene[487] have been studied.

Tautomerizations. The tautomeric behaviour of N-substituted 2-phenacylisothiouronium bromides,[488] imidazolin-5-ones,[489] indoline-2-thiones,[490] 2-chlorobenzimidazoles,[491] pyrimidines,[492] 4,8-dianilino-1,5-naphthoquinones,[493] 4-hydroxyazobenzenes,[494] xanthines,[495] 2-hydroxypyridines,[496] phthalimides and succinimides[497] have been examined.

Rearrangements Involving Ring Openings and Closures

A review dealing with the solvolytic, thermal and photochemical reorganizations of spiranes has been published.[498]

Three-membered Rings

When *trans*-1,2,3-triphenylcyclopropane is treated with Bu^{n}Li followed by D$_2$O *trans*-1,2,3-triphenyl[3-^{2}H$_1$]prop-1-ene is the major product; ring-opening does not occur in the presence of KOBu^t–[^{2}H$_6$]DMSO although the cyclopropyl-hydrogen atoms are exchanged for D.[499] Diradical intermediates are probably involved in the gas-phase

---

[478] J. Klein and S. Glily, *Tetrahedron*, **27**, 3477 (1971).
[479] H. W. Whitlock, Jr., C. Reich, and W. D. Woessner, *J. Am. Chem. Soc.*, **93**, 2483 (1971).
[480] R. Aumann, *Angew. Chem. Int. Ed.*, **10**, 190 (1971).
[481] R. Aumann, *Angew. Chem. Int. Ed.*, **10**, 560 (1971).
[482] R. J. Bushby, *Quart. Rev.*, **24**, 585 (1970).
[483] C. C. Chang and R. J. Kokes, *J. Am. Chem. Soc.*, **92**, 7517 (1970).
[484] J. Grimaldi and M. Bertrand, *Bull. Soc. Chim. France*, **1971**, 947.
[485] H. Hopf, *Chem. Ber.*, **104**, 3087 (1971).
[486] R. Walsh, *Trans. Faraday Soc.*, **67**, 2085 (1971).
[487] J.-L. Dumont, G. Pourcelot, and C. Georgoulis, *Bull. Soc. Chim. France*, **1971**, 1101.
[488] B. S. Shadbolt, *J. Chem. Soc.* (C), **1971**, 1667.
[489] R. Jacquier, J.-M. Lacombe, and G. Maury, *Bull. Soc. Chim. France*, **1971**, 1040.
[490] T. Hino, M. Nakagawa, T. Suzuki, S. Takeda, N. Kano, and Y. Ishii, *Chem. Comm.*, **1971**, 836.
[491] R. Benassi, P. Lazzeretti, L. Schenetti, F. Taddei, and P. Vivarelli, *Tetrahedron Letters*, **1971**, 3299.
[492] P. Lardenois, Marguerite Sélim, and Mohamed Sélim, *Bull. Soc. Chim. France*, **1971**, 1858.
[493] S. M. Bloom and G. O. Dudek, *J. Org. Chem.*, **36**, 235 (1971).
[494] E. Hofer and H. Uffmann, *Tetrahedron Letters*, **1971**, 3241; E. Manda, *Bull. Chem. Soc. Japan*, **44**, 1620 (1971).
[495] D. Lichtenberg, F. Bergmann, and Z. Neiman, *J. Chem. Soc.* (C), **1971**, 1676.
[496] E. Spinner and G. B. Yeoh, *J. Chem. Soc.* (B), **1971**, 279, 289, 296.
[497] J. Armand, S. Deswarte, J. Prinson, and H. Zamarlik, *Bull. Soc. Chim. France*, **1971**, 671.
[498] W. R. Dolbier, Jr., *Mech. Mol. Migr.*, **3**, 1 (1971).
[499] J. E. Mulvaney and D. Savage, *J. Org. Chem.*, **36**, 2592 (1971).

thermolyses of methylspiro[2.2]pentane to ethylidene- and 1-methyl-3-methylene-cyclobutanes,[500] and of (**272**) to hepta-1,2,6-triene and 3-methylenehexa-1,5-diene.[501] Trapping experiments suggest that the ylid (**273**) intervenes in the thermal rearrangement of methyl 2-azabicyclo[3.1.0]hex-3-ene-1-carboxylate to methyl 1,2-dihydropyridine-1-carboxylate.[502] The course of ring-opening of (**274**) depends upon the reaction conditions: with NaH in benzene followed by methanol (**275**) is formed, whereas with NaH in benzene–DMF followed by methanol (**277**) is produced, presumably via (**276**).[503] Alkali induces the rearrangement of 1-hydroxy-3-azabicyclo[4.1.0]hept-4-en-2-ones to 3-hydroxy-2-pyridones.[504] Cyclopropanediol intermediates are implicated in the Zn–acetic acid reduction of 1,3-diketones.[505] The thermal isomerization of alkylcyclopropenes has been studied and, for example, 2,3- and 3,4-dimethylpenta-1,3-dienes are formed from 1,2,3,3-tetramethylcyclopropene.[506]

The nucleophile-, acid-, base-induced and thermal rearrangements of aziridines have been reviewed.[507] 1-(3,4-Dihydro-4-oxoquinazolin-3-yl)-2-vinylaziridines undergo two competing thermal rearrangements: one to give the hydrazone and the other the pyrrolenine.[508] A hydrazone intermediate may intervene in the reorganization of (**278**) to (**279**), which occurs in hot toluene.[509]

There have been further studies of the base-induced rearrangements of epoxides to allylic alcohols.[510] In the presence of methanolic NaOH (**280**; R^1 = OH, R^2 = H) rearranges to (**281**) ca. 10 times faster than does (**280**; R^1 = H, R^2 = OH);[99] a small D isotope effect (k_H/k_D = 1.56) is observed with (**280**; R^1 = OH, R^2 = D). However, the proposal that a carbanion mechanism is involved requires substantiation.[511] Nortricyclanol is formed in the reaction of 2,3-epoxybicyclo[2.2.1]heptane with $LiNEt_2$.[512] A diradical intermediate is implicated in the ring-expansion of (**282**) to 2,2,5,5-tetramethylcyclopentane-1,3-dione at 650°.[513] Epoxysilanes readily undergo ring-opening in the presence of methanolic H_2SO_4, providing a new synthetic route to aldehydes.[514] Examples in which epoxides undergo intramolecular ring-openings include the base-induced isomerization of $\gamma\delta$-epoxy ketones[515] and the HCl-initiated rearrangement of *trans*-2,3-epoxy-3-(2-nitrophenyl)propiophenone.[516] The acid-catalysed isomerization of an oxazirane has been examined.[517]

---

[500] M. C. Flowers and A. R. Gibbons, *J. Chem. Soc.* (B), **1971**, 612.
[501] H. M. Frey, R. G. Hopkins, and L. Skattebøl, *J. Chem. Soc.* (B), **1971**, 539.
[502] F. W. Fowler, *Angew. Chem. Int. Ed.*, **10**, 135 (1971).
[503] P. S. Venkataramani, J. E. Karoglan, and W. Reusch, *J. Am. Chem. Soc.*, **93**, 269 (1971); K. Grimm, P. S. Venkataramani, and W. Reusch, *J. Am. Chem. Soc.*, **93**, 270 (1971).
[504] W. Ried and F. Batz, *Angew. Chem. Int. Ed.*, **10**, 735 (1971).
[505] K. Mori, *Tetrahedron*, **27**, 4907 (1971).
[506] R. Srinivasan, *Chem. Comm.*, **1971**, 1041.
[507] H. W. Heine, *Mech. Mol. Migr.*, **3**, 145 (1971).
[508] T. L. Gilchrist, C. W. Rees, and E. Stanton, *J. Chem. Soc.* (C), **1971**, 3036.
[509] D. W. Jones, *Chem. Comm.*, **1971**, 1130.
[510] R. P. Thummel and B. Rickborn, *J. Org. Chem.*, **36**, 1365 (1971); W. P. Cochrane and M. A. Forbes, *Can. J. Chem.*, **49**, 3569 (1971).
[511] J. M. Coxon, E. Dansted, R. P. Garland, M. P. Hartshorn, and W. B. Joss, *Tetrahedron*, **27**, 1287 (1971).
[512] J. K. Crandall, L. C. Crawley, D. B. Banks, and L. C. Lin, *J. Org. Chem.*, **36**, 510 (1971).
[513] N. J. Turro and D. R. Morton, *Tetrahedron Letters*, **1971**, 2535.
[514] G. Stork and E. Colvin, *J. Am. Chem. Soc.*, **93**, 2080 (1971).
[515] R. Brieman and M. Avramoff, *Israel J. Chem.*, **9**, 63 (1971).
[516] I. P. Sword, *J. Chem. Soc.* (C), **1971**, 820.
[517] P. Milliet and X. Lusinchi, *Tetrahedron Letters*, **1971**, 3763.

(272) 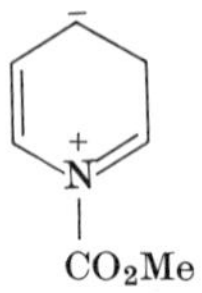 (273) 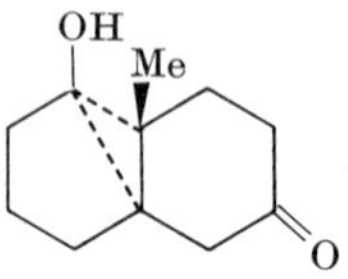 (274) 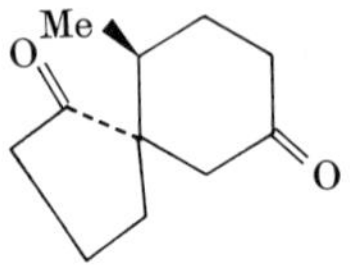 (275)

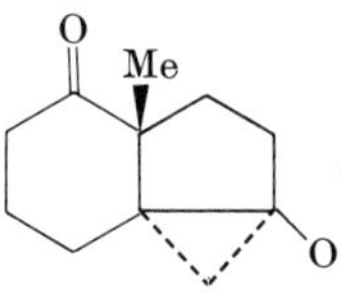

(278) (276) (277)

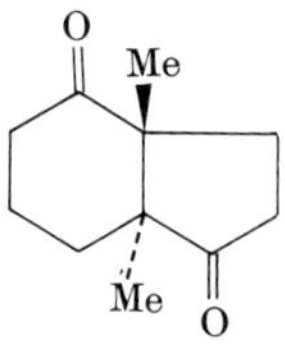

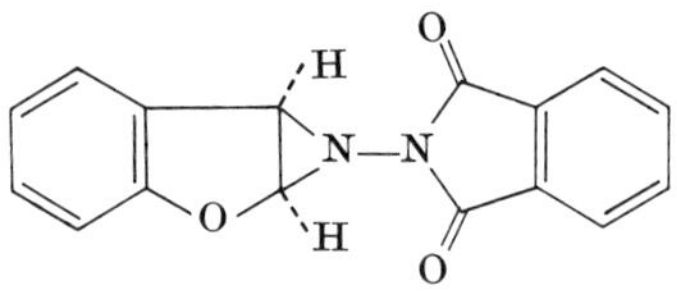

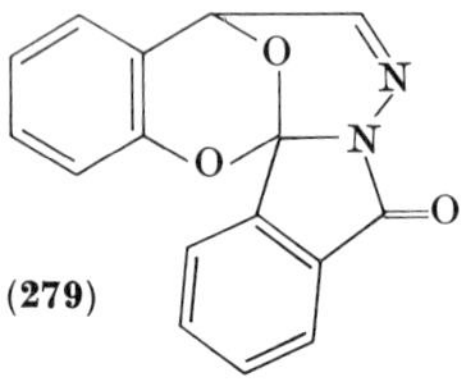

(279) 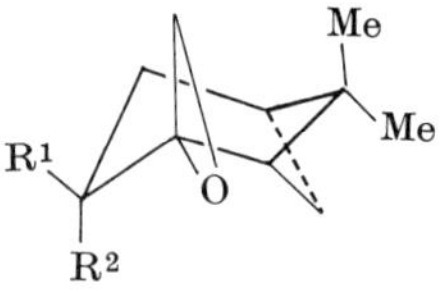 (280) 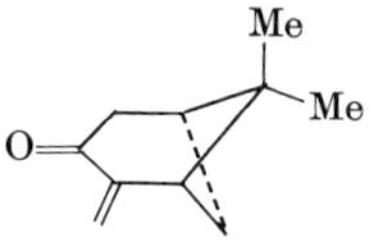 (281)

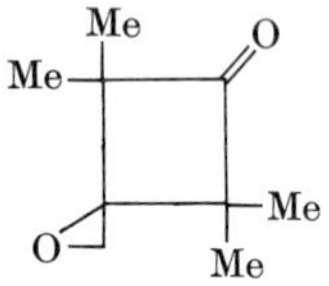

(282)

Four-membered Rings

The cyclobutanone ring of 7,7-dichlorobicyclo[3.2.0]hept-2-en-6-one is cleaved in the presence of aqueous Na_2CO_3 and 2-formylcyclopent-2-ene-1-carboxylic acid is formed.[518] Thermolysis of ethyl 1,4,6-triphenyl-2,3-diazabicyclo[3.2.0]hepta-3,6-diene-2-carboxylate yields the ylid (283).[519] The thermally induced rearrangement of acetals of 2-chlorothietan-3-one 1,1-dioxide is probably triggered by heterolysis of the 2—3 bond.[520]

[518] H. C. Stevens, J. K. Rinehart, J. M. Lavanish, and G. M. Trenta, *J. Org. Chem.*, **36**, 2780 (1971); P. R. Brook and A. J. Duke, *J. Chem. Soc.* (C), **1971**, 1764.
[519] G. Kan, M. T. Thomas, and V. Snieckus, *Chem. Comm.*, **1971**, 1022.
[520] L. A. Paquette and R. W. Houser, *J. Am. Chem. Soc.*, **93**, 944 (1971).

β-Lactams can epimerize when heated; for example, *cis*-4-methoxy-3-methyl-1-toluene-*p*-sulphonylazetidin-2-one equilibrates with the *trans*-isomer, probably via (**284**). The epimerizations are accompanied by slower irreversible rearrangements to the αβ-unsaturated amides.[521] An $E1cB$ mechanism has been postulated to account for the rearrangement of penicillanic acid derivatives to 1,4-thiazepines.[522] The cepham (**285**) is unstable in the presence of SiO_2 and undergoes an intramolecular rearrangement in which the β-lactam is attacked by the 3-hydroxyl group.[523] The major product resulting

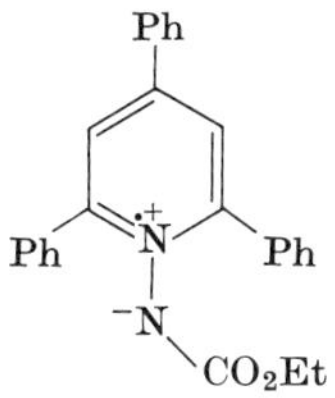

(**283**) 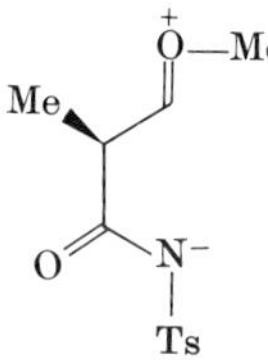　(**284**) 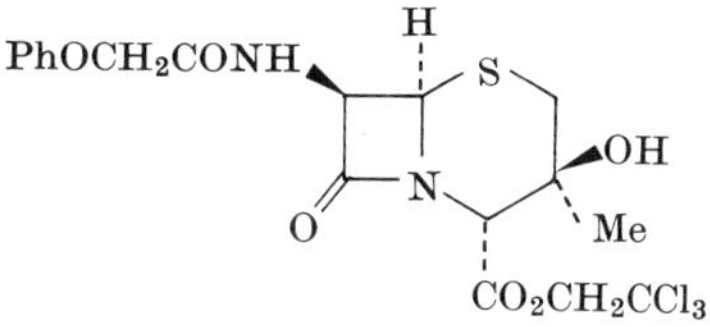　(**285**)

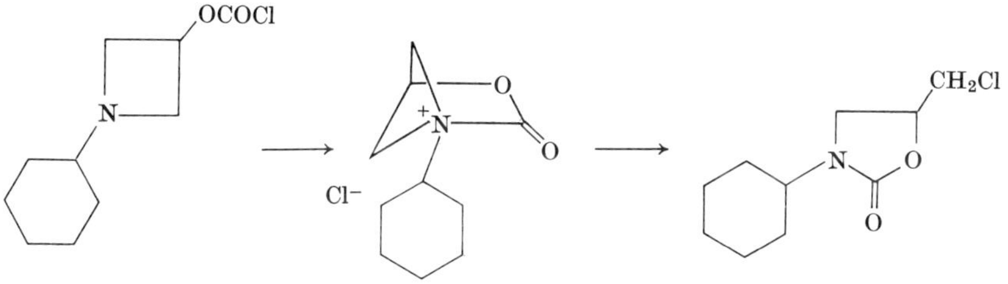

(**286**)　(**287**)

from the reaction of *trans*-3-benzamido-1,4-diphenylazetidin-2-one with acid is 3-benzoyl-1,2-diphenylimidazolid-5-one, in accord with a 3—4 bond cleavage; however, the minor products suggest that 1—2 and 1—4 bond ruptures also occur.[524] β-Lactams are likely intermediates in the thermal reorganizations of 2,3-bis(alkylimino)oxetanes.[525] When heated in toluene, 1-cyclohexylazetidin-2-ol chloroformate isomerizes to (**287**), probably via (**286**).[526] A [1,4]-shift of the sulphonamido group ensues when 1,6-dimethyl-7-toluene-*p*-sulphonyl-7-azabicyclo[4.2.0]oct-3-ene is brominated.[527]

Five-membered Rings

The base-catalysed Dimroth rearrangements of 5-amino-1-phenyl-1,2,3-triazoles,[528] 2-imino-5-nitro-4-thiazolines,[529] and *s*-triazolo[3,4-*a*]isoquinolines[530] have been studied.

[521] F. Effenberger, P. Fischer, G. Prossel, and G. Kiefer, *Chem. Ber.*, **104**, 1987 (1971).
[522] B. G. Ramsay and R. J. Stoodley, *Chem. Comm.*, **1971**, 450.
[523] G. E. Gutowski, C. M. Daniels, and R. D. G. Cooper, *Tetrahedron Letters*, **1971**, 3429.
[524] C. W. Bird and J. D. Twibell, *J. Chem. Soc.* (C), **1971**, 3155.
[525] T. Saegusa, N. Taka-ishi, and V. Ito, *Bull. Chim. Soc. Japan*, **44**, 1121 (1971).
[526] J. P. Li and J. H. Biel, *J. Org. Chem.*, **35**, 4110 (1970).
[527] L. A. Paquette and J. F. Kelly, *J. Org. Chem.*, **36**, 442 (1971).
[528] D. R. Sutherland and G. Tennant, *J. Chem. Soc.* (C), **1971**, 706.
[529] P. J. Islip and M. D. Closier, *Chem. Ind.* (*London*), **1971**, 95.
[530] C. Hoogzand, *Rec. Trav. Chim.*, **90**, 1225 (1971).

Related reorganizations are observed with mesoionic anhydro-2-arylamino-1,3,4-thiadiazolium hydroxides; thus, **(288**; R = S) affords **(290)** when heated in ethanol, via **(289)**.[531] A similar rearrangement occurs with **(288**; R = p-Cl-C$_6$H$_4$N) although **(288**; R = p-Me-C$_6$H$_4$N) equilibrates with its isomer.[532]

The degenerate thermal rearrangement of **(291)** can be observed by NMR spectroscopy; the isomerization is postulated to involve a [1,9]-sigmatropic shift since its rate is insensitive to the solvent polarity ($\Delta G^{\neq}$ 23 kcal mole^{-1}).[533]

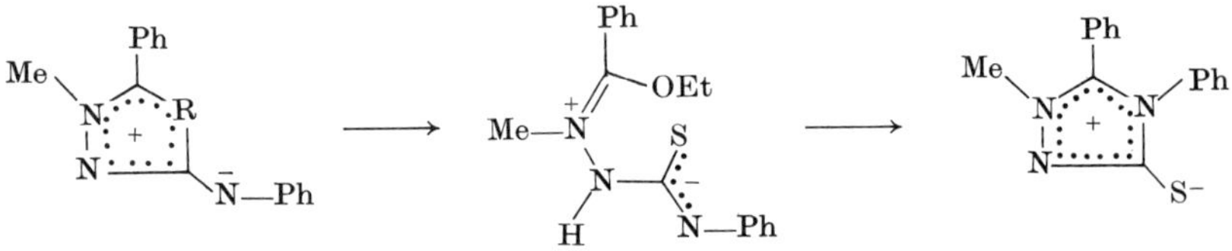

(288)   **(289)**   **(290)**

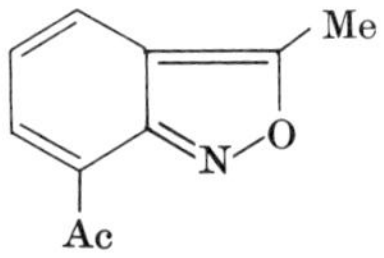

(291) 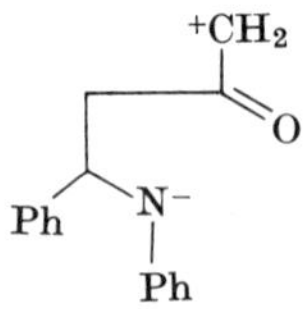   **(292)**

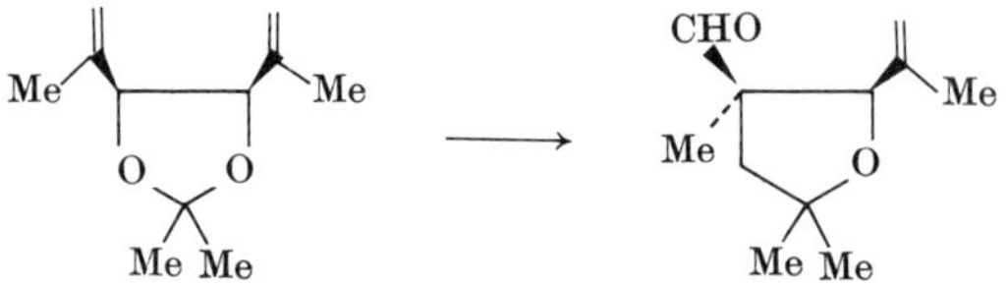

(293)   **(294)**

Several other isomerizations in which five-membered rings are opened and re-formed have been described. For example, the isolation of 1,5-diphenylpyrrolidin-3-one from the reaction of allene and C-benzoyl-N-phenylazomethine oxide suggests that the isoxazolidine, the expected cycloadduct, undergoes an intramolecular rearrangement via **(292)**.[534] On being heated, 5-amino-4-methyl-3-phenylisoxazole undergoes an interesting reorganization to 4-methyl-5-phenyl-4-imidazolin-2-one; 2-methyl-3-phenyl-2H-aziridine-2-carboxamide, which is also formed, is not an intermediate.[535] The conversions of N-(1,2,4-oxadiazol-3-yl)-N'-arylformamidines into 3-acylamino-1-aryl-1,2,4-triazoles,[536]

[531] W. D. Ollis and C. A. Ramsden, *Chem. Comm.*, **1971**, 1222.
[532] W. D. Ollis and C. A. Ramsden, *Chem. Comm.*, **1971**, 1224.
[533] K. P. Parry and C. W. Rees, *Chem. Comm.*, **1971**, 833.
[534] M. C. Aversa, G. Cum, and N. Uccella, *Chem. Comm.*, **1971**, 156.
[535] T. Nishiwaki, T. Saito, S. Onomura, and K. Kondo, *J. Chem. Soc.* (C), **1971**, 2644.
[536] M. Ruccia, N. Vivona, and G. Cusmano, *J. Heterocyclic Chem.*, 8, 137 (1971).

2-(2-aminophenyl)piperolidin-3-one into 3-(2-piperidylmethyl)indolin-2-one,[537] v-triazolo[1,5-a]pyridine-3-acraldehydes into 3-methyl-5-(2-pyridyl)pyrazoles,[538] and the isomerizations of acetoxonium salts of glycerol,[539] cyclopentane-1,2,3,4,5-pentaols,[540] and amino sugars,[541] have been examined. When heated in the presence of activated Al_2O_3, (**293**) rearranges to (**294**), while under more forcing conditions the *trans*-isomer of (**293**) affords a mixture of (**294**) and its epimer.[542] The reaction of (**293**) with a Grignard derivative also involves an initial isomerization to (**294**).[543] The acid-catalysed isomerizations of 5-(hydroxymethyl)-2-methyl-1,3-dioxolans[544] and the dimer of 3-hydroxy-2,3-dimethylindolenine[545] have been examined.

Furfural rearranges to 2,4-dianilinocyclopent-2-enone when treated with aniline;[546] cleavage of the furan ring also takes place in the reaction of 2-acetoxymethyl-3,4-di(azidocarbonyl)furan with ethanol to give $\alpha\beta$-dihydro-$\alpha\beta$-di(ethoxycarbonylamino)-γ-valerolactone.[547]

α-Thioacyl-γ-thiol-lactones rearrange in acidified ethanol to alkyl 2-alkyl-4,5-dihydrothiophen-3-carboxylates.[548] 4-Acetamido-5-phenylisothiazolinone 1,1-dioxide is converted into 4-benzylidene-2-methyl-2-oxazolin-5-one in the presence of acetic anhydride.[549] Ring-opening and -closing sequences are also implicated in the reactions of 4-arylazo-2-benzylthio-2-thiazolin-5-ones with aniline,[550] sydnones with ethoxide ion[551] and 7-aminobenzothiazoles with HNO_2.[552]

A number of reactions leading to the enlargement of five-membered rings have been reported. These include the conversions of 1,2-dimethylcyclopentene into cyclohexane and methylcyclohexane on a Pt-black electrode,[553] α-isocyano-γ-butyrolactone into ethyl 5,6-dihydro-4H-1,3-oxazine-4-carboxylate by NaOEt,[554] 2,3-dihydro-6-methylbenzofuran-2-carboxylic acid into 3,7-dimethylbenzopyran-2-one by $LiNPr^i_2$,[555] 2-amino-3-phenacyl-1,3,4-oxadiazolium bromides into *as*-triazines by hydrazine,[556] and penicillin sulphoxides into cephams and cephems in the presence of acid catalysts.[557] Ring-expansions of indole derivatives have also been reported.[558]

---

[537] P. A. Thio and M. J. Korner, *J. Heterocyclic Chem.*, **8**, 479 (1971).
[538] L. S. Davies and G. Jones, *J. Chem. Soc.* (C), **1971**, 759.
[539] H. Paulsen and H. Behre, *Chem. Ber.*, **104**, 1281 (1971).
[540] H. Paulsen and H. Behre, *Chem. Ber.*, **104**, 1299 (1971).
[541] H. Paulsen and C.-P. Herold, *Chem. Ber.*, **104**, 1311 (1971).
[542] P. Martinet and G. Mousset, *Bull. Soc. Chim. France*, **1971**, 4093.
[543] G. Mousset, *Bull. Soc. Chim. France*, **1971**, 4097.
[544] J. Gelas, *Bull. Soc. Chim. France*, **1970**, 4041.
[545] V. Dave and E. W. Warnhoff, *Can. J. Chem.*, **49**, 1921 (1971).
[546] K. G. Lewis and C. E. Mulquiney, *Austral. J. Chem.*, **23**, 2315 (1970).
[547] T. Yoshida, H. Katsura, and T. Kaneko, *Bull. Chem. Soc. Japan*, **44**, 1701 (1971).
[548] F. Duus and S.-O. Lawesson, *Tetrahedron*, **27**, 387 (1971).
[549] J. C. Howard, *J. Org. Chem.*, **36**, 1073 (1971).
[550] A. Mustafa, A. H. Harhash, M. H. Elnagdi, and F. A. El-All, *Ann. Chem.*, **748**, 79 (1971).
[551] Y. Saito, T. Teraji, and T. Kamiya, *Tetrahedron Letters*, **1971**, 2893.
[552] E. Haddock, P. Kirby, and A. W. Johnson, *J. Chem. Soc.* (C), **1971**, 3642.
[553] H. J. Barger, Jr., G. W. Walker, and R. J. York, *J. Am. Chem. Soc.*, **93**, 2800 (1971).
[554] U. Kraatz, H. Wamhoff, and F. Korte, *Ann. Chem.*, **744**, 33 (1971).
[555] B. Libis and E. Habicht, *Angew. Chem. Int. Ed.*, **10**, 748 (1971).
[556] A. Hetzheim and J. Singelmann, *Ann. Chem.*, **749**, 125 (1971).
[557] G. E. Gutowski, B. J. Foster, C. J. Daniels, L. D. H. Hatfield, and J. W. Fisher, *Tetrahedron Letters*, **1971**, 3433.
[558] R. M. Acheson and J. N. Bridson, *Chem. Comm.*, **1971**, 1225; F. Fried, J. B. Taylor, and R. Westwood, *Chem. Comm.*, **1971**, 1226; F. Linhart and S. Hünig, *Chem. Ber.*, **104**, 913 (1971); R. P. Ryan, W. G. Lobeck, Jr., C. M. Combs, and Y.-H. Wu, *Tetrahedron*, **27**, 2325 (1971).

The P(OEt)$_3$-induced ring-contraction of 5-amino-3-phenylisoxazole to an aziridine derivative has also been investigated.[559]

Kinetic studies suggest that cyclization of (**295**) represents the slow step of the acetic acid-catalysed conversion of $\alpha\beta$-unsaturated phenylhydrazones into pyrazolines.[560] Intramolecular rearrangements are implicated in the thermal cyclizations of 1-ethoxypropenyl esters of γ- and δ-keto acids,[561] of 1,4-diphenyl-1-thiocyanato-2,3-diazabuta-1,3-diene[562] and of N-cyclohexyl-o-nitroaniline.[563] In alkaline solution (**296**; R = Et) is converted into (**296**; R = H); UV spectroscopy provides evidence for the intervention of (**297**) and consequently an intramolecular cyclization is involved.[564]

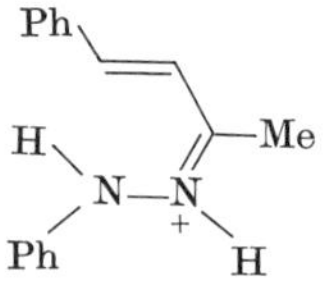

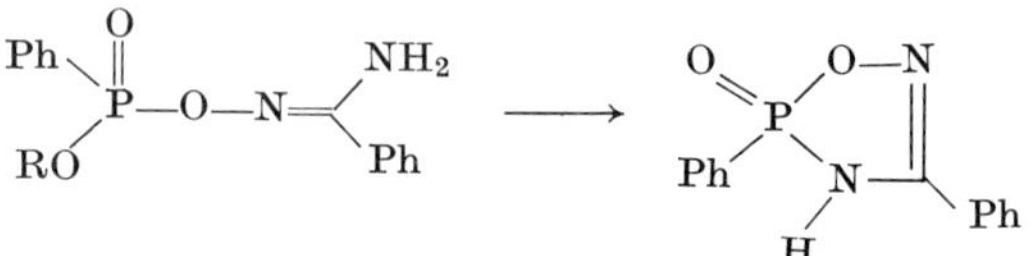

In the presence of KOBu^t–Bu^{t}OD (**298**) ring-opens to (**299**); the C—C bond cleavage occurs with >98% retention of configuration.[565] The ring-openings of 2-ethylbenzisoxazolium fluoroborate in the presence of dimedone[566] and bromolactonic acids in the presence of acid[567] have also been investigated.

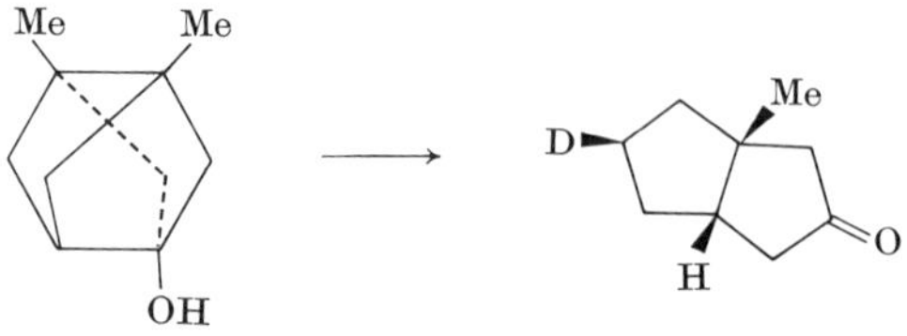

Six-membered Rings

N'-Alkoxy-5-formamidoimidazole-4-carboxamidines have been shown to be intermediates in the Dimroth rearrangement of 1-alkoxyadenines to 6-alkoxyaminopurines.[568] The rates of Dimroth rearrangement of 1,2-dihydro-2-imino-1-methyl-5-(p-substituted phenyl)pyrimidines decrease as the electron-donating properties of the *para*-substituents increase.[569] The influence of electronic and steric effects on the corresponding rearrangement of methyl-substituted imidazo[1,2-a]pyrimidines has also been reported.[570] The

---

[559] T. Nishiwaki and T. Saito, *J. Chem. Soc.* (C), **1971**, 3021.
[560] H. Ferres, M. S. Hamdam, and W. R. Jackson, *J. Chem. Soc.* (B), **1971**, 1892.
[561] M. S. Newman and Z. U. Din, *J. Org. Chem.*, **36**, 2740 (1971).
[562] W. T. Flowers, D. R. Taylor, A. E. Tipping, and C. N. Wright, *J. Chem. Soc.* (C), **1971**, 3097
[563] G. V. Garner and H. Suschitzky, *Tetrahedron Letters*, **1971**, 169.
[564] R. F. Hudson and R. Woodcock, *Chem. Comm.*, **1971**, 1050.
[565] W. T. Borden, V. Varma, M. Cabell, and T. Ravindranathan, *J. Am. Chem. Soc.*, **93**, 3800 (1971).
[566] G. Subrahmanyam and M. Jogibhukta, *Tetrahedron*, **27**, 5229 (1971).
[567] A. W. McCulloch, B. Stanovnik, D. G. Smith, and A. G. McInnes, *Can. J. Chem.*, **49**, 241 (1971).
[568] T. Fujii, T. Sato, and T. Itaya, *Chem. Pharm. Bull.* (*Japan*), **19**, 1731 (1971); T. Fujii, T. Haya, C. C. Wu, and F. Tanaka, *Tetrahedron*, **27**, 2415 (1971).
[569] D. J. Brown and B. T. England, *J. Chem. Soc.* (C), **1971**, 250.
[570] P. Guerret, R. Jacquier, and G. Maury, *J. Heterocyclic Chem.*, **8**, 643 (1971).

Dimroth rearrangement of 1,6-dihydro-6-imino-1-methylpyrimidine has been examined,[571] and a similar reorganization is implicated in the NaOH-induced conversion of 3-cyano-1-methylpyridinium iodide into 2-(N-methylamino)pyridine-3-carbaldehyde.[572] Hydrolysis of the imino linkage is probably responsible for the NaOH-promoted rearrangement of (300) to a mixture of (301) and (302).[573] The course of the reaction of 6-chloro-2-phenylpyrimidines and NaNH$_2$ is strongly influenced by the 5-substituent. For example, when an alkoxy group is present a mixture of (303) and (304) is formed; ^{14}C-labelling studies are in accord with nucleophilic attack at position 6 and rupture of the 5—6 bond.[574] 4-Bromo-6-phenylpyrimidine is converted into the 4-amino derivative by KNH$_2$; ^{15}N-labelling experiments suggest that 83% of the product is derived from an open-chain intermediate, probably formed by attack of the nucleophile at position 2 and fission of the 2—3 bond.[575] Pyrolysis of tetrafluoropyridazine at 800° yields a mixture of tetrafluoro-pyrimidine and -pyrazine.[576]

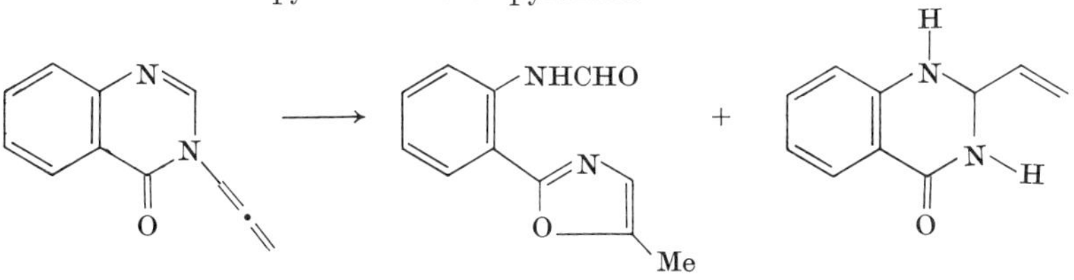

(300)   (301)   (302)

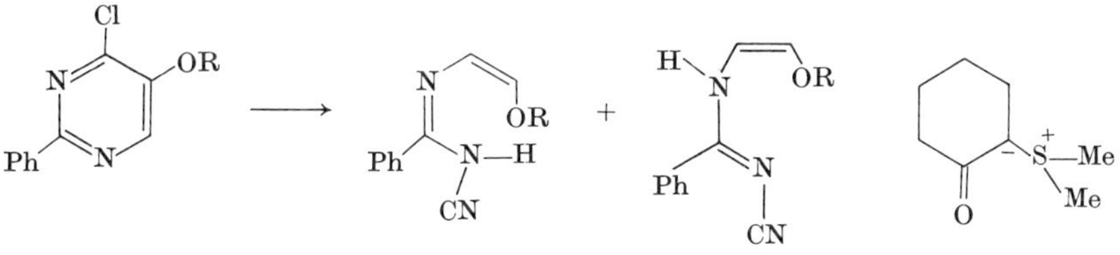

(303)   (304)   (305)

A number of reactions involving the ring-enlargement of six-membered rings have been described. For example, N-methylacridinium iodides react with hydroxylamine-O-sulphonic acid to give dibenzo-1,4-diazepines;[577] the sulphoxide of 8,3'-S-cycloadenosine is rearranged by NaOH to the 8,5'-O-cyclonucleoside;[578] and the ylid (305) undergoes a two-C-atom ring-expansion in the presence of dimethyl acetylenedicarboxylate.[579]

When 4,5-dichloro-2-phenyl-2H-pyridazin-3-one is heated with NaOH a novel ring-contraction to (306) ensues; D is incorporated into the pyrazole ring when NaOD is employed.[580] Cleavage of the 5—6 bond is possibly involved in the KNH$_2$-induced ring-contraction of 2-chloropyrazine into 2-cyanoimidazole.[581] The Grignard derivative

[571] D. J. Brown and B. T. England, J. Chem. Soc. (C), 1971, 2507.
[572] J. H. Blanch and K. Fretheim, J. Chem. Soc. (C), 1971, 1892.
[573] C. Bogentoft, O. Ericsson, and B. Danielsson, Acta Chem. Scand., 25, 551 (1971).
[574] H. W. van Meeteren and H. C. van der Plas, Rec. Trav. Chim., 90, 105 (1971).
[575] J. De Valk and H. C. van der Plas, Rec. Trav. Chim., 90, 1239 (1971).
[576] R. D. Chambers, J. A. H. MacBride, and W. K. R. Musgrave, J. Chem. Soc. (C), 1971, 3384.
[577] H. Hirobi and T. Ozawa, Tetrahedron Letters, 1971, 4493.
[578] M. Ikehara, Y. Ogiso, Y. Matsuda, and T. Morii, Tetrahedron Letters, 1971, 2965.
[579] T. Mukaiyama and M. Higo, Tetrahedron Letters, 1970, 5297.
[580] Y. Makai, G. P. Beardsley, and M. Takaya, Tetrahedron Letters, 1971, 1507.
[581] P. J. Lont, H. C. van der Plas, and A. Koudijs, Rec. Trav. Chim., 90, 207 (1971).

of 3-bromocyclohexene equilibrates with that of 3-bromomethylcyclopentene, presumably via bicyclo[3.1.0]hexylmagnesium bromide.[582] The Wolff–Kischner reduction of selenochromanone is accompanied by ring-contraction.[583] High yields of imidazolin-2-ones are formed when 1,2,4-triazin-3-ones are treated with hydroxylamine-O-sulphonic acid; in the case of cinnolin-3-one, **(307)** is an isolable intermediate which rearranges to oxindole, probably via **(308)**.[584] In contrast, 4,5-diphenyl-1,2,3-triazole is produced when 5,6-diphenyl-1,2,4-triazin-3($2H$)-one reacts with NH_2Cl.[585] An oxidative ring-contraction of 3,6-di-t-butyl-2-hydroxyl-1,4-benzoquinone to 2,4-di-t-butyl-2-chlorocyclopentene-1,3-dione is induced by $CuCl_2$ in acetic acid,[586] and a reductive ring-contraction of 1,4-benzoxazines to benzoxazoles occurs in the presence of phenylhydrazine.[587] Ring-contractions are also involved in the rearrangements of 1,4-dihydropyridine[588] and dihydropyran[589] derivatives.

An interesting cyclization of **(309)** to **(310)** occurs in the presence of I_2.[590] The cyclization of the mixed anhydride of 4-benzoylbutanoic and methyl hydrogen carbonate at 150° probably involves a bicyclic mechanism.[591] In the presence of NaH **(311)** gives **(312)** and not the expected 3-amino-1-benzylindazole; however, the latter derivative is an intermediate in the reaction.[592] Spectroscopic and kinetic evidence points to the intervention of **(313)** in the KOBu^t-induced rearrangement of N-benzoyl-O-glycylserinamide

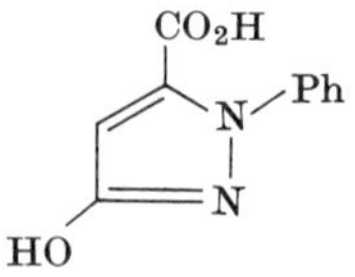

(306) 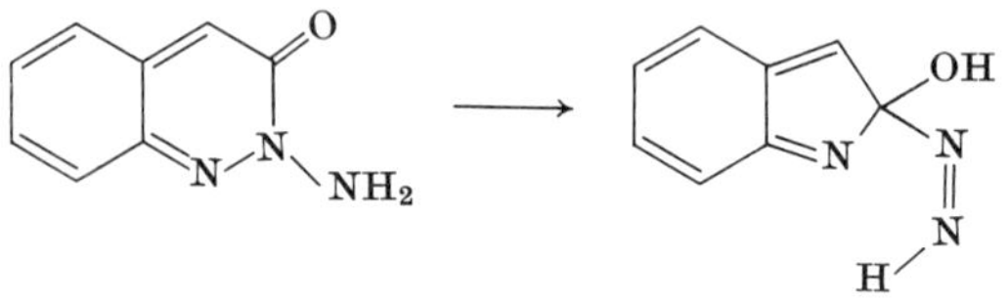   **(307)**   **(308)**

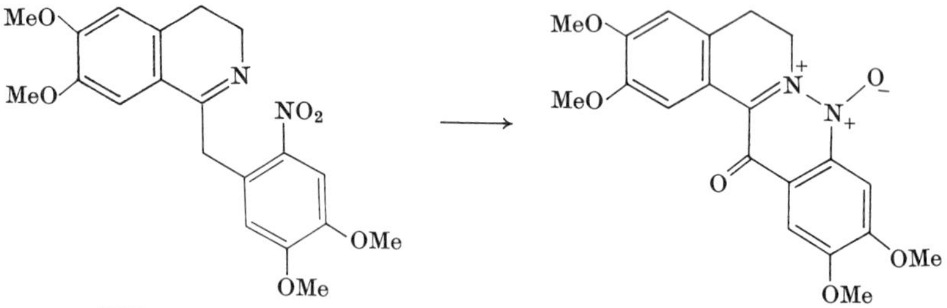

(309)   **(310)**

[582] A. Maercker and R. Geuss, *Angew. Chem. Int. Ed.*, **9**, 909 (1970).
[583] N. Bellinger, D. Cagniant, and P. Cagniant, *Tetrahedron Letters*, **1971**, 49.
[584] C. W. Rees and A. A. Sale, *Chem. Comm.*, **1971**, 531.
[585] C. W. Rees and A. A. Sale, *Chem. Comm.*, **1971**, 532.
[586] H. W. Moore and R. J. Wikholm, *Chem. Comm.*, **1971**, 1073.
[587] R. N. Henders, *J. Org. Chem.*, **36**, 2449 (1971).
[588] R. C. Allgrove, L. A. Cort, and J. A. Elvidge, *J. Chem. Soc.* (C), **1971**, 434.
[589] J. Brugidou, H. Christol, and Y. Langourieux, *Bull. Soc. Chim. France*. **1970**, 4062, 4067.
[590] M. P. Cava, M. J. Mitchell, and D. J. Hill, *Chem. Comm.*, **1970**, 1601.
[591] M. S. Newman and S. S. Gupte, *J. Org. Chem.*, **35**, 4176 (1970).
[592] N. Finch and H. W. Gschwend, *J. Org. Chem.*, **36**, 1463 (1971).

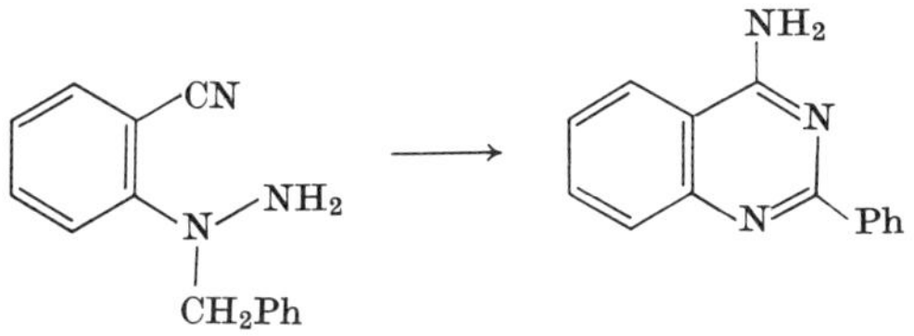

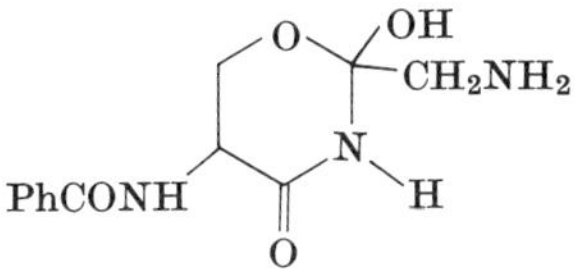

(311) (312) (313)

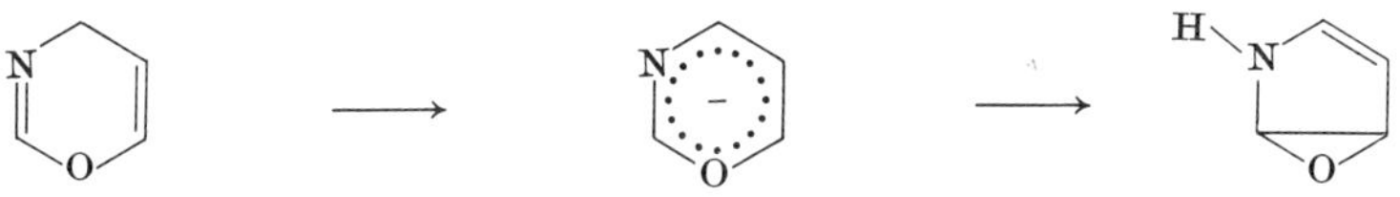

(314) (315) (316)

to *N*-benzoylserinylglycinamide.[593] When (314) is treated with Bu^{n}Li in THF at $-78°$ a deep blue solution of the anion (315) is formed; the solution rapidly decolourizes and yields (316) after acidification.[594]

Seven-membered and Larger Rings

The ring-contractions of 1-substituted 1*H*-azepines to 6-aminofulvenes by heat,[595] 4,5,6,7-tetrahydro-2-methoxy-1*H*-azepin-4-ones to 2-[2-(methoxycarbonyl)vinylidene]-pyrrolidines by acid,[596] 3-alkyl-8-nitro-*s*-triazolo[3,4-*b*]benzothiadiazepines to 1,3-benzothiazines in the presence of NaOEt,[597] and the base-induced rearrangements of 2,3-dihydro-5-methyl-6-phenyl-1,2-diazepin-4-ones[598] have been examined.

The adduct (317), derived from 1-nitrosocyclohexene and maleic anhydride, undergoes an unusual rearrangement in the presence of methanolic HCl to give (318).[599]

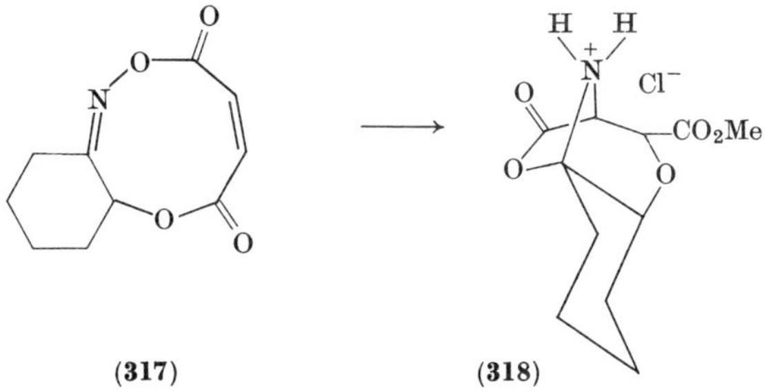

(317) (318)

---

[593] P. L. Russell, R. M. Topping, and D. E. Tutt, *J. Chem. Soc.* (B), **1971**, 657.
[594] R. R. Schmidt, *Angew. Chem. Int. Ed.*, **10**, 572 (1971).
[595] M. Mahendran and A. W. Johnson, *J. Chem. Soc.* (C), **1971**, 1237.
[596] Y. Tamura, Y. Yoshimura, and Y. Kita, *Chem. Pharm. Bull. (Japan)*, **19**, 1068 (1971).
[597] T. George and R. Tahilramani, *J. Org. Chem.*, **36**, 2190 (1971).
[598] J. A. Moore, E. J. Volker, and C. M. Kopay, *J. Org. Chem.*, **36**, 2676 (1971).
[599] G. Just and W. Zehetner, *Chem. Comm.*, **1971**, 81.

CHAPTER 9

Radical Reactions

A. LEDWITH and P. J. RUSSELL

Donnan Laboratories, University of Liverpool

Introduction	275
Structure and Stereochemistry	277
Decomposition of Peroxides	281
Decomposition of Azo-compounds	287
Diradicals	290
Atom-transfer Processes	296
Additions	304
Intermolecular Addition Reactions	304
Intramolecular Addition Reactions	310
Aromatic Substitution	313
Rearrangements	320
S_H2 **Reactions**	326
Reactions Involving Oxidation or Reduction by Metal Salts	331
Radical Ions and Electron-transfer Processes	336
Nitroxides	348
Autoxidation	353
Pyrolysis and Other Gas-phase Processes	355
Radiolysis, ESR Spectroscopy and Miscellaneous	359

Introduction

Workers in this field of chemistry are well served by two review volumes. The first[1] is a tribute to Professors D. H. Hey and W. A. Waters and consists of sixteen chapters giving timely surveys of the major areas of radical chemistry. The second[2] is a collection of special lectures given at two symposia of the IUPAC Congress (Boston, July 1971) under the titles "Short Lived Intermediates" and "Free Radicals and Homolytic Mechanisms".

The rapidly growing recognition of S_H2 processes in organic and organometallic chemistry—given a special section in this year's Report—is reflected in the appearance of a review monograph[3] which supplies a refreshing new look at many familiar free-radical reactions.

The technique of CIDNP[4] continues to grow in value for the free-radical chemist;[5] the

---

[1] R. O. C. Norman, ed., Chem. Soc. Special Publ., No. 24, 1970.
[2] XXIIIrd International Congress of Pure and Applied Chemistry, Vol. 4, Butterworths, London, 1971.
[3] K. U. Ingold and B. P. Roberts, "Free-Radical Substitution Reactions", Wiley-Interscience, John Wiley & Sons, New York, 1971.
[4] See *Org. Reaction Mech.*, **1970**, 309–311.
[5] R. Kaptein, F. W. Verheus, and L. J. Oosterhoff, *Chem. Comm.*, **1971**, 877; C. Walling and A. R. Lepley, *J. Am. Chem. Soc.*, **93**, 546 (1971).

early, tentative theories are being consolidated and have been fully described by Closs[6] and by Fischer.[7] Simple rules, devised by Kaptein,[8] facilitate qualitative predictions of the radical-pair mechanism of CIDNP and are likely to be of great value.

Rüchardt and his collaborators[9] have given a further survey of the tremendous importance of transition-state phenomena in controlling rates of free-radical reactions, while the more traditional polar and resonance effects—with treatments of Hammett, Taft, and Bamford and Jenkins—have been reviewed by Afanas'ev.[10]

Isotopic-exchange reactions of organic iodides with elementary iodine illustrate a remarkable variety of reaction types and the pioneering work of Noyes has been reviewed.[11]

Radical intermediates are involved in many gas-phase reactions of simple molecules with atoms (H, C, O and halogens), and their relevance to many technological applications such as explosion, combustion and halogenation has been surveyed.[12] Two related reviews[13,14] deal specifically with various aspects of chlorination by radical mechanisms. Thermodynamic and kinetic factors controlling free-radical cyclizations have been reviewed by Julia.[15] A comprehensive survey of kinetic-isotope, substituent and steric effects on hydrogen atom abstraction from (mainly phenolic) O—H bonds has been published,[16] and a related review[17] deals specifically with gas-phase abstraction of hydrogen and deuterium atoms by methyl and trifluoromethyl radicals.

Stable radicals have been surveyed from several different aspects including mechanistic aspects of oxidations by potassium nitrosodisulphonate (Fremy's radical),[18] structure and stability of aromatic hydrazyls,[19] and a general account of stable nitrogen free radicals.[20]

Other reviews appearing during the year include surveys of hydroxylation of aromatic compounds,[21] ESR spectra of fluorocarbon radicals[22] and of radicals (and radical ions) from thiophen derivatives,[23] reactions of the hydrated electron[24] and oxidation of organic nitrogen compounds by lead tetra-acetate.[25]

---

[6] G. L. Closs and D. R. Paulson, *J. Am. Chem. Soc.*, **92**, 7229 (1970); G. L. Closs and A. D. Trifunac, *J. Am. Chem. Soc.*, **92**, 7227 (1970); see also ref. 2, p. 19.
[7] H. Fischer and M. Lehnig, *J. Phys. Chem.*, **75**, 3410 (1971); see also ref. 2, p. 1.
[8] R. Kaptein, *Chem. Comm.*, **1971**, 732.
[9] See C. Rüchardt et al. in ref. 2, p. 223.
[10] I. B. Afanas'ev, *Russ. Chem. Rev.*, **40**, 216 (1971).
[11] R. M. Noyes and E. Körös, *Accounts Chem. Res.*, **4**, 233 (1971).
[12] H. G. Wagner and J. Wolfrum, *Angew. Chem. Int. Ed.*, **10**, 604 (1971).
[13] J. P. Soumillion, *Ind. Chim. Belge*, **35**, 851 (1970); *Chem. Abs.*, **73**, 130344 (1970).
[14] M. Kosugi, T. Migita, and Y. Nagai, *Nippon Kagaku Zasshi*, **92**, 477 (1971); *Chem. Abs.*, **75**, 7555 (1971).
[15] M. Julia, *Accounts Chem. Res.*, **4**, 386 (1971).
[16] M. Simonyi and F. Tüdös, *Adv. Phys. Org. Chem.*, **9**, 127 (1971).
[17] P. Gray, A. A. Herod, and A. Jones, *Chem. Rev.*, **71**, 247 (1971).
[18] H. Zimmer, D. C. Lankin, and S. W. Horgan, *Chem. Rev.*, **71**, 229 (1971).
[19] D. Braun, G. Peschk, and E. Hechler, *Chem.-Ztg., Chem. App.*, **94**, 703 (1970); *Chem. Abs.*, **74** 31317 (1971).
[20] A. T. Balaban, *Rev. Roum. Chim.*, **16**, 725 (1971).
[21] D. I. Metelitsa, *Russ. Chem. Rev.*, **40**, 563 (1971).
[22] V. I. Muromtsev, R. A. Asaturyan, and I. G. Akhvlediani, *Russ. Chem. Rev.*, **40**, 175 (1971).
[23] L. Lunazzi, A. Mangini, G. F. Pedulli, and M. Tiecco, *Gazzetta*, **101**, 10 (1971).
[24] J. Shankar, *J. Indian Chem. Soc.*, **48**, 97 (1971).
[25] J. B. Aylward, *Quart. Rev. (London)*, **25**, 407 (1971).

Structure and Stereochemistry

The combination of thermochemical and kinetic studies of reactions between iodine and organic molecules, as developed by Benson's group, continues to provide important and sometimes surprising information about the structures of organic radicals. Thus the gas-phase reaction between benzaldehyde and iodine has been shown[26] to have the following rate determining step:

$$PhCHO + I\cdot \rightarrow Ph\dot{C}O + HI$$

The carbonyl—hydrogen bond strength in benzaldehyde was found to be the same as for acetaldehyde and formaldehyde; it follows that stabilization of the radicals (RĊO) is independent of R and may be explained by conjugation of the odd electron with one of the oxygen lone-pair electrons. For the benzoyl radical there is no contribution to the radical stability from structures in which the unpaired electron is conjugated with the aromatic ring. Very similar conclusions may be drawn from ESR spectroscopic studies[27] of such radicals, trapped in a variety of matrices at 77°K by using the rotating cryostat technique.

Schleyer's group has shown[28] that the conformational-analysis programme, developed for carbonium ions, is applicable also for calculation of free-radical reactivities; bridgehead free-radical reactivities of a large number of systems have been predicted.

ESR evidence for the classical structure of 7-norbornenyl radical (1) was presented last year;[29] MINDO/2 calculations for (1) and its saturated homologue have now been reported[30] and lend support to the earlier conclusions.

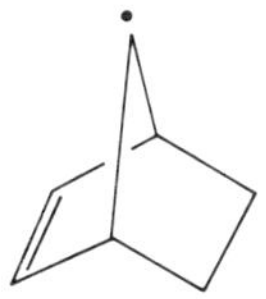

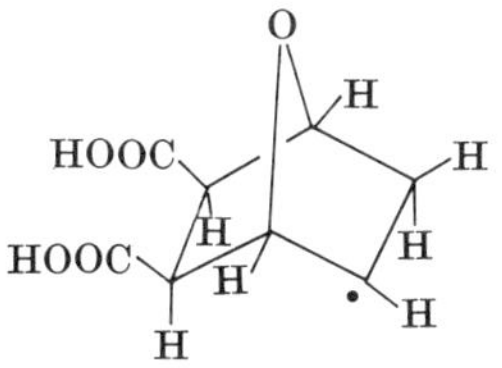

Radicals (2) and (3), formed by reactions of *exo-cis*-7-oxabicyclo[2.2.1]heptane-2,3-dicarboxylic acid and *exo-cis*-7-oxabicyclo[2.2.1]hept-2-ene-5,6-dicarboxylic acid, respectively, with H_2O_2–$TiCl_3$, are found by ESR to have non-planar (pyramidal) radical centre carbon atoms with the carbon—hydrogen bond bent in the *endo* direction.[31] Conceivably this might help to explain the well known preference[32] of free radicals of the 2-norbornyl type for reacting in atom-transfer reactions at the *exo*-position.

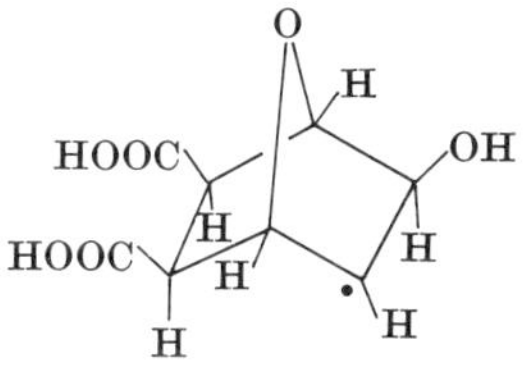

The 1-methyl-2,2-diphenylcyclopropyl radical (4) reacts in solution either to give (5), or to give (7) via the rearranged radical (6); apparently (4) is less selective in hydrogen abstraction than a chlorine atom.[33]

---

[26] R. K. Solly and S. W. Benson, *J. Am. Chem. Soc.*, **93**, 1592 (1971).
[27] J. E. Bennett and B. Mile, *Trans. Faraday Soc.*, **67**, 1587 (1971).
[28] R. C. Bingham and P. von R. Schleyer, *J. Am. Chem. Soc.*, **93**, 3189 (1971).
[29] See *Org. Reaction Mech.*, **1970**, 308.
[30] M. J. S. Dewar and W. W. Schoeller, *Tetrahedron*, **27**, 4401 (1971).
[31] T. Kawamura, T. Koyama, and T. Yonezawa, *J. Am. Chem. Soc.*, **92**, 7222 (1970).
[32] See *Org. Reaction Mech.*, **1970**, 315.
[33] H. M. Walborsky and J.-C. Chen, *J. Am. Chem. Soc.*, **92**, 7573 (1970).

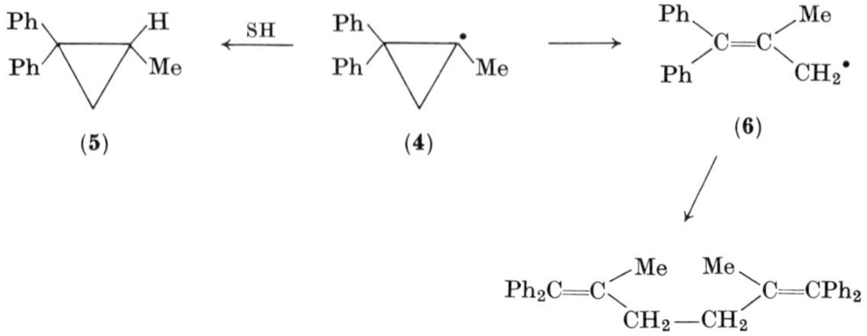

(5) ← SH — (4) → (6)

Ph₂C=C(Me)-CH₂-CH₂-C(Me)=CPh₂

(7)

Generation of (**4**) from a variety of optically active precursors yielded cyclopropyl products that were largely if not entirely racemized.[34] However, when the radical (**4**) disproportionates within the solvent cage, the 1-methyl-2,2-diphenylcyclopropane (**5**) obtained was found to be 31–37% optically pure (66–68% retention of configuration). This result adds to the growing body of evidence suggesting that cage reactions of radical pairs may occur with very high degrees of stereospecificity, and further emphasizes the point made last year[35] that retention of configuration in intramolecular rearrangements should not be taken as evidence *against* free-radical intermediates.

Whilst the energy barrier to inversion in cyclopropyl radicals having only hydrogen, alkyl or aryl substituents (e.g. **4**) appears to be quite small, halogen substitution at the radical centre causes a marked increase (CNDO/2 calculations); inversion barriers calculated[36] for a series of related radicals are as shown. In accord with these calculations,

Radical	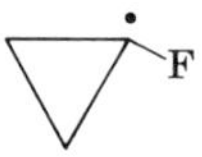 ▽·F	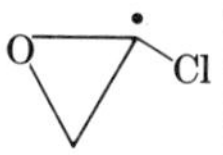 ▽·Cl	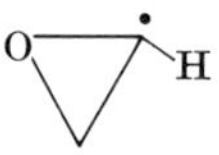 ▽·H	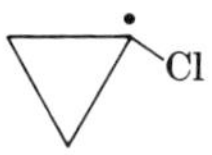 ▽·Cl	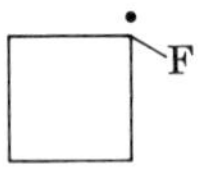 □·F	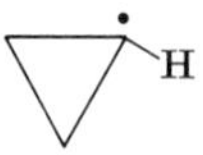 ▽·H
Inversion barrier (kcal mole⁻¹)	10.5	10.5	5.4	4.0	1.9	0.8

free-radical debromination of the two *endo–exo*-Cl/Br isomers of (**8**) by triphenyltin hydride occurs with very little equilibration of the intermediate radicals (e.g. **9**), whereas similar reductions of the *cis*- and *trans*-isomers of cyclobutane (**10**) gave rise to identical product ratios, indicating that rates of inversion of the α-fluorocyclobutyl radical

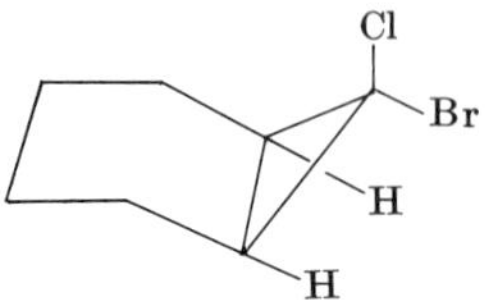

(8) 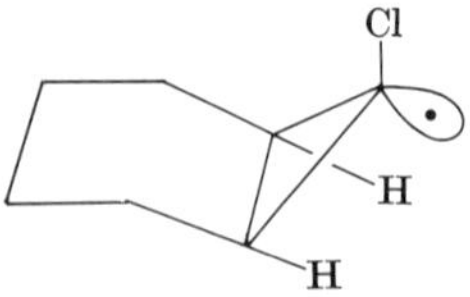   (9)

[34] H. M. Walborsky and J.-C. Chen, *J. Am. Chem. Soc.*, **93**, 671 (1971).
[35] See *Org. Reaction Mech.*, **1970**, 321.
[36] L. J. Altman and R. C. Baldwin, *Tetrahedron Letters*, **1971**, 2531.

(10)

(11)

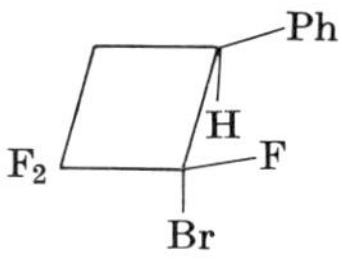

(e.g. **11**) were fast with respect to reaction with Ph_3SnH.[36] In a related study,[37] it is concluded that the identical ratios of products (**13**) and (**14**) obtained by hydrogen-abstraction reactions of the α-chlorocyclopropyl radical (**12**) indicate that the epimeric isomers (**12a**,**b**) interconvert more rapidly than they abstract hydrogen from cumene and toluene. MINDO/2 calculations indicate[38] that ring opening of the cyclopropyl

(13) **(14)**

radical should be disrotatory, in direct contradiction to earlier calculations based on the Extended Hückel method; an experimental test of this prediction would be of considerable interest.

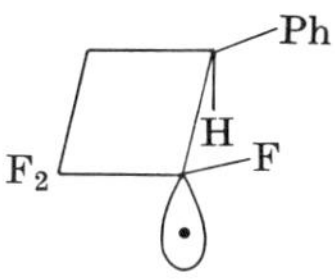

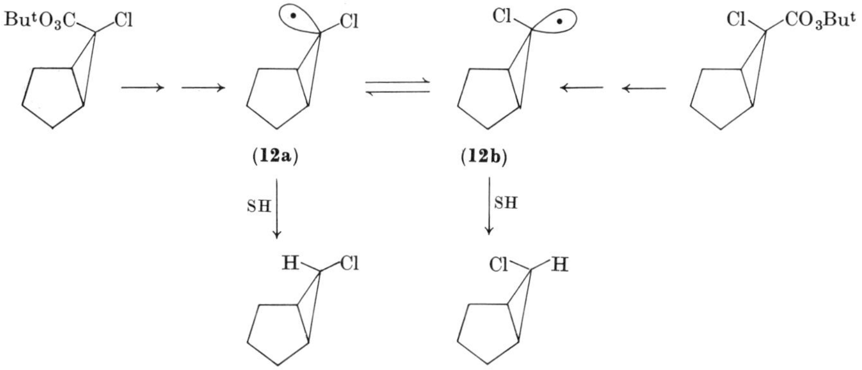

Further evidence for the comparatively high configurational stability of vinylic radicals has been provided[39] by studies of the sodium borohydride reduction of oxymercuration adducts of a series of C_{10} to C_{13} allenes. Thermal decomposition of *tert*-butyl *cis*- and *trans*-1-methoxypercrotonates (**15**) in cumene yielded olefinic products (**17**) with complete retention of the initial perester stereochemistry;[40] apparently hydrogen abstraction by the *cis*- and *trans*-isomeric radicals (**16**) occurs much more rapidly than loss of configurational identity (see also p. 308).

(15) *cis* or *trans* **(16)** *cis* or *trans* **(17)** *cis* or *trans*

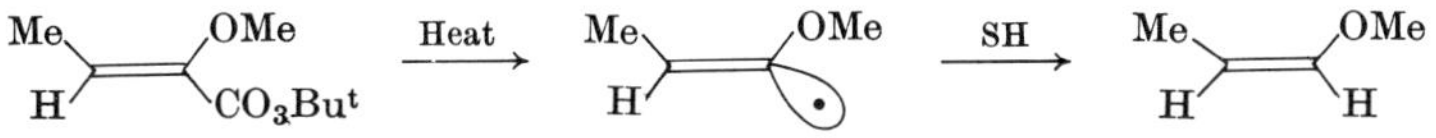

[37] L. A. Singer and J. Chen, *Tetrahedron Letters*, **1971**, 939.
[38] M. J. S. Dewar and S. Kirschner, *J. Am. Chem. Soc.*, **93**, 4290 (1971).
[39] R. Vaidyanathaswamy, D. Devaprabhakara, and V. V. Rao, *Tetrahedron Letters*, **1971**, 915.
[40] M. S. Liu, S. Soloway, D. K. Wedegaertner, and J. A. Kampmeier, *J. Am. Chem. Soc.*, **93**, 3809 (1971).

The elegant work of Norman and his collaborators in determining the geometry at the tervalent carbon of organic radicals was referred to last year.[41] Full details of this work have now been published[42] and isotropic ESR parameters are reported for a large number of cyclic and acyclic radicals in which the tervalent carbon is bonded to one or two oxygen substituents. Analysis of the data leads to characterization of two factors important in determining the degree of bonding at the tervalent carbon atom. These are, first, increase in the p-character of the carbon bonds, as induced, for example, by constraining these bonds within a small ring, and secondly, attachment to the tervalent carbon of a conjugatively electron-releasing group which serves to increase the electron density at the carbon atom; in the latter respect, two oxygen substituents have a far greater influence than one, and the changes in geometry that accompany the replacement of a hydrogen atom at the tervalent carbon atom by a methyl or trifluoromethyl group can be understood. In essence this work leads to a general conclusion that oxygen, and possibly other electronegative atoms, attached at a radical site cause the radical to adopt a non-planar conformation. Others have arrived at similar conclusions; for example, hyperfine splittings due to ^{13}C in natural abundance have been observed in several short-lived radicals produced photolytically in liquids, and the results indicate that whereas the radical $\dot{C}H_2COO^-$ is planar, the radical $Me_2\dot{C}-OH$ is not.[43] Related ESR studies of radicals $C-\dot{C}H-NH_2$, obtained by reactions of amines with the $Ti^{3+}-H_2O_2$ couple, indicate that the NH_2 group is not coplanar with the nodal plane of the π-system;[44] the experimental observations are in agreement with calculations.[45] A detailed study of the temperature-dependence of β-proton hyperfine splittings and line shapes indicates[46] that for the hydroxymethyl radical $\cdot CH_2OH$, there is a barrier of about 4 kcal/mole restricting rotation about the $C-O$ bond.

Theoretical calculations and a thorough experimental study of the ESR spectra of a series of alkyl radicals (**18**) substituted in the β-position with sulphur, silicon, germanium, or tin groups has been reported by Krusic and Kochi.[47] The results provide clear evidence of hindered rotation about the $C_\alpha-C_\beta$ bond and a preferred conformational orientation in these radicals in which the sulphur, silicon, germanium, and tin atoms are eclipsed with the p-orbital at the trigonal centre, as in (**19**).

$-\underset{|}{\overset{\cdot}{C}}-CH_2-X$

(**18**)

(**19**)

A structure involving symmetrically bridged radicals is apparently ruled out and it is suggested[47] that the conformational effects are due to incipient 1,3-bonding between unfilled $3d$ orbitals of the hetero atom and the p-orbital of the carbon radical centre.

---

[41] See *Org. Reaction Mech.*, **1970**, 315.
[42] A. J. Dobbs, B. C. Gilbert, and R. O. C. Norman, *J. Chem. Soc.* (A), **1971**, 124.
[43] R. Livingston, J. K. Dohrmann, and H. Zeldes, *J. Chem. Phys.*, **53**, 2448 (1970).
[44] R. Poupko, A. Loewenstein, and B. L. Silver, *J. Am. Chem. Soc.*, **93**, 580 (1971).
[45] R. Poupko and B. L. Silver, *J. Am. Chem. Soc.*, **93**, 575 (1971).
[46] P. J. Krusic, P. Meakin, and J. P. Jesson, *J. Phys. Chem.*, **75**, 3438 (1971).
[47] P. J. Krusic and J. K. Kochi, *J. Am. Chem. Soc.*, **93**, 846 (1971); see also T. Kawamura, M. Ushio, T. Fujimoto, and T. Yonezawa, *J. Am. Chem. Soc.*, **93**, 908 (1971).

Additional evidence for this type of conformational effect involving heteroatoms with filled d-orbitals has been reported[48] from similar studies of the ESR spectra of carbon radicals carrying tin, phosphorus or arsenic substituents in the β-position. In marked contrast, however, it is here concluded that the effects are due entirely to a hyperconjugative interaction of the β-substituent with σ-electrons linking it with the radical framework. A lively controversy should result from these conflicting interpretations, with a possible influence on interpretation of certain anomolous reactivities[49] of corresponding organometallic compounds.

It was reported last year[50] that the dibenzoyl peroxide-catalysed addition of menthyl methylphosphinate (**20a**) to alkenes proceeded with inversion of configuration at phosphorus. In apparent direct contradiction, it is now reported[51] that similar reactions of menthyl phenylphosphinate (**20b**) occur with retention of configuration at phosphorus.

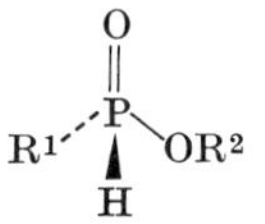

$$R^1 \underset{H}{\overset{O}{\underset{\|}{P}}} OR^2 \quad \begin{array}{l} a: R^1 = Me;\ R^2 = menthyl \\ b: R^1 = Ph;\ R^2 = menthyl \end{array}$$

(**20**)

Decomposition of Peroxides[52]

The range of peroxides, based on amino derivatives of cyclohexanone, referred to in last year's report[53a] has been extended[53b] to include interesting new sources of radicals such as (**21**) which decomposes[54] with homolysis of both the azo and the peroxide linkage.

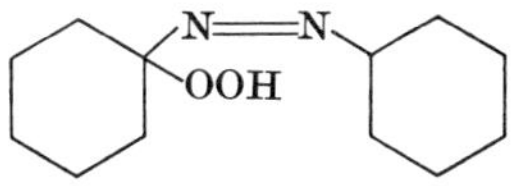

(**21**)

Details of the synthesis and thermal decomposition of the β-nitroalkyl peroxynitrate (**22**) have been reported;[55] the primary step in homolysis leads to formation of the β-nitroalkoxy radical (**23**) which then fragments (forming mainly acetone), dimerizes, and recombines with NO_2 and NO_3 to give a wide range of products.

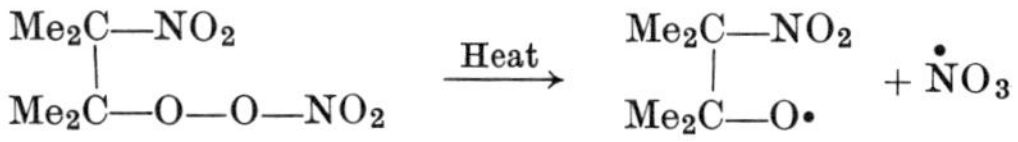

$$\begin{array}{c} Me_2C-NO_2 \\ | \\ Me_2C-O-O-NO_2 \end{array} \xrightarrow{\text{Heat}} \begin{array}{c} Me_2C-NO_2 \\ | \\ Me_2C-O\cdot \end{array} + \dot{N}O_3$$

(**22**)   (**23**)

---

[48] A. R. Lyons and M. C. R. Symons, *Chem. Comm.*, **1971**, 1068.
[49] W. Hanstein, H. J. Berwin, and T. G. Traylor, *J. Am. Chem. Soc.*, **92**, 7476 (1970).
[50] See *Org. Reaction Mech.*, **1970**, 317.
[51] W. B. Farnham, R. K. Murray, and K. Mislow, *Chem. Comm.*, **1971**, 146.
[52] D. Swern, ed., "Organic Peroxides", Vol. II, Wiley-Interscience, New York, 1971.
[53a] See *Org. Reaction Mech.*, **1970**, 322.
[53b] E. G. E. Hawkins, *J. Chem. Soc.* (C), **1971**, 160.
[54] E. G. E. Hawkins, *J. Chem. Soc.* (C), **1971**, 1474.
[55] E. F. J. Duynstee, J. L. J. P. Hennekens, W. van Raayen, and W. Voskuil, *Tetrahedron Letters*, **1971**, 3197.

Benzoyl toluene-p-sulphonyl peroxide (**24**) has been synthesized and its decomposition studied in a variety of solvents;[56] its decomposition in inert solvents accelerated rapidly as the reactions proceeded, owing to catalysis by developing toluene-p-sulphonic acid. This autocatalysis was prevented by adding magnesium oxide (suspension) to the reaction mixtures and, under these conditions, products were mainly those expected from formation of phenyl radicals. Related work with m-nitrobenzenesulphonyl peroxide (**25**) in the solvent range $CHCl_3$, $PhNO_2$, $PhCl$ and PhH established[57] a clear-cut solvent-dependence of reaction mechanism; for example, in $CHCl_3$ the thermolysis involves mainly homolysis of the $-O-O-$ linkage whereas in $PhNO_2$ the peroxide decomposes by electrophilic attack on the aromatic nuclei.

Formation and decomposition reactions of peroxides (**26**)[58] and (**27**)[59], containing heteroatoms, have been reported.

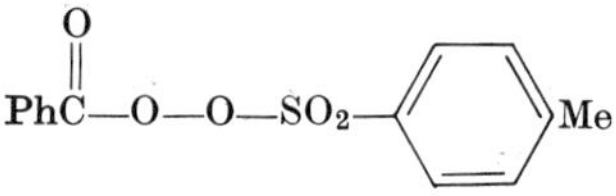

$$PhC(=O)-O-O-SO_2-C_6H_4-Me$$

(**24**)

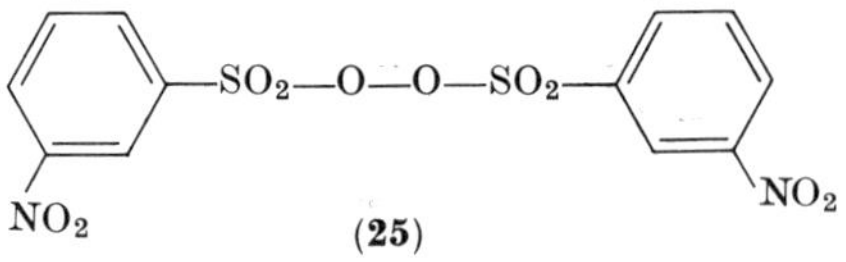

$$O_2N-C_6H_4-SO_2-O-O-SO_2-C_6H_4-NO_2$$

(**25**)

$$Ph_4SbO-OBu^t$$

(**26**)

$$Ph_3P-CHR^1R^2$$
$$|$$
$$O-OBu^t$$

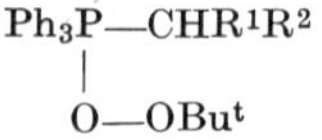

(**27**)

Rüchardt's group have published[60] details of the synthesis and decomposition processes for a wide range of α-substituted O-$tert$-butyl peroxyesters (**28**). Substituents Y, R^1, R^2 exert a polar effect on the homolysis of $tert$-butyl α-alkoxy- and α-acetoxyperoxycarboxylates (**28a,b**) and, because of the greater rate-enhancing effect of alkoxy substituents than of the acetoxy group, the former [(**28a**)] offer an interesting new range of low-temperature initiators. A range of eleven $tert$-butyl α-aryloxyperacetates and four $tert$-butyl α-arylthioperoxyacetates (**28c**) was found[61] to decompose homolytically, and again there is a marked effect of polar substituents on reaction rates: for substituents in the group Ar, Hammett ρ values were -1.1 and -1.3, respectively, for X = O and S.

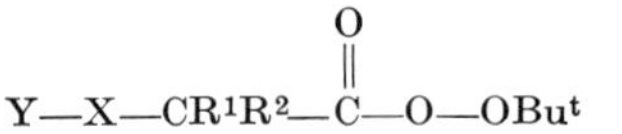

$$Y-X-CR^1R^2-C(=O)-O-OBu^t$$

(**28**)

a: Y = Me, Et or Pr^i; X = O;
   $R^1 = R^2 = H$
b: Y = CH_3CO; X = O;
   $R^1 = R^2 = H$ or Me
c: Y = Ar; X = O, S; $R^1 = R^2 = H$

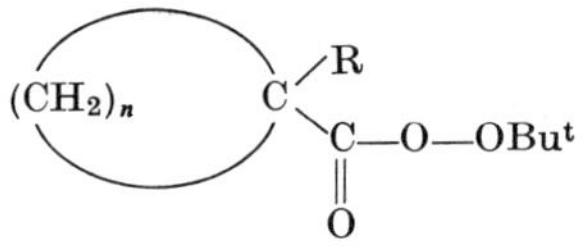

$$(CH_2)_n\begin{matrix}R\\C\\C(=O)-O-OBu^t\end{matrix}$$

(**29**)

$n = 2$–11
R = H, Me

---

[56] R. Hisada, H. Minato, and M. Kobayashi, *Bull. Chem. Soc. Japan*, **44**, 2541 (1971).
[57] Y. Yokoyama, H. Wada, M. Kobayashi, and H. Minato, *Bull. Chem. Soc. Japan*, **44**, 2479 (1971).
[58] G. A. Razuvaev, T. I. Zinov'eva, T. G. Brilkina, and E. P. Silkovskaya, *Dokl. Akad. Nauk SSSR*, **193**, 355 (1970); *Chem. Abs.*, **73**, 130436 (1971).
[59] K. Yamada, K. Akiba, and N. Inamoto, *Bull. Chem. Soc. Japan*, **44**, 2437 (1971).
[60] C. Rüchardt and I. Mayer-Ruthardt, *Chem. Ber.*, **104**, 593 (1971).
[61] C. Rüchardt and H. Böck, *Chem. Ber.*, **104**, 577 (1971).

When the α-carbon of the peroxyacetate component comprised part of a ring system (**29**), decomposition processes were again homolytic with transition states in which the C_α—CO bonds were relatively little stretched.[62] ^{18}O-Labelling experiments excluded a fast, reversible cleavage of the peroxide linkages with consecutive slow decarboxylation as a reaction mechanism, and the influence of ring size on the rate of formation of cycloalkyl free radicals was found to be a useful general criterion for investigation of the geometry of the transition states in these and related reactions.[63]

Study of the thermolysis of three O-[(*tert*-butylperoxy)carbonyl]ketoximes (**30a–c**) has led to the conclusion,[64] based on the very slow rates of decomposition, that, despite the known stability of iminoxy radicals, these reactions are not homolytic. It follows that homolytic perester fragmentations are not induced by the stability of the radicals being generated but depend mainly on favourable polar effects.

$$\begin{array}{cc}
R^1 \\
\diagdown \\
C=N-O-\overset{\overset{\displaystyle O}{\|}}{C}-OOBu^t & AcO-O-\overset{\overset{\displaystyle O}{\|}}{C}-OPr^i \\
R^2\diagup \\
(\mathbf{30}) & (\mathbf{31})
\end{array}$$

a: $R^1 = R^2 = Me$
b: $R^1 = Me, R^2 = Ph$
c: $R^1 = R^2 = Ph$

Decomposition of peracetyl isopropyl carbonate (**31**) at 75–90° yielded a mixture of products from which CH_4, Me_2CO, HOAc, Pr^iOH, Pr^iO_2Ac, and Pr^iOCO_2Me were isolated; nuclear dynamic polarization experiments established the major role of methyl and isopropoxy radical intermediates.[65]

Hydrogen gas is a significant product formed on photolysis of *tert*-butyl peroxyformate in solution, and it is suggested that the H_2 arises from reactions of hydrogen atoms produced in the primary photolytic process.[66] Attempts to isolate di-*sec*-butyl peroxyoxalate from reactions of *sec*-butyl hydroperoxide with oxalyl chloride were unsuccessful;[67] the products included di-*sec*-butyl peroxide, assumed to be formed by cage recombination of *sec*-butoxy radicals from homolysis of di-*sec*-butyl peroxyoxalate formed *in situ*. In contrast, and contrary to previous suggestions, it was found that interaction of two *sec*-butylperoxy radicals does not lead to formation of di-*sec*-butyl peroxide.

The CIDNP effects observed[68] during aroyl peroxide decomposition can be explained exclusively by singlet–triplet transitions in transient benzoyloxy/phenyl radical pairs; cross-relaxation in free phenyl radicals must contribute only to a minor extent, if at all, and the results lend strong support to current theories of CIDNP phenomena.

In agreement with existing results for acetyl peroxide and benzoyl peroxide, it has been shown that decomposition of [^{18}O]acetyl benzoyl peroxide and acetyl [^{18}O]benzoyl peroxide in benzene involves considerably more acetyl-^{18}O than benzoyl-^{18}O scrambling.[69]

---

[62] P. Lorenz, C. Rüchardt, and E. Schacht, *Chem. Ber.*, **104**, 3429 (1971).
[63] For further discussion of this point see also ref. 2, p. 223.
[64] C. Rüchardt and R. Pantke, *Chem. Ber.*, **104**, 3456 (1971).
[65] S. V. Rykov, A. L. Buchachenko, V. A. Dodonov, A. V. Kessenikh, and G. A. Razuvaev, *Dokl. Akad. Nauk SSSR*, **189**, 341 (1969); *Chem. Abs.*, **73**, 119929 (1970).
[66] W. A. Pryor and R. W. Henderson, *J. Am. Chem. Soc.*, **92**, 7234 (1970).
[67] R. Hiatt and L. Zigmund, *Can. J. Chem.*, **48**, 3967 (1970).
[68] B. Blank and H. Fischer, *Helv. Chim. Acta*, **54**, 905 (1971).
[69] M. Kobayashi, H. Minato, and R. Hisada, *Bull. Chem. Soc. Japan*, **44**, 2271 (1971).

High-pressure studies of homolytic processes by Neuman and his group[70-72] continue to provide valuable information, especially in respect of transition-state phenomena. In recent work[70] the pressure-dependences of the rates of thermolysis of m-Cl-, p-Cl-, p-Me- and p-MeO-substituted *tert*-butyl phenylperacetates (**32**) in cumene have been determined. Observed activation volumes for decomposition ($\Delta V^{\ddagger}$) are lower than expected for two-bond homolytic processes; they are also pressure-dependent and are thought to be a composite of contributions from bond-stretching and solvation of polar transition states. The data are in agreement with general conclusions previously outlined for the unsubstituted perester and show that transition-state polarity [e.g. (**33**)] varies with the nature of the substituent. Product studies of the decomposition of peresters (**32**) in cumene (79.6°) showed[71] formation of corresponding *tert*-butyl benzyl ethers, toluenes, bibenzyls, benzylcumyls, bicumyl, and *tert*-butyl alcohol. The effect of pressure on the ether/Bu^{t}OH ratios gives the pressure-dependence of the cage effect (combination versus diffusion); cage effects increased with pressure but not as rapidly as expected.

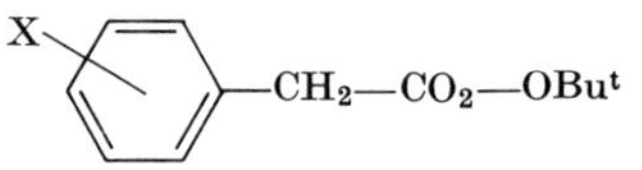

X—C$_6$H$_4$—CH$_2$—CO$_2$—OBu^t

(**32**)

X—C$_6$H$_4$—CH$_2$····CO$_2$····OBu^t $\delta+$ $\delta-$

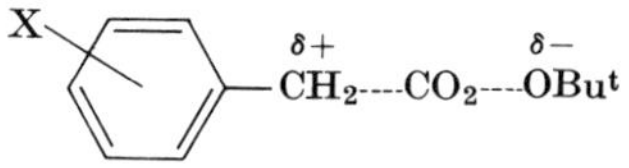

(**33**)

Effects of pressure on the rates and decomposition products of the *cis*- and *trans*-isomers of *tert*-butyl 2-propylperoxypent-2-enoate (**34a,b**) in cumene have been determined.[72] The results indicate that both isomers decompose by one-bond O—O scission to yield geminal vinylcarboxyl and *tert*-butoxy radicals which recombine to regenerate starting materials in competition with disproportionation and separative diffusion. Apparent activation volumes for the two isomers are different, suggesting that recombination of the radical pair derived from the *trans*-perester (**34b**) is more competitive with disproportionation and separative diffusion than is that of the other isomer. Disproportionation to give hept-3-yne and hepta-3,4-diene is more favourable from the

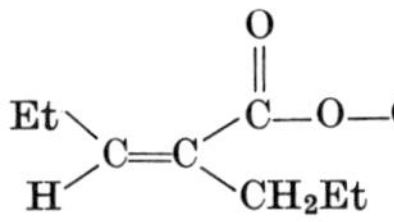

(**34a**) *cis*

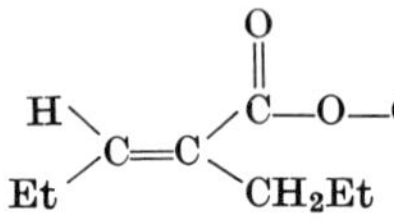

(**34b**) *trans*

cis-vinylcarboxyl radical (from **34a**) and it also appears that this radical undergoes unimolecular decarboxylation more rapidly than the *trans*-isomer; the isomeric vinyl radicals ultimately produced equilibrate rapidly at atmospheric pressure, but application of pressure permits pre-equilibrium trapping.

Decomposition of 3-methylbut-2-enoyl peroxide (**35**) in cumene or carbon tetrachloride gives low, but useful, yields of the appropriate vinyl radical products;[73] low efficiency in the production of vinyl radicals (**37**) is ascribed to competitive reactions of the precursor (**36**) as indicated.

[70] R. C. Neuman and J. V. Behar, *J. Org. Chem.*, **36**, 654 (1971).
[71] R. C. Neuman and J. V. Behar, *J. Org. Chem.*, **36**, 657 (1971).
[72] R. C. Neuman and G. D. Holmes, *J. Am. Chem. Soc.*, **93**, 4242 (1971).
[73] P. G. Webb and J. A. Kampmeier, *J. Am. Chem. Soc.*, **93**, 3730 (1971).

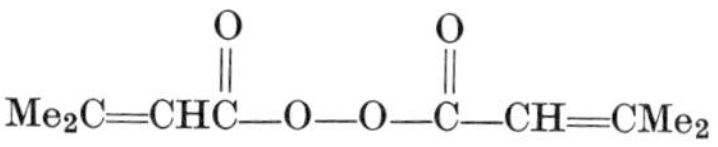

$$\text{Me}_2\text{C}=\text{CHC}(\text{O})-\text{O}-\text{O}-\text{C}(\text{O})-\text{CH}=\text{CMe}_2$$
(35)

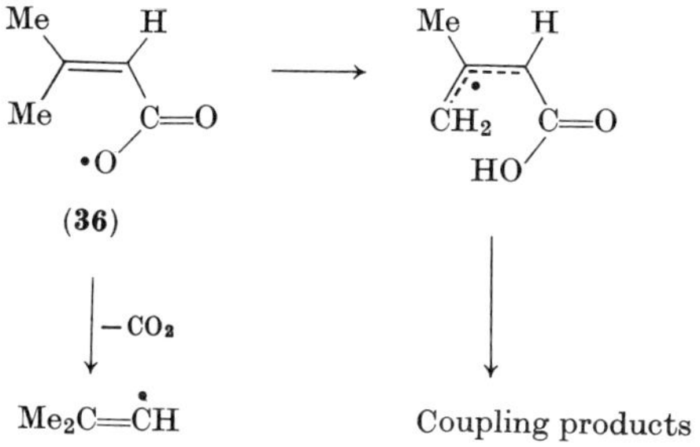

(36) →

(36) ↓ $-\text{CO}_2$

$\text{Me}_2\text{C}=\overset{\cdot}{\text{C}}\text{H}$
(37)

Coupling products

Although photolytic decomposition of aroyl peroxides is a well-established process leading to mixtures of products, including appropriate biaryls, photolysis of benzoyl peroxide in benzene yields a substantial amount of biphenyl which is apparently derived entirely from the solvent; reaction of solvent molecules with photoexcited peroxide are thought to be responsible.[74]

From a detailed re-investigation of the effect of solvent on the thermolysis of di-*tert*-butyl peroxide at 125° it is concluded[75] that solvent-induced decomposition does not make a significant contribution to the relative rates observed. Variations in rate of decomposition of di-*tert*-butyl peroxide with solvent are attributable to variations in reactant and transition-state solvation. The effects are comparable to, or smaller than, those previously observed with simple diacyl peroxides and peresters, which decompose by single-bond scission, and they are far smaller than those observed in peroxides that decompose rapidly by concerted multibond scission.

Evidence for a radical-induced mechanism of decomposition has been obtained[76] by a detailed kinetic study of the thermal decomposition of bis-(5-hexenoyl) peroxide in toluene; rather high concentrations of peroxide were required before the induced decomposition became important.

An interesting series of peroxide-containing free radicals (38) has been studied by Leffler's group.[77] The radicals were generated by reactions of the appropriate benzylic precursors with *N*-bromosuccinimide and in most cases the corresponding α-brominated products were isolated in good yield; the expected intramolecular radical-induced decomposition of the peroxide groups [e.g. via α-lactones such as (39)] was not observed.

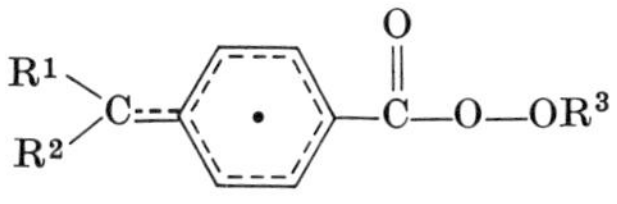

(38) 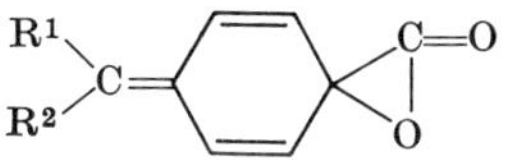   (39)

[74] J. D. Bradley and A. P. Roth, *Tetrahedron Letters*, **1971**, 3907.
[75] C. Walling and D. Bristol, *J. Org. Chem.*, **36**, 733 (1971).
[76] R. C. Lamb, W. E. McNew, J. R. Sanderson, and D. C. Lunney, *J. Org. Chem.*, **36**, 174 (1971).
[77] M. M. Schwartz and J. E. Leffler, *J. Am. Chem. Soc.*, **93**, 919 (1971).

In last year's report,[78] the effects of polar groups in radicals ArĊHOMe, obtained by hydrogen abstraction from substituted benzyl methyl ethers, in radical-induced decomposition of di-*tert*-butyl peroxide were noted. A further report[79] shows that, for similar radicals ArĊHOR, induced decomposition is favoured by chlorine substitution in the aromatic ring and is affected by steric effects of the group R (R = Me, Et, Pr^i, Bu^t or Ph).

Rate constant decomposition of a number of peroxides induced by the tri-*n*-butyltin radical (R$_3$Sn·) in benzene have been measured[80] at 10°C. The values range from ~ 100 mole^{-1} sec^{-1} for di-*tert*-butyl peroxide to 2.6×10^7 mole^{-1} sec^{-1} for di-*tert*-butyl diperoxyisophthalate. Di-*n*-alkyl sulphides are ten times more reactive towards R$_3$Sn· than are di-*n*-alkyl peroxides although the respective reaction exothermicities are ~ 15 and ~ 39 kcal mole^{-1}. Enhanced reactivity of the disulphides is attributed to easier formation of an intermediate or transition state with nine electrons around sulphur, compared with an analogous species with nine electrons around oxygen.[80]

Benzoyl and substituted benzoyl peroxides induce the decomposition of phenyliodine dibenzoate (**40a**) or dianisoate (**40b**) in chlorobenzene at 80°: rates of decomposition of the peroxides are slightly diminished, and yields of chlorobiphenyls from the peroxides slightly increased from those obtained in the absence of the iodine compounds.[81] Major products from the iodine compounds are iodobenzene and the corresponding carboxylic acid and it is suggested that phenylchlorocyclohexadienyl radicals (**41**) induce decomposition of the iodine compounds.

$$\text{PhI}\begin{matrix}\text{O—C—C}_6\text{H}_4\text{X} \\ \text{O—C—C}_6\text{H}_4\text{X}\end{matrix} + \text{Ph}\underset{\text{H}}{\diagdown}\!\!\!\!\overset{\text{Cl}}{\diagup} \longrightarrow \text{PhI—O—C—C}_6\text{H}_4\text{X}$$

(**40**)　　　　　(**41**)　　　　　　(**42**)

a, X = H
b, X = OMe

+ HOOCC$_6$H$_4$X
+ PhC$_6$H$_4$Cl

(**42**) + (**41**) ⟶ PhI + HOOCC$_6$H$_4$X + PhC$_6$H$_4$Cl

Reactions of hydroxylamines with peroxides may be important in inhibition of autoxidation by amine antioxidants; detailed studies[82] of the kinetics of reaction between benzoyl peroxide and diphenylhydroxylamine showed a first-order dependence on both reactants but failed to suggest a clear-cut mechanism. It is tentatively suggested that the first step in this reaction involves formation of a complex that collapses to radicals; a preferred mechanism is one in which this complex is formed by electron transfer from the highest occupied orbital of diphenylhydroxylamine to a low-lying vacant peroxide orbital. Photolysis of *tert*-butyl *N*,*N*-dialkylperoxycarbamates (**43**) gives rise to formation of corresponding *N*,*N*-dialkylamino radicals whereas photolysis of *tert*-butyl *N*-alkylperoxycarbamates (**44**) yields only *N*-alkylamino-*N*-*tert*-butoxy radicals (**45**).[83]

---

[78] See *Org. Reaction Mech.*, **1970**, 323.
[79] S. H. Goh, R. L. Huang, S. H. Ong, and I. Sieh, *J. Chem. Soc.* (C), **1971**, 2282.
[80] J. L. Brokenshire and K. U. Ingold, *Int. J. Chem. Kinetics*, **3**, 343 (1971).
[81] T. T. Wang and J. E. Leffler, *J. Org. Chem.*, **36**, 1531 (1971).
[82] G. R. Chalfont and M. J. Perkins, *J. Chem. Soc.* (B), **1971**, 245.
[83] W. C. Danen and C. T. West, *J. Am. Chem. Soc.*, **93**, 5582 (1971).

The identity of the radicals (**45**) was ascertained by generating them independently from the corresponding N,O-dialkylhydroxylamines either by direct photolysis or by photolysis in the presence of di-*tert*-butyl peroxide or of azomethane; ESR spectra of radicals (**45**) obtained by any of the methods were identical.

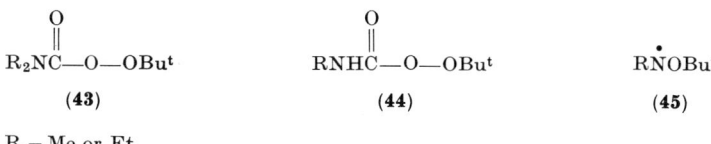

$$R_2NC(\!=\!\!O)\!-\!O\!-\!OBu^t \qquad RNHC(\!=\!\!O)\!-\!O\!-\!OBu^t \qquad R\dot{N}OBu^t$$
$$(\textbf{43}) \qquad\qquad (\textbf{44}) \qquad\qquad (\textbf{45})$$

R = Me or Et

The kinetics and mechanism of reactions of peroxy compounds with phosphites, sulphides and aromatic amines have been reviewed[84] and reactions of benzoyl peroxide in polymethylene oxide, $(CH_2O)_n$, have been compared with corresponding reactions with diethyl ether.[85] Decomposition of hydroperoxides, catalysed by a wide variety of metal chelates, has been studied by Russian workers.[86]

Decomposition of Azo-compounds

Azoalkanes are extremely convenient sources of alkyl radicals and may decompose by simultaneous two-bond scission or by successive one-bond scissions, i.e.:

$$R\!-\!N\!=\!N\!-\!R' \rightarrow R\cdot + N_2 + \cdot R'$$
$$R\!-\!N\!=\!N\!-\!R' \rightarrow R\cdot + \dot{N}\!=\!N\!-\!R' \rightarrow N_2 + \cdot R'$$

Detailed arguments in favour of both types of homolytic fragmentation have been presented,[87,88] and recent experimental evidence fails to resolve the matter completely. The elusive alkanediazonium radical $R-N\!=\!N\cdot$ has been shown[89] to have a significant lifetime by trapping experiments, and this result together with kinetic studies[90] of the (gas-phase) decomposition of a series of allylazoalkanes $RN\!=\!N\!-\!CH_2\!-\!CH\!=\!CH_2$ points to the stepwise decomposition route; in contrast, deuterium kinetic isotope effects on the decomposition of ring-deuteriated 1,1-diphenylazoethane are interpreted as favouring the two-bond scission route.[91] An essential feature of the arguments in favour of a concerted two-bond scission mechanism for azoalkanes is that the activation energy for homolysis of any particular azo compound may be readily approximated by additive treatment of data for others; careful re-investigation[92] of the decomposition of pyrazolines (**46**) and (**47**) leads to activation energies in close agreement with the approximated values and lends support to the concerted mechanism. Thermal

---

[84] D. G. Pobedimskii, *Russ. Chem. Rev.*, **40**, 142 (1971).
[85] E. G. Atovmyan and A. F. Lukovnikov, *Kinet. Katal.*, **11**, 1340 (1970); *Chem. Abs.*, **74**, 63765 (1971).
[86] L. I. Matienko, I. P. Skibida, and Z. K. Maizus, *Kinet. Katal.*, **12**, 595 (1971); B. G. Bal'kov, I. P. Skibida, and Z. K. Maizus, *Izv. Akad. Nauk SSSR, Ser. Khim.*, **1970**, 1780; G. M. Bulgakova, I. P. Skibida, E. G. Rukhadze, and G. P. Talyzenkova, *Kinet. Katal.*, **12**, 359 (1971); *Chem. Abs.*, **75**, 62798 (1971); G. M. Bulgakova, I. P. Skibida, and Z. K. Maizus, *Kinet. Katal.*, **12**, 76 (1971); *Chem. Abs.*, **74**, 111301 (1971).
[87] C. Rüchardt, *Angew. Chem. Int. Ed.*, **9**, 830 (1970).
[88] S. W. Benson and H. E. O'Neal, *Nat. Stand. Ref. Data Ser., Nat. Bur. Stand.*, No. 21, 31 (1970).
[89] N. A. Porter, M. E. Landis, and L. J. Marnett, *J. Am. Chem. Soc.*, **93**, 795 (1971).
[90] K. Takagi and R. J. Crawford, *J. Am. Chem. Soc.*, **93**, 5910 (1971).
[91] S. E. Scheppele, D. W. Miller, P. L. Grizzle, and F. A. Mauceri, *J. Am. Chem. Soc.*, **93**, 2549 (1971).
[92] J. W. Timberlake and B. K. Bandlish, *Tetrahedron Letters*, **1971**, 1393.

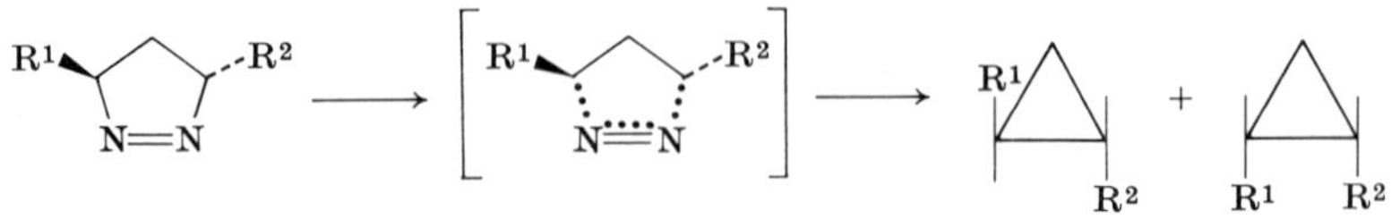

(46) $R^1 = R^2 = Ph$
(47) $R^1 = R^2 = p\text{-MeOC}_6H_4$

decomposition of (46) at 85° gives a mixture of *trans*-1,2-diphenylcyclopropane 90.8% and *cis*-1,2-diphenylcyclopropane 9.2%; at 50° the *trans*-isomer increases to 94.0% and the *cis*-isomer decreases to 6.0%; increase in stereoselectivity at the lower temperature is thought to arise as a result of competition between bond formation and bond rotation of the intermediate diradical.[92]

Disproportionation of cyanoisopropyl radicals obtained by photolysis of azobisisobutyronitrile accounts for only 5% of the radical reactions in fluid solution but for 95% in the crystal;[93] deuterium-labelling experiments now indicate that rates of disproportionation in the crystal are limited, not by atom transfer, but by rotational diffusion about the $C-C\equiv N$ axis. Difficulties to be expected with the various models used for computer simulation of cage processes have been discussed,[94] and the effect of viscosity of the medium on rates of radical formation during decomposition of azobisisobutyronitrile in solutions containing polystyrene has been determined.[95]

Apparent relative rates for homolysis of the strained azo compounds (48)–(50) are 9.2, 5.2×10^8, and 1.9×10^{11}, respectively; the differences are attributed[96] to varying strain in the transition states for (synchronous) loss of nitrogen and it is interesting that thermolysis of (48) leads only to norbornadiene and not to quadricyclene.

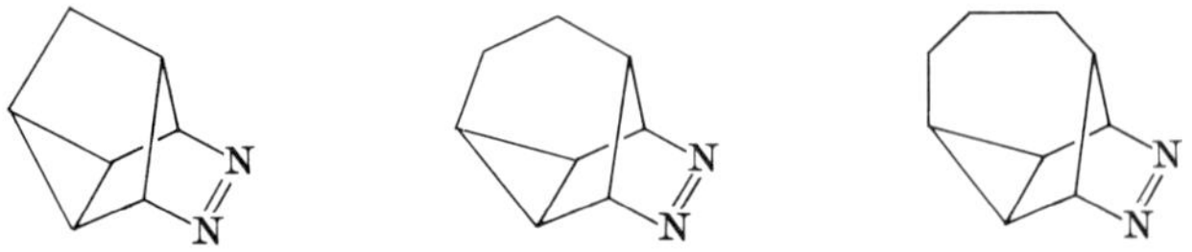

(48)   (49)   (50)

Products arising from the gas-phase thermolysis of azo-compounds (51) are entirely consistent with initial formation of a rapidly interconverting pair of butenyl radicals (52) and (53).[97]

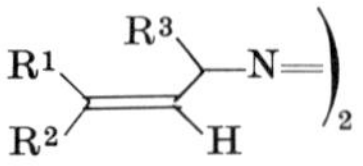

(51)   (52)   (53)

$R^1, R^2, R^3 = H$ or Me

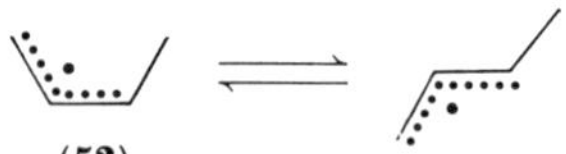

[93] J. M. McBride, *J. Am. Chem. Soc.*, **93**, 6302 (1971); see also K. Waki and T. Yamashita, *Kogyo Kagaku Zasshi*, **72**, 958 (1969); *Chem. Abs.*, **75**, 98018 (1971).
[94] C. Walling and A. R. Lepley, *Int. J. Chem. Kinet.*, **3**, 97 (1971).
[95] A. F. Guk and V. F. Tsepalov, *Kinet. Katal.*, **12**, 910 (1971).
[96] E. L. Allred and A. L. Johnson, *J. Am. Chem. Soc.*, **93**, 1300 (1971).
[97] R. J. Crawford, J. Hamelin, and B. Strehlke, *J. Am. Chem. Soc.*, **93**, 3810 (1971).

1,1'-Dimethyl-1,1'-diphenoxyazoethane (**54**; X = O)[98] and its sulphur anologue (**54**; X = S)[99] have been synthesized and subjected to thermal and photochemical homolysis; phenyl migration in the initial radical fragments is an important reaction preceding methyl-radical cleavage.

$$Me_2C-N=N-CMe_2$$
$$||$$
$$XPhXPh$$

(**54**) X = O or S

$$PhO-\overset{\bullet}{C}Me_2 \longrightarrow \overset{\bullet}{O}-\underset{|}{C}Me_2 \longrightarrow O=C\overset{Ph}{\underset{Me}{\diagdown}} + Me\bullet$$
$$\phantom{PhO-\overset{\bullet}{C}Me_2 \longrightarrow \overset{\bullet}{O}-}Ph$$

Phenylazo p-tolyl sulphone (**55**) provides an interesting new source of phenyl radicals but the presence of base (pyridine, CaO, MgO) is required to minimize competing, autocatalytic heterolysis;[100] under the best conditions for homolysis, orientations and partial rate factors for phenylation of substituted benzenes were similar to those obtained from phenyl radical produced by benzoyl peroxide/N-nitrosoacetanilide.

A series of tertiary alkyl hyponitrites (**56**) and (**57**) has been synthesized[101] and their decomposition was compared with that of the more common di-*tert*-butyl anologue;

Ph—N=N—S—⟨ ⟩—Me   Me$_2$CON=NOCMe$_2$ Ph$_2$CON=NOCPh$_2$
       ‖   R   R   Me   Me
       O | | | |
       ‖
       O

(**55**)   (**56**)   (**57**)

                R = Ph, Ar, C$_4$H$_9$, Me

relative reactivities of *tert*-butoxy and cumyloxy radicals with respect to hydrogen abstraction and β-scission in various solvents have been studied and it is clear that the latter radicals tend much more to β-scission.[101] For bis-(α,α-dimethylbenzyl) hyponitrite (**56**; R = Ph) the Eyring parameters were $\Delta H^{\ddagger} = 27.3$ kcal mole^{-1} and $\Delta S^{\ddagger} = 8.5$ eu (isooctane), and efficiency of radical production was 84% compared with 92% for the di-*tert*-butyl analogue;[102] rate constants for decomposition of substituted derivatives of this hyponitrite gave a Hammett correlation using σ^+ values with $\rho = +0.34$.

Hyponitrites are formed by thermal rearrangement of N-substituted derivatives of O-*tert*-butyl-N-nitrosohydroxylamine[103] and the range of compounds available has been further extended. N-Benzoyl-O-*tert*-butyl-N-nitrosohydroxylamine[104] gives *tert*-butyl perbenzoate as a major product of thermolysis, in yields that increase with increasing solvent viscosity; ^{18}O-labelling studies indicate complete randomization of the benzoyloxy oxygen atoms *subsequent* to initial homolysis.

Thermal decomposition of 2,2'-diphenyl-2,2'-azohexafluoropropane (**58**) in toluene[105]

---

[98] A. Ohno, N. Kito, and Y. Ohnishi, *Bull. Chem. Soc. Japan*, **44**, 467 (1971).
[99] A. Ohno, N. Kito, and Y. Ohnishi, *Bull. Chem. Soc. Japan*, **44**, 463 (1971).
[100] M. Kobayashi, H. Minato, M. Kojima, and N. Kamigata, *Bull. Chem. Soc. Japan*, **44**, 2501 (1971).
[101] L. Dulog and P. Klein, *Chem. Ber.*, **104**, 895 (1971).
[102] L. Dulog and P. Klein, *Chem. Ber.*, **104**, 902 (1971).
[103] See *Org. Reaction Mech.*, **1970**, 354; **1969**, 306–308.
[104] T. Koenig, M. Deinzer, and J. A. Hoobler, *J. Am. Chem. Soc.*, **93**, 938 (1971).
[105] J. B. Levy and E. J. Lehmann, *J. Am. Chem. Soc.*, **93**, 5790 (1971).

occurs with $\Delta H^{\ddagger} = 32.8$ kcal mole^{-1}, $\Delta S^{\ddagger} = 20$ eu; (cf. for the fully hydrogenated homologue $\Delta H^{\ddagger} = 29.0$ kcal mole^{-1}, $\Delta S^{\ddagger} = 11$ eu); products of interest include the *para*-coupled radical dimer (59) and coupling products of benzyl and $(CF_3)_2PhC\cdot$ radicals. It is noteworthy that the last-mentioned radicals can abstract a hydrogen from toluene at a velocity competitive with that of radical dimerization, while the non-fluorinated radicals cannot.

$Ph(CF_3)_2C-N=N-C(CF_3)_2Ph \longrightarrow Ph(CF_3)_2C\cdot\ N_2\ \cdot C(CF_3)_2Ph$
(58)

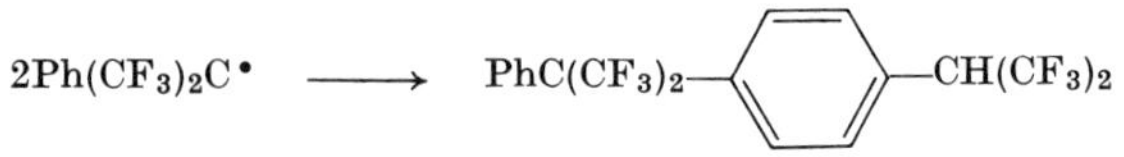

$2Ph(CF_3)_2C\cdot \longrightarrow PhC(CF_3)_2-\langle\ \rangle-CH(CF_3)_2$ (59)

$Ph(CF_3)_2C\cdot + PhCH_3 \longrightarrow PhCH(CF_3)_2 + PhCH_2\cdot$

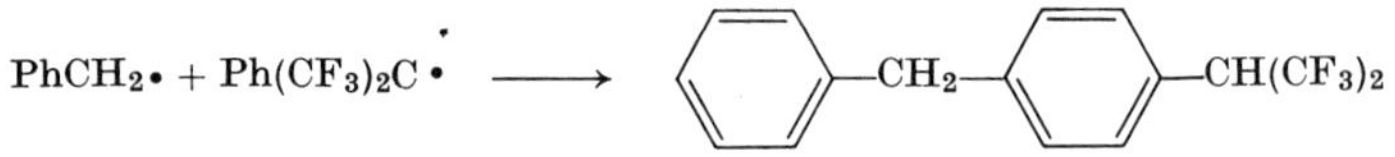

$PhCH_2\cdot + Ph(CF_3)_2C\cdot \longrightarrow \langle\ \rangle-CH_2-\langle\ \rangle-CH(CF_3)_2$

The pioneering work of Kosower in characterizing monosubstituted diazenes (diimides) $R-N=NH$ as reaction intermediates has been reviewed[106] and details of the preparation and reactions of alkyl-[107] and alkenyl-diazenes[108] have been reported; these reactive intermediates may be used as sources of corresponding hydrocarbon anions or radicals.

Diradicals

The tetramethylene diradical is a possible intermediate in the non-concerted fragmentation of cyclobutane to two ethylenes; recent calculations[109] indicate that no real intermediate tetramethylene biradical intervenes between cyclobutane and two ethylenes, rather, a new word "twixtyl" is suggested for a molecule or a range of molecular conformations that is not a minimum in a potential energy surface but that operationally behaves as a true intermediate. Experimental evaluation of this theoretical analysis has been sought[110] by studies of the pyrolysis of several [4.4.2]-propella-2,4-dienes, e.g. (60). Typically the *syn,cis*-isomer (60) pyrolysed cleanly to tetralin (61) and mainly *cis*-2-deuteriovinyl methyl ether (62); the *syn,trans*-isomer of (60) likewise cleaved to give mainly *trans*-2-deuteriovinyl methyl ether whereas the *anti,cis*-isomer of (60) gave approximately a 4:1 ratio of *cis*- to *trans*-vinyl ether. The high levels of preservation of geometrical relationships observed are attributed to increased rotational barriers due

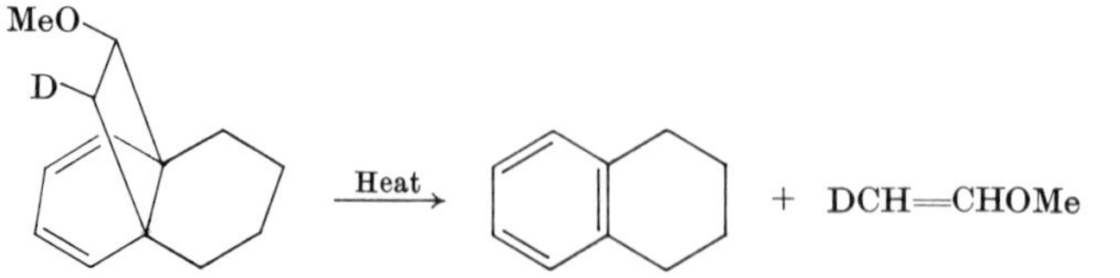

(60) $\xrightarrow{\text{Heat}}$ (61) + DCH=CHOMe (62) *cis* and *trans*

[106] E. M. Kosower, *Accounts Chem. Res.*, **4**, 193 (1971).
[107] T. Tsuji and E. M. Kosower, *J. Am. Chem. Soc.*, **93**, 1992 (1971).
[108] T. Tsuji and E. M. Kosower, *J. Am. Chem. Soc.*, **93**, 1999 (1971).
[109] R. Hoffmann, S. Swaminathan, B. G. Odell, and R. Gleiter, *J. Am. Chem. Soc.*, **92**, 7091 (1970).
[110] L. A. Paquette and G. L. Thompson, *J. Am. Chem. Soc.*, **93**, 4920 (1971).

to steric and ponderal factors. However, for the cases where loss of stereochemical integrity occurred [e.g. the *anti,cis*-isomer of (**60**)] it is concluded that intermediate singlet diradicals cannot be described as twixtyls, i.e. flat regions atop an energy profile.

Related work from the same group[111] has demonstrated that photochemical or thermal decomposition of the tricyclic azo compounds (**63**) readily forms the appropriate tricycloheptanes (**64**); the latter cleave thermally at 200° to give ethylenes and 5,5-dimethyl-1,4-diphenylcyclopentadiene (**65**), and a careful analysis of the products shows that opening of the *anti*-tricycloheptane nucleus occurs in highly stereoselective fashion, apparently ruling out the possibility of diradical intermediates. Full details

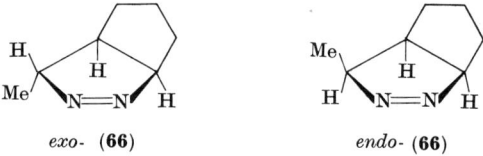

have been published[112] of intramolecular trapping of 1,3-diradical intermediates by a remote cyclobutene ring.[113]

Direct and sensitized photodecomposition of *exo*- and *endo*-4-methyl-2,3-diazobicyclo-[3.3.0]oct-3-enes (**66**) proceeds through singlet and triplet diradical intermediates, giving products with essential retention of configuration at the methyl-substituted carbon; in contrast, thermal decomposition of *exo*- and *endo*-(**66**) at 260° occurs with predominant inversion of configuration, demanding a more complex mechanism.[114]

Diradical intermediates (**68**) and (**70**) have been proposed[115] to explain the products (**69**), (**71**)–(**74**) formed by pyrolysis (580°) of *cis*-verbanone (**67**). Formation of acyclic

[111] L. A. Paquette and L. M. Leichter, *J. Am. Chem. Soc.*, **93**, 4922 (1971).
[112] L. A. Paquette and L. M. Leichter, *J. Am. Chem. Soc.*, **93**, 5128 (1971).
[113] See *Org. Reaction Mech.*, **1970**, 274, 328.
[114] P. B. Condit and R. G. Bergman, *Chem. Comm.*, **1971**, 4.
[115] J. M. Coxon, R. P. Garland, and M. P. Hartshorn, *Chem. Comm.*, **1971**, 1131.

products (**69**) and (**73**) involves cleavage of two C—C bonds of the cyclobutane ring and cleavage of the second bond will occur most readily when the singly occupied orbitals of the 1,4-diradical (**68**) are eclipsed with that bond. It is argued that, in contrast to (**68**), diradical (**70**) presents conformational constraints on π-delocalization for the secondary radical; consequently 1,5-hydrogen transfer from a methyl group becomes a competing reaction path leading to (**72**).

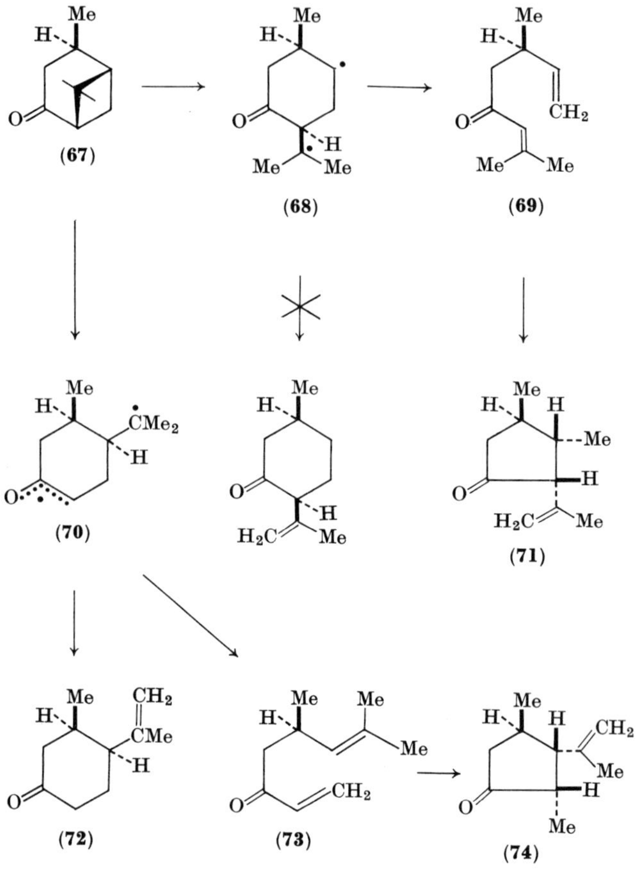

Recent studies[116] of the photolysis of *tert*-butylbenzoquinones (**75**) have pointed to the formation of diradical intermediates (**76**); it is now reported[117] that SO_2 is an efficient trap for the reaction intermediates, leading to products based on (**77**).

Unsubstituted trimethylenemethane has a triplet ground state,[118] and recent work from Berson's group[119] indicates that the substituted cyclic homologue (**78**), derived by three different methods including thermolysis and photolysis of the azo compound

[116] See *Org. Reaction Mech.*, **1970**, 529.
[117] S. Farid, *Chem. Comm.*, **1971**, 73.
[118] P. Dowd, A. Gold, and K. Sachdev, *J. Am. Chem. Soc.*, **90**, 2715 (1968).
[119] J. A. Berson, R. J. Bushby, J. M. McBride, and M. Tremelling, *J. Am. Chem. Soc.*, **93**, 1544 (1971).

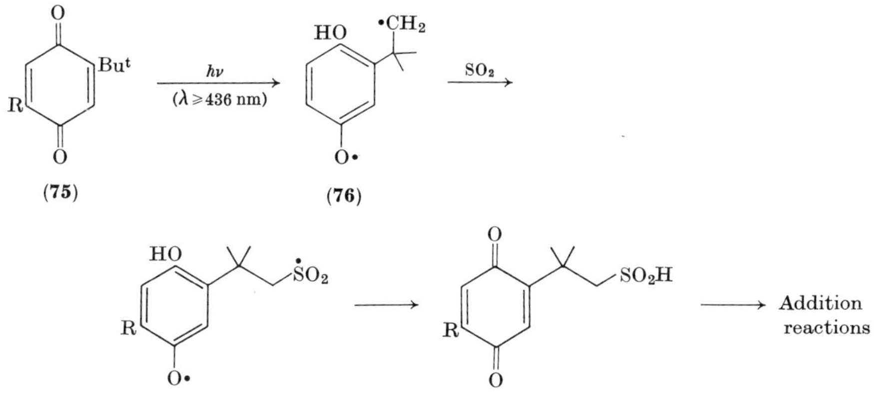

(79), gives product mixtures arising at least in part from a triplet species; theoretical aspects of CIDNP phenomena in triplet dimerizations have been discussed.[120]

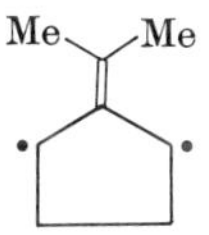

In the temperature range 310–428° ethylidenecyclobutane (80) undergoes a reversible, homogeneous, unimolecular isomerization to 1-methyl-2-methylenecyclobutane (81);[121] the isomerization and competing ring-opening fragmentation reactions occur by initial fission of an allylic bond of the cyclobutane to give a diradical (82).

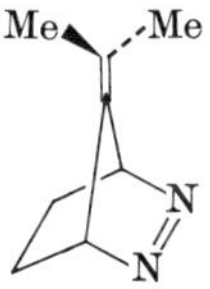

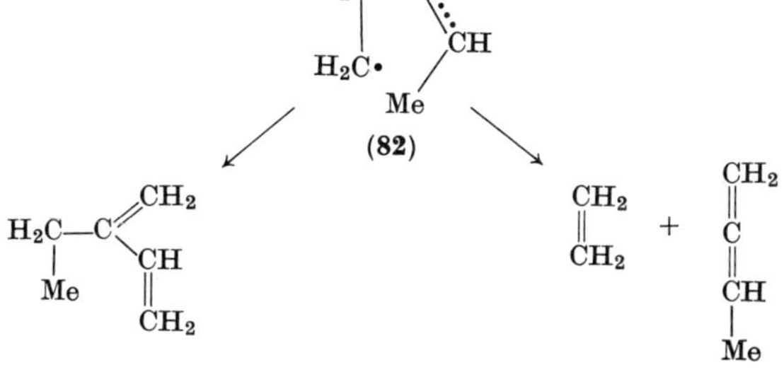

[120] G. L. Closs, J. Am. Chem. Soc., **93**, 1546 (1971).
[121] M. C. Flowers and A. R. Gibbons, J. Chem. Soc. (B), **1971**, 362.

The possible role of trimethylene diradicals in cyclopropane isomerizations continues to stimulate theoretical interest in their structure; *ab initio* SCF methods[122] now suggest a structure (**84**) with "crab-like" characteristics formed by collapse of the "open" cyclopropane (**83**) with trigonal terminal methylene groups. The singlet state of diradical (**84**) has a stabilization energy of 6.2 kcal mole^{-1}, accounted for by mixing of strong zwitterionic character into the diradical wave function.

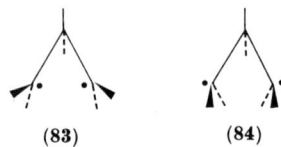

(83)   (84)

Trimethylene diradicals (**87**) and (**90**) are thought to be intermediates in the pyrolysis of the sulphone (**85**) and organolithium-induced decomposition of a sulphonium salt (**88**).[123] Both reactions show net cross-over, i.e. *cis*-2,4-dimethyl derivatives produce mainly *trans*-1,2-dimethylcyclopropanes and *trans*-2,4-dimethyl derivatives produce mainly *cis*-1,2-dimethylcyclopropanes; the inversions observed are understandable[124] on the basis of π-cyclopropane character for the diradicals.

[122] Y. Jean and L. Salem, *Chem. Comm.*, **1971**, 382; but see A. K. Q. Siu, W. M. St. John, and E. F. Hayes, *J. Am. Chem. Soc.*, **92**, 7249 (1970).
[123] B. M. Trost, W. L. Schinski, F. Chen, and I. B. Mantz, *J. Am. Chem. Soc.*, **93**, 676 (1971).
[124] R. G. Bergman and W. L. Carter, *J. Am. Chem. Soc.*, **91**, 7411 (1969); R. Hoffmann, *J. Am. Chem. Soc.*, **90**, 1475 (1968).

Thermolysis and photolysis of γ-methylperoxyvalero-γ-lactone (**91**) occurs by way of the diradical (**92**) although it is not yet clear whether oxetan formation occurs directly from (**92**) or from the decarboxylated diradical (**93**);[125] possible mechanisms for decomposition of cyclic peroxyesters have been discussed in previous reports.[126]

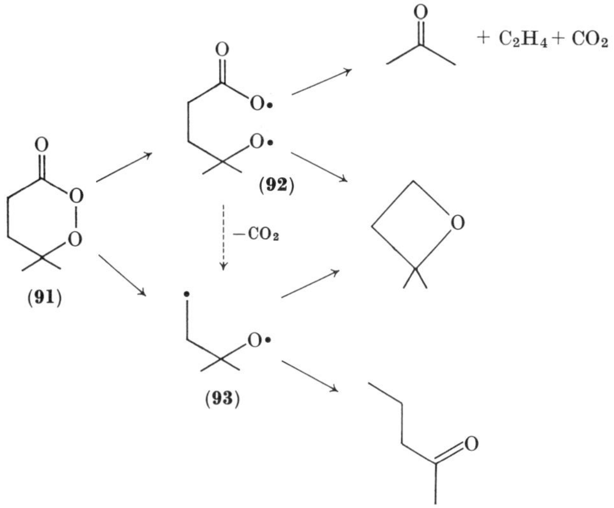

There has been a resurgence of interest in the nature of Tschitschibabin's hydrocarbon and similar molecules,[127] mostly concerned with the relative energies of singlet and triplet forms of the diradicals, and it has been shown, for example, that the Schlenk hydrocarbon (**94**) has a triplet ground state.

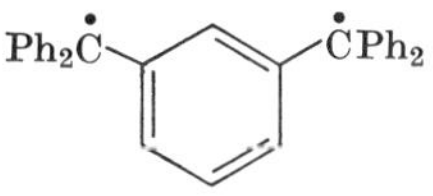

The chemistry of 1,3- and 1,4-diradical species derived from different sources (decomposition of azo-compounds, Norrish Type II photodecompositions, pyrolyses of small-ring compounds) has been discussed in terms of their origins and structures,[128] and diradical intermediates have been proposed for reactions of aromatic nitroso compounds with styrene,[129] photolytic reactions of 1,3-diphenylpropene,[130] and reactions

---

[125] W. Adam and L. Szendrey, *Chem. Comm.*, **1971**, 1299.
[126] See *Org. Reaction Mech.*, **1969**, 313.
[127] H. Stieger and H.-D. Brauer, *Chem. Ber.*, **103**, 3799 (1970); G. Kothe, K.-H. Denkel, and W. Sümmermann, *Angew. Chem. Int. Ed.*, **9**, 906 (1970); R. Schmidt and H.-D. Brauer, *Angew. Chem. Int. Ed.*, **10**, 506 (1971); G. R. Luckhurst, G. F. Pedulli and M. Tiecco, *J. Chem. Soc.* (B), **1971**, 329.
[128] L. M. Stephenson and J. I. Brauman, *J. Am. Chem. Soc.*, **93**, 1988 (1971).
[129] A. Yoneda, M. Tanaka, and N. Murata, *Kogyo Kagaku Zasshi*, **73**, 2185 (1970); *Chem. Abs.*, **74**, 87078 (1971).
[130] E. W. Valyocsik and P. Sigal, *J. Phys. Chem.*, **75**, 2079 (1971).

of diphenylmethylene with cyclopentadiene.[131] It has been concluded[132] that dimethylsilanone $Me_2Si=O$ is an intermediate in the gas-phase decomposition of octamethylcyclotetrasiloxane, but the silicon—oxygen bond is apparently to be regarded as a weak double bond rather than as having a diradical structure.

Atom-transfer Processes

Pryor's group have for some time been investigating the production and reactivity of hydrogen atoms by techniques not involving ionizing radiation and full details of these studies have now appeared.[133] Rate-constants k_H for the reactions of hydrogen atoms (generated by photolysis of thiols) with organic hydrogen donors: $QH + H\cdot \rightarrow Q\cdot + H_2$, are in excellent agreement with corresponding data from radiolysis experiments for all compounds studied except ethanol and propan-2-ol. The thiyl radical (RS·) is shown not to interfere with the analysis. Similar studies[134] with tritium-labelled thiols give correct values of k_H for alkanes where the Q—H bond being broken is much stronger than the RS—H bond. However, in this case compounds such as toluene which contain weak C—H bonds gave spuriously high values of k_H. Additional rate constants for reactions of hydrogen atoms with organic solutes in aqueous solution have also been determined[135] by radiation techniques, and there are several reports[136] of similar gas-phase reactions of hydrogen atoms.

The relationship between primary kinetic isotope effects and the nature of hydrogen-transfer transition states is a topic of increasing attention and importance. Arguments continue to appear[137,138] for and against the postulation that "the magnitude of the primary isotope effect in a hydrogen-transfer reaction varies with the symmetry of the transition state, and is a maximum when the hydrogen is symmetrically bonded to the atoms between which it is being transferred". Reactions of series of free radicals with thiols gave rise to a maximum ($k_H/k_D = 6.65$) in the isotope effect when the hydrogen transfer is approximately half-complete at the transition state,[137] supporting the postulate and entirely consistent with other recent observations of isotope-effect maxima in closely related transfer processes.[138]

Variation in the tritium kinetic isotope effects for additions of R^1SH to $R^2R^3C=CH_2$ are interpreted[139] in terms of the transition-state symmetry, as indicated by the difference in bond strengths to hydrogen in the initial state (R^1S—H) and final state (H—$CR^2R^3CH_2SR^1$). A large isotope effect ($k_H/k_T = 14.9$) was observed[140] in the reaction of trityl radicals with thiophenol, and further evidence of dependence of transition-state structure on the reactivity of free radicals has been noted for reactions with ethane and methane.[141] Kinetic isotope effects for the copper(I)-catalysed decomposition of ortho-

---

[131] D. J. Atkinson, M. J. Perkins, and P. Ward, *J. Chem. Soc.* (C), **1971**, 3247.
[132] I. M. T. Davidson and J. F. Thompson, *Chem. Comm.*, **1971**, 251.
[133] W. A. Pryor and J. P. Stanley, *J. Am. Chem. Soc.*, **93**, 1412 (1971).
[134] W. A. Pryor and M. G. Griffith, *J. Am. Chem. Soc.*, **93**, 1408 (1971).
[135] P. Neta, G. R. Holdren, and R. H. Schuler, *J. Phys. Chem.*, **75**, 449 (1971).
[136] D. W. Rathburn, K. Knarr, and H. C. Moser, *Trans. Faraday Soc.*, **67**, 2333 (1971); M. A. Contineanu, D. Mihelćić, R. N. Schindler, and P. Potzinger, *Ber. Bunsenges. Phys. Chem.*, **75**, 426 (1971); W.-K. Aders and H. G. Wagner, *Z. Phys. Chem.* (*Frankfurt*), **74**, 224 (1971); M. J. Lexton, R. M. Marshall, and J. H. Purnell, *Proc. Roy. Soc.* (*A*), **324**, 433 (1971); M. J. Lexton, R. M. Marshall, and J. H. Purnell, *Proc. Roy. Soc.* (A), **324**, 477 (1971).
[137] W. A. Pryor and K. G. Kneipp, *J. Am. Chem. Soc.*, **93**, 5584 (1971).
[138] See compilation of references in ref. 137.
[139] E. S. Lewis and M. M. Butler, *Chem. Comm.*, **1971**, 941.
[140] E. S. Lewis and M. M. Butler, *J. Org. Chem.*, **36**, 2582 (1971).
[141] I. B. Afanas'ev, *Kinet. Katal.*, **12**, 484 (1971); *Chem. Abs.*, **75**, 62797 (1971).

N,N-dimethylbenzamidoarenediazonium ions, reported previously,[142] have provided interesting information about relative rates of intramolecular hydrogen abstraction and rates of rotation around the carbonyl-C—N bond and the methyl-C—N bond; more recent work[143] details the probable effects of conformational barriers in the arene—carbonyl bonds on relative rates of intramolecular hydrogen-transfer and intermolecular radical-abstraction processes.

Results for studies of the reduction of optically active cyclopropyl bromides (95) by di-n-butyltin dihydride[144] are inconsistent with rapid non-cage reduction of a cyclopropyl radical whose inversion rate is of the same order as the rate of reduction; rather, net retention in the products (96) was observed[145] and ascribed to in-cage reduction of enantiomeric cyclopropyl radicals.

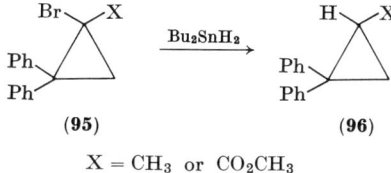

X = CH$_3$ or CO$_2$CH$_3$

Nitrogen cation radicals, R$_2$NH$^+$· have the potential to effect highly selective chlorinations and oxidations.[146] Minisci-type chlorination results in very selective (> 90%) ω-1 monochlorination of C$_6$ and C$_8$ alcohols, ethers, and carboxylic acids. The reactions are of obvious synthetic importance and occur in high yield.[147] Use of sterically hindered chlorinating agents[148] (achieved by increasing the size of R) results in a reversal of the usual tertiary > secondary > primary order of attack in chlorination reactions, and thus the possibility of selective attack on the terminal methyl group of a long alkane chain becomes feasible. Radical chlorination of propionic acid (with molecular chlorine) in the absence of enol-forming catalysts (H$_2$SO$_4$, HCl, FeCl$_3$) produces the β-chloro acid by a selective hydrogen-abstraction mechanism;[149] α-chlorination predominates in the presence of enolizing acids and chlorine-atom scavengers.

Examination of the rates of reaction of the stable pyridyl free radical (97) with benzyl bromide and a number of substituted benzyl chlorides in various solvents indicates[150] that atom-transfer processes best account for most of the results obtained. The halogen atom is transferred to the pyridyl radical, producing a halodihydropyridine and a benzyl radical which rapidly reacts with a second pyridyl radical. One striking exception[151] to the above generalization is the reaction of (97) with 4-nitrobenzyl halides, which is found to be much faster than with the other halides, extremely sensitive to solvent polarity, and probably involves an electron-transfer reaction as indicated in equations (1)–(4).

---

[142] See *Org. Reaction Mech.*, **1968**, 237.
[143] T. Cohen, K. W. Smith, and M. D. Swerdloff, *J. Am. Chem. Soc.*, **93**, 4303 (1971).
[144] See *Org. Reaction Mech.*, **1969**, 305; **1970**, 337.
[145] L. J. Altman and T. R. Erdman, *Tetrahedron Letters*, **1970**, 4891.
[146] For a review of this subject see N. C. Deno *et al.*, in ref. 2, p. 155; also see *Org. Reaction Mech.*, **1969**, 318–319; **1970**, 335–336.
[147] N. C. Deno, W. E. Billups, R. Fishbein, C. Pierson, R. Whalen, and J. C. Wyckoff, *J. Am. Chem. Soc.*, **93**, 438 (1971).
[148] N. C. Deno, R. Fishbein, and J. C. Wyckoff, *J. Am. Chem. Soc.*, **93**, 2065 (1971).
[149] Y. Ogata and K. Matsuyama, *Tetrahedron*, **26**, 5929 (1970).
[150] M. Mohammad and E. M. Kosower, *J. Am. Chem. Soc.*, **93**, 2709 (1971).
[151] M. Mohammad and E. M. Kosower, *J. Am. Chem. Soc.*, **93**, 2713 (1971).

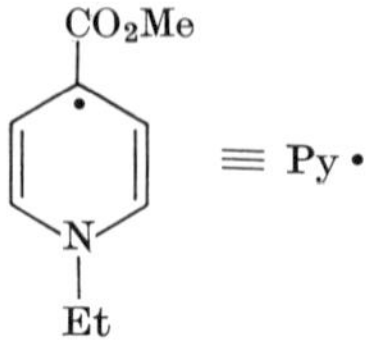

<img: pyridinyl radical structure with CO₂Me and N-Et, labeled (97), ≡ Py•>

$$\text{Py}^\bullet + \text{NBX} \rightleftharpoons \text{Py}^+, \text{NBX}^{\bar{\bullet}} \quad \ldots (1)$$

$$\text{Py}^+, \text{NBX}^{\bar{\bullet}} \rightleftharpoons \text{Py}^+ + \text{NBX}^{\bar{\bullet}} \quad \ldots (2)$$

$$\text{NBX}^{\bar{\bullet}} \rightarrow \text{NB}^\bullet + \text{X}^- \quad \ldots (3)$$

$$\text{Py}^\bullet + \text{NB}^\bullet \rightarrow \text{PyNB} \quad \ldots (4)$$

NBX = p-Nitrobenzyl halide

A detailed re-investigation of the photoreductions of benzophenone by propan-2-ol and of acetone by benzhydrol has shown that both benzpinacol and the mixed pinacol $\text{Ph}_2\text{C(OH)C(OH)Me}_2$ are produced.[152] Since in propan-2-ol solution the latter product can only arise from a cage reaction this work demonstrates that, contrary to previous reports, there are cage reactions in the benzophenone–isopropyl alcohol system. A value of 0.11 for the cage reaction indicates that electron spin flipping in a caged radical pair is fast compared to radical diffusion from the cage.

Homolysis of benzpinacol provides a thermal source of benzophenone ketyl radicals which, as when photochemically produced, efficiently reduce α,β-diketones (e.g. benzil) to α-hydroxy ketones (e.g. benzoin) by a hydrogen-transfer mechanism.[153]

Last year's report[154] showed that substituted tetrahydrofurans are obtained by photocyclization of the β-alkoxy ketone (98), results implying that the rate of δ-hydrogen abstraction was much greater than that of γ-hydrogen abstraction by the excited carbonyl group of (98). New results[155] based on studies of the ketones (99) and (100) show that such a conclusion is incorrect and instead it is argued that the rates of γ-hydrogen abstraction by excited triplet states would equal the rate of δ-hydrogen abstraction in compounds such as (98).

$$\underset{(98)}{\text{MeCCH}_2\text{CMe}_2} \quad \underset{(99)}{\text{PhCCH}_2\text{CH}_2\text{CH}_2\text{CH}_2\text{OMe}} \quad \underset{(100)}{\text{PhCCH}_2\text{CH}_2\text{OCH}_2\text{CH}_3}$$

(with O double-bonded to C, and OMe on CMe₂ for 98)

Gleicher has reported continued studies of hydrogen abstraction from non-aromatic polycyclic hydrocarbons.[156] Six such compounds were treated with trichloromethyl radicals, and the relative rates of hydrogen abstraction were correlated with a computer analysis of the change in strain energy in going from the ground state to the radical intermediate.[157] Relative rates of hydrogen abstraction from a series of arylmethanes

[152] S. A. Weiner, *J. Am. Chem. Soc.*, **93**, 425 (1971).
[153] M. B. Rubin and J. M. Ben-Bassat, *Tetrahedron Letters*, **1971**, 3403.
[154] See *Org. Reaction Mech.*, **1970**, 338–339.
[155] P. J. Wagner and R. G. Zepp, *J. Am. Chem. Soc.*, **93**, 4958 (1971).
[156] For previous studies see *Org. Reaction Mech.*, **1968**, 277; **1969**, 320; also **1970**, 338.
[157] V. R. Koch and G. J. Gleicher, *J. Am. Chem. Soc.*, **93**, 1657 (1971).

by trichloromethyl and *tert*-butoxy radicals at 70° have also been studied by the same group.[158] Whereas an excellent correlation between relative rates and changes in π-binding energies between incipient radicals and arylmethanes (SCF approach) was obtained with trichloromethyl, no such correlation was obtained with *tert*-butoxy radical. It is felt that the transition state for H-abstraction by the trichloromethyl radical must strongly resemble the intermediate free radical whilst that for abstraction by *tert*-butoxy radicals probably has a structure somewhere between the ground state and the intermediate radical. Related conclusions about the trichloromethyl radical have been drawn[159] from a study of the relative rates of Si-H hydrogen abstraction from substituted phenylsilanes, phenylmethylsilanes and phenyldimethylsilanes. Here the relative rates for the three sets of arylhydrosilanes are well correlated with Hammett constants for the ring substituents.

A correlation[160] with σ^+ for reactivities of toluenes towards *tert*-butoxy radicals giving a ρ value of -0.38 is in agreement with results of other workers.[161]

The rates of iodine abstraction from 1-iodoadamantane, 1-iodobicyclo[2.2.2]octane, 1-iodobicyclo[2.2.1]heptane, and *tert*-butyl iodide by the phenyl radical have been compared with rates of bromine abstraction from bromotrichloromethane.[162] All three bridgehead compounds were found to be less reactive than *tert*-butyl iodide, and reasons for the ease of formation of the 1-adamantyl and radicals (**101**) and (**102**) are discussed; the results permit determination of bond dissociation energies D(R–H) and D(R–I) for the bridgehead compounds. Similar studies[163] of the rates of iodine abstraction from

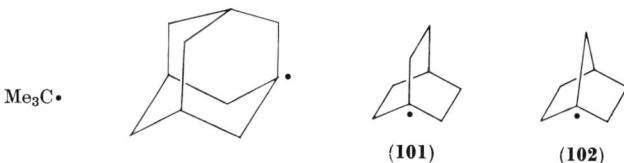

Me$_3$C•   (**101**)   (**102**)

a series of aliphatic iodides yielded a positive value in a Taft correlation and illustrate the generality of substituent effects on halogen abstraction reactions. The divergence from the correlation by aliphatic iodides with β-bromo and γ-iodo substituents suggests that anchimeric assistance to abstraction of iodine by these neighbouring groups is in operation. Neighbouring-group participation is also suggested to explain stereochemical aspects of hydrogen abstraction from halohydrins and hydroxy sulphides.[164]

The general problem of bridged radical intermediates is given further discussion[165] after a detailed re-investigation of the photobromination of optically active (+)-2-methylbutyl acetate, (−)-1-fluoro-2-methylbutane, and (+)-1-bromo-2-methylbutane with molecular bromine. Recovered, non-brominated fluoride and acetate had been racemized to a large extent, and recovered bromide only slightly during the reactions.

---

[158] J. D. Unruh and G. J. Gleicher, *J. Am. Chem. Soc.*, **93**, 2008 (1971).
[159] Y. Nagai, H. Matsumoto, M. Hayashi, E. Tajima, M. Ohtsuki, and N. Sekikawa, *J. Organomet. Chem.*, **29**, 209 (1971).
[160] T. Yamamoto, M. Hasegawa, and S. Nakamura, *Himeji Kogyo Daigaku Kenkyu Hokoku*, **1970**, No. 23A, 93; *Chem. Abs.*, **74**, 99322 (1971).
[161] H. Sakurai and A. Hosomi, *J. Am. Chem. Soc.*, **89**, 458 (1967); see also M. J. Perkins, P. Ward, and A. Horsfield, *J. Chem. Soc.* (B), **1970**, 395.
[162] W. C. Danen, T. J. Tipton, and D. G. Saunders, *J. Am. Chem. Soc.*, **93**, 5186 (1971).
[163] W. C. Danen and R. L. Winter, *J. Am. Chem. Soc.*, **93**, 716 (1971).
[164] E. S. Huyser and R. H. C. Feng, *J. Org. Chem.*, **36**, 731 (1971).
[165] D. D. Tanner, H. Yabuuchi, and E. V. Blackburn, *J. Am. Chem. Soc.*, **93**, 4802 (1971).

On the other hand, as reported by other workers,[166] the 1,2-dibromide obtained by bromination of (+)-1-bromo-2-methylbutane was found to be optically active. The results are interpreted as indicating bridging effects of a β-bromine substituent *after* formation of the radical and are given additional support by further studies[167] of the free-radical bromination of cyclohexyl bromide and 1-bromobutane with *N*-bromosuccinimide and molecular bromine. Mechanistic consequences of (*a*) the reversible reactions of alkyl radicals with hydrogen bromide and (*b*) direct reactions between hydrogen bromide and alkyl bromides, leading to olefins, were noted in last year's report;[168] further evidence of these phenomena in gas-phase brominations has been presented.[169]

Galvinoxyl (**103**) reacts with trihydrogalvinoxyl (**104**), producing hydrogalvinoxyl (**105**) exclusively in the stoichiometric proportions 2:1:3.[170] A kinetic study suggests an initial fast reversible oxygen–oxygen hydrogen transfer between galvinoxyl and trihydrogalvinoxyl, followed by a rate-controlled cross-disproportionation between galvinoxyl and its dihydro derivative (**106**). Other reports of the reactions of stable

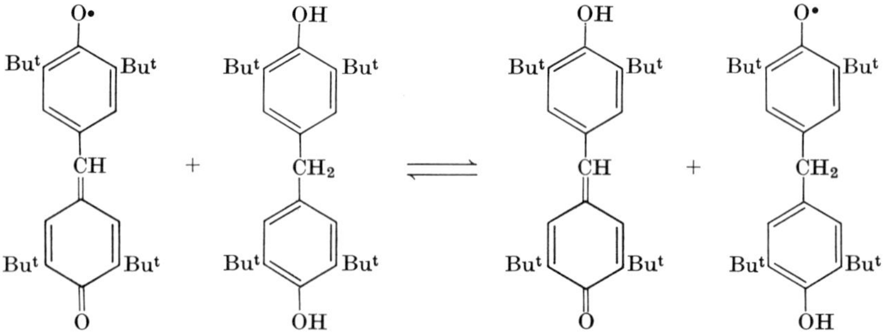

(**103**) + (**106**) $\xrightarrow{\text{Slow}}$ 2(**105**)

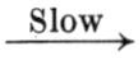

phenoxyl radicals include[171] those of 2,4,6-tri-*tert*-butoxy radical with HCl and HBr; spatially hindered bromocyclohexadienones react with 2,4,6-tri-*tert*-butylphenol, forming the corresponding radical.[172] Solvent effects on the kinetics of H-abstraction from this phenol by DPPH have also been studied.[173]

Although a ready initiator of radical polymerization, α-cyanoisopropyl radicals (Me$_2$ĊCN) do not normally participate in hydrogen-abstraction reactions. However such processes have been claimed to occur[174] from diphenylmethane, dibenzyl ether, mesitylene, and a number of other substrates. Hydrogen abstraction (by alkoxy radicals)

[166] P. S. Skell, D. L. Tuleen, and P. D. Readio, *J. Am. Chem. Soc.*, **85**, 2849 (1963).
[167] D. D. Tanner, M. W. Mosher, N. C. Das, and E. V. Blackburn, *J. Am. Chem. Soc.*, **93**, 5846 (1971).
[168] See *Org. Reaction Mech.*, **1970**, 333–335.
[169] D. S. Ashton, J. M. Tedder, and J. C. Walton, *Chem. Comm.*, **1971**, 1487.
[170] W. Adam and W. T. Chiu, *J. Am. Chem. Soc.*, **93**, 3687 (1971).
[171] A. A. Volod'kin, V. V. Ershov, A. I. Prokof'ev, S. P. Solodovnikov, and D. Kh. Rasuleva, *Izv. Akad. Nauk SSSR, Ser. Khim.*, **1971**, 856; *Chem. Abs.*, **75**, 62809 (1971).
[172] A. I. Prokof'ev, S. P. Solodovnikov, A. A. Volod'kin, and V. V. Ershov, *Izv. Akad. Nauk SSSR, Ser. Khim.*, **1971**, 174; A. A. Volod'kin, V. V. Ershov, A. I. Prokof'ev, S. P. Solodovnikov, and D. Kh. Rasuleva, *Izv. Akad. Nauk SSSR, Ser. Khim.*, **1971**, 859.
[173] J. C. Dearden, *J. Chem. Soc.* (B), **1971**, 2251.
[174] E. A. Tosman and M. V. Bazilevskii, *Kinet. Katal.*, **11**, 882 (1970); *Chem. Abs.*, **73**, 109019 (1970).

from benzaldimines[175] gives benzonitrile and products from the liberated radical R·
produced by the reaction:

$$ArCH=NR \xrightarrow{-H\cdot} Ar\overset{\bullet}{C}=NR \rightarrow ArCN + R\cdot$$

Ready abstractability of imidoyl hydrogen by alkoxy radicals contrasts with the lack of similar reactivity shown by vinylic hydrogen atoms.

The high reactivity of silyl radicals in halogen atom abstraction reactions from alkyl bromides and chlorides forms the basis of the Hudson–Jackson procedure for generating specific alkyl radicals, in concentrations adequate for ESR analysis, and a full report of of this technique has now been published.[176] Qualitative competition experiments indicate that ease of removal of halogen atoms may not follow the order of radical stability although electron-withdrawing groups in the alkyl halide favour attack by triethylsilyl radicals.

A detailed analysis of ESR spectra observed during photopinacolization of benzophenone with benzhydrol in benzene suggests[177] the intervention of a radical–ketone complex such as $Ph_2C=O \ldots H-O-\overset{\bullet}{C}Ph_2$ in the hydrogen-transfer equilibrium.

UV light-initiated reaction of 1,2,3,4,7,7-hexachloronorborn-2-ene (**107**) with oxalyl chloride does not result in chlorocarbonylation and affords instead[178] an almost quantitative yield of the heptachloronorborn-2-ene (**111**). Rearrangement of radical (**108**) must occur in preference to chain transfer with oxalyl chloride. The intermediate radical

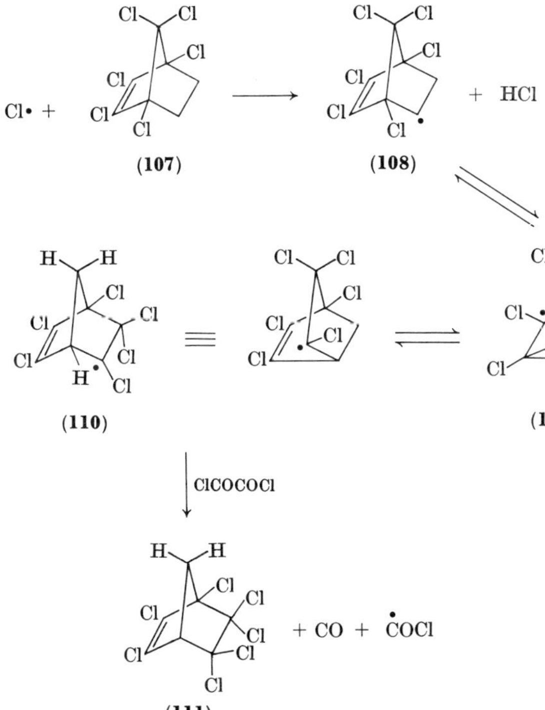

[175] H. Ohta and K. Tokumaru, *Chem. Comm.*, **1970**, 1601.
[176] A. J. Bowles, A. Hudson, and R. A. Jackson, *J. Chem. Soc.* (B), **1971**, 1947.
[177] G. O. Schenck, G. Behrens, and E. Roselius, *Tetrahedron Letters*, **1970**, 5185.
[178] D. I. Davies and P. Mason, *J. Chem. Soc.* (C), **1971**, 295.

(**109**) so formed is still sterically hindered for chain transfer, but such hindrance is reduced in structure (**110**) which has bridge and bridgehead hydrogens at C-7 and C-1, respectively, giving rise to chain transfer by chlorine abstraction.

The Cristol-Firth reaction (which provides a convenient way of studying free radicals R· generated from carboxylic acids RCO_2H) has been applied to several carboxylic acids containing a chloronorbornene ring system.[179] Reaction products suggest that chloronorborn-5-en-2-yl radicals are unlikely to be planar and are more probably represented by equilibrating *exo*- and *endo*-radicals, with the former favoured owing to minimized non-bonded interactions. Tedder's group have extended radical chlorination studies to chloro- and methyl-cyclopentanes,[180] chlorocycloheptane and fluorocyclohexane.[181] In each case the *cis*:*trans* ratio of the products derived from chlorine atom attack at the 2-position was evaluated and its significance was discussed.

Reactions of alkenes with trichloramine (Cl_3N) in non-polar solvents give good yields of *vic*-dichlorides; *meso*:*dl* ratios for chlorination of isomeric but-2-enes were similar to those obtained from molecular chlorine under radical conditions and, although a radical-chain mechanism is considered likely, the precise nature of the various chain steps is not yet clear.[182]

A historical and mechanistic review of the Wohl–Ziegler reaction (allylic bromination by *N*-bromo amides and imides) is presented[183] as an introduction to a re-examination of the reaction of *N*-bromoacetamide with a number of olefins. This reaction is found to lead, not to allylic bromination, but to 2,*N*-dibromoacetimidates, a new class of compound. Both free-radical and ionic reactions appear to be involved in a two-stage reaction; first, a radical reaction of *N*-bromoacetamide with itself, then ionic addition of the resulting adduct to the olefin.

Activation energies of the chlorination of THF have been evaluated,[184] and the fluorination of 1,4-dioxan and 1,4-oxathian by potassium tetrafluorocobaltate(III) proceeds through radical and radical-cation intermediates.[185] Photo-oximation of alkanes by nitrosyl chloride[186] probably proceeds by the non-chain radical process:

$$ClNO \xrightarrow{h\nu} Cl\cdot + NO$$
$$Cl\cdot + RH \rightarrow HCl + R\cdot$$
$$R\cdot + NO \rightarrow RNO \rightarrow R'{=}NOH$$

it is thus analogous to the recent proposal by de Boer's group[187] for photonitrosation by *tert*-butyl nitrite. The kinetics and mechanism of iodine exchange between monoiodoacetic acid and molecular iodine have been studied in benzene solution.[188]

A possible mechanism[189] accounting for the observed products [(**114**), (**115**), (**118**) and naphthalene] of the free-radical bromination of cycloprop[*a*]indane (**112**) is indicated.

[179] D. I. Davies and P. Mason, *J. Chem. Soc.* (C), **1971**, 288.
[180] D. S. Ashton and J. M. Tedder, *J. Chem. Soc.* (B), **1971**, 1719.
[181] D. S. Ashton and J. M. Tedder, *J. Chem. Soc.* (B), **1971**, 1723.
[182] K. W. Field and P. Kovacic, *J. Org. Chem.*, **36**, 3566 (1971); see also P. Kovacic and J.-H. C. Chang, *J. Org. Chem.*, **36**, 3138 (1971).
[183] S. Wolfe and D. V. C. Awang, *Can. J. Chem.*, **49**, 1384 (1971).
[184] J. Pichler and M. Kratochvil, *Coll. Czech. Chem. Comm.*, **36**, 2052 (1971).
[185] J. Burdon and I. W. Parsons, *Tetrahedron*, **27**, 4533 (1971).
[186] M. W. Mosher and N. J. Bunce, *Can. J. Chem.*, **49**, 28 (1971).
[187] A. Makor, J. U. Veenland, and T. J. de Boer, *Rec. Trav. Chim.*, **88**, 1249 (1969).
[188] E. Körös, M. Orban, and M. Burger, *Mag. Kém. Folyoirat*, **77**, 304 (1971).
[189] E. C. Friedrich and R. L. Holmstead, *J. Org. Chem.*, **36**, 971 (1971).

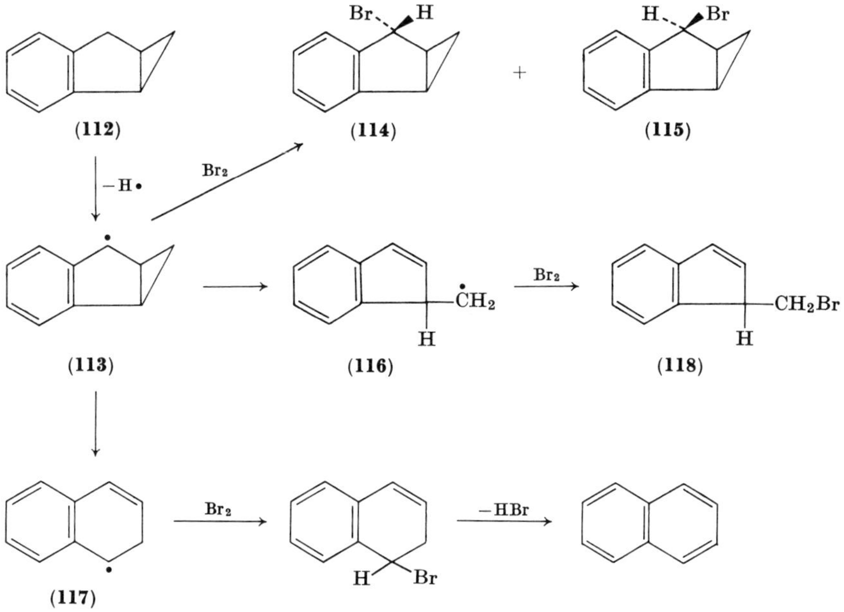

The reactions have been studied with a wide variety of brominating agents and conditions, and it is interesting that light-induced bromination with N-bromosuccinimide or molecular bromine in CCl_4 gives higher yields of (**114**) and (**115**) than are obtained at 77°. Further evidence for the radical intermediates proposed was obtained[189] by tin hydride reductions of the brominated products, and a notable result was the formation of (**112**) by tin hydride reduction of (**118**) which confirms the reversibility of the cyclopropyl–allyl rearrangement of the cycloprop[a]indanyl radical (**113**) to the inden-1-yl-methyl radical (**116**). 1,2-Dihydronaphthalene was obtained from tin hydride reductions of both (**114**) and (**115**), supporting the intermediacy of the 1,2-dihydronaphthyl radical (**117**) as the source of at least part of the naphthalene from free-radical bromination of (**112**).

Free-radical bromination of adamantanone and some 2-substituted adamantanes has been studied,[190] and a full report of a detailed kinetic study of the radical isomerization of 2-bromo-3,3,3-trichloropropene has been published.[191]

Afanas'ev has further extended investigation of the reactivity of alkyl free radicals in the liquid phase with reports on the reactions of sec-octyl radicals with chloromethanes,[192] bromo compounds,[193] acetone, toluene and anisole;[194] the reactivity of $ClCH_2CO_2H$ and $HO_2CCH_2CO_2H$ towards n-nonyl radicals has also been studied.[195]

[190] I. Tabushi, Y. Aoyama, and Z. Yoshida, *J. Am. Chem. Soc.*, **93**, 2077 (1971).
[191] R. G. Gasanov, T. T. Vasil'eva, and R. Kh. Freidlina, *Dokl. Akad. Nauk SSSR*, **193**, 1058 (1970); *Chem. Abs.*, **74**, 12395 (1971).
[192] I. B. Afanas'ev, I. V. Mamontova, and G. I. Samokhvalov, *Zhur. Org. Khim.*, **7**, 457 (1971); *Chem. Abs.*, **75**, 19504 (1971).
[193] I. B. Afanas'ev and I. V. Mamontova, *Zhur. Org. Khim.*, **7**, 678 (1971); *Chem. Abs.*, **75**, 34841 (1971).
[194] I. B. Afanas'ev and I. V. Mamontova, *Zhur. Org. Khim.*, **7**, 682 (1971); *Chem. Abs.*, **75**, 34842 (1971).
[195] I. B. Afanas'ev and E. D. Safronenko, *Zhur. Org. Khim.*, **7**, 453 (1971); *Chem. Abs.*, **75**, 19503 (1971).

Evidence has been adduced that methyl radicals abstract hydrogen from acetonitrile in the solid state; activation energies as low as 1.4 kcal mole^{-1} are claimed.[196] Attack of methyl radicals on amino acids,[197] and of dichloromethyl radicals on chloroform,[198] in solution has been studied. Hydrogen abstraction from organosilicon compounds by *tert*-butoxy radicals has been reported,[199] as have gas-phase reactions of methyl,[200] ·CD$_3$,[201] fluoromethyl,[202] trifluoromethyl,[203] perfluoroethyl[204] and other fluoroalkyl radicals.[205]

Numerous other gas-phase investigations include abstractions by trichloromethyl[206] and tribromomethyl radicals,[207] photochlorination of perfluorocyclobutene,[208] and other varied chlorination and bromination reactions.[209] Rates of reaction of sodium atoms with some haloalkyl-germanes, -stannanes and -silanes have been obtained,[210] and related studies[211] show the C—Cl bond in vinyl chloride to be stronger than that in cyclopropyl chloride. Relative rates of intramolecular H-abstraction by the sulphonamide radical,[212] and of intermolecular abstraction by F$_2$N·[213] and bromine atoms,[214] have also been reported.

An interesting report[215] describes genuine hydrogen abstraction from benzene by ·CF$_3$ radicals at high temperature, from which the bond-dissociation energy $D(C_6H_5-H)$ is calculated to be 110.0 ± 2.0 kcal mole^{-1}.

Additions

Intermolecular Addition Reactions

The photochemical reactions of trifluoroiodomethane and hydrogen bromide with 1,1,1-trifluorobut-2-ene have been investigated.[216] Because of the comparable steric effects of CH$_3$ and CF$_3$ and high degree of polarization, the olefin is well suited to a study

[196] E. D. Sprague and F. Williams, *J. Am. Chem. Soc.*, **93**, 787 (1971).
[197] I. A. Taha and R. R. Kuntz, *Photochem. Photobiol.*, **13**, 67 (1971).
[198] L. C. Dickey and R. F. Firestone, *J. Phys. Chem.*, **74**, 4310 (1970).
[199] H. Sakurai, A. Hosomi, and M. Kumada, *Bull. Chem. Soc. Japan*, **44**, 568 (1971).
[200] T. R. Donovan, W. Dorko, and A. G. Harrison, *Can. J. Chem.*, **49**, 828 (1971).
[201] T. N. Bell and A. E. Platt, *J. Phys. Chem.*, **75**, 603 (1971).
[202] J. A. Kerr and D. M. Timlin, *Int. J. Chem. Kinet.*, **3**, 1, 69 (1971).
[203] K. C. Ferguson and J. T. Pearson, *Trans. Faraday Soc.*, **67**, 754 (1971); L. M. Quick and E. Whittle, *Trans. Faraday Soc.*, **67**, 1727 (1971); S. H. Jones and E. Whittle, *Int. J. Chem. Kinet.*, **2**, 479 (1970).
[204] J. D. Clarke, C. Pearce, and D. A. Whytock, *Trans. Faraday Soc.*, **67**, 1049 (1971).
[205] L. O. Moore, *Can. J. Chem.*, **49**, 666 (1971); *J. Phys. Chem.*, **75**, 2075 (1971).
[206] F. B. Wampler and R. R. Kuntz, *Int. J. Chem. Kinet.* **3**, 283 (1971).
[207] S. Hautecloque and T. M. N. Nguyen, *Compt. Rend.* (C), **273**, 569 (1971).
[208] Z. R. Alberto, J. J. Cosa, C. A. Vallana, and E. H. Staricco, *Can. J. Chem.*, **49**, 1252 (1971).
[209] P. Smit and H. J. den Hertog, *Tetrahedron Letters*, **1971**, 595; J. P. Soumillion, P. Gouverneur, T. Burton, and A. Bruylants, *Bull. Soc. Chim. belges*, **80**, 233 (1971); N. N. Lebedev, V. F. Shvets, and V. A. Aver'yanov, *Kinet. Katal.*, **12**, 560 (1971); V. I. Stanko, A. I. Klimova, P. I. Belik, and K. P. Butin, *Zhur. Obshch. Khim.*, **41**, 338 (1971); *Chem. Abs.*, **75**, 20470 (1971); K. C. Ferguson and E. Whittle, *Trans. Faraday Soc.*, **67**, 2618 (1971).
[210] D. S. Boak, N. J. Friswell, and B. G. Gowenlock, *J. Organomet. Chem.*, **27**, 333 (1971); D. S. Boak and B. G. Gowenlock, *J. Organomet. Chem.*, **29**, 385 (1971).
[211] E. U. Emovon and J. F. Ojo, *Nigerian J. Sci.*, **3**, 163 (1969); *Chem. Abs.*, **73**, 130512 (1970).
[212] T. Ohashi, S. Takeda, M. Okahara, and S. Komori, *Bull. Chem. Soc. Japan*, **44**, 771 (1971).
[213] P. Cadman, C. Dodwell, A. J. White, and A. F. Trotman-Dickenson, *J. Chem. Soc.* (A), **1971**, 2967.
[214] K. D. King, D. M. Golden, and S. W. Benson, *Trans. Faraday Soc.*, **66**, 2794 (1970).
[215] G. A. Chamberlain and E. Whittle, *Trans. Faraday Soc.*, **67**, 2077 (1971).
[216] R. Gregory, R. N. Haszeldine, and A. E. Tipping, *J. Chem. Soc.* (C), **1971**, 1216.

of radical electrophilicity, variations of which are reflected in changes in the ratio of attack at each end of the double bond. Thus the direction of addition of the more electrophilic bromine atom is in marked contrast to that of trifluoromethyl radical and predominant attack occurs at the CF_3CH end of the olefin. For addition of a trifluoromethyl radical it is argued that a methyl group stabilizes a lone electron to approximately the same extent as a fluorine atom.

Addition of trichloromethyl radicals to nine nuclear-substituted styrenes has enabled the relative reactivities to be separated into polar and resonance effects.[217] Unlike the behaviour of the *meta*-substituted isomers, the relative rates of *para*-substituted styrenes did not follow the simple Hammett correlation and it is suggested that the polar term originates in charge-transfer interactions between olefin and trichloromethyl radical.

Free-radical addition of thiophenol to 3-methylenenortricyclene (**119**) gives the 1,2-addition product (**120**) as well as a variety of unsaturated thioethers that can be formulated as derivable, under reaction conditions, from the 1,5-homoconjugate addition product (**121**).[218] The search for π-bridged radicals as product-determining

       CH₂   CH₂SPh   CH₂SPh

    (**119**)   (**120**)   (**121**)

intermediates was unsuccessful, variation in product compositions with reagent concentrations demonstrating the existence of classical radicals rather than a single non-classical free radical.

Thermal and photochemical addition of diethyl azodicarboxylate to cycloheptene and cyclo-octene[219] is thought to proceed by a free-radical mechanism as for the addition of ethyl malonate and acetate to terminal olefins.[220] Preparative yields of adducts are obtained by photolysis of halogenated N-chloro- and N-bromo-acetamides in the presence of olefins:[221]

$$RCONHX + \; >C=C< \; \xrightarrow{h\nu} \; RCONH-\underset{|}{\overset{|}{C}}-\underset{|}{\overset{|}{C}}-X$$

X = Cl or Br
R = Me, CH_2Br, CH_2Cl, CH_2F or CCl_3

The α-halogen substituents have a marked effect on the tendency of acetylamino radicals to add to double bonds. Rates of addition of cyclopropyl radicals to vinyl halides have been compared with related data of other radicals.[222] Activation energies increase from the fluoride to the iodide, but the reactivity follows the same order because the pre-exponential factor increases faster than the activation energy; correlations between reactivity and ionization potentials of the olefins are discussed.[222]

Conditions for separate ionic and free-radical addition of hydrogen bromide to hexachloromethylenenorbornene (**122**) have been devised.[223] Radical addition affords a 60:40 mixture of (**126**) and (**124**) by way of the radical (**123**) which either undergoes

---

[217] H. Sakurai, S. Hayashi, and A. Hosomi, *Bull. Chem. Soc. Japan*, **44**, 1945 (1971).
[218] S. J. Cristol and R. Kellman, *J. Org. Chem.*, **36**, 1866 (1971).
[219] A. Shah and M. V. George, *Tetrahedron*, **27**, 1291 (1971).
[220] B. S. Kirkiacharian, *Bull. Soc. Chim. France*, **1971**, 1797, 1800.
[221] D. Touchard and J. Lessard, *Tetrahedron Letters*, **1971**, 4425.
[222] A. P. Stefani and H. E. Todd, *J. Am. Chem. Soc.*, **93**, 2982 (1971).
[223] R. Alexander and D. I. Davies, *J. Chem. Soc.* (C), **1971**, 896.

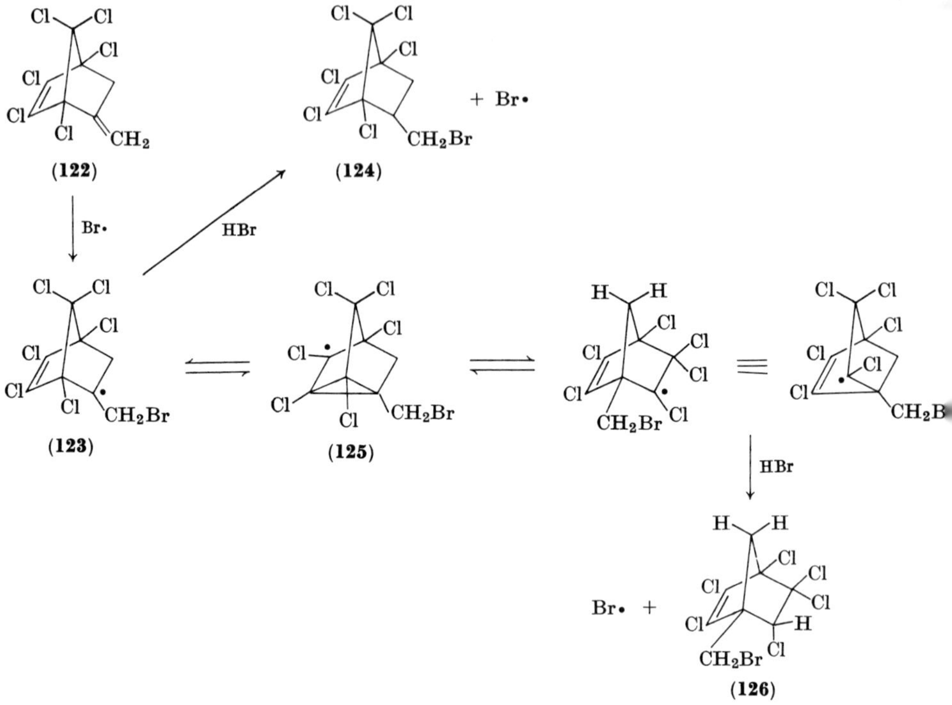

chain transfer or rearrangement to (125). No rearrangement products were observed in the ionic reaction. Synthetic procedures based on radical addition of hydrogen bromide to α- and β-pinene have been described,[224] the latter yielding p-menthane adducts by di-tert-butyl peroxide-induced addition of mono- and di-nitriles.[225]

Radical addition of methanethiol to bicyclo[3.1.0]hex-2-ene (127) proceeded smoothly upon irradiation to give a ca. 90% yield of a number of 1:1 addition products.[226] Results clearly favoured equilibria such as (128)⇌(129) rather than a single delocalized radical for the addition of thiyl radical to both faces of the double bond. Adducts indicating rearrangement of (129) to (130) by a cyclopropylcarbinyl–allylcarbinyl β-fission process are amongst the products isolated. The stereochemistry of condensation of organogermanium hydrides and halohydrides (R_3GeH, R_2ClGeH, RCl_2GeH and Cl_3GeH) with some substituted allenes has been studied[227] in terms of the structure of the hydro-

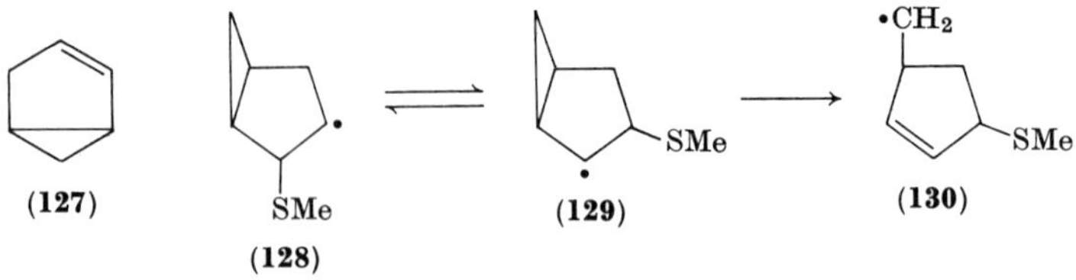

[224] A. Gaiffe and J. Castanet, Compt. Rend. (C), 272, 96 (1971).
[225] M. Cazaux and R. Lalande, Bull. Soc. Chim. France, 1971, 461.
[226] P. K. Freeman, M. F. Grostic, and F. A. Raymond, J. Org. Chem., 36, 905 (1971).
[227] M. Massol, Y. Cabadi, and J. Satgé, Bull. Soc. Chim. France, 1971, 3235.

germane and the initial allene; increase of electrophilic character of the germyl radical results in more terminal attack.

Nortricyclene derivatives are generally obtained from radical additions to norbornadiene, with only sulphur or silicon radicals leading to some norbornenyl derivatives. Addition of phosphorus radicals has now been examined[228] and, whilst the major products have a nortricyclene structure, formation of norbornenylphosphonate and other norbornene compounds does occur. Reaction of trimethyltin hydride and norbornadiene yields the norbornenes (131)–(133) and the nortricyclene (134); chlorodimethyl-, bromodimethyl-, iododimethyl-, and dichloromethyl-tin hydrides react similarly.[229] The *anti*-7-isomer of (131) is not formed and its absence is attributed to steric and torsional effects in the radical from which it would be formed. Norbornenes

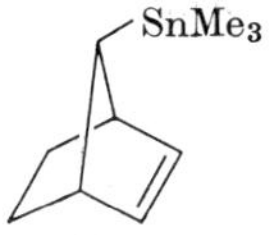

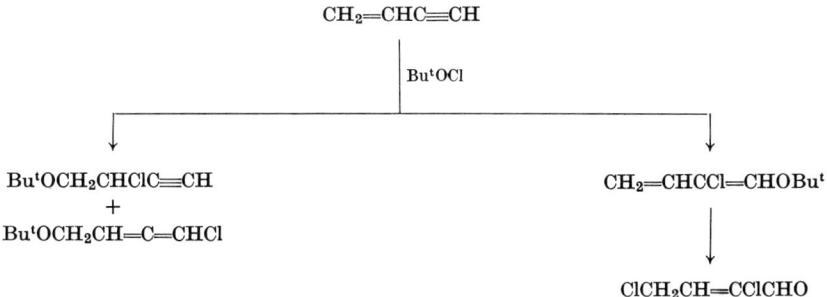

(131) and (132) result from initial *endo*-attack on the diene by the trimethyltin radical and amount to nearly 50% of the total product.

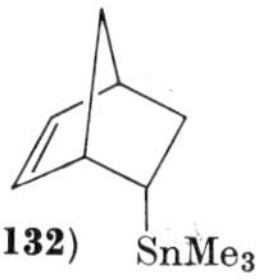

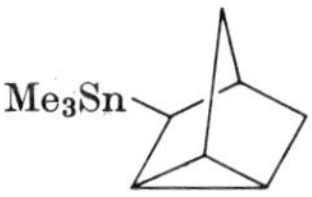

Relative rate comparisons with model hydrocarbons show that *tert*-butoxy radical attack at either terminus of enynes is enhanced in comparison with simple olefins and acetylenes.[230] Reaction of *tert*-butyl hypochlorite with but-1-en-3-yne proceeds[230] by competitive olefinic addition (to give 1,2- and 1,4-adducts) and acetylenic addition (to give a 4,3-adduct that suffers rapid, secondary electrophilic chlorination):

$$CH_2=CHC\equiv CH$$

$$\downarrow Bu^tOCl$$

Bu^{t}OCH$_2$CHClC≡CH
+
Bu^{t}OCH$_2$CH=C=CHCl

CH$_2$=CHCCl=CHOBu^t

↓

ClCH$_2$CH=CClCHO

---

[228] H. J. Callot and C. Benezra, *Can. J. Chem.*, **49**, 500 (1971).
[229] H. G. Kuivila, J. D. Kennedy, R. Y. Tien, I. J. Tyminski, F. L. Pelczar, and O. R. Khan, *J. Org. Chem.*, **36**, 2083 (1971).
[230] M. L. Poutsma and P. A. Ibarbia, *J. Org. Chem.*, **35**, 4038 (1970).

Olefin attack is preferred to acetylene attack by a factor of 4 at 25°. The addition of 1-chloropiperidine in acid solution to 2-methylbut-1-en-3-yne, pent-1-en-3-yne, and but-1-en-3-yne was also examined.[231]

Sulphonyl halides add readily and stereoselectively to acetylenes to form 1:1-adducts in good yields by free-radical mechanisms. Sulphonyl iodides add entirely in a *trans*-manner[232] and it is suggested that this remarkable stereoselectivity results because chain transfer is much faster than isomerization of the intermediate vinyl radical. In

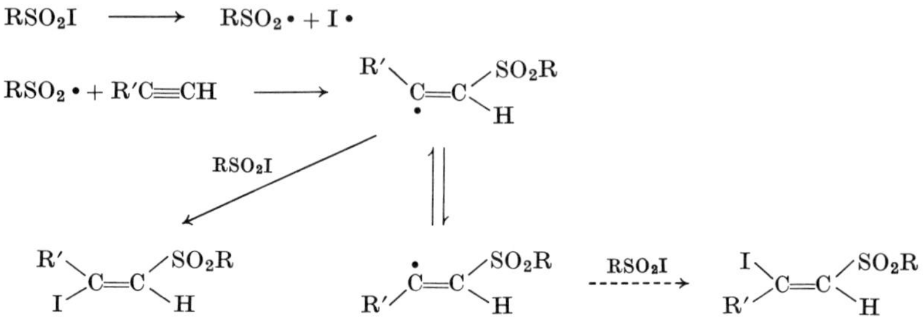

$$RSO_2I \longrightarrow RSO_2\cdot + I\cdot$$

$$RSO_2\cdot + R'C\equiv CH \longrightarrow \underset{\cdot}{\overset{R'}{\diagdown}}C=C\underset{H}{\overset{SO_2R}{\diagup}}$$

$$\underset{I}{\overset{R'}{\diagdown}}C=C\underset{H}{\overset{SO_2R}{\diagup}} \xleftarrow{RSO_2I} \underset{R'}{\overset{\cdot}{\diagdown}}C=C\underset{H}{\overset{SO_2R}{\diagup}} \xrightarrow{RSO_2I} \underset{R'}{\overset{I}{\diagdown}}C=C\underset{H}{\overset{SO_2R}{\diagup}}$$

the presence of a copper catalyst sulphonyl chlorides also add to terminal and non-terminal acetylenes,[233] although this reaction depended greatly on polar factors (e.g. solvent dielectric constant) and the course of addition could be controlled to give preferentially either *cis*- or *trans*-addition products. Radical addition of bromotrichloromethane with acetylene yields mainly *trans*-products.[234]

Kinetic data for the addition of iodine to pentene isomers[235] supports the view that stereospecific addition is the result of attack of an iodine atom or molecule on a charge-transfer complex between the olefin and the corresponding iodine molecule or atom.

A novel double radical-chain mechanism accounts for products observed from the thermal reaction of hexafluoroacetone azine (**135**) with cyclohexane.[236] The first chain reaction is initiated by the addition of a cyclohexyl radical to the azine, giving an adduct that decomposes to radical (**136**) and the diazo-compound (**137**); abstraction of hydrogen by (**136**) then gives the major isolated product and a new cyclohexyl radical to continue the chain. Addition of a cyclohexyl radical to (**137**) initiates the second chain, giving

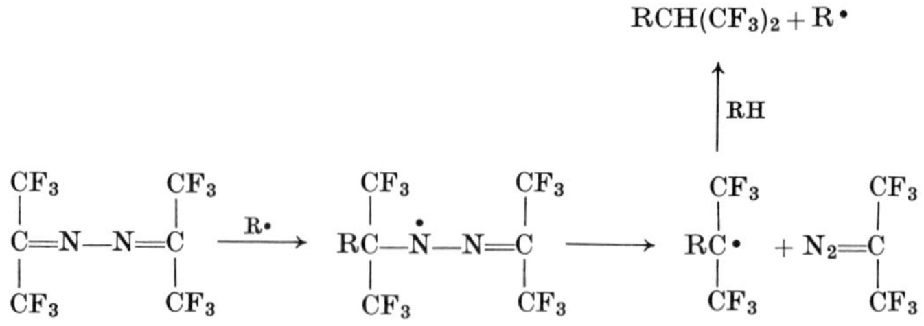

$$RCH(CF_3)_2 + R\cdot$$
$$\uparrow RH$$

$$\underset{CF_3}{\overset{CF_3}{|}}C=N-N=\underset{CF_3}{\overset{CF_3}{|}}C \xrightarrow{R\cdot} \underset{CF_3}{\overset{CF_3}{|}}RC-\overset{\cdot}{N}-N=\underset{CF_3}{\overset{CF_3}{|}}C \longrightarrow \underset{CF_3}{\overset{CF_3}{|}}RC\cdot + N_2=\underset{CF_3}{\overset{CF_3}{|}}C$$

(**135**)　　　　　　　　　　　　　　　　　　　(**136**)　(**137**)

R = Cyclohexyl

[231] M. L. Poutsma and P. A. Ibarbia, *J. Org. Chem.*, **36**, 2572 (1971).
[232] W. E. Truce and G. C. Wolf, *J. Org. Chem.*, **36**, 1727 (1971).
[233] Y. Amiel, *Tetrahedron Letters*, **1971**, 661.
[234] I. B. Afanas'ev, N. G. Baranova, and G. I. Samokhvalov, *Zhur. Org. Khim.*, **6**, 1338 (1970).
[235] R. L. Ayres, C. J. Michejda, and E. P. Rack, *J. Am. Chem. Soc.*, **93**, 1389 (1971).
[236] W. J. Middleton, *J. Am. Chem. Soc.*, **93**, 423 (1971).

$$(137) + R\cdot \begin{array}{c} \xrightarrow{} RN=N\overset{\overset{CF_3}{|}}{\underset{\underset{CF_3}{|}}{C}}\cdot \xrightarrow{RH} RN=N\overset{\overset{CF_3}{|}}{\underset{\underset{CF_3}{|}}{CH}} + R\cdot \\ (138) \\ \xrightarrow{} R\overset{\cdot}{N}N=C(CF_3)_2 \xrightarrow{RH} RNH-N=C(CF_3)_2 + R\cdot \\ (139) \end{array}$$

a resonance-stabilized radical that can abstract hydrogen (from cyclohexane) in two different ways, giving either the azo compound (138) or the hydrazone (139); the resulting cyclohexyl radical completes the chain cycle.

Radical-initiated reactions of isocyanides (RNC) with disubstituted phosphines takes two courses depending upon the nature of the isocyanide but not upon that of the phosphine.[237] The reaction may reasonably be explained by the annexed scheme involving imidoyl radicals (140) as key intermediates. The partitioning of (140) is determined mainly by the resonance stability of the liberated radical R·.

$$Et_2PH \xrightarrow{-H\cdot} Et_2P\cdot$$
$$Et_2P\cdot + RNC \longrightarrow Et_2P\overset{\cdot}{C}=NR \quad (140)$$

$$(140) \begin{array}{c} \xrightarrow{Et_2PH} Et_2PCH=NR + Et_2P\cdot \\ \searrow Et_2PCN + R\cdot \end{array}$$

$$R\cdot + Et_2PH \longrightarrow RH + Et_2P\cdot$$

Quantitative addition of ·CMe$_2$CN to biscyclopentadienylcobalt, Co(C$_5$H$_5$)$_2$, is claimed to result when an excess of [Me$_2$C(CN)—N=]$_2$ is decomposed in a boiling toluene solution of the cobalt complex.[238] Radical additions of dimethylisopropylamine,[239] ethanol and propan-1-ol[240] to trifluorochloroethylene and hydrobromic acid to cyclic allylic compounds[241] have been reported.

ESR studies[242] indicate that the principal adduct of trialkylstannyl radicals to butadiene is the *trans*-α-substituted allylic radical (141). No evidence for the presence of the *cis*-isomer (142) or the tin-centred radical (143) or (144) was obtained, and so the suggestion by Kuivila[243] (to explain the observed products) that the latter radicals are in equilibrium with the open-chain allylic isomers (141) and (142) remains open to question.

(141) (142)

[237] T. Saegusa, Y. Ito, N. Yasuda, and T. Hotaka, *J. Org. Chem.*, **35**, 4238 (1970).
[238] G. E. Herberich and J. Schwarzer, *Angew. Chem. Int. Ed.*, **9**, 897 (1970).
[239] F. Liška, *Coll. Czech. Chem. Comm.*, **36**, 1853 (1971).
[240] F. Liška and S. Šimek, *Coll. Czech. Chem. Comm.*, **36**, 3463 (1971).
[241] J.-M. Pabiot and R. Pallaud, *Compt. Rend.* (C), **273**, 475 (1971).
[242] T. Kawamura and J. K. Kochi, *J. Organomet. Chem.*, **30**, C8 (1971).
[243] H. G. Kuivila, *Accounts Chem. Res.*, **1**, 299 (1968).

(143)   (144)

The rate of trichloromethyl radical addition to a chlorine-substituted site is slower than the rate of addition at =CF$_2$ and much slower than addition to =CH$_2$.[244] Orientation of methyl radical addition to three fluoroethylenes shows substantial differences from orientations previously observed for the addition of trichloromethyl and heptafluoropropyl radicals to the same olefins.[245] Gas-phase additions of methyl radicals to hexafluoropropene,[246] methyl and CF$_3$· radicals to acetone,[247] and liquid-phase addition of CCl$_3$· radicals to isoprene,[248] have been reported.

1,3-Dioxolans undergo radical-catalysed addition to oct-1-ene; formation of all the products is accounted for by β-scission of heterocyclic radicals, the stability of which depends on the temperature at which reactions are carried out.[249] Radical addition of methyl dichloroacetate to dec-1-ene gives methyl 2,2-dichloro- and 2-chloro-dodecanoate,[250] the latter (C—Cl cleavage product) increasing as the temperature is raised. Relative activation energies of aroyloxy-radical addition to benzene are compared with elimination of carbon dioxide,[251] and rates of addition of atomic hydrogen to acetylene,[252] propene[253] and other olefins[254] have been measured.

Intramolecular Addition Reactions

Peroxide-initiated dimerization of 3,3,4,4-tetrafluoro-4-iodobut-1-ene (145) gives the isomers (146)[255] by a cyclization route analogous to that of the simple hex-5-enyl radical which is known to cyclize almost exclusively to the cyclopentylmethyl radical.

$$CH_2=CHCF_2CF_2I \xrightarrow[120°, 5\ hr.]{(Bu^tO)_2} CH_2=CHCF_2\dot{C}F_2 \xrightarrow[(R=CF_2CF_2I)]{RCH=CH_2} R\dot{C}HCH_2CF_2CF_2CH=CH_2$$

cis- and trans- (146)

[244] D. P. Johari, H. W. Sidebottom, J. M. Tedder, and J. C. Walton, *J. Chem. Soc.* (B), **1971**, 95.
[245] J. M. Tedder, J. C. Walton, and K. D. R. Winton, *Chem. Comm.*, **1971**, 1046.
[246] J. C. J. Thynne, *Int. J. Chem. Kinet.* **3**, 155 (1971).
[247] J. D. Reardon and C. E. Waring, *J. Phys. Chem.*, **75**, 735 (1971).
[248] I. B. Afanas'ev, I. V. Mamontova, T. M. Filippova, and G. I. Samokhalov, *Zhur. Org. Khim.*, **7**, 3 (1971); *Chem. Abs.*, **74**, 99352 (1971).
[249] B. Maillard, M. Cazaux, and R. Lalande, *Bull. Soc. Chim. France*, **1971**, 467; see also *Org. Reaction Mech.*, **1969**, 330.
[250] P. Guerrini, J. Sorba, and D. Lefort, *Compt. Rend.* (C), **272**, 1690 (1971).
[251] T. Suehiro and M. Ishida, *Bull. Chem. Soc. Japan*, **44**, 1692 (1971).
[252] K. Hoyermann, H. G. Wagner, J. Wolfrum, and R. Zellner, *Ber. Bunsenges. Phys. Chem.*, **75**, 22 (1971).
[253] M. J. Kurylo, N. C. Peterson, and W. Braun, *J. Phys. Chem.*, **54**, 4662 (1971).
[254] R. D. Kelley, R. Klein, and M. D. Scheer, *J. Phys. Chem.*, **74**, 4301 (1970).
[255] P. Piccardi and M. Modena, *Chem. Comm.*, **1971**, 1041.

The pent-4-ynyloxy radical formed on photolysis of the nitrite (**147**) surprisingly failed to cyclize in the various solvents employed;[256] no product from radical (**148**) was detected. Reasons for the failure are not well understood, but attention is drawn to the

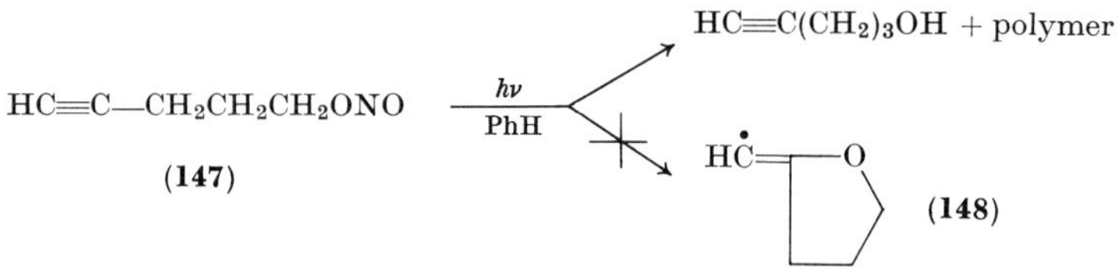

$$HC{\equiv}C-CH_2CH_2CH_2ONO \xrightarrow[PhH]{h\nu} HC{\equiv}C(CH_2)_3OH + polymer$$

(**147**) → HC=⟨O⟩ (**148**)

results of Surzur and his co-workers[257] who have shown that the corresponding sulphur analogues of (**148**) are formed when the terminal alkyne group is methyl or phenyl, but not when, as in this study, it is hydrogen.

Surzur's group have extended studies on cyclization by nitrogen- and sulphur-centred radicals. Effects of temperature and substituents on the reversibility of intramolecular addition of thiyl radicals similar to (**150**; X = S) have been elegantly delineated.[258] From these results control can be established in synthetic applications; thus the ratio of (**151**) to (**152**) is easily controlled by photolysis of the thiol (**149**) at different temperatures.[259] Syntheses of thiabicyclo-[2.2.2]- and -[3.2.1]-octane are controlled in a similar fashion by varying the substituents in the starting thiol.[260]

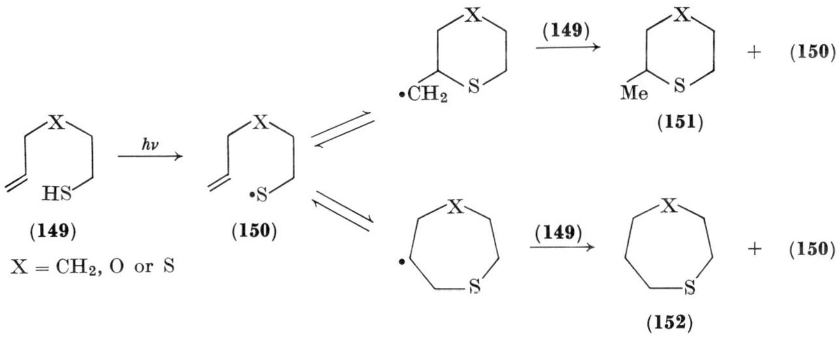

(**149**) X = CH$_2$, O or S

Intramolecular addition of (**154**), formed by treatment of (**153**) with TiCl$_4$, to radical (**155**) results in a stereoselective synthesis of the [3.2.1]-bridged azabicyclic compound (**156**; X = Cl), in 77% yield.[261] Lower yields were obtained from (**153**) by photolysis, and the products then indicated substantial attack on solvent by radical (**155**).

During the reduction of some *gem*-dihalogenocyclopropanes by an organotin hydride, a cyclization between the intermediate cyclopropyl radical and a double bond in a convenient position of the starting molecule has been observed.[262] Bicyclic compounds related to the [3.1.0]bicyclohexane system, with an *endo*- or *exo*-alkyl group in the 2-position, are the main products.

---

[256] R. D. Rieke and B. J. A. Cooke, *J. Org. Chem.*, **36**, 2674 (1971).
[257] See *Org. Reaction Mech.*, **1969**, 325.
[258] M.-P. Crozet, J.-M. Surzur, and C. Dupuy, *Tetrahedron Letters*, **1971**, 2031.
[259] J.-M. Surzur, M.-P. Crozet, and C. Dupuy, *Tetrahedron Letters*, **1971**, 2025.
[260] J.-M. Surzur, R. Nouguier, M.-P. Crozet, and C. Dupuy, *Tetrahedron Letters*, **1971**, 2035.
[261] J.-M. Surzur, L. Stella, and R. Nouguier, *Tetrahedron Letters*, **1971**, 903.
[262] C. Descoins, M. Julia, and H. V. Sang, *Bull. Soc. Chim. France*, **1971**, 4087.

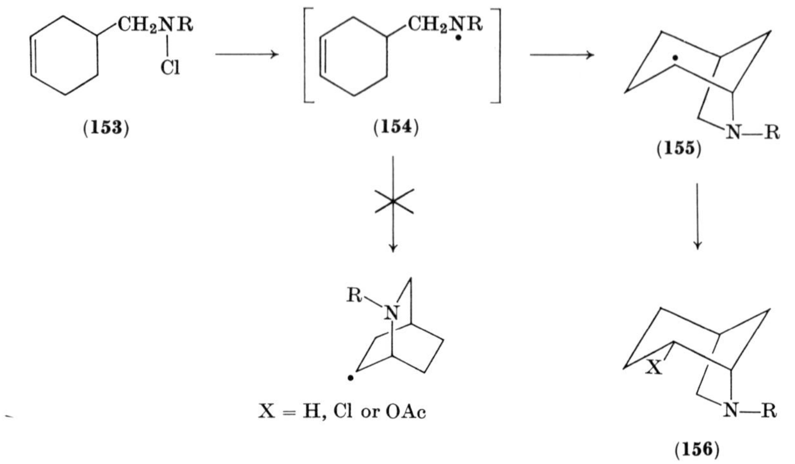

Reaction of the dienes (**157**) with iodobenzene dichloride gave the cyclized product (**158**) in 100% yield when R = C(CO$_2$Et)$_2$ but only 24% when R = CH$_2$.[263] In the latter case 62% of (**160**) and 14% of (**159**) are also produced.

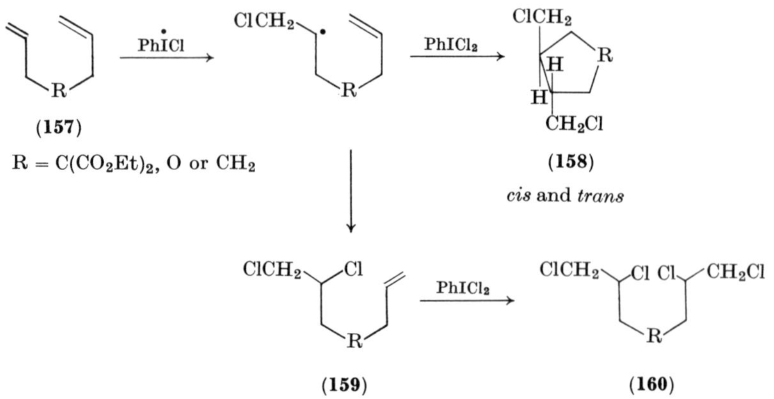

A series of N-substituted diallylamine derivatives, Y—N(CH$_2$CH=CH$_2$)$_2$, where Y = CH$_3$CO, CF$_3$CO, PhCO, CN, or CH$_2$CH$_2$CN, cyclized during the radical addition of nonafluoro-1-iodobutane.[264] The cyclic adducts were shown to possess a 3-iodomethyl-4-(2,2,3,3,4,4,5,5,5-nonafluoropentyl)pyrrolidine structure by spectroscopic and chemical methods. Diallylamine (Y = H) and nonafluoro-1-iodobutane, in the presence of a radical-generating agent, gave only a polymeric salt.

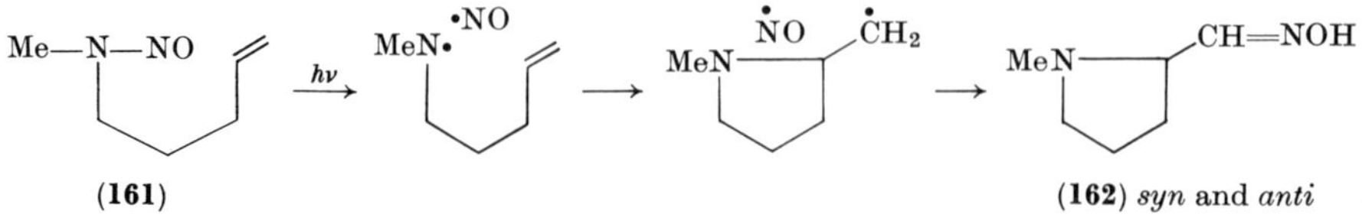

[263] M.-C. Lasne and A. Thuillier, *Compt. Rend.* (C), **273**, 1258 (1971).
[264] N. O. Brace, *J. Org. Chem.*, **36**, 3187 (1971).

Intramolecular photoadditions of a number of nitrosamines have been shown to be clearly specific in forming five-membered azacyclics and a radical pathway is indicated.[265] Thus, for example, photolysis of (**161**) in methanol containing 0.06N-HCl gave a mixture of *syn*- and *anti*-oximes (**162**) in 82% yield. Specific syntheses of analogous azabicyclic compounds are also described.[266]

Aromatic Substitution

In view of the rapid developments of recent years,[267] it would be rash to suppose that the existing theories regarding intermediates and reaction mechanisms for decomposition of acylarylnitrosamines represent the complete story. However, except for a report[268] of an enigmatic side-chain phenylation of 2,5-dimethylfuran during room-temperature decomposition of *N*-nitrosoacetanilide in a mixture of 2,5-dimethylfuran and benzene, and additional confirmation[269] of the electron-accepting character of arenediazonium ions, there is no new mechanistic feature to report. Nevertheless, two comprehensive surveys of the historical development of theories relating to acylarylnitrosamine decomposition have been published by Cadogan;[270,271] his group has also published a series of papers[272] giving detailed evidence for radical intermediates (**163**)–(**165**), the most recent of which[273] gives a useful comprehensive mechanism (including radical and aryne intermediates[274]) accounting for most of the products so far reported.

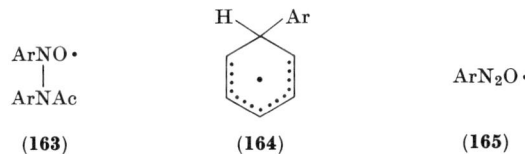

(**163**)   (**164**)   (**165**)

Homolytic substitution in aromatic and heteroaromatic compounds has also been reviewed.[275]

For some time it has been known that addition of small quantities of nitrobenzene to the benzoyl peroxide–benzene reaction enhances the yields of biphenyl and benzoic acid to greater than 0.8 mole per mole of peroxide, conferring considerable synthetic utility.[276] A full report[277] includes an attempted mechanistic rationalization of this "nitro-group effect" with essential features as shown in the annexed scheme. An im-

---

[265] Y. L. Chow, R. A. Perry, B. C. Menon, and S. C. Chen, *Tetrahedron Letters*, **1971**, 1545.
[266] Y. L. Chow, R. A. Perry, and B. C. Menon, *Tetrahedron Letters*, **1971**, 1549.
[267] See *Org. Reaction Mech.*, **1970**, 345; **1969**, 332.
[268] J. I. G. Cadogan, M. J. P. Harger, J. R. Mitchell, and J. T. Sharp, *Chem. Comm.*, **1971**, 1432.
[269] V. Habmann, C. Rüchardt, and C. C. Tan, *Tetrahedron Letters*, **1971**, 3885.
[270] J. I. G. Cadogan, *Accounts Chem. Res.*, **4**, 186 (1971).
[271] J. I. G. Cadogan, "Essays on Free-Radical Chemistry", *Chem. Soc. Special Publ.*, No. 24, 71 (1970).
[272] J. I. G. Cadogan, R. M. Paton, and C. Thomson, *J. Chem. Soc.* (B), **1971**, 583; J. I. G. Cadogan, J. Cook, M. J. P. Harger, P. G. Hibbert, and J. T. Sharp, *J. Chem. Soc.* (B), **1971**, 595; J. I. G. Cadogan, M. J. P. Harger, and J. T. Sharp, *J. Chem. Soc.* (B), **1971**, 602.
[273] D. L. Brydon, J. I. G. Cadogan, J. Cook, M. J. P. Harger, and J. T. Sharp, *J. Chem. Soc.* (B), **1971**, 1996.
[274] See also Chapter 6.
[275] H. J.-M. Dou, G. Vernin, and J. Metzger, *Bull. Soc. Chim. France*, **1971**, 4189.
[276] See *Org. Reaction Mech.*, **1965**, 156.
[277] G. R. Chalfont, D. H. Hey, K. S. Y. Liang, and M. J. Perkins, *J. Chem. Soc.* (B), **1971**, 233.

$(PhCO_2)_2 \rightarrow 2PhCO_2\cdot$

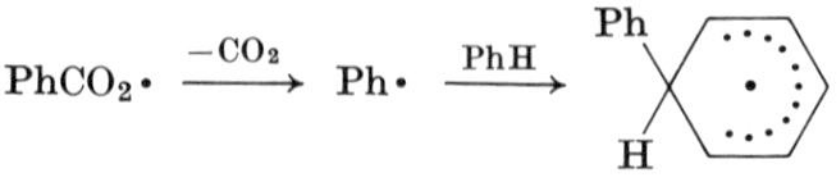

$PhCO_2\cdot \xrightarrow{-CO_2} Ph\cdot \xrightarrow{PhH}$ (166)

(166) is phenylcyclohexadienyl radical (Ph, H on sp3 carbon of cyclohexadienyl ring)

$PhNO_2 \rightarrow PhN{=}O \xrightarrow{Ph\cdot} Ph_2N{-}O\cdot$
(167)

$(166) + (167) \rightarrow Ph_2 + Ph_2NOH$

$Ph_2NOH + (PhCO_2)_2 \rightarrow Ph_2N{-}O\cdot + PhCO_2\cdot + PhCO_2H$

$PhCO_2\cdot + Ph_2NOH \rightarrow PhCO_2H + Ph_2N{-}O\cdot$

portant suggestion is that a small amount of the nitro-compound is reduced to nitrosobenzene, which scavenges phenyl radicals to form diphenylnitroxide (167). These nitroxide radicals oxidize phenylcyclohexadienyl radicals (166) efficiently to biphenyl, thus preventing the side-reactions that are thought to give products of high molecular weight. The nitroxide is regenerated by oxidation of diphenylhydroxylamine by molecular benzoyl peroxide.[278]

Reaction of benzoyl peroxide with *para*-disubstituted benzenes leads to substitution by phenyl and/or benzoyloxy radicals.[279] Product formation depends on the nature of the substituent groups and may be correlated with Hammett σ_p values although neither phenylation nor benzoyloxylation of the aromatic ring is observed in the reactions of benzoyl peroxide with sulphur-containing compounds. Total and partial rate factors for the phenylation of some of these *para*-disubstituted derivatives have been obtained.[280] Pentafluorophenylation of substituted benzenes has established the electrophilic character of the pentafluorophenyl radical.[281] Interesting CIDNP effects obtained during pyrolysis of bis(pentafluorobenzoyl) peroxide indicate that the related pentafluorobenzoate radical is also electrophilic.[282]

Photolysis of 2-iodo-*N*-methylbenzanilide (168) gives products arising from cyclization of the radical (170)[283] and are similar to those formed on copper-catalysed decomposition of the corresponding diazonium fluoroborate (169). The success of internuclear

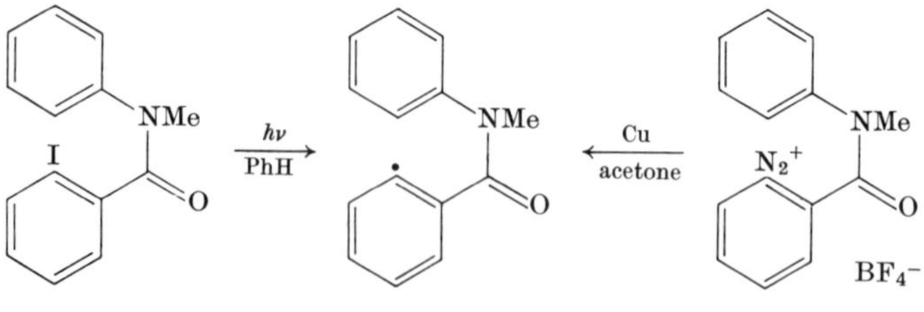

(168) → (170) via $h\nu$/PhH ; (169) → (170) via Cu/acetone

(168): 2-iodo-N-methylbenzanilide
(170): aryl radical
(169): diazonium fluoroborate, N_2^+ BF_4^-

[278] See also ref. 82.
[279] D. I. Davies, D. H. Hey, and B. Summers, *J. Chem. Soc.* (C), **1970**, 2653.
[280] D. I. Davies, D. H. Hey, and B. Summers, *J. Chem. Soc.* (C), **1971**, 2681.
[281] P. H. Oldham, G. H. Williams, and B. A. Wilson, *J. Chem. Soc.* (C), **1971**, 1094.
[282] J. Bargon, *J. Am. Chem. Soc.*, **93**, 4630 (1971).
[283] See *Org. Reaction Mech.*, **1970**, 346–347.

cyclization in these systems appears to depend on the conformational preference of the aryl groups for a *cis*-relationship.[284]

Phenyl radicals generated by photolysis of diphenylmercury react with anthracene at the 9,10-positions, giving 9-phenyl- and 9,10-diphenyl-anthracene.[285] These products are essentially the same as those obtained by generating phenyl radicals by the action of zinc powder on a suspension of benzenediazonium zinc chloride in dry acetone.[286] Photolysis of bis(trimethylsilyl)- and bis(trimethylgermyl)-mercury in benzene, toluene, and anisole gave rise to free-radical aromatic silylation and germylation; expected and unexpected (particularly in anisole) products were obtained and their origins have been discussed.[287] Hydrogen abstraction from the corresponding hydrosilanes has also been employed[288] in the study of homolytic aromatic substitution of silyl radicals with particular emphasis on the reactivity of the pentamethyldisilanyl radical (Me$_3$SiṠiMe$_2$); partial rate factors for pentamethyldisilanylation are correlated with Hammett σ constants resulting in a ρ value of $+1.4$; the total rate factors, relative to benzene, are 0.45 for anisole, 0.62 for toluene and 2.19 for benzotrifluoride. The relative nucleophilicity of the organosilyl radical is discussed in the light of these results and compared with that of other free radicals.

In a search[289] for intramolecular aryl migration in arylcyclohexadienyl radicals, 1-chloro-6-phenylcyclohexadienyl and 6,6-diphenylcyclohexadienyl (**172**) radicals have been generated, but the reactions of both radicals failed to give any evidence for intramolecular aryl migration. However, a major high-temperature reaction pathway of the 6,6-diphenylcyclohexadienyl radical is fragmentation, giving biphenyl and phenyl radicals, which react further with the precursor dimer (**171**) to produce benzene and

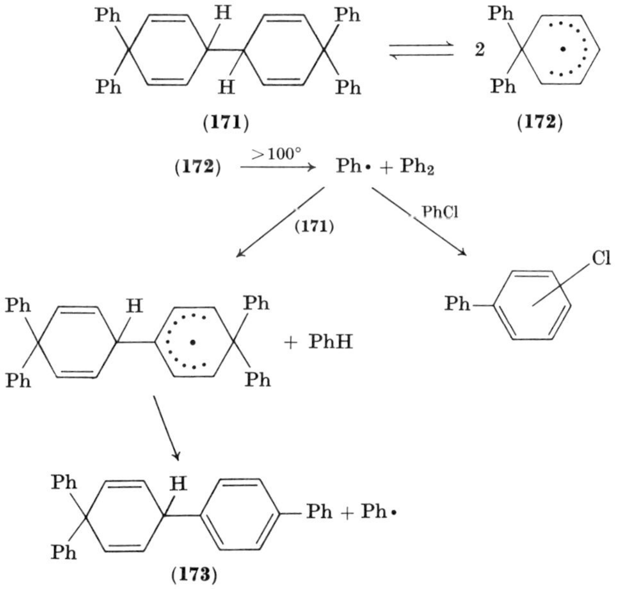

[284] D. H. Hey, G. H. Jones, and M. J. Perkins, *J. Chem. Soc.* (C), **1971**, 116.
[285] K. C. Bass and G. M. Taylor, *J. Chem. Soc.* (C), **1971**, 1.
[286] R. O. C. Norman and W. A. Waters, *J. Chem. Soc.*, **1958**, 167.
[287] S. W. Bennett, C. Eaborn, R. A. Jackson, and R. Pearce, *J. Organomet. Chem.*, **28**, 59 (1971).
[288] H. Sakurai and A. Hosomi, *J. Am. Chem. Soc.*, **93**, 1709 (1971).
[289] M. J. Perkins and P. Ward, *Tetrahedron Letters*, **1971**, 2379.

the new hydrocarbon (**173**). Similar products were obtained from a reaction in chlorobenzene at 210°, together with the three monochlorobiphenyls.[290]

Related work from Julia's group[291] on the pyrolysis of spirodienyl dimers supports the assumption that the spirodienyl radical (**174**) rearranges to both radicals (**175**) and (**176**), and similar conclusions are reached from a comparison of the lead tetraacetate oxidation of 5-(1-naphthyl)valeric acid and thermal decomposition of deuteriated 5-(1-naphthyl)valeryl peroxides.[292]

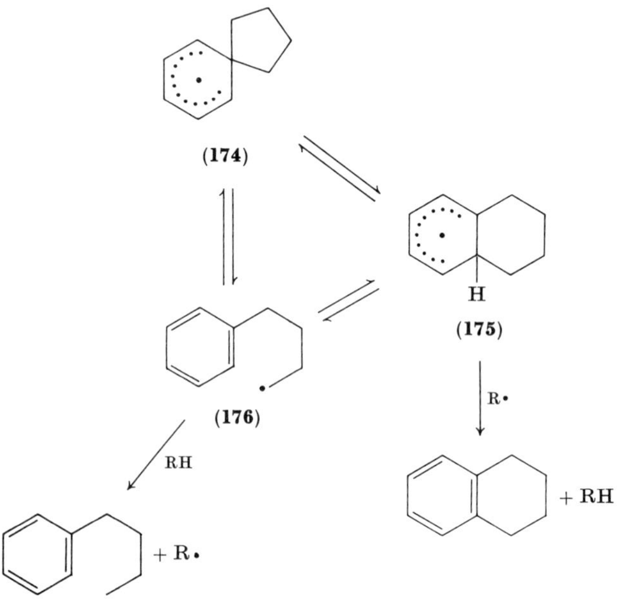

Cyano-radicals produced by photolysis of cyanogen iodide react with aromatic compounds, affording nitriles; isomer ratios and relative reactivities were determined for a number of monosubstituted benzenes, and the results indicate that the cyano radical preferably attacks relatively electronegative sites.[293]

Homolytic substitution in heteroaromatics continues to be widely studied. From competitive arylation the radical reactivity of quinoline has been determined and free valencies and localization energies for the different ring positions are shown to follow the order predicted by calculation.[294] Radical cyclohexylation in 4-methylpyridine occurs mostly at the 2- and 6-positions,[295] and similar results were noted for free-radical benzylation of 4-methyl- and 3,5-dimethyl-pyridine.[296] The distribution of isomers arising from radical methylation of some substituted heteroaromatics has been surveyed;[297] in the furan series an unusual displacement of the nitro group in (**177**) by methyl radicals was discovered.

[290] D. J. Atkinson, M. J. Perkins, and P. Ward, *J. Chem. Soc.* (C), **1971**, 3240.
[291] M. Julia and B. Malassine, *Tetrahedron Letters*, **1971**, 987.
[292] J. C. Chottard and M. Julia, *Tetrahedron Letters*, **1971**, 2561.
[293] P. Spagnolo, L. Testaferri, and M. Tiecco, *J. Chem. Soc.* (B), **1971**, 2006.
[294] G. Vernin, H. J.-M. Dou, and J. Metzger, *Bull. Soc. Chim. France*, **1971**, 2612.
[295] H. J.-M. Dou, G. Vernin, and J. Metzger, *Bull. Soc. Chim. France*, **1971**, 3553.
[296] H. J.-M. Dou, G. Vernin, M. Dufour, and J. Metzger, *Bull. Soc. Chim. France*, **1971**, 111.
[297] U. Rudqvist and K. Torssell, *Acta Chem. Scand.*, **25**, 2183 (1971).

Me• + [structure (177): O₂N-furan-CO₂Me] ⟶ [O₂N, Me, furan•, CO₂Me] $\xrightarrow{-NO_2}$ [Me-furan-CO₂Me]

(177)   15%

Arylation of substituted indoles and acridines has been reported,[298] and phenylation of pyridine and its N-oxide by electrolytic reduction of benzenediazonium tetrafluoroborate has enabled partial rate factors and total rate ratios to be estimated for pyridine N-oxide.[299]

Minisci and his co-workers have employed a number of new, selective methods for homolytic substitution of heteroaromatic bases.[300] As shown originally by Kochi,[301] the silver(I)-catalysed oxidative decarboxylation of acids by peroxydisulphate produces alkyl radicals; the reaction is now shown to be particularly useful as it permits the introduction of a large variety of primary, secondary and tertiary alkyl groups into heteroaromatic bases with good yield and high selectivity. Similar substitution results were obtained[302] with nucleophilic α-oxyalkyl radicals ($-\text{O}-\overset{|}{\underset{|}{\text{C}}}\cdot$) generated by oxidation of alcohols and ethers with a wide variety of electrophilic radicals. Both the radicals (**178**) and (**179**) attack protonated heteroaromatic bases selectively, in positions with high nucleophilic reactivity, to form the corresponding substitution products.[303] Oxidation of dimethylformamide with the sulphate radical anion generates (**179**) almost exclusively, in marked contrast to the *tert*-butoxy radical which gives rise to (**178**); an electron transfer rather than direct hydrogen abstraction is suggested to account for

        Me₂N—Ċ=O $\overset{\text{H}_2\dot{\text{C}}}{\underset{\text{Me}}{\diagdown}}$N—CH=O

         (**178**)   (**179**)

$$\text{Me}_2\text{NCHO} + \text{SO}_4^{-\cdot} \rightarrow \text{Me}_2\overset{+\cdot}{\text{N}}\text{CHO} + \text{SO}_4^{2-}$$
$$\downarrow$$
$$(\textbf{179}) + \text{H}^+$$

this difference.[303] Further studies of radical substitution of heterocyclic compounds by Minisci's group have demonstrated the nucleophilic character of methyl[304] and other simple alkyl radicals.[305]

An unusual reaction[306] involving decomposition of benzoyl peroxide in a benzene solution of 2,5-dimethylfuran results in almost exclusive (> 97%) reaction with the

---

[298] A. K. Sheinkman, V. A. Ivanov, and S. N. Baranov, *Dopov. Akad. Nauk Ukr. RSR, Ser. B*, **32**, 619 (1970); *Chem. Abs.*, **74**, 76298 (1971).
[299] R. M. Elofson, F. F. Gadallah, and K. F. Schulz, *J. Org. Chem.*, **36**, 1526 (1971).
[300] F. Minisci, R. Bernardi, F. Bertini, R. Galli, and M. Perchinummo, *Tetrahedron*, **27**, 3575 (1971).
[301] See *Org. Reaction Mech.*, **1970**, 358.
[302] W. Buratti, G. P. Gardini, F. Minisci, F. Bertini, R. Galli, and M. Perchinummo, *Tetrahedron*, **27**, 3655 (1971).
[303] G. P. Gardini, F. Minisci, G. Palla, A. Arnone, and R. Galli, *Tetrahedron Letters*, **1971**, 59.
[304] G. P. Gardini, F. Minisci, and G. Palla, *Chem. Ind. (Milan)*, **53**, 263 (1971).
[305] G. P. Gardini and F. Minisci, *Ann. Chim. (Rome)*, **60**, 746 (1970); *Chem. Abs.*, **74**, 87128 (1971).
[306] J. I. G. Cadogan, J. R. Mitchell, and J. T. Sharp, *Chem. Comm.*, **1971**, 1433.

furan, giving benzoic acid and 2-benzoyloxymethyl-5-methylfuran (**180**). This remarkably efficient scavenging reaction can be rationalized in terms of an induced decomposition of the peroxide following fast hydrogen abstraction by an initiating benzoyloxy

$$\text{Me}\underset{O}{\diagup\!\!\!\diagup}\text{Me} \xrightarrow{\text{PhCO}_2\cdot} \text{Me}\underset{O}{\diagup\!\!\!\diagup}\text{CH}_2\cdot$$

$$\downarrow (\text{PhCO}_2)_2$$

$$\text{Me}\underset{O}{\diagup\!\!\!\diagup}\text{CH}_2\text{OCOPh} + \text{PhCO}_2\cdot$$

(**180**)

radical as shown. The reaction is markedly different from that in furan which gives *cis*- and *trans*-2,5-dibenzoyloxy-2,5-dihydrofuran in high yields.[307]

Further evidence[308] substantiating the reversible addition of benzoyloxy radicals to benzene[309] is obtained from photolysis of benzoyl peroxide in the presence of oxygen in 1:1 mixtures of C_6H_6 and C_6D_6. Reversible addition of isopropoxycarbonyloxy radicals ($Pr^iOCOO\cdot$) to benzene is suggested by similar experimental results.[310]

Relative rate constants for the reaction of aqueous tritium atoms with a series of aromatic compounds ($ArH + T\cdot \rightarrow ArHT\cdot$) have been deduced from the effectiveness of the cupric ion-inhibition of β-radiation-induced aromatic tritium exchange in degassed aqueous solutions.[311] It is considered that the values obtained represent the relative reactivities of the aromatic compounds towards the sterically least demanding free radical, the hydrogen atom. Interestingly, the reactivity ratios found are numerically similar to results previously reported for free-radical phenylation of the corresponding compounds. From the same workers comes the finding[312] that, in addition to the replacement of H by T, replacement of halogen atoms by aqueous tritium atoms also occurs ($ArX \rightarrow ArT$). The two reactions, H-replacement and X-replacement, are in competition, and their relative rates were determined by GLC separation of the products and tritium assay. Both reactions are reduced in extent by addition of copper(II) sulphate, although dehalogenation is affected considerably more than H-exchange. The following mechanism accounts for all the essential observations. Isotopic exchange and dehalogenation are initiated by attack at different positions of the aromatic ring by tritium atoms to give radicals (**181**) and (**182**). Oxidation of these radicals by copper(II) ions leads to the corresponding cations (**183**) and (**184**), the former preferentially losing a triton in aqueous solution to regenerate inactive bromobenzene. However with few copper ions present halogen loss from radical (**181**) occurs, leading to significant amounts of tritium-labelled benzene.

Radical cyclohexylation of nitrobenzene is complicated owing to side reactions at the nitro group,[313] but corresponding reactions with alkyl-substituted thiazoles further

---

[307] See *Org. Reaction Mech.*, **1969**, 335–336.
[308] J. Saltiel and H. C. Curtis, *J. Am. Chem. Soc.*, **93**, 2056 (1971).
[309] See *Org. Reaction Mech.*, **1969**, 336.
[310] T. Nakata, K. Tokumaru, and O. Simamura, *Bull. Chem. Soc. Japan*, **43**, 3590 (1970).
[311] C. L. Brett and V. Gold, *Chem. Comm.*, **1971**, 148.
[312] C. L. Brett and V. Gold, *Chem. Comm.*, **1971**, 1426.
[313] M. Baule, G. Vernin, H. J.-M. Dou, and J. Metzger, *Bull. Soc. Chim. France*, **1971**, 2083.

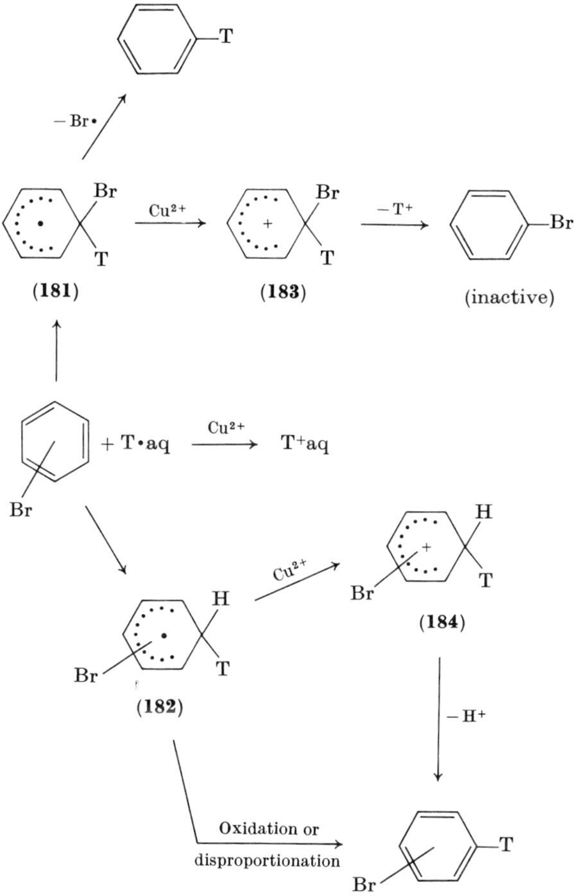

demonstrate the nucleophilic character of the cyclohexyl radical.[314] Aromatic *meta*-nitration,[315] aromatic fluorination,[316] and methylation of pyridine and its monomethyl derivatives[317] have been achieved by radical pathways. Benzothiazole (**185**) is acylated selectively in the 2-position, a result that may be used for the detection of acyl radicals in the oxidation of aldehydes.[318] Hydroxylation of thyropropionic acid (**186**; R = $CH_2CH_2CO_2H$) serves as a non-enzymic model for the metabolism of thyronines.[319]

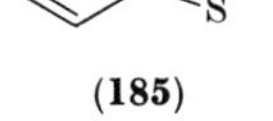

(**185**)

(**186**)

[314] See *Org. Reaction Mech.*, **1969**, 338.
[315] T. Suehiro, M. Hirai, and T. Kaneko, *Bull. Chem. Soc. Japan.*, **44**, 1402 (1971).
[316] N. B. Kaz'mina, L. S. German, I. D. Rubin, and I. L. Knunyants, *Dokl. Akad. Nauk SSSR*, **194**, 1329 (1970); *Chem. Abs.*, **74**, 124631 (1971).
[317] H. J.-M. Dou, G. Vernin, and J. Metzger, *Bull. Soc. Chim. France*, **1971**, 1021.
[318] T. Caronna, R. Galli, V. Malatesta, and F. Minisci, *J. Chem. Soc.* (C), **1971**, 1747.
[319] T. Matsuura, T. Nagamachi, and A. Nishinaga, *J. Org. Chem.*, **36**, 2016 (1971).

Hydroxyl-radical oxidation of pentaamminebenzoatocobalt(III) proceeds by addition to the aromatic ring.[320]

Rearrangements

The first example of a radical to undergo rearrangement by aryl migration from silicon to carbon was discussed last year;[321] the reverse migration has now been observed.[322] Thus radical (**187a**) gave rise to both the rearranged (**191a**) and cyclization (**190a**) products. Similar results were obtained with radicals (**187b**) and (**187c**) and, from studies of related compounds having the general formula $PhCHY(CH_2)_nSiMe_2H$, the mechanism outlined was adduced; factors governing the conversion of the spirocyclohexadienyl radical (**188**) into (**189**) are discussed by the Japanese workers.[322]

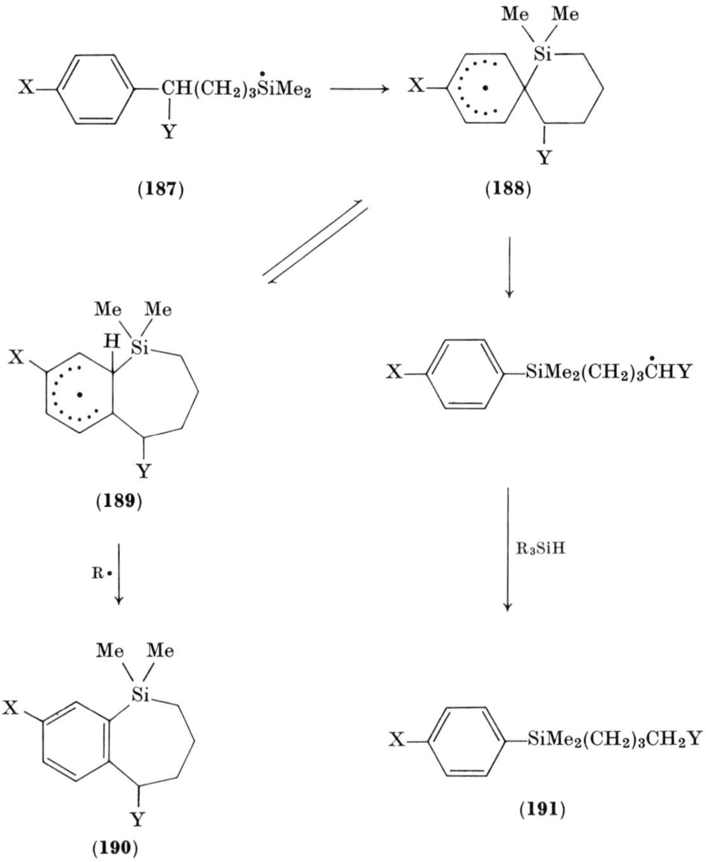

a: X = Y = H;   b: X = Me, Y = H;   c: X = H, Y = Me

Peroxide-induced decarbonylation of (**192**; X = H) gives a radical that undergoes 100% rearrangement with exclusive migration of the phenylene ring.[323] This system

[320] H. Cohen and D. Meyerstein, *J. Am. Chem. Soc.*, **93**, 4179 (1971).
[321] See *Org. Reaction Mech.*, **1970**, 351.
[322] H. Sakurai and A. Hosomi, *J. Am. Chem. Soc.*, **92**, 7507 (1970).
[323] B. M. Vittimberga, *Tetrahedron Letters*, **1965**, 2383.

thus presents an opportunity to study the effects of various substituted phenyl groups at the migration origin of free-radical rearrangement reactions and avoids the possibility that such a group would compete in the rearrangement. Radical-induced decarbonylation of the series of aldehydes (**192**) occurred, under standard conditions, at approximately the same rates and proceeded to the same extent;[324] it is concluded therefore that

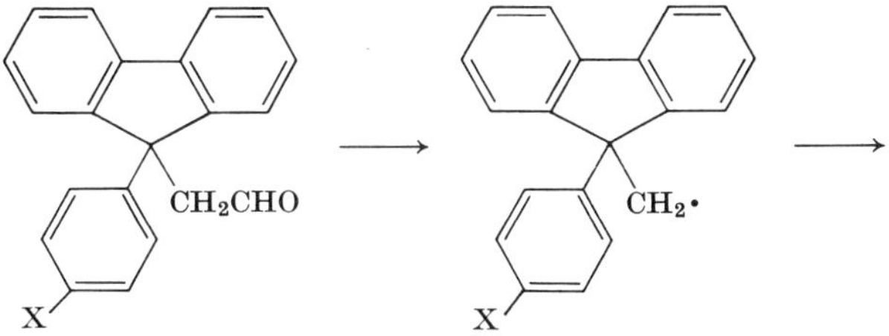

X = H, OMe, Me or Cl

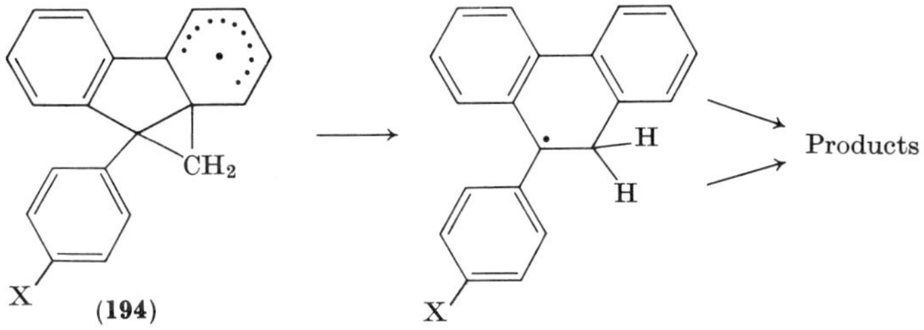

they decarbonylate by the same mechanism. In related work,[325] it was concluded that the rearrangement transition state was a semipolar hybrid with partial positive character at the migration origin. However, a careful analysis of the ratios of rearranged to unrearranged products, formed in the presence of toluene-α-thiol as a scavenger for radicals (**193**) and (**195**), leads to the conclusion[324] that the transition state for a free-radical aryl migration is purely radical in character with contributions from structures such as (**194**).

A stereoelectronic requirement in a radical fragmentation reaction is demonstrated[326] by the specific fission of the 4,5-bond in the 3β,5-cyclocholestan-6-yl radical: (**197**) → (**199**). Fragmentation of the 3,5-bond to give the more stable cholesteryl radical (**198**) did not occur. On the other hand, the structurally similar radical (**196**) is known to undergo specific fragmentation to the more stable radical. Thus, as in intramolecular radical cyclizations (the reverse of ring fragmentation), the nature and relative yields of products in rigid systems, or systems showing a strong conformational preference, are determined primarily by the stereochemistry of the molecule and not by the relative stabilities of possible intermediate radicals.

[324] P. N. Cote and B. M. Vittimberga, *J. Am. Chem. Soc.*, **93**, 276 (1971).
[325] C. Rüchardt and R. Hecht, *Chem. Ber.*, **98**, 2471 (1965).
[326] A. L. J. Beckwith and G. Phillipou, *Chem. Comm.*, **1971**, 658.

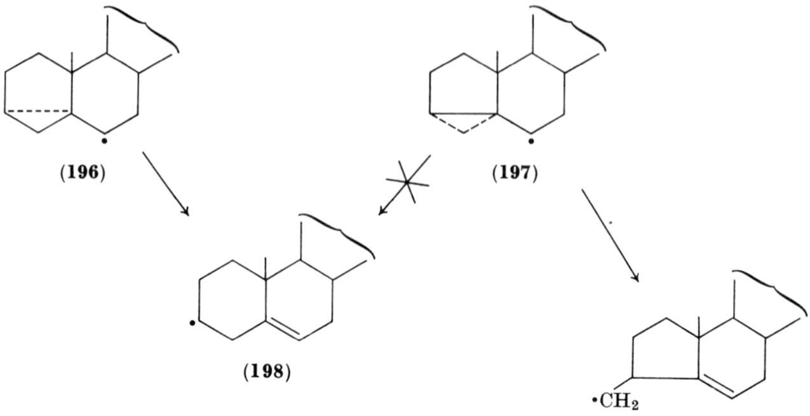

(196) (197) (198) (199) •CH₂

Free-radical additions of carbon tetrahalides to olefins (**200**) gives both unrearranged and rearranged products; the latter increase with increasing reaction temperature and decreasing concentration of chain-transfer agent.[327] As the stereochemistry of rearrangement is also dependent upon the nature and concentration of chain-transfer reagent it is concluded that radicals (**202**) and (**203**) are formed reversibly from the initial radical (**201**).

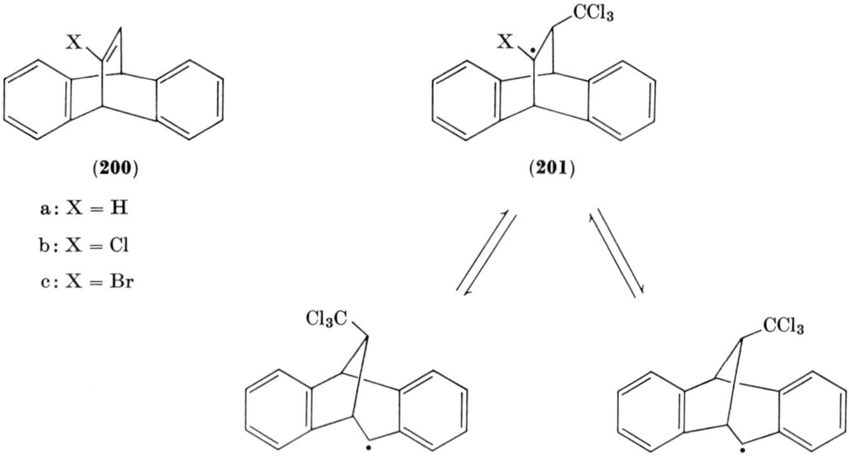

(200)
a: X = H
b: X = Cl
c: X = Br

(201) (202) (203)

Rearrangement of 1-methyl-2,2-diphenylcyclopropyl radicals (**4**) has already been discussed;[328] studies of the related radicals *trans*-2,3-diphenylcyclopropyl and 2-phenylcyclopropyl led to the conclusion[329] that formation of highly resonance-stabilized allylic radicals is the main factor in lowering the activation energy in rearrangement of phenyl-substituted cyclopropyl radicals.

[327] B. B. Jarvis, J. P. Govoni, and P. J. Zell, *J. Am. Chem. Soc.*, **93**, 913 (1971).
[328] See report on ref. 33.
[329] J. C. Chen, *Tetrahedron Letters*, **1971**, 3669.

Benzophenone-photosensitized conversion of cyclic acetals (**204**) to lactones (**206**) and esters (**207**) occurs through radical (**205**) which can undergo either *exo-* or *endo-* β-cleavage.[330] Relatively higher yields of the lactone may result from the enhanced

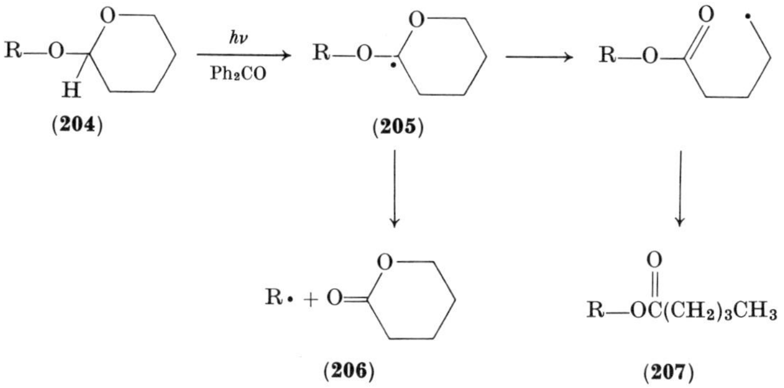

resonance stability of radical R·. Free-radical rearrangements via cyclopropylcarbinyl radicals have been observed during reduction of bullvalene bromides by tri-*n*-butyltin hydride.[331] Sodium naphthalene reduction of the organomercurial (**208**) has been shown to result in complete retention of the carbon skeleton when the reduction is carried out

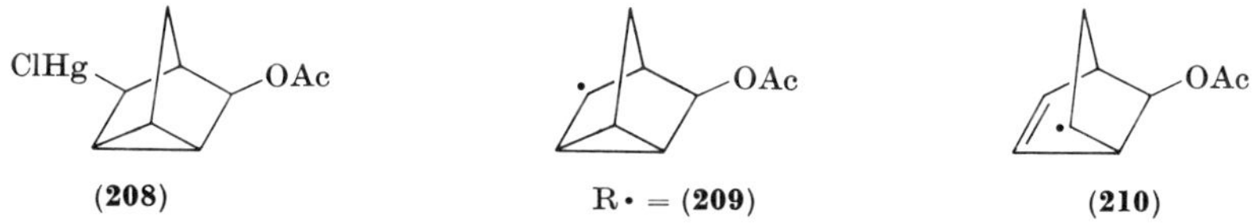

with radical anion in excess and at low temperatures.[332] Thus, under these conditions rearrangement of radical (**209**) to (**210**) is suppressed by the competitive electron-transfer process:

$$R\cdot + C_{10}H_8^{-} \rightarrow R:^- + C_{10}H_8 \quad [R\cdot = (\mathbf{209})]$$

From studies of the aromatization of the methylenecyclohexadiene (**211**), it is concluded[333] that the major rearrangement is a free-radical chain process. When the terminal methylene group was labelled with deuterium, complete equilibration of the ends of the

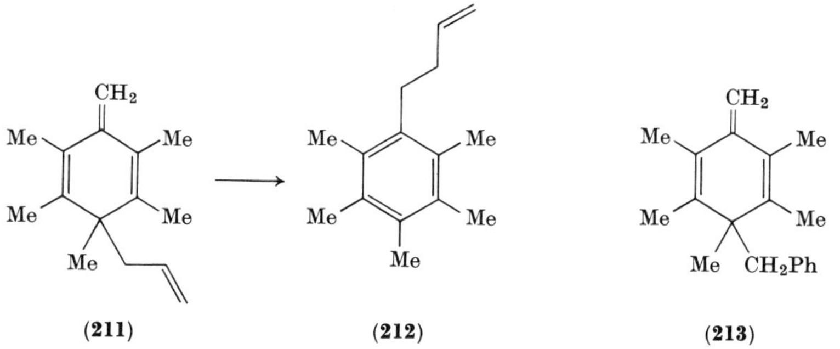

---

[330] T. Yamagishi, T. Yoshimoto, and K. Minami, *Tetrahedron Letters*, **1971**, 2795.
[331] H.-P. Löffler, *Chem. Ber.*, **104**, 1981 (1971).
[332] T. C. Morrill and F. L. Vandemark, *Tetrahedron Letters*, **1971**, 1811.
[333] B. Miller and K.-H. Lai, *Tetrahedron Letters*, **1971**, 1617.

allyl group in (**212**) was observed. A radical chain involving allyl radicals as the chain carriers is the only mechanism that fits the data. A similar mechanism probably accounts for the aromatization of the 4-benzyl derivative (**213**).

Acetoxy group migration in the rearrangement of radical (**214**) to (**215**) has been detected during an ESR study of possible rearrangements in a number of cyclic and acyclic radicals.[334] A novel photochemical rearrangement–elimination of an allylic alcohol having a di-π-methane structure has been observed.[335]

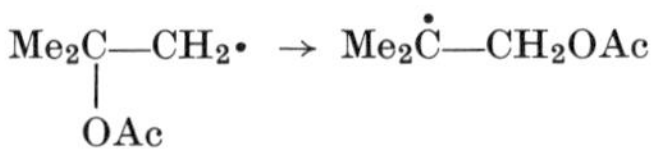

$$\underset{(214)}{\underset{\mathrm{OAc}}{\mathrm{Me_2\overset{|}{C}-CH_2\cdot}}} \rightarrow \underset{(215)}{\mathrm{Me_2\overset{\bullet}{C}-CH_2OAc}}$$

A rearrangement type which may prove to be of some generality has been detected during ESR studies on the oxidation of enols and enol ethers with the hydroxyl radical.[336] Adducts from enols and the ·OH radical (e.g. **216**) undergo acid-catalysed elimination of water to give the radical (**217**). Similarly the two adducts formed by reaction of the

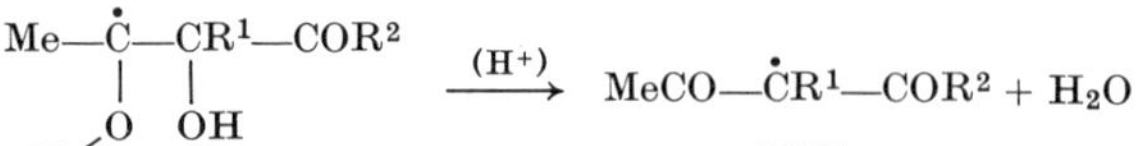

$$\underset{(216)}{\mathrm{Me-\overset{\bullet}{\underset{H\diagup O}{C}}-\overset{|}{\underset{OH}{CR^1}}-COR^2}} \xrightarrow{(H^+)} \underset{(217)}{\mathrm{MeCO-\overset{\bullet}{C}R^1-COR^2}} + H_2O$$

$$\underset{(218)}{\mathrm{R\overset{\cdot\cdot}{O}-\overset{\bullet}{C}H-CH_2-OH}} + H^+ \underset{}{\overset{-H_2O}{\rightleftharpoons}} \mathrm{R\overset{+}{O}=CH-\overset{\bullet}{C}H_2} \underset{}{\overset{H_2O}{\rightleftharpoons}} \underset{(219)}{\underset{\underset{+H^+}{\mathrm{OH}}}{\mathrm{RO-CH-\overset{\bullet}{C}H_2}}}$$

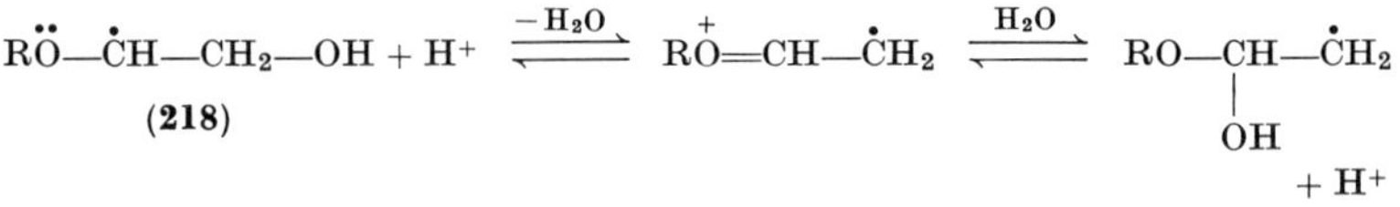

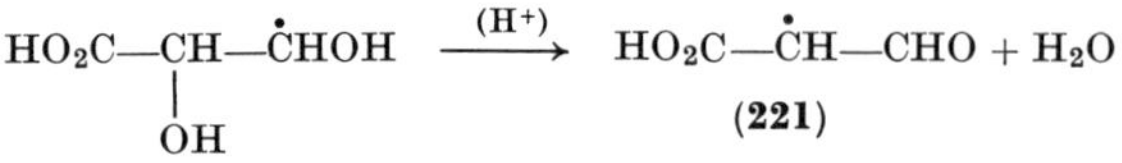

$$\underset{(220)}{\underset{\mathrm{OH}}{\mathrm{HO_2C-CH-\overset{\bullet}{C}HOH}}} \xrightarrow{(H^+)} \underset{(221)}{\mathrm{HO_2C-\overset{\bullet}{C}H-CHO}} + H_2O$$

·OH radical with enol ethers undergo acid-catalysed interconversion by way of an oxonium ion. Support for the above mechanism comes from oxidation of 2-methoxyethanol with the hydroxyl radical. Radical (**218**; R = Me) is produced by hydrogen abstraction but rearranges to (**219**; R = Me) when the pH is lowered below 2.5. Similar observations during photolysis of aqueous tartaric acid solutions have been made,[337] radical (**220**) being converted into (**221**) by acid-catalysis.

Last year's review summarized numerous reports that the Stevens (and related) rearrangements proceed by a radical-pair mechanism.[338] Rearrangement studies of the

[334] A. L. J. Beckwith and P. K. Tindal, *Austral. J. Chem.*, **24**, 2099 (1971).
[335] W. G. Dauben, W. A. Spitzer, and R. M. Boden, *J. Org. Chem.*, **36**, 2384 (1971).
[336] D. J. Edge, B. C. Gilbert, R. O. C. Norman, and P. R. West, *J. Chem. Soc.* (B), **1971**, 189.
[337] R. Livingston and H. Zeldes, *J. Chem. Phys.*, **53**, 1406 (1970).
[338] See *Org. Reaction Mech.*, **1970**, 354–356.

aminimides (**222**) and (**223**) also lead to the conclusion that products (**224**) and (**225**) are formed by an entirely radical pathway.[339] Radical intermediates are thought to be

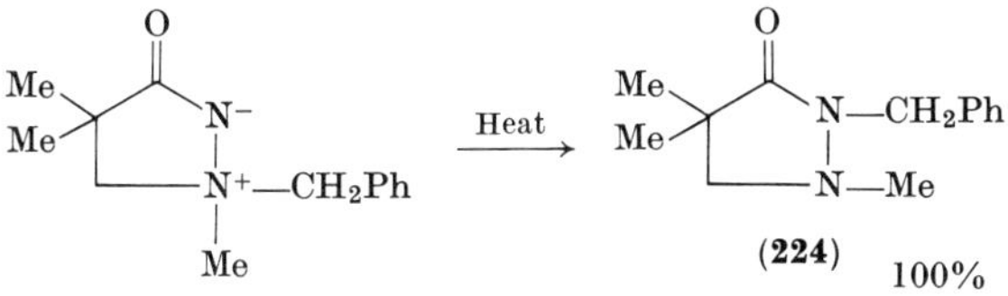

involved in the Stevens rearrangement of α-arylneopentylammonium salts in non-polar solvents.[340] CIDNP effects in the formation of oxime ether (**227**) during the thermolysis of (**226**) are, it is suggested,[341] due to a caged benzhydryl-iminoxyl radical pair, and ESR measurements taken before, during and after the thermal rearrangement of the

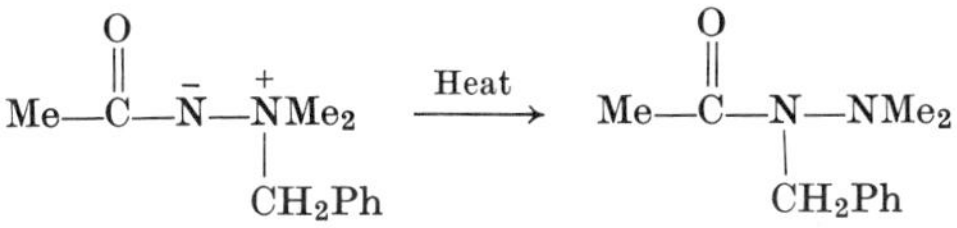

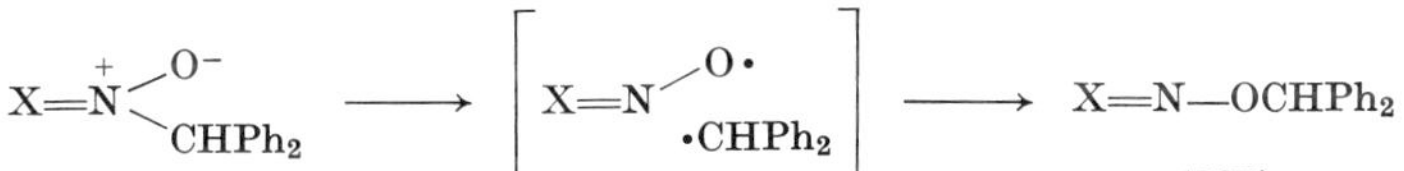

X = Fluoren-9-ylidene

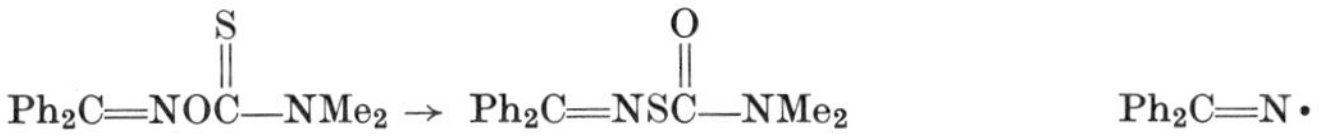

oxime thionocarbamate (**228**) to (**229**) showed the appearance and subsequent disappearance of a signal consistent with radical (**230**).[342]

It is known that migration of nitric oxide in the Barton reaction is intermolecular, and recent work[343] indicate a similar conclusion for photochemical formation of cyclic nitrones from nitrites of fused (steroidal) 5-membered ring alcohols. The same workers report a full study of this novel production of cyclic nitrones.[344] Synthesis of A-nor-B-homo-steroids from nitrite esters of some 11α-hydroxy-steroidal 4-en-3-ones has been achieved by conversion of a C-1 into a C-4 radical.[345]

---

[339] H. P. Benecke and J. H. Wikel, *Tetrahedron Letters*, **1971**, 3479.
[340] S. H. Pine, E. M. Munemo, T. R. Phillips, G. Bartolini, W. D. Cotton, and G. C. Andrews, *J. Org. Chem.*, **36**, 984 (1971).
[341] D. G. Morris, *Chem. Comm.*, **1971**, 221.
[342] R. F. Hudson, A. J. Lawson, and E. A. C. Lucken, *Chem. Comm.*, **1971**, 807; see also B. Cross, R. J. G. Searle, and R. E. Woodall, *J. Chem. Soc.* (C), **1971**, 1833.
[343] H. Suginome, T. Mizuguchi, and T. Masamune, *Tetrahedron Letters*, **1971**, 4723.
[344] H. Suginome, N. Sato, and T. Masamune, *Tetrahedron*, **27**, 4863 (1971).
[345] H. Reimann and O. Z. Sarre, *Can. J. Chem.* **49**, 344 (1971).

Re-investigation[346] of the mechanism of 1-phenylcyclobutene formation in the reaction of 4-bromo-1-chloro-1-phenylbut-1-ene (231; X = Br) with magnesium has shown previous proposals to be unacceptable. Amongst several possibilities considered, the radical cyclization mechanism shown here seems the most likely. Product (232) results

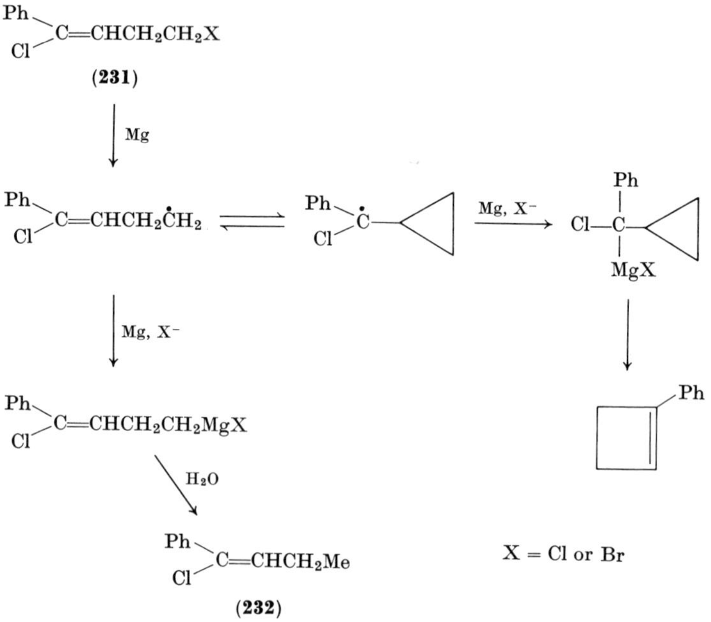

Ph\
   C=CHCH$_2$CH$_2$X\
Cl/

(231)

↓ Mg

Ph\C=CHCH$_2$ĊH$_2$ ⇌ Ph\Ċ—△ → Ph—C(Cl)(MgX)—△
Cl/   Cl/

↓ Mg, X⁻

Ph\C=CHCH$_2$CH$_2$MgX
Cl/

↘ H$_2$O

Ph\C=CHCH$_2$Me   X = Cl or Br
Cl/

(232)

from hydrolysis of the corresponding Grignard reagent which is shown not to be an intermediate in the formation of 1-phenylcyclobutene, as was previously suggested.

Homolytic rearrangement of 2,2,2-trichloroethyl and 1,2,2,2-tetrachlorethyl into 1,1,2-trichloroethyl and 1,1,2,2-tetrachloroethyl radicals, respectively, has been reported.[347]

S_H2 Reactions

It is now apparent that bimolecular homolytic substitution (S_H2) reactions occur very commonly at multivalent metallic centres as well as at univalent hydrogen or halogen atoms.[348] Most of the examples that have been recognized so far lie in the field of organo-derivatives of non-transition metals (M) where the displaced group is an alkyl radical (R·).[349]

$$X\cdot + MR_n \rightarrow XMR_{n-1} + R\cdot$$

Quantitative data are now available for the specific rates of several of these reactions and it is clear that they are often very much greater than those of equivalent reactions at hydrogen or halogen. This has important consequences for the interpretation of

[346] E. A. Hill and M. R. Engel, *J. Org. Chem.*, **36**, 1356 (1971).
[347] U. A. Ol'derkop and R. V. Kaberdin, *Zhur. Org. Khim.*, **6**, 1114 (1970).
[348] Attention is drawn to ref. 3 for a review of this topic in free-radical chemistry.
[349] See *Org. Reaction Mech.*, **1970**, 311–313.

several familiar organometallic reactions and to the prediction of new ones;[350] the involvement of boron has received most attention.

The kinetics of autoxidation of a series of n-, sec-, and $tert$-butylboron compounds in iso-octane have been investigated at 30°. In every case[351] the kinetics were compatible with a homolytic chain reaction involving the following steps:

Initiation $\quad R_3B + O_2 \longrightarrow R\cdot + R_2BO_2\cdot$

Propagation $\quad R\cdot + O_2 \longrightarrow ROO\cdot$

$\quad\quad\quad\quad\quad ROO\cdot + {>}BR \xrightarrow{k_p} ROOB{<} + R\cdot$

Termination $\quad 2ROO\cdot \longrightarrow$ Inactive products

Values for the "oxidizability" of different boranes at the same rates of initiation, and of the rate constants (k_p) for the bimolecular homolytic substitution by the alkylperoxy radicals at the boron centre have been listed. In certain cases the latter processes are considerably faster than the equivalent reaction at hydrogen, which is the corresponding propagation step in the autoxidation of a hydrocarbon. Increased crowding around the boron atom has a marked effect on the rate of initiation.[351,352]

The very rapid reaction of trialkylboranes with oxygen is strongly inhibited by the presence of small amounts of elemental iodine.[353] Indeed 0.2M-iodine prevents the uptake of oxygen by 0.5M-solutions of representative organoboranes over periods as long as several days. It is suggested that the oxidation must involve a relatively slow radical formation, followed by a very fast chain propagation. Thus alkyl radicals resulting from the initiation step are competitively scavenged by iodine, producing iodine atoms which are incapable of propagating the chain, i.e.:

$$R_3B + O_2 \to R\cdot + R_2BO_2\cdot$$

$$R\cdot + O_2 \to RO_2\cdot \to \text{Chain}$$

$$R\cdot + I_2 \to RI + I\cdot \to \text{No chain}$$

and

$$R_2BO_2\cdot + I_2 \to R_2BI + O_2 + I\cdot$$

However, an alternative mechanism that includes a fast autocatalytic initiation reaction has been suggested[354] to explain the inhibiting effect of iodine on the oxidation of triethylborane.

Typical organic iodides did not alter the rate of reaction of trialkylboranes with oxygen,[355] although the products are very different and suggest that a chain transfer occurs during the free-radical oxidation, i.e.:

$$R^1{}_3B + O_2 \to R^1{}_2BO_2\cdot + R^1\cdot$$

$$R^1\cdot + R^2I \to R^1I + R^2\cdot$$

$$2R^2\cdot \to R^2\text{—}R^2$$

---

[350] A. G. Davies and B. P. Roberts, *Nature (London), Phys. Sci.*, **229**, 221 (1971).
[351] A. G. Davies, K. U. Ingold, B. P. Roberts, and R. Tudor, *J. Chem. Soc. (B)*, **1971**, 698.
[352] H. C. Brown and M. M. Midland, *Chem. Comm.*, **1971**, 699.
[353] M. M. Midland and H. C. Brown, *J. Am. Chem. Soc.*, **93**, 1506 (1971).
[354] J. Grotewold, J. Hernandez, and E. A. Lissi, *J. Chem. Soc. (B)*, **1971**, 182.
[355] A. Suzuki, S. Nozawa, M. Harada, M. Itoh, H. C. Brown, and M. M. Midland *J. Am. Chem. Soc.*, **93**, 1508 (1971).

Thus products R^1I and dimer R^2–R^2 may be obtained by suitable choice of reaction conditions. Indeed, benzylic and allylic iodides are readily coupled in excellent yields by this technique.

Alkylthiyl radicals (RS·) also readily undergo S_H2 reactions. A full report describes the reaction of trialkylboranes with butane-1-thiol which gives dialkylbutylthioboranes, R_2BSBu^n by a free-radical chain mechanism involving bimolecular homolytic substitution by the butylthiyl radical at boron.[356] Competition experiments involving oct-1-ene permitted estimates of the relative rates of reaction of butylthiyl radical in S_H2 attack on boron, and addition to the olefin. Occurrence of the S_H2 process was confirmed by the observation of the ESR spectrum of the displaced alkyl radical on photolysis of a dialkyl or a diaryl disulphide in the presence of the trialkylborane.[356] Absolute rate coefficients were of the same order of magnitude (10^5–10^7 mole^{-1} sec^{-1}) as those found for the S_H2 reactions of alkylperoxy and alkoxy radicals with organoboranes. Synthetic applications of the radical chain reactions of trialkylboranes with organic disulphides have also been established.[357]

ESR has also been employed in studies of homolytic bimolecular substitution at boron by alkoxy radicals and the triplet state of ketones. Rate constants for the attack of *tert*-butoxy radicals on organoboranes can be estimated by monitoring the relative concentration of displaced alkyl radicals that are observed.[358a] The technique is equally applicable to many other organometallic compounds; reactions are very fast and are characterized by low (0–5 kcal mole^{-1}) activation energies. Triplet states of ketones behave in many ways like alkoxy radicals and it has now been established that they also react rapidly by an S_H2 process at the metallic centre of organometallic compounds such as organoboranes. Thus the photochemical reaction between ketones and trialkylboranes in the absence of a quencher is described by the following reactions:[358b]

$$R^1R^2CO \xrightarrow{h\nu} R^1R^2CO^*$$

$$R^1R^2CO^* + BR_3^3 \rightarrow R^1R^2\overset{\cdot}{C}OBR_2^3 + R^3\cdot$$
$$(233)$$

$$2(233) \rightarrow (R^1R^2COBR_2^3)_2$$

$$R^3\cdot + (233) \rightarrow R^1R^2R^3COBR_2^3 + \text{disproportionation products}$$

$$2R^3\cdot \rightarrow R^3\text{—}R^3$$

Excitation to the triplet state precedes an S_H2 displacement at the boron centre to give the radicals (**233**) and R^3·, which are observed and identified by ESR. The ultimate products of the reaction result from combination processes between these radicals as shown in the scheme. Rate constants for these reactions are very similar to those for the equivalent S_H2 reaction of the *tert*-butoxy radical.

The same group of workers have also demonstrated the intermediacy of the phosphoranyl radical during the oxidation of triethyl phosphite by *tert*-butoxy radicals:[359]

$$Bu^tO\cdot + P(OEt)_3 \rightleftharpoons Bu^tO\overset{\cdot}{P}(OEt)_3 \rightarrow Bu^t\cdot + OP(OEt)_3$$

---

[356] A. G. Davies and B. P. Roberts, *J. Chem. Soc.* (B), **1971**, 1830.
[357] H. C. Brown and M. M. Midland, *J. Am. Chem. Soc.*, **93**, 3291 (1971).
[358a] A. G. Davies, D. Griller, and B. P. Roberts, *J. Chem. Soc.* (B), **1971**, 1823.
[358b] A. G. Davies, D. Griller, B. P. Roberts, and J. C. Scaiano, *Chem. Comm.*, **1971**, 196; A. G. Davies, B. P. Roberts, and J. C. Scaiano, *J. Chem. Soc.* (B), **1971**, 2171.
[359] A. G. Davies, D. Griller, and B. P. Roberts, *Angew. Chem. Int. Ed.*, **10**, 738 (1971).

From quantitative measurements of the phosphorus-centred radical ($a_p = 891$ G, $g = 2.008$) and the liberated *tert*-butyl radical, Arrhenius parameters for the overall reaction and breakdown of the phosphoranyl radical were estimated.

A novel and useful S_H2-type process has come to light during investigation of the reaction of bromine with tri-*n*-butylborane.[360] Results are incompatible with a mechanism involving the direct rupture of the carbon−boron bond by bromine, but are accounted for by a free-radical chain bromination of the α-position of the organoborane (reactions 6 and 7). Initiation involves slow bromine attack on boron (reaction 5) and the hydrogen bromide produced in the substitution stage can then react preferentially to give the alkyl bromide (reaction 8) or, competitively, either with the intermediate or unchanged organoborane, to give an alkane (reaction 9). Application of this mechanism

$$Br_2 + BR_3 \rightarrow Br\cdot + BrBR_2 + R\cdot \qquad \ldots(5)$$

$$R_2B-\underset{H}{\overset{|}{C}}- + Br\cdot \rightarrow R_2B-\underset{\cdot}{\overset{|}{C}}- + HBr \qquad \ldots(6)$$

$$R_2B-\underset{\cdot}{\overset{|}{C}}- + Br_2 \rightarrow R_2B-\underset{Br}{\overset{|}{C}}- + Br\cdot \qquad \ldots(7)$$

$$R_2B-\underset{Br}{\overset{|}{C}}- + HBr \rightarrow R_2BBr + H\underset{Br}{\overset{|}{C}}- \qquad \ldots(8)$$

$$R-B\mathopen{<} + HBr \rightarrow RH + BrB\mathopen{<} \qquad \ldots(9)$$

has provided a simple procedure for the production of highly substituted tertiary alcohols.[361] Convenient general syntheses of alcohols[362] and alkyl hydroperoxides[363] by autoxidation of organoboranes have also been outlined, and recognition of S_H2 processes during the reaction of *N*-chlorodimethylamine with tributylborane has helped to clarify previous uncertainty.[364]

Buta-1,3-diene monoxide reacts readily with trialkylboranes in the presence of catalytic amounts of oxygen or other free-radical initiators, producing the corresponding 4-alkylbut-2-en-1-ols in relatively high stereochemical purity.[365] The radical-chain mechanism suggested is as outlined; the alkoxy radical (**235**) formed by rearrangement of (**234**) reacts with the trialkylborane, displacing an alkyl radical and forming a borinate (**236**). Subsequent hydrolysis yields the 4-alkylbut-2-en-1-ol.

Homolytic substitution at boron, and hydrogen-abstraction reactions of methyl radicals with triethylborane, all in the gas phase,[366] have been studied and it has been

[360] C. F. Lane and H. C. Brown, *J. Am. Chem. Soc.*, **92**, 7212 (1970).
[361] C. F. Lane and H. C. Brown, *J. Am. Chem. Soc.*, **93**, 1025 (1971).
[362] H. C. Brown, M. M. Midland, and G. W. Kabalka, *J. Am. Chem. Soc.*, **93**, 1024 (1971).
[363] H. C. Brown and M. M. Midland, *J. Am. Chem. Soc.*, **93**, 4078 (1971).
[364] A. G. Davies, S. C. W. Hook, and B. P. Roberts, *J. Organomet. Chem.*, **23**, C11 (1970).
[365] A. Suzuki, N. Miyaura, M. Itoh, H. C. Brown, G. W. Holland, and E. Negishi, *J. Am. Chem. Soc.*, **93**, 2792 (1971); but see *Org. Reaction Mech.*, **1970**, 312.
[366] J. Grotewold, E. A. Lissi, and J. C. Scaiano, *J. Chem. Soc.* (B), **1971**, 1187; T. N. Bell and A. E. Platt, *Int. J. Chem. Kinet.*, **3**, 307 (1971).

$$R_3B + O_2 \rightarrow R\cdot + R_2BO_2\cdot$$

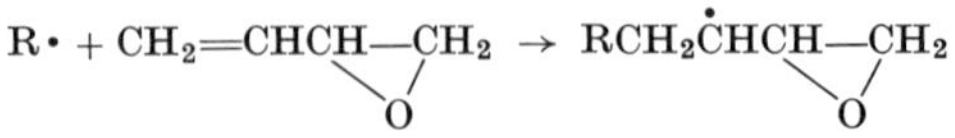

$$R\cdot + CH_2=CHCH-CH_2 \rightarrow RCH_2\overset{\cdot}{C}HCH-CH_2$$
$$\underset{O}{\diagdown\diagup} \phantom{\rightarrow RCH_2\overset{\cdot}{C}HCH-}\underset{O}{\diagdown\diagup}$$
$$(234)$$

$$(234) \rightarrow RCH_2CH=CHCH_2O\cdot$$
$$(235)$$

$$(235) + R_3B \rightarrow RCH_2CH=CHCH_2OBR_2 + R\cdot$$
$$(236)$$

$$(236) + H_2O \rightarrow RCH_2CH=CHCH_2OH + R_2BOH$$

suggested that S_H2 reactions are involved during cleavage of dialkyl peroxides and disulphides by Grignard and organolithium reagents.[367]

A complex series of reaction pathways can be envisaged for peroxide-initiated hydroxylation of aromatic compounds by trialkyl phosphite and oxygen.[368] Although radical (238) has been suggested as the hydroxylating species, other radicals (e.g. $RO_2\cdot$) may be directly involved and the actual existence of (238) must be questioned in view of the very rapid fragmentation of phosphorus radicals similar to (237) and (239).

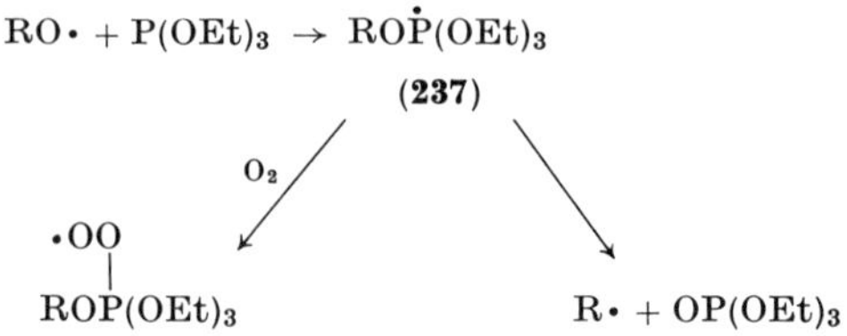

$$RO\cdot + P(OEt)_3 \rightarrow RO\overset{\cdot}{P}(OEt)_3$$
$$(237)$$

$$\cdot OO$$
$$|$$
$$RO\overset{}{P}(OEt)_3 R\cdot + OP(OEt)_3$$
$$(238)$$

$$(238) + ArH \rightarrow ArOH + OP(OEt)_3 + RO\cdot$$

$$R\cdot + O_2 \rightarrow RO_2\cdot + P(OEt)_3 \rightarrow ROO\overset{\cdot}{P}(OEt)_3$$
$$(239)$$

The conversion of alcohols into phosphate esters is an extremely important reaction and in general is achieved by nucleophilic reactions at phosphorus. A new and potentially valuable route[369] for photophosphorylation involves S_H2 attack of alkoxy radicals on trivalent phosphorus. Thus cholestan-3β-yl nitrite, on irradiation in the presence of tri-isopropyl phosphite, affords 5α-cholestan-3β-yl di-isopropyl phosphate (241; $R^2 = Pr^i$) by the route illustrated. No product indicating ejection of cholestane radicals

$$R^1ONO \rightarrow R^1O\cdot + NO$$

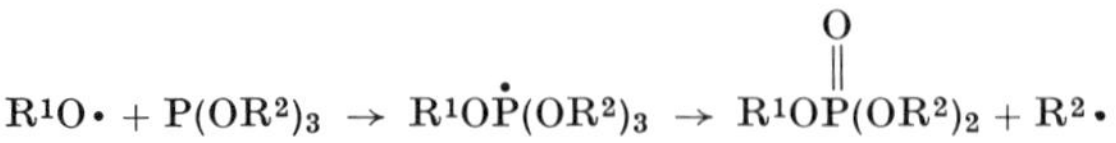

$$ O$$
$$ \|$$
$$R^1O\cdot + P(OR^2)_3 \rightarrow R^1O\overset{\cdot}{P}(OR^2)_3 \rightarrow R^1OP(OR^2)_2 + R^2\cdot$$
$$(240)(241)$$

$$R^1ONO = 3β\text{-hydroxycholestane nitrite}$$

[367] A. W. P. Jarvie and D. Skelton, *J. Organomet. Chem.*, **30**, 145 (1971).
[368] R. Higgins, K. M. Kitson, and J. R. L. Smith, *J. Chem. Soc.* (B), **1971**, 430.
[369] D. H. R. Barton, T. J. Bentley, R. H. Hesse, F. Mutterer, and M. M. Pechet, *Chem. Comm.*, **1971**, 912.

from the phosphoranyl intermediate (**240**) was detected and it appears that massive ligands are preferentially retained.

Reactions Involving Oxidation or Reduction by Metal Salts

Mihailović and his co-workers have continued their studies of the lead tetra-acetate oxidation of cyclic and acyclic aliphatic alcohols.[370] Production of ketones (e.g. **244**) from the corresponding secondary alcohol (**242**) has generally been assumed to arise by proton loss in a heterolytic pathway involving the alkoxylead(IV) acetate (**243**). However, by use of α-deuteriated substrates, it has now been established that ketone is formed, in part, by the homolytic route outlined.[371] 1,4-Transfer of hydrogen from the alcohol (α) carbon atom to the δ-carbon atom, (**246**)→(**247**), has been demonstrated and this can only follow intramolecular homolytic 1,5-hydrogen abstraction by alkoxy radical (**245**), since α-deuterated alcohols not undergoing this rearrangement afford only ketones without deuterium (**244**); significant involvement of alkoxy radicals during the (thermal) lead tetra-acetate oxidation of alcohols is thus further established.

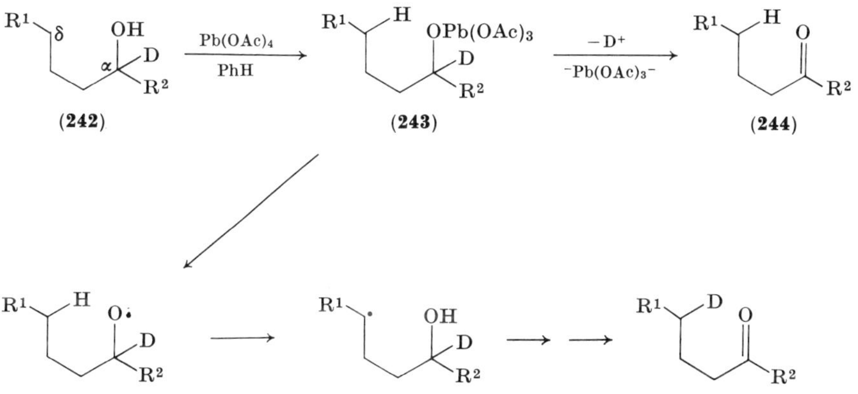

A full report of oxidative cleavage of 1,2-diarylethanols by ceric ammonium nitrate has been published.[372] The process involves primary one-electron oxidation, and similar conclusions are reached for certain chromic acid oxidation of related substrates. Relative rates of cleavage of substituted benzyl radicals were correlated with σ^+ for the substituent and give ρ values significantly different for the cerium(IV) and chromic acid oxidations; this difference is thought to exclude the intermediacy of free alkoxy radicals in both types of oxidation as it implies differences in the transition states for decomposition of the alcohol–metal complexes. Independent work supports chromium(IV) as the species leading to oxidative cleavage during chromic acid oxidation.[373] Relative rates of oxidative cleavage of methyl-, ethyl-, isopropyl,- and *tert*-butyl-phenylcarbinols by cerium(IV) are 0.04, 3.30, 184, and 195, reflecting the relative stability of the resulting alkyl radicals

---

[370] See *Org. Reaction Mech.*, **1970**, 356.
[371] D. Jeremić, S. Milosavljević, V. Andrejević, M. Jakovljević-Marinković, Ž. Čeković, and M. Lj. Mihailović, *Chem. Comm.*, **1971**, 1612.
[372] P. M. Nave and W. S. Trahanovsky, *J. Am. Chem. Soc.*, **93**, 4536 (1971).
[373] M. Rahman and J. Roček, *J. Am. Chem. Soc.*, **93**, 5455, 5462 (1971).

and showing that this is the key factor (as in alkoxy radical scission) in promoting the cleavage.[374]

Similar conclusions come from the results of a systematic ESR investigation of Ce(IV) photo-oxidation of primary, secondary and tertiary alcohols at 77°K.[375] Spectra of the corresponding hydroxyalkyl radicals are obtained with simple primary alcohols whereas alkyl radicals (from C—C cleavage) are obtained from tertiary alcohols; both types of radical are obtained from secondary alcohols. The same technique has been employed to study the radicals generated by the Ce(IV) oxidative decarboxylation of a number of carboxylic acids.[376]

Heiba and Dessau showed[377] that carboxyalkyl radicals ($R^1R^2\dot{C}$—COOH) are formed directly in the thermal decomposition of manganic carboxylates. These workers have now made the notable observation that similar radicals are also formed during thermal decomposition of ceric carboxylates; previously it had been thought that one general decarboxylative mechanism accounted for the thermal and photochemical decomposition of a number of higher-valent metal, including ceric, carboxylates. Elucidation of these mechanisms was achieved[378] by decomposing the ceric carboxylates in the presence of various olefins (e.g. styrene) and aromatic hydrocarbons (e.g. toluene). In the presence of an olefin, a decarboxylative pathway (path A) results in formation of both ester (**248**) and lactone (**249**) via the intermediate alkyl

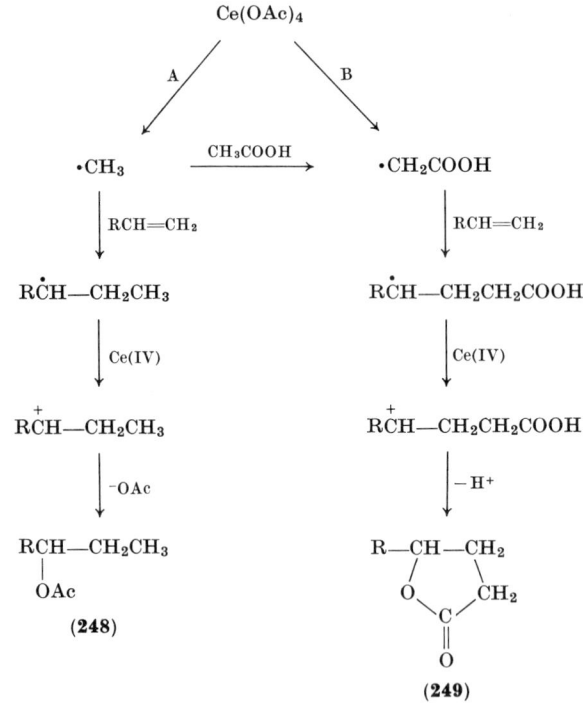

[374] W. S. Trahanovsky and J. Cramer, *J. Org. Chem.*, **36**, 1890 (1971).
[375] D. Greatorex and T. J. Kemp, *Trans. Faraday Soc.*, **67**, 56 (1971).
[376] D. Greatorex and T. J. Kemp, *Trans. Faraday Soc.*, **67**, 1576 (1971).
[377] *Org. Reaction Mech.*, **1969**, 346.
[378] E. I. Heiba and R. M. Dessau, *J. Am. Chem. Soc.*, **93**, 995 (1971).

radical, whereas a non-decarboxylative pathway (path B) leads, via carboxyalklyl radicals, to lactones only. Results of this work point to the latter pathway as the predominant, if not exclusive, mode of thermal decomposition of primary and secondary carboxylates. As well as providing useful tools for mechanistic studies, these oxidation reactions of olefins and aromatic hydrocarbons have broad synthetic utility, particularly for the preparation of γ-lactones and arylacetic acids.[378] Oxidation of substituted toluenes[379] and olefins[380] by manganese(III) acetates, catalysed by potassium bromide, has led to similar mechanisms being proposed by independent groups.

Further novel synthetic reactions employing manganese(III) and cerium(IV) acetates have been exploited; thus enolizable ketones undergo oxidative addition to olefins, leading to the formation of γ-keto esters:[381a]

e.g. $Me_2C=O + C_6H_{13}CH=CH_2 \xrightarrow[\text{or Ce(IV)}]{Mn(III)}$

$C_9H_{19}COMe + C_6H_{13}CH=CHCH_2COMe + C_6H_{13}CH(OAc)CH_2CH_2COMe$

Success of the reaction is attributed to the very selective oxidation of intermediate radicals by the metal salt, the initial α-keto radical ($MeCOCH_2\cdot$) is not easily oxidized, whereas the secondary alkyl radical formed by addition to the olefin is readily oxidized by the metal ion. Relative reactivity of the oxidation of secondary alkyl radicals by Mn(III), Ce(IV) and Cu(II) is of the order 1:12:350, supporting conclusions previously reached by Kochi.[381b]

Kochi has reviewed his recent work on the oxidation of alkyl radicals by metal complexes,[382] emphasizing the different mechanisms of copper(II) oxidations, the latter being particularly reactive toward alkyl radicals both as electron-transfer and ligand-transfer oxidants. Mechanisms of radical oxidations by various copper(II) complexes (halide, pseudohalide, acetate, etc.) have been studied in detail and it is shown that the ligands associated with the copper(II) nucleus and the solvent play critical roles in determining the relative importance of the various pathways for oxidation.

Although Grignard reagents do not generally undergo intermolecular additions to non-conjugated alkenes, intramolecular additions do occur and a study has been made of the rates and products of addition of the Grignard function to the multiple bond in a series of reagents based on (**250**).[383] Particularly unusual was the large rate-retarding effect by a methyl group when placed at either end of the double bond; the results are rationalized by a mechanism involving a reversible electron transfer to the double bond as shown.

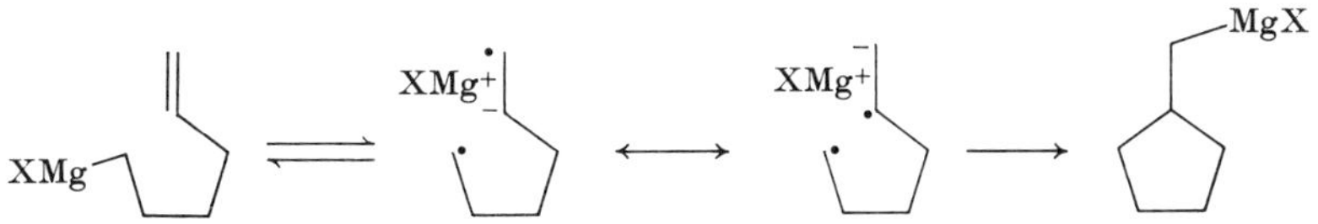

(**250**)

[379] J. R. Gilmore and J. M. Mellor, *Chem. Comm.*, **1970**, 507.
[380] J. R. Gilmore and J. M. Mellor, *J. Chem. Soc.* (C), **1971**, 2355.
[381a] E. I. Heiba and R. M. Dessau, *J. Am. Chem. Soc.*, **93**, 524 (1971).
[381b] See *Org. Reaction Mech.*, **1968**, 295.
[382] J. K. Kochi in ref. 2, p. 377; see also C. L. Jenkins and J. K. Kochi, *J. Org. Chem.*, **36**, 3095, 3103 (1971).
[383] W. C. Kossa, T. C. Rees, and H. G. Richey, *Tetrahedron Letters*, **1971**, 3455.

Products from the cobalt(II) chloride-catalysed reaction of methylmagnesium iodide with 5-*endo*-bromo-1,2,3,4,7,7-hexachloronorborn-2-ene are envisaged as arising from intermediate radicals produced by removal from the starting halide of the 5-*endo*-bromine and bridge-chlorine atoms.[384] Reactions of related halogenated-bridged polycyclic compounds are basically similar. Addition of both lithium di-*cis*- and di-*trans*-propenyl cuprate to cyclohex-2-enone is found to be completely stereospecific,[385] thus providing further evidence against the intermediacy of free radicals in conjugate additions of organocopper(I) compounds to α,β-unsaturated ketones. Furthermore, it places an upper limit of 10^{-11} sec on the life of a possible caged radical pair since the propenyl radical has an estimated life of 10^{-9} seconds for inversion.[386]

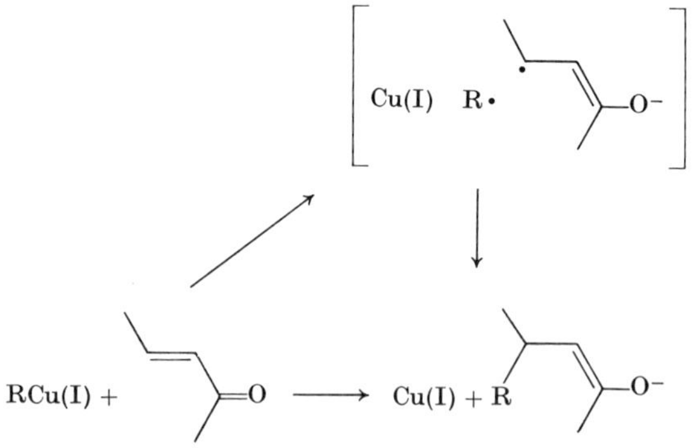

$$\begin{bmatrix} \text{Cu(I)} \quad R \cdot \diagdown\diagup\diagdown O^- \end{bmatrix}$$

$$RCu(I) + \diagdown\diagup\mathord=O \longrightarrow Cu(I) + R\diagdown\diagup\diagdown O^-$$

A fundamental investigation of the thermal decomposition of *cis*- and *trans*-propenyl copper(I) and *cis*- and *trans*-propenyl(tri-*n*-butylphosphine)copper(I) has shown that the hexa-2,4-dienes (dimers) obtained have complete retention of stereochemistry at the vinylic double bond.[387] This stereochemical result, taken with the known rate of inversion of configuration of the vinyl radical,[386] is sufficient to establish that free propenyl radicals are not intermediates in these thermal coupling reactions. Similar conclusions are drawn from the observed retention of configuration on decomposition of the analogous but-2-enyl-copper(I) and -silver(I) compounds. A survey of the yield and stereochemistry of the conversion of *trans*-propenyl-lithium into hexa-2,4-dienes on oxidation by a variety of transition-metal salts suggests that free propenyl radicals, likewise, are not involved.

Kochi and Tamura[388–391] have made related investigations based on the coupling of Grignard reagents with alkyl halides, reactions well known to be catalysed by transition-metal salts (Kharasch reaction); although the role of these salts has been extensively investigated, no general description of the elementary steps involved has previously been provided. Catalyses of the coupling reaction by silver and copper salts, despite their similarity, occur by fundamentally different mechanisms; the latter proceeds via alkyl

---

[384] R. Alexander, D. I. Davies, D. H. Hey, and J. N. Done, *J. Chem. Soc.* (C), **1971**, 2367.
[385] C. P. Casey and R. A. Boggs, *Tetrahedron Letters*, **1971**, 2455.
[386] G. M. Whitesides and C. P. Casey, *J. Am. Chem. Soc.*, **88**, 4541 (1966).
[387] G. M. Whitesides, C. P. Casey, and J. K. Krieger, *J. Am. Chem. Soc.*, **93**, 1379 (1971).
[388] M. Tamura and J. K. Kochi, *J. Am. Chem. Soc.*, **93**, 1485 (1971).
[389] M. Tamura and J. K. Kochi, *J. Am. Chem. Soc.*, **93**, 1483 (1971).
[390] M. Tamura and J. K. Kochi, *J. Am. Chem. Soc.*, **93**, 1487 (1971).
[391] M. Tamura and J. K. Kochi, *J. Organomet. Chem.*, **31**, 289 (1971).

copper(I) species and does not involve radicals[388] whereas it is proposed that silver(I) salts catalyse the reaction by the following steps:[389]

$$R'MgX + AgX \rightarrow R'Ag + MgX_2$$
$$RAg, R'Ag \rightarrow [R-R, R'-R, R'-R'] + 2Ag$$
$$Ag + RX \xrightarrow{Slow} R\cdot + AgX$$
$$R\cdot + Ag \rightarrow RAg, \text{ etc.}$$

Thus, whilst radicals from the halide are formed, the Grignard alkyl component, in agreement with Whitesides' investigations, does not become free. A third mechanistic type of catalytic activity is demonstrated by iron,[390] one of the most effective in these reactions. Full details of iron-catalysed reactions between various Grignard reagents and alkyl halides in THF have been published.[391]

Formation of alkyl halides, RX, from carboxylic acids, RCOOH, by decarboxylation with lead tetra-acetate and lithium halide (first described by Kochi[392]) has proved highly successful in the synthesis of numerous alkyl halides. The stereochemistry of the products has now been studied; on chlorodecarboxylation of *cis*- and *trans*-4-*tert*-butylcyclohexanecarboxylic acid (**251**) each isomer gave the same product composition: 67% of *cis*- and 33% of *trans*-4-*tert*-butylcyclohexyl chloride.[393] These results are very similar to those reported for the reaction of 4-*tert*-butylcyclohexyl radical, generated, for example, by thermolysis of the hypochlorites (**252**),[394] with chlorine donors such as carbon tetrachloride, and they thus confirm the existence of the cyclohexyl free radical during the decarboxylations. The Kochi reaction is also useful in demonstrating sub-

(**251**) *cis* and *trans*   (**252**) *cis* and *trans*

stantial polar effects of remote substituents upon free-radical stereoselectivities. Thus pairs of acids related to (**251**) give significantly different axial–equatorial chloride product ratios when everything but the *para*-substituent is the same.[395]

The Simonini reaction (reactions of silver carboxylates with iodine) has been shown to be a variant of the Hunsdiecker reaction and limitations of the synthetic applicability of the latter process have been discussed.[396]

ESR and product studies of the reaction between titanium(III) ions and some *N*-alkyl-hydroxylamines (**253**) suggest the annexed mechanistic scheme.[397] Alkylamino radicals (**254**) are formed first but undergo rapid reactions, including hydrogen abstraction to produce nitroxide (**255**) and rearrangement to α-aminoalkyl radicals (**256**). The latter radicals are effective one-electron reducing agents, and cations derived from them in this process undergo solvolysis to carbonyl compounds. Spectra of the radical intermediates are reported and discussed.

[392] See *Org. Reaction Mech.*, **1965**, 196.
[393] R. D. Stolow and T. W. Giants, *Tetrahedron Letters*, **1971**, 695.
[394] F. D. Greene, C. Chu, and J. Walia, *J. Am. Chem. Soc.*, 84, 2463 (1962); *J. Org. Chem.*, 29, 1285 (1964).
[395] R. D. Stolow and T. W. Giants, *J. Am. Chem. Soc.*, 93, 3536 (1971).
[396] N. J. Bunce and N. G. Murray, *Tetrahedron*, 27, 5323 (1971).
[397] N. H. Anderson and R. O. C. Norman, *J. Chem. Soc.* (B), **1971**, 993.

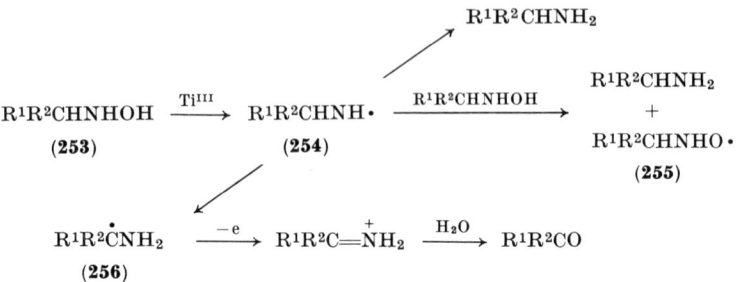

Walling and Kato re-investigated the oxidation of alcohols with Fenton's reagent and developed a new kinetic analysis.[398] Hydroxyl radicals attack both α- and β-hydrogen atoms of ethyl and isopropyl alcohol; β-hydroxyalkyl radicals are not oxidized by Fe(III) ions but are effectively scavenged by added copper(II) ions. Related studies with deuteriated alcohols have been made,[399] and oxidation of L-ascorbic acid with the Ti(III)/H_2O_2 couple has been investigated by flow-ESR techniques.[400] Detection of radical intermediates by optical and ESR techniques during the photo-oxidation of a number of alcohols, and benzilic acid, by transition-metal ions has been discussed in a preliminary report,[401] and substituent effects on the rate of photo-oxidation of alcohols by uranyl ions have been established;[402] the authors prefer a primary α-hydrogen-abstraction process despite considerable evidence suggesting an electron-transfer mechanism.

Radical mechanisms are thought to be operative in the cerium(IV)-catalysed decomposition of aryldiazomethanes to cis- and trans-stilbenes[403] and, in the presence of olefins, oxidation of azide ion with the same oxidant affords high yields of azido adducts,[404] possibly by initial addition of the azido radical. Oxidative cleavage competes effectively with α-hydrogen abstraction during reactions of nickel peroxide with 1,2-glycols; meso-hydrobenzoin gives a mixture of benzaldehyde and benzil.[405] Radicals generated by the lead(II) oxide oxidation of formazans,[406] metal-ion oxidation of sec-butyl-lithium,[407] and photoinduced ferricyanide decomposition of carboxylic acids[408] have also been studied.

Radical Ions and Electron-transfer Processes

Growing interest in ion radicals as reaction intermediates is reflected in the publication of several reviews during the year,[409-411] and an undoubted stimulus has been provided

[398] C. Walling and S. Kato, *J. Am. Chem. Soc.*, **93**, 4275 (1971).
[399] C. E. Burchill and G. F. Thompson, *Can. J. Chem.*, **49**, 1305 (1971).
[400] Y. Kirino and T. Kwan, *Chem. Pharm. Bull.*, **19**, 718 (1971).
[401] H. D. Burrows, D. Greatorex, and T. J. Kemp, *J. Am. Chem. Soc.*, **93**, 2539 (1971).
[402] R. Matsushima and S. Sakuraba, *J. Am. Chem. Soc.*, **93**, 5421 (1971).
[403] W. S. Trahanovsky, M. D. Robbins, and D. Smick, *J. Am. Chem. Soc.*, **93**, 2086 (1971).
[404] W. S. Trahanovsky and M. D. Robbins, *J. Am. Chem. Soc.*, **93**, 5256 (1971).
[405] R. Konaka and K. Kuruma, *J. Org. Chem.*, **36**, 1703 (1971).
[406] N. Azuma, K. Mukai, and K. Ishizu, *Bull. Chem. Soc. Japan*, **43**, 3960 (1971).
[407] H. J. M. Bartelink, H. K. Ostendorf, B. C. Roest, and H. A. J. Schepers, *Chem. Comm.*, **1971**, 878.
[408] A. L. Pozdnyak, G. A. Shagisultanova, and S. I. Arzhankov, *Zhur. Fiz. Khim.*, **44**, 2391 (1970); *Chem. Abs.*, **74**, 22380 (1971).
[409] N. L. Holy and J. D. Marcum, *Angew. Chem. Int. Ed.*, **10**, 115 (1971).
[410] L. L. Miller, *J. Chem. Educ.*, **48**, 168 (1971).
[411] J. F. Garst, *Accounts Chem. Res.*, **4**, 400 (1971).

by the recognition of a competing radical-chain mechanism for aromatic and nucleophilic substitutions previously thought to occur by only an aryne mechanism. Thus a very careful analysis of the amino compounds produced by reactions of 5- and 6-halopseudocumenes (**257**) and (**258**) with KNH_2 in liquid ammonia led Bunnett and his collaborators[412] to propose the following chain reaction:

$$\text{Electron donor} + ArI \rightarrow [ArI]^{\overline{\cdot}} + \text{Residue}$$
$$[ArI]^{\overline{\cdot}} \rightarrow Ar\cdot + I^-$$
$$Ar\cdot + NH_2^- \rightarrow ArNH_2^{\overline{\cdot}}$$
$$ArNH_2^{\overline{\cdot}} + ArI \rightarrow ArNH_2 + [ArI]^{\overline{\cdot}}, \text{ etc.}$$

(**257**) (**258**)

a, X = Cl;   b, X = Br;   c, X = I

Although the precise nature of the initiating electron-donor remains obscure, the mechanism is clearly substantiated by the observation[413] that addition of a good electron-donor (potassium metal) enhanced the radical pathway, sometimes to the exclusion of the aryne route. These observations, together with earlier related work of Russell[414] and Kornblum,[415] point clearly to a more general re-interpretation of some existing "nucleophilic" substitutions, and a wider recognition of chain reactions involving propagating ion radicals seems inevitable. The results of Russell and of Kornblum referred to above were obtained essentially from reactions of nitroalkanes with nitrobenzyl and nitrocumyl derivatives ($XRNO_2$), especially the halides (X = halogen), and an important and essential intermediate in all of these reactions is the anion radical of the nitroaralkyl compound [$XRNO_2^{\overline{\cdot}}$]. There has been much speculation on the nature of this species. A definitive polarographic study of the cathodic reduction of p-nitrobenzyl halides by Kosower and his collaborators[416] provides very good evidence for the existence of discrete intermediates $NO_2C_6H_4CH_2X^{\overline{\cdot}}$ (half-life for decomposition $\sim$ 30 msec) and indicates that rates of decomposition of the radical anions roughly parallel the carbon–halogen bond energies, with solvation differences making a smaller but unknown contribution. Another example[417] of a chain reaction involving propagating ion radicals is supplied by the photochemical dehalogenation of halogenobenzenes in alkaline alcoholic solutions:

$$PhI \xrightarrow{h\nu} Ph\cdot + I\cdot$$
$$Ph\cdot + Me_2CHOH \rightarrow PhH + Me_2\overset{\cdot}{C}-OH$$
$$Me_2\overset{\cdot}{C}-OH + OH^- \rightarrow Me_2\overset{\cdot}{C}-O^- + H_2O$$
$$Me_2\overset{\cdot}{C}-O^- + PhI \rightarrow PhI^{\overline{\cdot}} + Me_2CO$$
$$PhI^{\overline{\cdot}} \rightarrow Ph\cdot + I^-$$

[412] J. K. Kim and J. F. Bunnett, *J. Am. Chem. Soc.*, **92**, 7463 (1970).
[413] J. K. Kim and J. F. Bunnett, *J. Am. Chem. Soc.*, **92**, 7464 (1970).
[414] G. A. Russell, R. K. Norris, and E. J. Panek, *J. Am. Chem. Soc.*, **93**, 5839 (1971).
[415] See, for example, N. Kornblum, R. T. Swiger, G. W. Earl, H. W. Pinnick, and F. W. Stuchal, *J. Am. Chem. Soc.*, **92**, 5513 (1970).
[416] M. Mohammad, J. Hajdu, and E. M. Kosower, *J. Am. Chem. Soc.*, **93**, 1792 (1971).
[417] R. Backlin and W. V. Sherman, *Chem. Comm.*, **1971**, 453.

Aryl bromides and iodides may be dehalogenated in this way but the corresponding chlorides are unreactive.

The first report[418] of an observable complex in bimolecular reactions of olefinic cation radicals with neutral olefins comes from an elegant study, using ion cyclotron resonance techniques, of the reaction between styrene cation radical and neutral styrene. A nonconjugative "anchimeric" stabilization for anion radicals is suggested[419] by the observation that diphenylmethane anion radical is thermodynamically more stable than benzene anion radical by approximately 1 kcal mole^{-1}; this is unexpected since the PhCH$_2$ group is electron-donating and should destabilize, as indicated by the lower stability of anion radicals of toluene or *tert*-butylbenzene.

Reactions of sodium naphthalene with organic halogen compounds[420] continue to provoke lively dispute as to reaction mechanisms. Thus reactions of sodium naphthalene with aryl halides in tetrahydrofuran yield benzene as a major product, together with much smaller amounts of polyphenyls and phenylnaphthalene. The original suggestion,[421] that formation of benzene involves hydrogen abstraction from solvent tetrahydrofuran by phenyl radicals, has been disputed[422] and it is suggested, rather, that benzene results from abstraction of a proton by phenyl anion.

The reactions of triphenylsilyl halides with sodium naphthalenide yield mainly hexaphenyldisilane, together with smaller amounts of products including 1,4-dihydro-1,4-bis(triphenylsilyl)naphthalene (**259**); the yield of (**259**) is increased if an excess of free naphthalene is present and an electron-transfer mechanism, producing triphenylsilyl radical, is proposed for the first step.[423] However, when a similar type of reaction with chlorotrimethylsilane is carried out in the presence of reactive organic halides

$$Na^+, C_{10}H_8^{-} + Ph_3SiX \rightarrow Ph_3Si\cdot + C_{10}H_8 + NaX$$

$$2Ph_3Si\cdot \rightarrow Ph_6Si_2$$

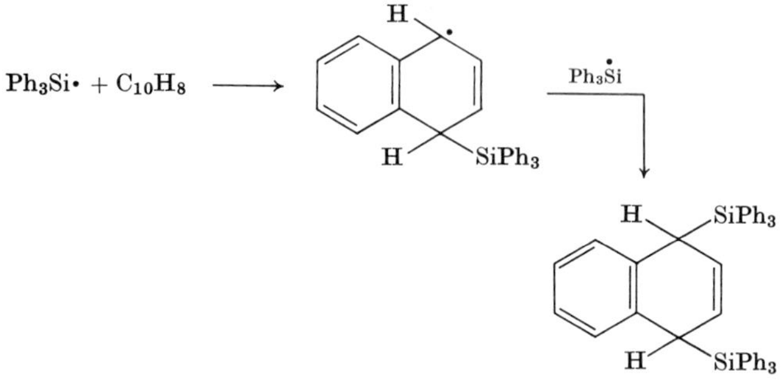

(**259**)

(e.g. PhCH$_2$Cl), the primary process is electron transfer to the organic halide forming, e.g., benzyl anion which then alkylates the silane.[424] In related work it has been shown

[418] C. L. Wilkins and M. L. Gross, *J. Am. Chem. Soc.*, **93**, 895 (1971).
[419] J. D. Young and N. L. Bauld, *Tetrahedron Letters*, **1971**, 2251.
[420] For a review see ref. 411.
[421] T. C. Cheng, L. Headley, and A. F. Halasa, *J. Am. Chem. Soc.*, **93**, 1502 (1971).
[422] G. D. Sargent, *Tetrahedron Letters*, **1971**, 3279.
[423] F. W. G. Fearon and J. C. Young, *J. Chem. Soc.* (B), **1971**, 272.
[424] S. Bank and J. F. Bank, *Tetrahedron Letters*, **1971**, 4581.

that (trimethylsilyl)sodium reacts with naphthalene, by electron transfer, to produce sodium naphthalenide and trimethylsilyl radicals.[425] Evidence for the intermediacy

$$Me_3Si^- + C_{10}H_8 \rightarrow Me_3Si\cdot + C_{10}H_8^{\bar{\cdot}}$$

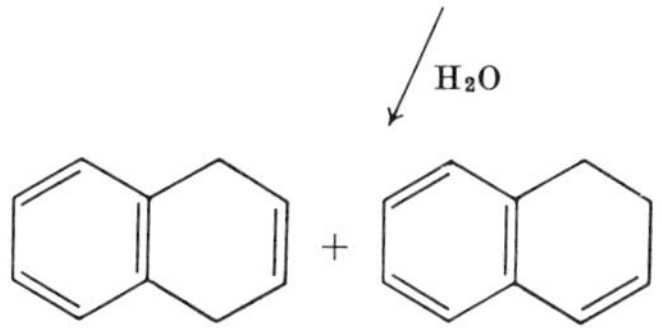

of benzylic anions in reactions of p-fluorobenzyl halides with sodium naphthalenide has been adduced[426] from ^{19}F-CIDNP studies. Trifluoroethyl ethers are efficiently cleaved by reaction with sodium naphthalenide[427] in dimethoxyethane:

$$C_{10}H_8^{\bar{\cdot}} + ROCH_2CF_3 \rightarrow C_{10}H_8 + ROCH_2\overset{\cdot}{C}F_2 + F^-$$

$$ROCH_2\overset{\cdot}{C}F_2 \xrightarrow{C_{10}H_8^{\bar{\cdot}}} RO^- + CH_2{=}CF_2$$

Further studies of the rates and mechanism of reactions of sodium naphthalenide with water have been reported.[428,429] It appears that the reactivity towards a proton source is significantly less than that of a structurally related anion and, quite surprisingly, the reactivity of tight ion-pairs is greater than that of solvent-separated ion-pairs.[430] Reaction of sodium naphthalenide with hydrogen yields a form of sodium hydride much more reactive than the usual commercial supplies.[431] The active form of sodium hydride reacts readily with benzyl chloride to give hydrogen, toluene, bibenzyl and stilbene.

The formation of both cis- and trans-9-alkyl-9,10-dihydro-10-methylanthracene when alkyl-lithium reagents are added to 9-methylanthracene suggests that alkyl-lithium addition is neither concerted not stereospecific.[432] It is suggested that the actual reaction pathway involves homolysis of the carbon—lithium bond, i.e.:

$$RLi + ArH \rightleftharpoons (Complex) \rightarrow ArH^{\bar{\cdot}} Li^+ + R\cdot$$
$$R\cdot + ArH \rightleftharpoons RArH\cdot$$
$$RArH\cdot + RLi \rightarrow RArH^- Li^+ + R\cdot$$

Radical intermediates are also thought to play a role in the rather more complex reactions of alkyl-lithiums with chloroformamidinium salts.[433]

Anion radicals of [2.2]paracyclophanes and 1,2-diarylethanes are extremely unstable,[434] unlike their simple aromatic counterparts. The anion-radical species readily accept another electron, either by disproportionation or direct reduction, to form the unstable two-electron reduced product which decomposes by scission of the ethane

---

[425] H. Sakurai, A. Okada, M. Kira, and K. Yonezawa, *Tetrahedron Letters*, **1971**, 1511.
[426] J. W. Rakshys, *Tetrahedron Letters*, **1971**, 4745.
[427] G. D. Sargent, *J. Am. Chem. Soc.*, **93**, 5268 (1971).
[428] See *Org. Reaction Mech.*, **1970**, 368.
[429] S. Hayano and M. Fujihira, *Bull. Chem. Soc. Japan*, **44**, 1496, 2046 (1971).
[430] S. Bank and B. Bockrath, *J. Am. Chem. Soc.*, **93**, 430 (1971).
[431] S. Bank and M. C. Prislopski, *Chem. Comm.*, **1970**, 1624.
[432] D. J. Schaeffer, R. Litman, and H. E. Zieger, *Chem. Comm.*, **1971**, 483.
[433] C. F. Hobbs and H. Weingarten, *J. Org. Chem.*, **36**, 2881 (1971).
[434] J. M. Pearson, D. J. Williams, and M. Levy, *J. Am. Chem. Soc.*, **93**, 5478 (1971); D. J. Williams, J. M. Pearson, and M. Levy, *J. Am. Chem. Soc.*, **93**, 5483 (1971).

linkage to produce two arylmethide anions. Detailed kinetic studies[435] of the reduction of 1,2-di-(1-naphthyl)ethane (NN) have established that free anion radicals (NN$\bar{\cdot}$) are stable in hexamethylphosphorotriamide, while corresponding ion pairs with Li$^+$, Na$^+$ or Cs$^+$ dissolved in ethers undergo decomposition as follows:

$$2\text{NN}\bar{\cdot}\text{ Metal}^+ \rightleftharpoons \text{NN}^{2-}, 2\text{ Metal}^+ + \text{NN}$$
$$\text{NN}^{2-}, 2\text{ Metal}^+ \rightarrow 2(\text{N}^-\text{ Metal}^+)$$

In agreement with the proposed mechanism, rates of bond cleavage were inversely proportional to the concentration of unreduced ethane derivative. Full details of the electron-transfer reactions between diphenylacetylene and sodium derivatives of biphenyl and naphthalene[436] have now been published.[437] In an important preliminary communication, Szwarc and his collaborators[438] give notice of evidence substantiating their hypothesis that electron-transfer and proton-transfer equilibria for a given type of ion-pair are independent of the solvent, and that equilibrium constants for the conversion tight pair–loose pair may be calculated from measurements of rates of electron and proton transfer to these species. ESR studies of thermodynamic equilibria between types of ion-pairs have been reported.[439]

In contrast to biacetyl, perfluorobiacetyl is readily and cleanly reduced to the semidione by alkali metals in tetrahydrofuran; ESR studies of the effect of counter-ion on ion-pair equilibria and rates of *cis–trans* interconversion for the perfluoro semidione have been reported.[440] Further extensive reports of the formation and reactions of semidiones from bicyclo[n.1.0]alkanes have been published by Russell's group,[441] who have also shown[442] that, contrary to previous reports for unsubstituted compounds, ketyls may be prepared only from cyclic α,β-unsaturated ketones when all hydrogen atoms α to the π-system are substituted by alkyl or aryl groups. Ketyls of cyclic α,β-unsaturated ketones are thought to undergo rapid oxidation giving rise to semidione precursors. Similar results were observed[443] during reduction of norbornen-7-one **(260)** and 9-benzonorbornenone **(261)** by alkali metals in ether solvents; the expected ketyls could not be detected; instead, the ESR spectra indicated formation of semidiones **(262)** and **(263)**.

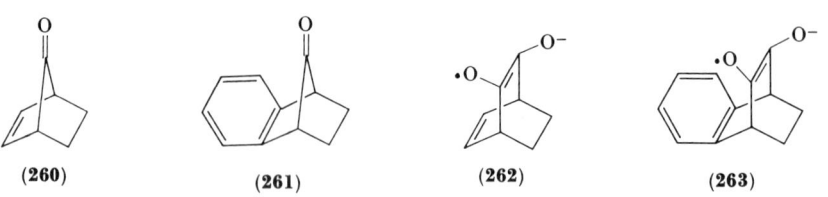

(260) **(261)** **(262)** **(263)**

---

[435] A. Lagendijk and M. Szwarc, *J. Am. Chem. Soc.*, **93**, 5359 (1971).
[436] See *Org. Reaction Mech.*, **1970**, 367.
[437] G. Levin, J. Jagur-Grodzinski, and M. Szwarc, *Trans. Faraday Soc.*, **67**, 768 (1971).
[438] Y. Karasawa, G. Levin, and M. Szwarc, *J. Am. Chem. Soc.*, **93**, 4614 (1971).
[439] K. S. Chen, S. W. Mao, K. Nakamura, and N. Hirota, *J. Am. Chem. Soc.*, **93**, 6004 (1971); L. Lee, R. Adams, J. Jagur-Grodzinski, and M. Szwarc, *J. Am. Chem. Soc.*, **93**, 4149 (1971).
[440] G. A. Russell, J. L. Gerlock, and D. F. Lawson, *J. Am. Chem. Soc.*, **93**, 4088 (1971).
[441] G. A. Russell, J. J. McDonnell, P. R. Whittle, R. S. Givens, and R. G. Keske, *J. Am. Chem. Soc.*, **93**, 1452 (1971); G. A. Russell, P. R. Whittle, and R. G. Keske, *J. Am. Chem. Soc.*, **93**, 1467 (1971).
[442] G. A. Russell and G. R. Stevenson, *J. Am. Chem. Soc.*, **93**, 2432 (1971); G. A. Russell and R. L Blankespoor, *Tetrahedron Letters*, **1971**, 4573.
[443] J. P. Dirlam and S. Winstein, *J. Org. Chem.*, **36**, 1559 (1971).

Thianthrenium cation radical readily substitutes activated benzenes to form aryl-sulphonium salts, but detailed kinetic studies suggest[444] that the active reagent is the thianthrene dication, formed in low equilibrium concentration by rapid disproportionation of the cation radical. Xenon difluoride in the presence of hydrogen fluoride is an efficient oxidant for benzene derivatives; aromatic cation radicals are primary intermediates and react to give aryl fluorides, biphenyls and polyphenyls.[445] Cation radicals are formed by reactions of benzidine, o-toluidine, and o-dianisidine with $Ce(SO_4)_2$ or with bromine; a possible correlation with carcinogenic activity of these amines is discussed.[446]

An interesting reaction of aromatic cation radicals is the nitration of perylene (and pyrene), readily effected by treating preformed cation radicals with sodium nitrite suspended in acetonitrile; 3-nitroperylene (**265**) is thought to arise by reaction of intermediate radical (**264**) with perylene cation radical.[447]

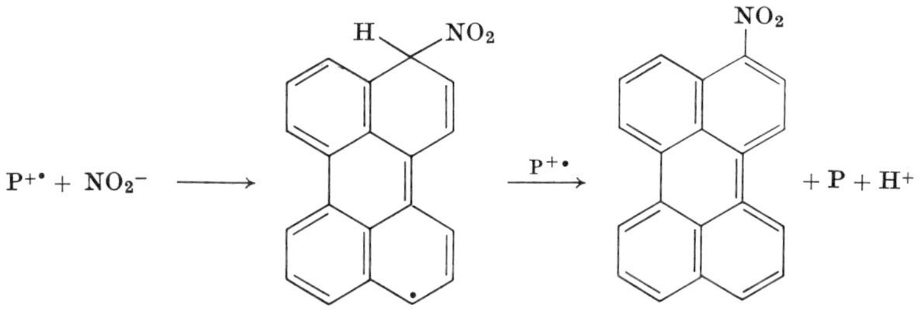

$$P^{+\cdot} + NO_2^- \longrightarrow (264) \xrightarrow{P^{+\cdot}} (265) + P + H^+$$

P = perylene

ESR and light-absorption studies[448] of some pyridinyl cation radicals $P\dot{y}(CH_2)_nPy^+$ (in which Py is 4-methoxycarbonylpyridine) reveal that intramolecular association of the pyridyl and pyridinium ends, with $n = 3$ or 4, occurs even at room temperature. "Paraquat" cation radical is produced[449] by an enigmatic reaction of aqueous hydroxide ion with di-(4-pyridyl) ketone dimethiodide.[450] Electron-transfer equilibria for redox systems of the violene type continue to be studied by Hünig and his group;[451] the most recent papers[452] give details of equilibria for the thiazolan-2-one azines, i.e. (**266**) $\rightleftharpoons$ (**267**) $\rightleftharpoons$ (**268**). Temperature-jump studies[453] of related violene series indicate that these equilibria may be rather more complex than shown and that more complex equilibria involving electron transfers between protonated species are important.

---

[444] J. J. Silber and H. J. Shine, *J. Org. Chem.*, **36**, 2923 (1971).
[445] M. J. Shaw, H. H. Hyman, and R. Filler, *J. Org. Chem.*, **36**, 2917 (1971).
[446] M. Matrka, J. Pipalova, Z. Sagner, and J. Marhold, *Chem. Prum.*, **21**, 14 (1971); *Chem. Abs.*, **74**, 87145 (1971).
[447] C. V. Ristagno and H. J. Shine, *J. Am. Chem. Soc.*, **93**, 1811 (1971).
[448] M. Itoh, *J. Am. Chem. Soc.*, **93**, 4750 (1971).
[449] F. E. Geiger, C. L. Trichilo, F. L. Minn, and N. Filipescu, *J. Org. Chem.*, **36**, 357 (1971); N. Filipescu, F. E. Geiger, C. L. Trichilo, and F. L. Minn, *J. Phys. Chem.*, **74**, 4344 (1970).
[450] But see also J. A. Farrington, A. Ledwith, and M. F. Stam, *Chem. Comm.*, **1969**, 259.
[451] S. Hünig, F. Linhart, and D. Scheutzow, *Angew. Chem. Int. Ed.*, **10**, 275 (1971).
[452] S. Hünig and G. Sauer, *Ann. Chem.*, **748**, 173, 189 (1971).
[453] C. F. Bernasconi, R. G. Bergstrom, and S. Hünig, *Chem. Comm.*, **1971**, 1485.

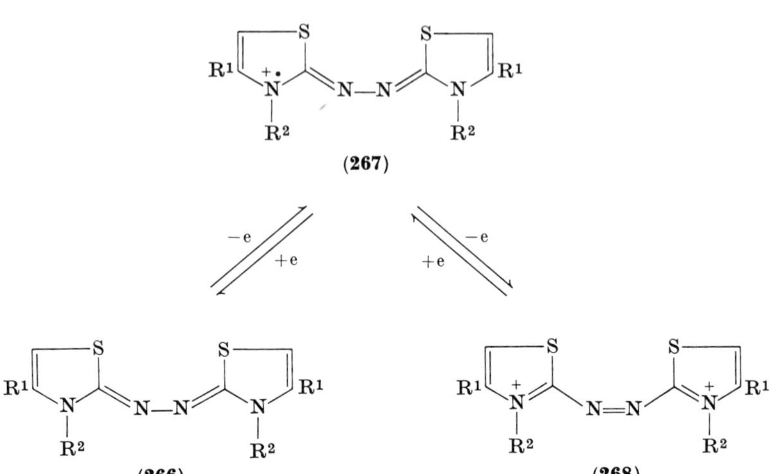

(267)

(266) (268)

Benzylic ethers (**269**) undergo electro-oxidative cleavage via cation radical formation, as indicated;[454] related oxidations of benzylic aldehydes and ketones proceed similarly,[455] with products arising by subsequent formation of benzoyl cations.

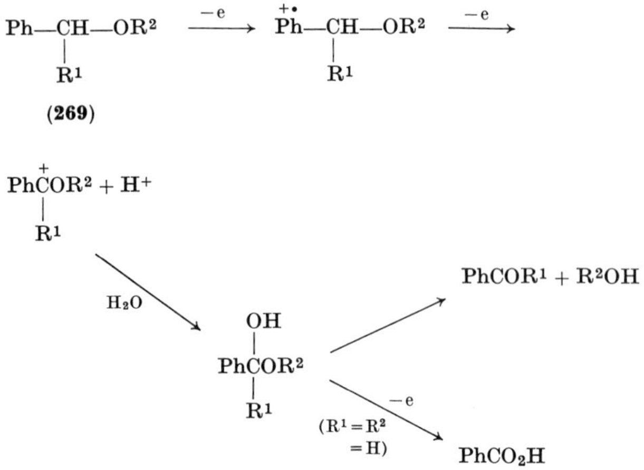

$$Ph-\underset{R^1}{\overset{|}{C}H}-OR^2 \xrightarrow{-e} Ph-\underset{R^1}{\overset{+\bullet|}{C}H}-OR^2 \xrightarrow{-e}$$

(269)

$$\underset{R^1}{\overset{|}{Ph\overset{+}{C}OR^2}} + H^+$$

$\downarrow H_2O$

$$\underset{R^1}{\overset{|}{Ph\overset{|}{C}OR^2}}\text{—OH}$$

→ $PhCOR^1 + R^2OH$

$\xrightarrow{-e}_{(R^1=R^2=H)}$ $PhCO_2H$

R^1 = H, alkyl, aryl;
R^2 = alkyl, H, acyl.

Anodic oxidation of aryl alkyl sulphides (**270**) in water-containing media gives rise to products that vary according to the nature of the alkyl group as indicated; similar products were observed after oxidation with bromine in 70% acetic acid.[456]

ESR studies continue to provide evidence of interesting new reactions and reacting species; evidence for the existence of SO_3^-, $S_2O_3^-$, S^-, and $\cdot SH$ is adduced[457] from

---

[454] L. L. Miller, J. F. Wolf, and E. A. Mayeda, *J. Am. Chem. Soc.*, **93**, 3306 (1971).
[455] L. L. Miller, V. R. Koch, M. E. Larscheid, and J. F. Wolf, *Tetrahedron Letters*, **1971**, 1389.
[456] K. Uneyama and S. Torii, *Tetrahedron Letters*, **1971**, 329.
[457] R. O. C. Norman and P. M. Storey, *J. Chem. Soc.* (B), **1971**, 1009.

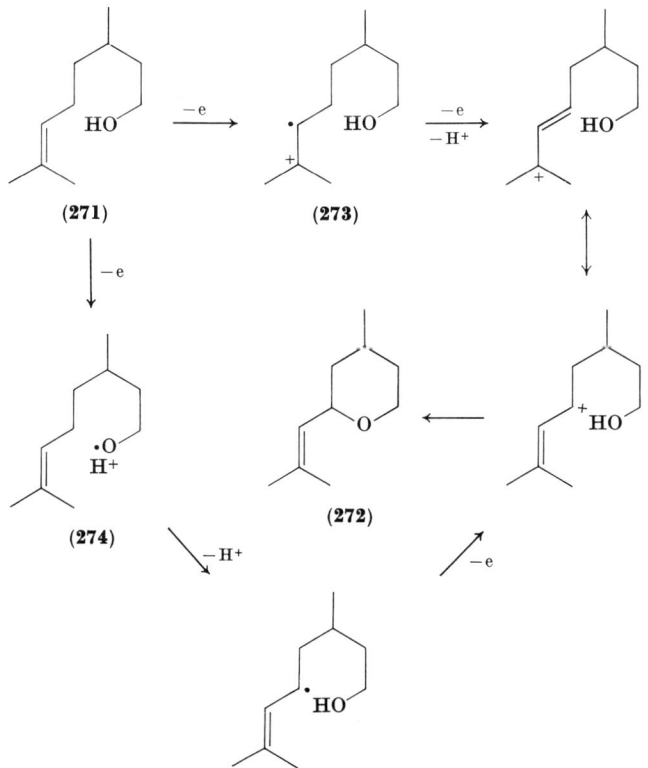

oxidations of SO_3^{2-}, $S_2O_3^{2-}$, S^{2-} by ·OH, ·NH$_2$ or Ce(IV), and the radicals ·COOH, ·COO$^-$, and ·CMe$_2$NH$_2$ are shown[458] to reduce a variety of carbonyl compounds.

Radical intermediates have been proposed for the alkali-metal reduction of disubstituted malonates and reactions of esters with base in the presence of chlorotrimethylsilane.[459] Further details of the cathodic reduction of arylsulphonamides have been

---

[458] N. H. Anderson, A. J. Dobbs, D. J. Edge, R. O. C. Norman, and P. R. West, *J. Chem. Soc.* (B), **1971**, 1004.
[459] Y.-N. Kuo, F. Chen, C. Ainsworth, and J. J. Bloomfield, *Chem. Comm.*, **1971**, 136.

presented[460] together with a full discussion of the (originally misinterpreted[461]) ESR spectrum of the radical anion of benzocyclobutene.[462] The nature and reactivity of radical anions formed by reductions of pyrimidine,[463] arylazo compounds[464] and naphthaquinone derivatives[465] have been studied.

The electrode reaction of organic compounds is a growing and interesting subject exemplified by the novel method of synthesis of cyclic ethers such as (**272**), produced from citronellol (**271**) in acetonitrile.[466] This example is interesting mechanistically as two possible routes are available, via cation radical (**273**) or (**274**), formed by primary electron transfer from (**271**). Electrolytic oxidation of some simple 1,2,3,4-tetrahydroisoquinoline phenols further demonstrates the potential use of this technique.[467]

A careful investigation of the anodic oxidation of phenylacetate ions suggests that the results are best interpreted by involving the electrode adsorption of key intermediates; expressions are derived relating product distribution to structural, electrochemical and adsorption parameters.[468] Particular attention is given to the competition between dimerization and cation formation for the intermediate benzyl radicals (see Scheme).

$$ArCH_2CO_2^- \rightleftharpoons ArCH_2CO_2^-\text{(ads)} \xrightarrow{-e} ArCH_2CO_2\cdot\text{(ads)} \xrightarrow[-CO_2]{\text{fast}}$$

$$ArCH_2\cdot \xrightarrow{O_2} ArCH_2OO\cdot \longrightarrow \longrightarrow ArCHO$$

$$\updownarrow \text{fast}$$

$$ArCH_2\cdot\text{(ads)} \longrightarrow ArCH_2CH_2Ar$$

$$\searrow -e$$

$$ArCH_2^+ \xrightarrow{MeOH} ArCH_2OMe$$

The reaction referred to above proceeds by electron removal from the carboxylate group (Kolbe reaction), but an example of the alternative pathway—electron removal from the aromatic ring—has been shown to occur during anodic oxidation of the caesium salt (**275**).[469] No Kolbe dimer is obtained from the electrolysis of a number of α-chloro and α-fluoro carboxylic acids.[470]

An alternative mechanism for the anodic substitution reaction between pyridine and 9,10-diphenylanthracene has been put forward,[471] namely that the initial anthracene cation radical, formed by electrode oxidation, disproportionates before reacting with pyridine.

[460] P. T. Cottrell and C. K. Mann, *J. Am. Chem. Soc.*, **93**, 3579 (1971).
[461] See *Org. Reaction Mech.*, **1970**, 368.
[462] R. D. Rieke, S. E. Bales, P. M. Hudnall, and C. F. Meares, *J. Am. Chem. Soc.*, **93**, 697 (1971).
[463] J. E. O'Reilly and P. J. Elving, *J. Am. Chem. Soc.*, **93**, 1871 (1971).
[464] A. G. Evans, J. C. Evans, P. J. Emes, C. L. James, and P. J. Pomery, *J. Chem. Soc.* (B), **1971**, 1484.
[465] H. Brockmann, H. Greve, and K. Hoyermann, *Tetrahedron Letters*, **1971**, 1493.
[466] T. Shono, A. Ikeda, and Y. Kimura, *Tetrahedron Letters*, **1971**, 3599.
[467] J. M. Bobbitt, H. Yagi, S. Shibuya, and J. T. Stock, *J. Org. Chem.*, **36**, 3006 (1971).
[468] J. P. Coleman, J. H. P. Utley, and B. C. L. Weedon, *Chem. Comm.*, **1971**, 438.
[469] J. P. Coleman and L. Eberson, *Chem. Comm.*, **1971**, 1300.
[470] P. C. Arora and R. G. Woolford, *Can. J. Chem.*, **49**, 2681 (1971).
[471] L. Marcoux, *J. Am. Chem. Soc.*, **93**, 537 (1971).

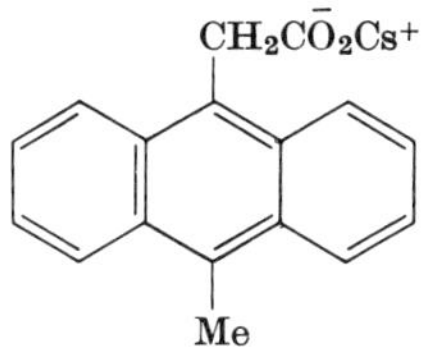

(275) [anthracene with CH2CO2-Cs+ at 9-position and Me at 10-position]

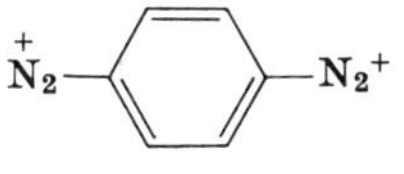

(276) [p-benzene bisdiazonium: +N2—C6H4—N2+]

A number of investigations of the electron-transfer reduction of arenediazonium ions have been reported. In acidic aqueous solution, benzene-p-bisdiazonium ion (276) reacts with alcohols more rapidly than it does with water and a free-radical chain process is proposed,[472] initiated either by electron transfer from diazotate to diazonium ion or by homolysis of a diazoanhydride; oxygen is necessary for the conversion of (276), promoted by iodide ion, into p-hydroxybenzenediazonium ion. The actual chain-propagation reactions involve electron transfer from alcohol-derived hydroxyalkyl radicals to arenediazonium ion, and more evidence of this type of oxidation is provided by the ^{60}Co-γ-radiation-initiated free-radical reaction of toluene-p-diazonium ions with methanol in aqueous solution:[473]

Initiation: $H_2O \xrightarrow{\sim\!\sim\!\sim} \cdot OH + H\cdot + e^-_{aq}$

$e^-_{aq} + ArN_2^+ \rightarrow Ar\cdot + N_2$

Propagation: $Ar\cdot + MeOH \rightarrow ArH + \cdot CH_2OH$

$ArN_2^+ + \cdot CH_2OH \rightarrow Ar\cdot + CH_2O + H^+ + N_2$

Termination: $Ar\cdot + ArN_2^+ \rightarrow$ Products

Product-distribution and the significance of the terminating step with respect to related ESR flow studies[474] are discussed.

Diethyl peroxydicarbonate initiates the reductive elimination of nitrogen from the diazo ketone (277) and related substrates;[475] the azo compound (279) is shown to be an intermediate in the reaction mechanism outlined.

$$Me_2CHOH \xrightarrow[40-70°]{(EtOCO_2)_2} Me_2\dot{C}OH$$

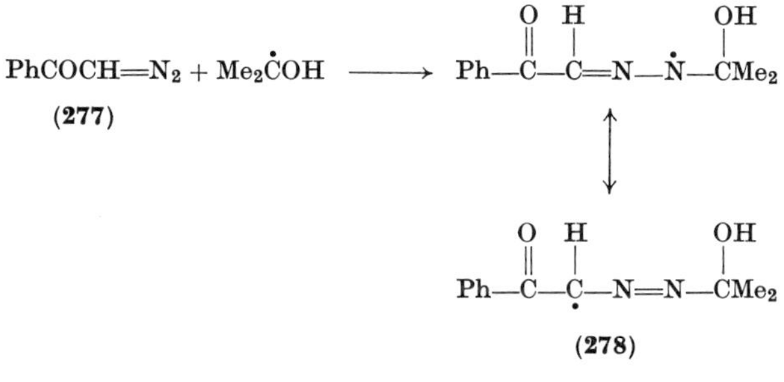

$$PhCOCH=N_2 + Me_2\dot{C}OH \longrightarrow Ph-\underset{\|}{C}-\underset{|}{C}=N-\dot{N}-\underset{|}{C}Me_2$$
$$\phantom{PhCOCH=N_2 + Me_2\dot{C}OH \longrightarrow Ph-}OH\phantom{=N-\dot{N}-}OH$$

(277)

$\updownarrow$

$$Ph-\underset{\|}{C}-\underset{\cdot}{C}-N=N-\underset{|}{C}Me_2$$
$$OHOH$$

(278)

[472] E. S. Lewis and D. J. Chalmers, *J. Am. Chem. Soc.*, **93**, 3267 (1971).
[473] J. E. Packer, D. B. House, and E. J. Rasburn, *J. Chem. Soc.* (B), **1971**, 1574.
[474] A. L. J. Beckwith and R. O. C. Norman, *J. Chem. Soc.* (B), **1969**, 403.
[475] L. Horner and H. Schwarz, *Ann. Chem.*, **747**, 1 (1971).

$$(278) + Me_2CHOH \longrightarrow PhCOCH_2N\!\!=\!\!NCMe_2 + Me_2\overset{\bullet}{C}OH$$
$$\overset{\displaystyle OH}{}$$
(279)

$$(279) \rightarrow PhCOMe + N_2 + Me_2CO$$

ESR has been employed to demonstrate the radical nature of the reduction of arenediazonium salts with substituted phenoxide (ArO$^-$) and alkoxide ions;[476] with 2,4,6-tri-*tert*-butylphenoxide the corresponding phenoxy radical is observed, i.e.:

$$ArO^- + PhN_2^+ \rightleftharpoons ArO\cdot + PhN_2\cdot$$

and, by using the spin trapping technique, phenyl radicals are shown to be produced by reaction of the diazonium salt with alkoxide ions, although the exact nature of this electron-transfer process is not made clear. Related work by Bunnett's group[477,478] shows that *p*-nitrobenzenediazonium ion combines rapidly with methoxide ion to form azomethyl *cis-p*-nitrophenyl ether, which then reacts further to form nitrobenzene (by a radical mechanism) and the *trans*-azo ether in approximately equal amounts.

Electron-transfer mechanisms also account for the arylation of quinones by arenediazonium[479] salts, and the reaction of the latter with *p*-Me$_2$NC$_6$H$_4$NMe$_2$.[480] Aryl radicals, from the hydroxide ion/ether reduction (Meerwein) of the corresponding diazonium salt react with phenyldiimine (PhN=NH) to give 1-aryl-2-phenylhydrazine.[481] An unusual ring closure leading predominantly to (281) has been observed during the decomposition of the diazonium salt (280); a radical mechanism is thought to operate.[482]

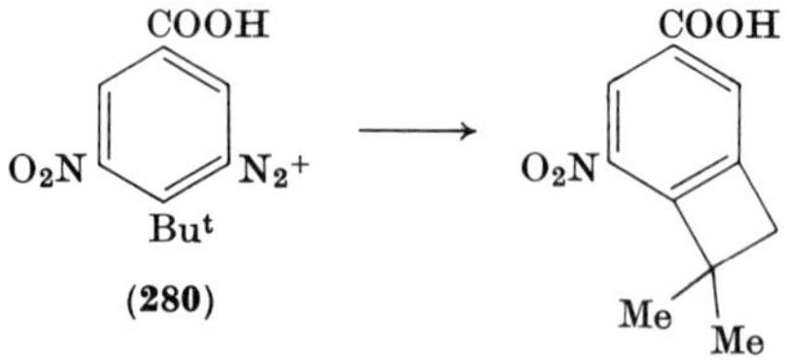

Tracer studies on the reductive coupling of naphthalene-1-diazonium ions containing deuterium at the 2- or 4-position show that the original orientations of deuterium relative to nitrogen are preserved in the 1,1′-azonaphthalene products.[483]

tert-Butoxy radicals, generated by photolysis of di-*tert*-butyl peroxide or thermolysis of di-*tert*-butyl hyponitrite in 1:1 *tert*-BuOH/H$_2$O, are efficiently scavenged by a simple one-electron transfer from the "Paraquat" cation radical[484] (see Scheme). The reaction

---

[476] A. Rieker, P. Niederer, and H. B. Stegmann, *Tetrahedron Letters*, **1971**, 3873.
[477] W. J. Boyle, T. J. Broxton, and J. F. Bunnett, *Chem. Comm.*, **1971**, 1469.
[478] J. F. Bunnett and H. Takayama, *J. Org. Chem.*, **33**, 1924 (1968).
[479] A. A. Matnishyan, G. V. Fomin, E. V. Prut, B. I. Liogon'kii, and A. A. Berlin, *Zhur. Fiz. Khim.*, **45**, 1308 (1971); *Chem. Abs.*, **75**, 48087 (1971).
[480] K. A. Bilevich, N. N. Bubnov, B. Ya. Medvedev, O. Yu. Okhlobystin, and L. V. Ermanson, *Dokl. Akad. Nauk SSSR*, **193**, 583 (1970); *Chem. Abs.*, **74**, 31401 (1971).
[481] A. Heesing and B.-U. Kaiser, *Tetrahedron Letters*, **1971**, 1307.
[482] M. H. Knight, T. Putkey, and H. S. Mosher, *J. Org. Chem.* **36**, 1483 (1971).
[483] D. V. Banthorpe and J. A. Thomas, *J. Chem. Soc.* (B), **1971**, 365.
[484] A. S. Hopkins and A. Ledwith, *Chem. Comm.*, **1971**, 830.

$$\text{Bu}^t\text{ON}=\text{NOBu}^t \xrightarrow[\text{Heat}]{\text{Slow}} \text{Bu}^t\text{O}\cdot + \text{N}_2 + \cdot\text{OBu}^t \xrightarrow[-\text{N}_2]{\text{Fast}} \text{Bu}^t\text{OOBu}^t$$

$$\text{Bu}^t\text{O}\cdot + \text{PQ}^{+\cdot} \xrightarrow{\text{Fast}} \text{Bu}^t\text{O}^- + \text{PQ}^{++}$$

$$(\text{PQ}^{++} = \text{Me}\overset{+}{\text{N}}\!\!\!\underset{}{\bigcirc}\!\!\!\underset{}{\bigcirc}\!\!\overset{+}{\text{N}}\text{Me})$$

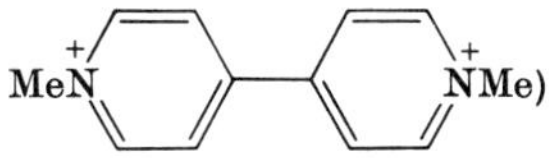

was predicted[485] on the basis of the mechanism proposed for photo-oxidation of alcohols by "Paraquat", and points the way to a new type of radical-scavenging process leading to evaluation of redox potentials for organic radicals. Intramolecular electron-transfer reactions of "Paraquat" salts of the 7,7,8,8-tetracyanoquinodimethane anion radical have been described,[486] along with electron-transfer reduction of new quaternary derivatives of 2-(2-pyridyl)quinoline.[487]

Hydrogen abstraction is widely assumed to be the primary step in oxidation of alcohols by peroxydisulphate ion, but recent findings support an electron-abstraction reaction resulting in initial formation of alkoxy radicals rather than the more usual hydroxyalkyl radicals;[488] the latter are produced by subsequent intermolecular hydrogen abstraction, as shown in the annexed Scheme for methanol.

$$\text{S}_2\text{O}_8{}^{2-} \xrightarrow[\text{or Heat}]{h\nu} 2\text{SO}_4{}^{-}$$

$$\text{SO}_4{}^{-} + \text{MeOH} \longrightarrow [\text{MeOH}]^{+\cdot} + \text{SO}_4{}^{2-}$$

$$\downarrow \text{MeOH}$$

$$\cdot\text{CH}_2\text{OH} \xleftarrow{\text{MeOH}} \text{MeO}\cdot + \text{MeOH}_2{}^+$$

Persulphate also appears to act as an electron-transfer reagent during oxidation of a number of N-methylbiphenyl-2-carboxamides in the presence of sodium carbonate.[489] Reaction of the p-methoxy derivative (**282**) leads to high yield of the dienone (**284**) as well as the corresponding N-methylphenanthridone.[490] Carboxamido-radicals are thought not to be intermediates in these reactions; it is suggested, rather, that cyclization occurs by intramolecular capture of the radical cation (**283**) produced by one-electron oxidation of the aromatic component.

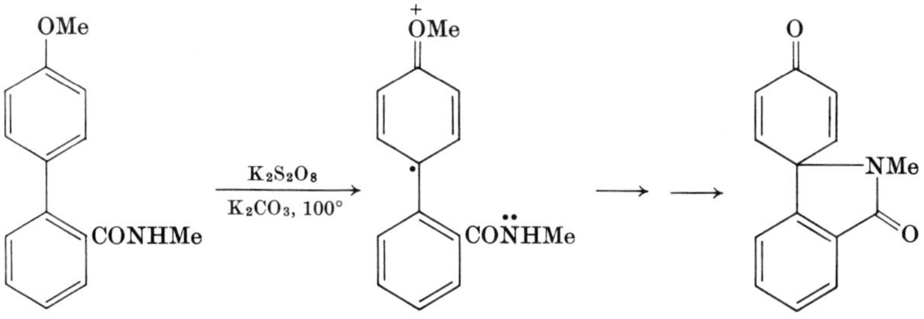

(**282**) → $\xrightarrow[\text{K}_2\text{CO}_3, 100°]{\text{K}_2\text{S}_2\text{O}_8}$ → (**283**) → → (**284**) 70%

[485] See *Org. Reaction Mech.*, **1970**, 374.
[486] A. Rembaum, V. Hadek, and S. P. S. Yen, *J. Am. Chem. Soc.*, **93**, 2532 (1971).
[487] A. L. Black and L. A. Summers, *J. Chem. Soc.* (C), **1971**, 2271.
[488] A. Ledwith, P. J. Russell, and L. H. Sutcliffe, *Chem. Comm.*, **1971**, 964.
[489] D. H. Hey, G. H. Jones, and M. J. Perkins, *Chem. Comm.*, **1971**, 998.
[490] See *Org. Reaction Mech.*, **1970**, 351.

Kinetics of oxidation of 2-piperidylethanol by persulphate,[491] and studies of the related peroxydiphosphate ($P_2O_8^{4-}$) species[492] have been reported. The relative importance of aryloxy radical versus aryloxy cation formation in oxidative coupling of plant phenols has been discussed in the light of recent data for electron-transfer reactions of phenols.[493] Phenoxy radicals may be formed by one-electron reduction of cyclohexadienones,[494] and certain electron-transfer reactions between cobalt(III) and europium(II) may be catalysed by organic molecules.[495] An electron-transfer mechanism has been proposed[496] as an alternative to an $E2$ pathway for the conversion of vic-dibromides into olefins by stabilized carbanions, and the one-electron reduction of a series of 1,4-diarylbicyclo[2.2.2]octanes has been studied.[497]

Nitroxides

The general use of nitrones and C-nitroso compounds as mechanistic probes for radical reactions in solution is now firmly established with the appearance of three independent reviews.[498-500] The three authors have each made significant contributions to this new development in free-radical chemistry and together the reviews represent a very full account of the uses and limitations of this radical scavenging (spin trapping) technique.

The C-nitroso-compound employed is usually 2-methyl-2-nitrosopropane ("nitrosobutane", Bu^{t}NO) which has proved particularly useful in the identification of radical intermediates. The fully deuteriated analogue has now been synthesized and an outline of its advantages given.[501] The deuteriated trap has also been employed in scavenging 2,6-dimethoxyphenyl radicals[502] and has facilitated identification of halocarbonyl nitroxides **(285)** formed as by-products in the reactions involving trihalogenomethyl radicals and nitrosobutane.[503] Further applications of the nitroso-compound include

$$\text{Bu}^t\text{N}=\text{O} + \overset{\bullet}{\text{C}}\text{Cl}_3 \longrightarrow \text{Bu}^t-\overset{\overset{\text{O}\bullet}{|}}{\text{N}}-\text{CCl}_3 \xrightarrow{\text{R}\bullet}$$

$$\text{Bu}^t-\overset{\overset{\text{O}\uparrow}{|}}{\text{N}}=\text{CCl}_2 \xrightarrow{\text{R}_2\text{NO}\bullet} \text{Bu}^t-\overset{\overset{\text{O}\bullet}{|}}{\underset{\text{ONR}_2}{\text{N}}}-\text{CCl}_2 \longrightarrow \text{Bu}^t-\overset{\overset{\text{O}\bullet}{|}}{\text{N}}-\overset{\overset{\text{O}}{\|}}{\text{C}}\text{Cl} \quad \textbf{(285)}$$

trapping of phosphorus-centred radicals[504] [$\cdot\text{PO}_3^{2-}$, Ph$\overset{\bullet}{\text{P}}H, Ph_2P\cdot$, Me$_2\overset{\bullet}{\text{P}}$S, (EtO)$_2\overset{\bullet}{\text{P}}$O, etc.] and radical intermediates produced during the photoreduction of a number of p-aminobenzophenones.[505]

---

[491] R. P. Meliksetyan, N. M. Beileryan, and O. A. Chaltykyan, *Arm. Khim. Zhur.*, **24**, 108 (1971); *Chem. Abs.*, **75**, 62830 (1971).
[492] R. J. Luzzier, W. M. Risen, and J. O. Edwards, *J. Phys. Chem.*, **74**, 4039 (1970).
[493] W. A. Waters, *J. Chem. Soc.* (B), **1971**, 2026.
[494] W. D. Pokhodenko, N. N. Kalibabtschuk, and W. A. Khizny, *Ann. Chem.*, **743**, 192 (1971).
[495] C. Norris and F. R. Nordmeyer, *J. Am. Chem. Soc.*, **93**, 4044 (1971).
[496] D. G. Korzan, F. Chen, and C. Ainsworth, *Chem. Comm.*, **1971**, 1053.
[497] H. E. Zimmerman and R. D. McKelvey, *J. Am. Chem. Soc.*, **93**, 3638 (1971).
[498] E. G. Janzen, *Accounts Chem. Res.*, **4**, 31 (1971).
[499] M. J. Perkins in ref. 1, p. 97.
[500] C. Lagercrantz, *J. Phys. Chem.*, **75**, 3466 (1971).
[501] R. J. Holman and M. J. Perkins, *J. Chem. Soc.* (C), **1971**, 2324.
[502] H. J. Jakobsen and K. Torssell, *Tetrahedron Letters*, **1970**, 5003.
[503] R. J. Holman and M. J. Perkins, *Chem. Comm.*, **1971**, 244; see also J. W. Hartgerink, J. B. F. N. Engberts, and T. J. de Boer, *Tetrahedron Letters*, **1971**, 2709.
[504] H. Karlsson and C. Lagercrantz, *Acta Chem. Scand.*, **24**, 3411 (1970).
[505] I. H. Leaver, *Tetrahedron Letters*, **1971**, 2333.

Trapping experiments with the benzylidene nitrone PhCH=N(→O)Bu^t have demonstrated that methoxy radicals are important intermediates in photochemical oxidations of methanol by "Paraquat" dichloride,[484] uranyl nitrate, potassium peroxydisulphate, and lead tetra-acetate.[488] In the thermal oxidation by peroxydisulphate ion both methoxy and hydroxymethyl radicals were trapped depending on the concentration of nitrone employed. Both nitrone and nitroso-compounds have shown that free phenyl radicals are produced during the reaction of N-aroyl-N'-arylhydrazines (**286**) with sodium hydride.[506] Nitroxide (**287**) was also obtained when the hydrazine was oxidized with

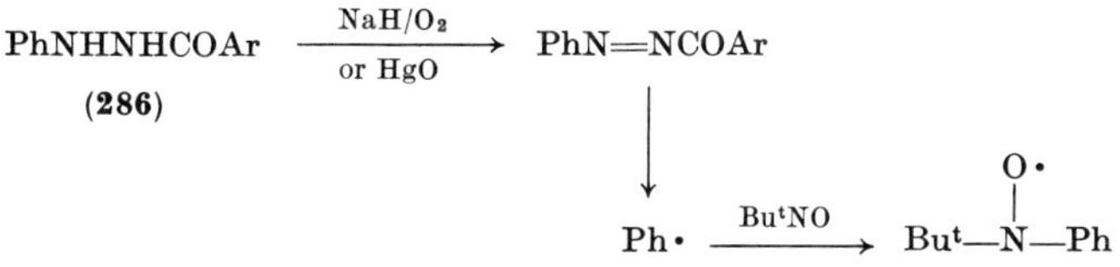

$$\text{PhNHNHCOAr} \xrightarrow[\text{or HgO}]{\text{NaH/O}_2} \text{PhN=NCOAr}$$
(**286**)

$$\text{Ph} \cdot \xrightarrow{\text{Bu}^t\text{NO}} \text{Bu}^t\text{—}\overset{\overset{\text{O} \cdot}{|}}{\text{N}}\text{—Ph}$$
(**287**)

mercuric oxide. Oxidation of 3-hydroxy-1,3-diphenyltriazene with silver oxide[507] yields nitrogen and diphenylnitroxide according to the annexed Scheme. Similar results occurred when (**288**) reacted with the 2,4,6-tri-*tert*-butylphenoxy radical.

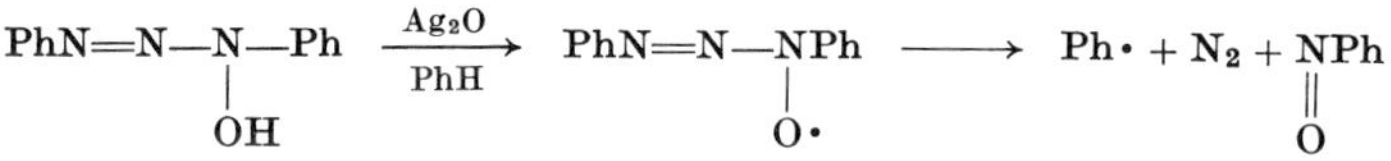

$$\underset{(\textbf{288})}{\text{PhN=N—N(OH)—Ph}} \xrightarrow[\text{PhH}]{\text{Ag}_2\text{O}} \text{PhN=N—N(O}\cdot\text{)Ph} \longrightarrow \text{Ph} \cdot + \text{N}_2 + \text{NPh(=O)}$$

$$\text{Ph} \cdot + \text{PhN=O} \rightarrow \text{Ph}_2\text{N—O} \cdot$$

Unlike nitrones, nitroso-compounds are poor scavengers of alkoxy radicals and this can be used to advantage. Thus the carbon radical formed by intramolecular addition of alkoxy radical (**289**) can be trapped and positively identified by carrying out the nitrite photolysis in the presence of nitrosobutane.[508]

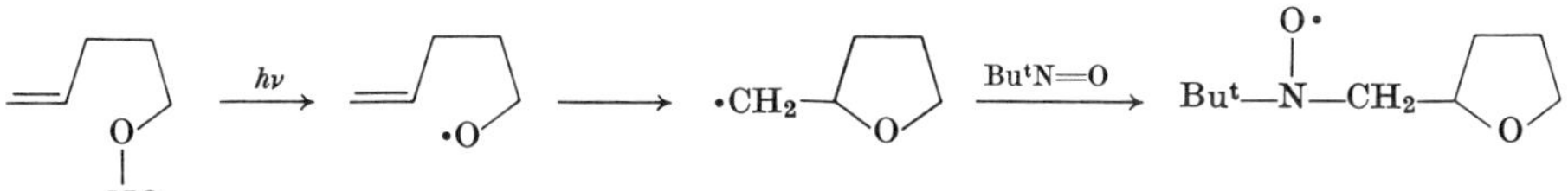

Self-formation of symmetrical nitroxides is one of the disadvantages encountered in photochemical reactions with nitroso compounds and may result from either reaction (11) or reaction (12).

$$\text{RNO} \xrightarrow{h\nu} \text{RNO}^* \xrightarrow{\text{RNO}} \text{R}_2\text{N—O} \cdot + \text{NO} \qquad \ldots (11)$$

$$\text{RNO} \xrightarrow{h\nu} \text{RNO}^* \longrightarrow \text{R} \cdot + \text{NO}$$
$$\text{R} \cdot + \text{RNO} \rightarrow \text{R}_2\text{N—O} \cdot \qquad \ldots (12)$$

---

[506] R. A. W. Johnstone, A. F. Neville, and P. J. Russell, *J. Chem. Soc.* (B), **1971**, 1183.
[507] G. A. Abakumov, V. K. Cherkasov, and G. A. Razuvaev, *Dokl. Akad. Nauk SSSR*, **197**, 823 (1971); *Chem. Abs.*, **75**, 34823 (1971).
[508] P. Tordo, M. P. Bertrand, and J.-M. Surzur, *Tetrahedron Letters*, **1970**, 3399.

Photolysis of 1-cyclopropyl-1-nitrosoethane (**290**) produces nitroxide (**292**), indicating that reaction (12) is operative;[509] radical (**291**) would have been formed if nitroxide formation took place by reaction (11) but, instead, the free cyclopropylmethyl radical undergoes rapid ring opening before addition to (**290**).

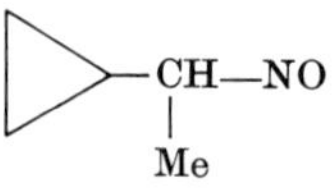

Synthesis and radical addition to nitrosocyclopropane has been carried out by the same group,[510] who have also used the trapping technique to identify intermediate radicals produced during photochemically initiated reactions of substituted 1,3-dioxolans and 1,3-oxathiolans (**293**) in $CFCl_3$.[511] Whilst cyclic radicals are trapped, no ring-opened (thio)ester radicals were identified, and the latter process is thought to occur by heterolysis.

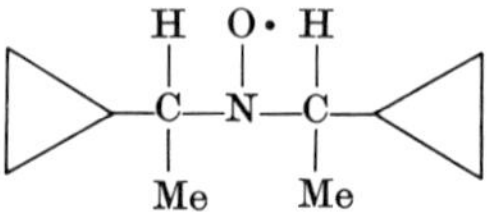

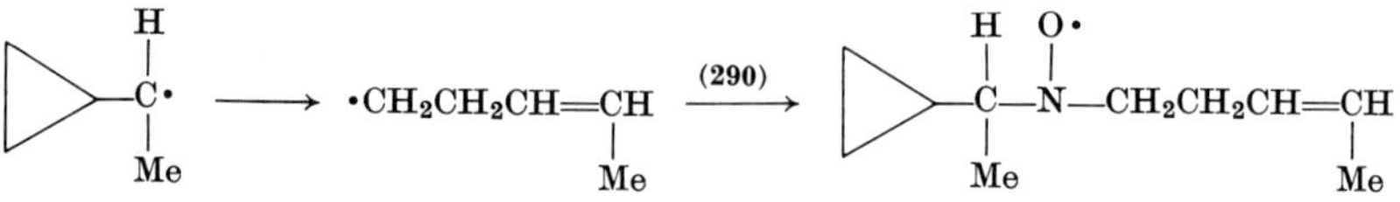

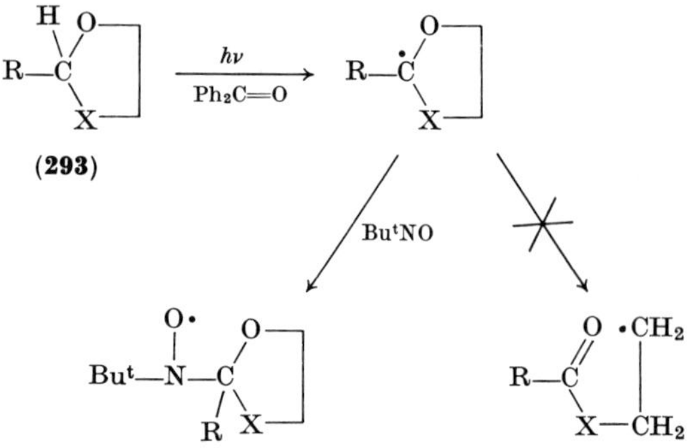

X = O or S

2,4,6-Tri-*tert*-butylnitrosobenzene is shown to offer considerable advantages over nitrosobenzene as a radical probe.[512] However, although primary alkyl radicals attack the nitrogen atom producing the normal type of addition, secondary alkyl radicals attack both the oxygen and the nitrogen atom of (**294**), and spectra of a mixture of nitroxide (**295**) and anilino (**296**) radicals are observed. CIDNP effects have been observed during addition of ·CMe_2CN radicals to nitrosobenzene.[513]

[509] J. A. Maassen, H. Hittenhausen, and T. J. de Boer, *Tetrahedron Letters*, **1971**, 3213.
[510] R. Stammer, J. B. F. N. Engberts, and T. J. de Boer, *Rec. Trav. Chim.*, **89**, 169 (1970).
[511] J. W. Hartgerink, L. C. J. van der Laan, J. B. F. N. Engberts, and T. J. de Boer, *Tetrahedron*, **27**, 4323 (1971).
[512] S. Terabe and R. Konaka, *J. Am. Chem. Soc.*, **93**, 4306 (1971).
[513] H. Iwamura, M. Iwamura, M. Tamura, and K. Shiomi, *Bull. Chem. Soc. Japan*, **43**, 3638 (1970).

$$\text{ArN}=\text{O} + \text{Me}_2\text{CH}\cdot \rightarrow \text{ArN}-\text{CHMe}_2 + \text{Ar}\dot{\text{N}}-\text{O}-\text{CHMe}_2$$
$$(294) \qquad\qquad\qquad\;\; | \qquad\qquad\qquad (296)$$
$$\qquad\qquad\qquad\qquad\;\; \text{O}\cdot$$
$$\qquad\qquad\qquad\quad (295)$$

Ar = 2,4,6-tri-t-butylphenyl

A full report of the stable cobalt nitroxides (**297**) obtained by reduction of aromatic nitro-compounds by the pentacyanocobalt(II) anion [·CO(CN)$_5$]$^{3-}$ or by addition of the latter to aromatic nitroso-compounds has been published.[514] Other complex ions of

$$\left[(\text{NC})_5\text{Co}-\text{N}\begin{array}{c}\text{O}\cdot\\ \diagdown\\ \text{Ar}\end{array}\right]^{3-} \qquad\qquad \text{CCl}_3-\text{N}=\text{CCl}_2$$
$$\qquad\qquad\qquad\qquad\qquad\qquad\qquad\qquad\quad \downarrow$$
$$\qquad\qquad\qquad\qquad\qquad\qquad\qquad\qquad\quad \text{O}$$
$$\qquad\quad (297) \qquad\qquad\qquad\qquad\qquad\qquad (298)$$

cobalt(II) have also been used.[515] Addition of chlorine atoms or phenyl radicals to the nitrone (**298**) does not occur but, instead, chlorine abstraction is observed.[516]

Whilst nitrones and nitroso compounds are efficient radical scavengers they also react with anions to give amino-oxy anions which are invariably oxidized to some extent to nitroxides.[517] These findings are important and demonstrate that results obtained from spin-trapping experiments should be used with caution, particularly when anionic species are present in the reaction mixture since nitrones and nitroso compounds are chemically reactive compounds. In certain cases other classes of radical scavenger may be of greater use; for example, *aci*-anions of nitroalkanes (R^{1}R^{2}C=NO$_2^-$) have been employed to identify transient radicals produced in flow systems.[518] Recent examples include the trapping of alkylamino radicals (**299**) during the interaction of Ti^{3+} ions and some N-alkylhydroxylamines; the amino radicals are too reactive to be observed directly.[397] In a similar manner the primary radicals produced on photolysis of alkaline aqueous thiosulphate solutions have been identified.[519]

$$\text{R}^1\text{R}^2\text{CHNHOH} \xrightarrow{\text{Ti}^{\text{III}}} \text{R}^1\text{R}^2\text{CHNH}\cdot \xrightarrow{\text{CH}_2=\text{NO}_2^-} \text{R}^1\text{R}^2\text{CHNHCH}_2-\text{NO}_2^{-\cdot}$$
$$\qquad\qquad\qquad\qquad\qquad (299)$$

The utility of aromatic nitrile N-oxides as spin-trapping reagents has also been outlined[520]—in this case iminoxy radicals are obtained—and in spite of spectral complications, the trapped radical can be identified, making nitrile N-oxides a useful addition to available spin traps. Fumarate ions have also been employed in radical trapping.[521]

A more detailed account of nitroxide formation in the reaction of NO$_2$ and NO with styrenes has been published.[522]

Apart from their role in mechanistic studies nitroxide radicals are of value synthetically. Progress in the chemistry of nitroxide radicals has been reviewed by the leading

---

[514] M. G. Swanwick and W. A. Waters, *J. Chem. Soc.* (B), **1971**, 1059.
[515] See *Org. Reaction Mech.*, **1970**, 365.
[516] V. Astley and H. Sutcliffe, *Chem. Comm.*, **1971**, 1303.
[517] A. R. Forrester and S. P. Hepburn, *J. Chem. Soc.* (C), **1971**, 701.
[518] See *Org. Reaction Mech.*, **1970**, 372.
[519] D. Behar and R. W. Fessenden, *J. Phys. Chem.*, **75**, 2752 (1971).
[520] B. C. Gilbert, V. Malatesta, and R. O. C. Norman, *J. Am. Chem. Soc.*, **93**, 3290 (1971).
[521] P. Neta, *J. Phys. Chem.*, **75**, 2570 (1971).
[522] L. Jonkman, H. Muller, and J. Kommandeur, *J. Am. Chem. Soc.*, **93**, 5833 (1971).

Russian worker[523] who has also reported[524] the quantitative oxidation of a stable nitroxide with $SbCl_6^-$. A full study on the reaction of bistrifluoromethyl nitroxide, $(CF_3)_2NO\cdot$, with some alkanes and alkenes illustrates its reactivity and usefulness. The alkane nitroxide reactions are thought to involve olefin formation by hydrogen abstraction from alkyl radicals formed initially; thus, (**301**) was produced from isobutane; coupling products, e.g. (**300**), were also obtained and reaction of the nitroxide with alkenes gave products from both addition and hydrogen-abstraction processes.[525a] A full study of the reactions of $(CF_3)_2NO\cdot$ with acetylenes has also been reported.[525b]

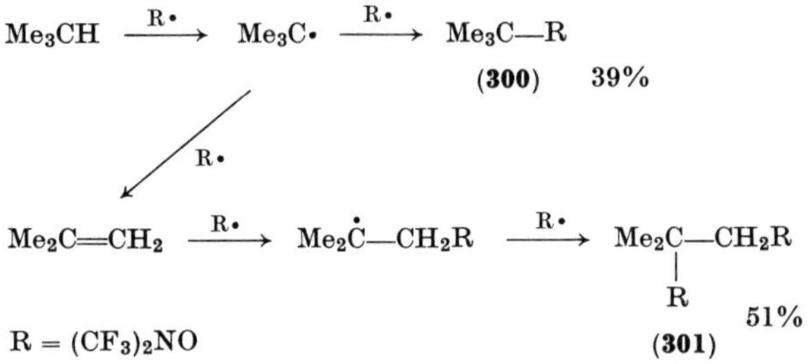

$$Me_3CH \xrightarrow{R\cdot} Me_3C\cdot \xrightarrow{R\cdot} Me_3C\text{---}R$$
$$(\textbf{300}) \quad 39\%$$

$$Me_2C\text{=}CH_2 \xrightarrow{R\cdot} Me_2\overset{\cdot}{C}\text{---}CH_2R \xrightarrow{R\cdot} Me_2C\text{---}CH_2R$$
$$\qquad\qquad\qquad\qquad\qquad\qquad\qquad\qquad | \\ \qquad\qquad\qquad\qquad\qquad\qquad\qquad\quad R \quad 51\%$$
$$R = (CF_3)_2NO \qquad\qquad\qquad\qquad\qquad (\textbf{301})$$

Measurement of the bimolecular rate constants for the self-reaction of diethyl nitroxide radicals in various solvents has suggested the reaction mechanism:[526]

$$2Et_2NO\cdot \rightleftharpoons [Et_2NO]_2 \rightarrow Et_2NOH + MeCH\text{=}\overset{\overset{O}{\uparrow}}{N}CH_2Me$$

A number of papers deal with the synthesis and chemical and spectral properties of nitroxides; these include steroidal,[527] [2.2]paracyclophane,[528] bis(organosilyl),[529] and some ^{13}C-labelled nitroxide radicals.[530] Photochemical denitrosation of a dinitroso compound has been employed to synthesize a stable isoquinuclidine nitroxide,[531] but claims[532] for the first three-membered-ring nitroxide (**303**) by similar treatment of (**302**) have been vigorously refuted.[533] The synthesis of some *tert*-butyl vinyl nitroxides has

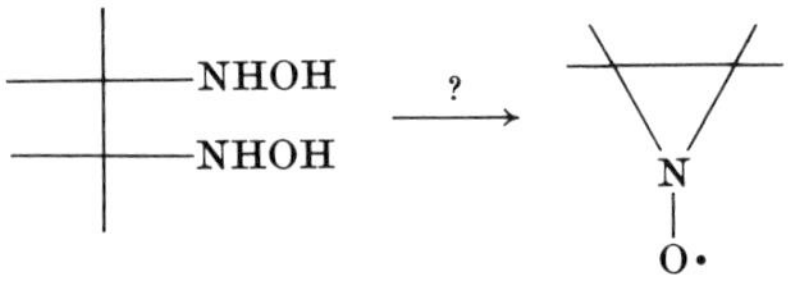

(**302**) → ? → (**303**)

---

[523] E. G. Rozantsev and V. D. Sholle, *Russ. Chem. Rev.*, **40**, 233 (1971).
[524] V. A. Golubev, R. I. Zhdanov, V. M. Gida, and E. G. Rozantsev, *Izv. Akad. Nauk SSSR, Ser. Khim.*, **1970**, 2815.
[525a] R. E. Banks, R. N. Haszeldine, and B. Justin, *J. Chem. Soc.* (C), **1971**, 2777.
[525b] R. E. Banks, R. N. Haszeldine, and T. Myerscough, *J. Chem. Soc.* (C), **1971**, 1951.
[526] K. Adamic, D. F. Bowman, T. Gillan, and K. U. Ingold, *J. Am. Chem. Soc.*, **93**, 902 (1971).
[527] R. Ramasseul and A. Rassat, *Tetrahedron Letters*, **1971**, 4623.
[528] A. R. Forrester and R. Ramasseul, *J. Chem. Soc.* (B), **1971**, 1638, 1645.
[529] R. West and P. Boudjouk, *J. Am. Chem. Soc.*, **93**, 5901 (1971).
[530] G. Chapelet-Letourneux and A. Rassat, *Bull. Soc. Chim. France*, **1971**, 3216.
[531] A. Rassat and P. Rey, *Chem. Comm.*, **1971**, 1161.
[532] G. R. Luckhurst and F. Sundholm, *Tetrahedron Letters*, **1971**, 675.
[533] P. Singh, D. G. B. Boocock, and E. F. Ullman, *Tetrahedron Letters*, **1971**, 3935.

been achieved and, although the monomeric radicals could not be homopolymerized, paramagnetic copolymers with styrene were obtained.[534] Isolation of an iminoxy radical, $Bu^t_2C=N-O\cdot$ has been reported.[535] Spectra of some bicyclic iminoxy radicals provide structural information on the parent bicyclic oxime.[536]

Nitroxides have been obtained during the chemical oxidation of N-(p-nitrosophenyl)-N-phenylhydroxylamine[537] and from Mannich-type condensation products of sulphinic acids with aldehydes and hydroxylamines.[538] Electron-donating effects of substituents have been found by UV photolysis of aromatic and heterocyclic nitro compounds to affect critically the formation of alkoxy nitroxide radicals.[539] Spectroscopic studies of phenyl N-nitrosonitroxide,[540] tetraphenylpyrrole N-oxide,[541] oxazolidine nitroxides,[542] acyl alkyl nitroxides[543] and tert-butyl phenyl nitroxides[544] have been published. Magnetic resonance has also been employed in studies of adamantane and bicyclic nitroxides.[545] Oxidation of bi-(N-arylnaphthylamines) produces stable nitroxides if the reactive naphthalene positions are blocked.[546] The reactivities of 2,2,6,6-tetramethyl-4-oxopiperidine N-oxide[547] and hindered pyrrole nitroxides[548] have been discussed. Complexes of nitroxides with Lewis acids,[549] aromatic solvents and monomers[550] have been described, and further uses of nitroxides as spin probes and spin labels have been recorded.[551]

Autoxidation

The kinetics of hydrocarbon autoxidation in the liquid phase[552] and more general aspects of liquid-phase thermal and photochemical oxidations[553] have been reviewed.

Uncatalysed autoxidation of anthranol to anthraquinone occurs readily in alkaline conditions; a kinetic study[554] suggests that the rate-determining step involves electron transfer from anthranol anion to molecular oxygen. It has been noted[555] that strongly

---

[534] A. R. Forrester and S. P. Hepburn, *J. Chem. Soc.* (C), **1971**, 3322.
[535] J. L. Brokenshire, G. D. Mendenhall, and K. U Ingold, *J. Am. Chem. Soc.*, **93**, 5278 (1971).
[536] H. Căldăraru and M. Moraru, *Tetrahedron Letters*, **1971**, 3183.
[537] G. Cauquis, A. Rassat, J.-P. Ravet and D. Serve, *Tetrahedron Letters*, **1971**, 971.
[538] G. Rawson and J. B. F. N. Engberts, *Tetrahedron*, **26**, 5653 (1970).
[539] R. B. Sleight and L. H. Sutcliffe, *Trans. Faraday Soc.*, **67**, 2195 (1971).
[540] A. T. Balaban, N. Negoiță, and I. Pascaru, *Rev. Roum. Chim.*, **16**, 721 (1971).
[541] R. Ramasseul, A. Rassat, G. Rio, and M.-J. Scholl, *Bull. Soc. Chim. France*, **1971**, 215.
[542] P. Michon and A. Rassat, *Bull. Soc. Chim. France*, **1971**, 3561.
[543] H. G. Aurich and J. Trösken, *Ann. Chem.*, **745**, 159 (1971).
[544] H. J. Jakobsen, T. E. Petersen, and K. Torssell, *Tetrahedron Letters*, **1971**, 2913.
[545] C. Morat and A. Rassat, *Bull. Soc. Chim. France*, **1971**, 893; A. Rassat and J. Ronzaud, *J. Am. Chem. Soc.*, **93**, 5041 (1971).
[546] R. F. Bridger and E. T. Strom, *J. Org. Chem.*, **36**, 560 (1971).
[547] T. Yoshioka, S. Higashida, S. Morimura, and K. Murayama, *Bull. Chem. Soc. Japan*, **44**, 2207 (1971).
[548] R. Ramasseul and A. Rassat, *Bull. Soc. Chim. France*, **1970**, 4330.
[549] T. B. Eames and B. M. Hoffman, *J. Am. Chem. Soc.*, **93**, 3141 (1971).
[550] L. Batt, G. M. Burnett, G. G. Cameron, and J. Cameron, *Chem. Comm.*, **1971**, 29.
[551] J. Oakes, *Nature*, **231**, 38 (1971); J. F. W. Keana and R. J. Dinerstein, *J. Am. Chem. Soc.*, **93**, 2808 (1971); I. Morishima, K. Endo, and T. Yonezawa, *J. Am. Chem. Soc.*, **93**, 2048 (1971).
[552] J. Betts, *Quart. Rev. (London)*, **25**, 265 (1971).
[553] S. V. Anantakrishnan and H. Jayaraman, *J. Sci. Ind. Res.*, **29**, 323 (1970); *Chem. Abs.*, **74**, 63599 (1971).
[554] Y. Ogata, Y. Kosugi, and K. Nate, *Tetrahedron*, **27**, 2705 (1971).
[555] P. Ashworth and W. T. Dixon, *Chem. Comm.*, **1971**, 1150.

alkaline solutions of hydroquinones, shaken in air, may give ESR spectra corresponding to radicals other than the expected semiquinones.

The kinetics of inhibition of cumene autoxidation at 60° by 1,1'-bi-(N-phenyl-2-naphthylamine), a sterically hindered amine, have been studied.[556] Two peroxy radicals are consumed per amino group and the rate-determining hydrogen transfer from amine to alkylperoxy radical ($k_H/k_D = 3.6$) is unusually slow, suggesting an abnormally fast reverse reaction. Experiments with tetralin[556] confirm the view that the inhibitor radicals abstract hydrogen atoms from the hydrocarbon, and the failure of steric hindrance about the amino group to reduce hydrogen abstraction stands in strong contrast to experience with hindered phenols. ESR studies[557] confirm the finding that oxidation of cumene on vanadium pentoxide and platinum occurs by a surface-initiated radical-chain process.

The report last year[558] that autoxidation of cumene at 35° catalysed by $(Ph_3P)_4Pd$ occurs by an oxygen-activation mechanism has been shown to be in error;[559] a detailed kinetic study of the reaction shows that chain initiation occurs on decomposition of traces of hydroperoxide impurities in the cumene.

Under certain conditions thiodipropionates $S(CH_2CH_2CO_2R)_2$ and especially the derived sulphinyldipropionates $O:S(CH_2CH_2CO_2R)_2$ have been found to be effective pro-oxidants; thermal decomposition of the latter (R = Me) has been studied[560] alone and in the presence of galvinoxyl, the results showing that radical intermediates capable of initiating oxidations are not produced under these conditions but are formed in the presence of hydroperoxides. It is suggested[560] that this type of reaction accounts for the pro-oxidant effects of sulphoxides and the thiodipropionate esters, and probably involves intermediate sulphenic acids, i.e.:

$$RO_2CCH_2CH_2SOH + R'OOH \rightarrow R'O\cdot + H_2O + RO_2CCH_2CH_2SO\cdot$$

Vitamin A, dissolved in liquid paraffin, is stable below room temperature but suffers oxidative decomposition at 80°, giving the epoxide as main product; a kinetic study[561] shows that cobalt(II) stearate is an effective catalyst.

Autoxidation of ketones and esters in aprotic solvents containing strong bases gives α-hydroperoxides as primary products; the initial step in the reaction involves formation of a resonance-stabilized anion, and arguments are advanced[562] for a non-radical mechanism involving interaction of substrate anion and oxygen to yield hydroperoxide anion in one step.

Autoxidation of phenols catalysed by complexes of copper salts and amines gives products arising from both one- and two-electron abstraction; reaction products depend markedly on reaction conditions but not on the structure of the catalyst.[563]

Autoxidation of triethylbismuth in cyclohexane at 25° gives major products EtOH and MeCHO, with minor products including EtOOEt, EtOEt, and MeCOOH; the reaction is strongly inhibited by diphenylamine and a radical chain reaction is proposed.[564]

[556] R. F. Bridger, *J. Org. Chem.*, **36**, 1214 (1971).
[557] N. P. Evmenko, Ya. B. Gorokhovatskii, and B. Ya. Dragin, *Neftekhimiya*, **10**, 546 (1970); *Chem. Abs.*, **73**, 109025 (1971).
[558] See *Org. Reaction Mech.*, **1970**, 379.
[559] R. A. Sheldon, *Chem. Comm.*, **1971**, 788.
[560] C. Armstrong and G. Scott, *J. Chem. Soc.* (B), **1971**, 1747.
[561] Y. Ogata, Y. Kosugi, and K. Tomizawa, *Tetrahedron*, **26**, 5939 (1970).
[562] H. R. Gersmann and A. F. Bickel, *J. Chem. Soc.* (B), **1971**, 2230.
[563] D. G. Hewitt, *J. Chem. Soc.* (C), **1971**, 2967.
[564] R. V. Winchester, *Can. J. Chem.*, **49**, 1747 (1971).

Slow combustion of keten in the gas phase occurs by two distinct mechanisms depending on the temperature although the products (CH_2O, CO, CO_2, H_2O) are the same in the two cases. Between 380° and 500° the reaction was autocatalytic,[565] with a low temperature coefficient, and formaldehyde appears to be the intermediate responsible for degenerate branching, i.e.:

$$CH_2O + O_2 \rightarrow HCOOH + O$$
$$O + CH_2O \rightarrow \cdot CHO + \cdot OH$$

In the lower temperature range 280–350°, the reaction differs chiefly because of the presence of peroxides, and a reaction scheme is proposed[566] that involves degenerate branching through decomposition of methyl hydroperoxide.

Pyrolysis and Other Gas-phase Processes

Products from the low-pressure gas-phase pyrolysis of diallyl, dicrotyl, dicinnamyl and related oxalates indicate that these compounds are convenient sources of substituted allyl radicals.[567] Unlike allyl and crotyl radicals, cinnamyl radicals (**304**; R = H) gave no coupling products, but instead formed indene (**305**) in a good yield. The observed isotope effect k_H/k_D of 2.92 for cyclization of the *o*-deuteriocinnamyl radical suggests that formation of the cyclic radical is reversible, and thus cyclization of an *ortho*-

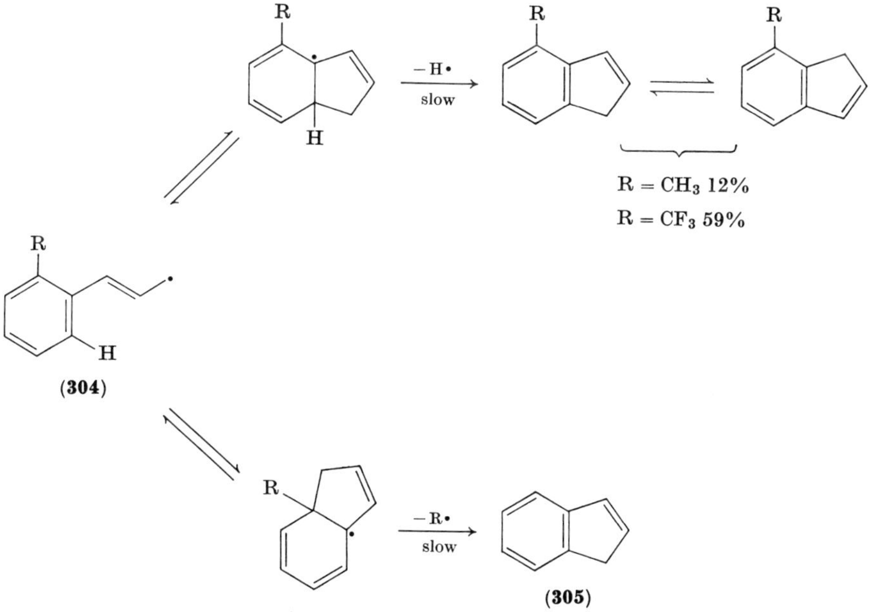

R = CH_3 12%
R = CF_3 59%

(**304**)

(**305**)

R = CH_3 23%
R = CF_3 <0.5%

[565] P. Michaud and C. Ouellet, *Can. J. Chem.*, **49**, 294 (1971).
[566] P. Michaud and C. Ouellet, *Can. J. Chem.*, **49**, 303 (1971).
[567] W. S. Trahanovsky and C. C. Ong, *J. Am. Chem. Soc.*, **92**, 7174 (1970); see also N. Wada, K. Tokumaru and O. Simamura, *Bull. Chem. Soc. Japan*, **44**, 1112 (1971).

substituted cinnamyl radical can lead to indene or a substituted indene by the mechanism outlined in the Scheme. From a study of the o-methyl- and o-trifluoromethylcinnamyl radicals a striking difference in the relative rates of the two cyclization pathways has been observed[568] (see Scheme) and clearly shows that the cinnamyl radical cyclization is very sensitive to the nature of the ring substituents. Pyrolysis of di-(α-substituted benzyl) oxalates has also been reported;[569] such compounds containing β-hydrogen atoms undergo elimination in addition to homolytic decomposition to α-substituted benzyl radicals.

Allyl radicals, from decomposition of diallyl oxalate in toluene, either dimerize or abstract hydrogen, and the thermochemistry of R· + R'H → RH + R'· where R, R' are allylic or benzylic, is outlined.[570]

Conversion of the cyclic azine (**306**) into (**307**) is thought to be a free-radical process since there is an induction period and oxygen has a marked adverse effect.[571]

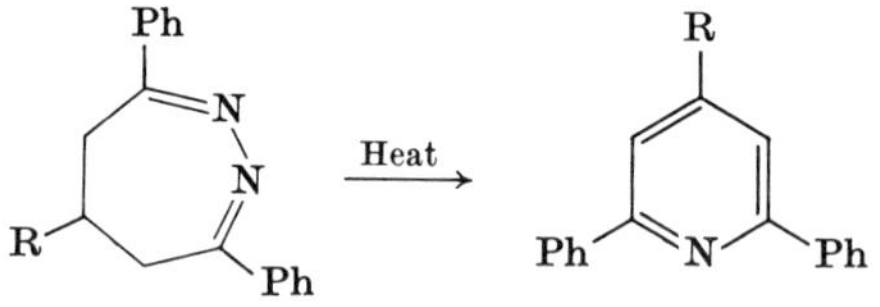

(**306**) → (**307**)

R = H or Ph

Thermal decomposition of acetaldehyde azine (**308**) yields N_2, HCN and MeCN, as well as a complex mixture of hydrocarbons;[572] participation of methyl and isopropyl radicals is proposed as shown.

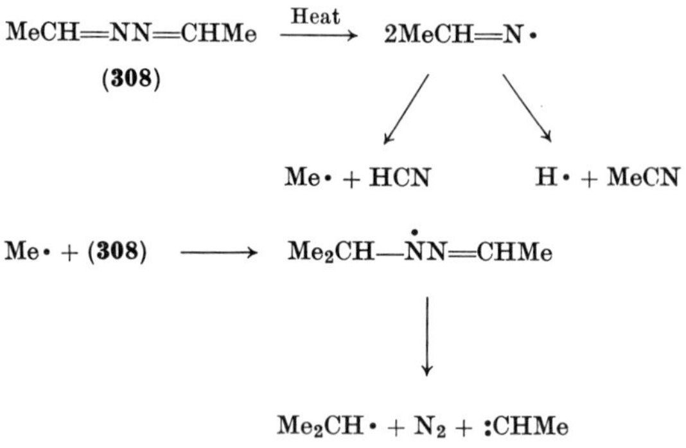

MeCH=NN=CHMe $\xrightarrow{\text{Heat}}$ 2MeCH=N·

(**308**)

Me· + HCN   H· + MeCN

Me· + (**308**) ⟶ Me$_2$CH—ṄN=CHMe

↓

Me$_2$CH· + N$_2$ + :CHMe

The conclusion from product studies of the photolysis and pyrolysis of camphor nitrimine (**309**) is that initial reaction occurs through homolysis of the $N-NO_2$ bond.[573] Mechanisms for formation of the two main products are outlined.

[568] W. S. Trahanovsky and P. W. Mullen, *Chem. Comm.*, **1971**, 102.
[569] W. S. Trahanovsky, C. C. Ong, J. G. Pataky, F. L. Weitl, P. W. Mullen, J. C. Clardy, and R. S. Hansen, *J. Org. Chem.*, **36**, 3575 (1971).
[570] R. Louw, *Rec. Trav. Chim.*, **90**, 469 (1971).
[571] G. Hasenheuttl, C. Opalka, and J. G. Krause, *Chem. Ind.*, **1971**, 1356.
[572] B. G. Gowenlock, R. M. Haynes, and C. A. F. Johnson, *J. Chem. Soc.* (B), **1971**, 1098.
[573] L. J. Winters, J. F. Fischer, and E. R. Ryan, *Tetrahedron Letters*, **1971**, 129.

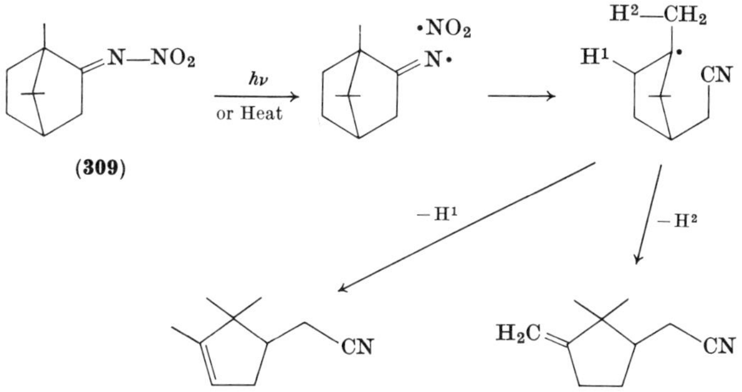

1-Nitroadamantane[574] decomposes at 500° to give products characteristic of intermediate adamantyl radicals, including fragmentation of the latter to alkylbenzenes and C_1–C_4 hydrocarbons. Pyrolysis of 2-nitropropane is compared with the decomposition of nitromethane and [$^2H_3$]nitromethane; an intramolecular mechanism is the major route to propene.[575]

Vapour-phase nitration of cyclohexane and cyclohexyl nitrite produces primary nitroalkanes; a mechanistic rationalization of the results is summarized in the annexed Scheme, as expected, the yield of the nitroalkanes decreases with decreasing number of carbon atoms.[576]

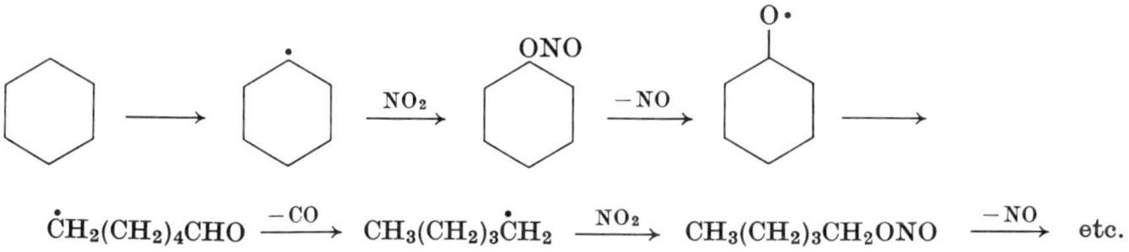

Standard enthalpies of formation (ΔH_f°) of diphenyl oxalate and benzoic anhydride, and values of bond-dissociation energies, have been derived for these and related oxygenated compounds by studies of thermolysis.[577]

Flash-vacuum thermolysis of phenyl-p-benzoquinone yields naphthalene, 2-hydroxydibenzofuran, and phenyl-1,4-hydroquinone; naphthalene may result from intramolecular trapping of a butadiene diradical by the phenyl ring.[578] Anisole derivatives[579] and p-benzoquinone[580] have been studied by the same technique.

A detailed study of the pyrolysis of benzohydroxamoyl chloride derivatives (310) has established that two reaction pathways are involved, depending on the substituents, R.[581] Benzohydroxamoyl chloride (R = H) and its 3-chloro-4-methoxy derivative gave

---

[574] A. I. Feinstein, E. K. Fields, P. J. Ihrig, and S. Meyerson, *J. Org. Chem.*, **36**, 996 (1971).
[575] D. J. Waddington and M. A. Warriss, *J. Phys. Chem.*, **75**, 2427 (1971).
[576] C. Matasa and H. B. Hass, *Can. J. Chem.*, **49**, 1284 (1971).
[577] A. S. Carson, D. H. Fine, P. Gray, and P. G. Laye, *J. Chem. Soc. (B)*, **1971**, 1611.
[578] H. J. Hageman and U. E. Wiersum, *Chem. Comm.*, **1971**, 497.
[579] R. A. Marty and P. de Mayo, *Chem. Comm.*, **1971**, 127.
[580] H. J. Hageman and U. E. Wiersum, *Tetrahedron Letters*, **1971**, 4329.
[581] Y. H. Chiang, *J. Org. Chem.*, **36**, 2155 (1971).

isocyanate derivatives, whereas nitro (R = p-NO$_2$ or m-NO$_2$) and chloro (R = p-Cl) derivatives gave (**311**) as the major product. An iminoxy-radical-addition mechanism may be operative for the formation of the latter type of compound.

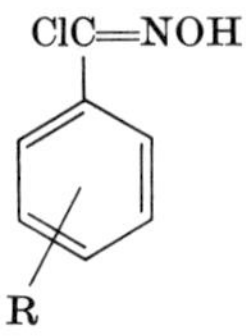

(**310**) 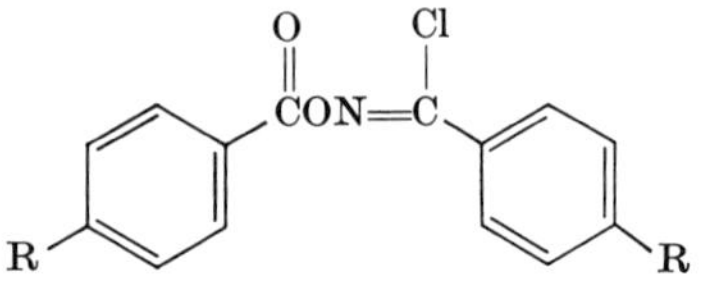   (**311**)

Depending on the nature of the *para*-substituent, the thermal decomposition of *para*-substituted tetra-arylhydrazines yields varying amounts of diarylamine, o- and m-semidine, and dihydrophenazine (as well as products of high molecular weight); evidently the Ar$_2$N• radicals either abstract H from the solvent or substitute into the aromatic ring of the starting material at either the *ortho*- or *meta*-position depending on the *para*-substituent.[582]

Gas-phase pyrolyses of *cis*-pinocarveol (**312**), *trans*-pinocarveol (**313**) and *trans*-pinocarvyl acetate (**314**) have been reported;[583] deuterium-labelling demonstrated the radical pathway involved in formation of the aldehyde (**315**) from (**312**) and (**313**).

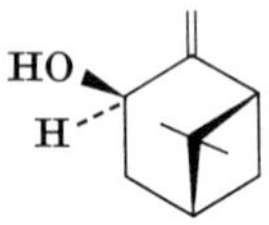

(**312**) 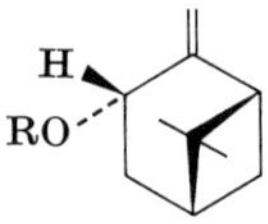   (**313**) R = H   (**314**) R = Ac   (**315**)

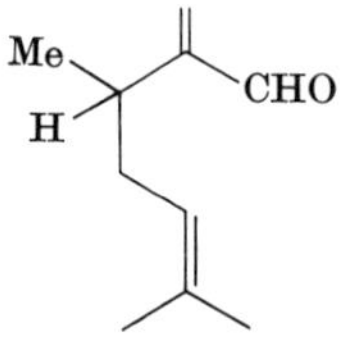

Pyrolysis of the tricyclic anhydride (**316**) involves an unusual decarboxylation to benzocyclobutenone (**317**) which subsequently decarbonylates to three C$_7$H$_6$ isomers.[584]

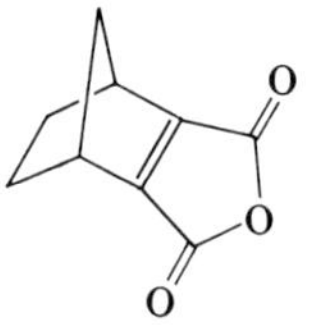

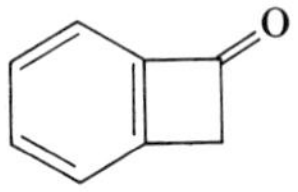

(**317**)

(**316**)

Other gas-phase investigations include the pyrolysis of 1,1,3,3-tetrafluoroacetone,[585] trimethylarsine,[586] ethyl bromide,[587] neopentane,[588] and 3-bromotropolone.[589] Kinetics

---

[582] F. A. Neugebauer and H. Fischer, *Chem. Ber.*, **104**, 886 (1971).
[583] J. M. Coxon, R. P. Garland, and M. P. Hartshorn, *Austral. J. Chem.*, **24**, 1481 (1971).
[584] O. A. Mamer, F. P. Lossing, E. Hedaya, and M. E. Kent, *Can. J. Chem.*, **48**, 3606 (1970).
[585] N. C. Craig, C. D. Jonah, J. T. Lemley, and W. E. Steinmetz, *J. Org. Chem.*, **36**, 3572 (1971).
[586] S. J. W. Price and J. P. Richard, *Can. J. Chem.*, **48**, 3209 (1970).
[587] D. A. Kairaitis and V. R. Stimson, *Austral. J. Chem.*, **24**, 2031 (1971).
[588] F. Baronnet, M. Dzierzynski, G. M. Côme, R. Martin, and M. Niclause, *Int. J. Chem. Kinet.*, **3**, 197 (1971).
[589] H. F. Gruetzmacher and J. Huebner, *Tetrahedron Letters*, **1971**, 1455.

Radical Reactions 359

of the pyrolysis of propene,[590] 2,3-epoxybutane isomers,[591] and mechanisms of acetylene[592] and trimethylsilane[593] pyrolyses have been discussed; so has the decomposition of benzoic acid[594] and of benzyl and furfuryl acetoacetate.[595] Further results[596] for the formation and reactions of $(PhS)_3C\cdot$ from the corresponding C—C dimer come from ^{13}C-labelling experiments.[597]

Kinetics of the gas-phase decomposition of *tert*-butoxy[598] and benzoyl[599] radicals are in good agreement with thermochemical data and transition-state models. Heats of formation and/or stabilization energies of radicals continue to be amassed and results for cyclopropylcarbinyl,[600] norbornyl,[601] cyclopentadienyl,[602] allyl, methylallyl, and dimethylallyl radicals[603] have been published.

The recombination rate of trimethylsilyl radicals has been measured in the gas-phase[604] and found to agree with the value obtained in solution.[605] Disproportionation–combination ratios of isopropyl and *tert*-butyl radicals with difluoroamino radicals have been determined.[606] Reactions of the OH radical with acetylene[607] and olefins,[608] of methyl and CF_3 radicals with ethylene oxide,[609] and of atomic hydrogen with vinyl chloride[610] have also been investigated kinetically. Arrhenius parameters for the nitrogen dioxide-catalysed *cis–trans* isomerization of *cis*-but-2-ene are consistent with a free-radical isomerization, and the results have relevance to olefin consumption in smog, by a non-photolytic pathway.[611]

Radiolysis, ESR Spectroscopy and Miscellaneous

UV photolyses (185 nm) of ethylene glycol[612] and liquid *tert*-butyl alcohol[613] have been extensively investigated; a detailed decomposition scheme is presented for the latter case, illustrating both molecular fragmentation and radical processes in product formation. Yields of products formed by γ-radiolysis of liquid *tert*-butyl alcohol[614] provide insight into the contribution of *tert*-butoxy radicals to product formation. Uracil radicals

---

[590] M. Simon and M. H. Back, *Can. J. Chem.*, **48**, 3313 (1970).
[591] M. C. Flowers and R. M. Parker, *J. Chem. Soc.* (B), **1971**, 1980.
[592] M. H. Back, *Can. J. Chem.*, **49**, 2199 (1971).
[593] I. M. T. Davidson and C. A. Lambert, *J. Chem. Soc.* (A), **1971**, 882.
[594] K. Winter and D. Barton, *Can. J. Chem.*, **48**, 3797 (1970).
[595] S. Ito and H. Imai, *Kogyo Kagaku Zasshi*, **74**, 526 (1971); *Chem. Abs.*, **74**, 124575 (1971).
[596] See *Org. Reaction Mech.*, **1970**, 313.
[597] D. Seebach, H. B. Stegmann, and A. K. Beck, *Angew. Chem. Int. Ed.*, **10**, 500 (1971).
[598] P. Cadman, A. F. Trotman-Dickenson, and A. J. White, *J. Chem. Soc.* (A), **1971**, 2296.
[599] R. K. Solly and S. W. Benson, *J. Am. Chem. Soc.*, **93**, 2127 (1971).
[600] D. F. McMillen, D. M. Golden, and S. W. Benson, *Int. J. Chem. Kinet.*, **3**, 359 (1971).
[601] H. E. O'Neal, J. W. Bagg, and W. H. Richardson, *Int. J. Chem. Kinet.*, **2**, 493 (1970).
[602] S. Furuyama, D. M. Golden, and S. W. Benson, *Int. J. Chem. Kinet.*, **3**, 237 (1971).
[603] A. B. Trenwith, *Trans. Faraday Soc.*, **66**, 2805 (1970).
[604] P. Cadman, G. M. Tilsley, and A. F. Trotman-Dickenson, *Chem. Comm.*, **1970**, 1721.
[605] See *Org. Reaction Mech.*, **1970**, 388.
[606] P. Cadman, Y. N. Inel, and A. F. Trotman-Dickenson, *J. Chem. Soc.*, (A), **1971**, 2859.
[607] J. E. Breen and G. P. Glass, *Int. J. Chem. Kinet.*, **3**, 145 (1971).
[608] E. D. Morris, D. H. Stedman, and H. Niki, *J. Am. Chem. Soc.*, **93**, 3570 (1971).
[609] S. H. Jones and E. Whittle, *Can. J. Chem.*, **48**, 3601 (1970).
[610] J. S. Tanner and J. W. S. Jamieson, *Can. J. Chem.*, **49**, 1023 (1971).
[611] J. L. Sprung, H. Akimoto, and J. N. Pitts, *J. Am. Chem. Soc.*, **93**, 4358 (1971).
[612] H. J. van der Linde and C. von Sonntag, *Photochem. Photobiol.*, **13**, 147 (1971).
[613] D. Sänger and C. von Sonntag, *Tetrahedron*, **26**, 5489 (1970).
[614] D. Verdin, *Int. J. Radiat. Phys. Chem.*, **2**, 201 (1970); *Chem. Abs.*, **75**, 48118 (1971).

(from 5-iodouracil) react 24 times faster with oxygen than with methanol.[615] Phenylthio and trichloromethyl radicals cyclize the cyclodeca-1,5-diene system of germacrene (**318**) to selinane derivatives (**319**) and (**320**), respectively.[616]

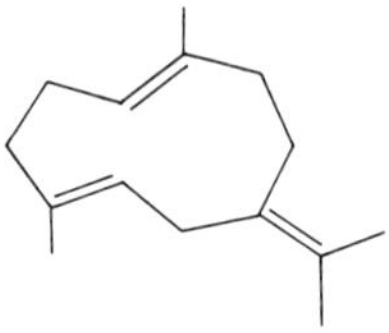

(**318**)

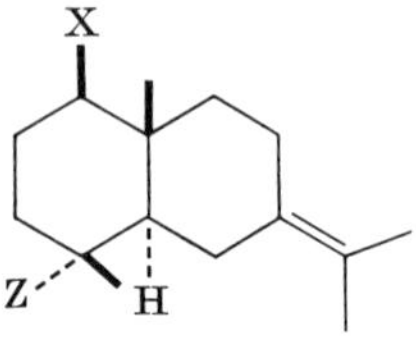

(**319**) X = SPh, Z = H
(**320**) X = CCl$_3$, Z = Cl

Photolysis of a benzene solution of *aldehydo*-D-glucose penta-acetate (**321**) produces three crystalline photoproducts, all of which are derived by way of radical (**322**), realized through α-bond cleavage.[617]

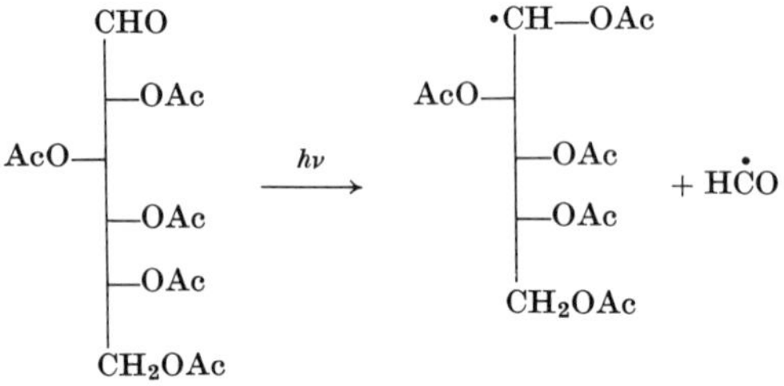

(**321**) $\xrightarrow{h\nu}$ (**322**) + H$\dot{\text{C}}$O

Radical mechanisms are envisaged for some of the 23 radiolysis products from aerated aqueous solutions of thymine.[618] The role of thiyl radicals in the radiolysis of mixed aqueous solutions of L-cysteine and formate have been elucidated,[619] and radical mechanisms are discussed for the radiolysis of liquid ethanol,[620] ethanol–ethylene systems,[621] chloroform vapour,[622] liquid (air-free) carbon tetrachloride,[623] clathrate hydrates of three-membered rings,[624] and propan-1-ol glass containing naphthalene or biphenyl.[625] Mechanisms of decay of *n*-alkyl radicals in γ-irradiated alkyl iodides have been discussed;[626] a chain reaction explains the extremely high radiation yields obtained during the induced oxidation of trichloroethylene.[627]

[615] E. Gilbert, G. Wagner, and D. Schulte-Frohlinde, *Z. Naturforsch.*, **26b**, 209 (1971).
[616] T. W. Sam and J. K. Sutherland, *Chem. Comm.*, **1971**, 970.
[617] R. L. Whistler and K.-S. Ong, *J. Org. Chem.*, **36**, 2575 (1971).
[618] R. Teoule and J. Cadet, *Chem. Comm.*, **1971**, 1269.
[619] M. Morita, K. Sasai, M. Tajima, and M. Fujimaki, *Bull. Chem. Soc. Japan*, **44**, 2257 (1971).
[620] S. M. S. Akhtar and G. R. Freeman, *J. Phys. Chem.*, **75**, 2756 (1971).
[621] R. A. Basson and L. van Wyk, *J. Chem. Soc.* (B), **1971**, 809.
[622] L. C. Dickey and R. F. Firestone, *J. Phys. Chem.*, **74**, 4310 (1970).
[623] N. E. Bibler, *J. Phys. Chem.*, **75**, 24 (1971).
[624] V. I. Trofimov, A. L. Blumenfeld, R. G. Kostyanovsky, and I. I. Chkheidze, *Tetrahedron Letters*, **1971**, 3267.
[625] F. S. Dainton, G. A. Salmon, and U. F. Zucker, *Proc. Roy. Soc.* (A), **325**, 23 (1971).
[626] H. W. Fenrick, N. B. Nazhat, P. J. Ogren, and J. E. Willard, *J. Phys. Chem.*, **75**, 472 (1971).
[627] A. R. Kazanjian and D. R. Horrell, *J. Phys. Chem.*, **75**, 613 (1971).

Hydrogen abstraction by radiolytically generated cyclohexyl radicals from pentachloroethane and *sym*-tetrachloroethane has been studied,[628] and comparison of the photochemical behaviour of but-2-yne with its radiolysis suggests that both free radicals and excited states play a role in the radiolysis reaction.[629] Evidence is presented that radiolysis of 2-methylpentane, 3-methylpentane and isopentane leads to loss of only tertiary hydrogen atoms.[630]

From an investigation of the radiolysis and photolysis of binary mixtures of alkyl halides it is demonstrated that non-geminate recombination does occur in this liquid-phase system.[631]

The reactivity and sites of attack of OH radicals on simple amides[632] has been extended[633] to explore the effects of branching at the carbon atoms bound to carbonyl or to the nitrogen atom. Pulse-radiolysis studies of aqueous solutions of luminol,[634] oxalic acid and oxalates,[635] indole and its derivatives,[636] thiocyanate,[637] and azide ions[638] have been reported. The technique has also been used to study electron-transfer reactions to oxidized lipoate,[639] e.g., from $Me_2\overset{\bullet}{C}OH$ radicals; and the spectra and reactivity of phenyl and hydroxyphenyl radicals.[640] Reaction kinetics of the radicals formed during pulse radiolysis of alkaline liquid ethanol indicates[641] that $Me\overset{\bullet}{C}HO^-$ is formed by the reaction

$$EtO\cdot + EtO^- \rightarrow Me\overset{\bullet}{C}HO^- + EtOH$$

and not by

$$Me\overset{\bullet}{C}HOH + EtO^- \rightarrow Me\overset{\bullet}{C}HO^- + EtOH$$

ESR spectra of the aziridino (**323**) and azetidino (**324**) radicals[642] indicate both are π-radicals similar to the simple dialkylamino radicals reported last year.[643] New data

(**323**)   (**324**)

on the chemistry of alkylamino radicals[644] indicates that they are unaffected by O_2 or NO, and their stability in oxidation systems helps to explain the inhibiting action of aliphatic amines. Diarylamino (including $Ph_2N\cdot$) radicals[645] have been observed during

---

[628] M. G. Katz, A. Horowitz, and L. A. Rajbenbach, *Trans. Faraday Soc.*, **67**, 2354 (1971).
[629] D. G. Whitten and W. Berngruber, *J. Am. Chem. Soc.*, **93**, 3204 (1971).
[630] S. W. Kanick, R. E. Linder, and A. C. Ling, *J. Chem. Soc.* (A), **1971**, 2971.
[631] R. E. Linder and A. C. Ling, *Chem. Comm.*, **1971**, 343.
[632] See *Org. Reaction Mech.*, **1970**, 385.
[633] E. Hayon, T. Ibata, N. N. Lichtin, and M. Simic, *J. Am. Chem. Soc.*, **93**, 5388 (1971).
[634] J. H. Baxendale, *Chem. Comm.*, **1971**, 1489.
[635] N. Getoff, F. Schwörer, V. M. Markovic, K. Sehested, and S. O. Nielsen, *J. Phys. Chem.*, **75**, 749 (1971).
[636] B. Iddon, G. O. Phillips, K. E. Robbins, and J. V. Davies, *J. Chem. Soc.* (B), **1971**, 1887.
[637] D. Behar, P. L. T. Bevan, and G. Scholes, *Chem. Comm.*, **1971**, 1486.
[638] E. Hayon and M. Simic, *J. Am. Chem. Soc.*, **92**, 7486 (1970).
[639] R. L. Wilson, *Chem. Comm.*, **1970**, 1425.
[640] B. Cercek and M. Kongshaug, *J. Phys. Chem.*, **74**, 4319 (1970).
[641] J. W. Fletcher, P. J. Richards, and W. A. Seddon, *Can. J. Chem.*, **48**, 3765 (1970).
[642] W. C. Danen and T. T. Kensler, *Tetrahedron Letters*, **1971**, 2247.
[643] See *Org. Reaction Mech.*, **1970**, 318.
[644] P. W. Jones and H. D. Gesser, *J. Chem. Soc.* (B), **1971**, 1873.
[645] F. A. Neugebauer and S. Bamberger, *Angew. Chem. Int. Ed.*, **10**, 71 (1971).

intense UV irradiation of the corresponding tetra-arylhydrazines and have been compared[646] with the corresponding diarylamine radical cations, $Ar_2\overset{\bullet}{N}\overset{+}{H}$.

Several members of the previously unknown class of isocyanatoalkyl (**325**) and isothiocyanatoalkyl (**326**) radicals have been prepared and studied by ESR,[647] using the adamantane matrix technique described last year;[648] spectroscopic results are consistent with a planar structure for these radicals. A full report on aliphatic amino-

$$R^1_{}\!\!\diagdown\overset{\bullet}{C}\!\!-\!\!N\!\!=\!\!C\!\!=\!\!O \qquad\qquad R^1_{}\!\!\diagdown\overset{\bullet}{C}\!\!-\!\!N\!\!=\!\!C\!\!=\!\!S$$
$$R^2\diagup \qquad\qquad\qquad\qquad\qquad R^2\diagup$$

(**325**)   (**326**)

alkyl radicals $R^1R^2\overset{\bullet}{C}\!-\!NR^3R^4$ generated by the same technique has appeared,[649] and it seems that the latter radicals may be formed in the primary photochemical process of the parent amines.[650]

Limitations of the use of flow methods in ESR studies of reaction mechanism have been described in a detailed survey of this much used and useful technique,[651] and an example of possible misinterpretation is provided by the dispute concerning the nature of radical species formed from propan-2-ol and hydroxyl radical.[652]

Radicals from Norrish Type I cleavage, photoreduction, and other reactions of excited ketones, are observed during UV-irradiation of aliphatic ketones in solution.[653] Photolysis of di-*tert*-butyl peroxide in the presence of *tert*-butyl formate[654] enables study of the *tert*-butoxycarbonyl radical to be made, thus:

$$Bu^tO\cdot \; + \; Bu^tOCH\!\!=\!\!O \;\rightarrow\; Bu^tO\overset{\bullet}{C}\!\!=\!\!O \;\rightarrow\; Bu^t\cdot \; + \; CO_2 + Bu^tOH$$

The rate constant for the oxidation of carbon monoxide by *tert*-butoxy radicals has also been evaluated[655] ($Bu^tO\cdot + CO \rightarrow Bu^t\cdot + CO_2$) as $(2.3 \pm 0.8) \times 10^5$ l m^{-1} sec^{-1}. By monitoring the concentration of intermediate radicals [·CMe(OH)COR] the kinetics of the photoreduction of biacetyl and pyruvic acid in various solvents have been established.[656] Photoreduction of some xanthene dyes affords the intermediate semiquinone radicals.[657] Radicals $HO\overset{\bullet}{S}O_2$, $Me\overset{\bullet}{S}O_2$ and $Bu^tO\overset{\bullet}{S}O_2$ are observed as intermediates in reactions between hydroperoxides and sulphur dioxide; adducts of these radicals with alkenes have been characterized.[658]

Photolysis of citric acid produces radicals resulting from the loss of the central and end carboxyl groups, together with one resulting from loss of a methylene hydrogen atom.[659] Short-lived radicals from photoreduction of aqueous solutions of alloxan,

---

[646] F. A. Neugebauer and S. Bamberger, *Angew. Chem. Int. Ed.*, **10**, 71 (1971).
[647] D. E. Wood, R. V. Lloyd, and W. A. Lathan, *J. Am. Chem. Soc.*, **93**, 4145 (1971).
[648] See *Org. Reaction Mech.*, **1970**, 382.
[649] D. E. Wood and R. V. Lloyd, *J. Chem. Phys.*, **53**, 3932 (1970).
[650] T. Richerzhagen and D. H. Volman, *J. Am. Chem. Soc.*, **93**, 2062 (1971).
[651] G. Czapski, *J. Phys. Chem.*, **75**, 2957 (1971).
[652] C. E. Burchill, *J. Phys. Chem.*, **75**, 167 (1971); R. E. James and F. Sicilio, *J. Phys. Chem.*, **75**, 1326 (1971).
[653] H. Paul and H. Fischer, *Chem. Comm.*, **1971**, 1038.
[654] D. Griller and B. P. Roberts, *Chem. Comm.*, **1971**, 1035.
[655] E. A. Lissi, J. C. Scaiano, and A. E. Villa, *Chem. Comm.*, **1971**, 457.
[656] P. B. Ayscough and M. C. Brice, *J. Chem. Soc.* (B), **1971**, 491.
[657] I. H. Leaver, *Austral. J. Chem.*, **24**, 753 (1971).
[658] B. D. Flockhart, K. J. Ivin, R. C. Pink, and B. D. Sharma, *Chem. Comm.*, **1971**, 339.
[659] H. Zeldes and R. Livingston, *J. Am. Chem. Soc.*, **93**, 1082 (1971).

parabanic acid and related compounds,[660] together with those from uracil and its derivatives,[661] have been investigated. The spectra and chemical behaviour of the radicals depend in many cases upon the pH; such observations are reported for the radicals observed during reaction of hydroxyl radicals with glycine.[662] Fessenden and Schuler[663] have produced a very comprehensive review of the ESR spectra of radiation-produced radicals that deals in particular with their own pioneering work. This technique has now been extended[664] to examination of aqueous solutions of solutes during steady irradiation with an electron beam, and the results demonstrate the practicality of studying, by ESR, radicals produced by reaction of the primary species of water radiolysis with various solutes. In particular, radicals from reaction of OH radicals with some amines, amino acids and related compounds,[665] and with thiols,[666] have been identified and investigated.

ESR spectroscopy continues to be used to identify the radicals and radical ions formed by radiation damage and, where possible, to elucidate mechanisms of their formation. A full report of such a study on eleven carboxylic acids has been published.[667] Radicals from tetrahydro-2-methylfuran,[668] methyl and ethyl isocyanate,[669] urea,[670] formamide and malonamide[671] have also been observed.

Attention is drawn to papers given at the Second International Symposium on ESR Spectroscopy.[672]

A series of papers reporting extensive studies on the properties and reactions of triarylimidazolyl free radicals (**327**) has been published. The effect of substitution on the

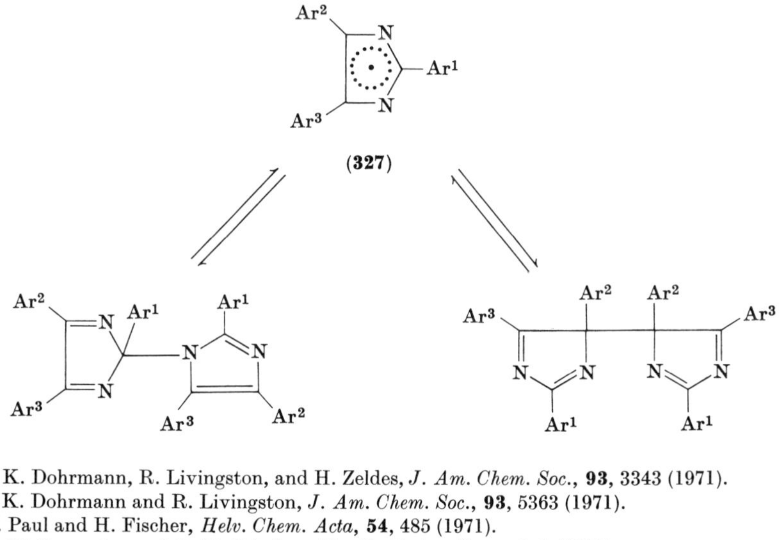

---

[660] J. K. Dohrmann, R. Livingston, and H. Zeldes, *J. Am. Chem. Soc.*, **93**, 3343 (1971).
[661] J. K. Dohrmann and R. Livingston, *J. Am. Chem. Soc.*, **93**, 5363 (1971).
[662] H. Paul and H. Fischer, *Helv. Chem. Acta*, **54**, 485 (1971).
[663] R. W. Fessenden and R. H. Schuler, *Adv. Radiation Chem.*, **2**, 1 (1971).
[664] K. Eiben and R. W. Fessenden, *J. Phys. Chem.*, **75**, 1186 (1971).
[665] P. Neta and R. W. Fessenden, *J. Phys. Chem.*, **75**, 738 (1971).
[666] P. Neta and R. W. Fessenden, *J. Phys. Chem.*, **75**, 2277 (1971).
[667] P. B. Ayscough and J. P. Oversby, *Trans. Faraday Soc.*, **67**, 1365 (1971).
[668] F. P. Sargent, *Can. J. Chem.*, **48**, 3453 (1970).
[669] Y. J. Chung and F. Williams, *J. Phys. Chem.*, **75**, 1893 (1971).
[670] H. Bower, J. McRae, and M. C. R. Symons, *J. Chem. Soc.* (A), **1971**, 2400.
[671] M. C. R. Symons, *J. Chem. Soc.* (A), **1971**, 3205.
[672] *J. Phys. Chem.*, **75**, 3383–3489 (1971).

rate of radical dimerization shows that any *ortho*-substituent in the aryl group increases the rate constant relative to position isomers, a fact consistent with radical destabilization by *ortho*-substituents through steric disruption of ring coplanarity.[673] The radicals (**327**) oxidize electron-rich substances by rapid electron abstraction from tertiary amines, iodide ion and metal ions, and by hydrogen-atom abstraction from phenols, thiols, primary and secondary amines and activated C—H compounds.[674] However, the radicals do *not* react with aliphatic alcohols, aromatic hydrocarbons or vinyl monomers, and they are insensitive to oxygen. Flash-photolysis studies indicate that imidazolyl radicals oxidize aromatic amines[675] and leuco-triphenylmethane dyes[676] by an electron-exchange reaction at the amino-nitrogen atom. Rates of the dye reaction were retarded by electron-donating and enhanced by electron-withdrawing groups on the imidazolyl radical.[677] An improved ESR spectrum of the 2,4,5-triphenylimidazolyl radical has been obtained.[678]

The synthesis of perchloro-9-phenylfluorenyl radical has been achieved in a number of ways and it is shown to be an extremely stable carbon radical.[679] A full report of other stable and highly inert chlorocarbon radicals derived from diphenylmethyl and triphenylmethyl has been published.[680] X-ray analysis of perchlorodiphenylmethyl radical,[681] and spectroscopic studies of diphenyl- and triphenyl-verdazyls[682] and disproportionation of the latter,[683] have been reported. Other studies of relatively stable free radicals include the dimerization of *tert*-butyl-substituted diphenylmethyls,[684] the use of diaryl-*p*-tolylmethyl radicals as initiators in polymerization of α,α-diaryl-1,4-quinodimethanes,[685] and the conversion of radicals such as triphenylmethyl and (**328**) into their carbanion conjugate species by butyl-lithium, a reaction which is Lewis-base catalysed and analogous to an "inner-sphere" electron transfer.[686]

(**328**)

---

[673] L. A. Cescon, G. R. Coraor, R. Dessauer, E. F. Silversmith, and E. J. Urban, *J. Org. Chem.*, **36**, 2262 (1971); B. S. Tanaseichuk and L. G. Rezepova, *Zhur. Org. Khim.*, **6**, 1065 (1970).
[674] L. A. Cescon, G. R. Coraor, R. Dessauer, A. S. Deutsch, H. L. Jackson, A. MacLachlan, K. Marcali, E. M. Potrafke, R. E. Read, E. F. Silversmith, and E. J. Urban, *J. Org. Chem.*, **36**, 2267 (1971).
[675] R. H. Riem, A. MacLachlan, G. R. Coraor, and E. J. Urban, *J. Org. Chem.*, **36**, 2272 (1971).
[676] A. MacLachlan and R. H. Riem, *J. Org. Chem.*, **36**, 2275 (1971).
[677] R. L. Cohen, *J. Org. Chem.*, **36**, 2280 (1971).
[678] N. Cyr, M. A. J. Wilks, and M. R. Willis, *J. Chem. Soc.* (B), **1971**, 404.
[679] M. Ballester, J. Castañer, and J. Pujadas, *Tetrahedron Letters*, **1971**, 1699.
[680] M. Ballester, J. Riera, J. Castañer, C. Badía, and J. M. Monsó, *J. Am. Chem. Soc.*, **93**, 2215 (1971).
[681] J. Silverman, L. J. Soltzberg, N. F. Yannoni, and A. P. Krukonis, *J. Phys. Chem.*, **75**, 1246 (1971).
[682] H. Brunner, K. H. Hausser, and F. A. Neugebauer, *Tetrahedron*, **27**, 3611 (1971); F. A. Neugebauer, H. Brunner, and K. H. Hausser, *Tetrahedron*, **27**, 3623 (1971).
[683] O. M. Polumbrik and G. F. Dvorko, *Kinet. Katal.*, **12**, 304 (1971).
[684] F. Bölsing and K.-D. Korn, *Tetrahedron Letters*, **1971**, 3865.
[685] D. Braun, U. Platzek, and H. J. Hefter, *Chem. Ber.*, **104**, 2581 (1971).
[686] C. G. Screttas, *Chem. Comm.*, **1971**, 406.

2-Thiazolylhydrazine and all three thiazolecarbonyl peroxides have been synthesized and examined as radical precursors to the 2-, 4- and 5-thiazolyl radicals (**329**–**331**) in benzene, bromobenzene and cumene.[687] Oxidation of the hydrazine by silver oxide or

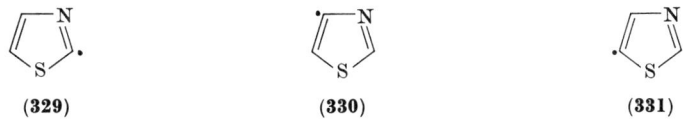

(**329**)   (**330**)   (**331**)

thermal decomposition of the 2-peroxide gives products fully supporting the involvement of 2-thiazolyl radicals. The 5-peroxide, on the other hand, gives no product clearly diagnostic of the corresponding radical (**331**).

Useful results are reported from a study of effects of viscosity and other physicochemical factors on the dimerization of cyclohexyl and substituted cyclohexyl radicals.[688] The observed high reactivity of the benzoylperoxy radical has relevance to inhibition of aldehyde oxidation.[689] Tropolone can undergo radical reactions with benzoyl peroxide at the 3-, 5- and 7-positions, as in other electrophilic substitutions.[690]

Phenylthiomethyl radical is produced by anodic oxidation of sodium phenylthioacetate and thermal decomposition of *tert*-butylphenyl thioperacetate.[691] Free radicals have been detected when methylphenylphosphinate is added to tetraphenylcyclopentadienone (tetracyclone) in the presence of base,[692] and anthranoxy radicals are observed in solutions of $\Delta^{10,10'}$-bianthrone;[693] reactions of carbon tetrachloride with aldehydes,[694] esters,[695] and organomercury compounds[696] are thought to involve radical mechanisms. The spectra of some organolead[697] and germanium-containing radicals,[698] as well as reactions of radicals with palladium and platinum complexes,[699] have been reported.

[687] A. L. Lee, D. Mackay, and E. L. Manery, *Can. J. Chem.*, **48**, 3554 (1970).
[688] M. Tessier and R. Pallaud, *Compt. Rend.* (C), **272**, 549 (1971).
[689] E. Niki, K. Ukegawa, and Y. Kamiya, *Kogyo Kagaku Zasshi*, **74**, 1354 (1971); *Chem. Abs.*, **75**, 76328 (1971).
[690] K. Doi and N. Chiba, *Tetrahedron Letters*, **1971**, 2891.
[691] K. Uneyama, S. Torii, and S. Oae, *Bull. Chem. Soc. Japan*, **44**, 815 (1971).
[692] M. J. Gallagher and I. D. Jenkins, *J. Chem. Soc.* (C), **1971**, 210.
[693] L. S. Singer, I. C. Lewis, T. Richerzhagen, and G. Vincow, *J. Phys. Chem.*, **75**, 290 (1971).
[694] V. I. Kovalenko and V. N. Dubchenko, *Ukr. Khim. Zhur.*, **37**, 462 (1971).
[695] V. N. Dubchenko and V. I. Kovalenko, *Zh. Prikl. Khim.* (*Leningrad*), **43**, 2049 (1970); *Chem. Abs.*, **73**, 130420 (1970).
[696] S. F. Zhil'tsov, L. F. Kudryavtsev, O. N. Druzhkov, and G. G. Petukhov, *Zhur. Obshch. Khim.*, **40**, 1537 (1970); *Chem. Abs.*, **75**, 34832 (1971).
[697] C. L. Cook and I. M. Napier, *Austral. J. Chem.*, **24**, 179 (1971).
[698] H. B. Stegmann, K. Scheffler, and F. Stöcker, *Angew. Chem. Int. Ed.*, **10**, 499 (1971).
[699] W. Beck, K. Schorpp, and K. H. Stetter, *Z. Naturforsch.*, **26b**, 684 (1971).

CHAPTER 10

Carbenes and Nitrenes

T. L. GILCHRIST

Department of Organic Chemistry, University of Liverpool

Structure	367
Methods of Generation	368
Carbenes	368
Nitrenes	371
Cycloadditions	373
Insertions and Abstractions	376
Rearrangements and Fragmentations	381
Reactions with Nucleophiles and Electrophiles	384
Carbenoids	387
Transition-metal Complexes	389

A new edition of Kirmse's definitive monograph on carbene chemistry has appeared,[1] and a book on the chemistry of the azide group contains several chapters where nitrene intermediates are discussed.[2] The chemistry of sulphonylnitrenes has been reviewed,[3] and there are two new introductory reviews on nitrenes.[4]

Structure

The geometries and energies of methylene[5,6] and of mono- and di-fluorocarbene[6] have been calculated by *ab initio* methods. The electrophilicity of the carbenes is in the expected order, $:CH_2 > :CHF > :CF_2$. Both the calculations and a study of the electron paramagnetic resonance (EPR) spectra in a solid matrix[7] indicate that triplet methylene is bent, the HCH angle being estimated as 125.5° or as 132° by calculation, and as 136° from the EPR study. The EPR spectrum of 9,9′-dianthrylcarbene (**1**) reveals that it is linear, however, with the two aromatic systems lying orthogonal.[8] This structure permits extensive delocalization through allenic resonance forms such as (**1a**). EPR spectra of nitrenes have been reviewed.[9]

The IR spectrum of dichlorocarbene has been obtained after pyrolysis of $PhHgCCl_3$ in the gas phase in argon, and rapid cooling of the pyrolysate to 8–10°K.[10]

---

[1] W. Kirmse, "Carbene Chemistry," 2nd Edition, Academic Press, New York, 1971.
[2] "The Chemistry of the Azide Group" (Ed. S. Patai), Interscience, 1971.
[3] R. A. Abramovitch and R. G. Sutherland, *Fortschr. Chem. Forsch.*, **16**, 1 (1970).
[4] R. Belloli, *J. Chem. Ed.*, **48**, 422 (1971); S. Hünig, *Helv. Chim. Acta*, **54**, 1721 (1971).
[5] W. A. Latham, W. J. Hehre, and J. A. Pople, *J. Am. Chem. Soc.*, **93**, 808 (1971).
[6] J. F. Harrison, *J. Am. Chem. Soc.*, **93**, 4112 (1971).
[7] E. Wasserman, V. J. Kuck, R. S. Hutton, and W. A. Yager, *J. Am. Chem. Soc.*, **92**, 7491 (1970).
[8] E. Wasserman, V. J. Kuck, W. A. Yager, R. S. Hutton, F. D. Greene, V. P. Abegg, and N. M. Weinshenker, *J. Am. Chem. Soc.*, **93**, 6335 (1971).
[9] E. Wasserman, in "Progress in Physical Organic Chemistry", Vol. 8 (Eds. A. Streitwieser and R. W. Taft), Wiley-Interscience, New York, 1971, p. 319.
[10] O. M. Nefedov, A. K. Mal'tsev, and R. G. Mikaelyan, *Tetrahedron Letters*, **1971**, 4125; A. K. Mal'tsev, R. G. Mikaelyan, and O. M. Nefedov, *Izv. Akad. Nauk. SSSR, Ser. Khim.*, **1971**, 199.

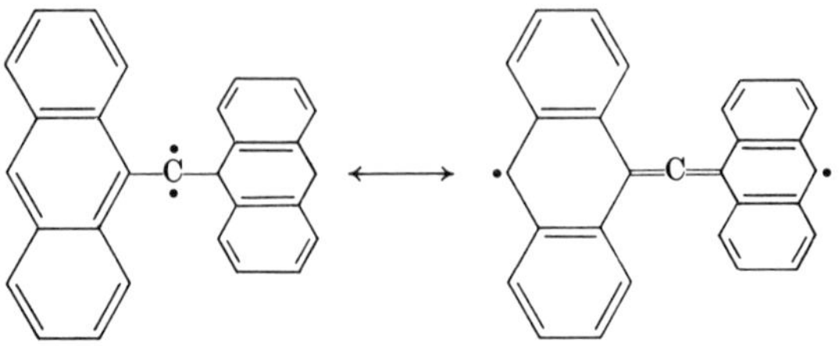

(1)   (1a)

Carbenes (**2**)[11] and (**4**)[12] have been generated and their rearrangement products studied in the hope of finding evidence for non-classical "foiled methylene" structures (in which the carbene centre is bent over towards the double bond). The results are not conclusive; for example, the major product from 7-norbornylidene (**2**) is bicyclo[3.2.0]-hepta-1,6-diene (**3**) which could be formed through a dipolar intermediate generated by attack of the carbene on one end of the double bond; on the other hand, an analogous rearrangement occurs with the saturated carbene (**5**) where there is no possibility of a non-classical structure.[13]

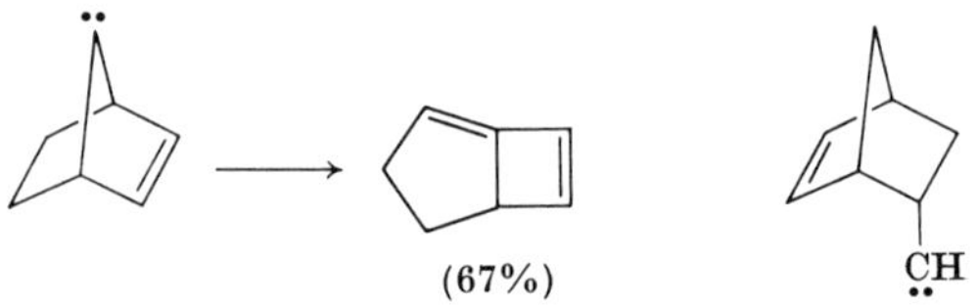

(2)   (3) (67%)   (4)

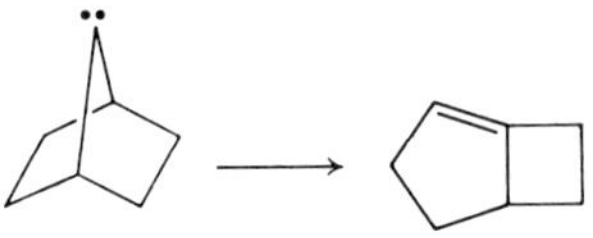

(5)

Methods of Generation

Carbenes

Two useful routes to dichlorocarbene have been developed and extended to the generation of other carbenes. One is the generation of dichlorocarbene from chloroform and aqueous alkali in the presence of catalytic amounts of tetra-alkylammonium salts—an example of "phase-transfer catalysis".[14] The method has now been used to generate dibromocarbene (from $CHBr_3$),[15] bromofluorocarbene (from $CHBr_2F$),[16] chlorofluoro-

---

[11] R. A. Moss, U.-H. Dolling, and J. R. Whittle, *Tetrahedron Letters*, **1971**, 931.
[12] P. K. Freeman and K. B. Desai, *J. Org. Chem.*, **36**, 1554 (1971).
[13] R. A. Moss and J. R. Whittle, *Chem. Comm.*, **1969**, 1298.
[14] C. M. Starks, *J. Am. Chem. Soc.*, **93**, 195 (1971).
[15] E. V. Dehmlow and J. Schönefeld, *Ann. Chem.*, **744**, 42 (1971); M. Makosza and M. Fedoryński, *Bull. Acad. Polon. Sci., Ser. Sci. Chim.*, **29**, 105 (1971).
[16] P. Weyerstahl, G. Blume, and C. Müller, *Tetrahedron Letters*, **1971**, 3869.

Carbenes and Nitrenes 369

carbene (from $CHCl_2F$)[16] and chloro(phenylthio)carbene (from $PhSCHCl_2$).[17] In each case the carbene was intercepted by olefins to give cyclopropanes in good yield.

The other general method to be extended is the thermal decomposition of organomercury derivatives. New precursors of dihalogenocarbenes have been described: $PhHgCBrClI$,[18] $PhHgCBr_2F$[18] and $PhHgCCl_2I$[19] are all unstable solids which can transfer :CBrCl, :CBrF and :CCl$_2$ respectively, even at room temperature or below. An improved route to difluorocarbene uses either $PhHgCF_3$ or CF_3HgI with sodium iodide.[20] Other organomercury derivatives which have been investigated are $PhHgCCl_2Me$[21] (for :CMeCl), $PhHgCHCl_2$[22] and $PhHgCHBrCl$[22] (for :CHCl), $PhHgCCl_2CF_3$[23] and $PhHgCBrClCF_3$[23] (for :CClCF$_3$), and $(Me_3SiCX_2)_2Hg$[24] (X = Cl or Br) (for :CXSiMe$_3$). A synthesis of (+)-methyl 6-bromopenicillinate (**6**) uses this method to generate a carbene which then undergoes an intramolecular insertion to form the four-membered ring.[25]

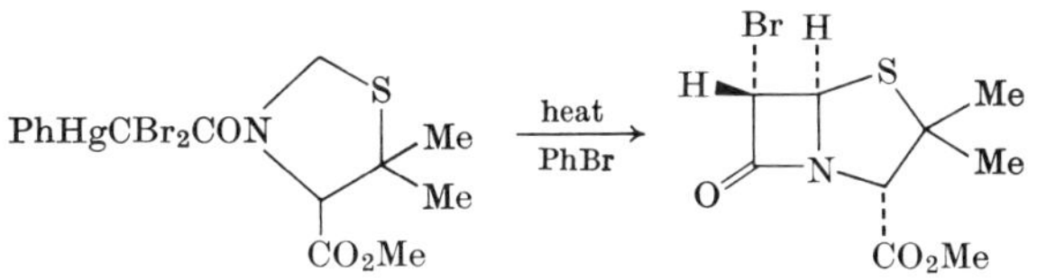

Methods of generating carbenes by thermal and photochemical cycloelimination have been reviewed,[26] and several new examples of these types of reaction have been published. Vapour-phase pyrolysis of *o*-phenylene carbonate gives the ketene (**7**), which can be trapped with methanol or which undergoes further fragmentation; a keto-carbene is a likely intermediate.[27] The tetrachloro derivative reacts in a similar way.[28] In an analogous reaction, it is suggested that vapour-phase pyrolysis of 1,2,3-triazoles (**8**) gives imino-

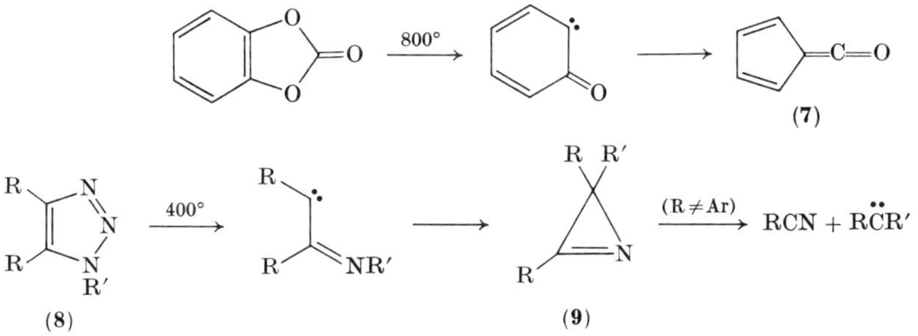

(R' = phthalimido)

[17] M. Makosza and E. Bialecka, *Tetrahedron Letters*, **1971**, 4517.
[18] D. Seyferth, C. K. Haas, and S. P. Hopper, *J. Organomet. Chem.*, **33**, C1 (1971).
[19] D. Seyferth and C. K. Haas, *J. Organomet. Chem.*, **30**, C38 (1971).
[20] D. Seyferth and S. P. Hopper, *J. Organomet. Chem.*, **26**, C62 (1971).
[21] D. Seyferth and D. C. Mueller, *J. Organomet. Chem.*, **28**, 325 (1971).
[22] D. Seyferth, H. D. Simmons, and H.-M. Shih, *J. Organomet. Chem.*, **29**, 359 (1971).
[23] D. Seyferth and D. C. Mueller, *J. Am. Chem. Soc.*, **93**, 3714 (1971).
[24] D. Seyferth and E. M. Hanson, *J. Organomet. Chem.*, **27**, 19 (1971).
[25] N. G. Johansson and B. Åkermark, *Tetrahedron Letters*, **1971**, 4785.
[26] R. W. Hoffmann, *Angew. Chem. Int. Ed.*, **10**, 529 (1971); G. W. Griffin, *Angew. Chem. Int. Ed.*, **10**, 537 (1971).
[27] D. C. DeJongh and D. A. Brent, *J. Org. Chem.*, **35**, 4204 (1970).
[28] D. C. DeJongh, D. A. Brent, and R. Y. Van Fossen, *J. Org. Chem.*, **36**, 1469 (1971).

carbenes by loss of nitrogen; the carbenes then rearrange to 2H-azirines (9), probably through 1H-azirine intermediates.[29] The 2H-azirines also undergo further thermal fragmentation to carbenes and nitriles, unless there is an aryl substituent at the 2-position, in which case a rearrangement to give an indole derivative takes precedence.

Extrusion of dimethoxycarbene from the benzonorbornadiene derivative (10) is much slower than from the corresponding norbornadiene (11);[30] the stepwise mechanism for the extrusion has therefore been questioned, on the basis that any intermediates would be expected to be better stabilized in the benzo derivative (10). This type of extrusion has also been investigated as a route to silylenes.[31]

The photochemical fragmentation of cyclopropanes to give carbenes and olefins has been extended to methylenecyclopropanes,[32] e.g. (12), and to the norcaradiene (13).[33] The reaction has also been observed with cyclopropane itself.[34] Carbenes are proposed as intermediates in the photolysis of tetrazole anions[35] and of the fluorene derivative (14),[36] but EPR spectra obtained after photolysis of 1,2,3-thiadiazoles in a solid matrix

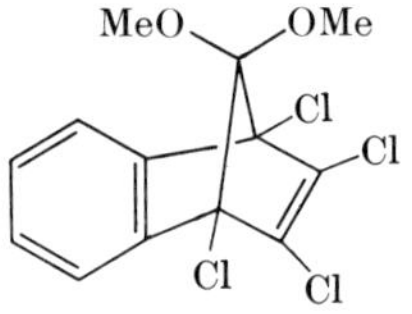

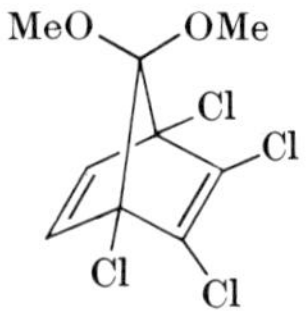

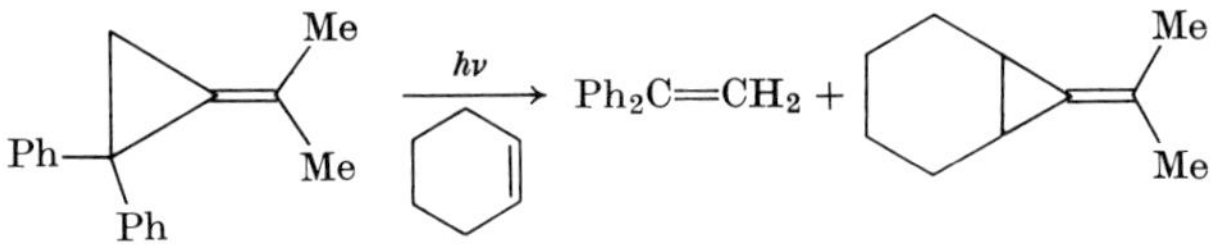

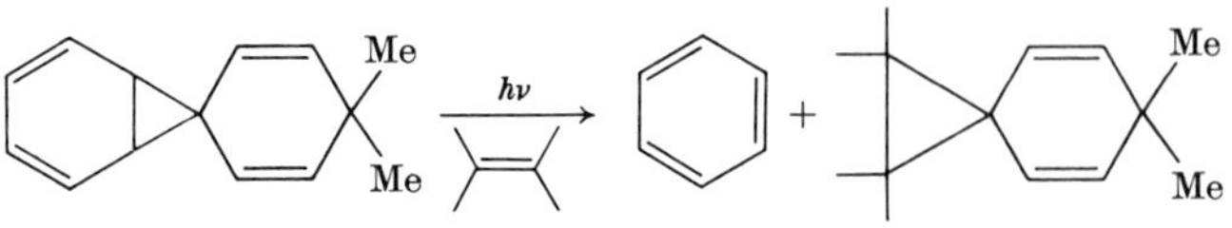

[29] D. J. Anderson, T. L. Gilchrist, G. E. Gymer, and C. W. Rees, *Chem. Comm.*, **1971**, 1518; T. L. Gilchrist, G. E. Gymer, and C. W. Rees, *Chem. Comm.*, **1971**, 1519.
[30] P. F. Ranken and M. A. Battiste, *J. Org. Chem.*, **36**, 1996 (1971).
[31] R. Maruca, *J. Org. Chem.*, **36**, 1626 (1971).
[32] J. C. Gilbert and J. R. Butler, *J. Am. Chem. Soc.*, **92**, 7493 (1970).
[33] R. H. Levin and M. Jones, *Tetrahedron*, **27**, 2031 (1971).
[34] A. K. Dhingra and R. D. Koob, *J. Phys. Chem.*, **74**, 4490 (1970).
[35] P. Scheiner, *Tetrahedron Letters*, **1971**, 4489.
[36] W. A. Henderson and A. Zweig, *Tetrahedron*, **27**, 5307 (1971).

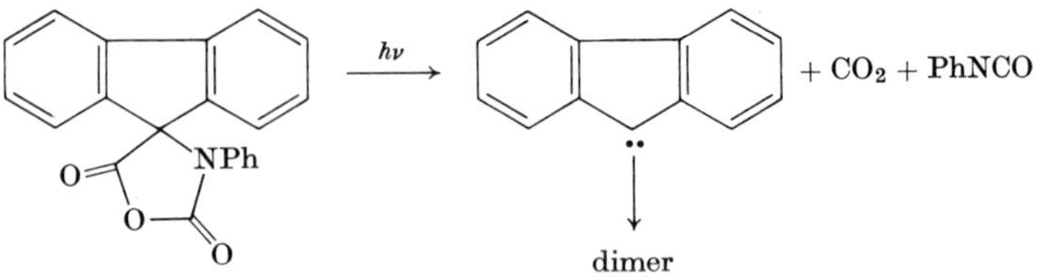

indicate that the extrusion of nitrogen is stepwise, and that thioketo-carbenes are not intermediates.[37]

Kinetic studies of the thermal decomposition of bromo- and chloro-difluoromethane show that the reactions are of the first order in the early stages, giving difluorocarbene.[38] Dichlorocarbene is suggested as an intermediate in the thermolysis of trichloronitrosomethane, Cl_3CNO;[39] pyrolysis of some oxalates may also involve the generation of carbenes, by an α-elimination:[40]

$$Ar_2CHO\text{—}CO\text{—}CO\text{—}OCHAr_2 \rightarrow 2Ar_2C: + 2CO_2 + H_2$$

The methoxy- and methine-protons of dimethylformamide dimethyl acetal can all be exchanged for deuterium in CD_3OD; to explain the observed rates of exchange it is proposed that the carbonium ion (15) undergoes competitive reaction with the solvent and fragmentation to give a carbene, which then reacts with the solvent:[41]

$$(MeO)_2CHNMe_2 \rightarrow MeO^- + Me\overset{+}{O}CHNMe_2 \rightarrow H^+ + Me O\overset{..}{C}NMe_2$$
$$(15)$$

Carbenes are also proposed as intermediates in the reaction of 1-bromoalkynes with alkoxides,[42] in the electrochemical reduction of 2,2-dichloronorbornane,[43] and in the reactions of fluorenone and of tetracyclone derivatives with tervalent phosphorus compounds.[44] A brief review of methods of generating dithiocarbenes appears in a paper on tetrathioethylenes.[45] The mechanism of the photochemical ring expansion of cyclic ketones to alkoxycarbenes has been investigated,[46] and another example of the reaction, involving a silacyclohexanone derivative, is reported.[47]

Nitrenes

Evidence has been presented that the photo-decomposition of some alkyl azides in solution does not involve free nitrenes: a study of migratory aptitudes in the photolysis of azides $R^1R^2R^3CN_3$ showed that the smallest group always migrates most readily.[48] The explanation offered is that in the preferred conformation of the ground-state azide the smallest group lies roughly orthogonal to the azide group. $\pi \rightarrow \pi^*$ Excitation of the azide leaves the p-orbital parallel to the bond joining the smallest group vacant; this group then migrates into this vacant orbital.

---

[37] P. Krauss, K.-P. Zeller, H. Meier, and E. Müller, *Tetrahedron*, **27**, 5953 (1971).
[38] G. R. Barnes, R. A. Cox, and R. F. Simmons, *J. Chem. Soc.* (B), **1971**, 1176; R. A. Cox and R. F. Simmons, *J. Chem. Soc.* (B), **1971**, 1625.
[39] V. Astley and H. Sutcliffe, *Tetrahedron Letters*, **1971**, 2707.
[40] R. E. Lehr and J. M. Wilson, *Chem. Comm.*, **1971**, 666.
[41] J. M. Brown and B. D. Place, *Chem. Comm.*, **1971**, 533.
[42] J. Cymerman-Craig and C. D. Beard, *Chem. Comm.*, **1971**, 691.
[43] A. J. Fry and R. G. Reed, *J. Am. Chem. Soc.*, **93**, 553 (1971).
[44] I. J. Borowitz, M. Anschel, and P. D. Readio, *J. Org. Chem.*, **36**, 553 (1971).
[45] D. L. Coffen, J. Q. Chambers, D. R. Williams, P. E. Garrett, and N. D. Canfield, *J. Am. Chem. Soc.*, **93**, 2258 (1971).
[46] N. J. Turro and D. R. Morton, *J. Am. Chem. Soc.*, **93**, 2569 (1971).
[47] A. G. Brook, H. W. Kucera, and R. Pearce, *Can. J. Chem.*, **49**, 1618 (1971).
[48] R. A. Abramovitch and E. P. Kyba, *J. Am. Chem. Soc.*, **93**, 1537 (1971).

New routes to nitrenes include the pyrolysis and photolysis of the azimine (**16**), which give products derived from ethoxycarbonylnitrene;[49] the pyrolysis of phenylsulphinylamine, PhNSO, at 1000°, which gives 1-cyanocyclopentadiene through an intermediate phenylnitrene;[50] the pyrolysis of 7-phenyloxindole (**17**);[51] and the photolysis of benzofurazan (**18**).[52] Benzofurazan, when photolysed in benzene, gives the azepine (**19**) and the isocyanate (**20**) which can be intercepted with methanol. The sequence shown has been proposed to explain these observations.

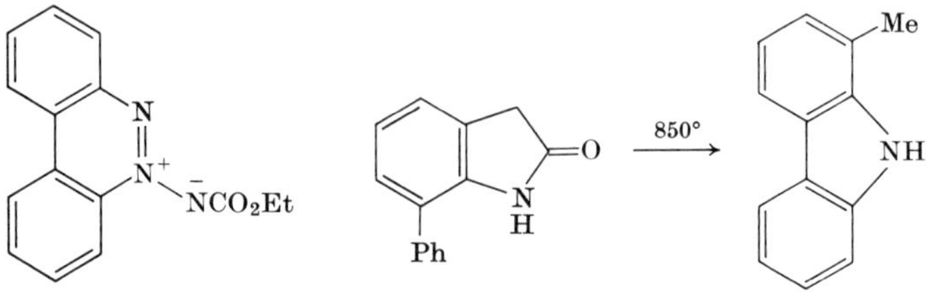

The tetrazole (**21**), generated by photolysis of a diazide, gives different products (**22**) and (**23**) on further photolysis and on pyrolysis, though a common intermediate imino-nitrene is suggested for the reactions.[53] Dinitrenes are formed by photolysis of p-diazidobenzene and of 4,4'-diazidoazobenzene in rigid matrices,[54] but methoxycarbonylnitrene was not detected in the photolysis of methyl azidoformate in a rare-gas matrix.[55]

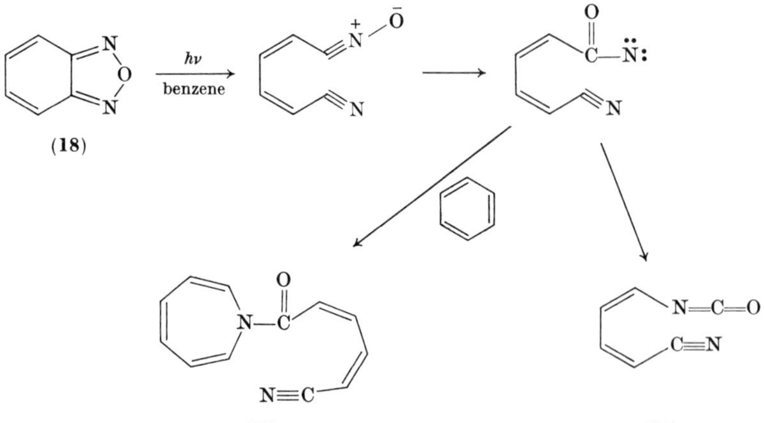

Fluoronitrene is proposed as an intermediate in the basic hydrolysis of difluoroamine.[56] A study of the relative rates of deoxygenation of substituted nitroso- and

[49] S. F. Gait, C. W. Rees, and R. C. Storr, *Chem. Comm.*, **1971**, 1545.
[50] C. Wentrup, *Tetrahedron*, **27**, 1027 (1971).
[51] R. F. C. Brown and M. Butcher, *Tetrahedron Letters*, **1971**, 667.
[52] M. Georgarakis, H. J. Rosenkranz, and H. Schmid, *Helv. Chim. Acta*, **54**, 819 (1971).
[53] R. M. Moriarty and P. Serridge, *J. Am. Chem. Soc.*, **93**, 1534 (1971).
[54] B. Singh and J. S. Brinen, *J. Am. Chem. Soc.*, **93**, 540 (1971).
[55] R. E. Wilde, T. K. K. Srinivasan, and W. Lwowski, *J. Am. Chem. Soc.*, **93**, 860 (1971).
[56] W. J. le Noble, E. M. Schulman, and D. N. Skulnik, *J. Am. Chem. Soc.*, **93**, 4710 (1971).

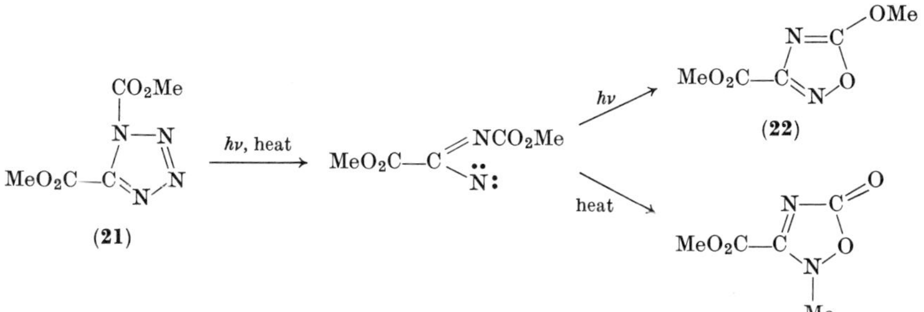

nitro-benzenes by triethyl phosphite shows that the rate is accelerated by electron-withdrawing groups; nucleophilic attack by phosphorus on oxygen is suggested.[57]

Cycloadditions

Photolysis of phenyldiazomethane in a solid *cis*-butene matrix allows reactions of triplet phenylcarbene to be observed; for example, it adds non-stereospecifically to *cis*-butene.[58] This is in contrast to solution photolyses in which most of the reactions can be ascribed to singlet phenylcarbene. Irradiation of the diazoalkanes (24) and (25) in solution with

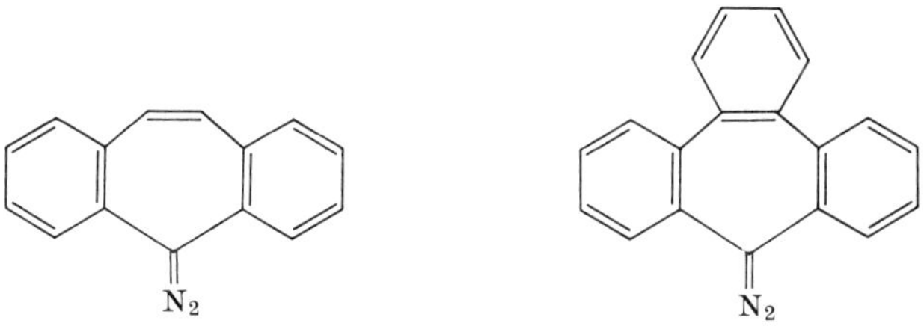

cis- or *trans*-but-2-ene gives cyclopropanes stereospecifically,[59] though the corresponding carbenes (like phenylcarbene) are known to have triplet ground states. In this case, however, the authors propose a new explanation for the stereospecific addition—that reaction occurs through a charge-transfer complex of the delocalized triplet carbene and the olefin. The results of irradiating these diazoalkanes in a solid matrix could help to solve the problem of whether the singlet or the triplet carbenes are responsible for the stereospecific addition.

Triplet carbenes are also implicated in reactions of diphenyldiazomethane[60] and of dimethyl diazomalonate[61] with acetylenes. The isolation of 5-membered ring products, e.g. (26), is ascribed to ring closure of intermediate diradicals formed by addition of the triplet carbenes; cyclopropenes and pyrazoles are ruled out as alternative intermediates. Pyrazoles may be intermediates in similar reactions of arylcarbenes with

[57] R. J. Sundberg and C.-C. Lang, *J. Org. Chem.*, **36**, 300 (1971).
[58] R. A. Moss and U.-H. Dolling, *J. Am. Chem. Soc.*, **93**, 954 (1971).
[59] S.-I. Murahashi, I. Moritani, and M. Nishino, *Tetrahedron*, **27**, 5131 (1971).
[60] M. E. Hendrick, W. J. Baron, and M. Jones, *J. Am. Chem. Soc.*, **93**, 1554 (1971).
[61] M. E. Hendrick, *J. Am. Chem. Soc.*, **93**, 6337 (1971).

acetylenedicarboxylic ester, however.[62] Reaction of m-nitrophenyldiazomethane with olefins has been reported;[63] photolysis gives cyclopropanes non-stereospecifically, but the thermal addition is stereospecific in the presence of zinc halides. The thermolysis of the ferrocene derivative (27) in the presence of 1,1-diphenylethylene yields the corresponding cyclopropane as well as the product formed by a 1,2-hydrogen shift.[64] The cycloadduct is said to be formed from the triplet carbene, since dec-1-ene does not react, and in this case a diradical intermediate would be much less stable. Oxygen also partially suppresses the intramolecular rearrangement, as would be expected if a singlet–triplet equilibrium existed.

A revised structure for the product of addition of dicyanocarbene to cyclo-octatetraene shows it to be a 1,4-adduct, probably formed from the triplet carbene.[65]

Synthetic uses of carbene addition include: addition of dichlorocarbene to 1,2,3-trimethylcyclopropene,[66] and to pinene and other terpenes;[67] the cuprous chloride-catalysed addition of methylene to methyl pyrrole-1-carboxylate[68] (where the mono- and di-adducts can be isolated) and to the amino acid derivative (28)[69] [which gives the naturally occurring amino acid (29) on saponification]; and the intramolecular cycloaddition of the carbene derived from the fulvene (30).[70]

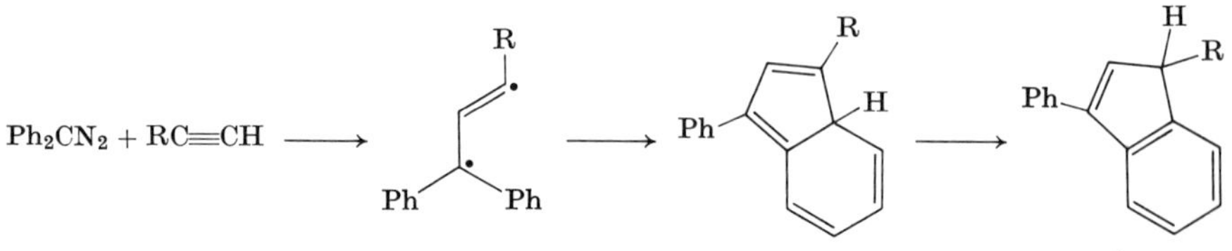

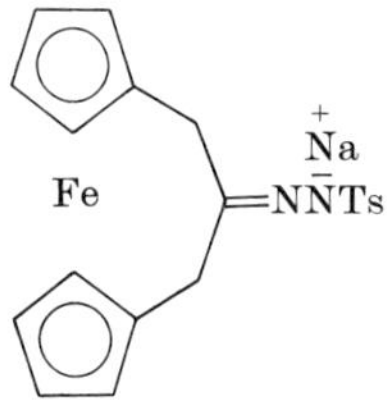

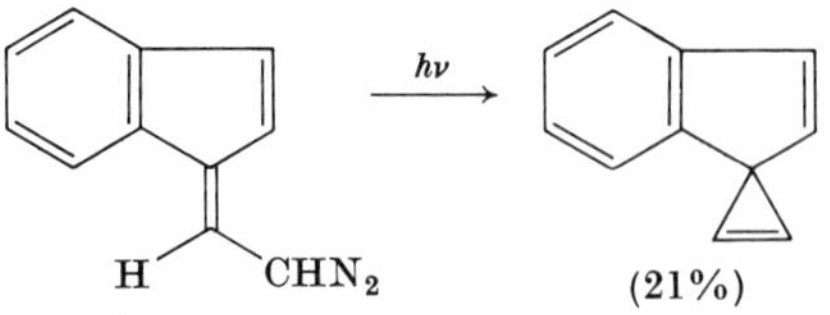

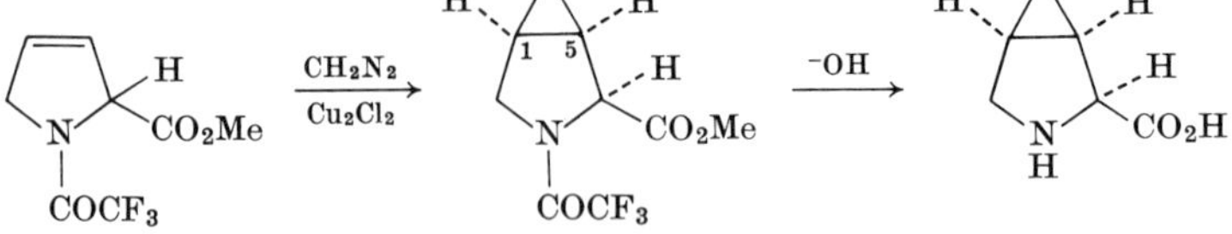

[62] H. Dürr, L. Schrader, and H. Seidl, *Chem. Ber.*, **104**, 391 (1971); L. Schrader, *Chem. Ber.*, **104**, 941 (1971).
[63] S. H. Goh, *J. Chem. Soc.* (C), **1971**, 2275.
[64] A. Sonoda and I. Moritani, *J. Organomet. Chem.*, **26**, 133 (1971).
[65] A. G. Anastassiou, R. P. Cellura, and E. Ciganek, *Tetrahedron Letters*, **1970**, 5267.
[66] B. M. Trost and R. C. Atkins, *Chem. Comm.*, **1971**, 1254.
[67] J. Hatem and B. Waegell, *Tetrahedron Letters*, **1971**, 2069.
[68] F. W. Fowler, *Angew. Chem. Int. Ed.*, **10**, 135 (1971).
[69] Y. Fujimoto, F. Irreverre, J. M. Karle, I. L. Karle, and B. Witkop, *J. Am. Chem. Soc.*, **93**, 3471 (1971).
[70] T. Severin, H. Krämer, and P. Adhikary, *Chem. Ber.*, **104**, 972 (1971).

The carbene $Ph_2C=C=C:$, formed through the diazoalkane, has been trapped (in very low yield) with 2,3-dimethylbut-2-ene,[71] and the corresponding carbene $(Me_3C)_2C=C=C:$, generated by base-catalysed elimination, has also been intercepted by olefins.[72] The diazoalkanes $PhC(N_2)PO(OMe)_2$ and $N_2CHPO(OMe)_2$ give cyclopropanes with olefins in the presence of copper catalysts.[73] Difluorocarbene adds to the carbon–carbon double bond of 2-benzylidenecyclohexanones,[74] and dichlorocarbene to the substituted double bond of phenylallene.[75] Copper-catalysed intramolecular cycloadditions of keto-carbenes are also reported.[76]

The isolation of cyclopropenone imines, e.g., (**31**), from the reaction of isocyanides with cycloalkynes and with electron-rich acetylenes,[77] represents a remarkable illustration of "carbene-like" properties of isocyanides; similar reactions of carbon monoxide itself might not be impossible to achieve.

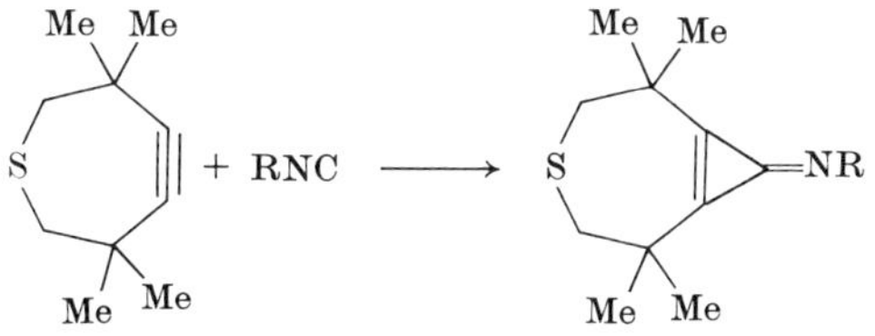

The first examples of addition of phenylnitrene to benzene[78] and of an arylnitrene to an olefin[79] are claimed, in the deoxygenation of nitrosobenzene in benzene–trifluoroethanol and in the decomposition of ferrocenyl azide in cyclohexene, respectively. The low decomposition temperature of ferrocenyl azide leads the authors to suggest that the metal participates in the reaction. Addition of ethoxycarbonylnitrene to porphins[80] and to o-, m- and p-xylenes[81] results in ring expansion. With 1-methylcyclopropene, the initial adduct, the azabicyclobutane (**32**), could not be isolated, the product obtained being the imine (**33**), but with 1,2-dimethylcyclobutene the adduct (**34**) was isolated and was found to be surprisingly stable.[82] Addition of phthalimidonitrene to benzofuran also gives an isolable aziridine, but this then rearranges when heated.[83]

Thermolysis of the azide (**35**) (R = 1-naphthyl) gives the anthracene derivative (**36**) and HCN;[84] the reaction sequence proposed to explain this involves an intramolecular cycloaddition followed by a 1,2-aryl shift and then loss of HCN in a retro-Diels–Alder

---

[71] D. J. Northington and W. M. Jones, *Tetrahedron Letters*, **1971**, 317.
[72] H. D. Hartzler, *J. Am. Chem. Soc.*, **93**, 4527 (1971).
[73] D. Seyferth, R. S. Marmor, and P. Hilbert, *J. Org. Chem.*, **36**, 1379 (1971).
[74] M. Derenberg and P. Hodge, *Chem. Comm.*, **1971**, 233.
[75] Y. M. Slobodin and I. Z. Egenburg, *Zhur. Org. Chim.*, **6**, 2629 (1970); R. R. Kostikov and A. I. Ioffe, *Zhur. Org. Khim.*, **6**, 2630 (1970).
[76] P. M. McCurry, *Tetrahedron Letters*, **1971**, 1845.
[77] A. Krebs and H. Kimling, *Angew. Chem. Int. Ed.*, **10**, 409 (1971).
[78] R. J. Sundberg and R. H. Smith, *Tetrahedron Letters*, **1971**, 267.
[79] R. A. Abramovitch, C. I. Azogu, and R. G. Sutherland, *Chem. Comm.*, **1971**, 134.
[80] R. Grigg, *J. Chem. Soc.* (C), **1971**, 3664.
[81] J. M. Photis, *J. Heterocyclic Chem.*, **7**, 1249 (1970).
[82] J. N. Labows and D. Swern, *Tetrahedron Letters*, **1971**, 4523.
[83] D. W. Jones, *Chem. Comm.*, **1971**, 1130.
[84] J. J. Looker, *J. Org. Chem.*, **36**, 1045 (1971).

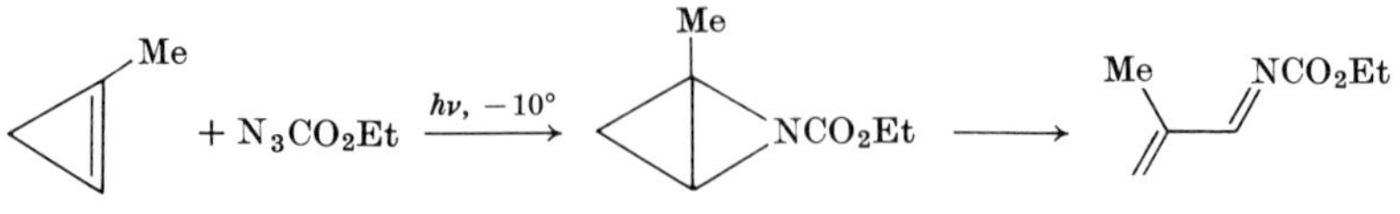

reaction. When R = Ph in structure (**35**), rearrangement of the nitrene is the major reaction; photolysis of the azides also promotes rearrangement.[85]

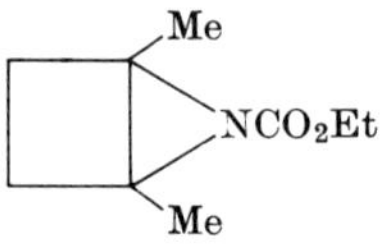

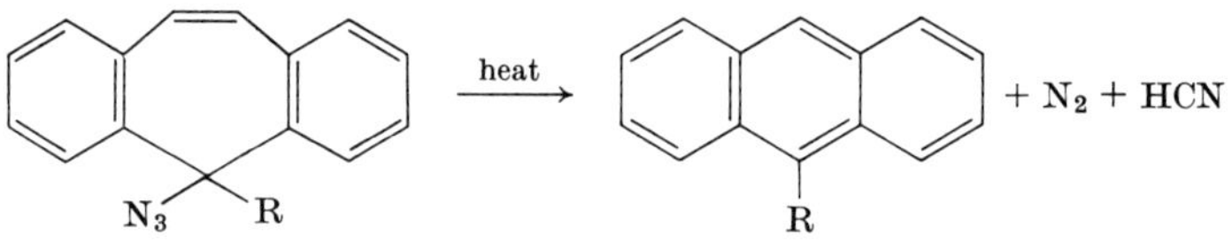

Insertions and Abstractions

Calculations of the potential energy surface for the concerted insertion of singlet methylene into a C—H bond of methane have been carried out by the extended Hückel method.[86] These indicate a nearly linear transition state for the process, the methylene approaching the C—H bond in the direction shown in (**37**). The hydrogen is transferred from methane to methylene while the C—C distance (2.5 Å) changes little; this is then followed by collapse to give ethane, with no evidence for discrete radical intermediates in the process. The mechanistic picture provided by these calculations contrasts with earlier suggestions by Doering and by Skell that the transition state is triangular. On the other hand, experimental work reported this year seems to favour the Doering–Skell picture, in that carbene (**38**) gives the *exo*-insertion product (**39**) (45%) by an intramolecular insertion for which a triangular transition state is the most reasonable.[87] The *endo*-product (**40**) (36%) is also formed, however, and further experimental tests are clearly desirable. The linear transition state does have the advantage that it explains the observed relative rates of insertion into different types of C—H bonds, in that it is radical-like, but involves transfer of electron density to the carbene carbon.

Several other pieces of work have strengthened the view that singlet carbenes react with C—H bonds in a concerted manner, whereas triplet carbenes react by abstraction of hydrogen. Calculations carried out on the reaction of singlet and triplet oxygen with hydrogen show that the singlet and triplet energy surfaces are strikingly different; singlet oxygen inserts by choice and may abstract by chance, whereas triplet oxygen

---

[85] J. J. Looker, *J. Org. Chem.*, **36**, 2681 (1971).
[86] R. C. Dobson, D. M. Hayes, and R. Hoffmann, *J. Am. Chem. Soc.*, **93**, 6188 (1971).
[87] C. D. Gutsche, G. L. Bachman, W. Udell, and S. Bäuerlein, *J. Am. Chem. Soc.*, **93**, 5172 (1971).

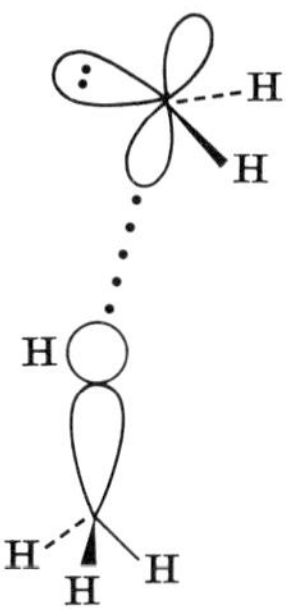

necessarily abstracts.[88] These results should also apply to carbene and nitrene reactions. The assertion[89] that triplet methylene (from diazomethane) can insert has been questioned[90] on the basis of some new results obtained by photolysis of diazomethane–propane mixtures. These show that triplet methylene is preferentially scavenged by diazomethane, so that even a small proportion of singlet methylene in a singlet–triplet mixture would be responsible for most of the products formed by reaction of the methylene with the substrate; insertion reactions observed with "deactivated" methylene could thus have come from a small percentage of singlet methylene.

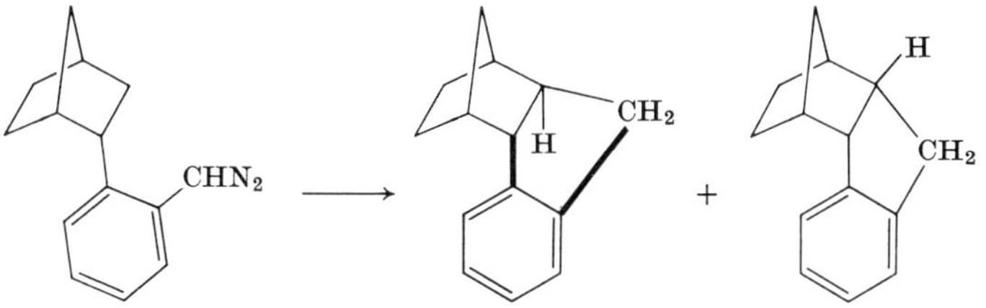

Evidence from CIDNP experiments[91] has shown that triplet methylene reacts with deuteriochloroform exclusively by abstraction of deuterium, the product (CH_2DCCl_3) then being formed by collapse of the triplet radical pair, whereas singlet methylene reacts exclusively at chlorine, the product ($CH_2ClCDCl_2$) being formed by abstraction of chlorine and collapse of the singlet radical pair (Scheme 1, p. 378). Similar mechanisms operate with tetrachloromethane[92] and with 1,2-dichloroethane.[93]

The reaction of dichlorocarbene with a variety of chlorinated C_3 hydrocarbons gives products resulting from dechlorination or dehydrochlorination;[94] for example, octachloropropane gives hexachloropropene and the dimer (41) of tetrachloroallene, the relative amounts depending on how much dichlorocarbene precursor is used. These reactions can again be rationalized as involving abstraction of a chlorine atom by the

---

[88] R. F. W. Bader and R. A. Gangi, *J. Am. Chem. Soc.*, **93**, 1831 (1971).
[89] D. F. Ring and B. S. Rabinovitch, *Can. J. Chem.*, **46**, 2435 (1968); see *Org. Reaction Mech.* **1968**, 329.
[90] J. A. Bell, *J. Phys. Chem.*, **75**, 1537 (1971).
[91] H. D. Roth, *J. Am. Chem. Soc.*, **93**, 4935 (1971).
[92] H. D. Roth, *J. Am. Chem. Soc.*, **93**, 1527 (1971).
[93] K. Dees, D. W. Setser, and W. G. Clark, *J. Phys. Chem.*, **75**, 2231 (1971).
[94] H. Khalaf, *Tetrahedron Letters*, **1971**, 4229.

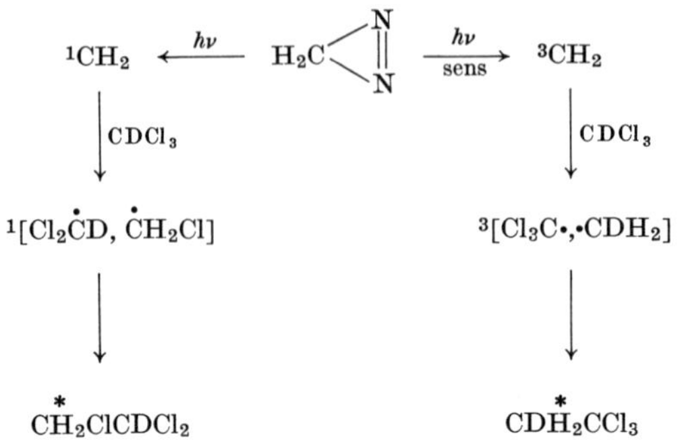

Scheme 1

$$^1CH_2 \xleftarrow{h\nu} H_2C\underset{N}{\overset{N}{\lVert}} \xrightarrow[\text{sens}]{h\nu} {}^3CH_2$$

$$\downarrow CDCl_3 \qquad\qquad \downarrow CDCl_3$$

$$^1[Cl_2\dot{C}D, \dot{C}H_2Cl] \qquad {}^3[Cl_3C\cdot, \cdot CDH_2]$$

$$\downarrow \qquad\qquad\qquad \downarrow$$

$$CH_2Cl\overset{*}{C}DCl_2 \qquad CD\overset{*}{H}_2CCl_3$$

carbene, this then being followed by the loss of a second chlorine atom rather than by recombination.

The kinetics of the abstraction of hydrogen from neopentane by triplet methylene have been measured,[95] and a radical-pair mechanism has been established for the allylic insertion of the triplet carbene fluorenylidene into cyclohexene.[96]

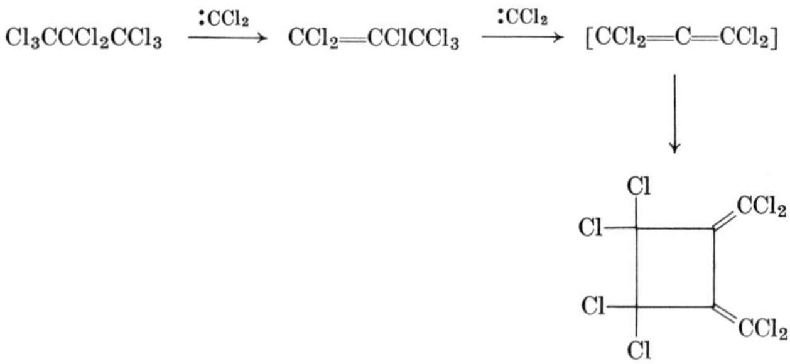

$$Cl_3CCCl_2CCl_3 \xrightarrow{:CCl_2} CCl_2{=}CClCCl_3 \xrightarrow{:CCl_2} [CCl_2{=}C{=}CCl_2]$$

$$\downarrow$$

(41)

The products obtained from carbenes (**42**)[97] and (**44**)[98] can be explained by proposing an equilibrium between singlet and triplet states. The carbene (**42**) gives identical products whether it is generated by direct or by sensitized photolysis, so it is proposed that the singlet and triplet carbenes interconvert more rapidly than they give products. A minor product is the olefin (**43**), and a concerted double-insertion mechanism is tentatively proposed to account for it. The carbene 9-anthronylidene (**44**) reacts with pyridine N-oxide in benzene to give anthraquinone; bianthrone is also formed by abstraction from the solvent. It is suggested that the triplet carbene is inefficiently trapped by the benzene, so that it can undergo partial conversion into the singlet state, which is intercepted by pyridine N-oxide. In the reactions of the carbene with olefins, the products (of abstraction and cycloaddition) can, however, all be rationalized as reactions of the triplet carbene.[99]

[95] J. R. McNesby and R. V. Kelly, *Int. J. Chem. Kinet.*, **3**, 293 (1971).
[96] J. E. Baldwin and A. H. Andrist, *Chem. Comm.*, **1971**, 1512.
[97] T. A. Baer and C. D. Gutsche, *J. Am. Chem. Soc.*, **93**, 5180 (1971).
[98] G. Cauquis and G. Reverdy, *Tetrahedron Letters*, **1971**, 3771.
[99] G. Cauquis and G. Reverdy, *Tetrahedron Letters*, **1971**, 4289.

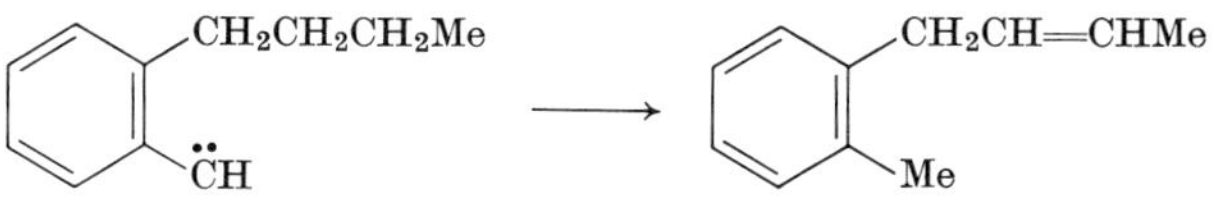

(42) — structure: benzene with CH₂CH₂CH₂Me and ĊH substituents
(43) — structure: benzene with CH₂CH=CHMe and Me substituents
(*cis* and *trans*: 5%)

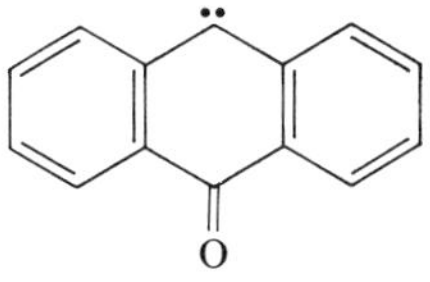

(44) — anthrone-like structure with carbene

Dichlorocarbene has for the first time been found to insert into C—Si bonds, namely, in its reaction with silacyclobutanes;[100] insertion has also been observed into a variety of tertiary C—H bonds,[101] into the benzylic C—H bond of (+)-2-phenylbutane (with complete retention of configuration),[102] into the β-C—H bonds of tetra-alkylgermanes,[103] and into allylic C—H bonds of cyclopentadiene, indene and fluorene.[104] Dibromocarbene inserts into Si—H and Ge—H bonds with complete retention of configuration.[105] The insertion of ethoxycarbonylcarbene into Si—H bonds[105] and into C—H bonds of n-alkanes[106] has also been studied, but attempts to insert difluorocarbene into a variety of metal–metal bonds were unsuccessful.[107] Insertion reactions of vinylidene carbenes have been observed with Si—H bonds and with C—H bonds α to oxygen.[108] Metastable singlet carbon atoms insert into the allylic C—H bonds of propene and butenes, as well as adding to the double bonds;[109] thus, propene reacts to give methylallene (by cycloaddition and rearrangement) and butadiene (by insertion and rearrangement) (Scheme 2).

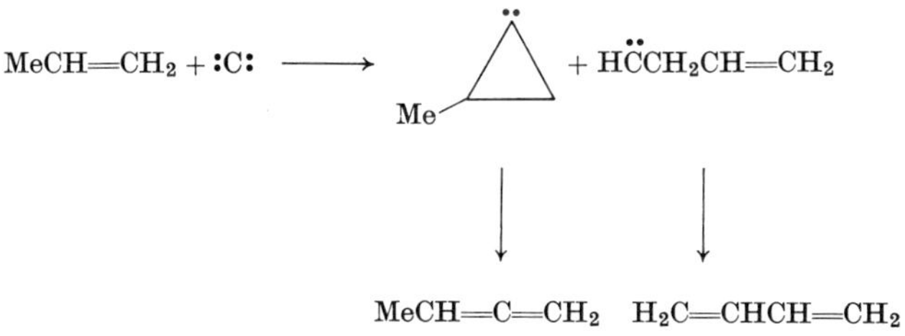

$$MeCH=CH_2 + :C: \longrightarrow \underset{Me}{\triangle} + H\ddot{C}CH_2CH=CH_2$$

$$MeCH=C=CH_2 \qquad H_2C=CHCH=CH_2$$

Scheme 2

[100] D. Seyferth, R. Damrauer, S. B. Andrews, and S. S. Washburne, *J. Am. Chem. Soc.*, **93**, 3709 (1971).
[101] E. V. Dehmlow, *Tetrahedron*, **27**, 4071 (1971).
[102] D. Seyferth and Y. M. Cheng, *J. Am. Chem. Soc.*, **93**, 4072 (1971).
[103] D. Seyferth, H.-M. Shih, P. Mazerolles, M. Lesbre, and M. Joanny, *J. Organomet. Chem.*, **29**, 371 (1971).
[104] M. B. D'Amore and R. G. Bergman, *Chem. Comm.*, **1971**, 461; R. E. Busby, M. Iqbal, R. J. Langston, J. Parrick, and C. J. G. Shaw, *Chem. Comm.*, **1971**, 1293.
[105] A. G. Brook, J. M. Duff, and D. G. Anderson, *J. Am. Chem. Soc.*, **92**, 7567 (1970).
[106] A. D. Forbes and J. Wood, *J. Chem. Soc.* (B), **1971**, 646.
[107] H. C. Clark and B. K. Hunter, *J. Organomet. Chem.*, **31**, 227 (1971).
[108] J. Cymerman-Craig and C. D. Beard, *Chem. Comm.*, **1971**, 692.
[109] P. S. Skell, J. E. Villaume, J. H. Plonka, and F. A. Fagone, *J. Am. Chem. Soc.*, **93**, 2699 (1971).

Intramolecular insertions of carbenes (**45**)[110] and (**46**)[111] have been observed. The copper-catalysed decomposition of diazo compounds (**47**) (where R′ is a chiral group) in the presence of amines, provides synthesis of alanine derivatives.[112]

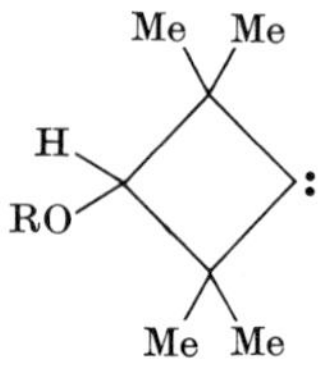

(**45**) 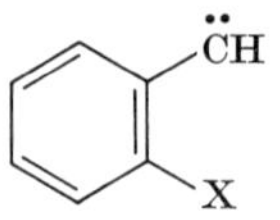   (**46**)   (X = OR, SR, NR$_2$)

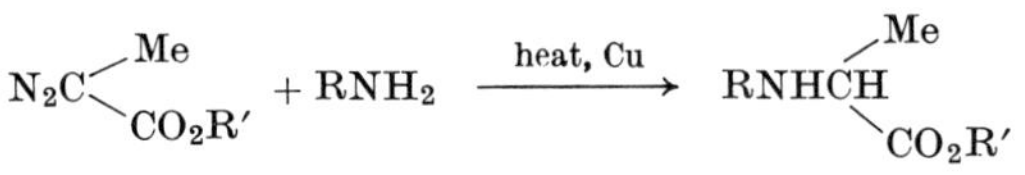

$$N_2C(Me)(CO_2R') + RNH_2 \xrightarrow{\text{heat, Cu}} RNHCH(Me)(CO_2R')$$

(**47**)

Aspects of the chemistry of arylnitrenes which have hitherto been in a rather unsatisfactory state have been greatly clarified by some newly published work on the photolysis of aromatic azides. It is shown, for example, that the polymer that is often a major product of photolysis of the azides in solution is almost certainly derived from the singlet nitrene; it is formed by insertion of the nitrene into the aromatic ring of another molecule of the azide, followed by decomposition of the azide and futher insertion.[113] Sensitized photolysis suppresses the formation of the polymer and, in the case of phenyl azide, gives a high yield (97%) of azobenzene. No polymer is formed in methanol, most of the product being derived from insertion of the nitrene into the C—H bonds of the solvent. It is concluded that direct photolysis of phenyl azide gives mainly singlet nitrene, only a small proportion (12–13%) being generated as the triplet. Photolysis of aryl azides in rigid polystyrene matrices, where rates of abstraction can be determined, allows a comparison between the lifetimes of arylnitrenes and the charge on the nitrogen atom.[114] Arylnitrenes appear to have the paradoxical character of electron-seeking reagents that are nevertheless repelled by sites of high electron density; hence their lower reactivity compared with carbenes or with acylnitrenes. The products of photolysis of aromatic azides in rigid matrices are primary and secondary amines;[115] the primary amines are formed by successive hydrogen abstraction from the medium, and the secondary amines by an abstraction–recombination mechanism. The proportion of primary to secondary amine thus depends markedly on the rigidity of the medium; addition of even a small percentage of toluene to a polystyrene matrix decreases the lifetime of the nitrenes drastically and increases the proportion of primary amine in the product.

The photolysis of the azide (**48**) has been reported.[116] The major product is the corresponding amine; a little of the azo compound and the "nitrene dimer" (**49**) (10%) are also formed.

[110] A. H. Rees and M. C. Whiting, *J. Org. Chem.*, **35**, 4167 (1970).
[111] G. V. Garner, D. B. Mobbs, H. Suschitzky, and J. S. Millership, *J. Chem. Soc.* (C), **1971**, 3693.
[112] J.-F. Nicoud and H. B. Kagan, *Tetrahedron Letters*, **1971**, 2065.
[113] A. Reiser and L. J. Leyshon, *J. Am. Chem. Soc.*, **93**, 4051 (1971).
[114] A. Reiser and L. J. Leyshon, *J. Am. Chem. Soc.*, **92**, 7487 (1970).
[115] A. Reiser, L. J. Leyshon, and L. Johnston, *Trans. Faraday Soc.*, **67**, 2389 (1971).
[116] B. Stanovnik, *Tetrahedron Letters*, **1971**, 3211.

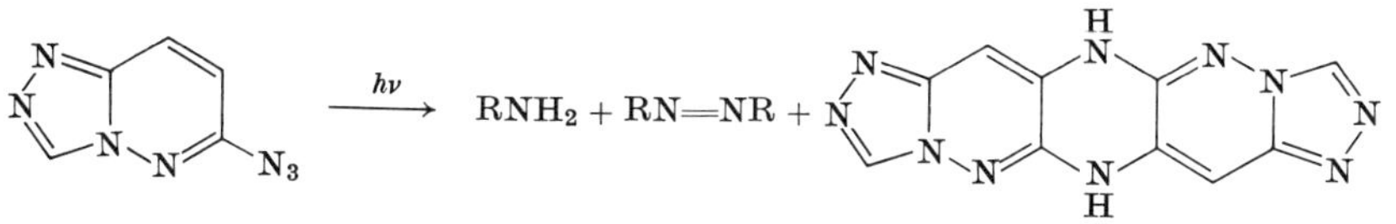

Rearrangements and Fragmentations

A ^{13}C-labelling experiment has elucidated the mechanism by which p-tolylcarbene is thermally converted into benzocyclobutene and into styrene.[117] The results are in accord with a mechanism suggested earlier;[118] namely, a series of equilibria between bicyclo-[4.1.0]heptatrienes, arylcarbenes and cycloheptatrienylidenes. The label is found at the positions required by this mechanism (Scheme 3).

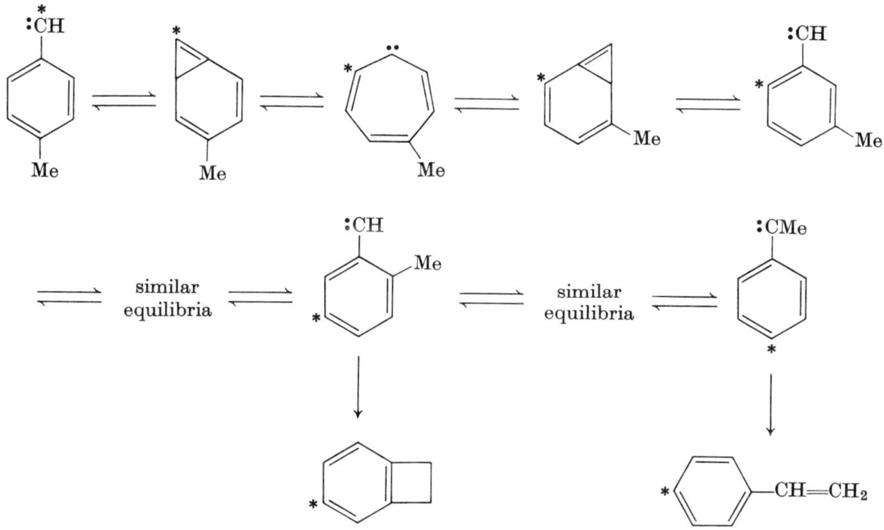

Scheme 3

The rearrangement of an optically active carbene of the cyclopropylidene series (**50**) gives an optically active allene, even when the substituents X and Y are of approximately equal size.[119] Preferential rotation of one carbon atom is suggested, this being the atom which furnishes the electron pair to the vacant p-orbital of the carbene centre. Thus, the influence of X and Y on the ring opening is electronic rather than steric.

Evidence for thermal Wolff rearrangement of a dialkoxycarbene is obtained from the gas-phase pyrolysis of dimethyl diazomalonate.[120] One of the products obtained is methyl vinyl ether, the formation of which is explained by proposing a rearrangement to the ketene which then loses CO to give the carbene (**51**). An independent study of

---

[117] E. Hedaya and M. E. Kent, *J. Am. Chem. Soc.*, **93**, 3283 (1971).
[118] W. J. Barron, M. Jones, and P. P. Gaspar, *J. Am. Chem. Soc.*, **93**, 4739 (1971).
[119] W. M. Jones and D. L. Krause, *J. Am. Chem. Soc.*, **93**, 551 (1971).
[120] D. C. Richardson, M. E. Hendrick, and M. Jones, *J. Am. Chem. Soc.*, **93**, 3790 (1971).

the carbene (**51**) in solution has been made, but only carbene dimers were isolated.[121] Wolff rearrangement of phosphinylcarbenes to give phosphenes (**52**), a new class of reactive intermediate, has also been proposed.[122] No comparable rearrangement was observed with the related carbenes (**53**);[123] intramolecular hydrogen or alkyl shifts (depending on the nature of the group R) occur instead.

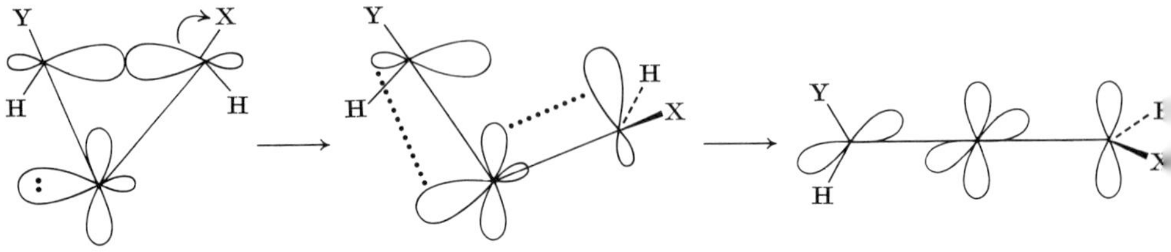

(**50**)

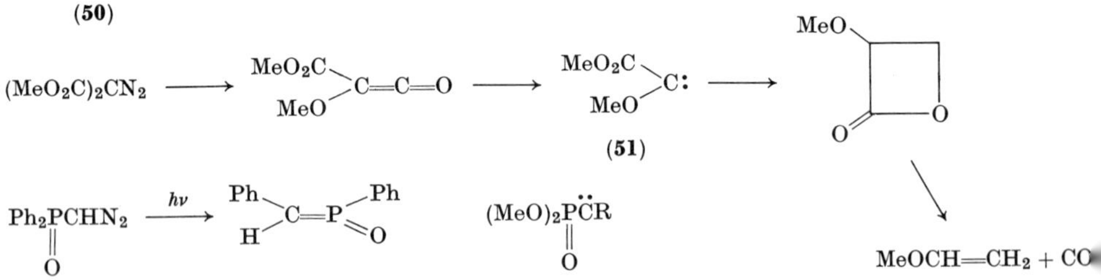

$(MeO_2C)_2CN_2 \longrightarrow \begin{matrix} MeO_2C \\ MeO \end{matrix} C=C=O \longrightarrow \begin{matrix} MeO_2C \\ MeO \end{matrix} C: \longrightarrow$

(**51**)

$\underset{\underset{O}{\parallel}}{Ph_2PCHN_2} \xrightarrow{h\nu} \underset{H}{\overset{Ph}{\diagdown}} C = \underset{O}{\overset{Ph}{P}} \quad \underset{\underset{O}{\parallel}}{(MeO)_2\ddot{P}CR}$

(**52**)   (**53**)

$MeOCH=CH_2 + CO$

Pyrolysis of the tosylhydrazone salt (**54**) gives 2,2-dimethylcyclobutanone as the major product:[124] this reaction represents the reverse of the photochemical ring expansion of cyclic ketones to alkoxycarbenes. Bicyclo[2.2.0]hexene has been generated by vacuum-pyrolysis of the tosylhydrazone salt (**55**), and its NMR spectrum has been recorded at low temperature.[125] Hydrogen shift is shown to be favoured over fluorine or fluoroalkyl shift in the rearrangement of the carbene (**56**).[126] A new type of skeletal rearrangement of 2-furylcarbenes (**57**) competes favourably with 1,2-hydrogen shift.[127]

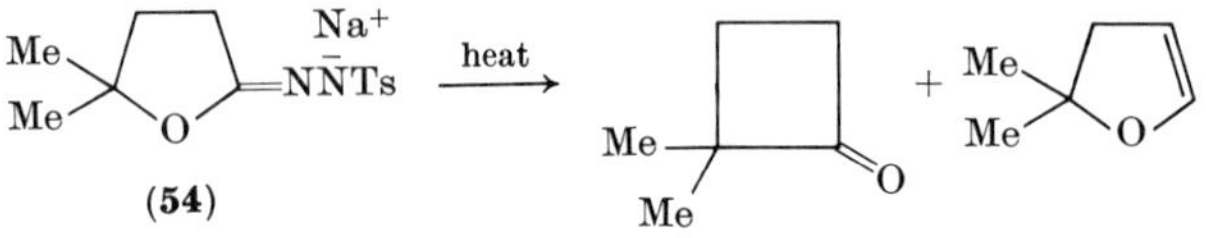

(**54**)   (70%)   (30%)

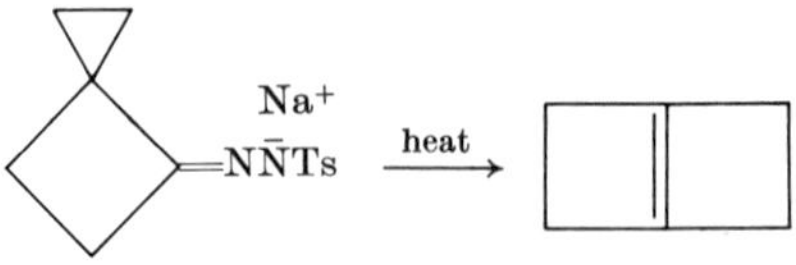

(**55**)

---

[121] J. Gehlhaus and R. W. Hoffmann, *Tetrahedron*, **26**, 5901 (1970).
[122] M. Regitz, A. Liedhegener, W. Anschütz, and H. Eckes, *Chem. Ber.*, **104**, 2177 (1971).
[123] R. S. Marmor and D. Seyferth, *J. Org. Chem.*, **36**, 128 (1971).
[124] W. C. Agosta and A. M. Foster, *Chem. Comm.*, **1971**, 433.
[125] K. B. Wiberg, G. J. Burgmaier, and P. Warner, *J. Am. Chem. Soc.*, **93**, 246 (1971).
[126] J. H. Atherton, R. Fields, and R. N. Haszeldine, *J. Chem. Soc.* (C), **1971**, 366.
[127] R. V. Hoffman and H. Shechter, *J. Am. Chem. Soc.*, **93**, 5940 (1971).

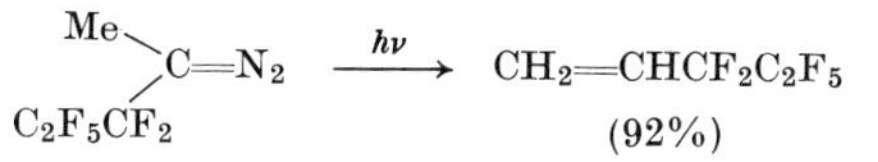

$$\underset{C_2F_5CF_2}{\overset{Me}{>}}C=N_2 \xrightarrow{h\nu} CH_2=CHCF_2C_2F_5$$
$$(92\%)$$
(56)

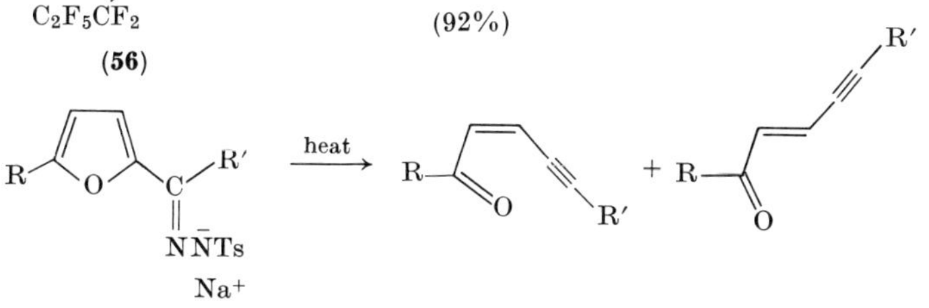

(57)

When heated, the tosylhydrazone salt (58) gives the corresponding diazoalkane which on further heating gives the products shown.[128] Non-carbene mechanisms were proposed to explain the formation of these products, but as the reaction conditions are typical of those in which carbenes are formed, the mechanisms are best regarded as tentative.

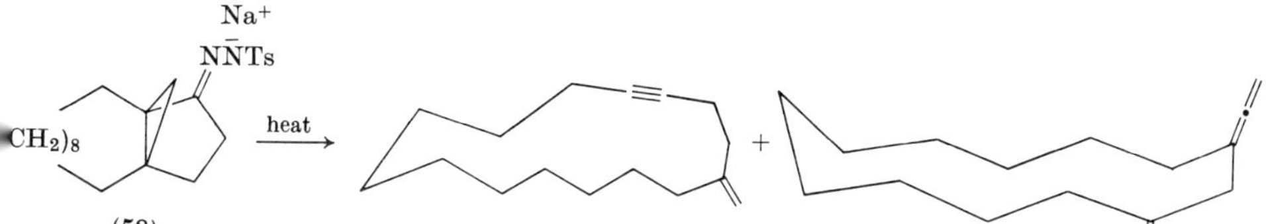

(58)

Photochemical fragmentation of dimethylcarbene to give methyl and vinyl radicals is reported.[129]

Further examples of skeletal rearrangements of imino-nitrenes, generated by vacuum-pyrolysis of fused tetrazoles, have been recorded.[130] An acyl shift in an intermediate nitrene is proposed for the thermolysis of the azide (59).[131] Examples of 1,2 oxygen-to-nitrogen migration have been observed in the rearrangement of alkoxynitrenes;[132] for example, oxidation of O-benzylhydroxylamine gives benzaldehyde oxime:

$$PhCH_2ONH_2 \xrightarrow[benzene]{Pb(OAc)_4} PhCH=NOH$$

The rearrangement of allylic amino-nitrenes (60)[133] provides further examples of allowed [2,3] sigmatropic shifts. Ring expansion is the major reaction of indazolinyl-nitrenes (61) and (62).[134] Further evidence is provided for the conversion of (phenyl-ethynyl)nitrene, $PhC{\equiv}C\ddot{N}{:}$, into α-cyano-α-phenylcarbene.[135]

[128] G. Ohloff and W. Pickenhagen, *Helv. Chim. Acta*, **54**, 1789 (1971).
[129] L. S. N. Lim and G. W. Norrish, *Nature, Phys. Sci.*, **229**, 42 (1971).
[130] C. Wentrup and W. D. Crow, *Tetrahedron*, **27**, 361 (1971); C. Wentrup, *Tetrahedron*, **27**, 367, 1281 (1971).
[131] J. S. Millership and H. Suschitzky, *Chem. Comm.*, **1971**, 1496; cf. H. W. Moore and D. S. Pearce, *Tetrahedron Letters*, **1971**, 1621.
[132] F. A. Carey and L. J. Hayes, *J. Am. Chem. Soc.*, **92**, 7613 (1970); R. Partch, B. Stokes, D. Bergman, and M. Budnik, *Chem. Comm.*, **1971**, 1504.
[133] J. E. Baldwin, J. E. Brown, and G. Höfle, *J. Am. Chem. Soc.*, **93**, 788 (1971).
[134] D. J. C. Adams, S. Bradbury, D. C. Horwell, M. Keating, C. W. Rees, and R. C. Storr, *Chem. Comm.*, **1971**, 828.
[135] R. Selvarajan and J. H. Boyer, *J. Org. Chem.*, **36**, 1679 (1971).

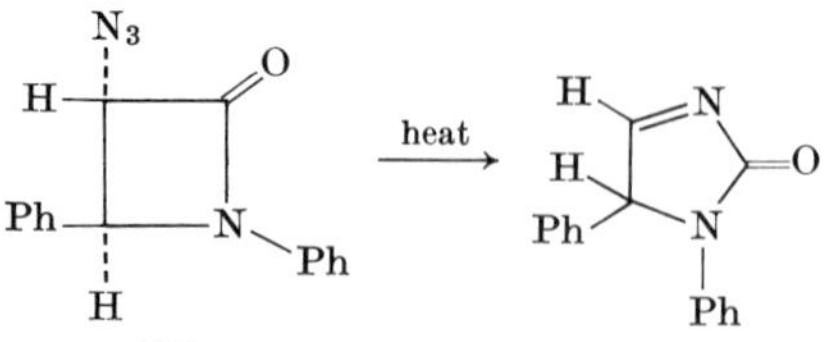

(59) 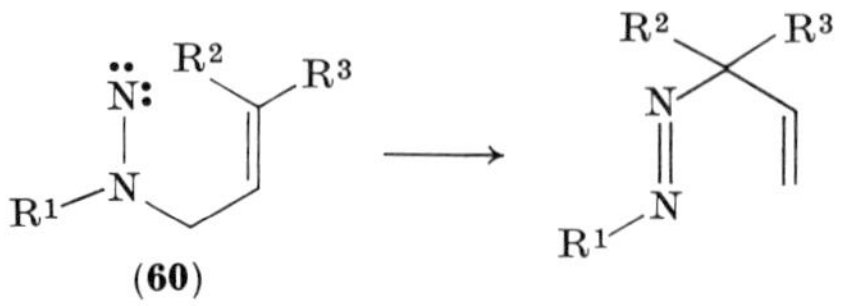 (60) 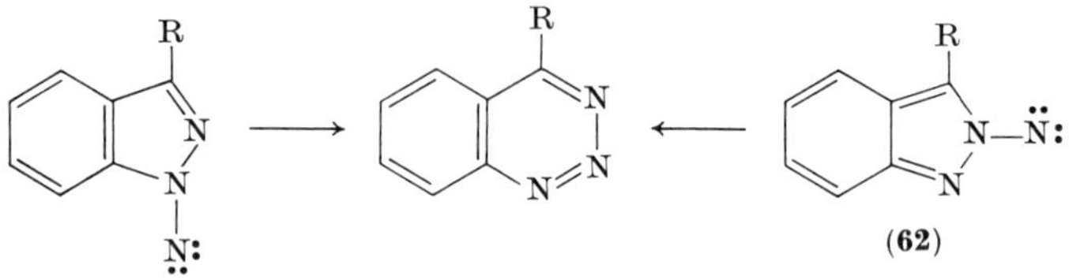 (61) (62)

Reactions with Nucleophiles and Electrophiles

The reaction of the organomercurial dichlorocarbene precursor $PhHgCCl_2Br$ is reported with a variety of nucleophiles, including cyclic allylic alcohols,[136] benzaldehyde,[137,138] benzophenone,[137] α-diketones,[139] azidobenzene,[140] and heterocumulenes.[141] It is, however, unlikely that free dichlorocarbene is involved in all these reactions; with benzaldehyde, for example, carbon monoxide is evolved at 25° and there is evidence of complex formation between the starting materials. Dichloro-oxirans are proposed as intermediates in the reactions with carbonyl compounds. The carbene adds to the double bonds of the allylic alcohols and there is no evidence that the hydroxyl group assists the approach of the carbene; on the other hand, ylids are proposed as intermediates in the reaction of ethoxycarbonylcarbene with allylic ethers, thioethers, and chlorides.[142] Allyl chloride reacts with di(methoxycarbonyl)carbene to give both addition and allylic insertion products, in the presence of methylene halides.[143] As the concentration of methylene halide is increased, the ratio of insertion to addition product decreases, showing that the (ground-state) triplet carbene is responsible for addition, and the singlet for insertion. In the reaction of *p*-halonitrobenzenes with trichloromethyl-lithium an intermediate zwitterion (63) is proposed as explaining the very high yields

[136] D. Seyferth and V. A. Mai, *J. Am. Chem. Soc.*, **92**, 7412 (1970).
[137] C. W. Martin and J. A. Landgrebe, *Chem. Comm.*, **1971**, 15.
[138] C. W. Martin, J. A. Landgrebe, and E. Rapp, *Chem. Comm.*, **1971**, 1438.
[139] D. Seyferth and W. E. Smith, *J. Organomet. Chem.*, **26**, C55 (1971).
[140] H. H. Gibson, J. R. Cast, J. Henderson, C. W. Jones, B. F. Cook, and J. B. Hunt, *Tetrahedron Letters*, **1971**, 1825.
[141] D. Seyferth, R. Damrauer, H.-M. Shih, W. Tronich, W. E. Smith, and J. Y.-P. Mui, *J. Org. Chem.*, **36**, 1786 (1971).
[142] W. Ando, T. Yagihara, S. Kondo, K. Nakayama, H. Yamato, S. Nakaido, and T. Migita, *J. Org. Chem.*, **36**, 1732 (1971).
[143] W. Ando, S. Kondo, and T. Migita, *Bull. Chem. Soc. Japan*, **44**, 571 (1971).

Carbenes and Nitrenes

of insertion products formed;[144] this reaction appears to be synthetically valuable. Reimer–Tiemann reaction of the phosphonium ylid (**64**) has also been observed,[145] and the usual assumption, that the proton transfer step from the *ortho* to the α-carbon in the Reimer–Tiemann reaction is an intermolecular process, has been experimentally established.[146]

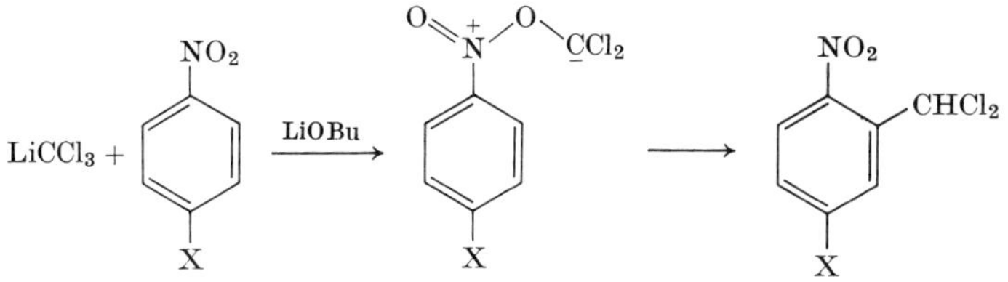

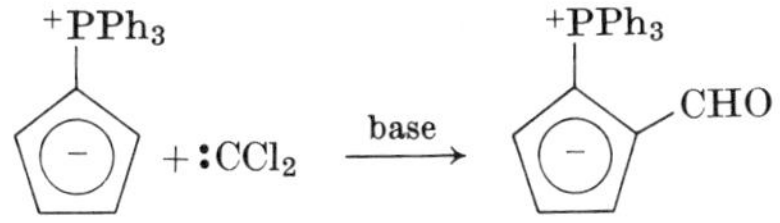

The reaction of alcohols with dichlorocarbene gives the corresponding alkyl chlorides with predominant, but not complete, retention of configuration; it is suggested that an ylid is formed that can then give the chloride either by a carbonium ion mechanism or by an $S_N i$ mechanism (Scheme 4).[147] Ylids are also proposed for the reaction of diarylcarbenes with alcohols; here the final products are alkyl diarylmethyl ethers (**65**).[148]

Atomic carbon is found to abstract sulphur from CS_2 and from thioethers; with episulphides, e.g. (**66**), the process is stereoselective.[149] It has been suggested that certain

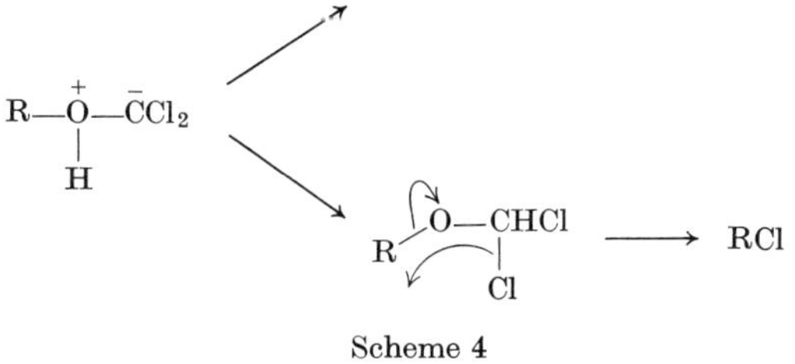

Scheme 4

$$Ph_2CN_2 + ROH \longrightarrow Ph_2CHOR$$
(**65**)

---

[144] E. T. McBee, E. P. Wesseler, and T. Hodgins, *J. Org. Chem.*, **36**, 2907 (1971).
[145] Z. Yoshida, S. Yoneda, and T. Yato, *Tetrahedron Letters*, **1971**, 2973.
[146] D. S. Kemp, *J. Org. Chem.*, **36**, 202 (1971).
[147] I. Tabushi, Z. Yoshida, and N. Takahashi, *J. Am. Chem. Soc.*, **93**, 1820 (1971).
[148] D. Bethell, A. R. Newall, and D. Whittaker, *J. Chem. Soc.* (B), **1971**, 23.
[149] K. J. Klabunde and P. S. Skell, *J. Am. Chem. Soc.*, **93**, 3807 (1971).

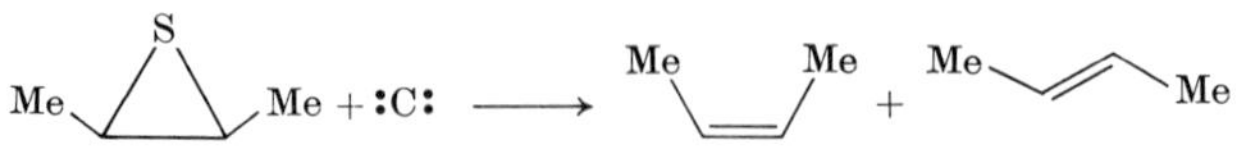

carbenes act as nucleophiles in their reactions with carbonyl compounds; for example, in the reaction of the carbene (67) with acetaldehyde,[150] and in the catalysis of the benzoin condensation by nucleophilic carbene precursors[151] (where the carbene takes the place of the usual cyanide catalyst). Other reactions of nucleophilic carbenes are also reported.[152]

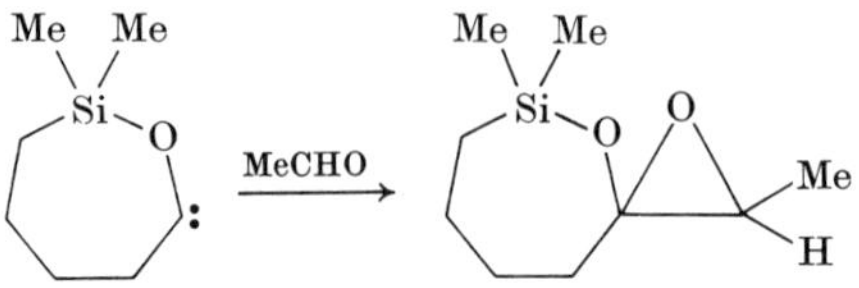

Further examples have appeared of intramolecular arylnitrene cyclizations of systems of the general type (68), where R is an electron-releasing group and X is O,[153] S,[154] NAc[155] or CH$_2$.[156] Mechanisms suggested are electrophilic attack to form a spiro intermediate,[153,154] and cycloaddition followed by rearrangement.[156] An intramolecular attack on a carbonyl group may be involved in the deoxygenation of o-nitrosostyryl ketones to give quinolines.[157] Deoxygenation of nitrosobenzene by triethyl phosphite in methanol gives o- and p-anisidine and, in this case, free nitrenes may not be involved: an intermediate zwitterion (69) would be subject to ready protonation, this being

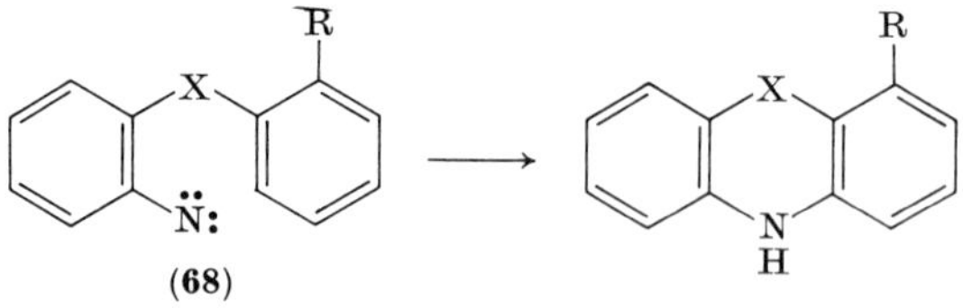

---

[150] A. G. Brook, R. Pearce, and J. B. Pierce, *Can. J. Chem.*, **49**, 1622 (1971).
[151] B. Lachmann, H. Steinmaus, and H.-W. Wanzlick, *Tetrahedron*, **27**, 4085 (1971).
[152] S. I. Burmistrov and S. E. Kondrat'eva, *Zhur. Org. Khim.*, **7**, 616 (1971); M. Regitz, J. Hocker, W. Schlössler, B. Weber, and A. Liedhegener, *Ann. Chem.* **748**, 1 (1971); J. Hocker and R. Merten, *Ann. Chem.*, **751**, 145 (1971).
[153] J. I. G. Cadogan and P. K. K. Lim, *Chem. Comm.*, **1971**, 1431.
[154] J. I. G. Cadogan and S. Kulik, *J. Chem. Soc.* (C), **1971**, 2621.
[155] Y. Maki, T. Hosokami, and M. Suzuki, *Tetrahedron Letters*, **1971**, 3509.
[156] G. R. Cliff and G. Jones, *Chem. Comm.*, **1970**, 1705.
[157] T. Kametani, K. Nyu, and T. Yamanaka, *Chem. Pharm. Bull.*, **19**, 1321 (1971).

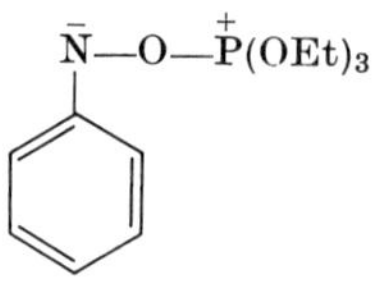

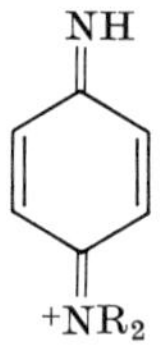

$$\text{RSO}_2\text{NH}_2 + \text{Pb(OAc)}_4 \xrightarrow{\text{Me}_2\text{SO}} \text{RSO}_2\text{N}{=}\text{SOMe}_2$$
$$(71)$$

$$\text{ArSON}_3 + \text{Me}_2\text{SO} \longrightarrow \underset{\underset{\text{O—SMe}_2}{|\ \ |}}{\text{Ar}\overset{\overset{\text{O}}{\|}}{\text{S}}{=}\text{N}} \longrightarrow \text{ArSO}_2\text{N}{=}\text{SMe}_2$$
$$(72)$$

followed by nucleophilic substitution by the solvent.[158] Quinone di-imine derivatives (**70**) are produced by photolysis of *p*-azidoaniline derivatives in water.[159]

Oxidation of sulphonamides in dimethyl sulphoxide gives sulphoximides (**71**), and sulphonylnitrenes are suggested as intermediates.[160] Decomposition of sulphinyl azides in dimethyl sulphoxide did not give the expected sulphoximides, however; instead, the isomeric sulphonylsulphimides (**72**) were isolated. A mechanism involving 2 + 2 addition of the sulphinylnitrene to dimethyl sulphoxide is proposed.[161]

Carbenoids

The reactions of diazoalkanes with derivatives of metals and metalloids have been reviewed.[162] Several metal-catalysed rearrangements of bicyclobutane derivatives have been interpreted as carbene retro-additions but, as other mechanisms have been proposed, these reactions are discussed with molecular rearrangements in Chapter 8.

A comparison of the reactivities of olefins towards the diethylzinc–di-iodomethane carbenoid system shows that the carbenoid is more electrophilic than dichlorocarbene.[163] The carbenoid inserts into the Si—H bonds of trialkylsilanes in good yield.[164] The reactions of olefins with zinc carbenoids derived from diethylzinc and PhCHI_2[165] and from diethylzinc and CHXYI[166] (where X and Y are F, Cl, Br or I) have also been studied. The phenylcarbenoid shows *syn*-selectivity in the cycloadditions, this selectivity being enhanced by electron-donating groups in the ring and by the use of ether as a solvent. It has also been found that in the diethylzinc–di-iodomethane system, the cyclopropanation is greatly accelerated by oxygen;[167] the yields are impressively high in these modified reaction conditions. Peroxides and other radical initiators also enhance the catalytic activity of the $(\text{MeO})_3\text{P–CuZ}$ system in the decomposition of dimethyl diazomalonate.[168] The insertion to addition ratio with cyclohexenes depends on the nature of Z:, the more insertion being observed the better its ability as a leaving group.[169] Mechanisms for these effects remain to be established.

The products formed from *cis*- and *trans*-7,7-dibromo-4-t-butylnorcarane with methyllithium (mainly those of transannular C—H insertion) are quite different for the two

[158] R. J. Sundberg and R. H. Smith, *J. Org. Chem.*, **36**, 295 (1971).
[159] R. C. Baetzold and L. K. J. Tong, *J. Am. Chem. Soc.*, **93**, 1347 (1971).
[160] T. Ohashi, K. Matsunaga, M. Okahara, and S. Komori, *Synthesis*, **1971**, 96.
[161] T. J. Maricich and V. L. Hoffman, *Tetrahedron Letters*, **1971**, 729.
[162] M. F. Lappert and J. S. Poland, *Adv. Organomet. Chem.*, **9**, 397 (1970).
[163] J. Nishimura, J. Furukawa, N. Kawabata, and M. Kitayama, *Tetrahedron*, **27**, 1799 (1971).
[164] J. Nishimura, J. Furukawa, and N. Kawabata, *J. Organomet. Chem.*, **29**, 237 (1971).
[165] J. Nishimura, J. Furukawa, N. Kawabata, and H. Koyama, *Bull. Chem. Soc. Japan*, **44**, 1127 (1971).
[166] J. Nishimura and J. Furukawa, *Chem. Comm.*, **1971**, 1375.
[167] S. Miyano and H. Hashimoto, *Chem. Comm.*, **1971**, 1418.
[168] B. W. Peace and D. S. Wulfman, *Chem. Comm.*, **1971**, 1179; *Tetrahedron Letters*, **1971**, 3799.
[169] D. S. Wulfman, B. W. Peace, and E. K. Steffen, *Chem. Comm.*, **1971**, 1360.

isomers, showing that a cyclic allene is not a common intermediate for the carbene reaction.[170] Intramolecular C—H insertion is also observed with the dibromocyclopropane (73) and methyl-lithium.[171] Lithium carbenoids are involved in the expansion of 2-dichloromethyl-2H-pyrroles, e.g. (74), to pyridines with butyl-lithium,[172] in the addition of CHFBrLi,[173] CHF$_2$Li,[174] and CF$_2$ClLi[174] to olefins, and in the reaction of hexachlorocyclobutene (75) with but-2-yne and butyl-lithium to give (76), the first spiro[2.3]hexa-1,4-diene.[175] The *endo*-halogen atom in dihalogeno-oxabicyclo[x.1.0]-alkanes, e.g. (77), is selectively displaced by lithium to give carbenoids that are apparently stabilized by the neighbouring oxygen.[176] Details of the LiNEt$_2$-catalysed isomerization of epoxide (78), for which a carbenoid mechanism is proposed, have appeared.[177]

Carbenoid mechanisms are discussed for the cupric-catalysed decomposition of diphenyldiazomethane,[178] for the reaction of chloromethylsilanes with sodium,[179] and for the oxidative cleavage [by mercuric, thallic, and lead(IV) acetates] of 1,3,3-trimethylcyclopropene.[180]

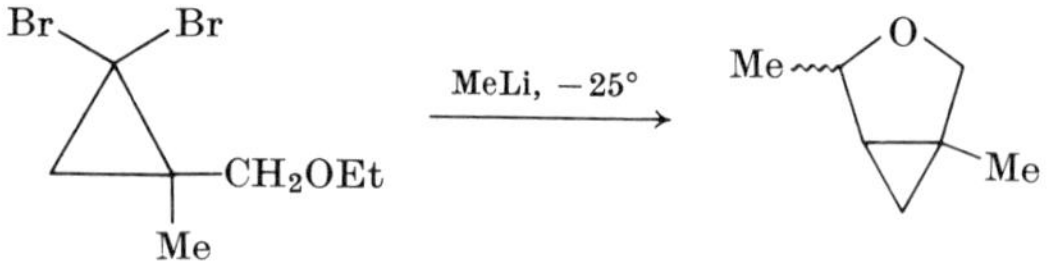

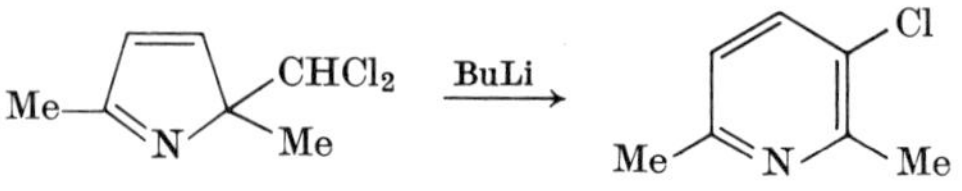

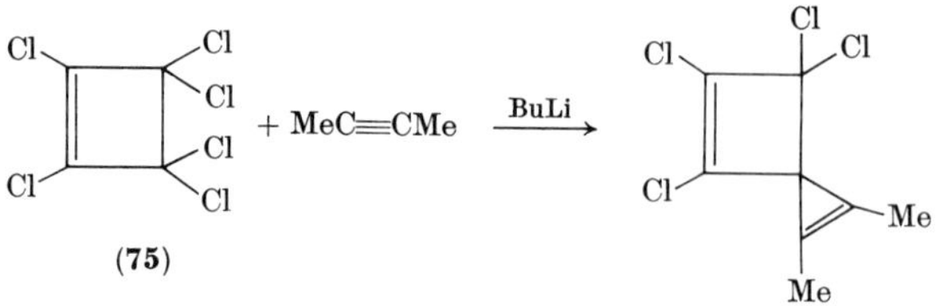

[170] W. R. Moore and B. J. King, *J. Org. Chem.*, **36**, 1877 (1971).
[171] M. S. Baird, *Chem. Comm.*, **1971**, 1145.
[172] A. Gambacorta, R. Nicoletti, and M. L. Forcellese, *Tetrahedron*, **27**, 985 (1971).
[173] M. Schlosser and G. Heinz, *Chem. Ber.*, **104**, 1934 (1971).
[174] M. Schlosser and L. Van Chau, *Angew. Chem. Int. Ed.*, **10**, 138 (1971).
[175] M. F. Semmelhack and R. J. DeFranco, *Tetrahedron Letters*, **1971**, 1061.
[176] K. G. Taylor, W. E. Hobbs, and M. Saquet, *J. Org. Chem.*, **36**, 369 (1971).
[177] J. K. Crandall, L. C. Crawley, D. B. Banks, and L. C. Lin, *J. Org. Chem.*, **36**, 510 (1971).
[178] T. Shirafuji, Y. Yamamoto, and H. Nozaki, *Tetrahedron*, **27**, 5353 (1971).
[179] J. W. Connolly and P. F. Fryer, *J. Organomet. Chem.*, **30**, 315 (1971).
[180] T. Shirafuji, Y. Yamamoto, and H. Nozaki, *Tetrahedron Letters*, **1971**, 4713.

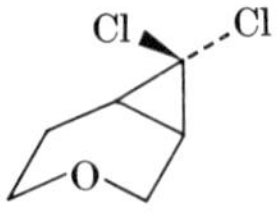

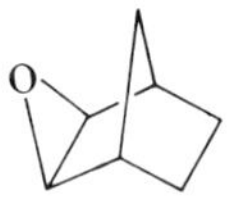

Transition-metal Complexes

The following discussion is restricted to the more important synthetic and mechanistic discoveries because annual surveys of inorganic aspects of carbene complexes are now available.[181]

The most important general method of synthesis of carbene–metal complexes, namely, the reaction of metal carbonyls with organolithium reagents, has been extended to several more systems.[182-184] Potentially the most promising new synthetic method is the transfer of a carbene ligand from one metal complex to another:[184] irradiation of complexes (**79**) in the presence of pentacarbonyliron gave the corresponding iron–carbene complexes. Other new syntheses include the reaction of electron-rich olefins, e.g., (**80**), with the platinum complex (**81**),[185] protonation and alkylation of acyl–metal complexes (**82**),[186] and several intramolecular cyclizations involving the formation of the carbene (**83**) as a ligand.[187] Chugaev's salt (**84**), originally prepared in 1915, has been recognized as probably the first carbene complex.[188]

Several examples of the reaction of methoxycarbene complexes with amines to give aminocarbene complexes are reported,[183,189] and the kinetics of the reaction of

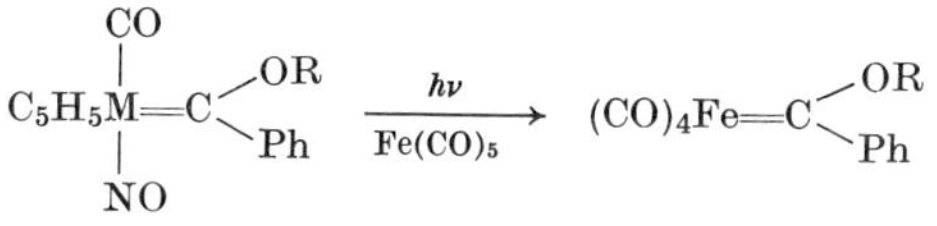

$$C_5H_5M(CO)(NO)=C(OR)Ph \xrightarrow[Fe(CO)_5]{h\nu} (CO)_4Fe=C(OR)Ph$$

(M = Cr, Mo, W)

(**79**)

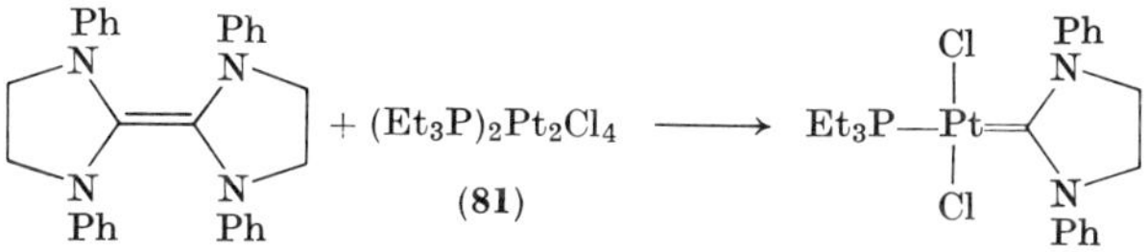

(**80**) + $(Et_3P)_2Pt_2Cl_4 \longrightarrow$ Et_3P—Pt(Cl)(Cl)=C(N-Ph)(N-Ph) cyclic

(**81**)

---

[181] *Organomet. Chem. Rev.* (B), Elsevier, Lausanne; *Ann. Rep. Chem. Soc.* (A), Chemical Society, London; "Spectroscopic Properties of Inorganic and Organometallic Compounds", Vols. 1–4, Chemical Society, London.
[182] G. A. Moser, E. O. Fischer, and M. D. Rausch, *J. Organomet. Chem.*, **27**, 379 (1971); E. O. Fischer and V. Kiener, *J. Organomet. Chem.*, **27**, C56 (1971); E. O. Fischer, C. G. Kreiter, H. J. Kollmeier, J. Müller, and R. D. Fischer, *J. Organomet. Chem.*, **28**, 237 (1971); J. A. Connor and E. M. Jones, *Chem. Comm.*, **1971**, 570; *J. Chem. Soc.* (A), **1971**, 1974; Y. Sawa, M. Ryang, and S. Tsutsumi, *J. Org. Chem.*, **35**, 4183 (1970); S. Fukuoka, M. Ryang, and S. Tsutsumi, *J. Org. Chem.*, **36**, 2721 (1971).
[183] J. A. Connor and E. M. Jones, *J. Chem. Soc.* (A), **1971**, 3368.
[184] E. O. Fischer and H.-J. Beck, *Chem. Ber.*, **104**, 3101 (1971).
[185] D. J. Cardin, B. Cetinkaya, M. F. Lappert, L. Manojlović-Muir, and K. W. Muir, *Chem. Comm.*, **1971**, 400.
[186] M. L. H. Green, L. C. Mitchard, and M. G. Swanwick, *J. Chem. Soc.* (A), **1971**, 794.
[187] F. A. Cotton and C. M. Lukehart, *J. Am. Chem. Soc.*, **93**, 2672 (1971); C. P. Casey and R. L. Anderson, *J. Am. Chem. Soc.*, **93**, 3554 (1971); M. H. Chisholm and H. C. Clark, *Chem. Comm.*, **1971**, 1484.
[188] G. Rouschias and B. L. Shaw, *J. Chem. Soc.* (A), **1971**, 2097.
[189] E. O. Fischer and H. J. Kollmeier, *Chem. Ber.*, **104**, 1339 (1971); E. O. Fischer, B. Heckl, and H. Werner, *J. Organomet. Chem.*, **28**, 359 (1971); L. Knauss and E. O. Fischer, *J. Organomet. Chem.*, **31**, C68 (1971).

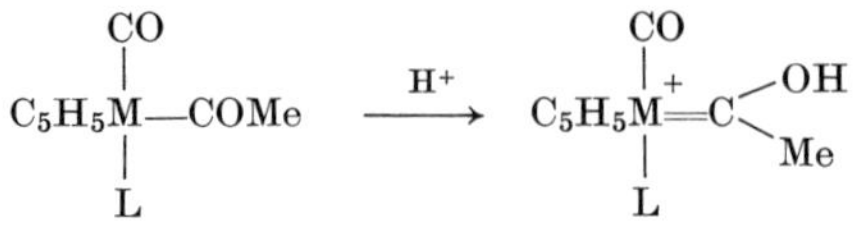

$(CO)_5CrC(OMe)Ph$ with primary amines have been studied.[190] The reaction follows a fourth-order rate law, requiring participation of a proton donor (in the initial step) and a proton acceptor (to activate the attacking amine) as well as the carbene complex and the amine; the mechanism is compared with that of ester aminolysis.

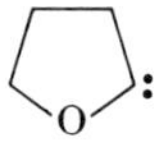

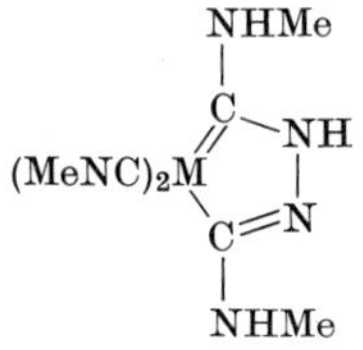

An extraordinary and unexplained reaction of the complex $(CO)_5CrC(OMe)Me$ occurs with lithium tributoxyhydridoaluminate in tetrahydrofuran; a new complex, assigned structure (**85**), is isolated in low yield.[191]

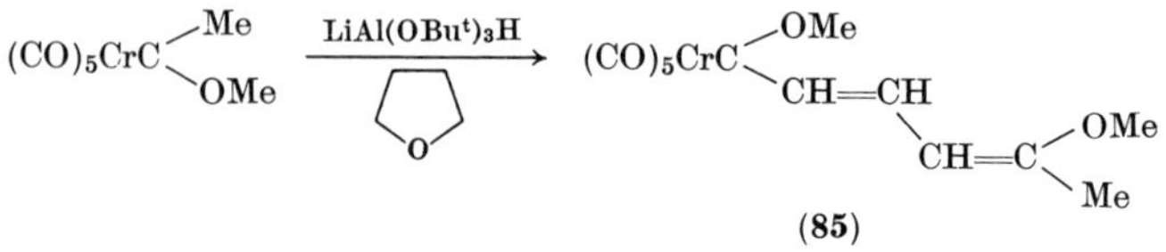

Certain carbene complexes have been found to be active catalysts for olefin disproportionation: for example, pent-1-ene gives a mixture of ethylene and *cis*- and *trans*-oct-4-ene in good yield at room temperature.[192] Various structural studies of carbene complexes based on spectra have been reported.[193] A vinylidenecarbene–cobalt complex has been postulated as an intermediate in the reaction of diphenylketene with cobalt octacarbonyl, in which the product isolated is the fulvene derivative (**86**).[194]

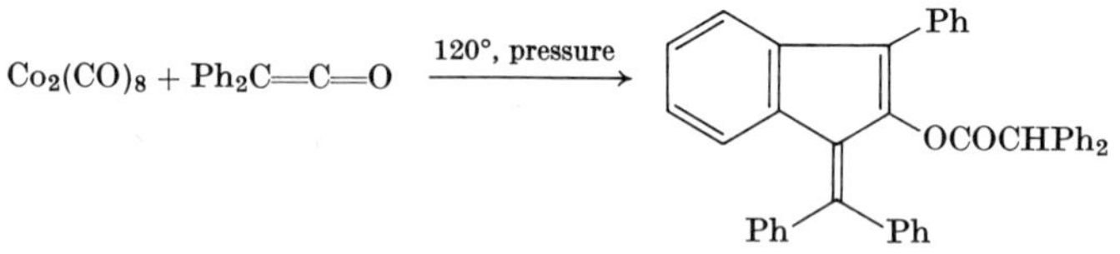

---

[190] H. Werner, E. O. Fischer, B. Heckl, and C. G. Kreiter, *J. Organomet. Chem.*, **28**, 367 (1971).
[191] L. Knauss and E. O. Fischer, *J. Organomet. Chem.*, **31**, C71 (1971).
[192] W. R. Kroll and G. Doyle, *Chem. Comm.*, **1971**, 839.
[193] H.-J. Beck, E. O. Fischer, and C. G. Kreiter, *J. Organomet. Chem.*, **26**, C41 (1971); E. M. Badley, B. J. L. Kilby, and R. L. Richards, *J. Organomet. Chem.*, **27**, C37 (1971); H. C. Clark and L. E. Manzer, *J. Organomet. Chem.*, **30**, C89 (1971); J. A. Connor and E. M. Jones, *J. Organomet. Chem.*, **31**, 389 (1971); L. F. Farnell, E. W. Randall, and E. Rosenberg, *Chem. Comm.*, **1971**, 1078.
[194] P. Hong, K. Sonogashira, and N. Hagihara, *Tetrahedron Letters*, **1971**, 1105.

Metal–nitrene complexes are proposed for the thermolysis of ferrocenesulphonyl azide[195] and the deoxygenation of dibenzylnitrosamine with pentacarbonyliron.[196] Reaction of the iridium–nitrene complex (**87**) with hydrochloric and sulphuric acid gives chloramine and hydroxylamine complexes, respectively.[197]

$$[Ir(NH_3)_5N_3]^{2+} \xrightarrow{H^+} N_2 + [Ir(NH_3)_5NH]^{3+}$$
(**87**)
$$\downarrow HX$$
$$[Ir(NH_3)_5NH_2X]^{3+}$$

[195] R. A. Abramovitch, C. I. Azogu, and R. G. Sutherland, *Tetrahedron Letters*, **1971**, 1637.
[196] A. Tanaka and J.-P. Anselme, *Tetrahedron Letters*, **1971**, 3567.
[197] B. C. Lane, J. W. McDonald, V. G. Myers, F. Basolo, and R. G. Pearson, *J. Am. Chem. Soc.*, **93**, 4934 (1971).

CHAPTER 11

Reactions of Aldehydes and Ketones and their Derivatives

B. CAPON

Chemistry Department, Glasgow University

Formation and Reactions of Acetals and Ketals	393
Hydrolysis and Formation of Glycosides	397
Non-enzymic Reactions	397
Enzymic Reactions	398
Hydration of Aldehydes and Ketones and Related Reactions	402
Reactions with Nitrogen Bases	404
Schiff Bases	404
Semicarbazones, Oximes, Hydrazones and Related Compounds	407
Hydrolysis of Enol Ethers and Esters	408
Enolization and Related Reactions	409
Aldol and Related Reactions	413
Other Reactions	414

Formation and Reactions of Acetals and Ketals

The hydrolysis of benzaldehyde di-t-butyl acetal is general acid-catalysed.[1] This acetal had previously been shown to undergo hydrolysis with aldehyde–oxygen fission.[2] It seems that steric overcrowding weakens the carbon—oxygen bond sufficiently for acid-catalysis with partial proton-transfer to be effective. The isotope effects $k(\text{AcOH})/k(\text{AcOD})$ and $k(\text{H}_3\text{O}^+)/k(\text{D}_3\text{O}^+)$ are 2.52 and 1.1, respectively, and the α-value is 0.6. The rate constants for the acetic acid-catalysed hydrolysis of a series of *para*-substituted benzaldehyde di-t-butyl acetals were correlated by the σ-constants to yield a ρ-value of −2.0 while those for the hydronium ion-catalysed reaction yielded a ρ-value of −4.0.

Fife and Anderson have explored the range of carbonium ion stabilities which lead the hydrolysis of unstrained dialkyl acetals to be general acid-catalysed. As shown previously,[3] the hydrolysis of tropone diethyl ketal is general acid-catalysed but those of 2,3-diphenylcyclopropenone diethyl ketal and ferrocenecarbaldehyde dimethyl acetal are not.[4]

Apparent general acid-catalysis by chloroacetic acid of the hydrolysis of triethyl orthobenzoate in 67.4:32.6 w/w dioxan–water probably arises from a specific effect of the salt (NaCl) used to maintain the ionic strength constant. It is clear that care is needed in interpreting investigations of general acid-catalysis in mixtures of water and organic solvents.[5,6]

[1] E. Anderson and T. H. Fife, *J. Am. Chem. Soc.*, **93**, 1701 (1971).
[2] J. J. Cawley and F. H. Westheimer, *Chem. Ind. (London)*, **1960**, 656.
[3] See *Org. Reaction Mech.*, **1969**, 399–400.
[4] T. H. Fife and E. Anderson, *J. Org. Chem.*, **36**, 2357 (1971).
[5] P. Salomaa, A. Kankaanperä, and M. Lahti, *J. Am. Chem. Soc.*, **93**, 2084 (1971).
[6] Review of salt effects: C. H. Rochester, *Progr. Reaction Kinetics*, **6**, 144 (1971).

Cordes and his co-workers have published details of their investigation of α-deuterium isotope effects in the hydrolysis of acetals and orthoesters. In addition to the results previously reported,[7] α-deuterium isotope effects for the hydronium ion-catalysed hydrolysis of p-methoxy-, p-chloro- and p-nitro-phenoxytetrahydropyran were also determined. That for the p-nitro-compound [$k(H)/k(D) = 1.06$] was significantly smaller than that for the other compounds (1.10). The hydrolysis of the p-nitro-compound shows general acid-catalysis[8] and the smaller α-deuterium isotope effect suggests that the transition state occurs earlier along the reaction co-ordinate.[9] Salt effects on the hydronium ion-catalysed hydrolysis of acetals and orthoesters have been studied. The quantities f^*/f_{R^+} and f^*/f_{BH^+} were determined where f^*, f_{R^+} and f_{BH^+} are, respectively, the activity coefficients of the transition state, the tris-(p-methoxyphenyl)methyl cation, and the anilinium ion. The ratio f^*/f_{R^+} increased in the order $MeCH(OEt)_2 < Me_2C(OEt)_2 < HC(OEt)_3 < MeC(OEt)_3$, but the ratio f^*/f_{BH^+} was always close to 1. It was proposed that the carbonium ion character of the transition state decreased in the order acetal > ketal > orthoformate > orthoacetate.[10]

Intramolecular catalysis in the hydrolysis of 2-(methoxymethoxy)-benzoic and -naphthoic acids has been investigated further. The rate of hydrolysis of 2-(methoxymethoxy)-1-naphthoic acid is about one-fifth that of 2-(methoxymethoxy)benzoic acid, probably as a result of steric hindrance to the carboxyl group lying in the same plane as the naphthalene ring.[11]

Dunn and Bruice[12] have investigated the hydrolysis of a series of (alkoxymethoxy)-benzoic acids (1). The spontaneous hydrolysis of these compounds could be a specific hydronium ion-catalysis of the forms with the carboxyl groups ionized with species (2) an intermediate. The plot of log k for this process against the σ*-values of the group R yielded a ρ*-value of −3.0, identical within experimental error with the ρ*-value for the hydronium ion-catalysed hydrolysis of the un-ionized form (1) and of acetals (3). It was therefore concluded that the spontaneous hydrolysis of (1) is probably a specific hydronium ion-catalysed hydrolysis of the ionized form rather than an intramolecularly catalysed hydrolysis of the un-ionized form. The high rate was attributed to electrostatic stabilization of species (2).[12]

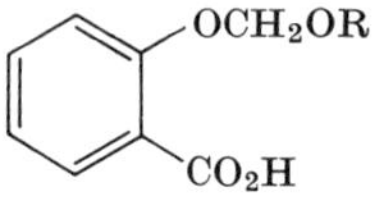

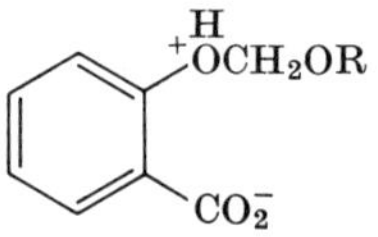

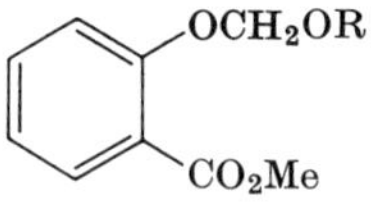

The benzaldehyde acetal (4) reacts in aqueous dioxan to yield the acylal (5). The pH–rate profile is bell-shaped, indicating that the most rapid reaction is the spontaneous hydrolysis of the monoionized form. The rate of this reaction was estimated to

---

[7] Org. Reaction Mech., **1970**, 417.
[8] Org. Reaction Mech., **1968**, 349–350; **1970**, 417–418.
[9] H. G. Bull, K. Koehler, T. C. Pletcher, J. J. Ortiz, and E. H. Cordes, J. Am. Chem. Soc., **93**, 3002 (1971).
[10] C. A. Bunton and J. D. Reinheimer, J. Phys. Chem., **74**, 4457 (1970).
[11] B. Capon, E. Anderson, N. S. Anderson, R. H. Dahm, and M. C. Smith, J. Chem. Soc. (B), **1971**, 1963.
[12] B. M. Dunn and T. C. Bruice, J. Am. Chem. Soc., **93**, 5725 (1971).

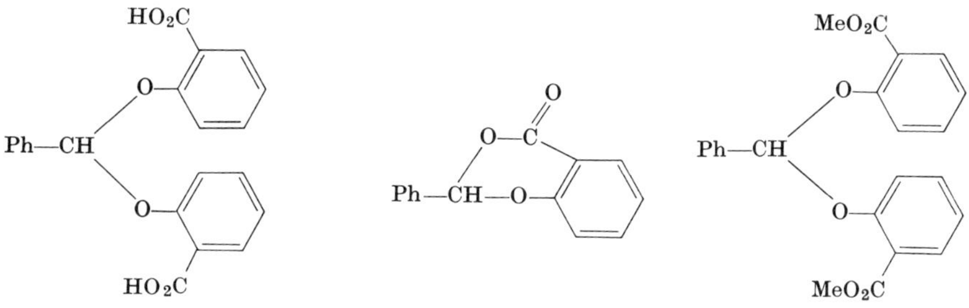

be ca. 10^9 times greater than the specific hydronium ion-catalysed reaction of (6) under the same conditions.[13]

Hydrolysis of the 5,6-isopropylidene residue of compound (7) which is present mainly as the hydrate (8), is reported to occur more rapidly at pH 5 than that of the corresponding group of 1,2:5,6-diisopropylideneglucose. This was attributed to "some kind of electrophilic assistance" as symbolized by (9).[14]

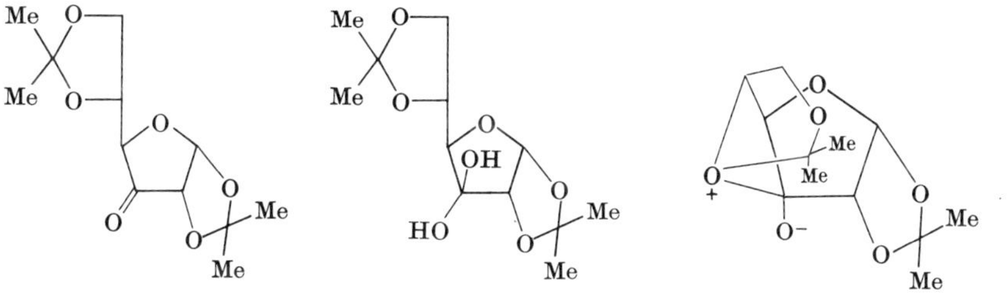

The NMR spectrum of 2,2-dimethyl-1,3-oxathiolan (11) in FSO_3H–SbF_5 has been compared with that of 2,2-dimethyl-1,3-dioxan and 2,2-dimethyl-1,3-dithiolan. It was concluded from the presence of an $-OH_2^+$ peak and the absence of an $-SH_2^+$ peak that the species present was (10) and not (12) and that the oxathiolan opened preferentially with acid-catalysis to form a sulphonium rather than an oxonium ion.[15] This is the opposite of the conclusion of Fife and Jao[16] who thought that the rate-limiting step in the hydrolysis of 2-aryl-1,3-oxathiolans was breaking of the carbon–sulphur bond. However, the present results are not necessarily incompatible with this conclusion since the preferential formation of ion (10) from (11) may be thermodynamically controlled, the important factor being that it is an O-protonated species.

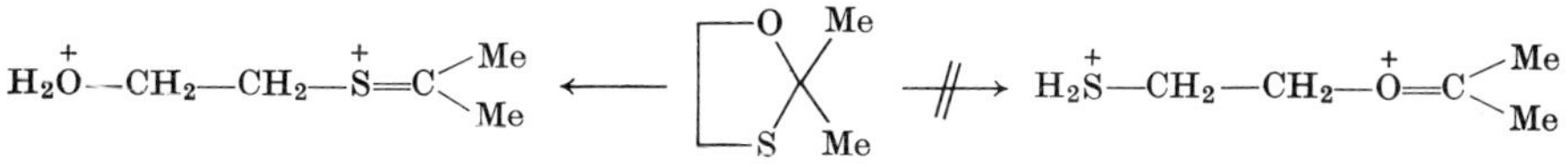

[13] E. Anderson and T. H. Fife, *Chem. Comm.*, **1971**, 1470.
[14] J. M. J. Tronchet and J. M. Bourgeois, *Helv. Chim. Acta*, **54**, 1580 (1971).
[15] F. Guinot, G. Lamaty, and H. Munsch, *Bull. Soc. Chim. France*, **1971**, 541.
[16] *Org. Reaction Mech.*, **1969**, 401.

The hydrolysis of 2,5-dihydroxy-2,5-dimethyl-1,4-dioxan dipyrophosphate (13) occurs with fission of the dioxan–oxygen bond. It was proposed that reaction involves protonation of the ring-oxygen followed by a rate-limiting bond-fission (a).[17a] However, a more likely pathway would involve protonation of the pyrophosphate group followed by a rate-limiting bond-fission (b).

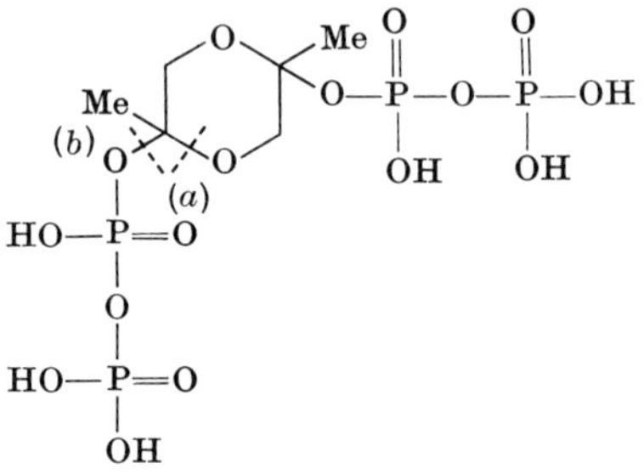

The effect of glycerol on the rate of hydrolysis of dimethoxymethane has been studied.[17b]

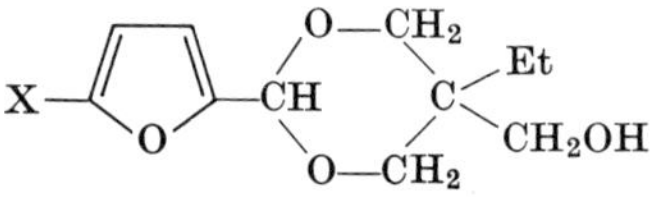

Other reactions studied include: hydrolysis of alkyl-substituted 1,3-dioxans,[18] the 2-furyl-1,3-dioxans (14),[19] and the polyacetal formed by periodate oxidation and reduction of cellulose;[20] methanolysis of 4-alkyl-1,3-dioxans,[21] cis-trans-equilibration of 2-alkyl-4-t-butyl-1,3-dioxolans;[22] and the acid-catalysed formation of cyclic acetals from n-butyraldehyde and 1-deoxy-D-glucitol, 2-deoxy-D-glucitol and 3-O-methyl-D-glucitol.[23]

The equilibrium constant for the formation of acetals from propionaldehyde, cyclohexanone, and benzaldehyde with methanol increases with increasing pressure.[24] The standard free energy, enthalpy, and entropy changes on formation of dioxolans and dioxans from acetone and a series of 1,2- and 1,3-diols have been measured.[25]

There have been numerous investigations of the conformations of cyclic acetals. The classes of compound studied include 1,3-dioxolans,[26,27] 1,3-dioxans,[27–36] spiro-1,3-dioxolans,[37] spiro-oxazolidines,[38] tetrahydro-1,3-oxazines,[39] and 1,3-oxathians.[40]

[17a] I. I. Semenyuk, N. V. Volkova, and A. A. Yasnikov, *Ukr. Khim. Zhur.*, **37**, 675 (1971).
[17b] C. Kalidas, K. U. Raman, and N. Chattanathan, *J. Indian Chem. Soc.*, **48**, 423 (1971); cf. *Org. Reaction Mech.*, **1970**, 420.
[18] K. Pihlaja and K. J. Teinonen, *Acta Chem. Scand.*, **25**, 323 (1971).
[19] Z. I. Zelikman, A. L. Chekhun, V. L. Pojrebnaya, and V. G. Kulnevich, *Khim. Geterosikl Soedin*, **6**, 152 (1971); *Chem. Abs.*, **75**, 62762 (1971).
[20] V. I. Ivanov, N. Y. Koznetsova, Z. P. Kovalenok, and G. M. Morozova, *Dokl. Akad. Nauk SSSR*, **192**, 341 (1970).
[21] B. N. Bobylev, M. I. Ferberov, and E. P. Tepenitsyna, *Kinet. Katal.*, **12**, 89 (1971); *Chem. Abs.*, **74**, 111204 (1971).
[22] G. Lemière and M. Anteunis, *Bull. Soc. Chim. Belges*, **80**, 215 (1971).
[23] T. G. Bonner, E. J. Bourne, P. J. V. Cleare, R. F. J. Cole, and D. Lewis, *J. Chem. Soc.* (B), **1971**, 957.
[24] D. G. Kubler and M. W. Young, *J. Org. Chem.*, **36**, 200 (1971).
[25] M. Anteunis and Y. Rommelaere, *Bull. Soc. Chim. Belges*, **79**, 523 (1970).
[26] M. Anteunis, G. Swaelens, and J. Gelan, *Tetrahedron*, **27**, 1917 (1971).
[27] W. J. Baumann, *J. Org. Chem.*, **36**, 2743 (1971).
[28] A. J. Jones, E. L. Eliel, D. M. Grant, M. C. Knoeber, and W. F. Bailey, *J. Am. Chem. Soc.*, **93**, 4772 (1971).

Hydrolysis and Formation of Glycosides

Non-enzymic Reactions

The alkaline fission of *p*-nitrophenyl α-D-glucoside (**15**) involves migration of the *p*-nitrophenyl group to position 2 through an intramolecular nucleophilic substitution.

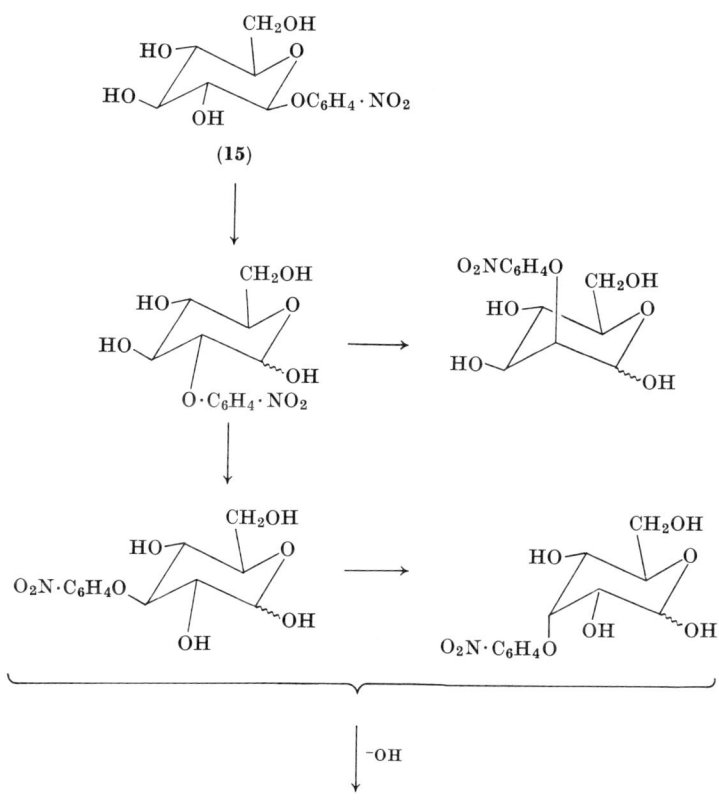

saccharinic acids + $O_2N \cdot C_6H_4O^-$

[29] G. Eccleston, E. Wyn-Jones, and W. J. Orville-Thomas, *J. Chem. Soc.* (B), **1971**, 1551.
[30] M. Anteunis, *Bull. Soc. Chim. Belges*, **80**, 3 (1971).
[31] M. Anteunis, G. Swaelens, F. Anteunis-De Ketelaere, and P. Dirinck, *Bull. Soc. Chim. Belges*, **80**, 409 (1971).
[32] D. Tavernier and M. Anteunis, *Bull. Soc. Chim. Belges*, **80**, 219 (1971).
[33] D. Tavernier and M. Anteunis, *Tetrahedron*, **27**, 1677 (1971).
[34] G. M. Kellie and F. G. Riddell, *J. Chem. Soc.* (B), **1971**, 1030.
[35] V. I. P. Jones and J. A. Ladd, *J. Chem. Soc.* (B), **1971**, 567.
[36] M. Anteunis and G. Swaelens, *Org. Mag. Resonance*, **2**, 389 (1970).
[37] R. A. Y. Jones, A. R. Katritzky, P. G. Lehman, K. A. F. Record, and B. B. Shapiro, *J. Chem. Soc.* (B), **1971**, 1302.
[38] R. A. Y. Jones, A. R. Katritzky, and P. G. Lehman, *J. Chem. Soc.* (B), **1971**, 1316.
[39] R. A. Y. Jones, A. R. Katritzky, and D. L. Trepanier, *J. Chem. Soc.* (B), **1971**, 1300.
[40] P. Pasanen and K. Pihlaja, *Tetrahedron Letters*, **1971**, 4515.

The resulting 2-O-p-nitrophenyl-D-glucose undergoes a further migration of the p-nitrophenyl group and epimerization as shown and is finally converted into saccharinic acid and p-nitrophenoxide.[41]

An extensive investigation of the acid-catalysed hydrolysis of alkyl α-D-glucopyranosides has been described. There is no simple relation between the rate and the structure of the alkyl group.[42]

Details of Capon and Ghosh's investigation of the kinetics of hydrolysis of 2-naphthyl β-D-glucuronide have been published.[43]

Acid-catalysed hydrolysis of the following glucosides has also been studied: sucrose,[44-46] raffinose,[44] cellobiose,[47] (1→4)α-D-glucans,[48] and α- and β-D-glucopyranose 1-phosphate.[49]

γ-Radiolysis of aryl glycosides has been investigated.[50]

Enzymic Reactions[51]

(a) *Galactosidases.* The pH-dependence of hydrolyses catalysed by the β-galactosidase from *E. coli* has been studied. The activities of the magnesium-activated and magnesium-free enzyme depend on an acidic group with pK_a less than 6.5, which was tentatively identified as a carboxyl group. The activity of the magnesium-free enzyme depends on basic group with pK_a ca. 6.5, which changes to 8.4 in the magnesium-activated enzyme. Nucleophilic competition experiments with methanol in the hydrolysis of o-nitrophenyl β-D-galactoside catalysed by the magnesium-free enzyme suggest that formation of the galactosyl enzyme is rate-limiting at all pH's. With the magnesium-activated enzyme, formation of the galactosyl enzyme is rate-limiting at low pH's, but at pH 7 the rate constants for formation and decomposition of the galactosyl enzyme are similar.[52]

N-Bromoacetyl-β-D-galactopyranosylamine inactivates the β-galactosidase from *E. coli* K12 by alkylating a methionine residue.[53]

The effect of sodium and magnesium ions on the β-galactosidase-catalysed hydrolysis of o-nitrophenyl β-D-galactoside has been investigated.[54]

The hydrolysis of β-D-galactosyl azide is catalysed by the β-galactosidase from *E. coli*.[55]

[41] D. Horton and A. E. Luetzow, *Chem. Comm.*, **1971**, 79.
[42] J. N. BeMiller and E. R. Doyle, *Carbohydrate Res.*, **20**, 23 (1971).
[43] B. Capon and B. C. Ghosh, *J. Chem. Soc.* (B), **1971**, 739; *Org. Reaction Mech.*, **1965**, 241.
[44] J. Szejtli, R. D. Henriques, and M. Castineira, *Acta Chim. Acad. Sci. Hung.*, **66**, 213 (1970); *Chem. Abs.*, **74**, 112345 (1971).
[45a] J. W. Barnett and C. J. O'Connor, *J. Chem. Soc.* (B), **1971**, 1163.
[45b] T. Moriyoshi, *Bull. Chem. Soc. Japan*, **44**, 2582 (1971).
[46] E. R. Gilliland, H. J. Bixler, and J. E. O'Connell, *Ind. Eng. Chem. Fundam.*, **10**, 185 (1971).
[47] N. A. Khalturinskii, Y. V. Moiseev, M. I. Vinnik and G. E. Zankov, *Dokl. Akad. Nauk. SSSR*, **198**, 149 (1971); *Chem. Abs.*, **75**, 49469 (1971).
[48] M. S. Weintraub and D. French, *Carbohydrate Res.*, **15**, 241, 251 (1970).
[49] C. Degani and M. Halmann, *J. Chem. Soc.* (C), **1971**, 1459; see also ref. 89.
[50] J. S. Moore and G. O. Phillips, *Carbohydrate Res.*, **16**, 79 (1971); G. O. Phillips, W. G. Filby, and J. S. Moore, *Carbohydrate Res.*, **16**, 89 (1971); G. O. Phillips, W. G. Filby, and J. S. Moore, *Carbohydrate Res.*, **16**, 105 (1971).
[51] B. Capon, *Biochimie*, **53**, 145 (1971).
[52] J.-P. Tenu, O. M. Viratelle, J. Garnier, and J. Yon, *Eur. J. Biochem.*, **20**, 363 (1971); see *Org. Reaction Mech.*, **1970**, 424–425.
[53] J. Yariv, K. J. Wilson, J. Hildesheim, and S. Blumberg, *FEBS Letters*, **15**, 24 (1971).
[54] R. Strom, D. G. Attardi, S. Forsén, P. Turini, F. Celada, and E. Antonini, *Eur. J. Biochem.*, **23**, 118 (1971).
[55] M. I. Sinnott, *Biochem. J.*, **125**, 717 (1971); cf. ref. 84.

(b) *Lysozymes.* The sequence of human leukaemia lysozyme is similar to that of hen-egg-white lysozyme but that of goose-egg-white is different.[56] The three-dimensional structures of human-leukaemia lysozyme and hen-egg-white lysozyme are also similar.[57] The hydrolysis of the cell wall of *M. lysodeikticus* catalysed by goose-egg-white lysozyme is inhibited by the cell wall tetra- and di-saccharides but not by NAG-2 or NAG-4.[58] Chemical modification of human lysozyme by acetylation and nitration of the tyrosine residues has been carried out.[59] Also the CD spectrum of human lysozyme has been reported.[60]

The sequence of turnip lysozyme has been determined.[61]

A standard method for the assay of lysozyme using the cell walls of *M. lysodeikticus* has been proposed.[62]

The lactone (16) derived from NAG-4 is bound 36 times more strongly by hen-egg-white lysozyme than is NAG-4. On the assumption that NAG-4 is bound mainly in subsites A to C and that the lactone (16) is bound mainly in subsites A to D it was estimated that the latter was bound 3600 times more strongly in this mode than NAG-4. This is consistent with the hypothesis that binding in the D-site requires distortion to a half-chair conformation.[63]

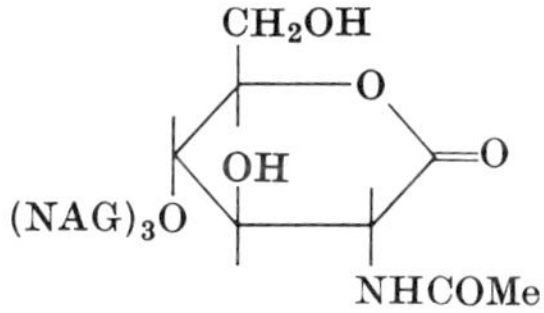

The coupling constant of the anomeric proton of the glucose residue of compound (17) is not changed in the presence of lysozyme and it was considered that there is no conformational distortion when this compound is bound by the enzyme.[64]

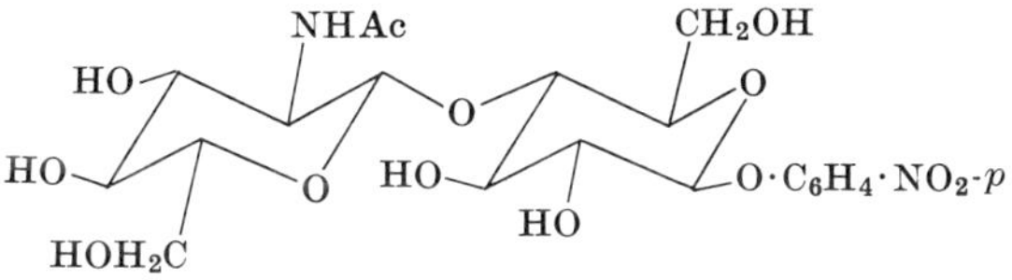

The interaction of lysozyme with *N*-trifluoroacetyl-D-glucosamine has been investigated by ^{19}F-magnetic resonance spectroscopy. The ^{19}F-signal of the α-anomer moves upfield in the presence of lysozyme, and the difference in shift between the unbound and bound *N*-trifluoroacetyl-α-D-glucosamine is similar to the difference in the proton shift of the acetyl group of unbound and bound *N*-acetyl-α-D-glucosamine. This suggests that the fluoro-compound is bound differently from the proton compound. *N*-Trifluoroacetyl-β-D-glucosamine and the methyl glycosides do not appear to be bound by lysozyme.[65] In contrast to these results the ^{19}F resonances of *N*-fluoroacetyl-D-glucosamine move

---

[56] R. E. Canfield, S. Kammerman, J. H. Sobel, and F. J. Morgan, *Nature New Biology*, **232**, 16 (1971).
[57] C. C. F. Blake and I. D. A. Swan, *Nature New Biology*, **232**, 12 (1971).
[58] A. K. Allen and A. Neuberger, *Biochim. Biophys. Acta*, **235**, 539 (1971).
[59] R. L. Fawcett, T. J. Limbird, S. L. Olivier, and C. L. Borders, *Can. J. Biochem.*, **49**, 816 (1971).
[60] J. P. Halper, N. Latovitzki, H. Bernstein, and S. Beychok, *Proc. Nat. Acad. Sci.*, **68**, 517 (1971).
[61] I. Bernier, E. Van Leemputten, M. Horisberger, D. A. Bush, and P. Jolles, *FEBS Letters*, **14**, 100 (1971).
[62] G. Gorin, S.-F. Wang, and L. Papapavlou, *Anal. Biochem.*, **39**, 113 (1971).
[63] I. I. Secemski and G. E. Lienhard, *J. Am. Chem. Soc.*, **93**, 3549 (1971).
[64] B. D. Sykes and D. Dolphin, *Nature*, **233**, 421 (1971); cf. *Org. Reaction Mech.*, **1968**, 355; **1969**, 405–406.
[65] H. Ashton, B. Capon, and R. L. Foster, *Chem. Comm.*, **1971**, 512.

downfield in the presence of lysozyme when trifluoroacetone is used as *internal* standard.[66,67] The pH-dependence of the binding of N-acetyl-D-glucosamine to lysozyme has been studied by NMR spectroscopy.[68]

A kinetic analysis of the reactions of the oligosaccharides from bacterial cell wall catalysed by lysozyme has been reported.[69]

The signals in the 220 MHz NMR spectrum of five of the tryptophan-NH groups of lysozyme have been assigned,[70] and the reversible denaturation of lysozyme has been studied by NMR spectroscopy.[71]

Other investigations on lysozyme are described in ref. 72.

(c) *Amylases* A general method for distinguishing between *endo* and *exo* activity of carbohydrates and the basis of their ability to hydrolyse periodate-oxidized pullullan has been proposed.[73]

An attempt to detect a "rapid initial burst" in the hydrolysis of *o*-nitrophenyl α-maltoside catalysed by saccharifying α-amylase was unsuccessful.[74] It has been reported that the α-amylase-catalysed hydrolysis of *p*-nitrophenyl α-maltoside and *p-t*-butylphenyl α-maltoside produces β-maltose whereas that of phenyl α-maltoside produces α-maltose.[75] The hydrolysis of 6-deoxy-6-fluoro-α-maltoside is catalysed by Taka amylase but the reaction is slower than that of phenyl maltoside.[76] The hydrolyses of phenyl 2-O-methyl-α-maltoside, phenyl 2-deoxy-α-maltoside and phenyl 4-O-α-glucosyl-α-mannoside are not catalysed by Taka amylase A.[77] Inhibition of the hydrolysis of phenyl and *p*-nitrophenyl α-maltoside catalysed by Taka amylase A has been studied.[78]

The turnover number for the hydrolysis of maltodextrins catalysed by Taka amylase A increases with increasing degree of polymerization until this is 7. It was suggested that the specificity region of this enzyme spans about 7 glucose units.[79]

The activity of Taka amylase A is increased on introduction of mercaptosuccinoyl groups.[80]

Other investigations on Taka amylase A are reported in ref. 81.

---

[66] R. A. Dwek, P. W. Kent, and A. V. Xavier, *Eur. J. Biochem.*, **23**, 343 (1971).
[67] P. W. Kent and R. A. Dwek, *Biochem. J.*, **121**, 11P (1971).
[68] J. F. Studebaker, B. D. Sykes, and R. Wien, *J. Am. Chem. Soc.*, **93**, 4579 (1971).
[69] D. M. Chipman, *Biochemistry*, **10**, 1714 (1971).
[70] J. D. Glickson, W. D. Phillips, and J. A. Rupley, *J. Am. Chem. Soc.*, **93**, 4031 (1971).
[71] C. C. McDonald, W. D. Phillips, and J. D. Glickson, *J. Am. Chem. Soc.*, **93**, 235 (1971).
[72] H.-C. Chien and L. C. Craig, *Bioorganic Chemistry*, **1**, 51 (1971); P. W. Linder and W. P. Bryan, *J. Am. Chem. Soc.*, **93**, 3061 (1971); L. J. Berliner, *J. Mol. Biol.*, **61**, 189 (1971); A. F. S. A. Habeeb and M. Z. Atassi, *Biochemistry*, **9**, 4939 (1970); E. Maron, C. Shiozawa, R. Arnon, and M. Sela, *Biochemistry*, **10**, 763 (1971); E. K. Achter and I. D. A. Swan, *Biochemistry*, **10**, 2976 (1971); S. Katz and J. E. Miller, *Biochemistry*, **10**, 3569 (1971); R. Roxby and C. Tanford, *Biochemistry*, **10**, 3348 (1971); S. S. Lehrer, *Biochemistry*, **10**, 3254 (1971); J. H. Bradbury and N. L. R. King, *Austral. J. Chem.*, **24**, 1703 (1971).
[73] G. S. Drummond, E. E. Smith, and W. J. Whelan, *FEBS Letters*, **15**, 302 (1971).
[74] N. Suetsugu, K. Hiromi, and S. Ono, *J. Biochem. (Tokyo)*, **69**, 421 (1971).
[75] T. Shibaoka, N. Suetsugu, K. Hiromi, and S. Ono, *FEBS Letters*, **16**, 33 (1971).
[76] H. Arita and Y. Matshushima, *J. Biochem. (Tokyo)*, **69**, 409 (1971).
[77] H. Arita, T. Ikenaka, and Y. Matsushima, *J. Biochem. (Tokyo)*, **69**, 401 (1971).
[78] Y. Nitta, K. Hiromi, and S. Ono, *J. Biochem. (Tokyo)*, **69**, 577 (1971).
[79] Y. Nitta, M. Mizushima, K. Hiromi, and S. Ono, *J. Biochem. (Tokyo)*, **69**, 567 (1971).
[80] S. Suzuki, Y. Hachimori, and R. Matoba, *Bull. Chem. Soc. Japan*, **43**, 3849 (1970).
[81] H. Yamaguchi, T. Ikenaka, and Y. Matsushima, *J. Biochem. (Tokyo)*, **68**, 843 (1970); M. Arita and Y. Masushima, *J. Biochem. (Tokyo)*, **68**, 717 (1970); M. Ohnishi, *J. Biochem. (Tokyo)*, **69**, 181 (1971).

The β-amylase from sweet potato has a molecular weight of 206,000 ± 10,000; it is tetrameric.[82] Another investigation on this enzyme is reported in ref. 83.

(d) *Other enzymes*. The hydrolysis of α-D-glucosyl fluoride catalysed by fungal amyloglucosidase occurs with inversion of configuration, and that catalysed by yeast and rabbit intestinal α-glucosidase occurs with retention of configuration. The hydrolysis of α-D-galactosyl fluoride catalysed by coffee-bean α-galactosidase and that of β-D-glucosyl fluoride catalysed by emulsin occur with retention of configuration.[84]

β-D-Glucopyranosyl azide is not a substrate for the β-glucosidase of almond emulsin (cf. ref. 55 above).[84] α-D-Glucosyl fluoride is a substrate for sucrose phosphorylase from *Pseudomonas saccarophila*.[85]

Yeast α-glucosidase is irreversibly inhibited by L-1,2-anhydro-*myo*-inositol (18).[86a]

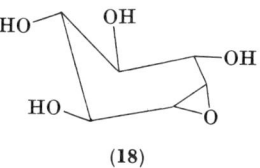

(18)

The reaction of phenyl β-D-glucopyranoside with methanol catalysed by emulsin β-glucosidase has been studied.[86b]

The *N*-acetyl-β-D-glucosaminidase of pig epidermis catalyses the hydrolysis of *p*-nitrophenyl 2-acetyl-2-deoxy-β-D-glucoside and -galactoside.[87]

The acid-catalysed hydrolysis of α-D-glucose 1-phosphate has an α-deuterium isotope effect $k_H/k_D = 1.13$ at 24°, consistent with a mechanism involving a carbonium ion. The hydrolyses catalysed by *E. coli* maltodextrin phosphorylase and rabbit-muscle phosphorylase b at 10° show isotope effects $k_H/k_D = 1.09$ and 1.10, respectively, and it was considered that the reactions also involved carbonium ions. 1,5-D-Gluconolactone was shown to be a competitive inhibitor for these enzymes. This compound was thought to be a transition-state analogue for a reaction going through a carbonium ion and hence this was considered to be additional evidence for reaction proceeding through a carbonium ion. The possibility that the inhibition was caused by 1,4-gluconolactone which is formed rapidly from the 1,5-lactone at neutral pH's[88] was not tested.[89]

Other glycosidases studied include α-glucosidase from rabbit muscle,[90] glucamylase from *A. awamori*,[91] yeast invertase,[92] and the β-glucosidase from emulsin.[93]

[82] P. M. Colman and B. W. Matthews, *J. Mol. Biol.*, **60**, 163 (1971).
[83] K. Uehara, T. Mizoguchi, K. Kishida, and S. Mannen, *J. Biochem. (Tokyo)*, **69**, 27 (1971).
[84] J. E. G. Barnett, *Biochem. J.*, **123**, 607 (1971).
[85] A. M. Gold and M. P. Osber, *Biochem. Biophys. Res. Comm.*, **42**, 469 (1971).
[86a] J. E. G. Barnett, D. Mercier, and S. D. Gero, *FEBS Letters*, **16**, 37 (1971).
[86b] K. Yoshida and N. Iino, *Chem. Pharm. Bull. (Japan)*, **19**, 6 (1971).
[87] G. V. Vikha, E. D. Kaverzneva, and A. Y. Khorlin, *Biokhimya*, **36**, 33 (1971).
[88] M. A. Jermyn, *Biochim. Biophys. Acta*, **37**, 78 (1960); T. Takahashi and M. Mitsumoto, *Nature*, **199**, 765 (1963).
[89] J.-I. Tu, G. R. Jacobson, and D. J. Graves, *Biochemistry*, **10**, 1229 (1971).
[90] T. N. Palmer, *Biochem. J.* **124**, 701, 713 (1971).
[91] K. M. Bendetskii, V. L. Yarovenko, L. N. Ermakova, and L. N. Luk'yanova, *Biokhimya*, **35**, 1099 (1970); K. M. Bendetskii, V. L. Yarovenko, and L. N. Luk'yanova, *Biokhimya*, **36**, 525 (1971).
[92] A. Waheed and S. Shall, *Enzymologia*, **41**, 291 (1971).
[93] T. Kleinschmidt and J. Horst, *Z. Physiol. Chem.*, **352**, 603 (1971).

Hydration of Aldehydes and Ketones and Related Reactions

The equilibrium constants for the hydration of a series of aryl trifluoromethyl ketones in aqueous sulpholan and aqueous dimethyl sulphoxide (DMSO) have been determined, and a hydration function W_0 has been defined:

$$W_0 = pK_D + \log Z/ZH_2O$$
$$= -\log f_Z \cdot a_{H_2O}/f_{ZH_2O}$$

where K_D is the dehydration constant,

$$K_D = Z \cdot f_Z a_{H_2O}/ZH_2O \cdot f_{ZH_2O}$$

and Z and ZH_2O refer to the ketone and ketone hydrate, respectively. W_0 is greater in the sulpholan mixtures than in the DMSO mixtures with the same mole fraction of water. Therefore sulpholan has a greater dehydrating effect than DMSO despite the fact that the activity of water in the DMSO–water mixtures decreases faster with increasing DMSO concentration than it does in sulpholan–water mixtures with increasing sulpholan concentration. It was proposed that "the ketone hydrate because of its *gem*-diol structure fits into the water structure without greatly perturbing it whereas the presence of sulpholan tends to exclude the ketone hydrate from the association pattern, causing f_{ZH_2O} to rise. DMSO, on the other hand, appears to form stronger association with the hydrate than does water causing f_{ZH_2O} in these solutions to fall." From 0 to 85 mole-% of DMSO, W_0 is negative as the ketones are hydrated more strongly than in water alone.[94]

The hydration of methyl and ethyl pyruvate is general base-catalysed but not acid-catalysed. The spontaneous hydration of these esters occurs at a similar rate to that of pyruvic acid, and so it seems that hydration of the latter is not intramolecularly catalysed.[95]

The formation and decomposition of the acetone–hydrogen peroxide adduct are general acid- and general base-catalysed. The α and β values for the decomposition were $\alpha = 0.43$, $\beta = 0.53$ and for the formation $\alpha = 0.53$ and $\beta = 0.55$.[96]

The kinetics of the tautomerization of phthalaldehydic acid have been measured by temperature-jump measurement. The reaction is catalysed by the phthalaldehydate anion and by fluoride ion.[97]

The kinetics of the addition of water to 1,3-dichloroacetone in dioxan have been investigated.[98] The heats of addition of water and methanol to hexafluoroacetone in the gas phase and in solution have been measured.[99]

The species present in aqueous solutions of glyoxal[100,101] and of acetaldehyde[102] have been studied by NMR spectroscopy.

[94] R. Stewart and J. D. Van Dyke, *Can. J. Chem.*, **48**, 3961 (1970).
[95] Y. Pocker, J. E. Meany, and C. Zadorojny, *J. Phys. Chem.*, **75**, 792 (1971).
[96] M. C. V. Sauer and J. O. Edwards, *J. Phys. Chem.*, **75**, 3004 (1971).
[97] R. P. Bell, B. G. Cox, and B. A. Timimi, *J. Chem. Soc.* (B), **1971**, 2247; see also K. Bowden and G. R. Taylor, *J. Chem. Soc.* (B), **1971**, 1390, 1395.
[98] R. P. Bell and J. E. Critchlow, *Proc. Roy. Soc.*, *A*, **325**, 35 (1971); cf. *Org. Reaction Mech.*, **1968**, 356–357.
[99] F. E. Rogers and R. J. Rapiejko, *J. Am. Chem. Soc.*, **93**, 4596 (1971).
[100] E. B. Whipple, *J. Am. Chem. Soc.*, **92**, 7183 (1970).
[101] G. C. S. Collins and W. O. George, *J. Chem. Soc.* (B), **1971**, 1352.
[102] F. Podo and V. Viti, *Org. Mag. Res.*, **3**, 259 (1971).

The pH–rate profile for the mutarotation of 6-deoxyglucohepturonic acid (**19**) is sigmoid and the form with the carboxyl group ionized reacts about 8 times faster than expected from a linear free-energy relationship for the mutarotation of other 6-substituted glucoses. The value of the catalytic constant for catalysis by hydroxyl ions for the mutarotation of 6-O-(o-hydroxyphenyl)-D-glucose (**20**) is 2800 times greater than that for the mutarotation of 6-O-phenyl-D-glucose, probably because the ionized form reacts very rapidly. It seems likely that the mutarotation of both (**19**) and (**20**) are intramolecularly catalysed, the mechanisms possibly being those symbolized by (**21**) and (**22**).[103]

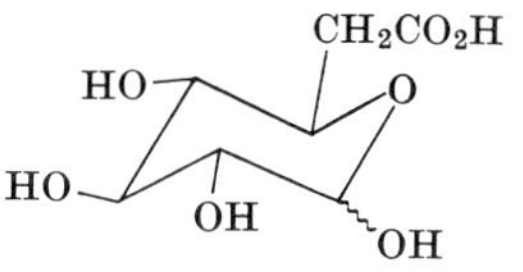

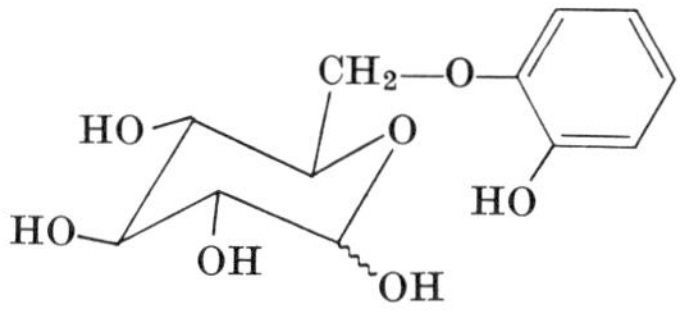

The mutarotation of tetra-O-methyl-D-glucose in benzene is enhanced by micellar dodecylammonium propionate and benzoate. The rate increases proportionally to the concentration of the long-chain salt at low concentrations, 10^{-5} to 10^{-3}M but reaches a plateau at higher concentrations $>10^{-2}$M.[104]

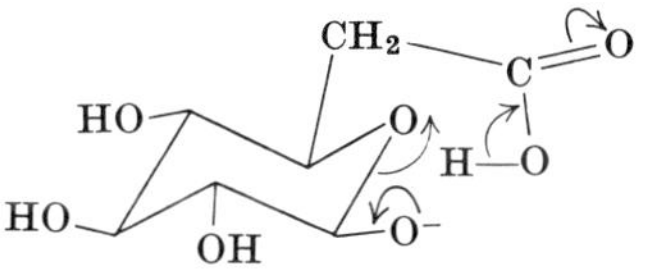

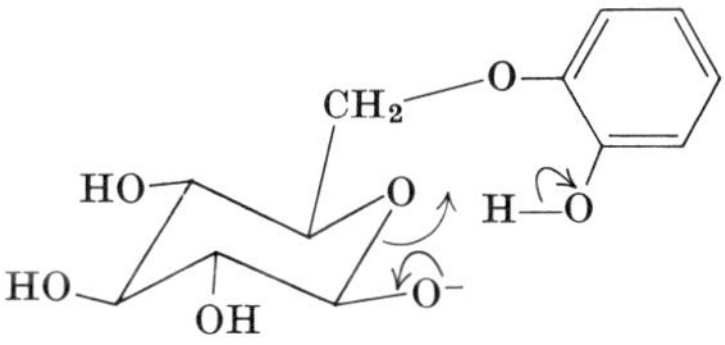

Mutarotatase from *E. coli* K12 catalyses the mutarotation of D-galactose, D-glucose, D-fucose, and D-xylose, but not that of D-mannose or 2-deoxy-D-glucose, although the last two sugars are inhibitors. The V_{max} depends on the concentration of one functional group, pK_a 5.5, in its basic form. K_m depends on the concentration of a functional group of pK_a 7.6 present in its acidic form.[105]

The kinetics of the complex mutarotation of 2-deoxy-β-D-ribose have been studied. It seems likely that reaction proceeds through an *aldehydo*-form which cyclizes more rapidly to form furanose than pyranose forms.[106,107] The *aldehydo* and *keto* content of several sugars and their phosphates have been determined by IR spectroscopy.[108] Contrary to an earlier report, it appears that the anomeric equilibrium of glucose does not shift towards the β-anomer in alkaline solution.[109]

---

[103] B. Capon and R. B. Walker, *Chem. Comm.*, **1971**, 1323.
[104] E. J. Fendler, J. H. Fendler, T. R. Medary, and V. A. Woods, *Chem. Comm.*, **1971**, 1497.
[105] F. Hucho and K. Wallenfels, *Eur. J. Biochem.*, **23**, 489 (1971).
[106] R. U. Lemieux, L. Anderson, and A. H. Conner, *Carbohydrate Res.*, **20**, 59 (1971).
[107] Discussion: B. Capon, *Chem. Rev.*, **69**, 454 (1969).
[108] C. A. Swenson and R. Barker, *Biochemistry*, **10**, 3151 (1971).
[109] D. E. Dorman and J. D. Roberts, *J. Am. Chem. Soc.*, **93**, 4463 (1971).

The following topics have also been studied: mutarotation of sugars in the presence of metal ions,[110] of lactose in the presence of sucrose,[111] and of xylose in the solid and molten state;[112] the effect of temperature on the rates of some mutarotation reactions;[113,114] and the hydration of the acetates of *aldehydo*-hexoses.[115]

Reactions with Nitrogen Bases

Schiff Bases

A detailed investigation of the reaction of cysteine with formaldehyde has been reported. Thiazolidine-4-carboxylic acid (**26**) is formed via a Schiff base (**24**); in acid solution the rate-limiting step is formation of the carbinolamine (**23**) but this changes at pH's above ca. 6 to dehydration of the carbinolamine which is general acid-catalysed with $\alpha = 0.66$; the hemithioacetal (**25**) is also formed rapidly in a parasitic equilibrium, but this reaction becomes less significant at higher pH's where the thiol group is dissociated.[116]

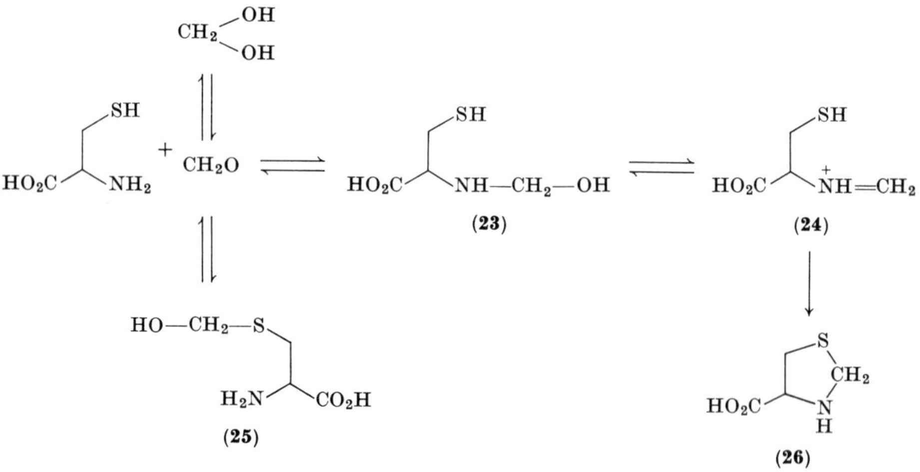

2,3-Diaminopropionic acid reacts rapidly with pyridoxal in deuterium oxide having pD > 6 to form two Schiff bases (**27**) and (**28**). The PMR signals of the CH=N groups of these compounds disappear slowly and a new signal appears which was attributed to the *gem*-diamino-species (**29**). It appears from the spectrum that there is only one resonance attributable to the CHNN group, which is surprising since two diastereoisomers of (**29**) are possible. There appeared to be no cyclic diamino-compound formed from 2,6-diaminohexanoic acid (lysine), but the seven-membered ring structure (**30**) was formed from 2,5-diaminopentanoic acid (ornithine) although the equilibrium was less favourable than that for formation of (**29**). There also appeared to be no cyclic diamino-compound

[110] R. Mitzner and E. Behrenwald, *Z. Phys. Chem.* (*Leipzig*), **247**, 78 (1971); *Z. Chem.*, **11**, 64 (1971).
[111] K. W. Patel and T. A. Nickerson, *J. Dairy Sci.*, **53**, 1654 (1970); *Chem. Abs.*, **74**, 88249 (1971).
[112] F. Shafizadeh, G. D. McGinnis, R. A. Susott, and H. W. Tatton, *J. Org. Chem.*, **36**, 2813 (1971).
[113] H. Schmid, *Monatsh*, **102**, 292 (1971).
[114] R. Mitzner and E. Behrenwald, *Z. Phys. Chem.* (*Leipzig*), **246**, 25 (1971).
[115] D. Horton and J. D. Wander, *Carbohydrate Res.*, **16**, 477 (1971).
[116] R. G. Kallen, *J. Am. Chem. Soc.*, **93**, 6227, 6236 (1971).

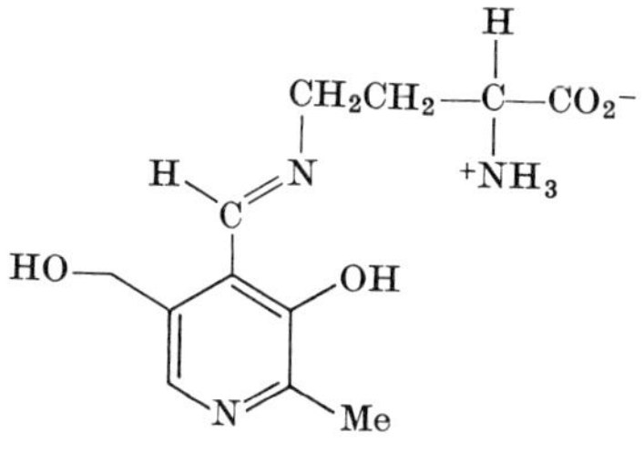

formed from 2,3-diaminopropionic acid but one of the carbinolamines (31) or (32) was thought to be especially stable, possibly as a result of hydrogen-bonding.[117] Ring-chain tautomerism of Schiff bases of pyridoxal has been studied.[118]

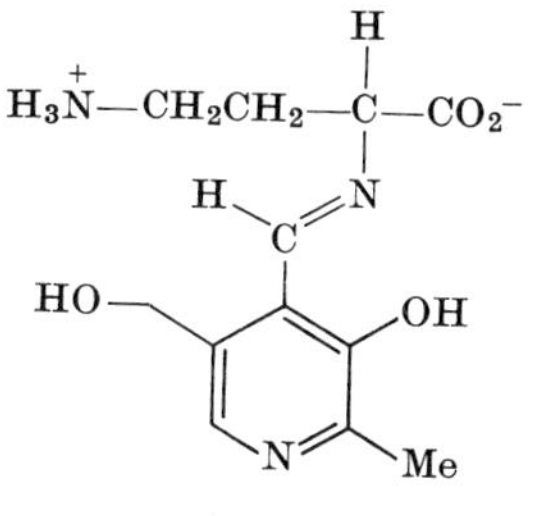

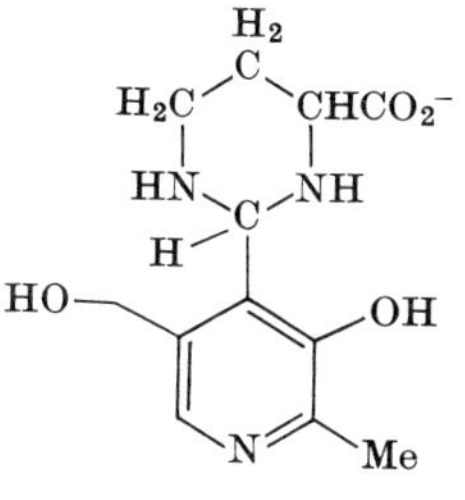

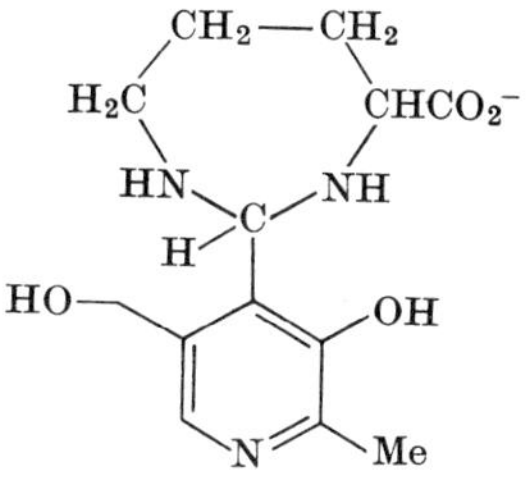

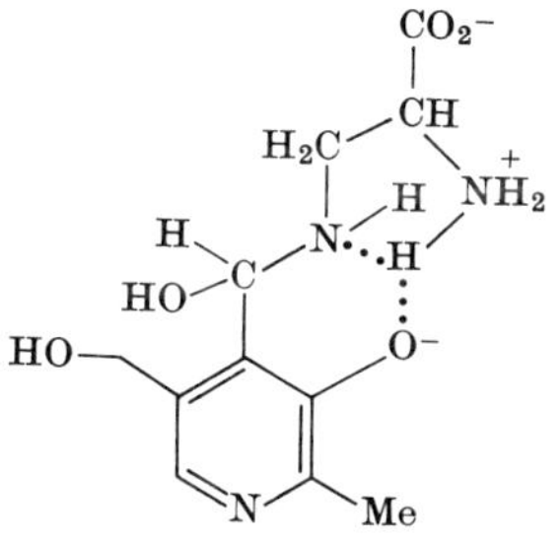

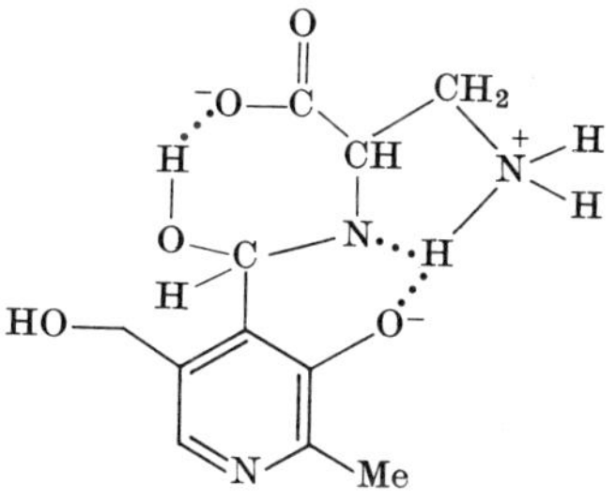

A detailed investigation of the formation of Schiff bases from salicylaldehyde and ethylamine, α-alaninate, β-alaninate and glycinate in the presence and absence of zinc has been described; the transimination reactions were also studied and shown to be promoted by zinc; these reactions may involve pre-equilibrium formation of a complex in which the imine and the amine are both bound to the zinc.[119]

The reaction of N-benzylideneaniline with HCN in methanolic acetonitrile, a step of the Strecker synthesis, has been investigated. The reaction shows Hammett ρ-values of -1.11 and -1.39 for substitution in the benzylidene and aniline rings, respectively. A mechanism symbolized by (33) was proposed.[120]

[117] E. M. Abbott and A. E. Martell, *J. Am. Chem. Soc.*, **93**, 5852 (1971).
[118] W. Korytnyk, H. Ahrens, and N. Angelino, *Tetrahedron*, **26**, 5415 (1970).
[119] B. E. Leach and D. L. Leussing, *J. Am. Chem. Soc.*, **93**, 3377 (1971).
[120] Y. Ogata and A. Kawasaki, *J. Chem. Soc.* (B), **1971**, 325.

The rate constants of the reaction of dehydroacetic acid with 8 aliphatic primary amines and 13 aromatic amines in ethanol have been determined. Those for the reaction with aromatic amines were correlated by the Hammett equation with $\rho = -2.32$, but those for the reactions of the aliphatic amines were not correlated by Taft's equation. It is uncertain which step is rate-limiting.[121a]

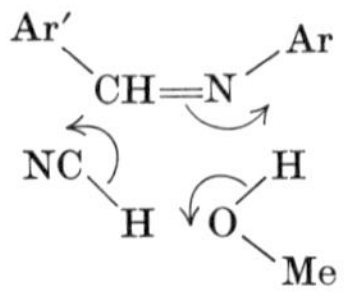

$$\begin{array}{c} Ar' \\ \diagdown \\ NC \\ \diagup \\ H \end{array} CH=N \begin{array}{c} Ar \\ \diagup \\ \diagdown \\ O \\ \diagdown \\ Me \end{array} \begin{array}{c} \\ H \end{array}$$

(33)

The α-deuterium isotope, k_D/k_H, for the attack of hydroxide ion on N-benzylidene-t-butylamine is ca. 1.22. The ρ-value for the attack of hydroxide ion on N-substituted benzylideneanilines is 2.7. These results were interpreted as indicating that the transition state is adduct-like.[121b]

The reaction of nitrosobenzene with substituted anilines in acetate buffers yields a ρ-value of -2.41.[122]

The following reactions have also been studied: condensation of 5-dialkylamino- and 5-arylalkylamino-furfuraldehyde with primary aromatic amines;[123] addition of alcohol and water to Schiff bases co-ordinated to copper(II);[124a] and hydrolysis of benzothiadiazines.[124b]

The tautomeric equilibria of Schiff bases derived from o-hydroxyaromatic aldehydes and ketones have been investigated further.[125]

The conformation of benzylideneaniline has been studied by NMR spectroscopy[126] and by X-ray crystallography.[127]

The pK_a's of some diaryl ketiminium ions have been determined by a continuous double-flow method.[128]

There have been numerous investigations of the *syn–anti* isomerization of Schiff bases.[129]

[121a] S. Goto and T. Toi, *Chem. Pharm. Bull.*, **19**, 632 (1971).
[121b] J. Archila, H. Bull, C. Lagenaur, and E. H. Cordes, *J. Org. Chem.*, **36**, 1345 (1971).
[122] R. A. Yunes, M. M. Meyer, A. J. Terenzani, O. D. Andrich, and C. A. Scarabino, *Rev. Fac. Ing. Quím. Univ. Nac. Litoral*, **38**, 239 (1969); *Chem. Abs.*, **75**, 87965 (1971).
[123] V. S. Pustovarov and Z. N. Nazarova, *Organic Reactivity (Tartu)*, **8**, 175 (1971).
[124a] M. Cressey, E. D. McKenzie, and S. Yates, *J. Chem. Soc.* (A), **1971**, 2677.
[124b] J. A. Mollica, C. R. Rehm, J. B. Smith, and H. K. Govan, *J. Pharm. Sci.*, **60**, 1380 (1971).
[125] G. Dudek and E. P. Dudek, *J. Chem. Soc.* (B), **1971**, 1356; P. Nagy and E. Kövér, *Mag. Kém. Folyorat*, **77**, 100 (1971); G. C. Percy and D. A. Thornton, *Chimia*, **25**, 195 (1971); cf. *Org. Reactio Mech.*, **1965**, 243–244; **1966**, 318–319; **1967**, 315; **1968**, 360; **1969**, 415; **1970**, 433.
[126] A. van Putten and J. W. Pavlik, *Tetrahedron*, **27**, 3007 (1971); J. W. Pavlik and A. van Putte *Tetrahedron*, **27**, 3301 (1971); V. M. S. Gil and M. E. L. Saraiva, *Tetrahedron*, **27**, 1309 (1971
[127] H. B. Bürgi and J. D. Dunitz, *Helv. Chim. Acta*, **54**, 1255 (1971).
[128] J. Taillades, A. Commeyras, A. Casadevall, and C. Bouchoule, *Bull. Soc. Chim France*, **1971**, 171
[129] C. H. Warren, G. Wettermark, and K. Weiss, *J. Am. Chem. Soc.*, **93**, 4658 (1971); G. E. Ha W. J. Middleton and J. D. Roberts, *J. Am. Chem. Soc.*, **93**, 4778 (1971); C. H. Bushweller, J. W O'Neil, and H. S. Bilofsky, *J. Am. Chem. Soc.*, **93**, 542 (1971); F. Kerek, G. Ostrogovich, an Z. Simon, *J. Chem. Soc.* (B), **1971**, 541; R. Damrauer and T. E. Rutledge, *J. Organomet. Chem* **29**, C9 (1971); G. Ostrogovich, Z. Simon, and F. Kerek, *Rev. Roum. Chim.*, **15**, 1453 (1970); Krebs and H. Kimling, *Angew. Chem. Int. Ed.*, **10**, 409 (1971).

The formation,[130] transglycosylation,[131] mutarotation,[132] and basic strengths[133a] of glycosylamines have been studied.

3'-Fluoro- and 3'-chloro-3'-deoxythymidine are hydrolysed faster than thymidine in 1M hydrochloric acid at 80°.[133b] Thymidine phosphorylase from *E. coli* has been studied.[133c]

Semicarbazones, Oximes, Hydrazones and Related Compounds

The rate constants for the reaction of semicarbazide with substituted acetophenones at pH 5.25 have been determined. Under these conditions the rate-limiting step was thought to be nucleophilic attack of the semicarbazide on the ketone and the ρ-value was 0.45.[134] The reactions of the following compounds with semicarbazide have also been studied: 5-nitrofurfuraldehyde,[135] pyruvic acid,[136] cycloalkyl methyl ketones,[137] and acyloins.[138]

The α-deuterium isotope effects for the reaction of ketones with hydroxylamine have been determined under acidic conditions where the rate-limiting step for oxime formation is nucleophilic attack by hydroxylamine. The ratios k_H/k_D are generally less than one but for cyclohexanone and 4-t-butylcyclohexanone they were 1.05. It was thought that in addition to the lower hyperconjugative stabilization of the ketone in the initial state of the deuteriated compound there was a steric isotope effect arising from the shortness of the C—N bond in the transition state. It was considered that this consisted partly of an "approach effect" with $k_H/k_D < 1$ and partly of a torsional effect with $k_H/k_D > 1$.[139]

The reaction of arylidenesemicarbazones with bromine to form oxadiazoles [reaction (1)] is of the first order in semicarbazone and of zero order in bromine. The interconversion

$$\text{ArCH=N-NHCONH}_2 + \text{Br}_2 \longrightarrow \underset{\underset{\text{Br}}{|}}{\text{ArC=N-NHCONH}_2} \longrightarrow \text{ArC}\underset{O}{\overset{N-N}{\underset{\|}{\|}}}\text{C-NH}_2$$

... (1)

of the less reactive form (*syn* or *anti*) into the more reactive (*anti* or *syn*) was thought to be the rate-limiting step. The minor component is present to the extent of about 10%, and so, if this proposal is correct, there should be an initial burst corresponding to the

---

[130] S. Kolka and J. Sokolowski, *Zesz. Nauk Wyzsz. Szk. Pedagog. Gdanskii: Mat. Fiz. Chem.*, **10**, 183 (1970); *Chem. Abs.*, **74**, 31925 (1971).
[131] F. V. Pishchiegen and V. A. Afanas'ev, *Zhur. Fiz. Khim.*, **44**, 2085 (1970).
[132] K. Smiataczowa, *Wiad. Chem.*, **25**, 343 (1971); *Chem. Abs.*, **75**, 88848 (1971); K. Smiataczowa, T. Jasiński, and J. Sokolowski, *Rocz. Chem.*, **45**, 329 (1971); K. Smiataczowa, T. Jasiński, and J. Sokolowski, *Rocz. Chem.*, **44**, 2405 (1970).
[133a] J. Jasińska and J. Sokolowski, *Rocz. Chem.*, **44**, 1913 (1970).
[133b] G. Etzold, R. Hintsche, G. Kowollik, and P. Langen, *Tetrahedron*, **27**, 2463 (1971).
[133c] M. Schwartz, *Eur. J. Biochem.*, **21**, 191 (1971).
[134] V. Baliah and V. N. V. Desikan, *Indian J. Chem.*, **8**, 902 (1970).
[135] L. DoAmaral and M. P. Bastos, *Rev. Brasil Tecnol.* **1970**, 23; *Chem. Abs.*, **74**, 124425 (1971).
[136] T. Pino and E. H. Cordes, *J. Org. Chem.*, **36**, 1668 (1971).
[137] N. Takeno, S. Kozuka, and N. Takano, *Muroran Koyyo Daigaku Kenkyu Hokotu*, **7**, 153 (1970); *Chem. Abs.*, **75**, 87801 (1971).
[138] D. Fleury and H. B. Fleury, *Compt. Rend.* (C), **271**, 406 (1970); cf. *Org. Reaction Mech.*, **1970**, 438.
[139] P. Geneste, G. Lamaty, and J.-P. Roque, *Tetrahedron Letters*, **1970**, 5015; *Tetrahedron*, **27**, 5561 (1971).

rapid reaction of this isomer. It was not reported whether this occurs.[140] The rate-limiting step in the bromination of 5-(arylmethylenehydrazino)tetrazoles is also thought to be a *syn–anti* isomerization.[141]

There have been several investigations of *syn–anti* isomers with carbon—nitrogen double bonds.[142]

The isomers of phenylglyoxime have been separated and characterized.[143] The conformations of sugar hydrazones have been studied.[144]

Hydrolysis of Enol Ethers and Esters

A detailed investigation of the general acid-catalysed hydrolysis of a series of vinyl ethers has been reported. The points for catalysis by pivalic and methoxyacetic acid showed positive deviations from the Brønsted plots, and those for cyanoacetic and chloroacetic acid showed negative deviations. The range of reactivities studied encompassed five powers of ten from phenyl vinyl ether to ethyl isopropenyl ether, but the α-values change by only about 0.2.[145] Ethyl styryl ether is hydrolysed about 500 times more slowly than ethyl vinyl ether. This must arise from the greater stability of the ethyl styryl ether due to the delocalization of the double bond with the benzene ring. The ρ-values for the hydrolysis of substituted ethyl styryl ethers are −1.07 and −0.70 for the *cis*- and *trans*-isomers, respectively. The *cis*-isomers are the slightly more reactive.[146]

The hydration of *trans*-cyclo-octene and of 2,3-dimethylbut-2-ene is general acid-catalysed in phosphate and sulphate buffers. Therefore, these reactions do in fact resemble the hydrolysis of enol ethers.[147]

There has been further discussion of solvent-isotope effects in H_2O–D_2O mixtures with special reference to the hydrolysis of vinyl ethers.[148]

The hydrolysis of silicon-containing vinyl ethers and vinyl sulphides has been studied. The vinyl sulphides are hydrolysed more rapidly than the corresponding vinyl sulphides not containing silicon.[149]

The isotope effect, $k(H_2O)/k(D_2O)$, in the hydrolysis of furan in 5.0M hydrochloric acid is 1.69. There is rapid hydrogen exchange at the α-position before hydrolysis; an $A2$ mechanism was proposed.[150]

---

[140] F. L. Scott, T. M. Lambe, and R. N. Butler, *Tetrahedron Letters*, **1971**, 2909.

[141] J. C. Tobin, A. F. Hegarty, and F. L. Scott, *J. Chem. Soc.* (B), **1971**, 2198.

[142] H. Kessler, P. F. Bley, and D. Leibfritz, *Tetrahedron*, **27**, 1687 (1971); M. Raban and E. Carlson, *J. Am. Chem. Soc.*, **93**, 685 (1971); M. Shanshal, *Z. Naturforsch.*, B, **25**, 1063 (1970), K. D. Berlin and S. Rengaraju, *J. Org. Chem.*, **36**, 2912 (1971); see also ref. 129.

[143] J. V. Burakevich, A. M. Lore, and G. P. Volpp, *J. Org. Chem.*, **36**, 1 (1971).

[144] J. M. J. Tronchet, B. Baehler, A. Jotterand, and F. Perret, *Helv. Chim. Acta*, **54**, 1660 (1971).

[145] A. J. Kresge, H. L. Chen, Y. Chiang, E. Murrill, M. A. Payne, and D. S. Sagatys, *J. Am. Chem. Soc.*, **93**, 413 (1971).

[146] T. Okuyama, F. Fueno, and J. Furukawa, *Bull. Chem. Soc. Japan*, **43**, 3256 (1970).

[147] A. J. Kresge, Y. Chiang, P. H. Fitzgerald, R. S. McDonald, and G. H. Schmid, *J. Am. Chem. Soc.*, **93**, 4907 (1971).

[148] R. A. More O'Ferrall, G. W. Koeppl, and A. J. Kresge, *J. Am. Chem. Soc.*, **93**, 1, 9 (1971).

[149] M. F. Shostakovskii, N. N. Vlasova, I. I. Tsykhanskaya, and I. S. Emel'yanov, *Zhur. Obshch. Khim.*, **40**, 1897 (1970); *Chem. Abs.*, **74**, 52681 (1971); M. F. Shostakovskii, F. P. L'vova, I. S. Emel'yanov, N. N. Vlasova, and B. V. Prokop'ev, *Izv. Akad. Nauk. SSSR, Ser. Khim.*, **1970**, 2147; *Chem. Abs.*, **74**, 52682 (1971).

[150] K. Unverferth and K. Schwetlick, *J. Prakt. Chem.*, **312**, 882 (1970).

The rates of hydrolysis of vinyl ethers have been correlated with the ^{13}C shifts of the α- and β-carbon atoms of the vinyl group[151] and with the geminal coupling constant of the $H_2C=$ protons.[152]

Other investigations of the hydrolysis of vinyl ethers are described in ref. 153.

The hydrolysis of the mixed vinyl ether acetal (34) has been studied. The solvent isotope effect on the hydrolysis is $k(HCl)/k(DCl) = 4.2$ and the reaction is general acid-catalysed. A mechanism involving a rate-limiting protonation of the double bond by the hydronium ion was proposed. However, the plot of k_{obs} against the concentration of acetic acid was not linear and it was suggested that the protonation step was reversible when carried out by acetic acid. In a deuterioacetic acid buffer there was no incorporation of deuterium at position 3 and it was therefore proposed that the two hydrogens H_a and H_b of carbonium ion (35) were eliminated at significantly different rates.[154] The authors of this paper appear to have overlooked previous work on the hydrolysis of mixed vinyl ether acetals.[155]

(34)   (35)

The methanolysis of cyanoketene dimethyl acetal is general acid-catalysed by dichloroacetic acid. The dependence of the rate on the stoichiometric concentration of dichloroacetic acid is non-linear owing to association. The solvent isotope effect $k(HCl)/k(DCl)$ is 3.0.[156]

The kinetics of the addition of ethanol to alkyl vinyl ethers[157,158] and the hydrolysis of vinyl esters in sulphuric acid[159a] have been studied.

The stereochemistry of the proton addition to phosphoenol α-oxobutyrate catalysed by pyruvate kinase has been determined.[159b]

Enolization and Related Reactions

When the logarithms of the second-order rate constants for the dedeuteriation of [$^2H_6$]-acetone catalysed by a series of diamines $Me_2N(CH_2)_nNH_2$ are plotted against the

---

[151] K. Hatada, K. Nagata, and H. Yaki, *Bull. Chem. Soc., Japan* **43**, 3195 (1970).
[152] A. F. Rekasheva, L. F. Kulish, and L. A. Kiprianova, *Teor. Eksp. Khim.*, **7**, 218 (1971); *Chem. Abs.*, **75**, 48014 (1971).
[153] B. A. Trofimov and I. S. Yemelyanov, *Organic Reactivity (Tartu)*, **7**, 564 (1970); J. P. H. Boyer, R. J. P. Corriu, and R. J. M. Perz, *Tetrahedron*, **27**, 4335, 5255 (1971); J. Toullec and J. E. Dubois, *Tetrahedron Letters*, **1971**, 3377.
[154] J. D. Cooper, V. P. Vitullo, and D. L. Whalen, *J. Am. Chem. Soc.*, **93**, 6294 (1971).
[155] See *Org. Reaction Mech.*, **1968**, 351; **1969**; 419.
[156] V. Gold and S. Grist, *J. Chem. Soc.* (B), **1971**, 2272.
[157] B. A. Trofimov, O. N. Vylegzhanin, and N. A. Nedolya, *Organic Reactivity (Tartu)*, **7**, 588 (1970).
[158] O. N. Vylegzhanin and B. A. Trofimov, *Izv. Akad. Nauk. SSSR, Ser. Khim.*, **1971**, 424; *Chem. Abs.*, **75**, 34966 (1971).
[159a] L. A. Kiprianova and A. F. Rekasheva, *Teor. Eksp. Khim.*, **6**, 413 (1970); *Chem. Abs.*, **73**, 119842 (1970).
[159b] J. A. Stubbe and G. L. Kenyon, *Biochemistry*, **10**, 2669 (1971).

logarithms of the dissociation constants of the diprotonated forms of the diamines the point for the amine with $n = 3$ lies above the straight line for the other (three) amines studied, suggesting that it is about seven times more negative than expected. The amines (36) and (37) are likewise about 100 times more reactive than expected. These rate-enhancements were attributed to bifunctional catalysis involving formation of an imine intermediate which reacts with intramolecular proton abstraction. The point for catalysis by the amine with $n = 3$ falls on the plot for the deprotonation of [^{2}H$_1$]isobutyraldehyde; with this compound the Schiff bases probably have a *trans*-structure and so intramolecular deprotonation is not possible.[160]

(36)   (37)

The rate of spontaneous iodination of pyruvic acid is slower than that of methyl and ethyl pyruvate and so the rate-limiting enolization is probably not intramolecularly catalysed by the carboxyl group.[161] This conclusion has been drawn previously.[162]

The solvent isotope effect for the methoxide-catalysed racemization of (+)-3-methyl-1-phenylbutyl phenyl ketone is 2.04. An isotope effect of this magnitude would be expected if the methoxide were hydrogen-bonded to three methanol molecules.[163]

The relative rates of the methoxide-catalysed deuterium-exchange of ketones (38)–(40) are as shown. The slow rate of exchange of (38) compared to *exo*-exchange of (40) was attributed to unfavourable angle strain in the enolate ion of (38) and the transition state for its formation. The difference in the rate of *endo*-exchange of (40) and of (39) was attributed to a similar effect. The difference in the rate of *endo*- and *exo*-exchange of (40) was attributed to the greater non-bonding interaction in the transition state with an adjacent two-carbon bridge than in that with an adjacent one-carbon bridge. The fact that the rate of exchange of (39) and *exo*-exchange of (40) are about equal appears to arise from a cancellation of two factors, unfavourable angle strain on formation of the enolate from (40) and the greater steric strain of the two-carbon bridge of (39) compared to the one-carbon bridge of (40).[164] The relative rates of exchange of (41) and (42) are as shown. This result seems to exclude the difference in the eclipsing interactions in the

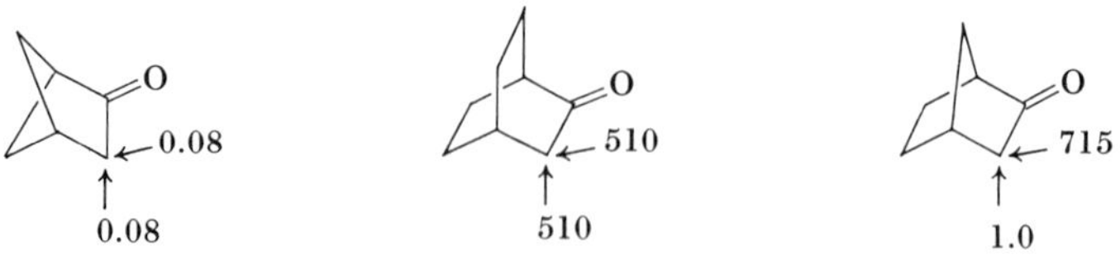

(38) 0.08 / 0.08   (39) 510 / 510   (40) 715 / 1.0

[160] J. Hine, M. S. Cholod, and J. H. Jensen, *J. Am. Chem. Soc.*, **93**, 2321 (1971); cf. *Org. Reaction Mech.*, **1970**, 443.
[161] J. E. Meany, *J. Phys. Chem.*, **75**, 150 (1971).
[162] See *Org. Reaction Mech.*, **1967**, 319.
[163] V. Gold and S. Grist, *J. Chem. Soc.* (B), **1971**, 2282.
[164] S. P. Jindal, S. S. Sohoni, and T. T. Tidwell, *Tetrahedron Letters*, **1971**, 779.

transition states for *exo*- and *endo*-abstraction as significant factors in determining the difference in the rates of *exo*- and *endo*-proton abstraction.[165]

The fraction of acid-catalysed bromination of methyl ketones, $RCH_2-CO-CH_3$ that occurs at the methyl group is higher in methanol than in carbon tetrachloride solution. The possibility that this arises from formation of a ketal was considered but

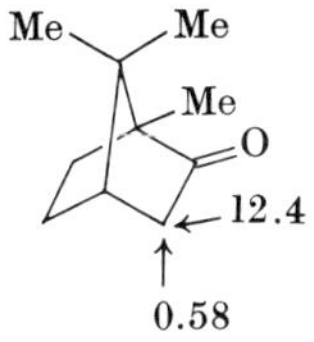

(41) 12.4 0.58

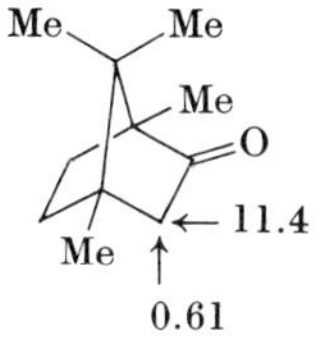

(42) 11.4 0.61

shown to be incorrect. It was suggested instead that methanol acts as a base catalyst and is more effective in facilitating removal of a proton from a methyl group than from a methylene group.[166a]

The kinetics of deuterium exchange in each branch of unsymmetrical ketones has been investigated further.[166b]

The kinetics of the conversion of mandelaldehyde dimer into 2-hydroxyacetophenone in pyridine and aqueous pyridine has been studied. It was thought that reaction proceeds through deprotonation of the open-chain dimer (43). The reaction is catalysed by hydroxylic species (including the substrate itself) and very efficiently by benzamidine.[166c]

The effect of lithium perchlorate on the isomerization of (−)-menthone into (+)-isomenthone in ethereal hydrogen chloride has been studied.[167]

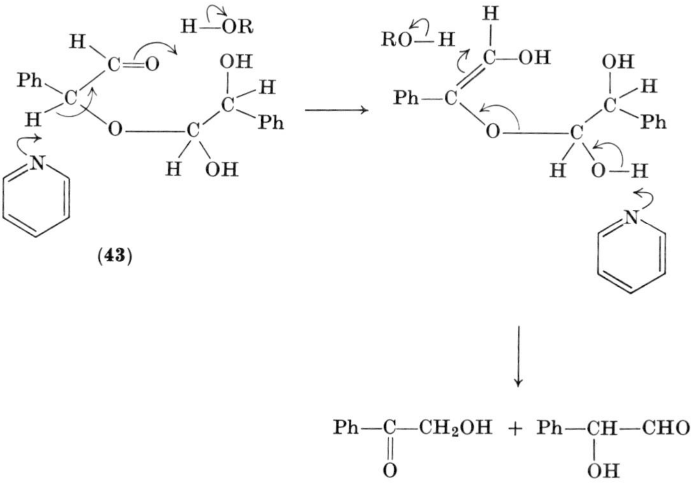

(43)

$Ph-\underset{\underset{O}{\|}}{C}-CH_2OH + Ph-\underset{\underset{OH}{|}}{CH}-CHO$

---

[165] S. P. Jindal and T. T. Tidwell, *Tetrahedron Letters*, **1971**, 783.
[166a] M. Gaudry and A. Morquet, *Tetrahedron*, **26**, 5611, 5617 (1970).
[166b] W. H. Sachs, *Acta Chem. Scand.*, **25**, 2643 (1971); cf. *Org. Reaction Mech.*, **1970**, 441.
[166c] D. W. Griffiths and C. D. Gutsche, *J. Am. Chem. Soc.*, **93**, 4788 (1971).
[167] Y. Pocker and R. F. Buchholz, *J. Am. Chem. Soc.*, **93**, 2905 (1971).

The bromination and iodination of acetone, diethyl ketone, and di-isopropyl ketone have been measured under conditions where enolization and halogenation occur at comparable rates. The second-order rate constants for bromination and iodination of each ketone are almost identical but vary with the ketone.[168]

The reaction of di-isopropyl ketone with hypobromite in aqueous sodium hydroxide shows third-order kinetic as the enolate ion does not all react with hypobromite but partly reverts to ketone. The concentrations of OCl⁻ and OBr⁻ at which the enolate ion reacts at the same rate with these species and with water are ca. 1M for OCl^- and 3.2×10^{-3}M for OBr^-.[169]

The effect of glycerol on the hydrogen chloride-catalysed iodination of acetone,[170] the reaction of β-ketoenol ethers with bromine,[171] the gas-phase equilibrium between bromoacetone and hydrogen bromide with acetone and bromine,[172a] and H–D exchange of ketones on a copper catalyst[172b] have been studied. Other investigations of enolization reactions are described in ref. 173. The rates of reactions of the anions of disulphones with halogens have also been studied.[174]

Epimerization of 5-t-butyl-2-halogenocyclohexanone occurs with exchange or migration of the halogen, the enol being an intermediate.[175] The rearrangement of ethyl α-methyl α-bromoacetoacetate into ethyl α-methyl γ-bromoacetoacetate has also been studied.[176]

In contrast to the homoketonization of nortricyclanol which proceeds with inversion of configuration,[177] the analogous cleavage shown in reaction (2) proceeds with retention of configuration.[178]

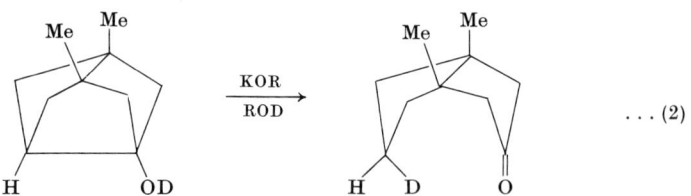

... (2)

R = Buᵗ or DOCH₂CH₂

1-Phenylindan-2-one (**44**) undergoes a methoxide-catalysed reaction with oxygen to yield o-benzoylphenylacetic acid. The reaction was thought to involve formation of

---

[168] J.-E. Dubois and J. Toullec, *Tetrahedron Letters*, **1971**, 3373, 3377.
[169] R. R. Lii and S. I. Miller, *J. Chem. Soc.* (B), **1971**, 2269, 2271.
[170] C. Kalidas, K. V. Raman, and N. Chattanathan, *J. Indian Chem. Soc.*, **48**, 423 (1971).
[171] D. R. Marshall and T. R. Roberts, *J. Chem. Soc.* (B), **1971**, 797.
[172a] K. D. King, D. M. Golden, and S. W. Benson, *J. Chem. Therm.*, **3**, 129 (1971).
[172b] W. R. Patterson and R. L. Burwell, *J. Am. Chem. Soc.*, **93**, 833 (1971); W. R. Patterson, J. A. Roth, and R. L. Burwell, *J. Am. Chem. Soc.*, **93**, 839 (1971).
[173] E. Haslam, M. J. Turner, D. Sargent, and R. S. Thompson, *J. Chem. Soc.* (C), **1971**, 1489; G. Lamaty, A. Roques, and L. Fonzes, *Compt. Rend.* (C), **273**, 521 (1971).
[174] R. P. Bell and B. G. Cox, *J. Chem. Soc.* (B), **1971**, 652; cf. Chapter 3, p. 118.
[175] P. Moreau, A. Casadevall, and E. Casadevall, *Bull. Soc. Chim. France*, **1971**, 3973; M. Lasperas, A. Casadevall, and E. Casadevall, *Bull. Soc. Chim. France*, **1971**, 1814; P. Moreau and E. Casadevall, *Compt. Rend.* (C), **272**, 801 (1971).
[176] M.-M. Mhala and P. S. Bhujang, *Indian J. Chem.*, **8**, 1109 (1970).
[177] See *Org. Reaction Mech.*, **1966**, 326.
[178] W. T. Borden, V. Varma, M. Cabel, and T. Ravindranathan, *J. Am. Chem. Soc.*, **93**, 3800 (1971).

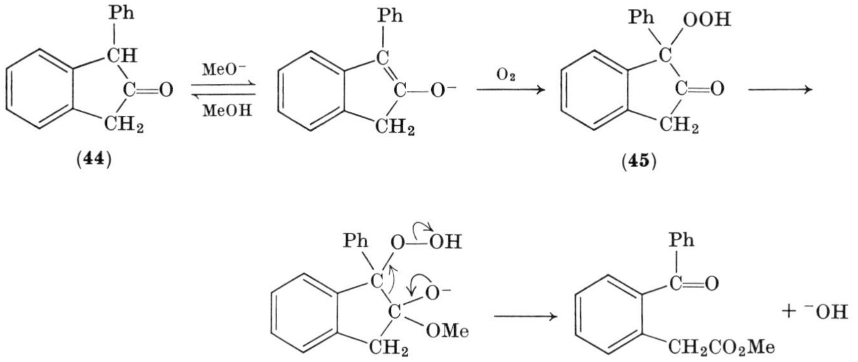

the hydroperoxide (**45**) from the enolate ion and oxygen followed by a slow base-catalysed cleavage.[179]

The alkylation[180] and protonation[181] of enolate ions and the protonation of enamines[182] has been studied.

Measurements of the equilibrium constants for keto–enol equilibria are noted in ref. 183.

More work on the reaction of triose phosphate isomerase with irreversible active-site-directed inhibitors has been reported.[184] Investigations of yeast enolase are noted in ref. 185.

Aldol and Related Reactions

The acid- and base-catalysed condensation of ketones with chloral generally occurs preferentially at the least highly substituted carbon atom.[186]

Schiff bases are intermediates in the Knoevenagel reaction of benzaldehyde and diethyl malonate catalysed by piperidine and amino acids,[187a] and an iminium ion was thought to be an intermediate in the aldolization and crotonization of acetaldehyde catalysed by morpholine.[187b]

[179] F. G. Bordwell and A. C. Knipe, *J. Am. Chem. Soc.*, **93**, 3416 (1971).
[180] H. D. Zook and J. A. Miller, *J. Org. Chem.*, **36**, 1112 (1971); H. O. House, M. Gall, and H. D. Olmstead, *J. Org. Chem.*, **36**, 2361 (1971).
[181] M. Charpentier, J. Sansoulet, and B. Tchoubar, *Compt. Rend.* (C), **273**, 554 (1971).
[182] L. Alais, R. Michelot, and B. Tchoubar, *Compt. Rend.* (C), **273**, 261 (1971).
[183] D. A. Erastov and S. N. Ignat'va, *Izv. Akad. Nauk SSSR, Ser. Khim.*, **1971**, 977; *Chem. Abs.*, **75**, 62900 (1971); Z. Bańkowska and I. Zadrozna, *Rocz. Chem.*, **44**, 2161 (1970); **45**, 183 (1971); A. D. Taneja and K. P. Srivastava, *Chimia*, **25**, 92 (1971); M. Bergon and J. P. Calmon, *Compt. Rend.* (C), **273**, 181 (1971); D. W. Thompson and A. L. Allred, *J. Phys. Chem.*, **75**, 433 (1971); L. Hevesi and A. Bruylants, *Bull. Soc. Chim. France*, **1971**, 4066; R. Haller and W. Hänsel, *Arch. Pharm.*, **304**, 140 (1971); *Chem. Abs.*, **74**, 111376 (1971).
[184] S. G. Waley, J. C. Miller, I. A. Rose, and E. L. O'Connell, *Nature*, **227**, 181 (1970); J. C. Miller and S. G. Waley, *Biochem. J.* **123**, 163 (1971); F. C. Hartman, *Biochemistry*, **10**, 146 (1971); *Org. Reaction Mech.*, **1970**, 446–447.
[185] S. Keresztes-Nayej and R. Orman, *Biochemistry*, **10**, 2506 (1971); J. C. W. Chien and E. W. Westhead, *Biochemistry*, **10**, 3198 (1971).
[186] E. Kiehlmann and P.-W. Loo, *Can. J. Chem.*, **49**, 1588 (1971).
[187a] I. V. Mel'nichenko, T. S. Boiko, and A. A. Yasnikov, *Ukr. Khim. Zhur.*, **37**, 561 (1971); *Chem. Abs.*, **75**, 87759 (1971).
[187b] B. Gaux and P. Le Hénaff, *Compt. Rend* (C), **272**, 1331 (1971).

The following reactions have also been studied: aldol condensation of formaldehyde with acetaldehyde,[188] and condensation of p-nitrobenzaldehyde with p-chlorophenylacetic acid,[189] and of p-dimethylaminobenzaldehyde with α- and γ-methyl-N-alkylpyridinium salts.[190]

The interaction of Mn^{2+} with the zinc-free apoaldolase of yeast has been investigated by ESR and NMR spectroscopy. The binding of the substrates, acetol phosphate, dihydroxyacetone phosphate, fructose diphosphate, and glyceraldehyde 3-phosphate was studied and it appears that they are directly co-ordinated to the manganese.[191a] The reaction of cobalt(II) with apoaldolase and the binding of fructose 1,6-diphosphate to the resulting active enzyme has also been studied.[191b] Other investigations on aldolase are described in ref. 192.

Other Reactions

A detailed investigation of the benzoin reaction in methanol strongly supports Lapworth's mechanism [reactions (3)–(6)].[193] Each step was investigated separately; the rate of addition of CN^- to benzaldehyde by broadening of the NMR signal of the aldehydic proton; the equilibrium constant for the addition by the rapid decrease in absorbance at 320 nm; the formation of the carbanion by measuring the rate of deuterium and tritium exchange of deuteriated and tritiated aldehyde and using the semiempirical equation $k_H/k_T = (k_H/k_D)^{1.44}$; and the formation of benzoin by UV spectroscopy.

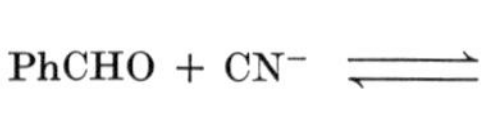

$$PhCHO + CN^- \rightleftharpoons PhCH\begin{matrix}O^-\\CN\end{matrix} \qquad \ldots (3)$$

$$PhCH\begin{matrix}O^-\\CN\end{matrix} \rightleftharpoons \left[PhC\begin{matrix}OH\\CN\end{matrix}\right]^- \qquad \ldots (4)$$

$$\left[PhC\begin{matrix}OH\\CN\end{matrix}\right]^- + PhCHO \longrightarrow Ph-\underset{CN}{\underset{|}{C}}(\overset{HO}{|})-\overset{O^-}{\underset{|}{C}}HPh \qquad \ldots (5)$$

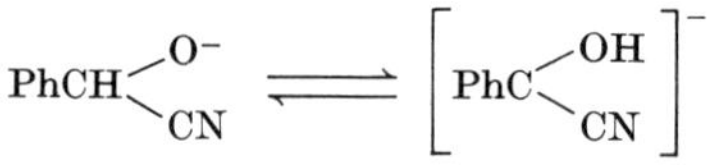

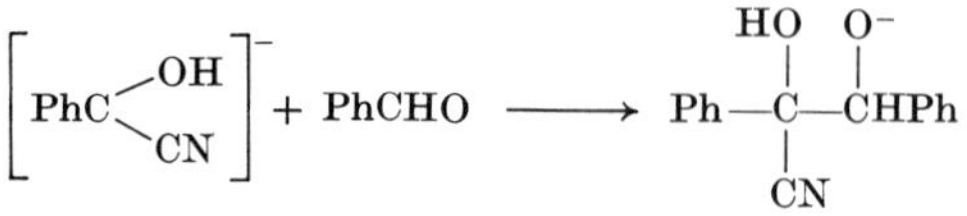

$$\underset{CN}{\underset{|}{Ph C}}(\overset{HO}{|})-\overset{O^-}{\underset{|}{C}}HPh \xrightarrow{fast} PhCOCHOHPh + CN^- \qquad \ldots (6)$$

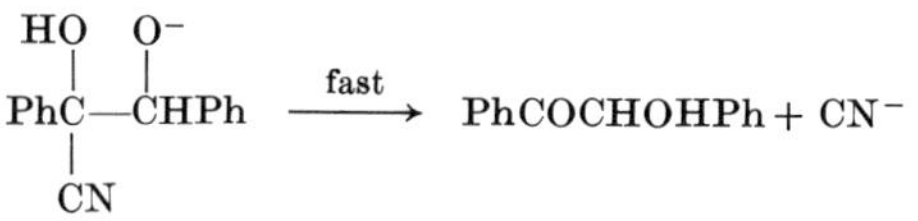

[188] H. Bertilsköld and J.-E. Vik, *Acta Chem. Scand.*, **25**, 2211 (1971).
[189] V. Chandra and V. B. Srivastava, *Indian J. Appl. Chem.*, **33**, 176 (1970); *Chem. Abs.*, **74**, 140449 (1971).
[190] V. E. Kononenko, T. N. Kashtonova, A. K. Sheinkman, and S. N. Baranov, *Organic Reactivity (Tartu)*, **8**, 185 (1971).
[191a] A. S. Mildvan, R. D. Kobes, and W. J. Rutter, *Biochemistry*, **10**, 1191 (1971).
[191b] R. T. Simpson, R. D. Kobes, R. W. Erbe, W. J. Rutter, and B. L. Vallee, *Biochemistry*, **10**, 2466 (1971).
[192] G. M. Lehrer and R. Barker, *Biochemistry*, **10**, 1705 (1971); H. J. Hinz, D. D. F. Shiao, and J. M. Sturtevant, *Biochemistry*, **10**, 1347 (1971); R. S. Lane, A. Shapley, and E. E. Dekker, *Biochemistry*, **10**, 1353 (1971); R. R. Marquardt, *Can. J. Biochem.*, **49**, 647, 658 (1971); C. J. Masters and D. J. Winzor, *Biochem. J.*, **121**, 735 (1971); C. Y. Lai and C. Chen, *Arch. Biochem. Biophys.* **144**, 467 (1971); L. C. Davis, G. Ribereau-Gayon, and B. L. Hurecker, *Proc. Nat. Acad. Sci.*, **68**, 416 (1971).
[193] J. P. Kuebrich, R. L. Schowen, M. Wang, and M. E. Lupes, *J. Am. Chem. Soc.*, **93**, 1214 (1971).

The reactions of anion (**46**) analogous to that proposed as an intermediate in the benzoin condensation were also studied. It was generated by the action of CN⁻ on benzoin in dimethylformamide or dimethyl sulphoxide, and in the absence of other electrophiles it yields (**50**) by reaction with another molecule of benzil. It reacts with aldehydes to yield (**49**), thought to arise via intermediate (**48**), and with ethanol it also yields (**49**) which is probably formed from benzaldehyde generated from (**47**). Competition experiments showed that the rates of reaction of anion (**46**) with these electrophiles were similar. It was suggested that there was little neutralization of the negative charge of the original carbanion in the transition states.[194]

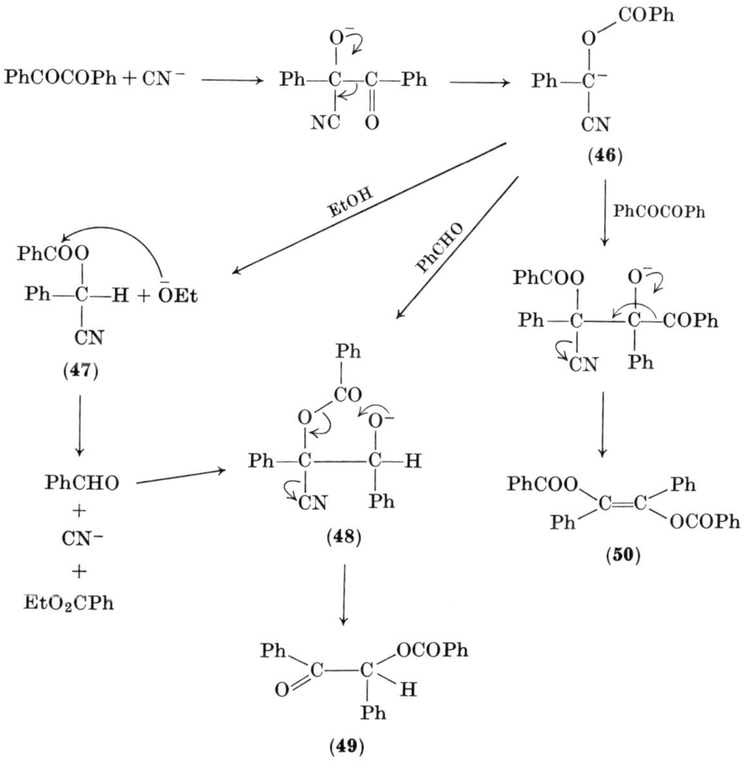

The dependence on R of the rate of reaction of trialkyl phosphites $(RO)_3P$ with benzil are correlated by the Taft equation $\log (k/k_0) = -3.38\sigma^+ + 0.40E_S + 0.03$. The importance of steric effects was interpreted as support for a mechanism involving a rate-determining nucleophilic attack by phosphorus on the carbonyl-carbon of the benzil.[195] The kinetics of the reaction of trimethyl phosphite with aliphatic diketones were also studied and thought to be consistent with a rate-limiting attack by phosphorus on the carbonyl-carbon. The rate constants were correlated by the Taft equation $\log (k/k_0) = 1.08 E_S + 0.012$ with steric effects dominant.[196]

---

[194] J. P. Kuebrich and R. L. Schowen, *J. Am. Chem. Soc.*, **93**, 1220 (1971).
[195] Y. Ogata and M. Yamashita, *J. Org. Chem.*, **36**, 2584 (1971); *Tetrahedron*, **27**, 2725 (1971); cf. *Org. Reaction Mech.*, **1970**, 450–451.
[196] Y. Ogata and M. Yamashita, *Tetrahedron*, **27**, 3395 (1971).

The variation in the ρ-value for the reactions of a series of aryl ketones with nucleophiles has been interpreted as indicating that there is a wide variation in transition-state structure from reactant-like for attack by hydroxylamine to product-like for attack by borohydride.[197] Consideration of the rates of attack of nucleophiles on 4-t-butylcyclohexanone and adamantanone was thought to support this proposal.[198]

The deuterium isotope effect on the rate constant for the reaction of ketones and the α-deuteriated analogues with bisulphite has values $k_H/k_D > 1$. This was thought to arise mainly because hyperconjugative stabilization of the deuteriated ketone was less than that of the proton-containing one.[199]

The addition of Grignard reagents to five-membered cyclic β-hydroxy ketones, which occurs stereospecifically *trans*, probably involves a chelate structure such as (**51**).[200a] The stereochemistries of other addition reactions to ketones are reported in ref. 200b.

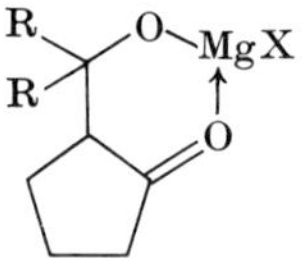

(**51**) 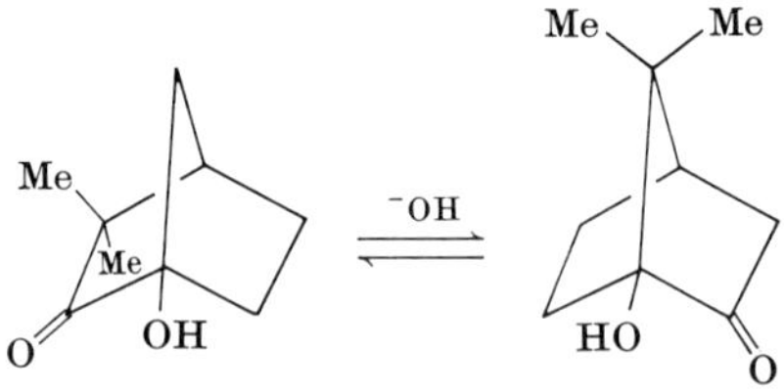   (**52**)   (**53**)

The Schmidt reactions of hexan-2-one[201] with substituted acetophenones[202] or hydrazoic acid have been studied. The dependence of the rate on the concentration of sulphuric acid and the deuterium isotope effects were considered to be consistent with the mechanism shown in reactions (7) and (8).

The interconversion of the ketols (**52**) and (**53**) has been studied.[203,204]

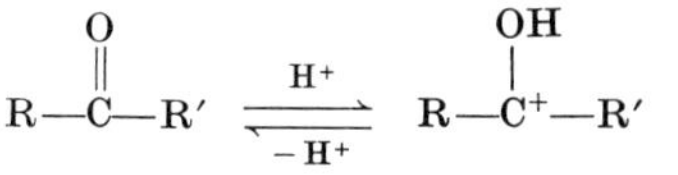

$$R-\overset{O}{\underset{\|}{C}}-R' \underset{-H^+}{\overset{H^+}{\rightleftharpoons}} R-\overset{OH}{\underset{|}{C^+}}-R' \qquad \ldots (7)$$

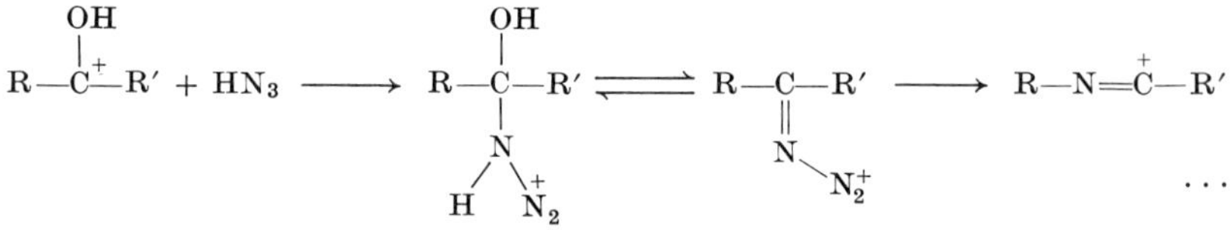

$$R-\overset{OH}{\underset{|}{C^+}}-R' + HN_3 \longrightarrow R-\overset{OH}{\underset{\underset{H}{\overset{|}{N}}\diagdown N_2^+}{\overset{|}{C}}}-R' \rightleftharpoons R-\overset{OH}{\underset{\underset{N_2^+}{\overset{\|}{N}}}{\overset{|}{C}}}-R' \longrightarrow R-N\overset{+}{=}C-R' \qquad \ldots (8)$$

[197] P. Geneste, G. Lamaty, and J. P. Roque, *Tetrahedron Letters*, **1970**, 5007.
[198] P. Geneste, G. Lamaty, C. Moreau, and J. P. Roque, *Tetrahedron Letters*, **1970**, 5011.
[199] P. Geneste, G. Lamaty, and J. P. Roque, *Tetrahedron*, **27**, 5539 (1971).
[200a] E. Ghera and S. Shoua, *Chem. Comm.*, **1971**, 398.
[200b] J. P. Battioni and W. Chodkiewicz, *Bull. Soc. Chim. France.*, **1971**, 1824; T. Matsumoto and K. Fukui, *Bull. Chem. Soc. Japan*, **44**, 1090 (1971); J. L. Namy, *Compt. Rend.* (C), **272**, 1334 (1971); F. Rocquet, A. Sevin, and W. Chodkiewicz, *Compt. Rend.* (C), **272**, 417 (1971); D. Guillerm-Dron, M. L. Capmau, and W. Chodkiewicz, *Compt. Rend.* (C), **272**, 486 (1971); M. S. Biernbaum and H. S. Mosher, *J. Org. Chem.*, **36**, 3168 (1971).
[201] G. I. Koldobskii, G. F. Tereshchenko, and L. I. Bagal, *Zhur. Org. Khim.*, **6**, 2395–2633 (1970); *J. Org. Chem. USSR*, **6**, 2407–2644 (1970).
[202] G. F. Tereshchenko, G. I. Koldobskii, A. S. Enin, and L. I. Bagal, *Organic Reactivity (Tartu)*, **7**, 1102 (1971); *Chem. Abs.*, **75**, 19413 (1971).
[203] A. Nickon, T. Nishida, J. Frank, and R. Muneyuki, *J. Org. Chem.*, **36**, 1075 (1971).
[204] J. V. Paukstelis and D. N. Stephens, *Tetrahedron Letters*, **1971**, 3549.

The following reactions have also been investigated: condensation of phenol and formaldehyde;[205] retro-Dieckman cleavage of some steroidal ketones;[206] hydrolysis of pyrylium salts;[207] decomposition of Mannich bases;[208] and reduction of ketones by magnesium.[209]

There have been several spectroscopic investigations of the reaction of aldehydes and ketones with Brønsted–Lowry acids[210] and Lewis acids.[211]

[205] M. I. Siling, B. Y. Akselrod, and I. V. Adorova, *Organic Reactivity (Tartu)*, **7**, 861, 888 (1971); M. I. Siling and B. Y. Akselrod, *Organic Reactivity (Tartu)*, **7**, 873 (1971).
[206] E. Fujita and Y. Nagao, *J. Chem. Soc.* (C), **1971**, 2902.
[207] G. Salvadori and A. Williams, *J. Am. Chem. Soc.*, **93**, 2727 (1971); A. Williams, *J. Am. Chem. Soc.*, **93**, 2733; see also Chapter 6, p. 193.
[208] J. A. Mollica, J. B. Smith, I. M. Nunes, and H. K. Govan, *J. Pharm. Sci.*, **59**, 1770 (1970).
[209] J.-F. Fauvarque, *Compt. Rend.* (C), **272**, 1053 (1971).
[210] D. M. Brouwer and J. A. van Doorn, *Rec. Trav. Chim.*, **90**, 535, 1010 (1971); E. Urbas and U. Haldna, *Organic Reactivity (Tartu)*, **7**, 1247 (1970); M. Rabinovitz and A. Elleneweig, *Tetrahedron Letters*, **1971**, 4439; W. V. Laurushin, N. N. Verkhovod, L. M. Grin, R. I. Pogonina, V. P. Izvehov, and L. M. Kutsenko, *Dokl. Akad. Nauk SSSR*, **192**, 1284 (1970); *Chem. Abs.*, **73**, 119819 (1970).
[211] F. Filippini and B.-P. Susz, *Helv. Chim. Acta*, **54**, 1175 (1971); G. Guiheneuf, C. Laurence, and B. Wojtkowiak, *Bull. Chim. Soc. France*, **1971**, 1157.

CHAPTER 12

Reactions of Acids and their Derivatives

B. CAPON

Chemistry Department, Glasgow University

Carboxylic Acids	419
Tetrahedral Intermediates	419
Intermolecular Catalysis	423
Reactions in hydroxylic solvents	423
Reactions in aprotic solvents	428
Intramolecular Catalysis and Neighbouring-group Participation	431
Association-prefaced Catalysis	441
Metal-ion Catalysis	442
Enzymic Catalysis	443
Serine proteinases	443
Thiol proteinases	450
Acid proteinases	450
Metallo-proteinases	452
Esterases	452
Other enzymes	453
Decarboxylation	453
Other Reactions	453
Non-carboxylic Acids	457
Phosphorus-containing Acids	457
Non-enzymic reactions	457
Enzymic reactions	461
Sulphur-containing Acids	463
Nitrous and Nitric Acid	465

Carboxylic Acids[1-3]

Tetrahedral Intermediates

There have been two interesting investigations of the hydrolysis of anilides in aqueous solution. Kershner and Schowen[4] have studied the hydrolysis of 2,2,2-trifluoro-N-methylacetanilides, and Pollack and Bender[5] studied that of p-formyl- and p-nitro-acetanilide. Under certain conditions breakdown of the tetrahedral intermediate is the rate-limiting step for both series. The first-order rate constant, k_0, for the hydrolysis of the trifluoro-N-methylacetanilides depends on hydroxide ion concentration according

---

[1] M. L. Bender, "Mechanisms of Homogeneous Catalysis from Protons to Proteins," Interscience, New York, 1971.
[2] K. Yates, "Kinetics of Ester Hydrolysis in Concentrated Acid," *Accounts Chem. Res.*, **4**, 136 (1971).
[3] Z. Rappoport (ed.), "The Chemistry of the Cyano Group," Interscience, London, 1970.
[4] L. D. Kershner and R. L. Schowen, *J. Am. Chem. Soc.*, **93**, 2014 (1971).
[5] R. M. Pollack and M. L. Bender, *J. Am. Chem. Soc.*, **92**, 7191 (1970).

to equation (1) and the mechanism may be written as shown in reaction (2) where the pathway having k_3' probably involves a dianionic tetrahedral intermediate. The ρ-values for the variation of $k_1'k_2'/k_{-1}'$ (written k_1 in reference 4) and $k_1'k_3'/k_{-1}'$ (written k_2 in reference 4) with structure change from ca. 0 when the expelled N-methylaniline has

$$k_0 = \frac{k_1'(k_1'k_2'/k_{-1}' + k_1'k_3'[^-\text{OH}]/k_{-1}')[^-\text{OH}]}{k_1' + k_1'k_2'/k_{-1}' + k_1'k_3'[^-\text{OH}]/k_{-1}'} \quad \ldots(1)$$

$$R-C\underset{NR'Ar}{\overset{O}{\diagdown}} \underset{k_{-1}'}{\overset{k_1'}{\rightleftharpoons}} R-\underset{OH}{\overset{O^-}{\underset{|}{C}}}-NR'Ar \overset{k_{-2}'}{\underset{k_3'[^-\text{OH}]}{\rightleftharpoons}} \text{PRODUCTS} \quad \ldots(2)$$

$$\begin{array}{c} \text{O}^-\;\text{Me} \\ |\;\;\;|\overset{\delta+}{}\;\;\;\overset{\delta-}{} \\ \text{CF}_3-\text{C}-\text{N}\cdots\text{H}\cdots\text{B} \\ |\;\;\;| \\ \text{O}^-\;\text{Ar} \\ (\mathbf{1}) \end{array} \qquad \begin{array}{c} \text{O}^-\;\text{Me} \\ |\;\;\;| \\ \text{CF}_3-\text{C}\cdots\text{N}\sim\!\!\sim\!\!\text{H}-\text{B} \\ |\;\;\;| \\ \text{O}^-\;\text{Ar} \\ (\mathbf{2}) \end{array} \qquad \begin{array}{c} \text{O}\;\;\;\text{H} \\ \overset{\delta-}{\|}\;\;\overset{\delta-}{|} \\ \text{CH}_3-\text{C}\cdots\text{N} \\ |\;\;\;| \\ \text{O}\;\;\;\text{Ar} \\ (\mathbf{3}) \end{array}$$

$pK_b < 9$ to 0.3, and 0.7 when it has $pK_b > 9$. It was suggested that the activated complex for breakdown of the tetrahedral intermediate to products involves simple proton-transfer to nitrogen from a general base, e.g. (**1**), when $pK_b < 9$, but heavy-atom reorganization with solvating catalyst, e.g. (**2**), with proton-transfer completed as a subsequent step when $pK_b > 9$. Support for this proposal comes from the sharp change in the solvent isotope effect on $k_1'k_3'/k_{-1}'$ which falls in the range $k_1'k_3'/k_{-1}'(\text{H}_2\text{O})/k_1'k_3'/k_{-1}'(\text{D}_2\text{O}) = 1.2$–$1.5$ when the leaving group has $pK_b < 9$ and 0.7–0.9 when $pK_b > 9$. The reaction is still (kinetically) general base-catalysed when the leaving N-methylaniline has electron-withdrawing substituents.[4] The first-order rate constant (corrected for ionization) for the hydrolysis of p-formyl- and p-nitro-acetanilide also depends on the concentration of hydroxide ion according to equation (1) but with k_2' zero. The ρ-value for the term $k'k_3'/k_{-1}'$ was estimated at ca. 11 and the isotope effect for this term for the nitroacetanilide was 0.65. It was suggested that the dianionic tetrahedral intermediate expels the anilide anion with transition state (**3**).[5]

The alkaline hydrolysis of N-methylnitroacetanilides,[6] N,N-diphenylformamide,[7] N,N-dimethylformamide,[8] amino-substituted-acid anilides,[9] and amides of aliphatic carboxylic acids[10] have been studied. The rate constants for the last of these reactions have been correlated by the full Taft equation with $\rho^* = 1.67$, $\delta = 0.985$, and $\lambda = -0.573$. There have been several investigations[11] of the alkaline hydrolysis of barbiturates.

[6] M. I. Vinnik and A. K. Pochikyan, *Izv. Akad. Nauk SSSR, Ser. Khim.*, **1971**, 1213, 1217; *Chem. Abs.*, **75**, 87780, 87802 (1971).
[7] S. Langlois and A. Broche, *Bull. Soc. Chim. France*, **1971**, 3432.
[8] E. Buncel, S. Kesmarky, and E. A. Symons, *Chem. Comm.*, **1971**, 120.
[9] S. O. Eriksson and U. Meresaar, *Acta Chem. Scand.*, **25**, 2688, 2697 (1971).
[10] P. D. Bolton and G. L. Jackson, *Austral. J. Chem.*, **24**, 969 (1971); see also ref. 45.
[11] M. Melzacka and W. Kahl, *Roczn. Chem.*, **44**, 2261, 2487 (1970); E. R. Garrett, J. T. Bojarski, and G. J. Yakatam, *J. Pharm. Sci.*, **60**, 1145 (1971); H. V. Maudling and M. A. Zoglio, *J. Pharm. Sci.*, **60**, 40 (1971).

The kinetics of the alkaline cleavage of 3-phenylpropynal to phenylacetylene and sodium formate also follow equation (1). Reaction was thought to proceed through the mono- and di-anionic forms of the hydrate (**4a**) and (**4b**), the former being present at high and the latter at low concentrations. Most of the reaction was thought[12] to proceed through the dianionic form.

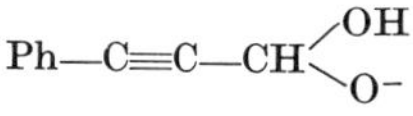

$$Ph-C\equiv C-CH\begin{pmatrix}OH\\O^-\end{pmatrix} \qquad Ph-C\equiv C-CH\begin{pmatrix}O^-\\O^-\end{pmatrix}$$

(**4a**) 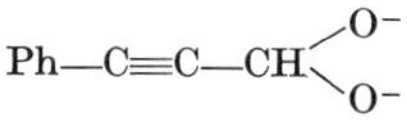 (**4b**)

Hydrolysis of the acyl–phosphorus compound (**5**) in phosphate buffers in aqueous dioxan has been studied. Kinetic general base-catalysis was observed and the reaction was thought to involve reversible formation of a tetrahedral intermediate that underwent a general base-catalysed expulsion of a phosphorus-containing anion;[13] in our opinion it is more likely that this group is expelled with general acid-catalysis.

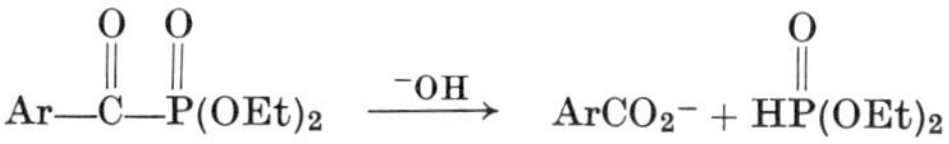

$$Ar-\overset{O}{\overset{\|}{C}}-\overset{O}{\overset{\|}{P}}(OEt)_2 \xrightarrow{^-OH} ArCO_2^- + H\overset{O}{\overset{\|}{P}}(OEt)_2$$

(**5**)

The acid-catalysed decarboxylation of phosphoroformic acid has a positive entropy of activation (+7.9 cal deg^{-1} mol^{-1}) and an isotope effect $k(D_2O)/k(H_2O) = 1.78$. The plot of log k against log $[HClO_4]$ has a slope of 1 and the Bunnett w-value is +10. The mechanism of reaction (3a) was proposed.[14]

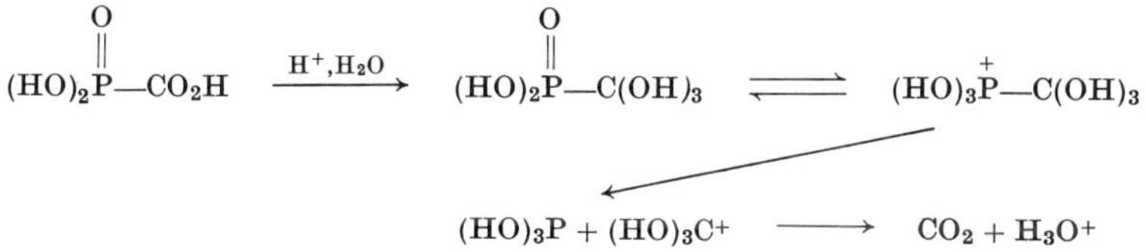

$$(HO)_2\overset{O}{\overset{\|}{P}}-CO_2H \xrightarrow{H^+, H_2O} (HO)_2\overset{O}{\overset{\|}{P}}-C(OH)_3 \rightleftharpoons (HO)_3\overset{+}{P}-C(OH)_3$$

$$(HO)_3P + (HO)_3C^+ \longrightarrow CO_2 + H_3O^+$$

...(3a)

The reactions of mercaptoethanol and mercaptoacetate with N-acetylimidazole are catalysed by imidazole; but the rate shows a non-linear dependence on the concentration of imidazole, as at high concentrations the reaction is slower than expected from linear extrapolation of the values at low concentration. This was explained by the mechanism (3b), the rate-limiting step being formation of the tetrahedral intermediate at high

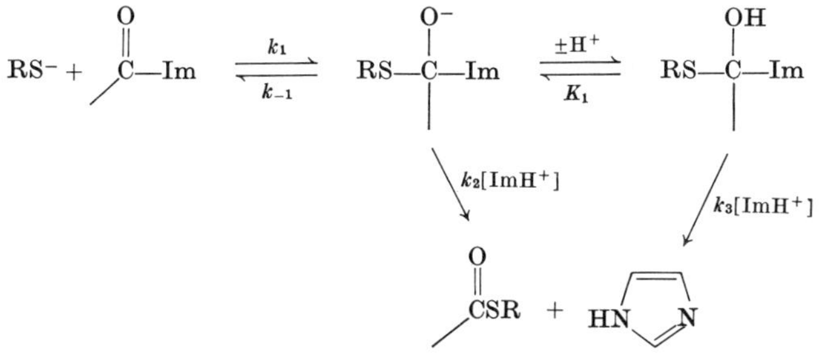

$$RS^- + \overset{O}{\overset{\|}{C}}-Im \underset{k_{-1}}{\overset{k_1}{\rightleftharpoons}} RS-\overset{O^-}{\underset{|}{C}}-Im \underset{K_1}{\overset{\pm H^+}{\rightleftharpoons}} RS-\overset{OH}{\underset{|}{C}}-Im$$

$$\searrow k_2[ImH^+] \qquad \swarrow k_3[ImH^+]$$

$$\overset{O}{\overset{\|}{C}}SR + HN\!\!\diagup\!\!\diagdown\!\!N$$

...(3b)

[12] J. Hine and G. F. Koser, *J. Org. Chem.*, **36**, 1348 (1971).
[13] W. Jugelt, S. Andreae, and G. Schubert, *J. Prakt. Chem.*, **313**, 83 (1971).
[14] S. Warren and M. R. Williams, *J. Chem. Soc.* (B), **1971**, 618.

concentrations and breakdown at low concentrations. No change in rate-limiting step could be detected in the uncatalysed reaction of acetylimidazole and ethanethiol and it was concluded that the catalysed and the uncatalysed reaction proceed through independent pathways. The latter was thought to involve formation of a tetrahedral intermediate from the thiol anion and the acetylimidazolinium cation, followed by rapid expulsion of imidazole.[15]

The aminolysis of the acetylimidazolinium ion has a β-value of 1.0. The plot is linear over 6 pK_a units and the points for amines with pK_a 3 units less than imidazole lie on the plot defined by the more basic amines. It was argued that if reaction proceeded through a tetrahedral intermediate a change in rate-limiting step would be expected when the pK_a of the attacking amine is approximately equal to the pK_a of imidazole as the less basic species would be expelled preferentially. This is not necessarily correct, however, since an important factor in deciding which group is expelled will be the ability of the remaining group to release electrons, as symbolized by the arrows in (6) and (7) and a

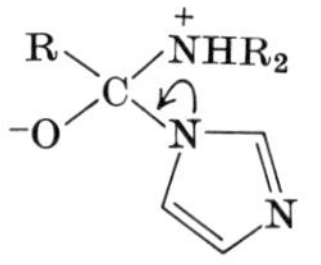

(6) 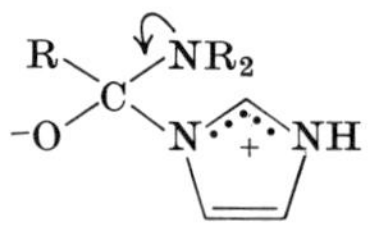   (7)

simple amino-group would be expected to do this more easily than an imidazole group, owing to the aromaticity of the imidazole ring. Also, if the amino-group in (6) leaves in its protonated form, as the electronic requirements for protonation are the reverse of those for an unprotonated leaving amino-group, the sensitivity of the rate of expulsion of R_2NH^+ to changes in R may be small. The reactions of acetylimidazole with oxygen nucleophiles were also studied.[16]

Details of their investigation of the aminolysis of the acetylimidazolinium ion and the 1-acetyl-3-methylimidazolinium ion have been published by Oakenfull, Salvesen and Jencks.[17]

Treatment of compound (8; R = tri-isopropylphenyl) in methylene dichloride solution with triethylamine leads to a yellow colour that was attributed to formation of a salt (9). It was therefore considered that the colour formed when (8) or (10) where R = Me is treated with triethylamine arose from (9) and not from a tetrahedral intermediate.[18]

The reaction of acetyl chloride with 2-methyl-2-thiazoline (11) in acetonitrile yields the N-acetylated ketene S,N-acetal [a 2-methylenethiazolidine] (12). Hydrolysis of this

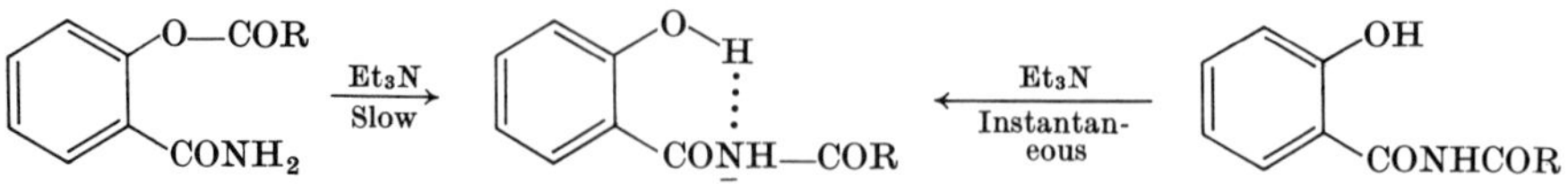

(8)   (9)   (10)

[15] W. P. Jencks and K. Salvesen, *J. Am. Chem. Soc.*, **93**, 1419 (1971).
[16] D. G. Oakenfull and W. P. Jencks, *J. Am. Chem. Soc.*, **93**, 178 (1971).
[17] D. G. Oakenfull, K. Salvesen, and W. P. Jencks, *J. Am. Chem. Soc.*, **93**, 188 (1971); see *Org. Reaction Mech.*, **1970**, 458.
[18] P. L. Russell, R. M. Topping, and D. E. Tutt, *J. Chem. Soc.* (B), **1971**, 657.

yields *N,S*-diacetylcysteine amine (**14**). There is an intermediate in this reaction that is probably the *N,N*-diacetyl derivative (**13**).[19]

The formation of the thioamides (**15b**) from the thiazolium salts (**15a**) has been studied.[20]

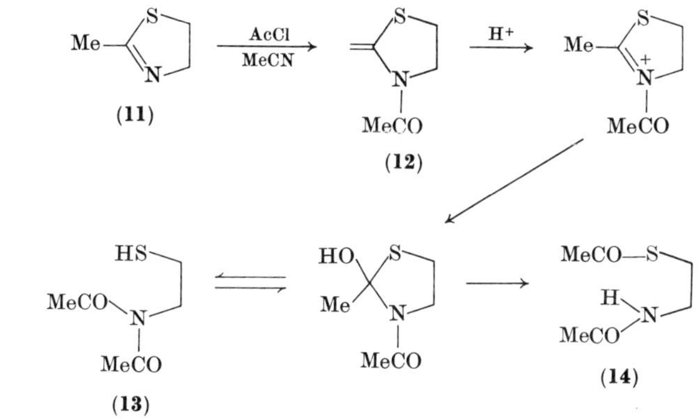

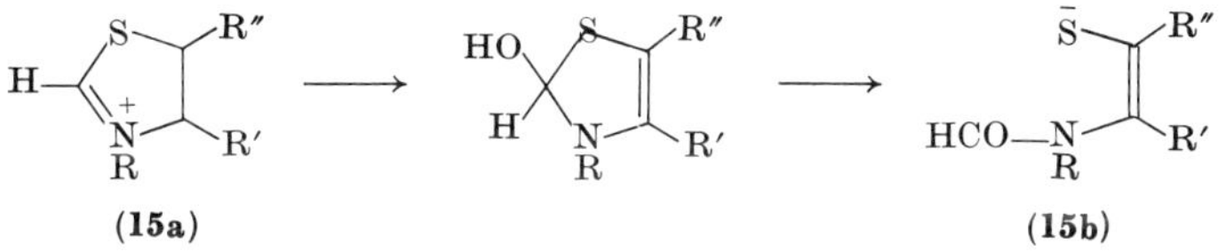

Substituent effects on the hydrolysis of *N*-aryl-*N'*-ethylformamidines[21] and amine-exchange of amidines[22a] have been investigated.

The rate of hydrolysis of the lysidinium cation (**16a**) in sulphuric acid is proportional to the concentration of acid up to 10M. It was proposed that the dication (**16b**) was the reactive species and that this underwent a rate-limiting attack by water.[22b]

$$CH_3\overset{O}{\overset{\|}{C}}NHCH_2CH_2\overset{+}{N}H_3 \longrightarrow CH_3CO_2H + H_3\overset{+}{N}CH_2CH_2\overset{+}{N}H_3$$

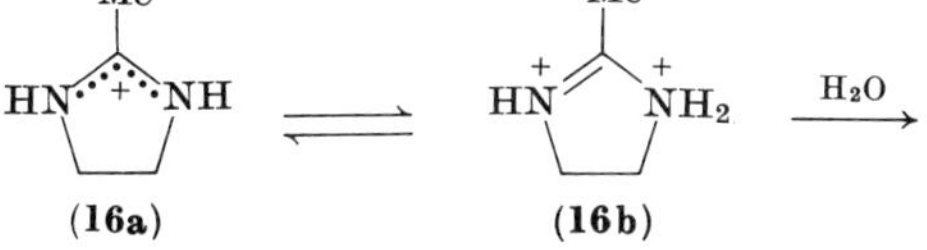

Intermolecular Catalysis

Reactions in hydroxylic solvents. The Brønsted coefficient for the reaction of substituted benzotriazole anions (**17**) with *p*-nitrophenyl acetate is 0.5. This suggests that there is

[19] L. V. Grobovsky and G. L. Schmir, *Tetrahedron*, **27**, 1185 (1971).
[20] Y. Asahi and M. Nagaoka, *Chem. Pharm. Bull.*, **19**, 1017 (1971).
[21] R. H. De Wolfe, *J. Org. Chem.*, **36**, 162 (1971).
[22a] J. Oszczapowicz and R. Orliński, *Roczn. Chem.*, **44**, 2327 (1970); **45**, 103 (1971).
[22b] P. Haake and J. W. Watson, *J. Org. Chem.*, **36**, 4063 (1970).

relatively little bond-forming in the transition state, as would be expected for a good nucleophile. From pH 2 to 9 the hydrolysis of acetylbenzotriazole follows the rate law:

$$-d[\text{Acetylbenzotriazole}]/dt = \{k_1 + k_2[^-\text{OH}]\}[\text{Acetylbenzotriazole}]$$

Acid-catalysis only becomes important below pH 2. This contrasts with the pH-dependence of the rate of hydrolysis of acetylimidazole, for which acid-catalysis is important at much higher pH's. The aminolysis of acetylbenzotriazole by glycine ethyl ester was also studied.[23]

The Brønsted coefficient for the reaction of the anions of a series of benzohydroxamic acids with p-nitrophenyl acetate is 0.7 and that for the reaction of a series of N-methylbenzohydroxamic acids is 0.23; it was thought that the nucleophilic species were (18) and (19) and not analogous.[24] The reaction of hydroxamic acids with p-nitrophenyl thiolacetate have been studied.[25a] A detailed analysis of substituent effects in the imidazole-catalysed hydrolysis of aryl acetates has been published.[25b]

(17)   (18)   (19)

The rate of hydrolysis of acetylimidazole is increased by the presence of formaldehyde owing, it was thought, to formation of N-acetyl-N'-(hydroxymethyl)imidazolinium ion. The rate of the imidazole-catalysed hydrolysis of p-nitrophenyl acetate is decreased by the presence of formaldehyde. It was proposed that the aldehyde converts the imidazole into its N-(hydroxymethyl) adduct[26] (see also ref. 171). The hydrolysis of N-alkyl-N'-acetylimidazolinium ions and 1-acetyl-4-(dialkyamino)pyridinium ions have been studied.[27a]

Hydrolysis of the nitrile group of 1-benzyl-3-cyanopyridinium bromide is catalysed by mercaptoethanol. It was shown that a parasitic equilibrium is set up [reaction (4)] and that the mercaptoethanol attacks the nitrile group to form the thioimide which yields the amide and thioester; the latter is finally hydrolysed to the acid [reaction (5)].[27b]

The hydrolysis of 1,3-diphenylurea in aqueous dioxan catalysed by triethylenediamine shows complex kinetics with terms that are of the second order in diphenylurea. The triethylenediamine was thought to act as a nucleophilic catalyst.[28]

Hydrolysis and aminolysis of a series of formate esters has been studied. It was concluded from the occurrence of a long-range coupling between the formyl-proton and the fluorine atoms of trifluoroethyl formate that this compound existed preferentially in the cis-conformation (20). Since this ester is not especially reactive it was questioned whether acyclic cis-esters are more reactive than acyclic $trans$-esters. The aminolysis

---

[23] M. Ravoux, M. Laloi-Diard, and M. Vilkas, *Tetrahedron Letters*, **1971**, 4015.
[24] M. Dessolin and M. Laloi-Diard, *Bull. Soc. Chim. France*, **1971**, 2946.
[25a] M. Dessolin, M. Laloi-Diard, and M. Vilkas, *Bull. Soc. Chim. France*, **1970**, 2573.
[25b] A. Williams and R. A. Naylor, *J. Chem. Soc.* (B), **1971**, 1967.
[26] C. J. Martin, N. B. Oza, and M. A. Marini, *Eur. J. Biochem.*, **20**, 276 (1971).
[27a] E. Guibe-Jampel and M. Wakselman, *Bull. Soc. Chim. France*, **1971**, 2554.
[27b] C. Zervos and E. H. Cordes, *J. Org. Chem.*, **36**, 1661 (1971).
[28] Y. Furuya, K. Itoho, and S. Fukutaka, *Bull. Chem. Soc. Japan*, **43**, 3846 (1970).

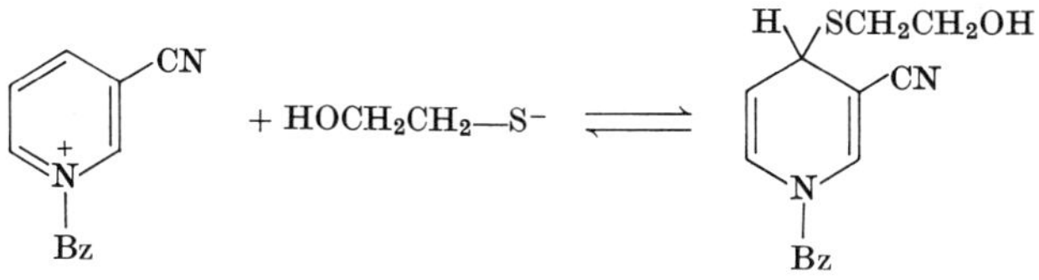

$$\text{(4)}$$

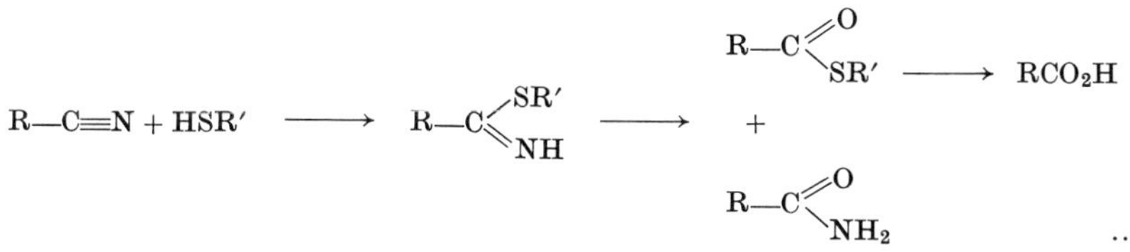

R—C≡N + HSR′ ⟶ R—C(SR′)(NH) ⟶ R—C(=O)SR′ + R—C(=O)NH$_2$ ⟶ RCO$_2$H   ...(5)

R = 3-(N-Benzylpyridinium)
R′ = 2-Hydroxyethyl

of N-acetyl-O-formylserine amide by glycine amide is general acid- and general base-catalysed by imidazole. Possible mechanisms are those symbolized by (**21**) and (**22**).[29a]

The aminolysis in methanol and the basic hydrolysis of the O-ethyl bromocarbonate ion has been studied.[29b]

There have been several measurements of the equilibrium constants for formation and for interconversion of amides. Jencks and his co-workers[30] have measured the free energies of acetyl-transfer from substituted acetanilides to aminoazobenzenesulphonic acid, using acetyl CoA-arylamine N-acetyltransferase to catalyse the reaction (cf. p. 453).

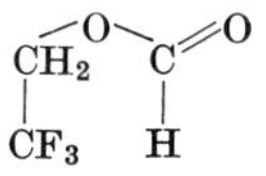

(**20**)

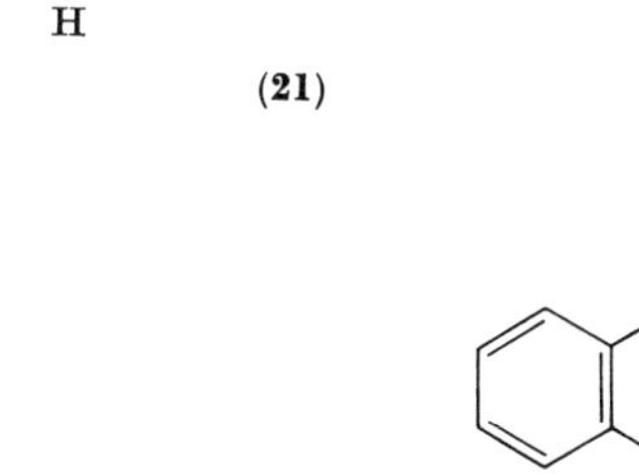

(**21**)

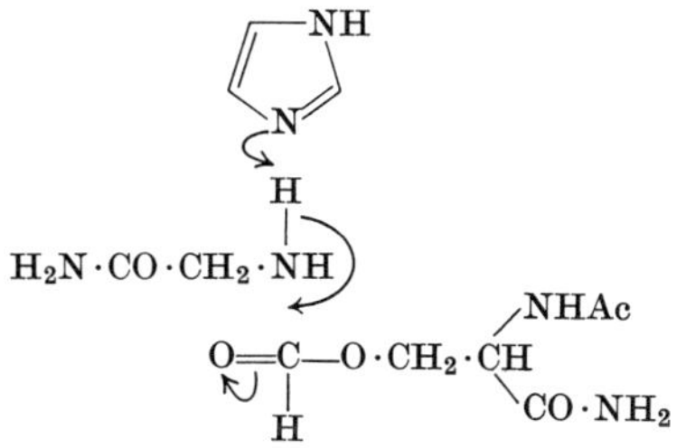

(**22**)

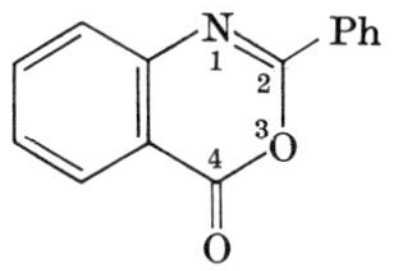

(**23**)

[29a] G. M. Blackburn and H. L. H. Dodds, *J. Chem. Soc.* (B), **1971**, 826; cf. M. Oki and N. Nakanishi, *Bull Chem. Soc. Japan*, **43**, 2558 (1970).
[29b] L. J. Malone, B. D. Hoewe, and R. M. Manley, *Inorg. Chem.*, **10**, 930 (1971).
[30] W. P. Jencks, B. Schaffhausen, K. Tornheim, and H. White, *J. Am. Chem. Soc.*, **93**, 3917 (1971).

Fersht and Requena[31] have measured the equilibrium constants for the formation of amides of formic acid derived from some good nucleophiles by measuring the rates of the forward and the reverse reactions; the equilibrium constants for the formation of amides derived from poorer nucleophiles were obtained from the equilibrium constants for amide-exchange with formohydroxamic acid. The equilibrium constant for the formation of formohydroxamic acid from hydroxylamine and formic acid is less than predicted from the pK_a of hydroxylamine and a linear plot of the logarithm of the formation constant against pK_a for other amines. This suggests that product stability is not an important factor in the origin of the α-effect. In contrast, Dixon and Bruice[32] have reported that the fast reaction of Malachite Green with hydrazine parallels a high formation-equilibrium constant for production of the Malachite Green–hydrazine adduct.

The equilibrium constant for the hydrolysis of the p-nitroanilide of N-acetyl-L-phenylalanine has been measured.[33]

Fersht[34] has calculated the rate constants for the alcoholysis, thiolysis, and phenolysis of amides from the equilibrium constants and the rate constants of the reverse reactions. In contrast to the alcoholysis of esters where the rate increases with decreasing pK_a of the leaving group, the rate of alcoholysis of amides increases with the pK_a of the leaving group. This arises from the need for acid-catalysis for departure of an amino group.

Tracer experiments[35] have shown that the hydrolysis of 2-phenyl-3,1-benzoxazin-4-one (**23**) occurs with attack at C-2 in acid solution and at C-4 in basic solution. The ρ-value of 0.71 for the reaction of 2-arylbenzoxazinones with hydroxide ions is consistent with this; a much larger value would have been expected if attack occurred at C-2. The plot of k_{obs} against concentration of acid passes through a maximum; this was attributed to complete protonation of the benzoxazinone and a decrease in the activity of water with increasing acid concentration.

A detailed investigation of the hydrolysis of substituted benzohydroxamic acids has been reported.[36] With these compounds, also, plots of k_{obs} against acid concentration pass through maxima when the hydroxamic acid is fully protonated. The ρ-value for the hydrolysis of the fully protonated forms is +0.64, consistent with an A2 mechanism, but Bunnett and Olsen's ϕ-value falls in the range 0.84–1.1, which suggests that water acts as a proton-transfer agent as well as a nucleophile.

O-Methylisourea reacts in sulphuric acid of concentration less than 40% by attack of water on the acyl- and methyl-carbon atoms of the monoprotonated form. In more concentrated acid, hydrolysis was thought to occur by attack of water on the acyl-carbon atom of the diprotonated form (**24**) and in 60–85% sulphuric acid O-methylisourea, N,N,N',N'-tetramethylurea and urea were thought to be hydrolysed by the same mechanism. In very concentrated acid urea is sulphonated before decomposition.[37] The decomposition of nitrourea in acidic solutions has also been studied.[38]

The acid-catalysed hydrolysis of methyl isocyanides yields N-methylformamide

[31] A. R. Fersht and Y. Requena, *J. Am. Chem. Soc.*, **93**, 3499 (1971).
[32] J. E. Dixon and T. C. Bruice, *J. Am. Chem. Soc.*, **93**, 3248 (1971).
[33] E. D. Djachenko, L. V. Kozlov, and V. K. Antonov, *Biokhimya*, **36**, 981 (1971).
[34] A. R. Fersht, *J. Am. Chem. Soc.*, **93**, 3504 (1971).
[35] A. Williams and G. Salvadori, *J. Chem. Soc.* (B), **1971**, 1105.
[36] A. J. Buglass, K. Hudson, and J. G. Tillett, *J. Chem. Soc.* (B), **1971**, 123.
[37] D. W. Farlow and R. B. Moodie, *J. Chem. Soc.* (B), **1971**, 407.
[38] F. Dewhurst and A. H. Lamberton, *J. Chem. Soc.* (B), **1971**, 788.

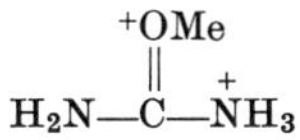

$$\underset{(24)}{\overset{+\text{OMe}}{\underset{|}{H_2N-C-\overset{+}{N}H_3}}}$$

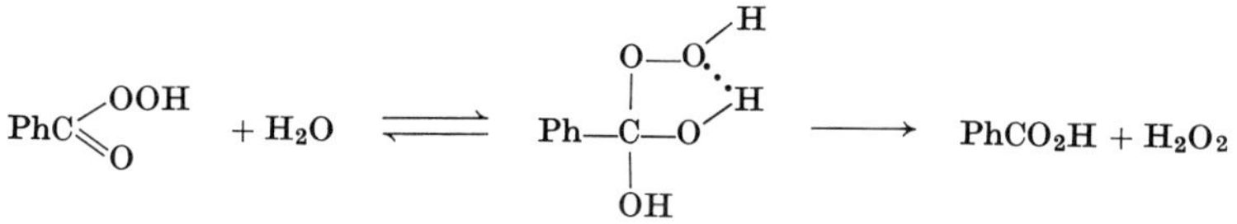

$$PhC\overset{OOH}{\underset{O}{\diagdown}} + H_2O \rightleftharpoons Ph-\underset{OH}{\overset{O-O\cdots H}{\underset{|}{C}-O}}\overset{H}{} \longrightarrow PhCO_2H + H_2O_2 \quad \ldots(6)$$

$$Ph\cdot CO_3H + H^+ \rightleftharpoons PhCO_3H_2^+ \quad \ldots(7)$$

$$PhC\overset{\overset{H}{\diagdown}O^+-O\overset{H}{\diagup}}{\underset{O}{\diagdown}} + H_2O \rightleftharpoons Ph\cdot\underset{OH}{\overset{H\diagdown\overset{+}{O}-O\cdots H}{\underset{|}{C}-O}}\overset{H}{} \longrightarrow PhCO_2H + H_2O_2 + H^+$$

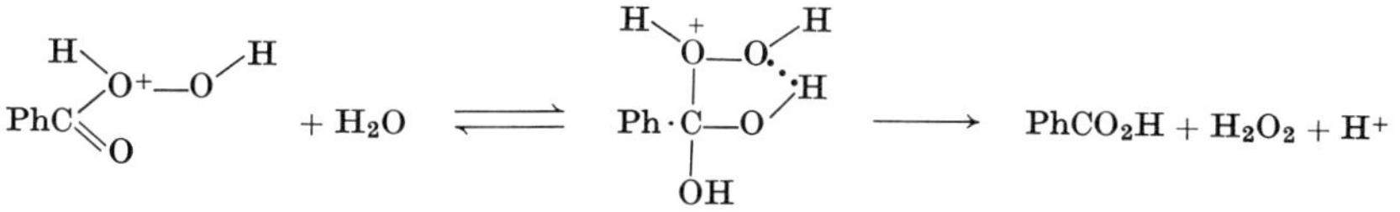

$$\ldots(8)$$

which is hydrolysed more slowly than it is formed. The reaction is general acid-catalysed and this was interpreted as indicating that the rate-limiting step is C-protonation.[39]

The hydrolysis of peroxybenzoic acid follows the rate law:

$$-d[PhCO_3H]/dt = k_1[PhCO_3H] + k_2[PhCO_3H][H_3O^+]$$

The plot of log $(k_{obs} - k_1)$ against H_0 in perchloric acid is curved, but the plot against log $[HClO_4]$ is a straight line with slope 1. The mechanism shown in reactions (6)–(8) was proposed.[40]

The rate of the acid-catalysed hydrolysis of benzyl acetate in acetone–water mixtures at 25° changes little with the composition of the solvent but that of t-butyl acetate decreases with increasing concentration of acetone.[41, 42]

The acid-catalysed hydrolyses of the following compounds have also been studied: benzoic anhydrides (in the presence of perchloric acid and lithium perchlorate);[43] lactams;[44] amides of simple carboxylic acids;[45] dinitroacetonitrile;[46] and ethyl acetate (in concentrated sulphuric acid).[47]

The signals of the methyl group of the *cis*- and *trans*-forms of N-methylformamide coalesce in sulphuric acid of concentration ca. 5%. In more concentrated acid (ca. 100%) two signals are seen again. This was interpreted as indicating that the N-protonated form predominates at moderate acid concentrations and the O-protonated form at high

[39] Y. Y. Lim and A. R. Stein, *Can. J. Chem.*, **49**, 2455 (1971).
[40] F. Secco and S. Celsi, *J. Chem. Soc.* (B), **1971**, 1792.
[41] F. Y. Khalil and H. Sadek, *Z. Phys. Chem.* (*Frankfurt*), **75**, 308 (1971).
[42] H. Sadek, F. Y. Khalil, and M. T. Hanna, *Z. Phys. Chem.* (*Frankfurt*), **73**, 77 (1970).
[43] G. Calvaruso and F. P. Cavasino, *J. Chem. Soc.* (B), **1971**, 483.
[44] T. Fujii, S. Yoshifuji, and A. Tamai, *Chem. Pharm. Bull.*, **19**, 369 (1971).
[45] P. D. Bolton and G. L. Jackson, *Austral. J. Chem.*, **24**, 471 (1971); see also ref. 10.
[46] E. S. Mints, R. S. Tesler, E. L. Golod, and C. I. Bagal, *Zhur. Org. Khim.*, **6**, 698 (1970); *J. Org. Chem. USSR*, **6**, 700 (1970).
[47] J. Siigur and U. Haldna, *Organic Reactivity* (*Tartu*), **7**, 431 (1970); *Chem. Abs.*, **73**, 130388 (1970); cf. *Org. Reaction Mech.*, **1967**, 336; **1968**, 382.

concentrations.[48] However, the results are also consistent with the presence of a low concentration of the N-protonated form in moderately concentrated acids which provides a pathway for the interconversion of the *cis*- and *trans*-forms. The NMR spectrum of [^{15}N]acetamide in fluorosulphonic acid at 172°K shows the presence of two nitrogen-bound protons and one oxygen-bound proton; the predominant form is therefore the O-protonated form.[49] Protonation of esters[50] and alcohols[51] has also been studied.

A detailed analysis of the effect of substituents in the alkaline hydrolysis of substituted-phenyl benzoates has been published.[52] The alkaline hydrolysis of benzoates of amino-alcohols[53, 54] and the acid- and base-catalysed hydrolysis of penicillenic acid[55] have been studied.

The hydrolysis of hydrazidic halides occurs via an uncatalysed pathway that involves a hydrazonocarbonium ion (**26**) and a base-catalysed pathway that involves the anion (**25**; X = Hal).[56]

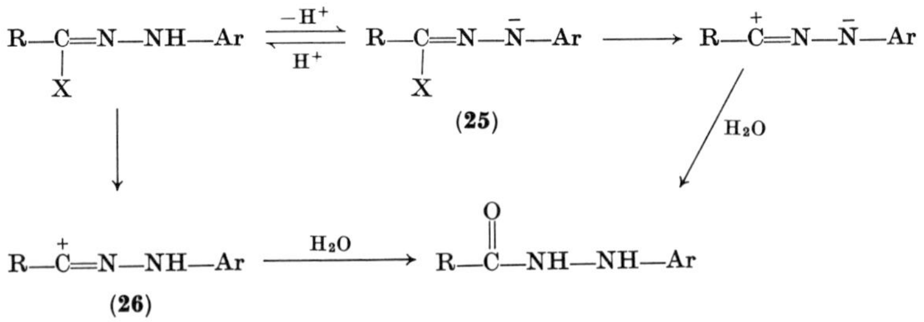

$$R-C(X)=N-NH-Ar \underset{H^+}{\overset{-H^+}{\rightleftharpoons}} R-C(X)=N-\bar{N}-Ar \longrightarrow R-\overset{+}{C}=N-\bar{N}-Ar$$

(**25**)

$$\downarrow \qquad \qquad \downarrow H_2O$$

$$R-\overset{+}{C}=N-NH-Ar \xrightarrow{H_2O} R-\overset{O}{\overset{\|}{C}}-NH-NH-Ar$$

(**26**)

Hypobromous acid, but not bromine, is a good catalyst for the dehydration of bicarbonate to carbon dioxide. The mechanism shown in (9) was proposed.[57]

$$HO-Br \quad O-C(=O)(O^-)H \longrightarrow HOBr + CO_2 + {}^-OH \qquad \ldots(9)$$

Reactions in aprotic solvents. The hydrolysis of *p*-nitrophenyl acetate in (wet ?) acetonitrile is catalysed by imidazole. The rate shows a mixed first- and second-order dependence on the concentration of imidazole, and benzoate is also a catalyst for the imidazole-catalysed reaction but has no effect alone. The mechanism symbolized by (**27**) was proposed.[58]

[48] M. Liler, *J. Chem. Soc.* (B), **1971**, 334; see also ref. 34.
[49] M. Liler, *Chem. Comm.*, **1971**, 115.
[50] M. I. Vinnik and N. B. Librovich, *Organic Reactivity (Tartu)*, **7**, 1221 (1970); J. Siigur, V. Toomes, E. Soonike, H. Kuura, and U. Haldna, *Organic Reactivity (Tartu)*, **7**, 412 (1970); J. Siigur, M. Magi, U. Haldna, and E. Lippmaa, *Organic Reactivity (Tartu)*, **8**, 267 (1971).
[51] E. M. Arnett and J. V. Carter, *J. Am. Chem. Soc.*, **93**, 1516 (1971).
[52] Z. S. Chaw, A. Fischer, and D. A. R. Happer, *J. Chem. Soc.* (B), **1971**, 1818.
[53] L. A. Kundryutskova, S. V. Karopy, S. V. Bogatkov, and E. M. Cherkassova, *Organic Reactivity (Tartu)*, **7**, 1050 (1970).
[54] L. A. Kundryutskova, S. V. Bogatkov, and E. M. Cherkassova, *Zhur. Org. Khim.*, **6**, 701 (1970); *J. Org. Chem. USSR*, **6**, 703 (1970).
[55] J. L. Longridge and D. Timms, *J. Chem. Soc.* (B), **1971**, 852.
[56] A. F. Hegarty, M. P. Cashman, and F. L. Scott, *Chem. Comm.*, **1971**, 684.
[57] M. Caplow, *J. Am. Chem. Soc.*, **93**, 230 (1971).
[58] G. Wallerberg, J. Boger, and P. Haake, *J. Am. Chem. Soc.*, **93**, 4938 (1971).

(27)

The aminolysis of methyl benzoate by ethanolamine in dimethylformamide is autocatalysed with the (2-hydroxyethyl)amide of benzoic acid a catalyst.[59a] The sodium salts of 2-pyridone and 4-pyridone are catalysts for the aminolysis of p-nitrophenyl esters of protected amino acids in aprotic solvents.[59b]

The aminolysis of p-isopropenylphenyl acetate in various solvents,[60] the acetic-acid catalysed reaction of 4,4-dimethyl-2-phenyl-2-oxazolin-5-one with the ethyl ester of DL-alanine in carbon tetrachloride,[61] and the formation of amides catalysed by di-imides[62a] have been studied. The ρ-value for the reaction of thioacids with substituted anilines in benzene is -2.40.[62b]

The Hammett plot for the reaction of substituted anilines with ethyl chloroformate in acetone is not a straight line.[63] It was proposed that the rate-limiting step was formation of the tetrahedral intermediate from amines containing electron-withdrawing groups and breakdown with amines containing electron-releasing groups. This is surprising. The possibility of third-order terms in the rate law for the reactions of the more basic amines does not appear to have been considered.

The acetic acid-catalysed reaction of p-chloroaniline with benzoyl chloride in toluene shows an isotope effect $k(\text{AcOH})/k(\text{AcOD}) = 1.91$ which was considered to indicate a cyclic synchronous mechanism.[64] The uncatalysed reactions of benzoyl bromide and benzoyl iodide with p-anisidine are faster than that of benzoyl fluoride; the acetic acid-catalysed reactions are also faster but by not such a large factor.[65] Pyridine N-oxides catalyse the reaction of primary amines with benzoyl chloride in benzene.[66] For other reactions of acyl halides with amines in aprotic solvents, see ref. 67. Solvent effects on the competitive reactivity of phenol and methanol with benzoyl chloride in the presence of triethylamine and pyridine have also been investigated.[68]

[59a] M. F. Sorokin and V. A. Yamskii, *Tr. Mosk. Khim.-Tekhnol. Inst.*, **1969**, No. 61, 107; *Chem. Abs.*, **73**, 108947 (1970).
[59b] N. Nakamizo, *Bull. Chem. Soc. Japan*, **44**, 2006 (1971).
[60] L. M. Kogan, A. I. Esrieler, A. B. Lebeder, and A. B. Peizner, *Zhur. Obshch. Khim.*, **40**, 2305 (1971); *Chem. Abs.*, **74**, 52669 (1971).
[61] H. Rodriguez, C. Chuaqui, S. Atala, and A. Marquez, *Tetrahedron*, **27**, 2425 (1971).
[62a] D. F. Mironova and G. F. Dvorko, *Ukr. Zhur. Khim.*, **37**, 458 (1971).
[62b] Y. Hirabayashi, M. Mizuta, M. Kojima, Y. Horio, and H. Ishihara, *Bull. Chem. Soc. Japan*, **44**, 791 (1971).
[63] G. Ostrogovich, C. Csunderlik, and R. Bacaloglu, *J. Chem. Soc.* (B), **1971**, 18.
[64] L. M. Litvinenko, G. V. Semenyuk, and N. M. Oleinik, *Zhur. Org. Khim.*, **6**, 2539 (1970); *J. Org. Chem. USSR*, **6**, 2552 (1970).
[65] L. M. Litvinenko and G. V. Semenyuk, *Organic Reactivity (Tartu)*, **7**, 1092 (1970).
[66] L. M. Litvinenko, G. D. Titskii, and I. V. Shpan'ko, *Zhur. Org. Khim.*, **7**, 107 (1971); *Chem. Abs.*, **74**, 99221 (1971).
[67] G. D. Titskii, I. V. Shpan'ko, L. M. Litvinenko, and V. M. Shuliko, *Organic Reactivity (Tartu)*, **8**, 39 (1971); L. M. Litvinenko and G. V. Semenyuk, *Organic Reactivity (Tartu)*, **7**, 1083 (1970); *Chem. Abs.*, **75**, 19435 (1971); G. D. Titskii and L. M. Litvinenko, *Zhur. Obshch. Khim.*, **40**, 2680 (1970); A. P. Grekov and V. Y. Veselov, *Zhur. Org. Khim.*, **7**, 323 (1971); *Chem. Abs.*, **75**, 48108 (1971); A. P. Grekov and V. Y. Veselov, *Zhur. Org. Khim.*, **6**, 1685 (1970); *J. Org. Chem. USSR*, **6**, 1691 (1970); Y. A. Strepikheev, A. L. Chimishkyan, and G. K. Goniharov, *Zhur. Org. Khim.*, **6**, 2290 (1970); *J. Org. Chem. USSR*, **6**, 2290 (1970).
[68] V. V. Korshak, S. S. Vinogradova, and V. A. Vasnev, *Dokl. Akad. Nauk SSSR*, **191**, 614 (1970).

The reaction of N-arylarenecarbimidoyl chlorides with amines in benzene [reaction (10)] follows the rate-law (11). The plots of log k_2 and log k_1 for the reaction with morpholine against the σ-constants of substituents in the Ar' group is curved. Substituents with $\sigma > 0.3$ give a positive slope and an addition–elimination mechanism (12) was proposed; substituents with $\sigma < 0.3$ give a negative slope and a mechanism (13) involving an ion pair was proposed.[69]

$$ArC(Cl)=N-Ar' + R_2NH \longrightarrow ArC(NR_2)=N-Ar' + HCl \qquad \ldots(10)$$

$$k_{obs} + k_2[BH] + k_3[BH]^2 \qquad \ldots(11)$$

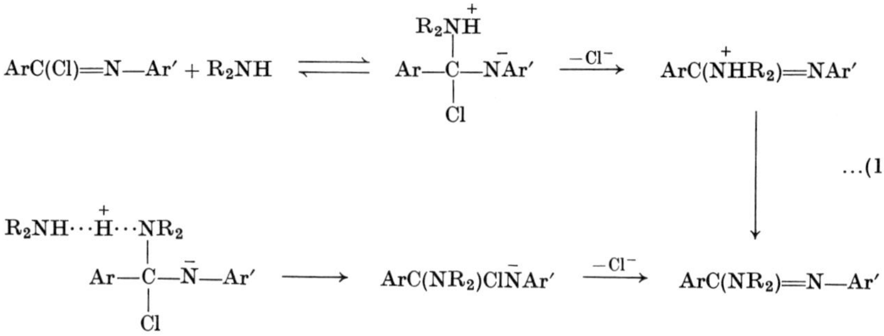

$$ArC(Cl)=N-Ar' + R_2NH \rightleftharpoons Ar-\underset{Cl}{\underset{|}{C}}(\overset{+}{R_2NH})-\bar{N}Ar' \xrightarrow{-Cl^-} ArC(\overset{+}{N}HR_2)=NAr' \qquad \ldots(12)$$

$$Ar-\underset{Cl}{\underset{|}{C}}(\overset{R_2NH\cdots\overset{+}{H}\cdots NR_2}{|})-\bar{N}-Ar' \longrightarrow ArC(NR_2)Cl\bar{N}Ar' \xrightarrow{-Cl^-} ArC(NR_2)=N-Ar'$$

$$ArC(Cl)=NAr' \rightleftharpoons Ar\overset{+}{C}=NAr'\,Cl^- \longrightarrow ArC(NR_2)=NAr' \qquad \ldots(13)$$

The reaction of tertiary trialkylamines with acetic anhydride results in formation of an amide and loss of the alkyl group that forms the most stable carbonium ion.[70] The mechanism (14) was proposed.

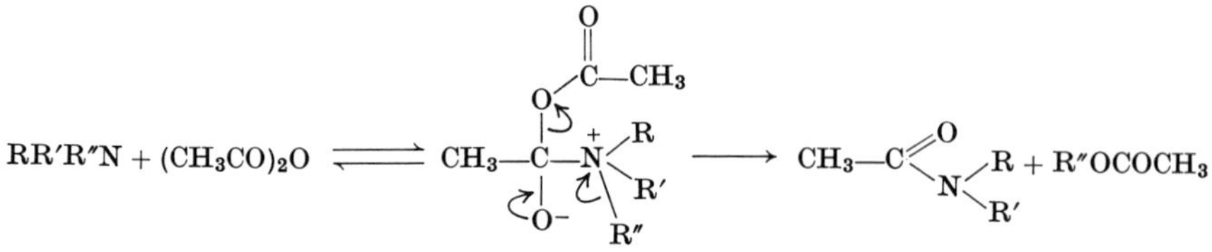

$$RR'R''N + (CH_3CO)_2O \rightleftharpoons CH_3-C\underset{O^-}{\overset{\overset{O}{\overset{\|}{C}-CH_3}}{|}}-\overset{+}{N}\begin{smallmatrix}R\\R'\\R''\end{smallmatrix} \longrightarrow CH_3-C\overset{O}{\underset{N}{\diagdown}}\begin{smallmatrix}R\\R'\end{smallmatrix} + R''OCOCH_3 \qquad \ldots(14)$$

The addition of phenol to aryl isocyanates in diethyl ketone is catalysed by N-methylaniline, and the addition of N-methylaniline is catalysed by phenols. The reactions were thought to proceed through six-membered cyclic transition states (**28**) and (**29**).[71]

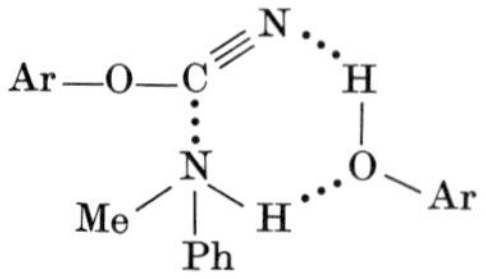

(**28**) 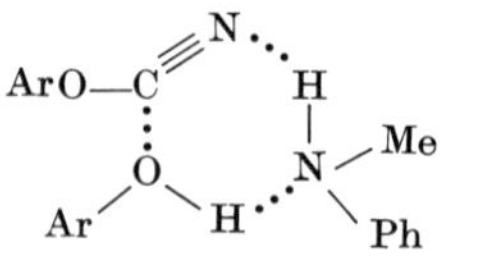 (**29**)

[69] Z. Rappoport and R. Ta-Shma, *Tetrahedron Letters*, **1971**, 3813.
[70] R. P. Mariella and K. H. Brown, *Can. J. Chem.*, **49**, 3348 (1971).
[71] D. Martin, K. Nadolski, R. Bacaloglu, and I. Bacaloglu, *J. Prakt. Chem.*, **313**, 58 (1971).

The reactions of amines with phenyl isocyanate are catalysed by amines, ureas (the products), amides, and carboxylic acids. Amides and carboxylic acids have similar catalytic activity. Bifunctional catalysis symbolized by (**30**) was proposed for the catalysis by ureas.[72] The reaction of aniline with *m*-chlorophenyl isocyanate in nitrobenzene has also been studied.[72]

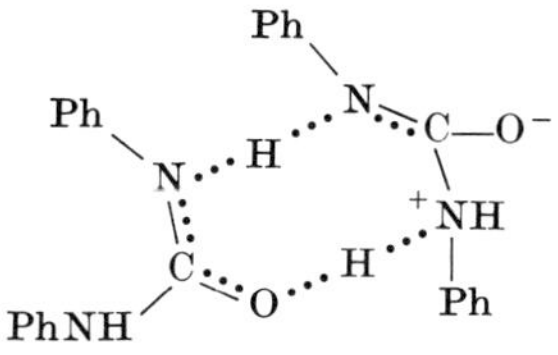

(**30**)

Intramolecular Catalysis and Neighbouring-group Participation[73a,b]

The hydrolysis of 8-quinolyl hydrogen succinate and 8- and 6-quinolyl hydrogen glutarate have bell-shaped pH–rate profiles. Succinic anhydride is an intermediate in the hydrolysis of the succinate ester, and glutaric anhydride is probably an intermediate in the hydrolysis of the glutarate esters as at the optimum pH these reactions are much faster than the hydrolyses of the corresponding quinolyl acetates. The rate constants for the hydrolysis of species (**31**) and (**32**) are similar and it was therefore proposed that the 8-quinolyl

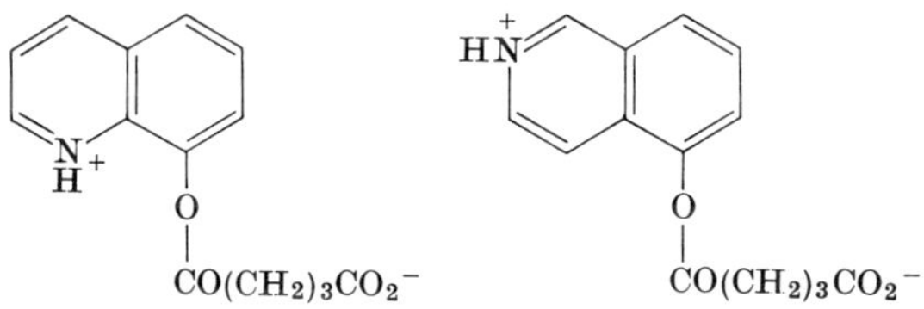

(**31**)   (**32**)

ester does not react with intramolecular bifunctional catalysis. It seems likely that the decrease in rate that occurs when the quinolyl ring is deprotonated arises not from the loss of an intramolecular acid catalyst but from the change in inductive and resonance effects. A similar conclusion was reached concerning the hydrolysis of 2-carboxyphenyl succinate although it is possible that this reaction involve series catalysis. It seems likely that the bell shaped pH–rate profile for the hydrolysis of methyl 2,6-dihydroxybenzoate[74] arises because hydrolysis of the monoionized form is accelerated by intramolecular catalysis and hydrolysis of the di-ionized form is decelerated by the unfavourable electronic effect of the second ionized hydroxyl group. The monoionized species of methyl 2,6-dihydroxybenzoate reacts much faster than that of methyl

---

[72] J. M. Briody and D. Narinesingh, *Tetrahedron Letters*, **1971**, 4143; N. K. Vorob'ev, and O. K. Shebanova, *Tr. Ivanov. Khim-Tekhnol. Inst.*, **1969**, No. 11, 16; *Chem. Abs.*, **74**, 140452 (1971).
[73a] A. J. Kirby and A. R. Fersht, *Prog. Bioorg. Chem.*, **1**, 1 (1971).
[73b] P. Salomaa, A. Kankaanperä, and K. Pihlaja, in 'The Chemistry of the Hydroxyl Group," S. Patai (Ed.), Interscience, London, 1971, p. 477.
[74] See *Org. Reaction Mech.*, **1969**, 447.

2,4-dihydroxybenzoate, but the predominant form of the latter is that in which the 4-hydroxyl group is ionized.[75]

The hydrolysis of the 2-methyl ester of benzene-1,2,3-tricarboxylic acid proceeds by two pathways that have enhanced rates. These correspond to terms in the rate law with the rate proportional severally to the concentrations of un-ionized and the monoionized forms. The mechanisms shown in (15a)[76] and (15b) were proposed.[77]

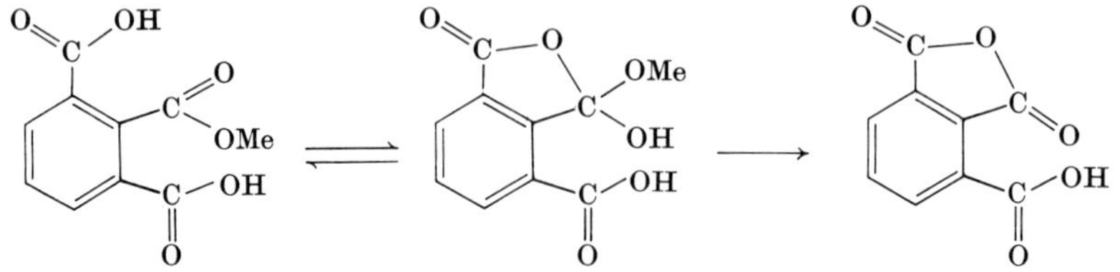

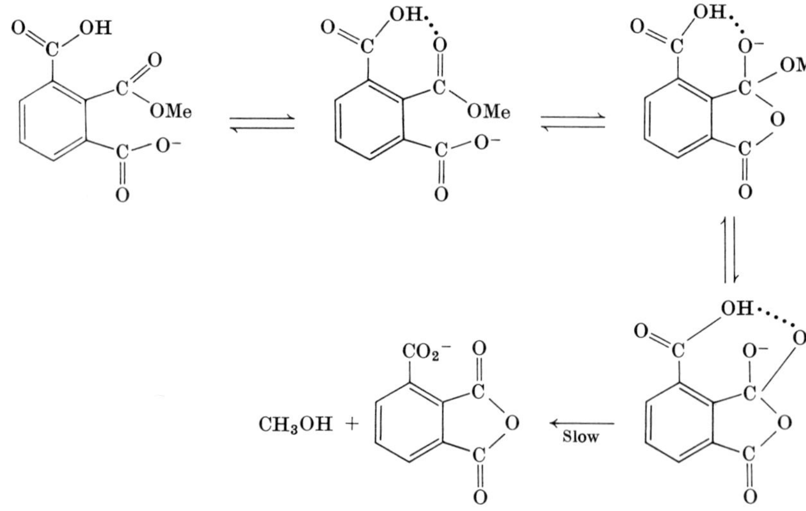

The pH–rate profiles for the hydrolysis of (**33**) and (**34**) are complex, depending on four equilibrium and four rate constants. The rate constant for the hydrolysis of the un-ionized form (**33**) is 8.4 times greater than that for the un-ionized form (**35**), which

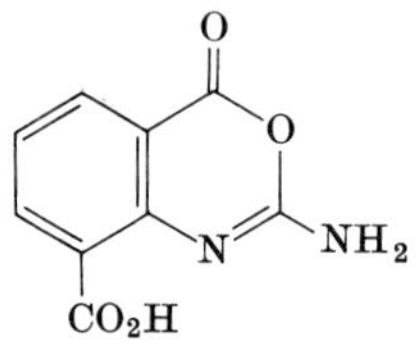

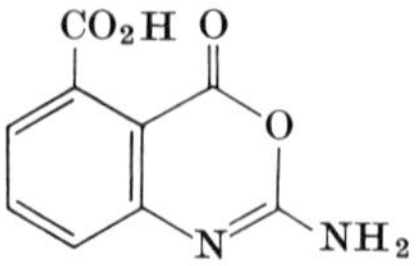

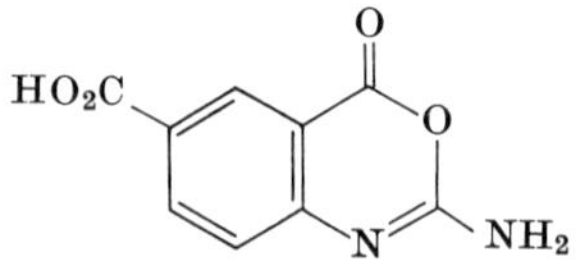

[75] T. Maugh and T. C. Bruice, *J. Am. Chem. Soc.*, **93**, 3237 (1971).
[76] Cf. *Org. Reaction Mech.*, **1966**, 342–343.
[77] G. H. Hurst and M. L. Bender, *J. Am. Chem. Soc.*, **93**, 704 (1971).

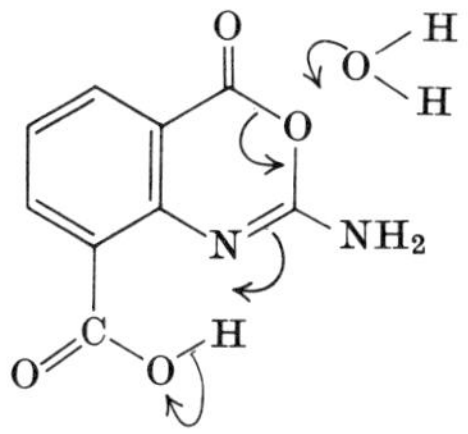

(36)

was attributed to intramolecular catalysis as symbolized by (**36**). There appears to be a larger rate enhancement in the morpholinolysis of (**33**). The hydrolysis of (**34**) is not intramolecularly catalysed.[78]

The pH–rate profile for the hydrolysis of hexachlorophene monosuccinate has a sigmoid portion showing that the monoanion is especially reactive. This was interpreted as arising from bifunctional catalysis symbolized by (**37**). The hydrolysis of hexachlorophene monoacetate is also enhanced and possible mechanisms involve intramolecular general acid- or general base-catalysis.[79]

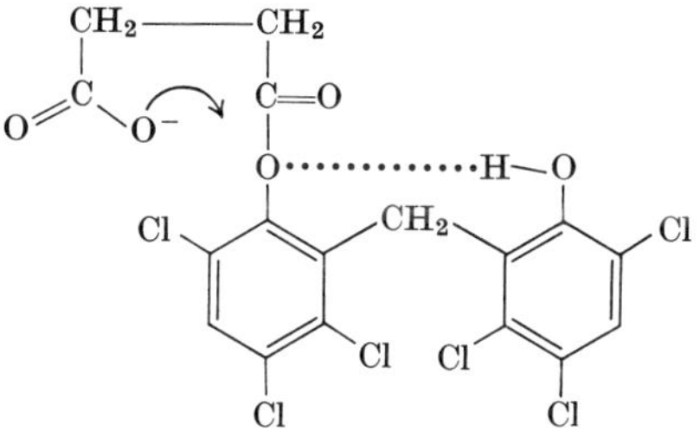

(37)

In contrast to hexachlorophene monosuccinate, pyrocatechol monosuccinate is hydrolysed with concurrent intramolecular nucleophilic catalysis by the carboxylate group and intramolecular general base-catalysis by the phenoxide group in buffers of pH 8–10.[80]

Aryl 2-hydroxyphenylacetates, 3-(2-hydroxyphenyl)propionates, 4-hydroxybutyrates, and 5-hydroxyvalerate react with neighbouring-group participation in aqueous solution to release the phenol and form a lactone. The second-order rate constant for the hydroxide-catalysed reaction of phenyl 4-hydroxybutyrate is about 3000 times greater than that for the hydroxide-catalysed hydrolysis of phenyl acetate. The hydroxybutyrate esters react 10–20 times faster than the hydroxyvalerate esters. 2-Naphthyl 2-hydroxyphenylacetate forms the lactone about five times faster than 2-naphthyl 3-(2-hydroxyphenyl)propionate. The hydroxyl group also participates in the acid-catalysed hydrolysis of the hydroxy-butyrates and -valerates but not in that of the hydroxyphenylacetates and hydroxyphenylpropionates.[81a] For other examples of hydroxyl-group participation in reactions of esters are described (see ref. 81b).

[78] A. F. Hegarty, R. F. Pratt, T. Giudici, and T. C. Bruice, *J. Am. Chem. Soc.*, **93**, 1428 (1971).
[79] T. Higuchi, M. Takechi, I. H. Pitman, and H. L. Fung, *J. Am. Chem. Soc.*, **93**, 539 (1971).
[80] L. E. Eberson and L.-A. Svenson, *J. Am. Chem. Soc.*, **93**, 3827 (1971).
[81a] B. Capon, S. T. McDowell, and W. V. Raftery, *Chem. Comm.*, **1971**, 389.
[81b] S. Ducher and J. Peyronnet, *Ann. Chim. (Paris)*, **5**, 415 (1970).

The conversion of amide (**38**) into lactone (**39**) is general base-catalysed. The pH-dependence of the spontaneous reaction is complex and was interpreted in terms of a change in rate-determining step at pH 8–9. Below this pH, $\beta = 1$ and the rate-limiting step was thought to be a diffusion-controlled proton-transfer between the catalyst and the tetrahedral intermediate. Above pH 9, $\beta =$ ca. 0.2 and the rate-determining step was thought to be conversion of the amide into the tetrahedral intermediate.[82]

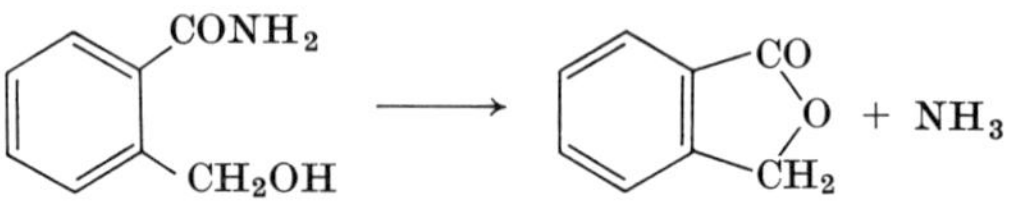

The lactonization of hydroxy-acid (**40**) is about 20 times faster than that of hydroxy-acid (**41**).[83] The lactonization of coumarinic acids[84, 85] and the opening of the lactone ring of coumaran-2,3-diones[86] have also been studied.

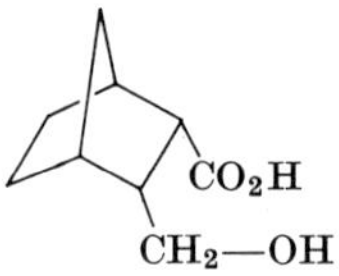

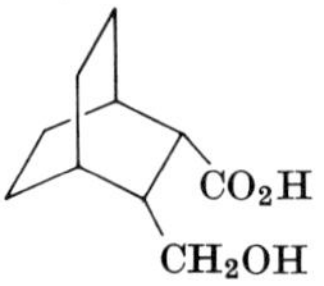

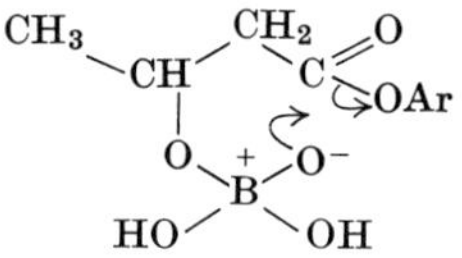

Ethyl and 2-naphthyl cis- and trans-2-hydroxycyclopentanecarboxylate are both hydrolysed faster than the corresponding cyclopentanecarboxylates in aqueous dioxan. The rate enhancement increases with increasing dioxan content of the solvent, an effect tentatively ascribed to solvent sorting. The hydrolyses of the 2-naphthyl 2-hydroxy-cyclopentanecarboxylates and 3-hydroxybutyrate are enhanced in borate buffers; this catalysis may involve formation of a small concentration of a borate ester and intramolecular catalysis as symbolized by (**42**).[87]

Hydrolysis of the carbonate (**43**) is about 10 times faster than that of the corresponding ortho-methoxy and para-hydroxy compounds. The entropy of activation is strongly negative and it was thought that the phenolic hydroxyl group acts as an intramolecular general-acid or general-base catalyst.[88]

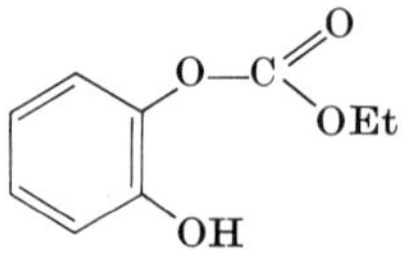

[82] C. J. Belke, S. C. K. Su, and J. A. Shafer, *J. Am. Chem. Soc.*, **93**, 4552 (1971).
[83] D. R. Storm, R. Tjian, and D. E. Koshland, *Chem. Comm.*, **1971**, 854.
[84] E. R. Garrett, B. C. Lippold, and J. B. Mielck, *J. Pharm. Sci.*, **60**, 396 (1971).
[85] B. C. Lippold and E. R. Garrett, *J. Pharm. Sci.*, **60**, 1019 (1971).
[86] T. Matsuura, M. Kawai, and Y. Butsugan, *Bull. Chem. Soc. Japan*, **47**, 3891 (1970).
[87] B. Capon and M. I. Page, *J. Chem. Soc. (B)*, **1971**, 741.
[88] J. G. Tillett and D. E. Wiggins, *Tetrahedron Letters*, **1971**, 911.

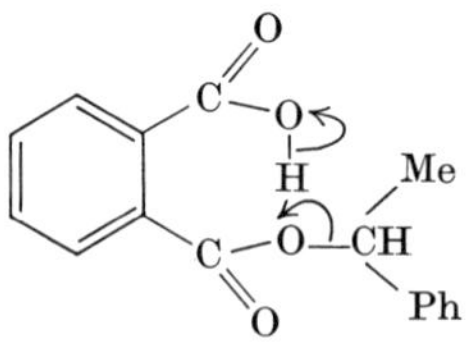

The hydrolyses in 75% aqueous acetone of optically active 1-phenylethyl hydrogen phthalate and hydrogen terephthalate yield racemic alcohol and hence occur with alkyl–oxygen fission. The phthalate ester reacts about 40 times faster than the terephthalic ester. A *p*-methoxy-substituent accelerates hydrolysis of both esters but that of the phthalate much more than that of the terephthalate so that the phthalate now reacts 20,000 times faster. These results suggest that the phthalate esters react with intramolecular general acid-catalysis as symbolized by (**44**).[89a]

The alkaline hydrolysis of dimethyl phthalate, isophthalate and terephthalate in aqueous dimethyl sulphoxide has been studied.[89b]

A detailed investigation of the equilibria between diacids and the corresponding anhydrides has been reported. Formation of the anhydrides of succinic acids is favoured on the introduction of alkyl substituents, which was attributed to release of steric strain on formation of the anhydride.[90]

The effect of substituents on the course of cyclization of polycarboxylic esters has been reviewed.[91]

When the groups R in (**45**) are changed from hydrogen to methyl the rate of the acid-catalysed lactonization is increased 24-fold, but a similar change in (**46**) increases it 140,000-fold.[92a]

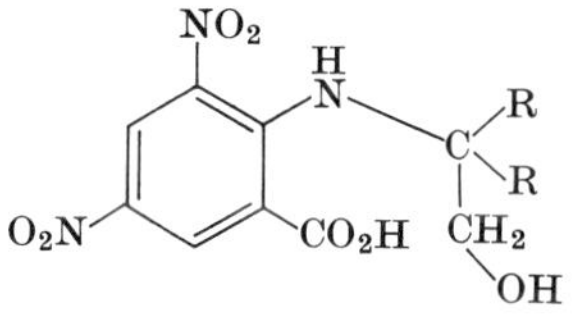

(**45**)

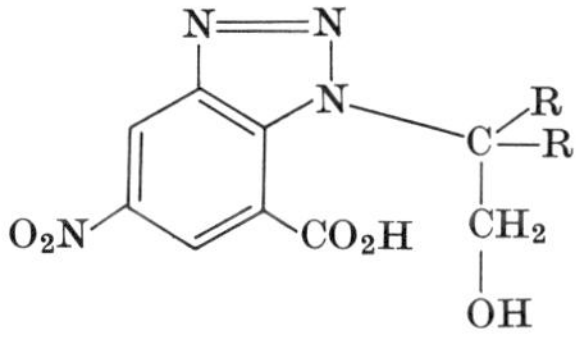

(**46**)

The effect of the leaving group on the rate of hydrolysis of a series of pseudo-esters of 2-benzoylbenzoic acid have been determined. The variation in rate is much less than with the corresponding benzoate esters.[92b]

Details of Bowden and Taylor's investigation of the hydrolysis of methyl 2-benzoylbenzoate have been published.[93] The interesting observation has been made that when methyl 2-benzoylbenzoate is allowed to hydrolyse in ^{18}O-enriched water there is only just over one atom of ^{18}O incorporated and this is mainly in the ketone group. On the assumption that reaction proceeds via an intermediate that breaks down as shown in (**47**), this observation shows that the reaction cannot involve a reversibly formed hydrate as shown in the sequence (16a) because this would lead to incorporation of two atoms of ^{18}O. It therefore seems either that attack of hydroxide on the ketone-carbonyl group is concerted with attack of the ketone-oxygen on the ester group (**48**) or that reaction involves an intermediate (**49**) that reacts by intramolecular attack faster than it captures a proton.[93] A detailed investigation of the hydrolysis of a series of methyl 2-acylbenzoates (**50**) was also reported. The rate decreased as the group R was changed

---

[89a] R. Anantaraman, T. D. R. Nair, and K. Saramma, *J. Chem. Soc.* (B), **1971**, 1142.
[89b] G. V. Rao and N. Venkatasubramanian, *Austral. J. Chem.*, **24**, 201 (1971).
[90] L. Eberson and H. Welinder, *J. Am. Chem. Soc.*, **93**, 5821 (1971).
[91] R. N. Chakravarti, *J. Indian Chem. Soc.*, **47**, 1025 (1970).
[92a] J. Turk, W. M. Haney, G. Heid, R. E. Barlow, and C. B. Clapp, *J. Heterocyclic Chem.*, **8**, 149 (1971).
[92b] M. V. Bhatt, G. V. Rao, and K. S. R. Rao, *Chem. Comm.*, **1971**, 822.
[93] K. Bowden and G. R. Taylor, *J. Chem. Soc.* (B), **1967**, 145, 149; cf. *Org. Reaction Mech.*, **1967**, 342.

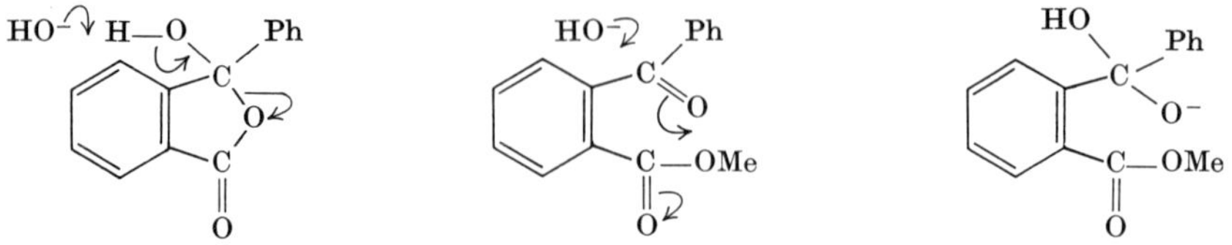

in the series H > Me > MeCH$_2$ > Me$_2$CH > Me$_3$C.[94] The rate of interconversion of these esters with the corresponding pseudo-esters in acid methanol and the proportion of the true ester present at equilibrium also decrease in this order. The mechanism shown in the sequence (16b) was proposed. Ring-chain tautomerism of the corresponding acids was also studied (cf. ref. 97 on p. 402).[94] An investigation of ring–chain tautomerism of keto-amides has been described.[95]

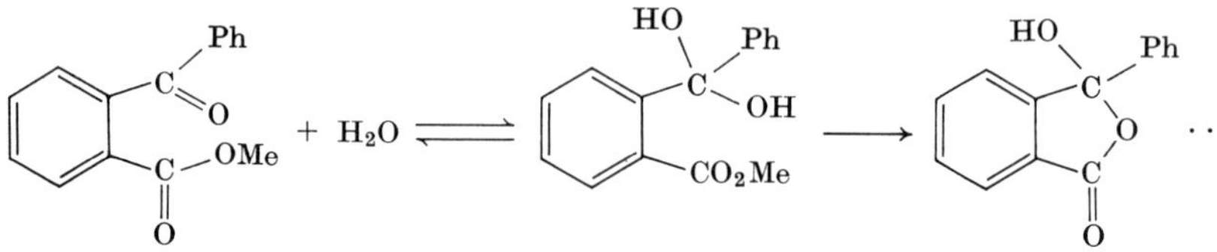

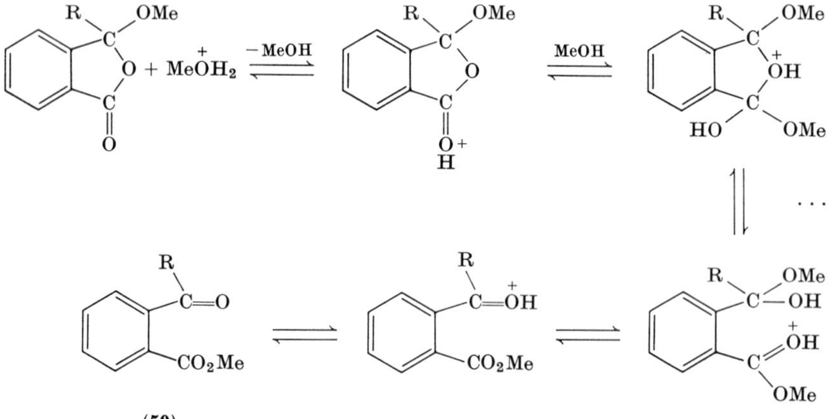

The ρ-value for the hydrolysis of methyl *cis*-3-benzoylacrylates is 2.56 compared to 0.685 for the *trans*-series. It was thought that the *cis*-series reacted with attack at the ketone-carbonyl group as shown in (**51**).[96]

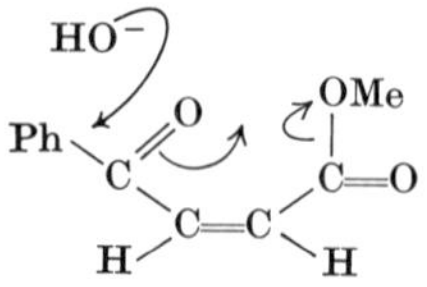

[94] K. Bowden and G. R. Taylor, *J. Chem. Soc.* (B), **1971**, 1390, 1395.
[95] R. Chiron and Y. Graff, *Bull. Soc. Chim. France*, **1971**, 2145.
[96] K. Bowden and M. P. Henry, *J. Chem. Soc.* (B), **1971**, 156.

Bersaldegenin 1,3,5-orthoacetate (partial structure, **52**) is converted into (**53**) under conditions where the analogous orthoacetate with the formyl group replaced by a methyl group is unchanged. The reaction was thought to involve participation by the aldehyde hydrate as shown.[97]

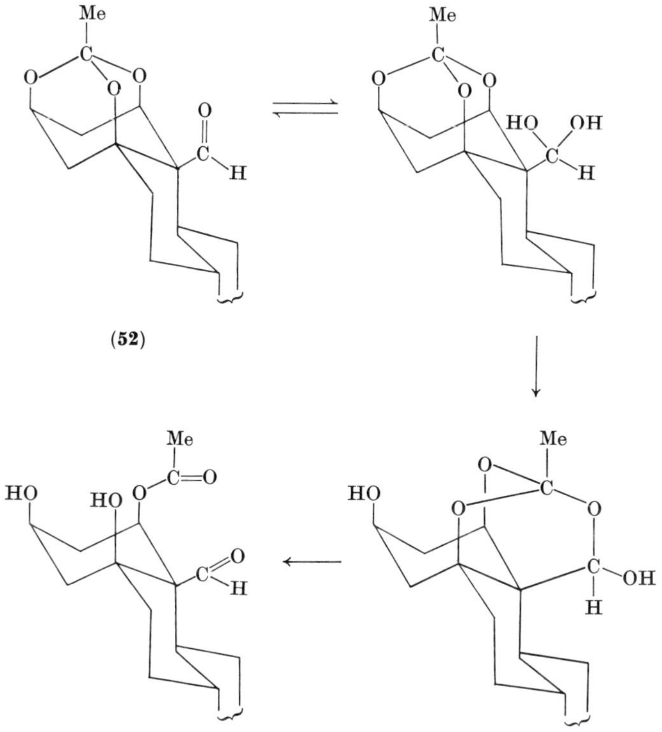

(**52**)

(**53**)

The ester group of compounds (**54**; R = H or Me) is hydrolysed in basic solution via the rearranged imide (**55**).[98]

Hydrolysis of MeNH(CH$_2$)$_3$CONHPh is 10–100 times faster than that of Me$_2$N(CH$_2$)$_3$-CONHPh at pH 5.5.[99a] Possibly MeNH(CH$_2$)$_3$CONHPh reacts with neighbouring-group

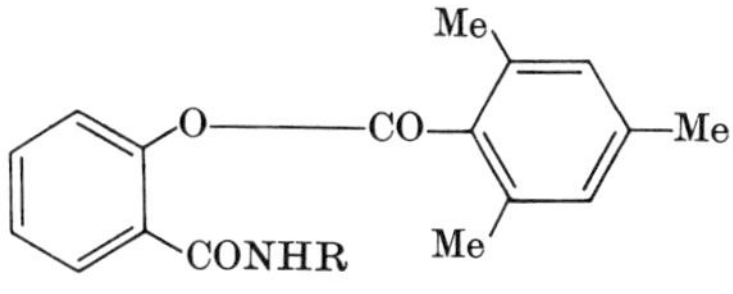

(**54**)

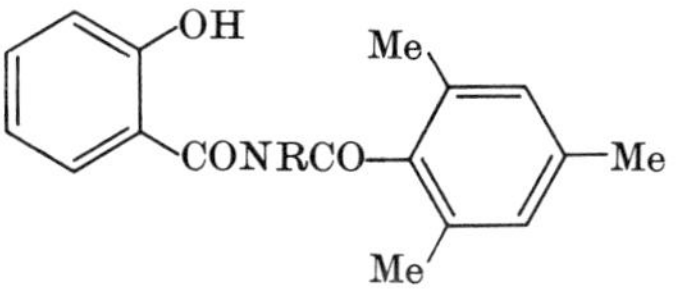

(**55**)

[7] S. M. Kupchan, I. Ognyanov, and J. L. Moniot, *Bioorg. Chem.*, **1**, 13 (1971).
[8] P. L. Russell, R. M. Topping, and D. E. Tutt, *J. Chem. Soc.* (B), **1971**, 657; cf. p. 422 and *Org. Reaction Mech.*, **1968**, 386 and **1969**, 448.
[9a] S. O. Eriksson and U. Meresaar, *Acta Chem. Scand.*, **26**, 2697 (1971).

participation. The hydrolysis of O-imidazolyl-4-propionylserine acetylamide has been studied.[99b]

When heated in diphenyl ether at 252° the Schiff base (**56**) is converted into the quinoline (**58**). Reaction was thought to proceed via the enamine (**57**). The reaction is autocatalysed and (**58**) may act as an acid catalyst for the cyclization step.[100]

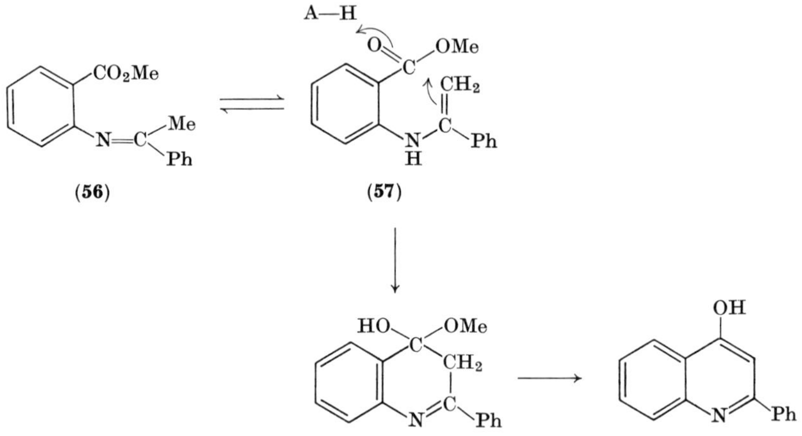

3-Methyluridine (**59**) undergoes a base-catalysed reaction to yield 3-methylurea riboside (**60**). Since methylation of the 5′-hydroxyl group inhibits this reaction it probably involves intramolecular nucleophilic attack as shown.[101]

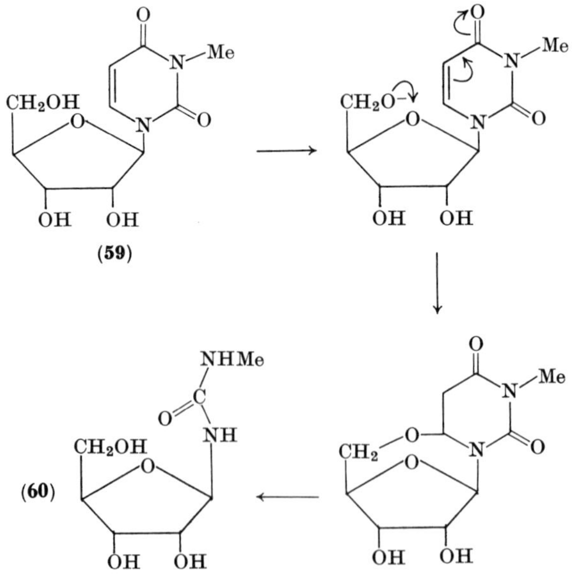

[99b] D. Nissen and F. Schneider, *Z. Naturforsch.*, **25b**, 1315 (1970).
[100] Y. Ogata, A. Kawasaki, and K. Tsujimura, *Tetrahedron*, **27**, 2765 (1971).
[101] Y. Kondo, J.-L. Fourney, and B. Witkop, *J. Am. Chem. Soc.*, **93**, 3527 (1971).

Compound (**62**) is an intermediate in the conversion of (**61**) into (**63**). Ring–chain tautomerism of (**62**) has been studied.[102]

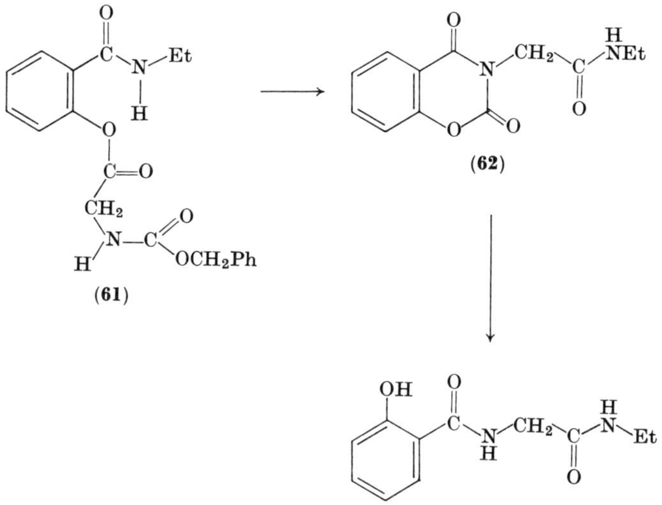

Methylamine is a convenient reagent for cleaving phthalimides. The reaction involves nucleophilic attack by the methylamine on the imide group to give a diamide, followed by intramolecular nucleophilic attack by the methylamido-group on the other amide group. It was demonstrated that diamide (**64**) is rapidly converted into *N*-methylphthalamic acid in deuterium oxide at 20°.[103a] It seems to the Reviewer that this reaction probably involves *O*-participation rather than *N*-participation since *N*-methylphthalimide would probably be stable under these conditions. The reaction of *N*-phenylmaleamic acid with acetic anhydride has also been studied.[103b]

Substituted 2-methyl-2-(*o*-phenylazophenoxy)propionyl groups, e.g. (**65**),[104] and analogous nitro-substituted groups, e.g. (**66**),[105] can be used as *N*-protecting groups in peptide synthesis. The phenylazo- or nitro-group is reduced to an amino-group which then carries out an intramolecular attack on the acyl-carbon atom to release the protected amino-group. The presence of the *gem*-dimethyl groups makes the intramolecular transamidation much faster.

---

[102] D. S. Kemp, J. M. Duclos, Z. Bernstein, and W. M. Welch, *J. Org. Chem.*, **36**, 157 (1971).
[103a] S. Wolfe and S. K. Hasan, *Can. J. Chem.*, **48**, 3572 (1970).
[103b] C. K. Sauers, C. L. Gould, and E. S. Ioannou, *J. Org. Chem.*, **36**, 1941 (1971).
[104] C. A. Panetta and A.-U. Rahman, *J. Org. Chem.*, **36**, 2250 (1971).
[105] F. Cuiban, *Tetrahedron Letters*, **1971**, 2471.

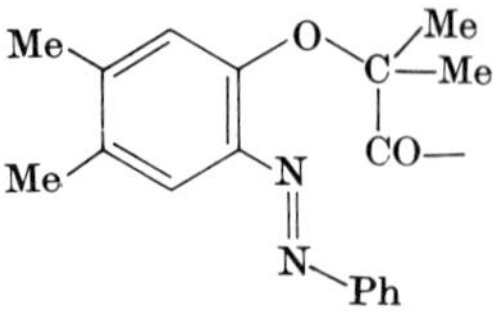

More work on *o*-hydroxyphenyl esters in peptide synthesis has been reported.[106] 2-Pyridyl thioesters have also been used in peptide coupling reactions. These reactions probably involve intramolecular catalysis.[107]

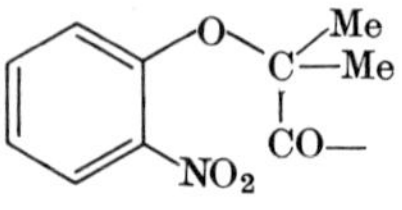

2-Pyridyl benzoate reacts with hydrogen chloride in acetonitrile to form an equilibrium mixture containing 2-pyridone and benzoyl chloride. 4-Pyridyl benzoate does not undergo an analogous reaction and the transition state (**67**) was proposed.[108]

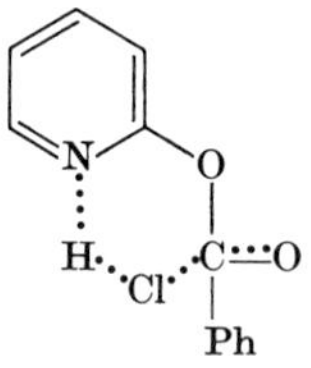

Aminolysis with benzylamine in ethyl acetate and racemization of aryl esters of *N*-benzoyl-leucine,[109] hydrolysis of cycloserine dimer,[110] and the Dakin–West reaction[111] have been studied.

The pH–rate profile for the conversion of *S*-benzoyl-1,1,3-trimethylisothiouronium bromide into 1-benzoyl-1,3,3-trimethylthiourea (**70**) is bell-shaped. It was thought[112] that reaction followed the course (17) with (**72**) and (**69**) present only at low concentra-

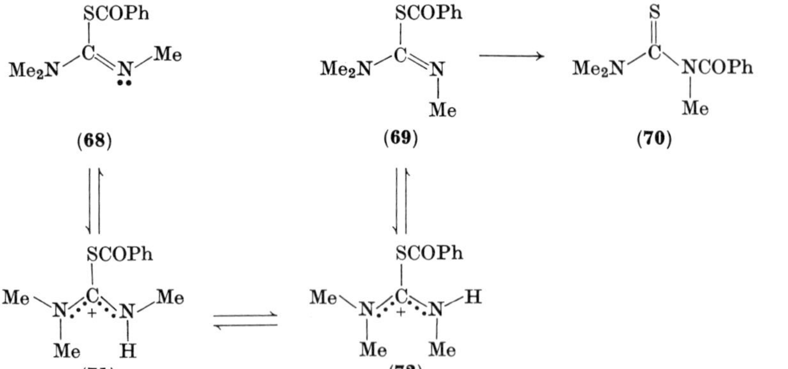

...(17)

[106] Y. Trudelle, *Chem. Comm.*, **1971**, 639; cf. *Org. Reaction Mech.*, **1967**, 344 and **1968**, 387.
[107] K. Lloyd and G. T. Young, *J. Chem. Soc.* (C), **1971**, 2890.
[108] P. A. Singgih and M. J. Janssen, *Tetrahedron Letters*, **1971**, 4223.
[109] J. Morawiec, D. Konopinska, and I. Z. Siemion, *Roczn. Chem.*, **45**, 771 (1971).
[110] F. O. Lassen and C. H. Stammer, *J. Org. Chem.*, **36**, 2631 (1971).
[111] R. Knorr, *Chem. Ber.*, **104**, 3633 (1971); R. Knorr and G. K. Staudinger, *Chem. Ber.*, **104**, 362 (1971); W. Steglich and G. Höfle, *Chem. Ber.*, **104**, 3644 (1971).
[112] T. C. Bruice and R. F. Pratt, *Chem. Comm.*, **1971**, 1259.

tions. The ascending limb of the bell on going to more alkaline solutions was thought to arise from the acyl-transfer occurring in the neutral species **(69)** and the descending limb from the need for acid-catalysis for conversion of **(68)**, the predominant form, into **(69)**, the unstable form.

The NMR signals of the aromatic and quinoid protons of naphthazarin monoacetate in $C_6D_5NO_2$ solution broaden at 120° and coalesce at 190°. This was attributed to intramolecular acetyl-transfer as shown in the equilibrium (18).[113]

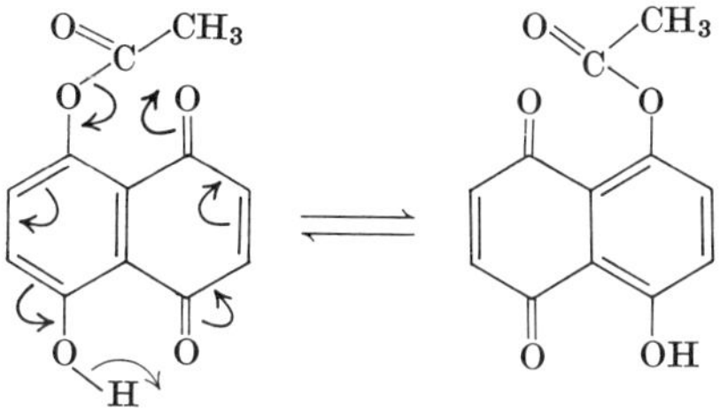

...(18)

Other examples of acyl migration are described by the authors of ref. 114, and the subject has been reviewed.[115]

It has been argued that a solvent isotope effect could arise from the fact that loss of translational entropy when a D_2O molecule becomes bound is greater than that when an H_2O molecule becomes bound. The difference in the entropy of activation for the hydrolysis of O-(chloroacetyl)salicyclic acid was $\Delta S^\ddagger(D_2O) - \Delta S^\ddagger(H_2O) = -0.9$ cal deg^{-1} mol^{-1}. It was considered that this result indicated that three water molecules were bound in the transition state. The reactions were studied in 10% v/v aqueous dioxan and the error in $\Delta S^\ddagger$ was given as ± 0.2 cal deg^{-1} mol^{-1}.[116]

Hydrolyses of the acetates of 2-(hydroxymethyl)benzimidazole, 2-(hydroxymethyl)-benzothiazole, and 2-(hydroxymethyl)-1-methylbenzimidazole are not intramolecularly catalysed in alkaline solutions.[117]

Orbital steering has been discussed.[118]

Association-prefaced Catalysis

Details of Blyth and Knowles' work on the reactions of p-nitrophenyl decanoate with decylamine and with long-chain N-alkylimidazoles have been described.[119] It has also been shown that the decylamine-catalysed reaction of decylamine with p-nitrophenyl acetate is 200 times faster than the ethylamine-catalysed reaction of ethylamine. There

[113] I. C. Calder, D. W. Cameron, and M. D. Sidell, *Chem. Comm.*, **1971**, 360.
[114] T. O. Oesterling, *Carbohydrate Res.*, **15**, 285 (1970); V. S. Mikhlin, M. A. Korshunov, Y. I. Turyan, and A. K. Kobyakov, *Organic Reactivity (Tartu)*, **7**, 1068 (1970); *Chem. Abs.*, **75**, 19564 (1971); G. Bernáth, K. Kovacs, and K. L. Lang, *Acta Chim. Acad. Sci. Hung.*, **65**, 347 (1970).
[115] R. M. Acheson, *Accounts Chem. Res.*, **4**, 177 (1971).
[116] L. Tseng and J. A. Stewart, *J. Am. Chem. Soc.*, **93**, 1273 (1971).
[117] R. D. Cook and S. Razmara, *Tetrahedron Letters*, **1971**, 2905.
[118] B. Capon, *J. Chem. Soc.* (B), **1971**, 1207; M. I. Page and W. P. Jencks, *Proc. Nat. Acad. Sci.*, **68**, 1678 (1971); T. C. Bruice, A. Brown, and D. O. Harris, *Proc. Nat. Acad. Sci.*, **68**, 658 (1971); J. Reuben, *Proc. Nat. Acad. Sci.*, **68**, 563 (1971); G. N. Port and W. G. Richards, *Nature*, **231**, 312 (1971); G. A. Dafforn and D. E. Koshland, *Biorg. Chem.*, **1**, 129 (1971).
[119] C. A. Blyth and J. A. Knowles, *J. Am. Chem. Soc.*, **93**, 3017, 3021 (1971); see *Org. Reaction Mech.*, **1967**, 344; **1969**, 454.

is probably a pre-association of the decylamine molecules. The rate constants for the uncatalysed reactions of decylamine and ethylamine with p-nitrophenyl acetate are similar.[120]

The hydrolysis of p-nitrophenyl laurate in aqueous hexane is catalysed by reverse micelles of cetyltrimethylammonium bromide.[121]

Contrary to the usual effect, some salts enhance the rate of decarboxylation of 6-nitrobenzisoxazole catalysed by cetyltrimethylammonium bromide.[122]

D-PhCH(OH)CHMeN$^+$Me$_2$C$_{12}$H$_{25}$ is a more effective catalyst for the hydrolysis of p-nitrophenyl O-methyl-D-mandelate than p-nitrophenyl O-methyl-L-mandelate.[123] The reactions of m-bromobenzaldoxime with p-nitrophenyl esters are accelerated by micelles of cetyltrimethylammonium bromide.[124] The rate of hydrolysis of acetylcholine esters is decreased by sodium lauryl sulphate.[125]

Cyclohepta-amylose is a catalyst for the hydrolysis of the β-lactam group of penicillins under mild alkaline conditions. Disappearance of the penicillin is faster than formation of the penicilloic acid, indicating that reaction proceeds via a penicilloyl cyclodextrin.[126]

Catalysis of the hydrolysis of p-acetoxybenzoic acid by the copolymer of N-(5-benzimidazolyl)acrylamide and vinylpyrrolidone is enhanced by the product p-hydroxybenzoic acid.[127]

Deoxyguanyldeoxyguanosine N-acetylhistidate catalyses the hydrolysis of deoxyguanyldeoxyguanosine p-nitrophenyl succinate in the presence of polycytidylic acid. Presumably, catalyst and substrate interact when they are bound of the polymer.[128]

There have been several investigations of catalysis of ester hydrolysis by polymers.[129]

Metal-ion Catalysis

In the presence of Cu^{2+} or Ni^{2+} tris reacts with the nitrile (**73**), but not with nitrile (**74**), to form an isoxazoline. The two mechanisms symbolized by (**75**) and (**76**) were

(73)   (74)

[120] D. G. Oakenfull, *Chem. Comm.*, **1970**, 1655.
[121] S. Friberg and S. I. Ahmad, *J. Phys. Chem.*, **75**, 2001 (1971).
[122] C. A. Bunton, M. Minch, and L. Sepulveda, *J. Phys. Chem.*, **75**, 2707 (1971); cf. *Org. Reaction Mech.*, **1970**, 487.
[123] C. A. Bunton, L. Robinson, and M. F. Stam, *Tetrahedron Letters*, **1971**, 121.
[124] A. K. Yatsimirski, K. Martinek, and I. V. Berezin, *Tetrahedron*, **27**, 2855 (1971).
[125] H. Nagami, J. Hasegawa, and M. Iwatsuru, *Chem. Pharm. Bull.*, **18**, 2297 (1970).
[126] D. E. Tutt and M. A. Schwartz, *J. Am. Chem. Soc.*, **93**, 767 (1971); cf. *Org. Reaction Mech.*, **1970**, 473.
[127] T. Kunitake and S. Shinkai, *J. Am. Chem. Soc.*, **93**, 4247, 4256 (1971).
[128] T. Shimidzu and R. L. Letsinger, *Bull. Chem. Soc. Japan*, **44**, 584 (1971).
[129] T. Komai and J. Noguchi, *J. Biochem.* (*Tokyo*), **70**, 467 (1971); J. Noguchi and H. Yamamoto, *J. Biochem.* (*Tokyo*), **69**, 119 (1971); J. Noguchi, S. Tokura, T. Komai, K. Kokazi, and T. Azuma, *J. Biochem.* (*Tokyo*), **69**, 1033 (1971); D. Lim, J. Zachoval, P. Strop, O. Prouzová, and J. Roda, *Chem. Ind.* (*London*), **1971**, 819; C. G. Overberger and M. Morimoto, *J. Am. Chem. Soc.*, **93**, 3222 (1971); C. G. Overberger, M. Morimoto, I. Cho, and J. C. Salamone, *J. Am. Chem. Soc.*, **93**, 3228 (1971); C. G. Overberger and C.-M. Shen, *Bioorg. Chem.*, **1**, 1 (1971).

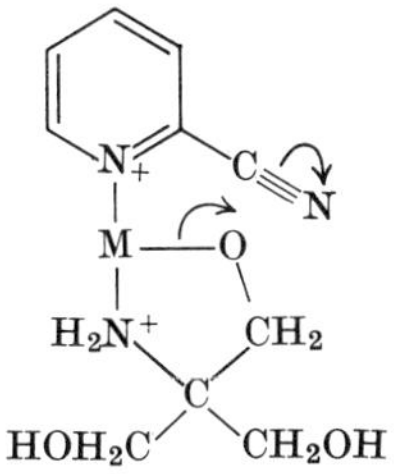

(75) 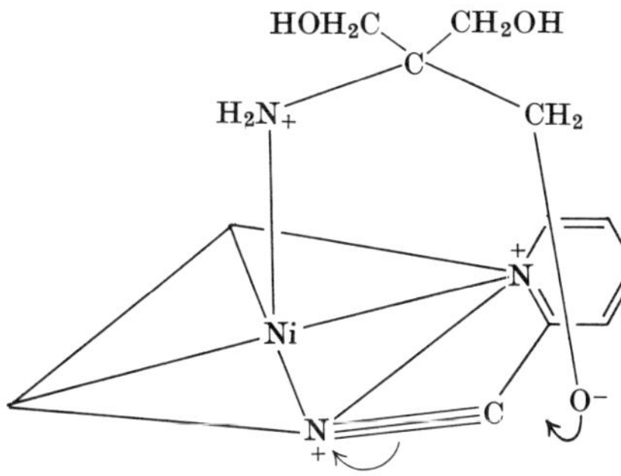   (76)

considered. It was thought that (**76**) was the more likely since (**75**) should be possible with nitrile (**74**).[130]

The hydrolysis of ethyl oxalate is catalysed by metal ions. The pathway shown in the reactions (19) was proposed.[131] The catalysis could involve attack of $^-$OH on the chelate structure (**77**) or intramolecular nucleophilic attack as symbolized by (**78**).

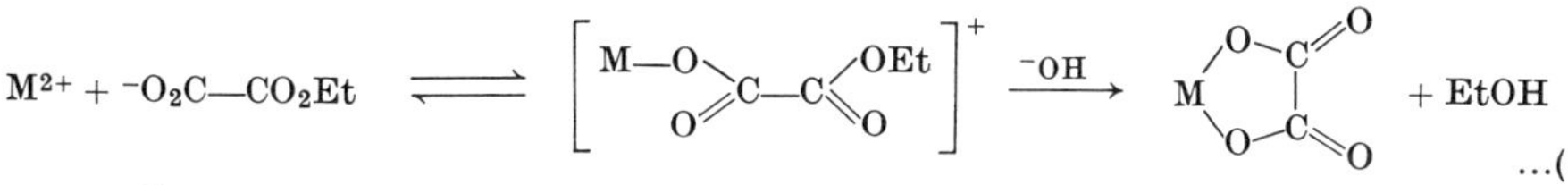

$$M^{2+} + {}^-O_2C\text{---}CO_2Et \rightleftharpoons \left[M\text{---}O\text{---}C\text{---}C\text{---}OEt \right]^+ \xrightarrow{{}^-OH} \text{chelate} + EtOH \quad \ldots(19)$$

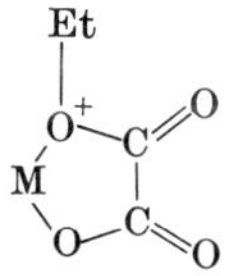

(77) 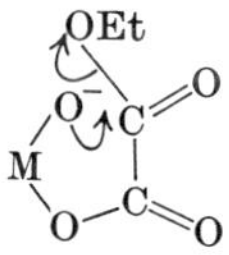   (78)

Other examples of catalysis by metal ions are reported in ref. 132.

Enzymic Catalysis[133–139]

Serine proteinases. Support for the view[140] that the loss of the ability of α-chymotrypsin to bind substrates at high pH's does not arise mainly from deprotonation of isoleucine-16,

---

[130] R. Breslow and M. Schmir, *J. Am. Chem. Soc.*, **93**, 4960 (1971); cf. *Org. Reaction Mech.*, **1967**, 346–347.
[131] G. L. Johnson and R. L. Angelici, *J. Am. Chem. Soc.*, **93**, 1106 (1971).
[132] M. A. Yampol'skaya, V. N. Paukov, S. L. Davydova, and A. V. Ablov, *Kinet. Katal.*, **12**, 759 (1971); *Chem. Abs.*, **75**, 87744 (1971); S. Suzuki and K. Watanabe, *Bull. Chem. Soc. Japan*, **43**, 3858 (1970); J. R. Cronin, D. A. Long, and T. G. Truscott, *Trans. Faraday Soc.*, **67**, 2096 (1971); D. A. Long, T. G. Truscott, J. R. Cronin, and R. G. Lee, *Trans. Faraday Soc.*, **67**, 1094 (1971); R. W. May and P. J. Morris, *J. Chem. Soc. (A)*, **1971**, 1518, 1524.
[133] P. D. Boyer (Ed.), "The Enzymes," Vols. 3 and 4, Academic Press, New York, 1971.
[134] R. G. W. Spickett, "Proteolytic Enzymes," *Chem. Ind. (London)*, **1971**, 82.
[135] G. J. H. Melrose, *Rev. Pure Appl. Chem.*, **21**, 83 (1971).
[136] G. P. Hess and J. A. Rupley, "Structure and Function of Proteins," *Ann. Rev. Biochem.*, **8**, 1013 (1971).
[137] A. S. Mildvan and M. Cohn, "Aspects of Enzyme Mechanisms Studied by Nuclear Relaxation Induced by Paramagnetic Probes," *Adv. Enzymol.*, **33**, 1 (1970).
[138] L. Polgar and M. L. Bender, "Simulated Mutation at the Active Site of Biologically Active Molecules," *Adv. Enzymol.*, **33**, 381 (1970).
[139] J. Turkova, "Mechanism of Action of Proteolytic Enzymes," *Chem. Listy*, **65**, 922 (1971).
[140] Cf. *Org. Reaction Mech.*, **1969**, 459.

but rather from deprotonation of alanine-149, has come from an investigation of the hydrolysis of N-acetyl-L-tryptophan methyl ester catalysed by α-chymotrypsin. K_m for the hydrolysis of this substrate catalysed by α-chymotrypsin which has N-terminal isoleucine-16 and alanine-149 shows a large increase at pH ca. 9.5, but that for the hydrolysis catalysed by α_1-chymotrypsin and δ-chymotrypsin, both of which have an N-terminal isoleucine-16 but not an N-terminal alanine-149, shows only a small increase at pH ca. 10. This suggests that deprotonation of both amino groups of α-chymotrypsin affects its ability to bind substrates in alkaline solution, but that the effect of alanine-149 is much greater than that of isoleucine-16.[141] Additional support for this view comes from the observation that when the amino-group of isoleucine-16 is succinoylated there is little loss of enzymic activity.[142]

Evidence that δ-chymotrypsin undergoes a conformational change on forming an acyl-enzyme has been obtained by Valenzuela and Bender[143] who have compared the reactivity of the amino-group of isoleucine-16 in succinoyl-δ-chymotrypsin and (indolylacryloyl)succinoyl-δ-chymotrypsin. In the former this amino-group reacts rapidly with acetic anhydride and nitrous acid, but in the latter hardly at all. Presumably, on formation of the acyl-enzyme the succinoyl-δ-chymotrypsin undergoes a conformational change that protects the amino-group of isoleucine-16.

The pH-dependence of the reactivity of the N-terminal isoleucine amino-group of α-chymotrypsin has been determined.[144] In 0.1M-potassium chloride at 10° the pK_a (app) is 8.9 and the reactivity is 20% of that of an amino-group with pK_a 8.9; but above pH 9.8 the reactivity increases sharply and at pH 12 the amino-group has normal reactivity. It was suggested that below pH 9.8 the state of protonation of this group controlled the activity of α-chymotrypsin but that above this pH additional structural changes occur in which alanine-149 is possibly involved.

The conformational transition of α-chymotrypsin at pH ca. 9 has been studied by UV and fluorescence spectroscopy. It seems that the amino group of isoleucine-16 has a normal pK_a in one conformation and a higher value (10.5) in the other in which it was thought to be buried inside the molecule. The interconversion of the two forms is relatively slow, with rate constants in the range 0.005–3.3 sec^{-1}. The equilibrium can be written as shown in equation (20).[145]

There have been several other investigations of conformational changes of chymotrypsin and its derivatives.[146]

$$\begin{array}{ccc} \mathrm{EH} & \xrightleftharpoons{K} & \mathrm{E} + \mathrm{H}^+ \\ k_1 \updownarrow k_{-1} & & k_2 \updownarrow k_{-2} \\ \mathrm{EH'} & \xrightleftharpoons{K^1} & \mathrm{E'} + \mathrm{H}^+ \end{array} \quad \ldots(20)$$

$K = 3 \times 10^{-11}$, $K^1 = 6 \times 10^{-8}$, $k_1 = 0.005$ s^{-1}, $k_{-1} = 0.235$ s^{-1}, $k_2 = 3.3$ s^{-1}, $k_{-2} = 0.07$ s^{-1}

[141] P. Valenzuela and M. L. Bender, *J. Am. Chem. Soc.*, **93**, 3783 (1971).
[142] T. T. Blair, M. A. Marini, S. P. Agarwal, and C. J. Martin, *FEBS Letters*, **14**, 86 (1971).
[143] P. Valenzuela and M. L. Bender, *Biochim. Biophys. Acta*, **235**, 411 (1971); cf. ref. 147.
[144] H. Kaplan, *Biochem. Biophys. Res. Comm.*, **42**, 1042 (1971).
[145] J. R. Gavel and B. Labouesse, *Biochimie*, **53**, 9 (1971).
[146] Y. D. Kim and R. Lumry, *J. Am. Chem. Soc.*, **93**, 1003 (1971); Y. Kim and R. Lumry, *J. Am. Chem. Soc.*, **93**, 5882 (1971); R. Biltonen and R. Lumry, *J. Am. Chem. Soc.*, **93**, 224 (1971); D. F. Shiao, R. Lumry, and J. Fahey, *J. Am. Chem. Soc.*, **93**, 2024 (1971); R. Lumry and S. Rajender, *J. Phys. Chem.*, **75**, 1387 (1971); R. Lumry, *Adv. Chem. Phys.*, **21**, 567 (1971).

The parameter k_2/K_s for the hydrolysis of the non-specific substrates, p-nitrophenyl m-nitrophenylacetate and 2-hydroxy-5-nitrotoluene-p-sulphonic acid sultone, catalysed by δ-chymotrypsin decreases at high pH's. This behaviour is different from that of specific substrates for which k_2/K_s remains unchanged. It was suggested that δ-chymotrypsin with the N-terminal amino-group of isoleucine-16 unprotonated can still bind specific substrates but not these non-specific substrates or that the complex between chymotrypsin and these substrates reacts slowly when the isoleucine-16 amino-group is not protonated.[147]

2-Phenylethaneboronic acid is an inhibitor for the hydrolysis of methyl hippurate catalysed by α-chymotrypsin. It is more effective at pH's where imidazole-57 is unprotonated and under these conditions it is bound to α-chymotrypsin 150 times more strongly than hydrocinnamide. It was proposed that there was a covalent bond formed between the boronic acid and the hydroxyl of serine-195 to yield a compound that would be analogous to a tetrahedral intermediate in a chymotrypsin-catalysed reaction.[148]

The hydrolysis of p-nitrophenyl thiolacetate catalysed by chymotrypsin shows an initial burst, indicating that reaction goes through an acyl-enzyme. The rate of acetylation is similar to that for the reaction with p-nitrophenyl acetate, suggesting that formation of the tetrahedral intermediate is the rate-determining step.[149]

Catechol cyclic phosphate reacts with chymotrypsin to form a phosphorylated enzyme that reacts much faster on intramolecular attack by the phenolic hydroxyl group than on attack by water.[150]

Substituent effects on the hydrolysis of anilides catalysed by α-chymotrypsin have been discussed further.[151] The second-order acylation rate constants, k_{cat}/K_m for the chymotrypsin-catalysed hydrolysis of aryl hippurates and aryl N-benzyloxycarbonyl-L-tryptophanates follow the Hammett equation with $\rho = +0.63$ and $+0.46$, respectively.[152]

The binding of N-trifluoroacetyl-D-tryptophan to α-chymotrypsin causes a downfield shift in the ^{19}F-NMR signal in the pH range 6.37–8.12. The values of K_I increase at pH 8, consistently with the presence of a group close to the binding site and ionizing at this pH.[153]

When N-[3-(2-furyl)acryloyl]tryptophan methyl ester is treated with α-chymotrypsin at pH 2.3 it is converted first into an oxazolinone and then into the acyl-enzyme. It seems that this species is also an intermediate at neutral pH's. It does not appear to be an intermediate in the hydrolysis of the acyl enzyme. The pathway for hydrolysis therefore appears to be that shown in (21) where ES* is the Michaelis complex of the

$$E + S \rightleftharpoons ES \rightarrow ES^* \rightarrow ES' \rightarrow E + P_2 \qquad \ldots(21)$$
$$+ P_1$$

oxazolinone and enzyme and ES' is the acyl-enzyme. (Phenylthiocarbamoyl)amino-acids are dehydrated by chymotrypsin.[154]

---

[147] S. F. Bosen and E. T. Kaiser, *J. Am. Chem. Soc.*, **93**, 1038 (1971); cf. ref. 143.
[148] K. A. Koehler and G. E. Lienhard, *Biochemistry*, **10**, 2477 (1971).
[149] A. Frankfater and F. J. Kézdy, *J. Am. Chem. Soc.*, **93**, 4039 (1971).
[150] E. T. Kaiser, T. W. S. Lee, and F. P. Boer, *J. Am. Chem. Soc.*, **93**, 2351 (1971).
[151] A. R. Fersht, *J. Am. Chem. Soc.*, **93**, 3504 (1971); W. P. Jencks, B. Schaffhausen, K. Tornheim, and H. White, *J. Am. Chem. Soc.*, **93**, 3917 (1971); see also A. R. Fersht and Y. Requena, *J. Am. Chem. Soc.*, **93**, 7079 (1971).
[152] R. E. Williams and M. L. Bender, *Can. J. Biochem.*, **49**, 210 (1971).
[153] H. Ashton and B. Capon, *Chem. Comm.*, **1971**, 513.
[154] M. A. Coletti-Previero, C. Axelrad-Cavadone, and A. Previero, *FEBS Letters*, **11**, 213, 218 (1970).

A kinetic investigation of the interaction of α-chymotrypsin with proflavin has been carried out. Above pH 9 the enzyme exists in a form that does not bind proflavin, but this form is present to the extent of ca. 20% at pH 6.9. This was demonstrated by rapid mixing of solutions of chymotrypsin and proflavin, both at pH 6.9; there was a rapid increase in absorbance corresponding to binding by the active form followed by a slow increase that was the result of binding by the active form after it had been formed slowly from the inactive form. When chymotrypsin at pH 12.0 was mixed with proflavin at pH 6.84 there was no rapid increase in absorbance but only a slow one corresponding to conversion of the inactive into the active form. The rate constant for the interconversion is 3.15 sec^{-1} at pH 6.84. It was considered that these results agreed with the proposal by Hess that in the active form the N-terminal isoleucine amino-group forms a salt bridge with aspartic acid-194 and that this is destroyed on deprotonation of isoleucine-16 when the enzyme undergoes a conformational change to yield the inactive form. The pK_a of isoleucine in the inactive form is 7.94 ± 0.1.[155] For other investigations of the binding of proflavin to chymotrypsin, see ref. 156. The binding of Biebrich Scarlet to chymotrypsin in the presence and absence of m-cresol has also been studied.[157]

The variation of refractive index, UV absorbance, and specific rotation of α-chymotrypsin in potassium chloride solution has been interpreted as indicating that the potassium chloride solution causes a conformational change in the enzyme.[158]

Photolysis of ^{14}C-diazoacetyl-α-chymotrypsin generates a carbene that was thought to react with the tyrosine-146 of another molecule of chymotrypsin.[159]

The rate of deacylation of *cis*-4-nitrocinnamoyl-α-chymotrypsin is 1000 times less than that of the *trans*-isomer. The *cis*-isomer can conveniently be converted into the *trans*-isomer by irradiation with light of 363 nm. Irradiation by light can therefore be made to release chymotrypsin.[160]

The quantum yield of the fluorescence of 3-benzyl-7-(diethylamino)-4-methylcoumarin is increased in the presence of chymotrypsin and there is a blue shift in the position of the fluorescence band. TPCK and DIP chymotrypsin have only a small effect. The changes in fluorescence were suppressed by 3-phenylpropionic acid. There is a change in the fluorescence band similar to that occurring when the medium is changed from water to alcohol or alcohol–water mixtures; this suggests that the binding site is less polar than water.[161]

The compound $PhCH_2CH(NHcbz)COCH_2Cl$ is a much more effective alkylating agent for chymotrypsin than is the corresponding formyl derivative.[162]

Although salicylaldoxime and acetoxime react at similar rates in non-enzymic reactions of esters, the former reacts 50 times faster with N-acetyl-L-tyrosinyl-chymotrypsin. Acetoxime appears to react with general base-catalysis by the imidazole group of histidine-57.[163]

---

[155] A. R. Fersht and Y. Requena, *J. Mol. Biol.*, **60**, 279 (1971).
[156] W. Chen and S. F. Russo, *Can. J. Biochem.*, **49**, 28 (1971); J. L. Marini and M. Caplow, *J. Am. Chem. Soc.*, **93**, 5560 (1971).
[157] R. Jayaram and I. D. Rattee, *Trans. Faraday Soc.*, **67**, 884 (1971).
[158] C. C. Cuppett, M. Resnick, and W. J. Canady, *J. Biol. Chem.*, **246**, 1135 (1971).
[159] C. S. Hexter and F. H. Westheimer, *J. Biol. Chem.*, **246**, 3928, 3934 (1971).
[160] S. D. Varflomeyev, A. M. Klibanov, K. Martinek, and I. V. Berezen, *FEBS Letters*, **15**, 118 (1971).
[161] Y. Nishimura, O. Takenaka, and K. Shibala, *J. Biochem.* (*Tokyo*), **70**, 293 (1971).
[162] E. Shaw and J. Rusaca, *Arch. Biochem. Biophys.*, **145**, 484 (1971).
[163] I. V. Berezen, N. F. Kazanskaya, and A. A. Klyosov, *Biokhimya*, **36**, 108 (1971).

Rate constants for the acylation and deacylation of chymotrypsin have been determined by a nucleophilic competition method in which butane-1,4-diol is used.[164]

The specificity of chymotrypsin with respect to the amino-acid residue one residue remote from the reaction centre of the substrate has been determined for the hydrolysis of the methyl ester of N-acetyl-leucyltyrosine and N-acetyltyrosyl-leucine.[165]

The chymotrypsin-catalysed hydrolysis of N-acylated L-tyrosines has been investigated. As the acylamino-group is enlarged the value of K_s is increased and k_2 is decreased but k_3 remains practically unchanged. It was proposed that this behaviour arose from non-productive binding.[166] Inhibition by N-acetyl-D-tryptophan of the hydrolysis of N-acetylglycine ethyl ester catalysed by chymotrypsin,[167] and the temperature-dependence of complex formation between α-chymotrypsin and competitive inhibitors, have also been studied.[168]

Compensation between $\Delta H^{\pm}$ and $\Delta S^{\pm}$, and between $\Delta H°$ and $\Delta S°$, in the chymotrypsin-catalysed hydrolysis of N-acetyl-L-tryptophan ethyl esters at various pH's has been studied.[169] There has been a discussion of structure–reactivity relationships in chymotrypsin-catalysed reactions.[170]

Formaldehyde causes a decrease in the activity of α-chymotrypsin that was attributed to formation of a derivative with the imidazole group of histidine-57.[171] The terminal and ϵ-amino groups also interact with formaldehyde.[172]

The kinetics of activation by trypsin of chymotrypsinogen modified by reaction with diazotized p-arsanilic acid have been studied by following the change in the CD spectrum.[173] When chymotrysinogen-A is activated by papain a chymotrypsin is formed that is identical with α-chymotrypsin except that it lacks serine-11, glycine-12 and leucine-13. The kinetic parameters are similar to those of chymotrypsin.[174]

Human chymotrypsin-II is composed of two chains with N-terminal half-cystine and isoleucine, and carboxyl-terminal serine and asparagine. It appears to resemble α-chymotrypsin.[175]

It was proposed[176] that dimerization of α-chymotrypsin arises from an electrostatic interaction between the imidazole of histidine-57 of one molecule and the carboxyl of tyrosine-146 of another.

The mechanism of the proton-transfer process in reactions catalysed by chymotrypsin has been discussed.[177]

---

[164] I. V. Berezen, N. F. Kazanskaya, and A. A. Klyosov, *FEBS Letters*, **15**, 121 (1971).
[165] N. F. Kazanskaya, E. M. Slobodyanskaya, V. I. Zetlin, E. N. Shepel, V. T. Ivanov, and Y. A. Ovchinnikov, *Biokhimya*, **35**, 1147 (1970).
[166] I. V. Berezen, N. F. Kazanskaya, and A. A. Klyosov, *Biokhimya*, **36**, 227 (1971).
[167] K. Martinek, A. V. Levashov, V. L. Rubajlo, S. D. Varfolomeyev, and I. V. Berezen, *Biokhimya*, **35**, 1207 (1970).
[168] K. Martinek, A. V. Levashov, and I. V. Berezen, *Biokhimya*, **36**, 167 (1971).
[169] S. Rajender, R. Lumry, and M. Han, *J. Phys. Chem.*, **75**, 1375 (1971).
[170] I. V. Berezen, N. F. Kazanskaya, A. A. Klyosov, and K. Martinek, *FEBS Letters*, **15**, 125 (1971).
[171] M. A. Marini and C. J. Martin, *Eur. J. Biochem.*, **19**, 153 (1971); see also ref. 26.
[172] M. A. Marini and C. J. Martin, *Eur. J. Biochem.*, **19**, 162 (1971).
[173] G. F. Fairclough and B. L. Vallee, *Biochemistry*, **10**, 2470 (1971).
[174] M. C. Shaw and D. Gratecos, *Biochim. Biophys. Acta*, **235**, 188 (1971).
[175] M. K. Coan, R. C. Roberts, and J. Travis, *Biochemistry*, **10**, 2711 (1971).
[176] K. C. Aune and S. N. Timasheff, *Biochemistry*, **10**, 1609 (1971); K. C. Aune, L. C. Goldsmith, and S. N. Timasheff, *Biochemistry*, **10**, 1617 (1971).
[177] L. Polgar, *J. Theor. Biol.*, **31**, 165 (1971).

For other investigations on α-chymotrypsin, see ref. 178.

Monoacetyl-trypsin was prepared from trypsin and p-nitrophenyl acetate. It was still catalytically active but k_{cat} was only 6% of that of natural trypsin for the hydrolysis of benzoyl-L-arginine ethyl ester and 37% of that of natural trypsin for the hydrolysis of benzyloxycarbonyl-L-arginine m-nitroanilide. K_m for the former reaction is unchanged but for the latter is increased three-fold. It was thought that the residual esterolytic activity arose from unacetylated trypsin present as an impurity but that the amidase activity arose from the acyl-trypsin. It was therefore claimed that trypsin can catalyse the hydrolysis of amides by a pathway other than through an acyl-enzyme.[179] On the other hand, when chymotrypsin is carbamylated by potassium cyanate the loss of esterolytic and amidase activity parallel each other.[180]

Trypsin modified by reaction with 5-(dimethylamino)naphthalene-1-sulphonyl chloride has enhanced activity towards N-benzoyl-DL-arginine p-nitroanilide but not towards N-benzoyl-L-arginine ethyl ester or N-benzoyl-L-arginine amide. The increased activity towards the anilide results from a decrease in K_m.[181]

Trypsin modified by coupling with quinoline-3-diazonium fluoroborate has enhanced activity towards N-benzoylarginine p-nitroanilide and N-toluene-p-sulphonyl-L-arginine methyl ester. The values of k_{cat} for both compounds are increased. Since acylation is rate-limiting for the anilide and deacylation for the ester it seems that the rate constants for both these processes are increased. Both substrates show substrate activation before and after modification of the enzyme.[182]

2-Alkylamines are more effective than 1-alkylamines in promoting the trypsin-catalysed hydrolysis of the p-nitrophenyl and ethyl esters of acetylglycine although the latter amines are more strongly bound to the enzyme. The hydrolysis of p-nitrophenyl acetate catalysed by trypsin is only slightly enhanced by amines.[183] There has been a discussion of the kinetics of trypsin-catalysed reactions in the presence of modifiers.[184]

The values of k_2, k_3 and K_s have been determined for the trypsin-catalysed hydrolyses of some specific (e.g. Bz—Arg—OEt) and non-specific (e.g. Ac—Phe—OMe) substrates. The main factor involved when specific substrates react faster is a higher value of k_2. Possibly specific substrates are bound to trypsin in a strained conformation.[185]

p-Nitrophenyl and ethyl p-amidinobenzoates are much better substrates for trypsin than are their *meta*-isomers, which are inhibitors.[186] p-Nitrophenyl p-amidobenzoate[187a]

[178] L. M. Del Castillo. Z. Nieto, E. Arce, G. Inei-Shizukawa, M. T. Cruz, and M. Castaneda-Agulló, *Biochim. Biophys. Acta*, **235**, 358 (1971); G. L. Rossi and S. A. Bernhard, *J. Mol. Biol.*, **55**, 215 (1971); D. N. Hague, J. S. Henshaw, V. A. John, M. J. Pooley, and P. B. Chock, *Nature*, **229**, 190 (1971); H. J. Gross and H. Albierty, *Z. Physiol. Chem.*, **352**, 1177 (1971); P. F. Sikk, A. A. Aaviksaar, N. N. Godovikov, N. A. Morozova, and V. A. Palm, *Organic Reactivity (Tartu)*, **7**, 986 (1970); J. P. Paris, A. A. Aaviksaar, A. A. Abduvakhavov, and P. F. Sikk, *Organic Reactivity (Tartu)*, **7**, 977 (1970); K. Martinek, I. E. Kirsh, A. A. Strongina, V. N. Dorovska, A. K. Iatsimirskii, I. V. Berezen, and V. A. Kabanov, *Doklady Akad. Nauk SSSR*, **199**, 148 (1971); S. A. Hawley, *Biochemistry*, **10**, 2436 (1971); R. Biltonen, A. T. Schwartz, and I. Wadso, *Biochemistry*, **10**, 3417 (1971).
[179] S. E. Bresler, V. M. Krutyakov, and G. P. Vlasov, *Eur. J. Biochem.*, **18**, 131 (1971).
[180] C. E. Stauffer, *FEBS Letters*, **16**, 45 (1971).
[181] J. G. Franklin and J. Leslie, *Can. J. Biochem.*, **49**, 516 (1971).
[182] M. Kanazawa, N. Yoshida, and Shin-ichi, *Biochim. Biophys. Acta*, **250**, 372 (1971).
[183] F. Seydoux, G. Coutouly, and J. Yon, *Biochemistry*, **10**, 2284 (1971).
[184] E. Holler, *J. Theor. Biol.*, **32**, 415 (1971).
[185] F. Seydoux and J. Yon, *Biochem. Biophys. Res. Comm.*, **44**, 745 (1971).
[186] K. Tanizawa, Shin-ichi, and Y. Kanaoka, *Chem. Pharm. Bull.*, **18**, 2247 (1970).
[187a] K. Tanizawa, Shin-ichi, K. Hamaguchi, Y. Kanaoka, and T. Ikenaka, *Chem. Pharm. Bull.*, **18**, 2571 (1970).

and p-nitrophenyl *trans*-4-aminomethylcyclohexane-1-carboxylate[187b] have been used to titrate the active site of trypsin. Inhibition of trypsin by aromatic amidines[188] and by silver ions[189] has been studied.

The pH-dependence of the hydrolysis of N-benzoyl-L-arginine amide catalysed by trypsin and N- and O-acetylated trypsin has been studied.[190]

There have been several other studies of trypsin.[191]

The effect of sodium chloride on hydrolyses catalysed by trypsin, chymotrypsin, papain, subtilopeptidase A, leucineaminopeptidase, or aminopeptidase B has been studied.[192a]

The active site of elastase consists of 6 or 7 sub-sites, some of which bind the amino-acids of the amide component (S_1', S_2', etc.) while others bind the amino-acids of the carboxyl component (S_1, S_2, etc.). S_2', S_1 and S_2 are hydrophobic in character and occupation of S_4 is important for effective hydrolysis.[192b] The value of k_{cat}/K_m for the elastase-catalysed hydrolysis of the esters shown in Table 1 increases strongly with chain length, largely as a result of a decrease in K_m. Elastase catalyses the hydrolysis of the amides of tripeptides but not those of smaller peptides. Amides of larger peptides are hydrolysed more rapidly, largely owing to an increase in k_{cat}. It is possible that the binding of peptides to elastase involves an induced fit.[193]

Table 1. The kinetics of hydrolysis of methyl esters of alanine oligomers at pH 9 and 37° (see ref. 193)

	k_{cat}/K_m (mole^{-1} sec^{-1})
Ac—Ala—OMe^a	49
Ac—Ala—Ala—OMe	2,200
Ala—Ala—Ala—OMe^b	980
Ala—Ala—Ala—OMe	2,700
Ac—Ala—Ala—Ala—OMe	300,000
Ac—Ala—Ala—Ala—Ala—OMe^c	1,800,000

^{a} pH 8. ^{b} pH 7.5. ^{c} 1% of trifluoroethanol.

1-Bromo-4-(2,4-dinitrophenyl)butan-2-one is a specific irreversible inhibitor for elastase. It probably alkylates the γ-carboxyl group of glutamic acid 6. This residue does not appear to be near the active site.[194]

[187b] K. Tanizawa, Shin-ichi, and Y. Kanaoka, *Chem. Pharm. Bull.*, **18**, 2346 (1970).
[188] K. Tanizawa, Shin-ichi, K. Hamaguchi, and Y. Kanaoka, *J. Biochem. (Tokyo)*, **69**, 893 (1971).
[189] K. Martinek, Y. V. Savin, and I. V. Berezen, *Biokhimya*, **36**, 806 (1971).
[190] W. E. Spomer and J. F. Wootton, *Biochim. Biophys. Acta*, **235**, 164 (1971).
[191] N. F. Kazanskaya and N. I. Larionova, *Biokhimya*, **36**, 187 (1971); M. Muramatu and S. Fizii, *Biochem. Biophys. Acta*, **242**, 203 (1971); A. A. Matsushima, K. Nakamura, K. Shibata, and Y. Inada, *J. Biochem. (Tokyo)*, **70**, 537 (1971); N. Yoshida, T. Kato, and N. Izumiya, *Bull. Chem. Soc. Japan*, **43**, 2912 (1970); R. A. Kenner and H. Neurath, *Biochemistry*, **10**, 551 (1971); R. A. Kenner, *Biochemistry*, **10**, 545 (1971); N. C. Robinson, R. W. Tye, H. Neurath, and K. A. Walsh, *Biochemistry*, **10**, 2743 (1971).
[192a] K. K. Makinen and P.-L. Makinen, *Acta Chem. Scand.*, **25**, 969 (1971).
[192b] D. Atlas, S. Levit, I. Schechter, and A. Berger, *FEBS Letters*, **11**, 281 (1970).
[193] R. C. Thompson and E. R. Blout, *Proc. Nat. Acad. Sci.*, **67**, 1734 (1970).
[194] L. Visser, D. S. Sigman, and E. R. Blout, *Biochemistry*, **10**, 735 (1971).

The optical properties of elastase have been studied.[195]

The following serine proteinases have also been studied: proteinase from *Aspergillus flavus*,[196] thermolysin,[197] and thrombin.[198]

Thiol proteinases. The specificity of papain to cleave peptide linkages one removed on the carboxyl side of an aromatic amino-acid has been shown to arise mainly from a high k_2 when the hydrolysis of N-acetyl-L-phenylalanylglycine p-nitroanilide is compared to the p-nitroanilide of hippuric acid. K_m is slightly smaller for the former reaction but a comparison of the kinetics of hydrolysis of the corresponding p-nitrophenyl esters shows that k_3 is unchanged. A detailed discussion of the specificity of papain was given.[199] The ρ-value of k_2 for the papain-catalysed hydrolysis of anilides of N-acetyl-L-phenylalanylglycine is -1.04. This is similar to the ρ-value for the hydrolysis of anilides of acylated amino-acids catalysed by chymotrypsin and it suggests that acid-catalysis is occurring in the rate-limiting step.[200]

The rate of release of nitrophenols from the nitrophenyl esters of N-(benzyloxycarbonyl)glycine catalysed by papain is increased in the presence of L-tryptophan amide. The increase in rate with concentration of this amide is linear at low concentrations but reaches a steady value at higher concentrations. This was interpreted as indicating a change in the rate-determining step from decomposition to formation of the acyl-enzyme at high concentrations. Similar results were obtained with ficin. It was concluded that the primary mode of binding of p-nitrophenyl esters to papain and ficin is non-productive.[201]

The pH-dependence of the reaction of bromoacetamide with papain shows a pK_a (app) of 7.3. The reaction is 200 times faster than the reaction of thiols with bromoacetamide.[202] It is reported that if inactive papain is treated with N-bromosuccinimide before activation with thioglycerol a more active enzyme is obtained.[203] Other work on papain is described in papers listed in ref. 204a.

The activity of stem-bromelin is not much affected by periodate oxidation of the carbohydrate component.[204b]

Acid proteinases.[205] The inhibition of the pepsin-catalysed hydrolysis of N-acetyl-L-phenylalanine at pH 2 is non-competitive and linear. This is consistent with a mechanism in which the N-acetyl-L-phenylalanine combines non-productively with the enzyme and the amino-enzyme intermediate. At pH 4.3 the inhibition is competitive, which is consistent with the inhibitor's combining productively with the intermediate. That this does in fact occur was confirmed by using labelled N-acetyl-L-phenylalanine which was incorporated into the substrate at pH 4.3 but not at pH 2.1. These results suggest that

---

[195] L. Visser and E. R. Blout, *Biochemistry*, **10**, 743 (1971).
[196] A. S. Tsyperovich and N. V. Lysenkov, *Biokhimya*, **36**, 937 (1971).
[197] T. Abe, K. Takahashi, and T. Avato, *J. Biochem. (Tokyo)*, **69**, 363 (1971).
[198] R. L. Lundblad, *Biochemistry*, **10**, 2501 (1971); R. K. H. Liem, R. H. Andreatta, and H. A. Scheraga, *Arch. Biochem. Biophys.*, **147**, 201 (1971); R. H. Andreatta, R. K. H. Liem, and H. A. Scheraga, *Proc. Nat. Acad. Sci.*, **68**, 253 (1971).
[199] G. Lowe and Y. Yuthavong, *Biochem. J.*, **124**, 107 (1971).
[200] G. Lowe and Y. Yuthavong, *Biochem. J.*, **124**, 117 (1971).
[201] P. M. Hinkle and J. F. Kirsch, *Biochemistry*, **10**, 2717 (1971).
[202] H. G. Löffler, F. Schneider, and M. Wenck, *Z. Naturforsch.*, **26b**, 43 (1971).
[203] D. M. Kirschenbaum, *Biochim. Biophys. Acta*, **235**, 159 (1971).
[204a] R. F. Steiner, *Biochemistry*, **10**, 771 (1971); J. L. Abernethy, E. Albano, and J. Comyno, *J. Org. Chem.*, **36**, 1580 (1971); G. Jori, G. Gennari, C. Tanolo, and E. Scoffine, *J. Mol. Biol.*, **59**, 151 (1971).
[204b] Y. Yasuda, N. Takahashi, and T. Murachi, *Biochemistry*, **12**, 2624 (1971).
[205] J. S. Fruton, *Adv. Enzymol.*, **33**, 401 (1970).

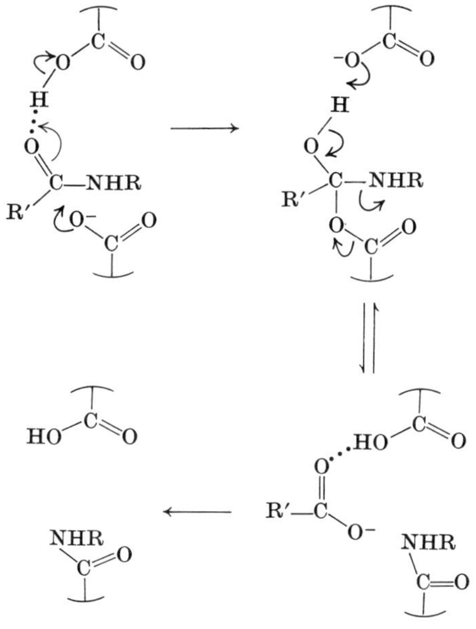

Scheme 1

the initial cleavage product of the peptide is the anion of N-acetyl-L-phenylalanine and the tentative mechanism shown in Scheme 1 was proposed.[206]

The possibility that the low kinetic pK_a of 1.0 for pepsin arose from electrostatic stabilization of the carboxylate group by a positively charged arginine residue has been examined by allowing pepsin to react with phenylglyoxal. At pH 1.3 the modified enzyme has an activity about 45% of that of the unmodified enzyme. This arose from a decrease in k_{cat}; K_m was unchanged. The kinetic pK_a changes from 1.0 to 1.5. It therefore seems that even in the modified enzyme the catalytic carboxyl group has a low pK_a.[207]

The hydrolysis of peptides Gly_n—Tyr—Tyr ($n = 1$, 2 or 3) catalysed by pepsin has been studied. The compounds with $n = 2$ or 3 are hydrolysed more rapidly than that with $n = 1$. Hydrolysis was accompanied by transpeptidation.[208]

The pH-dependence of k_{cat}/K_m for hydrolysis of bis-p-nitrophenyl sulphite with pepsin is bell-shaped with pK_a 0.82 and 5.17. These values are similar to those obtained with natural substrates. It was proposed that there are two catalytically active carboxyl groups, one of which is active in its acidic form and one in its basic form.[209]

Pepsin modified by reaction with p-bromophenacyl bromide has k_{cat} unchanged and K_m increased about two-fold compared to the unmodified enzyme for the hydrolysis of Ac—Phe—Tyr—OMe.[210]

Carboxylic acids are competitive inhibitors in the pepsin-catalysed hydrolysis of N-(benzyloxycarbonyl)-L-glutaryl-L-tyrosine.[211]

[206] T. M. Kitson and J. R. Knowles, *Biochem. J.*, **122**, 241, 249 (1971).
[207] T. M. Kitson and J. R. Knowles, *FEBS Letters*, **16**, 337 (1971).
[208] S. Terada, S. Yoshida, and N. Izumuja, *Biochem. J. (Tokyo)*, **70**, 133 (1971).
[209] S. W. May and E. T. Kaiser, *J. Am. Chem. Soc.*, **93**, 5567 (1971); see also C. D. Hubbard and T. P. Stein, *Biochem. Biophys. Res. Comm.*, **45**, 293 (1971).
[210] T. A. Valueva, L. M. Ginodman, and S. E. Terekhova, *Biokhimya*, **36**, 667 (1971).
[211] H. D. Shin, *Bull. Chem. Soc. Japan*, **43**, 3472 (1970).

Other work on pepsin is described by authors listed in ref. 212.

Diazoacetyl-DL-norleucine methyl ester is a reversible inhibitor for porcine pepsin C. It reacts with the β-carboxyl group of an aspartic acid residue in the sequence Ile—Val—Asp—Thr, which is identical with that found when the same inhibitor reacts with pepsin. The difference in specificity between pepsin and pepsin-C was discussed.[213] There has also been a discussion of the specificity of other acid proteinases.[214]

Metallo-proteinases. The hydrolysis of 3-(naphthyloxycarbonyl)-L-phenylalanine catalysed by carboxypeptidase-A has been studied by measuring the formation of the β-naphtholate ion spectrofluorimetrically. The binding of this substrate to the enzyme was also studied and quite good agreement was obtained between the value of K_s determined this way and kinetically.[215]

The pH-dependence of the rate of hydrolysis of tripeptides catalysed by carboxypeptidase-A shows that the value of K_s depends on the state of ionization of two carboxyl groups of pK_a 6 and 9 and that k_{cat} depends on the ionization of a group of pK_a 6–7.[216]

The kinetics of inactivation of carboxypeptidase-A_γ^{Leu} by N-bromoacetyl-N-methyl-L-phenylalanine suggests that reaction proceeds through a reversibly formed complex. The pH-dependence of the first-order rate constant for reaction of this complex suggests that a nucleophilic group of pK_a ca. 7 is involved. The site of modification is glutamic acid-270, thought to be important in the catalytic action. In addition, the α-amino-group of asparagine-8 and the imidazole group of histidine-13 were modified.[217] Inactivation by Woodward's reagent K has also been studied.[218] Cbz—Phe—Ala—Ala and other di- and tri-peptides are activators for carboxypeptidase-A.[219]

Other work on carboxypeptidase A[220] and on carboxypeptidase S from *Pseudomas stutzeri*[221] has been reported.

Esterases. The rate of disappearance of phenyl acetate catalysed by a pig-liver esterase preparation is increased 5.5 times by 0.5M-methanol, which was interpreted as indicating that reaction proceeded via an acyl-enzyme whose decomposition is rate-limiting. However, different acetate esters yielded different ratios of acetic acid to methyl acetate, and diethyl p-nitrophenyl phosphate has different effects on the ability of the preparation to catalyse hydrolysis and methanolysis. It was therefore proposed that two kinds of active site were present. The enhancing effect of alcohols, ROH, on the release of phenol from phenyl acetate in the presence of enzyme decreases as the electron-withdrawing power of R increases. This behaviour is different from that found with furoyl-chymotrypsin, where ethanol and trifluoroethanol have similar effects. It was proposed that in the transition state for the deacylation of the acyl-esterase the oxygen of the alcohol carries a partial positive charge and that O—C bond-forming has run ahead of proton transfer.[222]

---

[212] H. M. Lang and B. Kassell, *Biochemistry*, **10**, 2296 (1971); S. S. Husain, J. B. Ferguson, and J. S. Fruton, *Proc. Nat. Acad. Sci., N.Y.*, **68**, 2765 (1971); R. A. Badley and F. W. J. Teak, *J. Mol. Biol.*, **58**, 567 (1971); R. A. Valiulis and V. M. Stepanov, *Biokhimya*, **36**, 358 (1971).
[213] J. Kay and A. P. Ryle, *Biochem. J.*, **123**, 75 (1971).
[214] I. M. Voynick and J. S. Fruton, *Proc. Nat. Acad. Sci.*, **68**, 257 (1971).
[215] N. Lasser and J. Feitelson, *Biochemistry*, **10**, 307 (1971).
[216] D. S. Auld and B. L. Vallee, *Biochemistry*, **10**, 2892 (1971).
[217] G. M. Hass and H. Neurath, *Biochemistry*, **10**, 3535, 3541 (1971).
[218] P. H. Pétra, *Biochemistry*, **10**, 3163 (1971); P. H. Pétra and H. Neurath, *Biochemistry*, **10**, 3171 (1971).
[219] I. Schechter and E. Zazepizki, *Eur. J. Biochem.*, **18**, 469 (1971).
[220] C. A. Ryan, *Biochem. Biophys. Res. Comm.*, **44**, 1265 (1971).
[221] J. L. McCullough, B. A. Chabner, and J. R. Bertino, *J. Biol. Chem.*, **246**, 7207 (1971).
[222] P. Greenzaid and W. P. Jencks, *Biochemistry*, **10**, 1210 (1971).

Esterases from rat liver have been studied[223] and there have been several investigations on cholinesterases.[224]

Other enzymes. The maximum velocity of the reaction of *p*-nitroacetanilide with five substituted anilines, semicarbazide, hydroxylamine, and hydrazine catalysed by acetyl-coenzyme-A arylamine acetyl transferase are identical, which suggests that the rate-limiting step is formation of an acyl-enzyme. With *p*-nitrophenyl acetate as the acyl-donor this step becomes much more rapid and the deacylation of the more weakly basic anilines (pK_a 4) varied with the amine. Under these conditions the reactions show saturation kinetics and so there must be a binding site for the amine.[225]

Acylase,[226] aminopeptidase M,[227] and carbonic anhydrase[228] have also been studied.

Decarboxylation
Thermal decarboxylations of 2- and 4-pyridylacetic acid occur at similar rates, which suggests that the 2-isomer does not react via a cyclic mechanism. Photochemical decarboxylation was also studied.[229] The decarboxylation of pyridine-2-carboxylic acids has been investigated.[230]

Acetoacetate decarboxylase from *Clostridium acetobutylicum* reacts with 2,4-dinitrophenyl propionate which acylates the essential amino-group. The pH-dependence of the rate of this reaction shows an apparent pK_a of 5.9. It was thought that this value was the true pK_a although it is 4 pK_a units lower than that of an ordinary lysine-amino-group.[231] The α-ketoglutarate decarboxylase from *Rhodopseudomonas spheroides*[232] and the L-aspartate α-decarboxylase from *Escherichia coli*[233] have been studied.

Other Reactions
Tillett and his co-workers[234] have extended their investigation of the hydrolysis of sydnones to some 3-alkylsydnones. Unlike the hydrolysis of 3-phenylsydnones,[235] that of 3-methyl- and 3-isopropyl-sydnone is catalysed by perchloric acid as well as by hydrohalic acids, although the latter are more effective. Presumably, the conjugate acids are attacked by water and by halide ions (see Scheme 2, p. 454). The Bunnett *w*-values (3.45 and 3.88) and the entropies of activation (−19.9 and −10.2 cal deg^{-1} mol^{-1}) are also consistent with an $A2$ mechanism. 3-t-Butyl- and 3-furfuryl-sydnone are also hydrolysed by a pH-dependent pathway. The solvent isotope effect (k_D/k_H) for that of 3-furfurylsydnone is 1.0 and for the perchloric acid-catalysed pathway is 1.77. Perchloric acid is a more effective catalyst than hydrohalic acids. The Bunnett *w*-values are ca. 0

---

[223] A. Ljungquist and K.-B. Augustinsson, *Eur. J. Biochem.*, **23**, 303 (1971).
[224] J. B. Suszkiw, *Anal. Biochem.*, **44**, 321 (1971); A. P. Brestkin and T. U. Parkhomenko, *Biokhimya*, **36**, 950 (1971); A. P. Brestkin and D. L. Pevzner, *Biokhimya*, **36**, 81 (1971); J. Bajgar, *Coll. Czech. Chem. Comm.*, **36**, 1705 (1971); W. Leuzinger, *Biochem. J.*, **123**, 139 (1971); G. M. Bogolyubova, E. V. Karpinskaya, A. I. Kulikova, and V. J. Rozengart, *Biokhimya*, **36**, 1075 (1971).
[225] B. Riddle and W. P. Jencks, *J. Biol. Chem.*, **246**, 3250 (1971).
[226] L. Ötvös, E. Moravcsik, and G. Mády, *Biochem. Biophys. Res. Comm.*, **44**, 1056 (1971); A. Romeo, G. Lucente, D. Rossi, and G. Zanotti, *Tetrahedron Letters*, **1971**, 1799.
[227] U. Femert and G. Pfleiderer, *FEBS Letters*, **14**, 89 (1971); U. Femert, *FEBS Letters*, **14**, 92 (1971).
[228] R. G. Khalifah, *J. Biol. Chem.*, **246**, 2561 (1971); R. W. King and G. C. K. Roberts, *Biochemistry*, **10**, 558 (1971); Y. Pocker, M. W. Beug, and V. R. Ainardi, *Biochemistry*, **10**, 1390 (1971).
[229] F. R. Stermitz and W. H. Huang, *J. Am. Chem. Soc.*, **93**, 3427 (1971).
[230] E. V. Brown and R. J. Moser, *J. Org. Chem.*, **36**, 454 (1971).
[231] D. E. Schmidt and F. H. Westheimer, *Biochemistry*, **10**, 1249 (1971).
[232] T. Saito, S. Tuboi, Y. Nishimura, and G. Kikuchi, *J. Biochem. (Tokyo)*, **69**, 265 (1971).
[233] Y. Nakano and S. Kitaoka, *J. Biochem. (Tokyo)*, **70**, 327 (1971).
[234] S. Aziz, A. J. Buglass, and J. G. Tillett, *J. Chem. Soc.* (B), **1971**, 1912.
[235] See *Org. Reaction Mech.*, **1968**, 399.

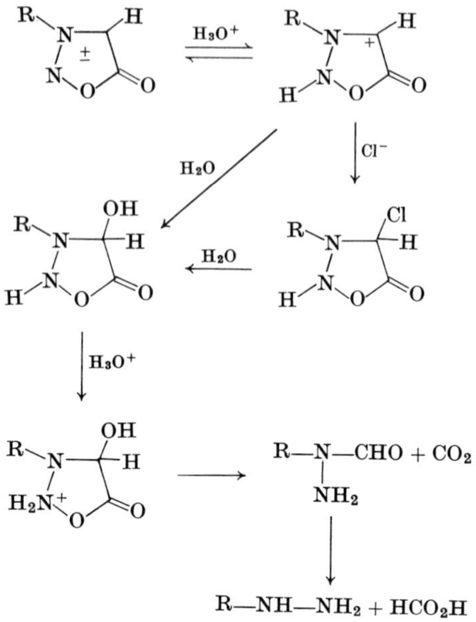

Scheme 2

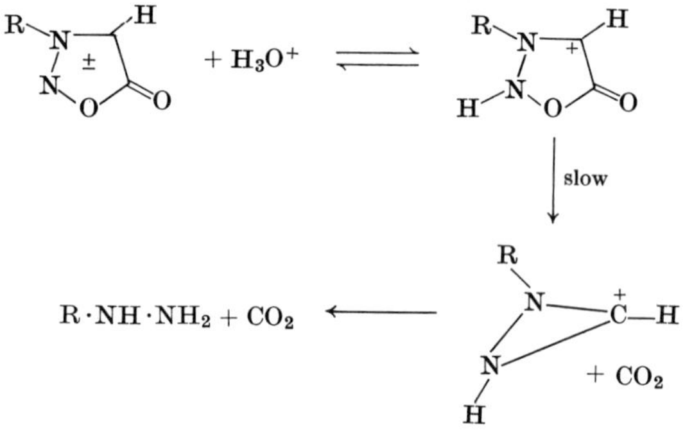

Scheme 3

and the entropies of activation positive (+10.1 and +5.6 cal deg^{-1}mol^{-1}). An A1 mechanism was proposed (Scheme 3).[235]

The conversion of phenylnitromethane into benzoic acid in concentrated sulphuric acid (Meyer reaction) has been studied and is thought to follow the sequence (22). The rate of hydrolysis of the intermediate benzonitrile oxide to hydroxamic acid increases with acid concentration and the reaction was thought to involve attack on the conjugate acid. The overall reaction shows a change in the rate-determining step from rearrangement-hydrolysis of the *aci*-nitro-compound in dilute acid to formation of the *aci*-form in concentrated acid.[236]

[236] J. T. Edward and P. H. Tremaine, *Can. J. Chem.*, **49**, 3484, 3489, 3493 (1971).

$$ArCH_2NO_2 \rightleftharpoons ArCH=N{\Large\diagdown}\!\!{}^{O^-}_{OH} \longrightarrow ArC\overset{+}{\equiv}N-O^- \longrightarrow$$

$$ArCONHOH \longrightarrow ArCO_2H \quad \ldots(22)$$

The rates of the acid and alkaline hydrolysis of a large number of cyclic carbonates have been measured. Generally the compounds with six-membered rings react faster than those with five-membered rings. Compound (**79**) reacts 50–70 times more rapidly than (**80**) in acid and alkaline solutions.[237, 238] The kinetics of hydrolysis of diethyl pyrocarbonate have also been studied.[239]

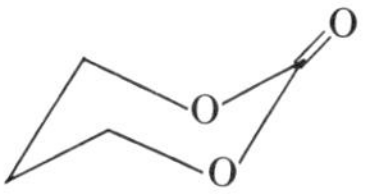

(**79**) 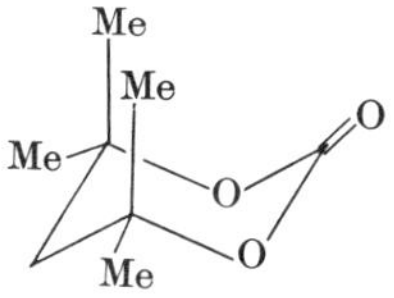 (**80**)

The rates of transesterification of ethyl acetoacetate and diethyl malonate with propan-1-ol in heptane are independent of the concentration of alcohol, and the reactions were thought to proceed through ketene intermediates.[240]

The rates of the alkaline hydrolysis of diethyl malonate and ethyl malonate in water–dimethyl sulphoxide mixtures decrease with increasing mole fraction of DMSO from 0 to 0.078 and thereafter increase up to mole fraction 0.2758.[241]

The reaction of ketene with acetic acid in the gas phase has been studied.[242, 243]

The reaction of benzyl alcohol with methyl 4,6-O-benzylidene-α-4-glucoside 2,3-dicarbonate catalysed by triethylamine in a variety of solvents has been studied. The reaction is first-order in substrate and in the amine–alcohol complex. The rate is enhanced by electron-withdrawing substituents in the benzyl alcohol.[244]

A detailed investigation of the alkaline hydrolysis of alkyl and aryl *para*-substituted benzoates and analysis of the results by means of linear-free-energy relationships have been reported.[245a]

The ρ-values for the alkaline hydrolysis of ethyl benzoates, cinnamates, and 5-phenylpenta-2,4-dienoates are 2.498, 1.329 and 0.83, respectively.[245b] Other investigations of the alkaline hydrolysis of cinnamate esters have also been reported.[246, 247] The ρ-value for the reaction of *trans*-2-arylcyclopropanecarboxylic acids with diphenyldiazomethane is 0.88 at 35°.[248]

---

[237] J. Katzhendler, L. A. Poles, and S. Sarel, *J. Chem. Soc.* (B), **1971**, 1847.
[238] J. Katzhendler, L. A. Poles, M. Dagan, and S. Sarel, *J. Chem. Soc.* (B), **1971**, 1035.
[239] R. Schelenz and E. Fischer, *Z. Lebensm.-Unters.-Forsch.*, **145**, 279 (1971); *Chem. Abs.*, **75**, 4940 (1971).
[240] D. S. Campbell and C. W. Laurie, *Chem. Comm.*, **1971**, 355.
[241] P. S. Radhakrishnamurti and P. C. Patro, *Tetrahedron*, **26**, 5503 (1970).
[242] P. G. Blake and H. H. Davies, *J. Chem. Soc.* (B), **1971**, 1727.
[243] P. G. Blake, H. H. Davies, and A. Speis, *J. Chem. Soc.* (B), **1971**, 2050.
[244] E. I. Strout, W. M. Doane, and K. E. Kolb, *J. Org. Chem.*, **36**, 3126 (1971).
[245a] J. R. Robinson, R. J. Washkuhn, and V. K. Patel, *J. Pharm. Sci.*, **60**, 736 (1971).
[245b] N. Wang and C.-H. Wang, *J. Org. Chem.*, **36**, 3178 (1971).
[246] K. Kamala, P. L. Nayak, and M. K. Rout, *Indian J. Chem.*, **9**, 680 (1971).
[247] P. L. Nayak and M. K. Rout, *J. Indian Chem. Soc.*, **48**, 141 (1971).
[248] R. R. Kostikov, N. P. Bobko, and I. A. Dyakonov, *Organic Reactivity* (*Tartu*), **8**, 97 (1971).

The hydrolysis of the following esters has been studied: methyl 6-substituted 2-naphthoates,[249] acetylmandelate ion,[250] ethyl 5,6-epoxybicyclo[2.2.2]octane-2-carboxylates,[251] ethyl acetate,[252] anthraquinone esters,[253] esters with double bonds in the alkyl and acyl portions,[254] methyl pyrimidine-, pyridazine- and pyrazine-carboxylates,[255] ethyl pyruvate,[256] bornyl acetate,[257] germine 3,16-diacetate,[258] glycerol monoleucinate,[259] glycol diacetate,[260] glucuronate esters,[261] and 1-phenylallyl esters (by alkyl–oxygen fission).[262, 263]

The alkaline hydrolysis of acetamide in the presence of methanol, glycol, or glycerol has been studied.[264] Other investigations of the hydrolyses of amides include those of N-acetylsulphanilic acid in sulphuric acid–water mixtures,[265] substituted phenylacetamides in alkaline solution,[266] and di-(p-formamido)phenyl sulphone over a wide pH range.[267]

The reaction of nitriles with hydrogen peroxide in alkaline solution[268] and the alkaline hydrolysis of nitriles[269a] have been studied.

The following reactions of anhydrides have been studied: methanolysis of acetic anhydride in mixtures of CH_3OH and CH_3OD,[269b] the conversion of acetic anhydride into acetyl sulphate in sulphuric acid,[270] the aluminium chloride-catalysed exchange between [$^{14}C_1$]acetic anhydride and alkyl thiolacetates in nitromethane,[271a] and the solvolysis of substituted 2-sulphobenzoic anhydrides.[271b]

The dependence of the proportion of C- to O-acylation of the 2-naphthyl oxide ion on the counterion, the solvent, and structure of the acyl halide has been interpreted as indicating that the transition state is similar to the corresponding reaction of alkyl

---

[249] V. Baliah and P. A. Nadar, *Indian J. Chem.*, **9**, 671 (1971).
[250] M. N. Das and A. K. Ray, *J. Chem. Soc.* (A), **1971**, 1831.
[251] C. L. Liotta, W. F. Fisher, and C. L. Harris, *Chem. Comm.*, **1971**, 1312.
[252] A. I. Morozov, I. A. Makolkin, and M. S. Sytilin, *Izv. Vyssh. Ucheb. Zaved. Khim. Khim. Tekhnol.*, **14**, 373 (1971); *Chem. Abs.*, **75**, 19432 (1971).
[253] P. H. Gore, A. Rahim, and D. N. Waters, *J. Chem. Soc.* (B), **1971**, 202.
[254] C. G. Evans and J. D. R. Thomas, *J. Chem. Soc.* (B), **1971**, 1502.
[255] L. W. Deady, D. J. Foskey, and R. A. Shanks, *J. Chem. Soc.* (B), **1971**, 1962.
[256] S. C. Rakshit and K. Sarkar, *J. Indian Chem. Soc.*, **48**, 605 (1971).
[257] A. A. Vereshchagina, O. N. Dolgopolou, A. F. Markova, and G. A. Rudakov, *Zhur. Org. Khim.*, **6**, 1379 (1970); *J. Org. Chem. USSR*, **6**, 1393 (1970).
[258] E. M. Cohen and R. Aizel, *J. Pharm. Sci.*, **60**, 193 (1971).
[259] M. Nakagaki and N. Funasaki, *Yakugaka Zasshi*, **90**, 1310 (1970); *Chem. Abs.*, **74**, 12335 (1971).
[260] C. Aubry, A. Zoulalian, and J. Villermaux, *Bull. Soc. Chim. France*, **1971**, 2483.
[261] V. G. Bukharov and L. N. Karneeva, *Izv. Akad. Nauk SSSR, Ser. Khim.*, **1970**, 1916; *Chem. Abs.*, **74**, 76627 (1971).
[262] G. Meyer and P. Viout, *Bull. Soc. Chim. France.*, **1971**, 2997.
[263] E. d'Incan and P. Viout, *Bull. Chem. Soc. France*, **1971**, 3312.
[264] H. Sadek, M. S. Abu Elamayem, and M. Elsemongy, *J. Chem. U.A.R.*, **12**, 469 (1969); *Chem. Abs.*, **74**, 12317 (1971).
[265] J. W. Barnett and C. J. O'Connor, *Tetrahedron Letters*, **1971**, 2161.
[266] D. Zavoianu, F. Cocu, and D. Pop, *Rev. Chim. (Bucharest)*, **21**, 527 (1970); *Chem. Abs.*, **74**, 12312 (1971); D. Zavolanu, F. Cocu, and D. Pop, *An. Univ. Bucuresti Chim.*, **19**, 53 (1970); *Chem. Abs.*, **75**, 87753 (1971).
[267] L. C. Garg, *J. Pharm. Sci.*, **60**, 606 (1971).
[268] J. E. McIsaac, R. E. Ball, and E. J. Behrman, *J. Org. Chem.*, **36**, 3048 (1971).
[269a] D. Zavoianu and F. Cocu, *An. Univ. Bucuresti Chim.*, **18**, 61 (1969); *Chem. Abs.*, **74**, 75770 (1971).
[269b] V. Gold and S. Grist, *J. Chem. Soc.* (B), **1971**, 2285.
[270] C. J. Clemett, *J. Chem. Soc.* (B), **1971**, 2202.
[271a] F. Dutka, A. F. Márton, and P. Vinkler, *Z. Naturforsch.*, **26b**, 703 (1971).
[271b] R. M. Laird and M. J. Spence, *J. Chem. Soc.* (B), **1971**, 454; cf. *Org. Reaction Mech.*, **1970**, 490

halides and that the reactions are direct displacements.[272] The hydrolysis of acetyl chloride[273] and the reaction of acetyl chloride with anisole catalysed by perchloric acid[274] have been studied.

The decomposition of carbanilates to isocyanates[275, 276] and the hydrolysis of isocyanates[277, 278] have been studied.

Other reactions investigated include dehydration of gaseous formic acid,[279] reactions catalysed by di-imides,[280] hydrolysis of 3-morpholinosydnone imine[281] and reaction of benzoic acid with diphenyldiazomethane.[282]

Non-carboxylic Acids

Phosphorus-containing Acids[283, 284]

Non-enzymic reactions. Hydrolyses of the ionized and un-ionized forms of esters (**81**) and (**82**) are enhanced by factors of ca. 10^7 over those for analogous esters without carboxyl groups. The ionized forms yield salicyclic acid and the dialkyl phosphate, but the un-ionized forms yield only 20% of these products and 80% of alcohol and monoalkyl 2-carboxylphenyl phosphate. Both forms were thought to react with intramolecular nucleophilic catalysis by the carboxyl group. Presumably, the pentacovalent intermediate can undergo pseudorotation to place the phenoxide group apical, unlike that from the corresponding monoaryl salicylyl phosphates which react to expel the aryloxy group exclusively.[285]

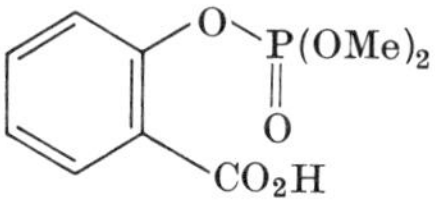

(**81**) 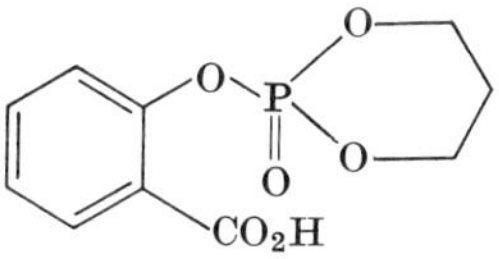   (**82**)

---

[272] P. Haberfield and R. B. Trattner, *Chem. Comm.*, **1971**, 1481.
[273] R. Vilcu and I. Ciocazanu, *Rev. Roum. Chem.*, **16**, 643 (1971).
[274] R. Corriu and G. Dabosi, *Compt. Rend. (C)*, **271**, 1404 (1971).
[275] A. B. Lateef, J. A. Reeder, and L. Rand, *J. Org. Chem.*, **36**, 2295 (1971).
[276] M. S. Fedoseev, G. N. Marchenko, and L. K. Kir'yanova, *Sin. Fiz.-Khim. Polim.*, **1970**, No. 7, 163; M. S. Fedoseev, G. N. Marchenko, and N. G. Rogov, *Sin. Fiz.-Khim. Polim.*, **1970**, No. 7, 158; *Chem. Abs.*, **75**, 48232, 48233 (1971).
[277] R. P. Tiger, L. S. Bekhli, and S. G. Entelis, *Kinet. Katal.*, **11**, 1347 (1970); *Chem. Abs.*, **74**, 63659 (1971).
[278] R. P. Tiger, L. S. Bekhli, and S. G. Entelis, *Kinet. Katal.*, **12**, 318 (1971).
[279] P. G. Blake, H. H. Davies, and G. E. Jackson, *J. Chem. Soc. (B)*, **1971**, 1923.
[280] D. F. Mironova and G. F. Kvorko, *Ukr. Khim. Zhur.*, **37**, 458 (1971).
[281] Y. Asahi, K. Shinozaki, and M. Nagaoka, *Chem. Pharm. Bull.*, **19**, 1079 (1971).
[282] N. B. Chapman, M. R. J. Dack, and J. Shorter, *J. Chem. Soc. (B)*, **1971**, 834.
[283] I. Ugi, D. Marquarding, H. Klusacek, P. Gillespie, and F. Ramirez, *Accounts Chem. Res.*, **4**, 288 (1971).
[284] P. Gillespie, P. Hoffman, H. Klusacek, D. Marquarding, S. Pfohl, F. Ramirez, E. A. Tsolis, and I. Ugi, "Non-rigid Molecular Skeletons—Berry Pseudorotation and Turnstile Rotation," *Angew. Chem. Int. Ed.*, **10**, 687 (1971).
[285] R. H. Bromilow, S. A. Khan, and A. J. Kirby, *J. Chem. Soc. (B)*, **1971**, 1091; cf. *Org. Reaction Mech.*, **1970**, 494–495.

Details of Schray and Benkovic's investigation of the hydrolysis of dibenzyl phosphoenolpyruvic acid have been reported.[286]

Cyanogen-promoted phosphorylation of reducing sugars is thought to involve reaction of the cyanogen with phosphate to produce an intermediate (**83**). When phosphate is allowed to react with cyanogen in the absence of sugar in ^{18}O-labelled water it incorporates the label, presumably by attack of water on (**83**).[287]

$$\text{NC-CN} + \text{HPO}_4{}^{2-} \longrightarrow \underset{\underset{\text{NH}}{\|}}{\text{NC-C-OPO}_3{}^{2-}}$$

(**83**)

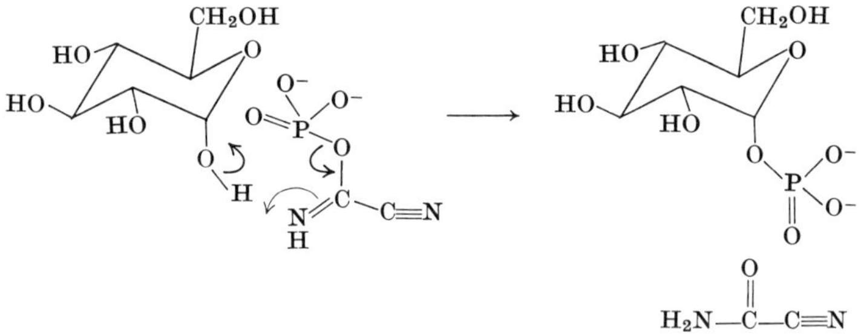

Hydrolysis of 2-imidazolyl phenyl phosphate does not appear to be intramolecularly catalysed by the imidazole group, but the cupric ion-catalysed hydrolysis is very rapid. The metal ion-catalysed and the uncatalysed reaction yield identical proportions of methyl phosphate and inorganic phosphate when the reactions are carried out in aqueous methanol. This suggests that the two reactions proceed through the same metaphosphate intermediate. Species (**84**) was thought to be reactive form in the metal ion-catalysed reaction.[288]

Yttrium(III) in the presence of pyridine-2-carboxaldehyde oxime is an efficient catalyst for the hydrolysis of p-nitrophenyl methylphosphonate. It was thought that the substrate became attached to one yttrium atom in a polymeric complex after which there was intramolecular nucleophilic attack by water or a hydroxyl group attached to another yttrium atom.[289]

The Zn(II)-pyridine-2-carboxaldehyde oxime anion reacts rapidly with phosphorylimidazole, transferring the phosphate group to the anionic oxygen. Saturation kinetics are observed and reaction was thought to involve a complex possibly as symbolized by (**85**).[290] The major pathway for the Mg^{2+}-catalysed hydrolysis of acetyl phosphate involves P—O cleavage but that for the Ca^{2+}-catalysed hydrolysis involves C—O cleavage.[291]

The metal ion-catalysed hydrolysis of 3-pyridyl and 8-quinolyl phosphate[292] and the copper ion-catalysed hydrolysis of disodium pyrophosphite[293] have been studied.

[286] K. J. Schray and S. J. Benkovic, *J. Am. Chem. Soc.*, **93**, 2522 (1971); cf. *Org. Reaction Mech.*, **1969**, 476.
[287] M. Halmann and C. Degani, *J. Chem. Soc.* (C), **1971**, 1459.
[288] S. J. Benkovic and L. K. Dunikoski, *J. Am. Chem. Soc.*, **93**, 1526 (1971).
[289] F. M. Blewett and P. Watts, *J. Chem. Soc.* (B), **1971**, 881.
[290] G. J. Lloyd and B. S. Cooperman, *J. Am. Chem. Soc.*, **93**, 4883 (1971).
[291] J. P. Klinman and D. Samuel, *Biochemistry*, **10**, 2126 (1971).
[292] Y. Murakami and J. Sunamoto, *Bull. Chem. Soc. Japan.*, **44**, 1827 (1971); see *Org. Reaction Mech.*, **1969**, 475.
[293] S. Ueda and Y. Sasaki, *Bull. Chem. Soc. Japan*, **44**, 1972 (1971).

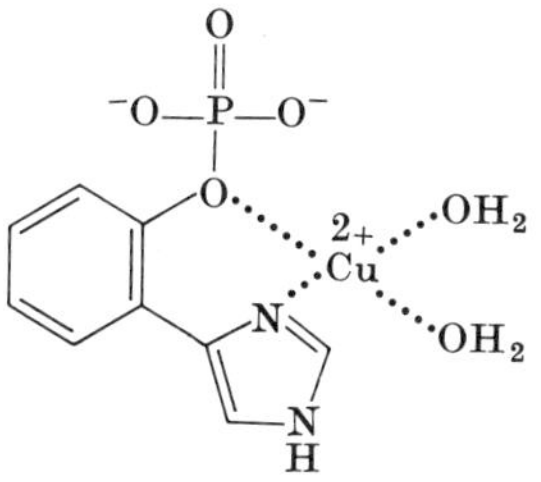

Micelles of decyl-(β-hydroxy-α-methylphenethyl)dimethylammonium bromide catalyse the hydrolysis of *p*-nitrophenyl diphenyl phosphate.[294] The hydrolysis of bis-2,4-dinitrophenyl phosphate in the presence of non-ionic detergents[295] and the reaction of α-cyclodextrin with di-isopropyl phosphofluoridate[296] have been studied.

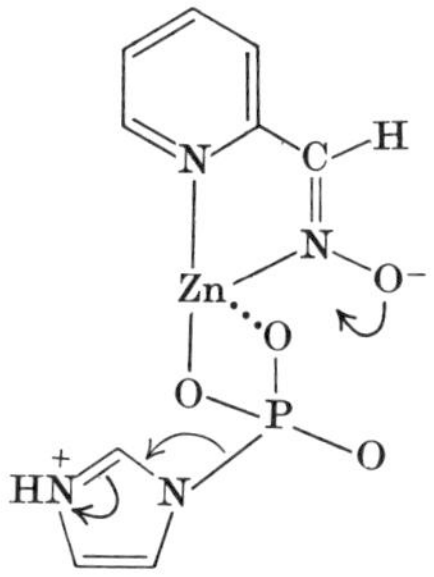

The monoanion is the most reactive species in the hydrolysis of 8-quinolyl phosphate. A mechanism involving intramolecular catalysis [see (**86**)] was proposed. However, the rate of hydrolysis is similar to that of the monoanion of 3-pyridyl phosphate and hence if this catalysis does take place it is not very effective.[297]

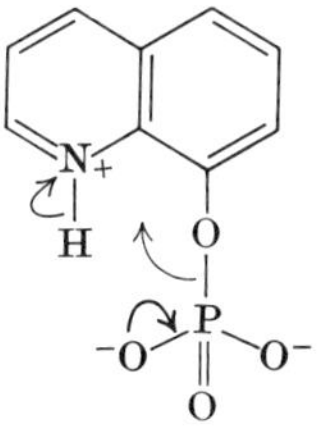

Details of the investigation by Cadogan and his co-workers of the hydrolysis of phosphonate esters with neighbouring oxime groups have been published.[298]

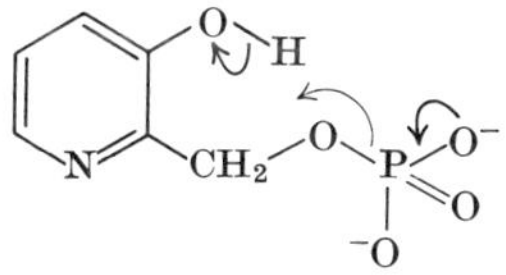

Hydrolysis of the dianionic form of (3-hydroxy-2-pyridyl)methyl phosphate is thought to involve intramolecular catalysis as symbolized by (**87**).[299]

Amines are general-base catalysts for the hydrolysis of isopropyl methylphosphonofluoridate. α-Effect amines do not show enhanced catalytic activity. The monoprotonated form of ethylenediamine is almost twice as reactive as would be expected from its pK_a, and the protonated amino-group may act as a general-acid catalyst.[300] Amines catalyse the hydrolysis of (dihydroxyphosphinyl)imidazole; pyridine is a catalyst but 2,6-lutidine is not, and so the catalysis was thought to be nucleophilic catalysis.[301]

---

[294] C. A. Bunton, L. Robinson, and M. Stam, *J. Am. Chem. Soc.*, **92**, 7393 (1970).
[295] C. A. Bunton, A. Kamego, and L. Sepulveda, *J. Org. Chem.*, **36**, 2571 (1917).
[296] C. van Hooidonk and J. C. A. E. Breehaart-Hansen, *Rec. Trav. Chim.*, **90**, 680 (1971).
[297] Y. Murakami and J. Sunamoto, *Bull. Chem. Soc. Japan*, **44**, 1939 (1971).
[298] J. I. G. Cadogan, J. A. Challis, and D. T. Eastlick, *J. Chem. Soc.* (B), **1971**, 1988; cf. *Org. Reaction Mech.*, **1967**, 362; **1970**, 495.
[299] Y. Murakami, J. Sunamoto, and H. Ishizu, *Chem. Comm.*, **1970**, 1665.
[300] J. Epstein, P. L. Cannon, and J. R. Sowa, *J. Am. Chem. Soc.*, **92**, 7390 (1970).
[301] G. J. Lloyd, C.-M. Hsu, and B. S. Cooperman, *J. Am. Chem. Soc.*, **93**, 4889 (1971).

Imidazole is a general-base catalyst for the hydrolysis of aryl diphenylphosphinates with ρ-value +2.88 at 25° compared to +1.55 for the hydroxide ion-catalysed reaction.[302]

The rate of hydrolysis of the monoanions of phosphoramidates, $^{2-}O_3PNH_2^+R$, is independent of the pK_a of the leaving amine when this is less than ca. 6 ($\beta = 0$), but the rate is highly dependent on the pK_a when this pK_a is greater than 6 ($\beta = 1$). To explain this it was proposed that phosphoramidates with strongly basic amino-groups exist wholly as the N-protonated form but that in those with weakly basic amino-groups the amino-group is unprotonated. Since the reaction probably involves the N-protonated form the experimental rate constant for compounds existing mainly in the unprotonated form must be a complex constant involving the equilibrium constant for protonation and the rate constant for reaction of the N-protonated form.[303]

The kinetics of hydrolysis of N,N-dimethyl-N'-phosphoguanidine (**88**), an analogue of creatine phosphate, has been studied. The pH–rate profile is bell-shaped and the neutral zwitterion is the reactive species. It was thought that the reaction proceeded through a metaphosphate formed as shown in (**89**).[304a]

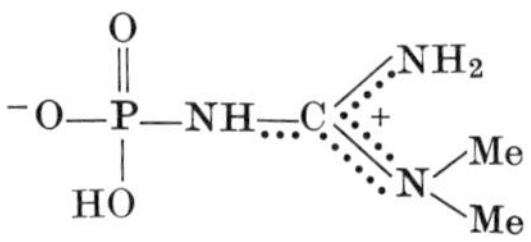

(**88**) 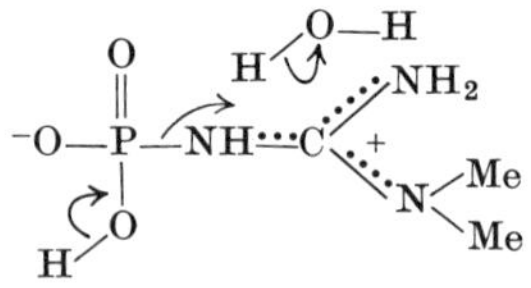   (**89**)

The dianions and monoanions of N-arylcarbamoyl phosphate are hydrolysed predominantly with phosphorus—oxygen cleavage. The ρ-value for the hydrolysis of the monoanion is less than that for the dianions, and the mechanisms (23) and (24) were proposed. Some carbon—oxygen bond fission occurs in the hydrolysis of N-p-nitrophenylcarbamoyl phosphate.[304b]

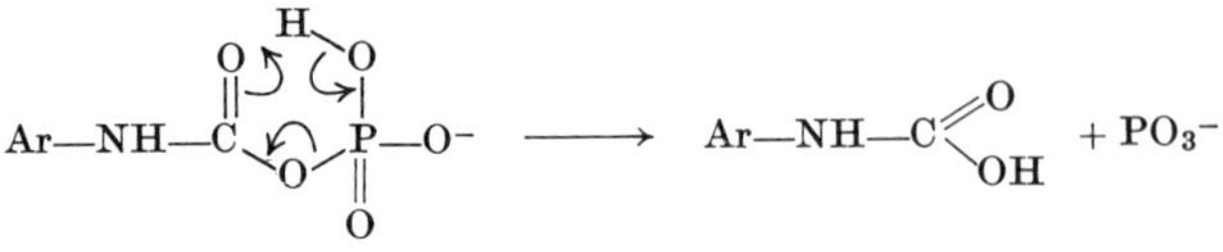

$$Ar-NH-C(O)(O)P-O^- \longrightarrow Ar-NH-C(O)OH + PO_3^- \quad \ldots(23)$$

$$Ar-NH-C(O)-O-P(O)(O^-)-O^- \longrightarrow ArNH-CO_2^- + PO_3^- \quad \ldots(24)$$

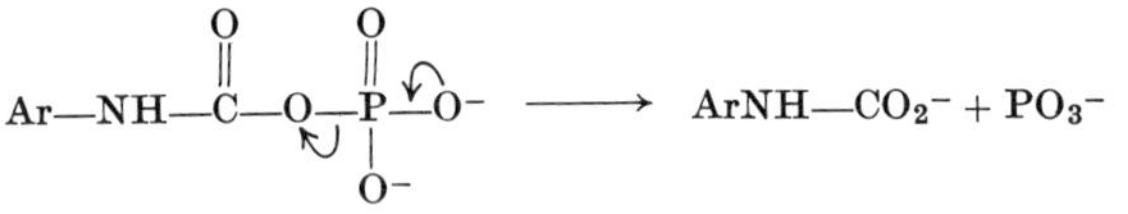

The hydrolysis of the phenoxyphosphole 1-oxide (**90**) is about four times faster than that of the phenoxyphospholene 1-oxide (**91**).[305]

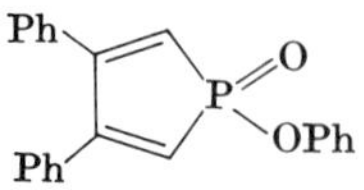

(**90**) 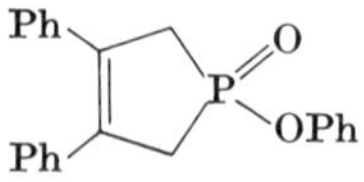   (**91**)

[302] A. Williams and R. A. Naylor, *J. Chem. Soc.* (B), **1971**, 1967.
[303] S. J. Benkovic and E. J. Sampson, *J. Am. Chem. Soc.*, **93**, 4009 (1971).
[304a] P. Haake and G. W. Allen, *Proc. Nat. Acad. Sci.*, **68**, 2691 (1971).
[304b] C. M. Allen and J. Jamieson, *J. Am. Chem. Soc.*, **93**, 1434 (1971).
[305] F. B. Clarke and F. H. Westheimer, *J. Am. Chem. Soc.*, **93**, 4541 (1971).

Acid-catalysed hydrolysis of isopentenyl phosphate occurs by hydration–dehydration to dimethylallyl dihydrogen phosphate which is hydrolysed unimolecularly with carbon—oxygen bond fission.[306]

Hydrolysis of $(PhO)_2PN(CH_2CH_2Cl)_2$ in a bicarbonate buffer of pH 8.3 proceeds with fission of the amide bond, and this is followed by hydrolysis of the chlorine atoms of the resulting chloroethylamine which involves neighbouring-group participation by the amino-group.[307]

The reaction of optically active O,S-di-isopropyl methylphosphonothioate with hydroxide ion yields racemic O-isopropylmethylphosphonothioic acid.[308]

Hydrolyses of the following phosphorus-containing esters have also been studied: phosphonates, phosphinates, and their sulphur-containing analogues;[309] 2,3-dimethoxyphenyl dihydrogen phosphate;[310] esters of phosphorous phosphoric anhydride;[311] pyrophosphates;[312] and 3β,17α-dihydroxy-6α-methylpregn-4-en-20-one 17-acetate 3-phosphate.[313]

The reaction of phosphorus esters with phenylmagnesium bromide[314] and the site of protonation of phosphinamides[315] have been investigated.

The effect of solvent on the rates and position of equilibrium in reactions of isopropyl methylphosphonofluoridate with oxygen nucleophiles has been studied.[316] The rate of hydrolysis of isopropyl methylphosphonofluoridate in sea-water has been estimated.[317] The Brønsted coefficient for the reaction of di-isopropyl phosphorochloridate with hydroxamate ions is 0.21.[318] The alcoholyses of diethylphosphinyl chloride, ethyl methylphosphonochloridate, and ethyl methylphosphonochloridothiolate have been studied.[319]

Enzymic reactions.[320a] There has been relatively little work on ribonuclease A this year. The circular dichroism spectrum of a mixture of ribonuclease and cytidine 2′-phosphate was thought to indicate that the cytidine has an *anti*-conformation when it is bound at the active site.[320b] The kinetics of the interaction of 2′-deoxyuridine 3′-phosphate with ribonuclease have been studied.[321] Transient and relaxation kinetics of enzymic reactions including those of ribonuclease have been reviewed.[322]

---

[306] B. K. Tidd, *J. Chem. Soc.* (B), **1971**, 1168.
[307] I. I. Kuz'menko and L. B. Rapp, *Khim.-Farm. Zhur.*, **5**, 7 (1971); *Chem. Abs.*, **75**, 62771 (1971).
[308] L. P. Reiff, L. J. Szafraniec, and H. S. Aaron, *Chem. Comm.*, **1971**, 366.
[309] V. E. Bel'skii, N. N. Bezzubova, V. D. Akamsin, V. N. Eliseenkov, N. I. Rizpolozhenskii, and A. N. Pudovik, *Dokl. Akad. Nauk SSSR*, **197**, 85 (1971); *Chem. Abs.*, **75**, 87762 (1971); V. E. Bel'skii, N. N. Bezzubova, V. N. Eliseenkov, and A. N. Pudovik, *Zhur. Obshch. Khim.*, **40**, 2557 (1970); *Chem. Abs.*, **74**, 124410 (1971); V. E. Bel'skii, L. S. Andreeva, I. A. Aleksandrova, and G. M. Vinokurova, *Izv. Akad. Nauk SSSR, Ser. Khim.*, **1970**, 1418; *Chem. Abs.*, **74**, 41563 (1971).
[310] M. M. Mhala and S. Prabha, *Indian J. Chem.*, **8**, 972 (1970).
[311] J. Mikolajczyk, J. Michalski, and A. Zwierzak, *Chem. Comm.*, **1971**, 1257.
[312] H. Berger, *Z. Naturforsch.*, **26b**, 694 (1971).
[313] T. O. Oesterling and J. H. Gustafson, *J. Pharm. Sci.*, **59**, 1612 (1970).
[314] H. R. Hays, *J. Org. Chem.*, **36**, 98 (1971).
[315] K. E. DeBruin, A. C. Padilla, and D. M. Johnson, *Tetrahedron Letters*, **1971**, 4279.
[316] G. T. Davis, M. M. Demek, J. R. Sowa, and J. Epstein, *J. Am. Chem. Soc.*, **93**, 4093 (1971).
[317] J. Epstein, *Science*, **170**, 1396 (1970).
[318] M. A. Weinberger, R. Greenhalgh, P. M. Lutley, and N. C. C. Gibson, *J. Chem. Soc.* (B), **1971**, 1950.
[319] A. A. Neimysheva, M. V. Ermolaeva, and I. L. Knunyants, *Zhur. Obshch. Khim.*, **40**, 2022 (1971); *Chem. Abs.*, **74**, 140429 (1971).
[320a] See "The Enzymes," P. D. Boyer (Ed.), Vol. 4, Academic Press, New York, 3rd ed., 1971.
[320b] T. Oshina and K. Imahori, *J. Biochem. (Tokyo)*, **70**, 193 (1971).
[321] F. G. Walz, *Biochemistry*, **10**, 2156 (1971).
[322] H. Gutfreund, *Ann. Rev. Biochem.*, **8**, 315 (1971).

The interaction of the S-peptide with the S-protein has been studied by observing the change in the ^{19}F-NMR spectrum when the S-protein is added to the S-peptide in which 80–90% of the lysine residues have been trifluoroacetylated.[323] The thermodynamics of the binding of S-peptide to S-protein have been studied.[324]

Ribonuclease A is inhibited if treated with pyridoxal phosphate and sodium borohydride. The pyridoxal was thought to interact with a lysine residue close to the active site.[325]

More work on the proton and carbon-13 NMR spectra of ribonuclease A has been reported.[326]

The partial sequence of ribonuclease T_1 has been described.[327] Ribonuclease T_1 is inactivated by phenylglyoxal or glyoxal; arginine-77 and the terminal alanine residue are lost; the loss of enzymic activity parallels the loss of arginine-77, and the reactivity of active-site glutamic acid-58 is also lost. This suggests that arginine-57 is present at or near the active site.[328]

The binding of guanosine 3′-phosphate to ribonuclease T_1,[329] the NMR spectrum of ribonuclease T_1,[330] and the photo-oxidation of ribonuclease T_1[331] have been investigated.

Ribonuclease III from *E. coli*,[332] ribonuclease from bovine seminal vesicles,[333] and ribonuclease from *Rhizopus oligosporus*[334] have been studied.

The value of k_{cat}/K_m for the phosphorylation of alkaline phosphatase from *E. coli* decreases about sixty-fold from methyl to isopropyl phosphate. Phenyl phosphoroamidate is hydrolysed more slowly than phenyl phosphate. These two pieces of evidence were taken to indicate that reaction involves an $S_N2(P)$ displacement by a serine-hydroxyl group of the enzyme. It was thought that this was assisted by general base-catalysis by an imidazole group and general acid-catalysis by a zinc ion.[335] Stopped-flow kinetic measurements of the formation of phosphoryl enzyme from alkaline phosphatase and ^{32}P-labelled phosphate esters suggest that there are two conformations of the enzyme, only one of which reacts with the esters, and that the slow step is the interconversion of these forms.[336] Evidence that alkaline phosphatase from *E. coli* has two, not one, active sites has been reported. A "flip-flop" mechanism was proposed.[337] Other papers on phosphatases are listed in ref. 338.

---

[323] W. H. Huestis and M. A. Raftery, *Biochemistry*, **10**, 1181 (1971).
[324] R. P. Hearn, F. M. Richards, J. M. Sturtevant, and G. D. Watt, *Biochemistry*, **10**, 806 (1971).
[325] G. E. Means and R. E. Feeney, *J. Biol. Chem.*, **246**, 5532 (1971).
[326] N. L. R. King and J. H. Bradbury, *Nature*, **229**, 404 (1971); J. P. Cohen-Addad, *Biochimie*, **53**, 173 (1971); A. Allerhand, D. Doddrell, V. Glushko, D. W. Cochran, E. Wenkert, P. J. Lawson, and F. R. N. Gurd, *J. Am. Chem. Soc.*, **93**, 544 (1971).
[327] K. Takahashi, *J. Biochem. (Tokyo)*, **70**, 477 (1971).
[328] K. Takahashi, *J. Biochem. (Tokyo)*, **68**, 659 (1970).
[329] T. Oshima and K. Imahori, *J. Biochem. (Tokyo)*, **70**, 197, 987 (1971); K. Takahashi, *J. Biochem. (Tokyo)*, **70**, 941 (1971).
[330] H. Rüterjans and O. Pongs, *Eur. J. Biochem.*, **18**, 313 (1971).
[331] K. Takahashi, *J. Biochem. (Tokyo)*, **69**, 331 (1971).
[332] H. Schweitz and J. P. Ebel, *Biochimie*, **53**, 585 (1971).
[333] M. Irie and S. Hosokawa, *J. Biochem. (Tokyo)*, **70**, 301 (1971).
[334] E. A. Woodroof and D. D. Glitz, *Biochemistry*, **10**, 1532 (1971).
[335] A. Williams and R. A. Naylor, *J. Chem. Soc. (B)*, **1971**, 1973.
[336] T. W. Reid and I. B. Wilson, *Biochemistry*, **10**, 380 (1971).
[337] M. Lazdunski, C. Petitclerc, D. Chappelet, and C. Lazdunski, *Eur. J. Biochem.*, **20**, 124 (1971).
[338] P. Christen, B. L. Vallee, and R. T. Simpson, *Biochemistry*, **10**, 1377 (1971); A. W. Hanson, M. L. Applebury, J. E. Coleman, and H. W. Wyckoff, *J. Biol. Chem.*, **245**, 4975 (1970); A. P. Brestin,

The following enzymes have also been investigated: staphylococcal nuclease;[339,340] myosin;[341] snake-venom phosphodiesterase;[342] yeast inorganic pyrophosphatase;[343] rabbit-liver fructose diphosphatase;[344] nucleoside diphosphatase;[345] a phosphodiesterase with high affinity for adenosine 3′,5′-cyclic phosphate;[346] potato phosphorylase;[347] DNA polymerase;[348] yeast pyruvate kinase;[349] creatine kinase;[350,351] carbamoyl phosphate hydrolase;[352] yeast hexokinase;[353,354] and N-acetyl-D-mannosamine kinase.[355]

Sulphur-containing Acids[356]

The activation parameters ($\Delta H^{\ddagger}$, $\Delta S^{\ddagger}$, $\Delta C_p^{\ddagger}$, and $\mathrm{d}\Delta C_p^{\ddagger}/\mathrm{d}T$) have been measured for the solvolysis of some *para*-substituted benzenesulphonyl chlorides. The solvent-isotope effect $k(\mathrm{H_2O})/k(\mathrm{D_2O})$ varied regularly through the series from 1.405 when the *para*-substituent was methoxyl to 1.822 when it was nitro.[357] The solvent isotope effect, $k(\mathrm{H_2O})/k(\mathrm{D_2O})$, for the hydrolysis of methanesulphonyl chloride is not changed significantly between 1 atmosphere and 2000 atmospheres at 25°; this result was considered to be inconsistent with Robertson's proposal that the isotope effect is due to a difference in the initial hydrogen-bonded structures of H_2O and D_2O, since much of this structure is lost at 2000 atmospheres.[358]

The ρ-values for reaction of substituted benzenesulphonyl chlorides with anilines vary from 1.14 for 3,4-dimethylaniline to 0.44 for 3-nitroaniline and the ρ-values for the reactions of benzenesulphonyl chlorides with substituted anilines vary from 0.93 for p-nitrobenzenesulphonyl chloride to 0.65 for p-methoxybenzenesulphonyl chloride. The results were interpreted in terms of a concerted nucleophilic displacement on the sulphonyl chloride.[359] The following aminolyses of sulphonyl chlorides have also been studied: p-nitrobenzenesulphonyl chloride with binuclear aromatic amines in nitro-

---

N. V. Novikova, and N. I. Rzhekhima, *Biokhimya*, **36**, 551 (1971); C.-W. Lin, H.-G. Su, and W. H. Fishman, *Biochem. J.*, **124**, 509 (1971); F. M. Hulett-Cowling and L. L. Campbell, *Biochemistry*, **10**, 1364, 1371 (1971); M. L. Blank and F. Snyder, *Biochemistry*, **9**, 5034 (1970).

[339] I. Parikh and G. S. Omenn, *Biochemistry*, **10**, 1173 (1971).
[340] H. F. Epstein, A. N. Schechter, and J. S. Cohen, *Proc. Nat. Acad. Sci.*, **68**, 2042 (1971).
[341] Y. Hayashi and Y. Tonomura, *J. Biochem.* (*Tokyo*), **68**, 665 (1970).
[342] L. B. Dolapchiev, *Biokhimya*, **35**, 1067 (1970).
[343] T. Negi, T. Samejima, and M. Irie, *J. Biochem.* (*Tokyo*), **70**, 359 (1971).
[344] T.-H. Chou and M. E. Kirtley, *Biochem. Biophys. Res. Comm.*, **45**, 98 (1971).
[345] V. L. Schramm and J. F. Morrison, *Biochemistry*, **10**, 2272 (1971).
[346] Y.-C. Huang and R. G. Kemp, *Biochemistry*, **10**, 2278 (1971).
[347] A. M. Gold, R. M. Johnson, and G. R. Sanchez, *J. Biol. Chem.*, **246**, 3444 (1971).
[348] T. R. Krugh, *Biochemistry*, **10**, 2549 (1971).
[349] R. T. Kuczenski and C. H. Suelter, *Biochemistry*, **10**, 2862, 2867 (1971).
[350] G. L. Rowley, A. C. Greenleaf, and G. L. Kenyon, *J. Am. Chem. Soc.*, **93**, 5542 (1971).
[351] E. P. Chetverikova, L. L. Alierskaya, and A. V. Krinskaya, *Biokhimya*, **36**, 717 (1971).
[352] D. Diederich, G. Ramponi, and S. Grisolia, *FEBS Letters*, **15**, 30 (1971).
[353] G. V. Titova, *Biokhimya*, **36**, 1083 (1971).
[354] C. T. Walsh and L. B. Spector, *Arch. Biochem. Biophys.*, **145**, 1 (1971).
[355] M. G. Uzbekov, *Biokhimya*, **36**, 174 (1971).
[356] E. Buncel, "Mechanistic Aspects of the Chemistry of Fluorosulphates," *Mech. React. Sulfur Compounds*, **5**, 71 (1970).
[357] R. E. Robertson and B. Rossall, *Can. J. Chem.*, **49**, 1441, 1451 (1971).
[358] M. L. Tonnet and A. N. Hambly, *Austral. J. Chem.*, **23**, 2427, 2435 (1970).
[359] O. Rogne, *J. Chem. Soc.* (B), **1971**, 1855.

benzene;[360] toluene-p-sulphonyl chloride with p-anisidine in various solvents;[361] toluene-p-sulphonyl bromide with t-butyl glycinate in benzene;[362] p-bromo- and p-nitro-benzenesulphonyl chloride with p-anisidine in nitrobenzene containing pyridine and pyridine N-oxide which act as nucleophilic catalysts.[363]

2-Methylpyridine is about 100 times less efficient as a catalyst for the methanolysis of benzenesulphonyl chloride than is pyridine, and the latter was therefore thought to act as a nucleophilic catalyst.[364]

The kinetics of hydrolysis of substituted benzenesulphonyl chlorides in aqueous dioxan have been studied. It was thought that 2,4,6-benzenesulphonyl chloride reacted by an S_N1 mechanism.[365]

The hydrolysis of isomeric dodecanesulphonyl chlorides in dodecane by aqueous sodium hydroxide with quaternary ammonium salts as "phase-transfer catalysts" has been studied (see p. 55).[366]

The acid and alkaline hydrolysis of *para*-substituted phenyl phosphosulphates has been studied.[367]

Ethane-1,2-disulphonic anhydride is methanolysed about 500 times faster than methanesulphonic anhydride at $-25°$. Propane-1,3-disulphonic anhydride is hydrolysed at approximately the same rate as methanesulphonic anhydride in aqueous dioxan. Benzene-1,2-disulphonic anhydride is methanolysed about ten times faster than benzenesulphonic anhydride.[368]

The imidazole- and N-methylimidazole-catalysed hydrolyses of 2-hydroxy-5-nitrotoluene-α-sulphonic acid sultone were thought to involve general base-catalysis on the grounds of the solvent-isotope effects $k(H_2O)/k(D_2O)$ which are 4.2 and 3.5, respectively.[369] The hydrolyses of 3-hydroxypropanesulphonic acid sultone and 4-hydroxybutanesulphonic acid sultone have also been studied.[370,371]

The ρ-value for the reaction of hydroxide ion with N-mesitylbenzenesulphinamides is $+1.3$.[372]

The following reactions have also been studied: hydrolysis of N-toluene-p-sulphonylimidazole and N-methyl-N'-toluene-p-sulphonylimidazole;[373,374] alkaline hydrolysis of

---

[360] R. S. Popova, A. F. Popov, and L. M. Litvinenko, *Zhur. Org. Khim.*, **6**, 1053 (1970); *J. Org. Chem. USSR*, **6**, 1053 (1970).
[361] V. A. Savyolova, L. M. Litvinenko, N. M. Chentsova, L. B. Sokolov, S. S. Medved, A. F. Popov, and V. I. Tokarev, *Organic Reactivity (Tartu)*, **7**, 847 (1971).
[362] L. M. Litvinenko, A. I. Bilobrova, Y. A. Sharanin, and A. F. Popov, *Zhur. Org. Khim.*, **7**, 795 (1971); *Chem. Abs.*, **75**, 36606 (1971).
[363] L. M. Litvinenko, V. A. Savelova, V. A. Shatskaya, and T. N. Sadovskaya, *Dokl. Akad. Nauk SSSR*, **198**, 844 (1971); *Chem. Abs.*, **75**, 62725 (1971).
[364] O. Rogne, *J. Chem. Soc.* (B), **1971**, 1334.
[365] M. L. Tonnet and A. N. Hambly, *Austral. J. Chem.*, **24**, 703 (1971); R. Foon and A. N. Hambly, *Austral. J. Chem.*, **24**, 713 (1971).
[366] C. M. Starks, *J. Am. Chem. Soc.*, **93**, 195 (1971).
[367] W. Tagaki, T. Eiki, and I. Tanaka, *Bull. Chem. Soc. Japan*, **44**, 1139 (1971); cf. *Org. Reaction Mech.*, **1970**, 501.
[368] R. M. Laird and M. J. Spence, *J. Chem. Soc.* (B), **1971**, 1434; cf. ref. 271*b*.
[369] E. T. Kaiser, K.-W. Lo, K. Kudo, and W. Berg, *Biorg. Chem.*, **1**, 32 (1971).
[370] A. Mori, M. Nagayama, and H. Mandai, *Bull. Chem. Soc. Japan*, **44**, 1669 (1971).
[371] A. Mori, M. Nagayama, and H. Mandai, *Kogyo Kagaku Zasshi*, **74**, 715 (1971); *Chem. Abs.*, **75**, 34794 (1971).
[372] J. B. Biasotti and K. K. Andersen, *J. Am. Chem. Soc.*, **93**, 1178 (1971).
[373] P. Voisin, P. Monjoint, M. Laloi-Diard, and M. Vilkas, *Compt. Rend.* (C), **272**, 322 (1970).
[374] P. Mointjoint, M. Laloi-Diard, and M. Vilkas, *Compt. Rend.* (C), **273**, 1177 (1971).

o-, m-, and p-trimethylammoniophenyl toluene-p-sulphonate;[375] hydrolysis of benzenesulphonamide and toluene-p-sulphonamide is concentrated sulphuric acid;[376,377] and hydrolysis of the triethylamine–sulphur trioxide complex.[378]

Nitrous and Nitric Acid

Investigations of the reaction of the benzenediazonium ion with hydroxide,[379] and of the N-nitrosation[380,381] and N-nitration[382,383] of amines, have been reported.

[375] V. M. Maremäe, T. O. Püsse, and V. A. Palm, *Organic Reactivity* (*Tartu*), **8**, 127 (1971).
[376] R. S. Ryabova, M. I. Vinnik, V. T. Lazareva, and R. D. Erlikh, *Zhur. Org. Khim.*, **6**, 797 (1970); *J. Org. Chem. USSR*, **6**, 800 (1970).
[377] M. I. Vinnik, R. S. Ryabova, and V. T. Lazareva, *Zhur. Org. Khim.*, **6**, 1434, 1438 (1970); *J. Org. Chem. USSR*, **6**, 1448, 1452 (1970).
[378] M. D. Bentley, S. E. Bowie, and R. D. Limoges, *J. Phys. Chem.*, **75**, 1763 (1971).
[379] C. D. Ritchie and D. J. Wright, *J. Am. Chem. Soc.*, **93**, 2425, 2429 (1971).
[380] J. Kroupa and M. Matrka, *Chem. Prum.*, **21**, 111 (1971); *Chem. Abs.*, **75**, 19452 (1971).
[381] M. Matrka, J. Kroupa, Z. Sagner, and V. Zverina, *Chem. Prum.*, **20**, 479 (1971); *Chem. Abs.*, **74**, 52785 (1971).
[382] B. V. Ghidaspov, I. M. Golubkov, and I. A. Stepanov, *Organic Reactivity* (*Tartu*), **7**, 931 (1970).
[383] B. V. Ghidaspov and I. M. Golubkov, *Organic Reactivity* (*Tartu*), **7**, 815 (1970).

CHAPTER 13

Photochemistry

R. S. DAVIDSON

Department of Chemistry, The University, Leicester

Physical Aspects	469
Emission Studies	469
The Triplet State	471
Energy Transfer	471
Energy Transfer and Reactions Occurring via Complex Formation	473
Carbonyl Compounds	475
Alkyl and Aryl Ketones, etc.	475
Enones	483
Diketones and Quinones	487
Aldehydes	489
Thioketones	490
Carboxylic Acids and Related Compounds	490
Olefins	493
Ionic Addition Reactions	493
Cycloaddition Reactions	494
Intramolecular Rearrangements	497
Miscellaneous Reactions of Alkenes and Alkynes	500
Aromatic Hydrocarbons	501
Heterocyclic Compounds	504
Nitrogen-containing Compounds	508
Azomethines and Related Compounds and Amines	508
N-Oxides and Related Compounds	512
Nitro- and Nitroso-compounds	513
Azo-, Diazo- and Diazonium Compounds and Azides	515
Halogen-containing Compounds	516
Carbonium Ions and Carbanions	517
Miscellaneous Compounds	518
Other Photoreactions	520
Photosensitized Oxidation	520
Chemiluminescence and Electrochemiluminescence	524
Solvolysis and Substitution Reactions	525
Photochromism	525

There have been further developments in physical techniques [1a-c,2a-c,3] during the year under review that allow a deeper probing of the processes leading to product formation in photochemical reactions. By the use of laser flash photolysis, excimer formation has been observed in benzene as well as a species, thought to be a biradical, that is formed by attack of triplet benzene on ground-state benzene.[1a] There has been

---

[1a] R. V. Bensasson, J. T. Richards, and J. K. Thomas, *Chem. Phys. Letters*, **9**, 13 (1971).
[1b] K. Sakurai, G. Capelle, and H. P. Broida, *J. Chem. Phys.*, **54**, 1412 (1971).
[1c] M. R. Topp, P. M. Rentzepis, and R. P. Jones, *Chem. Phys. Letters*, **9**, 1 (1971).

an explosive growth in the use of chemically induced dynamic nuclear polarization to study the mechanism of photochemical reactions.[2a-f] Use of this technique has shown that: (a) excited benzaldehyde can abstract hydrogen from ground-state aldehyde;[2a] (b) the radical formed by hydrogen abstraction from hydrocarbons by excited phenanthraquinone (**1**) is also produced by thermal decomposition of the photoadduct (**2**);[2b]

$$\text{(1)} + Ar_2CH_2 \xrightarrow{h\nu} \text{[intermediate]} + Ar_2\overset{\bullet}{C}H \underset{\Delta}{\rightleftharpoons} \text{(2)}$$

(c) that alkenes, alkanes and aldehydes generated by photolysis of dialkyl ketones arise by disproportionation of the radicals produced by a Type 1 reaction;[2c,d] and (d) that direct irradiation of dibenzoyl peroxide gives radicals by decomposition of its first excited singlet state.[2e] Transient species produced by flash photolysis have been detected mass-spectrometrically.[3] An interesting modification to the well used potassium ferrioxalate actinometer has been described which is claimed to allow a more rapid determination of light intensities.[4] The effects of orbital symmetry on photochemical transformations have been the subject of a recent excellent review.[5] Whole issues of journals have been devoted to reporting the 1970 IUPAC conference on photochemistry[6a] and a conference on singlet molecular oxygen.[6b] There have been further reviews of the photochemical literature[7] and several textbooks on the subject have been published.[8]

[2a] G. L. Closs and D. R. Paulson, *J. Am. Chem. Soc.*, **92**, 7229 (1970).
[2b] K. Maruyama, H. Shinod, and T. Maruyama, *Bull. Chem. Soc. Japan*, **44**, 585 (1971); K. Maruyama, T. Otsuki, H. Shinod, and T. Maruyama, *Bull. Chem. Soc. Japan*, **44**, 2000 (1971).
[2c] J. A. Den-Hollander, R. Kaptein, and P. A. T. M. Brand, *Chem. Phys. Letters*, **10**, 430 (1971).
[2d] M. Tomkiewicz, A. Groen, and M. Cocivera, *Chem. Phys. Letters*, **10**, 39 (1971).
[2e] R. Kaptein, J. A. Den-Hollander, D. Antheunis, and L. J. Oosterhoff, *Chem. Comm.*, **1970**, 1687.
[2f] H. D. Roth, *J. Am. Chem. Soc.*, **93**, 1527 (1971); G. L. Closs and A. D. Trifunac, *J. Am. Chem. Soc.*, **92**, 7227 (1970).
[3] O. P. Strausz, S. C. Barton, W. K. Duholke, H. E. Gunning, and P. Kebarle, *Can. J. Chem.*, **49**, 2048 (1971).
[4] K. C. Kurien, *J. Chem. Soc.* (B), **1971**, 2081.
[5] H. Katz, *J. Chem. Educ.*, **48**, 84 (1971).
[6a] A. A. Lamola, *Pure Appl. Chem.*, **24**, 599 (1970).
[6b] M. Kasha and A. U. Khan, *Ann. N.Y. Acad. Sci.*, **171**, 5 (1970).
[7] A. Gilbert, *Annual Reports (B)*, **67**, 195 (1970); N. J. Turro, G. S. Hammond, J. F. Endicott, J. C. Dalton, T. Kelly, J. E. Leonard, D. R. Morton, and D. M. Pond, *Ann. Survey Photochem.*, 1971. *Photochemistry (Specialist Periodical Report)*, ed. D. Bryce-Smith, Chemical Society, London, 1971; R. S. Davidson in *Organophosphorus Chemistry (Specialist Periodical Report)*, ed. S. Trippett, Chemical Society, London, 1971; *Adv. Photochemistry*, Vol. 8 (1971).
[8] A. Cox and T. J. Kemp, *Introductory Photochemistry*, McGraw-Hill, London, 1971; R. B. Cundall and A. Gilbert, *Photochemistry*, Nelson, London, 1970; R. P. Wayne, *Photochemistry*, Butterworths, London, 1970; J. P. Simons, *Photochemistry and Spectroscopy*, Wiley-Interscience, London, 1971; *Photochromism*, ed. G. H. Brown, in *Techniques of Chemistry*, Vol. 3, ed. A. Weissberger, Wiley-Interscience, 1971.

Physical Aspects

Emission Studies

A very exciting aspect of this field has been the development of equipment capable of detecting fluorescence from the second excited singlet state of aromatic hydrocarbons such as 3,4-benzopyrene[9a] and naphthalene.[9b] In a further study of the delayed fluorescence exhibited by benzophenone in polymer matrices, it was shown that the fluorescence had a similar lifetime to the normal phosphorescence.[10] Quantum yields of phosphorescence and phosphorescence lifetimes at room temperature and low temperature have been determined for several aromatic hydrocarbons.[11] The effects of hydrogen bonding on the emission spectra of a variety of compounds have been discussed.[12a,b,13] In the presence of triethylamine, proton transfer to the amine from 4-hydroxybenzophenone occurs and phosphorescence from the 4-benzoylphenoxide anion is observed.[12b] On the other hand, 1-naphthols form hydrogen-bonded complexes with the amine.[13] Aromatic azo-compounds of the type (**3**) that contain a 2-hydroxy- or amino-substituent exhibit fluorescence from the hydrazo-tautomer (**4**).[14] Attention has been drawn

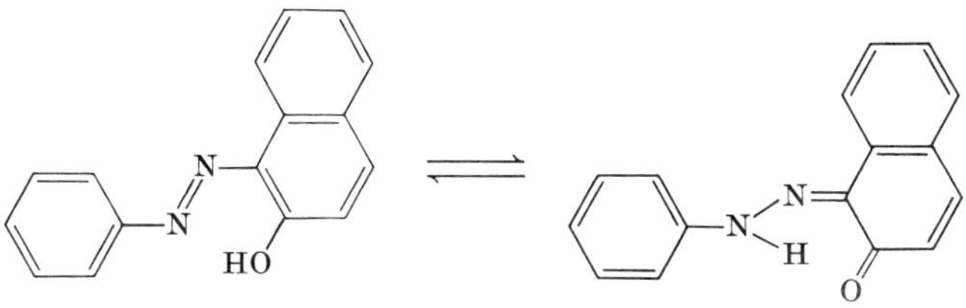

(**3**)   (**4**)

to the fact that compounds such as primary amines, that can form intramolecular hydrogen bonds, usually form aggregates in solid solutions of non-polar solvents;[15] in nearly all cases phosphorescence from the aggregate was very weak and much further to the red of the normal phosphorescence which was observed in polar matrices. It was shown that the phosphorescence exhibited by 4-aminobenzophenone in methylcyclohexane is from aggregates of the ketone, and consequently it is impossible to draw any conclusion about the type of transition that gives rise to the emission.[16] The phosphorescence spectrum of Michler's ketone has attracted further attention,[17] and evidence has been presented for the thesis that in protic solvents emission from the triplet $n\pi^*$ and $\pi\pi^*$ states can be observed. It has been found that *cis*-stilbene fluoresces in highly

---

[9a] P. Wannier, P. M. Rentzepis, and J. Jortner, *Chem. Phys. Letters*, **10**, 102 (1971).
[9b] P. Wannier, P. M. Rentzepis, and J. Jortner, *Chem. Phys. Letters*, **10**, 193 (1971).
[10] P. F. Jones and A. R. Calloway, *Chem. Phys. Letters*, **10**, 438 (1971).
[11] M. A. Slifkin and R. H. Walmsley, *Photochem. Photobiol.*, **13**, 57 (1971); J. Langelaar, R. P. H. Rettschnick, and G. J. Hoijtink, *J. Chem. Phys.*, **54**, 1 (1971).
[12a] R. E. Atkinson and F. E. Hardy, *J. Chem. Soc.* (B), **1971**, 357.
[12b] A. Matsuyama and H. Baba, *Bull. Chem. Soc. Japan*, **44**, 854 (1971).
[13] A. Matsuyama and H. Baba, *Bull. Chem. Soc. Japan*, **44**, 1162 (1971).
[14] D. Gegiou and E. Fischer, *Chem. Phys. Letters*, **10**, 99 (1971).
[15] R. S. Davidson and M. Santhanam, *J. Chem. Soc.* (B), **1971**, 1151.
[16] R. S. Davidson and M. Santhanam, *Chem. Comm.*, **1971**, 1114.
[17] W. Klöpffer, *Chem. Phys. Letters*, **11**, 482 (1971).

viscous media, and in a hydrocarbon glass a value of 0.75 for ϕ_F was reported.[18] From this observation it was concluded that isomerization of stilbenes by direct irradiation cannot occur from the first excited singlet state of the olefin. It has been found that metal alkyls of mercury, tin and lead promote intersystem crossing in anthracenes.[19]

Position of bromine substituent	ϕ_F	ϕ_P
1	0.070	0.45
2	0.004	0.56
3	0.010	0.32
4	0.010	0.47
None	0.41	0.057

(5)

The quantum yields of fluorescence and phosphorescence for some monobromo-derivatives of the naphtho-derivative (5) have been determined and it was found that when the substituent was in position 2 the heavy-atom effect was the most noticeable;[20] this efficiency was attributed to overlap of the back $p\pi$ orbital of the bromine atom with the naphthalene π-system. The quantum yield of fluorescence for 2-bromolysergic acid is much lower than for the unsubstituted acid and this also was attributed to the heavy-atom effect of the bromine substituent.[21] The suggestion has been made that molecules in excited states can undergo reorientation in rigid matrices and that, in the case of benzil, emission from more than one conformation is observed.[22a] This theory may also offer an explanation for the rather peculiar emission spectra of N-aroylbenzamides.[22b] It has been shown that in poly(methyl methacrylate) matrices solute molecules can exhibit appreciable mobility when there is 10–20% of residual monomer.[22c] There have been further investigations of the effects of solvent upon excited singlet and triplet levels[23] and acid–base equilibria of excited states.[24a,b] 2-Acetylnaphthalene is more basic in its first excited singlet state than in its ground state.[24b] Excitation of ketones dissolved in strongly acidic media results in emission from the protonated ketone.[24c]

Indan-1-one exhibits two phosphorescence bands when contained in polar protic matrices. From the fact that irradiation of the ketone in deuteriated methanol (MeOD) led to incorporation of deuterium, it was concluded that the ketone enolizes on excitation and that the long-wavelength phosphorescence band is due to the enolate anion.[24d]

---

[18] S. Sharafy and K. A. Muszkat, *J. Am. Chem. Soc.*, **93**, 4119 (1971).
[19] E. Vander-donckt and J. P. Van-Bellinghen, *Chem. Phys. Letters*, **7**, 630 (1970); E. Vander-donckt and J. P. Van-Bellinghen, *J. Chim. Phys. Physicochim. Biol.* **68**, 948 (1971).
[20] G. Kavarnos, T. Cole, P. Scribe, J. C. Dalton, and N. J. Turro, *J. Am. Chem. Soc.*, **93**, 1032 (1971).
[21] A. Bowd, J. B. Hudson, and J. A. Turnbull, *J. Chem. Soc.* (B), **1971**, 1509.
[22a] D. J. Morantz and A. J. C. Wright, *J. Chem. Phys.*, **54**, 692 (1971).
[22b] E. J. O'Connell, M. Delmauro, and J. Irwin, *Photochem. Photobiol.* **14**, 189 (1971).
[22c] R. J. Woods and J. F. Manville, *Can. J. Chem.*, **49**, 515 (1971).
[23] P. S. Song, M. L. Harter, T. A. Morris, and W. C. Herndon, *Photochem. Photobiol.*, **14**, 521 (1971); J. B. Gallivan and J. S. Brinen, *Chem. Phys. Letters*, **10**, 455 (1971); S. Hotchandani and A. C. Testa, *J. Chem. Phys.*, **54**, 4508 (1971).
[24a] W. R. Ware, P. R. Shukla, P. J. Sullivan, and R. V. Bramphis, *J. Chem. Phys.*, **55**, 4048 (1971).
[24b] A. R. Watkins, *Z. Phys. Chem. (Frankfurt)*, **75**, 327 (1971).
[24c] R. Rusakowicz, G. W. Byers, and P. A. Leermakers, *J. Am. Chem. Soc.*, **93**, 3263 (1971).
[24d] M. E. Long, B. Bergman, and E. C. Lim, *Mol. Photochem.*, **2**, 341 (1970).

The Triplet State

Use has been made of singlet–triplet absorption spectra in order to locate the triplet levels of several fluorinated benzenes[25a] and of all-*trans*-retinal.[25b] Further attempts have been made at an unequivocal assignment to the triplet energy of β-carotene;[26] this is of particular interest because of its relevance to the mechanism by which this polyene quenches singlet molecular oxygen. The triplet states of several dyes have been investigated by flash photolysis and in some cases triplet–triplet annihilation was observed.[27] Thymine triplets are deactivated by reaction with ground-state thymine and this process leads to dimer formation.[28] Evidence has been presented, from a study in which laser flash photolysis was used, that the triplet state of α-tocopherolquinone (**6**),

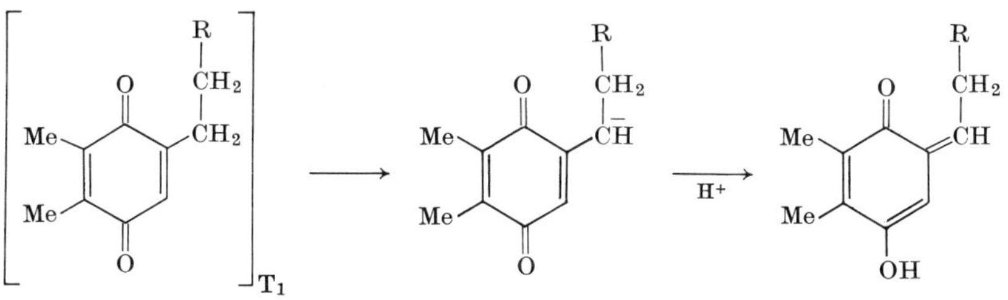

can deprotonate to give a species that ultimately gives the quinone methide (**7**).[29] The energy of the second triplet state of 1,5-dichloroanthracene has been determined as 70.0 kcal mol^{-1}.[30a]

The application of flash photolysis to study the triplet states of molecules has been reviewed;[30b] by this means the mechanism of decay of triplet acetone has been investigated.[30c]

Energy Transfer

An intriguing application of energy transfer has been to the determination of thickness of monolayers.[31] In the case of (**8**), the benzoyl group effectively deactivates the singlet

[25a] G. P. Semeluk and R. D. S. Stevens, *Can. J. Chem.*, **49**, 2452 (1971).
[25b] R. A. Raubach and A. B. Guzzo, *J. Phys. Chem.*, **75**, 983 (1971).
[26] E. J. Land, A. Sykes, and T. G. Truscott, *Photochem. Photobiol.*, **13**, 311 (1971).
[27] A. Kellmann, *Photochem. Photobiol.*, **14**, 85 (1971); J. P. Webb, W. C. McColgin, O. G. Peterson, D. L. Stockman, and J. H. Eberly, *J. Chem. Phys.*, **53**, 4227 (1970); Y. Kubota, *Bull. Chem. Soc. Japan*, **43**, 3121 (1970).
[28] D. Whillans and H. E. Johns, *J. Am. Chem. Soc.*, **93**, 1358 (1971).
[29] D. R. Kemp and G. Porter, *J. Chem. Soc.* (A), **1971**, 3510.
[30a] J. P. Roberts and R. S. Dixon, *J. Phys. Chem.*, **75**, 845 (1971).
[30b] A. K. Chibisov, *Uspekhi Khim.*, **39**, 1886 (1971); *Russ. Chem. Rev.*, **39**, 891 (1971).
[30c] G. Porter, R. W. Yip, J. M. Dunston, A. J. Cessna, and S. E. Sugamori, *Trans. Faraday Soc.*, **67**, 3149 (1971).
[31] H. Kuhn and D. Möbius, *Angew. Chem. Int. Ed.*, **10**, 620 (1971).

state of the naphthalene group and this group effectively deactivates the triplet state of the benzoyl group.[32] Of the several reported examples of quenching of phosphorescence of carbonyl compounds,[33,34] energy transfer to chromium complexes or terbic or europium salts seems to be of particular interest.[34] An investigation of the effect of oxygen on the singlet and triplet states of aromatic hydrocarbons has shown that the singlet state is deactivated because oxygen increases the efficiency of intersystem crossing.[35] By means of picosecond light pulses it has been possible to observe singlet–singlet energy transfer between Rhodamine 6 G and Malachite Green.[36] Examination of energy transfer between benzophenone and perylene has confirmed the dipole–dipole nature of the triplet–singlet energy-transfer process.[37] By using high-intensity light pulses it has been possible to observe singlet–singlet and singlet–triplet annihilation processes.[38] Deuterium isotope effects in energy-transfer processes have been the subject of further discussion.[39] The ability of porphyrins to sensitize the isomerization of some substituted ethylenes has been interpreted as occurring via the distorted triplet of the ethylene.[40a] Isomerization of vitamin A occurs by the same mechanism.[40b] An interesting example of energy transfer is the sensitized isomerization of stilbazole ligands of some ruthenium complexes;[41] irradiation in the charge transfer band of the complex leads to isomerization. Energy-transfer systems have been used to show that ϕ_{isc} for benzaldehyde is unity[42] and that the ϕ_{isc} for fluorenone is solvent-dependent.[43] In compounds of the type (**9**), it was shown that when structural factors were favourable, the Norrish Type II reaction could compete with intramolecular energy transfer,[44] e.g. in the case of (**10**).

[32] H. E. Zimmerman and R. D. McKelvey, *J. Am. Chem. Soc.*, **93**, 3638 (1971).
[33] S. E. Webber, *J. Phys. Chem.*, **75**, 1921 (1971); A. W. Jackson and A. J. Yarwood *J. Am. Chem. Soc.*, **93**, 2801 (1971).
[34] V. Balzani, R. Ballardini, M. T. Gandolfi, and L. Moggi, *J. Am. Chem. Soc.*, **93**, 339 (1971); G. W. Mushrusn, F. L. Minn, and N. Filipescu, *J. Chem. Soc.* (B), **1971**, 427.
[35] L. K. Patterson, G. Porter, and M. R. Topp, *Chem. Phys. Letters*, **7**, 612 (1970); R. Potashnik, C. R. Goldschmidt, and M. Ottolenghi, *Chem. Phys. Letters*, **9**, 424 (1971).
[36] D. Rehm and K. B. Eisenthal, *Chem. Phys. Letters*, **9**, 387 (1971).
[37] A. F. Vaudo and D. M. Hercules, *J. Am. Chem. Soc.*, **93**, 2599 (1971).
[38] S. D. Babanko, V. A. Benderskii, A. G. Levrushko, and V. P. Tychinskii, *Chem. Phys. Letters*, **8**, 598 (1971).
[39] R. A. Caldwell, G. W. Sovocool, and R. J. Peresie, *J. Am. Chem. Soc.*, **93**, 779 (1971).
[40a] P. D. Wildes and D. G. Whitten, *J. Am. Chem. Soc.*, **92**, 7609 (1970).
[40b] A. Sykes and T. G. Truscott, *Trans. Faraday Soc.*, **67**, 679 (1971).
[41] P. P. Zarnegar and D. G. Whitten, *J. Am. Chem. Soc.*, **93**, 3776 (1971).
[42] G. R. DeMaré, M. C. Fontaine, and M. Termonia, *Chem. Phys. Letters*, **11**, 617 (1971).
[43] R. A. Caldwell and R. P. Gajewski, *J. Am. Chem. Soc.*, **93**, 532 (1971).
[44] D. O. Cowan and A. A. Baum, *J. Am. Chem. Soc.*, **93**, 1153 (1971).

Energy Transfer and Reactions Occurring via Complex Formation

(a) *Excimers.* Theoretical calculations of the interplanar distance between the constituent molecules of excimers agree with the experimentally determined values.[45] The absorption and polarization spectra of the pyrene excimer have been obtained[46a] and the energy of the species has been determined by means of energy-transfer experiments;[46b] the excimer formation can result from triplet–triplet annihilation processes.[47] Excimers are also formed on excitation of crystalline aromatic hydrocarbons produced by cooling solutions of these compounds in hydrocarbon solvents.[48] Further examples of intramolecular excimer formation have also been reported.[49]

(b) *Exciplexes.* A particularly significant discovery is the finding that an exciplex $(D^+A^-)^*$ can be produced by interaction of excited D or A with the appropriate partner.[50] There is also a time delay in exciplex formation which increases as the lifetime of initially excited species (i.e. D* or A*) increases. Rate constants for exciplex formation and decay have been determined.[51] It has been claimed that absorption spectra of exciplexes can be observed by the application of low-temperature techniques.[52] Several new types of exciplex have been reported and these include complexes formed by aromatic hydrocarbons with olefins[53a,b] or benzoic acid.[53c] The naphthylalkylamines (**11**, **12**; $n = 2$ or 3)[54,55] and 9-anthrylalkylamines (**13**; $n = 3$) exhibit intramolecular exciplex formation;[55]

(**11**) (**12**) (**13**)

R = Me or Et

these compounds fail to undergo the normal reactions of the first excited states of naphthalene and anthracene, respectively, and this was attributed to deactivation through exciplex formation. Exciplex formation does not occur with (**11**), (**12**) or (**13**) when $n = 1$ and this was attributed to the fact that in these compounds the nitrogen lone pair cannot sit over the aromatic π-system. The triplet state of fluorenone is quenched by substituted anilines and the quenching efficiency is linearly related to σ^+ constants

---

[45] F. L. Minn, J. P. Pinion, and N. Filipescu, *J. Phys. Chem.*, **75**, 1794 (1971).
[46a] M. F. M. Post, J. Langelaar, and J. D. W. VanVoorst, *Chem. Phys. Letters*, **10**, 468 (1971).
[46b] O. L. J. Gijzeman, W. H. VanLeeuwen, J. Langelaar, and J. D. W. VanVoorst, *Chem. Phys. Letters*, **11**, 532 (1971).
[47] K. Kikuchi, H. Kokubun, and M. Koizumi, *Bull. Chem. Soc. Japan*, **44**, 1527 (1971).
[48] R. J. McDonald and B. K. Selinger, *Austral. J. Chem.*, **24**, 249 (1971).
[49] P. C. Johnson and H. W. Offen, *J. Chem. Phys.*, **55**, 2945 (1971).
[50] A. E. W. Knight and B. K. Selinger, *Chem. Phys. Letters*, **10**, 43 (1971); R. J. McDonald and B. K. Selinger, *Austral. J. Chem.*, **24**, 1797 (1971).
[51] K. Yoshihara, T. Kasuya, A. Inoue, and S. Nagakura, *Chem. Phys. Letters*, **9**, 469 (1971).
[52] H. Isheda and H. Tsubomura, *Chem. Phys. Letters*, **9**, 296 (1971).
[53a] G. N. Taylor, *Chem. Phys. Letters*, **10**, 355 (1971).
[53b] T. R. Evans, *J. Am. Chem. Soc.*, **93**, 2081 (1971).
[53c] D. W. Ellis, R. G. Hamel, and B. S. Solomon, *Chem. Comm.*, **1970**, 1697.
[54] E. A. Chandross and H. T. Thomas, *Chem. Phys. Letters*, **9**, 393, 397 (1971).
[55] D. R. G. Brimage and R. S. Davidson, *Chem. Comm.*, **1971**, 1385.

for the amines;[56] this lends credence to the suggestion that quenching is occurring via an exciplex mechanism.

(c) *Excited Charge-transfer Complexes, Electron Transfer and Ejection Reactions.* Further examples of intramolecular charge-transfer complex formation have been reported,[57,58] e.g. (14) and (15). In the case of (15), complex formation can take place because of the

$p\text{-}O_2NC_6H_4O(CH_2)_n NHNp$

(14)

Np = Naphthyl

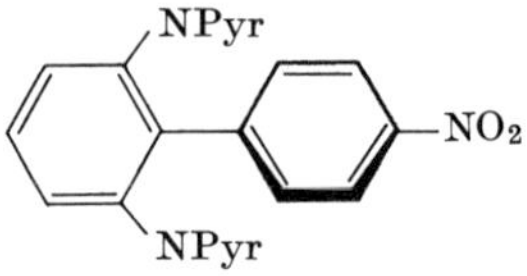

(15)

NPyr = Pyrrolidine

twisted biphenyl framework.[58] Much interest has been shown in determining the mechanism by which oxygen quenches the excited singlet state of aromatic hydrocarbons and it has been concluded that oxygen enhances intersystem crossing between the S_1 and T_1 states[35,59] and that this effect arises from initial complex formation between the hydrocarbon and oxygen.[59] There have been further reports of photoinduced electron transfer from Methylene Blue to zinc oxide anodes[60a,b] and this has been shown to cause oxidation of the dye.[60a] Radical-ion formation by excitation of charge-transfer complexes has also been detected.[61] A previous claim that excitation of tetracyanobenzene charge-transfer complexes produces emission from the complex has been refuted and the emission shown to be phosphorescence of the cyanobenzene.[62] The electronically excited species produced by electron transfer from radical anions to radical cations have been examined and in the case of the reaction of tri-p-tolylamine radical cation with aromatic hydrocarbon radical anions both exciplex and hydrocarbon fluorescence were detected:[63]

$$A^{\overline{\cdot}} + D^{\overset{+}{\cdot}} \to (A^-D^+)_{S_1} \to A_{S_1}{}^* + D_{S_0}$$

In polar solvents only hydrocarbon fluorescence is observed, and this is produced by a triplet–triplet annihilation reaction. This type of reaction also explains the occurrence of hydrocarbon fluorescence in the reaction of Würster's Blue cation with aromatic hydrocarbon radical anions:[64]

$$D^{\overset{+}{\cdot}}ClO_4^- + A^{\overline{\cdot}} \to D + ClO_4^- + A_{T_1}{}^*$$

$$A_{T_1}{}^* + A_{T_1}{}^* \to A_{S_0} + A_{S_1}{}^*$$

[56] S. G. Cohen and G. Parsons, *J. Am. Chem. Soc.*, **92**, 7603 (1970).
[57] K. Mutai, *Tetrahedron Letters*, **1971**, 1125; *Bull. Chem. Soc. Japan*, **44**, 2537 (1971); H. A. H. Craenen, J. W. Verhoeven, and T. J. de Boer, *Tetrahedron*, **27**, 2561 (1971).
[58] E. Daltrozzo, F. Effenberger, and P. Fischer, *Angew. Chem. Int. Ed.*, **10**, 567 (1971).
[59] T. Brewer, *J. Am. Chem. Soc.*, **93**, 775 (1971).
[60a] W. P. Gomes and F. Cardon, *Ber. Bunsengés., Phys. Chem.*, **75**, 914 (1971).
[60b] H. Tributsch and M. Calvin, *Photochem. Photobiol.* **14**, 95 (1971).
[61] H. Masuhara, M. Shimada, and N. Mataga, *Bull. Chem. Soc. Japan*, **43**, 3316 (1970).
[62] G. Briegleb and D. Wolf, *Z. Naturforsch.*, a, **25**, 1925 (1970).
[63] A. Weller and K. Zachariasse, *Chem. Phys. Letters*, **10**, 424, 590 (1971).
[64] A. Weller and K. Zachariasse, *Chem. Phys. Letters*, **10**, 197 (1971).

Further examples of similar reactions are to be found in the Section on chemiluminescence. Irradiation of perylene in polar solvents results in radical cation formation and this was shown to be a monophotonic process involving reaction of the excited singlet state of the hydrocarbon with the ground-state hydrocarbon.[65] Photoinduced electron ejection from various amines, e.g. tryptophan[66a] and tetramethyl-p-phenylenediamine[66b] has been studied.

(d) *Reactions Occurring by Exciplex and Excited Charge-transfer Complex Formation.* Irradiation of phthalate esters, e.g. (**16**), in the presence of mono-olefins produces oxetans and from the observation that the olefins quench the fluorescence of the phthalate

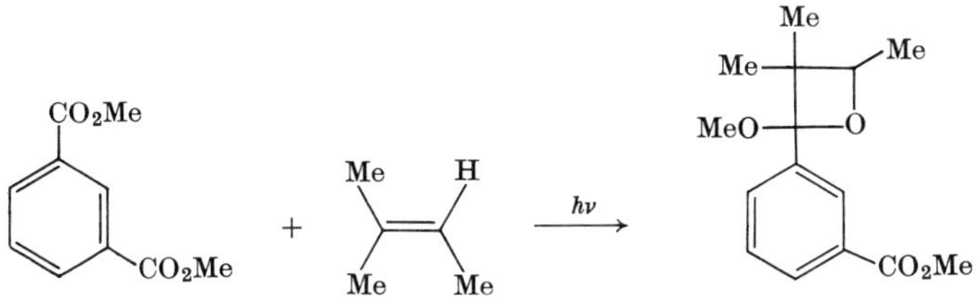

esters it is reasonable to assume that exciplex intermediates are involved.[67] Carboxylic acids of the type $ArXCH_2CO_2H$ (X = O, S or NH) are decarboxylated on irradiation in the presence of ketones,[68] quinones,[68] or aromatic nitro-compounds.[69] It was shown that the rate constants for reaction of the acids with benzophenone were similar to those for quenching triplet benzophenone by the substituted hydrocarbons $ArXCH_3$ (X = O, S or NH). This was taken as evidence that the reactions occur via exciplex intermediates of the type (**17**). Irradiation of the charge-transfer complex of toluene with 1,2,4,5-tetracyanobenzene produces 1-benzyl-2,4,5-tricyanobenzene via radical ion intermediates.[70]

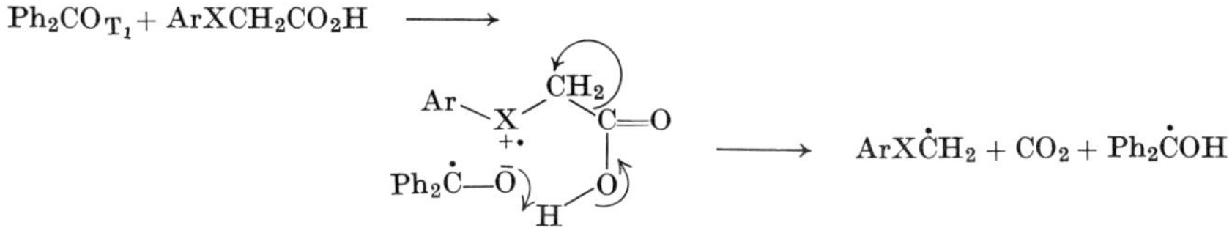

Carbonyl Compounds

Alkyl and Aryl Ketones, etc.

(a) *Intermolecular Reduction Reactions.* The triplet state of acetophenone has been detected in fluid solution by nanosecond flash photolysis.[71] When benzene was used as

---

[65] K. H. Grellmann and A. R. Watkins, *Chem. Phys. Letters*, **9**, 439 (1971).
[66a] J. Moan and H. B. Steen, *J. Phys. Chem.*, **75**, 2887, 2893 (1971); L. I. Grossweiner and Y. Usui, *Photochem. Photobiol.*, **13**, 195 (1971).
[66b] P. J. Bekowies and A. C. Albrecht, *J. Phys. Chem.*, **75**, 431 (1971); C. Chachaty, D. Shoemaker, and R. V. Bensasson, *Photochem. Photobiol.*, **12**, 317 (1970).
[67] Y. Shigemitsu, Y. Katsuhara, and Y. Odaira, *Tetrahedron Letters*, **1971**, 2887.
[68] R. S. Davidson and P. R. Steiner, *Chem. Comm.*, **1971**, 1115.
[69] R. S. Davidson, S. Korkut, and P. R. Steiner, *Chem. Comm.*, **1971**, 1052.
[70] A. Yoshino, M. Ohashi, and T. Yonezawa, *Chem. Comm.*, **1971**, 97.
[71] H. Lutz and L. Lindqvist, *Chem. Comm.*, **1971**, 493.

solvent formation of the acetophenone ketyl radical was observed. There have also been several studies of the photoreduction of benzophenone by benzene in which the benzophenone ketyl radical has been shown to be an intermediate species.[72,73] There appear to be three main mechanisms for ketyl radical formation: (i) hydrogen abstraction from benzene by the triplet ketone;[72] (ii) biradical (18) formation by addition of the triplet

$$Ph_2CO_{T_1} + C_6H_6 \longrightarrow Ph_2\dot{C}-O-\underset{H}{\underset{(18)}{\langle\cdot\rangle}} \xrightarrow{Ph_2CO} Ph_2\dot{C}OH + Ph_2\dot{C}OPh \longrightarrow Ph_2CO + \dot{P}h$$

$$\longrightarrow Dimer \longrightarrow Ph-Ph + 2Ph_2\dot{C}OH$$

ketone to benzene;[73,74] and (iii) initial formation of an exciplex between the ketone and benzene that ultimately collapses to the biradical (18). Strong evidence has been put forward in support of mechanisms (i) and (ii), and the recent finding[68,75,76] that carbonyl compounds can sensitize decarboxylations of acids of the type $ArXCH_2CO_2H$ (X = O, S or NH) suggests that mechanism (iii) should also be carefully considered. An unequivocal solution to this problem is awaited with interest. It has been found that the ketyl radical of acetone can react with other ketyl radicals, e.g. the ketyl radical of benzophenone[77a] and benzyl radicals;[77b] these reactions proceed more readily than dimerization of the radical. Further evidence has been presented for the photochemical decomposition of 1,2-diols to ketyl radicals,[78a] and these radicals have been used as reductants for 1,2-quinones.[78b] From rate-constant determinations it has been shown that the ability of tri-n-butyltin hydride to act as a reductant for triplet ketones is due to the weakness of the Sn—H bond.[79] Reductions with other good hydrogen donors, such as amines, have also been studied: when optically active amines are used, asymmetric reduction is observed and optically active 1,2-diols are obtained.[80] By the use of 2-methyl-2-nitrosopropane as a radical scavenger,[81] and also by low-temperature techniques[82] it has been shown that reaction of triplet benzophenone with N,N-dimethylaniline produces the $PhN(Me)\dot{C}H_2$ radical. Other aspects of reduction reactions have included the use of sulphides[83a] and amides[83b] as hydrogen donors, the formation of light-absorbing intermediates[83c] and determination of rate constants.[83d]

---

[72] J. Dedinas, *J. Phys. Chem.*, **75**, 181 (1971); A. V. Buettner and J. Dedinas, *J. Phys. Chem.*, **75**, 187 (1971).
[73] D. I. Schuster, T. M. Weil, and M. R. Topp, *Chem. Comm.*, **1971**, 1212.
[74] J. Saltiel and H. C. Curtis, *J. Am. Chem. Soc.*, **93**, 2056 (1971); J. Saltiel, H. C. Curtis, and B. Jones, *Mol. Photochem.*, **2**, 331 (1970).
[75] R. S. Davidson and P. R. Steiner, *J. Chem. Soc.* (C), **1971**, 1682.
[76] R. S. Davidson, K. Harrison, and P. R. Steiner, *J. Chem. Soc.* (C), **1971**, 3480.
[77a] S. A. Weiner, *J. Am. Chem. Soc.*, **93**, 425 (1971).
[77b] P. Singh, *J. Chem. Soc.* (C), **1971**, 714.
[78a] J. H. Stocker and D. H. Kern, *J. Org. Chem.*, **36**, 1095 (1971).
[78b] M. B. Rubin and J. M. Ben-Bassat, *Tetrahedron Letters*, **1971**, 3403.
[79] D. R. G. Brimage, R. S. Davidson, and P. F. Lambeth, *J. Chem. Soc.* (C), **1971**, 1241.
[80] D. Seebach and H. Daum, *J. Am. Chem. Soc.*, **93**, 2795 (1971).
[81] I. H. Leaver, *Tetrahedron Letters*, **1971**, 2333.
[82] S. Arimitsu and H. Tsubomura, *Bull. Chem. Soc. Japan*, **44**, 2288 (1971).
[83a] W. Ando, J. Suzuki, and T. Migita, *Bull. Chem. Soc. Japan*, **44**, 1987 (1971).
[83b] M. Nakashima and E. Hayon, *J. Phys. Chem.*, **75**, 1910 (1971).
[83c] N. Filipescu, L. M. Kindley, and F. L. Minn, *J. Org. Chem.*, **36**, 861 (1971).
[83d] G. C. Ramsay and S. G. Cohen, *J. Am. Chem. Soc.*, **93**, 1166 (1971).

By ESR spectroscopy it has been shown that the dimethylamino group in Michler's ketone acts as a hydrogen atom donor for the excited ketone.[81] A product that appears to be formed by such a route has been isolated.[84] In further discussion of the role of upper triplet states in the reactions of 9-anthraldehyde it has been concluded that dimerization and reduction occur from the same excited state and that the relative efficiency of the two processes depends on the concentration of the aldehyde.[85] A very interesting application of reduction reactions has been to the functionalization of the steroid nucleus[86a] and long-chain alkenes;[86b] thus the steroid (**19**) was transformed into (**20**) by irradiation

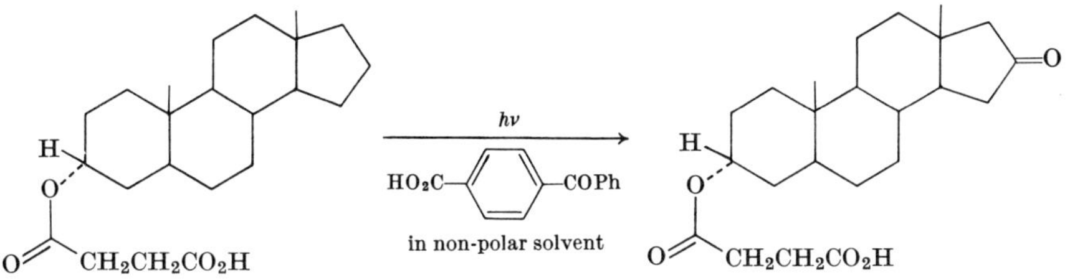

in the presence of 4-benzoylbenzoic acid; hydrogen-bonding between the carboxyl group of the acid and the ketone facilitates the reasonably regiospecific attack by the triplet ketone.

(b) *Intramolecular Reduction Reactions.* Conversion of acetophenones (**21**; R' = Me) into cyclobutanols was facilitated by increasing size of the *ortho*-substituents whereas

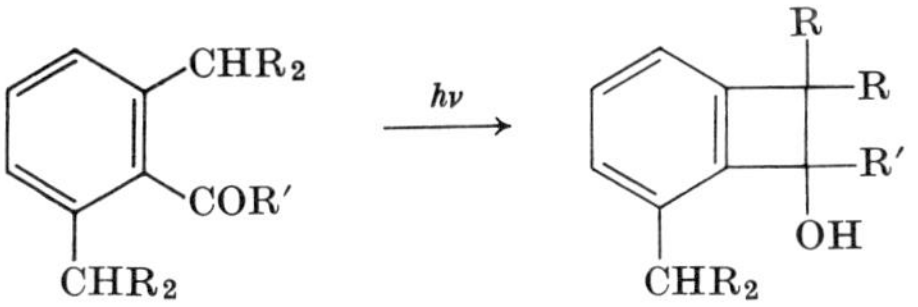

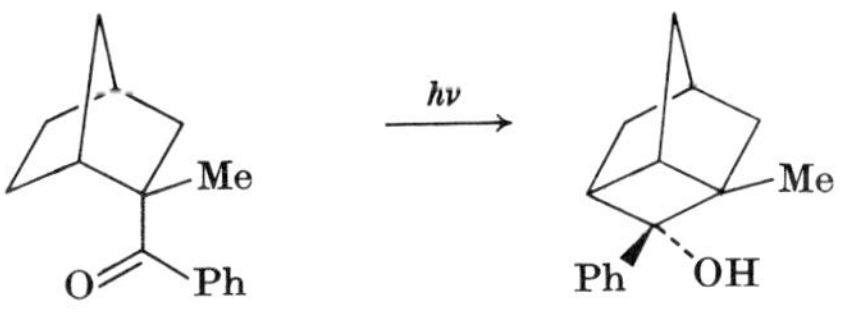

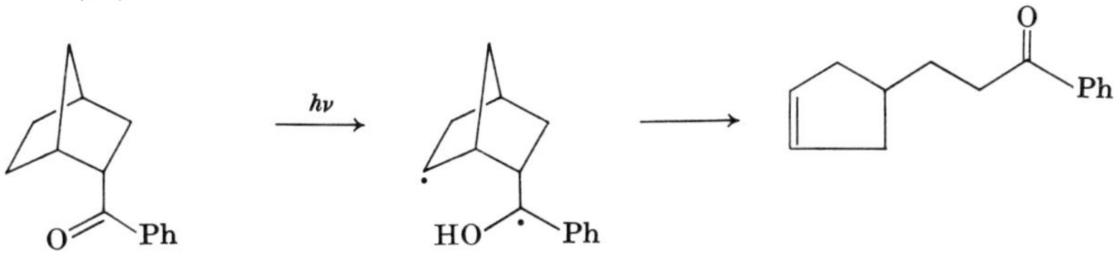

[84] T. H. Koch and A. H. Jones, *J. Am. Chem. Soc.*, **92**, 7503 (1970).
[85] P. Suppan, *Tetrahedron Letters*, **1971**, 4469.
[86a] R. Breslow and P. Kalicky, *J. Am. Chem. Soc.*, **93**, 3540 (1971); R. Breslow and P. C. Scholl, *J. Am. Chem. Soc.*, **93**, 2331 (1971).
[86b] L. Hoener and H. Schwarz, *Ann. Chem.*, **747**, 14 (1971).

such a change had the opposite effect on the benzophenones (**21**; R′ = Ph).[87] The triplet lifetimes of *o*-alkoxybenzophenones are very short and this has been attributed to deactivation by intramolecular hydrogen abstraction.[88] Triplet *o*-methylvalerophenone is also deactivated by a similar process but, as well, it undergoes the Norrish Type II reaction.[89] There have been a number of product studies on intramolecular reduction reactions[90a] and that of (**22**) is particularly interesting.[90b] When the benzoyl group is *exo* a Type I reaction occurs, whereas in the case of (**23**) a Type II reactions takes place. This is a further example of the subtle effects of substituents on the ratio of cyclobutanol formation to olefin formation in Type II reactions.[91] In the case of (**24**), its first excited

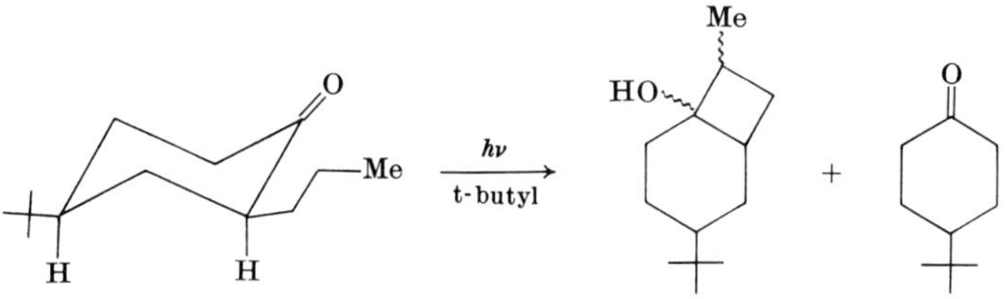

singlet and triplet state decomposed to give a cyclobutanol and Type II products.[92a,b] Cyclobutanol formation takes place more readily from the triplet than from the singlet state;[92a] it was suggested that, because of the greater energy of the singlet state, conformational effects played a smaller part in deciding the product ratios than in the case of the less energetic triplet state. There was greater stereospecificity in olefin formation from (**25**) when reaction took place from the singlet than from the triplet state;[93a] it was argued that the greater energy of the singlet state allowed fragmentation to occur from the initially formed biradical and that equilibration to a common biradical did not occur as it does in the case of the triplet state reaction.[93b] This argument was also applied to rationalize other reactions of 1,4-biradicals. There have been several other

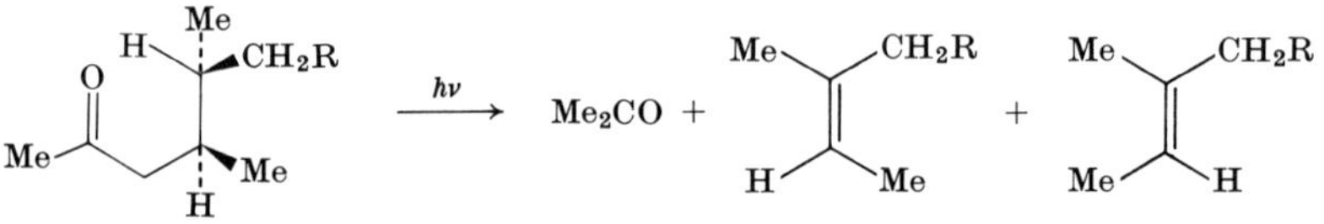

---

[87] Y. Kitaura and T. Matsura, *Tetrahedron*, **27**, 1597 (1971).
[88] G. R. Lappin and J. S. Zannucci, *J. Org. Chem.*, **36**, 1808 (1971).
[89] W. R. Bergmark, B. Beckmann, and W. Lindenberger, *Tetrahedron Letters*, **1971**, 2259.
[90a] N. Sugiyama, K. Yamada, and H. Aoyama, *J. Chem. Soc.* (C), **1971**, 830; L. M. Stephenson and J. L. Partlett, *J. Org. Chem.*, **36**, 1093 (1971); K. H. Schulte-Elte, B. Willhalm, A. F. Thomas, M. Stoll, and G. Ohloff, *Helv. Chim. Acta*, **54**, 1759 (1971); B. W. Finucane and J. B. Thomson, *J. Chem. Soc.* (C), **1971**, 1569; P. Gull, H. Wehrli, and O. Jeger, *Helv. Chim. Acta*, **54**, 2158 (1971).
[90b] F. D. Lewis and R. A. Ruden, *Tetrahedron Letters*, **1971**, 715.
[91] F. D. Lewis and J. A. Hilliard, *J. Am. Chem. Soc.*, **92**, 6672 (1970).
[92a] I. Flemming, A. V. Kemp-Jones, and E. J. Thomas, *Chem. Comm.*, **1971**, 1158.
[92b] K. Dawes, J. C. Dalton, and N. J. Turro, *Mol. Photochem.*, **3**, 71 (1971).
[93a] L. M. Stephenson, P. R. Cavigli, and J. L. Partlett, *J. Am. Chem. Soc.*, **93**, 1984 (1971).
[93b] L. M. Stephenson and J. I. Brauman, *J. Am. Chem. Soc.*, **93**, 1988 (1971).

examinations of Type II reactions[94] and a comprehensive review on the reaction has been published.[95] n-Butyl 2-naphthyl ketone has been shown to give Type II products from its $n\pi^*$ singlet state.[96] Cyclopropyl ketones (**26**) also undergo the Type II reaction, and the intermediate enol (**27**) was detected by IR spectroscopy.[97a] Fragmentation of

(**28**) occurs on irradiation in the vapour phase whereas no reaction is observed in unreactive solvents.[97a,b] There have been several reports of intramolecular hydrogen abstraction from C—H bonds adjacent to a nitrogen atom and the mechanism of the formation of the products of these reactions has been discussed.[98a] These reactions have also been the subject of a review.[98b] Intramolecular hydrogen abstraction from C—H bonds adjacent to sulphur has been observed and from a kinetic study it was concluded that reaction initially occurs by electron transfer from the sulphur atom to the excited carbonyl group.[98c]

(c) *Fragmentation Reactions.* There have been several determinations of the lifetimes of the first excited singlet states of aliphatic ketones[99a,b,100] and attempts have been made to correlate these lifetimes with the photoreactivity of the ketones. In the case of cyclobutanone, reducing the wavelength of irradiation decreases ϕ_F and increases cleavage of the ketone by a predissociation process.[99b] The lifetimes of the singlet states of many alkyl ketones are determined by the ease with which they undergo the Type II reactions; fluorescence lifetimes have been used to calculate rate constants for such reactions.[100] Examination of the hexahydroindanones (**29**) and (**30**) has shown that the *trans*-compound

---

[94] D. C. Neckers, R. M. Kellogg, W. L. Prins, and B. Schoustra, *J. Org. Chem.*, **36**, 1838 (1971); Y. Kubokawa, M. Kubo, and G. Nanjo, *Bull. Chem. Soc. Japan*, **43**, 3968 (1970); L. P. Y. Lee, B. McAneney, and J. E. Guillet, *Can. J. Chem.*, **49**, 1310 (1791); W. R. Oliver and L. R. Hamilton, *Tetrahedron Letters*, **1971**, 1837; P. J. Wagner and R. G. Zepp, *J. Am. Chem. Soc.*, **93**, 4958 (1971).
[95] P. J. Wagner, *Accounts Chem. Res.*, **4**, 168 (1971).
[96] N. C. Yang and A. Shani, *Chem. Comm.*, **1971**, 815.
[97a] D. G. Marsh, J. N. Pitts, K. Schaffner, and A. Tuinman, *J. Am. Chem. Soc.*, **93**, 333 (1971).
[97b] D. G. Marsh and J. N. Pitts, *J. Am. Chem. Soc.*, **93**, 326 (1971).
[98a] A. Padwa and W. Eisenhardt, *J. Am. Chem. Soc.*, **93**, 400 (1971); A. Padwa and A. Battisti, *J. Org. Chem.*, **36**, 230 (1971); A. Padwa, F. Albrecht, P. Singh, and E. Vega, *J. Am. Chem. Soc.*, **93**, 2928 (1971); E. H. Gold, *J. Am. Chem. Soc.*, **93**, 2793 (1971).
[98b] A. Padwa, *Accounts Chem. Res.*, **4**, 48 (1971).
[98c] A. Padwa and D. Pashayan, *J. Org. Chem.*, **36**, 3550 (1971).
[99a] A. M. Halpern and W. R. Ware, *J. Chem. Phys.*, **54**, 1271 (1971); G. M. Breuer and E. K. C. Lee, *J. Phys. Chem.*, **75**, 989 (1971).
[99b] J. C. Hemminger, C. F. Rusbult, and E. K. C. Lee, *J. Am. Chem. Soc.*, **93**, 1867 (1971); R. G. Shortridge, C. F. Rusbult, and E. K. C. Lee, *J. Am. Chem. Soc.*, **93**, 1863 (1971); J. C. Hemminger and E. K. C. Lee, *J. Chem. Phys.*, **54**, 1405 (1971); R. F. Klemm, *Can. J. Chem.*, **48**, 3320 (1970).
[100] J. C. Dalton and N. J. Turro, *J. Am. Chem. Soc.*, **93**, 3569 (1971); K. Dawes, N. J. Turro, and J. M. Conia, *Tetrahedron Letters*, **1971**, 1377.

(29) forms the aldehyde and isomerizes from both its singlet and its triplet state, whereas the *cis*-compound **(30)** isomerizes only from its singlet state and gives the aldehyde from both the singlet and the triplet state;[101] photoisomerization is much more efficient from

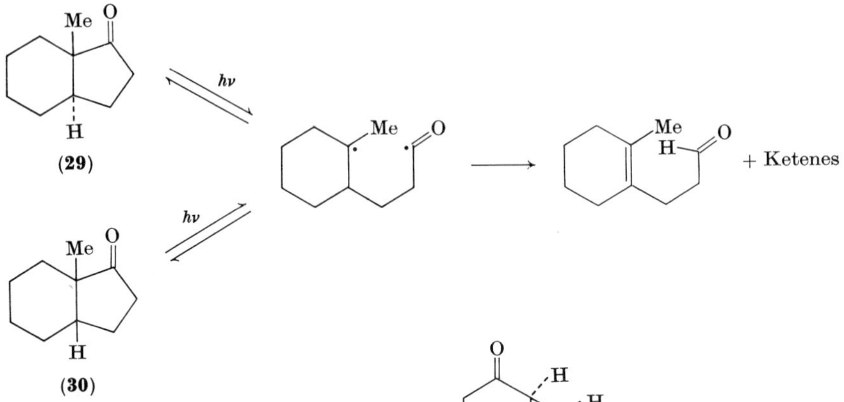

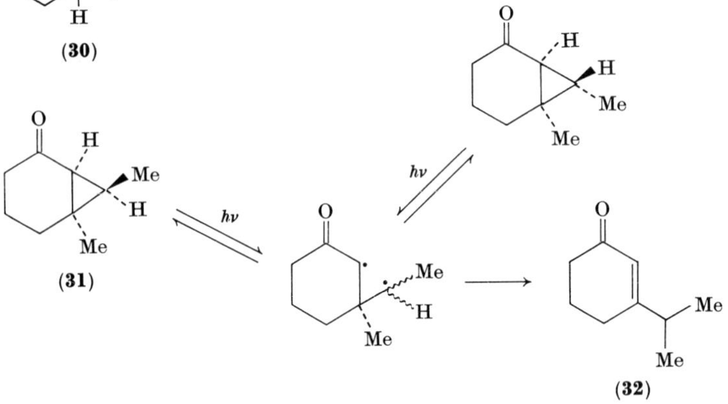

the *trans*- than from the *cis*-ketone. From this study it was concluded that the inefficiency of the Type I process for a number of aliphatic ketones is probably due to the cage recombination of the initially dissociated radicals.

Another interesting finding is that the ability of conjugated dienes to quench the the Type I reaction of t-butyl methyl ketone can be correlated with the ionization potential of the diene;[102a] this suggests that the mechanism of quenching may well involve an exciplex intermediate. A similar conclusion has been reached with respect to interaction of other carbonyl compounds with dienes.[102b] Irradiation of **(31)** leads not only to isomerization but also to the formation of **(32)** by migration of the methyl group.[102c] There have been several examples of the formation of carbenes from the biradicals formed by Type I fission of cycloalkanones.[103a-c] A particularly novel

[101] N. C. Yang and R. H. K. Chen, *J. Am. Chem. Soc.*, **93**, 530 (1971).
[102a] N. C. Yang, Man-Him-Hui, and S. A. Bellard, *J. Am. Chem. Soc.*, **93**, 4056 (1971).
[102b] R. R. Hautala and N. J. Turro, *J. Am. Chem. Soc.*, **93**, 5595 (1971).
[102c] W. G. Dauben and W. M. Welch, *Tetrahedron Letters*, **1971**, 4531.
[103a] N. J. Turro, D. R. Morton, E. Hedaya, M. E. Kent, P. F. D'Angelo, and P. Schissel, *Tetrahedron Letters*, **1971**, 2535.
[103b] N. J. Turro and D. R. Morton, *J. Am. Chem. Soc.*, **93**, 2569 (1971).
[103c] A. G. Brook, R. Pearce, and J. B. Pierce, *Can. J. Chem.*, **49**, 1622 (1971); A. G. Brook, H. W. Kucera, and R. Pearce, *Can. J. Chem.*, **49**, 1618 (1971).

(34)	11%	89% —in solution
	95%	5% —crystalline phase

reaction is that of the silacyclohexanone (33). There also have been a number of studies on aldehyde formation[103b, 104] and decarbonylation reactions[105, 106a, b] that ensue as a result of Type I cleavage. It was conclusively shown that decarbonylation of (34) occurs by a stepwise mechanism and that, in accord with the results previously discussed, the Type I cleavage is reversible.[106a] A result of real practical importance is that irradiation of the ketone in the crystalline phase increases the selectivity of the reaction.[106b] Further examples have been reported of "β cleavage" of phenacylsulphonium salts,[107] β-keto sulphoxides,[108a] β-keto sulphides[108b] and β-keto nitriles.[109] Several α-aryl ketones on irradiation in alcoholic solvents give ketals rather than products derived by fragmentation of the β bond.[110]

(d) *Addition Reactions*. Further interest has been shown in discovering how aliphatic ketones interact with olefins. From product studies, it was concluded that singlet excited acetone reacts with electron-rich olefins to give an exciplex that may collapse to give either starting compounds or an oxetan.[111a, b] For the reaction with electron-deficient olefins an initial charge-transfer interaction preceding 1,4-biradical formation[111a] was suggested. Exciplex intermediates were also postulated for the reaction of

---

[104] W. C. Agosta and W. L. Schreiber, *J. Am. Chem. Soc.*, **93**, 3947 (1971); J. Kagan, S. P. Singh, K. Warden, and D. A. Harrison, *Tetrahedron Letters*, **1971**, 1849; S. Moon and H. Bohm, *J. Org. Chem.*, **36**, 1434 (1971); J. K. Crandall, J. P. Arrington, and C. F. Mayer, *J. Org. Chem.*, **36**, 1428 (1971); H. Yoshioka, T. H. Porter, A. Higo, and T. J. Mabry, *J. Org. Chem.*, 1971, **36**, 229; J. D. Coyle, *J. Chem. Soc.* (B), **1971**, 1736.

[105] R. S. Cooke and G. D. Lyon, *J. Am. Chem. Soc.*, **93**, 3840 (1971); H. Ona, H. Yamaguchi, and S. Masamune, *J. Am. Chem. Soc.*, **92**, 7495 (1970); W. S. Wilson and R. N. Warrener, *Tetrahedron Letters*, **1970**, 5203; D. W. Jones and G. Kneen, *Chem. Comm.*, **1971**, 1356; P. M. Collins and P. Gupta, *J. Chem. Soc.* (C), **1971**, 1965; P. M. Collins, *J. Chem. Soc.* (C), **1971**, 1960; T. F. Thomas, and H. J. Rodriguez, *J. Am. Chem. Soc.*, **93**, 5918 (1971).

[106a] G. Quinkert, J. Palmowski, H. P. Lorenz, W. W. Wiersdorff, and M. Finke, *Angew. Chem. Int. Ed.*, **10**, 198 (1971).

[106b] G. Quinkert, T. Tabata, E. A. J. Hickmann, and W. Dobrat, *Angew. Chem. Int. Ed.*, **10**, 199 (1971).

[107] T. Laird and H. Williams, *J. Chem. Soc.* (C), **1971**, 1863, 3467.

[108a] S. Majeti, *Tetrahedron Letters*, **1971**, 2523; C. Ganter and J. F. Moser, *Helv. Chim. Acta*, **54**, 2228 (1971).

[108b] A. Padwa and A. Battisti, *J. Am. Chem. Soc.*, **93**, 1304 (1971).

[109] K. J. S. Arora, M. K. M. Dirania, and J. Hill, *J. Chem. Soc.* (C), **1971**, 2865.

[110] M. K. M. Dirania and J. Hill, *J. Chem. Soc.* (C), **1971**, 1213.

[111a] N. J. Turro, C. Lee, N. Schore, J. Barltrop, and H. A. J. Carless, *J. Am. Chem. Soc.*, **93**, 3079 (1971).

[111b] H. S. Samant and A. J. Yarwood, *Can. J. Chem.*, **49**, 2053 (1971).

aldehydes with olefins.[112a] Use has been made of the different modes of reaction of cis-1,2-diethoxyethylene and 1,2-dicyanoethylene with excited carbonyl groups to study the conformation of some alicyclic ketones.[112b] Quenching studies have confirmed that conjugated dienes react with ketones to give oxetans via the first excited singlet state of the ketone.[102a,113a] Benzophenone also reacts with conjugated dienes to give oxetans, and the triplet state of the ketone was shown to be the reaction species. The rate constants for these additions are quite low ($\sim 10^6$ mol^{-1} sec^{-1}) in comparison with those for energy transfer, and this explains the relative inefficiency of the addition reactions.[113b] Examples of intra-[114a] and inter-molecular[114b] oxetan formation have been reported.

(e) *Miscellaneous Reactions.* With regard to β,γ-unsaturated ketones, particular attention has been focused on the validity of the rationalization that 1,3-sigmatropic shifts occur from the S_1 state of the ketones and 1,2-shifts from the T_1 state. Theoretical justification of the statement has been attempted[115] and some further examples that fit the rule

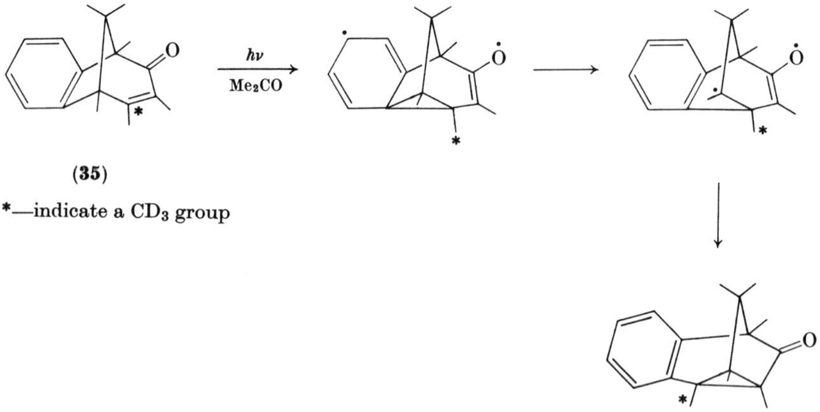

(35)

*—indicate a CD$_3$ group

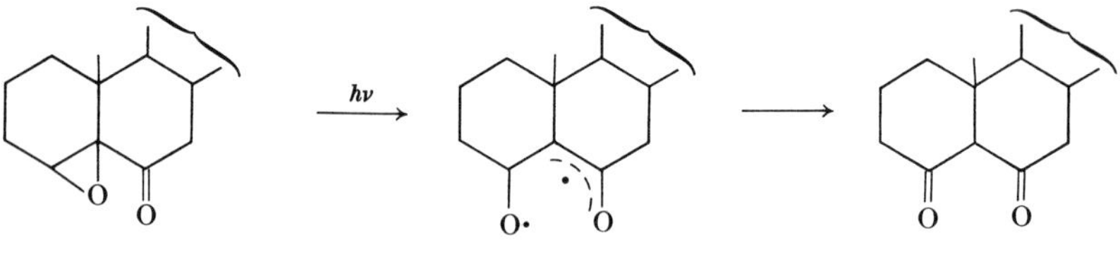

(36)

[112a] N. C. Yang and W. Eisenhardt, *J. Am. Chem. Soc.*, **93**, 1277 (1971).
[112b] N. J. Turro, M. Niemczyk, and D. M. Pond, *Mol. Photochem.*, **2**, 345 (1970).
[113a] J. A. Barltrop and H. A. J. Carless, *Chem. Comm.*, **1970**, 1637; K. Shima, Y. Sakai, and H. Sakurai, *Bull. Chem. Soc. Japan*, **44**, 215 (1971).
[113b] J. A. Barltrop and H. A. J. Carless, *J. Am. Chem. Soc.*, **93**, 4794 (1971).
[114a] M. Y. Mihailovic, Y. Lorenc, N. Popov, and J. Kalvoda, *Helv. Chim. Acta*, **54**, 2281 (1971); G. L. Lange and M. Bosch, *Tetrahedron Letters*, **1971**, 315; Y. Bahurel, F. Pautet, and G. Descotes, *Bull. Soc. Chim France*, **1971**, 2222.
[114b] C. Rivas, M. Vélez, and O. Crescente, *Chem. Comm.*, **1970**, 1474.
[115] D. I. Schuster, G. R. Underwood, and T. P. Knudsen, *J. Am. Chem. Soc.*, **93**, 4304 (1971).

have been reported.[116] Deuterium-labelling studies showed that (**35**) rearranges by a completely different mechanism.[117] Several steroidal α,β-epoxy ketones (e.g. **36**) have been shown to photoisomerize to 1,3-diketones, and the reactive state of the ketone was found to be the excited singlet state.[118] Several reactions have been reported in which triplet ketones have acted as radical initiators.[119] There have also been several reports of the decomposition of halohydrocarbons, e.g. carbon tetrachloride, when they are used as solvents for the irradiation of carbonyl compounds. The mechanism of these reactions is not clear and suggestions have included that the ketones act as true photosensitizers[120a] and that the aliphatic ketones undergo Type I cleavage, the radicals so formed attacking the solvent.[120b]

The triplet state of some ketones has been shown to attack trialkylboranes to give radical species that have been detected by ESR spectroscopy.[121]

$$R_2CO_{T_1} + BR'_3 \rightarrow R_2\dot{C}—O—BR'_2 + R^{1\cdot}$$

Enones

Calculations of the potential energy surfaces of the $n\pi^*$ and $\pi\pi^*$ triplet states of acrolein indicate that formation of *cis*-substituted cyclobutanes by their reaction with alkenes occurs from a planar triplet.[122] *trans*-Substituted cyclobutanes are formed from a twisted triplet state. The *cis–trans* photoisomerization of several linearly conjugated dienones has been studied;[123a,b] when the all-*cis*-dienone is not planar because of steric factors, the olefin photocyclizes to a dihydropyran.[123b]

There have been many reported examples of the photoaddition reactions of enones and the subject has been reviewed.[124] Addition reactions of cyclohex-2-enones substituted at the 3-position are very sensitive to the nature of the substituent, e.g. reactions occur readily with cyano and methoxyl groups, a chloro group retards reactions and the amino group completely stops it.[125] Examples of the addition of cyclohex-2-enone to alkenes[126a] and acetylenes,[126b] and of cyclopent-2-enones to fluoro-olefins[127a] and

---

[116] H.-D. Scharf and W. Küsters, *Chem. Ber.*, **104**, 3016 (1971); R. J. Chambers and B. A. Marples, *Tetrahedron Letters*, **1971**, 3747, 3751; R. S. Givens, W. F. Oettle, R. L. Coffin, and R. G. Carlson, *J. Am. Chem. Soc.*, **93**, 3957 (1971); R. S. Givens and W. F. Oettle, *J. Am. Chem. Soc.*, **93**, 3963 (1971).
[117] H. Hart and G. M. Love, *Tetrahedron Letters*, **1971**, 3563.
[118] J.-P. Pete and M. Viriot-Villaume, *Bull. Soc. Chim. France*, **1971**, 3699, 3709.
[119] J. W. Hartgerink, L. C. J. Van der Laan, J. B. F. N. Engberts, and T. J. de Boer, *Tetrahedron*, **27**, 4323 (1971); T. Yamagishi, T. Yoshimoto, and K. Minami, *Tetrahedron Letters*, **1971**, 2795; J. Sperling and D. Elad, *J. Am. Chem. Soc.*, **93**, 3839 (1971); B. S. Kirkiacharian, *Bull. Soc. Chim. France*, **1971**, 1797; K. Matsuura, S. Maeda, Y. Araki, and Y. Ishido, *Bull. Chem. Soc. Japan*, **44**, 292 (1971); J. Sperling and D. Elad, *J. Am. Chem. Soc.*, **93**, 967 (1971).
[120a] M. A. Golub, *J. Phys. Chem.*, **75**, 1168 (1971); H. Paul and H. Fischer, *Chem. Comm.*, **1971**, 1038.
[120b] A. G. Brook, P. J. Dillon, and R. Pearce, *Can. J. Chem.*, **49**, 133 (1971).
[121] A. G. Davies, D. Griller, B. P. Roberts, and J. C. Scaiano, *Chem. Comm.*, **1971**, 196.
[122] A. Devaquet and L. Salem, *Can. J. Chem.*, **49**, 977 (1971).
[123a] A. F. Kluge and C. P. Lillya, *J. Am. Chem. Soc.*, **93**, 4458 (1971).
[123b] A. F. Kluge and C. P. Lillya, *J. Org. Chem.*, **36**, 1988 (1971).
[124] P. DeMayo, *Accounts Chem. Res.*, **4**, 41 (1971).
[125] T. S. Cantrell, *Tetrahedron*, **27**, 1227 (1971).
[126a] J. J. McCullough and B. R. Ramachandran, *Chem. Comm.*, **1971**, 1180; P. Eaton and K. Nyi, *J. Am. Chem. Soc.*, **93**, 2786 (1971); D. C. Owsley and J. J. Bloomfield, *J. Chem. Soc. (C)*, **1971**, 3445; G. Adam, *Tetrahedron Letters*, **1971**, 1357.
[126b] J. W. Hanifin and E. Cohen, *J. Org. Chem.*, **36**, 910 (1971).
[127a] L. Tökés, A. Christensen, A. Cruz, and P. Crabbé, *J. Org. Chem.*, **36**, 2381 (1971).

allenes[127b] as well as the cyclodimerization of several enones[128] have been reported. The chlorinated enone (**37**) photoisomerizes to (**38**).[129a] This reaction follows a different course to that of the corresponding non-chlorinated compound.[129b] (2 + 2) Cycloaddition

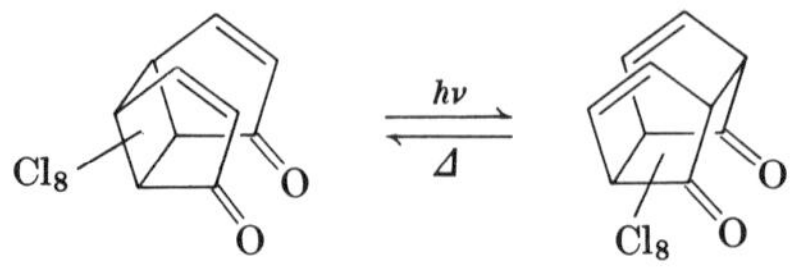

compounds (cyclobutanes) are formed on irradiation of cyclohex-2-enone and cyclopent-2-enone in the presence of an excess of a conjugated diene.[130a] Furan behaved as an abnormal diene in that it gave a (2 + 4) cycloaddition compound (**39**). Acyclic enones also form (2 + 2) cycloaddition compounds with conjugated 1,3-dienes and this reaction has been used in a synthesis of the boll-weevil sex attractant.[130b] The formation of these adducts emphasizes that care is necessary when conjugated dienes are used as triplet quenchers. Photoreactions of cyclo-octenones[131a,b] and cyclononadienone[131c] in which *trans*-enones are intermediates have been reported. The formation of (**41**) is another example of the way in which cuprous salts can alter the course of a photochemical reaction; the complex (**40**) was suggested as an intermediate.[131b] Tropone undergoes an

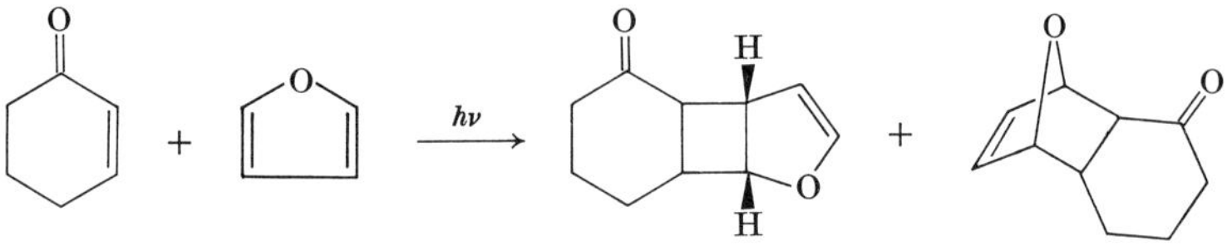

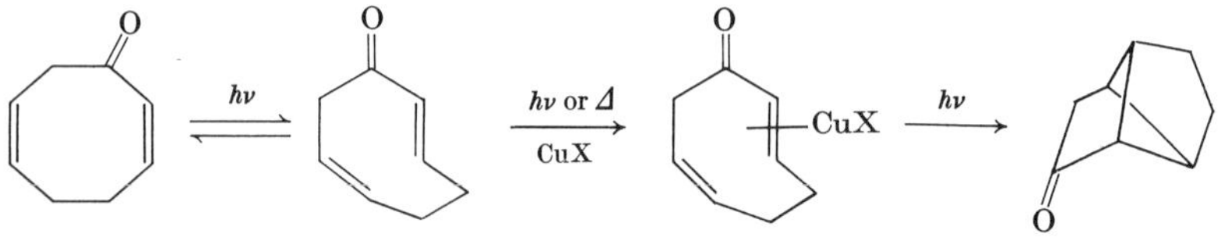

[127b] F. E. Ziegler and J. A. Kloek, *Tetrahedron Letters*, **1971**, 2201.
[128] N. E. Rowland, F. Sondheimer, G. A. Bullock, E. LeGoff, and K. Grohmann, *Tetrahedron Letters*, **1970**, 4769; H. George and H. J. Roth, *Tetrahedron Letters*, **1971**, 4057; G. Mark, H. Matthaeus, F. Mark, J. Leitich, D. Henneberg, G. Schomburg, I. Wilucki, and O. E. Polansky, *Monatsh. Chem.*, **102**, 37 (1971); P. H. Boyle, W. Cocker, D. H. Grayson, and P. V. R. Shannon, *J. Chem. Soc.* (C), **1971**, 1073; N. Sugiyama, T. Sato, H. Kataoka, and C. Kashima, *Bull. Chem. Soc. Japan*, **44**, 555 (1971).
[129a] A. Padwa, J. Masaracchia, and V. Mark, *Tetrahedron Letters*, **1971**, 3161.
[129b] *Org. Reaction Mech.*, **1970**, 528, refs. 142a, b.
[130a] T. S. Cantrell, *Chem. Comm.*, **1970**, 1656.
[130b] J. H. Tumlinson, R. C. Gueldner, D. D. Hardee, A. C. Thompson, P. A. Hedin, and J. P. Minyard, *J. Org. Chem.*, **36**, 2616 (1971).
[131a] G. L. Lange and E. Neidert, *Tetrahedron Letters*, **1971**, 4215.
[131b] R. Noyori, H. Inoue, and M. Katô, *Chem. Comm.*, **1970**, 1695.
[131c] R. Noyori, Y. Ohnishi, and M. Katô, *Tetrahedron Letters*, **1971**, 1515.

(8 + 2) cycloaddition reaction (new bonds formed with the oxygen atom and C-2) with olefins.[132] There have been some most interesting examples of intramolecular addition reactions,[133a,b] e.g. (**42**) → (**43**).[133a]

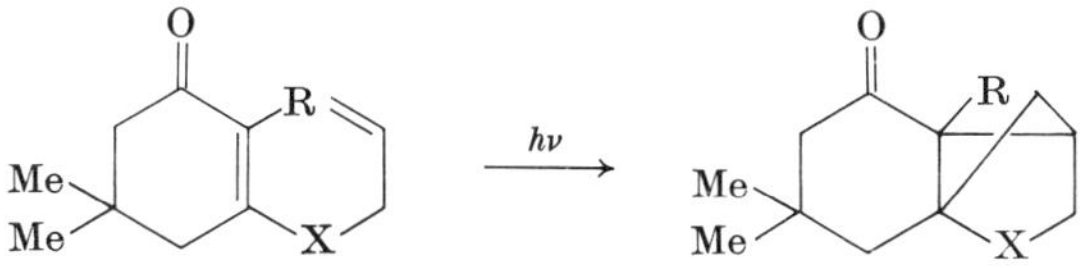

X = O, NMe, etc.

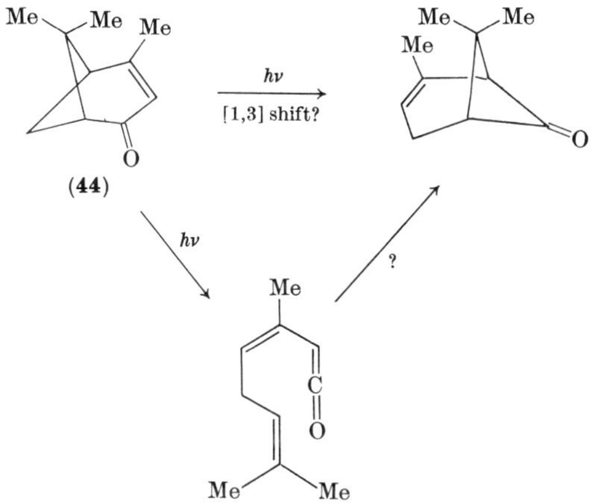

There have been further determinations of the energy levels of the excited states of enones by means of singlet–triplet absorption spectroscopy.[134] The fact that the triplet energies determined in this way are higher than those obtained by the energy-transfer experiments indicates that it is the energy of the twisted triplet that is determined by the latter method. Several further rearrangement reactions of enones have been reported;[135a,b] in the case of (**44**) it was shown that the triplet state was responsible for reaction;[135a] ketene (**45**) may be an intermediate in this reaction. Several reactions of cross-conjugated cyclohexadienones have been reported.[136,137] The relative yields of the two compounds (**47**) and (**48**), obtained by irradiation of (**46**), were found to be

---

[132] T. S. Cantrell, *J. Am. Chem. Soc.*, **93**, 2540 (1971).
[133a] Y. Tamura, Y. Kita, H. Ishibashi, and M. Ikeda, *Chem. Comm.*, **1971**, 1167.
[133b] R. Ramage and A. Sattar, *Tetrahedron Letters*, **1971**, 649; C. B. Hunt, D. F. MacSweeney, and R. Ramage, *Tetrahedron*, **27**, 1491 (1971); M. Yoshioka and M. Hoshino, *Tetrahedron Letters*, **1971**, 2413.
[134] G. Marsh, D. R. Kearns, and K. Schaffner, *J. Am. Chem. Soc.*, **93**, 3129 (1971).
[135a] D. I. Schuster and D. Widman, *Tetrahedron Letters*, **1971**, 3571.
[135b] W. G. Dauben, W. A. Spitzer, and M. S. Kellogg, *J. Am. Chem. Soc.*, **93**, 3674 (1971); R. C. Hahn and G. W. Jones, *J. Am. Chem. Soc.*, **93**, 4232 (1971); H. Hart and T. Takino, *J. Am. Chem. Soc.*, **93**, 720 (1971); D. I. Schuster and B. M. Resnick, *J. Am. Chem. Soc.*, **92**, 7502 (1970).
[136] D. I. Schuster and W. C. Barringer, *J. Am. Chem. Soc.*, **93**, 731 (1971); R. E. Harmon and B. L. Jensen, *J. Heterocycl. Chem.*, **7**, 1077 (1970); H. V. Secor, M. Bourlas, and J. F. Sebardeleben, *Experientia*, **27**, 18 (1971); K. Ogura and T. Matsuura, *Bull. Chem. Soc. Japan*, **43**, 3187 (1970); K. Ogura and T. Matsuura, *Bull. Chem. Soc. Japan*, **43**, 3181 (1970).
[137] D. I. Schuster and W. V. Curran, *J. Org. Chem.*, **35**, 4192 (1970).

solvent-dependent;[137] unfortunately it was impossible to tell whether the rearrangement occurred with retention or inversion at the migrating carbon atom. The previously reported[138a] rationale of the effect of steric factors upon the rearrangements of 4,4-dialkylcyclohexa-2,5-dienones has been questioned.[138b] Ring-opening of cyclohexa-2,4-dienones to give unsaturated ketenes occurs stereospecifically.[139a] The interconversion of this type of dienone with bicyclo[3.1.0]hexanones has been further examined.[139b]

Several examples of hydrogen abstraction reactions of α,β-unsaturated ketones have been reported,[140a-f] e.g. 5-ethoxycyclopent-2-enone undergoes a Norrish Type II reaction,[140a] and 3-t-butylcyclopent-2-enone is photoreduced by toluene.[140b] Enone (49) appears to give (50) by such a reaction;[140d] it was not established whether this reaction was inter- or intra-molecular. There have been a number of reported reactions in which the γ,δ-bond of enones undergoes homolytic cleavage,[141a,b] e.g. (51) → (52).[141b]

Rearrangement of (53) is thought to occur through the ketone (54).[142] Several 5,5-dialkylcyclopent-2-enones also undergo a Type I split to give a ketene, e.g. (55).[140f] Other reported rearrangements include those of 3-hydroxyflavones,[143a] thiopyran-4-ones[143b] and the *cis-trans* isomerization of benzylidene derivatives of cycloalkanones.[143c] By means of ^{14}C-labelling it has been possible to show that ketene gives the oxiren (56) on irradiation.[144]

Diketones and Quinones

The phosphorescence lifetimes of biacetyl[145a] and pentane-2,3-dione[145b] have been measured; in the case of the latter compound, deactivation of the triplet state occurs as a result of intramolecular hydrogen abstraction. The formation of enols of biacetyl and other 1,2-diketones[146] has attracted attention; determination of deuterium isotope effects[147a] and rate-constants [147b] confirmed the occurrence of this reaction; enolization takes place by 1,5-hydrogen abstraction. Biacetyl phosphorescence is quenched by

---

[138a] *Org. Reaction Mech.*, **1970**, 523, ref. 120.
[138b] D. I. Schuster, K. V. Prabhu, S. Adcock, J. Van-der-Veen, and H. Fujiwara, *J. Am. Chem. Soc.*, **93**, 1557 (1971).
[139a] M. R. Morris and A. J. Waring, *J. Chem. Soc.* (C), **1971**, 3266, 3269; A. J. Waring, M. R. Morris, and M. M. Islam, *J. Chem. Soc.* (C), **1971**, 3274.
[139b] B. Miller, *Chem. Comm.*, **1971**, 574.
[140a] A. B. Smith and W. C. Agosta, *Chem. Comm.*, **1971**, 343.
[140b] R. Reinfried, D. Bellus, and K. Schaffner, *Helv. Chim. Acta*, **54**, 1517 (1971).
[140c] W. L. Schreiber and W. C. Agosta, *J. Am. Chem. Soc.*, **93**, 3814 (1971).
[140d] J. Gloor, K. Schaffner, and O. Jeger, *Helv. Chim. Acta*, **54**, 1864 (1971).
[140e] W. L. Schreiber and W. C. Agosta, *J. Am. Chem. Soc.*, **93**, 6292 (1971).
[140f] W. C. Agosta and A. B. Smith, *J. Am. Chem. Soc.*, **93**, 5513 (1971).
[141a] A. E. Greene, J.-C. Muller, and G. Ourisson, *Tetrahedron Letters*, **1971**, 4147; E. Baggiolini, H. G. Berscheid, G. Bozzato, E. Cavalieri, K. Schaffner, and O. Jeger, *Helv. Chim. Acta*, **54**, 429 (1971); A. Balmain and G. Ourisson, *Chem. Comm.*, **1971**, 268.
[141b] G. Adam and B. Voigt, *Tetrahedron Letters*, **1971**, 4601.
[142] T. Sasaki, K. Kanematsu, and K. Hayakawa, *J. Chem. Soc.* (C), **1971**, 2142.
[143a] T. Matsuura, T. Takemoto, and R. Nakashima, *Tetrahedron Letters*, **1971**, 1539.
[143b] N. Ishibe and M. Odani, *Chem. Comm.*, **1971**, 702.
[143c] B. Furth, J. P. Morizur, and J. Kossanyi, *C. R. Acad. Sci. Paris*, (C), **271**, 691 (1970).
[144] R. L. Russell and F. S. Rowland, *J. Am. Chem. Soc.*, **92**, 7508 (1970).
[145a] R. F. Borkman, *Chem. Phys. Letters*, **9**, 77 (1971).
[145b] A. W. Jackson and A. J. Yarwood, *Can. J. Chem.*, **49**, 987 (1971).
[146] M. Bouchy, J. C. Andre, J. Lemaire, and M. Niclause, *C. R. Acad. Sci. Paris*, (C), **272**, 169 (1971).
[147a] N. J. Turro and T. J. Lee, *J. Am. Chem. Soc.*, **92**, 7467 (1970).
[147b] R. G. Zepp and P. J. Wagner, *J. Am. Chem. Soc.*, **92**, 7466 (1970).

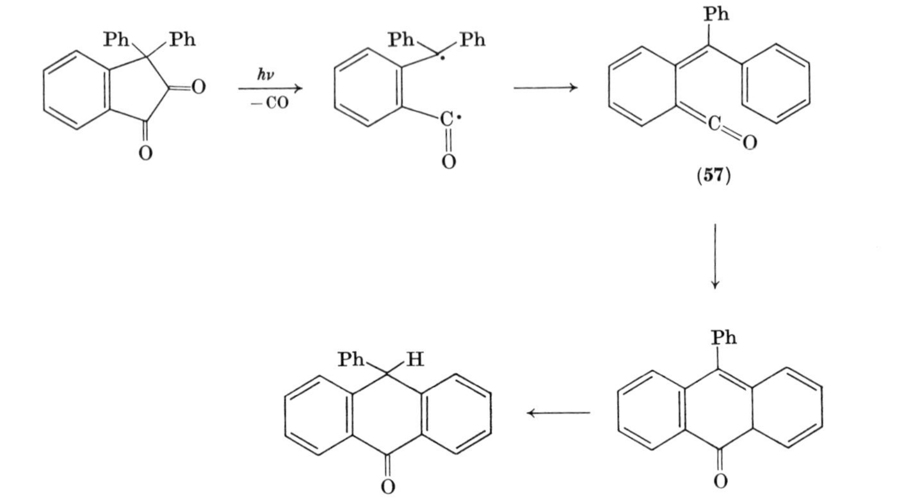

(57)

(59) (58)

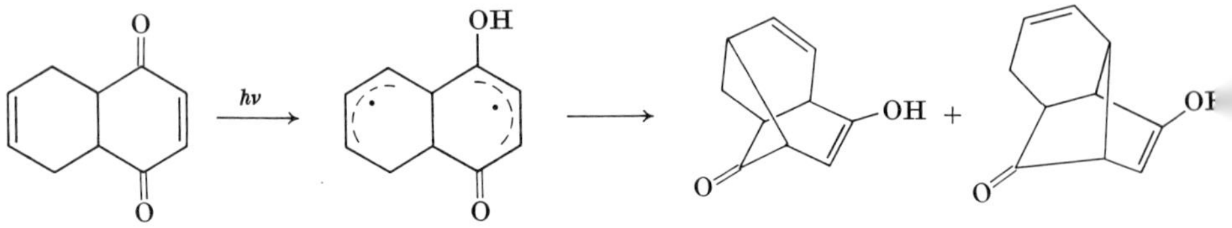

(60)

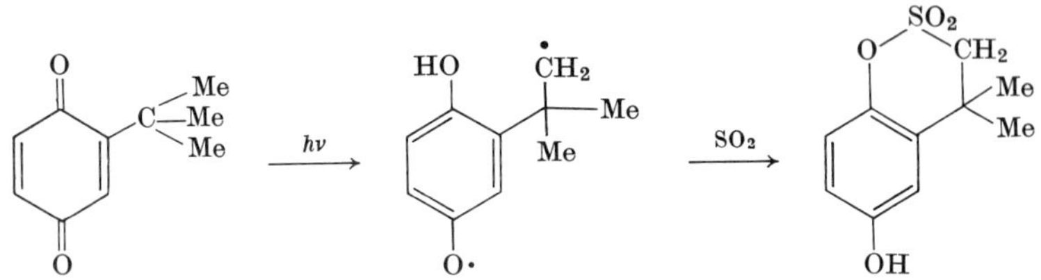

(61)

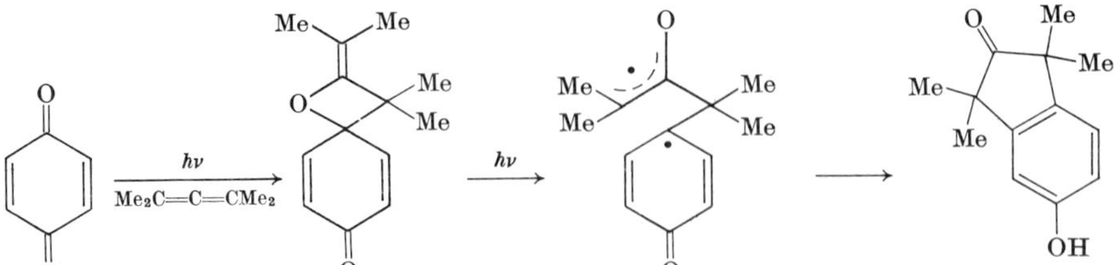

(62)

triethylborane as a result of chemical reaction.[121,148] Cyclobutene-1,2-diones[149a] and cyclobutane-1,2-diones[149b] have been shown, by low-temperature IR spectroscopy, to give bisketenes on irradiation; benzocyclobutane-1,2-dione fails to react in this way. Decarbonylation of naphthafuran-1,2-diones[150a] and indane-1,2-diones[150b] have been reported, and ketenes, e.g. (57), have been shown to be reaction products; in the presence of oxygen the indanediones gave anhydrides.[150c] Ketyl radicals produced by the photoreduction of biacetyl have been observed by ESR spectroscopy.[151]

The dihydroquinone (60) undergoes rearrangement by an intramolecular photoreduction reaction.[152] Biradicals formed in such reactions with other quinones (e.g. 61) have been trapped with sulphur dioxide.[153a] Biacetyl has also been shown to give adducts with olefins by electrophilic attack of the carbonyl group upon the olefin.[153b] The biradicals so formed undergo disproportionation to give the observed products. Semiquinone radicals have been detected by flash photolysis in the reactions of quinones,[154a] and biradicals have been postulated as intermediates in the reactions of a number of 2-alkylbenzoquinones[154b] and their mono-imines.[154c] The photoaddition of 1,2-quinones to alkenes,[155a] furans[155b] and allenes[155c,156] has been studied. Addition of tetramethylallene to p-benzoquinone gives an oxetan which subsequently undergoes rearrangement to (62).[156]

Aldehydes

Simple aliphatic aldehydes having a C_4 or longer chain undergo the Norrish Type II reaction from their S_1 and T_1 states.[157] The triplet lifetime of the aldehydes is governed by the ease of γ-hydrogen abstraction. A number of other aliphatic aldehydes decarbonylate from their excited singlet states and reaction is thought to involve initial α-cleavage:[158]

$$RCHO \xrightarrow{h\nu} R\cdot + \cdot CHO \rightarrow RH + CO$$

o-Tolualdehyde enolizes on excitation and a single adduct is formed between the enol and maleic anhydride.[159] An example of what appears to be intramolecular addition of

---

[148] E. Abuin, J. Grotewold, E. A. Lissi, and M. Umana, *J. Chem. Soc.* (A), **1971**, 516.
[149a] N. Obata and T. Takizawa, *Chem. Comm.*, **1971**, 587.
[149b] O. L. Chapman, C. L. McIntosh, and L. L. Barber, *Chem. Comm.*, **1971**, 1162.
[150a] W. M. Horspool and G. D. Khandelwal, *J. Chem. Soc.* (C), **1971**, 3328.
[150b] J. Rigaudy and N. Paillous, *Bull. Soc. Chim. France*, **1971**, 576.
[150c] J. Rigaudy and N. Paillous, *Bull. Soc. Chim France*, **1971**, 585.
[151] P. B. Ayscough and M. C. Brice, *J. Chem. Soc.* (B), **1971**, 491.
[152] J. R. Scheffer, J. Trotter, R. A. Wostradowski, C. S. Gibbons, and K. S. Bhandari, *J. Am. Chem. Soc.*, **93**, 3813 (1971).
[153a] S. Farid, *Chem. Comm.*, **1971**, 73.
[153b] H. S. Ryang, K. Shima, and H. Sakurai, *J. Am. Chem. Soc.*, **93**, 5270 (1971).
[154a] G. Leary, *J. Chem. Soc.* (A), **1971**, 2248.
[154b] J. M. Bruce, D. Creed, and K. Dawes, *J. Chem. Soc.* (C), **1971**, 2244.
[154c] I. Baxter and I. A. Mensah, *J. Chem. Soc.* (C), **1970**, 2604.
[155a] Y. L. Chow, T. C. Joseph, H. H. Quon, and J. N. S. Tam, *Can. J. Chem.*, **48**, 3045 (1970).
[155b] D. T. Anderson and W. M. Horspool, *Chem. Comm.*, **1971**, 615; W. M. Horspool, *J. Chem. Soc.* (C), **1971**, 400.
[155c] H. J. T. Bos, C. Slagt, and J. S. M. Boleij, *Rec. Trav. Chim.*, **89**, 1170 (1970); J. S. M. Boleij and H. J. T. Bos, *Tetrahedron Letters*, **1971**, 3201.
[156] N. Ishibe and I. Tanigushi, *Tetrahedron*, **27**, 4883 (1971).
[157] J. D. Coyle, *J. Chem. Soc.* (B), **1971**, 2254.
[158] H. Kuentzel, H. Wolf, and I. Schaffner, *Helv. Chim. Acta*, **54**, 868 (1971).
[159] S. M. Mellows and P. G. Sammes, *Chem. Comm.*, **1971**, 21.

an aldehyde group to an alkene was found with a substituted hex-5-enal[160]. The product, a methylcyclopentanone, could however be formed in a radical reaction.

Thioketones

Examples have been reported of the photoaddition of thioketones to acetylenes,[161a] allenes[161b] and imines.[161c] In all cases products are formed via intermediate biradicals. Addition to an imine is complicated by the fact that the biradical (e.g. **63**) can give rise

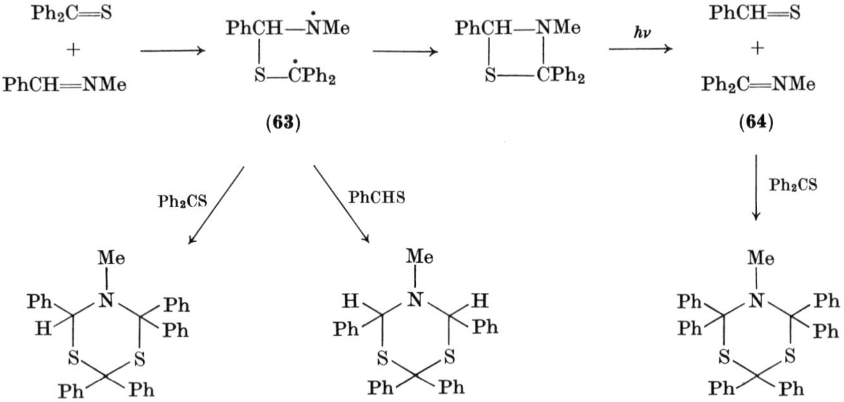

$$Ph_2C=S + PhCH=NMe \longrightarrow \underset{\underset{S-CPh_2}{|}}{PhCH-\overset{\cdot}{N}Me} \longrightarrow \underset{\underset{S-CPh_2}{|}}{PhCH-NMe} \xrightarrow{h\nu} PhCH=S + Ph_2C=NMe$$

$$(63) \qquad\qquad (64)$$

[Structures shown: products from Ph₂CS and PhCHS addition giving thiazine rings with Me, Ph, H substituents, and from Ph₂CS addition to (64).]

to the imine (**64**) as well as to addition products. Photoenolization of *o*-benzylthiobenzophenones has been shown to occur on irradiation.[162]

Carboxylic Acids and Related Compounds

Excitation of *o*-benzylphenyl glyoxalate causes an intramolecular hydrogen abstraction whose efficiency decreases as the solvent polarity increases;[163] this is attributed to the solvents affecting the energies of the $n\pi^*$ and $\pi\pi^*$ triplet states in different ways and consequently the efficiency of intersystem crossing is solvent-dependent. Photolysis of *O*-esters of thiobenzoic acid derived from allylic alcohols results in elimination of thiobenzoic acid and the formation of 1,3-dienes.[164] Deuterium-labelling studies showed that *cis*-elimination occurs. The photo-Fries rearrangement takes place on irradiation of alkyl aryl carbonates.[165a] In contrast, photolysis of *S*-phenyl thioacetate gives products derived by dimerization, etc., of the radicals produced by homolysis of the S—CO bond.[165b] Photolysis of carbamates derived from phenols results in "decarbamation", yielding aromatic hydrocarbons.[166a] The fragmentation of thiocarbamates has also been

[160] R. Aoyagi, T. Tsuyuki, and T. Takahashi, *Bull. Chem. Soc. Japan*, **43**, 3967 (1970).
[161a] A. Ohno, T. Koizumi, and Y. Ohnishi, *Bull. Chem. Soc., Japan*, **44**, 2511 (1971).
[161b] H. J. T. Bos, H. Schinkel, and T. C. M. Wijsman, *Tetrahedron Letters*, **1971**, 3905; H. Gotthardt, *Tetrahedron Letters*, **1971**, 2345.
[161c] A. Ohno, N. Kito, and T. Koizumi, *Tetrahedron Letters*, **1971**, 2421.
[162] N. Kito and A. Ohno, *Chem. Comm.*, **1971**, 1338.
[163] S. P. Pappas, R. D. Zehr, J. E. Alexander, and G. L. Long, *Chem. Comm.*, **1971**, 318.
[164] S. Achmatowicz, D. H. R. Barton, P. D. Magnus, G. A. Poulton, and P. J. West, *Chem. Comm.*, **1971**, 1014.
[165a] E. A. Caress and I. E. Rosenberg, *J. Org. Chem.*, **36**, 769 (1971).
[165b] E. L. Loveridge, B. R. Beck, and J. S. Bradshaw, *J. Org. Chem.*, **36**, 221 (1971).
[166a] E. F. Travecedo and V. I. Stenberg, *Tetrahedron Letters*, **1970**, 4539.

examined.[166b] Benzyl esters are decarboxylated on irradiation and benzyl radicals are produced;[167] quenching and sensitization studies have indicated that this reaction occurs from the triplet state of the ester. Decarboxylation also occurs on irradiation of γ-butyrolactone[168a] whereas, in contrast, benzobutyrolactones, e.g. (65) are decarbonyl-

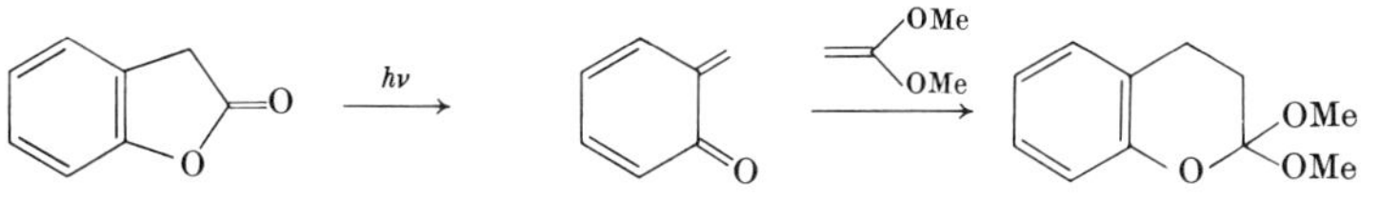

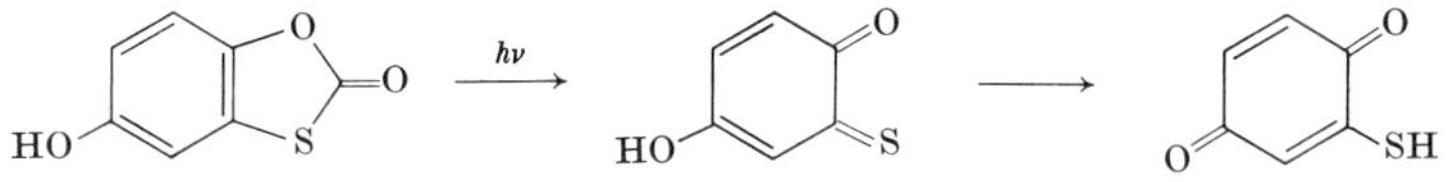

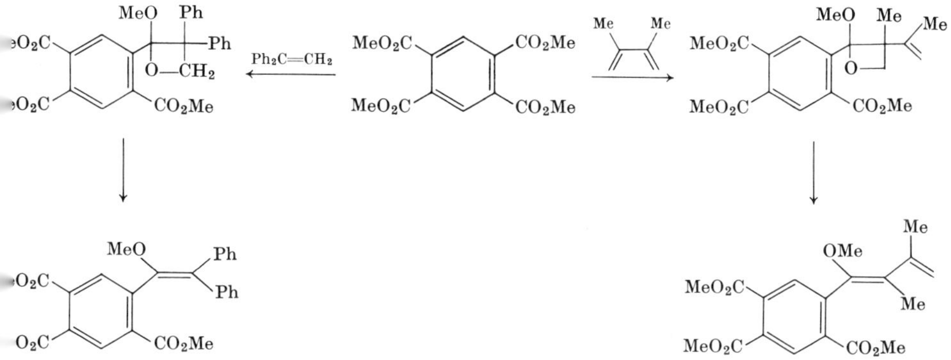

ated[168b] and quinone methides result. Many closely related compounds, e.g. (66), also suffer decarbonylation. There have been a number of reports of Type II reactions of aromatic[169a,b] and aliphatic esters;[169c,d,e] rate constants for reaction of alkyl benzoates from the singlet state were higher than those for reaction from the triplet state.[169a] Conjugated dienes[169d] and some mono-olefins[67,170] quench the fluorescence of aromatic esters; products have been isolated from the reactions that take place as a result of this

---

[166b] S. N. Singh and M. V. George, J. Org. Chem., **36**, 615 (1971).
[167] R. S. Givens and W. F. Oettle, J. Am. Chem. Soc., **93**, 3301 (1971).
[168a] R. Simonaitis and J. N. Pitts, J. Phys. Chem., **75**, 2733 (1971).
[168b] O. L. Chapman and C. L. McIntosh, Chem. Comm., **1971**, 383.
[169a] J. A. Barltrop and J. D. Coyle, J. Chem. Soc. (B), **1971**, 251.
[169b] M. Day and D. M. Wils, Can. J. Chem., **49**, 2916 (1971).
[169c] A. A. Scala and G. E. Hussey, J. Org. Chem., **36**, 598 (1971).
[169d] R. Brainard and H. Morrison, J. Am. Chem. Soc., **93**, 2685 (1971).
[169e] J. E. Gano, Mol. Photochem., **3**, 79 (1971).
[170] Y. Katsuhara, Y. Shigemitsu, and Y. Odaira, Bull. Chem. Soc. Japan, **44**, 1169 (1971).

interaction, e.g. tetramethyl pyromellitate gives (67) and (68) on reaction with 1,1-diphenylethylene and 2,3-dimethylbuta-1,3-diene, respectively. Products from the photolysis of anhydrides[171a] and peroxy-anhydrides[171b,c] have been investigated; photolysis of phthaloyl peroxide gives benzyne.[171c] Several amides have been shown to fragment on photolysis[172a,b] and, in the case of phthalimidoaziridines, cleavage of the aziridine ring results in nitrene formation.[172b] Irradiation of acylimines in hydrogen-donating solvents results in reduction from their triplet states.[173] The triplet states of α,β-epoxy-esters rearrange to β-keto-esters and, when the β-carbon atom is fully substituted, an alkyl group migrates to the α-carbon atom.[174]

Triplet sensitization of the esters (69a),[175a] (69b)[175b] and (70)[175c] results in cis–trans isomerization as well as the formation of intramolecular addition compounds. From

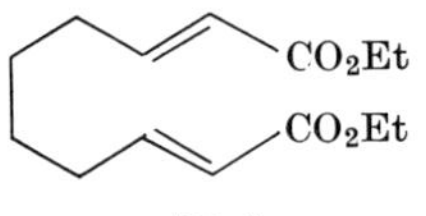

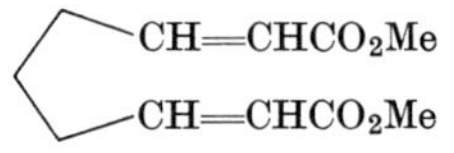

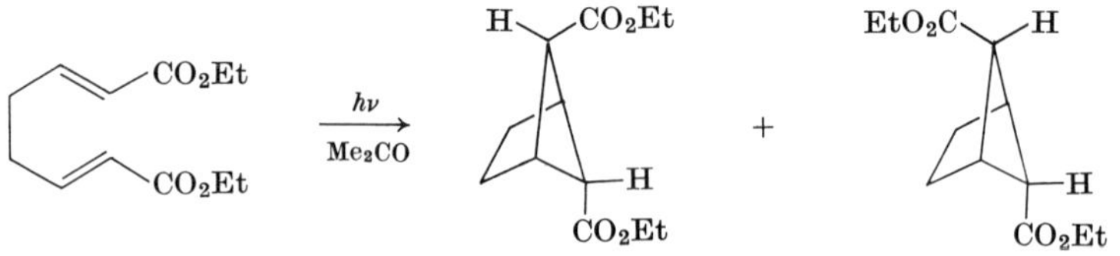

[171a] I. S. Krull, P. F. D'Angelo, D. R. Arnold, E. Hedaya, and P. O. Schissel, *Tetrahedron Letters*, **1971**, 771.
[171b] W. Adam and G. S. Aponte, *J. Am. Chem. Soc.*, **93**, 4300 (1971); W. Adam and R. Ruckstäschel, *J. Am. Chem. Soc.*, **93**, 557 (1971).
[171c] M. Jones and M. R. DeCamp, *J. Org. Chem.*, **36**, 1536 (1971).
[172a] W. A. Henderson and A. Zweig, *Tetrahedron*, **27**, 5307 (1971); D. Touchard and J. Lessard, *Tetrahedron Letters*, **1971**, 4425; J. Reisch and D. H. Niemeyer, *Tetrahedron*, **27**, 4637 (1971); L. J. Darlage, T. H. Kinstle, and C. L. McIntosh, *J. Org. Chem.*, **36**, 1088 (1971).
[172b] T. L. Gilchrist, C. W. Rees, and E. Stanton, *J. Chem. Soc. (C)*, **1971**, 988.
[173] N. Toshima, S. Asao, K. Takada, and H. Hirai, *Tetrahedron Letters*, **1970**, 5123.
[174] M. Tokuda, M. Hataya, J. Imai, M. Itoh, and A. Suzuki, *Tetrahedron Letters*, **1971**, 3133.
[175a] J. R. Scheffer and B. A. Boire, *J. Am. Chem. Soc.*, **93**, 5490 (1971).
[175b] J. R. Scheffer, R. A. Wostradowski, and K. C. Dooley, *Chem. Comm.*, **1971**, 1217.
[175c] J. R. Scheffer and R. A. Wostradowski, *Chem. Comm.*, **1971**, 144.

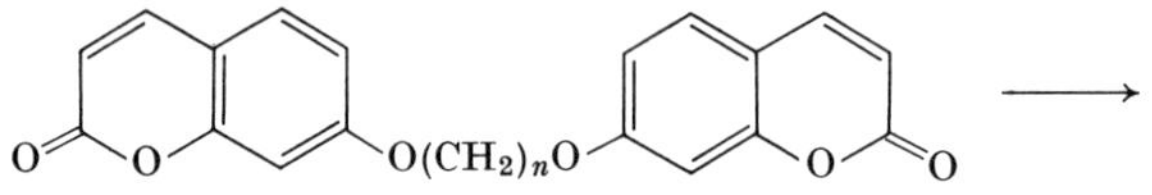

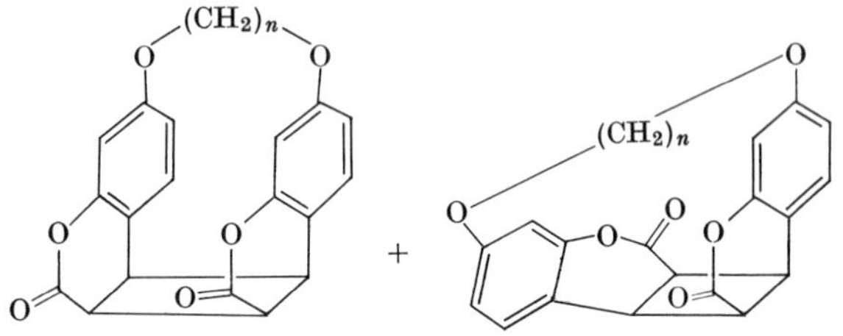

(69a) and (69b), cyclobutanes were formed, whereas (70) gave the "crossed" cycloaddition products (71a) and (71b). Direct excitation of (69a) caused *cis–trans* isomerization and deconjugation. The triplet state of the ester does not appear to be populated to any significant extent since intramolecular addition compounds were not produced. Dimerization of coumarin by reaction of its triplet state with ground-state coumarin is facilitated by the use of solvents containing heavy atoms (e.g. propyl bromide),[176a] i.e. these solvents increase the efficiency of intersystem crossing in the lactone. The biscoumarin derivative (72) gives intramolecular addition products (73a) and (73b) when $n = 3$, 4 or 5.[176b] From quenching and flash-photolysis studies it has been concluded that the photodimerization of carbostyril takes place from its triplet state.[177] Product studies have been made on the following cycloaddition reactions: addition of psoralen (a substituted coumarin) to DNA[178a] and of isocarbostyril to olefins;[178b] and dimerization of the γ-lactone of 4-hydroxycrotonic acid,[178c] cyclohexa-1,4-diene-1,2-dicarboxylic anhydride[178d] and 4,6-diphenyl-2-pyrone.[178e] Photocyclization of N-benzoyl-enamines has attracted attention because of the synthetic usefulness of the reaction.[179]

Olefins

Ionic Addition Reactions

A good review of this subject has been published.[180] Fragmentation of the octalin (74) occurs on irradiation in the presence of acetic acid.[181a,b] The reactions of the olefin (75) are also interesting in that in methanol solution an intermolecular addition takes place whereas in cyclohexane it is intramolecular.[182] Photocatalysed additions of acetic acid to 1-phenylcycloalkenes (ring size 6–8),[183a] to styrenes[183b] and to bicyclo[3.1.0]-hexanes[183c] have been reported; for the last of these reactions protonated cyclopropanes were suggested as intermediates. Trialkylboranes add to substituted cyclohexenes on irradiation in presence of triplet sensitizers.[184] Reactions of ferrocenylethylene and 2-ferrocenylpropene with alcohols occur by both ionic and free-radical mechanisms.[185a] Addition of alcohols to quaternary salts of pyridylethylenes is an ionic process whereas that of ethers involves free radicals.[185b]

[176a] R. Hoffman, P. Wells, and H. Morrison, *J. Org. Chem.*, **36**, 102 (1971).
[176b] L. Leenders and F. C. De Schryver, *Angew. Chem. Int. Ed.*, **10**, 338 (1971).
[177] T. Yamamuro, I. Tanaka, and N. Hata, *Bull. Chem. Soc. Japan*, **44**, 667 (1971).
[178a] D. M. Kramer and M. A. Pathak, *Photochem. Photobiol.*, **12**, 333 (1970).
[178b] G. R. Evanega and D. L. Fabiny, *Tetrahedron Letters*, **1971**, 1749.
[178c] K. Ohga and T. Matsu, *Bull. Chem. Soc. Japan*, **43**, 3505 (1970).
[178d] C. Ahlgrenm, B. Akermark, and R. Karlsson, *Acta Chem. Scand.*, **25**, 753 (1971).
[178e] R. D. Rieke and R. A. Copenhafer, *Tetrahedron Letters*, **1971**, 879.
[179] I. Ninomiya, T. Naito, and T. Kiguchi, *Tetrahedron Letters*, **1970**, 4451; I. Ninomiya, T. Naito, and S. Higuchi, *Chem. Comm.*, **1970**, 1662.
[180] J. A. Marshall, *Science*, **170**, 137 (1970).
[181a] J. A. Marshall and J. P. Arrington, *J. Org. Chem.*, **36**, 214 (1971).
[181b] *Org. Reaction Mech.*, **1970**, 536, ref. 194.
[182] A. Shani and R. Mechoulam, *Tetrahedron*, **27**, 601 (1971).
[183a] S. Fujita, Y. Hayashi, T. Nomi, and H. Nozaki, *Tetrahedron*, **27**, 1607 (1971).
[183b] N. Miyamoto, M. Kawanisi, and H. Nozaki, *Tetrahedron Letters*, **1971**, 2565.
[183c] S. Fujita, Y. Hayashi, and H. Nozaki, *Bull. Chem. Soc. Japan*, **44**, 1970 (1971).
[184] N. Miyamoto, S. Isiyama, K. Utimoto, and H. Nozaki, *Tetrahedron Letters*, **1971**, 4597.
[185a] C. Baker and W. M. Horspool, *Chem. Comm.*, **1971**, 615.
[185b] J. W. Happ, M. T. McCall, and D. G. Whitten, *J. Am. Chem. Soc.*, **93**, 5496 (1971).

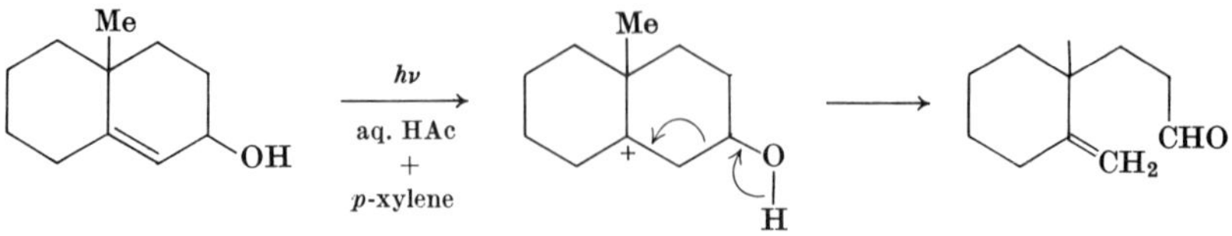

(74)

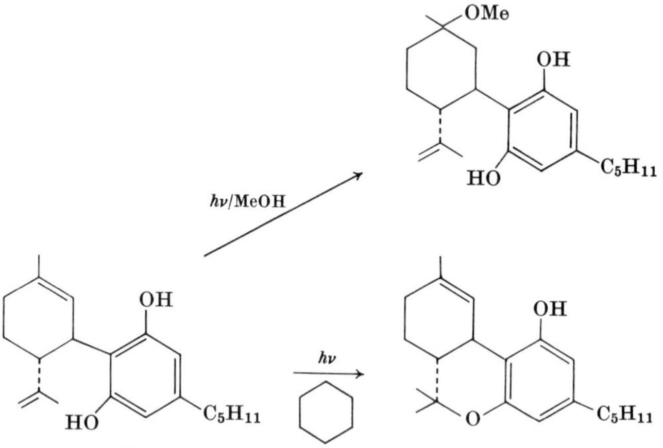

(75)

Cycloaddition Reactions

Calculations have been made of the effect of electron-donating and attracting-substituents on the energies of the highest occupied bonding molecular orbital and the lowest unoccupied antibonding orbital of alkenes, and the relation of these results to cycloaddition reactions has been discussed.[186] Numerous intermolecular (2 + 2) cycloadditions have been reported.[187-190] 1,1-Diphenylethylene forms (2 + 2) cycloaddition products with olefins but in some cases, e.g. with cyclopentene, competing hydrogen abstraction was observed.[188] From a study of the effect of temperature on the addition of stilbene to tetramethylethylene it was concluded that exciplex intermediates were involved.[189a] 4-Nitrostyrene photodimerizes on irradiation when in the form of frozen droplets.[189b] There has been much interest in the photoaddition of acenaphthylene to olefins and, in all cases, irradiations in solvents containing heavy atoms (e.g. PrBr) gave a different ratio of the stereoisomeric products than in non-halogenated solvents;[190a-d] the halogenated solvents appear to facilitate intersystem crossing. Hydrolysis, followed by decarboxylation, transformed (76a) into (76b); this product was completely stable on

---

[186] W. C. Herndon, *Tetrahedron Letters*, **1971**, 125.
[187] H. D. Schart and H. Seidler, *Chem. Ber.*, **104**, 2995 (1971); C. D. Tulloch and W. Kemp, *J. Chem. Soc. (C)*, **1971**, 2824; G. W. Griffin and U. Heep, *J. Org. Chem.*, **35**, 4222 (1970); T. Miyamoto, T. Mori, and Y. Odaira, *Chem. Comm.*, **1970**, 1598; J. C. Hinshaw, *Chem. Comm.*, **1971**, 630; T. Kubota, K. Shima, and H. Sakurai, *Chem. Comm.*, **1971**, 360.
[188] T. S. Cantrell, *Chem. Comm.*, **1971**, 1633.
[189a] J. Saltiel, J. T. D'Agostino, O. L. Chapman, and R. D. Lura, *J. Am. Chem. Soc.*, **93**, 2804 (1971).
[189b] J. H. Schauble, E. H. Freed, and M. D. Swerdloff, *J. Org. Chem.*, **36**, 1302 (1971).
[190a] J. Meinwald, G. E. Samuelson, and M. Ikedo, *J. Am. Chem. Soc.*, **92**, 7604 (1970).
[190b] W. Hartmann and H. G. Heine, *Angew. Chem. Int. Ed.*, **10**, 272 (1971).
[190c] B. F. Plummer and D. M. Chihal, *J. Am. Chem. Soc.*, **93**, 2071 (1971).
[190d] J. E. Shields, D. Gavrilovic, and J. Kopecky, *Tetrahedron Letters*, **1971**, 271.

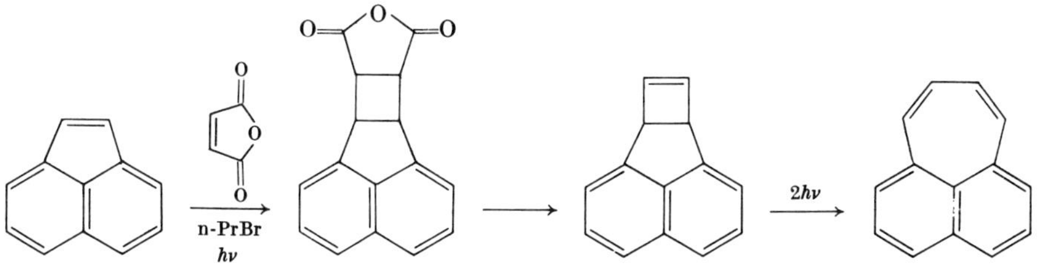

irradiation in solution whereas in a glass at low temperature it underwent a two-photon process to give (76c).[190a] Calculations of the potential-energy surfaces for this transformation to occur from singlet and triplet states have been made and these have shown that reaction from the lowest excited state is forbidden, i.e. reaction has to occur from an upper excited state.[191]

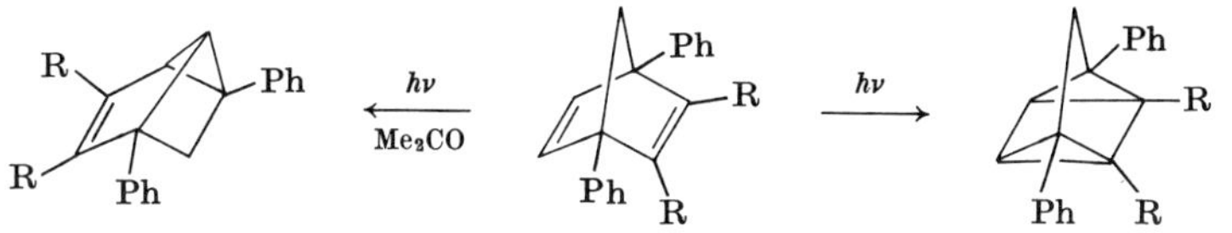

Several intramolecular (2 + 2) cycloaddition reactions have been reported.[192] 1,1,3-Trimethylbuta-1,3-diene isomerizes cleanly to 1,3,3-trimethylcyclobutene on irradiation in the vapour phase;[193] since cis–trans isomerization did not occur, it was possible to obtain the rate constant for the cyclization. "Crossed" (2 + 2) cycloaddition products have been obtained in a number of reactions,[194a,b] e.g. from 1,8-divinylnaphthalene.[194a] The formation of quadricyclene from norbornadiene has been shown to be a non-concerted reaction,[195] and the reactions of a number of norbornadienes have been discussed.[196a,b] The reaction of (77) takes a different course, giving (77a), when it is brought about by triplet sensitization to when direct excitation is used.[196a]

The photocatalysed interconversion of cyclohexa-1,3-dienes and hexatrienes has been studied extensively,[197a–d] including in many cases the role of thermal reactions therein. The ring opening of the cyclobutane (78) has been studied by flash photolysis and the

[191] J. Michl, *J. Am. Chem. Soc.*, **93**, 523 (1971).
[192] H. Hofmann and P. Hofmann, *Tetrahedron Letters*, **1971**, 4055; W. G. Dauben, C. H. Schallhorn, and D. L. Whalen, *J. Am. Chem. Soc.*, **93**, 1446 (1971).
[193] R. Srinivasan and S. Boue, *J. Am. Chem. Soc.*, **93**, 550 (1971).
[194a] J. Meinwald and J. W. Young, *J. Am. Chem. Soc.*, **93**, 725 (1971).
[194b] M. Miyashita and A. Yoshikoshi, *Chem. Comm.*, **1971**, 1091.
[195] G. Kaupp and H. Prinzbach, *Chem. Ber.*, **104**, 182 (1971); G. Kaupp, *Angew. Chem. Int. Ed.*, **10**, 340 (1971).
[196a] H. Prinzbach and M. Thyes, *Chem. Ber.*, **104**, 2489 (1971).
[196b] H. Prinzbach and J. Rivier, *Helv. Chim. Acta*, **53**, 2201 (1970); H. Prinzbach, W. Auge, and M. Basbudak, *Helv. Chim. Acta*, **54**, 759 (1971); M. Klaus and H. Prinzbach, *Angew. Chem. Int. Ed.*, **10**, 273 (1971).
[197a] K. H. Grellmann, J. Palmowski, and G. Quinkert, *Angew. Chem. Int. Ed.*, **10**, 196 (1971).
[197b] D. H. R. Barton, D. L. J. Clive, P. D. Magnus, and G. Smith, *J. Chem. Soc.* (C), **1971**, 2193; W. J. Gensler, Q. A. Ahmed, Z. Muljiani, and C. D. Gatsonis, *J. Am. Chem. Soc.*, **93**, 2515 (1971); E. E. Van Tamelen and R. H. Greeley, *Chem. Comm.*, **1971**, 601; A. Padwa, L. Brodsky, and S. C. Clough, *Chem. Comm.*, **1971**, 417; W. G. Dauben and M. S. Kellogg, *J. Am. Chem. Soc.*, **93**, 3805 (1971).

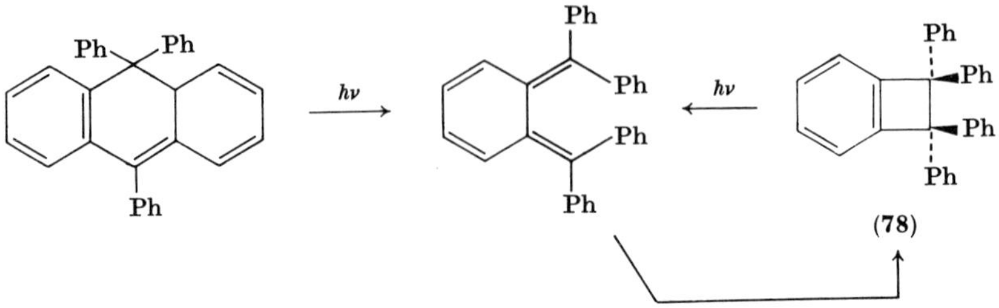

energy of activation for the thermal back-reaction determined ($t_{1/2} = 3.6 \times 10^{-5}$ sec^{-1}; $E_a = 15.3$ kcal mol^{-1}). The cycloreversion reaction of 1,2-dihydronaphthalenes, e.g. (**79**),[197c] has attracted attention.[197d] The cyclo-octatriene (**81**) isomerizes to (**82**) on direct irradiation, whereas on triplet sensitization it undergoes cycloreversion to (**80**);[198] by deuterium-labelling it was shown that (**81**) and (**82**) are not interconnected via (**80**).

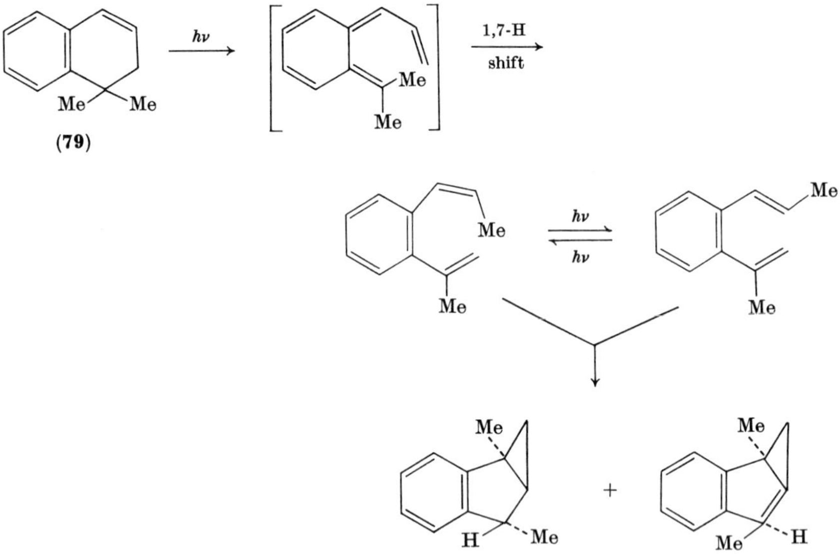

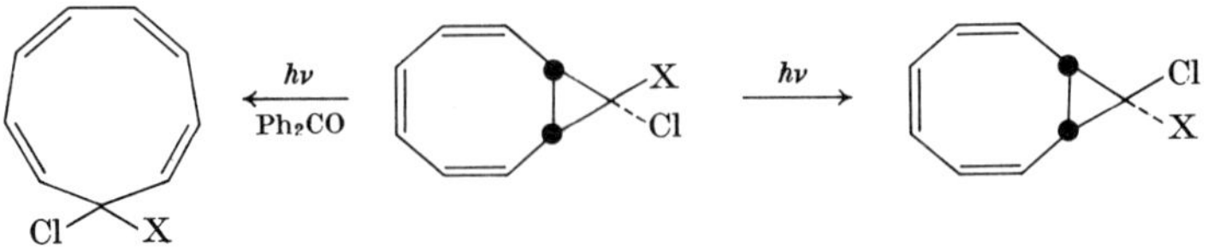

X = H or D

[197c] H. Heimgartner, H. Ulrich, H. J. Hansen, and H. Schmid, *Helv. Chim. Acta*, **54**, 2313 (1971).
[197d] K. Salisbury, *Tetrahedron Letters*, **1971**, 737; H. Kleinhuis, R. L. C. Wijting, and E. Havinga, *Tetrahedron Letters*, **1971**, 255; L. Ulrich, H. J. Hansen, and H. Schmid, *Helv. Chim. Acta*, **53**, 1323 (1970).
[198] A. G. Anastassiou and E. Yakali, *J. Am. Chem. Soc.*, **93**, 3803 (1971).

The cycloreversion of dihydropyrenes[199] and some intramolecular addition reactions of some iminoalkenes[200] have also been discussed.

As usual, the cyclization of stilbenes to dihydrophenanthrenes has attracted attention.[201-205] A useful simple rule has been published stating that "in those cases where cyclization may occur in more than one way, the dihydrophenanthrene which is produced is the one with the most number of benzene rings".[202] Optically active helicenes have been synthesized by irradiation of the appropriate stilbenes with circularly polarized light.[203] A number of substituted stilbenes are reported to give 9,10-dihydrophenanthrenes by way of radical intermediates.[204] 2-Biphenylyl isocyanide cyclizes to phenanthrene on irradiation.[205]

Intramolecular Rearrangements

Intramolecular energy-transfer from the aryl ring to the alkene in 1-phenylbut-2-ene occurs at a diffusion-controlled rate;[206a] here the double bond does not affect the excited singlet state of the aryl ring. In contrast this type of interaction is very strong in 6-phenylhex-2-ene.[206b] This interaction is thought to produce a singlet exciplex which can either decompose to give starting compounds or give rise to intramolecular addition products; the addition products have been identified. Quenching of the singlet state of the aryl ring also effectively reduces efficiency of that ring as a triplet sensitizer. There have been a number of reports of the interconversion of the allyl systems with cyclopropyl systems;[207a-c] the cyclopropyl systems are thought to open to give 1,3-biradicals that undergo a 1,2-hydrogen shift to give the allyl compounds,[207c] but it is much harder to rationalize the reaction in the opposite direction. Whilst with allylbenzene[207a] a di-π-methane rearrangement is possible, this cannot occur with simple allyl compounds.[207b] Several sigmatropic rearrangements of alkenes have been reported.[208,209a,b] The rearrangements of (**83**) and (**84**) are of interest because reaction of (**83**) can occur both thermally and photochemically, and in both cases a reasonable degree of retention at the migrating carbon atom is observed.[209a]

---

[199] H. R. Blattmann and W. Schmidt, *Tetrahedron*, **26**, 5885 (1970).
[200] G. Kan, M. T. Thomas, and V. Snieckus, *Chem. Comm.*, **1971**, 1022; T. H. Koch and D. A. Brown, *J. Org. Chem.*, **36**, 1934 (1971); A. Padwa and E. Glazer, *Chem. Comm.*, **1971**, 838.
[201] W. H. Laarhoven and T. J. H. M. Cuppen, *Tetrahedron Letters*, **1971**, 163; R. J. Hayward and C. C. Leznoff, *Tetrahedron*, **27**, 2085, 5115 (1971); N. Ishibe, M. Odani, and M. Sunami, *Chem. Comm.*, **1971**, 1034; M. B. Green, H. Schadenberg, and H. Wynberg, *J. Org. Chem.*, **36**, 2797 (1971); H. Wynberg, *Accounts Chem. Res.*, **4**, 65 (1971); G. Cauzzo, G. Galiazzo, P. Bortolus, and F. Coletta, *Photochem. Photobiol.*, **13**, 445 (1971); G. Rio and J. C. Hardy, *Bull. Soc. Chim. France*, **1970**, 3578.
[202] W. H. Laarhoven, T. J. H. M. Cuppen, and R. J. F. Nivard, *Tetrahedron*, **26**, 4865 (1970).
[203] H. Kagan, A. Moradpour, J. F. Nicoud, G. Balavoine, R. H. Martin, and J. P. Cosyn, *Tetrahedron Letters*, **1971**, 2479; A. Moradpour, J. F. Nicoud, G. Balavoine, H. Kagan, and G. Tsoucaris, *J. Am. Chem. Soc.*, **93**, 2353 (1971).
[204] R. Srinivasan and J. N. C. Hsu, *J. Am. Chem. Soc.*, **93**, 2816 (1971).
[205] J. De-Jong and J. H. Boyer, *Chem. Comm.*, **1971**, 961.
[206a] H. Morrison, J. Pajak, and R. Peiffer, *J. Am. Chem. Soc.*, **93**, 3978 (1971).
[206b] W. Ferree, J. B. Grutzner, and H. Morrison, *J. Am. Chem. Soc.*, **93**, 5502 (1971).
[207a] S. S. Hixson, *Tetrahedron Letters*, **1971**, 4211; E. W. Valyocsik, and P. Sigal *J. Phys. Chem.*, **75**, 2079 (1971); A. S. Kende, Z. Goldschmidt, and R. F. Smith, *J. Am. Chem. Soc.*, **92**, 7606 (1970).
[207b] K. Salisbury, *J. Chem. Soc. (B)*, **1971**, 931; K. H. Schulte-Elte and G. Ohloff, *Helv. Chim. Acta*, **54**, 370 (1971).
[207c] K. Salisbury, *Chem. Comm.*, **1971**, 934; E. W. Valyocsik and P. Sigal, *J. Org. Chem.*, **36**, 66 (1971).
[208] W. G. Dauben, C. D. Poulter, and C. Suter, *J. Am. Chem. Soc.*, **92**, 7408 (1970).
[209a] R. C. Cookson and J. E. Kemp, *Chem. Comm.*, **1971**, 385.
[209b] R. C. Cookson, J. Hudec, and M. Sharma, *Chem. Comm.*, **1971**, 107, 108.

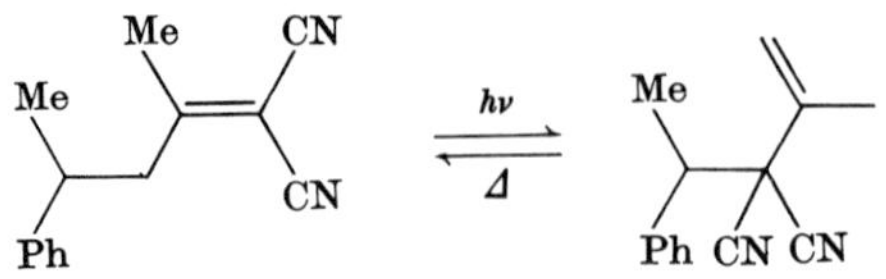

Several examples of 1,7 hydrogen shifts and of intramolecular (2 + 2) cycloaddition have been observed with cycloheptatrienes.[210] The rearrangement of norcaradienes and related compounds has also been studied. Usually, thermal disrotatory ring-opening occurs to give a cycloheptatriene,[211a–c] that then undergoes the photochemical reactions,[211b] e.g. (85).[211c] There have been several further examples of the di-π-methane rearrangement: one has been reported[212] that is in accord with the rationalization that "when more than one biradical intermediate can be postulated, reaction always occurs via the most stable biradical". Rearrangement of 6-methylene-3,3-diphenylcyclohexa-1,4-diene to 4-methylene-6,6-dimethylbicyclo[3.1.0]hex-2-ene has been shown to take place from the excited singlet state of the diene,[213] the triplet state of which is completely unreactive. Similar results were obtained with the corresponding 3,3-dimethyl compound. The bicyclo compounds produced in the singlet reactions are capable of undergoing further phototransformations. In the rearrangement of the singlet state of

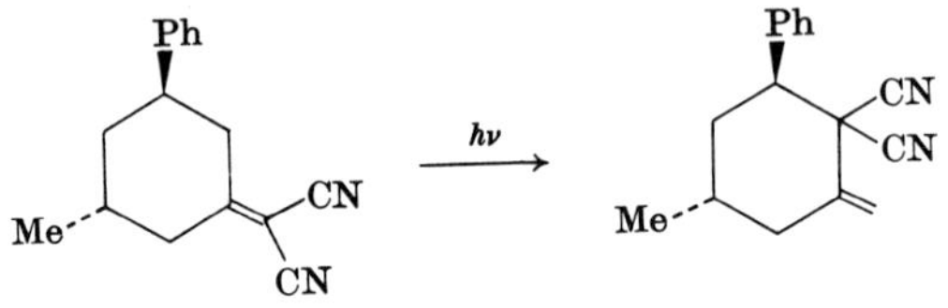

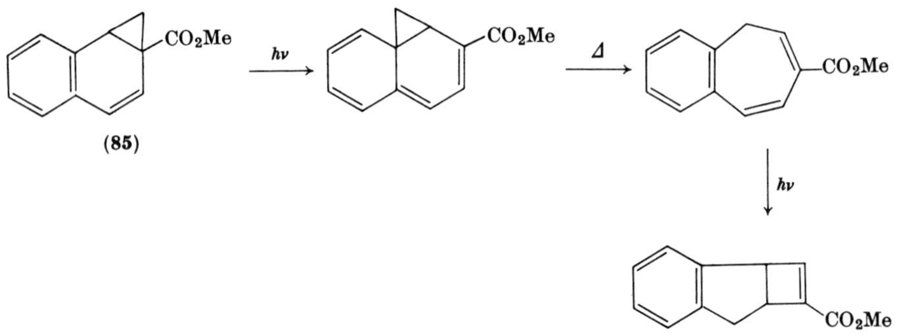

[210] A. R. Brember, A. A. Gorman, and J. B. Sheridan, *Tetrahedron Letters*, **1971**, 653; G. Linstrumelle, *Bull. Soc. Chim. France*, **1971**, 642.
[211a] H. Duerr and H. Kober, *Angew. Chem. Int. Ed.*, **10**, 342 (1971).
[211b] R. H. Levin and M. Jones, *Tetrahedron*, **27**, 2031 (1971); D. M. Madigan and J. S. Swenton, *J. Am. Chem. Soc.*, **92**, 7513 (1970).
[211c] D. M. Madigan and J. S. Swenton, *J. Am. Chem. Soc.*, **93**, 6316 (1971).
[212] H. E. Zimmerman and A. A. Baum, *J. Am. Chem. Soc.*, **93**, 3646 (1971).
[213] H. E. Zimmerman, P. Hackett, D. F. Juers, J. M. McCall, and B. Schröder, *J. Am. Chem. Soc.*, **93**, 3653 (1971).

(87) the cyclopropyl group was shown to move around the ring by a slither process;[214a,b] in contrast the triplet state of (87) only undergoes isomerization to (86). Irradiation of the diene (88) results in a di-π-methane rearrangement followed by ring opening of the bicyclo compound (89).[215] Irradiation of (90) brings about isomerization of the cyclopropyl system to give (91) as well as rearrangement to (92);[216] the rearrangement was shown to occur solely from the singlet state of (90) whereas the isomerization could occur from either the singlet or the triplet state. Rearrangement of (93) is an interesting example of a di-π-methane rearrangement since it occurs from the triplet

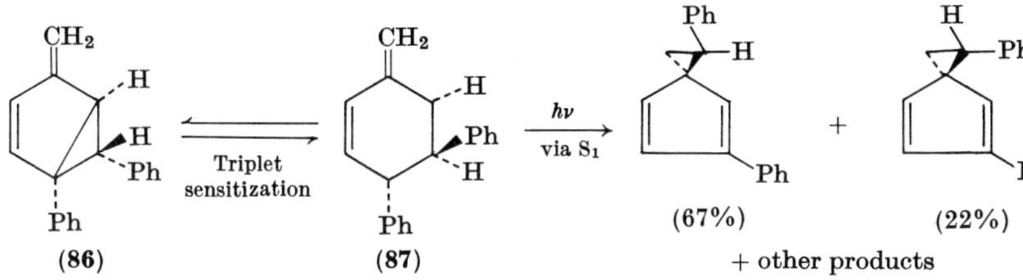

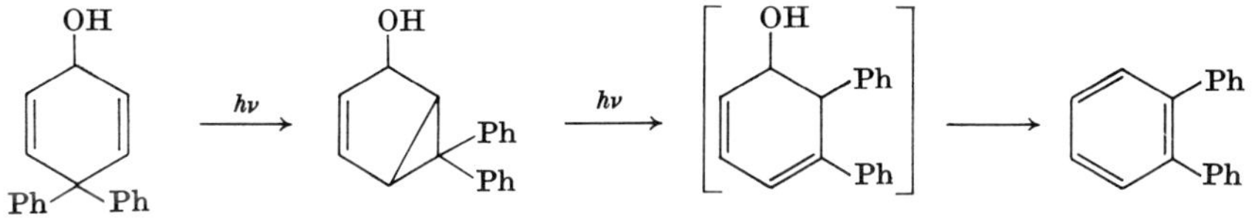

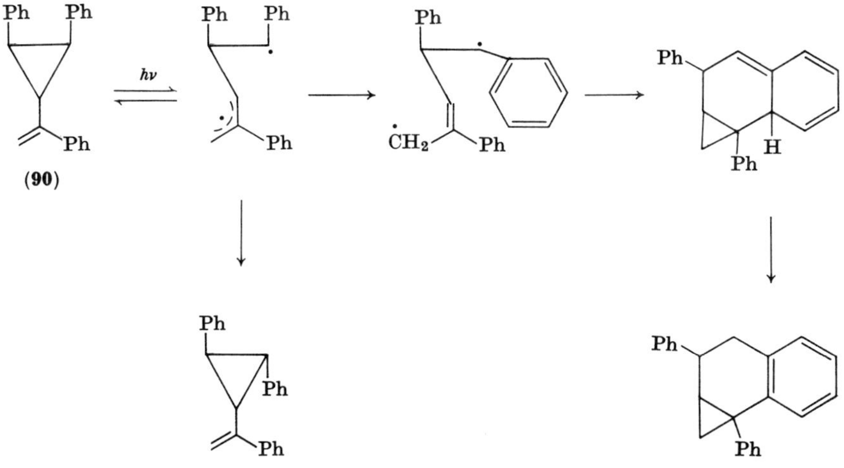

[214a] H. E. Zimmerman, D. F. Juers, J. M. McCall, and B. Schröder, *J. Am. Chem. Soc.*, **93**, 3662 (1971).
[214b] *Org. Reaction Mech.*, **1970**, 544, refs. 229a–c.
[215] W. G. Dauben, W. A. Spitzer, and R. M. Boden, *J. Org. Chem.*, **36**, 2384 (1971).
[216] H. E. Zimmerman and T. W. Flechtner, *J. Am. Chem. Soc.*, **92**, 7178 (1970).

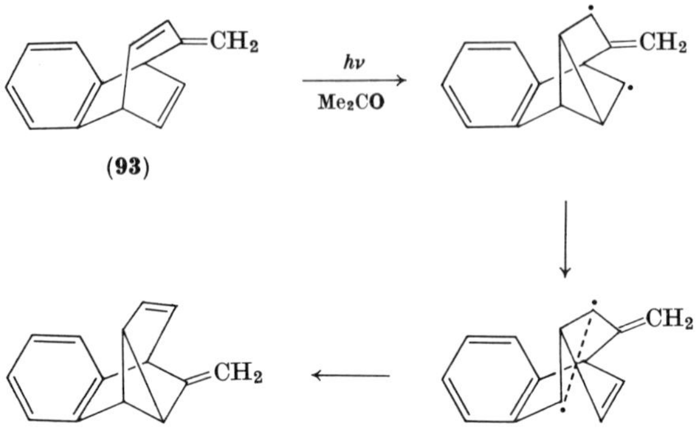

state.[217a,b] Presumably the rigidity of the molecule precludes dissipation of energy by the "free-rotor" mechanism.

Miscellaneous Reactions of Alkenes and Alkynes

There have been a number of studies of the *cis–trans* isomerization of alkenes.[218-221] For polyenes, calculations have shown that isomerization is more favourable about a central than about a terminal double bond.[219] Reference has already been made to the evidence supporting the conclusion that isomerization of stilbene under direct irradiation emanates from the triplet state of the olefin. Kinetic expressions have been developed that successfully describe the processes occurring in the photosensitized *cis–trans* isomerization of alkenes when energy-transfer to one of the isomers may be relatively inefficient.[220] The sensitized isomerization of allenes[221] and the rearrangement of allenes to buta-1,3-dienes have been discussed.[222] Convincing evidence has been presented to demonstrate that carbenes are produced on photolysis of vinylcyclopropanes.[223] The mercury-sensitized fragmentation of cyclo-octene[224a] and cycloocta-1,5-diene[224b] has been discussed.

Photoisomerization of the acetylene (**94**) to, *inter alia*, a cyclopropene has been reported.[225] *cis*-Stilbene has been suggested as an intermediate in the formation of phenanthrene from diphenylacetylene;[226] the triple bond of this compound, in an excited state, is susceptible to nucleophilic attack by, e.g. methanol. Examples of

[217a] Z. Goldschmidt and A. S. Kende, *Tetrahedron Letters*, **1971**, 4625.
[217b] P. D. Rosso, J. Oberdier, and J. S. Swenton, *Tetrahedron Letters*, **1971**, 3947.
[218] H. Guesten and D. Schulte-Frohlinde, *Chem. Ber.*, **104**, 402 (1971); R. S. H. Liu and Y. Butt, *J. Am. Chem. Soc.*, **93**, 1532 (1971); M. Wrighton, G. S. Hammond, and H. B. Gray, *J. Am. Chem. Soc.*, **93**, 3285 (1971); A. Gordon-Walker and G. K. Radda, *Biochem. J.*, **120**, 673 (1970); G. Cauzzo, M. Casagrande, and G. Galiazzo, *Mol. Photochem.*, **3**, 59 (1971); P. Bortolus, G. Favaro, and U. Mazzucato, *Mol. Photochem.*, **3**, 311 (1970).
[219] N. C. Baird and R. M. West, *J. Am. Chem. Soc.*, **93**, 4427 (1971).
[220] J. Saltiel, L. Metts, A. Sykes, and M. Wrighton, *J. Am. Chem. Soc.*, **93**, 5302 (1971).
[221] O. Rodriguez and H. Morrison, *Chem. Comm.*, **1971**, 679.
[222] J. Raffi and C. Troyanowsky, *C. R. Acad. Sci. Paris*, (*C*), **271**, 533 (1970).
[223] J. C. Gilbert and J. R. Butler, *J. Am. Chem. Soc.*, **92**, 7493 (1970).
[224a] S. Takamuku, K. Moritsugu, and H. Sakurai, *Bull. Chem. Soc. Japan*, **44**, 2562 (1971).
[224b] S. Takamuku and H. Sakurai, *Bull. Chem. Soc. Japan*, **44**, 569 (1971).
[225] B. Halton, M. Kulig, M. A. Battiste, J. Perreton, D. M. Gibson, and G. W. Griffin, *J. Am. Chem. Soc.*, **93**, 2327 (1971).
[226] T. D. Roberts, *Chem. Comm.*, **1971**, 362.

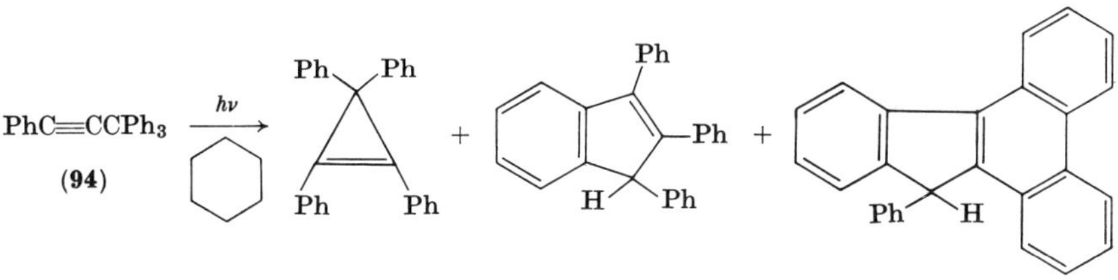

intermolecular[227a,b] and intramolecular[227c] addition have been discussed, e.g. addition of acetylene (95) to ethylene.[227a]

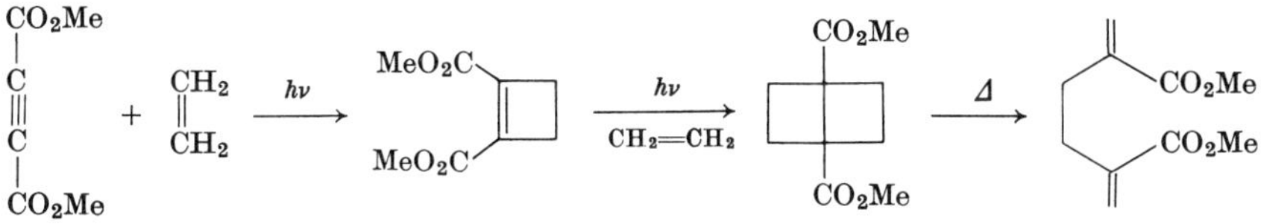

Aromatic Hydrocarbons

The dissipation of electronic energy in fluorinated benzenes[228a] and the photoisomerization of xylenes[228b] and 1,2,4-trifluorobenzenes[228c] have been discussed. A surprising and interesting result is that the "photo-oxidation" of benzene, in which cyclopenta-1,3-dienecarboxaldehyde is produced,[229a] does not in fact require oxygen but water. It was shown that the oxidation product was, not 4H-pyran-2-carboxaldehyde as previously suggested,[229b] also that benzvalene (96), produced by isomerization, is susceptible to nucleophilic attack.[230,231] This observation does not, however, settle how benzvalene is transformed into the aldehyde, and these details are awaited with interest.

A very useful non-photochemical synthesis that allows the preparation of benzvalene on a relatively large scale has been described.[231]

Addition of tetramethylethylene to benzene has been shown to give the three products (97), (98) and (100).[232] The last, newly identified, product is the most abundant. When the reaction was carried out in the presence of MeOD, the 4-deuterio-derivative of (100) was obtained. On the basis of this and other observations it was concluded that reaction occurred through the dipolar intermediate (99). Products derived by 1,2-, 1,3- and 1,4-addition have also been isolated from the reaction of benzene with cis- and trans-but-2-ene,[233] the configuration of the olefin being retained; the 1,4-adducts

---

[227a] D. C. Owsley and J. J. Bloomfield, *J. Am. Chem. Soc.*, **93**, 782 (1971).
[227b] W. Hartmann, *Chem. Ber.*, **104**, 2864 (1971).
[227c] L. Munchausen, I. Ookuni, and T. D. Roberts, *Tetrahedron Letters*, **1971**, 1917.
[228a] T. L. Brewer, *J. Phys. Chem.*, **75**, 1233 (1971); S. L. Lem, G. P. Semeluk, and I. Unger, *Can. J. Chem.*, **49**, 1567 (1971).
[228b] W. A. Noyes and D. A. Harter, *J. Phys. Chem.*, **75**, 2741 (1971).
[228c] G. P. Semeluk and R. D. S. Stevens, *Chem. Comm.*, **1970**, 1720.
[229a] L. Kaplan, L. A. Wendling, and K. E. Wilzbach, *J. Am. Chem. Soc.*, **93**, 3819 (1971); L. Kaplan, L. A. Wendling, and K. E. Wilzbach, *J. Am. Chem. Soc.*, **93**, 3821 (1971).
[229b] M. Luria and G. Stein, *Chem. Comm.*, **1970**, 1650.
[230] J. A. Berson and N. M. Hasty, *J. Am. Chem. Soc.*, **93**, 1549 (1971).
[231] T. J. Katz, E. J. Wang, and N. Acton, *J. Am. Chem. Soc.*, **93**, 3782 (1971).
[232] D. Bryce-Smith, B. E. Foulger, A. Gilbert, and P. J. Twitchett, *Chem. Comm.*, **1971**, 794.
[233] K. E. Wilzbach and L. Kaplan, *J. Am. Chem. Soc.*, **93**, 2073 (1971); R. Srinivasan, *Tetrahedron Letters*, **1971**, 4551.

were shown to be 7,8-dimethylbicyclo[2.2.2]octa-2,5-dienes. All three modes of addition are attributed to concerted reactions of the singlet (B_{2u}) state of benzene. Cyclobutene also gives a 1,3-adduct with benzene as well as with toluene and xylenes.[234] Addition is stereospecific and only two products were formed from xylenes, e.g. (**101**), and toluene. Dichlorovinylene carbonate (4,5-dichloro-1,3-dioxol-2-one) forms 1,4-addition products with benzene[235a] and mesitylene.[235b] Reference has already been made to the intramolecular addition products formed from 1-phenylhex-4-ene.[206b] Compound (**102**) forms the intramolecular 1,2-addition product (**103**).[236] Benzene has been shown to give addition products with ethers[237] and amines,[238] the former requiring the presence of acid. Addition of amines was facilitated by protic solvents and from this and other observations it was concluded that the initial reaction involved electron-transfer from the amine to the excited benzene. Excited benzene also reacts with trifluoroacetic acid, trifluoroacetylbenzene being produced;[239] this photo-Friedel–Crafts reaction was postulated as occurring through an excited charge-transfer complex.

Coniferyl alcohol has been shown to give a p-quinone methide on flash excitation.[240] A number of o-(aminophenoxy)-s-triazines rearrange to o-(hydroxyanilino)-s-triazines on irradiation, i.e. a photo-Smiles rearrangement takes place.[241] When compounds lacking the o-amino group are irradiated the photo-Fries rearrangement takes place. Optically active [2,2]para-[242a] and [2,3]meta-cyclophanes[242b] were racemized on irradiation, homolysis of the benzyl–benzyl bonds being suggested as the mechanism for this reaction.

Reported cycloaddition reactions of naphthalene and anthracene have included dimerization of 2-cyanonaphthalene[243a], addition of diphenylacetylene to numerous substituted naphthalenes[243b] and addition of 9,10-dimethylanthracene to tetracene.[243c] The rate constant for the dimerization of 9-anthroic acid has been determined.[244]

Several polycyclic aromatic hydrocarbons, e.g. acenaphthylene and phenanthrene, are reduced by tri-n-butyltin hydride to dihydro-compounds;[245] reaction occurs from the triplet state of the hydrocarbon and rate constants were determined. Further studies have been made of the quenching of aromatic hydrocarbon singlet states by amines in non-polar solvents.[246] The polarity of the solvent used for the photo-addition of pyrrole to naphthalene has been shown to affect the ratio of the products obtained;[247] an electron-transfer mechanism was proposed for this reaction.

---

[234] R. Srinivasan, *J. Am. Chem. Soc.*, **93**, 3555 (1971).
[235a] H. D. Scharf and R. Klar, *Tetrahedron Letters*, **1971**, 517.
[235b] G. Hesse and P. Letchken, *Angew. Chem. Int. Ed.*, **10**, 133 (1971).
[236] A. S. Kushner, *Tetrahedron Letters*, **1971**, 3275.
[237] D. Bryce-Smith and G. B. Cox, *Chem. Comm.*, **1971**, 915.
[238] D. Bryce-Smith, M. T. Clarke, A. Gilbert, G. Klunklin, and C. Manning, *Chem. Comm.*, **1971**, 916.
[239] D. Bryce-Smith, G. B. Cox, and A. Gilbert, *Chem. Comm.*, **1971**, 914.
[240] G. Leary, *Chem. Comm.*, **1971**, 688.
[241] H. Shizuka, T. Kanai, T. Morita, Y. Ohoto, and K. Matsui, *Tetrahedron*, **27**, 4021 (1971).
[242a] M. H. Delton and D. J. Cram, *J. Am. Chem. Soc.*, **92**, 7623 (1970).
[242b] M. H. Delton, R. E. Gilman, and D. J. Cram, *J. Am. Chem. Soc.*, **93**, 2329 (1971).
[243a] T. W. Mattingly, J. E. Lancaster, and A. Zweig, *Chem. Comm.*, **1971**, 595.
[243b] W. H. F. Sasse, P. J. Collin, D. B. Roberts, and G. Sugowdz, *Austral. J. Chem.*, **24**, 2151 (1971).
[243c] H. Bouas-Laurent and A. Castellan, *Chem. Comm.*, **1970**, 1648.
[244] D. O. Cowan and W. W. Schmiegel, *Angew. Chem. Int. Ed.*, **10**, 517 (1971).
[245] D. R. G. Brimage and R. S. Davidson, *Chem. Comm.*, **1971**, 281.
[246] V. R. Rao and V. Ramakrishnan, *Chem. Comm.*, **1971**, 971.
[247] J. J. McCullough and W. S. Wu, *Tetrahedron Letters*, **1971**, 3951.

Heterocyclic Compounds

The quantum yields of intersystem crossing of several heterocyclic compounds, e.g. quinoline and quinoxaline, have been determined,[248a] as have the pK values for deprotonation of singlet-state o-phenanthroline derivatives.[248b]

There has been further interest in the photoreduction of heterocyclic compounds. The products from the reduction of pyridine,[249a] quinoline[249b] and isoquinoline[249b] by cyclohexane have been elucidated and the intermediate heterocyclic radicals have been identified by ESR spectroscopy. 3-Cyanopyridine is reduced by methanol to the dihydro-compounds (104) and (105).[250a] 3-Cyanoquinoline and 4-cyanoisoquinoline react in a similar way. In contrast, 2-cyanoquinoline is photoreduced by isopropyl alcohol to (106).[250b] Photoreduction of quinine leads to loss of a 9-hydroxyl group.[251a] Dihydropyridinium salts, e.g. (107), are also photoreduced by methanol.[251b] Irradiation of 2- and 4-(hydroxymethyl)pyridines,[252a] (hydroxymethyl)purines[252b] and the dihydroisoquinoline (108)[252c] leads to fragmentation of the side chain. The last example appears to be an anologue of the Type II reaction of carbonyl compounds. There have been two reports of the reduction of halopyrimidines:[253a,b] in the case of (109), hydroxymethylation and loss of halogen occurred,[253a] whereas with the chloropyrimidine (110) a number of different reactions occurred,[253b] none of which corresponded to that of (109). Alkylation of purines[254a] and amination of thiophens[254b] take place when they are irradiated in the presence of amines, and imidazoles form unstable oxetans on irradiation in the presence of carbonyl compounds.[254c] Further details have been reported of the alkylation of purines caused by irradiation of the bases in the presence of alcohols.[254d] Ketyl radicals are produced on irradiation of alloxan[255a] and acyltriazoles[255b] in the presence of alcohols. Examples of the polar addition of methanol to some heterocyclic compounds have been reported.[256] Particular emphasis has been

---

[248a] S. G. Hadley, *J. Phys. Chem.*, **75**, 2083 (1971).
[248b] S. G. Schulman, R. T. Tidwell, J. J. Cetorelli, and J. D. Winefordner, *J. Am. Chem. Soc.*, **93**, 3179 (1971).
[249a] S. Caplain, J. P. Catteau, and A. Lablache-Combier, *Chem. Comm.*, **1970**, 1475; S. Caplain, A. Castellano, J. P. Catteau, and A. Lablache-Combier, *Tetrahedron*, **27**, 3541 (1971).
[249b] A. Castellano and A. Lablache-Combier, *Tetrahedron*, **27**, 2303 (1971); G. Allan, A. Castellano, J. P. Catteau, and A. Lablache-Combier, *Tetrahedron*, **27**, 4687 (1971).
[250a] M. Natsume and M. Wada, *Tetrahedron Letters*, **1971**, 4503.
[250b] N. Hata, I. Ono, and S. Ogawa, *Bull. Chem. Soc. Japan*, **44**, 2286 (1971).
[251a] V. I. Stenberg and E. F. Travecedo, *J. Org. Chem.*, **35**, 4131 (1970).
[251b] R. Gault and A. I. Meyers, *Chem. Comm.*, **1971**, 778.
[252a] V. I. Stenberg and E. F. Travecedo, *Tetrahedron*, **27**, 513 (1971).
[252b] D. E. I. Rosenthal, J. Salomon, and J. Sperling, *Chem. Comm.*, **1971**, 49.
[252c] Y. Ogata and K. Takagi, *Tetrahedron*, **27**, 2785 (1971).
[253a] R. A. F. Deeleman, H. C. van der Plas, A. Koudijs, and P. S. Darwinkel-Risseeuw, *Tetrahedron Letters*, **1971**, 4159.
[253b] L. R. Hamilton and P. J. Kropp, *Tetrahedron Letters*, **1971**, 1625.
[254a] N. C. Yang, L. S. Gorelic, and B. Kim, *Photochem. Photobiol.*, **13**, 275 (1971).
[254b] P. Grandclaudon and A. Lablache-Combier, *Chem. Comm.*, **1971**, 892.
[254c] T. Matsuura, A. Banba, and K. Ogura, *Tetrahedron*, **27**, 1211 (1971).
[254d] H. Steinmaus, I. Rosenthal, and D. Elad, *J. Org. Chem.*, **36**, 3594 (1971).
[255a] J. K. Dohrmann, R. Livingston, and H. Zeldes, *J. Am. Chem. Soc.*, **93**, 3343 (1971).
[255b] J. Van-Thielen, T. Van-Thien, and F. C. De Schryver, *Tetrahedron Letters*, **1971**, 3031.
[256] J. M. Patterson, R. L. Beine, and M. R. Boyd, *Tetrahedron Letters*, **1971**, 3923; M. F. Semmelhack, S. Kunkes, and C. S. Lee, *Chem. Comm.*, **1971**, 698; L. S. Davies and G. Jones, *J. Chem. Soc.* (C), **1971**, 2572.

put on identifying the excited states involved in the photoreduction of acridine, phenazine and related compounds; in the case of acridine, one report claims the $^3n\pi^*$ state to be the species,[257a] whereas another claims it to be the $^1n\pi^*$ state;[257b] kinetic and flash photolysis studies favour the latter. The reduction of phenazine by alkenes[258a] and alcohols[258b] and of 9,10-diazaphenanthrene by amines[258c] occurs from the $^1n\pi^*$ state. In contrast, the reduction of some benzophenazines by tri-n-butyltin hydride appears to occur via the $^3n\pi^*$ state.[258d] The reduction of acridine by acridane involves a charge-transfer complex.[259] Reduction of 9-chloroacridine to give 9,9′-biacridane appears to involve 9,9′-biacridine as an intermediate.[260] Further products have been isolated from the intra-[261a] and inter-molecular reduction[261b] of flavins. The radicals produced by the reduction of a variety of xanthene dyes, e.g. eosin, have been identified by ESR spectroscopy.[262] Several thiazine dyes and thionine are reduced by amine complexes of cobalt(III).[263] Tin(IV) porphyrins are reduced by tin(II) chloride to give tin(II) porphyrins by an electron-transfer reaction.[264] Electron-transfer processes of excited chlorophyll have been studied further;[265a,b] electron-transfer to p-benzoquinone was found to be quenched by β-carotene.[265a]

Full papers have now appeared in which the photoinduced decarboxylation of pyridyl-[266a] and pyrimidyl-acetic acid[266b] and nicotinic acids[266c] is discussed. Dimerization[267a] and trimerization[267b] of thymine, dimerization of 1,3-dimethyluracil,[267c] and formation of thymidine tetramers,[267d] dimerization of cytosine[267e] and thiouracils,[267f] and photosensitized splitting of pyrimidine dimers[267g] have all been discussed. Thymine[268a] and cytosine[268b] form adducts with benzo[a]pyrene; in the case of the cytosine adduct, the cytosyl residue was found substantially to reduce the ability of the hydrocarbon to sensitize carcinomas.

[257a] Y. Miyashita, S. Niizuma, H. Kokubun, and M. Koizumi, *Bull. Chem. Soc. Japan*, **43**, 3435 (1970).
[257b] D. G. Whitten and Y. J. Lee, *J. Am. Chem. Soc.*, **93**, 961 (1971).
[258a] S. M. Japar and E. W. Abrahamson, *J. Am. Chem. Soc.*, **93**, 4140 (1971).
[258b] G. A. Davies, J. D. Gresser, and P. A. Carapellucci, *J. Am. Chem. Soc.*, **93**, 2179 (1971).
[258c] G. A. Davies and S. G. Cohen, *Chem. Comm.*, **1971**, 675.
[258d] G. S. Davis, *Tetrahedron Letters*, **1971**, 3945.
[259] S. Niizuma, H. Kokubun, and M. Koizumi, *Chem. Phys. Letters*, **7**, 279 (1970); S. Niizuma, H. Kokubun, and M. Koizumi, *Bull. Chem. Soc. Japan*, **44**, 335 (1971).
[260] K. Nakamaru, S. Niizuma, and M. Koizumi, *Z. Phys. Chem. (Frankfurt)*, **73**, 113 (1970); K. Nakamaru, S. Niizuma, and M. Koizumi, *Bull. Chem. Soc. Japan*, **44**, 1256 (1971).
[261a] W. L. Cairns and D. E. Metzler, *J. Am. Chem. Soc.*, **93**, 2772 (1971).
[261b] M. Brülstein, W. R. Knappe, and P. Hemmerich, *Angew. Chem. Int. Ed.*, **10**, 804 (1971).
[262] I. H. Leaver, *Austral. J. Chem.*, **24**, 753 (1971).
[263] G. S. Singhal and E. Rabinowitch, *J. Chem. Phys.*, **53**, 4109 (1970).
[264] D. G. Whitten, J. C. Yau, and F. A. Carroll, *J. Am. Chem. Soc.*, **93**, 2291 (1971).
[265a] R. A. White and G. Tollin, *Photochem. Photobiol.* **14**, 15 (1971).
[265b] R. A. White and G. Tollin, *Photochem. Photobiol.* **14**, 43 (1971).
[266a] F. R. Stermitz and W. H. Huang, *J. Am. Chem. Soc.*, **93**, 3427 (1971).
[266b] S. Y. Wang, J. C. Nandi, and D. Greenfield, *Tetrahedron*, **26**, 5913 (1970).
[266c] F. Takeuchi, T. Fujimori, and A. Sugimori, *Bull. Chem. Soc. Japan*, **43**, 3637 (1970).
[267a] F. N. Hayes, D. L. Williams, R. L. Ratliff, A. J. Varghese, and C. S. Rupert, *J. Am. Chem. Soc.*, **93**, 4940 (1971); A. J. Varghese, *Photochem. Photobiol.*, **13**, 357 (1971).
[267b] S. Y. Wang, *J. Am. Chem. Soc.*, **93**, 2768 (1971).
[267c] D. Elad, I. Rosenthal, and S. Sasson, *J. Chem. Soc. (C)*, **1971**, 2053.
[267d] S. Y. Wang and D. F. Rhoades, *J. Am. Chem. Soc.*, **93**, 2554 (1971).
[267e] A. J. Varghese, *Biochemistry*, **10**, 2194 (1971).
[267f] J. B. Bremner, R. N. Warrener, E. Adman, and L. H. Jensen, *J. Am. Chem. Soc.*, **93**, 4574 (1971).
[267g] C. Helene and M. Charlier, *Biochem. Biophys. Res. Comm.*, **43**, 252 (1971).
[268a] C. Antonello and F. Carlassare, *Z. Naturforsch., b.*, **25**, 1268 (1970).
[268b] E. Cavalier and M. Calvin, *Photochem. Photobiol.*, **14**, 641 (1971).

There have been numerous papers on the photoisomerization of five- or six-membered heterocycles. The former series has been reviewed.[269] The stereochemistry of the photoisomerization of 2,3-dihydrofuran to cyclopropanecarboxaldehydes has again been discussed.[270a] 2,5-Dihydrothiophens, e.g. (111), undergo isomerization and ring-fragmentation on irradiation,[270b] a biradical intermediate formed by homolysis of the C—S bond being postulated. In a further discussion of the isomerization of furans to cyclopropenecarboxaldehydes, it has been suggested[271] that all the other rearrangements of five-membered heterocyclic compounds, (112), can be explained by postulating initial biradical formation on homolysis of the weakest bond; this is followed by cyclization

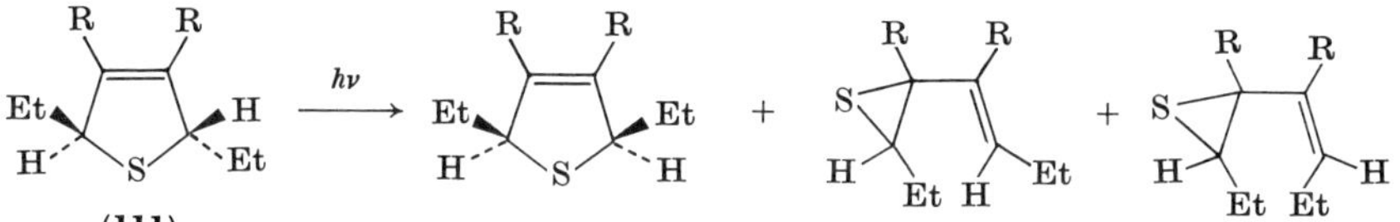

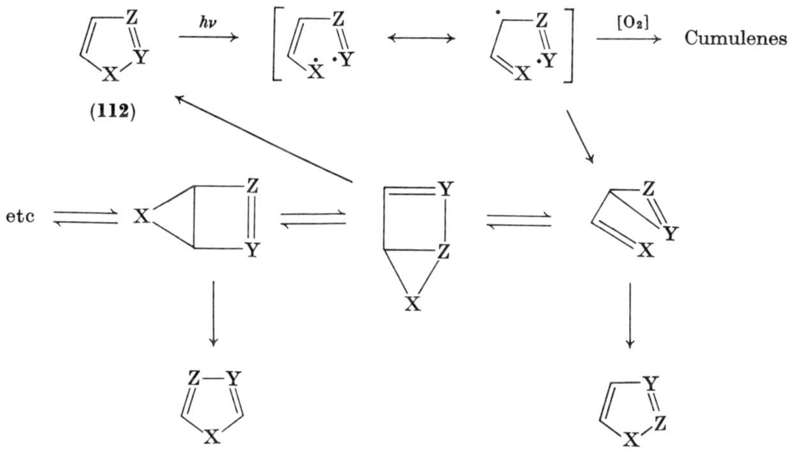

[269] A. Lablache-Combier and M. A. Remy, *Bull. Soc. Chim. France*, **1971**, 679.
[270a] P. Scribe, M. R. Monot, and J. Wiemann, *Bull. Soc. Chim France*, **1971**, 2268.
[270b] R. M. Kellogg, *J. Am. Chem. Soc.*, **93**, 2344 (1971).
[271] E. E. van Tamelen and T. H. Whitesides, *J. Am. Chem. Soc.*, **93**, 6129 (1971).

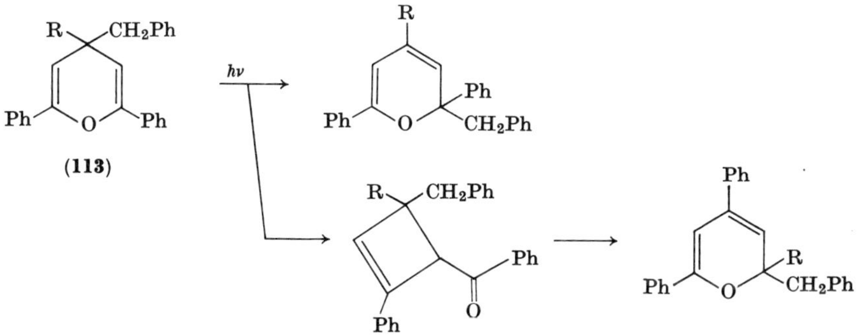

of the biradical to give cyclopropenes or bicyclo-compounds. Formation of cyclopropenecarboxaldehydes from furans[272a] and of cyclopropenethiocarboxaldehydes from thiophens[272b] has been confirmed by trapping experiments. The reported isomerization of isoxazoles,[273a] thiazoles[273b] and thiadiazoles[273c] may be understood in terms of the previously outlined mechanism. γ-Pyrans, e.g. (**113**), isomerize on irradiation; acylcyclobutenes appear to be intermediates.[274] Recent work has shown that the rearrangement of pyridazine to pyrazine does not involve a diazaprismane intermediate, as previously claimed.[275] Dihydro-pyrazines[276a] and -quinolizines[276b] undergo cycloreversion on irradiation, the former giving ring-contracted products, whereas the latter gives products that at first sight appear to be derived by a sigmatropic shift. Photolysis of benzofurazans has been used as a method for generating nitrenes,[277a,b] the initial photoreaction appearing to be a cycloreversion.[277a] The fragmentation of sulphur ylids as a means of generating carbenes has attracted further attention.[278] The fragmentation of sydnones and other mesoionic compounds has been found to be a fruitful source of reactive species;[279-284] the compounds studied include (**114**),[280] (**115**),[281] (**116**),[282] (**117**),[282] (**118**)[283] and (**119**),[284] and it will be interesting to see the application of these reactions in the coming year. There have been further examples of the homolysis of alloxan monohydrates and monoalcoholates,[285] the ring expansion of 5,6-methylenepyrimidines[286] and the homolysis of hexaarylbi-imidazoles.[287]

Nitrogen-containing Compounds

Azomethines and Related Compounds and Amines

Cleavage of the N—N bond of azines to give imino-radicals[288a,b] is thought, from quenching studies, to occur from more than one excited state.[288b] Azines form (2:1) adducts

[272a] A. Couture and A. Lablache-Combier, *Chem. Comm.*, **1971**, 891.
[272b] A. Couture and A. Lablache-Combier, *Tetrahedron*, **27**, 1059 (1971).
[273a] R. H. Good and G. Jones, *J. Chem. Soc.* (C), **1971**, 1196.
[273b] G. Vernin, J. C. Poite, J. Metzger, J. P. Aune, and H. M. Dou, *Bull. Soc. Chim. France*, **1971**, 1103.
[273c] K. P. Zeller, H. Meier, and E. Muller, *Tetrahedron Letters*, **1971**, 537.
[274] N. Kin-Cuong, F. Fournier, and J. J. Basselier, *C. R. Acad. Sci. Paris*, (C), **271**, 1626 (1970).
[275] D. W. Johnson, V. Austel, R. S. Feld, and D. M. Lemal, *J. Am. Chem. Soc.*, **92**, 7505 (1970).
[276a] D. R. Arnold, V. Y. Abraitys, and D. McLeod, *Can. J. Chem.*, **49**, 923 (1971).
[276b] R. M. Acheson and J. K. Stubbs, *J. Chem. Soc.* (C), **1971**, 3285.
[277a] M. Georgarakis, H. J. Rosenkranz, and H. Schmid, *Helv. Chim. Acta*, **54**, 819 (1971).
[277b] E. Giovannini, J. Rosales, and B. de Souza, *Helv. Chim. Acta*, **54**, 2111 (1971); M. A. Berwick, *J. Am. Chem. Soc.*, **93**, 5780 (1971).
[278] T. Kunieda and B. Witkop, *J. Am. Chem. Soc.*, **93**, 3487 (1971).
[279] C. W. Bird, D. Y. Wong, G. V. Boyd, and A. J. H. Summers, *Tetrahedron Letters*, **1971**, 3187; T. Shiba, K. Yanane, and H. Kato, *Chem. Comm.*, **1970**, 1592; R. Gleiter, D. Schmidt, and J. Streith, *Helv. Chim. Acta*, **54**, 1645 (1972).
[280] M. Maerky, H. J. Hansen, and H. Schmid, *Helv. Chim. Acta*, **54**, 1275 (1971); H. Gotthardt and F. Reiter, *Tetrahedron Letters*, **1971**, 2749; C. S. Angadiyavar and M. V. George, *J. Org. Chem.*, **36**, 1589 (1971); Y. Huseya, A. Chinone, and M. Ohta, *Bull. Chem. Soc. Japan*, **44**, 1667 (1971).
[281] H. Kato, T. Shiba, H. Yoshida, and S. Fujimori, *Chem. Comm.*, **1970**, 1591.
[282] H. Gotthardt, *Tetrahedron Letters*, **1971**, 1277.
[283] R. M. Moriarty, R. Mukherjee, O. L. Chapman, and D. R. Eckroth, *Tetrahedron Letters*, **1971**, 397.
[284] P. Scheiner, *Tetrahedron Letters*, **1971**, 4489.
[285] Y. Otsuji, S. Wake, and E. Imoto, *Tetrahedron*, **26**, 4139 (1970); S. Wake, T. Mawatari, Y. Otsuji, and E. Imoto, *Bull. Chem. Soc. Japan*, **44**, 2202 (1971); Y. Otsuji, S. Wake, and E. Imoto, *Tetrahedron*, **26**, 4293 (1970).
[286] T. Kunieda and B. Witkop, *J. Am. Chem. Soc.*, **93**, 3478 (1971).
[287] R. H. Riem, A. MacLachlan, G. R. Coraor, and E. J. Urban, *J. Org. Chem.*, **36**, 2272 (1971).
[288a] M. Kamachi, K. Kuwata, and S. Murahashi, *J. Phys. Chem.*, **75**, 164 (1971).
[288b] M. A. Fox and R. W. Binkley, *Tetrahedron Letters*, **1970**, 5257.

(114), (115), (116), (117), (118), (119)

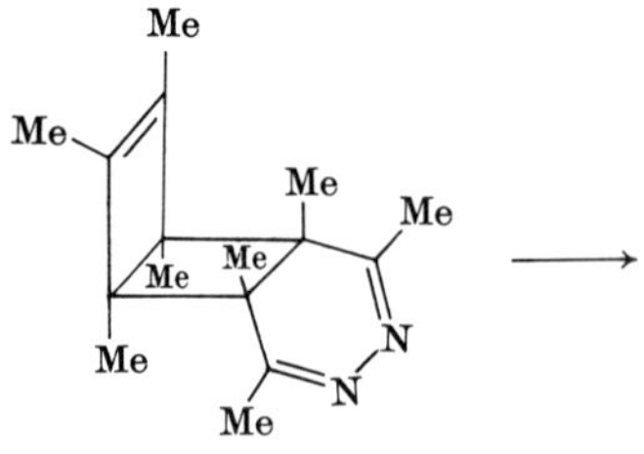

(120)

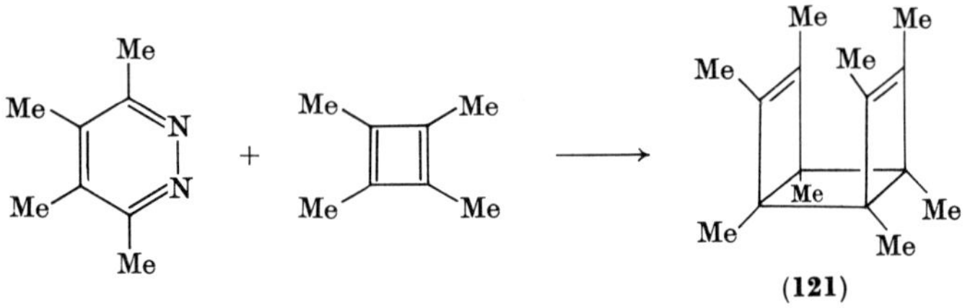

(121)

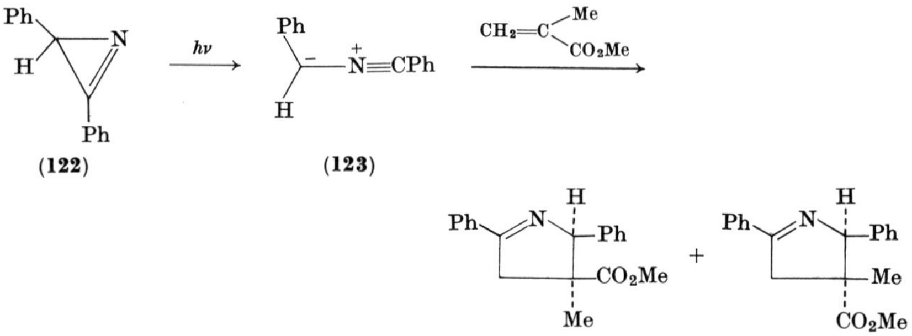

(122) → (123) →

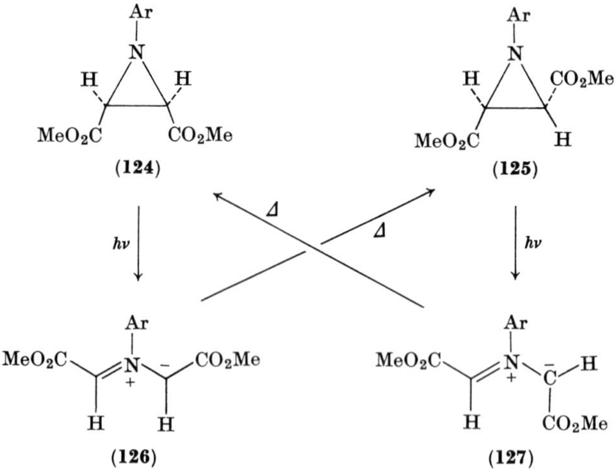

(124) (125)

| hν | | | hν |

(126)
$\tau_{1/2}$ 7.8 sec

(127)
$\tau_{1/2}$ 5.4 sec

with a number of olefins on irradiation.[289] Photolysis of the dihydropyrazine (**120**) results in tetramethylcyclobutadiene which dimerizes to give (**121**).[290] Some thiazepines undergo 1,3-hydrogen shifts on irradiation and the reaction is probably not concerted.[291] Photolysis of arylazirines, e.g. (**122**), produces dipolar species, e.g. (**123**), that can be trapped with dipolarophiles.[292] The kinetics of the interconversion of the aziridines (**124**) and (**125**) by photochemical disrotatory ring-opening and thermal conrotatory ring-closure of (**126**) and (**127**) have been examined by flash photolysis.[293]

Flash-photolysis studies of aniline[294a] and phenothiazine[294b] have shown that irradiation of these amines in non-polar solvents populates their triplet states, which cannot be seen in oxygenated solution. In polar solvents, radical cation formation takes place in either the presence or the absence of oxygen. Photoinduced electron-transfer from *leuco*-triphenylmethane dyes to imidazolyl radicals has been shown to occur.[295] Diquaternary salts of 4,4'-bipyridine form radical cations on irradiation.[296] Electron-ejection from indole[297a,b] is rapidly followed by proton loss, so that a 1-indolyl radical is formed; this rapidly rearranges to a 3-indolyl radical.[297b]

The acid-catalysed rearrangement of hydrazobenzene gives numerous products whereas photo-induced rearrangement produces solely *o*-semidines.[298a] In another report it is claimed that irradiation of hydrazobenzenes produces azo-compounds and amines.[298b] It was concluded that reaction did not occur by homolysis of the N—N bond. A number of *N*-benzyl-[299a] and *N*-allyl-anilines[299b] rearrange to give *ortho*- and *para*-substituted anilines on irradiation. In the presence of an oxidizing agent, e.g. ferric chloride, the allylanilines cyclized to quinolines.[299b] There have been mechanistic[300a,b] and synthetic investigations[300a] of the cyclization of *N*-aryl-enamines and diphenylamines; the former compounds afford a convenient route to dihydro-indoles and -carbazoles. Deuterium-labelling studies with (**128**) showed product formation both by 1,4-hydrogen shift and by two consecutive 1,2-hydrogen shifts. In the cyclization of

[289] T. P. Forshaw and A. E. Tipping, *J. Chem. Soc.* (C), **1971**, 2404.
[290] G. Maier and M. Schneider, *Angew. Chem. Int. Ed.*, **10**, 809 (1971).
[291] M. F. Semmelhack and B. F. Gilman, *Chem. Comm.*, **1971**, 988.
[292] A. Padwa and J. Smolanoff, *J. Am. Chem. Soc.*, **93**, 548 (1971).
[293] H. Hermann, R. Huisgen, and H. Mäder, *J. Am. Chem. Soc.*, **93**, 1779 (1971).
[294a] M. Hori, H. Itoi, and H. Tsubomura, *Bull. Chem. Soc. Japan*, **43**, 3765 (1970).
[294b] T. Iwaoka, H. Kokubun, and M. Koizumi, *Bull. Chem. Soc. Japan*, **44**, 341 (1971).
[295] A. MacLachlan and R. H. Riem, *J. Org. Chem.*, **36**, 2275 (1971).
[296] J. F. McKellar and P. H. Turner, *Photochem. Photobiol.* **13**, 437 (1971).
[297a] J. Feitelson, *Photochem. Photobiol.*, **13**, 87 (1971).
[297b] M. I. Pailthorpe and C. H. Nicholls, *Photochem. Photobiol.* **14**, 135 (1971).
[298a] H. J. Shine and J. D. Cheng, *J. Org. Chem.*, **36**, 2787 (1971).
[298b] D. V. Banthorpe, A. Cooper, D. A. Pearce, and J. A. Thomas, *J. Chem. Soc.* (B), **1971**, 2057.
[299a] Y. Ogata and K. Takagi, *Bull. Chem. Soc. Japan*, **44**, 2186 (1971).
[299b] Y. Ogata and K. Takagi, *Tetrahedron*, **27**, 1573 (1971).
[300a] O. L. Chapman, G. L. Eian, A. Bloom, and J. Clardy, *J. Am. Chem. Soc.*, **93**, 2918 (1971).
[300b] H. Shizuka, Y. Takayama, I. Tanaka, and T. Morita, *J. Am. Chem. Soc.*, **92**, 7270 (1970); C. Wentrup and M. Gaugaz, *Helv. Chim. Acta*, **54**, 2108 (1971).

the enamines as well as of the diphenylamines, the triplet was proposed as the reactive state.

N-Oxides and Related Compounds

From further studies[301a-c] of the rearrangement of nitrones to oxaziridines it has been concluded[301a] that ring-closure is concerted and occurs in a disrotatory manner. Product formation from irradiation of quinoline N-oxides,[302a] benzimidazole N-oxides,[302b] 2-phenylisatogens[302c] and quinoxaline di-N-oxides[302d] has been studied; oxaziridine intermediates have been postulated; with (129), deuterium-labelling showed the fate of the oxaziridine intermediate.[302a] Pyridazine N-oxides rearrange to furans, e.g. (130), as well as other products.[303] Further attempts have been made to characterize the excited state of pyridine N-oxides that is responsible for their deoxygenation.[304] There have been some investigations of the photolysis of the dimers of nitroso-compounds;[305] cleavage to monomers occurs and these readily homolyse to give nitric oxide plus an alkyl radical.

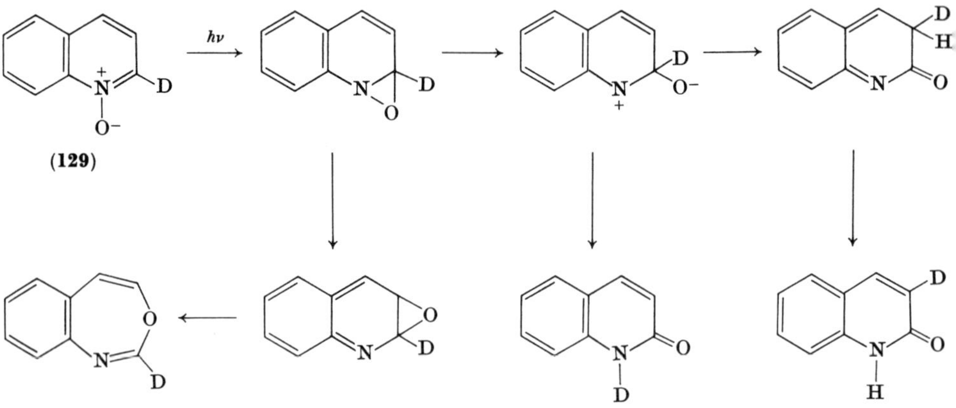

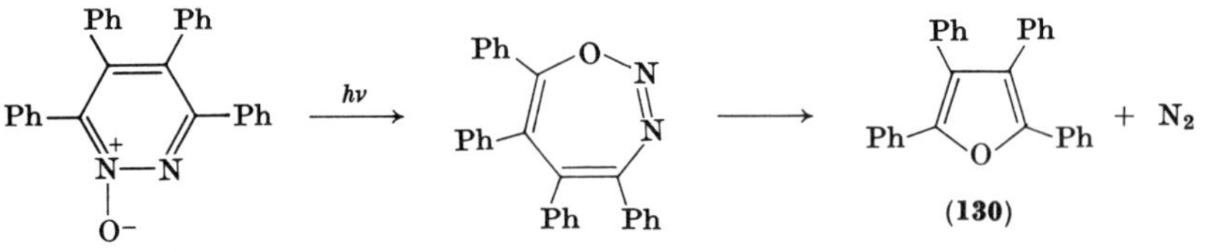

[301a] J. S. Splitter, T. M. Su, H. Ono, and M. Calvin, J. Am. Chem. Soc., **93**, 4075 (1971).
[301b] K. Koyana, H. Suzuki, Y. Mori, and I. Tanaka, Bull. Chem. Soc. Japan, **43**, 3582 (1970).
[301c] D. R. Eckroth, T. H. Kinstle, D. O. DeLa Cruz, and J. K. Sparacino, J. Org. Chem., **36**, 3619 (1971).
[302a] O. Buchardt, K. B. Tomer, and V. Madsen, Tetrahedron Letters, **1971**, 1311; G. Favaro, Mol. Photochem., **2**, 323 (1970).
[302b] R. Fielden, O. Meth-Cohn, and H. Suschitzky, Chem. Comm., **1970**, 1658.
[302c] D. R. Eckroth and R. H. Squire, J. Org. Chem., **36**, 224 (1971).
[302d] M. J. Haddadin, G. Agopian, and C. H. Issidorides, J. Org. Chem., **36**, 514 (1971).
[303] T. Tsuchiya, H. Arai, and H. Igeta, Tetrahedron Letters, **1971**, 2579.
[304] F. Bellamy, L. Guillermo, R. Barragan, and J. Streith, Chem. Comm., **1971**, 456.
[305] A. Rassat and P. Rey, Chem. Comm., **1971**, 1161; J. A. Maassen, H. Hittenhousen, and T. J. de Boer, Tetrahedron Letters, **1971**, 3213.

Nitro- and Nitroso-Compounds

There have been further examples of the use of nitrosyl chloride for the photo-oximation of alkanes.[306] The reactions of *gem*-chloronitroso-compounds have been interpreted as occurring through intermediate diazonium nitrates, e.g. (**131**).[307] Photolysis of *C*-nitroso-compounds give alkenes.[308] Investigations of the reactions of *N*-nitroso-compounds have included the decomposition of *N*-nitrososulphonamides,[309a] their reduction,[309b] and their addition to acetylene[309c] and alkenes.[309d,e] There have been

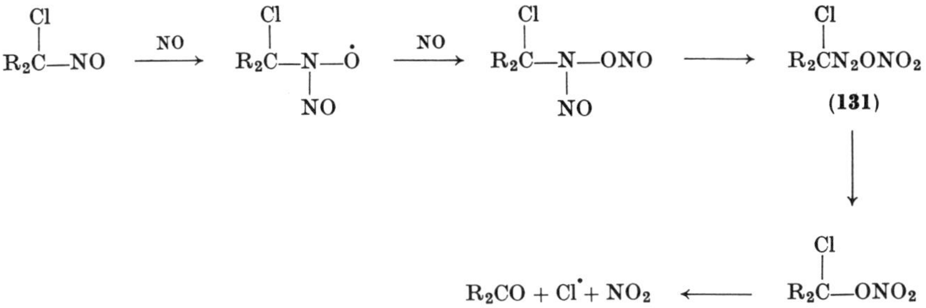

$$R_2C\text{--}NO \xrightarrow{NO} R_2C\text{--}\underset{NO}{\overset{Cl}{N}}\text{--}\dot{O} \xrightarrow{NO} R_2C\text{--}\underset{NO}{\overset{Cl}{N}}\text{--}ONO \longrightarrow R_2CN_2ONO_2 \quad (\mathbf{131})$$

$$R_2CO + Cl^{\bullet} + NO_2 \longleftarrow R_2\overset{Cl}{C}\text{--}ONO_2$$

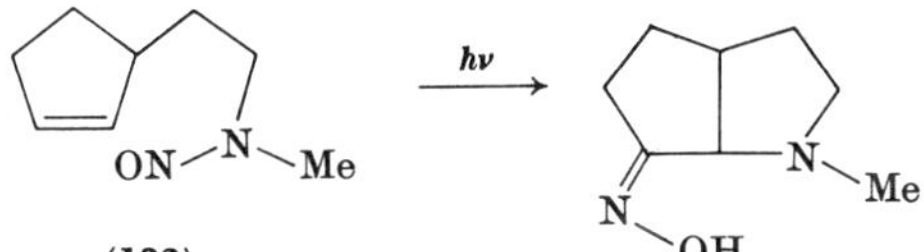

(**132**)

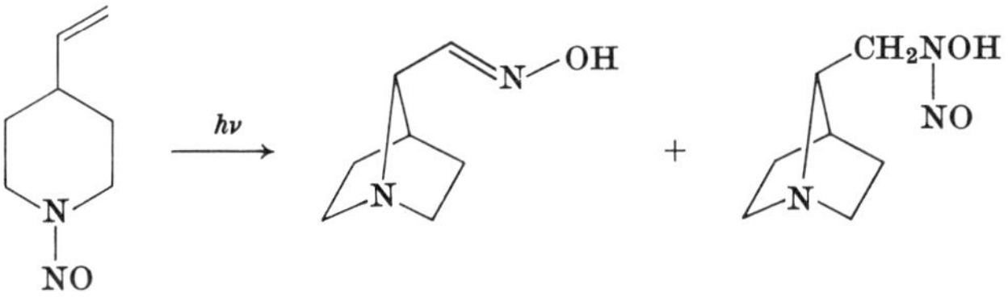

(**133**)

some very nice applications of the last-mentioned reaction, e.g. using (**132**) and (**133**), to the synthesis of bicyclic compounds;[309d] the reactions were found to be highly stereospecific, addition giving antidiaxial products. Photolysis of *N*-nitroamines gives many products, mostly derived by cleavage of the N—N bond.[310]

[306] M. W. Mosher and N. J. Bunce, *Can. J. Chem.*, **49**, 28 (1971); B. W. Tattersall, *Chem. Comm.*, **1970**, 1522.
[307] E. F. J. Duynstee and M. E. A. H. Meuis, *Rec. Trav. Chim.*, **90**, 932 (1971); A. H. M. Kayen, L. R. Subramanian, and T. J. de Boer, *Rec. Trav. Chim.*, **90**, 866 (1971).
[308] J. R. Dickson and B. G. Gowenlock, *Ann. Chem.*, **745**, 152 (1971).
[309a] J. B. F. N. Engberts, L. C. J. Vander Laan, and T. J. de Boer, *Rec. Trav. Chim.*, **90**, 901 (1971).
[309b] H. P. Haerter, M. Neuenschwander, and O. Schindler, *Helv. Chim. Acta*, **54**, 649 (1971); M. P. Lau, A. J. Cessna, Y. L. Chow, and R. W. Yip, *J. Am. Chem. Soc.*, **93**, 3808 (1971).
[309c] Y. L. Chow and D. W. L. Chang, *Chem. Comm.*, **1971**, 64.
[309d] Y. L. Chow, R. A. Perry, and B. C. Menon, *Tetrahedron Letters*, **1971**, 1549; Y. L. Chow, R. A. Perry, B. C. Menon, and S. C. Chen, *Tetrahedron Letters*, **1971**, 1545.
[309e] Y. L. Chow, S. C. Chen, and D. W. L. Chang, *Can. J. Chem.*, **49**, 3069 (1971).
[310] L. J. Winters, J. F. Fischer, and E. R. Ryan, *Tetrahedron Letters*, **1971**, 129.

There have been further studies of the addition of nitro-compounds to alkenes and reaction was shown to occur from the triplet state of the nitro-compound.[311] The rate constants for electrophilic addition of triplet nitrobenzene to alkenes were all quite high, ca. 10^7 mol^{-1} sec^{-1}. Studies of the photoreduction of nitro-compounds have included the reduction of 4-nitropyridine in acidified propan-2-ol[312a] and of the nitrobenzene–boron trichloride complex.[312b] For the latter reaction it was proposed that oxygen-transfer from the nitro-group to boron occurs with concomitant loss of a chlorine atom from the boron:

$$PhNO_2 \rightarrow BCl_3 \xrightarrow{h\nu} PhNO + OBCl_2 + Cl \cdot$$

There have been several studies of the intramolecular hydrogen abstraction which occurs on irradiation of o-(t-butyl)nitrobenzenes, e.g. (**134**);[313] a variety of products is formed, but in all cases reaction through the nitro-alcohol (**135**) seemed probable. When the reaction of these compounds were carried out in solvents that are good hydrogen-donors, e.g. aliphatic amines, reduction of the nitro to a hydroxylamino-group competed effectively with intramolecular reactions.[314] Other hindered nitro-compounds, e.g. nitromesitylene, are reduced by alcohols to amines.[315] Phenols are also produced and it was proposed that these compounds are derived by rearrangement of the nitro-group to a nitrite group and that this homolyses to give a phenoxyl radical. There have been further examples of the ability of nitro-compounds to sensitize the transformation of

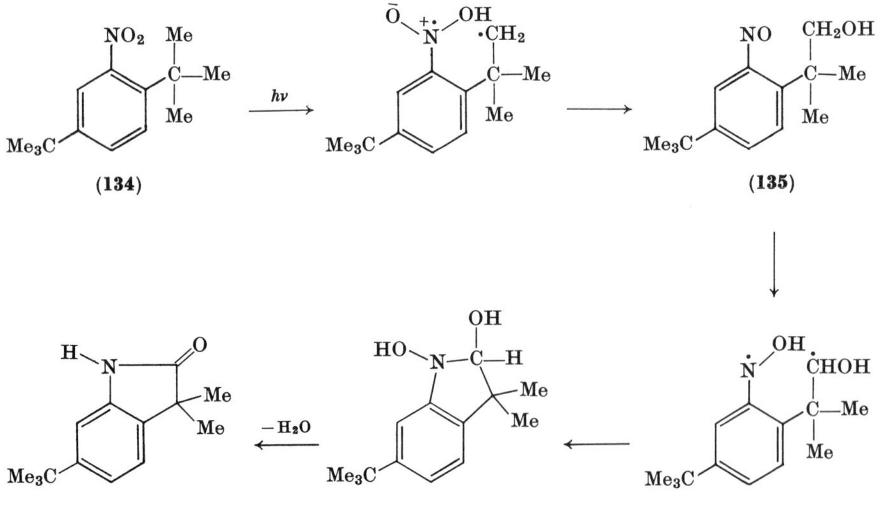

---

[311] J. L. Charlton, C. C. Liao, and P. deMayo, *J. Am. Chem. Soc.*, **93**, 2463 (1971).
[312a] S. Hashimoto, K. Kano, and K. Ueda, *Bull. Chem. Soc. Japan*, **44**, 1102 (1971).
[312b] W. Trotter and A. C. Testa, *J. Phys. Chem.*, **75**, 2415 (1971).
[313] D. Döpp, *Tetrahedron Letters*, **1971**, 2757; D. Döpp and K. H. Sailer, *Tetrahedron Letters*, **1971**, 2761; D. Döpp, *Chem. Ber.*, **104** 1043, 1035 (1971); L. R. C. Barclay and I. R. McMaster, *Can. J. Chem.*, **49**, 676 (1971).
[314] D. Döpp, *Chem. Ber.*, **104**, 1058 (1971).
[315] Y. Kitaura and T. Matsuura, *Tetrahedron*, **27**, 1583 (1971).

C—Me groups into C—CHO groups.[316] The observation that N-(p-nitrophenyl)valine is decarboxylated on irradiation[317] suggests that the decarboxylation of N-(nitrophenyl)-α-amino-acids may well be an intermolecular reaction. That such reactions can occur intermolecularly was substantiated by the finding that aromatic nitro-compounds, e.g. 1-nitronaphthalene, sensitize the decarboxylation of N-phenylglycines.[69] These reactions were suggested as occurring via excited charge-transfer complexes. Aromatic nitro-compounds also sensitize the decarboxylation of thioglycollic acids.[69] Similarly, S-(2-nitrophenyl)thioglycollic acid is decarboxylated on irradiation.[318].

Azo, Diazo- and Diazonium Compounds and Azides

Flash photolysis has been used to determine the rate constants for the thermal isomerization of *cis*- to *trans*-azobenzene.[319] Hydrogen-abstraction from cycloheptene by diethyl azodicarboxylate occurs on irradiation.[320a] By the use of optically active phenylazo-alkanes it has been shown that photoinduced elimination of the initially produced radicals can occur within the solvent cage;[320b] isomerization to the *cis*-azo-compound was also observed. Photochemical decomposition of cyclic azo-compounds[321a,b] involves typical biradicals, whereas thermal cleavage involves loss of nitrogen in a two-step process.[321a] The fate of the energy released by expulsion of nitrogen from electronically excited azo-compounds has been probed.[322a,b] Evidence that vibrationally excited 1,3-biradicals are formed in the decomposition of pyrazolines comes from the finding that in the mercury-sensitized decomposition a much higher yield of alkenes is obtained than in the benzophenone-sensitized decomposition.[322a] The formation of cyclopropanes by decomposition of pyrazolines has found synthetic application.[323a,b] There have been a number of reports of the photochemical decomposition of diazo-compounds being used for the formation of carbenes.[324] Decomposition of phenyldiazomethane to give triplet carbene is facilitated by the use of solid matrices, i.e. the restricted movement in the matrix gives the carbene sufficient time to undergo intersystem crossing before reaction takes place.[325] Attempts to produce optically active cyclopropanes by photolysis of diazoalkenes in the presence of olefins, using circularly polarized light, failed.[326]

---

[316] D. A. Mudry and A. R. Frasca, *Chem. Ind.* (*London*), **1971**, 1038; H. B. Land and A. R. Frasca, *Tetrahedron*, **26**, 5793 (1970).
[317] P. H. MacFarlane and D. W. Russell, *Tetrahedron Letters*, **1971**, 725.
[318] R. S. Goudie and P. N. Preston, *J. Chem. Soc.* (C), **1971**, 3081.
[319] P. D. Wildes, J. G. Pacifici, G. Irick, and D. G. Whitten, *J. Am. Chem. Soc.*, **93**, 2004 (1971).
[320a] A. Shah and M. V. George, *Tetrahedron*, **27**, 1291 (1971).
[320b] N. A. Porter, M. E. Landis, and L. J. Marnett, *J. Am. Chem. Soc.*, **93**, 795 (1971).
[321a] P. B. Conduit and R. G. Bergman, *Chem. Comm.*, **1971**, 4.
[321b] H. Tanida and S. Teratake, *Tetrahedron Letters*, **1970**, 4991.
[322a] E. B. Klunder and R. W. Carr, *Chem. Comm.*, **1971**, 742.
[322b] F. H. Dorer, E. Brown, J. Do, and R. Rees, *J. Phys. Chem.*, **65**, 1640 (1971).
[323a] H. Prinzbach and D. Stusche, *Angew. Chem. Int. Ed.*, **9**, 799 (1970).
[323b] A. G. Brook and P. F. Jones, *Can. J. Chem.*, **49**, 1841 (1971).
[324] L. Schrader, *Chem. Ber.*, **104**, 941 (1971); G. Lowe and J. Parker, *Chem. Comm.*, **1971**, 577; G. Cauquis, B. Divisia, M. Rastoldo, and G. Reverdy, *Bull. Soc. Chim. France*, **1971**, 3022; G. Cauquis, B. Divisia, and G. Reverdy, *Bull. Soc. Chim. France*, **1971**, 3027, 3031; T. Severin, H. Kraemer, and P. Adhikary, *Chem. Ber.*, **104**, 972 (1971); H. Duerr, L. Schrader, and H. Seidl, *Chem. Ber.*, **104**, 391 (1971).
[325] R. A. Moss and U. H. Dolling, *J. Am. Chem. Soc.*, **93**, 954 (1971).
[326] P. Boldt, W. Thielecke, and H. Luthe, *Chem. Ber.*, **104**, 353 (1971).

Ketyl radicals have been shown to induce decomposition of diazoalkanes.[327a,b] There have been a number of studies of the decomposition of azides.[327b,328a–f,329,330]

Photolysis of di-p-azidobenzene (136) gives the biradical (137).[328a] The same biradical is produced on photolysis of 4,4'-diazidoazobenzene. There is evidence

$$N_3-\langle\underline{}\rangle-N_3$$
(136)
$$\downarrow h\nu$$
$$\overset{\bullet}{N}=\langle\underline{}\rangle=N-N=\langle\underline{}\rangle=\overset{\bullet}{N} \quad \xleftarrow{h\nu} \quad N_3-\langle\underline{}\rangle-N=N-\langle\underline{}\rangle-N_3$$
(137)

that a number of azides decompose to give products that at first sight appear to have been formed via nitrenes; however, in a number of rearrangement reactions[328d,e] the migratory aptitudes of the moving group are very different from the expected values and therefore it was concluded[328d] that discrete nitrene intermediates are not involved. There has been a number of reports of the photolysis of diazonium compounds as a means of preparing aryl ra´dicals,[331a,b] and the reaction has found application in the synthesis of alkaloids.[331a]

Halogen-containing Compounds

Halomethyl radicals, produced by photolysis of tri- and tetra-halomethanes, have been trapped with t-nitrosobutanes.[332] Two competing processes occur on photolysis of hexahaloacetone: (a) the Type I reaction and (b) cleavage of a carbon–halogen bond.[333] Phenethyl bromide undergoes C—Br bond homolysis on irradiation with light of wavelength 313 and 254 nm; with the lower-wavelength light, energy-transfer from the aryl

[327a] L. Horner and H. Schwarz, *Ann. Chem.*, **747**, 1 (1971).
[327b] L. Horner and H. Schwarz, *Ann. Chem.*, **747**, 21 (1971).
[328a] B. Singh and J. S. Brinen, *J. Am. Chem. Soc.*, **93**, 540 (1971).
[328b] A. Reiser and L. J. Leyshon, *J. Am. Chem. Soc.*, **93**, 4051 (1971).
[328c] R. C. Baetzold and L. K. J. Tong, *J. Am. Chem. Soc.*, **93**, 1394 (1971).
[328d] R. A. Abramovitch and E. P. Kyba, *J. Am. Chem. Soc.*, **93**, 1537 (1971).
[328e] M. J. P. Harger, *Chem. Comm.*, **1971**, 442.
[328f] J. Ciabattoni and M. Cabell, *J. Am. Chem. Soc.*, **93**, 1482 (1971).
[329] H. W. Moore and W. Weyler, *J. Am. Chem. Soc.*, **93**, 2812 (1971); A. Pancrazi, Q. Khuong-Huu, and R. Goutarel, *Bull. Soc. Chim. France*, **1970**, 4446; Q. Khuong-Huu and A. Pancrazi, *Tetrahedron Letters*, **1971**, 37; M. F. G. Stevens, A. C. Mail, and J. Reisch, *Photochem. Photobiol.* **13**, 441 (1971); T. Kametani and M. Shio, *J. Heterocycl. Chem.*, **8**, 545 (1971); D. M. Clode and D. Horton, *Carbohydrate Res.*, **14**, 405 (1970); B. Stanovnik, *Tetrahedron Letters*, **1971**, 3211; R. A. Abramovitch, C. I. Azogu, and R. G. Sutherland, *Chem. Comm.*, **1971**, 134.
[330] D. H. R. Barton, P. G. Sammes, and G. G. Weingarten, *J. Chem. Soc.* (C), **1971**, 721.
[331a] T. Kametani, M. Koizumi, K. Shishido, and K. Fukumoto, *J. Chem. Soc.* (C), **1971**, 1923; T. Kametani, M. Koizumi, and K. Fukumoto, *J. Chem. Soc.* (C), **1971**, 1792; T. Kametani, R. Charubala, M. Ihari, M. Koizumi, K. Takahashi, and K. Fukumoto, *J. Chem. Soc.* (C), **1971**, 3315.
[331b] K. L. Kirk and L. A. Cohen, *J. Am. Chem. Soc.*, **93**, 3060 (1971); R. C. Petterson, A. Di Maggio, A. L. Hebert, T. J. Haley, J. P. Mykytka, and I. M. Sarkar, *J. Org. Chem.*, **36**, 631 (1971).
[332] J. W. Hartgerink, J. B. F. N. Engberts, and T. J. de Boer, *Tetrahedron Letters*, **1971** 2709.
[333] J. R. Majer, C. Olavesen, and J. C. Robb, *J. Chem. Soc.* (B), **1971**, 48.

ring to the C—Br bond was postulated.[334] Alkylation of aromatic hydrocarbons by their irradiation in the presence of halomethyl groups has been further investigated.[335a-c] A flash-photolysis study failed to give any positive evidence for the formation of exciplex intermediates.[335c] Irradiation of benzonitrile in aqueous solution containing iodide ion converts the cyano group into an aldehyde group.[336] It is thought that electron-transfer from the iodide ion to the cyano-group occurs; protonation of the species so formed, followed by hydrolysis, gives the aldehyde group. Numerous inter-[337a,b] and intra-molecular[338a,b] arylations have been reported, in which aryl radicals were generated by photolysis of halobenzenes; photolysis of dihalobenzenes always results in the rupture of the weakest carbon—halogen bond.[337a] In the case of 1,2-di-iodotetrafluorobenzene, bond homolysis gives tetrafluorobenzyne.[338b] There have been further examples reported of the isomerization of allylic chlorides, e.g. (138), to chlorocyclopropane derivatives;[339a,b] reaction was suggested as occurring via olefin diradicals[339a] although

previously evidence in favour of carbonium ion–chloride ion ion-pairs had been presented.[339c]

Carbonium Ions and Carbanions

Further details of the photoisomerization of the tropylium ion have been reported;[340a,b] the types of product formed are sensitive to the pH of the system.[340a] Irradiation of

---

[334] J. J. Dannenberg, K. Dill, and H. P. Waits, *Chem. Comm.*, **1971**, 1348.
[335a] S. Naruto and O. Yonemitsu, *Tetrahedron Letters*, **1971**, 2297.
[335b] C. M. Foltz, *J. Org. Chem.*, **36**, 24 (1971).
[335c] S. Naruto, O. Yonemitsu, N. Kanamaru, and K. Kimura, *J. Am. Chem. Soc.*, **93**, 4053 (1971).
[336] J. P. Ferris and F. R. Antonucci, *Chem. Comm.*, **1971**, 1294.
[337a] G. E. Robinson and J. M. Vernon, *J. Chem. Soc.* (C), **1971**, 3363.
[337b] K. Omura and T. Matsuura, *Tetrahedron*, **27**, 3101 (1971); R. F. C. Brown and M. Butcher, *Tetrahedron Letters*, **1971**, 667; G. E. Robinson and J. M. Vernon, *J. Chem. Soc.* (C), **1970**, 2586.
[338a] U. Weiss, H. Ziffer, and J. M. Edwards, *Austral. J. Chem.*, **24**, 657 (1971); G. DeLuca, G. Martelli, P. Spagnolo, and M. Tiecco, *J. Chem. Soc.* (C), **1970**, 2504; T. Sato, S. Shimada, and K. Hata, *Bull. Chem. Soc. Japan*, **44**, 2484 (1971); Z. Horii, Y. Nakashita, and C. Iwata, *Tetrahedron Letters*, **1971**, 1167; T. Kametani, H. Sugi, S. Shibuya, and K. Fukumoto, *Chem. Ind. (London)*, **1971**, 818; T. Kametani, T. Kohno, S. Shibuya, and K. Fukumoto, *Tetrahedron*, **27**, 5441 (1971); *Chem. Comm.*, **1971**, 774; T. Kametani, T. Sugahara, and K. Fukumoto, *Tetrahedron*, **27**, 5367 (1971); D. H. Hey, G. H. Jones, and M. J. Perkins, *Chem. Comm.*, **1971**, 47; *J. Chem. Soc.* (C), **1971**, 116; T. Kametani, S. Shibuya, H. Sugi, O. Kusama, and K. Fukumoto, *J. Chem. Soc.* (C), **1971**, 2446; S. M. Kupchan, J. L. Moniot, R. M. Kanojoa, and J. B. O'Brien, *J. Org. Chem.*, **36**, 2413 (1971); T. Kametani, T. Sugahara, H. Sugi, S. Shibuya, and K. Fukumoto, *Chem. Comm.*, **1971**, 724.
[338b] J. P. N. Brewer, I. F. Eckhard, H. Heaney, M. G. Johnson, B. A. Marples, and T. J. Ward, *J. Chem. Soc.* (C), **1970**, 2569.
[339a] F. Bohlmann, W. Skubulla, C. Zdero, T. Kühle, and P. Steirl, *Ann. Chem.*, **745**, 176 (1971).
[339b] S. J. Cristol, G. A. Lee, and A. L. Noreen, *Tetrahedron Letters*, **1971**, 4175.
[339c] *Org. Reaction Mech.*, **1971**, Chapter 13, ref. 224a.
[340a] E. E. van Tamelen, R. H. Greeley, and H. Schumacher, *J. Am. Chem. Soc.*, **93**, 6151 (1971).
[340b] H. Hogeveen and C. J. Gaasbeek, *Rec. Trav. Chim.*, **89**, 1079 (1970).

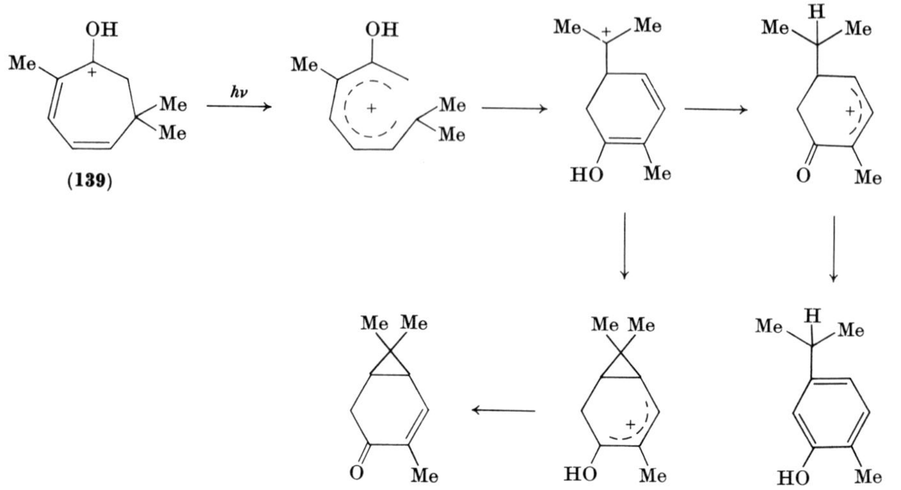

protonated eucarvone (**139**) gives (**140**) and (**141**),[341] which are thought to be formed by initial cycloreversion of (**139**). The products formed by irradiation of triphenylmethyl carbonium ions are also dependent on the pH of the system; thus, in 96% sulphuric acid cyclization to give the 9-phenylfluorenyl carbonium ion occurs,[342a,b] whereas in solutions of lower acidity substituted biphenyls are formed.[342b]

Irradiation of the cyclopentadienyl carbanion results in hydrogen-abstraction to give a cyclopentenyl radical.[343] In contrast, irradiation of cyclopentadiene brings about an electrocyclic reaction that yields bicyclo[2.1.0]pent-1-ene. Aryl-lithiums give biaryls on photolysis, and radical anions were suggested as intermediates.[344] The radical anion of diethyl maleate is isomerized on irradiation.[345]

Miscellaneous Compounds

The photoinduced Beckmann rearrangement of cycloalkanones oximes (ring size 4–15) has been studied;[346] unsymmetrical oximes were found to give equal mixtures of the two possible lactams. Optically active, α-substituted oximes give lactams with retention of configuration. Concerted breakdown of the intermediate oxazirane was postulated.

The decomposition of carbonyl sulphide, sensitized by aromatic hydrocarbons, produces triplet sulphur atoms.[347] The decay of triplet sulphur dioxide[348a] and reaction

[341] K. E. Hine and R. F. Childs, *J. Am. Chem. Soc.*, **93**, 2323 (1971).
[342a] D. M. Allen and E. D. Owen, *Chem. Comm.*, **1971**, 848 (1971).
[342b] E. E. van Tamelen and T. M. Cole, *J. Am. Chem. Soc.*, **93**, 6158 (1971).
[343] E. E. van Tamelen, J. I. Brauman, and L. E. Ellis, *J. Am. Chem. Soc.*, **93**, 6145 (1971).
[344] E. E. van Tamelen, J. I. Brauman, and L. E. Ellis, *J. Am. Chem. Soc.*, **93**, 6141 (1971).
[345] A. Torikai, T. Suzuki, T. Miyazaki, K. Fueki and, Z. Kuri, *J. Phys. Chem.*, **75**, 482 (1971).
[346] M. Cunningham, L. S. N. Lim, and G. Just, *Can. J. Chem.*, **49**, 2891 (1971).
[347] E. Leppin and K. Gollnick, *J. Am. Chem. Soc.*, **93**, 2848 (1971).
[348a] K. Otsuka and J. G. Calvert, *J. Am. Chem. Soc.*, **93**, 2581 (1971).

of excited sulphur dioxide with alkanes[348b] and alkenes[348c] have been investigated. The formation of a keto-sulphene by photolysis of the sulphonate has been reported.[349]

The direct[350a] and sensitized[350b] decomposition of dibenzoyl peroxide has been studied; in the case of sensitized decomposition by aromatic hydrocarbons, exciplex intermediates were postulated.[350b]

Reaction of triplet acetone with triethylborane causes quenching of the excited ketone and formation of ethyl radicals.[121,351] The photoisomerization of (**142**) to (**144**) is thought to involve (**143**).[352] There have been further investigations of biaryl formation

(**142**) (**143**) (**144**)

from the photolysis of tetra-aryl borates[353a] and triarylaluminiums.[353b] 1,1-Dimethyl-2,5-diphenyl-1-silacyclopentadiene dimerizes on irradiation.[354] Photolysis of (methoxycarbonyl)triphenylphosphonium bromide results in homolysis to phenyl and methoxycarbonyl radicals.[355] Flash-photolysis studies of tri- and di-phenylphosphine and of tetraphenyldiphosphine showed that diphenylphosphino radicals are produced on irradiation.[356] The oxidation of methanol, sensitized by diquaternary salts of 4,4'-bipyridine, results in formation of methoxyl radicals.[357] Photolysis of diarylsulphonamides produces azobenzenes, the reaction appearing to be intramolecular.[358] Irradiation of nitroxide radicals in toluene results in their reduction to hydroxylamines.[359] There have been a number of applications of the Barton and related reactions.[360]

[348b] C. C. Badcock, H. W. Sidebottom, J. C. Calvert, G. W. Reinhardt, and E. K. Damon, *J. Am. Chem. Soc.*, **93**, 3115 (1971).
[348c] H. W. Sidebottom, C. C. Badcock, J. G. Calvert, B. R. Rabe, and E. K. Damon, *J. Am. Chem. Soc.* **93**, 3121 (1971).
[349] S. T. Weintraub and B. F. Plummer, *J. Org. Chem.*, **36**, 361 (1971).
[350a] J. D. Bradley and A. P. Roth, *Tetrahedron Letters*, **1971**, 3907.
[350b] T. Nakata and K. Tokumaru, *Bull. Chem. Soc. Japan*, **43**, 3315 (1970).
[351] M. V. Encina and E. A. Lissi, *J. Organomet. Chem.*, **29**, 21 (1971).
[352] G. M. Clark, K. G. Hancock, and G. Zweifel, *J. Am. Chem. Soc.*, **93**, 1308 (1971).
[353a] P. J. Grisdale, J. L. R. Williams, M. E. Glogowski, and B. E. Babb, *J. Org. Chem.*, **36**, 544 (1971).
[353b] J. J. Eisch and J. L. Considine, *J. Organomet. Chem.*, **26**, 51 (1971).
[354] Y. Nakadaira and H. Sakurai, *Tetrahedron Letters*, **1971**, 1183.
[355] Y. Nagao, K. Shima, and H. Sakurai, *Tetrahedron Letters*, **1971**, 1101.
[356] S. K. Wong, W. Sytnyk, and J. K. S. Wan, *Can. J. Chem.*, **49**, 994 (1971).
[357] A. Ledwith, P. J. Russell, and L. H. Sutcliffe, *Chem. Comm.*, **1971**, 964.
[358] D. L. Forster, T. L. Gilchrist, and C. W. Rees, *J. Chem. Soc.* (C), **1971**, 993.
[359] J. F. Keana, R. J. Dinerstein, and F. Baitis, *J. Org. Chem.*, **36**, 209 (1971).
[360] G. Adam, D. Voigt, and K. Schreiber, *Tetrahedron*, **27**, 2181 (1971); H. Suginome, T. Mizuguchi, and T. Masamune, *Tetrahedron Letters*, **1971**, 4723; H. Reimann and O. Z. Sarre, *Can. J. Chem.*, **49**, 344 (1971); D. H. R. Barton, R. G. Sammes, and G. G. Weingarten, *J. Chem. Soc.* (C), **1971**, 729; H. Suginome, T. Kojima, K. Orito, and T. Masamune, *Tetrahedron*, **27**, 291 (1971); H. Suginome, N. Sato, and T. Masamune, *Tetrahedron*, **27**, 4863 (1971).

Other Photoreactions

Photosensitized Oxidation

Two helpful reviews dealing with the physical[361a] and chemical[361a,b] properties of singlet molecular oxygen have been published. Emission from the ($^1\Sigma_g^+$) state of oxygen has been detected after flash photolysis of a 1-fluoronaphthalene–oxygen mixture.[362] Absolute rate constants for the reaction of singlet molecular oxygen with a number of substrates have been determined;[363] these showed that endoperoxide formation from furans involves a much greater solvent effect than does hydroperoxide formation from alkenes, and that the lifetime of singlet oxygen was dependent on the viscosity of the solvent. Interest has been shown in the role of singlet oxygen in biological systems, e.g. in enzyme-catalysed oxygenation;[364] however, some doubt has been cast on the validity of these results,[365a] as also on the report that certain chromium complexes can act as sources of singlet oxygen.[365b]

Oxygen is known to quench the excited singlet and triplet states of aromatic hydrocarbons. By the use of high pressures of oxygen it has been shown that aromatic hydrocarbons and oxygen form charge-transfer complexes[366a] and that this causes static quenching of the excited states. Polymethoxybenzenes are oxidized to quinones on irradiation in the presence of oxygen.[366b] Alkanes[367a] and alkenes[367b] can also be photo-oxidized and in the former case a charge-transfer complex between the alkane and oxygen was postulated as an intermediate.[367a] Several examples of the oxidation of allylic[367b] and benzylic[368a,b] C—H bonds on direct irradiation has been reported.

In the case of (**145**), oxidation of the methyl in preference to a methylene group occurs whereas the latter group is readily oxidized in (**146**).[368b] Oxidation of ferrocene to the ferrocenium cation occurs on irradiation in carbon tetrachloride or chloroform solution, and reaction was shown to occur via a charge-transfer complex.[369]

Radical intermediates in the eosin-sensitized oxidation of phenols have been detected by ESR spectroscopy.[370a] It appears that oxidation of phenols can occur both by reaction with singlet oxygen and by radical reactions.[370b] Several examples of the dye-sensitized

---

[361a] D. R. Kearns, *Chem. Rev.*, **71**, 395 (1971).
[361b] E. F. J. Duynstee, *Chem. Weekbl.*, **67**, I, 37, 21 (1971).
[362] L. J. Andrews and E. W. Abrahamson, *Chem. Phys. Letters*, **10**, 113 (1971).
[363] R. H. Young, K. Wehrly, and R. L. Martin, *J. Am. Chem. Soc.*, **93**, 5774 (1971).
[364] H. W. S. Chan, *J. Am. Chem. Soc.*, **93**, 2357, 4632 (1971).
[365a] J. E. Baldwin, J. C. Swallow, and H. W. S. Chan, *Chem. Comm.*, **1971**, 1407.
[365b] H. W. S. Chan, *Chem. Comm.*, **1970**, 1550.
[366a] H. Ishida, H. Takahashi, and H. Tsubomura, *Bull. Chem. Soc. Japan*, **43**, 3130 (1970).
[366b] I. Saito and T. Matsuura, *Tetrahedron Letters*, **1970**, 4987.
[367a] N. Kulevsky, P. V. Sneeringer, L. D. Grina, and V. I. Stenberg, *Photochem. Photobiol.*, **12**, 395 (1970).
[367b] N. Friedman, M. Gorodetsky, and Y. Mazur, *Chem. Comm.*, **1971**, 874.
[368a] H. H. Wasserman, P. S. Mariano, and P. M. Keehn, *J. Org. Chem.*, **36**, 1765 (1971); H. B. Land and A. R. Frasca, *Tetrahedron*, **27**, 835 (1971); M. Chubachi and M. Hamada, *Tetrahedron Letters*, **1971**, 3537.
[368b] H. Shizuka, K. Sorimachi, T. Morita, K. Nishiyama, and T. Sato, *Bull. Chem. Soc. Japan*, **44**, 1983 (1971).
[369] O. Traverso and F. Scandola, *Inorg. Chim. Acta*, **4**, 493 (1970).
[370a] I. H. Leaver, *Austral. J. Chem.*, **24**, 891 (1971).
[370b] K. Pfoertner and D. Boese, *Helv. Chim. Acta*, **53**, 1553 (1970).

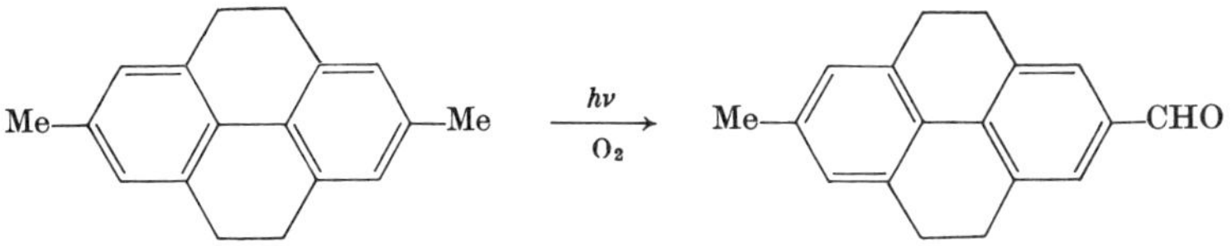

[Structure (145): dimethylpyrene → Me-pyrene-CHO under hν/O₂]

(145)

[Structure (146): pyrene derivative → two pyrene isomers under hν/O₂]

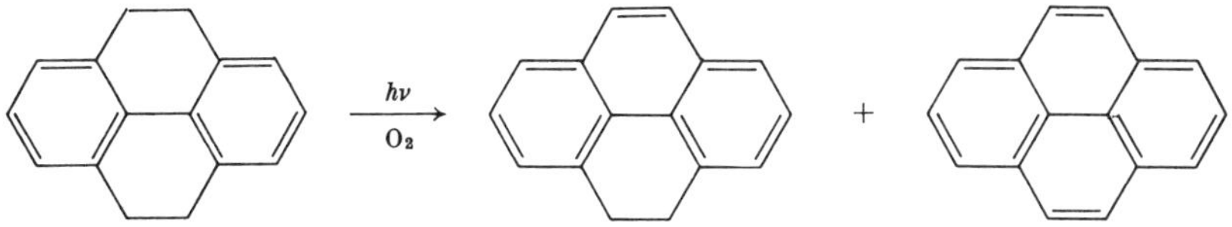

(146)

oxidation of alkenes to hydroperoxides have been reported[371a,b] and interest has been shown in the reactions that are of biochemical significance.[371b] Plastaquinone gives a variety of products on direct oxidation[371c] and it is of interest to know the role of the photoenol. Enol acetates, e.g. (147), are oxidized to enones via hydroperoxide inter-

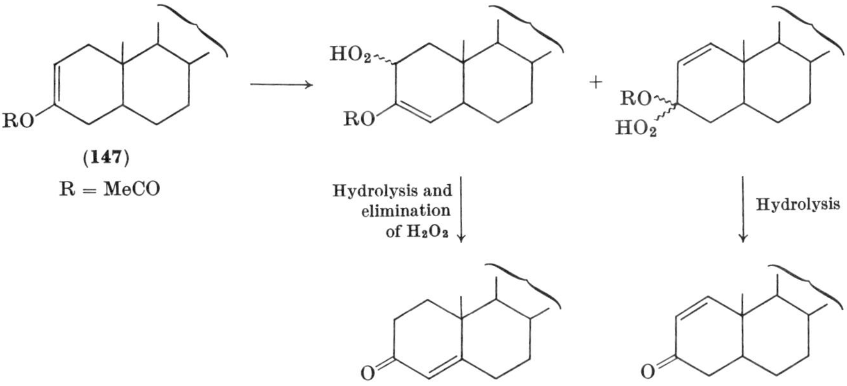

(147)
R = MeCO

Hydrolysis and elimination of H_2O_2

Hydrolysis

mediates.[371d] Dienol acetates are oxidized to a number of products via an endoperoxide. Systems of biological significance that have been studied include the oxidation of β-ionol derivatives to allenes[372a] and of β-carotene sensitized by hypercin.[372b]

Once again there has been much interest in the role of 1,2-dioxetans in oxygenation reactions. One intriguing observation related to the effect of added azide ion upon the course of oxygenation[373a] is that 1,2-dioxetans do not react with the azide anion to give

[371a] T. Matsuura, A. Horinaka, H. Yoshida, and Y. Butsugan, *Tetrahedron*, **27**, 3095 (1971); W. S. Gleason, I. Rosenthal, and J. N. Pitts, *J. Am. Chem. Soc.*, **92**, 7042 (1970); G. Rio and D. Bricout, *Bull. Soc. Chim. France*, **1971**, 3557; G. Rio and M. Charifi, *Bull. Soc. Chim France*, **1970**, 3598; G. W. K. Cavill and I. N. Coggiola, *Austral. J. Chem.*, **24**, 135 (1971); K. H. Schulte-Elte and B. L. Müller, *Helv. Chim. Acta*, **54**, 1870 (1971).
[371b] P. J. Dunphy, *Chem. Ind. (London)*, **1971**, 731.
[371c] D. Creed, H. Werbin, and E. T. Strom, *J. Am. Chem. Soc.*, **93**, 502 (1971).
[371d] J. Pusset, D. Guenard, and R. Beugelmans, *Tetrahedron*, **27**, 2939 (1971).
[372a] S. Isoe, S. Katsumura, S. Be Hyeon, and T. Sakan, *Tetrahedron Letters*, **1971**, 1089.
[372b] G. R. Seely and T. H. Meyer, *Photochem. Photobiol.*, **13**, 27 (1971).
[373a] *Org. Reaction Mech.*, **1970**, 566, ref. 408.

β-azido hydroperoxides but instead decompose to give carbonyl compounds;[373b] this raises the question whether the azide ion affects the course of the photo-oxygenation by by reacting with dye in its ground state. 1,2-Dioxetans have been isolated after some dye-sensitized photo-oxygenations, and studies have been made of their thermal decompositions.[374a,b] 1,2-Dioxetans appear to be intermediates in the oxidation of thujopsene,[375a] aryldihydropyrans[375b] and coumarones.[371c]

In the oxidation of 9,10-diphenylanthracenes having substituents in the 1- and the 4-position it was found that as the electron-donating power of the substituent increased so did the tendency to form an endoperoxide across the 1,4-positions.[376] Thermal decomposition of the endoperoxide was studied. Cycloheptatriene forms an endoperoxide across the 1,6-positions, i.e. a $6\pi + 2\pi$ cycloaddition appears to take place.[377] Endoperoxides from cyclopentadienes,[378a] buta-1,3-dienes[378b] and furans[378c] have been observed. The oxidation of some pyranoindolones to indoles appears to involve an endoperoxide intermediate.[378d] Photolysis of ascaridole causes rearrangement to a 1,2,3,4-diepoxycyclohexane.[379]

Interest has been shown in the oxidation of sulphides by singlet oxygen. Although efficient reaction between the two species occurs,[380a,b] only relatively few of the sulphide molecules are oxidized, i.e. the interaction of singlet oxygen with the sulphides results in quenching of singlet oxygen;[380a,b] it was shown very convincingly that a species such as (148) is formed in the reactions.[380a] The triphenyl phosphite–ozone adduct oxidizes disulphides, but singlet oxygen does not appear to be an intermediate.[381]

Irradiation of dimethyl sulphoxide causes an oxygen-transfer reaction that produces dimethyl sulphide and dimethyl sulphone.[382] Photo-oxidation of allylthiourea was shown to involve the conversion of a thiol into a sulphinic acid group.[383] Oxidation of thiopyran-4(4H)-thiones, e.g. (149),[384a] and sulphines [384b] is thought to occur by addition of singlet oxygen to the thiocarbonyl group; in contrast, oxidation of thiopyran-4(4H)-ones leads to ring-cleavage, probably via a 1,2-dioxetan formed across C-2 and C-3.[384c]

The mechanism of the photosensitized oxidation of amines has aroused interest

[373b] W. H. Richardson and V. F. Hodge, *Tetrahedron Letters*, **1971**, 749.
[374a] A. P. Schaap, *Tetrahedron Letters*, **1971**, 1757.
[374b] G. Rio and J. Berthelot, *Bull. Soc. Chim. France*, **1971**, 3555.
[375a] S. Ito, H. Takeshita, and M. Hirama, *Tetrahedron Letters*, **1971**, 1181.
[375b] R. S. Atkinson, *J. Chem. Soc.* (C), **1971**, 784.
[375c] G. Rio and J. Berthelot, *Bull. Soc. Chim. France*, **1971**, 1705.
[376] J. Rigaudy, J. Guillaume, and D. Maurette, *Bull. Soc. Chim. France*, **1971**, 144; J. Rigaudy, R. Dupont, and N. K. Cuong, *C. R. Acad. Sci. Paris, C*, **272**, 1678 (1971); J. Rigaudy, A. Defoin, and N. K. Cuong, *C. R. Acad. Sci. Paris, C*, **271**, 1258 (1970).
[377] A. S. Kende and J. Y. C. Chu, *Tetrahedron Letters*, **1970**, 4837.
[378a] W. R. Adams and D. J. Trecker, *Tetrahedron*, **27**, 2631 (1971); G. Rio and M. Charifi, *Bull. Soc. Chim. France*, **1970**, 3585.
[378b] G. Rio and J. Berthelot, *Bull. Soc. Chim France*, **1971**, 2938.
[378c] J. J. Basselier, J. Cherton, and J. Caille, *C. R. Acad. Sci. Paris, C*, **273**, 514, 1971.
[378d] J. Rokach, D. McNeill, and C. S. Rooney, *Chem. Comm.*, **1971**, 1085.
[379] K. K. Maheshwari, P. DeMayo, and D. Wiegand, *Can. J. Chem.*, **48**, 3265 (1970).
[380a] C. S. Foote and J. W. Peters, *J. Am. Chem. Soc.*, **93**, 3795 (1971).
[380b] R. A. Ackerman, I. Rosenthal, and J. N. Pitts, *J. Chem. Phys.*, **54**, 4960 (1971).
[381] R. W. Murray, R. D. Smetana, and E. Block, *Tetrahedron Letters*, **1971**, 299.
[382] K. Gollnick and H. U. Stracke, *Tetrahedron Letters*, **1971**, 203, 207.
[383] J. P. Dubosc, C. Mercier, and J. Bourdon, *Bull. Soc. Chim. France*, **1971**, 3286.
[384a] N. Ishibe, M. Odani, and M. Sunami, *Chem. Comm.*, **1971**, 118; *J. Chem. Soc.* (B), **1971**, 1837.
[384b] B. Zwanenburg, A. Wagenaar, and J. Strating, *Tetrahedron Letters*, **1970**, 4683.
[384c] N. Sugiyama, Y. Sato, and C. Kashima, *Bull. Chem. Soc. Japan*, **43**, 3205 (1970).

$$R_2S + {}^1\Delta O_2 \longrightarrow R_2\overset{+}{S}\!-\!O\!-\!\overset{-}{O} \xrightarrow{Ar_2S} R_2\overset{+}{S}\!-\!\overset{-}{O} + Ar_2\overset{+}{S}\!-\!\overset{-}{O}$$

(148)

$$\downarrow$$

$$R_2S + O_2$$

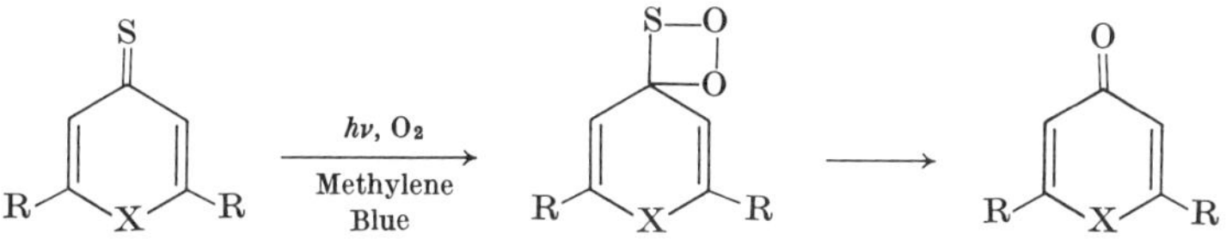

(149)

X = O or S

because two different mechanisms have been proposed. The oxidation reaction, which is sensitized by carbonyl compounds, dyes and aromatic hydrocarbons, has been proposed as occurring via singlet oxygen.[385] On the other hand, evidence has been presented[386a-c] that is consistent with a reaction occurring between the amines and the sensitizer, followed by reaction of the radicals so produced with oxygen. Evidence in favour of the latter mechanism includes the following: (a) only those amines are oxidized that can act as reductants for the excited sensitizer;[386a-c] (b) flash photolysis shows that amines react with dyes and aromatic hydrocarbons in the presence of oxygen;[386b,c] (c) the reactions sensitized by aromatic hydrocarbons showed a marked solvent effect;[386c] and (d) amines are known to be physical quenchers of singlet oxygen.[386d] The direct oxidation of aniline has been proposed as involving singlet oxygen[387] although it is well known that aromatic amines and oxygen form charge-transfer complexes. The oxidation taking place when amines are irradiated in chloroform solution is interpreted as occurring via radicals generated from the complex of the amine with solvent.[388] The eosin-sensitized oxidation of tyrosine and tryptophan has been shown to involve reaction of excited dye with the amino-acids;[389] oxidation of tryptophan results in cleavage of the pyrrole ring, probably via a 1,2-dioxetan formed across C-2 and C-3;[390a,b] the type of products formed in these reactions is pH-dependent. Oxidations of a number of pyrroles[391a] and

---

[385] M. H. Fisch, J. C. Gramain, and J. A. Olesen, *Chem. Comm.*, **1971**, 663.
[386a] R. F. Bartholomew and R. S. Davidson, *J. Chem. Soc.* (C), **1971**, 2342; R. F. Bartholomew, R. S. Davidson, and M. J. Howell, *J. Chem. Soc.* (C), **1971**, 2804.
[386b] R. F. Bartholomew and R. S. Davidson, *J. Chem. Soc.* (C), **1971**, 2347.
[386c] R. F. Bartholomew, D. R. G. Brimage, and R. S. Davidson, *J. Chem. Soc.* (C), **1971**, 3482.
[386d] *Org. Reaction Mech.*, **1970**, 566.
[387] B. Pouyet and R. Chapelon, *C. R. Acad. Sci. Paris, C*, **272**, 1753 (1971).
[388] E. A. Fitzgerald, P. Wuelfing, and H. H. Richtol, *J. Phys. Chem.*, **75**, 2737 (1971).
[389] A. G. Kepka and L. I. Grossweiner, *Photochem. Photobiol.*, **14**, 621 (1971).
[390a] G. Jori and G. Galiazzo, *Photochem. Photobiol.*, **14**, 607 (1971).
[390b] W. E. Savige, *Austral. J. Chem.*, **24**, 1285 (1971).
[391a] G. B. Quistad and D. A. Lightner, *Tetrahedron Letters*, **1971**, 4417; G. Rio and A. Nawrocka-Lecas, *Bull. Soc. Chim. France*, **1971**, 1723; A. Ranjon, *Bull. Soc. Chim. France*, **1971**, 2068; R. W. Franck and J. Auerbach, *J. Org. Chem.*, **36**, 31 (1971); G. B. Quistad and D. A. Lightner, *Chem. Comm.*, **1971**, 1099.

pyrazolines[391b] have been studied; the former are thought to involve endoperoxides as intermediates. In view of the interest in the mechanism of oxidation of purines, pyrimidines[392a] and nucleotides[392a,c] it is interesting that flavines [392b] and Methylene Blue[392c] sensitize the oxidation of nucleotides by a mutual reaction with the substrates and not by serving as sensitizers for singlet oxygen formations. Porphyrins sensitize the methionine residues in lysozyme.[393] Oxidation of phenylalanylglycine results in cleavage of the glycine residue.[394] Carboxylic acids and alcohols are oxidized on irradiation in the presence of a number of transition-metal ions.[395]

Chemiluminescence and Electrochemiluminescence

Utilization of the electronic energy released by thermal decomposition of 1,2-dioxetans continues to attract attention[374a,b] A number of lanthanide chelates have proved efficient acceptors for the energy released on decomposition of trimethyl-1,2-dioxetan.[396] Decomposition of the 1,2-dioxetan derived from 1,2-diethoxyethylene produces ethyl formate, the energy of activation for this process being 24 kcal mol^{-1}. Decomposition in the presence of 9,10-diphenylanthracene, pyrene or 9,10-dibromoanthracene leads to fluorescence of the energy-acceptors.[397] From oxygen-quenching and other studies it was concluded that triplet–singlet energy-transfer was involved; it was conclusively shown that triplet–triplet annihilation was not involved. Thermal decomposition of phthaloyl peroxide in phthalate esters containing dissolved 9,10-diphenylanthracene causes the latter to fluoresce;[398] emission was not observed when other solvents were used.

The electronic energy released by electron-transfer between a radical cation and a radical anion has been the subject of a number of thorough studies.[399,400a–c] Reference has already been made to those systems in which exciplex emission was observed.[63,64] In the electrochemiluminescent reaction of amines with aromatic hydrocarbons fluorescence of the hydrocarbon could be observed only if sufficient energy was released in the redox reaction to populate the triplet state of the hydrocarbon.[399] A similar conclusion was reached from a study of the reaction of radical ions generated by chemical methods.[64] Further confirmatory evidence for the theory that excited singlet states of the aromatic hydrocarbons are generated by triplet–triplet annihilation comes from the observation *trans*-stilbene is able to intercept the latter process and as a result the stilbene isomerized to *cis*-stilbene.[400a] Use of this reaction permitted calculation of the quantum yield of triplet formation for a number of electron-transfer reactions. Other investigated chemiluminescent reactions include oxidation of substituted phthalhydrazides[401a] and

[391b] L. Schrader, *Tetrahedron Letters*, **1971**, 2977.
[392a] M. Rippa, C. Picco, and S. Pontremoli, *J. Biol. Chem.*, **245**, 4977 (1970).
[392b] A. Knowles, *Photochem. Photobiol.*, **13**, 225 (1971).
[392c] A. Knowles, *Photochem. Photobiol.*, **13**, 473 (1971).
[393] G. Jori, G. Galiazzo, and E. Scoffone, *Experientia*, **27**, 379 (1971).
[394] O. H. Wheeler, D. A. Julian, and R. A. Ribot, *Photochem. Photobiol.*, **12**, 505 (1970).
[395] H. D. Burrows, D. Greatorex, and T. J. Kemp, *J. Am. Chem. Soc.*, **93**, 2539 (1971).
[396] P. D. Wildes and E. H. White, *J. Am. Chem. Soc.*, **93**, 6286 (1971).
[397] T. Wilson and A. P. Schaap, *J. Am. Chem. Soc.*, **93**, 4126 (1971).
[398] K. D. Gundermann, M. Steinfatt, and H. Fiege, *Angew. Chem. Int. Ed.*, **10**, 67 (1971).
[399] D. J. Freed and L. R. Faulkner, *J. Am. Chem. Soc.*, **93**, 2097 (1971).
[400a] D. J. Freed and L. R. Faulkner, *J. Am. Chem. Soc.*, **93**, 3565 (1971).
[400b] F. E. Lytle and D. M. Hercules, *Photochem. Photobiol.*, **13**, 123 (1971).
[400c] T. Matsumoto, M. Sato, S. Hirayam, and S. Uemura, *Bull. Chem. Soc. Japan*, **44**, 1450 (1971).
[401a] C. C. Wei and E. H. White, *Tetrahedron Letters*, **1971**, 3559.

related compounds,[401b] bioluminescent systems,[401c] and oxidation of purpurogallin[401d] and dimedone.[401e]

Solvolysis and Substitution Reactions

Irradiation of *p*-quinol in deuterium oxide leads to replacement of the ring-protons by deuterium,[402] reaction being mainly from the triplet state of the quinol; it was postulated that electrophilic attack by D^+ gives a Wheland-type intermediate that can deprotonate to give the deuteriated quinol. *p*-Nitroanisole (**150**) undergoes substitution by the

[structures: (**150**) 4-nitroanisole → KCNO → (**151**) 2-(NHCO$_2$H)-4-nitroanisole → 2-amino-4-nitroanisole]

cyanate ion to give a carbamic acid (**151**);[403] with methanol as solvent, a methyl carbamate was formed. This reaction is analogous to that with the cyanide ion. Nitrobenzene is reduced to aniline on irradiation in the presence of borohydride anion, whereas 1-nitronaphthalene undergoes replacement to give naphthalene.[404] 1-Amino-4-bromo-9,10-anthraquinone-2-sulphonate undergoes an interesting replacement reaction with ammonia or primary alkylamines to give 4-aminoanthraquinones.[405] Trifluoromethylnaphthols are readily hydrolysed to naphthoic acids on irradiation,[406] reaction occurring from the excited singlet state of the trifluoromethyl compound; the rate constants for reaction of a number of closely related compounds fell between 10^7 and 10^8 sec^{-1}. 5-Dimethylaminonaphthalene-1-sulphonamides derived from α-amino-acids and polypeptides undergo ready hydrolysis on irradiation.[407]

Photochromism

Reference has already been made to one of the reported[408] photochromic systems that involves a proton-transfer reaction of an excited state. The emission properties of some photochromic spiropyrans have also been studied.[409] The phosphorescence observed on excitation of photochromic chromenes emanates from the chromene and not from the coloured cycloreversion product.[410] The structures of some photochromic isomers of bianthrones have been determined.[411]

[401b] K. D. Gundermann and H. Fiege, *Ann. Chem.*, **743**, 200 (1971).
[401c] N. Suzuki and T. Goto, *Tetrahedron Letters*, **1971**, 2021.
[401d] J. Slawinski, *Photochem. Photobiol.*, **13**, 489 (1971).
[401e] J. Beutel, *J. Am. Chem. Soc.*, **93**, 2615 (1971).
[402] G. F. Vesley, *J. Phys. Chem.*, **75**, 1775 (1971).
[403] J. Hartsuiker, S. DeVries, J. Cornlisse and E. Havinga, *Rec. Trav. Chim.*, **90**, 611 (1971).
[404] W. C. Petersen and R. L. Letsinger, *Tetrahedron Letters*, **1971**, 2197.
[405] H. Inoue, T. D. Tuong, M. Hida, and T. Murata, *Chem. Comm.*, **1971**, 1347.
[406] P. Seiler and J. Wirz, *Tetrahedron Letters*, **1971**, 1683.
[407] L. DeSouza, K. Bhatt, B. M. Hadaiah, and R. A. Day, *Arch. Biochem. Biophys.*, **141**, 690 (1970).
[408] G. M. Wyman, *Chem. Comm.*, **1971**, 1332; D. L. Williams and A. Heller, *J. Phys. Chem.*, **74**, 4473 (1971); R. Pater, *J. Heterocyclic Chem.*, **7**, 1113 (1970).
[409] C. Balny, M. Mosse, C. Audic, and A. Hinnen, *J. Chim. Phys. Physicochim. Biol.*, **68**, 1078 (1971).
[410] J. Kolc and R. S. Becker, *Photochem. Photobiol.*, **12**, 383 (1970).
[411] R. Korenstein, K. A. Muszkat, and E. Fischer, *Helv. Chim. Acta*, **53**, 2102 (1970).

CHAPTER 14

Oxidation and Reduction

M. J. P. HARGER

Department of Chemistry, The University, Leicester

Ozonation and Ozonolysis	527
Oxidation by Metallic Ions	529
Oxidation by Molecular Oxygen	537
Other Oxidations	541
Reductions	546
Hydrogenation and Hydrogenolysis	555

Ozonation and Ozonolysis

A π-complex has frequently been postulated as the first intermediate in the reaction of an olefin with ozone. Such a complex, apparently involving the olefinic π-system, has now been detected spectroscopically at $-150°$ with α-mesitylstyrene (a hindered olefin) and ozone.[1]

In both the modified Criegee fragmentation mechanism[2] and the primary ozonide–aldehyde hypothesis[3] a 1,2,3-trioxolane primary ozonide is considered to be an intermediate in the conversion of an (unhindered) olefin into its normal ozonide. Although the primary ozonide might be formed directly from the olefin (or its π-complex with ozone) by 1,3-dipolar cycloaddition, new results suggest that, in some cases at least, this is not so. Ozonation of ethylidenecyclohexane (**1**) in pentane at $-78°$ gives normal ozonide (**3**) in 85% yield. In the presence of an equimolar quantity of propionaldehyde the expected mixture of (**3**) and crossed ozonide (**3**; Et in place of Me) is obtained in good yield, but with increased concentrations of aldehyde the total yield of ozonide is reduced. No ozonide is obtained with a 4-molar excess of propionaldehyde, the products then being cyclohexanone (85%) and acetaldehyde (62%), together with propionic acid (80%). A suggested mechanism is in Scheme 1 (p. 528). Initial addition of ozone gives the peroxy epoxide (**4**) (possibly by way of a π-complex) which rearranges to the Staudinger ozonide (**5**). This leads to the normal ozonide (**3**) by way of 1,2,3-trioxolane primary ozonide (**2**) and possibly, to some extent, also by other mechanisms. In any case, formation of the normal ozonide must compete with Baeyer–Villiger reaction of ozonide (**5**) with propionaldehyde, leading to a dioxetane (**6**) which would rapidly fragment to cyclohexanone and acetaldehyde. The high yield of normal ozonide obtained when acetaldehyde or acetone is present in place of propionaldehyde is attributed to the inefficiency of those compounds as traps for (**5**), as a result, respectively, of a high degree of association and relatively low reactivity. These results, which can be explained

---

[1] P. S. Bailey, J. W. Ward, and R. E. Hornish, *J. Am. Chem. Soc.*, **93**, 3552 (1971).
[2] N. L. Bauld, J. A. Thompson, C. E. Hudson, and P. S. Bailey, *J. Am. Chem. Soc.*, **90**, 1822 (1968); *Org. Reaction Mech.*, **1968**, 466–467.
[3] P. R. Story, R. W. Murray, and R. D. Youssefyeh, *J. Am. Chem. Soc.*, **88**, 3144 (1966); *Org. Reaction Mech.*, **1968**, 399–401.

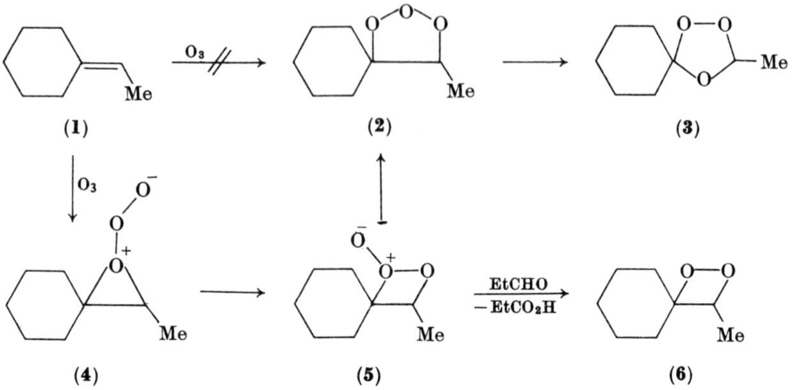

Scheme 1

equally well in terms of the reaction of peroxy epoxide (**4**) with propionaldehyde, seem incompatible with direct formation of the primary ozonide (**2**).[4]

Ozonolysis of *cis*- or *trans*-di-isopropylethylene in pentane gives a mixture of *cis*- and *trans*-ozonide (**7**). Variations with temperature of the isomer ratio of (**7**), and of crossed ozonide (**8**) formed in the presence of propionaldehyde, are consistent with competing mechanisms for the formation of normal ozonide. Moreover, the ozonide (**7**) obtained in the presence of [^{18}O]isobutyraldehyde contains some ^{18}O in the ether linkage, as predicted by the modified Criegee fragmentation mechanism, and some in the peroxide linkage as expected from the primary ozonide–aldehyde hypothesis. Significantly, at lower temperatures, where the reaction of primary ozonide with aldehyde would be expected to compete more efficiently with its fragmentation, the proportion of the label in the peroxide linkage is greater.[5] Several other aspects of the ozonation of olefins,[6] aromatic hydrocarbons,[7] and acetylenes,[8] and the reactions of ozonides,[9] have been discussed, including the formation of crossed ozonides from tetraphenylethylene and various ketones,[10] and the chromatographic resolution of the enantiomers of *trans*-ozonides.[11]

An example of ozonide formation in the absence of ozone is provided by the photo-oxidation of diphenyldiazomethane in acetaldehyde, which yields ozonide (**9**; 7.8%) in place of the benzophenone diperoxide (**10**) formed in the absence of aldehyde. Doubtless the Criegee zwitterion ($Ph_2C^+-O-O^- \leftrightarrow Ph_2C=O^+-O^-$) is responsible for both ozonide and diperoxide; their low yields might be a consequence of initial formation of carbonyl oxide as the diradical ($Ph_2\dot{C}-O-\dot{O}$), most of which reacts by some path other than conversion into zwitterion.[12]

[4] P. R. Story, J. A. Alford, J. R. Burgess, and W. C. Ray, *J. Am. Chem. Soc.*, **93**, 3042, 3044 (1971).
[5] R. W. Murray and R. Hagen, *J. Org. Chem.*, **36**, 1098, 1103 (1971).
[6] S. McLean and U. O. Trotz, *Can. J. Chem.*, **49**, 863 (1971); C. A. Grob and H. R. Pfaendler, *Helv. Chim. Acta*, **53**, 2156 (1970); H. J. Storesund and E. Bernatek, *Acta Chem. Scand.*, **24**, 3237 (1970).
[7] M. G. Sturrock, B. J. Cravy, and V. A. Wing, *Can. J. Chem.*, **49**, 3047 (1971).
[8] W. B. DeMore, *Int. J. Chem. Kinet.*, **3**, 161 (1971); *Chem. Abs.*, **74**, 99260 (1971).
[9] R. M. Ellam and J. M. Padbury, *Chem. Comm.*, **1971**, 1094; R. Criegee and H. Korber, *Chem. Eng. News*, **49** (No. 15), 31 (1971).
[10] R. Criegee and H. Korber, *Chem. Ber.*, **104**, 1812 (1971).
[11] R. Criegee and H. Korber, *Chem. Ber.*, **104**, 1807 (1971).
[12] R. W. Murray and A. Suzui, *J. Am. Chem. Soc.*, **93**, 4963 (1971).

Oxidation and Reduction

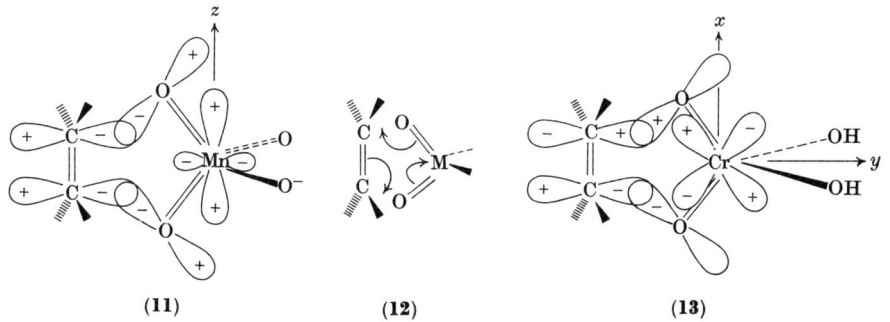

Insertion of ozone into the $C_\alpha-H$ bond is probably the first step in the oxidation of an acetal to the corresponding ester in CH_2Cl_2 [reaction (1)].[13]

Studies of the ozonation of aldehydes,[14] tertiary aromatic amines,[15] and organic derivatives of Group IV metals[16] have been reported. A dialkyl disulphide (RS—SR) is oxidized to the corresponding thiolsulphinate (RSO—SR) by the adduct of ozone and $(PhO)_3P$ in CH_2Cl_2. Since oxidation occurs below the normal temperature at which $(PhO)_3PO_3$ decomposes with liberation of $^1O_2$, it seems that the disulphide catalyses the decomposition, or that oxidation occurs by bimolecular reaction of the disulphide and $(PhO)_3PO_3$.[17]

Oxidation by Metallic Ions

The oxidation of several classes of organic compound by metal ions and other oxidants has been discussed with reference to orbital symmetry. For example, it has been argued that the concerted mechanism (12) for the oxidation of an olefin is "allowed" when permanganate is the oxidant but "forbidden" for oxidation by chromate. If we accept that reduction of Cr(VI) gives octahedral Cr(IV) the two electrons will enter a low energy t_{2g} orbital (d_{xy}) of the metal. On the other hand, when permanganate is reduced to tetrahedral Mn(V), the electrons will enter the d_{z^2} orbital. An orbital diagram (11) for Mn(VII) shows that (whether the d_{z^2} orbital is parallel, as shown, or perpendicular to

---

[13] P. Deslongchamps and C. Moreau, *Can. J. Chem.*, **49**, 2465 (1971).
[14] A. A. Syrov and V. K. Tsyskovskii, *Zhur. Org. Khim.*, **6**, 1392 (1970); *Chem. Abs.*, **73**, 87242 (1970).
[15] G. H. Kerr and O. Meth-Cohn, *J. Chem. Soc.* (C), **1971**, 1369.
[16] Y. A. Alexandrov, *Organomet. Chem. Rev. A*, **6**, 209 (1970); Y. A. Alexandrov and N. G. Sheyanov, *Zhur. Obshch. Khim.*, **40**, 1664 (1970); *Chem. Abs.*, **74**, 140565 (1971).
[17] R. W. Murray, R. D. Smetana, and E. Block, *Tetrahedron Letters*, **1971**, 299.

the olefinic bond) the reacting orbitals overlap in a bonding sense (Hückel-like transition state). By contrast with Cr(VI) (**13**) there is one antibonding overlap of reactant orbitals (Möbius-like transition state). Since the concerted mechanism (**12**) involves 6 electrons (i.e. $4n + 2$), it is "allowed" for Mn(VII), with a Hückel transition state, but is "forbidden" for Cr(VI) with a Möbius transition state.[18]

Equations (2) and (3) show the initial stages in the oxidation of a secondary alcohol by chromic acid, in which Cr(VI) is reduced to Cr(IV). The subsequent steps may be represented by reactions (4) and (5), in which Cr(IV) does not react directly with alcohol or by equations (6)–(8), in which it does. In the former case, formation of 2 equivalents of Cr(V) is accompanied by the formation of 1 equivalent of acetone while, in the latter, formation of 1 equivalent of Cr(V) is accompanied by formation of 2 equivalents of acetone. Using spectrometrically determined rate constants for the disappearance of Cr(VI) and Cr(V) during oxidation of Pr^{i}OH in 97% AcOH, the predicted concentration of

$$R_2CHOH + HCrO_4^- + H^+ \rightleftharpoons R_2CHOCrO_3H + H_2O \quad \ldots (2)$$

$$R_2CHOCrO_3H \rightarrow R_2CO + Cr(IV) \quad \ldots (3)$$

$$Cr(IV) + Cr(VI) \rightarrow 2Cr(V) \quad \ldots (4)$$

$$2Cr(V) + 2R_2CHOH \rightarrow 2Cr(III) + 2R_2CO \quad \ldots (5)$$

$$R_2CHOH + Cr(IV) \rightarrow R_2\dot{C}OH + Cr(III) \quad \ldots (6)$$

$$R_2\dot{C}OH + Cr(VI) \rightarrow R_2CO + Cr(V) \quad \ldots (7)$$

$$R_2CHOH + Cr(V) \rightarrow R_2CO + Cr(III) \quad \ldots (8)$$

$$Cr(IV) + 2Ce(IV) \rightarrow Cr(VI) + 2Ce(III) \quad \ldots (9)$$

$$Cr(IV) + Ce(III) \rightarrow Cr(III) + Ce(IV) \quad \ldots (10)$$

acetone as a function of time has been deduced for the two schemes. The observed accumulation of acetone is in accord with the predictions of equations (6)–(8).[19] This mechanism can also account for the formation of polymer during the oxidation (in aqueous solution) when acrylamide is present, and, on the assumption that chain termination is analogous to reaction (7) in that a radical is oxidized by Cr(VI) giving Cr(V), the decreasing yield of acetone approaching a limiting value of 67% as the concentration of monomer is increased.[20] Further direct participation of the Cr(IV) formed in the initial stages can be almost totally suppressed by addition of very small quantities of either Ce(III) or Ce(IV), presumably because both reactions (9) and (10) are operative.[21]

The interaction of Cr(IV) and alcohols [cf. reaction (6)] has been studied in detail by using the reduction of Cr(VI) by V(IV) [reactions (11) and (12)] to generate the oxidant. Rate-determining transfer of a H atom from the substrate (one-electron oxidation) within a Cr(IV)–alcohol complex [reaction (13)] is consistent with the observed lack of reactivity of Bu^{t}OH, the variation in rate in a series of primary alcohols (Taft $\rho^* = -0.85$), and the small value of the primary isotope effect ($k_H/k_D = 1.9$ for Me$_2$CDOH) which is similar to the values found with other one-electron oxidants. In this system the resulting radical is further oxidized to product by V(V) [reaction (14)] rather than by Cr(VI) [reaction (7)].[22]

[18] J. S. Littler, *Tetrahedron*, **27**, 81 (1971).
[19] K. B. Wiberg and S. K. Mukherjee, *J. Am. Chem. Soc.*, **93**, 2543 (1971).
[20] M. Rahman and J. Roček, *J. Am. Chem. Soc.*, **93**, 5462 (1971).
[21] M. P. Doyle, R. J. Swedo, and J. Roček, *J. Am. Chem. Soc.*, **92**, 7599 (1970).
[22] M. Rahman and J. Roček, *J. Am. Chem. Soc.*, **93**, 5455 (1971).

$$Cr(VI) + V(IV) \rightleftharpoons Cr(V) + V(V) \quad \ldots (11)$$

$$Cr(V) + V(IV) \rightarrow Cr(IV) + V(V) \quad \ldots (12)$$

$$\begin{array}{c} R \diagdown H \\ C \\ R \diagup O \\ | \\ H \end{array} + Cr^{IV} \rightleftharpoons \begin{array}{c} R \diagdown H \diagup OH \\ C \diagdown Cr^{IV} \\ R \diagup O \\ | \\ H \end{array} \xrightarrow{slow} \begin{array}{c} R \diagdown OH_2 \\ \overset{\bullet}{C} \diagdown Cr^{III} \\ R \diagup O \\ | \\ H \end{array} \quad \ldots (13)$$

$$R_2\overset{\bullet}{C}OH + V(V) \rightarrow R_2CO + V(IV) \quad \ldots (14)$$

Other aspects of the chromic acid oxidation of alcohols,[23] including α-hydroxy acids[24] and esters,[25] have been examined; the anomalously slow oxidation of benzocyclobutenol has been attributed to increased angle strain in the transition state, associated with the change in hybridization $sp^3 \rightarrow sp^2$ at C_α.[26] Further studies of the oxidation of aldehydes,[27] ketones,[28] carboxylic acids,[29] carbohydrates,[30] and hydrocarbons,[31] including olefins[32] and acetylenes,[33] by Cr(VI) have also been described. Oxidation of sodium toluene-p-sulphonate by aqueous $Na_2Cr_2O_7$ (pH 5.3) exhibits a primary isotope effect ($k_H/k_D = 4.6$ for $CD_3C_6H_4SO_3Na$ at 25°). A mechanism has been proposed in which the active oxidant is $HCrO_4^-$ and the rate-determining C—H bond cleavage is homolytic.[34]

Chromyl acetate, in common with mixed-function oxygenases, has the ability to hydroxylate aliphatic hydrocarbons with (partial) retention of configuration and to epoxidize olefins stereospecifically. A further analogy, hydroxylation of aromatic substrates with concomitant NIH shift, has now been demonstrated. Reaction of [1-³H,1-¹⁴C]naphthalene (³H/¹⁴C = 1.00) with $CrO_2(OAc)_2$ in CCl_4 gives naphthaquinone

---

[23] A. S. Perlin, *Can. J. Chem.*, **49**, 1972 (1971); R. Baker and T. J. Mason, *J. Chem. Soc. (B)*, **1971**, 988; V. S. Srinivasan, S. Sundaram, and N. Venkatasubramanian, *Proc. Indian Acad. Sci., Sect. A*, **72**, 261 (1970); *Chem. Abs.*, **74**, 140560 (1971); K.-H. Heckner, K.-H. Grupe, and R. Landsberg, *J. Prakt. Chem.*, **313**, 161 (1971); *Chem. Abs.*, **75**, 129026 (1971).

[24] K. K. Sengupta, A. K. Chatterjee, and S. P. Moulik, *Bull. Chem. Soc. Japan*, **43**, 3841 (1970); K. K. Sengupta, A. K. Chatterjee, B. B. Pal, and N. Sasmal, *Z. Phys. Chem. (Frankfurt)*, **72**, 330 (1970); S. Sundaram and N. Venkatasubramanian, *Indian J. Chem.*, **8**, 1104 (1970); K.-H. Heckner, K.-H. Grupe, and R. Landsberg, *Z. Phys. Chem. (Leipzig)*, **247**, 91 (1971).

[25] G. V. Bakore and C. L. Jain, *Z. Phys. Chem. (Leipzig)*, **245**, 1 (1970); D. S. Jha and G. V. Bakore, *J. Chem. Soc. (B)*, **1971**, 1166; *J. Indian Chem. Soc.*, **48**, 645 (1971); *Chem. Abs.*, **75**, 117759 (1971).

[26] P. Müller, *Helv. Chim. Acta*, **53**, 1869 (1970); **54**, 2000 (1971).

[27] K. K. Banerji and C. Goswami, *Tetrahedron Letters*, **1970**, 5039; C. Goswami and K. K. Banerji, *Z. Naturforsch.*, **26b**, 383 (1971).

[28] L. R. Subramanian and G. S. K. Rao, *J. Indian Inst. Sci.*, **52**, 112 (1970); *Chem. Abs.*, **74**, 75849, (1971); S. K. Tandon, K. K. Banerji, and G. V. Bakore, *Current Sci.*, **40**, 84 (1971); *Chem. Abs.*, **74**, 99266 (1971); *Indian J. Chem.*, **9**, 677 (1971); *Chem. Abs.*, **75**, 87852 (1971).

[29] M. A. Beg, A. A. Khan, and S. M. Khan, *Z. Phys. Chem. (Leipzig)*, **245**, 17 (1970); K. K. Sengupta and A. K. Chatterjee, *Z. Phys. Chem. (Frankfurt)*, **75**, 15 (1971).

[30] M. Bertolini and C. P. J. Glaudemans, *Carbohydrate Res.*, **17**, 449 (1971); S. Chandra and R. K. Mittal, *Carbohydrate Res.*, **19**, 123 (1971).

[31] N. Narayanankutty and M. V. Bhatt, *Tetrahedron Letters*, **1971**, 2121; H. H. Szmant, J. Colón, and J. Castrillón, *J. Org. Chem.*, **36**, 573 (1971).

[32] P. S. Kalsi, K. S. Kumar, and M. S. Wadia, *Chem. Ind. (London)*, **1971**, 31; U. Schwieter, W. Arnold, W. E. Oberhänsli, N. Rigassi, and W. Vetter, *Helv. Chim. Acta*, **54**, 2447 (1971).

[33] J. E. Shaw and J. J. Sherry, *Tetrahedron Letters*, **1971**, 4379.

[34] D. G. Lee and U. A. Spitzer, *Can. J. Chem.*, **49**, 2763 (1971).

with a ^{3}H/^{14}C ratio (0.70) substantially greater than 0.50. Clearly, the product is not an equimolar mixture of (14) and (15; X = H) but also contains some (15; X = T), formed most probably by way of the epoxide (16) and naphthol (17).[35]

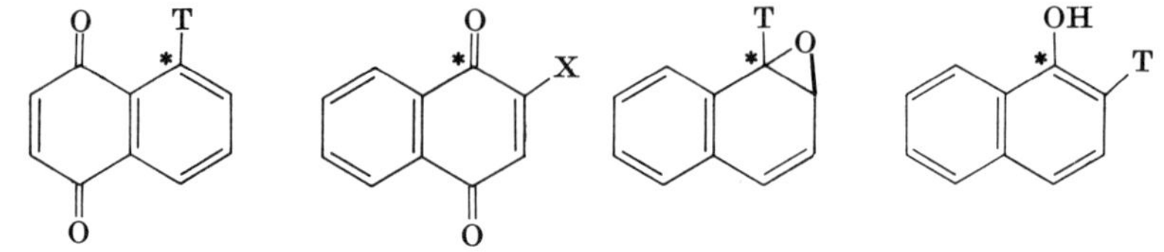

(14) (15) (16) (17)

Several permanganate oxidations have been investigated,[36] including those of diols,[37] fumaric acid, where an intermediate Mn(III) oxalato complex has been detected,[38] and alkenes (RCH=CHR), which in Ac$_2$O give good yields of α-diketones directly.[39] cis-2-Hydroxycyclohexanecarboxylic acid (18) is oxidized to the keto-acid by permanganate at pH 3–9 much faster than is the *trans*-isomer, and exhibits a bell-shaped pH–rate profile with a maximum at pH ~6. A mechanism consistent with the detailed

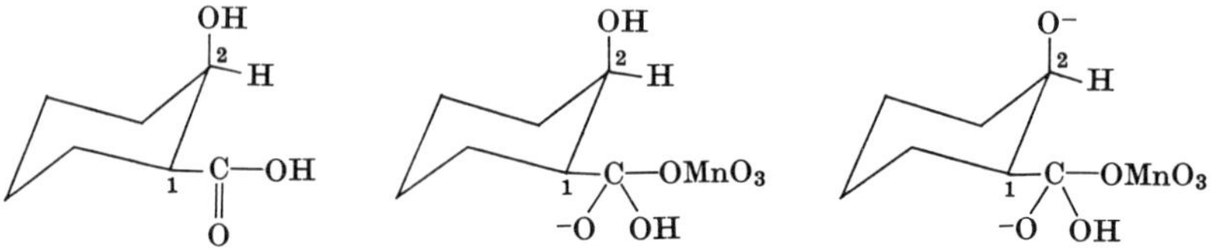

(18) (19) (20)

kinetic analysis involves reaction of the un-ionized substrate (18) with MnO$_4^-$ to give anion (19), in low concentration, which then dissociates to dianion (20). The rate-determining step, in agreement with the observed primary isotope effect [$k_H/k_D = 7$–8 for (18) with D at C-2] is the decomposition of (20) with intramolecular removal of H from C-2. Similar decomposition of the corresponding dianion from the *trans*-isomer is thought to be sterically unfavourable.[40]

The mechanism of the oxidation of benzenehexol on the surface of active MnO$_2$ has been discussed,[41] as have the oxidations of alkyl aryl sulphides,[42] alcohols,[43] alde-

---

[35] K. B. Sharpless and T. C. Flood, *J. Am. Chem. Soc.*, **93**, 2316 (1971).
[36] F. Freeman and D. K. Lin, *J. Org. Chem.*, **36**, 1335 (1971); P. V. Subba Rao, *Z. Phys. Chem.* (*Leipzig*), **246**, 352 (1971); R. S. Goody, A. S. Jones, and R. T. Walker, *Tetrahedron*, **27**, 65 (1971).
[37] P. Nath, K. K. Banerji, and G. V. Bakore, *Indian J. Chem.*, **8**, 1113 (1970); *Chem. Abs.*, **74**, 124517 (1971); *J. Indian Chem Soc.*, **48**, 17 (1971); *Chem. Abs.*, **74**, 99251 (1971).
[38] L. I. Simándi and M. Jáky, *Inorg. Nuclear Chem. Letters*, **7**, 605 (1971).
[39] K. B. Sharpless, R. F. Lauer, O. Repič, A. Y. Teranishi, and D. R. Williams, *J. Am. Chem. Soc.*, **93**, 3303 (1971).
[40] R. Stewart and J. A. MacPhee, *J. Am. Chem. Soc.*, **93**, 4271 (1971).
[41] A. J. Fatiadi, *J. Chem. Soc.* (B), **1971**, 889.
[42] J. R. Gilmore and J. M. Mellor, *Tetrahedron Letters*, **1971**, 3977.
[43] C. F. Wells and C. Barnes, *J. Chem. Soc.* (A), **1971**, 430, 1405; *Trans. Faraday Soc.*, **67**, 3297 (1971).

hydes,[44] and ketones[45] by Mn(III). Acetone reacts with an olefin and Mn(III) acetate in AcOH–KOAc to give a γ-keto-acetate. Unlike the α-keto radical (21), which is apparently resistant to further oxidation by virtue of the electron-withdrawing effect of the carbonyl group, the secondary alkyl radical (22) does form a carbonium ion (23), and thence the γ-keto-acetate. As would be expected, elimination of a proton from (23), leading to unsaturated ketone, competes with its reaction with acetate.[46] Formation of an allyl radical followed by further oxidation to the carbonium ion is thought to be the route to allylic acetate in the KBr-catalysed reaction of an olefin with Mn(III) acetate in AcOH.[47]

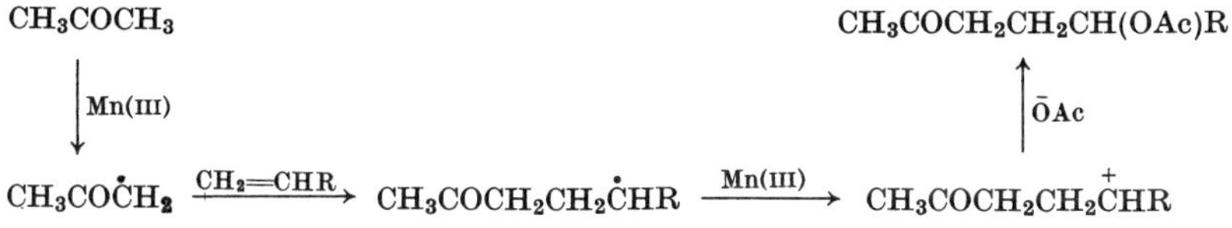

$$CH_3COCH_3 \xrightarrow{Mn(III)} CH_3CO\dot{C}H_2 \xrightarrow{CH_2=CHR} CH_3COCH_2CH_2\dot{C}HR \xrightarrow{Mn(III)} CH_3COCH_2CH_2\overset{+}{C}HR \xrightarrow{\bar{O}Ac} CH_3COCH_2CH_2CH(OAc)R$$

(21)　　　　　　(22)　　　　　　(23)

Oxidations of carboxylic acids by Co(III)[48] and of alcohols by Ce(IV)[49] continue to attract attention. Ceric ammonium nitrate (CAN) reacts with alkylphenylcarbinols [PhCH(OH)R] in aqueous MeCN to give the ketones [PhCOR] and the products of oxidative cleavage [PhCHO + products derived from R⁺] in proportions dependent on the nature of the alkyl group. The ratios of cleavage to ketone formation for R = Me (0.04), Et (3.30), Pr^i (184) and Bu^t (195) are reasonable for a mechanism, analogous to that proposed for 1,2-diarylethanols,[50] in which cleavage occurs at the one-electron

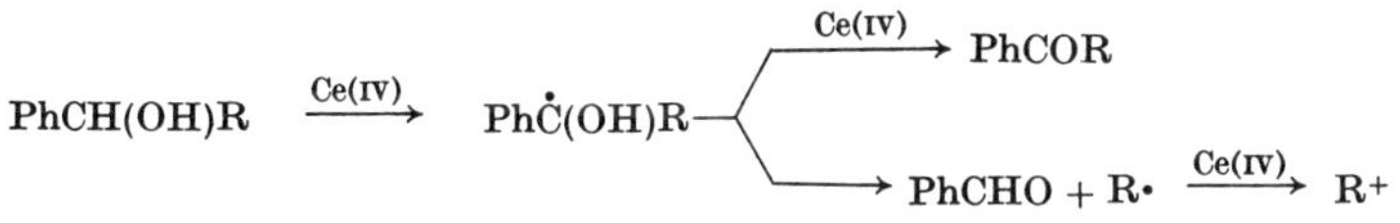

$$PhCH(OH)R \xrightarrow{Ce(IV)} Ph\dot{C}(OH)R \begin{cases} \xrightarrow{Ce(IV)} PhCOR \\ \longrightarrow PhCHO + R\cdot \xrightarrow{Ce(IV)} R^+ \end{cases} \quad \ldots (15)$$

$$PhCHN_2 + Ce(IV) \rightarrow PhCHN_2^{+\cdot} + Ce(III) \quad \ldots (16)$$

$$PhCHN_2^{+\cdot} + PhCHN_2 \rightarrow PhCHCHPh_2^{+\cdot} + 2N_2 \quad \ldots (17)$$

$$PhCHCHPh_2^{+\cdot} + PhCHN_2 \rightarrow PhCH=CHPh + PhCHN_2^{+\cdot} \quad \ldots (18)$$

---

[44] M. G. Vinogradov, S. P. Verenchikov, and G. I. Nikishin, *Kinet. Katal.*, **12**, 45 (1971); *Chem. Abs.*, **74**, 111340 (1971).
[45] K. K. Banerji, P. Nath, and G. V. Bakore, *Z. Naturforsch.*, **26b**, 30, 318 (1971); *J. Indian Chem. Soc.*, **48**, 535 (1971).
[46] E. I. Heiba and R. M. Dessau, *J. Am. Chem. Soc.*, **93**, 524 (1971).
[47] J. R. Gilmore and J. M. Mellor, *J. Chem. Soc.* (C), **1971**, 2355.
[48] J. K. Sthapak and S. Ghosh, *J. Indian Chem. Soc.*, **48**, 331 (1971).
[49] C. F. Wells and M. Husain, *Trans. Faraday Soc.*, **67**, 1086 (1971); V. I. Kurlyankina, N. V. Sarana, and O. P. Koz'mina, *Kinet. Katal.*, **11**, 1159 (1970); *Chem. Abs.*, **74**, 63754 (1971); T. R. Balasubramanian and N. Venkatasubramanian, *Indian J. Chem.*, **9**, 36 (1971); *Chem. Abs.*, **74**, 99265 (1971); D. L. Mathur and G. V. Bakore, *J. Indian Chem. Soc.*, **48**, 363 (1971); *Chem. Abs.*, **75**, 62817 (1971); *Bull. Chem. Soc. Japan*, **44**, 2595 (1971).
[50] P. M. Nave and W. S. Trahanovsky, *J. Am. Chem. Soc.*, **93**, 4536 (1971); see also *Org. Reaction Mech.*, **1968**, 472.

oxidation stage giving the alkyl radical R· (equation 15).[51] Other investigations of Ce(IV) oxidation have focused on substituted toluenes,[52] styrene[53] and anthracene; the last-named reacts with 2 equivalents of CAN in MeCN to give 9-anthryl nitrite which is further oxidized to anthraquinone on addition of an excess of CAN.[54] Phenyldiazomethane decomposes rapidly at $-5°$ in pentane in the presence of a catalytic amount of CAN. An oxidative chain reaction [equations (16)–(18)] can account for the observed formation of stilbene, although it is not obvious why the *cis*-isomer (85%) should be so dominant.[55]

While work on the oxidation of alcohols continues,[56] there have been several interesting reports of the reactions of Tl(III) with olefins and dienes.[57] 3-t-Butylcyclohexene reacts with acidic aqueous Tl(III) to give *trans*-diols (**25**; 73%) and (**27**; 10%) accompanied by less than 1% of either *cis*-diol. Such specificity suggests participation by the OH

group in the heterolysis of C—Tl bonds in intermediates such as (**24**) and (**26**).[58] Under comparable conditions 1-methylcyclobutene gives cyclopropyl methyl ketone,[59] and such oxidative rearrangements appear to be general for olefins with Tl(III) nitrate in MeOH; for example, cyclohexene gives cyclopentanecarboxaldehyde in high yield.[60] Thallium–carbene complexes are considered to be intermediates in the reaction of 1,3,3-trimethylcyclopropene with Tl(III) acetate in CH_2Cl_2.[61]

Olefins[62] and cyclopropanes[63] have figured prominently in investigations of oxidation by Pd(II) and Hg(II). Reaction of cyclohexene with Pd(II) chloride in AcOH [containing NaOAc, and Hg(II) acetate as reoxidant for Pd] at $25°$ gives a mixture of the allylic (92%) and homoallylic (8%) acetates. With tetradeuteriocyclohexene (**29**) the allylic acetate is obtained in reduced yield (70%) and consists of a 1:1 mixture of (**32**) and (**33**), while the homoallylic acetate, formed in increased yield, is a mixture of (**34**) and (**35**).[64] A symmetrical intermediate, probably a π-allylic complex[65] (**28**, partial structure),

---

[51] W. S. Trahanovsky and J. Cramer, *J. Org. Chem.*, **36**, 1890 (1971).
[52] P. S. Radhakrishnamurti and M. K. Mahanti, *J. Indian Chem. Soc.*, **47**, 1006 (1970); *Chem. Abs.*, **74**, 99325 (1971).
[53] E. I. Heiba and R. M. Dessau, *J. Am. Chem. Soc.*, **93**, 995 (1971).
[54] B. Rindone and C. Scolastico, *J. Chem. Soc.* (B), **1971**, 2238.
[55] W. S. Trahanovsky, M. D. Robbins, and D. Smick, *J. Am. Chem. Soc.*, **93**, 2086 (1971).
[56] V. P. Kudesia, *Bull. Soc. Chim. Belges*, **80**, 211 (1971); V. S. Srinivasan and N. Venkatasubramanian, *Indian J. Chem.*, **8**, 849 (1970); *Chem. Abs.*, **73**, 134324 (1970).
[57] S. Uemura, A. Tabata, M. Okano, and K. Ichikawa, *Chem. Comm.*, **1970**, 1630.
[58] C. Freppel, R. Favier, J.-C. Richer, and M. Zador, *Can. J. Chem.*, **49**, 2586 (1971).
[59] J. E. Byrd, L. Cassar, P. E. Eaton, and J. Halpern, *Chem. Comm.*, **1971**, 40.
[60] A. McKillop, J. D. Hunt, E. C. Taylor, and F. Kienzle, *Tetrahedron Letters*, **1970**, 5275; see also A. McKillop, B. P. Swann, and E. C. Taylor, *Tetrahedron Letters*, **1970**, 5281.
[61] T. Shirafuji, Y. Yamamoto, and H. Nozaki, *Tetrahedron Letters*, **1971**, 4713.
[62] H. B. Tinker, *J. Organomet. Chem.*, **32**, C25 (1971); see also R. Jira, *Tetrahedron Letters*, **1971**, 1225.
[63] R. J. Ouellette and C. Levin, *J. Am. Chem. Soc.*, **93**, 471 (1971).
[64] S. Wolfe and P. G. C. Campbell, *J. Am. Chem. Soc.*, **93**, 1497, 1499 (1971).
[65] See also F. Conti, M. Donati, G. F. Pregaglia, and R. Ugo, *J. Organomet. Chem.*, **30**, 421 (1971).

seems implicated in the formation of allylic acetate. Its formation from deuteriocyclohexene (29), involving C—D bond cleavage, would occur less readily than from undeuteriated substrate, and its solvolysis would afford allylic acetates (32) and (33). Acetoxypalladation [whether *cis* to give (30) or, as now seems probable,[66] *trans* to give (31)] is only a minor competing pathway leading to homoallylic acetates (34) and (35). Both of these have the acetoxyl group attached to one of the original olefinic C atoms and must have been formed by a mechanism (possibilities are discussed by the authors) involving stereospecific migration of deuterium.[64]

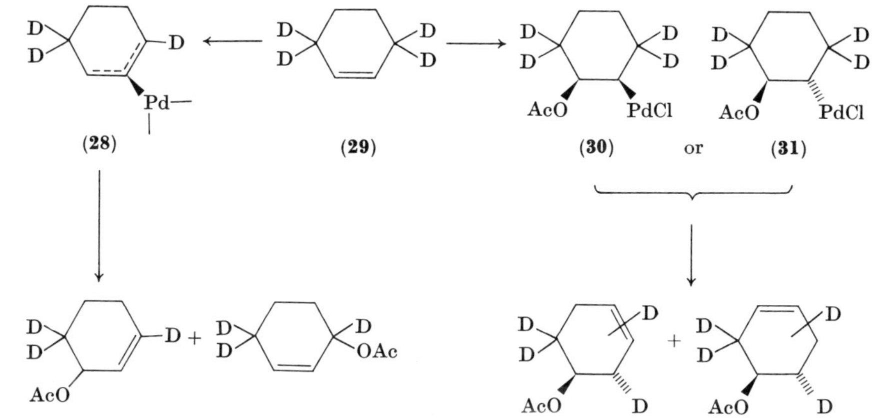

Anisole reacts with Pb(IV) acetate in CCl_3CO_2H to give the diaryl-lead bis(trichloroacetate) (36; 48%). This is decomposed by CF_3CO_2H in anisole to products which include 2,4'- and 4,4'-dimethoxybiphenyl. Possible arylating species are the *p*-methoxyphenyl

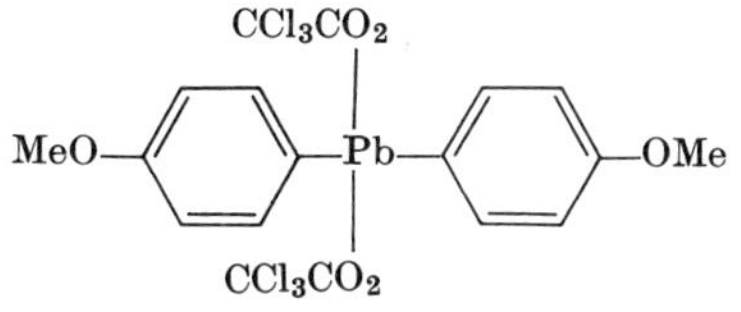

$$(CCl_3CO_2)_2PbAr_2 + CF_3CO_2H \rightleftharpoons (CF_3CO_2)(CCl_3CO_2)_2PbAr + ArH \quad \ldots (19)$$

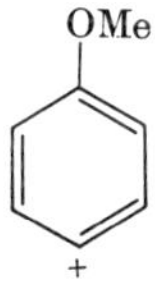

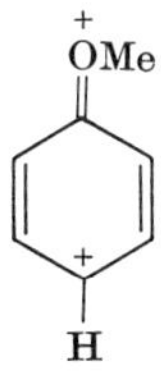

$$(CF_3CO_2)(CCl_3CO_2)_2PbAr \rightleftharpoons CF_3CO_2^- + (CCl_3CO_2)_2Pb + Ar^+ \quad \ldots (20)$$

cation (37), formed as in equations (19) and (20) (Ar = *p*-methoxyphenyl) and the dication (38), resulting from CF_3CO_2H-assisted oxidation of anisole by (36) or some derived Pb(IV) species.[67]

[66] P. M. Henry and G. A. Ward, *J. Am. Chem. Soc.*, **93**, 1494 (1971).
[67] R. O. C. Norman, C. B. Thomas, and J. S. Willson, *J. Chem. Soc.* (B), **1971**, 518.

Investigation of the Pb(IV) acetate oxidation of phenols,[68] α-diketones[69] and organic nitrogen compounds continues, the last topic having been reviewed.[70] Aminophenyl ketone hydrazones (**39**; R = Me or Ph) are oxidized to the 1,2,3-benzotriazines (**41**; 50%). Presumably diazo compounds (**40**) are first formed and cyclize to the dihydrotriazines which are then oxidized.[71] N-Aminoindazoles are also oxidized to 1,2,3-benzotriazines.[72]

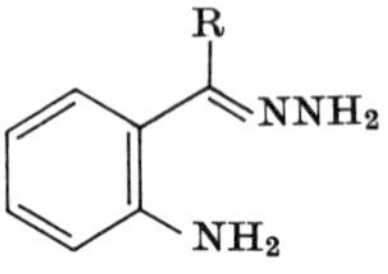

(**39**) 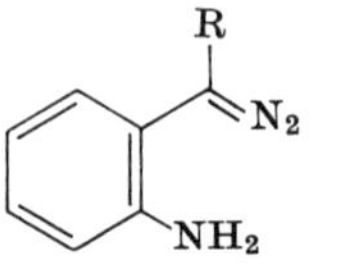   (**40**) 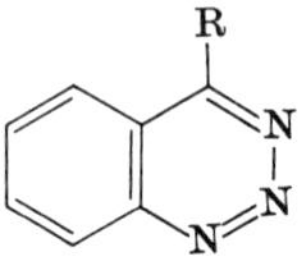   (**41**)

Oxidation of phenols by hexacyanoferrate(III) has been further investigated.[73] Simple one-electron oxidants, as well as enzymes, produce dimers resulting from either C—O or C—C coupling. Although both types of product can be formally rationalized in terms of dimerization of aryloxy radicals (ArO·), Waters[74] has surveyed evidence suggesting that C—C coupled products are more likely to result from reactions of aryloxy cations with phenol molecules (cf. ref. 68). He points out that, below a certain pH, ArO· is metastable with respect to ArO$^+$ and ArO$^-$ and, rather than dimerize, will disproportionate:

$$2ArO\cdot + H_3O^+ \rightleftharpoons ArO^+ + ArOH + H_2O$$

That C—O as well as C—C coupling may occur by way of aryloxy cations is emphasized by the demonstrated intermediacy of (**43**) in the oxidation of the bisphenol (**42**) to the spiran (**44**).[75]

Hexacyanoferrate(III) oxidations of p-hydroxyphenyl ethers,[76] hydroxamic acids,[77] tertiary amines,[78] thiols[79] and indole derivatives[80] have been considered. The influence of 3-substituents on the rates and products of oxidation of pyridinium salts to N-alkylpyridones by alkaline hexacyanoferrate(III) is not compatible with addition of hydroxide as the rate-determining factor.[81] The mechanism of the Cu(II) chloride oxidative dimeri-

---

[68] D. G. Hewitt, *J. Chem. Soc.* (C), **1971**, 1750.
[69] L. Canonica, B. Danieli, P. Manitto, and G. Russo, *Gazz. Chim. Ital.*, **100**, 1026 (1970).
[70] J. B. Aylward, *Quart. Rev.*, **25**, 407 (1971).
[71] S. Bradbury, M. Keating, C. W. Rees, and R. C. Storr, *Chem. Comm.*, **1971**, 827.
[72] D. J. C. Adams, S. Bradbury, D. C. Horwell, M. Keating, C. W. Rees, and R. C. Storr, *Chem. Comm.*, **1971**, 828.
[73] J. L. G. Nilsson and H. Selander, *Acta Chem. Scand.*, **24**, 2885 (1970); L. Taimr and J. Pospíšil, *Tetrahedron Letters*, **1971**, 2809.
[74] W. A. Waters, *J. Chem. Soc.* (B), **1971**, 2026.
[75] M. Chauhan, F. M. Dean, K. Hindley, and M. Robinson, *Chem. Comm.*, **1971**, 1141.
[76] J. Petránek, *Tetrahedron*, **27**, 5201 (1971).
[77] T. R. Oliver and W. A. Waters, *J. Chem. Soc.* (B), **1971**, 677.
[78] C. A. Audeh and J. R. L. Smith, *J. Chem. Soc.* (B), **1971**, 1741, 1745.
[79] R. C. Kapoor, R. K. Chohan, and B. P. Sinha, *J. Phys. Chem.*, **75**, 2036 (1971); O. P. Kachwaha, B. P. Sinha, and R. C. Kapoor, *Indian J. Chem.*, **8**, 806 (1970); *Chem. Abs.*, **74**, 3217 (1971).
[80] S. Hünig and F. Linhart, *Tetrahedron Letters*, **1971**, 1273.
[81] R. A. Abramovitch and A. R. Vinutha, *J. Chem. Soc.* (B), **1971**, 131; cf. H. Möhrle and H. Weber, *Chem. Ber.*, **104**, 1478 (1971).

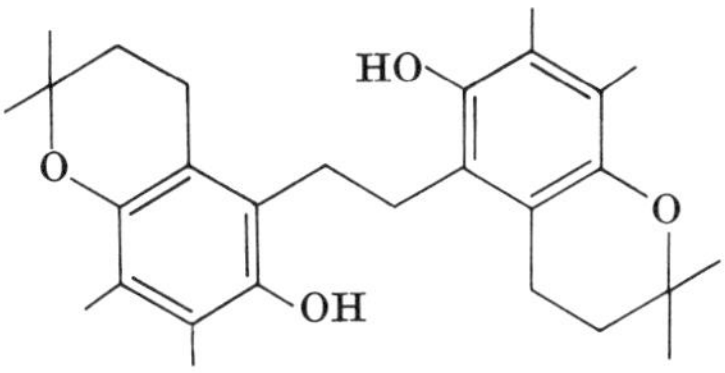

(42)

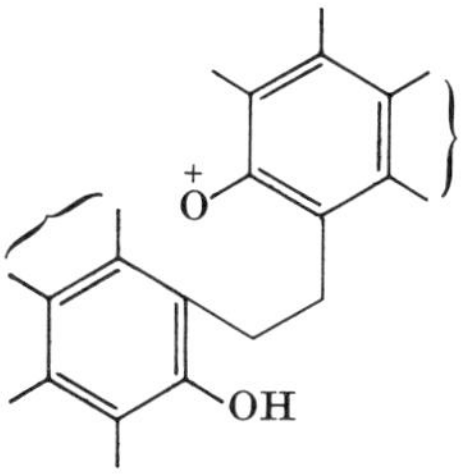

(43) 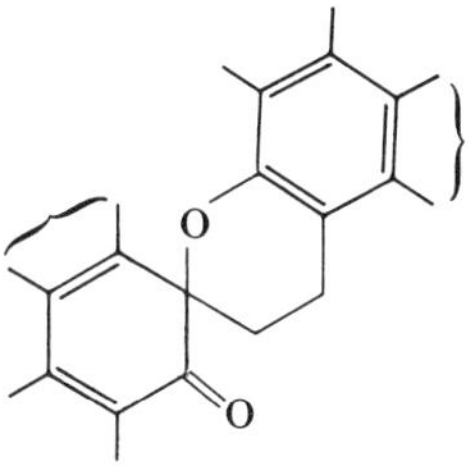 (44)

zation of aromatic hydrocarbons has been discussed,[82] and oxidation of cyclic semicarbazides by Cu(II) has proved a useful route to cyclic azoalkanes.[83]

Oxidation by Molecular Oxygen[84]

Although autoxidation of ^{14}C-labelled 6-(aminomethyl)tetrahydro-6-methylpterin [(45), partial structure] at pH 6.8 results in loss of the aminomethyl group, the 6-methylpterin produced still contains some of the original activity. Initial hydroxylation can occur at both the labelled CH$_2$ group, leading eventually to H^{14}CHO and inactive pterin, and at the ring-CH$_2$ group giving (46), which by ring opening to (47) and reclosure to (48) provides a precursor of active pterin.[85]

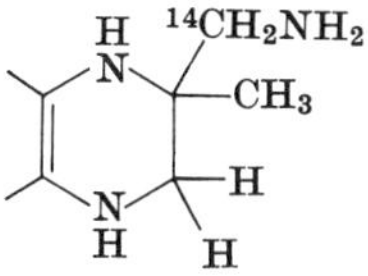

(45) 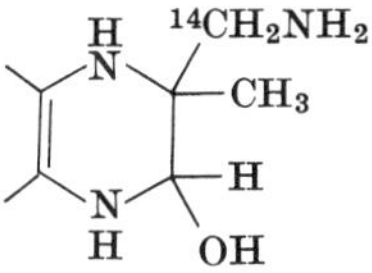 (46) 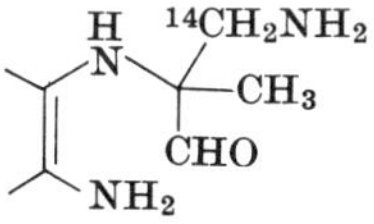 (47) 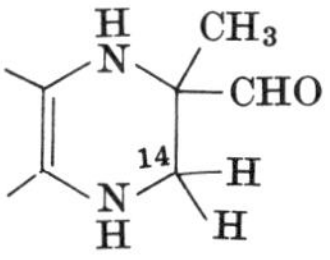 (48)

Monoprotonated N,N,N',N'-tetramethyl-p-phenylenediamine (TMPD) is an efficient catalyst for the oxidation (at pH ~ 7 and 25°) of 1-alkyl-1,4-dihydronicotinamides (PyH), including NADH and NADPH, to pyridinium salts (Py$^+$). Uptake of O$_2$ is of the first order in substrate concentration and one-half order in each of catalyst and O$_2$, and is

---

[82] C. A. Cummings and D. J. Milner, *J. Chem. Soc.* (C), **1971**, 1571.
[83] M. Heyman, V. T. Bandurco, and J. P. Snyder, *Chem. Comm.*, **1971**, 297.
[84] Review: S. V. Anantakrishnan, *J. Sci. Ind. Res.*, **29**, 323 (1970); *Chem. Abs.*, **74**, 63599 (1971).
[85] M. Viscontini and M. Argentini, *Helv. Chim. Acta*, **54**, 2287 (1971); see also M. Viscontini and M. Argentini, *Ann. Chim.*, **745**, 109 (1971); M. Viscontini and M. Cogoli-Greuter, *Helv. Chim. Acta*, **54**, 1125 (1971).

subject to only a small isotope effect ($k_H/k_D = 2.3$ with D at C-4 in the substrate). A radical-chain mechanism based on reactions (21)–(23) seems probable.[86]

$$TMPD + H^+ + O_2 \xrightarrow{initiation} TMPD^{+\cdot} + HO_2^{\cdot} \qquad \ldots (21)$$

$$HO_2^{\cdot} + PyH \longrightarrow H_2O_2 + Py^{\cdot} \qquad \ldots (22)$$

$$Py^{\cdot} + H^+ + O_2 \longrightarrow Py^+ + HO_2^{\cdot} \qquad \ldots (23)$$

The kinetics of hydrocarbon autoxidation[87] and its inhibition by 1,1'-bis-(N-phenyl-2-naphthylamine)[88] have been discussed, as have the oxidations of carbohydrates,[89] ketones,[90] dihydropyrans,[91] eleostearic acids,[92] trihydroxytoluenes,[93] 2,3-dialkylindoles,[94a] anthranol,[94b] trialkylboranes,[95] and diarylperfluoroacylphosphines.[96] Homocyclo-octatetraene dianion (50), formed from bicyclo[6.1.0]nonatriene (49) with Li–NH$_3$, reverts to (49), with the C$_1$—C$_8$ bond intact, on reaction with traces of O$_2$. However, on rapid exposure to larger amounts of O$_2$ the dianion (50) gives dimer (51; 51%), perhaps by one-electron oxidation followed by selective coupling of the radical anion (at C-3 or C-4) and protonation of the dimeric dianion.[97]

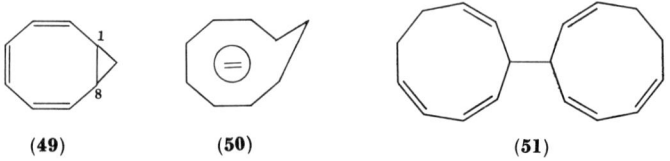

(49)   (50)   (51)

The catalytic effect of (Ph$_3$P)$_4$Pd in the liquid-phase autoxidation of cumene has now been ascribed to its catalysis of the initiation by traces of peroxide in the cumene,[98] and not to the formation of an oxygen complex [(Ph$_3$P)$_2$PdO$_2$] capable of hydrogen

---

[86] E. J. H. Bechara and G. Cilento, *Biochemistry*, **10**, 1831, 1837 (1971); see also M. da Silva Araujo, E. J. H. Bechara, K. Zinner, and G. Cilento, *Record Chem. Progr.*, **31**, 189 (1970).
[87] J. Betts, *Quart. Rev.*, **1971**, 265.
[88] R. F. Bridger, *J. Org. Chem.*, **36**, 1214 (1971).
[89] W. B. Gleason and R. Barker, *Can. J. Chem.*, **49**, 1425 (1971); H. G. J. De Wilt and B. F. M. Kuster, *Carbohydrate Res.*, **19**, 5 (1971).
[90] H. R. Gersmann and A. F. Bickel, *J. Chem. Soc.* (B), **1971**, 2230; F. G. Bordwell and A. C. Knipe, *J. Am. Chem. Soc.*, **93**, 3416 (1971); J. J. McDonnell and D. J. Pochopien, *J. Org. Chem.*, **36**, 2092 (1971).
[91] R. S. Atkinson, *Chem. Comm.*, **1971**, 585.
[92] L. Pekkarinen and M. Levomäki, *Suomen Kemistilehti*, **B44**, 110 (1971); L. Pekkarinen, *Suomen Kemistilehti*, **B44**, 273, 277 (1971).
[93] W. Flaig and H. Riemer, *Ann. Chem.*, **746**, 81 (1971).
[94a] S. McLean and G. I. Dmitrienko, *Can. J. Chem.*, **49**, 3642 (1971).
[94b] Y. Ogata, Y. Kosugi, and K. Nate, *Tetrahedron*, **27**, 2705 (1971).
[95] H. C. Brown, M. M. Midland, and G. W. Kabalka, *J. Am. Chem. Soc.*, **93**, 1024 (1971); M. M. Midland and H. C. Brown, *J. Am. Chem. Soc.*, **93**, 1506 (1971); H. C. Brown and M. M. Midland, *Chem. Comm.*, **1971**, 699; J. Grotewold, J. Hernandez, and E. A. Lissi, *J. Chem. Soc.* (B), **1971**, 182.
[96] E. Lindner, H.-D. Ebert, K. Geibel, and A. Haag, *Chem. Ber.*, **104**, 3121 (1971).
[97] T. I. Ito, F. C. Baldwin, and W. H. Okamura, *Chem. Comm.*, **1971**, 1440.
[98] R. A. Sheldon, *Chem. Comm.*, **1971**, 788; see also A. Fusi, R. Ugo, F. Fox, A. Pasini, and C. Cenini, *J. Organomet. Chem.*, **26**, 417 (1971).

abstraction as claimed by Stern.[99] Against this, the report[100] that [RhCl(cyclo-octene)$_2$]$_2$ forms an oxygen complex which can abstract hydrogen seems to support Stern's claim.

Co-oxidation of p-cymene and butane in AcOH catalysed by Co(III) acetate gives predominantly p-isopropylbenzoic acid (90%). Oxidation of the methyl group in preference to isopropyl is obviously incompatible with the usual radical mechanism and may result from "an electron transfer mechanism with radical cation intermediates" (no details are given).[101]

The Fe(III)-catalysed oxidation of acetoin (at 25° and pH ~2) produces 2 moles of biacetyl for each mole of O_2 consumed. If the sequence of equations (24) and (25) is correct the observed formation of only one mole of biacetyl per mole of O_2 in the presence of a high concentration of Fe(II) is readily explained by rapid reduction of H_2O_2. A radical mechanism for the reaction with O_2 [equation (24)] involving reduction of the catalyst by substrate and reoxidation by O_2, is ruled out by the stability of Fe(II) towards O_2

$$CH_3CH(OH)COCH_3 + O_2 \rightarrow CH_3COCOCH_3 + H_2O_2 \quad \ldots (24)$$

$$CH_3CH(OH)COCH_3 + H_2O_2 \rightarrow CH_3COCOCH_3 + 2H_2O \quad \ldots (25)$$

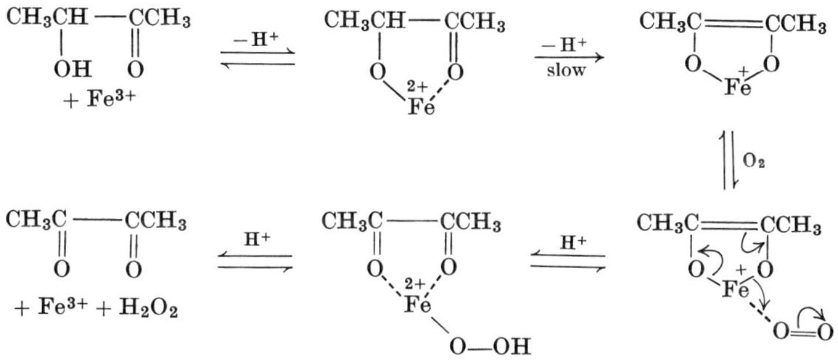

Scheme 2

under the reaction conditions. In the mechanism shown in Scheme 2 the metal ion complexes with both substrate and oxidant, thereby permitting electron transfer without the necessity for radical intermediates. The inverse dependence of rate on [H$^+$] and the primary isotope effect [k_H/k_D ~ 5 for CH$_3$CD(OH)COCH$_3$] are consistent with the loss of the second proton as rate-determining factor.[102] Oxidation of the second mole of acetoin by H_2O_2 can proceed by two different mechanisms, which have also been investigated.[102]

The metal-catalysed oxidation and dehydrogenation of olefins[103] and dienes,[104]

---

[99] E. W. Stern, *Chem. Comm.*, **1970**, 736.
[100] B. R. James and E. Ochiai, *Can. J. Chem.*, **49**, 975 (1971).
[101] A. Onopchenko, J. G. D. Schulz, and R. Seekircher, *Chem. Comm.*, **1971**, 939.
[102] P. K. Adolf and G. A. Hamilton, *J. Am. Chem. Soc.*, **93**, 3420 (1971).
[103] K. Takao, M. Wayaku, Y. Fujiwara, T. Imanaka, and S. Teranishi, *Bull. Chem. Soc. Japan*, **43**, 3898 (1970); Y. Moro-oka, Y. Takita, and A. Ozaki, *Bull. Chem. Soc. Japan*, **44**, 293 (1971); J.-C. Delgrange and M. Blanchard, *Bull. Soc. Chim. France*, **1971**, 1093; A. A. Potatuev, N. I. Deinikina, and G. I. Moskvina, *Neftekhimiya*, **11**, 391 (1971); *Chem. Abs.*, **75**, 87848 (1971).
[104] S. S. Hotanahalli and S. B. Chandalia, *J. Appl. Chem.*, **20**, 323 (1970); *Chem. Abs.*, **74**, 3220 (1971).

dihydrofurans,[105] ketones (to α,β-unsaturated ketones,)[106] p-xylene,[107] 3,3-dimethyl-1-phenyltriazene,[108] ascorbic acid[109] and vitamin A[110] have been examined.

Irreversible one-turnover reaction of [1-³H]glyceraldehyde 3-phosphate with the glyceraldehyde 3-phosphate dehydrogenase–NAD⁺ complex proceeds with incorporation of tritium into NADH but not into the enzyme. This is in agreement with a mechanism in which hydrogen is transferred directly from the substrate to NAD⁺ rather than one involving alternate oxidation and reduction of the methylene group of tryptophan side-chains in the enzyme.[111] Many other papers have been concerned with the action of oxidases[112] and dehydrogenases,[113] and a model reaction for flavoenzyme-catalysed dehydrogenation has been described. Thus 10-phenylisoalloxazine, in DMF–Bu^{t}OH–Bu^{t}OK, is rapidly reduced when either methyl mandelate [PhCH(OH)CO$_2$Me] or phenylglycine methyl ester [PhCH(NH$_2$)CO$_2$Me] is added under anaerobic conditions, but the initial isoalloxazine is regenerated on addition of air. Treatment of the anaerobic reaction mixture with aqueous base gives dehydrogenated substrate (isolated as PhCOCO$_2$H) in high yield. A mechanism in accord with the observed substrate specificity and the fact that structurally related alcohols and amines react at comparable rates involves formation of a covalent adduct (Scheme 3, with mandelate as substrate) and removal of a proton from C_α of the substrate. An analogous mechanism for all flavoenzyme-catalysed dehydrogenations is considered likely.[114]

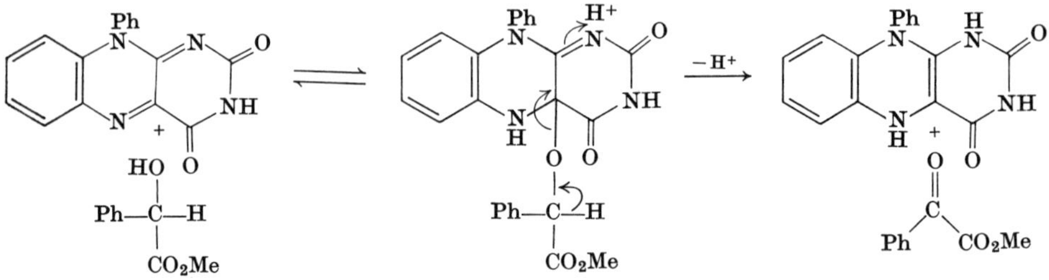

Scheme 3

[105] J.-P. Girault, P. Scribe, and G. Dana, *Bull. Soc. Chim. France*, **1971**, 2279.
[106] R. J. Theissen, *J. Org. Chem.*, **36**, 752 (1971).
[107] A. G. Caloyannis and W. F. Graydon, *J. Catal.*, **22**, 287 (1971).
[108] V. Zvěřina, J. Diviš, J. Marhold, and M. Matrka, *Coll. Czech. Chem. Comm.*, **36**, 1598 (1971).
[109] L. Pekkarinen and A.-L. Kyllönen, *Suomen Kemistilehti*, **B44**, 67 (1971); Y. Kurimura and E. Tsuchida, *Kagaku No Ryoiki*, **24**, 595 (1970); *Chem. Abs.*, **73**, 108904 (1970) (Review).
[110] Y. Ogata, Y. Kosugi, and K. Tomizawa, *Tetrahedron*, **26**, 5939 (1970).
[111] W. S. Allison, H. B. White, and M. J. Connors, *Biochemistry*, **10**, 2290 (1971).
[112] V. R. Holland and B. C. Saunders, *Tetrahedron*, **27**, 2851 (1971); R. Roman, H. B. Dunford, and M. Evett, *Can. J. Chem.*, **49**, 3059 (1971); A. Maradufu, G. M. Cree, and A. S. Perlin, *Can. J. Chem.*, **49**, 3429 (1971); M. Nishikimi, M. Osamura, and K. Yagi, *J. Biochem. (Tokyo)*, **70**, 457 (1971); W. Heimann and P. Schreier, *Helv. Chim. Acta*, **54**, 2794, 2803 (1971).
[113] H. Gutfreund, *Ann. Rev. Biochem.*, **40**, 315 (1971); H. Eisenberg, *Accounts Chem. Res.*, **4**, 379 (1971); R. A. MacQuarrie and S. A. Bernhard, *J. Mol. Biol.*, **55**, 181 (1971); *Biochemistry*, **10**, 2456 (1971); J. S. Barton and J. R. Fisher, *Biochemistry*, **10**, 577 (1971); G. di Prisco, *Biochemistry*, **10**, 585 (1971); S. F. Velick, J. P. Baggott, and J. M. Sturtevant, *Biochemistry*, **10**, 779 (1971); D. Balinsky, A. W. Dennis, and W. W. Cleland, *Biochemistry*, **10**, 1947 (1971); W. G. Hanstein, K. A. Davis, M. A. Ghalambor, and Y. Hatefi, *Biochemistry*, **10**, 2517 (1971); J.-L. Risler, *Biochemistry*, **10**, 2664 (1971); H. Theorell and K. Tatemoto, *Acta Chem. Scand.*, **24**, 3069 (1970); F. H. Bodley and A. H. Blair, *Can. J. Biochem.*, **49**, 1 (1971); A. Di Franco and M. Iwatsubo, *Biochemie*, **53**, 153 (1971); B. Wurster and B. Hess, *Z. Physiol. Chem.*, **351**, 1537 (1970).
[114] L. E. Brown and G. A. Hamilton, *J. Am. Chem. Soc.*, **92**, 7225 (1970).

Although the oxidations of 1,3-dienes catalysed by horseradish peroxidase and soybean lipoxidase may indeed by analogous to the reactions of $^1O_2$,[115] it now seems[116] that the product of the coupled oxygenation of tetracyclone by lipoxidase is not in fact the endoperoxide of tetracyclone as was thought.

Other Oxidations

Unlike simple cyclopropenes, which form α,β-unsaturated aldehydes or ketones by rearrangement of the initial epoxide, with m-chloroperoxybenzoic acid (MCPBA) in CH_2Cl_2,[117] the hindered tri-t-butylcyclopropene (**52**) gives mixed anhydride (**55**) and di-t-butylacetylene (or its oxidation products). Cyclopropenium ion (**53**), which can be isolated as the perchlorate when the oxidation is carried out in the presence of $HClO_4$, and a peroxyester (**54**) are likely intermediates.[118] Epoxidation is similarly unimportant in the (slow) reaction of di-t-butylcyclopropenone (**56**) with an excess of MCPBA to give the same mixtures of ketones[119] as is obtained by oxidation of di-t-butylacetylene. The concomitant evolution of CO_2 suggests that reaction occurs via (**57**) which collapses to di-t-butylacetylene and CO_2, possibly by way of a β-lactone (**58**).[120] Although the epoxide may be formed when cyclonona-1,2-diene reacts with CH_3CO_3H, the final products are cyclononane-1,2-dione and epoxycyclo-octane.[121]

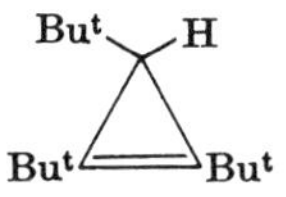

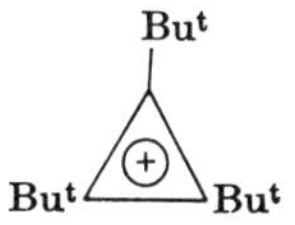

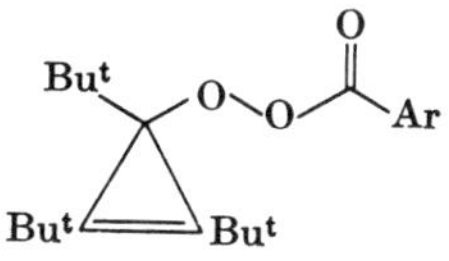

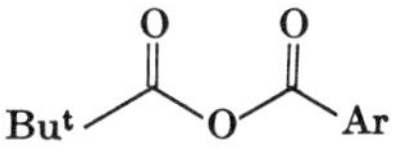

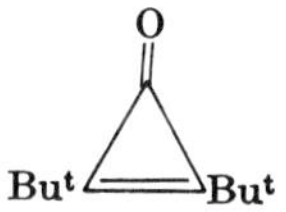

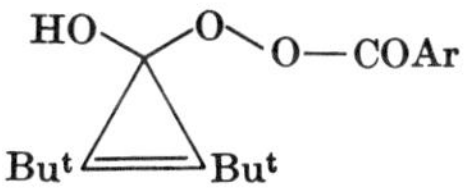

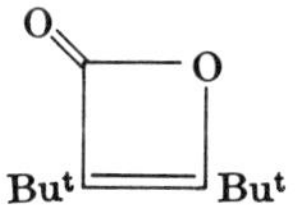

Baeyer–Villiger oxidation of cyclobutyl cyclohexyl ketone proceeds with equal migration of the two alkyl groups. However, tricyclo-octanone (**59**) gives almost exclusively the lactone (**61**) with CF_3CO_3H. The preferential migration of cyclobutyl in this rigid system may be a consequence of assisted heterolysis of the C-6—C-7 bond, participation by the C-1—C-8 (or C-5—C-8) bond delocalizing positive charge at the migrating centre (as in **60**).[122] Other investigations with peroxyacids have been concerned

---

[115] H. W.-S. Chan, *J. Am. Chem. Soc.*, **93**, 2357, 4632 (1971).
[116] J. E. Baldwin, J. C. Swallow, and H. W.-S. Chan, *Chem. Comm.*, **1971**, 1407.
[117] L. E. Friedrich and R. A. Cormier, *Tetrahedron Letters*, **1971**, 4761; see also *Org. Reaction Mech.*, **1969**, 555.
[118] J. Ciabattoni and J. P. Kocienski, *J. Am. Chem. Soc.*, **93**, 4902 (1971).
[119] *Org. Reaction Mech.*, **1970**, 590–591.
[120] J. K. Crandall and W. W. Conover, *Tetrahedron Letters*, **1971**, 583.
[121] W. P. Reeves and G. G. Stroebel, *Tetrahedron Letters*, **1971**, 2945; see also J. Grimaldi and M. Bertrand, *Bull Soc. Chim. France*, **1971**, 957.
[122] S. A. Monti and C. K. Ward, *Tetrahedron Letters*, **1971**, 697.

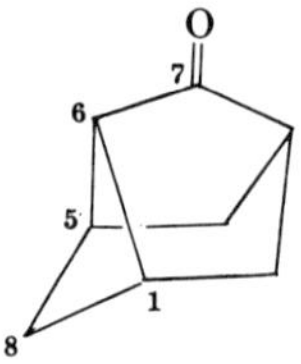

(59) 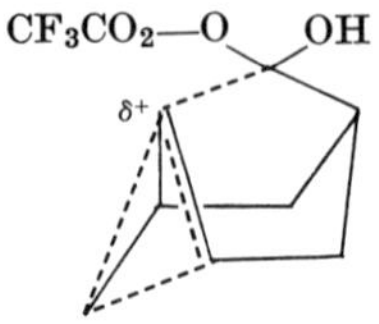 (60) 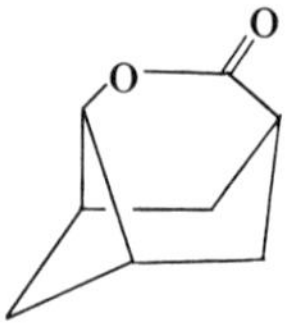 (61)

with CF_3CO_3H–BF_3 oxidations,[123a] deuterium isotope effects in the oxidation of diarylphosphine oxides,[123b] the formation of benzimidazoles from o-acylamino-N,N-dialkylanilines,[124] and the generation of sulphines from thioketones[125] and dithiocarboxylic esters.[126]

The pH–rate profile for the oxidation of uracil by H_2O_2 is bell-shaped in the region pH 8–11, with a maximum at pH 9.8–9.9. Together with the insensitivity of the reaction to radical traps, this points to nucleophilic attack of peroxy-anion on the neutral substrate, although competing radical reactions may become important at lower pH.[127] The mechanism of the oxidation of xanthopterin peroxide to melanurenic acid by H_2O_2 has also been studied,[128] as have the peroxide oxidation of α-amino-α,β-unsaturated ketones (where Baeyer–Villiger rearrangement can be observed),[129] the metal-ion catalysed oxidations of alcohols,[130] phenols,[131] and aromatic amines[132] and the cleavage of diols by nickel peroxide.[133]

Carboxamide (62) gives the spiro-lactam (64) in good yield when heated with alkaline aqueous peroxydisulphate. Carboxamido radicals are considered unlikely intermediates,

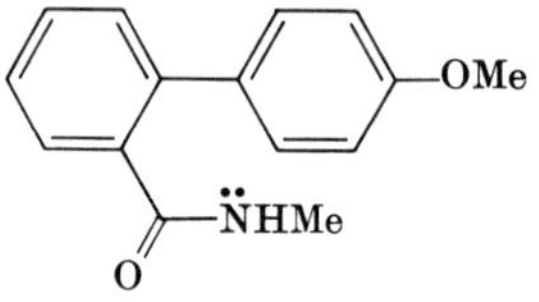

(62) 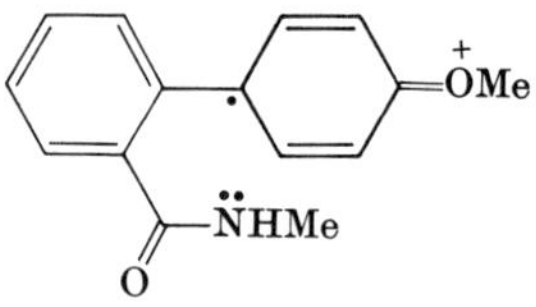 (63) 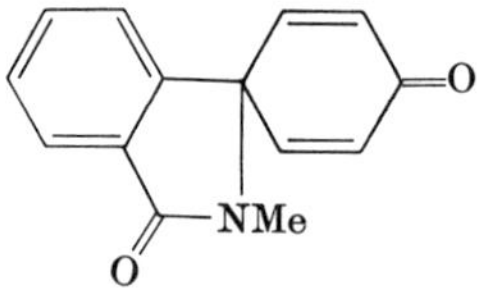 (64)

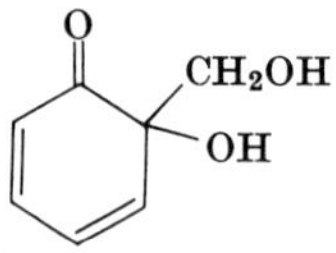

(65) 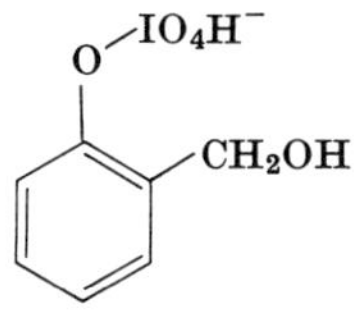 (66) 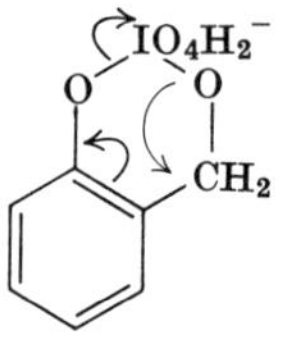 (67) 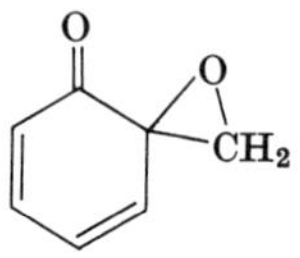 (68)

[123a] H. Hart, *Accounts Chem. Res.*, **4**, 337 (1971).
[123b] R. Curci and F. Di Furia, *Tetrahedron*, **27**, 4601 (1971).
[124] O. Meth-Cohn, *J. Chem. Soc.* (C), **1971**, 1356.
[125] A. Battaglia, A. Dondoni, P. Giorgianni, G. Maccagnani, and G. Mazzanti, *J. Chem. Soc.* (B), **1971**, 1547.
[126] B. Zwanenburg, L. Thijs, and J. Strating, *Rec. Trav. Chim.*, **90**, 614 (1971).
[127] L. R. Subbaraman, J. Subbaraman, and E. J. Behrman, *J. Org. Chem.*, **36**, 1256 (1971).
[128] G. B. Barlin and W. Pfleiderer, *Chem. Ber.*, **104**, 3069 (1971).
[129] D. L. Coffen and D. G. Korzan, *J. Org. Chem.*, **36**, 390 (1971).
[130] C. Walling and S. Kato, *J. Am. Chem. Soc.*, **93**, 4275 (1971).
[131] K. Günther, W. G. Filby, and K. Eiben, *Tetrahedron Letters*, **1971**, 251; R. G. R. Bacon and L. C. Kuan, *Tetrahedron Letters*, **1971**, 3397.
[132] G. R. Howe and R. R. Hiatt, *J. Org. Chem.*, **35**, 4007 (1970).
[133] R. Konaka and K. Kuruma, *J. Org. Chem.*, **36**, 1703 (1971).

and a mechanism involving intramolecular capture of the aromatic radical cation (63) is preferred.[134] Kinetic studies of the peroxydisulphate oxidations of aromatic amines,[135] phenols,[136] and diols[137] have been described, and the reactions of phosphines with di-isopropyl peroxydicarbonate[138] and of azines[139] and diols[140] with periodic acid have been discussed. By analogy with the reactions of 2-methylphenols with aqueous periodate it might be supposed that salicylyl alcohol would give o-quinol (65) which, being a 1,2-diol, would be cleaved to o-benzoquinone; this is not the case, however, the Diels–Alder dimer of spiroepoxycyclohexadienone (68) being obtained in high yield; presumably periodate ester (66) is first formed; this could give (68) in a number of ways, including ring-closure to, and decomposition of, periodate ester (67).[141]

Investigations of the Br_2 oxidation of arylidene semicarbazones,[142] alcohols[143] and aldehydes[144] continues. Intramolecular general base catalysis by the carboxylate group in the oxidation of o-formylbenzoic acid in aqueous solution supports the view that the reactive species is the aldehyde hydrate for aromatic, as well as aliphatic, aldehydes.[145]

α-Hydroxy carboxylic acids suffer oxidative decarboxylation with Br_2 in aqueous solution. The pH–rate profile for mandelic acid in neutral or basic solution (pH >5), where the substrate is completely ionized, is bell-shaped with a maximum at pH ~8. Addition of Br^-, while reducing the concentration of HOBr, increases the rate of oxidation. Apparently Br_2 is the most effective oxidant at pH >5, and a mechanism [equation (26)] involving its reaction with mandelate anion and base-catalysed decomposition of the resulting acyl hypobromite seems reasonable. Although relatively unimportant at pH >5, oxidation by HOBr becomes dominant at lower pH, where the active form of the substrate is undissociated acid.[146]

$$\text{PhCH}-\underset{\overset{|}{O^-}}{\overset{\overset{H}{O}}{C}}\overset{O}{\diagup} + Br_2 \underset{-Br^-}{\rightleftharpoons} \text{PhCH}-\underset{O-Br}{\overset{\overset{H\frown OH}{O}}{C}}\overset{O}{\diagup} \xrightarrow{-Br^-} \text{PhCH}\overset{O}{\overset{\|}{}} + \begin{array}{l}H_2O\\+CO_2\end{array} \quad \ldots (26)$$

Further evidence has appeared for the intermediacy of halocarbonium ions ($R_2\overset{+}{C}$-Hal) in the reaction of hydrazones ($R_2C=NNH_2$) with halogens in the presence of base,[147] and for sulphenyl iodides (RSI) in the oxidation of thiols (RSH) by I_2.[148] The following

[134] D. H. Hey, G. H. Jones, and M. J. Perkins, *Chem. Comm.*, **1971**, 998.
[135] A. Sabesan and N. Venkatasubramanian, *Austral. J. Chem.*, **24**, 1633 (1971).
[136] Y. Ogata and T. Akada, *Tetrahedron*, **26**, 5945 (1970).
[137] G. D. Menghani and G. V. Bakore, *Proc. Indian Nat. Sci. Acad. Part A*, **36**, 164 (1970); *Chem. Abs.*, **74**, 31413 (1971).
[138] W. Adam and A. Rios, *J. Org. Chem.*, **36**, 407 (1971).
[139] A. J. Fatiadi, *Chem. Ind. (London)*, **1971**, 64.
[140] G. J. Buist and C. A. Bunton, *J. Chem. Soc. (B)*, **1971**, 2117; G. J. Buist, C. A. Bunton, and W. C. P. Hipperson, *J. Chem. Soc. (B)*, **1971**, 2128.
[141] E. Adler, S. Brasen, and H. Miyake, *Acta Chem. Scand.*, **25**, 2055 (1971); see also E. Alder and K. Holmberg, *Acta Chem. Scand.*, **25**, 2775 (1971).
[142] F. L. Scott, T. M. Lambe, and R. N. Butler, *Tetrahedron Letters*, **1971**, 2669.
[143] V. P. Kudesia, *Bull. Soc. Chim. Belges*, **80**, 59, 213 (1971).
[144] V. P. Kudesia, *J. Sci. Res. (Hardwar, India)*, **2**, 33 (1970); *Chem. Abs.*, **74**, 99254 (1971).
[145] B. G. Cox, *J. Chem. Soc. (B)*, **1971**, 1704; see also *Org. Reaction Mech.*, **1970**, 592.
[146] J. M. Pink and R. Stewart, *Can. J. Chem.*, **49**, 649, 654 (1971).
[147] J. R. Campbell, A. Pross, and S. Sternhell, *Austral. J. Chem.*, **24**, 1425 (1971); A. Pross and S. Sternhell, *Austral. J. Chem.*, **24**, 1437 (1971).
[148] J. P. Danehy, B. T. Doherty, and C. P. Egan, *J. Org. Chem.*, **36**, 2525 (1971); J. P. Danehy, C. P. Egan, and J. Switalski, *J. Org. Chem.*, **36**, 2530 (1971); S. Beveridge and R. L. N. Harris, *Austral. J. Chem.*, **24**, 1229 (1971).

have also been examined: bromate oxidation of aldehydes;[149] the formation of azobenzene from aniline with $ClNH_2$;[150] hypohalite oxidation of adamantan-2-ols[151] and diarylsulphamides;[152] the conversion of ethers into esters by trichloroisocyanuric acid;[153] and the oxidation of 2-mercaptoethanol by o-iodosobenzoic acid.[154] Tetrazanes ($>$NNHNHN$<$), postulated intermediates in the Pb(IV) acetate oxidation of N-amino lactams ($>$N—NH$_2$) to *trans*-tetrazenes ($>$NN=NN$<$), can be isolated from the reaction of amino lactams with iodosobenzene diacetate at 0°; they decompose thermally to lactams ($>$NH) and can be oxidized to tetrazenes.[155]

Torssell's mechanism[156] for the dicyclohexylcarbodiimide–DMSO oxidation of alcohols to carbonyl compounds includes intramolecular proton abstraction by nitrogen in the protonated DCC–DMSO adduct (69). The dicyclohexylurea produced by using [$^2H_6$]-DMSO should therefore contain deuterium. Although it has been claimed that this is not the case,[157] new results support Torssell's suggestion. When testosterone is oxidized to androst-4-ene-3,17-dione by DCC and [$^2H_6$]DMSO, monodeuteriodicyclohexylurea is obtained. A plausible mechanism[158] consists of addition of the alcohol to (69) followed by cyclic decomposition of the adduct (70) with proton abstraction by nitrogen; the oxysulphonium ylid (71) so formed would collapse to Me_2S and the ketone. The first step in the NBS-induced DMSO oxidation of diphenylacetylene to benzil may be formation of a brominated vinyl cation (PhC^+=$CBrPh$).[159]

2-Methylhept-2-ene (72) is stereospecifically (>98%) oxidized by SeO_2 in 95% EtOH to *trans*-allylic alcohol (74). Such specificity is considered unlikely for S_N2' solvolysis

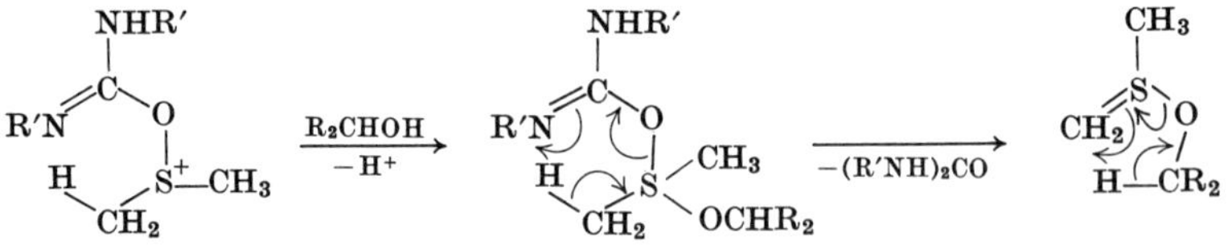

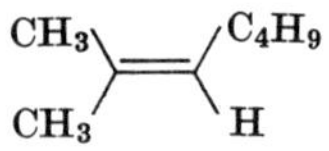

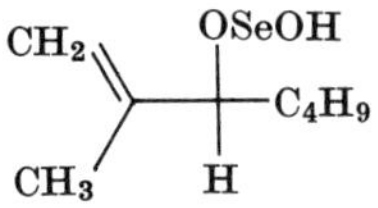

[149] V. Avasthi and A. C. Chatterji, *Z. Phys. Chem.* (*Leipzig*), **245**, 154, 161 (1970).
[150] G. A. Jaffari and A. J. Nunn, *J. Chem. Soc.* (C), **1971**, 823.
[151] R. M. Black and G. B. Gill, *Chem. Comm.*, **1971**, 172.
[152] D. L. Forster, T. L. Gilchrist, and C. W. Rees, *J. Chem. Soc.* (C), **1971**, 993.
[153] E. C. Jeunge, M. D. Corey, and D. A. Beal, *Tetrahedron*, **27**, 2671 (1971).
[154] J. Leslie, *Can. J. Chem.*, **48**, 3104 (1970).
[155] D. J. Anderson, T. L. Gilchrist, and C. W. Rees, *Chem. Comm.*, **1971**, 800.
[156] *Org. Reaction Mech.*, **1966**, 406–407; **1967**, 426.
[157] R. E. Harmon, C. V. Zenarosa, and S. K. Gupta, *Tetrahedron Letters*, **1969**, 3781; but see also Abstracts, 154th National Meeting of the American Chemical Society, Chicago, 1967, No. D3.
[158] J. G. Moffatt, *J. Org. Chem.*, **36**, 1909 (1971).
[159] S. Wolfe, W. R. Pilgrim, T. F. Garrard, and P. Chamberlain, *Can. J. Chem.*, **49**, 1099 (1971).

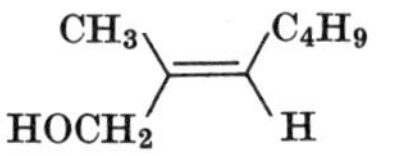

of the intermediate selenite ester (73), and an $S_N i'$ mechanism is preferred.[160] The influence of H_2SO_4 on the course of SeO_2 oxidation of olefins in AcOH has also been examined.[161]

Several studies of HNO_3,[162] H_2SO_4,[163] and SO_3[164] oxidation have been described, as have the oxidations of p-phenylenediamine derivatives by quinone di-imines,[165] epoxides by amine oxides,[166] tetrahydronaphthalene by DDQ,[167] and various substrates by potassium nitrosodisulphonate (Fremy's radical).[168] A cyclic ketal (75) reacts readily with trityl fluoroborate in CH_2Cl_2 (followed by aqueous work-up) to give the α-ketol (78) derived from the diol as well as the ketone (77) and Ph_3CH. A primary isotope effect ($k_H/k_D = 4$–5) is consistent with rate-determining hydride abstraction giving oxonium ion (76).[169]

Many electrochemical studies have again been concerned with the oxidation of amines[170] and phenols.[171] The formation of benzoin (82) and dideuteriodeoxybenzoin (83) from the enamine (79; R_2N = morpholino) in Bu^tOH–H_2O may be explained by

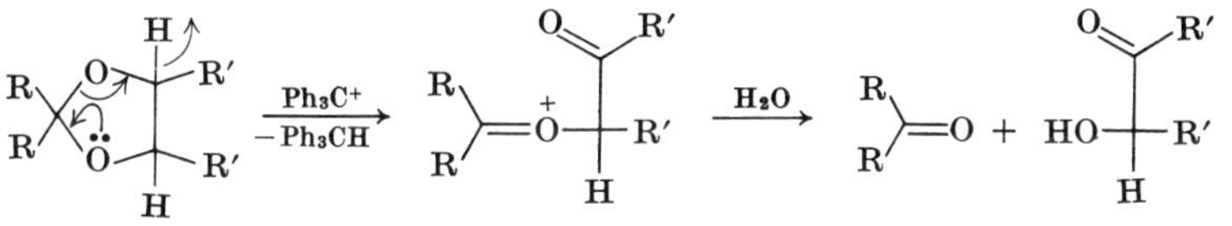

[160] U. T. Bhalerao and H. Rapoport, *J. Am. Chem. Soc.*, **93**, 4835 (1971).
[161] K. A. Javaid, N. Sonoda, and S. Tsutsumi, *Bull. Chem. Soc. Japan*, **43**, 3475 (1970).
[162] Y. Ogata and H. Tezuka, *Bull. Chem. Soc. Japan*, **43**, 3285 (1970); *Tetrahedron*, **26**, 5593 (1970); Y. Ogata and T. Kamei, *Tetrahedron*, **26**, 5667 (1970); E. J. Strojny, R. T. Iwamasa, and L. K. Frevel, *J. Am. Chem. Soc.*, **93**, 1171 (1971); W. Stec, A. Okruszek, and M. Mikolajczyk, *Z. Naturforsch.*, **26b**, 855 (1971).
[163] H. W. Geluk and J. L. M. A. Schlatmann, *Rec. Trav. Chim.*, **90**, 516 (1971).
[164] V. Mark, L. Zengierski, V. A. Pattison, and L. E. Walker, *J. Am. Chem. Soc.*, **93**, 3538 (1971).
[165] R. C. Baetzold and L. K. J. Tong, *J. Am. Chem. Soc.*, **93**, 1347 (1971).
[166] W. N. Marmer and D. Swern, *J. Am. Chem. Soc.*, **93**, 2719 (1971).
[167] P. J. van der Jagt, H. K. de Haan, and B. van Zanten, *Tetrahedron*, **27**, 3207 (1971).
[168] H. Zimmer, D. C. Lankin, and S. W. Horgan, *Chem. Rev.*, **71**, 229 (1971); K. Maruyama and T. Otsuki, *Bull. Chem. Soc., Japan*, **44**, 2873 (1971).
[169] D. H. R. Barton, P. D. Magnus, G. Smith, and D. Zurr, *Chem. Comm.*, **1971**, 861.
[170] G. Barbey, D. Delahaye, and C. Caullet, *Bull. Soc. Chim. France*, **1971**, 3377; G. Cauquis and J.-L. Cros, *Bull. Soc. Chim. France*, **1971**, 3760; G. Cauquis, J.-L. Cros, and M. Genies, *Bull. Soc. Chim. France*, **1971**, 3765; G. Cauquis, H. Delhomme, and D. Serve, *Tetrahedron Letters*, **1971**, 4113; G. Cauquis, J. Cognard, and D. Serve, *Tetrahedron Letters*, **1971**, 4645; G. Cauquis and M. Genies, *Tetrahedron Letters*, **1971**, 4677; R. Hand, M. Melicharek, D. I. Scoggin, R. Stotz, A. K. Carpenter, and R. F. Nelson, *Coll. Czech. Chem. Comm.*, **36**, 842 (1971); M. Masui and H. Sayo, *J. Chem. Soc. (B)*, **1971**, 1593.
[171] A. Ronlán and V. D. Parker, *J. Chem. Soc. (C)*, **1971**, 3214; V. D. Parker and A. Ronlán, *J. Electroanal. Chem. Interfacial Electrochem.*, **30**, 502 (1971); *Chem. Abs.*, **75**, 19541 (1971).

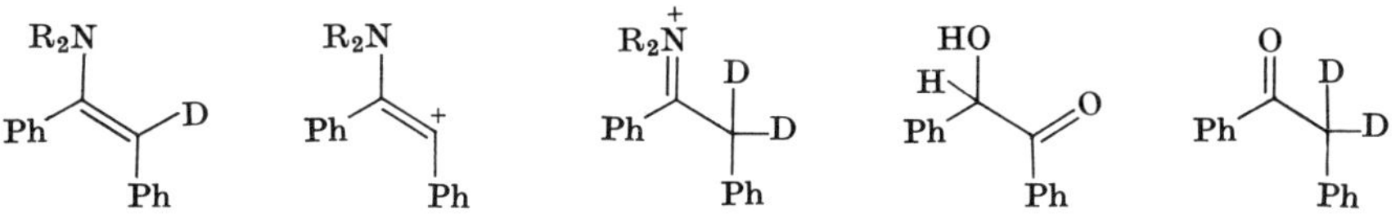

postulating that the radical cation formed by one-electron oxidation disproportionates to cations (**80**) and (**81**) which are then hydrolysed.[172] Simple arylethylenes have also been studied;[173] in MeCN, tetraphenylethylene gives 9,10-diphenylphenanthrene, probably (in part) via the dication.[174]

Anodic oxidation of phenylcyclopropane results in the cyclopropane ring being opened, although reaction is initiated by electron transfer from the aromatic nucleus.[175] A similar initiation is obviously not possible with bicyclo[4.1.0]heptane, although here too the products result from opening of the cyclopropane ring, seemingly by direct transfer of an electron from a C—C single bond to the anode.[176] Electrochemical oxidation of cyclopropanecarboxylic acids has been further investigated,[177] and voltammetric studies have revealed an arylacetate (caesium 9-methyl-10-anthraceneacetate) which suffers oxidative decarboxylation by a "pseudo-Kolbe" mechanism, i.e. electron removal from the aryl rather than from the carboxylate group.[178] Amongst other electrochemical oxidations reported are those of alkanes in HSO_3F,[179] anthracenes,[180] azines,[181] benzylic ether and carbonyl compounds,[182] phenyl sulphides,[183] ethylene glycol,[184] and 2,5-dimethylfuran.[185]

Reductions

The stereochemistry of reduction of cyclic ketones by metal hydrides continues to be actively investigated.[186,187] Although $LiAlH(OBu^t)_3$ is clearly more bulky than $LiAlH(OMe)_3$, it exhibits less selectivity (in THF) for attack on the carbonyl group of an asymmetric ketone from the less hindered side. It might be that the actual reducing species is $AlH(OBu^t)_2$ formed by dissociation, but new results argue against this. In a THF solution of $LiAlH(OBu^t)_3$ the equilibrium concentration of $AlH(OBu^t)_2$ is less than 1% as determined by IR spectroscopy. Moreover, the rates of reduction of cyclic ketones by $LiAlH(OBu^t)_3$ and pre-formed $AlH(OBu^t)_2$ are similar, although the stereochemistry of the products differ. Clearly, therefore, the product obtained from a ketone with $LiAlH(OBu^t)_3$ is not significantly derived from $AlH(OBu^t)_2$. An alternative

---

[172] S. J. Huang and E. T. Hsu, *Tetrahedron Letters*, **1971**, 1385.
[173] L. Eberson and V. D. Parker, *Acta Chem. Scand.*, **24**, 3553 (1970).
[174] J. D. Stuart and W. E. Ohnesorge, *J. Am. Chem. Soc.*, **93**, 4531 (1971).
[175] T. Shono and Y. Matsumura, *J. Org. Chem.*, **35**, 4157 (1970).
[176] T. Shono, Y. Matsumura, and Y. Nakagawa, *J. Org. Chem.*, **36**, 1771 (1971).
[177] L. B. Rodewald and M. C. Lewis, *Tetrahedron*, **27**, 5273 (1971); A. Takeda, S. Wada, and Y. Murakami, *Bull. Chem. Soc. Japan*, **44**, 2729 (1971).
[178] J. P. Coleman and L. Eberson, *Chem. Comm.*, **1971**, 1300.
[179] J. Bertram, M. Fleischmann, and D. Pletcher, *Tetrahedron Letters*, **1971**, 349.
[180] V. D. Parker, *Acta Chem. Scand.*, **24**, 2757, 2775, 3151, 3162, 3171, 3455 (1970).
[181] S. Hünig, G. Kiesslich, F. Linhart, and H. Schlaf, *Ann. Chem.*, **752**, 196 (1971).
[182] L. L. Miller, J. F. Wolf, and E. A. Mayeda, *J. Am. Chem. Soc.*, **93**, 3306 (1971); L. L. Miller, V. R. Koch, M. E. Larscheid, and J. F. Wolf, *Tetrahedron Letters*, **1971**, 1389.
[183] K. Uneyama and S. Torii, *Tetrahedron Letters*, **1971**, 329.
[184] A. K. Vijh, *Can. J. Chem.*, **49**, 78 (1971).
[185] K. Yoshida and T. Fueno, *J. Org. Chem.*, **36**, 1523 (1971).
[186] J.-M. Cense, A. Sevin, and W. Chodkiewicz, *Compt. Rend.* (C), **272**, 229 (1971); F. Rocquet, A. Sevin, and W. Chodkiewicz, *Compt. Rend.* (C), **272**, 417 (1971); F. Rocquet, A. Sevin, and W. Chodkiewicz, *Tetrahedron Letters*, **1971**, 1049; J.-P. Battioni and W. Chodkiewicz, *Bull. Soc. Chim. France*, **1971**, 1824.
[187] A. Daniel and A. A. Pavia, *Bull. Soc. Chim. France*, **1971**, 1060; C Bénard, M.-T. Maurette, and A. Lattes, *Compt. Rend.* (C), **273**, 426 (1971); O. Štrouf, *Coll. Czech. Chem. Comm.*, **36**, 2707 (1971); Y. Senda, S. Mitsui, R. Ono, and S. Hosokawa, *Bull. Chem. Soc. Japan*, **44**, 2737 (1971).

explanation of the greater selectivity of LiAlH(OMe)$_3$ is prompted by the discovery that it is increasingly associated in THF solutions of increasing concentration (approx. dimeric at 0.1M), whereas LiAlH(OBu^t)$_3$ remains monomeric. That the high selectivity of LiAlH(OMe)$_3$ is indeed largely a result of its concentration-dependent association is supported by the observation that the preference for attack on the less hindered side of 2-methylcyclohexanone, giving the less stable cis-alcohol, is greater in 0.1M (61%) than in 0.01M solution (28%).[188]

Reduction of benzophenone in ether containing an excess of an equimolar mixture of Me$_2$CHCH$_2$MgCl and Me$_2$CDCH$_2$MgCl preferentially gives undeuteriated diphenylcarbinol ($k_H/k_D = 2.7$*). However, the absolute reactivity of Me$_2$CDCH$_2$MgCl towards benzophenone is not appreciably less than that of Me$_2$CHCH$_2$MgCl. These two observations can be reconciled by a mechanism [reactions (27)–(29); R and R' = Me$_2$CHCH$_2$ or Me$_2$CDCH$_2$] in which the rate-determining step is electron transfer to the ketone [reaction (28)], and the diradical so formed can exchange its alkyl groups [reaction (29)] before decomposing to alcohol into which H rather than D is incorporated.[189]

$$2\text{RMgX} \underset{}{\overset{\text{fast}}{\rightleftharpoons}} \text{R}_2\text{Mg} + \text{MgX}_2 \qquad \ldots (27)$$

$$\text{Ph}_2\text{C}=\text{O} + \text{Mg}{<}^{\text{R}}_{\text{R}} \xrightarrow{\text{slow}} \text{Ph}_2\dot{\text{C}}-\text{O}-\dot{\text{Mg}}{<}^{\text{R}}_{\text{R}} \qquad \ldots (28)$$

$$\text{Ph}_2\dot{\text{C}}-\text{O}-\dot{\text{Mg}}{<}^{\text{R}}_{\text{R}} \underset{-\text{RMgX}}{\overset{+\text{R'MgX}}{\rightleftharpoons}} \text{Ph}_2\dot{\text{C}}-\text{O}-\dot{\text{Mg}}{<}^{\text{R'}}_{\text{R}} \qquad \ldots (29)$$

There has been further consideration of the factors that influence the proportions of addition and reduction, and the stereochemistry of the products, in the reactions of aldehydes and ketones with Grignard reagents[190] as well as with zinc, cadmium, beryllium and aluminium alkyls.[191] Results obtained from a detailed kinetic analysis of the reduction of benzophenone by lithium N-benzylanilide (in ether) are consistent with a mechanism [equation (30)] in which the reactants are in equilibrium with a 1:1 complex, which decomposes irreversibly with hydride transfer to the ketone in the rate-determining step.[192]

$$\underset{\text{O}}{\overset{\text{Ph}\quad\text{Ph}}{\text{C}}} + :\underset{\underset{\text{Li}}{|}}{\text{N}}{-}\text{CH}_2\text{Ph} \rightleftharpoons \underset{\text{O}\text{-----}\text{Li}}{\overset{\text{Ph}\quad\text{Ph}\quad\text{Ph}}{\text{C}\quad\quad\text{N}}}{-}\text{CH}_2\text{Ph} \longrightarrow \underset{\text{OLi}}{\overset{\text{Ph}\quad\text{Ph}}{\text{CH}}} + \underset{\text{CHPh}}{\overset{\text{Ph}}{\text{N}}} \quad \ldots (30)$$

* The implication that $k_D/k_H = 2.7$ in ref. 189 is assumed to be a misprint.

[188] E. C. Ashby, J. P. Sevenair, and F. R. Dobbs, *J. Org. Chem.*, **36**, 197 (1971).
[189] T. Holm, *J. Organomet. Chem.*, **29**, C45 (1971); see also T. Holm and I. Crossland, *Acta Chem. Scand.*, **25**, 59 (1971).
[190] M. Chastrette and R. Amouroux, *Bull. Soc. Chim. France*, **1970**, 4348; M. Chérest, H. Felkin, and C. Frajerman, *Tetrahedron Letters*, **1971**, 379; M. Chérest and H. Felkin, *Tetrahedron Letters*, **1971**, 383; D. Nasipuri, C. K. Ghosh, P. R. Mukherjee, and S. Venkataraman, *Tetrahedron Letters*, **1971**, 1587; see also J.-F. Fauvarque, *Compt. Rend.* (C), **272**, 1053 (1971).
[191] P. R. Jones, W. J. Kauffman, and E. J. Goller, *J. Org. Chem.*, **36**, 186 (1971); M. Chastrette and R. Amouroux, *Tetrahedron Letters*, **1970**, 5165; G. P. Giacomelli, R. Menicagli, and L. Lardicci, *Tetrahedron Letters*, **1971**, 4135.
[192] G. Wittig, H. F. Ebel, and G. Häusler, *Ann. Chem.*, **743**, 120 (1971).

Fluorenone (R_2CO) reacts with NaH in HMPT to give, after aqueous work-up, the pinacol (70%) together with a little fluorenol (10%). By contrast, the corresponding reaction in pyridine affords largely fluorenol (90%). In either case, quenching with D_2O produces 9-deuteriofluorenol, whereas use of NaD in place of NaH gives undeuteriated product. Clearly, formation of the alcoholate (R_2CHONa) by addition of NaH is not important. An alternative mechanism could involve electron transfer from a hydride ion to fluorenone to give the (caged) radical anion (84); this has been detected (in HMPT) by ESR spectroscopy. The radical anion could dimerize, leading to pinacol [reaction (31)], or be further reduced to the dianion, leading to fluorenol [reaction (32)]. Although it is not obvious why the solvent should be so influential in determining the reaction path, the fact that two equivalents of NaH are required to obtain a good yield of fluorenol (in pyridine) supports its formation via the dianion. Benzophenone gives largely diphenylcarbinol in either pyridine or HMPT. Although now the results with D_2O and NaD imply that reaction does occur via the alcoholate, detection of the ketone radical anion suggests that it may be formed, in part at least, by a mechanism involving initial electron transfer.[193]

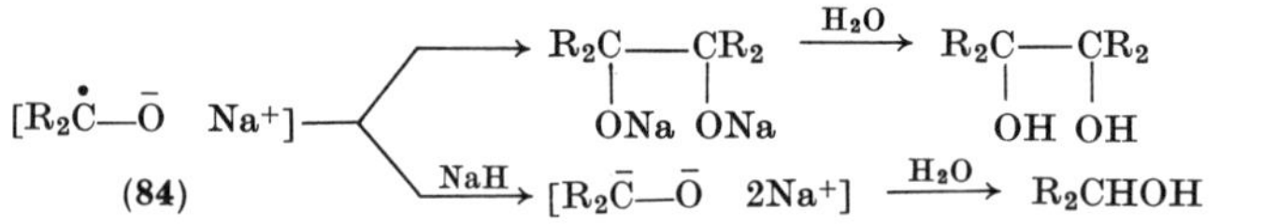

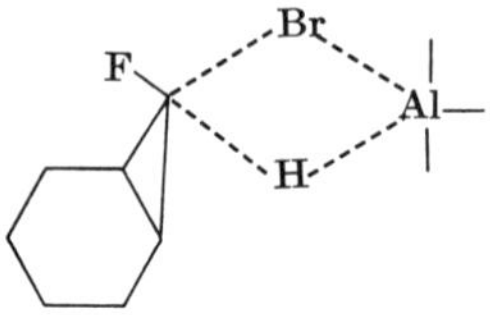

There have been further reports of the reactions of β-diketones,[194] acyl halides,[195] cyclic anhydrides,[196] cyclic acetals and ketals,[197] allenic epoxides,[198] and alkyl halides[199] with hydride reducing agents. Cyclonona-1,2-diene can result from the reduction of 9,9-dibromobicyclo[6.1.0]nonane by NaH in HMPT.[200] Retention of configuration in the LiAlH$_4$ reduction of the isomeric 7-bromo-7-fluorobicyclo[4.1.0]heptanes to the 7-fluorobicyclo[4.1.0]heptanes is consistent with a four-centre mechanism, e.g. (85).[201] LiAlH$_4$ not only replaces the halogen atoms in the dichlorocyclobutene (86) but also hydrogenates the ring olefinic bond giving (87; 41%).[202] The reduction of allylic alcohols has been further investigated,[203] as has the hydrogenation of acetylenes by NaBH$_4$ in

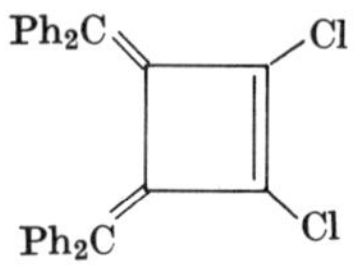

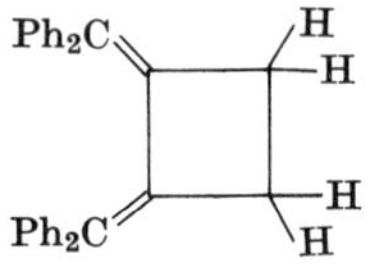

[193] P. Caubère and J. Moreau, *Bull Soc. Chim. France*, **1971**, 3270, 3276.
[194] J. W. Frankenfeld and W. E. Tyler, *J. Org. Chem.*, **36**, 2110 (1971).
[195] S. F. Sun and N. O. del Rosario, *J. Org. Chem.*, **35**, 4025 (1970).
[196] D. E. Burke and P. W. Le Quesne, *J. Org. Chem.*, **36**, 2397 (1971).
[197] M. S. Ahmad and S. C. Logani, *Austral. J. Chem.*, **24**, 143 (1971); H. A. Davis and R. K. Brown, *Can. J. Chem.*, **49**, 2166, 2563 (1971).
[198] J. Grimaldi and M. Bertrand, *Bull. Soc. Chim. France*, **1971**, 973.
[199] N. W. Gilman and L. H. Sternbach, *Chem. Comm.*, **1971**, 465; R. O. Hutchins, B. E. Maryanoff, and C. A. Milewski, *Chem. Comm.*, **1971**, 1097.
[200] J. Moreau and P. Caubère, *Tetrahedron*, **27**, 5741 (1971).
[201] H. Yamanaka, T. Yagi, K. Teramura, and T. Ando, *Chem. Comm.*, **1971**, 380.
[202] F. Toda, K. Kumada, N. Ishiguro, and K. Akagi, *Bull. Chem. Soc. Japan*, **43**, 3535 (1970).
[203] W. T. Borden and M. Scott, *Chem. Comm.*, **1971**, 381.

Oxidation and Reduction

aqueous solution, catalysed by molybdenum complexes of cysteine. A mechanism for the latter reaction, which accounts for the observed kinetics and for the stereospecific *cis*-hydrogenation, involves reduction of the original complex to a Mo(IV) species (**88**) having two *cis*-positions available for co-ordination of the substrate in the rate-determining step. Rapid hydrolysis of (**89**) will liberate the olefin, and a Mo(VI) species (**90**) which can be reduced back to (**88**).[204]

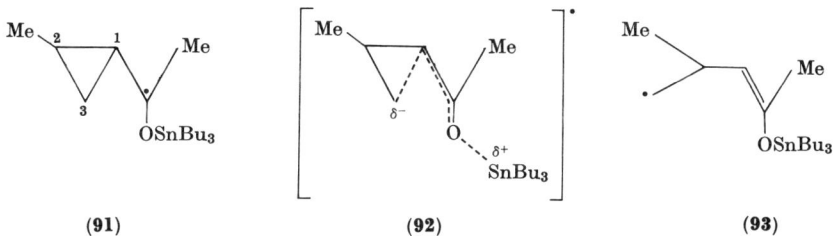

(88)   (89)   (90)

Use has been made of LiAlH$_4$ and NaBH$_4$ in investigations of the reduction of pyrazolones,[205] pyridinium salts,[206] and nitrobenzene,[207] the stereospecific desulphurization of phosphine sulphides[208] and the hydrogenolysis of organomercury compounds.[209] Phenyldiazene (PhN=NH) has been detected spectroscopically as an intermediate in the NaBH$_4$ reduction of benzenediazonium fluoroborate to benzene.[210] Reductions of various compounds by boranes,[211] silanes,[212] and stannanes[213] have been discussed. AIBN-initiated reduction of *trans*-2-methylcyclopropyl methyl ketone gives (after work-up) Bu^{i}COMe (85%) in preference to Bu^{n}COMe (15%). On grounds of radical stability alone, the observed preferential cleavage of the C-1–C-3 bond in intermediate (**91**) would not be expected, since it leads to a primary alkyl radical (**93**). If, on the other hand, the transition state (**92**) has polar character, as has been suggested,[214] the observed cleavage would not seem unreasonable.

(91)   (92)   (93)

[204] G. N. Schrauzer and P. A. Doemeny, *J. Am. Chem. Soc.*, **93**, 1608 (1971).
[205] J. Elguero, R. Jacquier, and D. Tizané, *Tetrahedron*, **27**, 133 (1971).
[206] F. Liberatore, A. Casini, V. Carelli, A. Arnone, and R. Mondelli, *Tetrahedron Letters*, **1971**, 2381, 3829.
[207] R. O. Hutchins, D. W. Lamson, L. Rua, C. Milewski, and B. Maryanoff, *J. Org. Chem.*, **36**, 803 (1971).
[208] R. Luckenbach, *Tetrahedron Letters*, **1971**, 2177.
[209] J. J. Perie and A. Lattes, *Bull. Soc. Chim. France*, **1971**, 1378; V. M. A. Chambers, W. R. Jackson, and G. W. Young, *J. Chem. Soc.* (C), **1971**, 2075.
[210] C. E. McKenna and T. G. Traylor, *J. Am. Chem. Soc.*, **93**, 2313 (1971).
[211] H. C. Brown, D. B. Bigley, S. A. Arora, and N. M. Yoon, *J. Am. Chem. Soc.*, **92**, 7161 (1970); S. A. Monti and R. R. Schmidt, *Tetrahedron*, **27**, 3331 (1971).
[212] R. A. Benkeser, *Accounts Chem. Res.*, **4**, 94 (1971); S. Spialter, W. A. Swansiger, L. Pazdernik, and M. E. Freeburger, *J. Organomet. Chem.*, **27**, C25 (1971).
[213] L. J. Altman and T. R. Erdman, *Tetrahedron Letters*, **1970**, 4891; H. Patin, L. Roullier, and R. Dabard, *Compt. Rend.* (C), **272**, 675 (1971).
[214] J.-Y. Godet and M. Pereyre, *Compt Rend.* (C), **273**, 1183 (1971).

From the reports of alkali-metal reduction of aromatic[215] and heterocyclic[216] systems comes a further illustration of the importance of ion-pair phenomena in determining the products and their stereochemistry. Table 1 shows the isomer composition of the 9,10-diethyl-9,10-dihydroanthracene obtained from 9,10-diethylanthracene with different metals and solvent compositions (in all cases H_2O was added slowly after 1 hr).

Table 1. Reduction of 9,10-diethylanthracene at 20°

Reducing metal	Na	Na	Li	Li
Solvent HMPT:THF ratio	3:2	1:4	3:2	1:4
Product cis:trans ratio	100:0	25:75	80:20	0:100

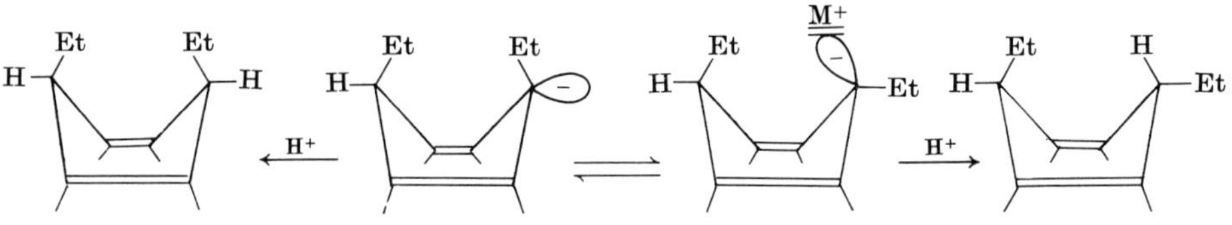

The authors[217] believe that the sequence reduction–protonation–reduction leads to a mixture of the free monoanion (**94**) and the solvent separated ion-pair (**95**), having preferred conformations as shown; and that the product cis:trans ratio reflects the position of equilibrium between (**94**) and (**95**). This being so, Li rather than Na as counter-ion, and a small proportion of HMPT in the solvent, encourage ion-pair formation.

Differences in ion-pairing are doubtless also responsible for the different behaviour of homocyclo-octatetraene dianion formed by two-electron reduction of bicyclo[6.1.0]-nona-2,4,6-triene in dimethoxyethane (DME) and NH_3. Protonation (with MeOH) in DME gives bicyclo[6.1.0]nona-2,4-diene, whereas in NH_3 (with NH_4Cl) it leads to cleavage of the cyclopropane ring and formation of cyclonona-1,3,6-triene.[218] Reductive cleavage of aryl-[219] and acyl-cyclopropanes[220] has also been further examined. The ring remains intact when allenic cyclopropane (**96**) reacts with Na–NH_3. Intramolecular protonation of an intermediate cyclopropyl carbanion by the hydroxyl group in (**96**) can explain the highly stereoselective formation of ($\pm$)-*trans*-chrysanthemyl alcohol (**97**).[221] Cyclic allenes from C_{11} to C_{14} give a higher proportion of *trans*-olefin if Na–NH_3 reduction is conducted in the presence of a proton donor, allyl radicals then being intermediates.[222]

---

[215] J. L. Marshall and T. K. Folsom, *Tetrahedron Letters*, **1971**, 757; H. van Bekkum, C. B. van den Bosch, G. van Minnen-Pathuis, J. C. de Mos, and A. M. van Wijk. *Rec. Trav. Chim.*, **90**, 137 (1971); L. Birkofer and N. Ramadan, *Chem. Ber.*, **104**, 138 (1971).
[216] W. A. Remers, G. J. Gibs, C. Pidacks, and M. J. Weiss, *J. Org. Chem.*, **36**, 279 (1971).
[217] R. Lapouyade, P. Labandibar, and H. Bouas-Laurent, *Tetrahedron Letters*, **1971**, 979.
[218] W. H. Okamura, T. I. Ito, and P. M. Kellett, *Chem. Comm.*, **1971**, 1317.
[219] H. M. Walborsky, M. S. Aronoff, and M. F. Schulman, *J. Org. Chem.*, **36**, 1036 (1971).
[220] Y. Bessière-Chrétien and M. M. El Gaied, *Bull. Soc. Chim. France*, **1971**, 2189.
[221] R. W. Mills, R. D. H. Murray, and R. A. Raphael, *Chem. Comm.*, **1971**, 555.
[222] R. Vaidyanathaswamy, G. C. Joshi, and D. Devaprabhakara, *Tetrahedron Letters*, **1971**, 2075.

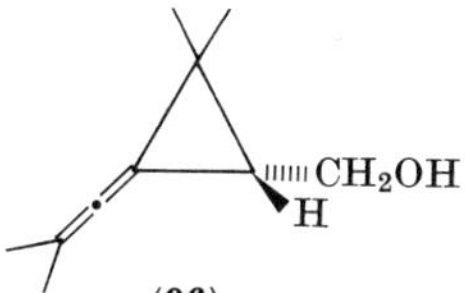

Reaction of cis-1,2-divinylcyclobutane (**98**) with Na–NH$_3$ is remarkably stereospecific, giving octa-cis-2,trans-6-diene (**99**; 69%) (by conrotatory ring opening) in much higher yield than the cis,cis- and trans,trans-isomers.[223]

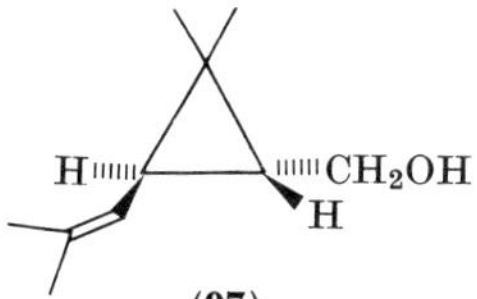

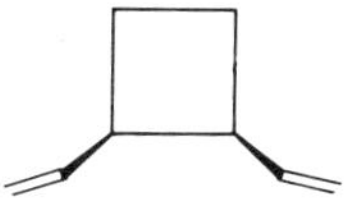

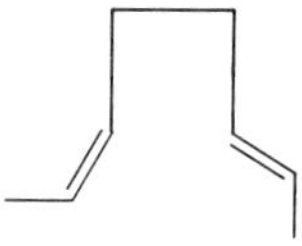

As might be expected, azocines, e.g. (**100**), readily form aromatic 10π-electron dianions on reduction with K–NH$_3$.[224] Although oxepins, e.g. (**101**), like azocines, have an 8π-electron system, their fifth MO is of such high energy that analogous behaviour would not be predicted. In fact, addition of K to (**101**) in NH$_3$–THF at −70° gives (after quenching) acyclic dienone (**104**). This probably arises from ring-opening at the one-electron reduction stage (**102**), giving the vinyl radical (**103**) which is further reduced to the anion.[225]

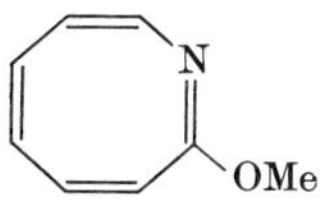

Metal–NH$_3$ reduction of camphor can give both alcohol (isoborneol + borneol) and pinacol products. The influence of the nature of the metal and of the presence of added salts, both on the stereochemistry of the alcohol and on the extent of pinacol formation, has been examined.[226,227] Table 2 contains representative results obtained in the

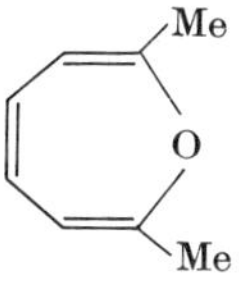

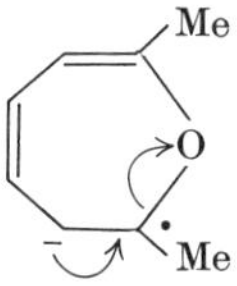

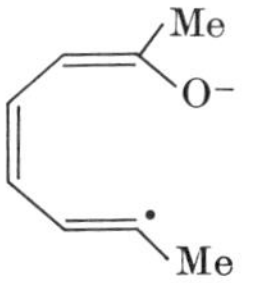

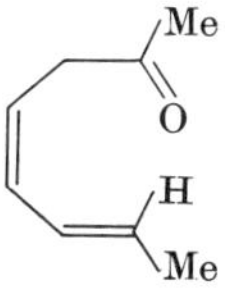

Table 2. Reduction of camphor in liquid ammonia

Reducing metal	Added salt	Metal:salt ratio	% Isoborneol in alcohol	Yield % pinacol
Li	None		20	70
K	None		58	0
Li	KBr	1:1	33	40
K	LiBr	1:1	34	40
K	LiBr	1:5	24	60

$$Li + KBr \text{ or } K + LiBr \rightarrow Li^+ + K^+ + e^- + Br^- \quad \ldots (33)$$
$$R_2C{=}O + e^- \rightleftharpoons R_2\overset{\bullet}{C}{-}\bar{O} \quad \ldots (34)$$
$$R_2\overset{\bullet}{C}{-}\bar{O} + Li^+ \text{ or } K^+ \rightarrow R_2\overset{\bullet}{C}{-}\bar{O}Li^+ \text{ or } R_2\overset{\bullet}{C}{-}\bar{O}K^+ \quad \ldots (35)$$

[223] H. Hey, *Angew. Chem. Int. Ed.*, **10**, 132 (1971).
[224] L. A. Paquette, J. F. Hansen, and T. Kakihana, *J. Am. Chem. Soc.*, **93**, 168 (1971).
[225] L. A. Paquette and T. McCreadie, *J. Org. Chem.*, **36**, 1402 (1971).
[226] A. Coulombeau and A. Rassat, *Bull. Soc. Chim. France*, **1970**, 4399, 4404; A. Coulombeau, *Bull. Soc. Chim. France*, **1970**, 4407.
[227] W. S. Murphy and D. F. Sullivan, *Tetrahedron Letters*, **1971**, 3707.

absence of an added proton donor. The first two entries emphasize the different course of the reaction with Li and with K while the other results show that a mixture of a metal and a salt of a different metal behaves as a "weighted average" of the individual metals. These and similar results lead to the conclusion that the ketone does not associate preferentially with the dissolving metal, but rather that ionization of the metal [reaction (33)] is complete before reduction [reaction (34)] and association to ion-pairs [reaction (35)] takes place.[227]

Aromatic ketones (ArCOR) are reduced by Li–NH_3 to hydrocarbons (ArCH$_2$R) when the reaction is quenched with NH_4Cl, but to alcohols [ArCH(OH)R] if $PhCO_2Na$ is used to destroy the excess of lithium. The alkoxide [ArCH(OLi)R] is first formed, and addition of NH_4Cl converts this into the alcohol, which can be further reduced before all the unchanged Li has been destroyed. When $PhCO_2Na$ is used, however, the Li is destroyed without the alkoxide being protonated.[228]

Semidiones have been detected in the reduction of norbornen-7-one with K–DME,[229] while other reports have dealt with the alkali-metal reductions of allyl ethers,[230] epoxides and ketals,[231] disubstituted malonic esters[232] and benzylphosphonium salts.[233] The influence of solvent and temperature on the stereochemistry of dehalogenation of halogenocyclopropanes by sodium naphthalenide has been examined,[234] and details of the reduction of propynal,[235] thioxanthylium salts[236] and ketones[237] by zinc or magnesium amalgam have been published.

Nitrosobenzene is rapidly reduced by dihydro-3-methyl-lumiflavin and is a likely intermediate in the (slower) reduction of nitrobenzene to N-phenylhydroxylamine in DMF. The effect of substituents on the second-order rate constants for the latter reaction ($\rho^- = +3.6$) is similar to that for the half-wave reduction potential of nitrobenzene. By analogy, a mechanism involving rate-determining one-electron transfer to nitrobenzene from the dihydroflavin has been suggested.[238] Reduction of nitro and nitroso compounds by alcohols,[239] by Na_2S (which gives olefins from vicinal dinitro compounds),[240] by vitamin B_{12S}[241] and by $(EtO)_3P$[242] has also been investigated, as have the reactions of Ph_3P with hydroperoxides[243] and tetracyanoethylene.[244]

[228] S. S. Hall, S. D. Lipsky, F. J. McEnroe, and A. P. Bartels, *J. Org. Chem.*, **36**, 2588 (1971); S. S. Hall, S. D. Lipsky, and G. H. Small, *Tetrahedron Letters*, **1971**, 1853.
[229] J. P. Dirlam and S. Winstein, *J. Org. Chem.*, **36**, 1559 (1971).
[230] D. H. Hunter and D. W. Moore, *Can. J. Chem.*, **49**, 1665 (1971).
[231] E. M. Kaiser, C. G. Edmonds, S. D. Grubb, J. W. Smith, and D. Tramp, *J. Org. Chem.*, **36**, 330 (1971).
[232] Y.-N. Kuo, F. Chen, C. Ainsworth, and J. J. Bloomfield, *Chem. Comm.*, **1971**, 136.
[233] A. W. Herriott, *Tetrahedron Letters*, **1971**, 2547.
[234] D. B. Ledlie, R. L. Thorne, and G. Weiss, *J. Org. Chem.*, **36**, 2186 (1971).
[235] Y. L. Pascal and S. Galaj, *Compt. Rend.* (C), **273**, 282, 1104 (1971).
[236] C. C. Price, M. Siskin, and C. K. Miao, *J. Org. Chem.*, **36**, 794 (1971).
[237] S. I. Goldberg, W. D. Bailey, and M. L. McGregor, *J. Org. Chem.*, **36**, 761 (1971); J. Binks and D. Lloyd, *J. Chem. Soc.*, (C), **1971**, 2641.
[238] M. J. Gibian and A. L. Baumstark, *J. Org. Chem.*, **36**, 1389 (1971).
[239] H. Toivonen, S. A. Laurema, and P. J. Ilvonen, *Tetrahedron Letters*, **1971**, 3203; S. L. Walters and T. C. Bruice, *J. Am. Chem. Soc.*, **93**, 2269 (1971).
[240] N. Kornblum, S. D. Boyd, H. W. Pinnick, and R. G. Smith, *J. Am. Chem. Soc.*, **93**, 4316 (1971).
[241] A. E. Brearley, H. Gott, H. A. O. Hill, M. O'Riordan, J. M. Pratt, and R. J. P. Williams, *J. Chem. Soc.* (A), **1971**, 612.
[242] J. I. G. Cadogan and S. Kulik, *J. Chem. Soc.* (C), **1971**, 2621; R. J. Sundberg and R. H. Smith, *J. Org. Chem.*, **36**, 295 (1971); R. J. Sundberg and C.-C. Lang, *J. Org. Chem.*, **36**, 300 (1971); Y. Maki, T. Hosokami, and M. Suzuki, *Tetrahedron Letters*, **1971**, 3509.
[243] R. Hiatt, R. J. Smythe, and C. McColeman, *Can. J. Chem.*, **49**, 1707 (1971); R. Hiatt and C. McColeman, *Can. J. Chem.*, **49**, 1712 (1971).
[244] M. P. Naan, R. L. Powell, and C. D. Hall, *J. Chem. Soc.* (B), **1971**, 1683.

Transformation of dialkyl disulphides into the monosulphides on reaction with $(Et_2N)_3P$ proceeds with inversion of configuration at one of the C atoms adjacent to S. Marked decreases in $\Delta H^{\neq}$ (and increases in rate) result from increased solvent polarity, consistently with formation of a charged intermediate in the rate-determining step. The mechanism shown in equation (36) is supported by the detection of the intermediate phosphonium salt in the desulphurization of bis(benzothiazol-2-yl) disulphide.[245] Reaction of thiolsulphonates (RSO_2–SR) with $(Et_2N)_3P$ can lead to both sulphones (RSO_2—R) and sulphinate esters (RSO–OR).[246]

$$\begin{array}{c} S-R \\ | H \\ | / \\ S-C{''''}R'' \\ \diagdown R' \end{array} \xrightleftharpoons[\text{slow}]{(Et_2N)_3P} (Et_2N)_3\overset{+}{P}-S-\overset{H}{\underset{R'}{C{''''}R''}} \bar{S}R \xrightarrow[\text{fast}]{-(Et_2N)_3PS} R''{''''}\underset{R'}{\overset{H}{C}}-SR \quad \ldots (36)$$

Deoxygenation of oxaziranes by thiourea and other sulphur-containing nucleophiles gives imines,[247] which are also obtained by reduction of oximes with Ti(III) chloride.[248] Further studies of the deoxygenation of sulphoxides[249] and seleninic acids[250] by iodide ion in aqueous acid have been described, as have the reductions of ketones by alcohols,[251] chloramine-T by hexacyanoferrate(II)[252] and diazoacetic esters by $Na_2S_2O_4$,[253] as well as the reductive coupling of naphthalene-1-diazonium ions.[254]

A model system has been found which underlines the importance of substrate activation by hydrogen bonding in NADH-mediated reduction of aldehydes. While both benzaldehyde and m-nitrobenzaldehyde are inert to reduction by the Hantzsch ester (**105**), salicylaldehyde (**106**; X = H) is slightly reduced in refluxing ethylene glycol, and the more acidic phenolic aldehyde (**106**; X = NO_2) is quantitatively reduced to the alcohol in refluxing dioxan or aqueous ethanol at 40°.[255]

(**105**)   (**106**)   (**107**)

[245] D. N. Harpp and J. G. Gleason, *J. Am. Chem. Soc.*, **93**, 2437 (1971).
[246] D. N. Harpp, J. G. Gleason, and D. K. Ash, *J. Org. Chem.*, **36**, 322 (1971).
[247] D. St.C. Black and K. G. Watson, *Angew. Chem. Int. Ed.*, **10**, 327 (1971).
[248] G. H. Timms and E. Wildsmith, *Tetrahedron Letters*, **1971**, 195.
[249] D. Landini, G. Modena, F. Montanari, and G. Scorrano, *J. Am. Chem. Soc.*, **92**, 7168 (1970); D. Landini, G. Modena, U. Quintily, and G. Scorrano, *J. Chem. Soc.* (B), **1971**, 2041; S. Allenmark and H. Johnsson, *Acta Chem. Scand.*, **25**, 1860 (1971); S. Tamagaki, M. Mizuno, H. Yoshida, H. Hirota, and S. Oae, *Bull. Chem. Soc. Japan*, **44**, 2456 (1971).
[250] F. Ferranti and D. De Filippo, *J. Chem. Soc.* (B), **1971**, 1925.
[251] T. Moriyoshi and M. Hirata, *Rev. Phys. Chem. Japan*, **40**, 59 (1970); *Chem. Abs.*, **74**, 46076 (1971).
[252] M. C. Agrawal and S. P. Mushran, *J. Phys. Chem.*, **75**, 838 (1971).
[253] N. T. Denisov, S. A. Solov'eva, and A. E. Shilov, *Kinet. Katal.*, **12**, 579 (1971); *Chem. Abs.*, **75**, 87882 (1971).
[254] D. V. Banthorpe and J. A. Thomas, *J. Chem. Soc.* (B), **1971**, 365.
[255] U. K. Pandit and F. R. M. Cabré, *Chem. Comm.*, **1971**, 552.

Analysis of polarographic half-wave potential data for a large number of substituted benzaldehydes, halogenobenzenes, and nitrobenzenes has led to the conclusion that *ortho*-substituents do not generally exert an important steric effect.[256]

Numerous investigations of the electrochemical reduction of carbonyl compounds, including hydroxy[257] and α,β-unsaturated[258] aldehydes and ketones, quinones,[259] β-diketones,[260] β-keto nitriles[261] and acetophenone,[262] have been described. The alcohol obtained from 4-t-butylcyclohexanone at a lead cathode in MeOH–NaOAc contains little (15%) of the less-stable *cis*-isomer, but this becomes the major product (56%) when AcOH is present. It may be that the protonated ketone becomes attached (in part) to the cathode by an equatorial C—Pb bond [as in (**107**)]; when discharged, the carbanion will initially have the OH group axial; provided protonation is rapid, the resulting alcohol will be rich in *cis*-isomer.[263]

Unlike cyclo-octatetraene, which is reduced in two separate one-electron steps,[264] and the π-equivalent 2-methoxyazocine, which is reduced by a single irreversible two-electron step,[265] 1,3,5,7-tetraphenylcyclo-octatetraene is electrochemically converted into the 10π-electron dianion in a single reversible two-electron step.[266]

Several authors have considered the cathodic reduction of halogen compounds.[267] It has been claimed that cyclopropane is formed from 1,3-dibromopropane by a concerted

$$\begin{array}{c}\delta^- \\ \text{Br} \cdots \text{CH}_2 \cdots \text{CH}_2 \cdots \text{Br} \\ \diagdown \quad \diagup \\ \text{CH}_2 \end{array}$$

(**108**)

(**109**)

(**110**)

process in which both C—Br bonds are breaking and the new C—C bond is forming in the transition state (**108**) for electron transfer.[268] This being so, *meso-* and ($\pm$)-2,4-dibromopentane should stereospecifically form *cis-* and *trans*-dimethylcyclopropane, respectively.

---

[256] M. Charton and B. I. Charton, *J. Org. Chem.*, **36**, 260 (1971).
[257] D. Barnes, P. C. Uden, and P. Zuman, *J. Chem. Soc.* (B), **1971**, 1114; M. Fedoroňko, E. Fülleová, and K. Linek, *Coll. Czech. Chem. Comm.*, **36**, 114 (1971).
[258] D. Barnes and P. Zuman, *J. Chem. Soc.* (B), **1971**, 1118; J.-M. Meunier and E. Laviron, *Bull. Soc. Chim. France*, **1971**, 3793; A. Albisson and J. Simonet, *Bull. Soc. Chim. France*, **1971**, 4213.
[259] E. R. Brown, K. T. Finley, and R. L. Reeves, *J. Org. Chem.*, **36**, 2849 (1971); R. D. Rieke, W. E. Rich, and T. H. Ridgway, *J. Am. Chem. Soc.*, **93**, 1962 (1971).
[260] E. Kariv, B. J. Cohen, and E. Gileadi, *Tetrahedron*, **27**, 805 (1971); E. Kariv, J. Hermolin, I. Rubinstein, and E. Gileadi, *Tetrahedron*, **27**, 1303 (1971); E. Kariv and E. Gileadi, *Coll. Czech. Chem. Comm.*, **36**, 476 (1971).
[261] A. Daver and G. Le Guillanton, *Bull. Soc. Chim. France*, **1971**, 312.
[262] J. H. Stocker and R. M. Jenevein, *Coll. Czech. Chem. Comm.*, **36**, 925 (1971).
[263] J. P. Coleman, R. J. Kobylecki, and J. H. P. Utley, *Chem. Comm.*, **1971**, 104.
[264] D. E. Smith and B. J. Huebert, *J. Electroanal. Chem. Interfacial Electrochem.*, **31**, 333 (1971); *Chem. Abs.*, **75**, 57997 (1971).
[265] L. B. Anderson, J. F. Hansen, T. Kakihana, and L. A. Paquette, *J. Am. Chem. Soc.*, **93**, 161 (1971).
[266] R. D. Rieke and R. A. Copenhafer, *Tetrahedron Letters*, **1971**, 4097.
[267] K. P. Butin, *Uspekhi Khim.*, **40**, 1058 (1971); *Russ. Chem. Rev.*, **40**, 525 (1971); Y. Matsui, T. Soga, and Y. Date, *Bull. Chem. Soc. Japan*, **44**, 513 (1971); H. Doupeux, P. Martinet, and J. Simonet, *Bull. Soc. Chim. France*, **1971**, 2299; H. A. O. Hill, J. M. Pratt, M. P. O'Riordan, F. R. Williams, and R. J. P. Williams, *J. Chem. Soc.* (A), **1971**, 1859.
[268] M. R. Rifi, *Coll. Czech. Chem. Comm.*, **36**, 932 (1971).

In fact, however, both compounds give a mixture of the isomeric dimethylcyclopropanes, presumably by cyclization of the following intermediate monobromocarbanion:[269]

$$(CH_3\bar{C}HCH_2CHBrCH_3).$$

Although benzyl groups are readily cleaved from acyclic phosphonium salts, controlled potential reduction of the cyclic diphosphonium salt (**109**) gives (**110**; 98%) in which the olefinic bonds have been hydrogenated without cleavage of any C—P bond.[270] Ammonium,[271] tropylium,[272] and dihydrodiazepinium[273] salts have also been studied, as have the electrogenerative hydrogenation of ethylene[274] and the cathodic reductions of nitro compounds,[275] amino acids,[276] benzidines,[277] azobenzenes,[278] azoxybenzenes,[279] diazoacetophenones,[280] imino compounds,[281] dihydropyridines,[282] ozonides,[283] and aromatic hydrocarbons.[284] The mixture of phthalide and o-(hydroxymethyl)benzanilide obtained from N-phenylphthalimide in DMF probably arises by further reduction of intermediate o-formylbenzanilide.[285]

Hydrogenation and Hydrogenolysis

Hydrogenation is included in a wide-ranging discussion of catalysis,[286] and reviews of the stereochemistry of hydrogenation of cyclic hydrocarbons,[287] asymmetric hydrogenation,[288] and homogeneous hydrogenation of C—C multiple bonds,[289] including selective hydrogenation of dienes and polyenes,[290] have been published.

[269] A. J. Fry and W. E. Britton, *Tetrahedron Letters*, **1971**, 4363.
[270] J. H. Stocker, R. M. Jenevein, A. Aguiar, G. W. Prejean, and N. A. Portnoy, *Chem. Comm.*, **1971**, 1478.
[271] A. Monvernay and P.-C. Lacaze, *Bull. Soc. Chim. France*, **1970**, 4206.
[272] P. H. Plesch and A. Stasko, *J. Chem. Soc.* (B), **1971**, 2052.
[273] H. P. Cleghorn, J. E. Gaskin, and D. Lloyd, *J. Chem. Soc.* (B), **1971**, 1615.
[274] S. H. Langer, I. Feiz, and C. P. Quinn, *J. Am. Chem. Soc.*, **93**, 1092 (1971).
[275] B. Lovreček, Z. Vajtner, and J. Hranilović, *Tetrahedron Letters*, **1971**, 3319; M. Heyrovský, S. Vavřička, and L. Holleck, *Coll. Czech. Chem. Comm.*, **36**, 971 (1971); C. Degrand, P. Brossier, and A. Darchen, *Bull. Soc. Chim France*, **1971**, 3798.
[276] S. Fujiwara, Y. Umezawa, and H. Ishizuka, *Bull. Chem. Soc. Japan*, **44**, 1984 (1971).
[277] D. Delahaye, G. Barbey, and C. Caullet, *Bull. Soc. Chim. France*, **1971**, 3082.
[278] K. G. Boto and F. G. Thomas, *Austral. J. Chem.*, **24**, 975 (1971).
[279] R. Hazard and A. Tallec, *Bull. Soc. Chim. France*, **1971**, 2917.
[280] M. Bailes and L. L. Leveson, *J. Chem. Soc.* (B), **1971**, 1957.
[281] J. Pinson and J. Armand, *Bull. Soc. Chim. France*, **1971**, 1764; J. Armand, L. Boularès, J. Pinson, and P. Souchay, *Bull. Soc. Chim. France*, **1971**, 1918; K. Kretzschmar and W. Jaenicke, *Z. Naturforsch.*, **26b**, 999 (1971); K. B. Wiberg and T. P. Lewis, *J. Am. Chem. Soc.*, **92**, 7154 (1970); H. Lund and E. T. Jensen, *Acta Chem. Scand.*, **25**, 2727 (1971).
[282] J. Kuthan, V. Simonek, V. Volkova, and J. Volke, *Z. Chem.*, **11**, 111 (1971); *Chem. Abs.*, **74**, 134139 (1971).
[283] J. Grignon and S. Fliszár, *Can. J. Chem.*, **49**, 3127 (1971).
[284] J. R. Jezorek and H. B. Mark, *J. Org. Chem.*, **36**, 666 (1971).
[285] D. W. Leedy and D. L. Muck, *J. Am. Chem. Soc.*, **93**, 4264 (1971).
[286] V. Gutmann and H. Noller, *Monatsh*, **102**, 22 (1971).
[287] O. V. Bragin and A. L. Liberman, *Uspekhi Khim.*, **39**, 2122 (1970); *Russ. Chem. Rev.*, **39**, 1017 (1970).
[288] E. I. Klabunovskii and E. S. Levitina, *Uspekhi Khim.*, **39**, 2154 (1970); *Russ. Chem. Rev.*, **39**, 1035 (1970).
[289] R. S. Coffey, in "Aspects of Homogeneous Catalysis", Vol. 1 (Ed. R. Ugo), C. Manfredi, Milan, 1970, p. 1.
[290] A. Andreetta, F. Conti, and G. F. Ferrari, in "Aspects of Homogeneous Catalysis", Vol. 1 (Ed. R. Ugo), C. Manfredi, Milan, 1970, p. 204.

A new type of hydrogenation catalyst selects olefins from solution on the basis of their size. The catalyst is prepared by chloromethylation of 10% of the aromatic rings in cross-linked polystyrene beads followed by replacement of 80% of the chlorine atoms with diphenylphosphine groups and equilibration of the product with $RhCl(PPh_3)_3$. The relationship between increasing molecular size of substrate and decreasing rate of hydrogenation is illustrated by the relative rates (in benzene, H_2 at 1 atm.) of hex-1-ene (1.00), cyclohexene (0.39), cyclo-octene (0.15) and cholest-2-ene (0.01). Such selectivity, which is not exhibited by $RhCl(PPh_3)_3$, suggests that hydrogenation occurs within the polymer beads, where the size of the solvent channels is restricted by cross-linking.[291]

Amongst results from studies of heterogeneous hydrogenation of olefins,[292] dienes[293] and acetylenes[294] with more conventional catalysts is one that emphasizes the directive effect of certain neighbouring groups. Thus, over Pd–C, olefin (**111**) gives almost exclusively the *cis*-isomer (**112**; 95%) resulting from preferential adsorption on the catalyst from the same side as the hydroxymethyl group. Evidently, specific interactions between the catalyst and the hydroxyl group more than compensate for steric hindrance to approach of the more hindered side of the olefin. When the hydroxymethyl substituent is replaced by methoxycarbonyl, hydrogenation gives predominantly the expected *trans*-product (85%).[295]

(**111**) (**112**)

Hydrogenation of a 1,4-diyne such as nonadeca-10,13-diynoic acid over Lindlar's catalyst in the presence of quinoline gives the corresponding *cis,cis*-1,4-diene. Although desorption of intermediate enynes shows that stepwise hydrogenation of the unsaturated centres is occurring, it is necessary, to account for the rapid formation of diene from the very beginning of the reaction, to postulate a competing pathway in which diyne is reduced directly to product.[296]

The similar stereochemistry of the ketone (**114**) (ca. 70% of *cis*-isomer) obtained by hydrogenation of α,β-unsaturated ketone (**113**) in neutral ethanolic solution over deactivated Raney Ni and by protonation of lithium enolate (**115**; X = Li) has been presented as evidence for the hydrogenation proceeding by 1,4-adsorption and -addition to give the enol (**115**; X = H).[297]

Hydrogenolysis of cyclopropyl carbonyl compounds (**116**; X = Me, OH or OMe) over Pd results in cleavage of the C-1—C-2 (or C-1—C-3) bond, rather than C-2—C-3,

---

[291] R. H. Grubbs and L. C. Kroll, *J. Am. Chem. Soc.*, **93**, 3062 (1971).
[292] O. V. Bragin, L. V. Pozdnova, and A. L. Liberman, *Izv. Akad. Nauk SSSR, Ser. Khim.*, **1970**, 1319; *Chem. Abs.*, **73**, 130545 (1970).
[293] P. F. Carr and J. K. A. Clarke, *J. Chem. Soc. (A)*, **1971**, 985.
[294] A. M. Sokol'skaya, V. A. Shóshenkova, V. F. Lokhmatova, S. A. Ryabinina, and K. K. Kuzembaev, *Zhur. Fiz. Khim.*, **45**, 610 (1971); *Chem. Abs.*, **74**, 140462 (1971).
[295] H. W. Thompson, *J. Org. Chem.*, **36**, 2577 (1971).
[296] A. Steenhoek, B. H. Van Wijngaarden, and H. J. J. Pabon, *Rec. Trav. Chim.*, **90**, 961 (1971).
[297] D. Baudry, J. P. Bégué, M. Bettahar, M. Charpentier-Morize, C. Pardo, and J. Sansoulet, *Bull. Soc. Chim. France*, **1971**, 1413.

(113)   (114)   (115)   (116)

irrespective of the other substituents (H, Me or Ph) on the cyclopropane ring. Whether this reflects the orientation of adsorbed substrate on the catalyst or simply the weakness of the C-1—C-2 bond is not clear.[298] Selectivity in the hydrogenolysis of unsymmetrical epoxides has been examined in detail; for example, 1,2-epoxybutane is selectively reduced by cleavage of the secondary C—O bond (giving butan-1-ol) over Ni and the primary C—O bond (giving butan-2-ol) over Pd; with either catalyst, 1-methylepoxycyclopentane gives largely 2-methylcyclopentanol (*trans* > *cis*) by fission of the more substituted C—O bond.[299]

Asymmetric hydrogenation of carbonyl,[300] olefin[301] and imine[302] functions in non-dissymmetric molecules has been further investigated by using catalysts modified with optically active compounds. Methyl acetoacetate gives (−)-methyl 3-hydroxybutyrate over Raney Ni modified with L-(S)-valine [Me$_2$CHCH(NH$_2$)CO$_2$H]. Introduction of a second asymmetric centre into the modifier [MeCHRCH(NH$_2$)CO$_2$H] increases the optical yield of the (−)-enantiomer. the increase being greater when the configuration at C$_\beta$ is R rather than S.[303] Further work on the alkali-metal-catalysed hydrogenation of aromatic hydrocarbons has been reported,[304] as have studies of the heterogeneous hydrogenation and hydrogenolysis of aniline,[305] benzyl alcohols,[306] cyclohexenyl ethers,[307] imines,[308] methylpyridines[309] and pyridinium salts.[310]

Assessment of the activity and selectivity of soluble transition-metal catalysts as a function of the metal and the ligands still attracts much attention.[311] For example,

---

[298] A. L. Schultz, *J. Org. Chem.*, **36**, 383 (1971).

[299] G. Sénéchal and D. Cornet, *Bull. Soc. Chim. France*, **1971**, 773; G. Sénéchal, J.-C. Duchet, and D. Cornet, *Bull. Soc. Chim. France*, **1971**, 783.

[300] F. Higashi, T. Ninomiya, and Y. Izumi, *Bull. Chem. Soc. Japan*, **44**, 1333 (1971); Y. Izumi, S. Yajima, K. Okubo, and K. K. Babievsky, *Bull. Chem. Soc. Japan*, **44**, 1416 (1971); Y. Izumi, T. Harada, T. Tanabe, and K. Okuda, *Bull. Chem. Soc. Japan*, **44**, 1418 (1971); T. Harada, M. Imaida, and Y. Izumi, *Bull. Chem. Soc. Japan*, **44**, 1419 (1971).

[301] T. Yoshida and K. Harada, *Bull. Chem. Soc. Japan*, **44**, 1062 (1971).

[302] K. Harada and K. Matsumoto, *Bull. Chem. Soc. Japan*, **44**, 1068 (1971).

[303] Y. Izumi and K. Ohkubo, *Bull. Chem. Soc. Japan*, **44**, 1330 (1971).

[304] S. Friedman, M. L. Kaufman, and I. Wender, *J. Org. Chem.*, **36**, 694 (1971); see also M. Ichikawa, M. Soma, T. Onishi, and K. Tamaru, *Bull. Chem. Soc. Japan*, **43**, 3672 (1970).

[305] K. Ikedate and S. Suzuki, *Bull. Chem. Soc. Japan*, **44**, 325 (1971).

[306] A. P. G. Kieboom, J. F. de Kreuk, and H. Van Bekkum, *J. Catal.*, **20**, 58 (1971).

[307] S. Nishimura, M. Katagiri, T. Watanabe, and M. Uramoto, *Bull. Chem. Soc. Japan*, **44**, 166 (1971).

[308] E. Frainnet, P. Braquet, and F. Moulines, *Compt. Rend.* (C), **272**, 1435 (1971).

[309] E. A. Mistryukov, E. L. Ilkova, and M. A. Ryashentseva, *Tetrahedron Letters*, **1971**, 1691.

[310] M. Tsuda and Y. Kawazoe, *Chem. Pharm. Bull.*, **18**, 2499 (1970); *Chem. Abs.*, **74**, 52799 (1971).

[311] G. Henrici-Olivé and S. Olivé, *Angew. Chem. Int. Ed.*, **10**, 105 (1971); L. Horner and H. Siegel, *Ann. Chem.*, **751**, 135 (1971); W. Strohmeier, R. Fleischmann, and T. Onoda, *J. Organomet. Chem.*, **28**, 281 (1971); W. Strohmeier and R. Fleischmann, *J. Organomet. Chem.*, **29**, C39 (1971); W. Strohmeier, *J. Organomet. Chem.*, **32**, 137 (1971); W. Strohmeier, W. Rehder-Stirnweiss, and R. Fleischmann, *Z. Naturforsch.*, **25b**, 1480, 1481 (1970); W. Strohmeier and W. Rehder-Stirnweiss, *Z. Naturforsch.*, **26b**, 193 (1971); W. Strohmeier and R. Endres, *Z. Naturforsch.*, **26b**, 730 (1971).

the rate of hydrogenation of each of the isomeric heptenes by $RhX(CO)L_2$ in toluene decreases as X changes Cl → Br → I → SCN (with L = PPh_3) and as L changes PPh_3 → $P(cyclohexyl)_3$ → $P(OPh)_3$ (with X = Cl).[312] A catalyst that is active for hydrogenation and isomerization of olefins is obtained by addition of PPh_3 to a benzene solution of $[Ir(cyclo-octene)_2Cl]_2$. Maximum activity is observed with a PPh_3:Ir ratio of 2:1, suggesting that the active species is $IrCl(PPh_3)_2$.[313]

Results reported last year clearly implicated O_2 in the isomerization of an olefin that accompanies its $RhCl(PPh_3)_3$-catalysed hydrogenation.[314] It has now been found that the isomerization of olefins catalysed by $RuCl_2(PPh_3)_3$ is enhanced by the presence of O_2. After isomerization of 4-vinylcyclohexene to ethylidenecyclohexene in the presence of O_2 the carbonyl complex $RuCl_2(CO)(PPh_3)_2(C_8H_{12})$ can be isolated, and this is an efficient catalyst for isomerization of pure olefin in the absence of O_2. Its formation presumably involves displacement of PPh_3 ligand by CO arising from catalysed autoxidation of the olefin. The enhanced activity of a metal complex in which CO has replaced PPh_3 may be of steric origin.[315]

Asymmetric hydrogenation of olefins catalysed by soluble Rh complexes containing phosphine ligands with asymmetry at P is well established.[316] New results show that complexes in which the asymmetry still originates in the phosphine but not at the P atom may be even more effective. For example, the catalyst obtained on addition of the chiral diphosphine (117) (2 moles) to a solution of $[Rh(cyclo-octene)_2Cl]_2$ causes α-acetamidocinnamic acid (118) to be reduced to (R)-N-acetylphenylalanine (119) with an optical yield of 72%.[317] Although in this instance the efficiency may result in part from the conformational rigidity of the diphosphine, this is not the case with $RhCl(neomenthyl-PPh_2)_3$, which gives a 61% enantiomeric excess of (S)-3-phenylbutanoic acid from (E)-β-methylcinnamic acid.[318] It may be significant that the substrates in these examples have carboxylate functions that could participate in the hydrogenation. The scope of catalysts (120) with an asymmetric carboxamide ligand has also been extended.[319]

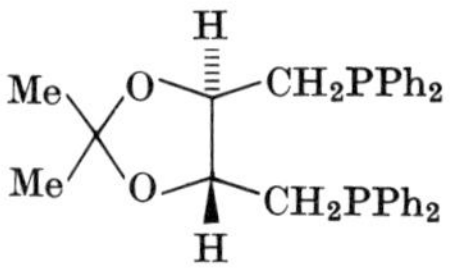

(117) 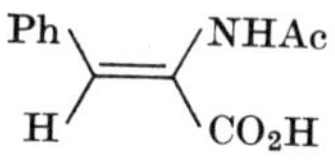   (118) 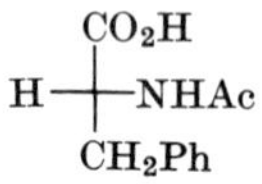   (119)

$py_2(carboxamide)RhCl_2(BH_4)$ $[Rh(CO)_2(O_2CEt)]_2$
(120)   (121)

---

[312] W. Strohmeier and W. Rehder-Stirnweiss, *Z. Naturforsch.*, **26b**, 61 (1971).
[313] H. van Gaal, H. G. A. M. Cuppers, and A. van der Ent, *Chem. Comm.*, **1970**, 1694.
[314] *Org. Reaction Mech.*, **1970**, 606; see also J. F. Biellmann, M. J. Jung, and W. R. Pilgrim, *Bull. Soc. Chim. France*, **1971**, 2720.
[315] J. E. Lyons, *Chem. Comm.*, **1971**, 562; *J. Org. Chem.*, **36**, 2497 (1971).
[316] *Org. Reaction Mech.*, **1968**, 495; **1969**, 571.
[317] T. P. Dang and H. B. Kagan, *Chem. Comm.*, **1971**, 481.
[318] J. D. Morrison, R. E. Burnett, A. M. Aguiar, C. J. Morrow, and C. Phillips, *J. Am. Chem. Soc.*, **93**, 1301 (1971).
[319] P. Abley, I. Jardine, and F. J. McQuillin, *J. Chem. Soc.* (C), **1971**, 840; P. Abley and F. J. McQuillin, *J. Chem. Soc.* (C), **1971**, 844.

Noteworthy amongst other investigations of the homogeneous catalytic reduction of olefins[320] and dienes[321] is the isolation of a carboxylate complex (**121**) from the hydroformylation of ethylene catalysed by $Rh_4(CO)_{12}$. By implication, this demonstrates the intermediacy in the hydroformylation of an acylrhodiumcarbonyl that has been oxidized by traces of O_2 to the complex isolated.[322] Soluble transition-metal catalysts have been employed in studies of the hydrogenation of olefins and ketones by alcohols[323] and by formic acid.[324] Simple n-alkanes exchange H/D with the solvent in CH_3CO_2D–D_2O containing K_2PtCl_4 at 100° at rates that increase with chain length and are inversely related to the ionization potentials of the alkanes.[325]

[320] M. G. Burnett and R. J. Morrison, *J. Chem. Soc.* (A), **1971**, 2325; H. A. Martin and R. O. de Jongh, *Rec. Trav. Chim.*, **90**, 713 (1971); L. A. Kheifits, A. E. Gol'dovskii, I. S. Kolomnikov, and M. E. Vol'pin, *Izvest. Akad. Nauk SSSR, Ser. Khim.*, **1970**, 2078; *Bull. Acad. Sci. USSR*, **1970**, 1951.

[321] T. Funabiki and K. Tarama, *Bull. Chem. Soc. Japan*, **44**, 945 (1971); *Tetrahedron Letters*, **1971**, 1111; C. K. Brown, W. Mowat, G. Yagupsky, and G. Wilkinson, *J. Chem. Soc.* (A), **1971**, 850.

[322] B. Heil, L. Markó, and G. Bor, *Chem. Ber.*, **104**, 3418 (1971).

[323] M. Gullotti, R. Ugo, and S. Colonna, *J. Chem. Soc.* (C), **1971**, 2652; Y. Sasson and J. Blum, *Tetrahedron Letters*, **1971**, 2167.

[324] M. E. Vol'pin, V. P. Kukolev, V. O. Chernyshev, and I. S. Kolomnikov, *Tetrahedron Letters*, **1971**, 4435.

[325] R. J. Hodges, D. E. Webster, and P. B. Wells, *Chem. Comm.*, **1971**, 462.

Author Index 1971

Aakano, Y., 453
Aaron, H. S., 88, 461
Aaron, J. J., 202
Aaviksaar, A. A., 448
Abakumov, G. A., 349
Abbott, D. J., 154, 163
Abbott, E. M., 405
Abduvakhavov, A. A., 448
Abe, K., 71, 157
Abe, T., 450
Abegg, V. P., 367
Abenhaim, D., 108
Abernethy, J. L., 450
Abley, P., 558
Ablov, A. V., 443
Abraham, M. H., 56, 124, 132
Abrahamson, E. W., 506, 520
Abraitys, V. Y., 508
Abramovich, L. D., 197
Abramovitch, R. A., 214, 367, 371, 375, 391, 516, 536
Absar, I., 87
Abu Elamayem, M. S., 456
Abuin, E., 489
Acheson, R. M., 67, 177, 239, 262, 269, 441, 508
Achmatowicz, S., 490
Achter, E. K., 400
Ackerman, R. A., 522
Acton, N., 501
Adam, G., 483, 487, 519
Adam, W., 95, 295, 300, 492, 543
Adamic, K., 352
Adams, D. G., 128
Adams, D. J. C., 383, 536
Adams, D. L., 234
Adams, R., 54, 340
Adams, W. R., 522
Adcock, S., 487
Adderley, C. J. R., 4, 161
Aders, W.-K., 296
Adhikary, P., 374, 515
Adler, E., 543
Adman, E., 506
Adolf, P. K., 539
Adolph, H. G., 106

Afanas'ev, I. B., 276, 296, 303, 308, 310
Afanas'ev, V. A., 407
Agarwal, S. P., 444
Agata, I., 191
Agopian, G., 512
Agosta, W. C., 382, 481, 487
Agrawal, M. C., 553
Aguiar, A., 555
Aguiar, A. M., 558
Ahlgrenm, C., 493
Ahmad, A., 147
Ahmad, M. S., 548
Ahmad, S. I., 442
Ahmed, Q. A., 495
Ahmed, S., 202
Ahonkhai, S. I., 145
Ahrens, H., 405
Aida, T., 92
Ainarde, V. R., 453
Ainsworth, C., 114, 144, 146, 228, 343, 348, 552
Aizel, R., 456
Akada, T., 543
Akagi, K., 208, 243, 548
Akamsin, V. D., 461
Åkermark, B., 369, 493
Akhrem, A. A., 241
Akhtar, S. M. S., 360
Akhvlediani, I. G., 276
Akiba, K., 226, 282
Akimoto, H., 359
Akselrod, B. Y., 417
Aksel'rod, Z. I., 197, 207
Al Holly, M. M., 217
Alais, L., 413
Albagli, A., 112
Albano, E., 450
Alberto, Z. R., 304
Albierty, H., 448
Albisson, A., 554
Albrecht, A. C., 475
Albrecht, F., 479
Alcais, P., 202, 251
Alder, R. W., 113, 175, 250
Aleksandrov, Y., 108
Aleksandrova, I. A., 461

Alekseeva, N. F., 156
Alexander, J. E., 490
Alexander, R., 305, 334
Alexandrov, Y. A., 529
Alford, J. A., 16, 168, 528
Alford, J. R., 14
Alierskaya, L. L., 463
Allan, G., 504
Allan, Z. J., 212
Allen, A. K., 399
Allen, C. M., 460
Allen, D. M., 48, 518
Allen, G. R., 125, 174
Allen, G. R., Jr., 11
Allen, G. W., 460
Allen, J. B. E., 7
Allen, P. E. M., 127
Allen, R. M., 71, 250
Allenmark, S., 553
Allerhand, A., 462
Allgrove, R. C., 272
Allinger, N. L., 107
Allison, W. S., 540
Allred, A. L., 413
Allred, E. L., 288
Alt, H., 131
Altman, L. J., 29, 278, 297, 549
Amano, K., 257
Ames, L. J., 247
Amiel, Y., 308
Amin, K., 187
Amir-Ud-Din, 147
Ammon, H. L., 205
Amouroux, R., 127, 547
Anand, N., 166
Anantakrishnan, S. V., 353, 537
Anantaraman, R., 54, 435
Anastassiou, A. G., 176, 232, 233, 374, 496
Andersen, K. K., 89, 92, 464
Anderson, A. G., 205
Anderson, D. G., 85, 379
Anderson, D. J., 230, 370, 544
Anderson, D. T., 489
Anderson, E., 393, 394, 395

Author Index

Anderson, H., 181
Anderson, H. J., 209
Anderson, L., 403
Anderson, L. B., 554
Anderson, N. H., 335, 343
Anderson, N. S., 394
Anderson, R. L., 389
Anderson, S. N., 208
Ando, T., 17, 101, 104, 253, 548
Ando, W., 106, 384, 476
Andre, J. C., 487
Andreae, S., 421
Andreatta, R. H., 450
Andreetta, A., 555
Andreeva, L. S., 461
Andrejević, V., 331
Andréu, P., 142
Andrews, G. C., 224, 239, 325
Andrews, G. D., 47, 160
Andrews, L. J., 520
Andrews, S. B., 379
Andrich, O. D., 406
Andrist, A. H., 229, 232, 378
Angadiyavar, C. S., 171, 508
Angelici, R. L., 443
Angelino, N., 405
Angeloni, A. S., 109
Anguibeaud, P., 96
Anh, N. T., 164, 165
Anschel, M., 371
Anschütz, W., 382
Ansell, M. F., 166, 217
Anselme, J.-P., 391
Anteunis, A., 397
Anteunis, M., 396, 397
Anteunis-De Ketelaere, F., 397
Antheunis, T., 468
Antonello, C., 506
Antonini, E., 398
Antonov, V. K., 426
Antonova, N. D., 160
Antonucci, F. R., 517
Aono, T., 191
Aoyagi, R., 490
Aoyama, H., 478
Aoyama, S., 245
Aoyama, T., 191
Aoyama, Y., 303
Aponti, G. S., 492
Applebury, M. L., 462
Arai, H., 512
Araki, Y., 214, 483
Aratani, T., 127
Arce, E., 448
Archie, W. C., Jr., 235
Archila, J., 406
Arcoria, A., 103

Argentini, M., 537
Arimitsu, S., 476
Arita, H., 400
Arita, M., 400
Arkell, A., 148
Armand, J., 264, 555
Armour, E. A., 256
Armstrong, C., 354
Arnaudo, T. F., 163
Arnett, E. M., 125, 202, 428
Arnold, D. R., 492, 508
Arnold, H., 79
Arnon, R., 400
Arnone, A., 317, 549
Arold, W., 531
Aronoff, M. S., 114, 550
Arora, K. J. S., 481
Arora, P. C., 344
Arora, S. A., 549
Arrington, J. P., 481, 493
Arzhankov, S. I., 336
Asahara, T., 229
Asahi, Y., 423, 457
Asami, R., 112
Asano, T., 202
Asao, S., 492
Asao, T., 102
Asaturyan, R. A., 276
Ash, D. K., 553
Ashbrook, C. W., 222
Ashby, E. C., 127, 128, 130, 547
Ashton, D. S., 300, 302
Ashton, H., 399, 445
Ashworth, J., 107
Ashworth, P., 353
Asmus, K.-D., 112
Astley, V., 351, 371
Atala, S., 429
Atassi, M. Z., 400
Atherton, J. H., 382
Atkins, R. C., 374
Atkins, T. J., 11, 254, 256
Atkinson, D. J., 296, 316
Atkinson, R. E., 469
Atkinson, R. S., 522, 538
Atlas, D., 449
Atovmyan, E. G., 287
Attardi, D. G., 398
Attridge, C. J., 257, 263
Aubry, C., 456
Audeh, C. A., 536
Audic, C., 525
Auge, W., 495
Augustinsson, K.-B., 453
Auld, D. S., 452
Aumann, R., 238, 264
Aumeer, P. S., 253
Aune, J. P., 508

Aune, K. C., 447
Aurich, H. G., 353
Austad, T., 97
Austel, V., 508
Avasthi, V., 544
Avato, T., 450
Avedikian, A.-M., 48, 107
Aversa, M. C., 170, 268
Averyanov, V. A., 304
Avraamides, J., 141
Avramoff, M., 265
Awang, D. V. C., 302
Axelrad-Cavadone, C., 445
Aylward, J. B., 45, 64, 276, 536
Ayres, R. L., 308
Ayscough, P. B., 362, 363, 489
Aziz, S., 453
Azogu, C. I., 391, 516
Azuma, N., 336
Azuma, T., 442

Baba, H., 469
Babanko, S. D., 472
Babb, B. E., 519
Babievsky, K. K., 557
Babior, B. M., 134
Bacaloglu, I., 430
Bacaloglu, K., 181
Bacaloglu, R., 183, 429, 430
Bach, R. D., 158
Bachhawat, J. M., 159
Bachman, G. L., 376
Baciocchi, E., 201
Back, M. H., 359
Backlin, R., 337
Bacon, R. G. R., 542
Bacon, W. E., 218
Badcock, C. C., 519
Bader, R. F. W., 377
Badía, C., 364
Badley, E. M., 390
Badley, R. A., 452
Baehler, B., 408
Baer, H. H., 122
Baer, T. A., 378
Bässler, T., 99
Baetzold, R. C., 387, 516, 545
Bäuerlein, S., 376
Bagal, C. I., 427
Bagal, I. L., 116
Bagal, L. I., 116, 258, 416
Bagg, J. W., 359
Baggiolini, E., 487
Baggott, J. P., 540
Bagol, L. I., 127
Bahn, A., 98
Bahurel, Y., 482
Bailes, M., 555
Bailey, D. M., 70

Author Index

Bailey, F. P., 196
Bailey, P. S., 527
Bailey, R. A., 79
Bailey, W. D., 150, 552
Bailey, W. F., 396
Baines, A. F. H., 247
Baiocchi, L., 12
Baird, M. S., 33, 74, 154, 229, 388
Baird, N. C., 500
Baird, W. C., 155, 224
Baitis, F., 519
Bajgar, J., 453
Bak, B. S., 46
Baker, A. J., 166
Baker, C., 493
Baker, D. J., 257
Baker, R., 2, 4, 21, 161, 531
Bakore, G. V., 531, 532, 533, 543
Bakuzis, P., 28, 65, 242, 250
Balaban, A. T., 130, 276, 353
Balasubrahmanyam, S. N., 156
Balasubramanian, A., 54
Balasubramanian, M., 156
Balasubramanian, T. R., 533
Balavoine, G., 497
Baldwin, F. C., 114, 538
Baldwin, J. E., 166, 167, 175, 222, 223, 224, 226, 227, 229, 232, 234, 378, 383, 407, 456, 541
Baldwin, R. C., 278
Bales, S. E., 344
Balf, J., 161
Balf, R. J., 25
Baliah, V., 107
Balinsky, D., 540
Bal'kov, B. G., 287
Ball, R. E., 456
Ballantyne, M. M., 218
Ballard, D. H., 208
Ballardini, R., 472
Ballester, M., 45, 364
Balmain, A., 487
Balny, C., 525
Balzani, V., 472
Bamberger, S., 361, 362
Bamkole, T. L., 180
Ban, Y., 261
Banba, A., 504
Bancroft, K. C. C., 196, 207
Bandlish, B. K., 287
Bandurco, V. T., 537
Banerji, K. K., 531, 532, 533
Banger, J., 146
Bank, J. F., 338
Bank, S., 338, 339

Bańkowska, Z., 413
Banks, D. B., 265, 388
Banks, R. E., 352
Banthorpe, D. V., 5, 212, 346, 511, 553
Banucci, E. G., 171
Barabas, A., 130
Baranov, S. N., 317, 414
Baranova, N. G., 308
Barbarella, G., 192
Barber, L. L., 489
Barbey, G., 545, 555
Barborak, J. C., 10, 13, 233
Bărbulescu, N., 108
Barclay, L. R. C., 514
Barends, R. J. P., 168
Barger, H. J., Jr., 269
Bargon, J., 314
Barili, P. L., 109
Bariou, B., 54
Barker, B. J., 58
Barker, R., 250, 403, 414, 538
Barlin, G. B., 182, 542
Barlow, M. G., 154, 167
Barlow, R. E., 435
Barltrop, J., 481
Barltrop, J. A., 482, 491
Barltrop, J. P., 482
Barnes, C., 532
Barnes, D., 554
Barnes, G. R., 371
Barnett, C., 202
Barnett, J. E. G., 401
Barnett, J. W., 398, 456
Barnett, W. E., 69, 150, 154
Baron, W. J., 373
Baronnet, F., 359
Barragan, R., 512
Barringer, W. C., 485
Barroeta, N., 143
Barron, W. J., 381
Bartelink, H. J. M., 336
Bartels, A. P., 552
Bartholemew, R. F., 523
Bartlett, P. D., 105, 166, 253
Bartolini, G., 239, 325
Bartollozzi, M., 192
Barton, D., 359
Barton, D. H. R., 46, 155, 219, 227, 258, 330, 490, 495, 516, 519, 545
Barton, J. S., 540
Barton, S. C., 468
Bartsch, R. A., 136, 150
Basbudak, M., 495
Basolo, F., 391
Bass, J. M. A., 199
Bass, K. C., 315
Basselier, J.-J., 232, 508, 522

Bassindale, A. R., 85, 204
Basson, R. A., 360
Basson, S. S., 108
Bastani, B., 224
Bastos, M. P., 407
Batalov, A. P., 127
Batt, L., 353
Battaglia, A., 169, 542
Battersby, A. R., 137
Battersby, J., 47
Battioni, J. P., 416, 546
Battiste, M. A., 4, 30, 161, 370, 500
Battisti, A., 479, 481
Batty, J. W., 67, 164
Batz, F., 265
Baudet, P., 185
Baudry, D., 556
Bauer, H. H., 236
Bauer, W., 235
Bauld, N. L., 338, 527
Baule, M., 318
Baum, A. A., 472, 498
Baumann, B. C., 217
Baumann, W. J., 396
Baumgarten, R. J., 65, 242
Baumstark, A. L., 552
Bawdry, D., 248
Baxendale, J. H., 361
Baxter, C. S., 175
Baxter, I., 489
Baylouny, R. A., 227
Bayne, W. F., 135
Bazilevskii, M. V., 300
Be Hyeon, S., 521
Beal, D. A., 544
Beale, J. H., 105
Beames, D. J., 20
Bean, G. P., 207
Beard, C. D., 45, 99, 371, 379
Beardsley, G. P., 191, 271
Beasley, G. H., 220
Beauchamp, J. L., 105
Beaute, C., 257
Bechara, E. J. H., 538
Beck, A. K., 359
Beck, B. R., 490
Beck, D., 228
Beck, H.-J., 389, 390
Beck, W., 365
Becker, R. H., 191
Becker, R. S., 525
Beckley, R. S., 254
Beckmann, B., 478
Beckwith, A. L. J., 47, 321, 324, 345
Beeby, P. J., 107
Beg, M. A., 531
Begtrup, M., 182

Bégué, J. P., 41, 102, 248, 556
Behar, D., 352, 361
Behar, J. V., 284
Behbahany, F., 132
Behre, H., 67, 269
Behrens, G., 301
Behrenwald, E., 404
Behrman, E. J., 456, 542
Beileryan, N. M., 348
Beine, R. L., 504
Bekhli, L. S., 457
Bekowies, P. J., 475
Belayev, V. D., 107
Beletskaya, I. P., 95, 116, 121, 131, 204
Belik, P. I., 304
Belke, C. J., 434
Bell, J. A., 377
Bell, R. P., 118, 122, 402, 412
Bell, T. N., 304, 329
Bellamy, F., 512
Bellard, S. A., 480
Belletire, J. L., 82
Bellinger, N., 72, 272
Belloli, R., 367
Bellucci, G., 109
Bellus, D., 487
Bel'skii, V. E., 64, 461
Beltrame, P., 169, 219
BeMiller, J. N., 398
Bénard, C., 546
Benassi, R., 264
Ben-Bassat, J. M., 298, 476
Bender, M. L., 419, 432, 443, 444, 445
Benderskii, V. A., 472
Bendetskii, K. M., 401
Benecke, H. P., 239, 325
Benezra, C., 307
Benezra, S. A., 195
Benjamin, B. M., 2
Benkeser, R. A., 549
Benkovic, S. J., 89, 458, 460
Benn, M. H., 70
Bennett, J. E., 277
Bennett, M. A., 154
Bennett, S. W., 315
Bennion, D. N., 57
Bensasson, R. V., 468, 475
Benschop, H. P., 88
Benson, S. W., 144, 277, 287, 304, 359, 412
Bentley, M. D., 43, 56, 465
Bentley, T. J., 330
Beránek, V., 200
Berbes, R., 198
Berezen, I. V., 442, 446, 447, 448, 449
Berezovskii, V. M., 197, 207

Berg, M., 168
Berg, W., 464
Berger, A., 449
Berger, D., 168
Berger, H., 461
Berger, S., 61
Bergman, B., 470
Bergman, D., 383
Bergman, N.-A., 123
Bergman, R. G., 29, 97, 99, 291, 292, 379, 515
Bergmann, F., 264
Bergmark, W. R., 478
Bergon, M., 413
Bergstrom, R. G., 187, 341
Berlin, A. A., 346
Berlin, K. D., 2, 408
Berliner, E., 155
Berliner, L. J., 400
Bernal, S., 219
Bernardi, R., 317
Bernasconi, C. F., 185, 187, 341
Bernatek, E., 528
Bernáth, G., 108, 441
Berngruber, W., 361
Bernhard, S. A., 448, 540
Bernier, I., 399
Bernstein, H., 399
Bernstein, Z., 439
Berrier, C., 77, 248
Berry, R. S., 192
Berscheid, H. G., 487
Berseck, L., 82
Berson, J. A., 5, 7, 9, 227, 235, 239, 292, 501
Berterello, H. E., 190
Berthelot, C., 522
Berthelot, J., 522
Berthelot, J.-P., 248
Berti, G., 247
Bertilsköld, H., 414
Bertini, F., 317
Bertini, J. R., 452
Bertolini, M., 531
Bertram, J., 77, 546
Bertrand, M., 23, 34, 108, 158, 167, 226, 264, 541, 548
Bertrand, M. P., 349
Berwin, H. J., 60, 73, 131, 281
Besserer, K., 48
Bessière-Chrétien, Y., 550
Bethell, D., 385
Betoux, J. M., 213
Bettahar, M., 556
Betts, J., 353, 538
Betz, W., 42
Beug, M., 200
Beug, M. W., 453

Beugelmans, R., 521
Beutel, J., 525
Bevan, P. L. T., 361
Beveridge, S., 543
Beverly, G. M., 159
Beychok, S., 399
Beyer, H., 212
Bezaguet, A., 226
Bezzubova, N. N., 461
Bhacca, N. S., 175
Bhalerao, V. T., 545
Bhandari, K. S., 489
Bhatmagar, A. K., 227
Bhatt, K., 525
Bhatt, M. V., 435, 531
Bhattacharyya, S. K., 156
Bhujang, P. S., 412
Biale, G., 141
Bialecka, E., 113, 369
Biasotti, J. B., 92, 464
Bibler, N. E., 360
Bickel, A. F., 354, 538
Biehler, J.-M., 54
Biel, J. H., 267
Biellmann, J. F., 116, 233, 558
Biernbaum, M. S., 240, 416
Biggi, G., 180, 181
Biggs, J., 108
Bigley, D. B., 549
Bilevich, K. A., 47, 346
Billups, W. E., 297
Bilobrova, A. I., 464
Bilofsky, H. S., 406
Biltonen, R., 444, 448
Bingham, R. C., 1, 13, 75, 277
Binkley, R. W., 508
Binks, J., 552
Binsch, G., 261
Binshtok, E. V., 118
Bird, C. W., 267, 508
Birkofer, L., 550
Bissell, R. L., 105
Bixler, H. J., 398
Black, A. L., 347
Black, D. St. C., 553
Black, R. M., 544
Blackburn, E. V., 299, 300
Blackburn, G. M., 425
Blackstock, D. J., 199
Blades, A. T., 144
Blair, A. H., 540
Blair, T. T., 444
Blake, C. C. F., 399
Blake, P. G., 145, 455, 457
Blanch, J. H., 271
Blanchard, M., 248, 250, 539
Blandamer, M. J., 60
Blank, B., 283

Author Index

Blank, M. L., 463
Blankespoor, R. L., 113, 340
Blattmann, H. R., 497
Blewett, F. M., 458
Bley, P. F., 261, 408
Blier, J. E., 230
Block, D. R., 189
Block, E., 522, 529
Bloom, A., 57, 511
Bloom, S. M., 264
Bloomfield, J. J., 343, 483, 501, 552
Blount, J., 251
Blout, E. R., 449, 450
Blum, J., 559
Blumberg, S., 398
Blume, G., 368
Blumenfeld, A. L., 360
Bly, R. S., 18
Blyth, C. A., 441
Boak, D. S., 130, 304
Bobbitt, J. M., 344
Bobko, N. P., 455
Bobylev, B. N., 396
Bocard, C., 157
Boch, G., 233
Boche, G., 13
Bochenkov, V. N., 70
Bocher, G., 98
Bock, H., 131
Bockrath, B., 339
Boden, R. M., 324, 499
Bodley, F. H., 540
Bodot, H., 75
Bodrikov, I. V., 155, 201
Böck, H., 282
Boeckman, R. K., 126
Boeckman, R. K., Jr., 36
Boelema, E., 15, 247
Bölsing, F., 364
Boer, F. P., 445
Boese, D., 520
Boettcher, R. R., 239
Bogathov, S. V., 166, 428
Bogentoft, C., 271
Boger, J., 428
Boggio, R. J., 219
Boggs, R. A., 334
Bogolyubova, G. M., 453
Bohen, J. M., 177
Bohlmann, F., 517
Bohm, H., 481
Bohme, D. K., 105
Boikess, R. S., 126
Boiko, T. S., 413
Boiko, V. N., 180
Boire, B. A., 492
Boireau, G., 108
Bojarski, J. T., 420

Bok, L. D. C., 108
Boldt, P., 515
Boleij, J. S. M., 489
Bolton, P. D., 420, 427
Bonazza, B. R., 64
Bonhomme, M., 203
Bonner, T. G., 396
Bonnier, J. M., 258
Bonnier, J.-M., 202, 205
Boocock, D. G. B., 352
Boone, D. E., 48, 205
Bor, G., 559
Borčić, S., 26, 52
Borden, W. T., 106, 116, 236, 270, 412, 548
Borders, C. L., 399
Bordner, J., 25, 248
Bordwell, F. G., 102, 118, 122, 138, 139, 144, 149, 242, 413, 538
Borer, M. C., 247
Borkman, R. F., 487
Bornais, J., 46
Borodkin, V. F., 70
Boron, E. R., 64
Borowitz, I. J., 96, 371
Bortolus, P., 497, 500
Bory, S., 119
Bos, H. J. T., 94, 489, 490
Bosch, M., 482
Boschetto, D. J., 106
Bosco, M., 182
Bose, A. K., 66, 245
Bosen, S. F., 445
Bosworth, N., 78, 247
Boto, K. G., 555
Bott, K., 48, 99
Bott, R. W., 204
Bottari, F., 247
Bouas-Laurent, H., 503, 550
Boubel, J. C., 57
Bouchoule, C., 406
Bouchy, M., 487
Boudjouk, P., 352
Boue, S., 495
Bouis, P. A., 43
Boularès, L., 555
Boulton, B. E., 202
Boulton, J. J. K., 183, 185
Bourdon, J., 522
Bourgeois, J. M., 395
Bourguignon, P., 248
Bourlas, M., 485
Bourne, E. J., 396
Bovington, C. H., 192
Bowd, A., 470
Bowden, K., 179, 193, 402, 435, 436
Bower, H., 363

Bowie, S. E., 465
Bowles, A. J., 301
Bowman, D. F., 352
Bowyer, P. M., 48
Boyce, R., 97, 116
Boyd, G. V., 508
Boyd, J., 15, 157, 245
Boyd, M. R., 504
Boyd, S. D., 552
Boyer, J. H., 383, 497
Boyer, J. P. H., 409
Boyer, P. D., 443, 461
Boyer, R. F., 228
Boyle, L. W., 169
Boyle, P. H., 83, 484
Boyle, W. J., 122, 346
Boyle, W. J., Jr , 144
Bozzato, G., 487
Brace, N. O., 137, 312
Bradamante, S., 113
Bradbury, J. H., 400, 462
Bradbury, S., 383, 536
Bradley, J. D., 285, 519
Bradshaw, J. S., 189, 490
Bradsher, C. K., 168
Brady, W. T., 172, 253
Bragin, O. V., 555, 556
Brainard, R., 491
Bramphis, R. V., 470
Brand, P. A. T. M., 468
Brand, W. W., 213
Brandenburg, C. F., 217
Brandt, E. V., 78
Braquet, P., 557
Brasen, S., 543
Brauer, H.-D., 108, 295
Brauman, J. I., 204, 235, 295, 478, 518
Braun, D., 215, 276, 364
Braun, W., 310
Bravo, P., 133
Braye, E. H., 132
Brearley, A. E., 552
Breehaart-Hansen,J.C.A.E., 459
Breen, J. E., 359
Brember, R. P., 498
Bremner, J. B., 506
Brener, L., 220, 254
Brenner, S., 115
Brent, D. A., 369
Bresler, S. E., 448
Breslow, R., 201, 443, 477
Bressan, G. B., 182
Bressel, U., 173, 207
Brestin, A. P., 462
Brestkin, A. P., 453
Brett, C. L., 318
Breuer, G. M., 479

Brewer, J. P. N., 190, 517
Brewer, T., 474
Brewer, T. L., 501
Bribes, J. L., 41
Brice, M. C., 489
Brickmann, J., 87
Bricout, D., 521
Bridger, R. F., 353, 354, 538
Bridon, J. N., 177
Bridson, J. N., 269
Briegleb, G., 474
Brieman, R., 265
Brieux, J. A., 179
Brilkina, T. G., 282
Brimacombe, J. S., 69
Brimage, D. R. G., 473, 476, 503, 523
Brinen, J. S., 372, 470, 516
Brinkman, B., 29, 81
Briody, J. M., 431
Briody, R. G., 51
Bristol, D., 285
Britton, W. E., 555
Broadhurst, M. J., 11, 125, 174
Broche, A., 420
Brockmann, H., 344
Brodowski, W., 229
Brodsky, L., 495
Broida, H. P., 467
Brokenshire, J. L., 286, 353
Bromilow, R. H., 69, 457
Brook, A. G., 47, 85, 240, 371, 379, 386, 480, 483, 515
Brook, A. J. W., 106
Brook, P. R., 101, 266
Brooke, G. M., 217
Brookhart, M., 20, 21, 39
Brophy, G. C., 5
Brossier, P., 555
Brouwer, D. M., 41, 45, 417
Brown, A., 441
Brown, C., 89, 261
Brown, C. K., 559
Brown, D. A., 497
Brown, D. B., 48
Brown, D. G., 134
Brown, D. J., 193, 270, 271
Brown, E., 515
Brown, E. R., 554
Brown, E. V., 192, 453
Brown, G. H., 218, 468
Brown, H. C., 17, 109, 130, 157, 158, 327, 328, 329, 538, 549
Brown, J. E., 222, 383
Brown, J. M., 121, 180, 371
Brown, J. N., 230
Brown, K. H., 107, 430

Brown, K. S., 200
Brown, L. E., 540
Brown, M. D., 119
Brown, R. F. C., 372, 517
Brown, R. K., 548
Brown, R. S., 60
Brown, R. T., 251
Brown, W., 215
Brownstein, S., 46
Broxton, T. J., 346
Bruce, J. M., 489
Bruce, R. L., 204
Bruck, D., 43
Brülstein, M., 506
Bruggink, A., 83, 113
Brugidou, J., 272
Bruice, T. C., 48, 109, 262, 394, 426, 432, 433, 440, 441, 552
Brummel, R. N., 158
Brune, H. A., 220
Brunet, J.-J., 101, 133
Brunn, E., 170
Brunner, H., 364
Bruylants, A., 179, 180, 304, 413
Bryan, W. P., 400
Bryce, M. C., 362
Bryce-Smith, D., 468, 501, 503
Brydon, D. L., 190, 313
Bryson, T. A., 25
Bubnov, N. N., 346
Buchachenko, A. L., 283
Buchanan, J. G., 69
Buchardt, O., 512
Buchholz, R. F., 243, 411
Buck, H. M., 43
Budde, W. L., 95
Buddrus, J., 229
Budnik, M., 383
Budylin, V. A., 197
Budziarek, R., 214
Budzis, M., 244
Buenker, R. J., 229
Bürgi, H. B., 406
Buettner, V., 476
Buglass, A. J., 426, 453
Buist, G. J., 543
Bukhaeva, V. T., 182
Bukharov, V. G., 456
Bulgakova, G. M., 287
Bull, H., 406
Bull, H. G., 394
Bullock, C., 61
Bullock, G. A., 484
Bullock, S., 246
Bunce, N. J., 302, 335, 513
Buncel, E., 212, 420, 463

Buncel, F., 198
Bundel, Y. G., 78, 79, 132
Bunnett, J. F., 105, 137, 189, 190, 191, 241, 337, 346
Bunton, C. A., 5, 55, 394, 442, 459, 543
Burakevich, J. V., 408
Burchill, C. E., 336, 362
Burdon, J., 302
Burger, K., 170
Burger, M., 302
Burger, V., 216
Burgess, J. R., 168, 528
Burgmaier, G. J., 382
Burighel, A., 100
Burka, L. T., 229
Burke, D. E., 548
Burke, P. L., 157
Burkinshaw, G. F., 13, 215
Burkoth, T. L., 231
Burley, T. W., 54, 111
Burmistrov, S. I., 386
Burnett, G. M., 353
Burnett, M. G., 559
Burnett, R. E., 558
Burnstein, I. J., 227
Burrows, H. D., 336, 524
Bursey, M. M., 195
Bursics, L., 251
Burt, D. W., 94
Burton, D. J., 101
Burton, T., 304
Burwell, R. L., 412
Busby, R. E., 379
Bush, D. A., 399
Bushby, R. J., 264, 292
Bushick, R. D., 260
Bushweller, C. H., 406
Buss, V., 26, 34, 74
Butcher, M., 372, 517
Butin, K. P., 121, 304, 554
Butler, A. R., 43, 199, 203, 207
Butler, G. B., 168
Butler, J. R., 370, 500
Butler, M. M., 296
Butler, R. N., 261, 408, 543
Butsugan, Y., 434, 521
Butt, Y., 500
Buu-Hoï, N. P., 205
Buza, M., 155
Byrd, J. E., 253, 534
Byrd, L. R., 154
Byers, A. E., 127
Byers, G. W., 470

Cabadi, Y., 306
Cabel, M., 412
Cabell, M., 116, 270, 516

Author Index

Cabell, M., Jr., 252
Cabré, F. R. M., 553
Cacace, F., 208
Cachaza, J. M., 108
Cadet, J., 360
Cadman, P., 304, 359
Cadogan, J. I. G., 190, 313, 317, 386, 459, 552
Cagniant, P., 72
Caille, J., 522
Cain, E. N., 121, 236
Cairns, W. L., 506
Calas, B., 251
Căldăraru, H., 353
Calder, I. C., 241, 441
Caldin, E. F., 122
Caldow, G. L., 177
Caldwell, R. A., 472
Callot, H. J., 307
Calloway, A. R., 469
Calmon, J. P., 413
Caloyannis, A. G., 540
Calvaruso, G., 427
Calvert, J. G., 518, 519
Calvin, M., 474, 506, 512
Cameron, A. F., 25
Cameron, D. W., 241, 441
Cameron, G. G., 353
Cameron, J., 353
Camioni, D. M., 184
Camp, R. L., 229
Campbell, D. S., 455
Campbell, G. A., 201, 213
Campbell, J. D., 92
Campbell, J. R., 543
Campbell, L. L., 463
Campbell, M. M., 235
Campbell, P., 201
Campbell, P. G. C., 534
Campbell, R. V. M., 72
Canady, W. J., 446
Canfield, N. D., 371
Canfield, R. E., 399
Cann, P. F., 47, 72, 244
Cannon, P. L., 459
Canonica, L., 536
Canselier, J. P., 203
Cantrell, T. S., 483, 484, 485, 494
Capelle, G., 467
Caplain, S., 504
Caple, R., 4, 152
Caplier, I., 132
Caplow, M., 428, 446
Capman, M. L., 416
Capon, B., 394, 398, 399, 403, 433, 434, 441, 445
Capozzi, G., 100, 245
Caprosu, M., 170

Caquis, G., 161
Caramella, P., 172
Carapellucci, P. A., 506
Cardaci, G., 95
Cardillo, B., 96
Cardin, D. J., 389
Cardon, F., 474
Carelli, V., 167, 549
Caress, E. A., 490
Carey, F. A., 47, 383
Carlassare, F., 506
Carless, H. A. J., 481, 482
Carlson, E., 261, 408
Carlson, R. G., 483
Carlson, R. M., 158
Carnduff, J., 159, 251
Caronna, T., 319
Carpenter, A. K., 545
Carr, J. B., 70
Carr, P. F., 556
Carr, R. W., 515
Carrasquillo, A., 34, 259
Carrié, R., 171
Carrington, D. E. L., 90
Carroll, F. A., 506
Carroll, J. T., 70
Carson, A. S., 357
Carter, J. V., 16, 428
Carter, W. L., 294
Cartledge, F. K., 47, 75
Carty, D., 160
Caruso, J. A., 58
Casadevall, A., 159, 243, 406, 412
Casadevall, E., 243, 412
Casagrande, M., 500
Caserio, M. C., 90, 153, 154
Casey, C. P., 334, 389
Cash, D. J., 2
Cashman, M., 45, 64
Cashman, M. P., 214, 428
Casilio, L. M., 184, 188
Casini, A., 549
Casiraghi, G., 97
Casnati, G., 96, 97
Casper, E. W. R., 96
Cassar, L., 253, 534
Cast, J. R., 384
Castañer, J., 45, 364
Castanet, J., 306
Castellan, A., 503
Castellano, A., 504
Castenada-Agulló, M., 448
Castineira, M., 398
Castrillón, J., 531
Catteau, J. P., 504
Caubère, P., 101, 133, 548
Caullet, C., 545, 555
Cauquis, G., 353, 378, 515, 545

Cauzzo, G., 497, 500
Cava, M. P., 272
Cavalier, E., 506
Cavalieri, E., 487
Cavasino, F. P., 427
Cavazza, M., 143
Cavender, C. J., 175, 235
Cavigli, P. R., 478
Cavill, G. W. K., 521
Cawley, J. J., 393
Cazaux, M., 306, 310
Celada, F., 398
Cellura, R. P., 176, 374
Celsi, S., 427
Cenini, C., 538
Cense, J.-M., 546
Cercek, B., 361
Cerefice, S. A., 256
Cerfontain, H., 197, 198
Černý, M., 109
Cescon, L. A., 364
Cessna, A. J., 29, 471, 513
Cetinkaya, B., 389
Cetorelli, J. J., 504
Čevokić, Ž., 331
Chabner, B. A., 452
Chabudzinski, Z., 159
Chachaty, C., 475
Chaeva, T. I., 95
Chakrabarti, J. K., 14, 98, 161, 250
Chakravarti, R. N., 435
Chalfont, G. R., 286, 313
Challis, B. C., 43, 200
Challis, J. A., 459
Chalmers, D. J., 345
Chaltykyan, O. A., 348
Chamberlain, G. A., 304
Chamberlain, P., 544
Chamberlain, T. R., 218
Chambers, J. Q., 371
Chambers, R. D., 241, 271
Chambers, R. J., 248, 483
Chambers, R. L., 74
Chambers, R. O., 188
Chambers, V. M. A., 549
Chan, H. W. S., 520, 541
Chan, S. C., 60
Chan, S. F., 60
Chan, T. H., 94
Chandalia, S. B., 539
Chandra, S., 531
Chandra, V., 414
Chandros, E. A., 473
Chang, B. C., 87
Chang, C. C., 264
Chang, D. W. L., 513
Chang, J.-H. C., 302
Chang, N. M. Y., 93

Chapelet-Letourneux, G., 352
Chapelon, R., 523
Chapman, N. B., 56, 108, 254, 457
Chapman, O. L., 489, 491, 494, 508, 511
Chappelet, D., 462
Charette, J. J., 150
Charifi, M., 521, 522
Charlier, M., 506
Charlton, J. L., 514
Charpentier, M., 413
Charpentier-Morize, M., 41, 102, 248, 556
Charton, B. I., 554
Charton, M., 145, 179, 554
Charubala, R., 516
Chassin, C., 10
Chastain, R. V., 107
Chastrette, M., 547
Chastrette, R., 127
Chattanathan, N., 396, 412
Chatterjee, A. K., 531
Chatterji, A. C., 544
Chaudhri, S. A., 112
Chauhan, M., 536
Chaw, Z. S., 428
Chawla, R. R., 82
Cheburkov, Yu. A., 241
Chegolya, T. N., 108
Chekhun, A. L., 396
Chekulaeva, V. N., 160
Chen, A., 175
Chen, C., 414
Chen, F., 114, 144, 146, 228, 294, 343, 348, 552
Chen, G. M. S., 4
Chen, G. M.-S., 152
Chen, H. J., 206
Chen, H. L., 408
Chen, J., 279
Chen, J. C., 322
Chen, J.-C., 277, 278
Chen, K. S., 340
Chen, R. H. K., 480
Chen, S. C., 313, 513
Chen, W., 446
Cheng, J. D., 511
Cheng, T. C., 338
Cheng, Y. M., 379
Chentsova, N. M., 464
Chérest, M., 129, 547
Cherkasov, V. K., 349
Cherkassova, E. M., 428
Chernik, C. L., 201
Chernova, T. N., 160
Chernyshev, V. O., 559
Cherton, J., 522
Chetverikova, E. P., 463

Chia, H. A., 177
Chiang, Y., 156, 206, 408
Chiang, Y. H., 134, 357
Chiba, N., 365
Chibisov, A. K., 471
Chien, H.-C., 400
Chien, J. C. W., 413
Chieu, N., 257
Chihal, D. M., 494
Childs, R. F., 48, 180, 254, 518
Chimishkyan, A. L., 429
Ching Puente, O. A., 62
Chinone, A., 244, 508
Chipman, D. M., 400
Chirkova, R. G., 181
Chiron, R., 436
Chisholm, M. H., 389
Chiu, W. T., 300
Chiyoda, Y., 108
Chkheidze, I. I., 360
Cho, I., 412
Choay, P., 77, 248
Chock, P. B., 448
Chodkiewicz, W., 416, 546
Chohan, R. K., 536
Choi, S. C., 227
Cholod, M. S., 410
Chottard, J. C., 316
Chou, T.-H., 463
Chow, Y. L., 230, 313, 489, 513
Christen, P., 462
Christensen, A., 483
Christensen, B. W., 90
Christol, H., 154, 272
Chu, C., 335
Chu, J. Y. C., 522
Chuaqui, C., 429
Chubachi, M., 520
Chuche, J., 234
Chun, M. C., 219
Chung, Y. J., 363
Ciabattoni, J., 48, 252, 516, 541
Ciganek, E., 176, 230, 374
Cilento, G., 538
Ciocazanu, I., 457
Cipollini, R., 208
Ciranni, G., 205, 208
Ciuffarin, E., 89
Clardy, J., 118, 511
Clardy, J. C., 356
Clark, F. R. S., 127
Clark, G. M., 157, 519
Clark, H. C., 379, 389, 390, 435
Clark, R. A., 225
Clark, W. G., 377

Clarke, F. B., 88, 460
Clarke, J. D., 304
Clarke, J. K. A., 556
Clarke, K., 90
Clarke, M. T., 503
Clarke, S. C., 30, 147
Claus, P., 224
Cleave, P. J. V., 396
Cleghorn, H. P., 555
Cleland, W. W., 540
Clementi, S., 197, 205
Clements, A. H., 166
Clemett, C. J., 456
Cliff, G. R., 386
Clifford, P. R., 5, 41, 73, 130, 158
Clinch, P. W., 54
Clinton, N. A., 60
Clive, D. L. J., 495
Clode, D. M., 516
Closier, M. D., 267
Closs, G. L., 276, 293, 468
Clough, J. C., 495
Clough, S. C., 233
Coan, M. K., 447
Coburn, M. D., 199
Cochran, D. W., 462
Cocivera, M., 468
Cocker, W., 83, 484
Cockerill, A. F., 121, 146
Cochrane, W. P., 265
Cocks, A. T., 145, 168, 224
Cocu, F., 456
Coffen, D. L., 72, 371, 542
Coffey, R. S., 555
Coffi Nketsia, S., 107
Coffin, R. L., 483
Coggiola, I. N., 521
Cognard, J., 545
Cogoli-Greuter, M., 537
Cohen, B. J., 554
Cohen, E., 217, 483
Cohen, E. M., 456
Cohen, H., 320
Cohen, J. S., 463
Cohen, L. A., 516
Cohen, R. L., 364
Cohen, S. G., 474, 476, 506
Cohen, T., 297
Cohen-Addad, J. P., 462
Cohn, M., 443
Cole, R. F. J., 396
Cole, T., 470
Cole, T. M., 518
Coleman, J. E., 462
Coleman, J. P., 344, 546, 554
Coletta, F., 497
Coletti-Previero, M. A., 445
Coller, B. A. W., 107, 202

Author Index

Collier, P. D., 181
Collin, P. J., 503
Collins, C. J., 2
Collins, G. C. S., 402
Collins, P. M., 481
Colman, P. M., 401
Colón, J., 531
Colonna, S., 163, 559
Colter, A. K., 46, 137
Colvin, E., 265
Combret, J.-C., 95
Combs, C. M., 269
Côme, G. M., 359
Comer, F., 227
Commeyras, A., 41, 406
Comyno, J., 450
Condon, E. V., 199
Conduit, P. B., 291, 515
Confer, A. H., 90
Congiu, L., 143
Conia, J. M., 245, 251, 479
Conn, J., 69
Conner, A. H., 403
Connolly, J. W., 388
Connor, J., 161
Connor, J. A., 389, 390
Connors, M. J., 540
Conover, W. W., 159, 541
Conrad, W. E., 87
Considine, J. L., 519
Conti, F., 534, 555
Contineanu, M. A., 296
Cook, B. F., 384
Cook, C. L., 365
Cook, D., 10, 141
Cook, J., 190, 313
Cook, K. L., 215
Cook, M. A., 75
Cook, M. J., 119
Cook, R. D., 441
Cook, R. E., 90
Cook, R. S., 179
Cooke, B. J. A., 311
Cooke, R. S., 481
Cooks, R. G., 17
Cooksey, C. J., 209
Cookson, R. C., 222, 224, 497
Cooper, A., 212, 511
Cooper, C. M., 227
Cooper, J. D., 409
Cooper, M. J., 107
Cooper, R. D. G., 267
Cooperman, B. S., 458, 459
Copenhafer, R. A., 493, 554
Coraor, G. R., 364, 508
Corbally, R. P., 188
Cordes, E. H., 394, 406, 407, 424

Core, S. K., 20, 62
Corey, E. J., 227, 245
Corey, M. D., 544
Corfield, J. R., 87, 89
Cornet, D., 557
Cornier, R. A., 541
Cornlisse, J., 525
Corre, E., 72
Corriu, R., 130, 205, 206, 457
Corriu, R. J. P., 83, 84, 409
Corson, F. P., 94
Cort, L. A., 272
Cortegiano, H., 123
Cosa, J. J., 304
Costa, G., 132
Costello, C. E., 154
Coste, J., 154
Cosyn, J. P., 497
Cote, P. N., 321
Cotton, F. A., 67, 389
Cotton, W. D., 239, 325
Cottrell, C. E., 128
Cottrell, P. T., 344
Coulombeau, A., 2, 5, 245, 551
Coulombeau, C., 243, 245
Coussemant, F., 157
Coutouly, G., 448
Couture, A., 508
Cover, R. E., 123
Cowan, D. O., 472, 503
Coward, J. K., 106
Cox, A., 468
Cox, B. G., 118, 122, 402, 412, 543
Cox, G. B., 503
Cox, R. A., 371
Cox, R. H., 112
Cox, W. G., 21
Coxon, J. M., 61, 77, 150, 161, 236, 247, 265, 291, 358
Coyle, J. D., 481, 489
Crabbe, P., 483
Craenen, H. A. H., 474
Craig, L. C., 400
Craig, N. C., 358
Cram, D. J., 19, 88, 90, 92, 120, 123, 195, 213, 503
Cramer, J., 332, 534
Crampton, M. R., 52, 187, 188
Crandall, J. K., 159, 265, 388, 481, 541
Cravy, B. J., 528
Crawford, R. J., 287, 288
Crawley, L. C., 265, 388
Cree, G. M., 540
Creed, D., 489, 521
Cremer, S. E., 89
Crescente, O., 482

Cressey, M., 406
Cresson, P., 217
Cretney, J., 85, 204
Criegee, R., 224, 528
Cristol, S. J., 19, 33, 46, 126, 156, 157, 161, 248, 305, 517
Critchlow, J. E., 402
Crombie, L., 72, 227
Cromwell, N. H., 107
Cronin, J. R., 443
Cros, J.-L., 545
Cross, B., 260, 325
Crossland, I., 128, 547
Crossland, R. K., 43, 106
Crouch, R. K., 96
Crout, D. H. G., 199
Crow, W. D., 383
Crowell, T. I., 138
Crozet, M.-P., 311
Cruz, A., 483
Cruz, M. T., 448
Csizmadia, I. G., 98
Csunderlik, C., 429
Cuddy, B. D., 14, 248, 250
Cuiban, F., 439
Cullen, W. R., 134
Cum, G., 170, 268
Cummings, C. A., 537
Cundall, R. B., 468
Cunningham, N., 518
Cuong, N. K., 522
Cupas, C. A., 14, 239, 245
Cuppen, T. J. H. M., 497
Cuppers, H. G. A. M., 558
Cuppett, C. C., 446
Curci, R., 542
Curran, W. V., 485
Curtis, H. C., 318
Curtis, J. R., 17
Cusmano, F., 95
Cusmano, G., 268
Cvitas, D., 206
Cymerman-Craig, J., 371, 379
Cyr, N., 364
Czapski, G., 362

da Silva Araujo, M., 538
Dabard, R., 549
Dabosi, G., 205, 457
Dack, M. R. J., 56, 457
Dacre, B., 192
Dafforn, G. A., 441
D'Agostino, J. T., 494
Dahlbom, R., 224
Dahm, R. H., 394
Dahman, A., 170, 234
Dahn, H., 81
Dai, S.-H., 177
Dain, J. G., 167

Dainton, F. S., 360
Dakubu, M., 145
Dal Monte, D., 183
Dalgliesh, D. T., 157
Dalton, D. R., 154
Dalton, J. C., 468, 470, 478, 479
Daltrozzo, E., 474
Daly, W. H., 127
Damon, E. K., 519
Damodaran, N. P., 247
D'Amore, M. B., 379
Damrauer, R., 379, 381, 406
Dana, G., 245, 540
Danehy, J. P., 72, 543
Danen, W. C., 286, 299, 361
Dang, T. P., 558
D'Angelo, P. F., 236, 480, 492
Daniel, A., 546
Danieli, B., 536
Daniels, C. J., 269
Daniels, C. M., 267
Danielsson, B., 271
Danilova, N. K., 179
Danishefsky, S., 61
Dannenberg, H., 215
Dannenberg, J. J., 517
Dannley, R. L., 260
Danno, S., 210
Dansted, E., 265
Darchen, A., 555
Darlage, L. J., 492
Darwinkel-Risseeuw, P. S., 504
Das, M. N., 456
Das, N. C., 300
Date, Y., 554
Datta, S. K., 261
Daub, J., 42
Dauben, W. G., 29, 254, 324, 480, 485, 495, 497, 499
Dauksas, V., 108
Daum, H., 476
Dave, V., 269
Daver, A., 554
Davidson, I. M. T., 296, 359
Davidson, M., 157
Davidson, R. S., 469, 473, 475, 476, 503, 523
Davies, A. G., 327, 328, 329, 483
Davies, D. I., 160, 301, 302, 305, 314, 334
Davies, G. A., 506
Davies, G. L. O., 146
Davies, H. H., 455, 457
Davies, J. V., 361
Davies, L. S., 269, 504

Davis, B. R., 13, 215
Davis, D. D., 74, 131
Davis, E. R., 39
Davis, F. A., 90, 215
Davis, G. S., 506
Davis, G. T., 461
Davis, H. A., 548
Davis, K. A., 540
Davis, L. C., 414
Davis, R. A., 239
Davydova, S. L., 443
Dawes, K., 478, 479, 489
Day, J., 90
Day, M., 491
Day, M. J., 258
Day, R. A., 525
de Bie, D. A., 192
de Boer, E., 57
de Boer, T. J., 193, 302, 348, 350, 474, 483, 512, 513, 516
De Bruin, K. E., 88
De Filippo, D., 553
De Grazia, C. G., 70
de Haan, H. K., 545
de Hoffmann, E., 150
de Jongh, R. O., 559
de Kreuk, J. F., 557
de la Mare, P. B. D., 106, 201, 203, 215
de Maindreville, M. D., 77, 252
De Maria, P., 109
de Mayo, P., 514
de Mos, J. C., 550
de Perez, C., 173
De Sarlo, F., 198, 199
De Schryver, F. C., 493, 504
De Shazo, M., 199
de Souza, B., 508
de Valk, J., 192, 271
De Wilt, H. G. J., 538
De Wolfe, R. H., 423
Deady, L. W., 456
Dean, F. M., 536
Dearden, J. C., 300
Dearing, C., 184
De'ath, N. J., 87
DeBruin, K. E., 461
DeCamp, M. R., 189, 492
Dedinas, J., 476
Deeleman, R. A. F., 504
Dees, K., 377
Defoin, A., 522
DeFranco, R. J., 388
Degani, C., 398, 458
Degrand, C., 555
Dehmlow, E. V., 229, 368, 379
Deinikina, N. I., 539

Deinzer, M., 289
De-Jong, J., 497
DeJongh, D. C., 369
Dekker, E. E., 414
Del Castillo, L. M., 448
Del Cima, F., 181, 192
Del Pesco, T. W., 5, 55
del Rosario, N. O., 548
Dela Cruz, D. O., 512
Delahaye, D., 545, 555
Delattre, S., 163
Delay, F., 107, 253
Delgrange, J.-C., 539
Delhomme, H., 545
Dell'Erba, C., 180
Delmauro, M., 470
Delton, M. H., 213, 503
DeLuca, G., 517
Delpuech, J. J., 57
DeMaré, G. R., 472
DeMayo, P., 235, 357, 483, 524
Dembech, P., 183
Dembinskiene, I., 108
Demek, M. M., 461
DeMember, J. R., 1, 41
Demetrescu, I., 54, 195, 198
Demole, E., 247
DeMore, W. B., 528
Dem'yanov, P. I., 95, 116
den Hertog, H. J., 304
Dence, J. B., 5
Den-Hollander, J. A., 468
Denis, J. M., 251
Denise, B., 113
Denisov, N. T., 553
Denkel, K.-H., 295
Denney, D. B., 87
Denney, D. Z., 87
Dennis, A. W., 540
Deno, N. C., 45, 297
Derenberg, M., 375
Derendyaev, B. G., 20, 43, 209
Desai, K. B., 368
Desbere, P.-L., 232
Deschamp, B., 116
Deshmane, S. S., 46, 77
Descoins, C., 311
Descotes, G., 482
Desikan, V. N. V., 407
Desimoni, G., 174
Deslongchamps, P., 13, 529
Desmaison-Brut, M., 78
DeSouza, L., 525
Dessau, R. M., 332, 333, 533, 534
Dessauer, R., 364
Dessolin, M., 424

Deswarte, S., 264
Detzer, N., 173
Deutsch, A. S., 364
Dev, S., 247
Devaprabhakara, D., 279, 550
Devaquet, A., 483
Devon, T. J., 215
DeVries, M. J., 229
DeVries, S., 525
Dewar, M. J. S., 25, 33, 34, 113, 151, 196, 216, 219, 226, 229, 277, 279
Dewhurst, F., 426
Deyrup, J. A., 172, 233
Dhingra, A. K., 370
Di Franco, A., 540
Di Furia, F., 542
Di Nunno, L., 183
di Prisco, G., 540
Diaz, A., 10, 52
Diaz, A. F., 10, 30, 262
Dickens, D., 235
Dickerson, R. E., 25, 248
Dickey, L. C., 304, 360
Dicko, D., 75
Dickson, D. R., 57
Dickson, J. R., 513
Dickson, S. J., 57
Diderich, G., 81
Diederich, D., 463
Dill, K., 517
Dilling, W. L., 16
Dillon, P. J., 483
DiMaggio, A., 516
Dimmel, D. R., 240
Din, Z. U., 270
d'Incan, E., 456
Diner, U. E., 155
Dinerstein, R. J., 353, 519
Dirania, M. K. M., 481
Dirinck, P., 397
Dirlam, J. P., 10, 340, 552
Divald, S., 90
Diviš, J., 540
Divisia, B., 515
Dix, D. T., 128
Dixneuf, P., 37
Dixon, J. E., 48, 109, 426
Dixon, R. S., 471
Dixon, W. T., 353
Djachenko, E. D., 426
Dmitrienko, G. I., 538
Dmytraczenko, A., 83, 246
Do, J., 515
DoAmaral, L., 407
Doane, W. M., 455
Dobbs, A. J., 280, 343
Dobbs, F. R., 547

Dobrat, W., 481
Dobson, R. C., 376
Dodd, D., 106, 124, 132, 209
Doddi, G., 186
Doddrell, D., 462
Dodds, H. L. H., 425
Dodonov, V. A., 283
Dodson, R. M., 239
Dodwell, C., 304
Doemeny, P. A., 549
Döpp, D., 514
Doering, W. von E., 220, 225
Doherty, B. T., 543
Doherty, C. F., 227
Dohrmann, J. K., 280, 363, 504
Doi, K., 365
Dolapchiev, L. B., 463
Dolbier, W. R., 1, 177
Dolbier. W. R., Jr., 226, 264
Dolby, J., 224
Dolenko, A., 212
Dolgina, T. I., 156
Dolgopolou, O. N., 456
Dolling, U.-H., 368, 373, 515
Dolphin, D., 208, 399
Dombroski, J. R., 173
Donald, D. S., 7
Donati, M., 534
Dondoni, A., 169, 542
Done, J. N., 334
Donk, L., 94
Donnay, R. H., 152
Donohue, J., 88
Donovan, T. R., 304
Dore, M., 206
Dorer, F. H., 515
Dorko, W., 304
Dorman, D. E., 403
Dorovska, V. N., 448
Dorsky, J., 246
Dou, H. J.-M., 313, 316, 318, 319
Dou, H. M., 508
Doupeux, H., 554
Dowd, P., 292
Dowd, W., 58
Doyle, E. R., 398
Doyle, G., 390
Doyle, M. P., 530
Dradi, E., 193
Draese, R., 114
Dragin, B. Ya., 354
Dreiding, A. S., 217
Drenth, W., 94
Dronov, V. N., 122
Drozd, V. N., 181
Drummond, G. S., 400
Drutâ, I., 170

Druzhkov, O. N., 365
Dryuk, V. G., 159
Dubchenko, V. N., 365
Dubois, G. E., 25, 161
Dubois, J. E., 44, 82, 103, 149
Dubois, J.-E., 152, 202, 412
Dubose, J. P., 522
Ducep, J. B., 116, 223
Duchamp, D. J., 90
Ducher, S., 433
Duchet, J.-C., 557
Duclos, J. M., 439
Ducom, J., 58, 113
Dudek, E. P., 406
Dudek, G., 406
Dudek, G. O., 264
Dueber, T. E., 99
Dürr, H., 230, 374
Duerr, H., 498, 515
Duff, J. M., 85, 379
Duff, R. E., 87
Duffy, M. J., 106
Dufour, M., 316
Duholke, W. K., 468
Duke, A. J., 101, 266
Dulog, L., 289
Dumont, J.-L., 264
Dunford, H. B., 540
Dunham, D. J., 248
Dunham, J., 75
Dunikoski, L. K., 458
Dunitz, J. D., 406
Dunkelblum, E., 154
Dunlop, A. M., 5, 55
Dunn, B. M., 394
Dunn, G. E., 204
Dunphy, P. J., 521
Dunston, J. M., 471
Dupont, R., 522
Dupuy, C., 311
Dupuy, W. E., 36
Durand, J.-P., 157
Durst, T., 119
Dutheil, M., 203
Dutka, F., 456
Dutta, V. P., 154
Duus, F., 269
Duynstee, E. F. J., 281, 513, 520
Dvorko, G. F., 364, 429
Dvorko, G. V., 163
Dwek, R. A., 400
Dyakonov, I. A., 455
Dyatkin, B. L., 132
Dybek, B., 78
Dzierzynski, M., 359

Eaborn, C., 75, 85, 131, 204, 207, 208, 315

Eames, T. B., 353
Eargle, D. H., 112
Earl, G. W., 337
Earl, R. A., 171
Easter, W. M., 246
Eastlick, D. T., 459
Eastwood, R., 131
Eaton, P., 483
Eaton, P. E., 253, 534
Ebel, H. F., 547
Ebel, J. P., 462
Eberhard, P., 170
Eberly, J. H., 471
Eberson, L., 210, 435, 546
Eberson, L. E., 433
Ebert, H.-D., 538
Eccleston, G., 397
Echigoya, E., 149
Eck, D. L., 137
Eckert, C. A., 166
Eckert-Maksić, M., 52
Eckes, H., 382
Eckhard, I. F., 190, 517
Eckroth, D. R., 508, 512
Edelman, R., 30, 87
Edgar, A. R., 69
Edge, D. J., 324, 343
Edmonds, A. C. F., 134
Edmunds, C. G., 552
Edward, J. T., 454
Edwards, J. M., 517
Edwards, J. O., 348, 402
Edwards, O. E., 70
Effenberger, F., 173, 202, 267, 474
Egan, C. P., 543
Egenburg, I. Z., 375
Egger, K. W., 145, 161, 227
Eguchi, S., 5, 25, 62, 161, 227, 243, 257, 258, 263
Eian, G. L., 511
Eiben, K., 363, 542
Eidenschink, R., 168
Eiki, T., 464
Eisch, J. J., 157, 240, 519
Eisenberg, H., 540
Eisenbraun, E. J., 48, 205
Eisenhardt, W., 479, 482
Eisenhut, M., 87
Eisenstadt, A., 40
Eisenstein, O., 164, 165
Eisenthal, K. B., 472
El Gaied, M. M., 550
El Ghariani, M., 187, 188
Elad, D., 483, 504, 506
Elakovich, S. D., 45
El-All, F. A., 269
Elguero, J., 228, 232, 549
Elia, V. J., 72

Eliel, E. L., 112, 261, 396
Eliseenkov, V. N., 461
Elix, J. A., 167
Ellam, R. M., 528
Elleneweig, A., 417
Ellis, D. W., 473
Ellis, K., 88
Ellis, L. E., 518
Elnagdi, M. H., 269
Elofson, R. M., 317
Elsemongy, M., 456
Elridge, J. A., 124, 272
Elving, P. J., 344
Emel'yanov, I. S., 408
Emes, P. J., 344
Emovon, E. U., 145, 304
Encina, M. V., 519
Endicott, J. F., 468
Endo, K., 78, 353
Endres, R., 557
Engberts, J. B. F. N., 83, 113, 348, 353, 483, 513, 516
Engel, M. R., 326
Engelbrecht, W. J., 229
Engelmann, T. R., 127
Engen, R. J., 97
Engewald, W., 207
Enggist, P., 247
England, B. D., 106
England, B. T., 270, 271
Engler, E. M., 14
Enin, A. S., 416
Entelis, S. G., 457
Epstein, H. F., 463
Epstein, J., 459, 461
Epstein, M. J., 233
Erastov, D. A., 413
Erbe, R. W., 414
Erdman, T. R., 297, 549
Erickson, W. F., 162, 223
Erickson, K. L., 240
Ericsson, O., 271
Eriksson, S. O., 420, 437
Erlikh, R. D., 465
Ermakova, L. N., 401
Erman, W. F., 28, 250
Ermanson, L. V., 346
Ermolaeva, M. V., 461
Ernsberger, M. V., 184
Ershov, V. V., 300
Eschenmoser, A., 104, 241
Esrieler, A. I., 429
Etzold, G., 407
Evanega, G. R., 493
Evans, A. G., 131, 344
Evans, C. G., 456
Evans, D. A., 224
Evans, E. A., 124
Evans, J. C., 344

Evans, T. R., 473
Evdokimov, V. F., 85
Evett, M., 540
Evmenko, N. P., 354
Evrard, M., 248
Ewing, S., 43
Exner, J. H., 64

Fabiny, D. L., 493
Fachinetti, G., 78
Fagan, P. J., 167
Fagone, F. A., 379
Fahey, J., 444
Fahey, R. C., 155
Fainzil'berg, A. A., 94, 95
Fainzil'berg, A. D., 122
Fair, R. W., 161
Fairclough, G. F., 447
Fakey, D. R., 201
Falk, H., 209
Fan, J. Y., 239
Fanta, P. E., 227
Faraone, F., 95
Fárcasiu, D., 81
Farid, S., 292, 489
Farlow, D. W., 426
Farnell, L. F., 390
Farnham, W. B., 88, 281
Farnum, D. G., 46, 236
Farona, M. F., 206
Farooq, S., 104, 241
Farrant, G. C., 175
Farrington, J. A., 341
Fatiadi, A. J., 532, 543
Faulkner, L. R., 524
Fauvarque, J.-F., 417, 547
Fava, A., 143, 192
Favaro, G., 500, 512
Favier, R., 534
Favorskaya, T. A., 241
Favre, H., 78, 263
Fawcett, R. L., 399
Fearon, F. W. G., 338
Feast, W. J., 167
Fedin, E. I., 207
Fedor, L. R., 138
Fedoroňko, M., 554
Fedoryński, M., 368
Fedoseev, M. S., 457
Feeney, R. E., 462
Fehn, J., 170
Feinstein, A. I., 357
Feit, E. D., 201
Feit, I. N., 241
Feitelson, J., 452, 511
Feiz, I., 555
Feld, R. S., 508
Felkin, H., 547
Feldman, M., 43

Author Index 573

Feldmann, R., 175
Felkin, H., 22, 23, 89, 129
Felkin, M., 129
Felty, R. E., 224
Femert, U., 453
Fendler, E. J., 184, 188, 403
Fendler, J. H., 184, 187, 188, 403
Feng, R. H. C., 299
Fenrick, H. W., 360
Fenton, D. F., 70, 164
Fenwick, J. D., 30
Ferberov, M. I., 396, 426, 431, 445, 446
Ferguson, J. B., 452
Ferguson, K. C., 304
Ferles, M., 195
Ferranti, F., 553
Ferrari, G. F., 555
Ferree, W., 497
Ferreira, D., 78
Ferres, H., 270
Ferrier, R. J., 219, 252
Ferris, J. P., 517
Fersht, A. R., 445
Fessenden, R. W., 351, 363
Feux, R. A., 133
Ficini, J., 227
Fiege, H., 524, 525
Field, K. W., 302
Field, L., 69
Fielden, R., 512
Fields, E. K., 190, 357
Fields, R., 382
Fife, T. H., 393, 395
Filby, W. G., 398, 542
Filipescu, N., 341, 472, 473, 476
Filippini, F., 417
Filippova, T. M., 310
Filler, R., 70, 341
Filleux, M. L., 206
Finch, A. F., 108
Finch, N., 272
Findlay, D. M., 227
Findlay, M. C., 153
Fine, D. H., 357
Finger, C., 196
Finke, M., 481
Fini, A., 109
Finley, K. T., 554
Finucane, B. W., 478
Firestone, R. A., 104
Firestone, R. F., 304, 360
Firl, J., 166
Fisch, M. H., 523
Fischer, A., 207, 428
Fischer, E., 455, 469, 525
Fischer, E. O., 389, 390

Fischer, H., 276, 283, 358, 362, 363, 483
Fischer, H. P., 123
Fischer, J. F., 356, 513
Fischer, P., 173, 267, 474
Fischer, R. D., 13, 389
Fishbein, R., 297
Fisher, J. R., 540
Fisher, J. W., 269
Fisher, R. D., 59
Fisher, W. F., 456
Fishman, W. H., 463
Fisichella, S., 103
Fitzgerald, E. A., 523
Fitzgerald, P. H., 63, 156, 160, 408
Fizii, S., 449
Flaig, W., 538
Flanagan, P. W., 48, 205
Flechtner, T. W., 499
Fleischhauer, J., 166
Fleischmann, M., 77, 546
Fleischmann, R., 557
Fleming, G. L., 154
Fleming, I., 32, 149, 229
Fleming, R. H., 226
Flemming, I., 478
Fletcher, J. W., 361
Fletcher, R., 69
Fletcher, V. R., 102, 172
Fleury, D., 407
Fleury, H. B., 407
Fleury, J.-P., 54
Flid, R. M., 156
Fliegl, E., 183
Fliszár, S., 555
Flockhart, B. D., 362
Flood, T. C., 532
Florio, S., 183
Flowers, M. C., 226, 265, 293, 359
Flowers, W. T., 270
Floyd, M. B., 244
Flynn, E. J., 70
Flythe, W. C., 43
Fodor, G., 107
Föhlisch, B., 230
Foglia, T. A., 70
Foley, J. W., 7
Folisch, B., 48
Folsom, T. K., 550
Foltz, C. M., 517
Fomin, G. V., 346
Fong, C. W., 204
Fong, W. C., 248
Fontaine, M. C., 472
Fonzes, L., 412
Foon, R., 464
Foote, C. S., 522

Forbes, A. D., 379
Forbes, M. A., 265
Forcellese, M. L., 388
Forkey, D. M., 205
Forlani, L., 182
Forney, L. S., 208
Forrester, A. R., 351, 352, 353
Forsén, S., 398
Forshaw, T. P., 511
Forster, D. L., 519, 544
Fort, R. C., 14, 253
Fort, R. C., Jr., 12
Foskey, D. J., 456
Foster, A. M., 382
Foster, A. W., 53
Foster, B. J., 269
Foster, R. L., 399
Foucaud, A., 72, 170
Foulger, B. E., 501
Fourney, J.-L., 438
Fournier, F., 508
Fowler, F. W., 55, 170, 175, 265, 374
Fowler, J. S., 203
Fox, F., 538
Fox, J. L., 218
Fox, M. A., 508
Fraenkel, G., 128
Frainnet, E., 557
Frajerman, C., 129
Frank, J., 251, 416
Frank, R., 242
Frankfater, A., 445
Frankenfeld, J. W., 548
Franklin, J. G., 448
Franz, H. J., 161
Franzus, B., 224
Frappier, F., 77, 248
Frasca, A. R., 515, 520
Fráter, G., 232
Frattini, P., 172
Frazee, W. J., 247
Freeburger, M. E., 549
Freed, D. J., 524
Freed, E. H., 494
Freed, J. H., 25
Freedman, H. H., 43
Freeman, F., 532
Freeman, G. R., 360
Freeman, J. P., 120, 171, 173
Freeman, P. K., 111, 306, 386
Frehel, D., 107
Freidlina, R. Kh., 303
French, D., 398
Fretheim, K., 271
Frey, H. M., 168, 224, 226, 235, 265
Frey, T. G., 163
Friberg, S., 442

Fried, F., 177, 269
Friedman, H. L., 58
Friedman, N., 520
Friedman, S., 557
Friedrich, E. C., 8, 302
Friedrich, H. J., 173
Friedrich, L. E., 22, 27, 235, 541
Freppel, C., 534
Frevel, L. K., 545
Fringuelli, F., 126
Friryelli, F., 197
Friswell, N. J., 304
Froehlich, R. A., 12
Froemsdorf, D. H., 137
Fronza, G., 133
Fruton, J. S., 450, 452
Fry, A. J., 371, 555
Fry, J. L., 44
Fryer, P. F., 388
Fryer, R. I., 251, 259
Fuchs, R., 105
Fueki, K., 518
Fülleová, E., 554
Fueno, F., 408
Fueno, T., 165, 546
Fujihara, M., 339
Fujii, A., 94
Fujii, T., 270, 427
Fujimaki, M., 360
Fujimori, S., 508
Fujimori, T., 506
Fujimoto, H., 103
Fujimoto, T., 280
Fujimoto, Y., 163, 374
Fujita, E., 417
Fujita, S., 109, 493
Fujiwara, H., 487
Fujiwara, S., 555
Fujiwara, Y., 210, 539
Fukui, K., 103, 151, 216, 416
Fukumoto, K., 516, 517
Fukuoka, S., 389
Fukutaka, S., 424
Fukumoto, T., 149
Fukunaga, T., 148
Funabiki, T., 559
Funamizu, M., 221
Funasaki, N., 456
Fung, H. L., 433
Funk, A. H., 158
Funke, B., 5
Furth, B., 487
Furukawa, J., 165, 387, 408
Furukawa, N., 61, 91, 246
Furuya, Y., 424
Furuyama, S., 359
Fusi, A., 538
Fyfe, C. A., 112

Gaal, W., 161
Gaasbeek, C. J., 517
Gaboriaud, R., 179
Gabrielsen, B., 18
Gadallah, F. F., 317
Gadola, M., 236
Gaertner, V. R., 71, 109
Gaiffe, A., 306
Gait, S. F., 372
Gajewski, J. J., 7, 175, 226, 229, 235
Gajewski, R. P., 472
Galaj, S., 552
Galiazzo, G., 497, 500, 523, 524
Gall, M., 96, 413
Gallagher, M. J., 365
Galli, R., 317, 319
Gallivan, J. B., 470
Gamba, A., 174, 219
Gambacorta, A., 388
Gan, L. G., 187
Gandolfi, M. T., 472
Ganesan, R., 202
Gangi, R. A., 377
Gano, J. E., 491
Ganter, C., 481
Garapon, J., 258
Garbarino, G., 180
Garbesi, A., 192
Gardini, G. P., 317
Garg, L. C., 456
Garin, D. L., 245
Garland, R. P., 236, 265, 291, 358
Garner, G. V., 270, 380
Garnier, F., 152
Garnier, J., 251, 398
Garrard, T. F., 544
Garratt, P. J., 175
Garrett, E. R., 420, 434
Garrett, P. E., 371
Garrett, P. J., 42
Garst, J. F., 112, 336
Garwood, D. C., 90, 162
Garwood, D. S., 90
Gasanov, R. G., 303
Gaskin, J. E., 555
Gaspar, P. P., 381
Gassman, P. G., 5, 7, 8, 11, 34, 92, 126, 174, 201, 213, 254, 256, 259
Gatsonis, C. D., 495
Gaudiano, G., 133
Gaudry, M., 411
Gaugaz, M., 511
Gault, R., 504
Gaux, B., 413
Gavel, J. R., 444

Gavrilovic, D., 494
Gaylor, J. R., 95
Gearen, M. M., 36, 126
Gedymin, V. V., 257
Gegiou, D., 469
Gehlhaus, J., 173, 382
Geibel, K., 538
Geiger, F. E., 341
Geiger, H., 5, 155, 156
Gein, N. V., 205
Geiseler, G., 108
Gelan, J., 396
Gelas, J., 269
Gelin, R., 154, 156
Geluk, H. W., 15, 545
Geneste, P., 129, 407, 416
Genies, M., 161, 545
Genkina, N. K., 95
Gennari, G., 450
Gensler, W. J., 244, 495
Georgarakis, M., 372, 508
George, A. D., 107
George, H., 484
George, M. V., 171, 305, 491, 508, 515
George, T., 273
George, W. O., 402
Georgoulis, C., 264
Gerdil, R., 62
Gerkin, R. M., 108, 245
Gerlock, J. L., 340
German, L. S., 121, 319
Gero, S. D., 401
Gershon, H., 201
Gersmann, H. R., 354, 538
Gerteisen, T. J., 5
Gesser, H. D., 361
Gesson, J. P., 79, 215
Getoff, N., 361
Geurtsen, G., 192
Geuss, R., 128, 272
Ghalambor, M. A., 540
Gharpure, S. B., 240
Ghera, E., 130, 416
Ghidaspov, B. V., 465
Ghosez, L., 172, 173
Ghosh, B. C., 398
Ghosh, C. K., 547
Ghosh, S., 533
Giacobbe, T. J., 252
Giacomelli, G. P., 547
Giannangeli, M., 12
Giants, T. W., 335
Giardi, I., 182
Gibbons, A. R., 226, 265, 293
Gibbons, C. S., 489
Gibian, M. J., 552
Gibs, G. J., 550
Gibson, D. H., 37

Author Index

Gibson, D. M., 500
Gibson, H. H., 384
Gibson, N. C. C., 461
Gida, V. M., 352
Giering, W. P., 180
Gijzeman, O. L. J., 473
Gil, V. M. S., 406
Gilbert, A., 468, 501, 503
Gilbert, B. C., 280, 324, 351
Gilbert, E., 360
Gilbert, J. C., 500
Gilchrist, J. C., 370
Gilchrist, T. L., 191, 265, 370, 492, 519, 544
Gileadi, E., 554
Giles, P. M., 69
Gill, G. B., 15, 247, 544
Gillan, T., 352
Gillespie, J. B., 126
Gillespie, P., 85, 457
Gillespie, P. D., 103
Gilliland, E. R., 398
Gillis, B. T., 167
Gilman, B. F., 511
Gilman, N. W., 548
Gilman, R. E., 213, 503
Gilmore, J. R., 333, 532, 533
Gilow, H. M., 199
Ginodman, L. M., 451
Ginsberg, H., 57
Ginzburg, A. G., 207
Giorgianni, P., 542
Giovannini, E., 508
Giral, L., 251
Giraudeau, P., 258
Girault J.-P., 540
Gitter, A., 208
Giudici, T., 433
Giumanini, A. G., 191, 239
Givens, R. S., 340, 483, 491
Gladstone, C. M., 252
Glaser, R., 246
Glass, G. P., 359
Glaudemans, C. P. J., 531
Glave, W. R., 138
Glazer, E., 497
Gleason, J. G., 90, 553
Gleason, W. B., 250, 538
Gleason, W. S., 521
Gleicher, G. J., 298, 299
Gleiter, R., 26, 38, 151, 290, 508
Glick, M. D., 90
Glickson, J. D., 400
Glily, S., 115, 264
Glitz, D. D., 462
Glogowski, M. E., 519
Gloor, B., 200
Gloor, J., 487

Glushko, V., 462
Goddart, W. A., Jr., 151
Godet, J.-Y., 549
Godovikov, N. N., 448
Goering, H. L., 27, 49, 50, 52
Goh, S. H., 286, 374
Gokel, G. W., 37
Golab, J., 162
Gold, A., 236, 292
Gold, A. M., 401, 463
Gold, E. H., 479
Gold, V., 123, 208, 318, 409, 410, 456
Goldberg, S. I., 150, 552
Golden, D. M., 304, 359, 412
Golden, H. J., 256, 257
Gol'dfarb, Y. L., 207
Golding, B. T., 36, 180
Gol'dovskii, A. E., 559
Goldschmidt, C. R., 472
Goldschmidt, J. M. E., 192
Goldschmidt, Z., 497, 500
Goldsmith, L. C., 447
Goldstein, M. J., 1
Golino, C. M., 62
Goller, E. J., 129, 547
Gollnick, K., 25, 518, 522
Golod, E. L., 127, 427
Golub, M. A., 483
Golubev, V. A., 352
Golubkov, I. M., 465
Gomes, W. P., 474
Gomez-Gonzales, L., 210
Gompper, R., 168, 172
Gonda, T., 127
Goniharov, G. K., 429
Gontarz, J. A., 158
Good, P. T., 145
Good, R. H., 508
Gooden, E. L., 107
Goody, R. S., 532
Gordon-Walker, A., 500
Gore, P. H., 205, 456
Gordon, J. E., 105
Gorelic, L. S., 504
Gorin, G., 399
Gorman, A. A., 498
Gorodetsky, M., 520
Gorokhova, Z. Y., 197, 205
Gorokhovatskii, Ya. B., 354
Gosser, L. W., 256
Goswami, C., 531
Goszczyński, S., 258
Gotô, R., 202
Goto, S., 406
Goto, T., 525
Gott, H., 552
Gotthardt, H., 171, 508
Goudie, A. J., 166

Goudie, R. S., 515
Gould, C. L., 439
Goutarel, R., 77, 248, 516
Gouverneur, P., 304
Govan, H. K., 406, 417
Govoni, J. P., 322
Gowenlock, B. G., 130, 304, 356, 513
Graff, Y., 436
Graham, D., 144
Graham, J. C., 107
Gramain, J. C., 523
Granapragsam, N. S., 202
Grandclaudon, P., 504
Grant, D., 14, 248, 250
Grant, D. M., 396
Grant, M. W., 43
Gratecos, D., 447
Gravel, D., 78
Graveling, F. J., 191
Graves, D. J., 401
Gravestoek, M. B., 25
Gray, G. A., 89
Gray, H. B., 500
Gray, P., 276, 357
Graydon, W. F., 540
Grayson, B. T., 261
Grayson, D. H., 83, 484
Graziano, M. L., 70
Gream, G. E., 47
Greatorex, D., 332, 336, 524
Grée, R., 171
Greeley, R. H., 495, 517
Green, B. S., 227
Green, H. E., 262
Green, M. B., 497
Green, M. L. H., 89, 389
Greene, A. E., 28, 62, 487
Greene, F. D., 229, 335, 367
Greenfield, D., 506
Greenhalgh, R., 461
Greenleaf, A. C., 463
Greenley, R. H., 231
Greenwald, B. E., 25, 33, 126, 156
Greenwood, G., 44, 175
Greenzaid, P., 452
Gregoriou, G. A., 17
Gregory, B. J., 50
Gregory, L. M., 70
Gregory, R., 304
Greig, D. G. T., 227
Greiss, G., 37, 209
Grekov, A. P., 429
Grellmann, K. H., 475, 495
Gresser, J. D., 506
Greve, H., 344
Gribble, G. W., 54
Gribble, M. Y., 188

Grieco, P. A., 25
Grieger, R. A., 166
Griffen, C. E., 69
Griffen, G. W., 369, 494, 500
Griffith, J. R., 108
Griffith, M. G., 296
Griffith, R. C., 232, 233
Griffiths, D. W., 243, 411
Grigg, R., 108, 208, 227, 233, 235, 257, 375
Grignon, J., 555
Grigsby, R. D., 48, 205
Griller, D., 328, 362, 483
Grimaldi, J., 108, 158, 167, 264, 541, 548
Grimm, K., 265
Grin, L. M., 417
Grina, L. D., 205, 520
Grisdale, P., 196
Grisdale, P. J., 519
Grisolia, S., 463
Grist, S., 123, 409, 410, 456
Grivas, J. S., 241
Grizzle, P. L., 287
Grob, C. A., 97, 99, 528
Grobovsky, L. V., 423
Groeger, F. A., 133, 261
Groen, A., 468
Grohmann, K., 484
Gross, H. J., 448
Gross, M. L., 46, 338
Grossweiner, L. I., 475, 523
Grostic, M. F., 306
Grotens, A. M., 57
Grotewold, J., 317, 329, 489, 538
Grubbs, E. J., 12
Grubb, S. D., 552
Grubbs, R. H., 556
Gruber, W., 79
Grünanger, P., 172
Gruetzmacher, H. F., 359
Grundmann, C., 261
Grundon, M. F., 215, 218
Grunwald, E., 52
Grunwell, J. R., 48, 116
Grupe, K.-H., 531
Grutzner, J. B., 5, 497
Gryaznov, A. P., 166
Grynkiewicz, G., 198, 199
Gschwend, H. W., 272
Guanti, G., 180
Gueldner, R. C., 484
Guenard, D., 521
Günther, K., 542
Guerret, P., 270
Guerrini, P., 310
Guest, I. G., 109, 247
Guesten, H., 500

Güsten, H., 262
Guibe-Jampel, E., 424
Guiheneuf, G., 417
Guilbault, L. J., 168
Guillaume, J., 522
Guillerm-Dron, D., 416
Guillermo, L., 512
Guillet, J. E., 479
Guinot, F., 395
Guk, A. F., 288
Gulyachkino, V. N., 204
Gull, P., 478
Gullotti, M., 559
Gund, T. M., 126
Gundermann, K. D., 524, 525
Gunn, P. A., 247
Gunning, H. E., 161, 468
Gunther, H., 230
Gupta, P., 481
Gupta, R., 113
Gupta, S. K., 157, 171, 544
Gupte, S. S., 272
Gurd, F. R. N., 462
Gurvich, L. G., 160
Guseinova, S. N., 205
Gushchina, E. G., 107
Gustafson, J. H., 461
Gustafsson, H., 109
Gutfreund, H., 461, 540
Guthrie, R. D., 58, 123
Guthrie, R. O., 116
Gutmann, V., 555
Gutowski, G. E., 267, 269
Gutsche, C. D., 243, 376, 376, 411
Guzzo, A. B., 471
Gymer, G. E., 370

Haag, A., 538
Haake, P., 423, 428, 460
Haas, C. K., 369
Halbeeb, A. F. S. A., 400
Haberfield, P., 57, 105, 457
Habicht, E., 269
Habmann, V., 313
Hachimori, Y., 400
Hackett, P., 498
Hadaiah, B. M., 525
Haddadin, M. J., 512
Haddock, E., 269
Haddon, V. R., 17
Hadek, V., 347
Hadley, S. G., 504
Hänsel, W., 413
Haerter, H. P., 513
Häusler, G., 547
Hagedorn, A. A., 46
Hagen, R., 528
Hagihara, N., 390

Hageman, H. J., 357
Hague, D. N., 448
Hahn, B.-S., 2
Hahn, R. C., 485
Haidukewych, D., 243
Hajdu, J., 337
Hakka, L. E., 206
Halasa, A. F., 97, 163, 338
Haldna, U., 417, 427, 428
Haldna, V., 206
Hales, N. J., 176
Hales, R. H., 189
Haley, T. J., 516
Hall, C. D., 101, 140, 162, 552
Hall, G. E., 230, 261, 406
Hall, J. H., 170, 233
Hall, R. E., 13, 50, 59
Hall, S. S., 552
Hallé, J.-C., 184
Hallensleben, M. L., 173
Haller, R., 413
Halliday, D. E., 4, 161
Halmann, M., 398, 458
Halper, J. P., 399
Halpern, A. M., 479
Halpern, J., 253, 534
Halpern, Y., 41, 42, 124
Halton, B., 500
Hamada, M., 520
Hamaguchi, K., 448, 449
Hamberger, H., 170, 234
Hambly, A. N., 463, 464
Hamdam, M. S., 270
Hamel, R. G., 473
Hamelin, J., 288
Hamid, M. A., 131
Hamill, H., 126
Hamilton, G. A., 539, 540
Hamilton, L. R., 479, 504
Hamlet, Z., 78
Hammen, P. D., 239
Hammer, C. F., 154
Hammond, G. S., 468, 500
Hamon, D. P. G., 172, 236
Han, M., 447
Hanach, M., 161
Hanack, M., 98, 99
Hancock, K. G., 519
Hand, R., 545
Handloser, L., 34
Hands, D., 15, 247
Haney, W. M., 435
Hang, E., 48
Hanifin, J. N., 217
Hanifin, J. W., 483
Hanna, M. T., 427
Hannan, B. N. B., 203
Hannaway, C., 25

Author Index

Hansen, H. J., 496, 508
Hansen, H.-J., 220, 227, 232
Hansen, J. F., 112, 231, 551, 554
Hansen, R. S., 356
Hanson, A. W., 462
Hanson, E. M., 369
Hanson, J. R., 109, 215
Hanson, K. R., 137
Hanson, M. P., 96
Hanssen, H. W., 197
Hanstein, W., 60, 73, 131, 281
Hanstein, W. G., 540
Happ, J. W., 493
Happer, D. A. R., 428
Harada, K., 557
Harada, M., 327
Harada, S., 150
Harada, T., 557
Hardec, D. D., 484
Harding, C. E., 2, 98, 161
Harding, D. R. K., 218
Hardy, F. E., 105, 469
Hardy, J. C., 497
Hardy, T. A., 111
Harfenist, M., 213
Harger, M. J. P., 89, 190, 313, 516
Hargrove, R. J., 99
Harhash, A. H., 269
Harlmann, A. A., 112
Harman, C. A., 226
Harmon, R. E., 171, 485, 544
Harness, J., 64
Harpp, D. N., 90, 94, 251, 553
Harrington, J. K., 33, 126, 156
Harris, C. L., 456
Harris, D. L., 10, 21, 30
Harris, D. O., 441
Harris, J. M., 13, 16, 50, 59
Harris, R. L. N., 543
Harris, R. O., 219
Harrison, A. G., 304
Harrison, D. A., 481
Harrison, J. F., 367
Harrison, K., 476
Harrison, L. W., 54
Harrison, R. M., 217
Harsthorn, M. P., 291, 358
Hart, H., 23, 28, 172, 250, 483, 485, 542
Harter, D. A., 501
Harter, M. L., 470
Hartgerink, J. W., 348, 350, 483, 516
Hartig, R., 145
Hartman, F. C., 413
Hartmann, W., 494, 501

Hartshorn, M. P., 61, 77, 150, 161, 199, 206, 236, 247, 265
Hartshorn, S. R., 198
Hartsuiker, J., 525
Harvey, R. G., 46, 215
Hartzler, H. D., 375
Hasan, S. K., 439
Hasegawa, J., 442
Hasegawa, M., 299
Haselbach, E., 228
Hasenheuttl, G., 356
Hashimoto, H., 118, 163, 197, 387
Hashimoto, S., 514
Haslam, E., 412
Hasokami, T., 386
Hass, G. M., 452
Hass, H. B., 357
Hasselgren, K.-H., 224
Hassner, A., 102, 103, 152, 171, 172, 230
Hasty, N. M., 501
Hasty, N. M., Jr., 9
Haszeldine, R. N., 144, 154, 167, 304, 352, 382
Hata, K., 517
Hata, N., 493, 504
Hatada, K., 409
Hataya, M., 492
Hatefi, Y., 540
Hatem, J., 33, 374
Hatanaka, N., 257
Hatfield, L. D. H., 269
Hattori, H., 263
Hauser, C. R., 70, 105, 127
Hauser, H., 173
Hausser, K. H., 364
Hautala, R. R., 480
Hautecloque, S., 304
Havinga, E., 496, 525
Hawkins, E. G. E., 281
Hawthorne, M. F., 95
Haya, T., 270
Hayakawa, K., 487
Hayakawa, T., 97
Hayami, J., 142
Hayano, S., 339
Hayasaha, T., 191
Hayashi, M., 299
Hayashi, N., 175
Hayashi, S., 191, 305
Hayashi, Y., 117, 463, 493
Hayatsu, H., 163
Hayes, D. M., 376
Hayes, E. F., 169, 294
Hayes, F. N., 506
Hayes, L. J., 383
Hayes, R., 108, 233
Haynes, R. M., 356

Haynie, E. C., 226
Hayon, E., 361, 476
Hays, H. R., 461
Hayward, R. J., 497
Hazard, R., 555
Headley, L., 338
Heaney, H., 13, 189, 190, 193, 248, 517
Hearn, R. P., 462
Hebert, A. L., 516
Hechler, E., 276
Hecht, R., 321
Heckl, B., 389, 390
Heckner, K.-H., 531
Hedaya, E., 236, 358, 381, 480, 492
Hedegaard, B., 218
Hedin, P. A., 484
Heep, U., 494
Heese, J., 79
Heesing, A., 346
Hefelfinger, D. T., 195
Hefter, H. J., 364
Hegarty, A. F., 45, 64, 133, 214, 261, 428, 433
Hehre, W. J., 48, 367
Heiba, E. I., 332, 333, 533, 534
Heid, G., 435
Heil, B., 559
Heimann, W., 540
Heimgartner, H., 496
Heinbach, P., 239
Heindel, N. D., 219
Heine, H. G., 251, 494
Heine, H. W., 265
Heinrick, B., 236
Heinz, G., 388
Heiszwolf, G. J., 96
Helene, C., 506
Helgeson, R. C., 19
Heller, A., 525
Hellerbach, J., 259
Hellin, M., 157
Helmick, L. S., 124
Helmy, E. E., 134
Hemmerich, P., 506
Hemminger, J. C., 479
Hemmington, J. A., 106
Hempelman, S., 36, 126
Henders, R. N., 272
Henderson, J., 384
Henderson, R. W., 283
Henderson, W. A., 370, 492
Hendrick, M. E., 373, 381
Hendrickson, J. B., 36, 126
Hendry, J. B., 199
Henneberg, D., 484
Hennekens, J. L. J. P., 281

Hennis, R. P., 263
Henrici-Olivé, G., 557
Henriques, R. D., 398
Henry, M. P., 436
Henry, P. M., 197, 252, 262, 535
Henry, T. J., 235
Henshaw, B. C., 5, 245
Henshaw, J. S., 448
Henson, W. L., 59
Hentschel, P., 229
Henzel, R. P., 11, 254
Hepburn, S. P., 351, 353
Herald, D. L., 108
Herberich, G. E., 37, 209, 244, 309
Hercules, D. M., 524
Hermann, H., 170, 511
Hermolin, J., 554
Hernandez, J., 327, 538
Herndon, W. C., 145, 470, 494
Herod, A. A., 276
Herold, C.-P., 67, 269
Herrāez, M. A., 108
Herriott, A. W., 88, 552
Herron, D. K., 227
Hess, B., 540
Hess, B. A., 189
Hess, B. A., Jr., 21
Hess, G. P., 443
Hesse, G., 48, 503
Hesse, R. H., 258, 330
Hetzheim, A., 269
Heublein, G., 154
Hevesi, L., 413
Hewitt, D. G., 354, 536
Hewitt, G., 227
Hexter, C. S., 446
Hey, D. H., 313, 314, 315, 334, 347, 517, 543, 551
Heyd, W. E., 14, 239, 245
Heyman, M., 537
Heyn, A. S., 162
Heyorly, A. F., 408
Heyrovský, M., 555
Hiatt, R., 283, 552
Hiatt, R. R., 159, 542
Hibbert, J., 123
Hibbert, P. G., 313
Hickmann, E. A. J., 481
Hickmott, P. W., 217
Hida, M., 192, 525
Hieble, J. P., 253
Higashi, F., 557
Higashida, S., 353
Higgins, R., 330
Higo, A., 481
Higo, M., 271

Higuchi, S., 493
Higuchi, T., 433
Hiiragi, M., 191
Hilbert, P., 375
Hildesheim, J., 398
Hill, D. J., 272
Hill, D. T., 107, 253
Hill, E. A., 326
Hill, H. A. O., 552, 554
Hill, J., 481
Hill, J. S., 251
Hill, M. J., 47, 262
Hilliard, J. A., 478
Hills, G. J., 57
Himmelreich, J., 259
Hindley, K., 536
Hine, J., 43, 410, 421
Hine, K. E., 48, 149, 518
Hinkle, P. M., 450
Hinnen, A., 525
Hino, T., 264
Hinshaw, J. C., 494
Hinshaw, W. B., Jr., 252
Hintsche, R., 407
Hintz, P. J., 126
Hinz, H. J., 414
Hipperson, W. C. P., 543
Hirabayashi, Y., 429
Hirai, H., 492
Hiraī, K., 133
Hirai, M., 319
Hirama, M., 522
Hirata, M., 553
Hirata, T., 245
Hirayam, S., 524
Hirobi, H., 271
Hiromi, K., 400
Hirota, H., 93, 553
Hirota, N., 340
Hirsch, L. K., 133, 163
Hirshmann, F. B., 46, 77
Hirshmann, H., 46, 77
Hisada, R., 282, 283
Hisatome, M., 38
Hirsjarvi, P., 5
Hirst, J., 180
Hitchcock, P. B., 154
Hittenhausen, H., 350
Hittenhousen, H., 512
Hixson, S. S., 497
Hiyama, T., 109
Hoare, M. J., 171
Hobbs, C. F., 339
Hobbs, W. E., 388
Hobson, J. D., 217
Hocker, J., 386
Hodge, P., 375
Hodge, V. F., 62, 522
Hodges, R., 13, 215

Hodges, R. J., 559
Hodgins, T., 385
Höbold, W., 161
Höfle, G., 222, 224, 227, 383, 440
Höhn, R., 161
Hoener, L., 477
Hoewe, B. D., 425
Hofer, E., 264
Hoffman, B. M., 353
Hoffman, M. K., 195
Hoffman, P., 85, 457
Hoffman, R., 1, 493
Hoffman, R. V., 382
Hoffman, V. L., 387
Hoffmann, H. M. R., 44, 175
Hoffmann, J., 108
Hoffmann, R., 25, 151, 290, 294, 376
Hoffmann, R. V., 260
Hoffmann, R. W., 173, 369, 382
Hofmann, H., 495
Hofmann, P., 495
Hogarth, M. J., 132
Hogett, J., 198
Hogeveen, H., 45, 98, 517
Hogg, D. R., 159
Hogg, V. A., 48
Hoijtink, G. J., 469
Hoizey, M.-J., 251
Holbrook, K. A., 145
Holcomb, A. G., 23
Holdren, G. R., 296
Holford, T. G., 162
Hollaender, J., 260
Holland, G. W., 329
Holland, H. L., 36
Holland, J. M., 175
Holland, V. R., 540
Holleck, L., 555
Holler, E., 448
Holm, T., 128, 547
Holman, R. J., 348
Holmberg, K., 543
Holmes, A. B., 232
Holmes, G. D., 284
Holmes, J. L., 145
Holmstead, R. L., 302
Holtz, D., 105, 121
Holy, N. L., 116, 336
Homer, G. D., 83
Honeck, H., 212
Honeycutt, S. C., 127
Hong, P., 390
Honig, M. L., 234
Honma, Y., 263
Hoobler, J. A., 289
Hoogzand, C., 126, 267

Hook, S. C. W., 329
Hopf, H., 51, 220, 264
Hopkins, A. S., 346
Hopkins, R. G., 265
Hopkinson, A. C., 43, 98
Hopper, S. P., 369
Hordis, C. K., 157
Horgan, S. W., 276, 545
Hori, M., 511
Horibe, I., 222
Horii, Z., 517
Horinaka, A., 521
Horio, Y., 429
Horisberger, M., 399
Horn, U., 36
Hornback, J. M., 5, 92
Horner, L., 345, 516, 557
Hornish, R. E., 12, 527
Horowitz, A., 361
Horrell, D. R., 360
Horsfield, A., 299
Horspool, W. M., 37, 229, 489, 493
Horst, J., 401
Horton, D., 398, 404, 516
Horvath, G., 117, 241
Horvath, V. M., 13
Horwell, D. C., 383, 536
Hoshina, M., 485
Hoskins, J. A., 193
Hoskins, K., 154
Hosokami, T., 215, 552
Hosokawa, S., 462, 546
Hosomi, A., 299, 304, 305, 315, 320
Hosoya, T., 157
Hotaka, T., 309
Hotamahalli, S. S., 539
Hotchandani, S., 470
Hough, L., 61, 246
Houk, K. N., 165, 175
Houlihan, S. A., 164
House, D. B., 345
House, H. O., 96, 413
Houser, R. W., 243, 266
Howard, J. C., 269
Howe, G. R., 159, 195, 196, 207, 542
Howell, B. A., 30
Howell, M. J., 523
Howells, D., 47, 72, 244
Howes, P. D., 67, 164
Hoyermann, K., 310, 344
Hoyos de Rossi, R., 190
Hranilović, J., 555
Hrivňák, J., 97
Hsieh, W. C., 2
Hsu, C.-M., 459
Hsu, E. T., 100, 546

Hsu, J. N. C., 497
Hsu, K., 229
Huang, R. L., 286
Huang, S. J., 100, 546
Huang, W. H., 454, 506
Huang, Y.-C., 463
Hubbard, C. D., 451
Hubbard, R., 167
Hucho, F., 403
Hudec, J., 497
Hudnall, P. M., 344
Hudrlik, A. M., 62
Hudrlik, P. F., 62
Hudson, A., 301
Hudson, C. E., 527
Hudson, H. R., 36, 54
Hudson, J. B., 470
Hudson, K., 426
Hudson, R. F., 89, 133, 260, 261, 270, 325
Huebert, B. J., 554
Huebner, J., 359
Hünig, S., 269, 341, 367, 536, 546
Huestis, W. H., 462
Huet, F., 263
Hüther, H., 220
Hug, R., 232
Hughes, A. N., 88
Hughes, C. R., 247
Hughes, R. D., 3, 156
Hughes, N. A., 64
Huisgen, R., 170, 171, 230, 231, 233, 234, 511
Huismann, H. O., 168
Hulett-Cowling, F. M., 463
Humffray, A. A., 204
Hummel, K., 98
Humski, K., 2
Hung, W. M., 169
Hunger, M., 142
Hunt, C. B., 485
Hunt, J. B., 384
Hunt, J. D., 203, 245, 534
Hunter, B. K., 379
Hunter, D. H., 139, 552
Hurecker, B. L., 414
Hurst, G. H., 432
Hursthouse, M. B., 155
Husain, M., 533
Husain, S. S., 452
Huseya, Y., 508
Hussey, G. E., 491
Hutchins, R. O., 248, 548, 549
Hutchinson, B. J., 119
Hutchinson, D. W., 123
Hutchinson, E. G., 13, 215
Hutchinson, J. J., 259
Hutton, R. S., 367

Huynh, X. Q., 152
Huyser, E. S., 299
Huyskens, P., 58
Hyman, H. H., 341
Hyne, J. B., 57
Hyvönen, M.-L., 182

Iatsimirskii, A. K., 448
Ibarbia, P. A., 48, 98, 156, 160, 307, 308
Ibata, T., 69, 361
Ibne-Rasa, K. M., 147
Ichikawa, K., 96, 203, 534
Ichikawa, M., 208, 557
Iddon, B., 361
Igeta, H., 512
Ignat'va, S. N., 413
Iguchi, M., 46
Ihari, M., 516
Ihrig, P. J., 357
Iida, M., 149
Iino, N., 401
Ikeda, A., 48, 344
Ikeda, M., 232, 485
Ikedate, K., 557
Ikedo, M., 494
Ikehara, M., 271
Ikekawa, N., 108, 263
Ikenaka, T., 400, 448
Ilkova, E. L., 557
Illuminati, G., 125, 182, 186, 187, 202
Ilvonen, P. J., 552
Imahori, K., 461, 462
Imai, H., 198, 359
Imai, J., 492
Imaida, M., 557
Imamoto, T., 258
Imanaka, T., 539
Imhoff, M. A., 46, 161, 248
Imoto, E., 192, 508
Inada, Y., 449
Inamoto, N., 176, 282
Inch, T. D., 250
Indelicato, J. M., 64
Inei-Shizukawa, G., 448
Inel, Y. N., 359
Ing, K. Y. W., 234
Inglis, D. B., 158
Ingold, C. K., 89, 198
Ingold, K. U., 275, 286, 327, 352, 353
Ingraham, L. L., 132
Inoue, A., 473
Inoue, H., 484, 525
Inoue, S., 218
Inoue, Y., 112
Inukai, T., 166
Ioannu, E. S., 439

Ioffe, A. I., 375
Iorio, M. A., 244
Ipaktschi, J., 30, 224
Iqbal, M., 379
Ireland, R. E., 25, 248
Irick, G., 515
Irick, G., Jr., 261
Irie, H., 248
Irie, M., 462, 463
Irie, T., 19
Irreverre, F., 374
Irving, H. M. N. H., 56, 89
Irwin, J., 470
Isaacs, N. S., 206
Ise, N., 150
Iseda, K., 206
Isheda, H., 473
Ishibashi, H., 485
Ishibe, N., 487, 489, 497, 522
Ishibi, N., 175
Ishida, H., 520
Ishida, M., 310
Ishido, Y., 483
Ishiguro, N., 548
Ishiguro, S., 72
Ishihara, H., 429
Ishii, S., 448
Ishii, Y., 264
Ishikawa, N., 191
Ishitobi, H., 221
Ishizu, H., 459
Ishizu, K., 336
Ishizuka, H., 555
Isiyama, S., 493
Islam, M. M., 487
Islip, P. J., 267
Isobe, K., 65, 242
Isoe, S., 521
Isola, M., 89
Issidorides, C. H., 512
Isvasyuk, N. V., 64
Ita, I., 45
Itani, H., 63, 245
Itaya, T., 270
Itier, J., 159
Ito, S., 45, 78, 359, 522
Itô, S., 167
Ito, T. I., 114, 538, 550
Ito, V., 267
Ito, Y., 263, 309
Itoh, I., 167
Itoh, M., 327, 329, 341, 492
Itoho, K., 424
Itoi, H., 511
Ivanov, V. A., 317
Ivanov, V. I., 396
Ivanov, V. T., 447
Iverson, P. E., 239
Ivin, K. J., 362

Iwamasa, R. T., 545
Iwamura, H., 213, 239, 350
Iwamura, M., 213, 239, 350
Iwaoka, T., 511
Iwata, C., 517
Iwata, K., 117
Iwata, Y., 106
Iwatsubo, M., 540
Iwatsuru, M., 442
Izumi, Y., 557
Izumiya, N., 449
Izumuja, N., 451
Izvehov, V. P., 417

Jablonski, C. R., 36
Jackman, L. M., 17
Jackson, A. W., 472, 487
Jackson, G. E., 457
Jackson, G. L., 420, 427
Jackson, H. L., 364
Jackson, J. R., 117
Jackson, R. A., 301, 315
Jackson, W. R., 270, 549
Jacobson, G. R., 401
Jacquesy, J. C., 47, 77, 79, 215, 248
Jacquesy, R., 47, 77, 79, 215, 248
Jacquier, R., 232, 264, 270, 549
Jacquignon, P., 205
Jaeger, D. A., 123
Jaenicke, W., 555
Jaffari, G. A., 544
Jaffé, M. H., 34, 59
Jagur-Grodzinski, J., 54, 340
Jagusztyn-Grochowska, J. M., 180
Jain, C. L., 531
Jain, P. C., 166
Jakobsen, H. J., 348, 353
Jakovljević-Marinković, Ž., 331
Jakubetz, W., 206
Jáky, M., 532
James, B. G., 113
James, B. R., 539
James, C. L., 344
James, R. E., 362
James, T. Ll., 227, 263
Jamieson, J., 460
Jamieson, J. W. S., 359
Jancis, E. H., 239
Janhoven, J., 181
Janssen, M. J., 440
Janzen, E. G., 348
Japar, S. M., 506
Jarczewski, A., 122
Jardine, I., 558

Jarreau, F.-X., 77, 248
Jarvie, A. W. P., 330
Jarvis, B. B., 322
Jasińska, J., 407
Jasiński, T., 407
Jaurdosiuk, M., 180
Jautelat, M., 5
Javaid, K. A., 545
Jayaram, R., 446
Jayaraman, H., 353
Jean, Y., 294
Jefford, C. W., 107, 216, 253
Jeger, O., 478, 487
Jencks, W. P., 422, 425, 441, 445, 452, 453
Jenevein, R. M., 554, 555
Jenkins, C. L., 333
Jenkins, I. D., 365
Jenkins, J. K., 22
Jennings, C. A., 127
Jenny, E. F., 176
Jensen, B. L., 485
Jensen, E. T., 555
Jensen, F. R., 131
Jensen, J. H., 410
Jensen, J. L., 156
Jensen, L. H., 506
Jeremić, D., 331
Jermyn, M. A., 401
Jesson, J. P., 280
Jeunge, E. C., 544
Jewess, P. J., 185
Jewett, J. G., 30
Jezorek, J. R., 555
Jha, D. S., 531
Jindal, S. P., 157, 410, 411
Jindall, S. P., 5
Jira, R., 534
Joanny, M., 379
Jogibhukta, M., 270
Johanson, R. G., 28, 41, 46, 79
Johansson, N. G., 369
Johari, D. P., 310
John, V. A., 448
Johns, H. E., 471
Johnson, A. L., 288
Johnson, A. P., 235
Johnson, A. W., 208, 227, 235, 269, 273
Johnson, B. F. G., 197
Johnson, B. L., 5, 30, 147, 245
Johnson, C. A. F., 356
Johnson, C. D., 199
Johnson, C. R., 90, 91, 108, 133
Johnson, D. M., 461
Johnson, D. W., 508
Johnson, G. L., 443

Johnson, H. T., 74
Johnson, H. W., 106
Johnson, M. D., 106, 124, 132, 208, 209
Johnson, M. G., 190, 517
Johnson, P. C., 473
Johnson, Q., 205
Johnson, R. M., 463
Johnson, W. S., 25
Johnsson, H., 553
Johnston, L., 380
Johnstone, R. A. W., 349
Jolles, P., 399
Jolly, P. W., 180
Jonah, C. D., 358
Jonas, J., 108
Jones, A., 276
Jones, A. H., 477
Jones, A. J., 396
Jones, A. S., 532
Jones, B., 476
Jones, C. W., 384
Jones, D. H., 191
Jones, D. N., 134
Jones, D. W., 175, 265, 375, 481
Jones, E. M., 389, 390
Jones, G., 269, 386, 504, 508
Jones, G. H., 315, 347, 517, 543
Jones, G. W., 485
Jones, H. A., 46
Jones, J. G. Ll., 109, 247
Jones, J. K. N., 83, 246
Jones, J. P., 47, 75
Jones, J. R., 121, 124
Jones, L. B., 13
Jones, M., 189, 370, 373, 381, 492, 498
Jones, P. F., 469, 515
Jones, P. R., 129
Jones, P. W., 361
Jones, R. A. Y., 397
Jones, R. P., 467, 547
Jones, S. H., 304, 359
Jones, V. I. P., 397
Jones, V. K., 13
Jones, W. M., 375, 381
Jonkman, L., 351
Jonsson, E. U., 91
Jordaan, A., 68
Jori, G., 450, 523, 524
Jortner, J., 469
Joseph, N., 202
Joseph, T. C., 489
Joshi, G. C., 550
Joshi, V. S., 247
Joss, W. B., 265
Jost, R., 41

Jotterand, A., 408
Joullié, M. M., 177
Javanovich, A. P., 22
Juers, D. F., 498, 499
Jugelt, W., 82, 421
Julia, M., 276, 311, 316
Julian, D. A., 524
Jung, F., 154
Jung, M. J., 558
Junge, H., 5, 7
Jurewicz, A. T., 208
Just, G., 175, 273, 518
Justin, B., 352

Kabachnik, M. I., 121
Kabalka, G. W., 329, 538
Kabanov, V. A., 448
Kaberdin, R. V., 326
Kachwaha, O. P., 536
Kadai, T., 157
Kagan, H., 497
Kagan, H. B., 380, 558
Kagan, J., 481
Kahl, W., 420
Kai, K., 163
Kairaitis, D. A., 359
Kaiser, B.-U., 346
Kaiser, E. M., 552
Kaiser, E. T., 445, 451, 464
Kaiser, G. V., 222
Kaiser, K. L., 180, 254
Kaiser, R. S., 56
Kaji, A., 142, 214
Kakehi, A., 171
Kakihana, T., 231, 551, 554
Kalamár, J.,97
Kalibabtschuk, N. N., 348
Kalicky, P., 477
Kalidas, C., 396, 412
Kalinin, V. N., 257
Kaliya, O. L., 156
Kallen, R. G., 404
Kalman, T. I., 123
Kalsi, P. S., 531
Kalvoda, J., 482
Kamachi, M., 508
Kamala, K., 455
Kamata, K., 245
Kamego, A., 459
Kamei, T., 545
Kametani, K., 516
Kametani, T., 191, 386, 516, 517
Kamigata, N., 289
Kamiya, T., 269
Kamiya, Y., 365
Kammerman, S., 399
Kampe, K.-D., 259
Kampel, V. T., 47

Kampmeier, J. A., 279, 284
Kan, G., 266, 497
Kanagasabapathy, V. M., 107
Kanai, T., 503
Kanamaru, N., 517
Kanaoka, Y., 448, 449
Kanazawa, M., 448
Kaneko, T., 269, 319
Kanematsu, K., 171, 232, 487
Kanick, S. W., 361
Kanitkar, K. B., 169
Kankaanperä, A., 393, 431
Kannan, S. V., 163, 205
Kano, K., 514
Kano, N., 264
Kanojoa, R. M., 517
Kantz, D. M., 46
Kaplan, H., 444
Kaplan, L., 501
Kaplan, L. A., 162, 187
Kaplan, M. S., 234
Kaptein, R., 275, 276
Karam, P. A., 36
Karasawa, Y., 54, 340
Karavan, V. S., 103
Karim, A., 14, 126
Kariv, E., 554
Karle, I. L., 374
Karle, J. M., 374
Karlin, A. V., 85
Karlsson, H., 348
Kapoor, R. C., 536
Kaptein, R., 468
Karneeva, L. N., 456
Karoglan, J. E., 265
Karopy, S. V., 428
Karpenko, T. F., 163
Karpinskya, E. V., 453
Kartashov, V. R., 201
Kasha, M., 468
Kashima, C., 484, 522
Kashin, A. P., 121
Kashtonova, T. N., 414
Kaspi, J., 99
Kassell, B., 452
Kasuya, T., 473
Katachandra, S., 202
Katagiri, M., 557
Katagiri, T., 161, 236
Katoaka, H., 484
Kato, H., 508
Katô, M., 175, 191, 484
Kato, S., 214, 336, 542
Kato, T., 449
Katritzky, A. R., 55, 119, 199, 207, 397
Katsamura, S., 521
Katsuhara, J., 247, 491

Katsuhara, Y., 475
Katsura, H., 269
Katz, H., 468
Katz, M. G., 361
Katz, S., 400
Katz, T. J., 256, 501
Katzhendler, J., 455
Kauffman, G. M., 124
Kauffman, W. J., 129, 547
Kauffmann, T., 168, 191
Kaufman, M. L., 557
Kaupp, G., 495
Kautz, D. M., 77
Kaválek, J., 179, 200
Kavarnos, G., 470
Kaverzneva, E. D., 401
Kawabata, N., 387
Kawabe, H., 106
Kawada, M., 17
Kawai, M., 434
Kawamura, T., 277, 280, 309
Kawanisi, M., 493
Kawasaki, A., 405, 438
Kawazoe, Y., 124, 207, 557
Kay, J., 452
Kay, P. S., 16
Kayama, Y., 175
Kayen, A. H. M., 513
Kazakova, L. I., 38
Kazanjian, A. R., 360
Kazanskaya, N. F., 446, 447, 449
Kazitsina, L. A., 38
Kazmaier, P., 70
Kaz'mina, N. B., 319
Keana, J. F. W., 353
Kearley, F. J., 189
Kearney, J. A., 214
Kearns, D. R., 485, 520
Keating, M., 383, 536
Kebarle, P., 468
Keehn, P. M., 520
Keena, J. F., 519
Keene, B. R. T., 108, 182
Keller, T., 230
Kellett, P. M., 550
Kelley, R. D., 310
Kelley, W. S., 202
Kellie, G. M., 397
Kellman, R., 305
Kellmann, A., 471
Kellog, R. M., 479, 507
Kellogg, M. S., 485, 495
Kelly, C. F., 136
Kelly, D. P., 41
Kelly, F. M., 143
Kelly, F. W., 142
Kelly, J. F., 70, 231, 267
Kelly, K. K., 258

Kelly, R. V., 378
Kelly, T., 468
Kelm, H., 108
Kelsey, D. R., 97, 99
Kemp, D. R., 471
Kemp, D. S., 385, 439
Kemp, J. E., 224, 497
Kemp, R. G., 463
Kemp, R. T., 138
Kemp, T. J., 332, 336, 468, 524
Kemp, W., 494
Kemp-Jones, A. V., 478
Kendall, F. H., 179
Kende, S., A. 497
Kende, A. S., 15, 22, 500, 522
Kennedy, J. D., 307
Kenner, R. A., 449
Kensler, T. T., 361
Kent, M. E., 236, 358, 381, 480
Kent, P. W., 400
Kenyon, G. L., 409, 463
Kenyon, W. G., 105
Kepka, A. G., 523
Kerek, F., 406
Keresztes-Nayej, S., 413
Kerfanto, M., 54
Kergomard, A., 48, 107
Kern, D. H., 476
Kernaghan, G. F. P., 44, 175
Kerr, G. H., 529
Kerr, J. A., 145, 304
Kershner, L. D., 64, 419
Keske, R. G., 340
Kesmarky, S., 420
Kessenikh, A. V., 122, 283
Kessler, H., 261, 262, 408
Kevek, F., 261
Kevill, D. N., 13, 54, 56
Key, J. M., 254
Kézdy, F. J., 445
Khaddar, M. R., 57
Khalaf, A. A., 18
Khalaf, H., 107, 262, 377
Khalifah, R. G., 453
Khalil, F. Y., 427
Khalturinskii, N. A., 398
Khan, A. A., 531
Khan, A. U., 468
Khan, H. A., 187
Khan, M. S., 244, 245
Khan, O. R., 307
Khan, S. A., 69, 457
Khan, S. M., 531
Khandelwal, G. D., 489
Kharitonov, V. V., 198
Khatomi, A. I., 207
Kheifits, L. A., 559

Khizny, W. A., 348
Khorlin, A. Y., 401
Khuddus, M. A., 108
Khuboretskii, V. M., 122
Khuong-Huu, Q., 77, 247, 248, 516
Khutoretskii, V. M., 94
Kiburis, J., 241
Kice, J. L., 92
Kieboom, A. P. G., 557
Kiefer, E. F., 253
Kiefer, G., 173, 267
Kiehlmann, E., 228, 413
Kiener, V., 389
Kienzle, F., 203, 245, 534
Kiesslich, G., 546
Kigasawa, K., 191
Kikuchi, O., 166
Kiguchi, T., 493
Kikuchi, G., 453
Kikuchi, K., 473
Kikukawa, O., 48
Kilby, B. J. L., 390
Killpack, D. R., 207
Kim, B., 504
Kim, C. J., 17
Kim, J. K., 190, 191, 337
Kim, K., 240
Kim, S. G., 17
Kim, Y. D., 444
Kim, Y.-H., 215
Kimling, H., 375, 406
Kimura, C., 108
Kimura, K., 517
Kimura, Y., 48, 344
Kin-Cuong, N., 508
Kindley, L. M., 476
King, B. J., 388
King, J. F., 218, 235
King, K. D., 304, 412
King, N. L. R., 400, 462
King, R. W., 453
Kingsland, M., 199
Kinstle, T. H., 492, 512
Kiprianova, L. A., 122, 409
Kira, M., 339
Kirby, A. J., 69, 431, 457
Kirby, G. W., 71, 250
Kirby, P., 183, 269
Kirchenbaum, D. M., 450
Kirchhoff, K., 262
Kirino, Y., 336
Kiriyama, T., 62
Kirk, B. E., 177
Kirk, D. M., 77
Kirk, D. N., 46, 248
Kirk, K. L., 516
Kirkiacharian, B. S., 305, 483
Kirmse, W., 29, 79, 81, 367

Kirch, G., 231
Kirsch, I. E., 448
Kirsch, J. F., 450
Kirschenbaum, D. M., 450
Kirschner, S., 33, 34, 113, 228, 229, 279
Kirtley, M. E., 463
Kishi, Y., 218
Kishida, K., 401
Kishida, Y., 39, 133
Kita, Y., 218, 257, 258, 273, 485
Kitahara, Y., 48, 102, 175, 221
Kitaoka, S., 453
Kitaura, Y., 478, 514
Kitayama, M., 387
Kitching, W., 204
Kito, N., 289, 490
Kitson, K. M., 330
Kitson, T. M., 451
Kiutamo, T., 5
Kivinen, A., 58
Kiwan, A. M., 56, 89
Klabunde, K. J., 385
Klabunovskii, E. I., 555
Klärner, F.-G., 224
Klar, R., 503
Klaus, M., 495
Klein, J., 115, 154, 264
Klein, P., 289
Klein, R., 310
Kleinfelter, D. C., 5
Kleinhuis, H., 496
Kleinschmidt, T., 401
Klemm, R. F., 479
Klibanov, A. M., 446
Klimova, A. I., 304
Klinman, J. P., 116, 117, 458
Kloek, J. A., 13, 484
Klöpffer, W., 469
Kloosterziel, H., 96
Klose, G., 239
Klose, T. R., 20
Kluge, A. F., 483
Klunder, E. B., 515
Klunklin, G., 503
Klusacek, H., 85, 451
Klyosov, A. A., 446, 447
Knappe, W. R., 506
Knarr, K., 296
Knauss, L., 389, 390
Kneen, G., 481
Kneen, W. R., 154
Kneipp, K. G., 296
Kniezo, L., 135
Knight, A. E. W., 473
Knight, M. H., 346
Knipe, A. C., 149, 413, 538

Knoeber, M. C., 396
Knorr, R., 440
Knowles, A., 524
Knowles, J. A., 441
Knowles, J. R., 451
Knudsen, T. P., 482
Knunyants, I. L., 132, 319, 461
Ko, E. C. F., 59
Kobayashi, H., 92
Kobayashi, M., 282, 283, 289
Kober, H., 230, 498
Kobes, R. D., 414
Koblowa, A. Z., 127
Kóbor, J., 108
Kobuke, Y., 165
Kobyakov, A. K., 441
Kobylecki, R. J., 554
Koch, T. H., 477, 497
Koch, V. R., 83, 298, 342, 546
Kochi, J., 132
Kochi, J. K., 280, 309, 333, 334
Kocienski, J. P., 48, 541
Kodama, M., 78
Koeberg-Telder, A., 197
Koehler, K., 394
Koehler, K. A., 445
Koenig, T., 289
Köhler, W., 107
Koehn, W., 173
Koenig, K. E., 133
Koeppl, G. W., 408
Koermer, G. S., 52
Körös, E., 276, 302
Kövér, E., 406
Kofman, V. A., 154
Koga, K., 17, 60
Kogan, L. M., 429
Kohen, F., 247
Kohno, T., 517
Kohnstam, G., 50
Koizumi, M., 473, 490, 506, 511, 516
Kojima, A., 161, 217
Kojima, M., 289, 429
Kojima, T., 166, 519
Kokazi, K., 442
Kokes, R. J., 264
Kokobun, H., 473, 506, 511
Kolb, K. E., 455
Koldobskii, G. I., 258, 416
Kole, J., 525
Kolka, S., 407
Kollmeier, H. J., 389
Kolodina, N. S., 121
Kolomnikov, I. S., 559
Kolwyck, K. C., 56
Koltai, E., 251

Komai, T., 442
Komeno, T., 63, 245
Kommandeur, J., 351
Komori, S., 304, 387
Konaka, R., 336, 350, 542
Konasewich, D. E., 82
Kondo, K., 268
Kondo, S., 384
Kondo, Y., 180, 438
Kondrat'eva, S. E., 386
Konecny, J., 107
Kong, M.-S., 239
Kongshaug, M., 361
Konishi, K., 112
Kononenko, V. E., 414
Konopinska, D., 440
Konz, W. E., 231
Koob, R. D., 370
Kopay, C. M., 273
Kopecky, J., 494
Koppel, G. A., 61
Koppel, J., 129
Koptyug, V. A., 20, 43, 209
Korber, H., 528
Korchagina, D. V., 20, 43, 209
Korenstein, R., 525
Korkut, S., 475
Korn, K.-D., 364
Kornblum, N., 337, 552
Korner, M. J., 269
Korneva, L. M., 197
Kornrumpf, B., 79
Korochkin, A. F., 159
Korshak, V. V., 429
Korshunov, I. A., 127
Korshunov, M. A., 441
Korswagen, A. R., 38
Korte, F., 269
Korvenranta, J., 5
Korytnyk, W., 405
Korzan, D. G., 72, 114, 146, 348, 542
Kosbahn, W., 166
Koser, G. F., 253, 421
Koshland, D. E., 434, 441
Koshmina, N. V., 195
Kosower, E. M., 290, 297, 337
Kossa, W. C., 333
Kossanyi, J., 487
Kust, A. N., 204
Koster, S., 174
Kostikov, R. R., 375, 455
Kostina, N. G., 197
Kostyanovsky, R. G., 360
Kosugi, M., 276
Kosugi, Y., 353, 354, 538, 540
Kothandaraman, G., 201
Kothe, G., 295

Kouar, J., 122
Kouba, J. E., 206
Koudijs, A., 271, 504
Kovacic, P., 16, 302
Kovacs, C. A., 240
Kovacs, J., 123
Kovacs, K., 441
Kovalenko, V. I., 365
Kovalenok, Z. P., 396
Kowerski, R. C., 29
Kowollik, G., 407
Koyama, H., 387
Koyama, T., 277
Koyana, K., 512
Kozak, I. V., 180
Kozhevnikov, I. V., 46, 107
Kozlov, L. V., 426
Kozlova, M. F., 116
Koz'mina, O. P., 533
Koznetsova, N. Y., 396
Kozuka, S., 212, 407
Kraatz, U., 269
Kraemer, H., 515
Krämer, H., 374
Kraiss, G., 71
Kramer, D. M., 493
Kramer, J. M., 192
Kramer, P. A., 43
Krantz, A., 167, 239
Krapcho, A. P., 28, 45, 79
Kratochvil, M., 108, 302
Kraus, W., 10
Krause, D. L., 381
Krause, J. G., 356
Krauss, P., 371
Krebs, A., 375, 406
Kreevoy, M. M., 82
Kreider, E. M., 213
Kreiter, C. G., 389, 390
Kresge, A. J., 129, 156, 206, 408
Kresze, G., 165
Kretchmer, R. A., 247
Kretzschmar, K., 555
Kricka, L. J., 167
Krieger, J. K., 334
Krimer, M. Z., 160
Krinskaya, A. V., 463
Krishnan, C. V., 58
Krivopalov, V. P., 182
Krokhina, I. N., 132
Kroll, L. C., 556
Kroll, W. R., 390
Kropp, P. J., 504
Kroupa, J., 465
Krow, G. R., 220
Krugh, T. R., 463
Kruk, C., 193
Krukonis, A. P., 364

Krull, I. S., 492
Krusic, P. J., 280
Krutsch, H. C., 101
Krutyakov, V. M., 448
Kruus, P., 57
Krylov, V. K., 116
Kryukov, S. I., 156
Kuan, L. C., 542
Kubin, D., 236
Kubler, D. G., 396
Kubo, M., 479
Kubokawa, Y., 479
Kubota, T., 494
Kubota, Y., 471
Kucera, H. W., 371, 480
Kucherov, V. F., 160
Kuck, V. J., 367
Kucsman, A., 95
Kuczenski, R. T., 463
Kudesia, V. P., 202, 534, 543
Kudo, K., 464
Kudryavtsev, L. F., 365
Kuduk-Jaworska, J., 159
Kuebrich, J. P., 414, 415
Kühle, T., 517
Kuentzel, H., 489
Küsters, W., 166, 483
Kuhlman, D. P., 191
Kuhn, H., 471
Kuivila, H. G., 307, 309
Kukawa, Y., 258
Kukolev, V. P., 559
Kukuyama, S., 117
Kulevsky, N., 520
Kulig, M., 500
Kulik, S., 386, 552
Kulikova, A. I., 453
Kulish, L. F., 409
Kulnevich, V. G., 396
Kumada, K., 548
Kumada, M., 72, 304
Kumagai, Y., 178
Kumamoto, T., 117
Kumar, K. S., 531
Kundryutskova, L. A., 428
Kundu, N. G., 7
Kuneida, T., 508
Kunichika, S., 161
Kunitake, T., 442
Kunkes, S., 504
Kuntz, R. R., 304
Kuo, S. C., 127
Kuo, Y.-N., 144, 228, 343, 552
Kupchan, S. M., 437, 517
Kuri, Z., 518
Kuriacose, J., 202
Kuriacose, J. C., 201
Kurien, K. C., 468

Kurihara, O., 206
Kurimura, Y., 540
Kuriyama, K., 248
Kurland, R. J., 46
Kurlyankina, V. I., 533
Kurnosova, N. P., 205
Kuroda, H., 189
Kursanov, D. N., 207
Kurts, A. L., 95, 116
Kuruma, K., 336, 542
Kurylo, M. J., 310
Kusama, O., 191, 517
Kushner, A. S., 503
Kuster, B. F. M., 538
Kusuda, K., 42, 230
Kuszmann, J., 117, 241
Kuthan, J., 195, 555
Kutney, J. P., 197
Kutschinski, J. L., 83
Kutsenko, L. M., 417
Kuura, H., 428
Kuwata, K., 508
Kuzembaev, K. K., 556
Kuz'menko, I. I., 461
Kuznetsova, N. I., 179
Kvasov, A. A., 127
Kvorko, G. F., 457
Kwan, T., 336
Kwart, K., 144
Kwok, W. K., 146
Kyba, E. P., 371, 516
Kyllönen, A.-L., 540

La Rochelle, R. W., 93
Laarhoven, W. H., 497
Labandibar, P., 550
L'abbe, G., 103, 169, 171, 240, 262
Lablache-Combier, A., 504, 507, 508
Labouesse, B., 444
Labows, J. N., 375
Labuschagne, A. J. H., 224
Lacadie, J. A., 43, 56
Lacaze, P.-C., 555
Lachmann, B., 386
Lacombe, J.-M., 264
Ladd, J. A., 397
Ladenberger, V., 149
Laemmle, J., 128
Lagenaur, C., 406
Lagendijk, A., 340
Lagercrantz, C., 348
Lagrange, M.-J., 96
Lahti, M., 393
Lai, C. Y., 414
Lai, K.-H., 44, 97, 215, 216, 323
Laidlaw, W. G., 229

Laird, R. M., 456, 464
Laird, T., 481
Lakshminarayana, P., 70
Lala, L. K., 247
Lalande, R., 306, 310
Laloi-Diard, M., 193, 424, 464
Lalor, F. J., 95
Lam, L. K. M., 2
Lamaty, G., 129, 395, 407, 412, 416
Lamb, R. C., 285
Lambe, T. M., 261, 408, 543
Lambert, C. A., 359
Lambert, J. B., 22, 23
Lambert, Y., 13
Lamberton, A. H., 426
Lamberts, L., 58
Lambeth, P. F., 476
Lamola, A. A., 468
Lamont, A. M., 226
Lamper, J. E., 121
Lamson, D. W., 549
Lancaster, J. E., 503
Land, E. J., 471
Land, H. B., 515, 520
Landgrebe, J. A., 234, 384
Landini, D., 92, 553
Landis, M. E., 287, 515
Landsberg, R., 531
Lane, B. C., 391
Lane, C. A., 43
Lane, C. F., 130, 329
Lane, R. S., 414
Lang, C.-C., 373, 552
Lang, H. M., 452
Lang, K. L., 441
Lange, G. L., 482, 484
Langelaar, J., 469, 473
Langen, H., 259
Langen, P., 407
Langer, S. H., 555
Langlois, S., 420
Langone, J. J., 244
Langourieux, Y., 272
Langston, R. J., 379
Lankin, D. C., 276, 545
Lanneau, G. F., 84
Lansbury, P. T., 25, 161
Lanthier, R., 78
Lapouyade, R., 550
Lappert, M. F., 387, 389
Lappin, G. R., 478
Lardenois, P., 264
Lardicci, L., 547
Larionova, N. I., 449
Larrabee, R. B., 228
Larscheid, M. E., 83, 342, 546
Larsen, E. R., 64
Larsen, J. W., 43, 187

Larson, H. O., 234
Lasne, M.-C., 312
Lasperas, M., 412
Lassen, F. O., 440
Lasser, N., 452
Lateef, A. B., 457
Latham, W. A., 367
Lathan, W. A., 48, 362
Lathrop, H., 12
Latimore, M. C., 144
Latovitzki, N., 399
Latters, A., 262, 263
Lattes, A., 158, 546, 549
Lau, M. P., 513
Lauer, R. F., 532
Laurema, S. A., 552
Laurence, C., 417
Lauria, F., 193
Laurie, C. W., 455
Laurishin, W. V., 417
Laval, J. P., 158
Lavanish, J. M., 101, 266
Lavie, D., 247
Lavielle, G., 95
Laviron, E., 554
Law, J., 36
Law, P. Y., 134
Lawesson, S.-O., 97, 218, 269
Lawler, R. G., 195
Lawson, A. J., 200, 260, 325
Lawson, D. F., 340
Lawson, P. J., 462
Lawton, R. G., 75, 248
Lay, W. P., 168, 231
Laye, P. G., 357
Lazar, H., 162
Lazareva, V. T., 465
Lazdunskii, C., 462
Lazdunskii, M., 462
Lazzeretti, P., 264
Le Demezet, M., 57
Le Guillanton, G., 554
Le Hénaff, P., 413
Le Men, J., 77, 251, 252
le Noble, W., 18
le Noble, W. J., 97, 372
Le Quesne, P. W., 548
Lea, J. R., 207
Leach, B. E., 405
Leard, M., 83, 84
Leary, G., 489, 503
Leaver, I. H., 348, 362, 476, 506, 520
Lebedev, A. B., 429
Lebedev, N. N., 304
LeBel, N. A., 171
LeBerre, A., 258
Lecourt, M.-J., 184
Ledlie, D. B., 552

Ledwith, A., 48, 341, 346, 347, 519
Lee, A. L., 365
Lee, B.-S., 106
Lee, C., 481
Lee, C. C., 2, 29, 36
Lee, C. S., 504
Lee, C. V., 99
Lee, D. G., 531
Lee, E. K. C., 479
Lee, G. A., 517
Lee, G. K. J., 204
Lee, I., 106
Lee, J. R., 208
Lee, L., 54, 236, 340
Lee, L. P. Y., 479
Lee, R. G., 443
Lee, T. B. K., 79
Lee, T. J., 487
Lee, T. W. S., 445
Lee, Y. T., 506
Leeder, W. R., 134
Leedy, D. H., 555
Leenders, L., 493
Leermakers, P. A., 470
Leffek, K. T., 59, 60, 122, 185
Leffler, J. E., 285, 286
Lefort, D., 310
Lefour, J.-M., 165
LeGoff, E., 484
Lehman, P. G., 397
Lehmann, E. J., 289
Lehn, J. M., 57
Lehner, H., 209
Lehnig, M., 276
Lehr, R. E., 371
Lehrer, G. M., 414
Lehrer, S. S., 400
Leibfritz, D., 261, 262, 408
Leicht, C. L., 65, 242
Leichter, L. M., 236, 291
Leipoldt, J. G., 108
Leitich, J., 484
Lelieuve, J., 179
Lenn, S. L., 501
Lemaire, J., 487
Lemal, D. M., 508
Lemière, G., 396
Lemieux, R. U., 403
Lemley, J. T., 358
Lempert, K., 69, 251
Lenoir, D., 14, 30, 245, 246
Leonard, J. E., 468
Leone, R. E., 1
Lepley, A. R., 191, 239, 275, 288
Leppard, D. G., 159, 251
Leppin, E., 518
LeRoux, J.-P., 232

Lesbre, M., 379
Leseticky, L., 263
Leshchev, V. P., 198
Leslie, J., 448, 544
Leslie, V. J., 217
Lessard, J., 305, 492
Letchken, P., 503
Letertre, G., 232
Letsinger, R. L., 442, 525
Lett, R., 119
Leussing, D. L., 405
Leuzinger, W., 453
Levashov, A. V., 447
Leveson, L. L., 555
Levi, E. M., 16
Levin, C., 534
Levin, G., 54, 114, 340
Levin, R. H., 370, 498
Levina, I. Y., 78
Levisalles, J., 248, 251
Levit, S., 449
Levitina, E. S., 555
Levomäki, M., 538
Levrushko, A. G., 472
Levy, E. C., 247
Lévy, J., 77, 251, 252
Levy, J. B., 289
Levy, M., 339
Lewandos, G. S., 257
Lewis, A., 195
Lewis, A. J., 61, 150, 161, 199, 247
Lewis, D., 396
Lewis, E. S., 124, 296, 345
Lewis, F. D., 478
Lewis, I. C., 365
Lewis, J., 197
Lewis, J. W., 244
Lewis, K. G., 269
Lewis, M. C., 33, 546
Lewis, P. H., 75
Lewis, T. P., 555
Lexton, M. J., 296
Ley, S. V., 13, 248
Leyshon, L. J., 380, 516
Leznoff, C. C., 497
Lhomme, J., 79, 250
Li, J. P., 267
Li Hsu, Y.-F., 165
Liang, G. A., 12
Liang, K. S. Y., 313
Liao, C. C., 514
Libbey, W. J., 7
Liberatore, F., 167, 549
Liberman, A. L., 555, 556
Libert, A., 97
Libis, B., 269
Libman, J., 77, 215
Librovich, N. B., 428

Licht, E., 192
Lichtenberg, D., 264
Lichtin, N. N., 361
Liden, A., 261
Lien, R. K. H., 450
Lien, E. L., 134
Lienhard, G. E., 399, 445
Liedhegener, A., 382, 386
Liggero, S. H., 14, 245
Lightner, D. A., 523
Lii, R. R., 116, 412
Liler, M., 428
Lill, C. F., 236
Lillien, I., 34
Lillya, C. P., 36, 483
Lim, D., 442
Lim, E. C., 470
Lim, L. S. N., 383, 518
Lim, P. K. K., 386
Lim, Y. Y., 427
Limbird, T. J., 399
Limoges, R. D., 465
Lin, C.-W., 463
Lin, C. Y., 167, 239
Lin, D. K., 532
Lin, H. C., 42, 124
Lin, L. C., 265, 388
Linda, P., 182, 197
Lindenberger, W., 478
Linder, P. W., 400
Linder, R. E., 361
Lindner, E., 538
Lindow, D. F., 46, 215
Lindqvist, L., 475
Linek, K., 554
Ling, A. C., 361
Linhart, F., 269, 341, 536, 546
Linsay, E. C., 52
Linstrumelle, G., 498
Lion, C., 22, 23, 103
Liogon'kii, B. I., 346
Liotta, C. L., 456
Lipmaa, E. T., 77, 177
Lippmaa, E., 428
Lippold, B. C., 434
Lipsky, S. D., 552
Lisitsyn, V. N., 181
Liška, F., 309
Lissi, E. A., 327, 329, 362, 489, 519, 538
Lister, J. H., 241
Litchfield, G. T., 204
Litinov, V. P., 207
Litman, R., 339
Littler, J. S., 530
Liturri, V., 182
Litvinenko, L. M., 103
Liu, J. H., 16

Liu, K. T., 158
Liu, M. S., 279
Liu, M. T. H., 166
Liu, R. S. H., 500
Litvinenko, L. M., 429, 464
Livingston, R., 280, 324, 362, 363, 504
Ljungquist, A., 453
Lloyd, D., 197, 202, 207, 552, 555
Lloyd, D. J., 141
Lloyd, G. J., 458, 459
Lloyd, K., 440
Lloyd, R. V., 362
Lo, K.-W., 464
Loader, C. E., 209
Lobeck, W. G., Jr., 269
Lodder, A. E., 43
Löffler, H. G., 450
Löffler, H.-P., 323
Loewenstein, A., 280
Logani, S. C., 548
Lokhmatova, V. F., 556
Lomas, J. S., 44, 82, 149
Lombardo, L., 189
Long, D. A., 443
Long, F. A., 123
Long, G. L., 490
Long, M. E., 470
Longevialle, P., 245
Longridge, J. L., 162, 428
Lont, P. J., 271
Loo, P.-W., 228, 413
Looker, J. J., 375, 376
Lord, E., 101, 140, 162
Lore, A. M., 408
Lorenc, Y., 482
Lorenz, H. P., 481
Lorenz, P., 283
Lossing, F. P., 358
Lotsch, W., 15
Lotspeich, F. J., 20, 62
Loukas, S. L., 17
Loupy, A., 106, 142
Lourens, G. J., 68
Louw, R., 356
Love, G. M., 23, 28, 172, 250, 483
Loveridge, E. L., 490
Lovreček, B., 555
Lovtsova, A. N., 77
Low, D. C. F., 165
Lowe, G., 450, 515
Lowe, J. N., 132
Lowe, J. P., 103
Lowe, R., 240
Lown, E. M., 161
Lown, J. W., 155, 233, 234
Lubinskaya, O. V., 160

Lucente, G., 453
Lucken, E. A. C., 260, 325
Luckenbach, R., 88, 549
Luckhurst, G. R., 295, 352
Luderer, J. R., 130
Ludorff, E., 168
Lüttke, W., 208
Luetzow, A. E., 398
Luft, R., 163
Lui, C. Y., 1
Luibrand, R. T., 7
Lukacs, G., 245
Lukas, J., 43, 45
Lukehart, C. M., 67, 389
Lukovnikov, A. F., 287
Luk'yanova, L. N., 401
Lumb, T. J., 256
Lumry, R., 444, 447
Lunazzi, L., 276
Lund, H., 555
Lundblad, R. L., 450
Lunney, D. C., 285
Lupes, M. E., 414
Lura, R. D., 494
Luria, M., 501
Lustgarten, R. K., 20, 21, 27, 52
Luskus, L. J., 165, 175
Lusinchi, X., 245, 265
Luthe, H., 515
Lutley, P. M., 461
Lutz, F. E., 253
Lutz, H., 475
Lutz, R. E., 138
Lutz, R. P., 219
Luzzier, R. J., 348
L'vova, F. P., 408
L'vova, M. S., 94, 95
Lwowski, W., 372
Lykov, Y. V., 108
Lyon, G. D., 481
Lyons, A. R., 281
Lyons, J. E., 263, 558
Lysenkov, N. V., 450
Lytle, F. E., 524
Lyushin, M. M., 205
Lyznicki, E. P., Jr., 23

Maassen, J. A., 350, 512
Mabry, T. J., 481
McAneny, B., 479
Macau, J., 58
McBee, E. T., 385
MacBride, J. A. H., 271
McBride, J. M., 288, 292
MacBridge, J. A. H., 241
McCabe, P. H., 62, 218
Maccagnani, G., 169, 542
McCall, J. M., 498, 499

McCall, M. T., 493
Maccarone, E., 199
Macchia, B., 109, 247
Macchia, F., 247
MacClarence, J. W., 105
McClory, M. R., 119
McColeman, C., 552
McColgin, W. C., 471
Maccoll, A., 143, 145
McCorkindale, N. J., 163
McCreadie, T., 254, 551
McCrindle, R., 247
McCulloch, A. W., 270
McCullough, J. J., 483, 503
McCullough, J. L., 452
McCurry, P. M., 158, 375
MacDiarmid, A. G., 204
McDonald, C. C., 400
Macdonald, D. D., 57
Macdonald, D. M., 57
McDonald, J. W., 391
McDonald, R. J., 473
McDonald, R. N., 17, 23, 72, 243
McDonald, R. S., 152, 156, 408
McDonnell, J. J., 340, 538
McDowell, S. T., 433
McEnroe, F. J., 552
McEwen, G. K., 89
McEwen, W. E., 169
McFarlane, N. R., 185
MacFarlane, P. H., 515
McGarry, B. E., 25
McGillivray, G., 203
McGillivray, N., 228
McGinnis, G. D., 404
McGregor, M. L., 150, 552
MacGregor, R. A., 177
Macháček, V., 200
Macháčková, A., 193, 200
McHales, D., 263
Machens, E., 220
Machiguchi, T., 102
Macias, A., 95, 116
Maciel, G. A., 48
McInnes, A. G., 270
McIntosh, C. L., 235, 489, 491, 492
McIntosh, J. M., 20
McIntyre, P. S., 108
McIsaac, J. E., 456
McIvor, M. C., 121
Mackay, D., 365
Mackay, M., 126
McKechnie, J. S., 46
McKellar, J. F., 511
McKelvey, R. D., 348, 472
McKenna, C. E., 549

McKenna, J. C., 69, 154
McKenna, J. M., 5, 7
McKennis, J. S., 220
McKenzie, E. D., 406
Mackenzie, K., 168, 231
McKervey, M. A., 14, 126, 248, 250
McKiernan, J. E., 37
McKillop, A., 203, 244, 245, 534
McKillop, T. F. W., 16
McKinley, S. V., 43
McKinney, M. A., 36, 126
MacLachlan, A., 364, 508, 511
McLafferty, F. W., 46
McLean, S., 227, 228, 528, 538
McLennan, D. J., 49, 142, 146, 241
McLeod, D. M., 508
McManus, S. P., 70
McMaster, I. R., 514
MacMillan, J. H., 222
McMillen, D. F., 359
McMurry, J. E., 16, 248
McMurry, T. B. H., 253
McNeil, D. W., 236
McNeil, M. W., 201
McNeill, D., 522
McNesby, J. R., 378
McNew, W. E., 285
McNicholas, M. W., 219
Macomber, R. S., 23, 49
MacPhee, J. A., 532
McPherson, C. A., 155
MacQuarrie, R. A., 540
McQuillin, F. J., 126, 558
McRae, J., 363
MacSweeney, D. F., 247, 485
Maddock, S. J., 257
Madigan, D. M., 498
Madsen, V., 512
Mády, G., 453
Maeder, H., 511
Mäder, H., 170, 233
Maeda, M., 124, 207
Maeda, S., 483
Maeda, T., 212
Maercker, A., 128, 272
Maerker, G., 70
Maerky, M., 508
Mageswaran, S., 222
Magi, M., 428
Magid, R. M., 114, 165, 263
Maglio, G., 37
Magnus, P. D., 46, 78, 219, 247, 490, 495, 545
Magrill, D. S., 163

Magyar, M., 198
Mahanti, M. K., 534
Mahendran, K., 105
Mahendran, M., 273
Maheshwari, K. K., 522
Mai, V. A., 384
Maia, H. L., 262
Maier, G., 511
Mail, A. C., 516
Maillard, R., 310
Main, L., 201
Maiorana, S., 113, 173, 222
Maiti, B. C., 70
Maitland, D. J., 215
Maitlis, M. M., 254
Maitlis, P. M., 180
Maizus, Z. K., 287
Majee, B., 113
Majer, J. R., 516
Majerski, Z., 14, 26, 52, 245
Majeti, S., 481
Majewski, P. J., 103
Makai, Y., 271
Makhaev, V. D., 133
Maki, Y., 191, 215, 386, 552
Makin, S. M., 166
Makinen, K. K., 449
Makinen, P.-L., 449
Makolkin, J. A., 456
Makor, A., 302
Makosza, M., 113, 368, 369
Makoszu, M., 180
Malassiné, B., 316
Malatesta, V., 96, 319, 351
Malinovskii, M. S., 159
Mallory, R. A., 247
Malo, H., 205
Malone, L. J., 425
Mal'tsev, A. K., 367
Mamatyuk, V. I., 20, 209
Mamayev, V. P., 182
Mamer, O. A., 358
Mamontova, I. V., 303, 310
Mancuso, A., 261
Manda, E., 264
Mandai, H., 464
Mandava, N., 107
Mandel, N., 88
Mandelbaum, A., 247
Mander, L. N., 20
Mandolini, C., 201, 202
Mandolini, L., 125
Manecke, G., 171
Manery, E. L., 365
Mangia, A., 113
Mangini, A., 169, 276
Mango, F. D., 177, 257
Manhas, M. S., 66

Man-Him-Hui, 480
Manitto, P., 536
Manley, R. M., 425
Mann, C. K., 344
Mannen, S., 401
Manning, C., 503
Manojlović-Muir, L., 389
Mantz, I. B., 294
Manville, J. F., 470
Manzer, L. E., 390
Mao, C.-L., 70
Mao, S. W., 340
Maradufu, A., 540
Marayama, T., 468
Marcali, K., 364
Marchenko, G. N., 457
Marchenko, V. A., 122
Marchese, G., 101, 150, 163
Marcopoulos, C. A., 192
Marcoux, L., 344
Marcum, J. D., 116, 336
Maremäe, V. M., 465
Margulis, T. N., 215
Marhold, J., 341, 540
Mariano, P. S., 520
Maricich, T. J., 387
Mariella, P., 107
Mariella, R. P., 430
Marini, J. L., 446
Marini, M. A., 424, 444, 447
Marino, G., 197, 205
Marinone, F., 174
Mark, F., 484
Mark, G., 484
Mark, H. B., 555
Mark, V., 154, 484, 545
Markett, J., 217
Markevich, V. S., 166
Markó, L., 559
Markova, A. F., 456
Markovic, V. M., 361
Markowski, V., 170, 234
Markstein, J., 240
Marmer, W. N., 108, 545
Marmor, R. S., 375, 382
Marnett, L. J., 287, 515
Maron, E., 400
Marples, B. A., 109, 190, 247, 248, 517
Marples, B. A., 483
Marquarding, D., 85, 457
Marquardt, R. R., 414
Marquet, A., 119
Marquez, A., 429
Marsh, D. G., 479
Marsh, G., 485
Marshall, D. R., 3, 152, 154, 202, 207, 412

Marshall, J. A., 28, 62, 82, 133, 163, 493
Marshall, J. L., 23, 66, 550
Marshall, R. M., 296
Marsili, A., 78, 247
Marsmann, H., 113, 126
Martell, A. E., 405
Martelli, G., 517
Martin, C. J., 424, 444, 447
Martin, C. W., 234, 384
Martin, D., 430
Martin, G. J., 206, 262
Martin, H. A., 559
Martin, J., 16
Martin, J. C., 89, 189
Martin, R., 213, 359
Martin, R. H., 497, 520
Martinek, K., 442, 446, 447, 448, 449
Martinet, P., 269, 554
Martin-Ramos, V., 170
Marton, A. F., 456
Marty, R. A., 357
Martynov, B. I., 132
Maruca, R., 370
Maruyama, K., 130, 468, 545
Maryanoff, B., 549
Maryanoff, B. E., 548
Marziano, N. C., 199
Marzin, C., 232
Masahura, H., 474
Masamune, S., 11, 12, 236, 254, 256, 481
Masamune, T., 325, 519
Masaracchia, J., 173, 484
Masiuke, T., 246
Maskornick, M. J., 121
Mason, P., 160, 301, 302
Mason, R., 154
Mason, T. J., 2, 531
Masse, J. P., 84
Massol, M., 306
Masters, C. J., 414
Masuda, T., 244
Masui, M., 545
Masuike, T., 61
Mataga, N., 474
Matasa, C., 357
Mathai, K. P., 218
Matheson, A. F., 60
Mathiaparam, P., 251
Mathur, D. L., 533
Mathur, N. K., 159
Matienko, L. I., 287
Matier, W. L., 215
Matnishyan, A. A., 346
Matoba, R., 400
Matrka, M., 341, 465, 540
Matsu, T., 493

Matsuda, H., 133
Matsuda, Y., 271
Matsui, K., 157, 503
Matsui, T., 56, 165
Matsui, Y., 554
Matsumoto, H., 299
Matsumoto, K., 107, 233, 234, 557
Matsumoto, T., 416, 524
Matsumura, Y., 48, 546
Matsunaga, K., 387
Matsura, T., 478
Matsushima, A. A., 449
Matsushima, R., 336
Matsushima, Y., 400
Matsuura, K., 483
Matsuura, T., 319, 434, 487, 504, 514, 517, 520, 521
Matsuyama, A., 469
Matsuyama, K., 297
Mattheus, H., 484
Matthews, B. W., 401
Matthews, J. S., 258
Mattingly, T. W., 503
Matuszak, A. J. B., 70
Mauceri, F. A., 287
Maudling, H. V., 420
Maugh, T., 432
Maundrell, D. F., 192
Maupérin, P., 77, 252
Maurette, D., 522
Maurette, M.-T., 546
Maurin, R., 34
Maury, G., 264, 270
Mawatari, T., 508
May, R. W., 443
May, S. W., 451
Mayeda, E. A., 342
Mayer, C. E., Jr., 241
Mayer, C. F., 481
Mayer-Ruthardt, I., 282
Mayers, G. L., 123
Mazarguil, H., 263
Mazerolles, P., 379
Mazet, M., 78, 244
Mazur, Y., 77, 215, 520
Mazzanti, G., 169, 542
Mazzochi, P. H., 225
Mazzucato, U., 500
Meakin, P., 280
Means, G. E., 462
Meany, J. E., 402, 410
Meares, C. F., 344
Mechoulam, R., 493
Medary, T. R., 403
Medved, S. S., 464
Medvedev, B. Ya., 346
Meek, A. G., 36
Meier, B., 215

Meier, H., 371, 508
Meinwald, J., 494, 495
Meisinger, R. H., 243
Meister, W., 123
Mekhtier, S. D., 205
Melander, L., 123
Melicharek, M., 545
Meliksetyan, R. P., 348
Melloni, G., 245
Mellor, J. M., 333, 532, 533
Mellor, J. W., 5
Mellows, S. M., 489
Mel'nichenko, I. V., 413
Mel'nikov, G. D., 154
Melrose, G. J. H., 443
Melzacka, M., 420
Ménard, H., 78
Mendenhall, G. D., 353
Meney, J., 215
Menger, F. M., 181
Menghani, G. D., 543
Menicagli, R., 547
Menon, B. C., 228, 313, 513
Mensah, I. A., 489
Menzel, P., 202
Mercier, C., 522
Mercier, D., 401
Meresaar, U., 420, 437
Merritt, M. V., 188
Messing, A. W., 83
Metcalfe, J., 235
Meth-Cohn, O., 512, 529, 542
Merten, H., 386
Metelitsa, D. I., 276
Metts, L., 500
Metzger, J., 313, 318, 319, 508
Motzgor, R., 316
Metzler, D. E., 506
Meuis, M. E. A. H., 513
Meunier, J.-M., 554
Meyer, G., 456
Meyer, G. R., 254
Meyer, M. M., 406
Meyer, T. H., 521
Meyer, W. C., 173
Meyers, A. I., 504
Meyerson, S., 190, 357
Meyerstein, D., 320
Mhala, M.-M., 412, 461
Miao, C. K., 552
Michael, K. W., 83
Michalski, J., 461
Michaud, P., 355
Michejda, C. J., 308
Michelot, R., 413
Michl, J., 495
Michon, P., 353
Middleton, S., 75

Middleton, W. J., 261, 308, 406
Midland, M. M., 327, 328, 329, 538
Mielek, J. B., 434
Migita, T., 106, 276, 384, 476
Mihailovič, M. Lj., 331
Mihailovic, M. Y., 482
Mihelćić, D., 296
Mikaelyan, R. G., 367
Mikhalevech, M. K., 107
Mikhlin, V. S., 441
Miki, T., 165
Mikolajczyk, J., 461
Mikolajczyk, M., 545
Mikula, J., 108
Mikuriya, Y., 21
Mildvan, A. S., 414, 443
Mile, B., 277
Miles, D. H., 25
Mileshkevich, V. P., 85
Milewski, C., 549
Milewski, C. A., 548
Miller, B., 44, 97, 215, 216, 217, 323, 487
Miller, D. W., 287
Miller, I. J., 30, 40, 263
Miller, J., 179
Miller, J. A., 96, 413
Miller, J. C., 413
Miller, J. E., 400
Miller, L. L., 83, 228, 336, 342, 546
Miller, M. J., 103
Miller, R. E., 137
Miller, R. G., 191, 256, 257
Miller, S. I., 94, 101, 116, 146, 412
Millership, J. S., 380, 383
Milliet, P., 265
Millot, F., 184
Mills, R. W., 109, 245, 550
Millward, G. E., 145
Milner, D. J., 537
Milosavljević, S., 331
Minami, K., 323, 483
Minato, H., 222, 282, 283, 289
Minch, M., 442
Minegishi, T., 157
Minisci, F., 317, 319
Minn, F. L., 341, 472, 473, 476
Mineo, I. C., 169
Minetti, R., 179, 180
Mints, E. S., 127, 427
Minyard, J. P., 484
Miotti, U., 54
Miraglia, M., 244

Mironova, D. F., 429, 457
Misiti, D., 258
Miskow, M. H., 20
Mislow, K., 88, 281
Mison, P., 247
Mistryukov, E. A., 557
Mitchard, L. C., 389
Mitchell, J. R., 190, 313, 317
Mitchell, M. J., 272
Mitchell, R. E., 70
Mitra, D. K., 158
Mitsui, S., 546
Mitsumoto, M., 401
Mittall, R. K., 531
Mitzner, R., 404
Miwa, T., 191
Miyake, H., 543
Miyamoto, N., 493
Miyamoto, T., 494
Miyano, S., 387
Miyashi, T., 234
Miyashita, M., 495
Miyashita, Y., 506
Miyaura, N., 329
Miyazaki, K., 214
Miyazaki, S., 18
Miyazaki, T., 518
Mizoguchi, T., 401
Mizuguchi, T., 325, 519
Mizuno, H., 252
Mizuno, M., 93, 553
Mizushima, M., 400
Mo, Y. K., 41, 42, 124
Moan, J., 475
Mobergans, C., 187
Mobbs, D. B., 380
Modena, G., 92, 93, 97, 99, 100, 245, 553
Modena, M., 310
Modro, T. A., 199
Möbius, D., 471
Möhrle, H., 536
Moffatt, J. G., 544
Mofti, A. M., 69
Moggi, L., 472
Mohammad, M., 297, 337
Moiseev, Y. V., 398
Mollet, P., 172
Mollica, J. A., 406, 417
Monack, L., 202
Mondelli, R., 549
Moniot, J. L., 437, 517
Monjoint, P., 193, 464
Monneret, C., 77, 247, 248
Monot, M. R., 507
Monsó, J. M. 364
Monson, R. S., 137, 149
Montaigne, R., 172
Montanari, F., 92, 553

Monti, L., 247
Monti, S. A., 259, 541, 549
Montillier, J. P., 94
Monvernay, A., 555
Moodie, R. B., 198, 199, 426
Moon, S., 481
Moore, D. W., 552
Moore, H. W., 272, 383, 516
Moore, J. A., 273
Moore, J. S., 398
Moore, L. O., 155, 304
Moore, R. E., 222
Moore, W. R., 388
Moradpour, A., 497
Morales, O., 243
Morantz, D. J., 470
Moraru, M., 353
Morat, C., 353
Moravcsik, E., 453
Morawiec, J., 440
Moreau, B., 119
Moreau, C., 129, 416, 529
Moreau, J., 548
Moreau, P., 243, 412
Moreau, S., 47
Morelli, I., 247
Morgan, F. J., 399
Morgan, T. D. B., 212
Mori, A., 464
Mori, K., 265
Mori, M., 261
Mori, T., 494
Mori, Y., 512
Moriarty, R. M., 175, 372, 508
Moriconi, E. J., 173
Morii, T., 271
Morimoto, M., 442
Morimura, S., 353
Morisaki, M., 108
Morisaki, N., 263
Morishima, I., 353
Morita, H., 511
Morita, M., 166, 360
Morita, T., 503, 520
Moritani, I., 173, 210, 373, 374
Moritsugu, K., 500
Moriyoshi, T., 245, 398, 553
Morizur, J. P., 487
Moro-oka, Y., 539
Morozov, A. I., 456
Morozova, G. M., 396
Morozova, N. A., 448
Morquet, A., 411
Morrill, J. G., 126
Morrill, T. C., 25, 33, 126, 156, 323
Morris, D. G., 5, 7, 239, 325
Morris, E. D., 359

Morris, M. R., 487
Morris, P. J., 443
Morris, T. A., 470
Morrison, H., 491, 493, 497, 500
Morrison, J. D., 558
Morrison, J. F., 463
Morrison, R. J., 559
Morrow, C. J., 558
Mortensen, J. Z., 218
Morton, D. R., 265, 371, 468, 480
Mortreux, A., 250
Moser, G. A., 191, 389
Moser, H. C., 296
Moser, J. F., 481
Moser, R. J., 453
Mosher, H. S., 240, 346, 416
Mosher, M. W., 300, 302, 513
Moskvina, G. I., 539
Moss, R. A., 79, 368, 373, 515
Mossa, G., 198
Mosse, M., 525
Mostashari, A., 46
Moulik, S. P., 531
Moulineau, C., 103
Moulines, F., 557
Mousset, G., 269
Mowat, W., 559
Mowery, P. C., 121, 195
Mravec, D., 97
Mroczek, A., 78
Muck, D. L., 555
Mudry, D. A., 515
Mühlstädt, M., 207
Müller, B. L., 521
Müller, C., 368
Mueller, D. C., 369
Müller, E., 371
Müller, H., 37, 244
Müller, J., 389
Müller, P., 531
Müller, R., 161
Müller-Hagen, G., 161
Mui, J. Y.-P., 384
Muir, C. N., 77, 247
Muir, D. M., 141
Muir, K. W., 389
Mukai, K., 336
Mukai, T., 234
Mukaiyama, T., 117, 271
Mukamal, H., 257
Mukerjee, Y. N., 166
Mukherjee, J., 248
Mukherjee, P. R., 547
Mukherjee, R., 508
Mukherjee, S. K., 530
Muldakhmetov, Z. M., 241
Muljiani, Z., 495

Mullen, P. W., 356
Muller, E., 508
Muller, H., 351
Muller, J.-C., 487
Mulligan, L. A., 202
Mulquiney, C. E., 269
Mulvaney, J. E., 264
Munchausen, L., 501
Munemo, E. M., 239, 325
Muneyuki, R., 251, 416
Munk, M. E., 243
Munsch, H., 395
Murachi, T., 450
Murahashi, S., 508
Murahashi, S.-I., 373
Murai, K., 108
Murakami, Y., 458, 459, 546
Muramatu, M., 449
Murase, I., 263
Murata, M., 232
Murata, N., 149, 295
Murata, T., 525
Murata, Y., 118, 163, 197
Murayama, K., 353
Murgia, S. M., 95
Murgulescu, G., 106
Murgulescu, I. G., 54
Murlo, P., 58
Muromtsev, V. I., 276
Murphy, J. W., 25
Murphy, W. S., 97, 116, 551
Murray, N. G., 335
Murray, R. D. H., 62, 109, 218, 245
Murray, R. K., 88, 281
Murray, R. W., 522, 527, 528, 529, 550
Murrill, E., 408
Murto, J., 181, 182
Museo, A., 37
Musgrave, W. K. R., 167, 188, 241, 271
Mushran, S. P., 553
Mushrusn, G. W., 472
Mustafa, A., 269
Muszkat, K. A., 470, 525
Mutai, K., 474
Mutterer, F., 330
Myers, R. F., 25
Myerscough, T., 352
Myhre, P. C., 47, 160, 200
Mykytka, J. P., 516
Mylonakis, S. G., 206
Mysov, E. I., 132

Naan, M. P., 101, 140, 162, 552
Nadar, P. A., 456
Nadolski, K., 430
Nádor, K., 71

Nagai, T., 215
Nagai, Y., 276, 299
Nagakura, S., 473
Nagamachi, T., 319
Nagami, H., 442
Nagao, Y., 417, 519
Nagaoka, M., 423, 457
Nagata, K., 409
Nagayama, M., 464
Nagy, P., 406
Náhlovská, Z., 219
Náhlovský, B. D., 219
Naiv, G. V., 197
Nair, T. D. R., 435
Naito, T., 493
Nakadaira, Y., 519
Nakagaki, M., 456
Nakagawa, M., 264
Nakagawa, Y., 546
Nakahira, T., 7
Nakai, M., 212
Nakai, T., 11, 48, 254
Nakamaru, H., 506
Nakamizo, N., 429
Nakamura, K., 340, 449
Nakamura, S., 299
Nakamura, Y., 71
Nakane, R., 206
Nakanishi, N., 425
Nakano, Y., 453
Nakao, R., 149
Nakao, Y., 96
Nakashima, M., 476
Nakashima, R., 487
Nakashita, Y., 517
Nakata, T., 318, 519
Nakayama, J., 239
Nakayama, S., 176
Namy, J. L., 108, 416
Nandi, J. C., 506
Nanjo, G., 479
Napier, I. M., 365
Narayanan, K. V., 252
Narayanankutty, N., 531
Narinesingh, D., 431
Naruto, S., 517
Nasipuri, D., 547
Naso, F., 101, 150, 163
Nate, K., 353, 538
Nath, P., 532, 533
Natsume, M., 504
Naulet, N., 262
Naumov, A. D., 122
Navada, K. C., 241
Nave, P. M., 331, 533
Nawrocka-Lecas, A., 523
Nayak, P. L., 158, 179, 455
Naylor, R. A., 424, 460, 462
Nazarova, Z. N., 406

Nazhat, N. B., 112, 360
Nebzydoski, J. W., 4, 161
Neckers, D. C., 479
Nedolya, N. A., 409
Nefedov, O. M., 367
Nefedova, M. N., 207
Negi, T., 463
Negishi, E., 157, 329
Negoiță, N., 353
Neidert, E., 484
Neidle, S., 155
Neiman, L. A., 177
Neiman, Z., 264
Neimysheva, A. A., 461
Nekrasov, Y. S., 177
Nelsen, S. F., 126
Nelson, D. A., 172
Nelson, R. F., 545
Nelson, J. D., 4, 152
Nerdel, F., 229
Nesmeyanov, A. N., 38
Nesmeyanov, N. A., 118
Neta, P., 296, 351, 363
Neuberger, A., 399
Neuenschwander, M., 513
Neugebauer, F. A., 361, 362, 358, 364
Neuheiser, L., 107
Neuman, R. C., 284
Neumann, C. L., 107
Neumann, H. M., 128
Neumann, W. P., 260
Neurath, H., 449, 452
Neville, A. F., 349
Newall, A. R., 385
Newman, M. S., 45, 99, 148, 270, 272
Newman, M. S., 270, 272
Ng Ying Kin, N. M. K., 78
Nguyen, C. H., 119
Nguyen, T. M. N., 304
Nicholls, C. H., 511
Nickerson, T. A., 404
Nickon, A., 251, 416
Niclause, M., 359, 487
Nicoletti, R., 388
Nicoud, J.-F., 380, 497
Niederer, P., 346
Nielsen, S. O., 361
Niemczyk, M., 482
Niemeyer, D. H., 492
Nierlich, F., 169
Nieto, Z., 448
Niiyama, H., 149
Niizuma, S., 506
Niki, E., 365
Niki, H., 359
Nikishin, G. I., 533
Nikolaev, G. A., 85

Nilles, G. P., 251
Nilsson, J. L. G., 202, 224, 536
Nilsson, M., 186
Ninomiya, I., 493
Ninomiya, T., 557
Nisbet, D. F., 177
Nishida, S., 173
Nishida, T., 213, 251, 416
Nishikata, K., 119
Nishikimi, M., 540
Nishimura, J., 387
Nishimura, S., 557
Nishimura, Y., 446, 453
Nishinaga, A., 319
Nishino, M., 373
Nishio, K., 119
Nishiwaki, T., 268, 270
Nishiyama, K., 520
Nissen, D., 438
Nitta, M., 234
Nitta, Y., 400
Nivard, R. J. F., 497
Niwa, M., 46
Noguchi, J., 442
Noguchi, S., 191
Noller, H., 142, 555
Nonhabel, D. C., 157
Nordmeyer, F. R., 348
Noreen, A. L., 517
Norin, T., 28
Norman, R. O. C., 275, 280, 315, 324, 335, 342, 343, 345, 351, 535
Norris, A. R., 187
Norris, C., 348
Norris, R. K., 337
Norrish, G. W., 383
Northington, D. J., 375
Nouguier, R., 311
Novikov, A. N., 203
Norikov, V. N., 182
Novikova, N. V., 463
Nowlan, V., 129
Noyes, R. M., 276
Noyes, W. A., 501
Noyori, R., 175, 180, 484
Nozaki, H., 109, 117, 127, 256, 388, 493, 534
Nozawa, S., 327
Nudelman, A., 57, 88
Nudelman, N. S., 179
Numata, T., 92
Numella, L., 182
Nunes, I. M., 417
Nunn, A. J., 544
Nye, M. J., 167, 245, 250
Nyholm, R. S., 154
Nyi, K., 483

Nyitrai, J., 251
Nyu, K., 386

Oae, S., 61, 92, 93, 146, 212, 213, 246, 365, 553
Oakenfull, D. G., 422, 442
Oakes, J., 353
Oancea, D., 106
Obata, N., 489
Oberdier, J., 500
Oberhänsli, W. E., 531
O'Brien, C., 124
O'Brien, J. B., 517
O'Callaghan, W. B., 161
Ochiai, S., 171
O'Connell, E. J., 470
O'Connell, E. L., 413
O'Connell, J. E., 398
O'Connor, C. J., 398, 456
Oda, M., 48, 175
Odaira, Y., 475, 491, 494
Odani, M., 487, 497, 522
Odell, B. G., 151, 290
Oesterling, T. O., 441
Östman, B., 200
Oettle, W. F., 483, 491
Ötvös, L., 453
O'Ferrall, R. A. M., 124, 408
Offen, H. W., 473
Ogata, Y., 155, 297, 354, 405, 415, 438, 504, 511, 538, 540, 543, 545
Ogawa, S., 504
Ogino, K., 213
Ogiso, Y., 271
Ogliaruso, M., 114
Ognyanov, I., 437
Ogren, P. J., 360
Ogura, K., 103, 485, 504
Ohashi, M., 475
Ohashi, T., 69, 304, 387
Ohga, K., 493
Ohkubo, K., 557
Ohloff, G., 236, 383, 478, 497
Ohmo, M., 243
Ohnesorge, W. E., 47, 546
Ohnishi, M., 400
Ohnishi, Y., 289, 484, 490
Ohno, A., 289, 490
Ohoto, Y., 503
Ohta, H., 257, 301
Ohta, M., 217, 244, 508
Ohtaka, H., 108
Ohtsuki, M., 299
Ohtsuru, M., 222
Oishi, T., 261
Ojo, J. F., 304
Okada, A., 339
Okahara, M., 304, 387

Okamoto, K., 53
Okamoto, T., 161
Okamoto, Y., 17, 57, 106, 191
Okamura, W. H., 114, 538, 550
Okano, M., 155, 203, 534
Okawara, M., 48
Okazaki, R., 176
Okeley, H. M., 218
Okhlobystin, O. Yu., 47, 346
Okhlobystina, L. V., 94, 122
Oki, M., 425
Okruszek, A., 545
Okubo, K., 557
Okubo, M., 130
Okuda, S., 71
Okuyama, T., 408
Olah, G. A., 1, 5, 17, 41, 42, 73, 124, 132, 158, 195
Olah, J. A., 124
Olavesen, C., 516
Ol'derkop, U. A., 326
Oleinik, N. M., 429
Olesen, J. A., 523
Olivé, S., 557
Oliver, T. R., 536
Oliver, W. R., 479
Olivier, L., 251
Olivier, S. L., 399
Ollis, W. D., 222, 268
Olmstead, H. D., 96, 413
Omenn, G. S., 463
Omura, K., 517
Ona, H., 236, 254, 481
Onda, M., 71
Ondvus, I., 62
O'Neal, H. E., 144, 287, 359
O'Neil, J. W., 406
Ong, C. C., 355, 356
Ong, K.-S., 360
Ong, S. H., 286
Onishi, I., 212
Onishi, T., 557
Ono, H., 512
Ono, I., 504
Ono, K., 229
Ono, N., 142
Ono, R., 546
Ono, S., 400
Onoda, T., 557
Onomura, S., 268
Onopchenko, A., 539
Ooi, N., 208, 243
Ookuni, I., 501
Oosterhoff, L. J., 43, 275, 468
Oosterling, T. O., 461
Opalka, C., 356
Opie, M. C. A., 207
Oppolzer, W., 235

Orahovats, A., 135
Oram, R. K., 87
Orban, M., 302
O'Rear, J. G., 108
O'Reilly, J. E., 344
Organ, T. D., 215
O'Riordan, E. A., 97, 116
O'Riordan, M., 552
O'Riordan, M. P., 554
Orito, K., 519
Orliński, R., 423
Orlova, E. Y., 181
Orman, R., 413
Orrell, K. G., 262
Orville-Thomas, W. J., 397
Ortiz, J. J., 394
Osamura, M., 540
Osber, M. P., 401
Osborn, C. L., 173
Oshima, T., 462
Oshina, T., 461
Osina, O. I., 197
Osokin, Y. G., 156
Ostendorf, H. K., 336
Ostrogovich, G., 183, 261, 406, 429
Osugi, J., 212
O'Sullivan, M., 212
Oszczapowicz, J., 423
Oth, J. F. M., 38, 112, 220, 231
Otsuji, Y., 192, 508
Otsuka, K., 518
Otsuki, K., 545
Otsuki, T., 468
Otter, B. A., 218
Otto, H.-H., 213
Ottolenghi, M., 472
Ouellet, C., 355
Ouellette, R. J., 534
Ourisson, G., 79, 250, 487
Ovchinnikov, Y. A., 447
Overberger, C. G., 442
Overmann, L. E., 78, 245
Oversby, J. P., 363
Overton, K. H., 15, 157, 245
Owen, C. R., 149
Owen, E. D., 48, 518
Owen, L. N., 244, 245
Owens, P. H., 113
Owsley, D. C., 483, 501
Oxer, G. W., 2, 69
Oza, N. B., 424
Ozaki, A., 539
Ozawa, K., 236
Ozawa, T., 271
Ozdrovskaya, I. M., 180
Ozdrovsky, E. N., 180
Ozolins, S., 207

Pabiot, J.-M., 309
Pabon, H. J. J., 556
Pacifici, J. G., 261, 515
Packer, J. E., 345
Padbury, J. M., 528
Paddock, N. L., 184
Padilla, A. C., 461
Padovan, M., 54
Padwa, A., 173, 479, 481, 484, 495, 511
Pagani, G., 113
Page, M. I., 434, 441
Paillous, N., 489
Pailthorpe, M. I., 511
Paine, J. B., 208
Pajak, J., 497
Pakter, M. K., 205
Paktev, M. K., 197, 205
Pal, B. B., 531
Paladini, J. C., 234
Palazzo, G., 12
Palla, G., 317
Pallaud, R., 309, 365
Palm, V. A., 448, 465
Palmer, J. S., 145
Palmer, M. H., 108
Palmer, T. N., 401
Palmowski, J., 481, 495
Palumbo, R., 37
Panchartek, J., 200
Pancrazi, A., 516
Pandit, U. K., 553
Panek, E. J., 337
Panetta, C. A., 439
Pánková, M., 135, 136
Pankratova, K. G., 79
Pannell, K. H., 47
Pantke, R., 283
Papapavlou, L., 399
Pappas, B. C. T., 231
Pappas, S. P., 490
Paquette, L. A., 11, 70, 112, 118, 120, 125, 173, 174, 180, 191, 220, 222, 228, 231, 233, 236, 243, 253, 254, 256, 266, 267, 290, 551, 554
Pardo, C., 41, 102, 556
Parikh, I., 463
Paris, C. W., 176
Paris, J. P., 448
Parker, A. J., 140, 141
Parker, J., 515
Parker, R. M., 359
Parker, V. D., 545, 546
Parker, W., 16
Parkhomenko, T. U., 453
Parmigiani, G., 109
Parrick, J., 379

Parris, G. E., 127
Parrott, M. J., 160
Parry, F. H., 172
Parry, K. A. W., 145
Parry, K. P., 268
Parry, R. J., 25
Parshall, G. W., 256
Parsons, G., 474
Parsons, I. W., 302
Partch, R., 383
Partlett, J. L., 478
Pasanen, P., 397
Pascal, Y. L., 552
Pascaru, I., 353
Pascoe, J. D., 240
Pascone, J. M., 92
Pashayan, D., 479
Pasini, A., 538
Pass, G., 106
Passerini, R. C., 199
Passet, B. V., 200
Pasto, D. J., 102, 158, 175
Patai, S., 367, 431
Pataky, J. G., 356
Patel, K. W., 404
Patel, V. K., 455
Pater, R., 525
Pathak, M. A., 493
Patin, H., 549
Paton, J. M., 70
Paton, R. M., 313
Patrick, J. E., 223
Patrick, T. B., 226
Patro, P. C., 455
Patrushkin, Y. A., 103
Pattenden, G., 72, 113, 227
Patterson, J. M., 504
Patterson, L. K., 472
Patterson, W. R., 412
Pattison, V. A., 545
Paukov, V. N., 443
Paukstelis, J. V., 251, 416
Paul, H., 362, 363, 483
Paul, I. C., 46
Paul, J., 209
Paulme, J.-P., 227
Paulsen, H., 67, 244, 269
Paulson, D. R., 229, 276, 468
Pauson, P. L., 157
Pautet, F., 482
Pavia, A. A., 546
Pavlik, J. W., 406
Pavlyuchenko, A. I., 204
Pawlak, W., 245
Paxson, T., 95
Payne, M. A., 408
Pazdernik, L., 549
Pazos, J. F., 229

Peace, B. W., 387
Peagram, M. J., 169
Pearce, C., 304
Pearce, D. A., 511
Pearce, D. S., 383
Pearce, R., 315, 371, 386, 480, 483
Pearson, B. U. M., 36
Pearson, J. M., 339
Pearson, J. T., 304
Pearson, L., 36, 126
Pearson, M. J., 219
Pearson, R. G., 34, 151, 216, 391
Peavy, R. E., 167, 175
Pechet, M. M., 258, 330
Pechhold, E., 128
Pedersen, E. B., 97
Pedulli, G. F., 276, 295
Peguy, A., 57
Pehk, T. I., 77
Pehk, T. T., 177
Peiffer, R., 497
Peizner, A. B., 429
Pekkarinen, L., 538, 540
Pelczar, F. L., 307
Pelletier, S. W., 108
Pendygraft, G. W., 116
Penev, P., 171
Penton, J. R., 198, 199, 200
Pepekina, L. V., 204
Perchinummo, M., 317
Percy, G. C., 406
Perel'man, L. A., 103
Perelova, E. G., 38
Peresie, R. J., 472
Peregyre, M., 85, 549
Perez, G., 208
Perie, J., 158
Perie, J. J., 549
Perin, F., 205
Perkins, M. J., 111, 187, 286, 296, 299, 313, 315, 316, 347, 348, 517, 543
Perlin, A. S., 531, 540
Perret, F., 408
Perreton, J., 500
Perrin, C., L., 55, 196
Perry, R. A., 313, 513
Perz, R. J. M., 409
Peschk, G., 276
Pete, J.-M., 483
Peters, J. A., 36
Peters, J. W., 522
Petersen, T. E., 353
Petersen, W. C., 525
Peterson, D. J., 127
Peterson, N. C., 310
Peterson, O. G., 471

Peterson, P. E., 64
Petit, F., 248
Petitclerc, C., 462
Pétra, P. H., 452
Petránek, J., 536
Petrenko, T. I., 180
Petrosyan, V. S., 133
Petrov, E. S., 121, 122
Petrova, S., 260
Petrovanu, M., 170
Petterson, R. C., 516
Pettiford, L. R., 205
Pettit, R., 220, 254, 257
Pettus, J. A., Jr., 222
Petty, H. E., 17
Petukhov, G. G., 365
Pevzner, D. L., 453
Pews, R. G., 94
Peyerimhoff, S. D., 229
Peyronnet, J., 433
Pfaendler, H. R., 99, 528
Pfeiffer, W. D., 98
Pfleiderer, G., 453
Pffeiderer, W., 542
Pfoertner, K., 520
Pfohl, S., 85, 457
Philips, J. C., 231, 243
Philips, T. R., 239
Phillipe, M., 133, 163
Phillips, C., 558
Phillips, G. O., 106, 361, 398
Phillips, J. C., 118
Phillips, T. R., 325
Phillips, W. D., 400
Phillipou, G., 321
Photis, J. M., 375
Piccardi, P., 310
Picco, C., 524
Pichler, J., 108, 302
Pickard, H. B., 162
Pickenhagen, W., 383
Pidacks, C., 550
Pierce, J. B., 386, 480
Pierron, C., 251
Piers, K., 235
Pierson, C., 297
Piescone, J. M., 5
Pietra, F., 78, 180, 181, 192
Pietropaolo, R., 95
Pigasse, D., 154, 156
Pignataro, S., 197
Pihlaja, K., 396, 397, 431
Pijselman, J., 85
Pilgrim, W. R., 544, 558
Pillai, P. M., 243
Pincock, J. A., 153
Pincock, R. E., 236
Pine, S. H., 239, 325
Pines, H., 162, 163, 205

Pinion, J. P., 473
Pink, J. M., 543
Pink, R. C., 362
Pinke, P. A., 256
Pinnick, H. W., 337, 552
Pino, T., 407
Pinschmidt, R. K., 166, 175
Pinson, J., 555
Pipalova, J., 341
Pishchiegen, F. V., 407
Pitman, I. H., 433
Pittman, C. U., 70
Pitts, J. N., 359, 479, 491, 521, 522
Place, B. D., 371
Plasz, A. C., 192
Platenburg, D. H. J. M., 88
Platt, A. E., 304, 329
Platzek, U., 364
Plénat, F., 154
Plesch, P. H., 47, 48, 555
Pletcher, D., 77, 546
Pletcher, T. C., 394
Plieninger, H., 218
Ploner, K.-J., 239
Plonka, J. H., 378
Plotkina, N. I., 205
Plummer, B. F., 494, 519
Plyusnin, V. S., 205
Pobedimskii, D. G., 287
Pochikyan, A. K., 420
Pochini, A., 96
Pochopien, D. J., 538
Pocker, Y., 47, 243, 262, 402, 411, 453
Podo, F., 402
Podstata, J., 212
Pogonina, R. I., 417
Poignant, S., 206
Poite, J. C., 508
Pojer, P. M., 244
Pojorlieff, I. G., 123
Pojrebnaya, V. L., 396
Pokhodenko, W. D., 348
Poland, J. S., 387
Polansky, O. E., 169, 484
Poles, L. A., 455
Polevaya, O. Y., 64, 81
Polgar, L., 443, 447
Polishchulk, V. R., 121
Pollack, R. M., 419
Polston, N. L., 157
Polumbrik, O. M., 364
Pomerantz, M., 224
Pomery, P. J., 344
Pommeret, J.-J., 170
Pond, D. M., 468, 482
Pongs, O., 462
Ponpipom, M. M., 252

Pontremoli, S., 524
Pooley, M. J., 448
Poonian, M. S., 7
Pop, D., 456
Pope, B. M., 235
Pople, J. A., 34, 48, 74, 367
Popov, A. F., 103, 464
Popov, N., 482
Popova, R. S., 464
Porai-Koshits, B. A., 200
Porfir'eva, Y. I., 154
Port, G. N., 441
Porter, G., 471, 472
Porter, N. A., 287, 515
Porter, R. D., 1, 17, 41
Porter, T. H., 481
Portnoy, N. A., 555
Portoghese, P. S., 262
Pospíšil, J., 536
Post, M. F. M., 473
Potashnik, R., 472
Potatuev, A. A., 539
Potmischil, F., 108
Potrafke, E. M., 364
Potzinger, P., 296
Pouliquen, J., 227
Poulter, C. D., 497
Poulton, G. A., 490
Poupko, R., 280
Pourcelot, G., 264
Poutsma, M. L., 48, 98, 156, 160, 307, 308
Pouyet, B., 523
Povarenkina, S. V., 64
Powell, K. G., 126
Powell, R. L., 87, 552
Poyser, J. P., 155
Pozdnova, L. V., 556
Pozdnyakovich, Y. V., 209
Pozdnyak, A. L., 336
Pozharskii, A. F., 108
Pozharskii, F. T., 182
Prabha, S., 461
Prabhu, K. V., 487
Prange, U., 38
Pratt, D. R., 189
Pratt, J. M., 552, 554
Pratt, R. F., 262, 433, 440
Pregaglia, G. F., 534
Prejean, G. W., 555
Pressing, J., 55
Preston, P. N., 515
Prest, S. F., 79
Pretsch, E., 176
Previero, A., 445
Price, A. P., 193
Price, C. C., 552
Price, M. J., 193
Price, S. J. W., 359

Prince, R. D., 263
Prins, R., 38
Prins, W. L., 479
Prinson, J., 264
Prinzbach, H., 217, 238, 495, 515
Prislopski, M. C., 339
Pritzkow, W., 103, 161
Priestley, H. M., 87
Probst, W. J., 226
Procházka, M., 263
Prokof'ev, A. I., 300
Prokop'ev, B. V., 408
Pross, A., 543
Prossel, G., 173, 267
Prouzová, O., 442
Proverb, R. J., 222
Pruss, G. M., 136
Prut, E. V., 346
Pryor, W. A., 283, 296
Prystaš, M., 109
Przheval'skii, N. M., 79
Pudovik, A. N., 461
Püsse, T. O., 465
Pujadas, J., 364
Purnell, J. H., 296
Purohit, G. B., 156
Pushkarev, V. P., 201
Puskas, J., 69
Pusset, J., 521
Pustovarov, V. S., 406
Putkey, T., 346
Putze, B., 38
Puxeddu, A., 132
Pyle, J. L., 130
Pyun, C., 191

Quast, H., 242
Queen, A., 50
Quemeneur, F., 54
Quemeneur, M. T., 206
Quick, L. M., 304
Quillinan, A. J., 217
Quin, L. D., 263
Quinkert, G., 481, 495
Quinn, C. P., 555
Quintily, U., 90, 92, 93, 553
Quistad, G. B., 523
Quon, H. H., 489

Raaen, V. F., 2
Raban, M., 261, 408
Rabe, B. R., 519
Raber, D. J., 13, 14, 16, 50, 253
Rabinovitch, B. S., 377
Rabinovitz, M., 43, 417
Rabinowitch, E., 506
Rabjohn, N., 46

Rack, E. P., 308
Radda, G. K., 500
Radhakrishnamurti, P. S., 179, 455, 534
Rad'ko, S. I., 107
Radom, L., 34, 74
Rae, I. D., 244
Raffi, J., 500
Raftery, M. A., 462
Raftery, W. V., 433
Ragonnet, B., 23
Rahim, A., 456
Rahman, A.-U., 439
Rahman, M., 331, 530
Rajagopalan, P., 171
Rajaram, J., 201, 202
Rajbenbach, L. A., 361
Rajender, S., 444, 447
Rajeswari, K., 40
Rakhimzhanova, N. A., 241
Rakshit, S. C., 456
Rakshys, J. W., 116, 339
Rakshys, J. W., Jr., 43
Ramachandran, B. R., 483
Ramage, R., 239, 247, 485
Ramadan, N., 550
Ramakrishnan, V., 503
Raman, K. U., 396
Raman, K. V., 412
Ramasseul, R., 352, 353
Ramey, K. C., 220
Ramirez, F., 85, 89, 457
Ramponi, G., 463
Ramsay, B. G., 140, 267
Ramsay, G. C., 476
Ramsden, C. A., 268
Rand, L., 457
Randall, E. W., 390
Randall, G. L. P., 197
Randall, J. K., 229
Randel, W., 48
Ranganathan, T. N., 184
Ranjon, A., 523
Rank, W., 122
Ranken, P. F., 30, 370
Ranneva, Y. I., 122
Rao, B., 25, 161
Rao, D. C., 202
Rao, G. S. K., 531
Rao, G. V., 435
Rao, K. S. R., 435
Rao, P. V. S., 532
Rao, V. N. M., 42, 230
Rao, V. R., 503
Rao, V. V., 279
Rao, Y. S., 70
Raphael, R. A., 163, 550
Rapiejko, R. J., 402

Rapoport, H., 545
Rapp, E., 234, 384
Rapp, L. B., 461
Rappoport, Z., 99, 100, 138, 142, 419, 430
Rasburn, E. J., 345
Rassat, A., 2, 5, 243, 245, 352, 353, 512, 551
Rastoldo, M., 515
Rasuleva, D. Kh., 300
Ratcliff, R. L., 506
Rathburn, D. W., 296
Rathke, M. W., 97
Rattee, I. D., 447
Rau, D., 218
Raubach, R. A., 471
Rauen, H. M., 107
Raunio, E. K., 163
Rausch, M. D., 191, 389
Ravet, J.-P., 353
Ravindranathan, T., 116, 245, 270, 412
Ravoux, M., 424
Rawson, G., 353
Ray, A. K., 456
Ray, G. J., 46
Ray, W. C., 168, 528
Raymond, F. A., 306
Rayner, D. R., 90
Razmara, S., 441
Razuvaev, G. A., 282, 283, 349
Read, L. K., 118
Read, R. E., 364
Readhead, M. J., 244
Readio, P. D., 300, 371
Reagan, M. T., 206
Reali, M., 79
Reardon, J. D., 310
Record, K. A. F., 397
Reed, G. W. B., 228
Reed, R. G., 371
Reeder, J. A., 457
Rees, A. H., 380
Rees, C. W., 191, 265, 268, 272, 370, 372, 383, 492, 519, 536, 544
Rees, N. H., 131
Rees, R., 515
Rees, T. C., 333
Reese, C. B., 33, 74, 154, 229
Reeves, R. L., 56, 554
Reeves, W. P., 159, 541
Regel, W., 173
Regitz, M., 230, 382, 386
Rehder-Stirnweiss, W., 557, 558
Rehm, C. R., 406
Rehm, D., 472

Reich, C., 264
Reich, I. L., 52
Reichardt, P. B., 78, 245
Reid, D. J., 50
Reid, T. W., 462
Reiff, L. P., 461
Reimann, H., 325, 519
Reines, S. A., 108
Reinfried, R., 487
Reinhardt, G. W., 519
Reinheimer, J. D., 394
Reisch, J., 492, 516
Reisenhofer, E., 132
Reiser, A., 380, 516
Reiter, F., 171, 508
Reitz, R. R., 72
Rekasheva, A. F., 409
Relkasheva, A. F., 122
Renault, C., 258
Rembaum, A., 347
Remers, W. A., 550
Remy, M. A., 507
Rengaraju, S., 2, 408
Rentzea, C. N., 189
Rentzea, M., 189
Rentzepis, P. M., 467, 469
Répásy, O., 198
Repič, O., 532
Requena, Y., 426, 445, 446
Reshelova, M. D., 38
Resnick, B. M., 485
Resnick, M., 446
Rettschnick, R. P. H., 469
Reuben, D. M. E., 121
Reuben, J., 441
Reusch, W., 265
Reutov, O. A., 64, 77, 78, 79, 81, 95, 116, 118, 131, 132, 133, 158, 204
Reverdy, G., 378, 515
Rey, M., 217
Rey, P., 352, 512
Reynolds-Warnhoff, P., 3, 152
Rezepova, L. G., 364
Rhoades, D. F., 506
Rhoads, S. J., 217
Ribereau-Gayon, G., 414
Ribot, R. A., 524
Ricci, A., 183, 198
Riccio, P., 182
Rich, W. E., 554
Richard, J. P., 359
Richards, E. M., 250
Richards, F. M., 462
Richards, J. T., 467
Richards, K. E., 199, 206
Richards, P. J., 361
Richards, R. L., 390

Richards, W. G., 441
Richardson, A. C., 61, 246
Richardson, D. C., 381
Richardson, W. H., 62, 359, 522, 552
Richer, J.-C., 534
Richerzhagen, T., 362, 365
Richey, H. G., 162
Richey, W. G., 333
Richter, R. F., 152
Richter, W., 217
Richtol, H. H., 523
Rickborn, B., 108, 126, 133, 265
Rickborn, R., 245
Ridd, J. H., 195, 198, 199
Riddell, F. G., 397
Riddle, B., 453
Ridway, T. H., 554
Ried, W., 265
Riedel, K., 33, 229
Rieder, W., 224
Rieff, L. P., 88
Riehl, J. J., 154
Rieke, R. D., 195, 311, 344, 493, 554
Rieker, A., 346
Rieker, R., 61
Riem, R. H., 364, 508, 511
Riemann, J. M., 226
Riemer, H., 538
Riera, J., 364
Riera-Figueras, J., 45
Rifi, M. R., 554
Rigassi, N., 531
Rigau, J. J., 90
Rigaudy, J., 489, 522
Rigby, C. W., 101
Rillings, H. C., 29
Rimatori, V., 258
Rimmelin, P., 41
Rinaudo, J., 202
Rindone, B., 534
Rinehart, J. K., 101, 266
Ring, D. F., 377
Rio, G.,353, 497, 521, 522, 523
Rios, A., 95, 543
Rippa, M., 524
Risen, W. M., 348
Risler, J.-L., 540
Ristagno, C. V., 341
Ritchie, C. D., 465
Ritchie, M. H., 252
Rittig, F. R., 206
Rivas, C., 482
Rivier, J., 495
Riviere, M., 262
Rizpolozhenskii, N. I., 461
Robb, J. C., 516

Author Index

Robba, M., 203
Robbins, K. E., 361
Robbins, M. D., 137, 336, 534
Robert, A., 170
Robert, J.-B., 113, 126
Roberts, B. P., 275, 327, 328, 329, 362, 483
Roberts, D. B., 503
Roberts, D. D., 28, 34
Roberts, G. C. K., 453
Roberts, J. D., 5, 230, 261, 403, 406
Roberts, J. P., 471
Roberts, J. S., 16
Roberts, M., 10
Roberts, R. C., 447
Roberts, R. M., 18
Roberts, R. M. G., 131
Roberts, T. D., 500, 501
Roberts, T. R., 154, 412
Robertson, G. B., 154
Robertson, J. M., 25
Robertson, R. E., 60, 463
Robertson, R. K., 106
Robey, R. L., 203
Robinson, G. E., 517
Robinson, J. R., 3, 152, 455
Robinson, L., 442, 459
Robinson, M., 536
Robinson, N. C., 449
Robinson, P. J., 144
Robinson, R. A., 70
Roblot, G., 78
Roček, J., 331, 530
Rochester, C. H., 393
Rockett, B. W., 127
Rocquet, F., 416, 546
Roda, J., 442
Rodewald, L. B., 33, 546
Rodgers, M., 101
Rodima, T. K., 206
Rodionov, P. P., 188
Rodriguez, H., 429
Rodriguez, H. J., 481
Rodriguez, O., 500
Rodriguez-Siurana, A., 45
Roe, R., 172
Roedig, A., 173, 236
Roest, B. C., 336
Röttele, H., 220
Rogers, F. E., 166, 402
Rogge, P. T., 202
Rogne, O., 463, 464
Rogov, N. G., 457
Rohr, C., 229
Roitman, J. N., 120
Rokach, J., 522
Roman, R., 540
Romanskii, I. A., 122
Rome, D. W., 5, 245
Romeo, A., 247, 453
Romm, R., 57
Rommelaere, Y., 396
Ronlán, A., 545
Ronzaud, J., 353
Roobeck, C. F., 45, 98
Rooney, C. S., 522
Roque, J., 129
Roque, J.-P., 407, 416
Roques, A., 412
Roques, B., 203
Rosales, J., 508
Rose, A. I., 116, 117
Rose, I. A., 413
Roselius, E., 301
Rosenberg, E., 390
Rosenberg, I. E., 490
Rosenberg, V. I., 132
Rosenblum, M., 180
Rosenkranz, H. J., 372, 508
Rosenthal, D. E. I., 504
Rosenthal, I., 504, 506, 521, 522
Roskos, P. D., 16
Rosnati, V., 102
Ross, C. H., 170, 233
Rossall, B., 463
Rosser, M. J., 75
Rossi, D., 453
Rossi, G. L., 448
Rossi, R. A., 190
Rosso, P. D., 500
Rostokin, G. A., 127
Roth, A. P., 285, 519
Roth, H. D., 377, 468
Roth, H. J., 484
Roth, J. A., 412
Roth, W. R., 226
Rothberg, I., 2, 106
Rothenberg, F., 202
Roullier, L., 549
Rouschias, G., 389
Roussel, A., 172
Roussel, J., 158
Roussi, G., 129
Rout, M. K., 158, 455
Roux, D. G., 78
Rowanskii, I. A., 122
Rowland, F. S., 487
Rowland, N. E., 484
Rowley, G. L., 463
Roxby, R., 400
Roy, R. G., 247
Royo, G., 84, 130
Rozantsev, E. G., 352
Rozengart, V. J., 453
Rozhdestvenskaya, L. M., 200
Rua, L., 549
Ruban, E., 33, 229
Rubenstein, K. E., 27
Rubin, I. D., 319
Rubin, M. B., 298, 476
Rubinstein, I., 554
Ruccia, M., 268
Ruckstäschel, R., 492
Rudakov, E. S., 46, 107
Rudakov, G. A., 456
Rudakova, R. I., 107
Ruden, R. A., 133, 163, 478
Rudqvist, U., 316
Rüchardt, C., 276, 282, 287, 313, 321
Rüterjans, H., 462
Ruff, F., 95
Rukhadze, E. Q., 287
Ruotsalainen, H., 205
Rupert, C. G., 506
Rupley, J. A., 400, 443
Rusaca, J., 446
Rusakowicz, R., 470
Rusbult, C. F., 479
Russell, D. W., 515
Russell, G. A., 113, 337, 340
Russell, J. W., Jr., 263
Russell, P. J., 347, 349, 422, 437, 519
Russell, P. L., 273
Russell, R. L., 487
Russo, G., 536
Russo, R. V., 2, 106
Russo, S. F., 446
Rutherford, R. J. D., 55
Rutledge, T. E., 406
Rutter, W. J., 414
Ryabinina, S. A., 556
Ryabova, R. S., 465
Ryan, C. A., 452
Ryan, E. R., 356, 513
Ryan, R. P., 269
Ryang, H. S., 489
Ryang, M., 389
Ryashentseva, M. A., 557
Rybinskaya, M. I., 197
Rydon, H. N., 262
Rykov, S. V., 283
Ryle, A. P., 452
Rynbrandt, R. H., 92
Rys, P., 30, 183
Rzhekhima, N. I., 463

Sabresan, A., 543
Sabnis, S. D., 166
Sachdev, K., 292
Sachs, W. H., 411
Sadek, H., 427, 456
Sadler, I. H., 226
Sadovskaya, T. N., 464

Saegusa, T., 263, 267, 309
Sänger, D., 359
Safronenko, E. D., 303
Sagatys, D. S., 44, 82, 149, 408
Sagner, J., 341
Sagner, Z., 465
Sagramora, L., 89
Sahotjian, R. A., 36
Sahu, J., 179
Sailer, K. H., 514
St. John, W. M., 294
Saitner, H., 166
Saito, I., 520
Saitô, S., 53
Saito, T., 268, 270, 453
Saito, Y., 269
Sakai, K., 263
Sakai, M., 1, 10, 11, 12, 30, 254, 256
Sakai, Y., 482
Sakakibara, Y., 161
Sakan, T., 521
Sakrikar, S., 36
Sakuraba, S., 336
Sakurai, H., 206, 299, 304, 305, 315, 320, 339, 482, 489, 494, 500, 519
Salaun, J., 245
Sale, A. A., 192, 272
Saleh, M. A., 8
Salem, L., 216, 294, 483
Salemnik, G., 212, 219
Salisbury, K., 496, 497
Salmon, G. A., 360
Salamone, J. C., 442
Salmon, J. R., 10
Salomaa, P., 393, 431
Salomon, J., 504
Salomon, R. G., 227
Salter, J. C., 21
Saltiel, J., 318, 476, 494, 500
Salvadori, G., 109, 193, 417, 426
Salvesen, K., 422
Salwińska, E., 258
Sam, T. W., 360
Samant, H. S., 481
Samchenko, I. P., 122
Samejima, T., 463
Samenyuk, I. I., 396
Sammes, P. G., 155, 227, 489, 516
Sammes, R. G., 519
Samokhvalov, G. I., 303, 308, 310
Sampson, E. J., 89, 460
Samuel, D., 458

Samuelson, G. E., 494
Sanchez, G. R., 463
Sanderson, A. P., 203
Sanderson, J. R., 285
Sandhu, H. S., 144
Sandri, E., 183
Sandrock, G., 51
Sandström, J., 261
Saneyoshi, M., 124
Sang, H. V., 311
Sannicolo, F., 102
Sansoulet, J., 413, 556
Santelli, M., 23, 226
Santhanam, M., 469
Santiago, C., 262
Saquet, M., 388
Saraiva, M. E. L., 406
Saramma, K., 54, 435
Sarana, N. V., 533
Sarel, S., 69, 455
Sargent, D., 412
Sargent, F. P., 363
Sargent, G. D., 338, 339
Sarkar, I. M., 516
Sarkar, K., 456
Sarre, O. Z., 325, 519
Sartirana, P., 169
Sartori, G., 97
Sasaki, K., 198, 360
Sasaki, M., 212
Sasaki, O., 155
Sasaki, T., 5, 25, 62, 161, 171, 217, 227, 232, 243, 257, 258, 487
Sasaki, Y., 458
Sasmal, N., 531
Sasse, W. H. F., 503
Sasson, S., 506
Sasson, Y., 559
Satgé, J., 306
Sato, M., 524
Sato, N., 325, 519
Sato, S., 213
Sato, T., 257, 270, 484, 517, 520
Sato, Y., 206, 522
Satoh, J. Y., 102
Satpathy, K. K., 179
Sattar, A., 239, 485
Saucy, G., 25, 248
Sauer, G., 341
Sauer, M. C. V., 402
Sauers, C. K., 439
Saunders, B. C., 540
Saunders, D. G., 299
Saunders, M., 9, 13, 34, 35, 36, 59
Saussez, R., 132
Savage, G., 264

Sauvage, J. P., 57
Savelova, V. A., 464
Savige, W. E., 523
Savin, Y. V., 449
Savinykh, L. V., 131, 204
Savyolova, V. A., 464
Sawa, Y., 389
Sayo, H., 545
Scaiano, J. C., 328, 329, 362, 483
Scala, A. A., 491
Scamehorn, R. G., 102, 242
Scandola, F., 520
Scarabino, C. A., 406
Scarpati, R., 70
Schaal, R., 179, 184
Schaap, A. P., 522, 524
Schacht, E., 283
Schachtschneider, J. H., 177
Schade, G., 25
Schadenberg, H., 497
Schäfer, D., 229
Schaeffer, D. J., 339
Schaffhausen, B., 425, 445
Schaffner, I., 489
Schaffner, K., 479, 485, 487
Schallhorn, C. H., 29, 254, 495
Schaper, K.-J., 206
Scharf, H. D., 503
Scharf, H.-D., 166, 483
Schart, H. D., 494
Schatz, P. F., 236
Schaulbe, J. H., 494
Schaumann, E., 262
Schechter, A. N., 463
Schechter, I., 449, 452
Scheer, I., 247
Scheer, M. D., 310
Scheer, W., 170, 233
Scheerer, B., 168
Scheffer, J. R., 489, 492
Scheffler, K., 365
Scheiber, P., 71
Scheidt, F., 29
Scheinbaum, M. L., 224
Scheiner, P., 370, 508
Scheinmann, F., 217
Schelenz, R., 455
Schelly, Z. A., 200
Schenck, G. O., 301
Schenck, H.-U., 171
Schenetti, L., 163, 264
Schenk, H. P., 65, 242
Schenker, K., 228
Schepers, H. A. J., 336
Schepperle, S. E., 2, 287
Scheraga, H. A., 450
Scherer, H., 230

Scherer, K. V., Jr., 248
Schested, K., 361
Scheuneman, E. C. W., 229
Scheutzow, D., 341
Schiess, P. W., 245
Schindler, O., 513
Schindler, R. N., 296
Schinkel, H., 490
Schinski, W. L., 294
Schissel, P., 480
Schissel, P. O., 236, 492
Schlaf, H., 546
Schlatmann, J. L. M. A., 15, 545
Schleyer, P. von R., 1, 10, 13, 14, 16, 26, 30, 34, 50, 59, 74, 75, 98, 126, 233, 245, 246, 253, 277
Schlierf, C., 257
Schlober, A., 11
Schlössler, W., 386
Schlosser, A., 126
Schlosser, M., 149, 150, 388
Schmid, G. H., 63, 156, 160, 408
Schmid, H., 220, 232, 372, 404, 496, 508
Schmid, M., 220
Schmidpeter, A., 169
Schmidt, C., 166
Schmidt, D., 508
Schmidt, D. E., 453
Schmidt, E., 166
Schmidt, E. K. G., 225
Schmidt, G., 218
Schmidt, H., 161
Schmidt, M., 206
Schmidt, R., 295
Schmidt, R. R., 191, 273, 549
Schmidt, T., 226
Schmidt, W., 497
Schmidt, W. F., 112
Schmiegel, W. W., 503
Schmir, G. L., 423
Schmir, M., 443
Schmit, J. P., 150
Schmitt, E., 242
Schmitz, A., 183
Schmutzler, R., 87
Schnegg, U., 231
Schneider, D. F., 224
Schneider, F., 438, 450
Schneider, G., 13, 233
Schneider, M., 511
Schneider, W., 191
Schoeller, W. W., 25, 113, 219, 277
Schöllkopf, U., 229
Schönefeld, J., 229, 368

Schofield, K., 198, 199
Scholes, G., 361
Scholl, M.-J., 353
Scholl, P. C., 45, 477
Schollkopf, U., 33
Schomburg, G., 484
Schood, L. J., 189
Schore, N., 481
Schorpp, K., 365
Schoustra, B., 479
Schowen, R. L., 414, 415, 419
Schrader, L., 374, 515, 524
Schramm, V. L., 463
Schran, H., 186
Schrauzer, G. N., 549
Schray, K. J., 458
Schreiber, K., 519
Schreiber, W. L., 481, 487
Schreier, P., 540
Schriewer, H., 107
Schröder, B., 498, 499
Schroeder, G., 208
Schröder, G., 38, 112, 220, 231
Schroff, A. P., 257
Schubert, B., 217
Schubert, G., 421
Schubert, W. M., 59
Schuetz, R. D., 251
Schuler, R. H., 296, 363
Schulman, E. M., 372
Schulman, M. F., 114, 550
Schulman, S. G., **201**, 504
Schulte-Elte, K. H., 236, 478, 497, 521
Schulte-Frohlinde, D., 262, 360, 500
Schultz, A. L., 557
Schulz, J. G. D., 539
Schulz, K. F., 317
Schumacher, H., 517
Schumann, W. C., 224
Schuster, D. I., 476, 482, 485, 487
Schuster, G. B., 27, 235
Schuster, P., 169, 206
Schwaiger, W., 205
Schwartz, A. T., 448
Schwartz, H., 477
Schwartz, L. H., 81
Schwartz, M., 407
Schwartz, M. A., 29, 442
Schwartz, M. M., 285
Schwartz, R. N., 202
Schwarz, H., 345, 516
Schwarzer, J., 309
Schweitz, H., 462
Schwetlick, K., 207, 408
Schwörer, F., 361

Schwieter, U., 531
Sciacovelli, V., 163
Sclove, D. B., 229
Scoffine, E., 450
Scoffone, E., 524
Scoggin, D. I., 545
Scolastico, C., 534
Scorrano, G., 90, 92, 93, 241, 553
Scott, F. L., 45, 64, 70, 133, 164, 214, 261, 408, 428, 543
Scott, G., 354
Scott, J. M. W., 50
Scott, M., 106, 548
Screttas, C. G., 364
Scribe, P., 470, 507, 540
Scrowston, R. M., 90
Searle, R. J. G., 260, 325
Sears, K. D., 97
Sebardeleben, J. F., 485
Sebastian, J. F., 48, 116
Secco, F., 427
Secemski, I. I., 399
Seconi, G., 183
Secor, H. V., 485
Seddon, W. A., 361
Sedor, E. A., 239
Sedov, A. M., 203
Seebach, D., 359, 476
Seeger, R., 38
Seekircher, R., 539
Seely, G. R., 521
Seguchi, K., 202
Seibl, J., 104, 241
Seidl, H., 374, 515
Seidl, P., 10
Seidler, H., 494
Seiler, P., 525
Sekiguchi, S., 157
Sekikawa, N., 299
Sekiya, M., 72
Sela, M., 400
Selander, H., 202, 536
Selby, I. A., 262
Sélim, Marguerite, 264
Sélim, Mohamed, 264
Selinger, B. K., 473
Selvarajan, R., 383
Seltzer, S., 262
Semenovsky, A. V., 160
Semenyuk, G. V., 429
Semmelhack, M. F., 388, 504, 511
Sen, A. K., 97
Sen, M., 248
Senatore, L., 89
Senda, Y., 546
Sendega, R. V., 107
Sénéchal, G., 557

Sengupta, K. K., 531
Sengupta, P., 248
Senov, C. V., 95
Sepulveda, L., 442, 459
Sera, A., 202
Sergeev, E. V., 107
Sergeev, G. B., 154
Serguchev, Y. A., 154
Serridge, P., 372
Serve, D., 353, 545
Sestako, I., 48
Seter, J., 8
Setkina, V. N., 207
Setser, D. W., 377
Sevenair, J. P., 102, 547
Severin, T., 374, 515
Sevin, A., 416, 546
Seyden-Penne, J., 106, 116, 142
Seydoux, F., 448
Seyferth, D., 369, 375, 379, 382, 384
Sgherza, V., 150
Shadbolt, R. S., 264
Shafer, J. A., 434
Shafizadeh, F., 404
Shagisultanova, G. A., 336
Shah, A., 305, 515
Shakhelgel'diev, M. A., 205
Shakhidayatov, Kh., 262
Shall, S., 401
Shanan-Atidi, H., 228
Shani, A., 479, 493
Shankar, J., 276
Shanks, R. A., 456
Shanmugam, P., 70
Shannon, P. V. R., 83, 484
Shanshal, M., 408
Shapiro, B. B., 397
Shapiro, I. O., 122
Shapley, A., 414
Sharafy, S., 470
Sharanin, Y. A., 464
Sharma, B. D., 362
Sharma, M., 497
Sharp, J. T., 190, 313, 317
Sharp, T., 313
Sharpless, K. B., 532
Shatenstein, A. I., 121, 122, 207
Shatkina, T. N., 77
Shatskaya, V. A., 464
Shaw, A., 154
Shaw, B. L., 389
Shaw, C. J., 257
Shaw, C. J. G., 379
Shaw, E., 446
Shaw, G., 204
Shaw, J. E., 531

Shaw, M. C., 447
Shaw, M. J., 341
Shaw, P. M., 46, 77, 248
Shchelkunov, A. V., 241
Shearing, D. J., 139
Shebanova, O. K., 431
Shechter, H., 382
Shein, S. M., 179, 180, 182, 188
Sheinkman, A. K., 317, 414
Sheldon, R. A., 354, 538
Shelton, G., 227, 257
Shemyakin, M. M., 177
Shen, C.-M., 442
Shen, J., 42, 124
Shen, K.-W., 236
Shen, Y. H., 169
Shepard, K. L., 97, 116
Shepel, E. N., 447
Sheppard, G., 217
Sheradsky, T., 212, 219
Sheridan, J. B., 498
Sherman, W. V., 337
Shermergorn, I. M., 64
Sherrington, D. C., 48
Sherrod, S. A., 29, 99
Sherry, J. J., 531
Shestakov, G. K., 156
Shevelev, S. A., 95
Sheyanov, N. G., 529
Shiao, D. D. F., 414
Shiao, D. F., 444
Shiba, H., 198
Shiba, T., 508
Shibala, K., 446
Shibaoka, T., 400
Shibasaki, M., 239
Shibata, K., 449
Shibuya, S., 344, 517
Shieh, T. C., 201
Shields, D. E., 494
Shields, T. C., 161
Shigemitsu, Y., 475, 491
Shih, H.-M., 369, 379, 384
Shilov, A. E., 553
Shilov, E. A., 152, 154
Shima, K., 482, 489, 494, 519
Shimada, M., 474
Shimada, S., 517
Shimagaki, M., 218
Shimidzu, T., 442
Shin, H. D., 451
Shine, H. J., 341, 511
Shiner, V. J., 58, 59, 106
Shiner, V. J., Jr., 13, 43
Shingu, H., 53
Shin-ichi, 448, 449
Shinkai, S., 442
Shinod, H., 468

Shinozaki, K., 457
Shio, M., 516
Shiomi, K., 350
Shiozawa, C., 400
Shirafugi, T., 256
Shirafuji, T., 388, 534
Shishido, K., 516
Shizuka, H., 503, 511, 520
Shklyaev, V. S., 127
Shmyreva, Z. V., 154
Shoemaker, D., 475
Shohamy, E., 138
Sholle, V. D., 352
Shono, T., 48, 344, 546
Shook, H. E., 263
Shorter, J., 56, 457
Shortridge, R. G., 479
Shortridge, T. J., 227
Shoshenkova, V. A., 556
Shostakovskii, M. F., 408
Shoua, S., 130, 416
Shpan'ko, I. V., 429
Shteingarts, V. D., 197
Shtivel, N.-K., 166
Shubin, V. G., 20, 43, 209
Shudo, K., 5, 259
Shukla, P. R., 470
Shuliko, V. M., 429
Shushunov, V. A., 108, 260
Shutov, G. M., 181
Shvets, A. I., 182
Shvets, V. F., 108, 304
Shvo, Y., 228
Sicher, J., 135, 136
Sicilio, F., 362
Sidebottom, H. W., 310, 519
Sidell, M. D., 241, 441
Sidorov, V. A., 200
Siebert, W., 206
Siedle, A. R., 187
Siegel, H., 557
Sieh, I., 286
Siemion, I. Z., 440
Sievertsson, H., 202
Sigal, P., 295, 497
Sigimura, Y., 39
Sigman, D. S., 449
Siigur, J., 427, 428
Sikk, P. F., 448
Silber, J. J., 341
Siling, M. I., 417
Silkovskaya, E. P., 282
Sillion, B., 258
Silver, B. L., 280
Silver, F. M., 78, 245
Silver, M. S., 36
Silverman, G., 251, 259
Silverman, J., 364
Silversmith, E. F., 364

Author Index

Simamura, O., 257, 318, 355
Simándi, L. I., 532
Šimek, S., 309
Simic, M., 361
Simmons, H. D., 369
Simmons, R. F., 371
Simon, M., 359
Simon, Z., 261, 406
Simonaitis, R., 101, 491
Simonek, V., 555
Simonet, J., 554
Simonetta, M., 219
Simonik, J., 162, 163
Simonnin, M.-P., 184
Simonov, A. M., 182
Simons, J. P., 468
Simonyi, M., 276
Simpson, G. E. F., 240
Simpson, P., 94
Simpson, R. T., 414, 462
Sims, C. L., 224
Singelmann, J., 269
Singer, G. M., 214
Singer, L. A., 279
Singer, L. S., 365
Singer, M. I. C., 197
Singgih, P. A., 440
Singh, A., 203, 215
Singh, B., 372, 516
Singh, P., 222, 352, 476, 479
Singh, S. N., 491
Singh, S. P., 481
Singhal, G. S., 506
Singler, R. E., 19
Sinha, B. P., 536
Sinnott, M. I., 398
Sinnott, M. L., 43
Sirowej, H., 218
Siskin, M., 552
Siu, A. K. Q., 169, 294
Skånberg, I., 202
Skattebøl, L., 265
Skell, P. S., 300, 379, 385
Skelton, D., 330
Skibida, I. P., 287
Skinner, G. A., 196
Skrabal, P., 30
Skubulla, W., 517
Skulnik, D. N., 372
Skvortova, M. A., 127
Slagt, C., 489
Slawinski, J., 525
Sleight, R. B., 353
Sleiter, G., 182, 187
Slifkin, M. A., 469
Sliva, W., 168
Slobodin, Y. M., 375
Slobodyanskaya, E. M., 447

Slocum, D. W., 127
Slopianka, M., 217
Slovetskii, V. I., 94, 95
Small, G. H., 552
Smallcombe, S. H., 90
Smedman, L. A., 28
Smetana, R. D., 522, 529
Smets, G., 169, 240, 262
Smiataczowa, K., 407
Smick, D., 534
Smid, J., 57, 108
Smirnov-Zamkov, I. V., 180
Smirnov, V. V., 154
Smissman, E. E., 70
Smit, P., 304
Smit, W. A., 160
Smith, A., 108
Smith, A. B., 487
Smith, C. L., 124
Smith, D. E., 554
Smith, D. G., 270
Smith, D. J. H., 88, 235
Smith, E. E., 400
Smith, G., 495, 545
Smith, G. E. P., 97, 163
Smith, G. F., 251
Smith, G. G., 142, 143, 144
Smith, H. A., 105
Smith, J. B., 406, 417
Smith, J. C., 224
Smith, J. G., 240
Smith, J. R. L., 330, 536
Smith, J. W., 552
Smith, K. W., 297
Smith, L., 172
Smith, M. C., 394
Smith, M. J., 89, 180, 235
Smith, M. S., 54
Smith, R. A., 5
Smith, R. F., 497
Smith, R. G., 552
Smith, R. H., 375, 387, 552
Smith, S. G., 96
Smith, S. H., 36, 126
Smith, W. E., 384
Smolanoff, J., 511
Smolina, T. A., 64, 81
Smolinski, S., 78
Smolyen, Z. S., 155
Smordina, N. Y., 78
Smoyer, R., 125, 202
Smythe, R. J., 552
Snatzke, G., 259
Sneen, R. A., 16
Sneeringer, P. V., 520
Snieckus, V., 266, 497
Snyder, E. I., 18, 109, 135, 172

Snyder, F., 463
Snyder, H. R., 186
Snyder, J. P., 87, 236, 537
Sobel, J. H., 399
Socha, J., 179
Soga, T., 554
Sohár, P., 117, 241
Sohma, K., 203
Sohn, W. H., 150
Sohoni, S. S., 5, 410
Sokolov, L. B., 464
Sokolov, V. I., 158
Sokolowski, J., 407
Sokol'skaya, A. M., 556
Solly, R. K., 277, 359
Solodova, K. V., 180
Solodovnikov, S. P., 300
Solomon, B. S., 473
Solov'eva, S. A., 553
Soloway, S., 279
Soltzberg, L. J., 364
Soma, M., 557
Soma, N., 39
Sommer, J. M., 41
Sommer, L. H., 83
Sondheimer, F., 232, 484
Song, P. S., 470
Songstad, J., 97
Sonoda, A., 374
Sonoda, N., 545
Sonogashira, K., 390
Soonike, E., 428
Sorba, J., 310
Sorensen, T. S., 36, 40, 263
Šorm, F., 109
Sorokin, M. F., 429
Sorimachi, K., 520
Souchay, P., 555
Soumillion, J. P., 276, 304
Sovelli, G., 197
Sovocool, G. W., 472
Sowa, J. R., 459, 461
Spaar, R., 99
Spagnolo, P., 316, 517
Spangler, C. W., 263
Sparacino, J. K., 512
Speckamp, W. N., 168
Spector, L. B., 463
Speis, A., 145, 455
Spence, M. J., 456, 464
Spencer, T. A., 94
Speranza, M., 187
Sperling, J., 483, 504
Spialter, S., 549
Spickett, R. G. W., 443
Spiegelman, G., 66
Spinner, E., 264
Spiridonova, S. V., 155
Spitzer, U. A., 531

Author Index

Spitzer, W. A., 324, 485, 499
Splitter, J. S., 512
Spomer, W. E., 449
Sprague, E. D., 303
Sprecher, R. F., 261
Sprenger, W. A., 252
Sprung, J. L., 359
Spryskov, A. A., 197, 198
Spurlock, L. A., 21
Squire, R. H., 512
Sridhar, N., 202
Srinivasan, R., 221, 265, 495, 497, 501, 503
Srinivasan, T. K. K., 372
Srinivasan, V. S., 531, 534
Srivanavit, C., 88
Srivastava, K. P., 413
Srivastava, V. B., 414
Stackhouse, T. F., 215
Stage, D. J., 176
Staley, W. S., 235
Stalick, W. M., 162, 163, 205
Stam, M. F., 341, 442, 459
Stammer, C. H., 440
Stammer, R., 350
Stang, P. J., 98, 99
Stangeland, L. J., 97
Stanko, V. I., 304
Staninets, V. I., 152
Stanley, J. P., 296
Stanovnik, B., 270, 380, 516
Stanton, E., 265, 492
Stapleford, K. S. J., 251
Staricco, E. H., 304
Stark, F. O., 83
Starks, C. M., 55, 368, 464
Starness, W. H., Jr., 253
Staroscik, J., 108, 133
Stasko, A., 47, 555
Staude, E., 109
Staude, U., 109
Staudinger, G. K., 440
Stauffer, C. E., 448
Stauffer, R. D., 256, 257
Staunton, J., 137
Stebles, M. R. D., 74
Stec, W., 545
Stedman, D. H., 359
Steen, H. B., 475
Steenhoek, A., 556
Stefani, A. P., 305
Stefanovsky, J. N., 123
Stegel, F., 186
Steglich, W., 440
Stegmann, H. B., 346, 359, 365
Stéhelin, L., 79
Stein, A. R., 427
Stein, G., 501

Stein, T. P., 451
Stein, Y., 69
Steinberg, N. G., 245
Steiner, P. R., 475, 476
Steiner, R. F., 450
Steinfatt, M., 524
Steingarts, V. D., 209
Steinmaus, H., 386, 504
Steinmetz, W. E., 358
Steirl, P., 517
Stella, L., 311
Stenberg, V. I., 490, 520
Stepanov, I. A., 465
Stepanov, V. M., 452
Stephens, D. N., 416
Stephens, R., 47
Stephenson, L. M., 295, 478
Štěrba, V., 193, 200
Sterlin, S. R., 132
Stermitz, F. R., 453, 506
Stern, E. W., 539
Stern, M. J., 58, 150
Sternbach, L. H., 251, 259, 548
Sternberg, V. I., 504
Sternhell, S., 5, 107, 543
Sterns, M., 167
Stetter, H., 14, 253
Stetter, K. H., 365
Stevens, C. L., 243
Stevens, D. N., 251
Stevens, H. C., 101, 266
Stevens, J. D. R., 141
Stevens, J. I., 97, 116
Stevens, M. F. G., 516
Stevens, R. D. S., 471, 501
Stevenson, G. R., 340
Stevenson, J. R., 16
Stevenson, R., 215
Stewart, C. A., 167
Stewart, H. F., 240
Stewart, J. A., 441
Stewart, J. A. G., 226
Stewart, R., 112, 402, 532, 543
Sthapak, J. K., 533
Stieger, H., 295
Stille, J. K., 3, 156
Stimson, V. R., 359
Stirling, C. J. M., 67, 154, 163, 164
Stock, J. T., 344
Stock, L. M., 202
Stocker, J. H., 476, 554, 555
Stockman, D. L., 471
Stockton, J. D., 172
Stodermann, D., 154
Stöcker, F., 365
Stokes, B., 383

Stoll, M., 478
Stolow, R. D., 335
Stoodley, R. J., 117, 140, 267
Storesund, H. J., 14, 528
Storey, P. M., 342
Stork, G., 25, 265
Storm, D. R., 434
Storr, R. C., 372, 383, 536
Story, P. R., 168, 527, 528
Stotz, R., 545
Stowell, J. C., 220, 254
Strachan, W. M. J., 198
Stracke, H. U., 522
Strandman, L., 58
Strating, J., 15, 247, 522, 542
Strauss, M. J., 48, 186
Strausz, O. P., 161, 468
Streckert, G., 46
Strehlke, B., 288
Streith, J., 508, 512
Streitwieser, A., 113, 121, 195
Strepikheev, Y. A., 429
Strickland, R. C., 18
Stroebel, G. G., 541
Stroebel, G. S., 159
Strohmeier, W., 557, 558
Strojny, E. J., 545
Strom, E. T., 353, 521
Strom, R., 398
Strongina, A. A., 448
Strop, P., 442
Štrouf, O., 546
Strout, E. I., 455
Struble, D. L., 47
Stuart, J. D., 47, 546
Stubbe, J. A., 409
Stubbs, J. K., 177, 508
Stuchal, F. W., 337
Studebaker, J. P., 400
Stütz, A., 161
Sturrock, M. G., 528
Sturtevant, J. M., 414, 462, 540
Stusche, D., 238, 515
Stutz, A., 4
Su, C., 181
Su, H.-G., 463
Su, S. C. K., 434
Su, T.-M., 13, 233, 512
Suama, M., 96
Subbaraman, J., 542
Subbaraman, L. R., 542
Subbotin, A. I., 155
Subrahmanyam, G., 270
Subramanian, L. R., 513, 531
Suchkova, L., 179
Suchkova, L. A., 179
Sucrow, N., 217
Suehiro, T., 310, 319

Author Index

Suelter, C. H., 463
Sümmermann, W., 295
Suetsugu, N., 400
Suga, T., 245
Sugahara, T., 517
Sugamori. S. E., 471, 506
Sugimoto, M., 5, 48, 161
Suginome, H., 325, 519
Sugita, T., 96
Sugiyama, H., 107
Sugiyama, N., 478, 484, 522
Sugowdz, G., 503
Sullivan, D. F., 551
Sullivan, J. M., 19, 145, 157
Sullivan, P. J., 470
Sullivan, T. F., 118, 139
Summers, A. J. H., 508
Summers, B., 314
Summers, L. A., 347
Sun, S. F., 548
Sunami, M., 497, 522
Sunamoto, J., 458, 459
Sundaram, S., 531
Sundberg, R. J., 373, 375, 387, 552
Sundholm, F., 352
Sunko, D. E., 26, 52
Suppan, P., 477
Surridge, J. H., 155
Surzur, J.-M., 311, 349
Suschitzky, H., 270, 380, 383, 512
Susott, R. A., 404
Sustmann, R., 164, 168
Susuki, A., 492
Susuki, H., 512
Susz, B.-P., 417
Suszkiw, J. B., 453
Sutcliffe, H., 351, 371
Sutcliffe, L. H., 347, 353, 519
Sutcliffe, M., 62
Suter, C., 497
Sutherland, D. R., 267
Sutherland, I. O., 222
Sutherland, J. K., 360
Sutherland, R. G., 37, 229, 367, 375, 391, 516
Sutton, J. R., 37
Suzui, A., 528
Suzuki, A., 327, 329
Suzuki, H., 215
Suzuki, J., 476
Suzuki, K., 166
Suzuki, M., 386, 552
Suzuki, N., 525
Suzuki, S., 400, 443, 557
Suzuki, T., 180, 264, 518
Svenson, L.-A., 433
Svensson, G.-G., 202

Swaelens, G., 396, 397
Swallow, J. C., 541
Swaminathan, S., 151, 252, 290
Swan, I. D. A., 399, 400
Swann, B. P., 244, 534
Swansiger, W. A., 549
Swanwick, M. G., 351, 389
Swedo, R. J., 530
Sweeney, A., 107, 108, 208, 233, 253
Sweet, W. D., 106
Swenson, C. A., 403
Swenton, J. S., 225, 498, 500
Swerdloff, M. D., 297, 494
Swern, D., 108, 281, 375, 545
Swierczewski, G., 89
Swiger, R. T., 337
Swisher, J. V., 243
Switalski, J., 543
Swiven, E. F. V., 199
Sword, I. P., 265
Sykes, A., 471, 472, 500
Sykes, B. D., 399, 400
Symons, E. A., 420
Symons, M. C. R., 281, 363
Syrov, A. A., 529
Sytilin, M. S., 456
Sytnyk, W., 519
Szacs, S. S., 162
Szafraniec, L. J., 88, 461
Szarek, W. A., 83, 246
Szeimies, G., 11, 126
Szejtli, J., 398
Szendrey, L., 295
Szilagyi, P. J., 41
Szmant, H. H., 230, 531
Szwarc, M., 54, 114, 340

Tabata, A., 534
Tabata, T., 481
Tabatskaya, A. A., 43
Tabushi, I., 189, 303, 385
Tacconi, G., 174
Taddei, F., 264
Taft, R. W., 46
Tagaki, W., 464
Taguchi, T., 157
Taha, I. A., 304
Tahilramani, R., 273
Tahnács, J., 198
Taillades, J., 406
Taimr, L., 536
Tajima, E., 299
Tajima, M., 360
Takabe, K., 161
Takada, K., 492
Takagi, K., 161, 287, 504, 511
Takahashi, J., 141

Takahashi, K., 450, 462, 516
Takahashi, M., 112, 520
Takahashi, N., 385, 450
Takahashi, T., 401, 490
Takahashi, T. T., 102
Taka-ishi, N., 267
Takaki, M., 112
Takamatsu, N., 218
Takamuku, S., 206, 500
Takano, N., 407
Takao, K., 539
Takaya, H., 157, 180
Takaya, M., 191, 271
Takayama, H., 346
Takayama, Y., 511
Takebayashi, M., 69
Takechi, M., 433
Takeda, A., 546
Takeda, K., 222
Takeda, S., 264, 304
Takehira, Y., 243
Takematsu, A., 206
Takemoto, T., 487
Takenaka, J., 157
Takemaka, O., 446
Takeno, N., 166, 407
Takeshita, H., 522
Takeuchi, F., 506
Takeuchi, H., 215
Takihana, T., 112
Takino, T., 485
Takita, Y., 539
Takizawa, T., 489
Talapatra, B., 70
Talapatra, S. K., 70
Talkowski, C. J., 79
Tallec, A., 555
Talvik, A., 116
Talyzenkova, G. P., 287
Tam, J. N. S., 230, 489
Tamagaki, S., 93, 553
Tamai, A., 427
Tamaka, A., 391
Tamao, K., 72
Tamaru, K., 208, 245, 557
Tamate, K., 48
Tamburin, H. J., 225
Tamura, M., 132, 334, 350
Tamura, T., 258
Tamura, Y., 218, 232, 257, 273, 485
Tan, C. C., 313
Tanabe, K., 263
Tanabe, T., 557
Tanaka, F., 270
Tanaka, I., 464, 493, 511, 512
Tanaka, J., 161, 236
Tanaka, K., 102, 229
Tanaka, M., 295

Tanaka, R., 101
Tanaseichuk, B. S., 364
Tandon, S. K., 531
Taneja, A. D., 413
Tanford, C., 400
Tang, W. P., 245, 250
Tanida, H., 18, 60, 63, 221, 245, 515
Tanigushi, I., 489
Tanimizu, Y., 198
Tanizawa, K., 448, 449
Tanner, D. D., 299, 300
Tanner, J. S., 359
Tanno, T., 65, 242
Tanolo, C., 450
Tarama, K., 559
Taranin, B. I., 108
Ta-Shma, R., 100, 430
Tatemoto, K., 540
Taticchi, A., 126, 197
Tatlow, J. C., 47
Tatsukami, Y., 97
Tattersall, B. W., 513
Tatton, H. W., 404
Tavernier, D., 397
Taube, A., 218
Tavares, R. F., 246
Tawara, T., 48
Tawara, Y., 43
Taylor, D. K., 166
Taylor, D. R., 173, 177, 270
Taylor, E. C., 203, 244, 245, 534
Taylor, G. M., 315
Taylor, G. N., 473
Taylor, G. R., 402, 435, 436
Taylor, J. B., 177, 269
Taylor, K. G., 388
Taylor, R., 85, 143, 195, 196, 204
Taylor, R. J. K., 134
Tchoubar, B., 413
Teak, F. W. J., 452
Tebby, J. C., 250
Tedder, J. M., 300, 302, 310
Teeter, J. S., 154
Teinonen, K. J., 396
Telang, V. G., 262
Telford, J. R., 168, 231
Temkin, O. N., 156
Temler, J., 78
Temnikova, T. I., 103
Tenhunen, E., 5
Tennant, G., 267
Tenno, T., 116
Tenu, J.-P., 398
Tenud, L., 104, 241
Teoule, R., 360
Tepenitsyna, E. P., 396

Terabe, S., 350
Terada, S., 451
Teraji, T., 173, 269
Teranura, K., 104, 548
Teranishi, A. Y., 532
Teranishi, S., 210, 539
Terashima, M., 218, 257
Terashima, S., 239, 252
Teratake, S., 515
Terekhova, M. I., 121, 122
Terekhova, S. E., 451
Terent'ev, V. A., 166
Terenzani, A. J., 406
Tereshchenko, G. F., 258, 416
Termonia, M., 472
Terrier, F., 184
Terry, H. W., 54
Teske, H., 108
Tesler, R. S., 427
Tessier, M., 365
Testa, A. C., 470, 514
Testaferri, L., 316
Tette, J. P., 72
Teufel, H., 176
Tezuka, H., 545
Thadani, C. K., 205
Thebtaranonth, Y., 222
Theissen, R. J., 540
Theorell, H., 540
Thielecke, W., 13, 515
Thiem, K.-W., 155
Thierry, J., 77, 248
Thies, R. W., 222
Thijs, L., 542
Thio, J., 38
Thio, P. A., 269
Thoai, N., 257
Thömel, F., 239
Thom, E., 213
Thomas, A. F., 245, 478
Thomas, C. B., 535
Thomas, E. J., 32, 229, 478
Thomas, F. G., 555
Thomas, H. T., 473
Thomas, J. A., 346, 511, 553
Thomas, J. D. R., 456
Thomas, J. K., 467
Thomas, M. T., 266, 497
Thomas, P. J., 143
Thomas, T. F., 481
Thomassin, R., 206, 248
Thompson, A. C., 484
Thompson, A. R., 85, 204
Thompson, B. J., 229
Thompson, D. W., 413
Thompson, G. F., 336
Thompson, G. L., 118, 152, 290
Thompson, H. W., 556

Thompson, J. A., 46, 527
Thomson, B. J., 37
Thomson, J. B., 478
Thompson, J. F., 296
Thompson, R. C., 449
Thompson, R. S., 412
Thomson, C., 313
Thorburn, S., 205
Thorne, R. L., 552
Thornton, D. A., 406
Thorstenson, J. H., 248
Thuillier, A., 312
Thummel, R. P., 265
Thyes, M., 495
Thynne, J. C. J., 310
Tibbetts, F. E., 191
Ticozzi, C., 133
Tidd, B. K., 461
Tidwell, R. T., 504
Tidwell, T. D., 5
Tidwell, T. T., 157, 410, 411
Tiecco, M., 276, 295, 316, 517
Tien, R. Y., 307
Tiger, R. P., 457
Tilgner, H., 108
Tillett, J. G., 203, 215, 426, 434, 453
Tilsley, G. M., 359
Timasheff, S. N., 447
Timberlake, J. W., 46, 287
Timimi, B. A., 402
Timlin, D. M., 145, 304
Timms, D., 162, 428
Timms, G. H., 553
Timms, R. E., 85
Timofeeva, L. A., 103
Timofeyeva, T. M., 200
Timuru, B. A., 402
Tindal, P. K., 324
Tinker, H. B., 534
Tipping, A. E., 154, 270, 304, 511
Tipton, T. J., 299
Tishchenko, A. D., 181
Titov, Yu. A., 241
Titova, G. V., 463
Titskii, G. D., 429
Tizané, D., 549
Tjian, R., 434
Tkatchenko, I., 180
Tobey, S. W., 165
Tobin, J. C., 261, 408
Tochtermann, W., 229
Toda, F., 208, 243, 548
Todd, A., 14, 98, 161, 250
Todd, H. E., 305
Todd, S. M., 184
Todesco, P. E., 182, 183
Todo, N., 157

Author Index

Tökés, L., 483
Toi, T., 406
Toivonen, H., 552
Tokarev, V. I., 464
Token, K., 17
Tokuda, M., 492
Tokumaru, K., 301, 318, 355, 519
Tokura, N., 56, 180, 215
Tokura, S., 442
Toldy, L., 261
Tollin, G., 506
Tomalia, D. A., 252
Tomer, K. B., 512
Tomizawa, K., 354, 540
Tomkiewicz, M., 468
Tomlinson, J. H., 484
Tonellato, U., 97, 99, 100
Tong, L. K. J., 387, 516, 545
Tong, M.-M., 236
Tonne, P., 33, 229
Tonnet, M. L., 463, 464
Tonomura, Y., 463
Toomes, V., 428
Topp, M. R., 467, 472, 476
Toppet, S., 262
Topping, R. M., 273, 422, 437
Tordo, P., 349
Tori, K., 222
Torii, S., 342, 365, 546
Torikai, A., 518
Tornheim, K., 425, 445
Torssell, K., 316, 348, 353
Tortorella, V., 247
Toru, T., 257, 258
Toscano, L., 247
Toscano, V. G., 220
Toshima, M., 492
Tosman, E. A., 300
Toth, G., 108, 261
Toru, T., 25
Touchard, D., 305, 492
Toullec, J., 412
Tournaire, M., 159
Towl, A. D. C., 154
Towns, R. L., 230
Townsend, J. M., 94
Toyama, T., 106
Toyne, K. J., 254
Trahanovsky, W. S., 18, 331, 332, 336, 355, 356, 533, 534
Tramp, D., 552
Trattner, R. B., 457
Travecedo, E. F., 490, 504
Traverso, O., 520
Travis, J., 447
Traylor, T. G., 60, 73, 130, 281, 549
Traynham, J. G., 45, 53

Trecker, D. J., 173, 522
Trefonas, L. M., 230
Tremaine, P. H., 185, 454
Tremelling, M., 292
Tremper, H. S., 47
Trenta, G. M., 101, 266
Trenwith, A. B., 359
Trepanier, D. L., 397
Treptow, R. S., 28, 250
Tret'yakov, V. P., 107
Tribonova, O. I., 181
Tributsch, H., 474
Trichilo, C. L., 341
Trifunac, A. D., 276, 468
Trinka, T., 109
Trippett, S., 87, 88, 89, 468
Trivedi, J. P., 199
Trösken, J., 353
Trofimov, B. A., 409
Trofimov, V. I., 360
Troisi, L., 182
Troitskaya, L. L., 158
Tronchet, J. M. J., 395, 408
Tronich, W., 384
Trost, B. M., 93, 148, 294, 374
Trotman-Dickenson, A. F., 304, 359
Trotter, J., 489
Trotter, W., 514
Trotz, U. O., 528
Troyanowsky, C., 500
Truce, W. E., 213, 308
Trudelle, Y., 440
Truscott, T. G., 443, 471, 472
Tschuikow-Roux, E., 145
Tselmskii, I. V., 116
Tselinsky, I. V., 122
Tseng, L., 441
Tsepalov, V. F., 288
Tsolis, E. A., 85, 89, 457
Tsoucaris, G., 497
Tsubomura, H., 473, 476, 511, 520
Tsuchida, E., 540
Tsuchihashi, G., 103
Tsuchiya, T., 512
Tsuda, M., 557
Tsuda, Y., 65, 242
Tsuji, T., 63, 221, 245, 290
Tsujihara, K., 92
Tsujimoto, N., 232
Tsujimura, K., 438
Tsukurimichi, E., 149
Tsuno, Y., 258
Tsurugi, J., 149
Tsushima, T., 18, 60
Tsutsumi, S., 389, 545
Tsuyuki, T., 490
Tsvetkov, E. N., 121

Tsykhanskaya, I. I., 408
Tsyperovich, A. S., 450
Tsyskovskii, V. K., 529
Tu, J.-I., 401
Tuboi, S., 453
Tudor, R., 327
Tüdös, F., 276
Tufariello, J. J., 72
Tuinman, A., 479
Tuleen, D. L., 69, 300
Tulloch, C. D., 494
Tunggal, B. D., 230
Tuong, T. D., 525
Tupitsyn, I. F., 121
Turbitt, T. D., 38, 180
Turini, P., 398
Turk, J., 435
Turkova, J., 443
Turnbull, J. A., 470
Turner, A. B., 215
Turner, G. L., 108, 182
Turner, J. O., 48
Turner, M. J., 412
Turner, P. H., 511
Turner, S. R., 168
Turro, N. J., 265, 371, 468, 470, 478, 479, 480, 481, 482, 487
Turyan, Y. I., 441
Tutt, D. E., 273, 437, 442
Tuong, T. O., 192
Tuulmets, A., 129
Twibell, J. D., 267
Twitchett, P. J., 501
Tychinskii, V. P., 472
Tye, R. W., 449
Tyler, W. E., 548
Tyminski, I. J., 307

Uccella, N., 170, 268
Udell, W., 376
Uden, P. C., 554
Ueda, K., 69, 514
Ueda, S., 458
Ueda, T., 150
Uehara, K., 401
Uemura, S., 155, 524, 534
Uffmann, H., 264
Ugi, I., 85, 103, 457
Ugi, I. K., 37
Uglova, E. V., 133
Ugo, R., 534, 538, 555, 559
Ukai, A., 65, 242
Ukegawa, K., 365
Ullenius, C., 186
Ullman, E. F., 352
Ulrich, H., 496
Umana, M., 489
Umemura, T., 243

Umezawa, Y., 555
Una, S. J., 180
Underwood, G. R., 482
Underwood, J. G., 127
Underwood, W. G. E., 227
Uneyama, K., 342, 365, 546
Unger, I., 501
Unruh, J. D., 299
Unverferth, K., 207, 408
Uosaki, K., 180
Uramoto, M., 557
Urasaki, I., 155
Urban, E. J., 364, 508
Urbánek, J., 179
Urbas, E., 417
Urness, C. M., 40, 263
Ushio, M., 112, 280
Ustynyuk, T. K., 241
Usui, Y., 475
Utimoto, K., 493
Utley, J. H. P., 108, 344, 554
Uyehara, T., 221
Uyeo, S., 248
Uzan, R., 202
Uzbekov, M. G., 463

Vaidyanathaswamy, R., 279, 550
Vaiga, S., 129
Vajtner, Z., 555
Valenzuela, P., 444
Valiulis, R. A., 452
Vallana, C. A., 304
Vallee, B. L., 414, 447, 452, 462
Valueva, T. A., 451
Valyocsik, E. W., 295, 497
Vamplew, D., 132
van Bekkum, H., 36, 550, 557
Van Chau, L., 388
Van Cleve, W. C., 199
Van den Berg, G. R., 88
van den Bosch, C. B., 550
Van Den Elzen, R., 119
van der Ent, A., 558
van der Jagt, P. J., 545
van der Laan, L. C. J., 350, 483
van der Linde, H. J., 359
van der Lugt, W. T. A. M., 177
van der Plas, H. C., 192, 271, 504
van Doorn, J. A., 41, 45, 417
Van Dyke, J. D., 402
Van Fossen, R. Y., 369
van Gaal, H., 558
van Hooidonk, C., 459
Van Horn, W. F., 94

Van Leemputten, E., 399
Van Loock, E., 262
van Meeteren, H. W., 271
van Minnen-Pathuis, G., 550
van Putten, A., 406
van Raayen, W., 281
Van Roodselaar, A., 161
van Tamalen, E. E., 25, 29, 160, 231, 495, 507, 517, 518
van Velzen, J. C., 193
Van Wazer, J. R., 87, 113, 126
van Wijk, A. M., 550
Van Wijngaarden, B. H., 556
van Wyk, L., 360
van Zanten, B., 545
Van-Bellinghen, J. P., 470
Vancas, I., 3, 152
Vandemark, F. L., 323
Vander Laan, L. C. J., 513
Vander-donckt, E., 470
Van-der-Veen, J., 487
VanLeeuwen, W. H., 473
Vanlierde, H., 172
Van-Thielen, J., 504
Van-Thien, T., 504
VanVoorst, J. D. W., 473
Varfalvy, L., 263
Varflomeyev, S. D., 446
Varfolomeyev, S. D., 447
Varghese, A. J., 506
Varimbi, S. P., 202
Varma, V., 116, 270, 412
Varughese, P., 105
Varveri, F. S., 17
Vasil'eva, T. T., 303
Vaslov, G. P., 448
Vasnev, V. A., 429
Vaudo, A. F., 472
Vaughan, J., 199, 206, 207
Vavřička, S., 555
Večeřa, M., 179, 200
Vechietti, V., 193
Vedejs, E., 220
Veenland, J. U., 302
Vega, E., 479
Vélez, M., 482
Velick, S. F., 540
Velkou, M. R., 17
Vemura, S., 203
Venkataraman, S., 547
Venkataramani, P. S., 265
Venkatasubramanian, N., 435, 531, 533, 534, 543
Verdïn, D., 359
Verenchikov, S. P., 533
Vereshchagina, A. A., 456
Vergoni, M., 197
Verheus, F. W., 275

Verhoeven, J. W., 474
Verkade, J. G., 89
Verkhovod, N. N., 417
Vernin, G., 313, 316, 318, 319, 508
Vernon, C. A., 106
Vernon, J. M., 167, 517
Verploegh, M. C., 94
Veselov, V. Y., 429
Vesley, G. F., 525
Vethaviyaser, N., 219
Vetter, W., 531
Vetushi, C., 247
Vezer, S., 69
Viana, C. A. N., 57
Viau, R., 119
Vijh, A. K., 546
Vik, J.-E., 414
Vikha, G. V., 401
Vilcu, R., 457
Vilen, M., 5
Vilkas, M., 193, 424, 464
Villa, A. E., 362
Villaume, J. E., 379
Villermaux, J., 456
Villieras, J., 95
Vincow, G., 365
Vinkler, P., 456
Vinnik, M. I., 197, 204, 398, 420, 428, 465
Vïnogradov, M. G., 533
Vinogradova, S. S., 429
Vinokurova, G. M., 461
Vintani, C., 169
Vinutha, A. R., 536
Viola, A., 222
Viout, P., 456
Viratelle, O. M., 398
Viriot-Villaume, M., 483
Virtanen, P. O. I., 205
Viscontini, M., 537
Visser, L., 449, 450
Viti, V., 402
Vittimberga, B. M., 320, 321
Vitullo, V. P., 206, 409
Vivarelli, P., 183, 264
Vivona, N., 268
Vizgert, R. V., 107, 180
Vlasov, G. P., 448
Vlasova, N. N., 408
Vodrazka, W., 230
Vogel, D. C., 150
Vogel, P., 9, 13, 34, 35, 36, 59
Vogel, P. C., 58
Vohra, K. N., 70
Voigt, B., 487
Voigt, D., 519
Voisin, P., 193, 464
Volford, J., 108

Author Index

Volke, V., 155
Volker, E. J., 273
Volkova, N. V., 396
Volkova, V., 555
Vollhardt, K. P. C., 42
Vollmer, J. J., 239
Volman, D. H., 362
Volod'kin, A. A., 300
Vol'pin, M. E., 559
Volpp, G. P., 408
Volz, H., 42, 114
Volz de Lecca, M., 42
Von Rein, F. W., 162
von Schriltz, D. M., 90
von Sonntag, C., 359
Vonnahme, R. L., 37
Voorhees, K. J., 143, 144
Vorob'ev, N. K., 431
Voroshilova, L. I., 103
Voskuil, W., 281
Voynick, I. M., 452
Vyeudilik, W., 224
Vyhidy, A., 198
Vylegzhanin, O. N., 409

Waber, A., 259
Wachs, R. H., 62
Wada, H., 282, 355
Wada, M., 504
Wada, S., 546
Waddington, D. J., 357
Wadia, M. S., 531
Wadso, I., 448
Waegell, B., 33, 374
Waegell, M., 167
Wagenaar, A., 522
Wagner, G., 360
Wagner, H. G., 276, 296, 310
Wagner, H.-U., 172
Wagner, P. J., 298, 479, 487
Wagner, U., 209
Wagner-Jauregg, T., 176
Wagnon, J., 248
Waheed, A., 401
Waits, H. P., 517
Wakatsuka, H., 257
Wake, S., 508
Waki, K., 288
Wakselman, M., 424
Walborsky, H. M., 114, 277, 278, 550
Waley, S. G., 413
Walia, J., 335
Walker, G. W., 269
Walker, J. A., 224
Walker, L. E., 167, 545
Walker, R. B., 403
Walker, R. T., 532
Walker, R. W., 145

Wall, A. A., 138
Wallenfels, K., 403
Waller, R. L., 260
Wallerberg, J., 428
Wallin, D. G., 150
Walling, C., 275, 285, 288, 336, 542
Walmsley, D. E., 95
Walmsley, R. H., 469
Walser, A., 251
Walsh, C. T., 463
Walsh, K. A., 449
Walsh, R., 226, 264
Walter, R. I., 202
Walter, W., 262
Walters, E. A., 123
Walters, S. L., 552
Walton, D. R. M., 75, 85, 131, 204
Walton, J. C., 300, 310
Walz, F. G., 461
Wamhoff, H., 269
Wampler, F. B., 304
Wan, J. K. S., 519
Wander, J. D., 404
Wang, C.-H., 455
Wang, E. J., 501
Wang, M., 414
Wang, N., 455
Wang, N. C., 479
Wang, S.-F., 399
Wang, S. Y., 506
Wang, T. T., 286
Wannier, P., 469
Wanzlick, H.-W., 386
Waples, D. A., 206
Warburton, M. R., 173
Ward, C. K., 259, 541
Ward, G. A., 535
Ward, J. P., 205
Ward, J. S., 220
Ward, J. W., 527
Ward, P., 111, 296, 299, 315, 316
Ward, T. J., 190, 517
Wardell, J. L., 202
Warden, K., 481
Ware, W. R., 470, 479
Waring, A. J., 215, 487
Waring, C. E., 310
Warkentin, J., 127, 149
Warnebolt, R. B., 102, 253
Warner, P., 38, 42, 180, 191, 382
Warnhoff, E. W., 3, 152, 269
Warren, C. H., 261, 406
Warren, S., 47, 72, 244, 421
Warrener, R. N., 167, 481, 506

Warriss, M. A., 357
Wartiovaara, I., 182
Washburne, S. S., 82, 379
Washkuhn, R. J., 455
Wasielewski, M. R., 202
Wasserman, E., 367
Wasserman, H. H., 520
Wasson, J. S., 226
Watanabe, H., 247
Watanabe, K., 443
Watanabe, T., 557
Watanatada, C., 70
Wataya, Y., 163
Waters, D. N., 456
Waters, W. A., 315, 348, 351, 536
Waters, W. L., 153
Watkins, A. R., 470, 475
Watson, J. M., 217
Watson, J. W., 181, 423
Watson, K. G., 553
Watson, W. R., 145
Watt, G. D., 462
Watts, P., 250, 458
Watts, P. H., 205
Watts, W. E., 38, 180
Wavtiovaara, I., 181
Wayaku, M., 539
Wayne, R. P., 468
Webb, C. F., 5
Webb, J. P., 471
Webb, P. G., 284
Webber, S. E., 472
Weber, B., 386
Weber, H., 536
Weber, J., 14, 253
Weber, W. P., 133
Webster, D. E., 559
Wedegaertner, D. K., 279
Weedon, B. C. L., 344
Wege, D., 2, 69, 167, 189
Wehrli, H., 478
Wehrly, K., 520
Wei, C. C., 524
Weil, E. D., 154
Weil, T. M., 476
Weiler, L., 25, 102, 161, 253
Weinberger, M. A., 461
Weiner, S. A., 298, 476
Weingarten, G. G., 516, 519
Weingarten, H., 339
Weinshenker, N. M., 367
Weinstock, J., 118, 138
Weintraub, M. S., 398
Weintraub, S. T., 519
Weiss, C., 207
Weiss, K., 261, 406
Weiss, M. J., 550
Weiss, R., 257

Weiss, R. G., 34, 109
Weiss, U., 517
Weissberger, A., 468
Weissman, B. A., 69
Weitl, F. L., 54, 56, 356
Welch, W. M., 439, 480
Welinder, H., 435
Weller, A., 474
Wells, C. F., 532, 533
Wells, C. H. J., 185
Wells, D. K., 18
Wells, P., 493
Wells, P. B., 559
Wells, W. E., 43, 106
Wenck, M., 450
Wender, I., 557
Wendling, L. A., 501
Wenkert, E., 28, 65, 242, 250, 462
Wennerström, O., 186
Wentrup, C., 372, 383, 511
Wepster, B. M., 199
Werbin, H., 521
Werner, H., 389, 390
Werstiuk, N. H., 3, 152, 157
Wesseler, E. P., 385
Wessels, G. F. S., 108
West, C. T., 286
West, P. J., 490
West, P. R., 324, 343
West, R., 42, 174, 230, 240, 352
West, R. M., 500
Westberg, H. H., 11, 254, 256
Westerman, I. J., 168
Westhead, E. W., 413
Westheimer, F. H., 88, 393, 446, 453, 460
Weston, R. G., 167
Westwood, R., 177, 269
Wettermark, G., 261, 406
Wetzel, B., 168
Wetzel, R. B., 215
Wexler, A., 225
Weyerstahl, P., 368
Weyler, W., 516
Whalen, D. L., 30, 254, 409, 495
Whalen, R., 297
Wheeler, O. H., 524
Whelan, W. J., 400
Wheland, R. C., 2
Whillans, D., 471
Whipple, E. B., 402
Whistler, R. L., 360
White, A. J., 304, 359
White, D. W., 87
White, E. H., 524

White, H., 425, 445
White, H. B., 540
White, J. F., 206
White, R. A., 506
White, W. N., 212
Whitesides, G. M., 106, 334
Whitesides, T. H., 507
Whitham, G. H., 157, 169, 262
Whiting, M. C., 43, 380
Whitlock, H. W., 78
Whitlock, H. W., Jr., 236, 245, 264
Whitney, T. A., 92
Whittaker, D., 10, 156, 248, 385
Whittaker, G., 175
Whitten, D. G., 261, 361, 472, 493, 506, 515
Whitten, J. L., 188
Whittle, E., 304, 359
Whittle, J. R., 368
Whittle, P. R., 340
Whytock, D. A., 304
Wiberg, K. B., 7, 382, 530, 555
Widman, D., 485
Wieland, D. M., 108, 133
Wiemann, J., 245, 257, 507
Wien, R., 400
Wiersdorff, W. W., 481
Wiersum, U. E., 357
Wiggins, D. E., 434
Wightman, R. H., 137
Wijsman, T. C. M., 490
Wijting, R. L. C., 496
Wikel, J. H., 239, 325
Wikholm, R. J., 272
Wilcox, C. F., 161
Wilde, R. E., 372
Wilder, P., 2
Wilder, P., Jr., 2
Wildes, P. D., 261, 472, 515, 524
Wildsmith, E., 553
Wilhite, D. L., 188
Wilke, G., 180
Wilkins, C. L., 338
Wilkinson, G., 559
Wilkinson, P. J., 263
Wilks, M. A. J., 364
Willard, A. K., 133
Willard, J. E., 360
Willhalm, B., 478
Williams, A., 193, 417, 424, 426, 460, 462
Williams, C. M., 10, 156, 248
Williams, D. J., 339
Williams, D. L., 506, 525

Williams, D. L. H., 212
Williams, D. R., 371, 532
Williams, E. A., 7
Williams, F., 303, 363
Williams, F. J., 7, 8, 126, 254
Williams, F. R., 554
Williams, H., 481
Williams, J. L. R., 519
Williams, J. M., 78
Williams, M. R., 421
Williams, N., 250
Williams, R. E., 445
Williams, R. H., 186
Williams, R. J. P., 552, 554
Williamson, K. L., 165
Willis, M. R., 364
Willson, J. S., 535
Wils, D. M., 491
Wilson, G. E., 93
Wilson, I. B., 462
Wilson, J. A., 185
Wilson, J. M., 371
Wilson, K. J., 398
Wilson, K. R., 236
Wilson, M. H., 155
Wilson, R. L., 361
Wilson, S. E., 11, 114, 165, 254, 256, 263
Wilson, T., 524
Wilson, W. S., 167, 481
Wilucki, I., 484
Wilzbach, K. E., 501
Winchester, R. V., 354
Winefordner, J. D., 504
Wing, V. A., 528
Wingard, R. E., 118
Wingard, R. E., Jr., 243
Winstein, S., 1, 5, 10, 20, 30, 38, 40, 52, 141, 340, 552
Winter, K., 359
Winter, R. E. K., 234
Winter, R. L., 299
Winterfeldt, E., 218
Winters, L. J., 356, 513
Winton, K. D. R., 310
Winzor, D. J., 414
Wirthwein, R., 191
Wirz, J., 525
Wiskott, E., 14, 353
Wisson, M., 245
Witherup, T. H., 224
Witkop, B., 374, 438, 508
Wittig, G., 189, 239, 259, 547
Woessner, W. D., 264
Wojtkowiak, B., 417
Wolf, A. P., 14, 245
Wolf, D., 474
Wolf, G. C., 308

Author Index

Wolf, H., 489
Wolf, J. F., 83, 342, 546
Wolfe, N. L., 17
Wolfe, S., 302, 439, 534, 544
Wolff, H. P., 220
Wolff, T., 215
Wolfinger, M. D., 118
Wolfrum, J., 276, 310
Wolinsky, J., 248
Woltermann, A., 168
Wong, D. Y., 508
Wong, J. L., 252
Wong, R., 179
Wong, S. M., 123, 519
Wood, D. E., 362
Wood, G., 20
Wood, J., 379
Wood, J. M., 134
Wood, S. E., 126
Woodall, J. E., 130
Woodall, R. E., 260, 325
Woodcock, D. J., 2
Woodcock, R., 270
Woodgate, P. D., 13, 215
Woodgate, S. D., 105
Woodroof, E. A., 462
Woods, D. K., 227
Woods, R. J., 470
Woods, V. A., 403
Woodworth, C. W., 14, 253
Woolard, J. McK., 177
Woolford, R. G., 344
Wootton, J. F., 449
Worsley, M., 155
Wostradowski, R. A., 489, 492
Wray, V., 108
Wright, A., 240
Wright, A. J. C., 470
Wright, C. N., 270
Wright, D. B., 173
Wright, D. J., 465
Wright, G. J., 85, 204, 206, 207, 208
Wright, G. W., 2
Wright, J. L. C., 163
Wright, M., 157, 262
Wright, N. D., 177
Wrighton, M., 500
Wristers, J., 254
Wu, C. C., 17, 69, 270
Wu, J.-H., 269
Wu, P.-W., 205
Wu, W. S., 503
Wucherpfennig, W., 173
Wudl, F., 79
Wuelfing, P., 523
Wulfman, D. S., 387
Wurster, B., 540

Wyckoff, H. W., 462
Wyckoff, J. C., 297
Wyman, G. M., 525
Wynberg, H., 15, 247, 497
Wyn-Jones, E., 397
Wyvratt, M. J., 118

Xavier, A. V., 400

Yablokov, V. A., 260
Yablonovskaya, S. D., 166
Yabuuchi, H., 299
Yager, W. A., 367
Yagi, H., 344
Yagi, K., 540
Yagi, T., 104, 548
Yagupol'skii, L. M., 180
Yagupsky, G., 559
Yajima, S., 557
Yakali, E., 496
Yakatam, G. J., 420
Yaki, H., 409
Yakobson, G. G., 197
Yakovleva, Z. A., 197, 198
Yakushin, F. S., 122
Yakushina, T. A., 207
Yamabe, S., 103
Yamada, F., 163
Yamada, H., 189, 227, 236, 254, 256
Yamada, K., 282, 478
Yamada, S., 17, 69, 252
Yamada, S.-I., 239
Yamagishi, T., 323, 483
Yamaguchi, H., 11, 400, 481
Yamahura, S., 46
Yamamoto, H., 442
Yamamoto, N., 247
Yamamoto, T., 299
Yamamoto, Y., 109, 256, 388, 534
Yamamuro, T., 493
Yamanaka, H., 104, 548
Yamanaka, T., 386
Yamashita, M., 415
Yamashita, T., 288
Yampol'skaya, M. A., 443
Yamskii, V. A., 429
Yanagita, M., 106
Yang, K.-U., 5, 55
Yang, N. C., 201, 480, 482, 504
Yannoni, N. F., 364
Yano, T., 17, 57, 106
Yano, Y., 146
Yanovskaya, L. A., 262
Yao, N.-P., 57
Yariv, J., 398

Yarovenko, V. L., 401
Yarwood, A. J., 472, 481, 487
Yasnikov, A. A., 396, 413
Yasuda, N., 309
Yasuda, Y., 450
Yatagai, M., 78
Yates, B. L., 222
Yates, K., 152, 153, 419
Yates, P., 30
Yates, S., 406
Yato, T., 385
Yatsimirski, A. K., 442
Yau, J. C., 506
Yawakawa, K., 38
Yeh, C.-L., 175
Yelvington, M. B., 62
Yemelyanov, I. S., 409
Yen, S. P. S., 347
Yenin, A. S., 258
Yeoh, G. B., 264
Yip, R. W., 471, 513
Ykman, P., 169, 240
Yokoyama, Y., 282
Yon, J., 398, 448
Yoneda, A., 295
Yoneda, S., 48, 117, 118, 163, 197, 385
Yonemitsu, O., 517
Yonezawa, K., 339
Yonezawa, T., 277, 280, 353, 475
Yoon, N. M., 549
York, R. J., 269
Yoshida, H., 92, 93, 508, 521, 553
Yoshida, K., 401, 546
Yoshida, M., 239, 257
Yoshida, N., 448, 449
Yoshida, S., 451
Yoshida, T., 146, 269, 557
Yoshida, Z., 43, 48, 117, 118, 163, 189, 197, 303, 385
Yoshifuji, M., 176
Yoshifuji, S., 427
Yoshihara, K., 473
Yoshii, N., 263
Yoshikoshi, A., 102, 495
Yoshimoto, T., 323, 483
Yoshimura, Y., 273
Yoshina, A., 475
Yoshioka, H., 481
Yoshioka, M., 485
Yoshioka, T., 353
Young, A. C., 182
Young, A. E., 43
Young, A. T., 116
Young, G. T., 440
Young, G. W., 549

Young, J. C., 338
Young, J. D., 338
Young, J. M., 109, 245
Young, J. W., 495
Young, L. B., 105
Young, M. W., 396
Young, R. H., 520
Young, R. N., 54, 111
Young, W. G., 262
Young, W. R., 121
Youssefyeh, R. D., 527
Yu, S., 130
Yudin, L. G., 197, 204
Yukawa, Y., 17
Yukimoto, Y., 171
Yunes, R. A., 406
Yusuf, M., 205
Yutani, K., 192
Yuthavong, Y., 450
Yvernault, T., 78, 244

Zachariasse, K., 474
Zachoval, J., 442
Zador, M., 534
Zadorojny, C., 402
Zadrozna, I., 413
Zagulayeva, O. A., 182
Zahorieva, K., 123
Zakharcheva, I. I., 204
Zakharkin, L. I., 257
Zalukaev, L. P., 154
Zaluski, M.-C., 203
Zamarlik, H., 264

Zamashchikov, V. V., 107
Zankov, G. E., 398
Zannucci, J. S., 478
Zanotti, G., 453
Zarnegar, P. P., 472
Zatsepina, N. N., 121
Závada, 135, 136
Zavoianu, D., 456
Zazepizki, E., 452
Zbarskii, V. L., 181
Zbiral, E., 4, 161
Zdero, C., 517
Zecchi, G., 102
Zefirov, N. S., 160
Zehetner, W., 175, 273
Zehr, R. D., 490
Zeiss, W., 169
Zeldes, H., 280, 324, 362, 363, 504
Zelesko, M. J., 203
Zelikman, Z. I., 396
Zell, P. J., 322
Zeller, K.-P., 371, 508
Zellner, R., 310
Zeltner, M., 203, 215
Zenarosa, C. V., 544
Zengierski, L., 545
Zepp, R. G., 298, 479, 487
Zervos, C., 424
Zetlin, V. I., 447
Zhdanov, R. I., 352
Zhilin, V. F., 181
Zhil'tsov, S. F., 365

Zhukova, S. V., 177
Ziderman, I., 69
Zieger, H. E., 339
Ziegler, F. E., 13, 484
Ziegler, G. R., 121, 195
Ziffer, H., 517
Zigmund, L., 283
Ziman, S. D., 148
Zimmer, H., 276, 545
Zimmerman, H. E., 216, 348, 472, 498, 499
Zimmerman, M., 150
Zinner, K., 538
Zinov'eva, T. I., 282
Zirngibl, L., 176
Zoglio, M. A., 420
Zollinger, H., 30, 183, 200
Zoltewicz, J., 192
Zoltewicz, J. A., 124
Zook, H. D., 96, 413
Zoulalian, A., 456
Zubiani, M. G., 133
Zucker, U. F., 360
Zugrăvescu, I., 170
Zuman, P., 554
Zurr, D., 46, 545
Zverina, V., 465, 540
Zvyagintseva, E. N., 207
Zwanenburg, B., 83, 113, 522, 542
Zweifel, G., 157, 519
Zweig, A., 370, 492, 503
Zwierzak, A., 461

Cumulative Subject Index 1970–71

Aceheptylene, tetramethyl, **70**, 237–238
Acenaphthene,
  hydrogen exchange, **71**, 207
  iodination, **70**, 230
  tetrahydro, acetylation, **71**, 205
Acetals,
  conformation of cyclic, **70**, 422; **71**, 396
  formation of, **70**, 420–421; **71**, 396
  hydrolysis of, **70**, 417–423; **71**, 393–396
  isomerization of, **70**, 419; **71**, 266
Acetic acid, additions, **70**, 10, 172, 180–181; **71**, 155, 157
Acetoxonium salts, rearrangement, **70**, 79, 300; **71**, 67, 269
Acetylenes,
  acidities, **71**, 121
  additions to, **70**, 107, 127, 176, 179, 187, 189, 193, 197, 201, 203, 206, 209; **71**, 98, 152, 153, 155, 156, 157, 160, 161, 162, 163, 169, 171, 172, 175, 176, 177, 307, 308, 310
  base-catalysed rearrangement of, **70**, 297; **71**, 264
  intramolecular radical addition, **71**, 311
  reduction of, **71**, 114
Acetylenic halides, **70**, 104
Acetyl hypohalites, additions, **70**, 177
Acetyl imidazole, **71**, 421, 424
Acetyl peroxide, scrambling of oxygens, **70**, 251
Acid halides,
  alcoholysis, **70**, 464, 490; **71**, 429
  aminolysis, **70**, 464, 465; **71**, 429
  hydrolysis, **70**, 490; **71**, 457
  reaction with anisole, **71**, 457
  reaction with catechol, **70**, 469
Acidity functions,
  effect of micelles on, **70**, 473
  for DMSO-water mixtures, **71**, 121
  in decarboxylation of phosphoformic acid, **71**, 421
  in deoxymetalation reactions, **70**, 166
  in hydrolysis of benzylidene diacetates, **70**, 422
  in hydrolysis of hydroxamic acids, **71**, 426
  in hydrolysis of sydnones, **71**, 453
  in protonation of anisole, **71**, 206
  in rearrangement of dienones, **71**, 203
  in rearrangement of 1,2,3,4,5-pentamethylcyclopentenyl cation, **71**, 40
  H_- and acidities of hydrocarbons, **70**, 128
  H_- and hydrogen exchange of sulphoxides, **70**, 135
  H_- and hydrolysis of chloroform, **71**, 107

Acidity functions—*continued*
  J_- and hydrolysis of substituted nitrobenzenes, **70**, 214; **71**, 179
Acridines, radical arylation, **71**, 317
Acyl migrations, **71**, 272, 441
Adamantane,
  anodic oxidation, **70**, 376
  1-(N,N'-dichloroamino-), **71**, 16
  hydrogen abstraction from, **70**, 388
  insertion of carbenes, **70**, 399
  isomerization, **71**, 245
  nitroxides, **71**, 253
  oxidation, **71**, 15
  π-route to, **71**, 25
  rearrangement, **71**, 14–15
Adamantanone, **71**, 126, 129, 258, 303
Adamantyl cations, **70**, 2, 21, 287; **71**, 13, 161
Adamantyl derivatives,
  pyrolysis, **71**, 15, 157, 245
  solvolysis, **70**, 16–19, 60, 67; **71**, 13, 14, 26, 50, 54, 58
Adamantylideneadamantane, **70**, 20, 185; **71**, 14, 247
2-(1-Adamantyl) propanol, **71**, 253
Adamantyl radicals, **71**, 299, 357
Adamantyl-vinyl cation, **70**, 20, 108
Addition–elimination reactions, **70**, 109, 145; **71**, 150
Additions to dienes, **70**, 174, 180; **71**, 154, 156
Additions, 1,4 to ketazines, **70**, 173
Additions to olefins,
  1,3-cheletropic, **71**, 172
  cis- or *syn*-, **70**, 176, 179; **71**, 157
  cyclo, **70**, 190–209; **71**, 164–178; *see* Cycloadditions
  electrophilic, **70**, 8–10, 16, 20, 59–61, 63, 173–186; **71**, 25, 28, 152–161
  intramolecular, **70**, 10, 31–32, 189; **71**, 152, 154, 158, 161, 162
  Markovnikov rule, **70**, 172, 177
  nucleophilic, **70**, 127, 186; **71**, 113, 116, 118, 123, 161–164
  ρ-values for, **70**, 174
  radical, **70**, 174, 177, 305, 314, 315, 340–345; **71**, 281, 304
  rearrangement tendencies in, **70**, 186
  to cyclohexene, **71**, 57
  to diethyl maleate and fumarate, **70**, 178–179
  to serratene derivatives, **70**, 186
Aldehydes,
  hydration of, **70**, 429–430
  protonated, **70**, 451; **71**, 41, 417

Aldol reaction, **70**, 447–448; **71**, 117, 413–414
  retro-aldol, **70**, 280; **71**, 117
Alkaloids,
  rearrangements, **70**, 300
  1,2-shifts in **70**, 288
D-Allal, 4,6-O-benzylidene-3-deoxy-3-C-(iodomethyl), hydrolysis of, **70**, 30, 33
Allantoin, hydrolysis, **70**, 492
Allenes,
  additions to, **70**, 108, 178, 179, 183, 197, 202, 340; **71**, 160, 161, 164, 172, 173, 268
  dimerization of, **70**, 202, 328; **71**, 173
Allyl azides, **71**, 219
Allylic ammonium ylids, **71**, 222
Allylic brominations, **70**, 335; **71**, 302
Allylic carbanions, **70**, 120, 195
Allylic cations,
  bond angle distortions, **70**, 30
  cycloadditions, **71**, 175
  NMR spectra, **70**, 31; **71**, 41
Allylic esters, solvolysis, **70**, 492; **71**, 456
Allylic fluorides, dimerization, **71**, 160
Allyl phenyl sulphide, rearrangement, **70**, 260
Allyl sulphenates, **70**, 256
Allyl sulphoxides, **70**, 256
Allyl thiocyanates, **71**, 219
α-Effect, **70**, 116, 491; **71**, 109, 180, 459
Aluminium alkyls
  additions to olefins, **70**, 183
  alkyl exchange, **71**, 127
  elimination reactions, **71**, 145
  in dimerization of olefins, **71**, 161
  reactions with hydrocarbons, **71**, 236
  reactions with ketones, **71**, 130
Aluminium triphenyl, addition to olefins, **71**, 157
Ambident nucleophiles, **70**, 104–105; **71**, 95–97, 116, 129, 163, 240, 456
Amides,
  acid hydrolysis, **70**, 462; **71**, 427, 439
  alcoholysis, **71**, 426
  alkaline hydrolysis, **70**, 456, 462; **71**, 419–420
  aminolysis of, **70**, 458, 463
  equilibrium constant for formation, **71**, 425, 426
  metal-ion catalysed hydrolysis, **70**, 474
  phenolysis, **71**, 426
  protonation, **70**, 463; **71**, 427–428
  rate of N-protonation, **70**, 462
  syn–anti interconversion, **71**, 262
  thiolysis, **71**, 426
Amidines, **71**, 181, 423
Amination, photolytic, **70**, 565
Amines, aromatic, halogenation, **71**, 202
Amines, tertiary, cleavage of, **71**, 107
Anchimeric retardation, **70**, 192–193
Anhydrides,
  acetic anhydride–sulphuric acid mixtures, **70**, 490

Anhydrides—*continued*
  aminolysis of, **70**, 463–464; **71**, 430
  carboxylic–sulphuric, **70**, 502
  conformation of, **70**, 491
  equilibrium constants for formation, **71**, 435
  hydrolysis of, **70**, 459; **71**, 427
  methanolysis of, **71**, 456
  reaction with pyridones, **70**, 464–465
  o-sulphobenzoic anhydride, **70**, 490
  sulphonic-, **71**, 464
Anilinium salts, rearrangement of, **71**, 243
Anisole,
  acylation of, **71**, 205
  chlorination, **71**, 201, 203
  methylation, **70**, 255
  nitrosation, **71**, 200
  protonation of, **71**, 200
  reaction with alkyl radicals, **71**, 303
  reaction with helium tritide ion, **71**, 208
[24]Annulene, dehydro-, **70**, 120
Anthracenes,
  dianion, **71**, 208
  Diels–Alder reaction, **70**, 190, 195
  halogenation, **70**, 232
  photochemistry of, **70**, 547; **71**, 503
  protonation, **70**, 238
  radical bromination, **70**, 336
Anthraquinones, nitration of, **70**, 229
Antimony, nucleophilic displacement at, **70**, 100
Arbusov reaction, **70**, 98
Arenonium ions, **70**, 53, 237; **71**, 43
Aromaticity of transition states in pericyclic reactions, **71**, 151
Arsenic,
  electrophilic substitution at, **71**, 134
  nucleophilic substitution at, **70**, 100
1,2-Aryl shifts,
  carbanionic, **70**, 124
  cationic, **70**, 21–25; **71**, 16–20
  radical, **70**, 352; **71**, 315, 320, 321
1,5-Aryl shifts, **71**, 227–228
Aspartic acid,
  enzymic decarboxylation, **70**, 489; **71**, 453
  non-enzymic deamination, **70**, 188
Aspirin, hydrolysis, **70**, 460
  -ate complexes, **70**, 140; **71**, 130
Autoxidation, **70**, 377–380, 586–589; **71**, 353–355, 537–541
  inhibition, **71**, 286
  of acetoin, **71**, 539
  of acetylenes, **70**, 379
  of alcohols, **70**, 379
  of amines, **70**, 378; **71**, 537
  of anthrol, **71**, 353, 538
  of ascorbic acid, **71**, 540
  of boranes, **70**, 380; **71**, 327, 538
  of carbohydrates, **71**, 538
  of chloroketones, **71**, 586
  of cholesterol, **71**, 587

Subject Index

Autoxidation—*continued*
  of cumene, **70**, 379; **71**, 354, 538
  of cyclooctatetraene dianion, **71**, 538
  of cymenes, **71**, 539
  of dibenzylhydroxylamine, **71**, 587
  of dihydrofurans, **71**, 540
  of dihyropyrans, **71**, 538
  of eleostearic acids, **71**, 538
  of ethers, **70**, 378
  of hydrazines, **70**, 587
  of hydrocarbons, **70**, 378, 380, 587, 588; **71**, 353, 578
  of hydroquinones, **71**, 354
  of indoles, **71**, 538
  of ketones, **71**, 354, 538, 540
  of olefins, **71**, 539
  of phenols, **70**, 378; **71**, 354, 538
  of phosphines, **70**, 588; **71**, 538
  of sulphides, **70**, 588
  of tetra(dimethylamino)ethylene, **70**, 587
  of triazines, **71**, 540
  of trichloroethylene, **71**, 587
  of triethylbismuth, **71**, 354
  of Vitamin A, **71**, 354, 540
1-Aza-3-arylbicyclobutanes, solvolytic cleavage, **70**, 117
Azabicyclo[2.2.1]hept-2-yl derivatives, **71**, 54
Azabicyclo[2.2.2]octyl derivatives, **70**, 91
Azepines,
  2,3-homo-1H-cycloadditions of, **70**, 194
  1-substituted-H, **70**, 302; **71**, 273
  tautomerism, **70**, 272
Azetidinones, **70**, 201, 268
Azides,
  cycloadditions, **70**, 198; **71**, 171
  nitrenes from, **70**, 387, 400, 407, 412; **71**, 367, 371, 375, 383
  photolysis of, **70**, 397–398, 562; **71**, 380, 516
  reactions with nitrosium ions, **70**, 57
Aziridines,
  acyl, **70**, 301
  additions to acetylenic sulphones and sulphoxides, **70**, 187
  aroyl, 1,3-dipolar additions, **70**, 197, 198
  rearrangement, **71**, 265
  rearrangement of 1-arylthiocarbonyl, **70**, 291
  ring-opening, **70**, 116, 125–126, 267; **71**, 109, 169–170, 233
Azlactones, racemization, **71**, 123
Azo-compounds,
  cyclo-additions, **70**, 201; **71**, 175
  decomposition, **70**, 310, 318–322; **71**, 287–290, 291
  electrophilic substitution, **71**, 198, 201
  photolysis, **70**, 329, 560–562; **71**, 291, 515
Azo-coupling, **70**, 229–230; **71**, 200
Azodicarboxylic ester, cycloaddition of, **70**, 203, 206
Azomethine dyes, **70**, 492

Azophenol cyanates, additions to, **70**, 198
Azulenes,
  nitrosation, **70**, 229
  protonation of, **70**, 238; **71**, 206
Azulenes, aza, Diels–Alder reactions, **70**, 193

Baeyer–Villiger oxidation, **70**, 293, 590; **71**, 259, 541
Barbaralone, homo, Diels–Alder reaction, **70**, 195
Barbaryl cations, **70**, 19, 50
Barbiturates, hydrolysis, **71**, 420
Barrelenes, **71**, 13
Barton reaction, **70**, 306, 564; **71**, 325, 519
Baudisch reaction, **70**, 229
Beckmann rearrangement, **70**, 291; **71**, 257–258
  photo-, **71**, 518
Benzene, triplet, **71**, 467
Benzene-1,3,5-triol, **71**, 206
Benzenonium ions, **70**, 53, 56, 237; **71**, 46, 206
Benzidine rearrangement, **70**, 244; **71**, 212, 511
Benzil, reaction with trimethyl phosphite, **70**, 450; **71**, 415
Benzilic acid rearrangement, **70**, 289; **71**, 251
Benzimidazole, hydrogen exchange, **71**, 124, 209
Benzobicyclo[4.2.1]nonenyl derivatives, **71**, 19
Benzobicyclo[3.2.1]octadiene, hydrogen exchange, **70**, 131; **71**, 121
Benzobicyclo[2.2.2]octanone, hexamethyl, rearrangement, **71**, 27–28
Benzobicyclo[4.2.1]nonenyl derivatives, **71**, 19
Benzobutadiene, dimer, **71**, 274
Benzofuran, **71**, 197
Benzoin reaction, **71**, 414
Benzonorbornadienes,
  additions, **71**, 19, 152, 156
  photolysis, **70**, 544
  thermolysis, **71**, 224
Benzonorbornene, **71**, 4
Benzonorbornenone hydrogen exchange, **70**, 132
Benzonorbornenyl derivatives, **70**, 22–23; **71**, 18
  chromium tricarbonyl complexes, **71**, 18
  cyclopropyl substituted, **71**, 30
Benzoquinuclidine, electrophilic substitution in, **70**, 226
Benzothiadiazines, **71**, 406
Benzothiazoles, **71**, 319
Benzothiophenes, **70**, 229; **71**, 197, 207, 208
Benzotrichloride nitration, **71**, 199
Benzotricyclo [4.2.1.0^{2,5}]nonadienes, **71**, 175
Benzoxazinones, 2-aryl-, hydrolysis, **71**, 426
N-Benzylaniline, nitration, **71**, 199
Benzyl halides, nucleophilic substitutions of, **70**, 65, 66, 68, 116; **71**, 54, 57, 116
Benzyl tosylates, solvolysis, **70**, 63

Benzynes,
  1,3-, **71**, 188, 189
  1,4-, **71**, 188
  as intermediates in decomposition of *N*-nitrosoacetanilides, **70**, 220; **71**, 190, 313
  cyclization of, **70**, 222
  from
    benzene diazonium carboxylate, **70**, 221
    dihalobenzenes, **71**, 189
    di-*p*-tolyl sulphoxide, **70**, 222
    halobenzenes, **71**, 337
    halonaphthalenes, **71**, 189
    oxidation of benzotriazoles, **70**, 220
    photolysis of phthaloyl peroxides, **71**, 189, 492
    photolysis of 1,2,3,4-tetrafluoro-5,6-diiodobenzene, **70**, 563
  hetarynes, **70**, 221; **71**, 191–192
  reaction with
    allenes, **70**, 202, 221
    bicyclo[6.1.0]nonatriene, **71**, 175
    biphenylene, **70**, 221
    carbon disulphide, **70**, 221, 395
    cinnamaldehyde, **70**, 221
    cyanoheptafulvene, **70**, 221
    cycloheptatriene, **71**, 188
    cyclohexadiene, **70**, 221
    cyclooctatetraene, **70**, 221; **71**, 178
    cyclopentadienyl magnesium bromide, **70**, 124, 195, 221
    diaryltellurides, **70**, 221
    1,2-di-deuterocyclohexene, **70**, 207
    5,5-dimethoxy-1,2,3,4-tetrachlorocyclopentadiene, **70**, 221
    furan, **71**, 191
    isoindoles, **71**, 167
    tropones, **71**, 190
Bicarbonate, dehydration, **71**, 428
Bibenzyl, nitration, **70**, 227
Bicyclic cations, gas-phase stabilities, **70**, 10
Bicyclic mechanism, **71**, 272
Bicyclic olefins, acid-catalysed isomerization, **70**, 296
Bicyclo[*m*,1,0]alkanes, oxidation by Tl^{3+}, **70**, 144
Bicyclo[*m.n.*1]alkanes, strain energies of, **70**, 15
Bicyclobutanes,
  additions to, **70**, 143; **71**, 11, 125, 174
  metal-ion-catalysed rearrangements, **71**, 11, 254, 387
  3-phenyl-1-aza-, 16
  pyrolysis, **71**, 235
  ring-opening of, **70**, 13, 143
Bicyclo[1.1.0]butylmethyl derivatives, **70**, 39
Bicyclo[4.2.2]decatetraene, **70**, 52, 181; **71**, 238
Bicyclo[6.2.0]decatriene, **71**, 235
Bicyclo[4.3.1]decatrienyl *p*-nitrobenzoate, **71**, 10
Bicyclo[3.3.2]decyl derivatives, **70**, 15, 87
Bicyclo[4.4.0]decyl derivatives, **71**, 12

Bicyclo[7.1.0]decyl derivatives, **70**, 36–37
Bicyclo[5.4.1]dodecapentaenylium ions, **70**, 55
Bicyclo[3.2.0]hepta-2,6-diene, octachloro, **71**, 42
Bicyclo[4.1.0]heptadiene, tetraphenyl-diaza, **71**, 230
Bicyclo[3.2.0]hepta-3,6-dien-4-one, 5-methyl-6-phenyl-1,2-diaza-, **70**, 270
Bicyclo[3.1.1]heptanes, **71**, 236
Bicyclo[4.1.0]heptanes, **71**, 256
Bicyclo[3.2.0]heptanol, reaction with triphenyl phosphine and carbon tetrachloride, **71**, 109
Bicyclo[3.2.0]hepta-1,3,6-triene, acidity of, **70**, 129
Bicyclo[3.2.0]heptene, **70**, 257, 276; **71**, 224
Bicyclo[3.2.0]hept-2-en-6-one, 7,7-dichloro, **71**, 253, 266
Bicyclo[3.2.0]hept-6-enyl derivatives, **71**, 13
Bicyclo[2.2.1]heptyl derivatives, *see* Norbornyl derivatives
Bicyclo[3.1.1]heptyl derivatives, **70**, 13, 92
Bicyclo[3.2.1]heptyl derivatives, **70**, 11, 92
Bicyclo[4.1.0]heptyl derivatives, **71**, 26
Bicyclohexadienes, **70**, 268; **71**, 257
Bicyclo[2.2.0]hexanes, **70**, 204, 274; **71**, 236
Bicyclo[3.1.0]hexane,
  6-formyl, **70**, 263
  endo-6-methyl, **70**, 16, 180
  reaction with bromine, **70**, 143–144, 175–176
Bicyclo[2.1.1]hexene, **70**, 257
Bicyclo[2.2.0]hexene, octafluoro, **70**, 268
Bicyclo[3.1.0]hexenes, **71**, 225, 227, 306
Bicyclo[3.1.0]hex-2-enones, photolysis, **70**, 522
Bicyclo[3.1.0]hex-2-en-6-yl derivatives, **71**, 29
Bicyclo[3.1.0]hex-3-en-2-yl derivatives, **71**, 9
Bicyclo[2.1.1]hexyl derivatives, **70**, 13
Bicyclo[3.1.0]hexyl derivatives, **71**, 8
Bicyclo[5.2.0]nonadiene, **70**, 252
Bicyclo[6.1.0]nonadiene, **70**, 252
Bicyclo[3.3.1]nonane-2,3-*exo*-oxide, **70**, 16
  2-*endo*-phenyl-, **70**, 16
Bicyclo[3.2.2]nonatriene, **71**, 221
Bicyclo[4.2.1]nonatriene,
  protonation, **71**, 52
  rearrangement, **71**, 239
Bicyclo[4.3.0]nonatrienes, **71**, 231, 232, 233
Bicyclo[6.1.0]nonatrienes, **70**, 204; **71**, 114, 175, 224, 231, 232, 234, 550
Bicyclo[3.2.2]nonatrienyl derivatives, **71**, 19
Bicyclo[6.1.0]nonatrienyl derivatives, **71**, 13
Bicyclo[3.3.1]non-1-ene, **70**, 170, 185
Bicyclo[6.1.0]non-2-ene,
  9-hydroxymethyl, **70**, 263
  9-methylene, **70**, 263
Bicyclo[6.1.0]non-4-ene, 9,9-dibromo, **70**, 37
Bicyclo[3.2.2]nonenyl derivatives, **70**, 16
Bicyclo[3.3.1]non-7-yl carbonium ion, 3-*exo*-methylene, **70**, 21
Bicyclo[3.2.2]nonyl derivatives, **70**, 16

Bicyclo[3.3.1]nonyl derivatives, **71**, 13
Bicyclo[4.2.1]nonyl derivatives, **70**, 15
Bicyclo[4.3.0]nonyl derivatives, **71**, 12
Bicyclo[6.1.0]nonyl derivatives, **70**, 36, 37–38, 43; **71**, 7
Bicyclo[3.2.1]octadiene, 3,4-dibromo, **70**, 132
Bicyclo[3.3.0]octadienes, **71**, 234
Bicyclo[4.2.0]octa-2,7-dienes, **70**, 269
Bicyclo[2.2.2]octadienones, **71**, 250
Bicyclo[5.2.0]octadienone, 7-azo, **71**, 231
Bicyclo[3.2.1]octadienyl anions, **70**, 131
Bicyclo[3.2.1]octa-2,6-dienyl derivatives, **71**, 10
Bicyclo[5.1.0]octanes, **70**, 42–43
Bicyclo[4.2.0]octan-7-one, 8-chloro-8-methyl-, **71**, 253
Bicyclo[2.2.2]octanones, hydrogen exchange, **71**, 410
Bicyclo[4.2.0]octatriene,
1-bromo, **70**, 273
1,6-dimethyl-7,8-diphenyl, **70**, 273
Bicyclo[2.2.2]octatrienylmethyl cation, tribenzo, **70**, 16
Bicyclo[2.2.2]octene, retro-Diels–Alder reaction, **71**, 168
Bicyclo[3.2.1]octene, trimethyl-7-oxa, **71**, 248
Bicyclo[5.1.0]oct-3-ene, **71**, 126
Bicyclo[2.2.2]octenyl derivatives, **71**, 72
Bicyclo[3.2.1]octenyl derivatives, **71**, 21
Bicyclo[5.1.0]octenyl derivatives, **70**, 27–28
Bicyclo[2.2.2]octyl derivatives, **70**, 14, 15, 92; **71**, 135
Bicyclo[3.2.1]octyl derivatives, **70**, 14
Bicyclo[3.2.1]oct-1-yl toluene-*p*-sulphonate, 7-methylene, **71**, 13
Bicyclo[3.3.0]octyl derivatives, **70**, 14; **71**, 12
Bicyclo[4.2.0]octyl derivatives, **70**, 10–12, 44
Bicyclo[5.1.0]octyl derivatives, **70**, 38, 43; **71**, 8
Bicyclo[4.2.0]octylmethyl derivatives, **70**, 15
Bicyclo[1.1.1]pentane, radical chlorination of, **70**, 336
Bicyclo[2.1.0]pentane, **70**, 207; **71**, 174, 178, 256
Bicyclo[2.1.0]pentane, oxa, cycloaddition, **70**, 196
Bicyclo[1.1.1]pentan-2-ol, 2-phenyl, **70**, 274
Bicyclo[2.1.0]pentene, **70**, 268
Bicyclo[1.1.1]pentyl derivatives, **70**, 13, 15
Bicyclo[2.1.0]pentyl derivatives, **70**, 12
Bicyclo[10.1.0]tridecadiene, *trans*, 13, 13-dibromo, **70**, 16, 38
Bicyclo[10.1.0]tridecane, *trans*-13,13-dibromo, **70**, 16, 38
Biphenylene, electrophilic substitution, **71**, 195
Biphenyls, nitration, **70**, 229
Birch reduction, **70**, 600; **71**, 550
Bischler–Napieralski reaction, **70**, 235
Bishomocubane, **70**, 20

Bishomocubyl derivatives, **71**, 15
Bishomotropylium ions, **70**, 52
Boranes,
α-bromo, **71**, 109
electrophilic displacements of, **70**, 140, **71**, 130
radical reactions of, **70**, 312–313; **71**, 327–330
Borates, trialkyl cyano-, conversion to ketones, **70**, 140
Boron, nucleophilic displacements at, **71**, 95
Bredt's rule, **70**, 170, 185
Bridgehead reactivities, **71**, 13, 15
Bromination of
amines, **71**, 202
2,6-dialkylphenyl acetate, **70**, 232
ethers, **71**, 232
phenols, **70**, 230; **71**, 203
reversibility of aromatic, **70**, 230; **71**, 202
sulphides, **71**, 232
Bromine,
addition to olefins, **70**, 174–176; **71**, 3–4, 152–155
nucleophilic displacement at, **70**, 104; **71**, 94
Bromine azide, addition to olefins, **70**, 177
Bromine monoflouride, additions, **70**, 176
Bromodeboronation, **70**, 140; **71**, 130
Bromodemercuration, **70**, 144
Bromodestannylation, **71**, 131
Bromonium ion, **70**, 173, 174, 176; **71**, 4, 152, 153, 154
N-Bromoacetamide, **71**, 302
N-Bromosuccinimide,
additions promoted by, **70**, 176; **71**, 153, 154
bromination by, **70**, 10, 145, 332; **71**, 303
oxidation by, **70**, 592
source of succinimidyl radicals, **71**, 312
Bromohemimellitene, **71**, 28
Brønsted equation for
aminolysis of acetylimidazolinium ion, **71**, 422
aminolysis of 2-amino-6-oxo-4,5-benzo-1,3-oxazine, **70**, 460
aminolysis of chelated glycine isopropyl ester, **70**, 475
bromination and detritiation of malononitrile, **71**, 123
elimination reactions of p-henyl 2-chloroethylsulphones, **70**, 161
hydrogen exchange of 1,3,5-trimethoxybenzene, **70**, 236
hydrolysis and formation of acetone hydrogen peroxide adduct, **71**, 402
hydrolysis of aryl diazomethanes, **71**, 81
hydrolysis of benzaldehyde di-t-butyl acetal, **71**, 393
hydrolysis of enol ethers, **71**, 408
nucleophilic substitution at sulphur, **70**, 101; **71**, 89

Brønsted equation for—*continued*
  proton abstraction from aliphatic nitro-compounds, **70**, 130, 131; **71**, 122
  reaction of di-isopropyl phosphochloridate with hydroxamate ions, **71**, 461
  reaction of *p*-nitrophenyl acetate with benzohydroxamic acids, **71**, 424
  reaction of *p*-nitrophenyl acetate with benzotriazole anions, **71**, 423
  reactions of phosphate esters, **70**, 493
Brønsted β-coefficient, correlation with isotope effect, **71**, 122
Bullvalenes,
  additions to, **70**, 181, 207, 253
  dihydro, homo-Diels–Alder reaction, **70**, 195
  reduction by tri-butyl tin hydride, **71**, 323
Butadienes, conversion into cyclobutenes, **70**, 268
t-Butyl anion, unsolvated, **70**, 122
t-Butyl cation, **71**, 41, 46
t-Butylcyclohexanols, dehydration, **71**, 149
t-Butylcyclohexanones, addition of Grignard reagents, **71**, 129
t-Butylcyclohexene, cycloadditions, **71**, 172
t-Butylcyclohexyl radicals, **71**, 335
t-Butylcyclohexyl toluene-*p*-sulphonates, **71**, 53
t-Butyl halides, solvolysis, **70**, 66, 67; **71**, 46

Cadmium alkyls, **71**, 129
Cadmium dimethyl, exchange of methyl groups, **70**, 141
Caesium cyclohexylamide as base in hydrogen exchange, **70**, 134
Cannabinoid-1,5-dienes, cyclization, **70**, 207
Caranones, cleavage, **71**, 126
Carbamates, dithio, **70**, 492
Carbamoyl phosphates, **71**, 460
Carbanilates, **71**, 457
Carbanions, **70**, 119–145; **71**, 111–134
  acetylenic, **70**, 297
  additions to olefins, **71**, 113, 162
  allylic, **70**, 120; **71**, 111, 116
  aromatic, **70**, 120, 213; **71**, 111
  azaallylic, **70**, 124; **71**, 123
  azapentalenyl, **70**, 121
  benzylic, **71**, 339
  basicities, **70**, 128–129; **71**, 121–122
    effect of counter ion on, **71**, 128–129
    effect of solvent on, **71**, 128–129
  bridgehead, **71**, 121
  cyclic dienyl, **70**, 120, 122
  cyclononatetraenyl dianion, **71**, 114–115, 538
  cyclopropyl, **70**, 144; **71**, 121, 550
  dimerization, **70**, 125
  dinitro, **70**, 130; **71**, 122
  electronic spectra, **70**, 119; **71**, 111
  ESR spectra, **71**, 112
  α-fluoro-, **70**, 126–127; **71**, 121, 188

Carbanions—*continued*
  fluorenyl, **70**, 128–129
  from cyclooctatetraene, **70**, 122
  from dienes, **70**, 296–297
  from 5,6-dihydro-oxepin, **70**, 121
  from dinitrobenzene, **70**, 213
  from 2-methoxy-azacyclooctatetraene, **70**, 121–122
  from naphthalene, **70**, 125
  from phenylpropynes, **70**, 120–121
  from sorbaldehyde, **70**, 121
  from styrenes, **70**, 126
  from tolan, **70**, 125
  from trinitrobenzene, **70**, 213
  heptatrienyl anion, cyclization, **70**, 123
  in addition reactions, **70**, 179–180, 187; **71**, 162
  isoinversion of, **70**, 123
  neighbouring group participation by, **70**, 81–83; **71**, 72
  NMR spectra, **70**, 65, 119–122, 129; **71**, 112
  non-classical, **70**, 131–132, 145, 282; **71**, 121
  7-oxaheptatrienyl, **70**, 282
  pentadienyl, cyclization, **70**, 122–123
  phenyl, **70**, 283
  photolysis of, **71**, 518
  reaction with *vic*-dibromides, **71**, 341
  1,2-shifts in, **70**, 124; **71**, 113
  sulphinyl, **70**, 135; **71**, 115, 119
  sulphonyl, **70**, 130, 135; **71**, 119
  α-thio, **70**, 122, 135; **71**, 112–113
  trapping by electron transfer, **71**, 341
Carbenes, **70**, 391–415; **71**, 367–391
  abstraction reactions, **70**, 399–401; **71**, 376–381
  anion–carbene pair, **70**, 397
  as intermediates in photolyses, **70**, 519
  cycloadditions, **70**, 401–403; **71**, 373–376
  cyclopropylidene, **71**, 381
  dicyano, **71**, 176
  dimethoxy, **71**, 370
  diphenyl, **71**, 296
  diphenylthio, **70**, 313–314
  dithio, **71**, 371
  electrophilicity of, **71**, 367
  EPR spectra, **71**, 367
  from azirines, **70**, 396
  from 1-bromoalkynes and alkoxides, **71**, 371
  from bromodifluoromethane, **71**, 371
  from carbon disulphide and acetylenes, **70**, 395
  from carbon tetrachloride, **70**, 394
  from chlorodifluoromethane, **71**, 371
  from chloroform, **71**, 369
  from deoxygenation of ketones, **70**, 521; **71**, 371
  from dimethylformamide dimethyl acetal, **71**, 371
  from ketenes, **71**, 522
  from mercurials, **70**, 394; **71**, 367, 368, 384

Subject Index

Carbenes—*continued*
  from phenyl carbonate, **71**, 369
  from photolysis of cycloketones, **71**, 371
  from photolysis of cyclopropanes, **71**, 370
  from photolysis of diazo-compounds, **70**, 562; **71**, 373, 374, 377, 515
  from photolysis of sulphurylids, **71**, 508
  from photolysis of tetrazolide anions, **70**, 553; **71**, 370
  from silicon compounds, **70**, 397
  from triazoles, **71**, 371
  from trichloronitrosomethane, **71**, 371
  insertion in adamantane, **70**, 390
  insertion reactions, **70**, 399–400; **71**, 376–381
  intramolecular reactions, **70**, 399; **71**, 369
  IR spectra, **71**, 367
  metal-complexes of, **70**, 412–415; **71**, 256, 389
  non-classical, **70**, 392; **71**, 368
  nucleophilic, **70**, 395; **71**, 386
  rearrangement of, **70**, 405–407; **71**, 381–384
  singlet, **70**, 391–392 393, 402; **71**, 376, 378, 384
  triplet, **70**, 391–392, 393, 402; **71**, 373, 376, 378, 384
Carbenoids, **70**, 410; **71**, 387–389
Carbohydrates,
  addition to glycals, **70**, 173, 343
  alcoholysis of methyl 4,6-O-benzylidene-α-D-glucoside-2,3-dicarbonate, **71**, 455
  catalysis by N-acetylmannosamine kinase, **71**, 463
  conformation of sugar hydrazones, **71**, 408
  conversion of 1,2-(N-aminoepimino)-3,4,5,6-diethylidene-1,2-dideoxy-L-iditol into 1-deoxy-3,5 : 4,6-diethylidene-L-xylo-hex-2-ulose N-acetylhydrazone, **71**, 244
  cyanogen-promoted phosphorylation of reducing sugars, **71**, 458
  cycloadditions of glycosyl halides, **71**, 171
  cyclization of 1-thio-3,4,6-tri-O-acetyl-S-(o-aminophenyl)-2-O-methanesulphonyl-thio-β-D-glucopyranoside, **71**, 72
  elimination reactions of tosylates, **70**, 155
  elimination reactions of uronic acids, **70**, 155
  epimerization of 4-amino-4-deoxy-L-xylose, **70**, 282
  equilibration of acetoxonium salts of sugar acetates, **70**, 300; **71**, 67, 269
  field ionization mass spectra of aryl glycosides, **70**, 424
  fragmentation of 4,6-O-benzylidene-2-O-toluene-p-sulphonyl α-D-ribo-hexopyranoside-3-ulose, **71**, 83
  glycosylamines,
    basic strengths, **71**, 407
    formation, **70**, 437; **71**, 407
    hydrolysis, **70**, 437; **71**, 407
    mutarotation, **70**, 437; **71**, 407
    rearrangement, **70**, 437

Carbohydrates—*continued*
  glycosyl bromides, bromide exchange, **71**, 106
  hydration of 1,2 : 5,6-diisopropylidene-D-hexofuranos-3-uloses, **70**, 430
  hydrolysis of
    arabinosylcytosine to arabinosyluracil, **70**, 469
    4,6-O-benzylidene-3-deoxy-3C-(iodomethyl)-D-allal, **70**, 30, 38
    cytidine, **70**, 490
    glucuronate esters, **71**, 456
    glycosides, **70**, 423–429; **71**, 397–401
    glycosylamines, **70**, 437; **71**, 407
    glycosyl phosphates, **70**, 424
    nucleosides, **70**, 435–437; **71**, 407
    sodium (N-acetyl-N-2-fluorenylhydroamino-β-D-glucoside)-uronate, **70**, 469
  interconversion of glucose, mannose, and fructose, **70**, 446
  mutarotation, **70**, 431; **71**, 403–406
  neighbouring group participation, **71**, 69
  osazones, **70**, 437, 439
  oxidation, **70**, 580, 583; **71**, 531, 538
  photolysis of *aldehydo*-D-glucose-pentaacetate, **71**, 360
  Pummerer rearrangement of thioanhydrohexitol sulphoxides, **71**, 241
  radiolysis of glycosides, **71**, 393
  reaction of ascorbic acid with hydrazines, **70**, 387
  reaction of α-D-glucopyranose-4-methanesulphonates with sodium azide, **71**, 246
  rearrangement of
    8,3′-S-cycloadenosine sulphoxide, **71**, 271
    ethyl 4,6-di-O-acetyl-dideoxy-α-D-erythro-hex-2-enopyranoside, **71**, 68
    hex-2-enopyranosylpurine nucleosides, **71**, 252
    isopropylidene derivatives, **70**, 410
    D-*threo*-pentulose, **70**, 282
  reduction of adenine nucleosides and nucleotides, **70**, 605
  1,2-shifts of, **70**, 288
  solvolysis of 1,6-di-O-acetyl-2,3-di-O-benzyl-4-O-mesylglucose, **71**, 61–62
  solvolysis of 5-O-tosyl-L-arabinose, **71**, 64
Carbon acids, acidity, **70**, 128; **71**, 121
Carbon, atomic, **70**, 393–394, 409; **71**, 385
Carbonium ions, **70**, 1–58; **71**, 1–49
  addition to olefins, **70**, 10, 175; **71**, 4, 48, 155, 157, 160, 161
  arylalkoxy, **71**, 43
  carbonylation, **70**, 2; **71**, 45
  heats of formation, **71**, 43
  in addition reactions, **70**, 9–10, 176, 177, 178, 179; **71**, 3, 155, 157, 160
  α-keto, **71**, 41
  NMR spectra, **70**, 1–2, 34, 50, 53, 56; **71**, 1, 9, 17, 34–36, 38, 48
  polarography, **71**, 47

Carbonium ions—continued
  polyenyl, **70**, 1; **71**, 1, 41
  stability, **70**, 10; **71**, 43
Carboranes, acidities, **70**, 129
Carboxylic acids, protonation of, **70**, 463
Caryophyllene nitrosite, **70**, 366
Catalysis,
  association-prefaced, **70**, 472–474, 487, 588; **71**, 201, 441–442
  bifunctional in,
    addition of amines to isocyanates, **71**, 413
    addition of methanol to ketenes, **70**, 464
    enolization, **71**, 410
    ester hydrolysis, **70**, 471; **71**, 431
  bisulphite, of deuterium exchange of cytidine-5-phosphate, **71**,
  copper in
    decomposition of azides, **70**, 412
    decomposition of diazo-compounds, **70**, 412; **71**, 296–297
    nucleophilic aromatic substitution, **70**, 215–216, 222
    reaction of amines with polyhalogenoalkanes, **70**, 359
  enzymic by
    acetoacetate decarboxylase, **71**, 453
    acetyl Co-A-arylamine-N-acetyl transferase, **71**, 425, 453
    N-acetyl-β-D-glucoaminidase, **70**, 429; **71**, 401
    N-acetylmannosamine kinase, **71**, 463
    acid phosphatase, **70**, 499–500
    aconitate isomerase, **71**, 116
    acylase, **71**, 453
    adenosine deaminase, **70**, 223
    adenosine triphosphate citrate lyase, **71**, 117
    aldolase, **70**, 448; **71**, 415
    alkaline phosphatase, **70**, 499; **71**, 462
    aminopeptidase-B, **70**, 485; **71**, 449
    aminopeptidase-M, **70**, 485; **71**, 453
    amylases, **70**, 426; **71**, 400–401
    amyloglucosidase, **71**, 401
    amylo-1,6-glucosidase-oligo-1,4→1,4-glucantransferase, **70**, 429
    aspartate decarboxylase, **70**, 489; **71**, 453
    bromelin, **70**, 483; **71**, 450
    carbamoylphosphate hydrolase, **71**, 463
    carbonic anhydrase, **70**, 485; **71**, 453
    carboxypeptidase-A, **70**, 484; **71**, 452
    carboxypeptidase-B, **70**, 484
    carboxypeptidase-S, **71**, 452
    ceruloplasmin, **70**, 588
    cholinesterase, **70**, 485; **71**, 453
    chymotrypsin, **70**, 476–480; **71**, 443–448, 449
    citrate lyase, **71**, 117
    creatine kinase, **71**, 443
    dehydrogenases, **70**, 588; **71**, 540
    DNA polymerase, **70**, 500

Catalysis—continued
  enzymic by—continued
    elastase, **70**, 481; **71**, 449–450
    enolase, **70**, 155; **71**, 413
    esterase, **70**, 485; **71**, 452
    β-fructofuranosidase, **70**, 429
    fructose-1,6-diphosphatase, **70**, 500; **71**, 463
    fucosidase, **70**, 429
    fumarase, **70**, 180
    galactosidase, **70**, 424–425; **71**, 398, 401
    glucoamylases, **70**, 429
    glucosidases, **70**, 425–426; **71**, 401
    glucuronidases, **70**, 429
    glyceraldehyde-3-phosphate dehydrogenase, **70**, 589; **71**, 540
    glycerol kinase, **70**, 500
    glycogen phosphorylase, **70**, 500
    invertase, **71**, 401
    isocitrate dehydrogenase, **70**, 589
    kallikrein, **70**, 482
    ketoglutarate decarboxylase, **71**, 453
    leucineaminopeptidase, **71**, 449
    lipase, **70**, 485
    lipoxidase, **71**, 541
    lysozymes, **70**, 426–428; **71**, 399–400
    α-mannosidase, **70**, 429
    methylmalonyl isomerase, **71**, 132
    mutarotatase, **71**, 403
    myosen, **70**, 500
    nucleoside diphosphatase, **71**, 467
    oxidases, **70**, 588
    papain, **70**, 482–483; **71**, 449, 450
    pepsin, **70**, 483–484; **71**, 450–452
    pepsin C, **71**, 452
    peroxidase, **71**, 540
    phenylalanine ammonia-lyase, **71**, 137
    potato phosphorylase, **71**, 463
    proteinase from *A. flavus*, **71**, 450
    pyruvate decarboxylase, **70**, 136
    pyruvate kinase, **70**, 500; **71**, 409, 463
    phosphoribosyl adenosine triphosphate-pyrophosphate phosphoribosyl transferase, **70**, 499
    ribonuclease, **70**, 479, 498–499; **71**, 461–462
    serine transhydroxymethylase, **70**, 133
    staphylococcalnuclease, **70**, 500; **71**, 463
    subtilopeptidase, **71**, 449
    succinate dehydrogenase, **70**, 133
    sucrose phosphorylase, **70**, 499; **71**, 401
    thermolysin, **70**, 484; **71**, 450
    triosephosphate isomerase, **70**, 447; **71**, 413
    trypsin, **70**, 480–481; **71**, 448–449
    trypsin-like enzyme from *Evasterias trochelii*, **70**, 482
    urease, **70**, 485
    xylosidases, **70**, 428
    yeast hexokinase, **70**, 500; **71**, 463
    yeast inorganic pyrophosphatase, **71**, 463

Catalysis—*continued*
  enzymic, of squalene oxide cyclization, **70**, 208
  general acid, in
    acetal hydrolysis, **70**, 317–318; **71**, 393–394
    aminolysis of 1-acetyl-3-methylimidazolinium ion, **70**, 458
    enol-ether hydrolysis, **71**, 408
    formation of thiazolidine-4-carboxylic acid, **71**, 404
    hydration of ketones, **70**, 430; **71**, 394–395
    hydration of olefins, **71**, 156
    hydrolysis of aryl diazoketones, **71**, 81
    orthoester hydrolysis, **70**, 485; **71**, 393
  general base, in
    additions to double bonds, **70**, 187
    aminolysis of N-acetyl imidazole, **71**, 422
    aminolysis of esters, **70**, 463; **71**, 442
    azo-coupling, **70**, 230
    Baeyer–Villiger oxidation, **70**, 293
    bromination of disulphones, **71**, 118
    elimination reactions of 4-aryloxybutanones, **71**, 138
    halogenation of nitroalkanes, **71**, 116
    hydration of ketones, **70**, 430; **71**, 402
    hydrolysis of dihydro-6- and -1-methyl uracil, **70**, 455
    hydrolysis of aryl diphenylphosphinates, **71**, 460
    hydrolysis of esters, **71**, 428
    hydrolysis of phosphates, **70**, 492; **71**, 421
    hydrolysis of phosphofluoridates, **71**, 459
    hydrolysis of sultones, **71**, 464
    nucleophilic aromatic substitution, **70**, 211
  hypobromous-acid-, in dehydration of bicarbonate, **71**, 428
  intramolecular in
    acetal hydrolysis, **70**, 431; **71**, 394–395
    alcoholysis of *o*-nitrobenzoyl chloride, **70**, 464
    amide hydrolysis, **70**, 458, 465–472; **71**, 434, 439, 440
    enol ether hydrolysis, **70**, 440
    enolization, **70**, 443
    ester hydrolysis, **70**, 465–472; **71**, 431–441
    mutarotation of aldoses, **70**, 431; **71**, 403
    oxidation, **71**, 532, 543
    phosphate hydrolysis, **71**, 494–496; **71**, 457, 459
    reductions, **70**, 81, 599; **71**, 553
    sulphate hydrolysis, **70**, 500
  mercury, in
    addition of HCl to acetylenes, **71**, 158
    ligand replacement at cobalt, **70**, 141–142
  metal-ion in,
    additions, **70**, 209; **71**, 161, 177–178
    autoxidation, **70**, 379, 588; **71**, 353–358, 359

Catalysis—*continued*
  metal-ion in—*continued*
    hydration of pyruvate, **70**, 430
    hydrolysis of alkyl halides, **71**, 107
    hydrolysis of amides, **70**, 474–476
    hydrolysis of esters, **70**, 471, 476; **71**, 443
    hydrolysis of glycosides, **70**, 423
    hydrolysis of nitriles, **71**, 442
    hydrolysis of phosphates, **70**, 497; **71**, 458
    hydrolysis of phosphonates, **70**, 496–497; **71**, 458
    hydrolysis of pyrophosphates, **71**, 458
    hydrolysis of Schiff bases, **70**, 432
    isomerization of olefins, **70**, 208
    mutarotation of sugars, **71**, 404
    oligomerization of olefins, **70**, 200
    reaction of Grignard reagents, **71**, 334–335
    rearrangement of hydrocarbons, **71**, 253–257
    transamination, **71**, 405
  nucleophilic in
    electrophilic aromatic substitution, **70**, 141
    ester hydrolysis, **70**, 459
    hydrolysis of acetic anhydride, **70**, 459
    hydrolysis of dihydroxyphosphinyl-(imidazole), **71**, 459
    hydrolysis of phosphates, **70**, 492
    methanolysis of benzenesulphonyl chloride, **71**, 464
  phase transfer, **71**, 55, 368, 464
  silver-ion in
    peroxydisulphate oxidations, **70**, 358
    reaction of 1-acyl-1-bromo-cyclohexanes, **71**, 102
    reactions of *trans*-13,13-dibromobicyclo-[10.1.0]trideca-4t,8t-ene and tridecane, **70**, 38
    reactions of hydrocarbons, **70**, 275–276; **71**, 11, 253–257
    solvolysis of 1-adamantyl halides, **71**, 13
    solvolysis of α-bromoisobutyrophenone, **71**, 102
    solvolysis of 2-chloro[1-^{14}C]cyclohexanone, **71**, 61
    solvolysis of 6-iodohexa-2,3-diene, **71**, 99
    transition metal and orbital symmetry, **70**, 208
    transition metal of valence isomerizations, **70**, 275
Caveol, **70**, 249
Chapman rearrangement, **71**, 214
Charge-transfer complexes,
  and "vertical" stabilization, **71**, 70, 84
  correlation of charge transfer bands with solvolytic reactivity, **71**, 84
  in additions, **71**, 173, 178
  in azo-coupling reactions, **70**, 230

Charge-transfer complexes—*continued*
  in chlorination by trichloroisocyanuric acid, **70**, 231
  in ion-pair return, **70**, 62
  in nucleophilic aliphatic substitution, **70**, 117
  in nucleophilic aromatic substitution, **71**, 180
  in radical substitution reactions, **70**, 349
  in reactions of tetranitromethane with olefins, **70**, 342
  intramolecular, **70**, 514; **71**, 474
  of trimethylsilyl benzene and tetracyaroethylenes, **71**, 131
  photochemistry of, **71**, 475
Chelation, effect on reactivity of Grignard reagents, **71**, 130
Chemiluminescence, **70**, 331, 386, 571–573; **71**, 524-525
Chloramines, rearrangement, **70**, 293; **71**, 16, 259
  solvolysis, **70**, 30, 80, 104, 293; **71**, 34
Chloraziridines, **70**, 43, 104; **71**, 34
Chlorination,
  electrophilic aliphatic, **70**, 145; **71**, 125
  electrophilic aromatic, **70**, 231; **71**, 201–203
  radical, **70**, 332–336; **71**, 276, 297, 302
Chlorine,
  addition to olefins, **70**, 173; **71**, 3, 152–155
  nucleophilic substitution at, **70**, 104, 394; **71**, 94
Chloroactic acid, hydrolysis, **71**, 107
Chloroformates, **71**, 54
Chloronium ions, **70**, 173
Chlorosulphones, **71**, 103
Chlorosulphonyl isocyanate, **70**, 201, 206, 207, 253, 265, 343; **71**, 125, 168, 173, 174
Chromium alkyls, **71**, 132
Chromium tricarbonyl complexes, **71**, 18
Chrysanthemone, **71**, 28
CIDNP, **70**, 127–128, 246, 256, 261, 276, 277, 309–311, 319, 354, 355; **71**, 213, 239, 275–276, 283, 293, 314, 325, 339, 350, 377, 468
Cinnolines, formation from phenylhydrazines of glycollic acid, **70**, 235
  nitration, **71**, 199
Cinnolin-3-one, **71**, 272
Claisen rearrangement, **70**, 248–251; **71**, 216–219
  abnormal, **71**, 227
  amino-Claisen, **70**, 250; **71**, 218
  thio-Claisen, **70**, 250; **71**, 218
Clemmensen reduction, **70**, 289, 603; **71**, 150
Cobalamines, metal replacement of, **70**, 142
Cobalt alkyls, **71**, 106
Cobinamide coenzymes, **71**, 134
π-Complexes, *see after* Pyrylium salts
"Conducted tour" mechanisms, **70**, 137

Conformational transmission, **70**, 166
Cope rearrangement, **70**, 251, 260, 275; **71**, 219–222
  oxy-Cope, **70**, 253–254; **71**, 222, 236
Copper(I) alkyls, **70**, 359; **71**, 133
Cristol–Firth rearrangement, **71**, 301
Crown ethers, effect on reaction rates, **70**, 131; **71**, 57, 120, 123, 139
Cubane-tricyclooctadiene rearrangement, **70**, 208
Cumyl chlorides, solvolysis, **70**, 56
Curtius reaction, **70**, 292
Cycloadditions,
  2 + 2, **70**, 198; **71**, 172–174, 177, 484, 494
  2 + 3, **70**, 195–198; **71**, 168
  2 + 4, **70**, 190–195; **71**, 38, 164–168, 484
  intra-, **71**, 239
  4 + 3, **71**, 44
  6 + 4, **71**, 38
  8 + 2, **71**, 175
  8 + 8, **71**, 175
  1,3-dipolar, **70**, 195–198, 205; **78**, 168–172
  intramolecular, **70**, 197; **71**, 171, 175
  isotope effects on, **70**, 196; **71**, 169
  metal-catalysed, **71**, 177
  stereochemistry, **70**, 197
  1,4-dipolar, **71**, 168
Cycloalkyl cations, NMR, **70**, 53
Cycloalkyl chlorides, α-aryl, **70**, 15
Cycloalkylmethyl derivatives, **71**, 28, 78
Cyclobutadienes, **70**, 203
Cyclobutanes,
  cleavage of, **70**, 203–204, 274; **71**, 151
  ring contraction, **70**, 287; **71**, 34
  ring expansion, **71**, 240
Cyclobutadienyl dications, **70**, 55
Cyclobutene-butadiene isomerization, **70**, 268; **71**, 234–235
Cyclobutenyl cations, **71**, 42, 174
Cyclobutenylethyl toluene-p-sulphonate, **71**, 23
Cyclobutyl derivatives, solvolysis and rearrangement, **70**, 10–14, 39, 43–45; **71**, 26, 34
Cyclodecapentaene, 1,6-methano, protonation, **70**, 52
Cyclodecyl arenesulphenates, **71**, 53
Cyclodecyl toluene-p-sulphonate elimination, **71**, 136
Cyclodec-5-ynyl toluene-p-sulphonate, **70**, 33
Cyclodextrins,
  catalysis by, **70**, 472–473, 487, 497; **71**, 201, 442, 459
  spin-labelled, **70**, 472–473
Cycloheptatriene,
  cycloadditions, **70**, 206
  7-cyano, isomerization, **70**, 134, 272
  [1,5]-H-shifts, **70**, 262
  valence tautomerism, **70**, 272; **71**, 230
Cyclohexa-1,2-diene, **70**, 202

Subject Index

Cyclohexadienes,
 additions to, **70**, 178, 205
 base-catalysed isomerization, **70**, 132
 elimination of hydrogen, **70**, 154
 reactions with bases, **70**, 125
Cyclohexenes, additions to, **70**, 175, 179, 181, 183; **71**, 157, 158
Cyclohexadienones, **70**, 247; **71**, 13, 163
Cyclohexyne, **71**, 101, 133, 161
Cyclohexyl cations, 4-t-butyl, **71**, 45, 53
Cyclooctadienes, dimerization of, **70**, 203; **71**, 173
Cyclooctatetraene,
 as dienophile, **70**, 194
 Diels–Alder reaction, **70**, 195; **71**, 230
 reaction with benzyne, **70**, 221; **71**, 178
 reaction with nitrile oxide, **70**, 198
 tricarbonyl iron complex, **71**, 39–41, 197
Cyclooctatetraene, dianion, carboxylation, **70**, 145
Cyclooctatetraene, dianion, homo, **71**, 550
Cyclooctene, addition, **71**, 158
Cyclooctene-diol-di toluene-p-sulphonate, **70**, 29
Cyclooctyl arenesulphenates, **71**, 53, 107
Cyclopentadienes,
 cycloadditions, **70**, 203, 205; **71**, 172, 178
 Diels–Alder reactions, **70**, 195, 203; **71**, 165
 photo-rearrangement, **71**, 478
Cyclopentadienide anions, **70**, 132, 195; **71**, 132, 163, 197, 518
 dehydro, **71**, 189
Cyclopentadienone, dimerization of, **70**, 195
Cyclopentenylethyl derivatives, **70**, 3–4
Cyclopentylmethyl cation, **71**, 77
Cyclophanes, electrophilic substitution, **71**, 195
Cyclopropane rings, conjugation, **70**, 41; **71**, 147, 220
Cyclopropanes,
 cis–trans-isomerization, **70**, 294
 hydrogenolysis, **70**, 608
 hydrogen exchange, **70**, 131, 133, 134; **71**, 264
 methylene, **71**, 175, 226
 protonated, **70**, 45–47, 87, 88–89, 143, 144, 180; **71**, 34–36, 74
 racemization of optically active, **70**, 144–145
 ring-opening, **70**, 16, 24, 37, 38, 42–43, 117, 143–144, 175, 180, 183, 186, 266, 295, 600–601; **71**, 8, 11, 25–26, 32–34, 36, 125–126, 228, 229, 243–244, 264–265, 294
 vinyl, rearrangement, **70**, 263; **71**, 222, 224
Cyclopropanols, reaction with mercuric acetate, **70**, 144
 deoxygenation of, **71**, 29
Cyclopropanones,
 cycloadditions of, **70**, 204
 ring opening of, **70**, 278
Cyclopropenes, **70**, 207, 301
 1-methylene, dimerization, **70**, 207

Cyclopropenones, **70**, 198
 reduction of, 301
Cyclopropenyl cations, **70**, 55; **71**, 541
Cyclopropenyl radicals, **70**, 376
Cyclopropylallenes, **71**, 175
Cyclopropylallyl cations, **71**, 29
Cyclopropylethyl derivatives, **70**, 41
Cyclopropyl ketones, **70**, 301
Cyclopropylmethyl cations,
 NMR of, **70**, 34
 rearrangement of, **70**, 13, 34; **71**, 7–9, 25–26, 28, 29, 78, 174
Cyclopropylmethyl radicals, **70**, 320; **71**, 303
Cyclopropyl radicals, **70**, 337, 341, 357, 519; **71**, 277–279, 297, 322
Cyclopropylvinyl cation, **71**, 29
Cycloreversion reactions, **70**, 274
Cyclization,
 1,5-dipolar, **70**, 208
 1,7-dipolar, **70**, 208
Cytidine, hydrolysis, **71**, 490
Cytosines,
 dihydro, hydrogen exchange, **70**, 136
 hydrolysis, **70**, 490

Darzens reaction, **71**, 116
Deamination reactions, **70**, 5–6, 43–44, 45–46, 73, 74, 86, 88–89, 188; **71**, 2, 5, 7, 36, 64, 69, 78–81, 244
Decarbonylation, **70**, 353, 488, 520, 522, 525, 558; **71**, 30, 147, 320–321, 481, 489
Decarboxylation,
 copper-catalysed in quinoline, **70**, 488
 of 5-amino-1-β-D-ribofuranosyl-imidazole--4-carboxylic acid, **71**, 204
 of anthranilic acid, **70**, 488
 of aspartic acid, **70**, 488
 of azulene-1-carboxylic acid, **70**, 233
 of 2-(1-carboxy-1-hydroxyethyl)-3,4-dimethylthiazolium ion, **70**, 486
 of 2-(o-chlorophenyl)-2-cyanoacetate ion, **70**, 487
 of 3,3-dialkyl-2-oxo-carboxylic acids, **70**, 289
 of fumaric and maleic acids, **70**, 361
 of glycidic acids, **70**, 488
 of lysine, **70**, 488
 of [^{2}H$_2$]maleic acid, **70**, 488
 of malonic acid, **70**, 488
 of 6-nitrobenzisoxazole, **71**, 442
 of pentafluorobenzoic acid, **70**, 233
 of phenylcyclopropane carboxylic acids, **70**, 585
 of phenylpropiolic acid, **70**, 488
 of phosphoroformic acid, **71**, 241
 of picolinic acid, **70**, 488
 of pyridine-2-carboxylic acid, **71**, 453
 of pyridylacetic acid, **71**, 453
 of pyrrole-2-carboxylic acid, **71**, 204
 of sarcosine, **70**, 488

Decarboxylation—continued
  of α,β-unsaturated acids, **70**, 488
  oxidative, **70**, 357, 358; **71**, 317, 337, 335
  photo-, **70**, 533; **71**, 475, 476, 481, 491, 494, 506, 515
Dehydrojanusene, **71**, 161
Deltacyclyl derivatives, **70**, 7–8
Demethoxycarbonylation of dimethyl halogenomalonates, **70**, 489
Deoxygenation of
  carbonyl compounds, **70**, 393, 521
  cyclopropanols, **71**, 29
  nitro-compounds, **70**, 398, 407; **71**, 372–373, 386
  nitroso-compounds, **70**, 398; **71**, 372–373, 386
  oxaziranes, **71**, 553
  oxetans, **70**, 409
  N-oxides, **70**, 557
  oxirans, **70**, 409
  selenic acids, **71**, 553
  sulphoxides, **71**, 552
Desulphurization, **70**, 603, 611
Dewar benzene,
  additions to, **71**, 160
  from 1,1′-bicyclopropenes, **71**, 257
Diamantane, **71**, 120
Diazabicyclo[2.2.1]heptenyl derivatives, **71**, 91
Diazepinium ion, dihydro, **71**, 202, 207
Diazo-compounds,
  copper-catalysed decomposition of, **70**, 412
  cycloadditions, **70**, 197, 198, 205; **71**, 170–171
  decomposition of, **70**, 40, 90; **71**, 81, 336, 346
  oxidation, **71**, 336
  photolysis of, **70**, 392, 395, 405, 562; **71**, 373–374, 377, 575
  pyrolysis of, **70**, 395; **71**, 374
  reaction with carboxylic acids, **71**, 450
  reaction with ketones, **70**, 86, 288
  reaction with trinitrobenzene, **71**, 193
Diazo-ketones,
  hydrolysis of, **70**, 89; **71**, 81
  reaction with peroxides, **71**, 345
Diazonium salts,
  decomposition of, **70**, 234, 346, 348, 381; **71**, 189, 190
  electron-transfer reactions, **71**, 345
  reaction with dihydrodiazepinium ions, **70**, 57
  reaction with ferrous chloride, **70**, 360
  reaction with hydroxide ion, **71**, 465
  reaction with phenoxide and alkoxide, **71**, 346
Diazotization, **70**, 503; **71**, 200
Dibenzobarrelene, **71**, 4
Dibenzocyclononatetraenide, **70**, 120
Dibenzopentalenyl dianion, **70**, 120
Dibenzothiepin oxide, **70**, 135

Dibenzotricyclo[3.3.0.0^{2,8}]octadiene, **70**, 186
Dicarbonium ions, **71**, 46
Dieckman reaction, retro, **71**, 417
Diels–Alder reactions, **70**, 190–195; **71**, 164–168
  catalysed, **70**, 193
  concerted nature of, **70**, 190, 191, 192; **71**, 168
  correlation of rates with *para*-localization energies, **70**, 190
  intramolecular, **71**, 239
  of 1-alkoxydienes, **71**, 166
  of bicyclopentene, **71**, 166
  of cyclopentadiene, **70**, 191, 195; **71**, 165
  of diarylmethylenecyclopropanes, **71**, 167
  of dihydroxyprazines, **70**, 194
  of 7,8-dimethylenecycloocta-1,3,5-trienes, **70**, 194
  of 2,5-dimethyl-3,4-diphenylcyclopentadienone, **71**, 165
  of 1,4-diphenylbutadiene, **71**, 166
  of 1,3-diphenylquinoxalino[2,3-c]furan, **70**, 193
  of furan, **71**, 165
  of hepta-3,5-dien-2-one, **71**, 166
  of hexafluorobicyclo[2.2.0]hexa-2,5-diene, **71**, 167
  of methyl (*p*-methoxyphenyl)penta-*trans*-2-*trans*-4-dienoate, **71**, 166
  of myrcene, **70**, 673
  of nitrosobenzenes, **71**, 406
  of pentachlorocyclopentadiene, **71**, 165
  of polyfluorobicyclo[2.2.0]hex-2-enes, **71**, 167
  of tetrabromo-5,5-dimethoxyloxycyclopentadiene, **70**, 194
  of tetrazines, **70**, 195
  of thebaine, **70**, 195
  of tropylium cation, **71**, 167
  of vinyl allenes, **71**, 171
  reactivity of *cis*- and *trans*-dienophiles, **71**, 165
  regiospecificity of, **71**, 165
  retardation of by internal hydroxyl groups, **70**, 192–193
  retro-, **70**, 167, 203, 275; **71**, 168, 220, 236, 375
  stereochemistry of, **70**, 191
  substituent effects, **71**, 164–165
  volume change in, **70**, 191
Dienone-phenol rearrangement, **70**, 247–248; **71**, 203, 215–216
Dihydrodiazepinium ions, **70**, 57
Dihydro-dimethylpyrroloquinoline, **71**, 197
Dihydropyridines, **71**, 302
Diimides,
  catalysis by, **71**, 452
  hydrolysis of, **70**, 491
Di-isopropylidenefuranose-3-toluene-*p*-sulphonate, S_N2 reactions, **70**, 114

Subject Index 623

Dimroth rearrangement, **70**, 298; **71**, 267–268, 270–271
Dinitroacetonitrile, nitration of, **70**, 229
Dioxans,
  conformation of, **70**, 422; **71**, 396
  formation of, **71**, 396
  hydrolysis of, **70**, 421; **71**, 396
Dioxolans,
  conformation of, **70**, 422; **71**, 396
  formation of, **70**, 396
  hydrolysis of, **70**, 419, 420; **71**, 395–396
  radical reactions, **71**, 310, 350
  rearrangement of, **71**, 269
1,2-Dioxetans, **70**, 331; **71**, 520-522, 524
Di-π-methane rearrangement, **71**, 497, 499
Dipolar aprotic solvents for
  nucleophilic additions, **70**, 186; **71**, 162–163
  nucleophilic aliphatic substitutions, **70**, 67, 68; **71**, 56–58
  nucleophilic aromatic substitutions, **71**, 179, 184
Diquinoethylene, dimerization, **71**, 174
Diradicals, **70**, 202, 204, 257, 258, 259, 260, 268, 327–331; **71**, 151, 221, 229, 236, 264, 290–296
Disulphides, nucleophilic substitution reactions, **71**, 90

Electrophilic substitution,
  aliphatic, **70**, 119–145; **71**, 41, 42, 111–134
    intramolecular, **71**, 126–127
  aromatic, **70**, 141, 142; **71**, 143, 195, 210
    effect of positive poles, **71**, 208
    intramolecular, **70**, 235; **71**, 208
    mode of breakdown of σ-complex, **71**, 196
    photo-induced, **71**, 525
    steric hindrance, **71**, 199, 202
  base-catalysed, **70**, 128–131; **71**, 119–125
  transition state for aliphatic, **71**, 124
Electrocyclic rearrangements, **70**, 266–274; **71**, 228–235
Electrons hydrated, **71**, 276
Elimination reactions, **70**, 147–170; **71**, 135–150
  acetylene-forming, **70**, 169; **71**, 149
  asymmetric selection in, **70**, 156
  Chugaev, **70**, 155; **71**, 150
  1,6-cycloelimination of SO_2; **70**, 165
  decarboxylative dehalogenations, **70**, 170
  dehydrations, **70**, 152–154; **71**, 148–149
  dehydrogenations, **70**, 154, 157
  deoxymetalations, **70**, 166
  double-bond character of transition state, **70**, 156, 163, 164; **71**, 136
  electron-impact-induced, **70**, 153–154
  1,3-eliminations, **70**, 82, 87
  1,4-eliminations, **70**, 167
  $E1cB$ mechanism, **70**, 158–162; **71**, 137–140, 146, 149, 264
  $E2C$ mechanism, **70**, 162; **71**, 140–142

Elimination reactions—*continued*
  gas phase, **70**, 167–169; **71**, 142–145
  halide-ion promoted, **70**, 151–152, 163–165; **71**, 136, 142
  heterogeneous, **70**, 167; **71**, 142
  Hofmann, **70**, 147, 157; **71**; 135, 136
  Hofmann rule, **70**, 157; **71**, 141
  ion-pair return in, **71**, 49–50
  lithium-perchlorate promoted, **70**, 170
  metal-ion promoted, **70**, 151
  of (2-acetoxyethyl)-tri-n-butyltin, **71**, 149
  of alkanesulphonyl chlorides, **70**, 170
  of 4-aryloxybutan-2-ones, **71**, 137–138
  of arylsulphonylmethyl nitrates, **71**, 113
  of aryl trialkylsilyl acetals, **71**, 144
  of bicyclo[2.2.2]octyl derivatives, **71**, 135
  of 9,9-bi-fluorenyl derivatives, **70**, 170
  of 4-t-butylcyclohexanols, **71**, 149
  of n-butyl(tri-n-butylphosphine)copper, **70**, 169
  of chlorotetralins, **70**, 158
  of cycloalkyl halides, **71**, 136, 145
  of dibromoalkanes, **70**, 151, 152; **71**, 113–114, 143, 146–147, 348
  of di-t-butyl phosphonium salts, **71**, 87
  of 2,2-difluoroethyltrimethoxysilane, **71**, 144
  of dihydrothiophenium hexafluorophosphates, **71**, 148
  of 2,6-dimethyl-4($\alpha,\alpha,\beta,\beta$-tetracyanoethyl)-aniline, **71**, 139
  of α-ethylbenzyl toluene-p-sulphonate, **71**, 106
  of fluorenylmethanol, **70**, 159
  of halogenocyclohexenes, **70**, 158
  of menthyl tosylate, **70**, 162
  of phenethyl derivatives, **70**, 170; **71**, 142, 146
  of phosphates, **71**, 150
  of steroidal selenoxides, **70**, 150
  of steroidal sulphenates, **70**, 151
  of steroidal sulphoxides, **70**, 150–151
  of stilbene dibromides, **70**, 151; **71**, 146
  of Streptovitacin A, **70**, 170
  of styryl chlorides, **71**, 149–150
  of tri-isobutyl aluminium, **71**, 145
  orientation in, **70**, 156–165; **71**, 135–137
  pyrolytic of
    esters, **70**, 155; **71**, 143, 144
    N-oxides, **70**, 155
    sulphoxides, **70**, 168
    xanthates, **70**, 155; **71**, 150
  ρ-values for, **70**, 149, 150, 152, 153, 158, 159, 164, 165, 168, 169, 170; **71**, 138, 139, 141, 142, 143, 144, 146, 149
Saytzeff rule, **70**, 157, 158
six-centred transition state, **71**, 143–144
syn-eliminations, **70**, 147–150, 153, 284; **71**, 136, 138, 139
tosylate-bromide rate ratios, **70**, 169; **71**, 142

Enamines,
  alkylation of, **70**, 450
  as intermediates in the Michael addition, **70**, 188–189
  cycloadditions, **70**, 198, 201, 203; **71**, 173
  hydrolysis of, **70**, 448–450
  isomerization of, **70**, 450, 492
  mercuration, **71**, 158
  protonation of, **71**, 413
  reaction with deuteroacetone, **71**, 127
  rearrangement, **71**, 228
  synthesis of, **71**, 127
Endo- and exo-cyclic intramolecular reactions, **71**, 104–105
Ene reactions, **70**, 207; **71**, 177
  intramolecular, **71**, 239
Energy transfer in photochemical reactions **70**, 510–515
  intramolecular, **70**, 510
Enolate anions, alkylation, **70**, 104, 126; **71**, 96, 116
  protonation, **71**, 413
Enol esters, hydrolysis of, **70**, 440, 441; **71**, 409
Enol ethers,
  addition of bromine, **71**, 154
  addition of N-sulphuryltosylamide, **71**, 173
  exchange reactions of, **70**, 440
  hydrolysis of, **70**, 439; **71**, 408–409
  methanolysis of, **71**, 409
  reaction with hydroxyl radicals, **71**, 324
Enolization, **70**, 132–133, 441–447; **71**, 409–413
Entropy of activation for
  chymotrypsin-catalysed hydrolyses, **71**, 447
  decarboxylation of phosphoroformic acid, **71**, 421
  decomposition of azo-compounds, **70**, 288–289
  elimination reactions, **70**, 164
  exchange of magnetically non-equivalent methyl groups in 9-chloro-9-durylfluorene, **70**, 65–60
  exchange of methyl groups in dimethyl cadmium, **70**, 141
  hydrolysis of 2-bromocyclopentanol and cyclohexanol, **71**, 87
  hydrolysis of enol ethers, **70**, 440
  hydrolysis of ketals, **70**, 418
  hydrolysis of sydnones, **71**, 454
  nucleophilic displacement reactions of vinyl halides, **71**, 100
  rearrangement of α-substituted allylic didisulphides, **71**, 224
  rearrangement of aryl thionobenzoates to aryl thiobenzoates, **71**, 213
  rearrangement of 1,1-dideuteriohexa-1,5-diene, **71**, 220
  rearrangement of cis-dienones, **70**, 271
  solvolysis of benzenesulphonyl chlorides, **71**, 463

Entropy of activation for—*continued*
  solvolysis of benzonorbornen-2-*endo*-yl p-bromobenzenesulphonate, **70**, 23
  solvolysis of benzyl chloride, **70**, 66
  solvolysis of t-butyl chloride, **70**, 67
  solvolysis of α-bromoarylacetic acids, **70**, 78
  solvolysis of cyclopropylmethyl 2-naphthalenesulphonate, **71**, 28
  suprafacial, 1,6-cycloelimination, **70**, 165
Enzymes, inactivation by hydrogen atoms, **70**, 341
Enzymic catalysis, *see* Catalysis, enzymic
Episulphonium ion, **71**, 63, 160
Epoxidation of olefins, **70**, 172, 184–185, 189; **71**, 158, 531
  intramolecular, **71**, 251
Epoxides,
  decomposition of, **70**, 330
  deoxygenation, **70**, 167, 330
  fragmentation of α-pinene epoxide, **71**, 83
  α-halo, **71**, 243
  photolysis of, **70**, 548
  radical reactions of, **70**, 305
  reaction with acetone, **70**, 420
  reaction with dialkyl copper, **71**, 133
  reaction with Grignard reagents, **70**, 139
  rearrangement of, **70**, 10, 284–285; **71**, 108, 233–234, 244, 247, 257, 265
  ring-opening of, **70**, 71, 81, 113, 116, 299; **71**, 5, 108, 170
Esterification, **70**, 492
Esters,
  acidities, **70**, 130
  alcoholysis of methyl formate, by NMR, **70**, 489
  alkyl-oxygen fission, **71**, 435, 456
  aminolysis, **70**, 456, 459, 465; **71**, 424
  conformation, **70**, 491; **71**, 424
  hydrolysis, **70**, 459, 461, 476, 491; **71**, 424, 427, 428, 431–441, 455–456
    elimination-addition mechanism, **70**, 489; **71**, 455
  of amino-acids, **70**, 489
  of thiochloroformates, **70**, 489
  oxygen exchange, **70**, 491
  protonation of, **70**, 463; **71**, 428
  reaction with amidines, **71**, 181
Ethers, aromatic,
  alkylation of, **70**, 235
  basicity, **71**, 48
  halogenation, **70**, 231, 232
  protonation of, **71**, 41
  rearrangement of, **70**, 242
Excimers, **70**, 512–513, 538, 573; **71**, 467
Exciplexes, **70**, 513–515, 536, 558, 560; **71**, 473, 474, 475, 476, 481, 494, 497, 524
  intramolecular, **71**, 473
Excited states, dipole moments, **70**, 507
Extended selectivity relationship, **70**, 226

Subject Index

Favorskii rearrangement, **70**, 278, 279, 586; **71**, 241–242
 homo-Favorskii, **71**, 242
Feist's ester, **70**, 258
Ferrocenes, lithiation of, **71**, 127
Ferrocenophan-1-ones, acylation of, **70**, 234
Ferrocenophanyl cations, **71**, 37
α-Ferrocenylmethyl cations, **70**, 49; **71**, 37, 38
Five-membered rings, enlargement of, **71**, 269
Fischer indole reaction, **70**, 244
Flash photolysis, **70**, 506, 512; **71**, 475, 511, 515, 519, 523
Fluoranthene, hydrogen exchange of, **70**, 237; **71**, 207
 dihydro, **71**, 196
 tetrahydro, nitration of, **70**, 229
Fluorenes,
 iodination of, **70**, 230
 sulphonation of, **70**, 227
Fluorenyl anions, **70**, 128–129
Fluorenyl cations, **70**, 58
Fluorescence spectra, **70**, 507–508; **71**, 469–470, 491
Fluorination of aromatic compounds, **70**, 231; **71**, 201, 319
Fragmentation reactions, **70**, 9, 91, 145, 170, 201, 408–409; **71**, 44, 45, 82–83
 photochemical, **71**, 479–481, 493–494
Friedel–Crafts reaction, **70**, 233; **71**, 197, 205–206
Fries rearrangement, **70**, 242; **71**, 213
 photo-, **70**, 532
Fulvene,
 8-cyano, **71**, 175
 dimethyl, reaction with diazomethane, **70**, 205, 206
 1,2,3,5,6-pentamethyl, formation and rearrangement, **70**, 57
 reaction with maleic anhydride, **71**, 166
 reaction with tropone, **71**, 175
 6-vinyl, **71**, 175, 235
Fulvenes,
 adducts with dichloroketene, cycloadditions, **71**, 166
 homo, cycloadditions, **70**, 206
Furans,
 halogenation of, **70**, 226; **71**, 197, 203, 205, 207
 hydrolysis of, **70**, 439; **71**, 408; *see* Errata
 oxidation, **71**, 546
 radical substitution, **71**, 316
 side-chain phenylation, **71**, 313, 317–318
Furfural, reaction with aniline, **71**, 269

Germacrone, **71**, 46
Germanes, base-catalysed cleavage of acetylenic, **71**, 131

Germanium, nucleophilic displacement at, **70**, 91
Glycosides,
 hydrolysis of, **70**, 423–429; **71**, 397–401
 synthesis, 424
Glycosylamines, **70**, 437; **71**, 407
Glycosyl halides, nucleophilic displacement, **70**, 116; **71**, 106
Goering–Schwene diagram, **70**, 2
Grignard reagents,
 alkyl magnesium fluorides, **70**, 138
 allyl, **70**, 139; **71**, 129
 anodic addition to olefins, **70**, 376
 autoxidation of, **70**, 380
 benzyl, alkylation of, **70**, 105
 cobalt-catalysed reactions of, **71**, 334
 cyclization of alkenyl, **71**, 333
 cyclohexenyl, **70**, 137
 cyclopentadienyl, reaction with benzyne, **70**, 137
 cyclopent-3-enylmethyl, **70**, 137
 hexamethylphosphortriamide solutions, **71**, 113
 isomerization, catalysed by Ti(IV), **70**, 139
 metal-ion catalysed reactions with alkyl halides, **71**, 334–335
 radical reactions of, **70**, 309; **71**, 320
 reaction with
  alkynols, **71**, 162
  acid chlorides, **70**, 360
  alkyl bromides and iodides, **70**, 309
  α-bromocrotonic acid, **70**, 189
  cinnamyl alcohol, **70**, 188
  dichlorobis (triphenylphosphene) nickel, **71**, 89
  epoxides, **70**, 139
  fluorosilanes, **71**, 84
  ketones, **70**, 139; **71**, 128, 130, 547
  norbornenones, **71**, 127
  orthoesters, **70**, 64
  peroxides, **71**, 130
  phosphorus esters, **71**, 461
  silyl halides, **71**, 130
  sulphoxides, **70**, 278
 rearrangement of, **71**, 272
 reduction by, **70**, 596–597; **71**, 547
 structure of, **70**, 137; **71**, 127–128
Guanidines, *syn–anti*, conversion, **70**, 438

Halide exchange reactions, **70**, 67–68, 109, 215–216; **71**, 106
Halogen, nucleophilic displacement at, **70**, 104; **71**, 94, 192
Halogenation of
 aromatic compounds, **70**, 230–232; **71**, 201–203
  in presence of thallium salts, **70**, 230; **71**, 203
 phenylacetylene, **70**, 126

Halogenaromatics,
  nitration, **70**, 228; **71**, 197
  reactivity to electrophilic substitution, **70**, 226; **71**, 196
α-Halogenocarbonyl compounds, **70**, 110–111; **71**, 101–103
Hammett ρ–σ relationship for
  addition of bromine to styrene, **70**, 174
  addition of trichloromethyl radicals to styrenes, **71**, 305
  alkaline hydrolysis of benzoate esters, **71**, 455
  alkaline hydrolysis of cinnamate esters, **71**, 455
  alkaline hydrolysis of 5-phenyl-penta-2,4-dienoate esters, **71**, 455
  alkaline hydrolysis of substituted phenyl benzoates, **71**, 428
  aminolysis of N-arylarenecarbimidoyl chlorides, **71**, 430
  aminolysis of esters, **70**, 456, 460
  attack of nucleophiles on methyl halides, **71**, 50–51
  azo-coupling, **71**, 200
  Baeyer–Villiger oxidation, **70**, 293
  base-catalysed cleavage of acetylenic silanes and germanes, **71**, 131
  charge-transfer complexing, **70**, 70
  chloride-exchange of aryl-(methyl)-1-naphthyl chlorosilanes, **71**, 83
  Claisen rearrangement, **70**, 248
  cleavage of 1-aza-3-arylbicyclobutanes, **70**, 117
  cleavage of phenylcyclopropanes by thallium acetate, **70**, 535
  decarboxylations, **70**, 487
  decomposition of bis(α,α-dimethylbenzyl) hyponitrite, **71**, 289
  decomposition of peroxides, **71**, 282
  diazonium coupling, **70**, 230
  Diels–Alder reaction of 6-substituted fulvenes, **71**, 166
  disilanylation of benzenes, **71**, 315
  electrophilic aromatic substitution, **70**, 226, 237; **71**, 196, 197, 205
  elimination reactions, **70**, 149, 158, 159, 161, 164; **71**, 138, 139, 142, 143, 144, 146, 149
  hydrogen exchange of N,N-dimethylanilines, **70**, 237
  hydrolysis of
    (alkoxymethoxy)benzoic acids, **71**, 394
    2-arylbenzoxazinones, **71**, 426
    N-arylcarbamoyl phosphates, **71**, 460
    2-arylcyclopropanecarboxylate, esters, **71**, 455
    aryl diazoketones, **71**, 81
    aryl diphenylphosphinates, **71**, 460
    2-aryloxytetrahydropyrans, **70**, 418
    aryltrimethylsilanes, **71**, 85
    aryltrimethylstannanes, **71**, 85

Hammett ρ–σ relationship for—*continued*
  hydrolysis of—*continued*
    benzaldehyde di-t-butyl acetals, **71**, 393
    benzohydroxamic acids, **71**, 426
    benzylidene diacetates, **70**, 422
    α-diazo-phosphonates and phosphine oxides, **70**, 90
    ethylstyryl ethers, **71**, 408
    methyl 3-benzoylacrylates, **71**, 436
    5-methylsalicylidene anilines, **71**, 432
    N-mesitylarenesulphinamides, **71**, 92, 464
    peroxybenzoic acids, **71**, 427
    Schiff bases, **71**, 406
    sultones, **70**, 502
    trifluoro-N-methylacetanilides, **71**, 420
    vinyl ethers, **70**, 439; **71**, 408
  iodination of acetophenones, **70**, 442
  ionization potentials, **71**, 197
  isomerization of α-N-diphenylnitrone, **70**, 290
  isotopic exchange of benzhydryl iodides with iodine, **70**, 67
  methanolysis of silanes, **70**, 94
  nucleophilic aromatic substitution, **70**, 211; **71**, 180, 182
  oxidation of N-aryl-2-naphthylamines, **71**, 359
  oxidation of 1,2-diarylethanols by cerium, **71**, 331
  ozonide formation, **70**, 575
  papain-catalysed hydrolysis of anilides of N-acetyl-L-phenylalanyl glycine, **71**, 450
  phenylation of substituted benzenes, **70**, 347; **71**, 314
  pyrolysis of 1-arylethylacetates, **71**, 196
  pyrolysis of aryl n-propyl sulphoxides, **70**, 168
  Pummerer rearrangement, **70**, 277
  radical abstraction from,
    benzaldehydes, **70**, 334
    neopentyl benzenes, **70**, 338
    toluenes, **70**, 319, 362
  radical abstraction of benzylic hydrogen atoms, **70**, 334
  radical reactions, **71**, 276
  reaction of
    acetophenones with semicarbazide, **71**, 407
    anilines with ethyl chloroformate, **71**, 424
    arenesulphenyl chlorides with benzhydryl aryl sulphides, **71**, 90
    aryl dimethylsilanes with tri-butyl tin alkoxides, **71**, 85
    1-aryl-2-haloacetylenes with alkanethiolates, **71**, 94
    aryl sulphenanilide with 2,4-dichloroaniline, **71**, 90
    benzenediazonium salts with acetoacetanilide, **70**, 229
    benzenesulphonyl chlorides with anilines, **71**, 463

Hammett ρ–σ relationship for—*continued*
  reaction of—*continued*
    benzylidene aniline with HCN, **71**, 405
    dehydroacetic acid with amines, **71**, 406,
    2,2-diarylvinyl bromides with t-butoxide **70**, 109
    diphenylsulphides with *trans*-dichlorodipyridine platinum (II), **71**, 95
    Grignard reagents with benzophenones, **71**, 128
    p-nitrophenyl benzenesulphonates with thiophenoxides, **70**, 501
    p-nitrophenyl sulphate with substituted thiophenoxides, **70**, 501
    nitrosobenzene with anilines, **71**, 406
    toluenes with t-butoxy radicals, **71**, 299
  rearrangement of aryl thionobenzoates, **71**, 213
  rearrangement of triazenes, **70**, 243
  reduction of nitro-compounds, **71**, 552
  solvolysis of
    β-arylalkyl derivatives, **70**, 21
    1-aryl-1-chloro-3,3-diphenylallene, **70**, 108
    2-[π-(aryl)chromium tricarbonyl]-2-methyl-1-propyl methanesulphonate, **70**, 48
    2-arylcyclopropylcarbinyl 3,5-dinitrobenzoates, **70**, 35
    arylthio-diarylvinyl sulphonates, **70**, 106; **71**, 100
    N-benzoyl-*N*-phenyldiimide, **70**, 491
    benzyl bromides, **70**, 65
    benzyl tosylates, **70**, 63–64
    α-bromoketones, **70**, 111
    cumyl chlorides, **70**, 56
  xylosidase-catalysed reactions, **71**, 428
Hammond's postulate, **70**, 142
Hard and soft acids and bases, **70**, 103, 105, 109; **71**, 94, 96, 103, 142, 165
Heat capacity of activation, **70**, 67, 87; **71**, 403
Helium tritiide ion, **71**, 208
Hell–Volhard–Zelinsky reaction, **70**, 333
Hemiacetals formation, **70**, 430
Hemimellitene, nitration, **70**, 228
  protonation, **71**, 206
Hemimercaptals, **70**, 430
Hofmann rearrangement, **71**, 258–259
Homoenolization, **70**, 126, 446; **71**, 412
Homopropargyl derivatives, **70**, 33
Horner–Emmons olefin synthesis, **70**, 128
Hudec's rules, **70**, 91
Hunsdiecker reaction, **71**, 335
Hydration of olefins, **70**, 180; **71**, 154
Hydrazides, hydrolysis of, **70**, 462
Hydrazidic halides, **71**, 428
Hydrazines,
  addition to acetylenes, **70**, 159
  addition to olefins, **70**, 190
  oxidation of, **70**, 405

Hydrazones,
  bromination, **71**, 133
  conformation of sugar hydrazones, **71**, 408
  cyclization of, **71**, 270
  decomposition of sulphonyl hydrazones, **70**, 404, 408; **71**, 382, 383
  formation from nitrilimines and aryl acetylenes, **70**, 195
  hydrolysis of, **70**, 437
  isomerization of, **70**, 295
  reaction with nitrous acid, **70**, 291
  syn–anti interconversion, **70**, 439
Hydrazonyl bromides, **71**, 45
Hydride-ion shift,
  1,2-, **70**, 46, 53, 284, 286, 288; **71**, 7, 14, 59, 80, 244–250
  1,3-, **71**, 45
  1,5-, **70**, 87; **71**, 79, 253
  in norbornyl derivatives, **70**, 2,3; **71**, 1,2
  intermolecular, **70**, 19, 55, 56, 57, 143, 185; **71**, 45, 46
  transannular, **70**, 15, 585; **71**, 53
Hydrindanyl derivatives, **70**, 15
Hydroboration, **70**, 10, 172, 182; **71**, 157
Hydrocarbons, acidities, **70**, 128, 129; **71**, 121
  hydrogen exchange, **70**, 132, **71**, 111, 122
Hydroformylation, **70**, 186
Hydrogen exchange,
  aromatic, **70**, 225, 236–238; **71**, 124, 202, 206–208
  base catalysed, **70**, 130–137; **71**, 119–125
  of aliphatic nitro compounds, **70**, 130, 134; **71**, 122
  of aromatic nitro compounds, **70**, 136, 213
  of ketones, **70**, 132–133, 441; **71**, 409–412
  of sulphones, **71**, 120
  of sulphoxides, **70**, 135
Hydrogen halide additions, **70**, 178–180; **71**, 155–156, 305–306
Hydrogen migrations,
  1,5-, **71**, 175, 221, 226, 227, 230
  in carbenes, **70**, 406
  in radical reactions, **70**, 356; **71**, 297, 298
Hydrogen peroxide, addition to acetone, **70**, 430
  reaction with 3,4-benzotropolone-1,2-quinone, **71**, 18
Hydrogenation, **70**, 605–611; **71**, 555–559
  heterogeneous, of
    acetylenes, **70**, 609; **71**, 556
    aromatic hydrocarbons, **71**, 559
    azo-compounds, **70**, 610
    benzylamines, **70**, 610
    benzyl alcohols, **70**, 610; **71**, 557
    carbonyl compounds, **70**, 610; **71**, 557
    chloro-aromatic compounds, **70**, 610
    cyclopropanes, **70**, 608; **71**, 556–557
    dienes, **71**, 555
    dimethoxybenzenes, **70**, 610
    epoxides, **71**, 557

Hydrogenation—*continued*
  heterogeneous, of—*continued*
    imines, **70**, 610; **71**, 557
    norbornenes, **70**, 172
    olefins, **70**, 607–609; **71**, 556–557
    polyenes, **71**, 555
    pyridines, **71**, 551
    unsaturated carbonyl compounds, **70**, 609; **71**, 556
  homogeneous of
    acetylenes, **70**, 607; **71**, 548–549
    benzenesulphonyl chloride, **70**, 607
    dimethyl sulphoxide, **70**, 607
    ketones, **70**, 607; **71**, 559
    nitrobenzenes, **70**, 607
    olefins, **70**, 605–606; **71**, 556–559
Hydroperoxides, acid-induced rearrangement, **70**, 294
Hydroxamic acid, hydrolysis, **71**, 426
Hydroxylation
  electrophilic, **70**, 238
  radical, **70**, 349, 565
Hyperconjugation, **71**, 75
Hypohalous acid additions, **70**, 173–174, 176

Imidazole,
  as catalyst for ester hydrolysis, **71**, 428, 460
  hydrogen exchange of, **70**, 136
  nitration, **70**, 229
Imides, aminolysis of, **70**, 465
Imines, *see* Schiff bases
Indenes,
  cycloaddition, **70**, 203
  hydrogen exchange, **70**, 131
  isomerization of, **70**, 296
  metalation, **70**, 232
  thermolysis, **71**, 228
  racemization, **70**, 131
Indenyl mercuric chloride, **71**, 142
Indoles,
  halogenation, **70**, 231
  nitrosation, **70**, 224
  protodemercuration, **71**, 204
  radical arylation, **71**, 317
  rearrangement of, **70**, 290, 291, 299, 300; **71**, 269
  sulphonation, **70**, 227
Indolizine, hydrogen exchange, **70**, 237; **71**, 207
Interfaces, reactions at, **70**, 473
Iodides, alkyl, exchange with iodine, **71**, 276
Iodination, aromatic, **70**, 230–231, **71**, 203
Iodine, addition to olefins, **70**, 177–178; **71**, 155
  nucleophilic displacement at, **70**, 104; **71**, 94
Iodine azide, addition to olefins, **70**, 177, 289; **71**, 154
Iodine isocyanate, addition to olefins, **70**, 177

Iodobenzene dichloride, **70**, 173
Iododeboronation, **71**, 204
Iododemercuration, **71**, 131, 204
Iodonium ions, **70**, 177; **71**, 155
Iodonium nitrate, **70**, 178; **71**, 155, 203
Ion-cyclotron resonance spectroscopy, **71**, 46, 105, 195, 338
Ion-pair exchange, **71**, 52
Ion-pair return, **70**, 12, 59–66, 119–145, 325; **71**, 49–54, 58, 60, 111
  steric course, **71**, 51
Ion pairs, **70**, 15, 65–66, 67, 367; **71**, 54, 121, 123, 142, 340, 399
  in additions, **70**, 59, 178, 179
  in elimination reactions, **70**, 148; **71**, 106, 135
  in rearrangement of alkyl dimethylallyl indoles, **70**, 291
  in reductions by alkali metals, **71**, 550, 552
Iron, nucleophilic displacement at, **70**, 104; **71**, 95
Iron carbonyl complexes, **71**, 132
  hydrogen exchange of, **70**, 136
  of dienyl-3,5-dinitrobenzoates, solvolysis, **70**, 47
Isobullvalene, **70**, 252
Isocyanates, **70**, 492; **71**, 430, 457
Isoimides, hydrolysis of, **70**, 457
Isonitriles, cycloadditions, **70**, 198
  hydrolysis, **71**, 426
  radical additions, **71**, 309
Isoquinolines, rearrangements to quinolines, **70**, 246
Isotope effects,
  carbon, **70**, 72, 488, 590; **71**, 17
  chlorine, **70**, 112–113
  deuterium,
    in energy transfer processes, **71**, 472
    primary, **70**, 8, 131, 133, 148, 158, 179, 229, 230, 243, 252, 284, 319, 324, 383, 399, 406, 443, 578–579, 593, 596; **71**, 75, 122, 123, 135, 140, 142, 143, 144, 150, 177, 206, 215, 256, 265, 276, 296, 354, 393, 429, 441, 487, 531, 538, 539, 545, 547
    secondary, **70**, 3, 22, 60, 69–70, 84, 85, 196, 199, 202, 417; **71**, 2, 13, 17, 18, 34–35, 58, 59, 91, 172, 177, 287, 394, 401, 406, 407, 416
    solvent, **70**, 70, 90, 99, 117, 142, 158, 417, 418, 440, 455, 592; **71**, 60, 81, 152, 179, 408, 410, 421, 453, 463, 464
    steric, **71**, 234
  nitrogen, **70**, 161, 214
  tritium,
    primary, **70**, 136, 609
    secondary, **70**, 69
Isoxazoles, hydrogen exchange of, **70**, 136
  rearrangement, **71**, 268
  ring contraction, **71**, 270
Isoxazolin-5-ones, ring opening, **70**, 492

Jacobsen rearrangement, **70**, 298
Jahn–Teller distortion, **70**, 120
Janusene, **71**, 46
  dehydro, **71**, 248
  dibromide, **71**, 248

Ketals, hydrolysis of, **70**, 418–423; **71**, 393–396
Ketene acetals, methanolysis, **71**, 409
Ketenes,
  addition of alcohols, **70**, 464
  combustion, **71**, 355
  cycloadditions, **70**, 198–201; **71**, 172, 176
  intermediates in ester hydrolysis and alcoholysis, **70**, 489; **71**, 455
  intermediates in photolysis of enones, **70**, 522
  photolysis of, **70**, 392, 522
  reaction with acetic acid, **71**, 455
Ketones,
  acidities, **70**, 130
  addition of lithium propenyl cuprate to unsaturated, **71**, 334
  bromination, **70**, 444; **71**, 411, 412
  hydration, **70**, 430; **71**, 402
  iodination, **70**, 442, 443; **71**, 410, 412
  photolysis of, **70**, 515–532; **71**, 475–489
  protonation of, **70**, 457; **71**, 41, 417
  reaction with bisulphite, **71**, 416
  reaction with Grignard, **70**, 139; **71**, 128–129, 410
  reduction of, **70**, 595–597, 602, 603, 605; **71**, 546–549, 552, 554
  reductive alkylation, **70**, 445
Kharasch reaction, **71**, 334
Knoevenagel reaction, **71**, 334
Koch–Haaf reaction, **70**, 19; **71**, 14, 36
Kochi reaction, **71**, 335
Kolbe reaction, **71**, 344

α-Lactams, **70**, 279
Lactams, hydrolysis, **71**, 427
α Lactones, **70**, 78
β-Lactones, **70**, 202, 492; **71**, 69
Lactones,
  acidities of, **70**, 130
  hydrolysis and formation, **70**, 468; **71**, 434, 435
Linolenyl ethers, **71**, 115
Liquid crystals, Claisen rearrangement in, **70**, 248
Lithium alkyls,
  carbenoids from, **70**, 410, 412; **71**, 388
  oxidation of, **71**, 336
  radical reactions of, **71**, 339
Lithium allyls, **71**, 93
Lithium α-aryl compounds, **70**, 124, 188; **71**, 127
Lithium benzhydryl, **70**, 126
Lithium benzyl, **70**, 126

Lithium butenyl, **70**, 124
Lithium butyl, **70**, 124, 137; **71**, 114–115, 127, 148
Lithium t-butyl, **70**, 125, 137
Lithium crotyl, **70**, 122
Lithium 1,3-diphenylallyl, **70**, 137
Lithium phenyl, **71**, 93
Lithium tolyl, **70**, 137; **71**, 112
Lithium trimethylsilyl, **70**, 137
Lithium trityl, **70**, 126
Lithium vinyl, **71**, 93
London dispersion forces, **71**, 165
Longicamphenilyl toluene-p-sulphonate, **70**, 71
Longifolene, rearrangement, **70**, 58; **71**, 79
Lumazine, hydrogen exchange of, **70**, 136

Malachite green, **71**, 48, 109, 426
Maleic anhydride, additions to
  anthracene, **70**, 190
  cyclohexadiene, **70**, 191
  fulvenes, **71**, 166
  hepta-3,5-dien-2-one, **71**, 166
  isoprene, **70**, 191
  methoxybutadiene, **70**, 191
  nitrosocyclohexene, **71**, 175
  tosylazocyclohexene, **70**, 193
Malononitrile, bromination and tritiation, **71**, 123
Manganese dialkyls, **71**, 132
Mannich bases, **71**, 123
  decomposition, **71**, 417
  reaction with hydrazines, **71**, 164
Mannich reaction, **70**, 189
Martynoff rearrangement, **71**, 239
Meerwein–Ponndorf–Verley reduction, **70**, 600
Meisenheimer complexes, **70**, 212, 214, 217–219; **71**, 184–188
Meisenheimer rearrangement, **70**, 276, 355, 356; **71**, 239
Meldrum's acid, **70**, 423
Memory effects, **71**, 5–7
Menschutkin reaction, reverse, **70**, 65; **71**, 54
Menthone,
  enolization of, **70**, 443
  epimerization of, **70**, 133
Mercuration, **70**, 183; **71**, 158, 204
Mercurideboronation, **70**, 140
Mercuridemetalations of 4-pyridiomethyl derivatives, **70**, 141
Mercurideprotonations, **70**, 141
Mercuridestannylation, **70**, 141; **71**, 131, 132
Mercurinium ions, **70**, 183; **71**, 5, 132, 158
Mercury alkyls, **71**, 133
Mercury aryls, **70**, 233; **71**, 197
Mesitylene, acylation of, **70**, 234
  chloromethylation of, **71**, 205
Mesoionic compounds, **70**, 196, 490, 553; **71**, 169, 171, 269, 453, 508

Metalation of
  allenes, **70**, 138
  arylalkanes, **71**, 121
  cumene, **70**, 138
  dicyclopropyl acetylene, **70**, 138
  m-dinitrobenzenes, **70**, 125
  hept-1-ene, **70**, 138
  4-methoxybenzyl dimethylamine, **70**, 138
  toluene, **70**, 138
  toluidines, **70**, 138
Metal carbonyls, electrophilic substitution, **71**, 206, 207, 208
Metal cleavage, **71**, 204
Metallocenes,
  hydrogen exchange of, **70**, 237; **71**, 207
  rearrangement, **70**, 260–261; **71**, 209, 244
Metalocenylmethyl cations, **70**, 47, 286–287; **71**, 36–38
Methoxymercuration, **70**, 184
Meyer reaction, **71**, 454
Meyer–Schuster rearrangement, **71**, 252
Micelles, **70**, 423
  in deaminations, **70**, 88; **71**, 79–80
  in decarboxylations, **70**, 487; **71**, 442
  in hydrolysis of *p*-nitrophenyl laurate, **71**, 442
  in hydrolysis of phosphates, **70**, 497; **71**, 459
  in hydrolysis of sulphates, **70**, 502
  in intramolecularly catalysed reactions of acetals, **70**, 419
  in mutarotation of tetramethylglucose in benzene, **71**, 403
  reactions at, **70**, 473
Michaelis–Arbusov reaction, **71**, 89
Michael reaction, **70**, 188, 189; **71**, 116, 123, 163
Migration of alkyl groups, **71**, 44
Mills–Nixon effect, **70**, 386; **71**, 195
Moffat oxidation, **71**, 544
Molecular-orbital calculations, on
  acidity of CH_4 and C_2H_6 **70**, 129
  addition of vinyl cation to olefins, **71**, 172
  addition reactions, **70**, 171, 172, 174, 194
  approach of singlet methylene, **70**, 410
  aziridinyl cation, **71**, 34
  benzenonium ions, **71**, 206
  benzynes, **71**, 188
  carbanions, **70**, 122
  carbenes, **71**, 367
  carbonium ions (*ab initio*), **71**, 34, 74
  conjugated ions, **71**, 48
  Cope rearrangement, **70**, 251; **71**, 219
  cyclobutane cleavage to ethylene, **71**, 151, 290
  cyclooctatetraene ions, **70**, 55
  cyclopropanes, **71**, 121
  cyclopropyl anion, **71**, 113, 228
  cyclopropyl cation, **71**, 33, 228
  cyclopropyl radical, **71**, 228
  cyclopropylvinyl cation, **71**, 97
  Diels–Alder reaction, **71**, 164, 166

Molecular-orbital calculations, on—*continued*
  enolate ions, **71**, 116
  electrophilic aromatic substitution, **70**, 225, 228, 231, 234; **71**, 195
  ferrocenylmethyl cation, **71**, 38
  fulminate–isocyanate isomerization, **70**, 293
  hydrocarbons, **71**, 48
  insertion reactions of carbenes, **71**, 376
  inversion barriers of cyclopropyl radicals, **71**, 278
  ionization of aryl carbinols, **70**, 56
  norbornadiene, norbornene and their cations, **70**, 27; **71**, 25
  norbornenyl and norbornyl radicals, **71**, 277
  oxaziridine, formation from *N*-oxides, **70**, 557
  protonated ethylene, **70**, 83, 84
  protonation of amides, **70**, 463
  reactivity of arylmethyl derivatives, **70**, 63
  rearrangement of cyclopropylcarbenes to allenes, **70**, 408
  rearrangement of isopropyl cation to protonated cyclopropane, **70**, 47
  ring closure of 3-*exo*-methylenebicyclo [3.3.1]non-7-yl cation, **70**, 21
  ring opening of cyclopropyl radical, **71**, 279
  solvolysis of,
    2-arylethyl toluene-*p*-sulphonates, **70**, 22
    cyclobutyl and cyclopropylmethyl derivatives, **70**, 35
  syn–anti-interconversion of benzylidene aniline, **70**, 439
  S_N2 transition states, **70**, 111–112; **71**, 103–105
  trimethylene, **71**, 226, 294
Munchnones, **71**, 169
Mutarotation of sugars, **70**, 431; **71**, 403–404

Naphthalenes,
  acylation of, **70**, 234; **71**, 205
  dilithio-, **70**, 125
  disodio-, **70**, 125
  halogenation, **70**, 231
  hydrogen exchange, **71**, 207
  iododeboronation, **71**, 204
  nitration, **70**, 228
  oxidation, **71**, 531
  photochemistry of, **70**, 547; **71**, 503
  sulphonation, **71**, 197
Naphthols,
  azo-coupling, **70**, 229, 230; **71**, 200
  hydrogen exchange, **71**, 207
  methylation of, **70**, 235
Narcissistic reactions, **70**, 257; **71**, 216
Neighbouring group participation by
  acetal group, **70**, 74–75
  allenic double bonds, **71**, 23
  amide groups, **70**, 78–80, 180, 301, 471, 472, 502; **71**, 69–70, 439

Subject Index 631

Neighbouring group participation by—*continued*
  amino groups, **70**, 80, 299, 470, 471; **71**, 70–72, 107, 158, 272, 437, 439, 440, 459
  aryl groups, **70**, 21–25; **71**, 16–20
  azulenyl group, **71**, 17
  boron, **70**, 81, 140
  carbanions, **70**, 81–83; **71**, 72, 105
  carbon, **70**, 83, 87; **71**, 28, 37, 46, 59, 74–79, 246
  carbon–metal bond, **71**, 73
  carbonyl group, **70**, 72, 77, 81, 82, 83, 469, 470; **71**, 61, 65–67, 435, 437, 441
  carboxyl group, **70**, 78, 458, 466–467, 494, 496; **71**, 69, 154, 431, 432, 433, 435, 457
  cyclobutyl group, **70**, 41–42
  cyclopropyl group, **70**, 7–8, 13, 27, 33–41, 57; **71**, 25–32
  double bonds, **70**, 25–32, 38, 40, 80, 185–186, 337–338; **71**, 2, 20–25, 28, 154
  enamine, **71**, 439
  enolate, **70**, 81, 82, 83; **71**, 66
  enol group, **71**, 23
  epoxide ring, **71**, 30–32
  ester group, **70**, 79, 81, 353; **71**, 67, 161
  ether group, **70**, 72–74, 174; **71**, 61
  guanidino group, **70**, 72–74, 174
  halogen, **70**, 76–77, 174, 180; **71**, 64
  hydrazone, **70**, 189
  hydrogen, **70**, 83–87; **71**, 28, 75
  hydroperoxyl group, **71**, 62
  hydroxyl group, **70**, 71, 73–74, 174, 184, 467–469, 495; **71**, 61–62, 150, 431, 433, 434, 435, 438, 439, 459
  imidazole group, **70**, 471, 500
  methoxyl, **71**, 62, 70
  nitrile, **71**, 161
  nitro group, **70**, 81, 299
  oxime group, **71**, 459
  phenyl groups, **70**, 21–25, 351–352; **71**, 57
  phosphinyl group, **71**, 72
  silyl group, **70**, 81; **71**, 72
  stannyl group, **70**, 81
  sulphonamido group, **71**, 267
  sulphoxide group, **70**, 81, 103, 177
  thioether group, **70**, 75–76, 107, 178; **71**, 62–64, 100, 245
  triple bond, **70**, 33; **71**, 25
  ureido group, **70**, 465–466
  urethano group, **70**, 79
  vinyl ether group, **71**, 22
Neighbouring group participation, in
  additions to acetylenes, **70**, 180; **71**, 162–163
  additions to olefins, **70**, 32, 174–175, 177, 178, 180, 184, 185–186, 188, 189; **71**, 69, 152, 154, 158, 161
  amide hydrolysis, **70**, 458, 465–472; **71**, 434, 439
  epoxide ring opening, **70**, 77, 78, 81, 82
  ester hydrolysis, **70**, 465–472; **71**, 431–441

Neighbouring group participation, in—*continued*
  hydrolysis of sulphonyl fluorides, **70**, 502
  metalation reactions, **70**, 138
  phosphate hydrolysis, **70**, 494–496; **71**, 457
  radical reactions, **70**, 320, 334–335, 337, 338, 351–352, 353; **71**, 299–300
  reactions of nitriles, **71**, 273
  reduction of sulphoxides, **70**, 603–604
  sulphate hydrolysis, **70**, 500
Nematic solvents, **71**, 150, 218
Neoclovene, **71**, 16
Neopentyl derivatives, **71**, 57, 106, 109
Nickel tetracarbonyl, reaction with bromonaphthalene, **70**, 222
NIH shift, **71**, 531
Nitramine rearrangement, **70**, 242; **71**, 212
Nitration, **70**, 225–229; **71**, 196, 198–200, 465
  effect of mixing, **70**, 227; **71**, 195
  of aliphatic hydrocarbons, **71**, 124–129
  of amines, **70**, 503
  of dinitroacetonitrile, **71**, 127
  of phenylene cation radical, **71**, 341
Nitrile oxides, **70**, 196, 197, 198, 293, 492; **71**, 169, 172, 454
Nitriles, hydrolysis, **70**, 491–492; **71**, 421, 427, 442, 454
  reaction with hydrogen peroxide, **71**, 456
Nitrenes, **70**, 392–415; **71**, 167, 367–391
  cycloadditions of, **70**, 404; **71**, 375
  EPR spectra, **71**, 454
  from photolysis of azides; **70**, 582; **71**, 371, 380, 516
  insertion of optically active, **70**, 400
  sulphonyl, **71**, 367, 387
  "stable", **70**, 392–393
Nitronium ions, **70**, 293; **71**, 201, 258
Nitroalkanes,
  addition to olefins, **71**, 161–162
  deprotonation of, **70**, 130
  equilibrium acidities, **70**, 128–129, 130
  halogenation, **71**, 116
  hydrogen exchange of, **70**, 134–135; **71**, 122
  pyrolysis of, **71**, 357
  radical anions, **71**, 337
Nitroalkenes,
  additions to, **70**, 189–190
  base-catalysed isomerization, **70**, 184, 291
Nitrobenzene, hydrogen exchange of, **70**, 136, 213
Nitro-compounds, reductive cyclization, **71**, 272
Nitrogen, nucleophilic displacement at, **70**, 104; **71**, 95
Nitrones, **70**, 198; **71**, 239, 258, 348–353
Nitronium ions, **70**, 142
Nitronyl nitroxides, **70**, 366
Nitrosamines, rearrangement, **70**, 243; **71**, 212
Nitrosation, **70**, 229; **71**, 200, 465
Nitrosoacetanilides, decomposition of, **70**, 220, 345; **71**, 190, 313

Nitroso-compounds, **70**, 503; **71**, 175, 295, 406
Nitrosonium ions, reaction with azides, **70**, 57
Nitrosyl chloride, **70**, 503
Nitroxides, **70**, 362–367, 515, 558, 588; **71**, 134, 348–353, 519
Nopinol, **71**, 236
Nopinone, pyrolysis, **70**, 275
Norandrostan-16β-ylamine, **70**, 43
Norandrostan-16β-yl toluene-p-sulphonate, **70**, 44
Norbornadienes,
  additions, **71**, 25, 156, 160
  conversion to quadricyclene, **70**, 207, 208; **71**, 495
  formation from quadricyclene, **70**, 275
  radical addition to, **70**, 342, 345; **71**, 307
  rearrangement, **71**, 224
Norbornadienyl cations, **70**, 32; **71**, 20
Norbornadienyl derivatives, **71**, 21
Norbornanone,
  hydrogen exchange of, **70**, 132, 441; **71**, 157, 410
  reaction with diazomethane, **71**, 78
  tosylhydrazone, **70**, 7
Norbornenes,
  cycloaddition, **71**, 169, 172
  electrophilic additions to, **71**, 8–10, 16, 124, 172, 175, 180, 183, 188; **72**, 3–4, 5, 152, 155, 157, 158, 160, 161
  perdeuteration, **71**, 157
  radical additions to, **70**, 315; **71**, 305–306
  5-vinyl, base-catalysed isomerization, **70**, 132
  tin derivatives, **71**, 131
Norbornenones,
  decarbonlation of, **70**, 165; **71**, 147
  hydrogen exchange of, **70**, 132
  reduction of, **71**, 340, 552
Norbornenyl derivatives, **70**, 26–28, 30, 64; **71**, 21, 235
Norbornenyl radicals, **70**, 308; **71**, 301–302
Norbornyl derivatives, **70**, 1–10, 20, 57, 60; **71**, 1–5
  amination of, **70**, 9
  arylene, **70**, 9; **71**, 2–4
  borane, bromination of, **71**, 130
  camphene-8-carboxylic acid, **70**, 10
  cyano-substituted, **70**, 4
  deamination of, **70**, 5–6; **71**, 5
  dimethyl substituted, **70**, 4; **71**, 1
  elimination reactions, **70**, 149–150, 160
  epoxides, **71**, 5
  equilibration of *exo-* and *endo-*, **70**, 2, 3; **71**, 5
  ferrocenyl, **70**, 49–50
  hydride shifts in, **70**, 2, 3
  hyperconjugative effects in, **70**, 3
  isotope effects in solvolysis of, **70**, 3
  7-keto, **70**, 9
  mass spectra, **70**, 10
  methoxycarbonyl-substituted, **71**, 2

Norbornyl derivatives—*continued*
  methoxyl substituted, **71**, 4, 6
  methyl substituted, **70**, 2; **71**, 23, 245
  2-norbornyl cations, **70**, 1; **71**, 1, 60, 155
    2-ferrocenyl, **70**, 49
    halogeno-substituted, **70**, 2
    heat of formation, **70**, 10
    NMR spectra of, **70**, 1–3; **71**, 1
    π-route to, **70**, 3–4
    Raman spectra of, **70**, 1; **71**, 1
    σ-delocalization energy of, **70**, 2
  7-norbornyl derivatives, **70**, 15, 20, 40, 42; **71**, 5, 25
  2-norbornyl oxocarbonium ions, **70**, 2
  nitrile oxides, **70**, 293; **71**, 454–455
  oxidation of, **70**, 8
  phenyl substituted, **70**, 2
  radical reactions of, **70**, 307, 314, 315, 598; **71**, 302, 334
  rearrangement of 1-bromo-7,7-dimethylnorbornanone, **71**, 248
  rearrangement of camphene, **71**, 248
  rearrangement of 1-hydroxy-3,3-dimethylnorbornanone, **71**, 250
  rearrangement of 2-*endo*-phenylisoborneol, **71**, 247
  1,2-shifts in, **70**, 288
  sulphur oxygen cleavage in reactions of norbornyl toluene-p-sulphonates, **71**, 92
  thiocyanates, **70**, 3
  valence force-field calculations for, **70**, 10
  X-ray diffraction of, **70**, 10
Norcaradienes, valence tautomerism, **70**, 272; **71**, 230
Norkawanyl derivatives, **70**, 16
Norphyllocladanyl derivatives, **70**, 16
Norrish type I process, **70**, 520, 521; **71**, 362, 478, 480, 481, 483, 487, 516
Norrish type II process, **70**, 517, 519; **71**, 295, 472, 478, 479, 487, 489, 491, 504
Nortricyclene protonated, **70**, 1
Nortricyclyl bromide, **71**, 152
Nucleophilic aliphatic substitution, **70**, 1–117; **71**, 1–109
  photoinduced, **71**, 525
Nucleophilic aromatic substitution, **70**, 211–223; **71**, 179–193
  base-catalysis in, **70**, 211, 212
  copper-catalysis in, **70**, 215, 222
  effect of micelles, **70**, 212; **71**, 188
  halide exchange, **71**, 179, 192
  Hammett equation, **70**, 211, 213
  in DMSO-water mixtures, **70**, 212, 214; **71**, 179–184
  intramolecular, **70**, 214, 220; **71**, 193
  of benzimidazoles, **70**, 217
  of benzofurazan, **71**, 182–183
  of benzothiazoles, **71**, 217
  of bromodinitrobenzene, **71**, 179
  of bromonitrobiphenyls, **70**, 213

Nucleophilic aromatic substitution—*continued*
of chloroacenaphthylene, **71**, 187
of chlorodinitrobenzene, **70**, 212; **71**, 179, 180, 181
of chlorodinitrophenylmethane, **70**, 215
of chloronitrobenzenes, **70**, 212
of 4-chloro-3-nitrobenzotrifluoride, **70**, 211
of chloroquinoline, **70**, 215
of diazonium compounds, **70**, 216
of dichloronitrobenzenes, **70**, 212
of difluoronitrobenzenes, **70**, 212; **71**, 180
of dihalogenobenzenes by cyclohexyl radicals, **70**, 305
of dinitroanisole, **70**, 213
of dinitrobenzenes, **70**, 213
of dinitrobenzenesulphonyl chlorides, **71**, 180
of diphenyl ethers, **70**, 211
of fluoro-2,4-dinitrobenzene, **70**, 212, 214; **71**, 185
of fluoro-2,6-dinitrobenzene, **71**, 180
of fluoronitrobenzene, **71**, 180
of 2-halo-3,5-dinitrobenzoic acids, **71**, 181
of 1-halo-2-naphthols, **70**, 213
of halonitrobenzenediazonium ions, **71**, 193
of 2-halotropones, **71**, 181
of imidazoles, **71**, 193
of iododinitrobenzene, **71**, 179, 180, 192
of nitroaniline, **70**, 214
of nitrophenyl trifluoromethylsulphones, **71**, 180
of phenanthridines, **71**, 182
of 9-phenoxyacridine, **71**, 223
of phosphonitrilic compounds, **71**, 184, 192
of picryl chloride, **70**, 214; **71**, 180
of polyhalogen-compounds, **70**, 215, 219–220; **71**, 188
of pyridines, **71**, 182
of pyrimidines, **70**, 216; **71**, 182, 193
of quinolines, **70**, 216
of quinoxalines, **71**, 182
of substituted phenyl trifluoromethylsulphones, **70**, 213
of thiazoles, **71**, 182
of thiophenes, **70**, 217; **71**, 186
of triazines, **70**, 217; **71**, 183
of triazolium salt, **71**, 172
of trinitroanisole, **71**, 179
of xanthines, **70**, 217
radical anions in, **70**, 214
photoinduced, **70**, 573; **71**, 525
solvent effects, **70**, 212; **71**, 179–180
steric effects in, **70**, 213
Nucleophilicity, **70**, 112; **71**, 105, 147, 179
Nucleosides, hydrolysis of, **70**, 435–437; **71**, 407

Olefinic alcohols, **71**, 48
Olefins,
acid-catalysed isomerization, **71**, 156, 263

Olefins—*continued*
base-catalysed isomerization, **70**, 131–132, 296–297; **71**, 263
carbonylation of, **71**, 45
cis–trans-isomerization, **70**, 294; **71**, 262, 359, 500
hydration, **70**, 180; **71**, 156, 408
isomerization, **71**, 47
metal-catalysed isomerization, **71**, 262–264
oxidation of cyclic, **70**, 305
ozonolysis of, **70**, 379, 575–578; **71**, 527
photochemistry of, **70**, 535-546; **71**, 493–501
protonation, **71**, 41
silver complexes, **71**, 161
thallium induced rearrangements, **71**, 244
Orbital orientation, **70**, 58
Orbital symmetry in photochemical reactions, **71**, 468
Orbital symmetry in transition-metal catalysed reactions **70**, 208
Orbital steering, **70**, 468; **71**, 441
Ortho effect, **70**, 115; **71**, 143, 145, 179
Ortho esters
hydrogen exchange of, **70**, 486
hydrolysis, **70**, 79, 485–486; **71**, 393, 394
reaction with Grignard reagents, **70**, 64
Orton rearrangement, **70**, 244
Osazones, **70**, 437, 439
6-Oxabicyclo[3.2.1]oct-1-ylmethyl *p*-bromobenzenesulphonate, **71**, 12
Oxathiazolones, decomposition, **70**, 198
Oxathiolans, ring opening, **71**, 395
Oxazines, aminolysis and hydrolysis, **70**, 469
Oxaziridine, 2-t-butyl, **71**, 43
Oxazolines, rearrangement to pyrroles, **70**, 267
Oxazolines, 2-benz-, hydrolysis, **70**, 492
Oxepin, 1,2-dihydro, **70**, 282
Oxetans, deoxygenation, **70**, 330
Oxibase scale, **71**, 94, 97, 163
Oxidation, anodic, **70**, 357, 376, 594–595; **71**, 83, 100, 342, 344, 365, 545–546
Oxidation, by
amine oxides, **71**, 545
ascorbic acid, **71**, 336
bromate, **71**, 544
bromine, **70**, 592; **71**, 543
N-bromosuccinimide, **70**, 592
t-butyl hypochlorite, **70**, 592
cerium(IV), **71**, 331, 332, 333, 336, 341, 343, 533–534
chlorites, **70**, 592
m-chloroperbenzoic acid, **71**, 488, 541
N-chlorosuccinimide, **70**, 592; **71**, 244
chromium(VI), **70**, 8, 361, 578–579; **71**, 331, 529–532
chromyl acetate, **71**, 531
chromyl chloride, **70**, 579
cobalt(III), **70**, 360, 361, 582
copper(II), **70**, 583; **71**, 333, 536–537
DDQ, **71**, 253, 545

Oxidation by—*continued*
- diethyl azodicarboxylate, **70**, 405
- dimethyl sulphoxide, **70**, 594; **71**, 544
- dinitrogen tetroxide, **70**, 593
- Fenton's reagent, **71**, 336
- ferricenium cations, **70**, 584
- ferricyanide, **70**, 362, 582, 583; **71**, 336, 536
- Fremy's radical, **71**, 545
- fuming sulphuric acid, **71**, 15
- hexachloroantimonate, **70**, 584
- hydrogen peroxide, **71**, 542
- hypohalite, **71**, 544
- iodate, **70**, 592
- iodine, **70**, 593
- iodosobenzene diacetate, **70**, 592
- iridium(IV), **70**, 359
- iron(III), **71**, 336
- lead oxide, **70**, 584; **71**, 336
- lead tetra-acetate, **70**, 356–357, 409, 584, 585; **71**, 4, 47, 222, 276, 331, 349, 535, 536, 544
- manganese(III), **70**, 358, 361, 581; **71**, 333, 533
- manganese dioxide, **71**, 532
- mercury(II), **70**, 584; **71**, 534
- molybdenum(VI), **70**, 584
- nickel peroxide, **71**, 336, 542
- nitric acid, **70**, 593; **71**, 545
- nitrosodisulphonate, **71**, 276
- osmium(VIII), **70**, 584
- oxygen, **70**, 377–380, 586–589; **71**, 353–355, 537–541
- ozone-triphenyl phosphite, **71**, 522
- palladium(II), **70**, 584; **71**, 534
- per-acids, **70**, 589–591; **71**, 541–543
- periodate, **70**, 278; **71**, 543
- permanganate, **70**, 359, 386, 579; **71**, 529–530, 532
- peroxydiphosphate, **71**, 349
- peroxydisulphate, **70**, 358; **71**, 347, 348, 349, 542, 543
- potassium bromate, **70**, 591
- quinone-di-imines, **71**, 545
- quinones, **70**, 593
- selenium dioxide, **70**, 593; **71**, 544
- silver(I), **70**, 583, 584; **71**, 349
- silver(II), **70**, 584
- sulphuric acid, **71**, 545
- sulphur trioxide, **71**, 543
- thallium(III), **70**, 584–586; **71**, 534
- titanium(III), **71**, 335
- trityl fluoroborate, **71**, 545
- uranyl nitrate, **71**, 349
- vanadium(V), **70**, 582
- xenon fluorides, **71**, 341

Oxidation, enzymic, **70**, 588–589; **71**, 540–541

Oxidation of
- acetylenedicarboxylic acid, **70**, 580
- acetylenes, **70**, 590
- adamantane, **71**, 15

Oxidation of—*continued*
- alcohols, **70**, 356, 578, 582, 584, 591–593; **71**, 331, 347, 348, 530, 532, 533, 534, 542, 543, 544, 545
- aldehydes, **70**, 361, 579, 582, 584, 590, 592, 593; **71**, 365, 531, 532, 543
- aliphatic amines, **70**, 582, 594
- alkylidenecyclopropanes, **70**, 585
- allyl benzene, **70**, 584
- amino acids, **70**, 584
- N-aminophthalimide, **70**, 409
- aminoazobenzenes, **70**, 582
- 1-aminodimethoxybenzotriazoles, **70**, 221
- 1-amino-2,3-diphenylaziridine, **70**, 581
- aminolactams, **71**, 544
- anisole, **71**, 535
- aromatic amines, **70**, 359, 580, 584, 593, 595; **71**, 341, 542, 543, 544, 545
- aromatic hydrocarbons, **71**, 536
- aryldiazomethanes, **71**, 336
- aryl methanes, **70**, 593
- aryl propenes, **71**, 253
- ascorbic acid, **70**, 584
- azines, **71**, 543, 546
- azo-compounds, **70**, 589
- 2-benzamido-2-(benzylthio)-propanoic ester **71**, 244
- benzenehexol, **71**, 532
- carbanions, **70**, 386
- carbohydrates, **70**, 583; **71**, 531
- carboxylic acids, **70**, 358, 581, 583, 595; **71**, 336
- cycloheptatriene, **70**, 581
- cyclohexane, **70**, 361
- cyclopropanes, **71**, 546
- cyclopropenes, **71**, 541
- diarylphosphine oxides, **70**, 591
- diketones, **71**, 536
- dimethylnitrosobenzene, **70**, 593
- dienes, **70**, 580; **71**, 534
- dimedone, **71**, 525
- dimethoxybenzenes, **70**, 585
- diols, **71**, 336, 532, 542, 543, 546
- disulphides, **71**, 522
- enamines, **71**, 545
- epoxides, **71**, 545
- ethers, **71**, 546
- formazans, **71**, 336
- fumaric acid, **71**, 532
- furfural, **70**, 580
- glycols, *see* diols
- hydrazines, **70**, 405
- hydrazones, **70**, 584, 593, 594; **71**, 222, 543
- hydrocarbons, **71**, 531, 532, 540
- hydroxamic acid, **71**, 536
- α-hydroxy acids, **70**, 579, 581; **71**, 531, 532, 543
- α-hydroxyketones, **70**, 584; **71**, 336
- hydroxylamines, **71**, 335
- hydroxyphenyl ethers, **71**, 536

Oxidation of—*continued*
  imines, **70**, 589
  indoles, **71**, 536
  ketones, **70**, 359, 584, 590; **71**, 521, 532, 541, 546
  lactose, **70**, 580
  maltose, **70**, 580
  mandelic acid, **70**, 580
  N-methylbiphenyl-2-carboxyamides, **71**, 347
  1-methylpiperidine-2,3-dione, **70**, 278
  naphthalene, **71**, 531
  olefins, **70**, 305, 579, 584, 593; **71**, 47, 333, 524, 532, 544
  oxalic acid, **70**, 579, 580
  oximes, **70**, 585
  phenols, **70**, 536, 584, 593; **71**, 344, 536, 542, 543, 545
  phenylacetate ion, **71**, 344
  phenylnitromethane, **70**, 580
  γ-phenylvaleric acid, **70**, 579
  phosphines, **70**, 593; **71**, 543
  phosphorous acids, **70**, 579
  phthalhydrazides, **71**, 524
  propenyl lithium, **71**, 334
  propylmercury chloride, **71**, 47
  purpurogallin, **71**, 525
  quinols, **70**, 582
  semicarbazides, **71**, 577
  semicarbazones, **71**, 543
  sodium toluene-p-sulphonate, **71**, 531
  spiro[4.2]heptane, **70**, 585
  spiro[5.2]octane, **70**, 585
  sulphides, **70**, 591, 592; **71**, 532, 546
  sulphonamides, **71**, 387
  sulphoxides, **70**, 591
  tertiary amines, **71**, 536
  2-thiabicyclo[2.2.1]heptane, **70**, 592
  thiols, **70**, 361, 584; **71**, 536, 543, 544
  thiophosphinate, **71**, 87
  toluenes, **70**, 581, 591; **71**, 333
  toluene-p-sulphonic acid, **70**, 580
  triazenes, **71**, 349
  tri-t-butylcyclopropene, **71**, 48
  triethylphosphite, **71**, 328
  1,1,3-triphenylindene, **70**, 595
  unsaturated alcohols, **70**, 591
  uracil, **71**, 542
  xanthopterin peroxide, **71**, 542
Oxides,
  photolysis of N-oxides, **70**, 556–557; **71**, 512
  reaction of N-oxides with acetic anhydride, **70**, 310
  rearrangement of N-oxides, **70**, 245–246, 255, 276, 355–356; **71**, 213, 227, 236
Oximes,
  formation of, **70**, 438; **71**, 407
  hydrolysis of, **70**, 437
  photolysis of, **70**, 555–556, **71**, 518
  syn–anti-interconversion, **70**, 439; **71**, 261

Oxocanyl derivatives, **70**, 28
Oxocinyl derivatives, **70**, 28
Oxonin, cycloaddition, **70**, 206
Oxygen, nucleophilic displacement from, **70**, 104; **71**, 95
Oxymercuration, **70**, 172, 183–184; **71**, 158, 160
Ozonolysis of
  aldehydes, **71**, 529
  amines, **71**, 529
  diphenylketene, **70**, 78
  disulphides, **71**, 529
  ethers, **70**, 566
  olefins, **70**, 379, 575–578; **71**, 168, 527–529
  organometallics, **71**, 529

Palladium acetate, **70**, 235; **71**, 210
Palladium chloride–copper chloride, **71**, 4, 11–12
Palladium complexes, **71**, 43
Palladium vinyl, **71**, 133
Paracyclophanyl derivatives, **71**, 19
Penicillenic acid, **71**, 162
Pentacyclo[5.3.0^{2,5}0^{3,9}0^{4,8}]deca-6,10-dione, octachloro, **70**, 20
Pentacyclo[4.4.0.0^{2,4}0^{3,8}0^{5,7}]dec-9-yl toluene-p-sulphonate, **71**, 29
Pentacyclo[4.3.0.0^{2,4}0^{3,8}0^{5,7})non-9-yl p-nitrobenzoate, **70**, 40
Peptides,
  hydrogen exchange of, **70**, 133
  hydrolysis of, **70**, 471
  protonation of, **70**, 463
  racemization of, **71**, 123
Peresters, decomposition of, **70**, 324; **71**, 282–283
Perinaphthane, **71**, 207
Perispecificity, **70**, 206
Perkow reaction, **71**, 103
Peroxides,
  decomposition of, **70**, 322–327; **71**, 281–287, 314, 318, 365, 524
  effect of sulphonium salts, **70**, 323
  in benzyl ethers, **70**, 323
  oxygen scrambling of, **70**, 326
  photolysis of, **71**, 318
  radicals, **71**, 285
Peroxybenzoate decomposition, **70**, 489
Peroxybenzoic acid, hydrolysis, **71**, 427
Phenanthrene, **71**, 205, 207
Phenanthrenonium ions, **71**, 209
Phenolenone, hydrogen exchange, **70**, 236–237
Phenolphthalein, monopositive ion, **70**, 55
Phenolsulphonphthalein, monopositive ion, **70**, 55
Phenols,
  acylation of, **70**, 234
  bromination of, **70**, 230, 232

Phenols—continued
 chlorination, 71, 203
 hydrogen exchange, 71, 208
 iodination of, 70, 230
 nitrosation, 71, 200
 oxidative coupling, 70, 361; 71, 341, 536
 oxidation, 70, 357; 71, 344
Phenonium, ions, 70, 21, 25
Phenylacetylene, halogenation of, 70, 126
Phenyl cations, 70, 234
Phenyl groups,
 migration, 70, 21–25, 286, 289; 71, 16–20, 154, 244, 245, 248
 1,3-migration, 70, 407
α-Phenethyl derivatives, nucleophilic substitution reactions of, 70, 64; 71, 51
Phenylenediamine, sulphonation, 71, 198
5-Phenyl-1,3,4-oxathiazol-2-one, nitration, 70, 228
N-Phenyl-5-N-phenylaminopenta-2,4-dienylidimine, 70, 271
4-Phenyl-1,2,4-triazoline-3,5-dione, 70, 206
Phosphates, hydrolysis, 70, 492–500; 71, 457–463
 photochemically induced, 70, 496, 573
Phosphinamides, protonation, 71, 461
Phosphinates, 71, 460, 461
Phosphinyl chlorides, 70, 497; 71, 461
Phosphites, reaction with diketones, 70, 450–451; 71, 415
Phosphoenolpyruvate, 71, 458
Phosphoguanidine, 71, 460
Phosphole-1-oxides, 71, 88
Phospholene-1-oxides, 71, 88
Phosphonates, thio, 70, 494; 71, 88, 461
Phosphonium-cyclopentadiene ylid, 71, 197
Phosphonium salts,
 alkaline hydrolysis, 70, 98; 71, 87
 nitration, 70, 229
Phosphoramidates, 71, 89, 460
Phosphoramide, 70, 496
Phosphorane, 70, 496
Phosphorescence, 71, 469, 487
Phosphothioate, 70, 496
Phosphorus, nucleophilic displacement at, 70, 94–100, 492–500; 71, 85–89, 457–463
 steric course of radical substitution, 71, 281
Phosphorus compounds, trivalent, addition to dienes, 70, 207
Photoadditions,
 cycloadditions of olefins, 70, 536–539; 71, 494–499
 intramolecular, 70, 526; 71, 482, 485, 495
 of α-acetoxyacrylonitrile to dienes, 70, 203
 of aromatic hydrocarbons, 70, 547; 71, 501–503
 of carbonyl compounds to olefins, 70, 521; 71, 481–485
Photochromism, 70, 574; 71, 525
Photo-Claisen rearrangement, 70, 549

Photocycloadditions to acetylenes, 70, 546; 71, 171, 483
 to olefins, 70, 536–539, 564–565; 71, 481–482, 493, 494–497
Photocyclization of β-alkoxyketones, 70, 338
Photodecarbonylations, 70, 520, 522, 525; 71, 481, 489, 491
Photodecarboxylation, 70, 533; 71, 475, 476, 481, 491, 506, 515
Photodimerization to thymine, 70, 554
Photoenolization, 70, 516; 71, 487, 490
Photo-esterification of cinnamic acid, 70, 534
Photo-Fries rearrangement, 70, 532; 71, 490
Photo-induced electron ejection reactions, 70, 515; 71, 474–475
Photolysis of
 acetylenes, 70, 546; 71, 500–501
 N-acetyldiphenylamine, 70, 244
 aldehydes, 70, 531; 71, 468, 489–490
 aldehydo-D-glucose pentaacetate, 71, 360
 aluminium trianyls, 71, 518
 amides, 71, 492
 amines, 70, 556; 71, 511
 anhydrides, 71, 492
 aromatic hydrocarbons, 70, 546; 71, 501–504
 aryloxirans, 70, 397
 arylthallium bistrifluoroacetates, 70, 348
 azides, 70, 397–398, 562; 71, 380, 516
 azine monoxide, 70, 557
 azines, 71, 508
 aziridines, 71, 492, 512
 azirines, 71, 511
 azo-compounds, 70, 560–562; 71, 291, 515
 azoxy compounds, 70, 556
 benzofurazons, 71, 372
 borates, 71, 519
 camphor nitrimine, 71, 356
 carbanions, 71, 518
 carbonates, 70, 396
 carbonium ions, 70, 563–564; 71, 518
 carboxylic acids, 70, 363, 532–535; 71, 490–493
 N-chloro-N-propylpent-4-enylamine, 70, 340
 cholest-4-en-3-ol, 70, 536
 cholesterol, 70, 536
 citric acid, 71, 362
 p-cyanophenol, 70, 573
 cyclobutanes, 71, 495–496
 cycloheptadienyl anion, 70, 123
 cycloheptatriene esters, 70, 402
 diazo-compounds, 70, 395, 405, 562; 71, 515, 516
 dibenzoyl peroxides, 70, 564
 dienes, 70, 536; 71, 494–495
 dihydrofurans, 71, 507
 dihydropyrazines, 71, 508, 511
 dihydroquinolizines, 71, 508
 dihydrothiophenes, 71, 507
 diketones, 70, 528–531; 71, 487–489

Subject Index

Photolysis of—*continued*
  dimethylenecyclobutane, **70**, 202
  diphenylpropene, **71**, 295
  enamides, **70**, 535
  enol acetates, **71**, 521
  enones, **70**, 522–528; **71**, 172, 482, 483, 487
  epoxides, **70**, 549
  epoxyesters, **71**, 492
  epoxyketones, **71**, 483
  esters, **70**, 534–535; **71**, 490–493
  ethers, **70**, 548
  ethylene glycol, **71**, 359
  eucarvone, **71**, 48, 518
  furans, **71**, 507, 508
  halogeno-compounds, **70**, 562–563; **71**, 337, 338, 516–517
  heterocyclic compounds, **70**, 550–554; **71**, 504–508
  hexaphenylditin, **70**, 382
  hydrazobenzene, **71**, 511
  imidazoles, **71**, 508
  indoles, **71**, 511
  isoxazoles, **71**, 508
  ketenes, **70**, 522
  keto-acids and esters, **70**, 531–532
  ketones, **70**, 381, 515–532; **71**, 468, 475–489, 516, 519
  limonene, **70**, 535
  mercurials, **71**, 315
  2-methylbenzophenone, **70**, 506
  naphthotriazines, **71**, 331
  nitriles, **71**, 517
  nitrite ion, **70**, 387
  nitrites, **71**, 349
  nitro-compounds, **70**, 558, 574; **71**, 514
  nitrones, **70**, 556; **71**, 512
  nitroso-compounds, **70**, 559–560; **71**, 349–350, 513
  olefins, **70**, 535–546; **71**, 493–501, 524
  oxadiazolones, **70**, 398
  oxaziridines, **70**, 557
  N-oxides, **70**, 556–557; **71**, 512
  oximes, **70**, 556; **71**, 518
  peroxides, **70**, 308, 564; **71**, 283, 285, 286, 468, 519
  peroxyanhydrides, **71**, 492
  phenothiazines, **71**, 511
  phosphines, **71**, 519
  phosphorus ylids, **70**, 565
  pyrans, **71**, 508
  pyrazolines, **70**, 329; **71**, 515
  pyridazines, **71**, 508
  pyrimidines, **71**, 508
  quinolines, **71**, 504
  quinones, **70**, 528–531; **71**, 291, 468, 489
  quinoxalines, **71**, 504
  Schiff bases, **70**, 555
  sulphides, **70**, 549
  sulphonamides, **71**, 519
  sulphonium salts, **70**, 550

Photolysis of—*continued*
  sulphoxides, **70**, 550
  tetraphenylmethane, **70**, 397
  tetrazenes, **70**, 318
  thiadiazoles, **71**, 508
  thiazepines, **71**, 511
  thiazoles, **71**, 508
  thiocarbamates, **71**, 490
  thioesters, **71**, 490
  thioketones, **70**, 565; **71**, 490
  trienes, **70**, 537–539; **71**, 495–496
  trimethylstannylethyl methyl ketone, **70**, 565
  triphenylmethane dyes, **71**, 511
  tropylium ion, **71**, 517
  unsaturated esters, **71**, 493
  ylids, **70**, 398; **71**, 508
Photo-oxidation of
  alcohols, **70**, 374; **71**, 336
  allylthiourea, **71**, 522
  amines, **70**, 570; **71**, 522–523
  aminoacids, **71**, 523, 525
  anthracenes, **71**, 522
  benzenes, **71**, 501
  benzilic acid, **71**, 336
  carboxylic acids, **71**, 524
  β-carotene, **71**, 521
  dienes, **71**, 522
  ferrocene, **71**, 520
  furans, **71**, 522
  hydrazones, **70**, 571
  hydrocarbons, **71**, 520
  imines, **70**, 571
  β-ionol, **71**, 521
  methanol, **71**, 349
  methionine, **70**, 571
  2-methylbenzophenone, **70**, 506
  nucleotides, **71**, 524
  olefins, **70**, 567–569; **71**, 520–522
  ψ-pelletierine, **70**, 570
  phenols, **71**, 520
  polymethoxybenzenes, **71**, 520
  porphyrins, **71**, 524
  purines, **71**, 524
  pyrazolines, **71**, 524
  pyrimidines, **71**, 524
  pyrroles, **71**, 523
  quinones, **71**, 521
  sulphides, **71**, 522
  sulphoxides, **71**, 522
  thiophenes, **70**, 570
  thiopyran-4(4H)-thiones, **71**, 522
  triphenylthiazole, **70**, 570
  tropinones, **70**, 570
  N-vinylcarbazole, **70**, 571
Photo-oximation, **70**, 564; **71**, 302, 513
Photorearrangement of
  acetylenes, **71**, 500
  alkenes, **71**, 499–500
  aromatic hydrocarbons, **70**, 546; **71**, 501

Photorearrangement of—*continued*
  cycloaekanone-oximes, **71**, 518
  cyclopentadiene, **71**, 518
  N,N-dimethylphenylethynylamine, **70**, 356
  heterocyclic compounds, **70**, 550–555; **71**, 507–508
  hexahydroindanones, **71**, 479–480
  hydrazobenzene, **71**, 511
  hydroxylbenzenonium ions, **70**, 56
  nitrones, **71**, 512
  N-oxides, **71**, 512
  protonated durene, **70**, 56
  triphenylmethyl cation, **71**, 518
  tropylium ion, **71**, 517
Photoreduction of
  aromatic hydrocarbons, **71**, 503
  acridines, **71**, 506
  benzophenazines, **71**, 500
  carbonyl compounds, **70**, 363, 515–519; **71**, 298, 301, 476
    intramolecular, **70**, 516
  enones, **70**, 523
  flavines, **71**, 500
  hydrazones, **70**, 555
  2-hydroxybenzophenone, **70**, 516
  isoquinoline, **71**, 504
  nitro-compounds, **70**, 558; **71**, 514
  olefins, **70**, 546
  oximes, **70**, 555
  phenazines, **70**, 550; **71**, 506
  porphyrins, **71**, 506
  pyridazines, **70**, 550
  pyridines, **70**, 550; **71**, 504
  pyrimidines, **71**, 504
  quinolines, **70**, 550; **71**, 504
  Schiff bases, **70**, 555
  trifluoroacetophenone, **70**, 514
  xanthine dyes, **71**, 506
Photo-Smiles rearrangement, **70**, 549
Picrates, alkyl, solvolysis, **71**, 43
Pinacol rearrangement, **70**, 25, 84, 283–284, 286; **71**, 15
Pinacols, thermal dissociation of, **70**, 315; **71**, 298
Pinacolyl arenesulphonates, **70**, 60; **71**, 46
Pinane, **71**, 246
α-Pinene, additions to, **70**, 61, 179; **71**, 10, 156, 248, 306
β-Pinene, additions to, **70**, 183, 345; **71**, 10, 156, 248, 306
Pinocarveol, **71**, 236
Piperazines, hydrogen exchange of, **70**, 136
Platinum,
  insertion in cyclopropanes, **71**, 126
  nucleophilic displacement at, **71**, 95
Polyalkylbenzenes,
  Friedel–Crafts reaction, **71**, 205
  halogenation, **70**, 230; **71**, 203
  nitration, **70**, 228–229
  sulphonation, **71**, 197

Polymers, catalysis by, **70**, 474; **71**, 442
Porphyrins, **71**, 208
Potassium methyl, structure, **70**, 137
Pressure, effect on reaction rate, **70**, 39, 66, 116, 423, 462, 468; **71**, 57, 108, 212, 284, 463
Principle of Least Motion, **70**, 84, 96, 156, 161, 268, 410, 442
Principle of Microscopic Reversibility, **71**, 124
Prins reaction, **70**, 181; **71**, 157
Propynyl phenyl sulphide, **70**, 250
Protoadamantanone, **71**, 250
Protoadamantyl derivatives, **70**, 20, 30; **71** 14, 157, 245, 246, 248
Protodearsonation, **70**, 233
Protodegermanation, **70**, 140
Protodemercuration of
  aromatic compounds, **70**, 233; **71**, 204
  benzylchloromercury, **71**, 131
  mercury dialkyls, **71**, 131
  vinyl mercuric halides, **71**, 142
Protodemetalation of pyridiomethyl derivatives, **70**, 142
Protodephosphonation, **70**, 142
Protodesilylation, **70**, 140, 232; **71**, 195, 196
Protodesulphation, **71**, 198
Protodestannylation, **70**, 140, 232; **71**, 204
Pseudorotation of
  arsenic, **71**, 87
  phosphorus, **70**, 94–98, 494; **71**, 85–87
  sulphur, **70**, 103
Pulse radiolysis, **71**, 112
Pummerer rearrangement, **70**, 227; **71**, 117, 241
Purines, hydrogen exchange, **71**, 125
Pyrazine, **71**, 205
Pyrazole,
  halogen, **71**, 202
  hydrogen exchange of, **70**, 136, 237
  nitration of, **70**, 226, 228; **71**, 199
  ring contraction, **71**, 271
Pyrazolines, azo-coupling of, **70**, 230
Pyrazolones, iodo, iodine exchange, **70**, 232
Pyridazines,
  acylation, **71**, 205
  halogeno, reaction will sodium in liquid ammonia, **70**, 239
  photoreduction of, **70**, 550
  pyrolysis of tetrafluoro, **71**, 271
Pyridazinones, ring contraction, **71**, 271
Pyridines,
  dehydrogenation of dihydro, **70**, 154
  Friedel–Crafts reaction, **71**, 205
  halogenation of, **70**, 231, 232
  nitration, **70**, 229; **71**, 199
  photo-rearrangement, **70**, 551
  photo-reduction, **70**, 550
  radical alkylation of, **70**, 350; **71**, 316
  radical phenylation of, **70**, 351; **71**, 317
Pyridine-*N*-oxides,
  hydrogen exchange of, **70**, 136, 237

Pyridine-N-oxides—*continued*
 radical phenylation, **71**, 317
 rearrangement of, **70**, 245–246
4-Pyridiomethyl organometallics, demetalation, **71**, 141
Pyridinium ions,
 additions to, **70**, 190, 197
 hydrogen exchange, **71**, 124
 reaction with bisulphite, **70**, 239
Pyridinones, nitration, **70**, 229
Pyrimidines,
 halogeno, reaction with sodium in liquid ammonia, **70**, 239
 nucleophilic substitution in, **70**, 216
 rearrangement of alkoxy, **70**, 246
 ring opening of, **71**, 271
Pyrimidones, **71**, 199
Pyrocarbonates, **71**, 455
Pyrolyses, gas-phase, **70**, 383; **71**, 353
Pyrolysis of
 acetaldehyde, **70**, 383
 acetoacetates, **71**, 359
 acetylene, **71**, 359
 5-azidopyrazoles, **70**, 393
 benzohydroxamoyl chloride, **71**, 357
 benzoic acid, **71**, 359
 camphor nitrimine, **71**, 356
 diallyl, **71**, 355
 diallyl oxalate, **71**, 356
 diazines, **70**, 396; **71**, 356
 dibenzyl oxalates, **71**, 356
 dicinnamyl, **71**, 355
 dicrotyl, **71**, 355
 2,3-epoxybutanes, **71**, 359
 hydrazines, **71**, 358
 nitroalkanes, **71**, 357
 olefins, **71**, 359
 phenyl-p-benzoquinone, **71**, 357
 pinocarveol, **71**, 358
 1,3,4-thiadiazolines, **70**, 196
 trimethylsilane, **71**, 359
Pyrones, electrocyclic ring-opening, **70**, 270
Pyrophosphates, **70**, 498; **71**, 461
Pyrroles, **71**, 197, 205, 207, 209
Pyrylium salts, **71**, 193
π-complexes in,
 addition reactions, **70**, 188; **71**, 162
 aromatic metalations, **70**, 232
 aromatic nitration, **70**, 225; **71**, 195
 Friedel–Crafts reactions, **70**, 233; **71**, 206
 ozonolysis, **71**, 527
 rearrangements, **70**, 242
 solvolysis of 2-arylethyl toluene-p-sulphonates, **70**, 22

Quadricyclenes,
 additions to, **71**, 156
 cleavage of, **71**, 126
 interconversion with norbornadiene, **70**, 207, 208, 275; **71**, 236–238

Quadricyclyl cations, **71**, 21
Quaternization reactions, **70**, 68, 115, 116–117; **71**, 107, 108
Quinolines,
 halogenation, **71**, 201
 hydrogen exchange, **71**, 207
 nitration, **71**, 199
 photochemistry of, **71**, 489
 photoreductions of, **70**, 550
 radical substitution, **70**, 350; **71**, 316
 rearrangement into isoquinoline, **70**, 246
Quinolizines, rearrangement, **70**, 262
Quinones, dehydrogenation by, **70**, 157

Radical anions, **70**, 367–373; **71**, 317, 323, 336–340, 424, 524, 548
 as reaction intermediates, **71**, 116
 ESR spectra, **70**, 317; **71**, 344
 in apparent S_N2 reactions, **70**, 117; **71**, 337
 in apparent S_N2-Ar reactions, **70**, 214; **71**, 337
 in reductions, **70**, 604; **71**, 339
 ion pairs of, **70**, 119, 367; **71**, 339, 340
 semidones, **71**, 340
Radical cations, **70**, 373–375; **71**, 341–348, 471, 511, 524, 546
 ESR spectra, **71**, 341
 in chlorinations by N-halogeno-amines, **70**, 335
 in oxidation of di-n-butylamine, **70**, 594
 in oxidation of olefins, **71**, 47
 in oxidation of toluenes, **70**, 581
 intramolecular association, **71**, 341
 paraquat, **71**, 341, 346–347
 semidones, **70**, 317; **71**, 340
 sigmatropic group migration in **70**, 352
 thianthrenium, **71**, 341
Radical reactions, **70**, 305–389; **71**, 275–365
 acetoxymigration, **71**, 324
 addition of carbon tetrachloride to dibenzobicyclo[2.2.2]octane, **71**, 322
 addition of diethyl azodicarboxylate, **71**, 305
 addition of organoboranes to α, β-unsaturated carbonyl compounds, **70**, 312–313
 addition of thiophenol to 3-methylenenortricyclene, **71**, 305
 alkyl–alkoxy, **70**, 309
 alkyl–alkyl, **70**, 309
 aromatic substitution, **70**, 345–351; **71**, 313–320
 aromatization of methylenecyclohexadiene, **71**, 323
 aryl migrations, **70**, 351–352
 bromination, **70**, 306, 331–338; **71**, 302
 chlorinations, **70**, 306, 332–338; **71**, 276
 cyclization of alkenyl radicals, **70**, 340, 380; **71**, 276, 310–313, 349, 353
 cyclization of cinnamyl radical, **71**, 355
 cyclization of germacrene, **71**, 360

Radical reactions—continued
  decomposition of acylarylnitrosamines, **70**, 220, 345; **71**, 313
  aryldiimines, **70**, 419
  azosulphones, **70**, 381; **71**, 289
  disulphones, **70**, 380
  dipole effect in, **70**, 333
  gas-phase radical abstraction, **70**, 339
  hydrogen abstraction from non-aromatic polycyclic hydrocarbons, **71**, 298
  hydroxylation of aromatic compounds, **71**, 270
  intermediates in, **70**, 306
  iodinations, **71**, 277
  isomerization of 2-bromo-3,3,3-trichloropropene, **71**, 303
  Martynoff rearrangement of nitrones to oxime ethers, **71**, 239
  of acetone with bistrimethylsilylmercury, **70**, 386
  of benzhydryl ethers, **70**, 334
  of benzyl ethers, **70**, 334; **71**, 286
  of carbon tetrachloride with aldehydes, **71**, 365
  of diazenes, **71**, 290
  of hydrogen atoms, **71**, 296
  of hyponitrites, **71**, 289
  of lithium in tetrahydrofuran with 2-(2-pyridyl)ethyl chloride, **71**, 240
  of γ-methylperoxyvalero-γ-lactone, **71**, 295
  of β-nitroalkyl peroxynitrate, **71**, 281
  of sodium naphthalene with alkyl halides, **70**, 309; **71**, 338
  of sulphur ylids, **70**, 127, 276, 277
  of tritium with aromatic compounds, **71**, 318
  oxidations by manganese(III), **71**, 533
  pyrolysis of anhydrides, **71**, 358
  pyrolysis of azines, **71**, 350
  pyrolysis of benzohydroxamoyl chloride, **71**, 357
  pyrolysis of hydrazines, **71**, 358
  pyrolysis of nitro-compounds, **71**, 357
  pyrolysis of phenyl-p-benzoquinone, **71**, 357
  pyrolysis of pinocarveol, **71**, 358
  rearrangement of amine oxides, **70**, 276
  reversibility of addition of benzyloxy radicals to benzene, **71**, 318
  solvent viscosity in, **70**, 322
  stereoselectivity of, **70**, 317; **71**, 308
  structure and reactivity, **70**, 306
  study by ESR spectroscopy, **70**, 306, 360, 378, 381, 386, 388; **71**, 301, 324, 328, 329, 332, 336, 340, 345, 346, 362–363
  tin hydride reactions, **70**, 312, 337, 344; **71**, 278, 279, 297
Radicals,
  acyl, **70**, 529
  adamantyl, **71**, 357
  alkyl, **70**, 315–316, 327; **71**, 303, 317
  alkyl diazonium, **71**, 287

Radicals—continued
  allenyl, **70**, 307
  amino, **70**, 318, 389; **71**, 361, 362
  aminoalkyl, **70**, 317, 319
  aryl, **70**, 345, 348, 363, 563; **71**, 282, 314–315, 516, 517
  aryloxy, **71**, 536
  benhydryl, **70**, 309
  benzyl, **70**, 309; **71**, 344
  bridged bromo-, **70**, 334–335; **71**, 114, 300
  bridgehead, **71**, 299
  t-butylcyclohexyl, **71**, 335
  t-butoxy, **70**, 307, 322, 332, 342; **71**, 284, 299, 307, 317, 328, 346, 359, 362
  cage-recombination of, **70**, 308, 321–322, 324, 327, 580; **71**, 278, 284, 288, 298
  cinnamyl, **71**, 355
  cyanoisopropyl, **71**, 288, 300
  cyclohexadienyl, **70**, 347–350; **71**, 315–316, 320
  cyclohexenyl, **71**, 314
  cyclohexyl, **70**, 315; **71**, 318–319, 341, 357, 519
  cyclopropyl, **70**, 337; **71**, 277–279
  cyclopropylmethyl, **70**, 326
  electrochemical generation, **70**, 375–377
  electrophilic character of, **70**, 341
  ESR spectra of, **70**, 307, 315, 317, 324, 336, 340, 362, 363, 365, 378; **71**, 276, 277, 280, 309, 343, 361–364
  fluorenyl, **71**, 364
  fluoroalkyl, **71**, 304
  Fremy's, **71**, 545
  galvinoxyl, **71**, 300
  geometry of, **71**, 280–281
  germanium containing, **71**, 365
  halogenomethyl, **70**, 332, 389; **71**, 516
  heats of formation, **71**, 359
  hexenyl, cyclization of, **70**, 380
  hydrazyls, **71**, 276
  hydroxyl, **71**, 318–319
  imidazolyl, **71**, 511
  imino, **71**, 508
  iminoxy, **71**, 317
  2-iodophenyl, **70**, 319
  ketyl, **71**, 476, 489, 504, 516
  methyl, **71**, 276, 304, 311, 329, 361, 504, 516
  nitroalkoxy, **71**, 281
  nitroxyl, **70**, 306
  non-classical, **70**, 337
  non-planarity, **70**, 315, 337
  norbornenyl, **70**, 308, 314; **71**, 271, 301
  norbornyl, **70**, 307, 308, 314, 315, 338; **71**, 299
  nortricyclyl, **71**, 323
  nucleophilic character of, **70**, 341, 344, 361
  oxanorbornyl, **71**, 277
  phenoxy, **71**, 300, 348
  phenyl, **70**, 345; **71**, 283, 289, 299, 314, 315, 349

Radicals—*continued*
  1-phenylethyl, dimerization of, **70**, 316
  phosphoramyl, **71**, 328, 330–331
  phosphinyl, **70**, 317, 344
  planarity of, **70**, 314–315
  propargyl, **70**, 307
  pyridyl, **71**, 297
  rates of combination, **70**, 382
  semiquinone, **70**, 529
  silyl, **70**, 313; **71**, 301, 315, 320, 338, 339
  spin trapping, **70**, 362–364
  stabilization by methoxy groups, **70**, 320
  stereoselectivity in, **70**, 316
  trichloromethyl, **71**, 299, 305
  trifluoromethyl, **70**, 363, 388; **71**, 276, 304, 310
  tri(phenylthio)methyl, **70**, 313
  uracil, **71**, 360
  vinyl, **70**, 345; **71**, 279, 308, 334, 551
  X-ray crystal analysis, **71**, 364
Radical substitution,
  aliphatic, **70**, 311–312, 325, 380; **71**, 275, 326–331
  aromatic, **70**, 305, 345–351; **71**, 313–320
  gas-phase, **70**, 330
  partial rate factors in, **70**, 347–348; **71**, 314, 315
  sigma complexes in, **70**, 349; **71**, 315, 316
  bimolecular at metal centres, **70**, 311; **71**, 326–327
  on oxygen, **70**, 323; **71**, 286
Radiolyses, **70**, 384–385; **71**, 360–361
Ramberg–Bäcklund reaction, **70**, 135, 279; **71**, 118, 243
Rearrangements, **70**, 241–303; **71**, 211–273
Rearrangements of
  4-acetamido-5-phenylisothiazolinone-1,1-dioxide, **71**, 69
  4β-acetoxy-1,2-dihydroxantonene, **71**, 253
  aldazines, **71**, 232
  allylic azides, **71**, 219
  allylic isocyanates, **71**, 219
  allylic sulphides, **71**, 223, 224
  α-amino-aldehydes, **70**, 289
  3-amino-3,4-dihydro-1-hydroxycarbostyril, **70**, 244
  α-amino-ketones, **71**, 243
  5-amino-4-methyl-3-phenylisoxazole, **71**, 268
  2-(aminophenyl)piperolidin-3-one, **71**, 269
  anhydropenicillins, **70**, 300
  annulenes, **71**, 231
  aroyl esters of 1-phenylpent-2-enol, **70**, 291
  aryl-2-azidoaryl sulphides, **70**, 293
  N-arylaziridine-carboximidoyl chlorides, **71**, 252
  aryl dimethylindoles, **70**, 290
  N-aryl-*S*,*S*-dimethylsulphimides, **71**, 224
  aryl hydrazonates, **71**, 214
  aryloxypyridinium salts, **71**, 214

Rearrangements of—*continued*
  arylthiocarbonylaziridines, **70**, 291
  aryl thionobenzoates, **71**, 213
  aryl thionocarboxylates, **71**, 214
  arylthiovinyl sulphonic esters, **70**, 288
  2-azabicyclo[2.2.1]heptanes, **71**, 259
  2-azidopyridines, **71**, 232
  aziridine-*N*-oxides, **71**, 227
  azomethine ylids, **71**, 232
  basketene, **71**, 220, 236
  trans-benzamido-1,4-diphenylazetidin-2-one, **71**, 267
  benzisoxazoles, **70**, 267, 271
  benzobarrelenes, **71**, 249
  benzocyclooctatetraenes, **70**, 272
  benzo-1,3-dioxans, **70**, 300
  benzofurobenzopyrans, **70**, 288
  1,2,3-benzothiazole-7-diazonium salts, **70**, 298
  N-benzoyl-*O*-glycylserinamide, **71**, 272
  3-benzylideneiosbornyl acetate, **70**, 291
  2-(benzyloxyphenyl)diphenylmethyl chloride, **71**, 253
  bicyclobutanes, **71**, 254
  bromoalkanes, **70**, 333
  bromocyclooctatetraenes, **70**, 272
  o-bromo-*N*,*N*-dimethylaniline, **71**, 202
  N-bromo-β-Lactams, **71**, 259
  bromomethylenecyclobutanes, **71**, 240
  β-bromo-β-nitrostyrene, **70**, 293
  9-bromo-1,3,5,7-tetramethyl-2,4,6,8-tetrathiaadamantane, **70**, 288
  but-2-enyl propionate, **71**, 252
  t-butyl-2-halocyclohexanones, **71**, 243
  1-t-butyl-2-(methylsulphonyloxymethyl)azetidine, **71**, 244
  cephams, **71**, 267
  3-chloro-3-ethyl-2-methyl-3H-azirines, **71**, 282
  N-chloro-*N*-methyl-(1-phenylcyclobutyl)amine, **71**, 259
  1-chloro-2-phenylazetidine, **71**, 259
  1-chloro-3-phenylthiopropan-2-one, **70**, 278
  chlorothietan-3-one-1,1-dioxide acetals, **71**, 260
  chrysanthenumdicarboxylic acid, **71**, 227
  cubane, **71**, 253
  cyclobutane-1,2-dione, **71**, 251
  cyclodeca-1,5-dienes, **71**, 222
  cyclodecatetraenes, **71**, 235
  1-cyclohexylazetidin-2-ol chloroformate, **71**, 267
  decalins, **70**, 288
  2,3-dibenzoyl-1-thiobenzoylglycerol, **70**, 291
  N,*N*-dichlorocyclohexylamine, **70**, 293
  dichloropropanes, **70**, 288
  3,3-dicyano-2-methyl-4-phenylpent-1-ene, **71**, 224
  dienes, **70**, 290; **71**, 215, 219, 226

Rearrangements of—*continued*
 dienones, **70**, 247–248; **71**, 203, 215–216
 3,6-dihydro-3,3,6,6-tetramethyl-2H-azepin-2-one, **70**, 252
 4,5-dihydro-4,6,6-trimethyl-1,3-oxazine, **70**, 282
 di-isopropylbenzene hydroperoxides, **71**, 260
 7,7-dimethoxycyclohepta-1,3,5-triene, **70**, 266
 2,2-dimethylbut-3-yne, **71**, 250
 5,5-dimethylcyclopentadiene, **71**, 227
 1,6-dimethyl-2,5-diphenyl-3,4-diazabicyclo[4.4.0]decatetraene, **70**, 265
 cis-2,2-dimethyl-3-isobutenylcyclopropyl isocyanate, **70**, 252
 2,3-diphenyl-2H-azirines, **70**, 293
 N,N-diphenylbenzylamine, **71**, 240
 α-N-diphenylnitrone, **70**, 290
 divinyl cyclopropanes, **71**, 222
 divinyl episulphones, **71**, 222
 6-ethoxy-4-thiouracils, **70**, 246
 ethylidenecyclobutane, **71**, 293
 ethyl 1,4,6-triphenyl-2,3-diazabicyclo[3.2.0]hepta-3,6-diene-2-carboxylate, **71**, 266
 α-haloepoxides, **71**, 243
 homocubanes, **71**, 254
 hydrazinium salts, **70**, 277
 1-hydroperoxy-1-isopropylnaphthalen-2(1H)-one, **71**, 251
 1-hydroxy-3-aza-bicyclo[4.1.0]hept-3-en-2-ones, **71**, 265
 2-hydroxy-3,4-bis(diphenylmethylene)cyclobutanone, **71**, 243
 α-hydroxy-α,α-diphenylacetaldehyde, **71**, 250
 hydroxyindoles, **71**, 212
 2-hydroxy-2-methylcyclobutanones, **70**, 289
 4-hydroxy-4-vinylcyclooctene, **70**, 257
 7β-iodoneopinone, **71**, 250
 isoxazolidine, **71**, 268
 ketols, **71**, 416
 mandelaldehyde, **71**, 243
 mandelaldehyde dimers, **71**, 411
 menthone, **71**, 243
 1-methoxy-6,7-bis(methoxycarbonyl)bicyclo[3.2.0]nona-3,6,8-triene-2-one, **71**, 221
 4-methoxychalcones, **70**, 288
 methyl 2-azabicyclo[3.1.0]hex-3-ene-1-carboxylate, **71**, 265
 methylenecyclobutanes, **71**, 226
 methylenecyclopropanes, **71**, 226
 methyl 6β-phthalimidopenic illanate, **70**, 280
 naphthazarin diacetate, **71**, 360
 nitrobenzenesulphenanilides, **71**, 215
 p-nitrobenzoyl-13-azabicyclo[10.1.0]tridecanes, **71**, 227

Rearrangements of—*continued*
 nitrones, **71**, 234, 258
 nitrosoanilines, **71**, 212
 oxadiaziridine, **70**, 267
 N-(1,2,4-oxadiazol-3-yl)-N′-arylformamidines, **71**, 268
 oxaziranes, **71**, 265
 oxaziridinobenzodiazepine, **71**, 293
 oxazolines, **70**, 267
 paraccylophanes, **71**, 213
 penicillanic acid derivatives, **71**, 267
 pentamethyl-3H-pyrrolenine, **71**, 252
 polyhalogenobenzenes, **71**, 241
 prismanes, **71**, 254
 2-(2-pyridyl)ethyl chloride, **71**, 240
 pyrimidines, **70**, 271
 quinolizinium-1-diazonium salts, **70**, 298
 salivene, **71**, 248
 silyl carbinols, **71**, 240
 spiro[4.5]dec-6-en-2-one, **70**, 290
 stilbazolium salts, **70**, 295
 tertiary alicyclic peroxides, **71**, 266
 tetrazenes, **70**, 300
 α-thioacyl-γ-thiol-lactones, **71**, 269
 thioethers, **70**, 244, 250, 260
 thiouronium salts, **70**, 295
 triarylpentaaza-1,4-dienes, **70**, 291
 triazolopyrazine, **70**, 298
 triazolo[1,5-a]pyridine-3-acraldehydes, **71**, 269
 triazolo[1,5-a]pyridines, **70**, 271
 trichloro-2-hydroxyalkan-4-ones, **71**, 228
 trichloroprop-1-enyl ketones, **71**, 291
 trimethylsilylcyclopentadiene, **70**, 265; **71**, 228
 trimethylsilyloxypent-3-en-2-one, **71**, 228
 triphenylpyrylium-3-oxide, **70**, 290
 triphenylstannylcycloheptatriene, **71**, 228
 vinylcycloheptatriene, **71**, 221
Rearrangements, radical, **70**, 351–356; **71**, 320–326
Reduction by
 alcohols, **71**, 552, 553
 alkali metals, **70**, 600–602; **71**, 240, 340, 343, 550, 551–552
 aluminium hydrides, **70**, 596–598; **71**, 245, 546–547
 aluminium isopropoxide, **70**, 600
 aluminium–mercury couple, **70**, 301
 borohydride, **70**, 595, 598; **71**, 549
 chromium(II), **70**, 289, 316, 358–359, 604
 cobalt(II), **70**, 604
 copper hydrides, **70**, 598
 dihydro-3-methyl-lumiflavin, **71**, 552
 di-imide, **70**, 604
 dithionite, **71**, 553
 ethyl phosphite, **71**, 552
 ferrocyanide, **71**, 553
 formic acid, **71**, 251
 Grignard reagents, **70**, 596

Subject Index 643

Reduction by—*continued*
  Hantzsch ester, **71**, 553
  hexachlorodisilane, **70**, 99, 599
  lithium, **71**, 112
  lithium aluminium hydride, **70**, 14, 73, 94, 100, 223, 291, 597, 599; **71**, 92, 106, 253, 549
  lithium *N*-benzanilide, **71**, 547
  magnesium, **71**, 552
  magnesium alkyls, **71**, 596
  silanes, **71**, 549
  sodium hydride, **71**, 339, 548
  sodium naphthalenide, **71**, 552
  sodium sulphide, **71**, 552
  tervalent phosphorus reagents, **70**, 603
  tin hydrides, **70**, 337, 352, 599; **71**, 297, 303, 311, 323, 549
  titanium(III), **71**, 553
  trichlorosilane, **70**, 599
  vanadium(II), **70**, 604
  zinc, **70**, 602, 603; **71**, 77, 252
  zinc–copper couple, **71**, 279
Reduction of
  acetals, **70**, 598; **71**, 548
  acetylenes, **71**, 548
  acyl halides, **71**, 548
  adenine nucleosides and nucleotides, **70**, 605
  akuammicine, **71**, 77
  aldehydes, **70**, 596; **71**, 547, 553
  alkyl halides, **70**, 598, 599; **71**, 548
  alkyl palladium complexes, **70**, 598
  allenes, **70**, 604
  allylbenzene, **70**, 604
  allyl ethers, **71**, 552
  allylic alcohols, **71**, 106, 548
  anhydrides, **70**, 598; **71**, 548
  annulenes, **71**, 112
  aromatic ethers, **70**, 602
  aromatic hydrocarbons, **70**, 600, 604; **71**, 550
  aryl halides, **70**, 223, 598
  arsonium salts, **70**, 599
  azocines, **71**, 551
  azo-compounds, **70**, 605; **71**, 344
  benzaldehyde, **70**, 604
  benzocyclobutene, **70**, 605
  benzoic anhydride, **70**, 599
  benzophenone anil, **71**, 240
  benzoyl cyanide, **70**, 603
  benzylphosphonium salts, **71**, 552
  benzyltriethylammonium nitrate, **70**, 604
  bicyclo[6.1.0]nonatriene, **71**, 550
  bicyclo[2.2.2]octyl tosylates, **70**, 14
  1-bromoethylbenzene, **70**, 316
  1-t-butyl-3-phenylpropargyl alcohol, **70**, 599
  chloramine-T, **71**, 553
  citronellol, **71**, 551
  cyclobutanes, **71**, 551
  cyclohexane-1,2-dione, **70**, 600
  cyclopentane-1,2-dione, **70**, 600

Reduction of—*continued*
  cyclopropanes, **70**, 600; **71**, 114, 500
  cyclopropylmethylketones, **70**, 601
  dialkylidenecyclobutane, **70**, 600
  diarylbicyclo[2.2.2]octanes, **71**, 348
  diazo-compounds, **71**, 553
  diazonium salts, **71**, 549, 553
  2,4-dibromo-2,4-dimethylpentan-3-one, **70**, 279
  di-t-butyldiaziridinone, **70**, 603
  dihalogeno-compounds, **70**, 599; **71**, 311
  diphenylcyclopropenone, **70**, 301
  disulphides, **71**, 553
  2-en-4-yn-1-ols, **70**, 599
  epoxides, **70**, 549; **71**, 548, 552
  esters, **70**, 598; **71**, 552
  halogeno-compounds, **70**, 599
  halogenofluorocyclopropanes, **70**, 337
  humulones, **70**, 605
  imines, **70**, 605
  indenes, **71**, 228
  p-iodonitrobenzene, **70**, 369
  ketals, **70**, 598; **71**, 552
  ketones, **70**, 595–597, 602, 603, 605; **71**, 546–549, 552, 553
  ketosulphones, **70**, 605
  lactams, **71**, 267
  lactones, **70**, 602
  1-methoxycyclohexa-1,3-diene, **70**, 602
  methyl-1-naphthylmenthoxyphenylgermane, **70**, 94
  naphthaquinones, **71**, 344
  nitrobenzyl halides, **71**, 337
  nitrocamphene, **71**, 289
  nitro-compounds, **70** 599, 603, 605; **71**, 549, 552
  nitroso-compounds, **71**, 552
  norbornenone, **71**, 340
  octadehydro[24]annulene, **70**, 602
  olefins, **70**, 602
  organomercurials, **70**, 352
  orthoformates, **70**, 597
  9-oxalbicyclo[4.2.1]non-2-yl iodides, **70**, 73
  oximes, **70**, 598
  oxepins, **71**, 551
  phenylacetylene, **71**, 114
  phosphine oxides, **70**, 99, 599
  phosphine disulphides, **71**, 549
  propynal, **71**, 552
  pyrazolones, **71**, 549
  pyridinium salts, **71**, 549
  pyrimidines, **71**, 344
  quinoids, **70**, 605
  sulphides, **70**, 599
  sulphonamides, **71**, 343
  sulphonate of hydroxy-12β-conanine, **71**, 245
  sulphones, **70**, 599; **71**, 92
  sulphoxides, **70**, 599, 604
  tabersonine, **71**, 77, 251–252

Reduction of—*continued*
  tetranitromethane, **71**, 112
  thioacetals, **70**, 602
  tosylates, **70**, 598
  tropylium cation, **71**, 47
  unsaturated ketones, **70**, 602
Reductions, electrochemical, **70**, 369, 604; **71**, 47, 337, 343–344, 554–555
Reimer–Tiemann reaction, **71**, 385
Ritter reaction, **71**, 107
Rupe rearrangement, **71**, 252

Sabinene, **71**, 28
Salt effects, **70**, 61, 67, 68; **71**, 5, 50, 55, 150, 427, 442
Sativene, **71**, 16, 188
Schiff bases,
  addition of alcohols, **71**, 406
    chloroketene, **71**, 172
    phenols, **70**, 434
  addition to maleimides, **71**, 176
  formation, **71**, 405
  hydrogen atom abstraction from, **71**, 301
  hydrolysis, **70**, 431–433; **71**, 406
  intermediates in Knoevenagel reaction, **71**, 413
  pK_a, **71**, 406
  reaction with HCN, **71**, 405
  ring-chain tautomerism, **70**, 433
  syn–anti-interconversion, **70**, 439; **71**, 261, 406
  transamination, **71**, 405
  tautomerism, **71**, 406
Schmidt reaction, **70**, 292; **71**, 258, 416
Selenophen, **70**, 226; **71**, 197
Selenoxides, elimination reactions of, **70**, 170
Semibullvalene, **70**, 257; **71**, 219
Semicarbazones
  formation of, **70**, 438; **71**, 407
  reaction with bromine, **71**, 407
  rearrangement of, **71**, 293
  syn–anti isomerization, **71**, 261
Serini reaction, **70**, 286
Sigmatropic rearrangements,
  [1,3], **70**, 246, 257–261; **71**, 224, 482
  [1,4], **70**, 246
  [1,5], **70**, 248, 261–265, 268, 269; **71**, 226–228
    stereochemistry of, **70**, 264
  [1,7], **70**, 248, 265–266
  [2,3], **70**, 254–257, 277; **71**, 222, 227
  [3,3], **70**, 246, 248–254; **71**, 216
  [3,5], **70**, 248
  [5,5], **70**, 249
Silanes, base-catalysed cleavage of acetylenic, **71**, 131
Silylenes, **71**, 370
Silyl groups, migration of, **70**, 265, 277
Simonini reaction, **71**, 335
Simmons–Smith reaction, **70**, 401, 411

Singlet oxygen, **70**, 507, 566–571; **71**, 520, 522
  additions to olefins, **70**, 203, 207; **71**, 520
  reactions with fluorenyl anions, **70**, 126
Singlet states of hydrocarbons, **70**, 506; **71**, 474
Smiles rearrangement, **70**, 124, 242, 243, 245; **71**, 213
S_E2' mechanism, **70**, 139, 140; **71**, 129
$S_E i$ mechanism, **70**, 139
S_H2 reactions, **70**, 311–312, 325, 380; **71**, 275, 326–331
S_N2' reactions, **70**, 115; **71**, 106–107
$S_N i$ reactions, **70**, 93; **71**, 54
S_N1 reactivity and carbonium ion stability, **71**, 43
Sodium hydride from sodium dihydronaphthalenyl and hydrogen, **70**, 145
Solvent assistance, **70**, 17, 61
Solvent effects, **70**, 18, 22, 60, 66–68, 111, 212, 466, 491; **71**, 55–58, 179–180, 553
  in hydrolysis of chlorosilanes, **71**, 85
  in hydrolysis of esters, **70**, 446, 491; **71**, 427, 455
Solvent participation, **70**, 17–18; **71**, 13, 59
Sommelet reaction, **70**, 124
Sommelet–Hauser rearrangement, **71**, 239
Spin-labelling, **70**, 366
Spin-trapping, **70**, 362–367; **71**, 348–351
Spiro[m.2]alkanes, oxidation by thallium, **70**, 585
Spiro-compounds, rearrangement, **70**, 258, 260, 265; **71**, 264
Spiroconjugation, **70**, 194
Spirocyclopropane-anthrones, **70**, 25
Spiro[2,4]heptyl derivatives, **70**, 11
Spiro[2,5]octyl derivatives, **70**, 11
Spiropentylamine, **70**, 43
Spiropentylcarbinols, **70**, 39
Squalene, **71**, 25
Squalene oxide, **71**, 25
Steroids,
  acetolysis of 4β-bromo-5β-androstan-3-one, **71**, 107
  acetolysis of cholestanyl toluene-*p*-sulphonates, **70**, 166
  3β-acetoxy-5α-pregnan-20α-yl toluene-*p*-sulphonate, **71**, 46, 77
  acid-catalysed rearrangement of, **70**, 290
  acyloin rearrangement of D-homosteroids, **70**, 289
  autoxidation of cholesterol, **70**, 379, 587
  backbone rearrangement, **71**, 77
  cholestan-3β-yl nitrite irradiation in the presence of isopropyl phosphite, **71**, 330
  cholestene, cycloaddition, **71**, 172
  cholest-14-en-7-ols, addition of HF to, **71**, 47
  deamination of 3-amino-4,4-dimethyl-5α-cholestane, **70**, 89
  deamination of 3-aminomethylcholestanol, **70**, 86

Steroids—*continued*
  deamination of norandrostan-16β-ylamine, **70**, 43
  dehydration of steroidal alcohols, **71**, 247
  dienone-phenol rearrangement, **70**, 247
  elimination reactions of 3-*p*-tolylsulphonoxy-4,5-epoxycholestanes, **70**, 155
    cholestanyl trimethylammonium hydroxides, **70**, 157
  ergosterol, additions to, **71**, 155
  esterification of steroidal alcohols with acetyl nitrate, **70**, 492
  formation of lanosterol, **71**, 25
  formation of *D*-ring by triple-bond participation, **71**, 25
  fragmentation of 3β,5-cyclocholestan-6-yl radical, **71**, 321
  Hofmann degradation of steroidal perhydroazepines, **70**, 207
  D-homoandrost-5-ene, **70**, 46
  hydrolysis of 2α,5-epithio-5α-cholestan-3β-yl bromide, **71**, 62
  isomerization of 2β-methylcholestane, **70**, 442
  photo-fragmentation of steroidal epoxides, **70**, 548
  photolysis, **70**, 517, 536; **71**, 483
  photolysis of steroidal nitrites, **71**, 325
  protonation of steroidal olefins, **71**, 248
  reaction of 20β-acetoxy-5α-pregnanes, **71**, 245
  reaction of 4-bromo-3-keto-5β-steroids with potassium acetate, **71**, 253
  reaction of 3-hologeno-cholestanes with organolithium and potassium, **70**, 113
  reaction of steroidal alcohols with PCl$_5$ and SOCl$_2$, **70**, 114
  rearrangement of androst-5-ene, **71**, 46
  rearrangement of α-halogenated 20-ketosteroids, **70**, 278
  rearrangement of 2α-hydroxy-3α-mesyloxy-2β-methylcholestane, **71**, 240
  rearrangement of oestrone, **71**, 79
  reconjugation of cholest-5-en-3-one, **70**, 442
  ring contractions of 5α-oestrane diones, **71**, 251
  ring opening of steroidal epoxides, **71**, 247
  selective functionalization, **71**, 477
  1,2-shifts, **70**, 288
  solvolysis of 5α-cholestan-3-yl toluene-*p*-sulphonates, **70**, 87
  solvolysis of 5α-cholesten-3-yl toluene-*p*-sulphonates, **70**, 87
  solvolysis of 2α,5-epithio-5α-cholestane derivatives, **71**, 245
  solvolysis of 3β-(1-hydroxyethyl)-A-norcholest-5-enyl toluene-*p*-sulphonate, **70**, 29
  solvolysis of 4β-methylcholesteryl toluene-*p*-sulphonates, **70**, 29

Steroids—*continued*
  solvolysis of norandrostan-16β-yl toluene-*p*-sulphonate, **70**, 44
  steroidal nitroxides, **71**, 352
Stevens rearrangement, **70**, 276, 307, 354–355; **71**, 239, 324–325
Stilbene dibromides, debromination of, **70**, 151
Stilbenes,
  additions to, **70**, 174; **71**, 173
  metalation of, **70**, 232
  photolysis of, **71**, 496–497, 524
Strecker synthesis, **71**, 405
Styrenes,
  additions, **70**, 174, 177, 181–182; **71**, 157, 305
  cycloadditions to, **70**, 197, 199; **71**, 166
  dimerization of, **70**, 186
  radical anion from, **70**, 370
  reaction with benzene in presence of palladium acetate, **70**, 235
  reaction with nitroso compounds, **71**, 294
Styryl cations, **71**, 41
Sulphate, hydrolysis, **70**, 500–503
Sulphenanilides, *o*-nitro-, intramolecular oxygen-transfer, **70**, 243
Sulphenates,
  hydrolysis, **70**, 103
  rearrangements, **70**, 257, 311
Sulphenes, **70**, 201; **71**, 173
Sulphenyl halides,
  addition reactions of, **70**, 10, 172, 178; **71**, 159–160
  conversion into thiophenes, **70**, 235
  reaction with amines, **70**, 103; **71**, 94
  reaction with sulphides, **71**, 90
Sulphides, protonation, **71**, 41
Sulphilimines, **71**, 92
Sulphinates, ethylbenzyloxy, **70**, 502
Sulphinic acids, protonation of, **70**, 502
Sulphinyl sulphones, **71**, 92
Sulphite, dimethyl, isomerization, **71**, 106
Sulphonamides, **70**, 502; **71**, 92, 465
Sulphonate, hydrolysis, **70**, 501; **71**, 465
Sulphonation, **70**, 226, 227; **71**, 197–198
Sulphones,
  acetylenic, additions to, **70**, 187
  acidities of, **70**, 130
  bromination of disulphones, **71**, 118
  hydrogen exchange of, **71**, 119–120
  pyrolysis, **71**, 294
  rearrangement of, **70**, 245
  vinyl, additions to, **70**, 186, 187; **71**, 163
Sulphonic acids,
  nitration of, **70**, 229
  protonation of, **70**, 502
  rearrangement, **70**, 245
Sulphonic anhydrides, **71**, 464
Sulphonium salts,
  nucleophilic substitution reactions, **70**, 223
  reaction with aryl lithium, **70**, 222

Sulphonyl chlorides, **70**, 502; **71**, 94, 463
Sulphonyl halides, radical addition, **71**, 388
Sulphonylsulphilimines, elimination reactions, **70**, 150
Sulphoxides,
  acidities, **70**, 131
  benzynes from, **70**, 322
  chlorination, **70**, 135
  conformation of 5-membered cyclic, **71**, 119
  elimination reactions of, **70**, 151, 168; **71**, 138
  halogenation of, **70**, 232
  hydrogen exchange, **70**, 135; **71**, 119
  α-lithio, **71**, 119
  nitration of, **71**, 199
  nucleophilic substitution reactions, **71**, 90, 92–93
  protonation of, **70**, 502
  racemization, **70**, 103
  rearrangement, **70**, 257, 264, 277–278; **71**, 227, 230
  vinyl, addition of bromine, **71**, 156
Sulphur,
  nucleophilic displacement at, **70**, 101–104, 500–503; **71**, 89–94, 553
  singlet and triplet, **70**, 565
Sulphur dichloride, additions, **70**, 63, 178
Sulphur diimides, **70**, 201
Sulphur dioxide, reaction with vinyldiazomethane, **70**, 253
Sulphur trioxide, complexes with amines, **70**, 502; **71**, 465
Sulphuryl chloride, **70**, 231, 333, 334
Sultones, hydrolysis, **70**, 502; **71**, 464
Swain–Scott equation for reaction of
  chloro-2,4-dinitrobenzene with amines, **71**, 179
  methyl halides with nucleophiles, **71**, 51
  N-methylpyridinium-3-sulphonate with anions, **70**, 112
  phenacyl bromide with amines, **70**, 111
Sydnones, **70**, 196, 490, 553; **71**, 171, 269, 453, 508

Taft equation for
  deacylation of acyl-chymotrypsins, **70**, 477
  dehydration, **71**, 149
  dehydrogenation reactions, **70**, 168–169; **71**, 149
  demethylation of tertiary amines by hexacyanoferrate, **70**, 582
  dissociation constants of acetylacetanilides, **70**, 130
  dissociation constants of dinitropropenes, **70**, 130
  hydrolysis of amides, **71**, 420
  hydrolysis of aryldiazoketones, **71**, 82
  hydrolysis of esters, **70**, 461
  iodine abstraction from alkyl iodides, **71**, 299
  radical reactions, **71**, 276

Taft equation for—*continued*
  reaction of dehydroacetic acid with amines, **71**, 406
  reaction of diketones with phosphites, **71**, 415
  silver-ion-catalysed rearrangement of cubanes, **71**, 253
  solvolysis of 2-substituted 1-adamantyl toluene-p-sulphonates, **70**, 33
Tautomerizations, **71**, 264
Tellurophen, **71**, 197
Terpenes, acid-catalysed rearrangements of, **70**, 290
Tetracyanoethylene oxide, cycloadditions, **70**, 196
Tetracyclododecanones, oxidation of, **70**, 7–8
Tetracyclodecyltosylates, **70**, 7
Tetracyclooctyl derivatives, **71**, 15
Tetrahedral intermediates in
  cleavage of 3-phenylpropynal, **71**, 421
  nucleophilic aromatic substitution, **70**, 211; **71**, 183, 187
  reactions of carboxylic acid derivatives, **70**, 453–458; **71**, 419–423
Thallation of aromatic compounds, **70**, 232; **71**, 203
Thalliodemetalations, **70**, 145
Thallium, alkyl, **70**, 114; **71**, 132
Thermolysis, *see* Pyrolysis
Thiazole,
  hydrogen exchange of, **70**, 136
  nitration of, **70**, 229
  radical substitution, **71**, 318
Thiazolin-5-ones, **71**, 269
Thiazolyl-propionate, bromophenyl, hydrogen exchange of, **70**, 136
Thienothiophens, **71**, 207
Thietans ring-opening, **70**, 102
Thiirans, ring-opening, **70**, 102
Thiironium ion *see* Episulphonium ion
Thioacetals, **70**, 418
Thioanisole, electrophilic substitution in, **70**, 226
Thiocarbonates, **70**, 62
Thiocyanates, **70**, 3, 15, 63, 105, 291; **71**, 219
Thiols,
  nucleophilic addition of, **70**, 190; **71**, 162, 163
  hydrogen atom abstraction from, **71**, 296
  radical addition, **70**, 305, 306; **71**, 305, 306
  intramolecular, **71**, 311
Thiophene,
  acylation, **71**, 197, 205
  electrophilic substitution, **71**, 197
  halogenation of, **70**, 226, 230
  hydrogen exchange, **70**, 237
  nitration, **71**, 199
  proton exchange, **71**, 207
  radical phenylation, **70**, 350, 351
  radicals from, **71**, 276
  Vilsmeier reaction of, **71**, 238

Thujene, **71**, 28
Thujopsene, protonation, **71**, 78
Thujopsene-widdrol interconversion, **70**, 38
Tin,
    nucleophilic displacement at, **70**, 93
    radical displacement at, **70**, 312
Tin alkyls, **71**, 113, 149
Toluene,
    Friedel–Crafts reaction of, **70**, 234; **71**, 205–206
    oxidation, **71**, 333
    oxygenation, **70**, 222
Topomerization, **71**, 261
Torsional effects, **70**, 314; **71**, 5, 152, 158–159
Transamination, **70**, 435, 443; **71**, 123
Transannular rearrangement, **70**, 302; **71**, 154, 161
Transition-metal complexes intramolecular substitution in, **70**, 227
Transition state, activity coefficient, **71**, 105
Triazines,
    chloro, nucleophilic displacement, **71**, 183
    halogeno, reaction with sodium in liquid ammonia, **70**, 239
    hydrolysis of triphenyl, **70**, 492
    Smiles rearrangement of, **70**, 244
Triazinones, **71**, 272
Tribenzobicyclo[2.2.2]octatrienyl methyl cations, **70**, 16
Tribenzocycloheptatriene, base-catalysed hydrogen exchange, **70**, 132
Tribenzo[c, i, o]triphenylene dianion, **70**, 120
Tri-t-butyl-2-nitrobenzene, nitration, **71**, 199–200
Tricycloheptenes, **71**, 236
Tricyclo[4.1.0.0^{2,7}]heptane, **70**, 13; **71**, 11
Tricyclo[3.2.0.0^{2,7}]heptyl derivatives, **71**, 27
Tricyclo[4.1.0.0^{2,4}]heptyl derivatives, **70**, 38
Tricyclo[3.3.0.0^{3,7}]octane, 2-chloro, **70**, 20
Tricyclo[3.3.0.0^{3,7}]octan-1-ol, 3,7-dimethyl, **71**, 116
Tricyclo[3.2.1.0^{2,4}]oct-6-en-8-one, decarbonylation, **70**, 30
Tricyclo[3.2.1.0^{2,4}]oct-6-en-8-yl derivative, **71**, 29
Tricyclo[3.3.0.0^{3,7}]oct-2-yl p-bromobenzenesulphonate, **70**, 20
Tricyclo[3.2.1.0^{2,4}]oct-8-yl anti-toluene-p-sulphonate, exo-3,3-diphenyl, **70**, 23
Trifluoroacetic acid additions, **70**, 10, 172, 180
Trifluoroethanesulphonates, **71**, 106
Trifluoroethanol, as limiting solvent, **71**, 43, 56
Trimethoxybenzene, hydrogen exchange, **70**, 236
Trimethylenemethane, **71**, 292
Triphenylamines, bromination, **71**, 202
Triphenylethyl toluene-p-sulphonate, **71**, 17
Triphenylmethylcations, **70**, 53, 55, 57–58, 61, 68, 234, 387, 563; **71**, 46, 48, 545
Triphenylmethyl radicals, **70**, 387; **71**, 296

Triphenylphosphine, reaction with alcohols, **71**, 109
Triplet state, **70**, 508–515; **71**, 161, 467, 471–472, 475, 477, 478, 489, 499, 512, 518, 524
Trishomocyclopropenyl cations, **70**, 40
Tropane, quaternization, **71**, 107
Tropanyl methanesulphonate, **71**, 71
Tropolones, radical substitution, **71**, 365
Tropone, tricarbonyl iron complex, **71**, 39
Tropones, cycloadditions, **70**, 205, 206; **71**, 175
Tropones, 2-substituted, nucleophilic displacement, **70**, 109
Tropylethyl p-nitrobenzenesulphonate, **70**, 25
Tropylium cations, **70**, 55; **71**, 43, 48, 167
    bishomo, **70**, 38
    photoisomerization of, **70**, 32, 564
Turnstile rotation, **71**, 85
Twixtylic intermediates, **71**, 152, 290

Ullmann reaction, **70**, 222; **71**, 192
Urea, hydrolysis, **71**, 424, 426
Uridine, hydrogen exchange, **71**, 123
Uronic acid derivatives, elimination reactions, **70**, 156

VARYTEMP, **70**, 131
Verbanone, **71**, 291
Vertical stabilization, **71**, 70, 208
Vilsmeier reaction, **70**, 235, 238; **71**, 206
Vinyl anions, **70**, 125, 127; **71**, 240
Vinyl cations, **70**, 39, 106–109, 172, 180, 288; **71**, 29, 41, 48, 97–101, 172, 248
Vinylic carbon, nucleophilic substitution at, **70**, 75, 106–110; **71**, 97–101
Vinyl ethers, *see* Enol ethers
Vinyl group, 1,2-shift, **70**, 285, 286, 290
Vitamin B12, **71**, 132
Volume of activation for
    benzidine rearrangement, **71**, 212
    decomposition of peroxides, **71**, 284
    hydrolysis of acetals, **70**, 423
        N-t-butylacetamide, **70**, 462
        γ-butyrolactone, **70**, 468
        cyclopropylcarbinyl and cyclobutyl chlorides, **70**, 89
        sulphonyl chlorides, **71**, 463
    quaternization of pyridine and 2,6-dialkylpyridines, **70**, 116; **71**, 108
    solvolysis of benzyl chloride, **70**, 66; **71**, 57
        1-phenyl-2-methyl-2-propyl chloride, **70**, 67

Wallach rearrangement, **70**, 243; **71**, 212
Winstein–Grunwald equation for
    decarboxylative desulphonation, **71**, 149
    solvolysis of adamantyl derivatives, **70**, 67; **71**, 56
    α-bromoketones, **70**, 111
    bromomethylenecyclopropane, **71**, 99

Winstein–Grunwald equation for—*continued*
    dimethyltrimethylsilylmethyl bromide, **71**, 75
    tropan-3α-yl methanesulphonate, **71**, 71
    substitution of 1,10-phenanthroline on to [Febipyr$_2$(CN)$_2$]^0, **70**, 104
Wittig reaction, **71**, 118
Wittig rearrangement, **70**, 276, 355; **71**, 239
    all carbon analogue, **70**, 124
    of allyl benzyl ethers, **70**, 123, 254–255
    of allylphenyl ether, **71**, 240
    butynyl fluorenyl ether, **70**, 124
    diallyl ethers, **70**, 124
Wolff rearrangement, **70**, 200, 400, 405; **71**, 381–382
Wolff–Kishner reduction, **70**, 283, 302; **71**, 272
Woodward–Hoffmann rules, **70**, 104, 190, 199, 266; **71**, 151
Wurtz reaction, **70**, 139, 311

Xenon fluorides, **71**, 201, 341
Xylenes,
    alkylation of, **70**, 235
    nitration, **70**, 228

Ylids, **71**, 133, 384
    arsenic, **71**, 118
    carbenes from, **70**, 385
    carbon, **71**, 171
    carbonyl, **70**, 584; **71**, 234
    nitrenes from, **70**, 398
    nitrogen, **70**, 127, 128, 136, 276; **71**, 124, 170–171, 222, 265–266
    oxygen, **71**, 385
    phosphorus, **70**, 127, 128, 136–137, 188, 255, 327, 565; **71**, 117, 118, 162, 163, 197, 385
    sulphonium, **71**, 106, 117, 133, 239
    sulphur, **70**, 127, 128, 256, 277; **71**, 122, 223. 224, 228, 271, 508
Ynamines, cyclic additions, **70**, 198, 203; **71**, 173
Yukawa–Tsuno relationship for
    electrophilic aromatic substitution, **70**, 226, 229
    hydrolysis of 2-alkoxytropones, **71**, 193

Zimmerman–Groevenstein rearrangement, **70**, 355
Zinc alkyl, **71**, 127, 129

Errata for Organic Reaction Mechanisms 1968

P. 399, line 1: *For* 2-phenylsydnone *read* 3-phenylsydnone.

Errata for Organic Reaction Mechanisms 1970

P. 26, line 11: *For* compounds (**65**) *read* 7-aryl-7-norbornyl p-nitrobenzoates.
P. 26, line 12: *For* series (**64**) *read* *syn*-7-aryl-*anti*-norbornenyl p-nitrobenzoates.
P. 27, line 2: *For* (**65**) *read* saturated.
P. 268, line 25: *For* $[\sigma^2 s + \sigma^2 s]$ *read* $[\sigma^2 s + \sigma^2 a]$.
P. 439 (last two lines) and P. 440 (first two lines): The comment in these lines arose from a misreading of reference 174 and should be deleted.

QD
258
O 82
1971

FEB 10 1975